# THE HISTORY OF THE THEORY OF STRUCTURES

KARL-EUGEN KURRER

# The History of the Theory of Structures

From Arch Analysis to Computational Mechanics

Author: Dr.-Ing Karl-Eugen Kurrer
Ernst & Sohn Verlag für Architektur und technische
Wissenschaften GmbH & Co. KG
Rotherstraße 21, D -10245 Berlin, Germany

This book contains 667 illustrations.

Bibliographic information published by the Deutsche Nationalbibliothek
The Deutsche Nationalbibliothek lists this publication in the Deutsche Nationalbibliografie;
detailed bibliographic data is available in the Internet at <http://dnb.ddb.de>.

Cover: Computer-generated drawing of an FEM model for the Göltzsch Viaduct
by Dr. Roger Schlegel (Dynardo GmbH, Wiemar) plus historical illustrations
(sources given in book).

ISBN  978-3-433-01838-5

© 2008 Ernst & Sohn Verlag für Architektur und
technische Wissenschaften GmbH & Co. KG, Berlin

All rights reserved (including those of translation into other languages). No part of this book may
be reproduced in any form – by photoprinting, microfilm, or any other means – nor transmitted
or translated into a machine language without written permission from the publishers. Registered
names, trademarks, etc. used in this book, even when not specifically marked as such, are not to be
considered unprotected by law.

English translation: Philip Thrift, Hannover
Typodesign: Sophie Bleifuß, Berlin
Typesetting: Uta-Beate Mutz, Leipzig
Drawings: Peter Palm, Berlin
Production: HillerMedien, Berlin
Printing: betz-druck, Darmstadt

Printed in Germany

# FOREWORD

The title of the book alone makes us curious: What is "theory of structures" anyway? Used cursorily, the term describes one of the most successful and most fascinating applied science disciplines. But actually, you can't use this term cursorily; for this is not just about theory, not just about methods of calculation, but rather those fields plus their application to real loadbearing structures, and in the first place to the constructions in civil engineering. Languages sometimes find it difficult to define such a wide field rigorously and, above all, briefly; in the author's country, the term *Baustatik* (literally "building statics") has acquired a widely accepted meaning, even though that meaning is also too narrow. And even the English expression "structural analysis" does not tell the whole story precisely because this is not just about analysis, but about synthesis, too, the overall picture in the creation of a loadbearing structure.

Right at the start we learn that the first conference on the history of theory of structures took place in Madrid in 2005. This theme, its parts dealt with many times, is simply crying out for a comprehensive treatment. However, this book is not a history book in which the contributions of our predecessors to this theme are listed chronologically and described systematically. No, this is "Kurrer's History of Theory of Structures" with his interpretations and classifications; luckily – because that makes it an exciting treatise, with highly subjective impressions, more thematic than chronological, and with a liking for definitions and scientific theory; indeed, a description of the evolution of an important fundamental engineering science discipline with its many facets in teaching, research and, first and foremost, practice.

The history of theory of structures is in the first place the history of mechanics and mathematics, which in earlier centuries were most definitely understood to be applied sciences. K.-E. Kurrer calls this period up to 1825 the preparatory period – times in which structural design was still dominated very clearly by empirical methods. Nevertheless, it is worth noting that the foundations of many structural theories were laid in this period. It is generally accepted that the structural report for the retrofitting works to St. Peter's Dome in Rome (1742/43) by the *tre mattematici* represents the first structural calculations as we understand them today. In other words, dealing with a constructional task by the application of scientific methods – accompanied, characteristically, by the eternal dispute between theory and practice (see section 11.2.5). These days, the centuries-old process of the theoretical abstraction of natural and technical processes in almost all scientific disciplines is called "modelling and simulation" – as though it had first been introduced with the invention of the computer and the world of IT, whereas in truth it has long since been the driving force behind mankind's ideas and actions. Mapping the loadbearing properties of building constructions in a theoretical model is a typical case. One classic example is the development of masonry and elastic arch theories (see chapter 4). It has become customary to add the term "computational" to these computer-oriented fields in the individual sciences, in this case "computational mechanics".

The year 1825 has been fittingly chosen as the starting point of the discipline-formation period in theory of structures (see chapter 6). Theory of structures is not just the solving of an equilibrium task, not just a computational process. Navier, whose importance as a mechanics theorist we still acknowledge today in the names of numerous theories (Navier stress distribution, Navier-Lamé and Navier-Stokes equations, etc.), was very definitely a practitioner. In his position as professor for applied mechanics at the École des Ponts et Chaussées, it was he who combined the subjects of applied mechanics and strength of materials in order to apply them to the practical tasks of building. For example, in his *Résumé des Leçons* of 1826 he describes the work of engineers thus: "… after the works have been designed and drawn, [the engineers] investigate them to see if all conditions have been satisfied and improve their design until this is the case. Economy is one of the most important conditions here; stability and durability are no less important …" (see section 2.1.2). Theory of structures as an independent scientific discipline had finally become established. Important structural theories and methods of calculation would be devised in the following years, linked with names like Clapeyron, Lamé, Saint-Venant, Rankine, Maxwell, Cremona, Castigliano, Mohr and Winkler, to name but a few. The graphical statics of Culmann and its gradual development into graphical analysis are milestones in the history of structural theory.

Already at this juncture it is worth pointing out that the development did not always proceed smoothly: controversies concerning the content of theories, or competition between disciplines, or priority disputes raised their heads along the way. This exciting theme is explored in detail in Chapter 11 by way of 12 examples.

In the following years, the evolution of methods in theory of structures became strongly associated with specific structural systems and hence, quite naturally, with the building materials employed, such as iron (steel) and later reinforced concrete (see chapters 7, 8 and 9). Independent materials-specific systems and methods were devised. Expressed in simple terms, structural steelwork, owing to its modularity and the fabrication methods, concentrated on assemblies of linear members, whereas reinforced concrete preferred two-dimensional structures such as slabs, plates and shells. The space frames dealt with in chapter 8 represent a fulcrum to some extent.

This materials-based split was also reflected in the teaching of structural theory in the form of separate studies. It was not until many years later that the parts were brought together in a homogeneous theory of structures, albeit frequently "neutralised", i. e. no longer related to the specific properties of the particular building material – an approach that must be criticised in retrospect. Of course, the methods of structural analysis can encompass any material in principle, but in a specific case they must take account of the particular characteristics of the material.

Kurrer places the transition from the discipline-formation period – with its great successes in the shape of graphical statics and the systematic approach to methods of calculation in member analysis – to the consolidation period around 1900. This latter period, which lasted until 1950, is characterised by refinements and extensions, e.g. a growing interest in shell structures, and the consideration of non-linear effects. Only after this does the "modern" age begin – designated the integration period in this instance and typified by the use of modern computers and powerful numerical methods. Theory of structures is integrated into the structural planning process of conceptual design – analysis – detailing – construction – manufacturing. Have we reached the end of the evolutionary road? Does this development mean that theory of structures, as an independent engineering science, is losing its profile and its justification? The developments of recent years indicate the opposite.

The history of yesterday and today is also the history of tomorrow. In the world of data processing and information technology, theory of structures has undergone rapid progress in conjunction with numerous paradigm changes. It is no longer the calculation process and method issues, but rather principles, modelling, realism, quality assurance and many other aspects that form the focal point. The remit includes dynamics alongside statics; in terms of the role they play, thin-walled structures like plates and shells are almost equal to trusses and frames, and taking account of true material behaviour is obligatory these days. During its history so far, theory of structures was always the trademark of structural engineering; it was never the discipline of "number crunchers", even if this was and still is occasionally proclaimed as such upon launching relevant computing programs. Theory of structures continues to play an important mediating role between mechanics on the one side and the conceptual and detailed design subjects on the other side in teaching, research and practice. Statics and dynamics have in the meantime advanced to what is known internationally as "computational structural mechanics", a modern application-related structural mechanics.

The author takes stock of this important development in chapter 10. He mentions the considerable rationalisation and formalisation, the foundations for the subsequent automation. It was no surprise when, as early as the 1930s, the structural engineer Konrad Zuse began to develop the first computer. However, the rapid development of numerical methods for structural calculations in later years could not be envisaged at that time. J. H. Argyris, one of the founding fathers of the modern finite element method, recognised this at an early stage in his visionary remark "the computer shapes the theory" (1965): besides theory and experimentation, there is a new pillar – numerical simulation (see section 10.4).

By their very nature, computers and programs have revolutionised the work of the structural engineer. Have we not finally reached the stage where we are liberated from the craftsman-like, recipe-based business so that we can concentrate on the essentials? The role of "modern theory of structures" is also discussed here, also in the context of the relationship between the structural engineer and the architect (see chapter 12). A new "graphical statics" has appeared, not in the sense of the automation and visual presentation of Culmann's graphical statics, but rather in the form of graphic displays and animated simulations of mechanical relationships and processes. This is a decisive step towards the evolution of constructions and to loadbearing structure synthesis, to a new type of structural doctrine. This potential as a living interpretation and design tool has not yet been fully exploited.

It is also worth mentioning that the boundaries to the other construction engineering disciplines (mechanical engineering, automotive engineering, shipbuilding, the aerospace industry, biomechanics) are becoming more and more blurred in the field of computational mechanics; the relevant conferences no longer make any distinctions. The concepts, methods and tools are likewise universal. And we are witnessing similar developments in teaching, too.

This "history of theory of structures" could only have been written by an expert, an engineer who knows the discipline inside out. Engineering scientists getting to grips with their own history is a rare thing. But this is one such lucky instance. This fully revised English edition, which explores international developments in greater depth, follows on from the highly successful German edition. We should be very grateful to Dr. Kurrer, and also "his" publisher, Ernst & Sohn, for this treatise.

Stuttgart, September 2007
Ekkehard Ramm
Professor of Structural Mechanics, University of Stuttgart

**Preface**

Encouraged by the engineering profession's positive response to the first edition of this book, which appeared in German only under the title of *Geschichte der Baustatik* in 2002, and the repeated requests for an English edition, two years ago I set myself the task of revising, expanding and updating the book. Although this new version still contains much of the original edition unaltered, the content now goes much further, in terms of quantity and quality. My aim was not only to take account of the research findings of the intervening years, but also to include the historical development of modern numerical methods of structural analysis and structural mechanics; further, I wanted to clarify more rigorously the relationship between the formation of structural analysis theories and progress in construction engineering. The history of the theory of spatial frameworks, plus plate, shell and stability theory, to name just a few examples, have therefore been given special attention because these theories played an important role in the evolution of the design language of lightweight steel, reinforced concrete, aircraft and ship structures. Without doubt, the finite element method (FEM) – a child of structural mechanics – is one of the most important intellectual technologies of the second half of the 20th century. I have therefore presented the historico-logical sources of FEM, their development and establishment in this new edition. Another addition is the chapter on scientific controversies in mechanics and theory of structures, which represents a "pocket guide" to the entire historical development from Galileo to the early 1960s and therefore allows an easy overview. There are now 175 brief biographies of prominent figures in theory of structures and structural mechanics, over 60 more than in the first edition, and the bibliography has been considerably enlarged.

Certainly the greatest pleasure during the preparation of this book was experiencing the support of friends and colleagues. I should like to thank Jennifer Beal (Chichester), Antonio Becchi (Berlin), Norbert Becker (Stuttgart), Alexandra R. Brown (Hoboken), José Calavera (Madrid), Christopher R. Calladine (Cambridge, UK), Kostas Chatzis (Paris), Mike Chrimes (London), Ilhan Citak (Lehigh), René de Borst (Delft), Giovanni Di Pasquale (Florence), Werner Dirschmid (Ingolstadt), Holger Eggemann (Aachen), Jorun Fahle (Gothenburg), Amy Flessert (Minneapolis), Hubert Flomenhoft (Palm Beach Gardens), Peter Groth (Pfullingen), Carl-Eric Hagentoft (Gothenburg), Torsten Hoffmeister (Berlin), Santiago Huerta (Madrid), Andreas Kahlow (Potsdam), Sándor Kaliszky (Budapest), Klaus Knothe (Berlin), Eike Lehmann (Lübeck), Werner Lorenz (Cottbus/Berlin), Andreas Luetjen (Braunschweig), Stephan Luther (Chemnitz), William J. Maher (Urbana), René Maquoi (Liège), Gleb Mikhailov (Moscow), Juliane Mikoletzky (Vienna), Klaus Nippert (Karlsruhe), John Ochsendorf (Cambridge, USA), Ines Prokop (Berlin), Patricia Radelet-de Grave (Louvain-la-Neuve), Ekkehard Ramm (Stuttgart), Anette Ruehlmann (London), Sabine Schroyen (Düsseldorf), Luigi Sorrentino (Rome), Valery T. Troshchenko (Kiev), Stephanie Van de Voorde (Ghent), Volker Wetzk (Cottbus), Jutta Wiese (Dresden), Erwin Wodarczak (Vancouver) and Ine Wouters (Brussels).

Philip Thrift (Hannover) is responsible for the English translation. This present edition has benefited from his particular dedication, his wealth of ideas based on his good knowledge of this subject, his sound pragmatism and his precision. I am therefore particularly indebted to him, not least owing to his friendly patience with this writer! At this point I should also like to pay tribute to the technical and design skills of Peter Palm (drawings), Sophie Bleifuß (typodesign), Uta-Beate Mutz (typesetting) and Siegmar Hiller (production), all of whom helped ensure a high-quality production. My dear wife and editor Claudia Ozimek initiated the project at the Ernst & Sohn publishing house and steered it safely to a successful conclusion. Finally, I would like to thank all my colleagues at Ernst & Sohn who have supported this project and who are involved in the distribution of my book.

I hope that you, dear reader, will be able to absorb some of the knowledge laid out in this book, and not only benefit from it, but also simply enjoy the learning experience.

Berlin, January 2008
Karl-Eugen Kurrer

**Preface to the first, German edition**

For more than 25 years, my interest in the history of structural analysis has been growing steadily – and this book is the result of that interest. Whereas my initial goal was to add substance to the unmasking and discovery of the logical nature of structural analysis, later I ventured to find the historical sources of that science. Gradually, my collection of data on the history of structural analysis – covering the didactics, theory of science, history of engineering science and construction engineering, cultural and historical aspects, aesthetics, biographical and bibliographical information – painted a picture of that history. The reader is invited to participate actively by considering, interpreting and forming his or her own picture of the theory of structures.

I encountered numerous personalities as that picture took shape and I would like to thank them for their attention, receptiveness and suggestions – they are too numerous to mention them all by name here. In writing this book I received generous assistance – also in the form of texts and illustrations – from the following:
- Dr. Bill Addis, London (biographies of British structural engineers),
- Dr. Antonio Becchi, Genoa (general assistance with the biographies and the bibliography),
- Emer. Prof. Dr. Zbigniew Cywiński, Gdańsk (biographies of Polish structural engineers),
- Prof. Dr. Ladislav Frýba, Prague (biographies of Czechoslovakian structural engineers),
- Prof. Dr. Santiago Huerta, Madrid (biography of Eduardo Saavedra),
- Prof. Dr. René Maquoi, Liège (biographies of Belgian structural engineers),
- Dr. Gleb Mikhailov, Moscow (biographies of Russian structural engineers),
- Prof. Dr. Ekkehard Ramm, Stuttgart (foreword),
- Prof. Dr. Enrico Straub, Berlin (biography of his father, Hans Straub),
- Emer. Prof. Dr. Minoru Yamada, Kyoto (biographies of Japanese structural engineers).

I would also like to thank Mike Chrimes, London, Prof. Dr. Massimo Corradi, Genoa, Dr. Federico Foce, Genoa, Prof. Dr. Mario Fontana, Zurich, Prof. Dr. Wolfgang Graße, Dresden, Prof. Dr. Werner Guggenberger, Graz, and Prof. Dr. Patricia Radelet-de Grave, Louvain-la-Neuve, who helped me with literature sources.

This book would not have been possible without the valued assistance of my very dearest friend Claudia Ozimek, who was responsible for the prudent supervision by the editorial staff. And I should also like to thank all my other colleagues at Ernst & Sohn for their help in the realisation of this book.

I very much hope that all the work that has gone into this book will prove worthwhile reading for you, the reader.

Berlin, September 2002
Dr.-Ing. Karl-Eugen Kurrer

# CONTENTS

| | | |
|---|---|---|
| 5 | | **Foreword** |
| 9 | | **Preface** |
| 11 | | **Preface to the first, German edition** |
| | | |
| 20 | **1** | **The tasks and aims of a historical study of theory of structures** |
| 21 | 1.1 | Internal scientific tasks |
| 25 | 1.2 | Practical engineering tasks |
| 26 | 1.3 | Didactic tasks |
| 27 | 1.4 | Cultural tasks |
| 28 | 1.5 | Aims |
| 28 | 1.6 | An invitation to a journey through the history of theory of structures |
| | | |
| 30 | **2** | **Learning from the history of structural analysis: 11 introductory essays** |
| 31 | 2.1 | What is structural analysis? |
| 31 | 2.1.1 | Preparatory period (1575–1825) |
| 34 | 2.1.2 | Discipline-formation period (1825–1900) |
| 37 | 2.1.3 | Consolidation period (1900–1950) |
| 39 | 2.1.4 | Integration period (1950 to date) |
| 41 | 2.2 | From the lever to the truss |
| 42 | 2.2.1 | Lever principle according to Archimedes |
| 43 | 2.2.2 | The principle of virtual displacements |
| 43 | 2.2.3 | The general law of work |
| 44 | 2.2.4 | The principle of virtual forces |
| 44 | 2.2.5 | The parallelogram of forces |
| 45 | 2.2.6 | From Newton to Lagrange |
| 46 | 2.2.7 | Kinematic or geometric view of statics? |
| 46 | 2.2.8 | Stable or unstable, determinate or indeterminate? |
| 47 | 2.2.9 | Syntheses in statics |
| 50 | 2.3 | The development of higher engineering education |
| 51 | 2.3.1 | The specialist and military schools of the *ancien régime* |

| | | |
|---|---|---|
| 52 | 2.3.2 | Science and enlightenment |
| 52 | 2.3.3 | Science and education during the French Revolution (1789–1794) |
| 53 | 2.3.4 | Monge's teaching plan for the École Polytechnique |
| 55 | 2.3.5 | Austria, Germany and Russia in the wake of the École Polytechnique |
| 58 | 2.3.6 | The education of engineers in the United States |
| 63 | 2.4 | Insights into bridge-building and theory of structures in the 19th century |
| 64 | 2.4.1 | Suspension bridges |
| 70 | 2.4.2 | Timber bridges |
| 72 | 2.4.3 | Composite systems |
| 73 | 2.4.4 | The Göltzsch and Elster viaducts (1845–1851) |
| 76 | 2.4.5 | The Britannia Bridge (1846–1850) |
| 79 | 2.4.6 | The first Dirschau Bridge over the River Weichsel (1850–1857) |
| 80 | 2.4.7 | The Garabit Viaduct (1880–1884) |
| 84 | 2.4.8 | Bridge engineering theories |
| 92 | 2.5 | The industrialisation of steel bridge-building between 1850 and 1900 |
| 92 | 2.5.1 | Germany and Great Britain |
| 94 | 2.5.2 | France |
| 95 | 2.5.3 | United States of America |
| 99 | 2.6 | Influence lines |
| 101 | 2.6.1 | Railway trains and bridge-building |
| 102 | 2.6.2 | Evolution of the influence line concept |
| 103 | 2.7 | The beam on elastic supports |
| 104 | 2.7.1 | The Winkler bedding |
| 105 | 2.7.2 | The theory of the permanent way |
| 107 | 2.7.3 | From permanent way theory to the theory of the beam on elastic supports |
| 108 | 2.8 | Displacement method |
| 109 | 2.8.1 | Analysis of a triangular frame |
| 112 | 2.8.2 | Comparing the displacement method and trussed framework theory for frame-type systems |
| 113 | 2.9 | Second-order theory |
| 113 | 2.9.1 | Josef Melan's contribution |
| 114 | 2.9.2 | Suspension bridges become stiffer |
| 115 | 2.9.3 | Arch bridges become more flexible |
| 116 | 2.9.4 | The differential equation for laterally loaded struts and ties |
| 116 | 2.9.5 | The integration of second-order theory into the displacement method |
| 117 | 2.9.6 | Why do we need fictitious forces? |
| 121 | 2.10 | Ultimate load method |
| 121 | 2.10.1 | First approaches |
| 123 | 2.10.2 | Foundation of the ultimate load method |
| 127 | 2.10.3 | The paradox of the plastic hinge method |
| 130 | 2.10.4 | The acceptance of the ultimate load method |
| 136 | 2.11 | Structural law – Static law – Formation law |
| 136 | 2.11.1 | The five Platonic bodies |
| 137 | 2.11.2 | Beauty and law |

| 142 | 3 | **The first fundamental engineering science disciplines: theory of structures and applied mechanics** |
| --- | --- | --- |
| 143 | 3.1 | What is engineering science? |
| 144 | 3.1.1 | First approximation |
| 146 | 3.1.2 | Raising the status of engineering sciences through philosophical discourse |
| 153 | 3.1.3 | Engineering and engineering sciences |
| 157 | 3.2 | Revoking the encyclopaedic in the system of classical engineering sciences: five case studies from applied mechanics and theory of structures |
| 158 | 3.2.1 | On the topicality of the encyclopaedic |
| 161 | 3.2.2 | Franz Joseph Ritter von Gerstner's contribution to the mathematisation of construction theories |
| 166 | 3.2.3 | Weisbach's encyclopaedia of applied mechanics |
| 173 | 3.2.4 | Rankine's *Manuals*, or the harmony between theory and practice |
| 177 | 3.2.5 | Föppl's *Vorlesungen über technische Mechanik* |
| 180 | 3.2.6 | The *Handbuch der Ingenieurwissenschaften* as an encyclopaedia of classical civil engineering theory |
| | | |
| 186 | **4** | **From masonry arch to elastic arch** |
| 189 | 4.1 | The geometrical thinking behind the theory of masonry arch bridges |
| 189 | 4.1.1 | The Ponte S. Trinità in Florence |
| 195 | 4.1.2 | Establishing the new thinking in bridge-building practice using the example of Nuremberg's Fleisch Bridge |
| 199 | 4.2 | From the wedge to the masonry arch – or: the addition theorem of wedge theory |
| 201 | 4.2.1 | Between mechanics and architecture: masonry arch theory at the Académie Royale d'Architecture de Paris (1687–1718) |
| 201 | 4.2.2 | La Hire and Bélidor |
| 203 | 4.2.3 | Epigones |
| 204 | 4.3 | From the analysis of masonry arch collapse mechanisms to voussoir rotation theory |
| 204 | 4.3.1 | Baldi |
| 206 | 4.3.2 | Fabri |
| 207 | 4.3.3 | La Hire |
| 208 | 4.3.4 | Couplet |
| 210 | 4.3.5 | Bridge-building – empiricism still reigns |
| 211 | 4.3.6 | Coulomb's voussoir rotation theory |
| 212 | 4.3.7 | Monasterio's *Nueva Teórica* |
| 213 | 4.4 | The line of thrust theory |
| 216 | 4.4.1 | Gerstner |
| 218 | 4.4.2 | The search for the true line of thrust |
| 219 | 4.5 | The breakthrough for elastic theory |
| 220 | 4.5.1 | The dualism of masonry arch and elastic arch theory under Navier |
| 221 | 4.5.2 | Two steps forwards, one back |
| 223 | 4.5.3 | From Poncelet to Winkler |
| 227 | 4.5.4 | A step back |

| | | |
|---|---|---|
| 227 | 4.5.5 | The masonry arch is nothing, the elastic arch is everything – the triumph of elastic arch theory over masonry arch theory |
| 232 | 4.6 | Ultimate load theory for masonry arches |
| 234 | 4.6.1 | Of cracks and the true line of thrust in the masonry arch |
| 235 | 4.6.2 | Masonry arch failures |
| 236 | 4.6.3 | The maximum load principles of the ultimate load theory for masonry arches |
| 236 | 4.6.4 | The safety of masonry arches |
| 238 | 4.6.5 | Analysis of a masonry arch railway bridge |
| 241 | 4.7 | The finite element method |
| 243 | 4.8 | On the epistemological status of masonry arch theories |
| 245 | 4.8.1 | Wedge theory |
| 245 | 4.8.2 | Collapse mechanism analysis and voussoir rotation theory |
| 246 | 4.8.3 | Line of thrust theory and elastic theory for masonry arches |
| 248 | 4.8.4 | Ultimate load theory for masonry arches as an object in the historical theory of structures |
| 248 | 4.8.5 | The finite element analysis of masonry arches |
| | | |
| 250 | **5** | **The beginnings of a theory of structures** |
| 252 | 5.1 | What is the theory of strength of materials? |
| 255 | 5.2 | On the state of development of structural design and strength of materials in the Renaissance |
| 260 | 5.3 | Galileo's *Dialogue* |
| 261 | 5.3.1 | First day |
| 264 | 5.3.2 | Second day |
| 270 | 5.4 | Developments in the strength of materials up to 1750 |
| 277 | 5.5 | Civil engineering at the close of the 18th century |
| 279 | 5.5.1 | Franz Joseph Ritter von Gerstner |
| 283 | 5.5.2 | Introduction to structural engineering |
| 289 | 5.5.3 | Four comments on the significance of Gerstner's *Einleitung in die statische Baukunst* for theory of structures |
| 290 | 5.6 | The formation of a theory of structures: Eytelwein and Navier |
| 291 | 5.6.1 | Navier |
| 294 | 5.6.2 | Eytelwein |
| 296 | 5.6.3 | The analysis of the continuous beam according to Eytelwein and Navier |
| | | |
| 306 | **6** | **The discipline-formation period of theory of structures** |
| 308 | 6.1 | Clapeyron's contribution to the formation of classical engineering sciences |
| 308 | 6.1.1 | *Les Polytechniciens*: the fascinating revolutionary élan in post-revolution France |
| 310 | 6.1.2 | Clapeyron and Lamé in St. Petersburg (1820–1831) |
| 313 | 6.1.3 | Clapeyron's formulation of the energy doctrine of classical engineering sciences |
| 314 | 6.1.4 | Bridge-building and the theorem of three moments |
| 317 | 6.2 | From graphical statics to graphical analysis |
| 318 | 6.2.1 | The founding of graphical statics by Culmann |

| | | |
|---|---|---|
| 320 | 6.2.2 | Rankine, Maxwell, Cremona and Bow |
| 322 | 6.2.3 | Differences between graphical statics and graphical analysis |
| 324 | 6.2.4 | The breakthrough for graphical analysis |
| 330 | 6.3 | The classical phase of theory of structures |
| 331 | 6.3.1 | Winkler's contribution |
| 340 | 6.3.2 | The beginnings of the force method |
| 350 | 6.3.3 | Loadbearing structure as kinematic machine |
| 358 | 6.4 | Theory of structures at the transition from the discipline-formation to the consolidation period |
| 358 | 6.4.1 | Castigliano |
| 362 | 6.4.2 | The foundation of classical theory of structures |
| 365 | 6.4.3 | The dispute about the fundamentals of classical theory of structures is resumed |
| 373 | 6.4.4 | The validity of Castigliano's theorems |
| 374 | 6.5 | Lord Rayleigh's *The Theory of Sound* and Kirpichev's foundation of classical theory of structures |
| 375 | 6.5.1 | Rayleigh coefficient and Ritz coefficient |
| 377 | 6.5.2 | Kirpichev's congenial adaptation |
| 379 | 6.6 | The Berlin school of structural theory |
| 380 | 6.6.1 | The notion of the scientific school |
| 381 | 6.6.2 | The completion of classical theory of structures by Heinrich Müller-Breslau |
| 383 | 6.6.3 | Classical theory of structures takes hold of engineering design |
| 387 | 6.6.4 | Müller-Breslau's students |
| 396 | **7** | **From construction with iron to modern structural steelwork** |
| 398 | 7.1 | Torsion theory in iron construction and theory of structures from 1850 to 1900 |
| 398 | 7.1.1 | Saint-Venant's torsion theory |
| 402 | 7.1.2 | The torsion problem in Weisbach's *Principles* |
| 405 | 7.1.3 | Bach's torsion tests |
| 408 | 7.1.4 | The adoption of torsion theory in classical theory of structures |
| 411 | 7.2 | Crane-building at the focus of mechanical and electrical engineering, structural steelwork and theory of structures |
| 412 | 7.2.1 | Rudolph Bredt – the familiar stranger |
| 412 | 7.2.2 | The Ludwig Stuckenholz company in Wetter a. d. Ruhr |
| 423 | 7.2.3 | Bredt's scientific-technical publications |
| 429 | 7.2.4 | The engineering industry adopts classical theory of structures |
| 433 | 7.3 | Torsion theory in the consolidation period of structural theory (1900–1950) |
| 433 | 7.3.1 | The introduction of an engineering science concept: the torsion constant |
| 435 | 7.3.2 | The discovery of the shear centre |
| 440 | 7.3.3 | Torsion theory in structural steelwork from 1925 to 1950 |
| 443 | 7.3.4 | Summary |
| 443 | 7.4 | Searching for the true buckling theory in steel construction |
| 443 | 7.4.1 | The buckling tests of the DStV |

| | | |
|---|---|---|
| 448 | 7.4.2 | German State Railways and the joint technical-scientific work in structural steelwork |
| 449 | 7.4.3 | Excursion: the Olympic Games for structural engineering |
| 451 | 7.4.4 | A paradigm change in buckling theory |
| 452 | 7.4.5 | The standardisation of the new buckling theory in the German stability standard DIN 4114 |
| 454 | 7.5 | Steelwork and steelwork science from 1950 to 1975 |
| 456 | 7.5.1 | From the truss to the plane frame: the orthotropic bridge deck |
| 463 | 7.5.2 | The rise of composite steel-concrete construction |
| 469 | 7.5.3 | Lightweight steel construction |
| 471 | 7.6 | Eccentric orbits – the disappearance of the centre |
| | | |
| 474 | **8** | **Member analysis conquers the third dimension: the spatial framework** |
| 475 | 8.1 | Development of the theory of spatial frameworks |
| 476 | 8.1.1 | The original dome to the Reichstag (German parliament building) |
| 478 | 8.1.2 | Foundation of the theory of spatial frameworks by August Föppl |
| 481 | 8.1.3 | Integration of spatial framework theory into classic structural theory |
| 485 | 8.2 | Spatial frameworks in an era of technical reproducibility |
| 486 | 8.2.1 | Alexander Graham Bell |
| 487 | 8.2.2 | Vladimir Grigorievich Shukhov |
| 487 | 8.2.3 | Walther Bauersfeld and Franz Dischinger |
| 489 | 8.2.4 | Richard Buckminster Fuller |
| 490 | 8.2.5 | Max Mengeringhausen |
| 491 | 8.3 | Dialectic synthesis of individual structural composition and large-scale production |
| 491 | 8.3.1 | The MERO system and the composition law for spatial frameworks |
| 494 | 8.3.2 | Spatial frameworks and computers |
| | | |
| 496 | **9** | **Reinforced concrete's influence on theory of structures** |
| 498 | 9.1 | The first design methods in reinforced concrete construction |
| 498 | 9.1.1 | The beginnings of reinforced concrete construction |
| 500 | 9.1.2 | From the German Monier patent to the *Monier-Broschüre* |
| 503 | 9.1.3 | The *Monier-Broschüre* |
| 511 | 9.2 | Reinforced concrete revolutionises the building industry |
| 512 | 9.2.1 | The fate of the Monier system |
| 514 | 9.2.2 | The end of the system period: steel reinforcement + concrete = reinforced concrete |
| 527 | 9.3 | Theory of structures and reinforced concrete |
| 528 | 9.3.1 | New types of loadbearing structures in reinforced concrete |
| 554 | 9.3.2 | Prestressed concrete: "Une révolution dans les techniques du béton" (Freyssinet) |
| 561 | 9.3.3 | The paradigm change in reinforced concrete design takes place in the Federal Republic of Germany too |
| 562 | 9.3.4 | Revealing the invisible: reinforced concrete design with truss models |

| | | |
|---|---|---|
| 570 | **10** | **From classical to modern theory of structures** |
| 571 | 10.1 | The relationship between text, image and symbol in theory of structures |
| 573 | 10.1.1 | The historical stages in the idea of formalisation |
| 580 | 10.1.2 | The structural engineer – a manipulator of symbols? |
| 581 | 10.2 | The development of the displacement method |
| 582 | 10.2.1 | The contribution of the mathematical elastic theory |
| 585 | 10.2.2 | From pin-jointed trussed framework to rigid-jointed frame |
| 589 | 10.2.3 | From trussed framework to rigid frame |
| 591 | 10.2.4 | The displacement method gains emancipation from trussed framework theory |
| 596 | 10.2.5 | The displacement method during the invention phase of structural theory |
| 597 | 10.3 | The groundwork for automation in structural calculations |
| 598 | 10.3.1 | Remarks on the practical use of symbols in structural analysis |
| 600 | 10.3.2 | Rationalisation of structural calculation in the consolidation period of structural theory |
| 606 | 10.3.3 | The dual nature of theory of structures |
| 608 | 10.3.4 | First steps in the automation of structural calculations |
| 610 | 10.3.5 | The diffusion of matrix formulation into the exact natural sciences and fundamental engineering science disciplines |
| 619 | 10.4 | "The computer shapes the theory" (Argyris): the historical roots of the finite element method and the development of computational mechanics |
| 622 | 10.4.1 | Truss models for elastic continua |
| 630 | 10.4.2 | Modularisation and discretisation of aircraft structures |
| 640 | 10.4.3 | The matrix algebra reformulation of structural mechanics |
| 648 | 10.4.4 | FEM – formation of a general technology engineering science theory |
| 654 | 10.4.5 | The founding of FEM through variational theorems |
| 671 | 10.4.6 | Computational mechanics – a broad field |
| 672 | 10.4.7 | A humorous plea |
| | | |
| 674 | **11** | **Twelve scientific controversies in mechanics and theory of structures** |
| 675 | 11.1 | The scientific controversy |
| 675 | 11.2 | Twelve disputes |
| 675 | 11.2.1 | Galileo's *Dialogo* |
| 676 | 11.2.2 | Galileo's *Discorsi* |
| 677 | 11.2.3 | The philosophical dispute about the true measure of force |
| 678 | 11.2.4 | The dispute about the principle of least action |
| 679 | 11.2.5 | The dome of St. Peter's in the dispute between theorists and practitioners |
| 681 | 11.2.6 | Discontinuum or continuum? |
| 682 | 11.2.7 | Graphical statics versus graphical analysis, or the defence of pure theory |
| 683 | 11.2.8 | Animosity creates two schools: Mohr versus Müller-Breslau |
| 684 | 11.2.9 | The war of positions |
| 685 | 11.2.10 | Until death do us part: Fillunger versus Terzaghi |
| 687 | 11.2.11 | "In principle, yes …": the dispute about principles |
| 689 | 11.2.12 | Elastic or plastic? That is the question. |
| 690 | 11.3 | Résumé |

| 692 | **12** | **Perspectives for theory of structures** |
| 694 | 12.1 | Theory of structures and aesthetics |
| 694 | 12.1.1 | The schism of architecture |
| 695 | 12.1.2 | Beauty and utility in architecture – a utopia? |
| 699 | 12.1.3 | Alfred Gotthold Meyer's *Eisenbauten. Ihre Geschichte und Ästhetik* |
| 702 | 12.1.4 | The aesthetics in the dialectic between building and calculation |
| 707 | 12.2 | A plea for the historico-genetic teaching of theory of structures |
| 708 | 12.2.1 | Historico-genetic methods for teaching of theory of structures |
| 709 | 12.2.2 | Content, aims, means and characteristics of the historico-genetic teaching of theory of structures |
| 709 | 12.2.3 | Outlook |
| 712 | | **Brief biographies** |
| 778 | | **Bibliography** |
| 831 | | **Name index** |
| 839 | | **Subject index** |

CHAPTER 1

# The tasks and aims of a historical study of theory of structures

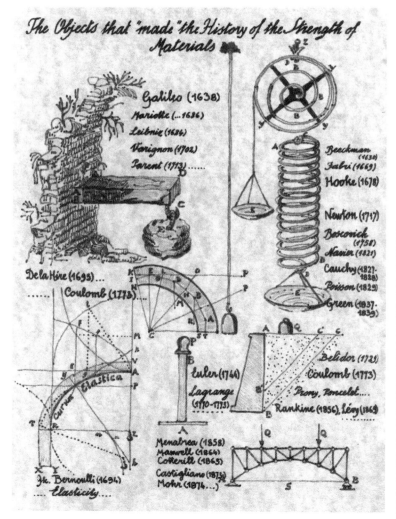

**FIGURE 1-1**
Drawing by Edoardo Benvenuto

Until the 1990s, the history of theory of structures attracted only marginal interest from historians. At conferences dealing with the history of science and technology, but also in relevant journals and compendiums, the interested reader could find only isolated papers investigating the origins, the chronology, the cultural involvement and the social significance of theory of structures. This gap in our awareness of the history of theory of structures has a passive character: most observers still assume that the stability of structures is guaranteed a priori, that, so to speak, structural analysis wisdom is naturally bonded to the structure, is absorbed by it, indeed disappears, never to be seen again. This is not a suppressive act on the part of the observer, but rather is due to the nature of building itself – theory of structures had appeared at the start of the Industrial Revolution, claiming to be a "mechanics derived from the nature of building itself" [Gerstner, 1789, p. 4].

Only in the event of failure are the formers of public opinion reminded of structural analysis. Therefore, the historical development of theory of structures followed in the historical footsteps of modern building, with the result that the historical contribution of theory of structures to the development of building was given more or less attention in the structural engineering-oriented history of building, and therefore was included in this.

The history of science, too, treated the history of theory of structures as a diversion. If indeed theory of structures as a whole strayed into the field of vision, it was only in the sense of one of the many applications of mechanics. Structural engineering, a profession that includes theory of structures as a fundamental engineering science discipline, only rarely finds listeners outside its own disciplinary borders.

Today, theory of structures is, on the one hand, more than ever before committed to formal operations with symbols, and is less apparent to many users of structural design programs. On the other hand, some attempts to introduce formal teaching into theory of structures fail because the knowledge about its historical development is not adequate to define the concrete object of theory of structures. Theory of structures is therefore a necessary but unpopular project.

Notwithstanding, a history of theory of structures has been gradually coming together from various directions since the early 1990s, the first highlight of which was the conference "Historical perspectives on structural analysis" – the world's first conference on the history of theory of structures – organised by Santiago Huerta and held in Madrid in December 2005. The book published on the occasion of the conference (Fig. 1-2) demonstrates that the history of theory of structures already possesses a number of the features important to an engineering science discipline and can be said to be experiencing its constitutional phase.

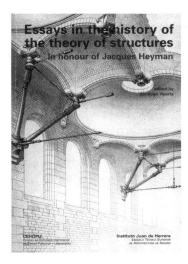

**FIGURE 1-2**
Cover of the book published to mark the first conference on the history of theory of structures (2005)

## 1.1 Internal scientific tasks

Like every scientific cognition process, the engineering science cognition process in theory of structures also embraces history insofar as the idealised reproduction of the scientific development supplanted by the status of knowledge of an object forms a necessary basis for new types of scientific ideas: science is truly historical. Reflecting on the genesis and development of the object of theory of structures always becomes an element in the engineering science cognition process when rival, or rather coexistent, theories are superseded by a more abstract theory – possibly by a basic theory of a fundamental engineering science discipline. Therefore, the question of the inner consistency of the more abstract theory, which is closely linked with this broadening of the object, is also a question of the historical evolution. This is how Saint-Venant proceeded in 1864 with his extensive historical and critical commentary of Navier's beam theory [Navier, 1864], in the middle of the establishment phase of structural theory (1850–75). Theory formation in structural analysis is the classification of the essential properties of technical artefacts or artefact classes reflected in theoretical models. This gives rise to the historically weighted comparison and the criticism of the theoretical approaches, the theoretical models and the theories, especially in those structural analysis theory formation processes that grew very sluggishly, e. g. masonry arch theory. One example of this is Winkler's 1879/80 historico-logical analysis of masonry arch theories in the classical phase of structural theory (1875–1900) [Winkler, 1879/1880].

In their monumental work on the history of strength of materials, Todhunter and Pearson had good reasons for focusing on elastic theory (see [Todhunter & Pearson, 1886 & 1893; Pearson, 1889]), which immediately became the foundation for materials theory in applied mechanics as well as theory of structures in its discipline-formation period (1825–1900), and was able to sustain its position as a fundamental theory in these two engineering science disciplines during the consolidation period (1900–50). The mathematical elastic theory first appeared in 1820 with Navier's *Mémoire sur la flexion des plans élastiques* (Fig. 1-3). It inspired Cauchy and others to contribute significantly to the establishment of the scientific structure of elastic theory and induced a paradigm change in the constitution phase of structural theory (1825–50), which was essentially completed by the middle of the establishment phase of structural theory (1850–75). One important outcome of the discipline-formation period of structural theory (1825–1900) was the constitution of the discipline's own conception of its epistemology – and elastic theory contributed substantially to this. Theory of structures thus created for itself the prerequisite to help define consciously the development of construction on the disciplinary scale. And looked at from the construction engineering side, Gustav Lang approached the subject in his evolutionary portrayal of the interaction between loadbearing construction and theory of structures in the 19th century [Lang, 1890] – the first monograph on the history of theory of structures.

Up until the consolidation period of structural theory (1900–50), the structural analysis theory formation processes anchored in the emerging specialist literature on construction theory contained a historical element that was more than mere references to works already in print. It appears,

**FIGURE 1-3**
Lithographic cover page of Navier's
*Mémoire sur la flexion des plans élastiques*
[Roberts & Trent, 1991, p. 234]

after all, to be a criterion of the discipline-formation period of structural theory that recording the relationship between the logical and historical was a necessary element in the emerging engineering science cognition process. If we understand the logical to be the theoretical knowledge reflecting the laws of the object concerned in abstract and systematic form, and the historical to be the knowledge and reproduction of the genesis and evolution of the object, then it can be shown that the knowledge of an object's chronology had to be a secondary component in the theoretical knowledge of the object. This is especially true when seen in terms of the leaps in development in the discipline-formation period of structural theory. Whereas Pierre Duhem pursued the thinking of natural philosophy from the theory of structures of the Middle Ages to the end of the 17th century in his two-volume work *Les origines de la Statique* [Duhem, 1905/06], the comprehensive contributions of Mehrtens [Mehrtens, 1900 & 1905], Hertwig [Hertwig, 1906 & 1941], Westergaard [Westergaard, 1930/1], Ramme [Ramme, 1939] and Hamilton [Hamilton, 1952] to the origins of the discipline of theory of structures provide reasons for the history of theory of structures in a narrower sense. The famous book by Timoshenko on the history of strength of materials (Fig. 1-4) contains sections on the history of structural theory [Timoshenko, 1953].

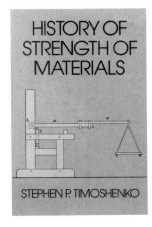

**FIGURE 1-4**
Cover of Timoshenko's *History of strength of materials* [Timoshenko, 1953]

In the former USSR, Rabinovich [1949, 1960, 1969] and Bernshtein [1957, 1961] contributed to the history of strength of materials and theory of structures in particular and structural mechanics in general. But of all those monographs, only one has appeared in English [Rabinovich, 1960], made available by George Herrmann in the wake of the Sputnik shock. In that book, Rabinovich describes the future task of a type of universal history of structural mechanics as follows: "[Up] to the present time [early 1957 – the author] no history of structural mechanics exists. Isolated excerpts and sketches which are the elements do not fill the place of one. There is [a] need for a history covering all divisions of the science with reasonable thoroughness and containing an analysis of ideas and methods, their mutual influences, economics, and the characteristics of different countries, their connection with the development of other sciences and, finally, their influence upon design and construction" [Rabinovich, 1960, p. 79]. Unfortunately, apart from this one exception, the Soviet contributions to the history of structural mechanics were not taken up in non-Communist countries – a fate also suffered by Rabinovich's monograph on the history of structural mechanics in the USSR from 1917 to 1967 (Fig. 1-5).

**FIGURE 1-5**
Dust cover of the monograph entitled *Structural Mechanics in the USSR 1917–67* [Rabinovich, 1969]

In his dissertation *The art of building and the science of mechanics*, Harold I. Dorn deals with the relationship between theory and practice in Great Britain during the preparatory period of structural theory (1575–1825) [Dorn, 1971]. T. M. Charlton concentrates on the discipline-formation period of structural theory in his book [Charlton, 1982]. He concludes the internal scientific view of the development of theory of structures as the history of structural theory enters its initial phase. And as early as 1972, Jacques Heyman's monograph *Coulomb's memoir on*

FIGURE 1-6
Dust cover of the Spanish edition of Heyman's *Structural analysis. A historical approach* [Heyman, 2004]

*statics: An essay in the history of civil engineering* [Heyman, 1972/1] was not only lending a new emphasis to the treatment and interpretation of historical sources, but was also showing how practical engineering can profit from historical knowledge. This was followed nine years later by Edoardo Benvenuto's universal work *La scienza delle costruzioni e il suo sviluppo storico* [Benvenuto, 1981], the English edition of which – in a much abridged form – did not appear until 10 years later [Benvenuto, 1991]. Heyman's later monographs [Heyman, 1982, 1995/1, 1998/1] in particular demonstrate that the history of theory of structures is able to advance the scientific development of structural analysis. Many of Heyman's books have been published in Spanish in the *Textos sobre teoría e historia de las construcciones* series founded and edited by Santiago Huerta (see, for example, Fig. 1-6).

In 1993 Benvenuto initiated the series of international conferences under the title of *Between Mechanics and Architecture* together with the Belgian science historian Patricia Radelet-de Grave. The conferences gradually became the programme for a school and after Benvenuto's early death were continued by the Edoardo Benvenuto Association headed by its honorary president Jacques Heyman. Only six results of this programme will be mentioned here:

– The first volume in this series edited by Benvenuto and Radelet-de Grave and entitled *Entre Mécanique et Architecture. Between Mechanics and Architecture* [Benvenuto & Radelet-de Grave, 1995].
– The compendium *Towards a History of Construction* edited by Becchi, Corradi, Foce and Pedemonte [Becchi et al., 2002].
– *Degli archi e delle volte* [Becchi & Foce, 2002], a bibliography of structural and geometrical analysis of masonry arches past and present with an expert commentary by Becchi and Foce.
– The volume of essays on the history of mechanics edited by Becchi, Corradi, Foce and Pedemonte (Fig. 1-7) [Becchi et al., 2003].
– The compendium on the status of the history of construction engineering edited by Becchi, Corradi, Foce and Pedemonte *Construction History. Research Perspectives in Europe* [Becchi et al., 2004/2].
– The reprint of Edoardo Benvenuto's principal work *La scienza delle costruzioni e il suo sviluppo storico* made available by Becchi, Corradi and Foce [Benvenuto, 2006].

Erhard Scholz has investigated the development of graphical statics in his habilitation thesis [Scholz, 1989] from the viewpoint of the mathematics historian. Dieter Herbert's dissertation [Herbert, 1991] analyses the origins of tensor calculus from the beginnings of elastic theory with Cauchy (1827) to its use in shell theory by Green and Zerna at the end of the consolidation period of structural theory (1900–50).

In the past two decades, we have seen a slowly accelerating upswing in working through the backlog in the history of modern structural mechanics by specialists. The development of modern numerical engineering methods was the subject of a conference held in Princeton by the Association for Computing Machinery (ACM) in May 1987 [Crane,

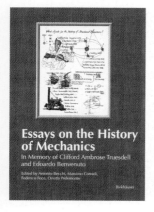

FIGURE 1-7
Cover of the volume of essays on the history of mechanics [Becchi et al., 2003]

1987]. Ekkehard Ramm provides a fine insight into the second half of the consolidation period (1900–50) and the subsequent integration period of structural theory (1950 to date) [Ramm, 2000]. As a professor at the Institute of Theory of Structures at the University of Stuttgart, Ramm supervised Bertram Maurer's dissertation *Karl Culmann und die graphische Statik* (Karl Culmann and graphical statics) [Maurer, 1998]. And Malinin's book *Kto jest' kto v soprotivlenii materialov* (who's who in strength of materials) [Malinin, 2000] continued the biographical tradition popular in the Soviet history of mechanics.

Publications by Samuelsson and Zienkiewicz [Samuelsson & Zienkiewicz, 2006] plus Kurrer [Kurrer, 2003] have appeared on the history of the displacement method. Carlos A. Felippa deals with the development of matrix methods in structural mechanics [Felippa, 2001] and the theory of the shear-flexible beam [Felippa, 2005]. On the other hand, the pioneers of the finite element method (FEM) Zienkiewicz [Zienkiewicz, 1995 & 2004] and Clough [Clough, 2004] concentrate on describing the history of FEM. It seems that a comprehensive presentation of the evolution of modern structural mechanics is necessary. Only then could the history of theory of structures make a contribution to a historical engineering science in general and a historical theory of structures in particular, both of which are still awaiting development.

## 1.2 Practical engineering tasks

Every structure moves in space and time. The question regarding the causes of this movement is the question regarding the history of the structure, its genesis, utilisation and nature. Whereas the first dimension of the historicity of structures consists of the planning and building process, the second dimension extends over the life of the structure and its interaction with the environment. The historicity of the knowledge about structures and their theories plus its influence on the history of the structure form the third dimension of the historicity of structures. In truth, the history of the genesis, usage and nature of the structure form a whole. Nevertheless, the historicity of structures is always broken down into its three dimensions. Whereas historicity in the first dimension is typically reduced to the timetable parameters of the participants in the case of new structures, understanding the second dimension is an object of history of building, preservation of monuments and construction research plus the evolving history of construction engineering and structural design. One vital task of the history of theory of structures would be to help develop the third dimension, e.g. through preparing, adapting and re-interpreting historical masonry arch theories. Its task in practical engineering is not limited to the province of the expanding volume of work among the historical building stock. The knowledge gleaned from the history of theory of structures could become a functional element in the modern construction process because the unity of the three-dimensionality in the historicity of structures is an intrinsic anticipation in this; for the engineering science theory formation and the research trials, the conception, the calculation and the design as well as the fabrication, erection and usage can no longer be

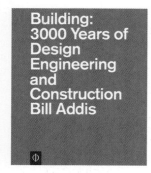

**FIGURE 1-8**
Cover of the new book by Bill Addis
*Building: 3000 Years of Design Engineering and Construction*
[Addis, 2007]

**Didactic tasks**

**FIGURE 1-9**
Cover to the collection of essays on columns *La colonne. Nouvelle histoire de la construction*
[Gargiani, 2008]

separated from the conversion, preservation and upkeep of the building stock. The task of the history of theory of structures lies not only in feeding the planning process with ideas from its historical knowledge database, but also in introducing its experiences from work on historical structures into the modern construction process. In this sense, the history of theory of structures could be further developed into a productive energy in engineering.

When engineers conceive a building, they have to be sure, even before the design process begins, that it will function exactly as envisaged and planned. That applies today and it also applied just the same to engineers in Roman times, in the Middle Ages, in the Renaissance and in the 19th century. All that has changed is the methods with which engineers achieve this peace of mind. Bill Addis has written a history of design engineering and construction which focuses on the development of design methods for buildings (Fig. 1-8).

Bill Addis looks into the development of graphical and numerical methods plus the use of models for analysing physical phenomena, but also shows which methods engineers employ to convey their designs. To illustrate this, he uses examples from structural engineering, building services, acoustics and lighting engineering drawn from 3000 years of construction engineering history. Consequently, the knowledge gleaned from the history of theory of structures serves as one of the cornerstones in his evolution of the design methods used by structural engineers.

Roberto Gargiani pursues an artefact-based approach in his collection of essays on columns [Gargiani, 2008] (Fig. 1-9), which are presented from the history of building, history of art, history of construction engineering, history of science and history of structural theory perspectives. The discipline-oriented straightforwardness of the history of theory of structures is especially evident here.

## 1.3

The work of the American Society for Engineering Education (ASEE), founded in 1893, brought professionalism to issues of engineers' education in the USA and led to the formation of engineering pedagogy as a subdiscipline of the pedagogic sciences. In the quarterly *Journal of Engineering Education*, the publication of the ASEE, scientists and practitioners have always reported on progress and discussions in the field of engineering teaching. For example, the journal reprinted the famous *Grinter Report* [Grinter, 1955; Harris et al., 1994, pp. 74–94], which can be classed as a classic of engineering pedagogy and which calls for the next generation of engineers to devote 20 % of their study time on social sciences and the humanities, e.g. history [Harris et al., 1994, p. 82]. Prior to L. E. Grinter, another prominent civil engineering professor who contributed to the debate about the education of engineers was G. F. Swain. In his book *The Young Man and Civil Engineering* (Fig. 1-10), Swain links the training of engineers with the history of civil engineering in the USA [Swain, 1922].

Nevertheless, students of the engineering sciences still experience the division of their courses of study into foundation studies, basic specialist studies and further studies as a separation between the basic subjects and the specific engineering science disciplines, and the latter are often presented only in the form of the applications of subjects such as mathematics and mechanics. Even the applied mechanics obligatory for many engineering science disciplines at the fundamental stage are understood by many students as general collections of unshakeable principles – illustrated by working through idealised technical artefacts. Closely related to this is the partition of the engineering sciences in in-depth studies; they are not studied as a scientific system comprised of specific internal relationships, for example, but rather as an amorphous assemblage of unconnected explicit disciplines whose object is only a narrow range of technical artefacts. The integrative character of the engineering sciences thus appears in the form of the additive assembly of the most diverse individual scientific facts, with the result that the fundamental engineering science disciplines are learned by the students essentially in the nature of formulas. The task of a history of theory of structures is to help eliminate the students' formula-like acquisition of theory of structures. In doing so, the separation of the teaching of theory of structures into structural analysis for civil and structural engineers and structural engineering studies for architects presents a challenge. Proposals for a historicised didactic approach to structural engineering studies have been made by Rolf Gerhardt [Gerhardt, 1989]. Introducing the historical context into the teaching material of theory of structures in the project studies in the form of a historic-genetic teaching of structural theory could help the methods of structural engineering to be understood, experienced and illustrated as a historico-logical development product, and hence made more popular. The history of theory of structures would thus expand significantly the knowledge database for a future historic-genetic method of teaching for all those involved in the building industry.

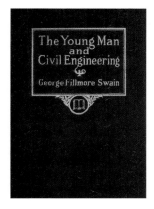

**FIGURE 1-10**
Cover of Swain's *The Young Man and Civil Engineering* [Swain, 1922]

### 1.4 Cultural tasks

There is an elementary form of the scientist's social responsibility: the democratising of scientific knowledge through popularising; that is the scientist's account of his work – and without it society as a whole would be impossible. Popular science presentations are not just there to provide readers outside the disciplinary boundaries with the results of scientific knowledge reflected in the social context of scientific work, but rather to stimulate the social discussion about the means and the aims of the sciences. Consequently, the history of theory of structures, too, possesses an inherent cultural value. The author Christine Lehmann, together with her partner the mathematics teacher Bertram Maurer, has written a biography of Karl Culmann (Fig. 1-11) based on Maurer's dissertation [Maurer, 1998] in which the results of research into the history of theory of structures are presented to the layman in an understandable, narrative fashion within an appealing literary framework.

The individual sciences physics, biology and even chemistry transcend again and again the boundaries of their scientific communities. This may

**FIGURE 1-11**
Cover of the biography of Karl Culmann [Lehmann & Maurer, 2006]

be due to their role as constituents of worldly conceptions and the close bond with philosophy and history. But the same does not apply to the engineering sciences; even fundamental engineering science disciplines find it difficult to explain their disciplinary intent in the social context. The fragmentation of the engineering sciences complicates the recognition of their objective coherence, their position and function within the ensemble of the scientific system and hence their relationship as a whole to the society that gave birth to them and which surrounds them. This is certainly the reason why the presentations, papers and newspaper articles of the emeritus professor of structural analysis Heinz Duddeck plead for a paradigm change in the engineering sciences, which in essence would result in a fusion between the engineering sciences and the humanities [Duddeck, 1996]. As the history of theory of structures forms a disciplinary union between structural analysis and applied mechanics with input from the humanities (philosophy, general history, sociology, histories of science, technology, industry and engineering), it is an element of that fusion. It can therefore also assist in overcoming the "speechlessness of the engineer" [Duddeck, 1999].

## Aims 1.5

The aim of a history of theory of structures therefore consists of solving the aforementioned scientific, practical engineering, didactic and cultural tasks. This book, written from the didactic, scientific theory, construction history, aesthetic, biographical and bibliographical perspectives (Fig. 1-12), aims to provide assistance.

## An invitation to a journey through the history of theory of structures 1.6

In Franz Kafka's parable of the gatekeeper from the chapter entitled "In the Cathedral" in his novel *The Trial* published in 1925, Josef K. searches in vain for a way to enter the law via a gate guarded by a gatekeeper. Kafka's protagonist Josef K. could be studying civil engineering or architecture, history of science or history of technology – for him the motives for acquiring the fundamentals of theory of structures were duly spoiled: he would sit in front of the gate or exit the stage like an actor in a theatre.

Dear Mr. Josef K.! There are various gates through which the laws of structural analysis can be learned with joy (Fig. 1-12). You can consider, dear Mr. Josef K., which phantasmagorical gatekeeper you can evade most easily – but let me tell you this: the gatekeepers don't exist! Please get up, open any gate and pass through it, and you will see the form in which theory of structures appears to you. If you are inquisitive and wish to open all seven gates, then you will be in possession of a picture of the history of structural analysis – your picture. But never guard your picture jealously as if it were your property because then at the final curtain the same will happen to you as happened to your Kafkaesque namesake: you'll be put on trial without knowing who is prosecuting and why – perhaps you'll even prosecute yourself! You would be sentenced to life imprisonment, sitting and waiting, hoping to be allowed in. The shadow cast by your property would seem like the cool draught of your approaching death. So choose

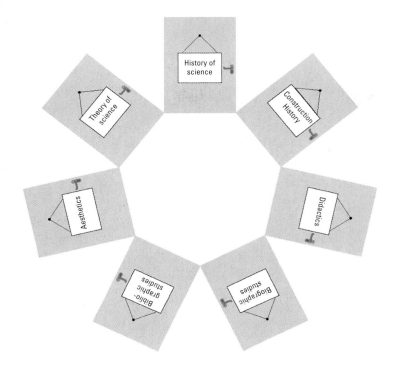

**FIGURE 1-12**
Seven gates to the knowledge of the history of theory of structures

instead – like Friedrich Hölderlin recommended to his friend Christian Landauer in 1801 [*Der Gang aufs Land*, Hölderlin, 1992, p. 102] – the path to freedom:

… *Und gefunden das Wort, und aufgegangen das Herz ist,*
    *Und von trunkener Stirn höher Besinnen entspringt,*
*Mit der unsern zugleich des Himmels Blüte beginnen,*
    *Und dem offenen Blick offen der Leuchtende sein.*

… That when found is the word, and joy releases our heartstrings,
    And from drunken excitement higher reflection is born:
At such time of our blossom will heaven's flowering begin too,
    And, to opened eyes here, open that radiance be.

    (*The Path by Land*, translation: Michael Loughridge)

If you take this path to freedom, then the gloomy shadows will disappear not only in springtime, but in autumn, too.
So: Open the Black Box
    Of the history of theory of structures,
    Craving for the knowledge.
    But I bid of you just one thing:
    Do not be afraid of formulas!

With this in mind, I would like to invite you, dear reader, to join me in a journey through the history of theory of structures. Experience the moment, make it your own and give it as a gift.

# CHAPTER 2

# Learning from the history of structural analysis: 11 introductory essays

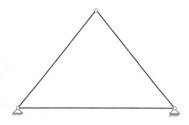

While a tutor between 1977 and 1981 in the theory of structures department at Berlin TU led by Prof. Gebhard Hees, one of the author's most important teaching and learning experiences was grasping the basic principles of structural analysis from the historical point of view. This journey into the past of theory of structures was the start of a long search that only later began to take proper shape. The intention of the handwritten introductory lectures on the history of each structural analysis method was to help the students understand that theory of structures, too, is the outcome of a socio-historical everyday process in which they themselves play a part and, in the end, help to mould. The goal was to create a deeper sense of the motivation for and enjoyment of the learning of structural analysis. The formula-type acquisition of the subject matter had to be overcome: a didactic approach to the fundamentals of structural theory through their historical appreciation. Since then, two more introductory lectures have been added to take account of the current level of knowledge. Hopefully, they will provide the reader with an easy introduction to the history of structural analysis.

It can certainly be claimed "that the history of science is science itself. One cannot properly appreciate what one has until one knows what others have possessed before us" [Goethe, 1808]. This quotation from the preface of Johann Wolfgang von Goethe's (1749–1832) *Theory of Colours* also applies to those engineering sciences that were first seeing the light of day as Goethe finally closed his eyes on the world. As structural analysis is a fundamental engineering science discipline, it follows that the history of structural analysis is structural analysis itself. On a pedagogical level, "learning from the history of structural analysis" means discovering the logic of structural analysis from its history, i.e. comprehending the principles, theorems, methods and terminology of structural analysis as an educational process in the literal sense.

The aim of this chapter is to introduce the reader to the historical elementary forms of structural analysis. Through this tactic of the practical discovery of examples of education processes in structural analysis, which relates to the theory of the foundation course in theory of structures common at universities these days, does it become possible to comprehend the evolution of the structural analysis disciplines in the connotations of the history of science; only then is that which structural engineers possess today truly explicable.

## 2.1 What is structural analysis?

In order to answer this question in the historical dimension, we must first divide the history of this subject into periods and break those down further into phases. The second step is to present selected and commented quotations from each development period. The quotations and comments are intended to illustrate not only the features of the individual development phases, but also to show specifically the historical progression of the nature of structural analysis as a whole.

### 2.1.1 Preparatory period (1575–1825)

This period stretching over some 250 years is characterised by the direct application of the mathematics and mechanics of the dawn of the modern world to simple loadbearing elements in structures. In terms of empirical knowledge and theory, it is the empirical knowledge that prevails in the design of buildings and structures; theory is evident primarily in the form of geometrical design and dimensioning rules. Not until the transition to the discipline-formation period of structural analysis, the initial phase (1775–1825), is the structural analysis of buildings and structures regarded as an independent branch of knowledge.

#### 2.1.1.1 Orientation phase (1575–1700)

Generally, this phase is characterised by the sciences (mathematics and mechanics) of this new age "discovering" the building industry. The theoretical basis for the design of structures is still dominated by geometry. Nevertheless, in the middle of the orientation phase, Galileo's *Dialogue* (1638) added elements of strength of materials to the menu in the form of the first beam theory, which Nicolas François Blondel (1618–86), Marchetti, Fabri and Grandi were able to make use of directly. Robert Hooke (1635–1703) took the next step in 1660 with the discovery of the law of elasticity – confirmed by Christiaan Huygens (1629 – 95) –, which later became known as Hooke's law. Although Simon Stevin (1548 – 1620) had already achieved progress in statics with his *Beghinselen der Weeghconst* (The Elements of the Art of Weighing) published in 1586, no evidence of the use of the parallelogram of forces in construction engineering before the late 17th century has yet come to light. The supremacy of geometry and the independent development of statics, strength of materials and elastic theory permitted the analysis of loadbearing elements only in isolated cases.

*Sagredo:* "While Simplicio and I were awaiting your arrival we were trying to recall that last consideration which you advanced as a principle and basis for the results you intended to obtain; this consideration dealt with the resistance which all solids offer to fracture and depended upon a certain cement which held the parts glued together so that they would yield and separate only under considerable pull. Later we tried to find the explanation of this coherence [= cohesion – the author] …; this was the occasion of our many digressions which occupied the entire day and led us far afield from the original question which … was the consideration of the resistance that solids offer to fracture."

*Salviati:* "… Resuming the thread of our discourse, whatever the nature of this resistance which solids offer to large tractive forces there can at least be no doubt of its existence; and though this resistance is very great in the case of a direct pull [= tensile strength – the author], it is found, as a rule, to be less in the case of bending forces [= bending strength – the author] … It is this second type of resistance which we must consider, seeking to discover in what proportion it is found in prisms and cylinders of the same material, whether alike or unlike in shape, length, and thickness. In this discussion I shall take for granted the well-known mechanical principle which has been shown to govern the behaviour of a bar, which we call a lever …" [Galileo, 1638/1964, pp. 93 – 94].

*Commentary:* Galileo organised his *Dialogue* as a discussion between his friend Francesco Sagredo (1571 – 1620), a senator of the Republic of Venice, Filipo Salviati (1582 – 1614), a wealthy Florentine and in real life a pupil of Galileo, and the dull Simplicio, a fictitious character introduced to represent the outdated Aristotelian doctrine. On the second day of the discussion, Galileo develops the principles of a new science – strength of materials; contributions to the analysis of loadbearing elements in the preparatory period concentrated on the structure of beam theory as the nucleus of strength of materials. Navier, with his practical bending theory, was the first to break away radically from the Galileo tradition.

**Application phase (1700 –1775)**

**2.1.1.2**

Differential and integral calculus appeared for the first time around 1700 and during the 18th century forced its way into applications in astronomy, theoretical mechanics, geodesy and construction engineering. Mathematicians and natural science researchers such as Gottfried Wilhelm von Leibniz (1646 –1716), the Bernoullis and Leonhard Euler (1707 – 83) brought progress to beam theory and the theory of the elastic line. In France, the first engineering schools developed in the first half of the 18th century from the corps of military engineers. These schools had a scientific self-conception that was based on the use of differential and integral calculus for the world of technical artefacts, a view that would not change significantly before the start of the consolidation period of structural theory around 1900. For example, Bernard Forest de Bélidor's (1697–1761) book *La Science des Ingénieurs*, much of which is based on differential and integral calculus, was already available by 1729. Bélidor dealt with earth pressure, arches and beams in great detail. Algebra and analysis seized the world of artefacts of builders and engineers in the form of applications. Contrasting with this, geometric methods still prevailed in the design of buildings and structures, although they could be increasingly interpreted in terms of statics.

"Although the advantages brought about by mathematical methods were great and important for science, the benefits the mathematical truths revealed to those artists is equally great …; we need only now mention civil and military engineering as such arts which are closer to our intentions and demonstrate modestly what an honour mathematics, so frequently mentioned, has already brought to this splendid art, and in future, we hope, will continue to bring" [Bélidor, 1729/1757, translator's foreword].

**Commentary:** The German translator of Bélidor's *La Science des Ingénieurs* understands mathematics as a direct application to construction engineering problems. Mathematics gets its justification from the benefits that it can endow on the useful arts of the budding civil engineer. Mathematics itself therefore appears as a useful art.

### 2.1.1.3 Initial phase (1775–1825)

Charles Augustin de Coulomb's (1736–1806) paper *Essai sur une application des maximis règles et de minimis à quelques problèmes de statique relatifs à l'architecture* presented to the Academy of Sciences in Paris in 1773 and published in 1776 was the first publication to apply differential and integral calculus to beam, arch and earth pressure theories in a coherent form. Coulomb's paper is not only a concentrated expression of the application phase, but also makes theory of structures its scientific object. Engineers such as Franz Joseph Ritter von Gerstner (1756–1832) and Johann Albert Eytelwein (1764–1849) also emphasized its independence as "structural engineering" [Gerstner, 1789] and the "statics of solid bodies" [Eytelwein, 1808]. Nevertheless, this branch of knowledge still did not have a coherent and theoretical foundation (fundamental theory).

"Among those parts of applied mathematics that are indispensable scientific aids for the builder, it is the statics of solid bodies that takes precedence. … It was not possible to express all those theories of statics required in architecture without higher analysis …" [Eytelwein, 1808, pp. III–IV].

**Commentary:** Statics of solid bodies is seen as an independent branch of knowledge of builders and architects. In contrast to the application phase, the statics of solid bodies is only indirectly applied mathematics. Differential and integral calculus advanced to become an integral component of higher education in engineering that started to develop after 1800.

**Discipline-formation period (1825–1900)**

**2.1.2**

The individual fragments of knowledge that accumulated during the preparatory period were brought together in the discipline-formation period of structural theory through the elastic theory that evolved in the first half of the 19th century in France. Of the three great strides forward in the discipline-formation period, Louis Henri Navier (1785–1836) took the greatest – in the shape of formulating a programme of structural analysis and its partial realisation through the practical bending theory in his *Résumé des Leçons* (1826). The second stride was that of Karl Culmann (1821–81), with the expansion of his trussed framework theory (1851) into graphical statics (1864/66) as an attempt to give structural analysis mathematical legitimacy through projective geometry. The third stride was the consequential assimilation of elements of elastic theory into the construction of a linear elastic theory of trusses by James Clerk Maxwell (1831–79), Emil Winkler (1835–88), Otto Mohr (1835–1918), Alberto Castigliano (1847–84), Heinrich Müller-Breslau (1851–1925) and Viktor Lvovich Kirpichev (1845–1913). And with his force method – a general method for calculating statically indeterminate trusses – Müller-Breslau rounded off the discipline-formation period of structural analysis.

**Constitution phase (1825–1850)**

**2.1.2.1**

Navier's practical bending theory [Navier, 1826] formed the very nucleus of structural analysis in the constitution phase and reflected the self-conception of this fundamental engineering science discipline. Navier used the practical bending theory to analyse numerous timber and iron constructions by setting up a structural model and integrating the linearised differential equation of the curvature of the deflection curve taking into account the boundary and transitional conditions of the curved member. Navier's practical bending theory thus became a reference point in the theory of structures. In Germany it was Moritz Rühlmann (1811–96) who adopted Navier's work comprehensively and in 1851 commissioned the first German translation entitled *Mechanik der Baukunst* (mechanics of architecture) [Navier, 1833/1878].

"The majority of design engineers determine the dimensions of parts of structures or machines according to the prevailing customs and the designs of works already completed; only rarely do they consider the compression that those parts have to withstand and the resistance with which those parts oppose said compression. This may have only a few disadvantages as long as the works to be built are similar to those others built at other times and they remain within conventional limits in terms of their dimensions and loads. But one should not use the same method if the circumstances require one to exceed those limits or if it is a whole new type of structure of which there is as yet no experience" [Navier, 1833/1878, pp. IX–X].

"When they produce designs for works for which they are responsible, engineers customarily follow a path that in mathematics is called the method of 'regula falsi'; i.e. after the works have been designed and drawn, they investigate them to see if all conditions have been satisfied and improve their design until this is the case. Economy is one of the most important conditions here; stability and durability are no less important. With the help of the rules that are developed in this book, it will be possible to establish all the limits that one may not exceed without exposing the structure to a lack of stability. However, one should not assume that one must always approach these limits in order to satisfy economy. The differences that prevail among materials have a role to play, and other reasons, too; the skill is to assess how close one may approach those limits" [Navier, 1826/1878, pp. XIII–XIV].

*Commentary:* In his book, Navier discusses the strength tests that various scientists and engineers had carried out on customary building materials during the 18th century. However, he also goes much further and synthesises the empirical data obtained – including his own – with the beam theory and the theory of the elastic line to create his practical bending theory. Civil and structural engineers now no longer have to rely solely on the handing-down of construction engineering knowledge for this branch of technical artefacts. They can create structural models of technical artefacts in an iterative design process based on engineering science theory with the help of quill, paper, calculating aids and tables of building materials. And more besides: they can anticipate technical artefacts ideally and optimise them in the model in order to construct economic loadbearing structures that fulfil their loadbearing functions.

### 2.1.2.2 Establishment phase (1850–1875)

As the building of iron bridges became customary after 1850, so structural analysis became established in continental Europe in the form of trussed framework theory and, later, graphical statics. The railway boom was the driving force behind the building of iron bridges, which resulted in an incessant demand for wrought iron with its good tensile strength as produced in the puddling furnace. And until the introduction of the Bessemer method after 1870, engineers tried to relieve the pressure on production volumes by using the material sparingly. Therefore, in the establishment phase, iron bridge-building and theory of structures go hand in hand.

"The purpose of all stability investigations, all determinations of the forces acting on the individual constructions, is to execute the intended construction with a minimum of material. It is certainly not difficult to establish all the dimensions for every bridge system such that they are certainly adequate, and it is not difficult to imagine a leap from the limits of the necessary into the superfluous. The English engineer, for example, does this with nearly every iron bridge he designs; characteristic of the English structures in particular is that they appear fattened and even the uninitiated gets the feeling: 'It will hold.' … What is befitting for the wealthy Englishman, who goes everywhere fully conscious of the idea 'I am in possession of the iron and do not need to worry myself about the statics', is less fitting for the poor devils on the continent; they have to fiddle and experiment, stake out and estimate many solutions for every railway to be planned in order to discover the cheapest, and draw various force diagrams for every bridge to be built in order that no material is wasted and only that which is essential is used … From the viewpoint of a national economy, it is the American who treads the right path: he uses no more than is absolutely necessary, and preferably a little less; the structure will probably just hold. Insurance companies of various kinds overcome the feeling of uncertainty that can occur in some instances. Like in everything, the middle way is the best …" [Culmann, 1866, pp. 527 – 528].

*Commentary:* Like Navier before him, Culmann sees the purpose of structural analysis in the economic use of materials for buildings and structures, a matter that is still part and parcel of structural analysis today. His remarks on British and American engineers were based on his travels to Great Britain and North America which he undertook on behalf of Bavarian State Railways in 1849 – 50 and which formed the themes of his famous reports published in 1851 and 1852. It was in these reports that Culmann developed his theory of statically determinate frameworks. For the first time, various force diagrams could be drawn for every bridge to be built; trussed framework theory and graphical statics became the incarnation of iron bridge-building.

### 2.1.2.3 Classical phase (1875 – 1900)

Culmann's graphical statics experienced unforeseen popularity in the classical phase. However, this method was less suitable for analysing statically indeterminate systems in everyday engineering. It was against this backdrop that Müller-Breslau's *Die neueren Methoden der Festigkeitslehre und der Statik der Baukonstruktionen* (the newer methods of strength of materials and the statics of building constructions) [Müller-Breslau, 1886] evolved, which were based on the principle of virtual forces and – in the form of a practical elastic theory – fused statics and strength of materials into a general theory of linear elastic trusses, i. e. into a classical theory of structures. In 1903 Kirpichev achieved a coherent and compact presentation of the theory of statically indeterminate systems [Kirpichev, 1903].

"This book discusses in context the methods of strength of materials founded principally by Mohr, Castigliano and Fränkel that are based on the laws of virtual displacements [= principle of virtual forces – the author]. The exercises selected for explaining the general relationships between the internal and external forces are for the most part drawn from the statics of building structures and those in turn from the theory of statically indeterminate beams; they relate to both more difficult and also to those simpler cases that can be dealt with equally briefly – and perhaps even more briefly – in another way. However, they will be included here because obtaining known results in a new way may be especially suitable for quickly acquainting the reader with the doubtful methods. The prime task of all exercises is to explain the given laws in the most informative way but not to hone the theory of a limited number of cases in detail. Therefore, the majority of exercises concerning statically indeterminate beams are carried out only as far as the static indeterminacy is eliminated because it is precisely the uniform calculation of the internal and external forces linked to elasticity equations plus a clear presentation of the deformations that form the area in which the discussion can be applied successfully" [Müller-Breslau, 1886, p. III].

*Commentary:* In the preface to his book *Die neueren Methoden der Festigkeitslehre und der Statik der Baukonstruktionen,* Müller-Breslau formulates the need for a methodical foundation to classical theory of structures: the focal point is not the solution of specific tasks concerning statically indeterminate systems, but rather the method derived from the principle of virtual forces. Müller-Breslau therefore places the idea of the operative use of symbols on the level of an individual science. Whereas Navier's practical bending theory advanced to become the model of structural analysis theory formation at the start of the discipline-formation period, the entire classical theory of structures after 1900 would form *the* model for other fundamental engineering science disciplines.

### 2.1.3 Consolidation period (1900–1950)

Structural analysis experienced a significant expansion of its scientific objective on a secure basis in the consolidation period. As early as 1915, the growth in reinforced concrete construction led to the development of a framework theory and 15 years later to a theory of shell structures. The displacement method quickly became a partner to the force method, but without disputing the leading position of the latter. On the other hand, during the 1930s structural analysis lost the innovative branch of aircraft construction, which in just a few years gave rise to the independent engineering science discipline of aviation engineering. In terms of everyday calculations, both the force method and the displacement method quickly reached their limits during the skyscraper boom of the 1920s. Relief initially came in the form of iterative methods such as those of Hardy Cross (1930), with which the internal forces of systems with a high degree of static indeterminacy could be quickly handled in a very simple way. Rationalisation of structural calculations thus became a scientific object in theory of structures.

#### 2.1.3.1 Accumulation phase (1900–1925)

Structural analysis spread to other technical fields during the accumulation phase: reinforced concrete construction, mechanical and plant engineering, crane-building and, finally, aircraft construction. Structural analysis therefore realised the outside world's universal demand for a theory of linear elastic trusses. But within the branch, it achieved its universal applicability by publishing linear algebra (the foundation of the force method) in the form of determinant theory. Alongside this there was the displacement method which had developed from the theory of secondary stresses in trusses. So by the end of the accumulation phase the contours of the dual nature of structural analysis had already been drawn. Another feature of this phase was the formulation of numerous special structural analysis methods for the quantitative control of systems with multiple degrees of static indeterminacy. At the end of the accumulation phase, the coherent and consistent arrangement of structural analysis arose out of the principle of virtual displacements. And that completed the rise in the status of theory of structures through applied mathematics and mechanics.

"I see the primary aim of the study of structural design as the scientific recognition and mastery of the theory that enables an independent treatment of the individual case, even an unusual one ... I use solely the principle of virtual displacements – with deliberate restrictions – for founding and developing the theory" [Grüning, 1925, p. III].

"The structural design of the loadbearing structure deals with ... two related tasks:
1. Calcultion of the magnitude of the resistance of each member considered as a force [internal forces and support reactions – the author] that withstands the application of the external actions in the equilibrium position: 'equilibrium task'.
2. Calculation of the magnitude of the displacements of the nodes from the unstressed initial position to the equilibrium position: 'deformation task'.

Both tasks are interlinked in such a way that when one is fully solved, the other can be regarded as solved" [Grüning, 1925; p. 7].

*Commentary:* Martin Grüning's (1869–1932) deductive structure of the entire theory of structures based on the principle of virtual displacements (which he considered subsidiary to the principle of virtual forces) led to knowledge of the internal relationship between the equilibrium and deformation tasks. This becomes visible in the tendency to equate force and displacement variables in practical structural calculations: whereas in the past only the force variables were interesting in structural systems, the structural engineer now had to pay more attention to calculating displacement variables as a result of ever more slender assemblies.

### 2.1.3.2 Invention phase (1925–1950)

The invention phase in structural analysis was characterised by several new developments: theory of shell structures, development of the displacement method to become the second main technique of structural analysis alongside the force method, recognition of non-linear phenomena (second-order theory, plasticity), formulation of numerical methods. The lack of a theory for the practical solution of systems with a high degree of static indeterminacy focused attention on the study of structural calculations. Resolving structural calculations into elementary arithmetical operations was the goal here; setting up algorithms was its internal feature, Taylorising the calculation work of the engineer the external one.

"The purpose of this paper is to explain briefly a method which has been found useful in analyzing frames which are statically indeterminate. The essential idea which the writer wishes to present involves no mathematical relations except the simplest arithmetic" [Cross, 1932/1, p. 1].

"A method of analysis has value if it is ultimately useful to the designer; not otherwise. There are apparently three schools of thought as to the value of analyses of continuous frames. Some say, 'Since these problems cannot be solved with exactness because of physical uncertainties, why try to solve them at all?' Others say, 'The values of the moments and shears cannot be found exactly; do not try to find them exactly; use a method of analysis which will combine reasonable precision with speed.' Still others say, 'It is best to be absolutely exact in the analysis and to introduce all elements of judgment after making the analysis.'

"The writer belongs to the second school; he respects but finds difficulty in understanding the viewpoint of the other two. Those who agree with his viewpoint will find the method herein explained a useful guide to judgment design.

"Members of the last named school of thought should note that the method here presented is absolutely exact if absolute exactness is desired. It is a method of successive approximations; not an approximate method" [Cross, 1932/1, p. 10].

*Commentary:* It was in the May 1930 edition of the *Proceedings of the American Society of Civil Engineers* that Hardy Cross published his 10-page paper on the iterative method of calculating statically indeterminate systems which was later to bear his name. Two years later, the paper was republished in the *Transactions of the American Society of Civil Engineers*, but this time accompanied by a 146-page discussion in which 38 respected engineers took part. Never has such a paper in the field of theory of structures triggered such a broad discussion. In his paper, Cross proposed abolishing exact structural solutions and replacing them with a step-by-step approximation of the reality. He preferred structural analysis methods that combined acceptable accuracy with quick calculations. The infinite progress (in the meaning of the limit state concept) superseded by the symbols of differential and integral calculus is replaced by the finite progress of the real work of the computer. It was only a question of time before this work would be mechanised. Just a few years later, Konrad Zuse would be using such a machine: the "engineer's computing machine" [Zuse, 1936]. Cross represents the Henry Ford-type manner of production in structural calculations at the transition to the integration period of structural analysis. No wonder countless publications on his method appeared until well into the 1960s.

### 2.1.4 Integration period (1950 to date)

Aviation engineering, too, soon reached its limits using the methods adapted from theory of structures supplemented by theories of lightweight construction. The calculation of systems with a high degree of static indeterminacy, i.e. the pressure to rationalise structural calculations, was joined by a further problem in aircraft construction: aeroplane structures consist of bars, plates and shells of low weight which as a whole are subjected to dynamic actions and therefore experience large deformations. What could have been more obvious than to divide the whole into elements, consider these separately in the mechanical sense and then put them back together again taking into account the jointing conditions? Which is exactly what the creators of the finite element method – Turner, Clough, Martin and Topp – did in 1956. What could have been more obvious than to use the formal elegance of the force and displacement methods in order to reformulate the entire theory of structures from the perspective of matrix analysis? Which is what Argyris did in 1956. The perspective was one of transferring the entire discipline to the computer in the form of a suite of programs! And that is where the traditional fundamental engineering science disciplines transcend their boundaries. In the first half of the integration period, the computer gave birth to our modern structural mechanics, of which theory of structures today appears to be a subdiscipline. Computational mechanics, which includes the objectives of structural mechanics, became established after the integration period. The content of the initial phase of the preparatory period of structural analysis some 250 years ago, i.e. the conscious application of *infinitesimal* calculus for investigating loadbearing structures, had been repeated at a higher level through the conscious application of a *finitesimal* calculation. Does computational mechanics, or rather structural mechanics, and the theory of structures it replaced represent the start of a different kind of discipline-formation period? Will the aforementioned disciplines be resolved into a universal technical physics? Will they become an integral part of engineering informatics (computational engineering) heavily based on mathematics [Pahl & Damrath, 2001]?

## 2.1.4.1 Innovation phase (1950–1975)

The innovation phase is characterised on a theoretical level by the emergence of modern structural mechanics, and on a practical level by the automation of structural calculations. Compared to other numerical engineering methods, the finite element method (FEM) gained more and more ground. Clough gave this method its name in 1960 und Zienkiewicz outlined it for the first time in a monograph published in 1967 [Zienkiewicz, 1967].

"A method is developed for calculating stiffness influence coefficients of complex shell-type structures. The object is to provide a method that will yield structural data of sufficient accuracy to be adequate for subsequent dynamic and aeroelastic analyses. Stiffness of the complete structure is obtained by summing stiffnesses of individual units. Stiffness of typical structural components are derived in the paper. Basic conditions of continuity and equilibrium are established at selected points (nodes) in the structure. Increasing the number of nodes increases the accuracy of results. Any physically possible support conditions can be taken into account. Details in setting up the analysis can be performed by nonengineering trained personnel; calculations are conveniently carried out on automatic digital computing equipment. … It is to be expected that modern developments in high-speed digital computing machines will make possible a more fundamental approach to the problems of structural analysis; we shall expect to base our analysis on a more realistic and detailed conceptual model of the real structure than has been used in the past. As indicated by the title, the present paper is exclusively concerned with methods of theoretical analysis; also it is our object to outline the development of a method that is well adapted to the use of high-speed digital computing machinery" [Turner et al., 1956].

*Commentary:* The authors, active in aircraft and construction engineering, proposed a fundamental concept for structural mechanics analysis which a few years later was named the "method of finite elements". It soon became clear that, in principle, this method could solve all the problems met with in practice. The *epistēmē* (epistemic) and the *tekhnē* (art, craft) were fused together in the method of finite elements: from now on, developments in theory would be directly related to computer-assisted practical calculations.

### 2.1.4.2 Diffusion phase (1975 to date)

The introduction of desktop computers, computer networks and, lastly, the Internet turned computer-assisted structural calculations into an everyday reality. The epistemological interest in theory of structures shifted from the automation of structural calculations and their theoretical background to contexts: structural engineering is increasingly understood as a process within a system. The superseding of structural analysis by modern structural mechanics was followed by the redefinition of structural calculations in the sequence from draft, calculate, dimension, design and draw via fabricate and erect to use, reuse, repair and dispose of. This had to involve a fundamental change in traditional engineering mathematics because the engineer was increasingly confronted with problems such as the coordination of planning processes right up to documentation management.

"The computer has brought about a fundamental change in theory of structures in recent years. On the practical side, the actual calculations – previously the mainstay of the structural engineer's workload – have now taken a back seat. Experience so far shows that even resourceful computer programs require a considerable basic knowledge about the underlying concepts and assumptions of the methods, and for assessing the results.

"In terms of methods and program engineering, trusses have progressed furthest. Expansions on the calculation side currently involve additional effects such as time-related and cyclic actions, non-linearity, local and global failure phenomena, interface problems such as soil-structure interaction and stochastic issues. The integration of structural analysis into the structural engineering setup (design, calculate, dimension, draw) is well advanced. Software has so far made few inroads into the actual conception process. It is therefore to be expected that future developments will take place precisely in this area, and will include structure optimisation, sensitivity analyses ('what-if' studies), automatic investigation of load-bearing structure and material alternatives plus detail planning and science-based systems. The mechanical model represents the heart of these investigations" [Ramm & Hofmann, 1995, p. 340].

*Commentary:* The authors reflect the consideration of physical effects which owing to the limited computer capacity had been fringe issues in the analysis of loadbearing structures hitherto; on the other hand, they see gaps in the research regarding the synthesis of the loadbearing structure in the conception stage. The creation of a concept-oriented theory of structures on the one hand and a construction-oriented theory of structures on the other could contribute to improved computer-aided cooperation between conception, calculation, dimensioning, design and drawing.

### 2.2 From the lever to the truss

Even Archimedes (267–212 BC) wanted to lever the world out of its hinges: "Take hold and you will move it" is the title of the vignette on the cover of Pierre Varignon's (1654–1722) 1687 paper entitled *Projet d'une nouvelle mécanique* (Fig. 2-1), which shows Archimedes doing just that.

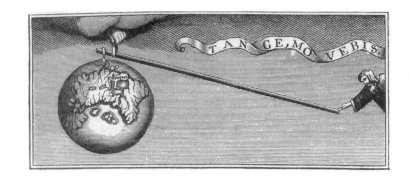

**FIGURE 2-1**
Archimedes levering the world out of its hinges [Rühlmann, 1885, p. 20]

Until the appearance of Galileo's strength of materials theory in 1638, the lever principle, the principle of virtual displacements and the parallelogram of forces dominated the historico-logical evolution of structural theory. Massimo Corradi [Corradi, 2005] has written a splendid, brief summary of how the scientists of the early days of the modern age adopted the principles of ancient structural theory.

On the level of the statics of solid bodies, the kinematical views of statics followed their own paths until their analytical synthesis by Lagrange in 1788. It was only through the fusing of statics and strength of materials to form theory of structures in the 19th century that the energy principle and the principle of virtual forces were able to take over the analysis of deformable loadbearing structures in the final three decades of that century. And it was during this period that the principle of virtual displacements lost its importance for theory of structures.

### 2.2.1 Lever principle according to Archimedes

Archimedes (267–212 BC), the outstanding mathematician of the Hellenistic scientific tradition, describes the lever principle in his book *De aequiponderantibus* by way of three axioms [Mach, 1912, p. 10]:

Bodies of equal weight on equal lever arms are in equilibrium (Fig. 2-2a):

$$\sum M_a \stackrel{!}{=} 0 \;\rightarrow\; F \cdot l = F \cdot l \quad \text{(equilibrium)}$$

Bodies of equal weight on unequal lever arms are not in equilibrium (Fig. 2-2b):

$$\sum M_a \stackrel{!}{=} 0 \;\rightarrow\; F \cdot l_1 \neq F \cdot l_2 \quad \text{(no equilibrium)}$$

If two bodies on given lever arms are in equilibrium, and $F_1$ or $F_2$ is changed, then the condition

$$\sum M_a \stackrel{!}{=} 0 \;\rightarrow\; F_1 \cdot l_1 = F_2 \cdot l_2 \quad \text{(equilibrium)} \tag{2-1}$$

must be satisfied (Fig. 2-2c).

Archimedes' precise formulation of the lever principle (2-1) – based on centuries of practical experience of such simple machines – was the first time that the phenomenon of equilibrium had been expressed mathematically.

**FIGURE 2-2**
Lever principle according to Archimedes

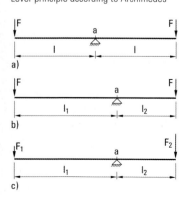

Almost 1500 years were to pass before Jordanus Nemorarius (?–1237) would add further fundamental knowledge about statics. He analysed the cranked lever, for example. With the help of the principle of virtual velocities, he tried to prove the lever principle. In doing so, he probably made use of the Aristotelian approach to the lever [Ramme, 1939, pp. 15–17]. At the start of the 16th century, the results of the statics of Nemorarius and his pupils were known in relevant scientific circles, which enabled various statics problems to be solved to a tolerable standard.

### 2.2.2 The principle of virtual displacements

Observations on the lever principle are already evident in the work of the greatest Greek natural philosopher, Aristotle of Stagira (384–322 BC), and his school. For example, idealising dynamics as statics can be found in the work *Quaestiones Mechanicae* which is attributed to him, or at least to his school [Dijksterhuis, 1956, pp. 34–36]. The following proof is taken from that work, merely translated here into the language of modern algebra. In order to show that eq. 2-1 applies, the following proof is carried out. As the movements take place simultaneously, it follows that (Fig. 2-3b)

$$F_1 \cdot v_1 = F_2 \cdot v_2 \qquad (2\text{-}2)$$

Eq. 2-2 is nothing more than the principle of virtual velocities, where

$\delta u_1 = v_1 \cdot t$ and $\delta u_2 = v_2 \cdot t$ plus

$\delta \varphi = \delta u_1 / l_1$ and $\delta \varphi = \delta u_2 / l_2$ (Fig. 2-3c) which results in

$$\delta \varphi \cdot (F_1 \cdot l_1 - F_2 \cdot l_2) = 0 \qquad (2\text{-}3)$$

Eq. 2-3 is the principle of virtual displacements. As in eq. 2-3 $\delta \varphi \neq 0$, the expression in brackets must vanish, i.e. the principle of virtual displacements is equivalent to the equilibrium statement, or rather the lever principle (eq. 2-1). So the principle of virtual displacements states that the virtual work done by a true equilibrium condition (Fig. 2-3a) on a virtual displacement condition (Fig. 2-3c) is zero. In doing so, the virtual displacement condition must be theoretically and geometrically possible and small in terms of the differential geometry. This general and final formulation of the principle of virtual displacements for rigid bodies was communicated to Varignon by Johann Bernoulli (1667–1746) in a letter dated 26 January 1717.

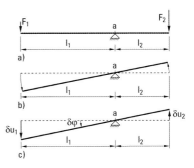

**FIGURE 2-3**
On the principle of virtual displacements according to Aristotle

### 2.2.3 The general law of work

The general law of work states that the work done by an equilibrium condition (Fig. 2-4a) on a geometrically (kinematically) possible displacement condition (Fig. 2-4b) is zero:

$$F_1 \cdot u_1 - F_2 \cdot u_2 = 0 \qquad (2\text{-}4)$$

Using the kinematic relationships (see Fig. 2-4b)

$u_1 = \varphi \cdot l_1$ and $u_2 = \varphi \cdot l_2$

eq. 2-4 becomes

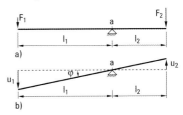

**FIGURE 2-4**
On the general law of work

$$\varphi \cdot (F_1 \cdot l_1 - F_2 \cdot l_2) = 0 \qquad (2\text{-}5)$$

Eq. 2-5 is the general law of work transformed into the equilibrium statement, or rather the lever principle (eq. 2-1).

## The principle of virtual forces

### 2.2.4

The principle of virtual forces states that the virtual work done by a virtual equilibrium condition (Fig. 2-5a) on a real displacement condition (Fig. 2-5b) is zero. In doing so, the virtual force condition must be possible in theoretical (virtual) and statical (equilibrium) terms.

The principle of virtual forces was formulated independently by Maxwell in 1864 and Mohr in 1874/75 for the calculation of elastic trusses.

The principle of virtual forces takes on the following form for rigid bodies (Fig. 2-5):

$$\delta F_1 \cdot u_1 - \delta F_2 \cdot u_2 = 0 \qquad (2\text{-}6)$$

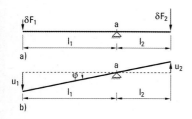

FIGURE 2-5
On the principle of virtual forces for rigid bodies

As in the principle of virtual forces the virtual force condition (Fig. 2-5a) must be in equilibrium, the following must apply according to the lever principle (eq. 2-1):

$$\delta F_2 \cdot l_2 = \delta F_1 \cdot l_1$$

If this virtual equilibrium statement is rewritten as

$$\delta F_2 = \delta F_1 \cdot l_1 / l_2$$

and entered into eq. 2-6, the result is

$$\delta F_1 \cdot u_1 - \delta F_1 \cdot (l_1 / l_2) \cdot u_2 = 0 \quad \text{or}$$

$$\delta F_1 \cdot [u_1 - (l_1 / l_2) \cdot u_2] = 0$$

As in the latter equation $\delta F_1 \neq 0$, the expression within square brackets must vanish, i.e.

$$u_1 = (l_1 / l_2) \cdot u_2 \qquad (2\text{-}7)$$

applies. This proves that the principle of virtual forces for rigid bodies (eq. 2-6) corresponds to the geometrical expression (eq. 2-7) (see Fig. 2-5b).

## The parallelogram of forces

### 2.2.5

In 1586 the Dutch mathematician Simon Stevin achieved a great step forward in the explanation of the equilibrium concept in his book *Beghinselen der Weeghconst* (The Elements of the Art of Weighing) by solving the problem of two inclined planes (Fig. 2-6).

Starting with the axiom that perpetual motion of the loop of chain $ABC$ is impossible, he discovered that the equilibrium of the two inclined planes was satisfied by

$$G_1 / G_2 = S_1 / S_2 \qquad (2\text{-}8)$$

Eq. 2-8 now has to be proved. If equilibrium prevails, then the following must be true:

$$S_1 = S_2 = G_1 \cdot \sin\beta = G_2 \cdot \sin\alpha \rightarrow G_1/G_2 = \sin\alpha/\sin\beta = S_1/S_2$$

(which had to be verified).

Stevin provided further examples for the correct breakdown of forces into components – albeit not yet systematically and not fully verified.

### 2.2.6 From Newton to Lagrange

We have to thank Roberval and Varignon for summarising and coordinating the three basic concepts of statics lever, virtual velocity and parallelogram of forces. In his book *Nouvelle mécanique* (published in 1687 and completed posthumously in 1725), Varignon – to a certain extent still in the Aristotelian tradition – spread the idea of the equivalence of lever principle, principal of virtual velocities and parallelogram of forces. The equivalence of these three is nothing other than the implied recognition of a single axiom – the equilibrium conditions, which for the *x-z*-plane take on the form

$$\sum F_x = 0 \quad \text{and} \quad \sum F_z = 0 \tag{2-9}$$

The equilibrium conditions of eq. 2-9 can also be easily derived from Newton's second law (mass × acceleration = sum of all applied forces)

$$m \cdot a_x = \sum F_x \quad \text{and} \quad m \cdot a_z = \sum F_z \tag{2-10}$$

as a special case of rest (or constant, straight-line motion: $a_z = a_x = 0$, i.e. the static case).

The third equilibrium condition of plane statics, $\sum M_y = 0$, could not be discovered based on Newton's *Principia* (1687) because Newton worked merely with point mechanics, and therefore the concept of the moment was unnecessary. It was not until 1775 that Euler discovered the twisting theorem [Truesdell, 1964]. This provided the chance to obtain all the equilibrium conditions completely from the specialisation of Newton's second law and the twisting theorem:

$$\sum F_x = 0, \quad \sum F_z = 0 \quad \text{and} \quad \sum M_y = 0 \tag{2-11}$$

In his pioneering work of 1788, *Mécanique analytique*, Lagrange based all mechanics, i.e. also statics, on a single principle: the principle of virtual velocities. Lagrange therefore recognised not only the equivalence of the three principles of statics (Fig. 2-7)
- lever principle: eq. 2-1,
- principle of virtual displacements: eq. 2-3, and
- parallelogram of forces: eq. 2-8,

but also stated clearly that the principle of virtual displacements can be converted mathematically into the equilibrium principle.

Statics in the meaning of the equilibrium of a body on the level of theoretical mechanics was therefore complete from the logic viewpoint. And that created the logic prerequisites for the historical development of theory of structures in the discipline-formation period.

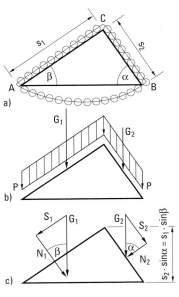

**FIGURE 2-6**
Two inclined planes after Stevin

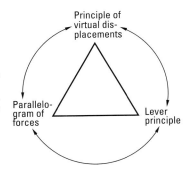

**FIGURE 2-7**
The equivalence of the principles of statics

### Kinematic or geometric view of statics?

**Kinematic view of statics**
Represented by:
**Aristotle**, Heron, Vitruvius, Thabit ibn Qurra, Nemorarius, Leonardo da Vinci, Tartaglia, Cardano, **Lagrange**, Mohr, Land, Müller-Breslau

**Geometric view of statics**
Represented by:
**Archimedes**, Heron, Pappus, Thabit ibn Qurra, Guidobaldo del Monte, Stevin, **Galileo**, Roberval, Varignon, Clapeyron, Müller-Breslau

| | |
|---|---|
| potential | current |
| virtual | real |
| idealised | material |
| what could be | what is |
| theoretical | practical |
| rationally oriented | technically oriented |
| motion | rest |
| machine | structure |
| disruption of equilibrium | equilibrium |
| principle of virtual displacements | lever principle, parallelogram of forces |

**FIGURE 2-8**
Comparison of the kinematic and geometric views of statics

### Stable or unstable, determinate or indeterminate?

#### 2.2.7

The nucleus of the theory behind the kinematic view of statics, founded by Aristotle and completed by Lagrange, is the principle of virtual displacements, which was applied successfully to simple machines such as the lever, the pulley or the inclined plane. Leonardo da Vinci, for example, understood the masonry arch as a kinematic machine (see Fig. 4-16). However, he did not calculate the thrust of the arch with the help of the principle of virtual displacements, but instead proposed how this could be determined with a kinematic experiment. Converting building structures into machine models in order to analyse them mechanically was a characteristic of the kinematic view of statics. The effect of the equilibrium on the machine model could then be determined indirectly through a geometrical disruption of the equilibrium condition in the meaning of the principle of virtual displacements.

The kinematic view of statics (Fig. 2-8/left) was an integral component of the Aristotelian motion theorem and natural philosophy and was first overthrown in the early part of the modern age by Galileo and others. This was associated with a significant rise in importance of the geometrical mechanics founded by Archimedes – and here again we have to mention Galileo's *Dialogue*. Galileo's cantilever beam, on which he demonstrated the bending failure problem (see Fig. 5-7), advanced to become a metaphor for the geometric view of statics.

Whereas the kinematic view of statics as pure theoretical work in the meaning of Plato enjoyed a high social standing in ancient times, the geometric view of statics (Fig. 2-8/right) belonged to architecture and was hence regarded as "lowbrow art". The geometric view of statics evolved on the basis of Euclidian geometry and from the rudimentary practical needs of building, where equilibrium was the natural condition that could not be eliminated without external disturbances. The basic concept of stability enabled the geometric view of statics to gain supremacy in the discipline-formation period of structural theory (1825–1900). Nevertheless, eminent structural engineers like Mohr, Land, Müller-Breslau and others made significant contributions to the ongoing development of the kinematic view of statics. The difference between the kinematic and geometric views of statics formed an important element in the controversies surrounding the foundations of structural analysis.

#### 2.2.8

The logic side of simple statics around the middle of the discipline-formation period – the establishment phase of structural analysis – is presented below in its historical context with the help of examples. Fig. 2-9 shows how simple statics (requiring only equilibrium conditions), as a theory of statically determinate systems, is positioned between kinematics and higher statics (requiring equilibrium and deformation conditions).

The clear establishment of the concepts shown in Fig. 2-9 and the theory of statically indeterminate systems and the kinematic methods in theory of structures are the outcome of the classical phase of structural theory. On the other hand, the establishment phase of structural theory

**What is kinematic, statically determinate and statically indeterminate?**

| Kinematics | Statics | Statics |
|---|---|---|
| Systems with one degree of freedom | as a pure equilibrium exercise | as a unison of the solution of equilibrium and deformation tasks |
| Kinematic | Statically determinate | Statically indeterminate |
| Motion possible | Equilibrium conditions are required for the statically determinate and indeterminate system, i.e. kinematically rigid (stable) | |
| Unequivocal state of motion | Equilibrium conditions are adequate to determine an unequivocal force state | Equilibrium conditions alone are inadequate here |
| No force state | One force state | Infinite number of force states possible |
| State of motion determinate | Force and deformation states unconnected | Force and deformation states connected |

FIGURE 2-9
Breakdown of the concepts of kinematically determinate, statically determinate and statically indeterminate

essentially consists of the development of the theory of statically determinate systems.

### 2.2.9 Syntheses in statics

Numerous scientists and engineers had already looked into the statically determinate basic systems of the cantilever beam and the simply supported beam during the preparatory period of structural theory. A third statically determinate basic system was added to these two during the establishment phase: the three-pin frame (Fig. 2-10).

FIGURE 2-10
Idealised three-pin system

The first three-pin frames were implemented by Johann Wilhelm Schwedler (1823–94) in the early 1860s specifically as iron constructions, whereas Claus Köpcke (1831–1911) described the three-pin frame theoretically in 1861 [Lorenz, 1990, p. 6]. Nevertheless, Rudolf Wiegmann (1804–65) had realised earlier – in a paper dating from 1836 [Schädlich, 1967, p. 84] – that three interconnected pin-jointed plates or bars also form a plate [Wiegmann, 1839]. Therefore, Wiegmann was the first to express, implicitly, the first formation law of trussed framework theory (Fig. 2-11), and even proposed executing the system entirely in wrought iron [Schädlich, 1967, p. 85]. In 1840 Camille Polonceau (1813–59), working independently of Wiegmann, published his paper *Notice sur un nouveau système de charpente en bois et en fer* [Polonceau, 1840/1], which in that same year was translated into German under the title of *Neues Dachkonstrukzionssystem aus Holz und Eisen* (new roof construction systems in timber and iron) [Polonceau, 1840/2]. Two years later, in the journal *Allgemeinen Bauzeitung*, Wiegmann accused Polonceau of copying his idea [Schädlich, 1967, p. 84]. This is certainly incorrect because Polonceau was inspired by Amand Rose Emy (1771–?); the latter exhibited a 1:10 scale model of his system at the Paris industry exhibition in 1839 (and was awarded a silver medal for it!) [Emy, 1841, p. 284], which he described in 1841 in the second volume of his book on carpentry. In structural terms

**FIGURE 2-11**
Plate made up of three pin-jointed bars

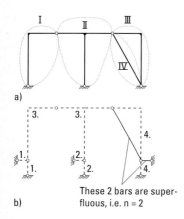

**FIGURE 2-12**
a) Given system, b) system with two degrees of static indeterminacy

this corresponds to the Wiegmann-Polonceau truss [Emy, 1841]. Christian Schädlich draws attention to Russian roof structures built prior to 1840 which in static-constructional terms are similar to the Wiegmann-Polonceau truss. "So we may assume that Wiegmann, the French and the Russians discovered the same system at the same time and certainly independently" [Schädlich, 1967, p. 84]. Nevertheless, Wiegmann explored the loadbearing quality of the system in greatest detail so that according to Max Förster (1867–1930) we are justified in designating the roof structure known in the engineering literature as a "Polonceau truss" as a "Wiegmann-Polonceau truss" [Förster, 1909, p. 323].

But back to the first formation law. As a hinged connection between two bars, it functions as a logical development principle in the analysis and synthesis of loadbearing structures. The question of whether a structural system is statically determinate or indeterminate (Fig. 2-9) can now be answered as follows: a system is statically determinate when it can be developed based on the three basic systems with the help of the first formation law using a minimum number of bars (members); if this minimum number is exceeded by $n$ bars, the system is statically indeterminate with $n$ degrees of static indeterminacy. The structural synthesis of members in a two-bay frame makes the difference between statically determinate and statically indeterminate very clear (Fig. 2-12).

The Wiegmann-Polonceau truss is analysed as an example in the historico-logical sense with the help of the three statically determinate basic systems (cantilever beam, simply supported beam and three-pin truss) and the first formation law in Fig. 2-13: the trussed beam as a structural system is designated plate I. This plate is joined to plate II, which is identical with plate I, via a hinge at the ridge. The resulting system would be kinematically determinate, i.e. unstable. Therefore, plate III has to be introduced as a tie so that lateral deflection to the right is prevented. According to the first formation law, all three plates in turn form one plate. The result is an externally statically determinately supported Wiegmann-Polonceau truss. However, the Wiegmann-Polonceau truss has one degree of static indeterminacy internally because the timber rafters of the frame plates I and II (Fig. 2-13) act as continuous two-span beams.

The continuity effect of the rafters in the Wiegmann-Polonceau truss was first correctly assessed quantitatively in German engineering literature by Gustav Adolf Breymann (1807–59) in 1854 [Breymann, 1854, pp. 76–77]. In this context, Stefan M. Holzer points out the difficulties in the structural modelling and calculation of the Wiegmann-Polonceau truss, which were first overcome in everyday structural calculations around 1890 as the pinned trussed framework model became established in structural engineering [Holzer, 2006, pp. 430–34]. But this had been started long before by Karl Culmann with the systematic static-constructional criticism of the existing truss systems, which reached its climax in the pinned trussed framework model and, based on that, the statically determinate truss theories of Culmann [Culmann, 1851] and Schwedler [Schwedler, 1851]. Prior to the appearance of Culmann's graphical statics

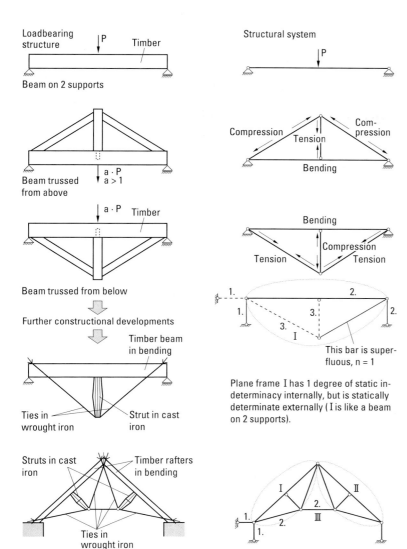

FIGURE 2-13
Historico-logical development of the Wiegmann-Polonceau truss

in 1864 and 1866, trussed framework theory developed in close cooperation with the transition from composite systems comprising timber, cast iron and wrought iron (Fig. 2-13) to truss systems consisting exclusively of wrought iron members. This process had already begun to take hold in practical building in the 1850s, e.g. in the roof trusses of the Gare de l'Ouest in Paris (Eugène Flachat, 1802–73), built in 1853 entirely from riveted wrought iron [Schädlich, 1967, p. 87]; from then on, the truss idea was implemented consequently as constructional reality. Trussed frameworks could now be modelled as pin-jointed trusses, although it was clear that riveted joints could not work as hinges. Nonetheless, trussed framework theory now permitted a satisfactory quantitative control of the structural design of this new category of loadbearing structures. The fact that it first gradually conquered the practical side of structural calculations was due not only to the elimination of composite systems from

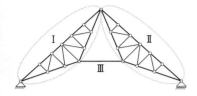

**FIGURE 2-14**
Pinned trussed framework model of the Wiegmann-Polonceau truss (roof to retort facilities after Schwedler)

framework construction, but also due to the rising formula-like acquisition of graphical statics after 1870 in the form of graphical analysis for the practical side of structural calculations, as was particularly widespread in building structures. In bridge-building, on the other hand, the practical side of structural calculations developed in close conjunction with the formation of structural theories like in building structures. Nevertheless, contrary examples can be found at an early date, also in building structures. For example, in the 1860s Schwedler implemented the trussed framework idea in static *and* constructional terms for the roofs to retort facilities at a Berlin gasworks employing Wiegmann-Polonceau trusses [Schwedler, 1869/1]. And for the roof over the council chamber in Berlin's new City Hall Schwedler went so far as to design the rafters of the Wiegmann-Polonceau truss as parabolic simply supported beams with the joints essentially pinned [Schwedler, 1869/2 & 1869/3]. His design therefore corresponds to the principle of the idealised, typical pinned trussed framework model shown in Fig. 2-14.

In the structural model, the strut of the trussed beam (Fig. 2-13) for plane frames I and II (Fig. 2-14) is pin-jointed at the chords, i.e. the chord of plane frames I and II no longer acts as a statically indeterminate continuous beam, but rather as the sum of several statically determinate simply supported beams. The elimination of the internal static indeterminacy allows the member forces of the pinned trussed framework model shown in Fig. 2-14 to be calculated with the help of the Cremona diagram or Ritter's method of sections: the equilibrium conditions are sufficient to determine all support reactions and member forces (force condition) unequivocally. Hence, major progress in theory of structures during the establishment phase has a practical example.

## The development of higher engineering education

### 2.3

In the 17th and 18th centuries, the repression of individual feudal powers by central powers led to the formation of absolutist states which enabled faster economic development. Besides the establishment of a permanent army for exerting dynastic interests, the absolutist states created an adequate transport infrastructure of roads and canals which fitted into the framework of their mercantile economic policies. Sébastien le Prêtre de Vauban (1633–1707), Commissioner-General of French Fortifications and Marshal of France, saw "making rivers navigable and the building of inland waterways primarily as an effective means of defending frontiers in times of war and encouraging trade and prosperity in times of peace" [Straub, 1992, p. 163]. The École Polytechnique, which was a direct outcome of the French Revolution of 1794, became the model for engineering education in the first half of the 19th century for countries such as Prussia and Austria as well as other European states and even the USA. In the USA after the civil war and in the German Empire proclaimed following Bismarck's top-down revolution of 1871, the demand for engineers to cope with the rapid rise in economic prosperity reached unprecedented proportions. It was during this period that the polytechnic schools in Germany changed to technical universities [König, 2006/1, pp. 198–211].

They advanced to become the most successful model for teaching and research in the engineering sciences in the first decades of the 20th century, but from the late 1930s onwards were superseded step by step by the US system of engineer education.

### 2.3.1 The specialist and military schools of the *ancien régime*

The first corps of engineers for the building of roads and fortifications was founded in France as early as 1604, and that was followed by Vauban's famous Corps des Ingénieurs militaires. Numerous specialist schools then appeared around the middle of the 18th century. A training facility for civil engineers set up in 1747 by Daniel Charles Trudaine (1703–69) and Jean-Rodolphe Perronet (1708–94) had by 1775 evolved into the École des Ponts et Chaussées. In addition to these there were shipbuilding and mining schools which represented elements for satisfying the technical needs of those branches of the evolving manufacture-based capitalism.

Special mention should be made of the military engineering school for the corps of military engineering in Mezières, which was founded in 1736. The two-year course of study was intended to prepare the students for later tasks such as the design of new fortifications and inspecting the magazine. However, the geometry and algebra skills taught to the students could not be used directly for solving specific tasks; instead, the trial-and-error method was preferred, which Wußing describes as follows: "They had to learn how to defilade a fortification taking into account the terrain and specific military aspects" [Wußing, 1958, p. 649]. Gaspard Monge (1746–1818), the creator of the curriculum at the later École Polytechnique, developed descriptive geometry into a means for achieving constructional solutions to such tasks. Monge, whose father was a knife and scissors grinder, was initially only allowed to attend the special institute (the "plasterers' school"). Fourier, too, wished to take the examinations at the artillery school, but – despite being recommended by the mathematician Legendre – was turned down by the minister responsible, who argued that "Fourier is not from the aristocracy and could not join the artillery even if he was a second Newton" [Wußing, 1958, p. 650].

Similar class privileges were decisive in gaining entry to the civil engineering schools as well: "Personal connections governed admissions; there were no entrance examinations, and hardly any final examinations" [Wußing, 1958, p. 649].

The schools of civil and military engineering were poorly furnished in terms of personnel and equipment. For example, apart from a few cannons, the artillery school had no visual aids, and neither a physics room nor a library, and there were no permanent teachers at the civil engineering schools.

Taking these facts into account, it is difficult to understand how Schnabel arrives at the conclusion that engineering sciences formed part of the curricula at those special schools [Schnabel, 1925, p. 5]. At best, we can speak of a period preparing the scientific foundation of those technical branches that formed the heart of mercantile aspirations (e. g. mining, building).

**Science and enlightenment**

**2.3.2**

Taton describes pre-revolutionary science as an exclusive privilege of enlightened amateurs who exchanged letters on a regular basis and met in academies [Taton, 1953]. In France, writers such as Voltaire and scientists such as Clairaut asserted the new Newtonian system against the prevailing Cartesianism and therefore played a not insignificant role in the mechanics boom in that country. Voltaire and Rousseau praised the educational value of exact and experimental sciences in their works. Turgot, Montesquieu, Condorcet and other compilers of encyclopaedias tried to base the political economics (physiocracy) on scientific principles in order to justify a radical social transformation.

The middle classes' criticism of the social conditions in the *ancien régime*, but also their demand for a universal and rational explanation of the world, culminated in the 35-volume *Encyclopédie ou Dictionnaire* (1751–81), which was published by the philosopher and writer Diderot (editor until 1765) and the physicist D'Alembert (editor until 1758). In his grand introduction to the *Encyclopédie* (1751), D'Alembert writes: "But when it is often difficult enough to define the sciences and the arts individually with a small number of rules or basic concepts, it is no less difficult to accommodate the infinite diversity of human knowledge in one unifying system. Our first step in this investigation is therefore to check the genealogy and the interlinking of our knowledge, the supposed causes of its genesis and the features of its differences; in a word we must return to the origin and to the creation of our thoughts" (cited in [Treue, 1989/90, p. 175]). The *Encyclopédie* was a vehicle for the greatest writers, philosophers and scientists of France to convey their knowledge of the world from its beginnings right up to the point at which they handed over their manuscripts for printing [Treue, 1989/90, p. 180]. Was it a surprise, then, when the French Revolution tried to implement – in a political sense – the universal demand of the encyclopaedists for a rational view of the world?

**Science and education during the French Revolution (1789–1794)**

**2.3.3**

Differences in foreign policy issues and the reaction to the king's flight led to a split in the middle classes in 1791. Brissat, the spokesman of the Girondins, therefore called for a great war against Prussia and Austria, who were opposed to the citizens' revolution in France. Robespierre, the spokesman of the Jacobins, rejected this because the army was disorganised and French industry lay in ruins.

Finally, on 20 April 1792, the Girondinian Ministry declared war on Austria. This brought about a worsening of the domestic situation, which gradually led to the Jacobin dictatorship. It was during this period that all the important social decisions were made which then led to the founding of the École Polytechnique in conjunction with the reorganisation of the military.

After its reorganisation on 10 July 1793, the Committee of Public Safety – the revolutionary government per se –, in particular its member Lazare Carnot (1753–1823), began to set up the Revolutionary Army. "The national needs of the young republic – the military engineering ones arising out of the Revolutionary Wars and also the civil engineering ones – encouraged the founding of the École Polytechnique, which represented the most important creation of the French Revolution in the field of natural science and technical education" [Klemm, 1977, p. 25]. Thomas Hänseroth and Klaus Mauersberger have acknowledged the work of Lazare Carnot from the viewpoint of structural and machine mechanics in 18th-century France [Hänseroth & Mauersberger, 1989].

All regulatory measures in the education sector were aimed at a nationalisation of this sector, although Condorcet had rejected this in his "draft of a decree on the general organisation of state education" (1792). Nevertheless, he influenced all the subsequent reforms in education.

Condorcet proposed a primary school for every village, which would be linked to a central school in towns with more than 4000 inhabitants. Every departement should have one higher education establishment. These central schools played an important role in the establishment of the École Polytechnique, especially with regard to the mathematics and natural science entrance requirements [Bradley, 1976, pp. 13/14]. Almost all the polytechnic schools that succeeded the École Polytechnique complained about the prior education of their candidates for admission.

Although the plans for establishing the École Polytechnique had already been drawn up before the fall of Robespierre and a corresponding commission had been set up, the École Polytechnique was not opened until the end of 1794, i.e. after the closure of the Club des Jacobins. It can therefore lead to misunderstandings when Klemm writes: "After the end of the Reign of Terror, the old times before Robespierre's dictatorship returned in many areas. The downtrodden education and training structure now had to be built up in the sense of a true revolutionary ideal" [Klemm, 1977, p. 17]. For in the end the organisers of the École Polytechnique were indeed members of the Club des Jacobins such as Fourcroy, Carnot and Monge. "In the period in which we see the first germ of the idea of the polytechnic schools, a permanent learned society had formed alongside the Committee of Public Safety which brought about many beneficial resolutions through the organ of this feared committee" [Jacobi, 1891, p. 359].

### 2.3.4 Monge's teaching plan for the École Polytechnique

The role played by descriptive geometry in Monge's teaching and study plans was drawn from the following [Belhoste, 2003, pp. 200–205]: after a three-year course of study, the students of the École Polytechnique would graduate as fully trained military engineering officers, artillery officers and civil engineers. It was not until a few years later that the programme was changed to a two-year mathematics/natural sciences foundation course followed by – depending on the standard achieved in the final examination – training at one of the various reactivated special schools (in order of priority): 1) École des Ponts et Chaussées, 2) École des Mines, 3) École du Génie, 4) École d'Artillerie, etc.

Descriptive geometry occupied half the period of study, with physics, chemistry, freehand drawing and mathematics sharing the other half.

Monge managed to elevate descriptive geometry to become the language of the engineer:

"1. Three-dimensional artefacts had to be represented and defined exactly by means of two-dimensional drawings. Looked at from this viewpoint, a person conceiving a project must be able to use a language that is understood by those executing the work. This is also valid for those who in turn carry out the various individual parts.

2. The object of descriptive geometry is to derive everything arising necessarily from its forms and associated positions based on the exact description of the bodies" [Monge, 1795, cited in Hänseroth & Mauersberger, 1989, p. 51].

Descriptive geometry advanced to become the most important fundamental discipline of the polytechnic schools to emerge in the first half of the 19th century, and legitimised their scientific character. Therefore, in the history of drawing, graphics and the use of images in engineering sciences and architecture, descriptive geometry represents a significant watershed in the phenomenon of visual communication of architects as well as construction and mechanical engineers [Kahlow & Kurrer, 1994]. A few insights into Monge's book *Géométrie descriptive* (Fig. 2-15) [Monge, 1794/95] show us that the applications of descriptive geometry were not merely academic, but had also been drawn from practice:

– An engineer travels to a mountainous country – he has his aids, and the altitude of his location is required.
– A general with a tethered balloon at his disposal requires a map of the terrain in which an enemy has taken up positions.
– Applications of involutes and evolutes in engineering, e. g. in the cam levers of rotating shafts, which raise stamping beams in crushers and stamping mills [Wußing, 1958, p. 655].

Monge was convinced that "these lessons [in descriptive geometry – the author] ... [would] certainly contribute to the ongoing improvements in national industry" [Wußing, 1958, p. 655].

It is precisely the example of Monge's descriptive geometry that makes it clear how certain areas of mathematics surfaced systematically with a "technical ferment" [Manegold, 1970]. The 18th century was a century of expansion in differential and integral calculus. Its advocates sought practical technical applications (application phase). These aspirations of individual scientists, mostly resident in academies, had evolved to such an extent by the end of the 18th century that a systematic compilation of the scattered applications of differential and integral calculus to form engineering science theories was almost within reach. This would be realised initially within the framework of the École Polytechnique and the special schools, and later at the polytechnic schools in Austria and Germany. The unfolding of the engineering sciences in the 19th century in Austria and Germany must therefore be equated with the rise of the polytechnic schools to become technical universities [Manegold, 1970, p. 8].

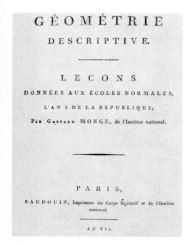

FIGURE 2-15
Cover of *Géométrie descriptive* by Gaspard Monge [Monge, 1794/95]

## 2.3.5 Austria, Germany and Russia in the wake of the École Polytechnique

The polytechnic schools in Prague, Vienna and St. Petersburg were the direct successors to the École Polytechnique.

The artillery attack at Valmy on 20 September 1792 was the first time that the French revolutionary troops managed to repel the army of the alliance of Austria, Prussia and a number of small German states. After that, the revolutionary troops switched from the defensive to the offensive in the first Coalition War, in which they gained victory in 1795 by taking Savoy, the entire left bank of the Rhine and the Netherlands. In the light of the important role played by higher technical education in the victory of the French revolutionary army, Austria's Royal Commission for the Reform of State Education started its work in 1795. Its primus inter pares was the professor of higher mathematics at the University of Prague, Franz Joseph Ritter von Gerstner (1756–1832).

An engineering professorship, dealing in particular with fortifications, bridge- and road-building plus river training works, had already been set up at the University of Prague on the initiative of the Bohemian Parliament. But it was the École Polytechnique model that showed that a polytechnic institute must be based on mathematics and natural science subjects, which up until then had been reserved exclusively for the faculty of philosophy. Recognition of this fact led to Prague's existing school of engineering being incorporated into the university. However, Gerstner was soon focusing on an independent polytechnic institute that would dedicate itself entirely to "raising the status of the Fatherland's industry through scientific teaching" [Gerstner, 1833, p. V]. Therefore, Gerstner's ideas about reforming natural science studies were happily accepted, even by Emperor Franz II, on 27 September 1797: "I expect that the reorganisation of the study of philosophy will also result in a detailed plan for a higher technical institute whose splendid benefits are already obvious from the sketch in my possession" [Kraus, 2004, pp. 125–126]. One year later, Gerstner presented this "detailed plan", which was enthusiastically received by the Bohemian Parliament. Apart from a few omissions from Gerstner's plan due to the uncertainties of the Napoleonic Wars, the Royal Engineering State Education Establishment (now Prague Technical University) opened its doors to students for the first time on 10 November 1806 – making it the second-oldest technical university in Europe. Gerstner was in charge of this establishment until 1822, responsible for the subjects of mathematics and mechanics. One major shortcoming in this polytechnic institute was the low standard of foregoing education among candidates for admission, which led to a secondary school being annexed to the institute (Fig. 2-16).

The Vienna Polytechnic (now Vienna Technical University) proposed by Prechtl in 1815 was attributed a university character by Schnabel [Schnabel, 1925, p. 19]. Lessons, according to Prechtl's underlying idea, "should not be taught, i.e. the sciences should not be a means in themselves, but instead serve only as the necessary vehicle with which to execute the various apposite duties of civil life properly and safely. But the

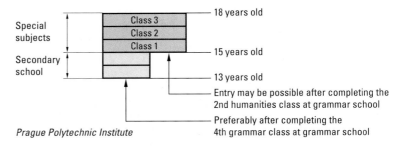

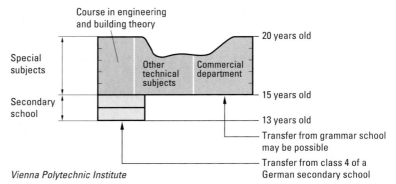

**FIGURE 2-16**
Organisation of the polytechnic schools in Prague and Vienna around 1850

sciences may also not strive to attain false popularity on the level of a completely untrained mental capacity because then the nature and dignity of science would be annihilated and the end not achieved!" [Schnabel, 1925, p. 18]. Fig. 2-16 illustrates the organisational structure of the polytechnic schools around 1850. Prechtl's school is a stark contrast to the École Polytechnique and Prague Polytechnic; it became the model for many European countries, also outside the German-speaking states [Hantschk, 1990, p. 488]. In particular, it was the attempt to mitigate Austria's backwardness through technical education without changing the feudal structure of society that attracted a number of German states to follow Austria's lead. Vienna Polytechnic remained the leading establishment for higher engineering education in the German-speaking countries until 1850, and produced such worthy graduates as Redtenbacher, the founder of scientifically based mechanical engineering, and Karmarsch, for many years the director of Hannover Polytechnic School.

The unified engineering established in Vienna Polytechnic by Prechtl anticipated the division into departments and faculties in the subsequent

**FIGURE 2-17**
The historical roots of the Polytechnic School in Karlsruhe

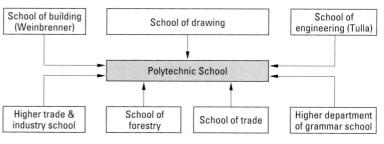

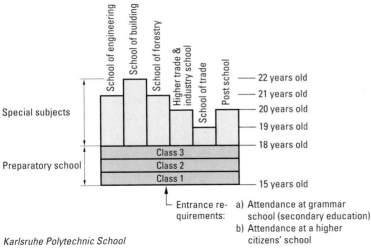

*Karlsruhe Polytechnic School*

**FIGURE 2-18**
Organisation of the Polytechnic School in Karlsruhe around 1850

technical universities. This principle played a role in the founding of the Polytechnic School in Karlsruhe, which in 1825 succeeded several technical schools (Fig. 2-17).

As early as 1832, Nebenius reorganised Karlsruhe Polytechnic and summarised in a book the discussion about engineering education, which had been raging for some time on a broader front. Nebenius clearly distinguishes between the training of manual workers and masters in the trade and industry schools and the polytechnic school, which had a university character. But in fact this status was not achieved until the 1850s when the Industrial Revolution reached its zenith in Germany (Fig. 2-18).

Like the foundation of the Building Academy in Berlin in 1799 can be understood as the pinnacle of the reform activities of the Prussian State Building Authority, an engineering school also marked the climax of the reforms in Russian highways and waterways construction by Count Nikolaj Petrovič Rumjancev (1754–1826) in the first decade of the 19th century: the Institute of Engineers of Ways of Communication (Institut du Corps des Ingénieurs des Voies de Communication) founded in 1809/10 in St. Petersburg. The years leading up to the founding of this institute are interesting, as Sergej G. Fedorov analyses in his monograph on the canal-, road- and bridge-building engineer Carl Friedrich Wiebeking (1767–1842) [Fedorov, 2005]. In June 1804, Wiebeking, at that time a Privy Counsellor in Vienna, sent a copy of his monograph on waterways engineering to the Russian Minister of Trade – Count Rumjancev. In the accompanying letter, Wiebeking explained that if his proposed road-building method were to be approved, he could contribute to the reforms carried out by Rumjancev and to improving transport routes in Russia [Fedorov, 2005, p. 47]. The improvement was essentially to use soldiers in times of peace to help with the building of roads, bridges and canals. Wiebeking's proposals were not followed up. Although shortly afterwards Wiebeking was promoted to director-general of canals, bridges and roads for the whole of Bavaria, he still maintained his contacts with Russia. At a

**FIGURE 2-19**
The first home of the Institute of Engineers of Ways of Communication and the Russian Highways Authority – Jusupov Palace on the River Fontanka, St. Petersburg [Fedorov, 2005, p. 57]

meeting in Erfurt between Napoleon and his German allies (27 September to 14 October 1808) to which Tsar Alexander I was also invited, "they discussed important military policy but also personnel issues which affected the future of the Russian Highways Authority. Apart from the Spanish engineer Augustin de Betancourt, who was later to become director of the Institute of the Engineers of Ways of Communication, Wiebeking also travelled to Erfurt for an audience with the Russian Tsar (or rather his entourage)" [Fedorov, 2005, p. 51]. Fedorov suspects that Wiebeking was considered as a possible alternative candidate to Betancourt for the post of director-general of the Russian Highways Authority (Fig. 2-19) [Fedorov, 2005, p. 51].

What is certain is that one of the reasons behind Wiebeking's visit to Erfurt was to discuss the potential training of the next generation of Russian engineers. In September 1808 in Erfurt, Wiebeking handed Count Rumjancev his conditions for such a venture: "In order to prepare them for their future 'scientific design activities', six to eight persons with profound geometry skills between the ages of 18 and 25 should be invited. Two to four persons should be available for the practical execution of canal-building works and they should complete a shipbuilding master's training course. Three years of training are proposed in which they will work under the supervision of Bavarian engineers" [Fedorov, 2005, p. 53]. Fedorov suspects that the Russian building students would be accommodated in the canal-building school built shortly before [Fedorov, 2005, p. 55]. In the end the project failed because the Institute of Engineers of Ways of Communication and the Russian Highways Authority were founded in St. Petersburg (see Fig. 2-19) in 1810. As compensation for the expenses on the Bavarian side, the first director of the Russian Highways Authority recommended purchasing 100 copies of the treatise on timber bridges edited by Wiebeking [Fedorov, 2005, p. 57].

### 2.3.6 The education of engineers in the United States

The institutionalisation of higher technical education in the United States stems from three roots. Firstly, the United States Military Academy (USMA) at West Point, which was founded in 1802 as a direct outcome of the War of Independence upon the decree of Thomas Jefferson (1743–1826) and which was closely allied to the model of higher technical education in continental Europe, of which the École Polytechnique was the best-known example. To study at the USMA you needed a personal recommendation from a member of the US Congress, or the US President himself! The students graduated with the academic degree of "bachelor" and the military rank of "second lieutenant" (Fig. 2-20). The second root is the Canals Administration, which operated in New York between 1816 and 1825 and was mainly concerned with the building of the Erie Canal between Buffalo on Lake Erie and Albany in New York state [Gerstner, 1842, 1997]. And thirdly, a small percentage of the engineers needed for the large canal-building projects gained their training in private schools. The most important of these was Rensselaer Polytechnic Institute, which had been founded in 1824 and quickly became a renowned school of civil engineering, and in 1835 became the first institution in the United States to award the academic degree "Civil Engineer" [Gispen, 2006, p. 155]. The American Society of Civil Engineers (ASCE) was founded in 1852.

By 1839 the Austrian engineer Franz Anton Ritter von Gerstner (1793–1840) – who from 1824 to 1829 supervised the building of the first (albeit horse-drawn) railway in continental Europe between Budweis and Pramhöf – had realised the strategic importance of the interaction between technical education, banking and communication systems such as railways, canals and steamships for the economic development of the USA: "The United States has three aspects to thank for its prosperity. The schools, which provide general, useful education and enable everyone to assess and calculate his undertakings; the banks, 800 in number, which enable everyone to borrow money with ease, according to his assets, and place them in a position of being able to participate in speculations of all kinds; and finally, railways, canals and steamships, which promote transport in this vast country in such a way that anyone who has not seen it cannot fully appreciate" [Gerstner, 1839, p. 1]. But Gerstner saw this infrastructure with his own eyes and had a much better picture than any of his contemporaries. After completing the first Russian railway between St. Petersburg and Zarskoe-Selo in 1838, Gerstner spent 12 months in the USA on behalf of the Russian government in order to find out about the new railway and canal networks. Gerstner's plan was to describe these infrastructure networks in detail and to write about the influence of engineering education and the banks on economic development in the USA in later reports. Unfortunately, Gerstner died in Philadelphia in 1840, which

**FIGURE 2-20**
Examination at West Point, 1868
[Gispen, 2006, p. 155]

left his assistant L. Klein to publish the monumental two-volume work on the early American railway and canal networks posthumously based on Gerstner's notes [Gerstner, 1842, 1843]. In the meantime, this unique document depicting the early history of American transport systems has been translated into English [Gerstner, 1997].

The founders and directors of the military academy at West Point were also aware of the importance of this school for civil engineering right from the outset. For instance, in 1830 the school management envisaged West Point as a sort of national school of engineering which would of course furnish the military with engineering competence, but also serve the progress of the country as a whole. For the latter, the academy would call on a corps of engineering "which is in a position to steer the entrepreneurial spirit blowing across this country in a healthy direction" [Gispen, 2006, p. 155]. Right up until the American Civil War, USMA graduates played an important part in creating the colossal infrastructure systems, like canal and railway networks, of the United States.

One example is Hermann Haupt (1817–1905). At the age of just 14, he entered the USMA on the recommendation of US President Andrew Jackson (1767–1845). He graduated in 1835 with the rank of second lieutenant and became involved in the building of bridges and tunnels for a railway company. After that, Haupt worked until 1847 as professor of mathematics and civil engineering at Pennsylvania College (now Gettysburg College) before becoming the leading railway engineer in the USA. The book he wrote in 1851, *General Theory of Bridge Construction* [Haupt, 1851], is an interim report of his creative activities in bridge-building and had a profound influence on the theory of bridge-building in the United States throughout the second half of the 19th century.

During the civil war, Haupt organised the office of construction works and military railways plus military engineering works and thus contributed to the Union's victory over the Confederate States. President Abraham Lincoln (1809–65) acknowledged the rebuilding of Potomac Creek Bridge within nine days during a visit on 28 May 1862 with the following words: "That man Haupt has built a bridge four hundred feet long and one hundred feet high across Potomac Creek, on which loaded trains are passing every hour, and upon my word, gentlemen, there is nothing in it but cornstalks and beanpoles" (cited in [Flower, 1905, p. 225]). Haupt was a master of the efficient use of materials in bridge-building.

The reunification between the Confederacy and the Union to create the USA allowed Lincoln to create the political conditions for dynamic industrial development on a scale that had never been seen before. The Massachusetts Institute of Technology (MIT) had been founded in 1861, but it was not until the Morrill Act of 1862 that the political framework conditions were established for a new stage of development in higher engineering education in the USA, which initiated a wave of new engineering courses and institutions [Gispen, 2006, p. 165]:

- Sibley College of Engineering at Cornell University (1868)
- Stevens Institute of Technology (1870)

- engineering courses of study at the University of Wisconsin (1870)
- engineering courses of study at Purdue University founded in 1869 (1874)
- Case School of Engineering (1880).

According to Kees Gispen, the number of education establishments for engineers grew from less than 10 before the civil war to 85 in 1880 [Gispen, 2006, p. 164]. Siegmund Müller counted 118 establishments with a total of 17,200 engineering students for the 1901/02 semester – and that did not include a number of less important institutions [Müller, 1908, pp. 20-21]. Fig. 2-21 shows clearly that around 1900 the main centres of education for engineers were still to be found in the states of New York, Massachusetts and Pennsylvania. At that time the State of California had only three technical universities, Berkeley (No. 6), Pasadena (No. 7) and Stanford (No. 8), but these did generate decisive momentum in the ongoing development of engineering sciences in general and theory of structures in particular in the second half of the 20th century.

In qualitative terms, too, engineering education in the USA underwent a significant change in the final decades of the 19th century. The empirical-practical/academic-theoretical mix in engineering education prior to the civil war "cleared the stage for the far stronger training- and science-oriented engineering culture of the post-civil war years" [Gispen, 2006, p. 156]. A trademark of American engineering education is the early integration of laboratory experiments into courses of study (Fig. 2-22). During his visit to the 1893 World Exposition in Chicago, Alois Riedler (1850–1936), a mechanical engineering professor from Berlin, toured American training establishments on behalf of the Prussian Ministry of Education; his subsequent report recommended that the practical lessons in the well-equipped engineering laboratories of the USA be copied by Germany's technical universities. By 1896 Riedler had managed to establish a mechanical laboratory at Berlin-Charlottenburg Technical University [Knobloch, 2004, pp. 134–137]; this was followed in 1901 by a testing facility for the structural design of building structures, where

**FIGURE 2-21**
Map of the USA showing of the locations of the 118 technical universities [Müller, 1908, p. 14]

Müller-Breslau carried out extensive investigations into earth pressure on retaining walls, the results of which were published in a monograph [Müller-Breslau, 1906]. For the protagonists of practical engineering training in Germany, the USA thus advanced to become a prototype. But the debate surrounding the reform of engineering studies in Germany went well beyond the laboratory issue and would involve the auxiliary sciences of engineering – mathematics in particular – and in 1894–97 would culminate in the anti-mathematics movement. On the other side of the coin, leading engineering scientists from the USA stressed the need "to overcome the lack of scientific training in American engineering education systems and to look to Germany for solutions" [Puchta, 2000, p. 133]. The synthesis of theory and practice in the teaching and research of engineering sciences as advocated by the Stuttgart-based mechanical engineering professor Carl von Bach (1847–1931) would finally become established in Germany after 1900.

Nevertheless, in scientific terms, America did not catch up with German engineering education until the late 1930s. In contrast to the technical universities of Germany, which demanded that all students first pass an entrance examination at grammar school level, in the USA anybody wishing to study at a technical or other kind of university was admitted provided he or she had a minimum level of knowledge, laid down in the entrance conditions, regardless of the type of previous education. The Society for the Promotion of Engineering Education founded in the wake of the 1893 World Exposition in Chicago classified universities according to five groups. Group A, the highest category, was reserved for those universities that required entrance candidates to possess the following knowledge:
– algebra, including quadratic equations
– planimetry
– stereometry or plane trigonometry

**FIGURE 2-22**
Materials laboratory at Sibley College of Engineering, Cornell University [Müller, 1908, p. 63]

- at least one year's study of a foreign language
- thorough knowledge of English.

Of the 116 American education establishments for engineers recognised by the Society for the Promotion of Engineering Education, 31 were assigned to group A, 33 to group B, 25 to group C and the rest to groups D and E [Müller, 1908, pp. 42 – 45].

The knowledge tested in entrance examinations by the group A universities corresponded roughly to the level of knowledge of an 18-year-old leaving the high schools that were evolving in the USA in the final decades of the 19th century, and that of a school-leaver two years younger in Germany. Decades later, Stepan P. Timoshenko (1878 – 1972) would complain about the low level of mathematics at US high schools, where although pupils could use a slide rule, they were unaware of the theory of logarithms on which the slide rule was based [Timoshenko, 2006, p. 216]. The entrance conditions for US universities were much lower than those in Germany and that led to the proportion of general sciences occupying only 16 – 20 % of the total teaching time, even at the MIT [Müller, 1908, p. 51]. The subjects were covered in the first (Freshman) and second (Sophomore) years of study. It was not until the third (Junior) and forth (Senior) years that engineering disciplines such as theory of structures, iron construction and bridge-building got their chance. The differences between universities in the USA, "which in some cases use almost advertising-type endeavours to highlight their own university as much as possible and through the honour of the degrees to be gained encourage attendance at that university, have led to a curious abundance of the most diverse degrees" [Müller, 1908, p. 51]. For example, the MIT awarded the academic degree "Bachelor of Science" (B. S.) after four years of study, and after a further year the degree "Master of Science" (M. S.); after a minimum of two years of study in the research department of the "graduate school" and the preparation of a written thesis, the MIT could award the successful student the title "Doctor of Philosophy" or "Doctor of Engineering". In essence, it is the university at which the academic degree is obtained that is more valuable than the academic degree itself. This structure of the studies and this underlying attitude has in the meantime spread from the American to the continental European sphere and has been consolidated through countless university ranking lists. In the course of the Bologna Process initiated in 1998, the American system of teaching and research, which gave the natural and engineering sciences decisive momentum in the second half of the 20th century, has been adapted and further developed by almost all European states in order to promote the mobility, international competitiveness and employability of university graduates.

## 2.4 Insights into bridge-building and theory of structures in the 19th century

Bridge-building in the first half of the 19th century was characterised by a rapid change in the use of various building materials and their corresponding loadbearing structures. Fig. 2-23 provides an overview. Ewert provides a more detailed insight into the logic behind the development of bridge systems [Ewert, 2002].

| PHASE | BUILDING MATERIAL | | | |
|---|---|---|---|---|
| | Stone/concrete (good in compression) | Timber (good in tension and compression) | Cast iron (good in compression) | Wrought iron (good in tension and compression) |
| up to 1780 | Arch bridges | Bridges as beams, beams trussed from above and below, plus combinations of these types | – | – |
| 1780 to 1840 | Arch bridges | As above, plus arch bridges, early forms of trusses (Howe, Town, Long) | Arch bridges Beam bridges for short spans | Suspension bridges (after about 1800) |
| 1840 to 1860 | Arch bridges | Howe truss Trussed frame | Numerous transitional forms: systems by Néville, Warren, Schifkorn and others – | Lattice girder, trussed frame |
| after 1860 | Concrete bridges | Trussed frame | – | Trussed frame |

Solid structures | Structures resolved into individual members

**FIGURE 2-23**
Overview of 19th-century bridge systems

The Industrial Revolution in Great Britain opened up a new chapter in the history of bridge-building: the bridge over the River Severn at Coalbrookdale (1779) was the first be built in cast iron (Fig. 2-24), but its loadbearing structure is strongly reminiscent of the stone arch bridges that prevailed at that time. Numerous cast-iron bridges were built in Great Britain in the following years, and prefabricated cast-iron parts were allegedly exported to America as well. Mehrtens even claims that "cast iron replaced its main rival, timber, in many facets of building" [Mehrtens, 1908, p. 56]. On the continent, it was not until 1796 that a – much more modest – cast-iron bridge was built over the Striegauer Wasser river at Laasan in Prussia.

**Suspension bridges**  **2.4.1**

The first chain suspension bridge built of wrought iron was erected over Jacob's Creek in Pennsylvania, USA, in 1796. James Finley, the designer of this 22 m-span bridge, was granted a patent for his invention in 1801. By 1808, about 40 bridges had been built in North America according to this patent, the largest of them being the chain suspension bridge with two spans of 47 m each over Schuylkill Falls, Philadelphia. This bridge failed in 1811 under the weight of a herd of cattle and after being rebuilt collapsed again due to an excessive load of snow and ice [Mehrtens, 1908, p. 234].

The first suspension bridge in Europe was erected in 1816 by the textiles manufacturer Richard Lees in the Scottish town of Galashiels; it was a footbridge spanning over the Gala Water river built to provide access to his factory [Ruddock, 1999/2000, p. 103].

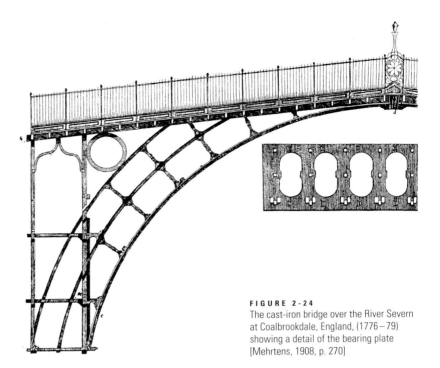

**FIGURE 2-24**
The cast-iron bridge over the River Severn at Coalbrookdale, England, (1776–79) showing a detail of the bearing plate [Mehrtens, 1908, p. 270]

One early highlight in the building of suspension bridges was the Menai Strait Bridge in North Wales, which still stands today. Designed by Thomas Telford (1757–1834) and built between 1818 and 1826, its span of 175 m was an astounding achievement for that time. Telford carried out countless strength tests on wrought-iron chains – he was apparently unaware of a coherent suspension bridge theory. According to the writings of Malberg, however, we can at best speak of a scanty preliminary design [Malberg, 1857/59, p. 561]. Nonetheless, Telford's suspension bridge (Fig. 2-25) served as the prototype for suspension bridge construction until well after the middle of the 19th century.

**FIGURE 2-25**
Suspension bridge over the Menai Strait near Bangor, Wales [Dietrich, 1998, p. 115]

By the mid-1820s, suspension bridges had been erected in other countries as well, e.g.
- in Switzerland to the designs of Guillaume Henri Dufour (1787–1875) after 1822 [Peters, 1987, p. 79],
- in France to the designs of Marc Seguin (1786–1875) after 1822 and Navier after 1823/24 [Wagner & Egermann, 1987, p. 68], and
- in Russia to the designs of Wilhelm von Traitteur (1788–1859) after 1823/24 [Fedorov, 2000, pp. 123–151].

### Austria

#### 2.4.1.1

Austria, too, achieved important progress in the building of suspension bridges at an early date. For instance, in 1823 the Emperor of Austria Franz I (known as Franz II, Emperor of the Holy Roman Empire of the German Nation until 1806) granted the company founded by Ignaz Edler von Mitis the right to build a suspension bridge in Vienna and to levy a bridge toll for 40 years [Pauser, 2005, pp. 92–93 & pp. 123–125]. Together with the designer of the bridge, Johann von Kudriaffsky, who Alfred Pauser quite rightly credits as the co-founder of the Viennese school of bridge-building [Pauser, 2005, p. 123], Mitis dedicated himself to the erection of this bridge and, above all, to the theory of structures principles established by Navier [Navier, 1823/1] for this new branch of bridge-building. Consequently, the 76 m-span Sophienbrücke footbridge in Vienna (Fig. 2-26), completed in 1825, was the first "suspension bridge erected following an elaborate structural analysis" [Pauser, 2005, p. 123], and did not have to be replaced by a new bridge until 1872. By contrast, the suspension bridge over the Seine in Paris, designed and calculated by Navier, had to be demolished due to a formal complaint before being put into service [Pauser, 2005, p. 124]. And by 1828 Austria's second chain suspension bridge, the Karl Footbridge over the Danube Canal, had been inaugurated. This footbridge, designed and calculated by Mitis spanned 95 m but the sag of the chains was just 6 m, which made it much bolder than the Sophienbrücke. However, its susceptibility to vibration due to its very low self-weight led again and again to closures and finally to its demolition in 1870 [Pauser, 2005, p. 124]. The problem of stiffening the deck of a suspension bridge was first solved in a clear structural and constructional way after the appearance of trussed girders (see also Fig. 2-81).

**FIGURE 2-26**
Sophienbrücke footbridge in Vienna, Austria's first (chain) suspension bridge [Pauser, 2005, p. 123]

### 2.4.1.2 Bohemia and Moravia

The first chain suspension bridge for traffic other than just pedestrians on the European mainland was opened on 8 June 1824. It spanned 29.70 m over a tributary of the River Morava near Strážnice in Moravia and was designed by Bedřich (Friedrich) Schnirch for Count Magnis [Hruban, 1982]. In 1827 Count Karel Chotek, the senior burgrave of Bohemia, nominated Schnirch Superintendent of Roads for the Bohemian Building Authority with the special task of designing a chain suspension bridge over the River Moldau (now Vltava) in Prague. From the five proposals submitted, the company responsible for bridge-building chose Schnirch's design at the end of 1828. At the suggestion of Franz Joseph Ritter von Gerstner (1756–1832), the company contacted William Tierney Clark, the engineer responsible for the Hammersmith Bridge in London, which had been completed in 1827. Clark sent his project documents for his bridge to Prague, but Schnirch came to the conclusion that the proposed positions of the towers were not possible in Prague, and that the deflection of Clark's design would be excessive; the vote went against him and the building of the bridge had to be postponed. Schnirch and Clark thus became "trusted but antagonistic friends" [Pauser, 2005, p. 64]. Not until 1836 did Count Chotek manage to revive his Prague chain suspension bridge project, and Schnirch was appointed to carry out the design work. Building work on the bridge began on 19 April 1839 and was completed in October 1841 [Hruban, 1982] (Fig. 2-27).

**FIGURE 2-27**
Emperor Franz Bridge, Prague
[photo: František Fridrich]

### 2.4.1.3 Germany

What was the situation regarding suspension bridges in Germany in the 1820s? The guyed bridge over the River Saale at Nienburg, which had been designed by Christian Gottfried Heinrich Bandhauer (1790–1837), the master-builder of Count Ferdinand von Anhalt-Köthen (1769–1830), and completed in 1824, collapsed on 6 December 1825 during a torchlight procession, resulting in the deaths of 55 people [Pelke et al., 2005, p. 33], [Birnstiel, 2005, p. 179]. Notwithstanding, the Count nominated Bandhauer Director of Building in 1826. Bandhauer was acquitted in 1827 by the commission set up to investigate the collapse of the bridge; two years later, he published his report on the hearing and the investigation. The subtitle of the report was remarkable: "published by the master-builder of this bridge himself following requests in public newspapers" [Bandhauer, 1829]. But as they say, "it never rains but it pours", during the building of the Catholic St. Mary's Church in Köthen, the scaffolding for the bell-tower collapsed, killing seven workers. Bandhauer was arrested, and just a few weeks before the death of his aristocratic patron, was dismissed without notice on 5 July 1830. He was accused of causing death by negligence but in the end was given a conditional discharge. Bandhauer published a book on the theory of masonry arches and catenaries in 1831 [Bandhauer, 1831], which deals with the arch form given by the inverted catenary. The book employs theory of structures as virtually a belated legitimation of his building works. He withdrew to Roßlau, the town of its birth, and died, impoverished, on 20 March 1837.

Another suspension bridge fared much better: the footbridge built by the mechanics theorist Johann Georg Kuppler over the River Pegnitz in Nuremberg, which was inaugurated on 31 December 1824 after three months of prefabrication and erection, still stands today. There is a model of this bridge in Nuremberg Museum, which was presumably used by Kuppler to erect the bridge [Petri & Kreutz, 2004, p. 308]. This two-span suspension bridge with spans of 33.78 and 32.72 m was extensively refurbished after the great floods of 1909. Vibration problems led to the addition of two additional trestles per span in the river in 1931. However, as the city authorities found temporary structures in the old quarter unacceptable during the rallies of the National Socialists, they started to plan the demolition of the bridge and its replacement with a new suspended footbridge in 1939: "The renewal of the footbridge is regarded as necessary by the mayor of the City of Nuremberg" [Petri & Kreutz, 2004, p. 310]. However, World War 2 prevented demolition, and today the Nuremberg society "BauLust" is dedicated to maintaining this important witness to the history of structural engineering.

The suspension bridge opened in 1827 in Malapane, Silesia (now Ozimek, Poland), over the Mala Panew river is still in use today. The building of the 31.56 m-span road bridge consumed 75 t of cast iron and 18.4 t of steel. The inspector of machines Schottelius was responsible for its design and building [Pasternak et al., 1996]. The value of this monument to bridge engineering is acknowledged in one way by the fact that the bridge graces the coat of arms of the Polish town of Ozimek.

**United States of America**   2.4.1.4

To mark the 200th anniversary of the birth of Johann August Röbling (born on 12 June 1806 in Mühlhausen, Germany, died on 22 July 1869 in New York, USA), two conferences on the life and work of John A. Roebling (the name he adopted in the USA) took place on 12 June 2006 in Potsdam and 27 October 2006 in New York [Green, 2006]. To accompany the exhibition in the place of his birth, a special edition of the *Mühlhäuser Beiträge* covering the work of J. A. Röbling appeared, edited by Nele Güntheroth (Berlin City Museum Foundation) and Andreas Kahlow (Potsdam Polytechnic) [Güntheroth & Kahlow, 2006]. Two essays by Nele Güntheroth contain new material about Röbling's early life, which paints a splendid picture of the times. Eberhard Grunsky writes about a spectacular chance find from 1998: Röbling's design for a suspension bridge over the River Ruhr at Freienohl, to which Grunsky assigns the date 1828. The design published by Grunsky permits a deep insight into the statics-constructional thinking of civil and structural engineers in the first third of the 19th century. Donald Sayenga analyses Röbling's lasting technological achievements in the field of wire cable production: "John A. Roebling was the most productive and most innovative wire rope pioneer of the 19th century. He was the founder of the wire rope industry in the USA. The company he set up in New Jersey in 1849 remained in business until 1973" [Güntheroth & Kahlow, 2006, p. 96]. Fig. 2-28 shows the John A. Roebling's Sons

FIGURE 2-28
Postcard showing Roebling's Wire Mills in Trenton, New Jersey
[courtesy of Mühlhausen city archives]

Company (JARSCO) wire rope factory as a postcard motif dating from 1910.

The unification of the statics-constructional and the technological in the thoughts and actions of J. A. Röbling was in the end the basis of his success. He therefore gained world fame with his two-tier Niagara railway bridge opened in 1855 (Fig. 2-29); Göran Werner describes the background to this unique project in his essay. Andreas Kahlow analyses magnificently Röbling's achievements as a design engineer in the context of the history of construction engineering. And the special edition concludes with extracts from Washington A. Roebling's (1837–1926) biography of his father and information about the building of the 487 m-span Brooklyn Bridge. After the death of Johann August Röbling, his son Washington and his daughter-in-law Emily Warren Roebling (1843–1903) took over responsibility for the Brooklyn Bridge, which was completed in 1883 and immediately became a symbol of technical progress in the 19th century.

Röbling's approach to design and his practical experience in construction were appropriately acknowledged by Stephen G. Buonopane with the following words: "Roebling's technical writings reveal his skills as an engineer – proficient with theory, but able to draw on experience and observation, Roebling relied on approximate design methods and

FIGURE 2-29
Röbling's Niagara Bridge
[Güntheroth & Kahlow, 2005, p. 135]

possessed a deep understanding of structural behaviour. Roebling's suspension bridges pushed the limits of the 19th century bridge design, and they inspired the development of more exact structural analysis methods. Roebling's stayed suspension bridge system represents an extremely safe and economical system for long span bridges. Although rarely built after Roebling's career, variations of the stayed suspension bridge continue to be proposed for very long spans" [Buonopane, 2006, p. 21].

### 2.4.2 Timber bridges

During the 18th century, the Grubenmanns, that Swiss carpentry dynasty, developed the customary covered trussed beam bridges to such a perfection that they achieved spans that pushed this bridge system to its limits. For example, Hans Ulrich Grubenmann's (1709–83) bridge over the Rhine at Schaffhausen (1757) was a two-span suspended trussed frame spanning a total of about 60 m [Killer, 1959, pp. 21–31].

The trussed beam has been known since at least the 6th century [Valeriani, 2006, p. 121]. In the trussed beam bridge, several trussed beams (see Fig. 2-13 for the basic form) are joined together to form a complete system. The most important joints are subjected to compression only and can therefore be built according to the rules of carpentry; however, the joints with the vertical hangers are at risk of slippage. Therefore, as early as 1621, Bernardino Baldi suggested connecting the vertical hangers of trussed beams to the horizontal beams solely by means of loops of metal so that the tensile force in the verticals could be used to full effect [Valeriani, 2006, p. 123]. The carpentry joints consequently remained the Achilles heel of timber bridge-building until well into the 19th century.

Soon after it came into being, the USA took the lead in timber bridge-building. The bridges designed in 1804 by Theodore Burr (1771–1820) are especially noteworthy. Fig. 2-30 shows how the construction of bridges had progressed compared to the long-span timber bridges of the Grubenmanns.

The Delaware Bridge built between 1804 and 1806 consists of a timber arch and a timber tie to counteract the thrust of the arch. With its span of 61 m, the Delaware Bridge was longer than the boldest timber bridges of the Grubenmanns. The bridge over the Connecticut River at Bellow Falls differs from the Delaware Bridge in that a lattice girder is used to stiffen the arch. Burr's lattice-type bridge systems would later give rise to semi-parabolic beams.

During the 1820s and 1830s, timber bridge-building in North America was enriched by the lattice girder bridges of Town (Fig. 2-31) and Long (Fig. 2-32). Ithiel Town (1784–1844) came from Connecticut and in 1820 was granted an American patent for his lattice girder, which was suitable for spans up to 60 m. The sides consisted of groups of criss-crossing planks, usually spruce and measuring 75 × 300 mm in section, which were connected to the chords and at their intersections by wooden pegs [Pottgießer, 1985, p. 67]. In structural terms, the lattice girder acted like a multiple trussed framework because the groups of planks crossed more than once: the system consists of several simple trussed frameworks all of

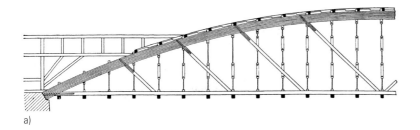

a)

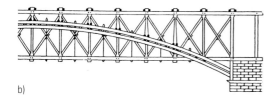

b)

FIGURE 2-30
Bridges after Burr (USA): a) Delaware Bridge at Trenton (top), and b) bridge over the Connecticut River at Bellow Falls (bottom) [Mehrtens, 1900, p. 10]

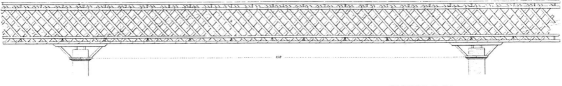

FIGURE 2-31
Town's lattice girder bridge over the River James at Richmond, USA [Pottgießer, 1985, p. 67]

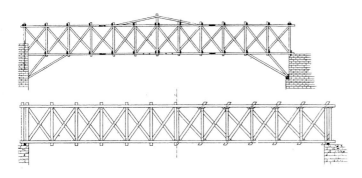

FIGURE 2-32
Lattice girders by Long [Pottgießer, 1985, p. 68]

which are stable in themselves. The structural analysis of such systems still formed part of theory of structures around 1900. Town's timber bridges became a model for the iron lattice girder bridges of continental Europe, the most distinguished of which is the Dirschau Bridge over the River Weichsel, completed in 1857.

Colonel Stephen H. Long (1784–1864) introduced the word "framework" into the English language; his bridges consisted of top and bottom chords linked by a system of diagonal and vertical members (Fig. 2-32). He was granted a patent for his bridge system and its many improvements in 1839 [Pottgießer, 1985, p. 68]. In Long's system, we can see the trussed framework model of the structural analysis: "Apparently, he [Long – the author] was one of the first to realise the magnitudes and effects of the various forces that act on the individual parts of the loadbearing construction" [Pottgießer, 1985, p. 67]. He published his report in the *Journal of the Franklin Institute.* Together with Squire Whipple (1804–88), Long is regarded as the father of structural engineering in the USA [Griggs & DeLuzio, 1995].

William Howe (1803–52) improved Long's system by using wrought-iron ties instead of pairs of timber collars and in 1840 was granted an American patent for this innovation. By prestressing the ties, a compressive stress could be induced in the individual struts such that no tensile stresses occur in the struts even under imposed loads. Fig. 2-33 shows the failure of the railway bridge at Unghvár in Austria-Hungary (now Hungary) where on 7 December 1877 a passing locomotive literally broke through the deck construction. However, the sides of the bridge, built as Howe trusses, remained intact. Strictly speaking, the Howe truss belongs to the composite systems.

### 2.4.3 Composite systems

Whereas up until the 1830s bridge-building could be broken down into stone bridges, timber bridges, cast-iron bridges and wrought-iron bridges (mainly suspension bridges), the following 25 years were characterised by a creative period of self-discovery among design engineers willing to experiment, and this resulted in diverse composite systems, but mainly combinations of cast and wrought iron. Using the example of the roof constructions built in the 1840s for the Walhalla Temple (Niederstauf), State Hermitage Museum (St. Petersburg) and the New Museum (Berlin), Werner Lorenz has introduced the category of "construction language" into the history of building: "The history of building can be described as the rise and fall of ever newer construction languages which are expressed in the continuous production of ever newer 'texts' built in these different languages" [Lorenz, 2005, pp. 172–173].

The history of bridge structures in the 19th century from composite systems to wrought-iron trussed framework systems can also be interpreted as a development towards a uniform construction language grammar. Whereas the construction language in the historical phase of composite systems is realised only in the form of diverse dialects, it later brings together its grammatical rules systematically to form a trussed framework theory. Since the end of the establishment phase of structural theory in the mid-1870s, the design, assembly and building of wrought-iron, later steel, frameworks has followed the grammatical rules of trussed framework theory. Prior to that, a fascinating diversity of composite systems, of dialects, evolved, just two of which will be examined here. For example, the footbridge over the St. Denis Canal at Aubervilliers designed by the Belgian engineer Néville in 1845 had top and bottom chords made from an $\mathrm{I}$-shaped cast-iron part with wrought iron "infills" on both sides. The triangulated sides consisted of wrought-iron diagonals braced horizontally by cast-iron members which in turn were secured by wrought-iron straps (Fig. 2-34). The details in Fig. 2-34 illustrate
- a section through the wrought-iron diagonals at the top chord (detail a),
- a longitudinal section through the top chord in the plane of the truss (detail b),

**FIGURE 2-33**
Failure of the railway bridge (Howe system) on 7 December 1877 at Unghvár, Austria-Hungary (now Hungary) [Pottgießer, 1985, p. 83]

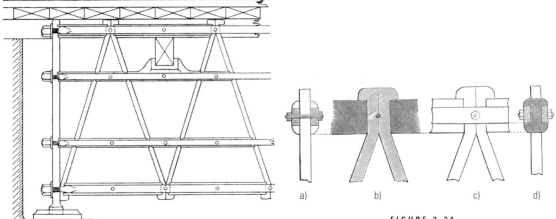

**FIGURE 2-34**
Truss by Néville [Mehrtens, 1908, p. 534]

- an elevation on a top chord joint (detail c), and
- a section through the cast/wrought-iron top chord (detail d).

James Warren and Willoughby Monzoni finally omitted the superfluous middle chords of the Néville truss and used cast iron for the ascending struts and wrought iron for the descending ties; they built the bottom chord – in tension – in wrought iron and the top chord – in compression – in cast iron. We can already see here the advance in the qualitative knowledge of the path of the forces in the resolved bridge girder in the sense of a trussed framework. Warren and Monzoni were granted a British patent for their truss system in 1848. The Warren truss was the basis for a number of further developments such as the subdivided Warren truss (a Warren truss with additional vertical bars), the double Warren truss and the quadrilateral Warren truss. In 1852 Joseph Cubitt used a Warren truss with struts of cast iron and ties of wrought iron for the 79 m-span Newark Dyke railway bridge – the first time this had been done – and 1857 saw the completion of the first larger trussed framework bridge entirely in wrought iron: Crumlin Viaduct in south Wales, which was designed by Thomas William Kennard and consists of Warren trusses with pin joints (Fig. 2-35). The standard wrought-iron bars were connected with bolts which allowed a high degree of prefabrication. "The ingenious building-kit system was exported as far as India" [Dietrich, 1998, p. 104]. The age of the composite system in bridge-building was at an end; the trussed framework bridges constructed entirely of wrought iron became more and more popular and helped to develop the uniform construction language of structural steelwork.

**FIGURE 2-35**
View of the Crumlin Viaduct in south Wales [Dietrich, 1998, p. 111]

### 2.4.4 The Göltzsch and Elster viaducts (1845–1851)

After a contract to build a railway line from Leipzig to Nürnberg via Plauen, Hof and Bamberg had been signed by the states of Saxony and Bavaria in January 1841, the two railway committees published an "offer to participate in the Saxon-Bavarian railway company" in April of that year. The ink was hardly dry on this offer before the applications for shares came rolling in for the new company, and within a few hours all the available shares (4.5 million thaler) had been bought. However, shortly after building work started on 1 July 1841, the permanent way engineer Robert Wilke (1804–89) realised it would be necessary to cross the deep valleys of the Göltzsch and Weiße Elster rivers. The speculators among the shareholders were bitterly opposed to this view because the work had already consumed 6 million thaler prior to this, the most difficult section of the line. The government of Saxony therefore advanced considerable sums to the railway company so that the bridge-building work could begin at all.

The bridge tender was advertised on 27 January 1845. A total of 81 designs were submitted for assessment to the six-man commission chaired by Johann Andreas Schubert (1808–70), but none of them satisfied the engineering principles of the commission because none of the designs had any scientific basis. The commission published its report on the competing designs on 31 July 1845. Schubert appended a masonry arch theory tailored to bridge-building entitled *Kurzgefaßte Theorie der Kreisrundbogen-Brückengewölbe* (brief theory of circular masonry arch bridges) [Conrad & Hänseroth, 1995, p. 759] to the commission's report. In that same year, Schubert published a paper on free and "prescribed thrust lines" [Schubert, 1845]. Even if his ideas about the prescribed thrust line he introduced did not lend momentum to masonry arch theory, as a bridge-builder he could already foresee the nature of the influence line, at least in outline. In order to quantify how the travelling load $Q$ influences the horizontal thrust, Schubert modelled the masonry arch implicitly as a pin-jointed rigid arch plate (Fig. 2-36) and specified an equation for calculating the horizontal thrust $H(\alpha)$ due to the travelling load $Q$ for the curved line of the intrados. Here, he understood "prescribed thrust line" as nothing more than the intrados. This is valid for the two hinges only; all the other points on the line of thrust depend on the load and therefore cannot coincide with the intrados. Schubert improved his masonry arch theory in close conjunction with bridge-building [Schubert, 1847/1848]. He supposedly completed his structural analysis for the commission project within a few weeks; a number of senior students at Dresden Polytechnic produced the working drawings [Weichhold, 1968, p. 277].

Work on site began in the summer of 1845. In some of the excavations for the piers they encountered a stratum of alum shale. These subsoil difficulties caused some shareholders to continue their obstructive tactics because they saw their railway company on the brink of bankruptcy. Once again, the state of Saxony came to the rescue and effectively nationalised the company by buying up the shares on 1 July 1847. Wilke wanted to overcome the subsoil difficulties in the Göltzsch Valley by omitting the piers affected and spanning the ensuing large gap in the second tier with two strainer arches and the gap in the fourth tier with a loadbearing arch. Each was to be built as a circular arch with a span of about 31 m [Conrad & Hänseroth, 1995, p. 762]. Schubert agreed to Wilke's proposal in principle but suggested using one arch per tier instead of two; furthermore, he wanted an elliptical surmounted arch. Again, Schubert underpinned his

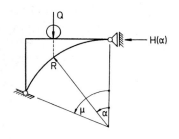

**FIGURE 2-36**
Schubert's approach to calculating the horizontal thrust in masonry arches due to travelling loads

FIGURE 2-37
Göltzsch Viaduct around 1850
[Conrad & Hänseroth, 1995, p. 762]

proposals with a masonry arch theory analysis. A report commissioned by the Saxony government, to which the highly regarded Bavarian railway builder F. A. v. Pauli contributed, pleaded for a compromise: two many-centred surmounted arches per tier. The recommendations of the report were implemented in practice (Fig. 2-37). Schubert and Wilke agreed on a totally different design for the Elster Viaduct [Conrad & Hänseroth, 1995, p. 762]. In historical terms, Schubert's contributions to masonry arch theory still belong to the constitution phase of structural theory (1825–50) because they already form part of the design activities but still have a characteristic justifying flavour.

The railway managers placed large orders for building materials in 1845, e.g. one single order for the supply of 25 million clay bricks; thereupon, the company swiftly established its own brickworks. The managers could not find any private quarry owners and stonemasons willing and able to supply the dressed granite stones required, so the company had to operate more than 40 stone quarries and bear the costs itself. The scaffolding for the Göltzsch Viaduct consumed nearly 230 000 tree trunks. Of the 10 000 + workers on the Leipzig–Plauen–Hof–Bamberg–Nuremberg line in the Vogtland, 20 % of them were at times employed on the two bridge sites. The opening of the bridges for traffic in 1851 was a triumph for stone and masonry bridge-building: the Göltzsch Viaduct, 78 m high and 574 m long, was at the time the highest railway bridge in the world. Models of the Göltzsch and Elster viaducts were proudly displayed at the Great Exhibition in London in 1851.

**The Britannia Bridge (1846–1850)**

**2.4.5**

The Conway and Britannia bridges built for the twin-track railway line between Chester and Holyhead are two structures that became symbols of technical progress in the 19th century and illustrate the heart of the process that Tom F. Peters has called "Building the Nineteenth Century", which he describes in a fascinating way in his monograph of the same name by way of case studies [Peters, 1996]. The complete railway line was intended to lop six hours off the journey time from London to Dublin. After Chester, the line was to run parallel to the north coast of Wales, cross the River Conway (Conway Bridge) and reach the Menai Strait at Bangor. Here, the railway would cross the water by way of the 465 m-long Britannia Bridge before continuing on to the ferry terminal at Holyhead on the island of Anglesey, from where travellers had only to cross the Irish Sea (105 km wide at this point) in order to reach Dublin.

Robert Stephenson (1803–59), the engineer entrusted with the design of the Conway and Britannia bridges for the Chester Holyhead Railway Company, initially planned a cast-iron arch bridge with spans of 110 m for the Menai Strait crossing but this was rejected by the British Admiralty who required a clear opening measuring 137 m wide by 32 m high above water level. The solution worked out was a beam-type bridge with two box-like tubes of wrought iron (Fig. 2-38). This was the first time a continuous beam had been selected as the structural system for a large bridge project. The two middle spans measure 141.73 m and the two end spans 71.90 m.

After the British Parliament had passed an act for the erection of the Britannia Bridge on 30 June 1845, it was decided to determine the final dimensions after carrying out tests. Robert Stephenson turned to the British Association for the Advancement of Science, which in turn appointed William Fairbairn to carry out the tests, and he in turn called on the services of his friend Eaton Hodgkinson to work out the theoretical basis. The railway company approved the tests in August 1846. Fairbairn and Hodgkinson ordered the building of beam models with circular, elliptical and rectangular cross-sections at a scale of 1 : 6 and discovered the ratio between their flexural stiffnesses to be 13.03 : 15.30 : 21.50, and therefore the rectangular cross-section represented the most efficient form. The trials also revealed that reinforcing the top and bottom of the hollow box (Fig. 2-39) gave the continuous beam such a high load-carrying capacity

**FIGURE 2-38**
The Britannia Bridge over the Menai Strait
[Pottgießer, 1985, p. 61]

that the chain suspension bridge-type suspension arrangement originally envisaged was no longer necessary. Robert Stephenson provided two hollow tubes with overall heights that varied from 7.00 m at the supports to 9.15 m at mid-span. Fairbairn developed this new type of box construction for bridges on the basis of his wealth of experience in the design and building of iron ships. The new form of construction inspired George Biddell Airy to draw up the theory of elastic plates and publish his findings in 1863 [Airy, 1863].

Of direct practical use for the building of the Britannia Bridge was Navier's theory of continuous beams in the form presented by Henry Moseley (1801–72). It was William Pole (1814–1900) and Edwin Clark (1814–94) in particular who produced a comprehensive presentation of continuous beam theory as applied to the Britannia and Conway bridges, backed up by experiments, within the scope of a two-volume book about both bridges [Clark, 1850]; on the cover it read: "Published with the sanction and under the supervision of Robert Stephenson". However, Stephenson and Fairbairn quarrelled before the Britannia Bridge was finished in 1850. The latter rushed his story of the Britannia and Conway bridges into print, giving it the title *An account of the construction of the Britannia and Conway tubular bridges with a complete history of their progress, from the conception of the original idea to the conclusion of the elaborate experiments which determined the exact form and mode of construction ultimately adopted* [Fairbairn, 1849/2]. This lengthy book title leads us to suspect that Fairbairn was keen to describe his role in the design process of both bridges, but particularly the Britannia Bridge. In that same year, a publication about the erection of the tunnel tubes of the Britannia Bridge appeared, with the subtitle *The stupendous tubular bridge was projected by R. Stephenson* (Fig. 2-40). The bridge superstructure constructed from sheets of wrought iron riveted together to form multi-cellular hollow tubes and thus a gigantic continuous beam was spectacular and innovative. But this was only one aspect; equally spectacular and innovative was the prefabrication and the span-by-span erection of the tubes plus the statics-constructional penetration of this highly complex technological process. We can thus speak of a triumph of science (see Fig. 2-40) in this synthesis of construction and technology.

Edwin Clark devised the innovative erection procedure for the two middle spans (Figs. 2-41 and 2-42):

1. Conventional erection of the two end spans $AB$ and $DE$ on timber scaffolding.
2. Building of beam $CD$ at the prefabrication yard.
3. Upon completion of beam $CD$, removal of the temporary timber supports so that the tube can be lowered onto two walls built for the purpose; at this point the tube functions structurally as a beam on two supports, whereby the loading case 'self-weight during erection' (see span $CD$ in Fig. 2-41) can be tried out on the ground.
4. Following successful checking of the load-carrying capacity, four pontoons are positioned under the each end of tube $CD$ at low tide so that

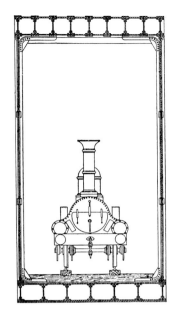

**FIGURE 2-39**
Section through one of the tubes of the Britannia Bridge [Raack, 1977, p. 4]

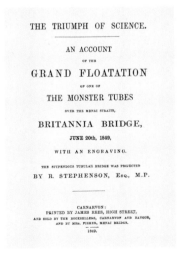

**FIGURE 2-40**
Cover of the publication on the erection of the tubes for the Britannia Bridge [n.n., 1849/2]

**FIGURE 2-41**
On the erection of the tubes for the Britannia Bridge after Clark [Clark, 1850]

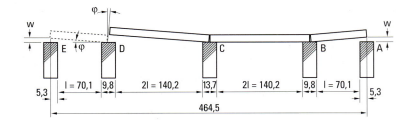

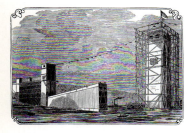

**FIGURE 2-42**
Floating one of the middle spans of the Britannia Bridge into position [n.n., 1849/2]

as the tide rises the tube is raised and its weight fully supported on the pontoons.

5. By using ropes, winches and anchors, the tube is floated out to the appropriate middle span (Fig. 2-42).
6. As the tide recedes, the tube can be lowered onto the bearings on the piers.
7. Lifting procedure with the help of hydraulic jacks and lowering onto the bearings at $C$ and $D$; the middle span $CD$ functions structurally as a beam on two supports under the self-weight loading case; at the ends $C$ and $D$, the end tangent and hence the cross-section is inclined at an angle $\varphi$.
8. Raising the bearing $E$ of beam $CD$ by the amount $w = \varphi \cdot l$ (Fig. 2-41).
9. As the two cross-sections of beam $CD$ at $D$ and beam $CD$ at $D$ exhibit the same angle $\varphi$, they can now be riveted together at $D$ to produce a two-span continuous beam $CDE$.
10. Lowering the bearing $E$ of the two-span continuous beam $CDE$ by the amount $w = \varphi \cdot l$ to its final level.
11. The fabrication and erection of middle beam $BC$ proceeds similarly to steps 2 to 7.
12. Raising the bearing $A$ of beam $AB$ by the amount $w = \varphi \cdot l$ (Fig. 2-41).
13. Beams $AB$ and $BC$ are riveted together at $B$ to produce a two-span continuous beam $ABC$.
14. Partial riveting together of the pair of two-span continuous beams $ABC$ and $CDE$ at $C$ to produce a four-span continuous beam $ABCDE$ with partial continuity at $C$.

The erection steps 1 to 14 are repeated for the superstructure of the second railway track.

Intrinsic to the whole erection process was the measurement of the deformations and angles of inclination of the tubes at important stages of the work, e.g. steps 3 and 7 to 10; the measurements were compared with the corresponding theoretical values taken from the structural calculations. The creation of partial continuity over the central pier $C$ in step 14 avoided an excessively large support moment at $C$; the span moment of the simply supported beam $BC$ or $CD$ in the erection condition governed the design. William Pole was responsible for the structural calculations for the erection conditions, and it was he who wrote the corresponding chapters on beam theory for Clark's book [Clark, 1850]: "This is an excellent example for all time of the application of scientific principles for practical purposes: it has hardly been surpassed for its elegance through ingenuity founded upon a thorough understanding of elastic theory of beams" [Charlton, 1976, p. 175].

After a construction time of five years, the tubes for the northern track were opened for traffic on 18 March 1850. The interaction of empirical findings and theory in the design, calculation, fabrication and erection of the Britannia Bridge became the prototype for the building of large bridges based on engineering science principles, as is in essence still normal today. The building of the Britannia Bridge therefore initiated the establishment phase of structural theory (1850–75).

### 2.4.6 The first Dirschau Bridge over the River Weichsel (1850–1857)

Whereas the first railway bridges in Germany were built of timber and stone, the River Weichsel crossing at Dirschau represents a departure because it was the first iron railway bridge, with six spans each of 131 m. A committee of experts was set up by the Prussian Building Authority in 1844 to handle the preliminary planning of what was at the time the largest beam-type bridge in continental Europe; the committee also advised on and specified the associated river and dyke works. Carl Lentze (1801–83), inspector of waterways and a member of this committee, was entrusted with the design of the bridge and was therefore sent to Great Britain and France by the Ministry of Finance in order to study the building of iron bridges in those countries.

In particular, the Britannia Bridge and its completion in 1850 supplied, in Lentze's own words, "by way of the thickness of the rolled iron connected with rivets [about 1.5 million rivets had to be inserted – the author], the necessary experience that was lacking for the building of a permanent wrought-iron bridge with an unsupported span exceeding 200 feet [= 62.77 m]" [Mehrtens, 1893, p. 104].

Lentze designed a suspension bridge according to Stephenson's method because it "was the sole proven means of achieving a large clear span" [Mehrtens, 1893, p. 101]. In 1847 the Prussian king instructed Lentze to investigate the option of reducing the cost of building the bridge by designing it not for carrying heavy locomotives, but instead only railway wagons, e. g. drawn by horses; something similar had been considered for the Britannia Bridge but had been immediately rejected. If it had not been for the subsequent economic crisis and the revolution, a suspension bridge would have been built which would have given the Prussian Building Authority no joy at all!

By the time the king stopped the work in 1847, a large brickworks employing 200 workers had been set up and a mechanical engineering factory with iron foundry had been established by a private company, and furthermore, construction plant had been procured and contracts concluded for the supply of stone, timber and other building materials. The work on the bridge did not resume until 1850. The royal postponement "offered a favourable opportunity for the workshops to acquire and train a skilful crew of permanent workers" [Mehrtens, 1893, p. 104].

After Lentze and Mellin had studied the Britannia Bridge in situ, they abandoned the suspension bridge idea. Lentze was now in favour of build-

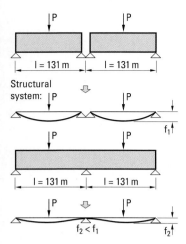

FIGURE 2-43
Comparison of the structural action of two simply supported beams and one continuous beam with one degree of static indeterminacy

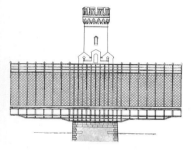

FIGURE 2-44
How the continuity effect was realised in practice for the Dirschau Bridge over the River Weichsel [Hertwig, 1930, p. 17]

**The Garabit Viaduct (1880–1884)**

ing a close-mesh lattice girder bridge and initially preferred the idea of setting up a full-size trial span! However, as news of the lecture on the Britannia Bridge given by Edwin Clark in London on 15 March 1850 reached him, he decided that the trial span was superfluous because of the structurally favourable continuity effect (Figs. 2-43 and 2-44).

Following the ceremonial laying of the foundation stone by the Prussian king in 1851, the building work proceeded relatively briskly. Railway engineer Rudolph Eduard Schinz (1812–55), in charge of the technical office, was responsible for setting up the building site complete with plant and machinery for erecting the iron superstructure. "The calculations and execution of the design were essentially his duties" [Mehrtens, 1893, p. 117]. Schinz divided the six spans of the bridge into three pairs of two spans (see Fig. 2-43). Shortly before the scaffolding to the first two spans was struck in 1855, Schinz suffered a fatal stroke as a result of overwork; unfortunately, he was unable to see his calculations confirmed by the trussed framework theory that evolved in the late 1850s in the hands of Culmann and Schwedler. In that same year the journal *Zeitschrift für Bauwesen* published a long paper by Schinz describing the bridges over the Weichsel at Dirschau and the Nogat at Marienburg, which were still under construction. The paper also includes details of his structural calculations for the two-span beams with their one degree of static indeterminacy. He models the two-span beam as a simply supported beam fixed at the central support; Schinz determines the bending moment diagram, the deflection and the forces in the lattice girder members for the deck loaded and deck unloaded cases.

A further bridge was built between 1888 and 1891 in direct proximity to the old Dirschau bridge over the Weichsel. Both bridges are described by Trautz [Trautz, 1991, pp. 63–67], also Ramm [Ramm, 1999] and Groh [Groh, 1999]. Wieland Ramm was responsible for organising an extensive exhibition, in Polish, German and English, entitled "The old Vistula Bridge at Tczew (1850–57)" at the University of Kaiserslautern in close cooperation with Gdańsk Technical University. It toured several German and Polish cities and could also be seen at the "1st International Congress on Construction History" from 20 to 24 January 2003 in Madrid. One year later, Wieland Ramm published a book [Ramm, 2004] that depicts the old Weichsel Bridge in Dirschau in the contexts of history of building and writings on the subject; the monograph also includes the exhibition catalogue. In the opinion of that outstanding connoisseur of the history of bridge-building theory and practice Georg Christoph Mehrtens (1843–1917), the building of the lattice girder bridges at Dirschau and Marienburg "was a magnificent achievement for the theory and practice of the time" (cited in [Ramm, 2004, exhibition catalogue, p. 18]).

**2.4.7**

The successful loading test of the Garabit Viaduct in 1888 undoubtedly marked the zenith of the creative works of Gustave Eiffel (1832–1923) in the field of bridge-building. As the Garabit Viaduct went into service in July 1888, Eiffel presented his *Mémoire* on this bridge to the Société des

**FIGURE 2-45**
The Garabit Viaduct shortly after completion [Eiffel, 1889]

Ingénieurs Civils, of which he was president at that time, and one year later published this as a monograph on the occasion of the World Exposition in Paris [Eiffel, 1889]. The year 1889 was to become the most famous year of Eiffel's life due to the tower erected by his company for the Exposition – and named after him but actually designed by his chief assistant Maurice Koechlin (1856–1946). But back to the Garabit Viaduct (Fig. 2-45).

The Garabit Viaduct spans the gorge cut by the River Truyère about 12 km south of the small town of Saint-Flour, enabling the single-track line from Marvejols to Neussargues to cross the deep divide at a height of 122 m (difference between top level of rails and water level before the river was dammed, measured at the centre of the arch). The Garabit Viaduct therefore held the height record for arch bridges for 92 years [Pottgießer, 1985, p. 227]. The total length of the bridge is 564.65 m and the crescent-shaped trussed arch (Fig. 2-46) has the following dimensions [Eiffel, 1889, p. 71 ff; Stiglat, 1997, p. 86]:

- span: 165 m
- rise: approx. 57 m
- depth of arch cross-section at crown: 10 m
- width of arch cross-section at crown: 6.28 m
- width of arch cross-section at springing: 20 m.

As the trussed arch is designed as pinned at the springings (acts like a cylindrical bearing with its axis in the $y$-direction), the structurally effective depth of the cross-section at this point is zero – in the longitudinal direction the system is an elastic two-pin arch with one degree of static indeterminacy with respect to the self-weight $g_z$ and the vertical imposed load due to railway operations $q_z$ ($x$–$z$-plane = vertical plane) (Fig. 2-46a). The trussed arch (the outline of its elevation is shown as a dotted line in Fig. 2-46a) has to withstand bending and axial forces. On the other hand, the system with regard to the wind loads acting horizontally $q_y$ ($x$–$y$-plane = horizontal plane) on the crescent-shaped trussed arch is a curved elastic trussed girder fixed at the supports, i.e. three degrees of static indeter-

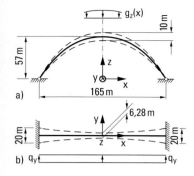

**FIGURE 2-46**
Details of the arch of the Garabit Viaduct: structural system a) on elevation, and b) on plan

minacy (Fig. 2-46b). The trussed girder (the outline of its plan shape is shown as a dotted line in Fig. 2-46b) is subjected to bending and torsion. The main arch of the Garabit Viaduct can therefore be classified as an externally statically indeterminate space frame.

The Garabit Viaduct is based on the Maria Pia Bridge over the River Douro at Porto, Portugal, which Eiffel completed in 1877. Eiffel's price submitted to the Royal Portuguese Railway Company was about 40 % lower (per linear meter of bridge) than the next cheapest tenderer [Walbrach, 2006, p. 278]. It was this structure that placed Eiffel on the international stage for the first time. Eiffel's assistant Théophile Seyrig played the leading role in the planning and execution of the work. The 160 m span of the arch of the Maria Pia Bridge is semicircular, and the arch takes on the function of the beam-like superstructure in the vicinity of the crown. Seyrig's structural calculations are based on the following actions:
- imposed load due to railway operations $q_z$ = 40 kN/m
- wind pressure on unloaded construction $q_y$ = 2.75 kN/m$^2$
- wind pressure on loaded construction $q_y$ = 1.50 kN/m$^2$.

Seyrig investigated only three loading cases: maximum load, imposed load on one half of the arch with $q_y$ (asymmetrical loading case), and imposed load on both halves of the arch with $q_y$ acting on a length extending 40 m left and right of the crown (symmetrical loading case).

Eiffel's success led to him being appointed by the Minister of Public Works to build the Garabit Viaduct – along the lines of the Maria Pia Bridge – on 14 June 1879 at the suggestion of the state engineers Bauby and Léon Boyer (1851–86) – without issuing a tender [Eiffel, 1889, pp. 135–140]. Shortly afterwards, Eiffel sacked his assistant Seyrig because he had asked for a share in the revenue; Karl Culmann recommended his pupil Maurice Koechlin as a replacement. Koechlin substantially revised Boyer's preliminary design for the Garabit Viaduct, which had taken the Maria Pia Bridge as its starting point:
- arch axis follows a quadratic parabola
- distribution of mass in trussed arch adapted to suit local loading conditions
- separation of trussed girder of superstructure from trussed arch
- replacement of cast-iron tubular sections for the columns by box-like riveted rectangular sections made from steel plates.

Koechlin obtained the rise/span ratio by solving an optimisation exercise in which he minimised the weight of the arch. The deviations of the parabolic arch axis from the line of thrust are considerably smaller than for a circular arch axis.

The structural calculations for the Garabit Viaduct which stem from Koechlin's pen form the main part of the book published by Eiffel in 1889 [Eiffel, 1889]. In the calculations, Koechlin achieves an independent combination of Jaques Antoine Charles Bresse's (1822–83) elastic arch theory [Bresse, 1854] and Culmann's graphical statics [Culmann, 1864/66] accompanying the elastic arch theory. Both thermal effects and the influence of moving loads were considered in the static indeterminacy calcula-

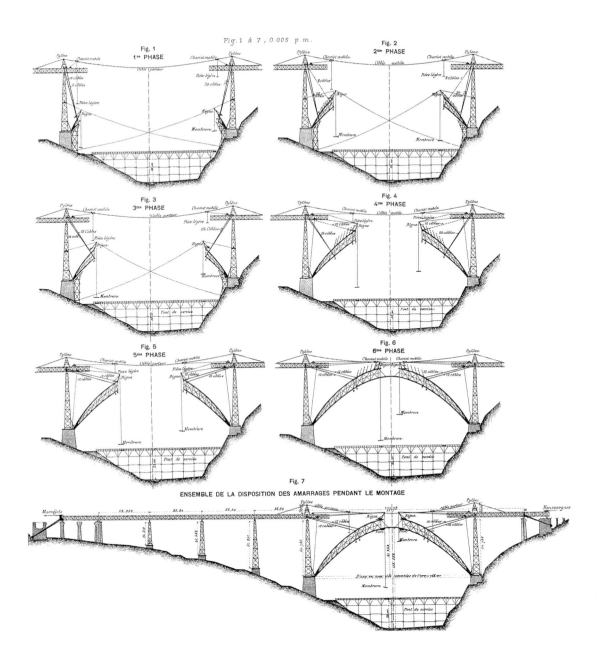

FIGURE 2-47
How the arch of the Garabit Viaduct was erected [Eiffel, 1889]

tions. The numbers were evaluated in graphic and tabular form. For instance, when determining the arch displacements, Koechlin evaluated the integrals graphically, a method that Otto Mohr introduced into graphical statics and which would later be called "Mohr's analogy" in the engineering literature.

But it was not only the conception, calculations and design that posed new challenges – the building of the Garabit Viaduct also tested Eiffel's company. The necessary infrastructure had to be created first in this uninhabited region of the Auvergne: accommodation, canteen, offices, workshops, stores, cattle stalls and even a school for the children of the workers, who were expected to remain on site for several years and so brought their families with them. At times there were up to 500 workers on site, mostly Italians [Stiglat, 1997, p. 87]. Construction work began in the spring of 1880 with the foundations for the arch abutments. Fig. 2-47 shows how the arch was erected between 24 June 1883 and 6 April 1884. Eiffel's long-serving assistant Émile Nouguier (1840–98) was responsible for the entire erection procedure.

Like with the Maria Pia Bridge, the arch was built up from both springings like guyed cantilevers: the guy ropes were draped over the tops of the piers rising from the springings and continued to the abutments, where they were anchored [Stiglat, 1997, p. 87]. Some 4000 t of iron and 20 000 m³ of granite had to be transported the 34 km from the railway station at Neussargues using teams of horses, oxen and cattle. The iron members were then carried across the timber temporary bridge on rail-borne wagons (see Fig. 2-48) before each piece, weighing about 2 t, could be heaved into position with manually operated cranes. The guyed cantilever construction employed such "mathematical precision in the calculation, fabrication and erection that reworking of the rivets inserted on site – about half of the total of 500 000 – was seldom necessary" [Walbrach, 2006, p. 279]. Although the deck was completed in June 1884, the loading test could not be carried out for another four years because the railway line was not ready!

The Garabit Viaduct project achieved a new level of quality in the relationship between practice and theory in bridge-building in terms of design, calculation, building and published works because statics-constructional and technological theory and practice were moulded into a superior unison. It is especially Koechlin's rational forms of structural calculations and drawings based on structural engineering theory that express the new technological and scientific footing of engineering work. During the construction of the Eiffel Tower in 1889, this process would be literally taken to a new height by Eiffel's company. Koechlin therefore symbolises the historico-logical middle of the classical phase of structural theory (1875–1900).

**Bridge engineering theories**

## 2.4.8

The formation of theory of structures into a fundamental engineering science discipline of civil and structural engineers (1825–1900) is closely linked with the development of bridge-building. The overview in Fig. 2-48 conveys a rough impression of developments in masonry, monolithic and iron bridge-building during the 19th century; the relationships between practice and theory are summarised schematically here. The growing recognition of continuity effects in stone and masonry bridges from the arch of large-format, dressed natural stones to the reinforced concrete arch matched the trend towards continuum mechanics on the level of structural engineering theory. On the other hand, the development of loadbearing structures in iron bridge-building tended towards resolving the structure into bar-like arrangements, which led to trussed framework

**MASONRY & MONOLITHIC BRIDGES**

*Tendency: developing continuity of the structure*

| | Construction | Theory |
|---|---|---|
| Many theories | Arches built with large-format dressed stones | French arch theories of La Hire (1695), Bélidor (1729), etc., completed by Coulomb (1773) |
| Many theories | Arches built from small-format clay bricks | Eytelwein (1808) etc. |
| Many theories | Arches built from rubble stones in mortar bed | Young (1817), Gerstner (1831), Moseley (1835), Schubert (1847) – Thrust line theories |
| Harmonised theories | Concrete arches | Winkler (1868) was the first to design the arch based on elastic theory |
| Harmonised theories | Reinforced concrete arches | Stages in modelling:<br>– arch as elastic bar<br>– arch as elastic plate<br>– arch as elastic barrel shell<br>→ Theory of shell structures |

**IRON BRIDGES**

*Tendency: resolving structure into individual members*

| | Construction | Theory |
|---|---|---|
| Many theories | Cast-iron arches | Individual arch theories, e.g. Reichenbach (1811), Young (1817) |
| Many theories | Riveted solid-web beams | Navier's bending theory (1826) |
| Many theories | Close-mesh lattice girders | Navier's bending theory, tests, later calculations according to trussed framework theories of Culmann and Schwedler |
| Harmonised theories | Trussed frameworks (statically determinate) | Culmann's graphical statics (1866) |
| Harmonised theories | Trussed frameworks in general | General theory of statically indeterminate systems. Mohr, Müller-Breslau, Maxwell, Engesser, etc.<br>→ Member analysis |

**FIGURE 2-48**
Relationship between theory and practice in iron and masonry/monolithic bridge-building

theory and graphical statics, and at the end of the discipline-formation period of structural theory (1825–1900) was crowned by the general theory of elastic bar trusses. From the mid-19th century onwards, the distinctive bridge engineering theories were absorbed into the scientific canon of structural analysis.

### 2.4.8.1 Reichenbach's arch theory

In his introduction to the first edition of his paper *Theorie der Brückenbögen und Vorschläge zu eisernen Brücken in jeder beliebigen Größe* (theory of bridge arches and proposals for iron bridges of any size) published in 1811, Georg von Reichenbach (1772–1826) calls bridge-building "one branch of applied mechanics" (cited in [Mehrtens, 1908, p. 301]). As the Baden-based mechanics theorist could only conceive his structures in terms of stone and cast iron, it should be no surprise to learn that Reichenbach's bridges made from prefabricated cast-iron tubes owe much to stone bridge-building in terms of their construction (Fig. 2-49).

And that corresponds to his theory: starting with the masonry arch theories used in stone bridge-building which were popular at that time, especially in France, Reichenbach developed his theory to such an extent that he could calculate a vertical load $Q$ under which the cast-iron arch would fail at two predefined points. Here, $Q$ depends on a "rupture moment" which in turn is a function of the tensile strength of the cast iron; Reichenbach's equation for the "rupture moment" is correct for the case of an axial force acting on the circumference of the tube (= static moment of an axial force applied to the circumference of the tube in relation to

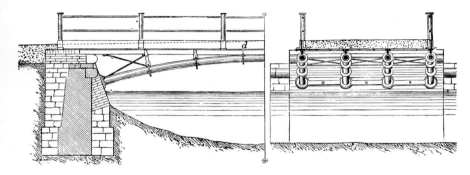

the centre-of-gravity axis of the tube). Taking the aforementioned restrictions into account, Reichenbach's "rupture moment" can be interpreted as the failure moment. His bridge theory is based on his own tensile strength tests on cast iron. He does not calculate the compressive stresses occurring in the tube cross-section. After Reichenbach has determined the failure moments of the assembled arch (i. e. several tubes in the bridge elevation), he first considers the effect of a travelling load $Q$ on a single arch and establishes the relationship between this effect and the failure moment plus the arch dimensions required to prevent the arch from collapsing [Mehrtens, 1908, p. 309].

Reichenbach specifies an equation for $Q$ in which he enters the position of load $Q$ and discovers through trial and error that the most unfavourable position of the load is at the first quarter-point of the arch. Finally, he specifies the corresponding equation for the assembled arch. Unfortunately, Reichenbach overlooks the significance of the self-weight for the loadbearing behaviour of tubular bridges. Nevertheless, his theory must be seen as an attempt to provide a scientific footing for a special branch of iron bridge-building, although Reichenbach's work does not lead to any dimensions. Reichenbach himself did not design any tubular bridges that were actually built; the few that were built by others disappeared a few decades later. The goal for which Reichenbach's theory and construction proposals were intended was the building of iron bridges "whose clear span is, so to speak, unlimited, whose cast components can be produced in any iron foundry, and which, with sufficient thickness, can be built well below the price of solid stone arches with the same arch length and span and would be not too much more expensive, perhaps in some circumstances only slightly more expensive, than the newest and best timber bridges" (cited in [Mehrtens, 1908, p. 302]).

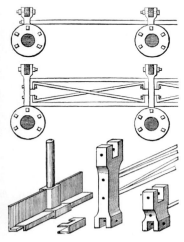

**FIGURE 2-49**
Tubular bridge over the River Hammerstrom at Peitz [Mehrtens, 1908, p. 306]

### 2.4.8.2
**Young's masonry arch theory**

Thomas Young's (1773–1829) deliberations concerning masonry arch theory fall in the period between 1801 and 1816: they begin with his agreement to hold lectures for the Royal Institution in London, which at the same time initiated his interest in the mechanical arts, and close with the completion of his article on bridges which appeared in 1817 in the *Supplement to the fourth edition of the Encyclopaedia Britannica* [Young, 1817]. George Peacock's *Miscellaneous works of the late Thomas Young*

[Young, 1855] contains an abridged version of the article. In analysing Young's masonry arch theory, Santiago Huerta discovered its anticipatory nature and allocated it to the history of the development of masonry arch theory [Huerta, 2005].

Young's article on bridges is divided into six parts:
1. Resistance of materials
2. The equilibrium of arches
3. The effects of friction
4. Earlier historical details
5. An account of the discussion which have taken place respecting the improvement of the Port of London
6. A description of some of the most remarkable bridges which have been erected in modern times

As the fifth part shows, Young gets involved in the discussions surrounding the redesign of the Port of London, the central aspect of which was the design of a new bridge to replace the old London Bridge (see [Dorn, 1970; Ruddock, 1979]). Thomas Telford supplied a forward-looking design in the shape of a 183 m-span, cast-iron arch bridge with a rise of 19.52 m (Fig. 2-50). Unfortunately, the committee entrusted with selecting the new bridge chose a conventional stone arch bridge with three spans.

After Young has developed his masonry arch theory in the first three parts, he uses the knowledge gained to answer in detail the 21 questions the committee posed in relation to Telford's design (see [Huerta, 2005, pp. 227–229]) in the fifth part. Young introduces his article as follows: "The mathematical theory of the structure of bridges has been a favourite subject with mechanical philosophers; it gives scope to some of the most refined and elegant applications of science to practical utility; and at the same time that its progressive improvement exhibits an example of the very slow steps by which speculation has sometimes followed execution, it enables us to look forwards with perfect confidence to that more desirable state of human knowledge, in which the calculations of the mathematician are authorised to direct the operations of the artificer with security,

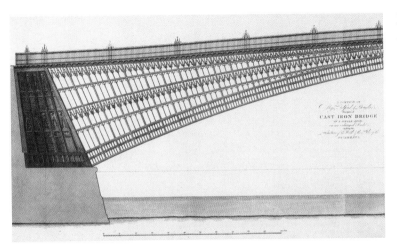

**FIGURE 2-50**
Telford's design for a cast-iron arch bridge to replace the old London Bridge [Ruddock, 1979, p. 156]

instead of watching with servility the progress of his labours" [Young, 1855, p. 194]. Despite his criticism of bridge engineering theory at that time, Young does not plead for practice driven by theory, but instead for scientific observation of the progress of the building work. He therefore highlights knowledge as the cognitive and design as the practical goal of engineering sciences. Does his masonry arch theory fulfil these aims of engineering science theory formation?

In the first part of his article, "Resistance of materials", Young formulates the law of distribution of deformations over the arch cross-section (arch depth $d$, arch width $b$) when the axial force $N$ is applied eccentrically. As he presumes proportionality between force and deformation, Huerta converts the relationships described by Young in text form into the algebraic language of the stress concept (Fig. 2-51).

Young describes the position of the neutral axis in the arch cross-section as follows: "The distance of the neutral point from the axis is to the depth, as the depth to twelve times the distance of the force, measured in the transverse section" [Huerta, 2005, p. 202], which can be expressed algebraically in the following form

$$z = \frac{d^2}{12 \cdot y} \qquad (2\text{-}12)$$

For the axial stresses $\sigma$ for the fibres on the extreme right (Fig. 2-51), Huerta translates Young's passage of text into

$$\sigma = \sigma_m \cdot \left( \frac{d + 6 \cdot y}{d} \right) \qquad (2\text{-}13)$$

where

$\sigma_m$ average stress when axial force $N$ is applied concentrically, i.e.

$$\sigma_m = \frac{N}{d \cdot b} \qquad (2\text{-}14)$$

If $N$ is applied on the right edge of the kern $y = d/6$, then according to eq. 2-12 the neutral axis is on the extreme left of the cross-section, i.e. the axial stress is zero there, but the axial stress according to eq. 2-13 assumes the value $2\sigma_m$ on the extreme right of the cross-section. Young has therefore indirectly specified the middle-third rule – years ahead of Navier: if the axial force $N$ is applied in the middle-third of the depth $d$ of the arch, i.e. in the kern of the arch cross-section, then all the stresses are compressive. After that, Young analyses the thermal influences on the internal

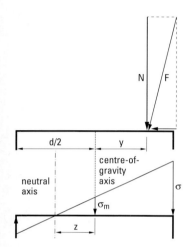

FIGURE 2-51
Stress distribution over the cross-section after Young [Huerta, 2005, p. 202]

FIGURE 2-52
Equilibrium of Telford's cast-iron arch bridge (see Fig. 2-50) for self-weight after Young [Huerta, 2005, p. 214]

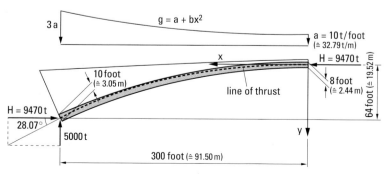

forces of iron arches [Huerta, 2006]; here again, he is, in principle, ahead of the corresponding contributions from Bresse [Bresse, 1854], Rankine [Rankine, 1862], Winkler [Winkler, 1868/69] and Castigliano [Castigliano, 1879].

In the second part of his article on bridges, Young introduces the concept of the line of thrust and develops the line of thrust theory for masonry arches. He derives the equation of the line of thrust for a symmetrical masonry arch subjected to asymmetrical vertical load distribution and applies this to Telford's design for a cast-iron arch bridge (Fig. 2-52)

$$y = \frac{a \cdot x^2}{2 \cdot H} \cdot \left(1 + \frac{x^2}{25\,117}\right) \tag{2-15}$$

where
- $y$    ordinate of line of thrust in [m]
- $x$    abscissa of line of thrust in [m]
- $H$    horizontal thrust in [t] ($H$ = 9,470 t in this example)
- $a$    ordinate of parabolic uniformly distributed self-weight $g$ in the axis of symmetry in [t/m] ($a$ = 31.79 t/m in this example)

Parameter $b$ in the parabolic uniformly distributed self-weight $g$ (see Fig. 2-52) is $a/4186.25$. Young specifies the ordinates of the line of thrust according to eq. 2-15 in a table (see [Huerta, 2005, p. 212]). The line of thrust according to eq. 2-15 is shown in Fig. 2-52 as a dotted line; it runs within the arch profile, which means that the arch is stable. However, Young assumes that the line of thrust passes through the centres of gravity of the springers and the keystone. This fixing of the line of thrust enables Young to reduce a problem with three degrees of static indeterminacy to the solution of a pure equilibrium issue. However, an infinite number of lines of thrust can be found, each of which represents an equilibrium condition. The question is: Which is the true line of thrust? The later elastic arch theories of Bresse, Winkler, Castigliano and others tried to find the answer to this question. Finally, Young calculates the line of thrust for self-weight and a point load acting at an arbitrary point. Huerta evaluates Young's masonry arch theory and Telford's design as follows: "Young's analysis of Telford's design is completely correct, combining statements of equilibrium (curves of equilibrium for given loads) with statements about the material (cast iron must work in compression; therefore the curve must lie within the arch). ... Telford's design, with some modifications, would have been a completely safe structure." Had it been built, it would today be "a symbol of London, in the same way as the Eiffel tower is a symbol of Paris. It is to regret that ignorance, fear and parsimony stopped Telford's grand design" [Huerta, 2005, p. 227]. Young's masonry arch theory lay dormant for decades; logically, it belongs to the establishment phase of structural theory (1850 – 75), even though it was published in 1817. The masonry arch theory of Thomas Young is therefore an ingenious anticipation of an engineering science theory in which cognition and design are fused dialectically into a higher accord.

**Navier's suspension bridge theory**

#### 2.4.8.3

Whereas Galileo, the Bernoullis, Leibniz, Huygens, Gregory and Euler had already dealt with the equilibrium form of the catenary – Hooke and Gregory the inverted catenary, the line of thrust – from a purely mathematical viewpoint, Navier used this work as a basis for the first practical suspension bridge theory in 1823 [Navier, 1823/1] (see also [Wagner & Egermann, 1987]). Navier's suspension bridge theory deals with important problems such as

- the influence of bridge loads on the design and span of the bridge,
- the equilibrium of the towers for one or more spans,
- the calculation of the chain dimensions using the most reliable test results available at that time and the elastic elongation of the iron (Gauthey, Rondelet, Duleau, Barlow, Pictet),
- the dynamic influences of loads on the vibration of the bridge deck, and
- the equilibrium of chains and hangers.

Navier's theory, a fortunate synthesis of mathematical theory and evaluation of tests, was to determine the calculation of suspension bridges for the next 50 years.

What are the differences between the bridge theories of Reichenbach and Navier?

- In contrast to Reichenbach, Navier solves the detailed design task, which was only possible because of his comprehensive evaluation of tests.
- Reichenbach justifies his construction system theoretically, Navier achieves the theoretical analysis of suspension bridge construction hitherto.
- Compared to the colossal suspension bridges, the cast-iron tubular bridges, albeit reliable, are very modest. It was therefore necessary to switch from empirical findings to engineering theory in the practical design of suspension bridges. Before the cast-iron arch bridge could undergo this qualitative change, it had already disappeared.
- In its approach to and treatment of the problem, Reichenbach's theory followed the tradition of the masonry arch theories of the 18th century.

Reichenbach's arch theory must therefore be assigned to the initial phase of structural theory (1775–1825), whereas Navier's suspension bridge theory belongs to the constitution phase (1825–50).

**Navier's *Résumé des Leçons***

#### 2.4.8.4

Whereas Galileo, Mariotte, Leibniz, Parent, Jakob Bernoulli (1655–1705), Euler, Coulomb and others had already dealt with beam theory and the theory of the elastic line, Navier brought together these two, often merely co-existent, threads of theory into a practical bending theory in his *Résumé des Leçons* [Navier, 1826, 1833/1838 & 1839]. Peter Gold speaks of the first coherent theory of structural analysis from a scientific theory perspective [Gold, 2006 & 2007].

However, it took some time before the practical bending theory actually became established in practice. "In 1855, i.e. about 30 years after Navier published his bending theory, opinions as to its usability differed considerably among experts" [Werner, 1979, p. 49]. Navier's view of beams resolved into individual members certainly contributed to this (Fig. 2-53).

Navier uses only the cross-sections of the top and bottom chords in his calculation, but the "parallel positioning of the two bars [top and bottom chords – the author] can only be achieved if the bars are joined together with a series of transverse pieces and St. Andrews crosses" [crossing diagonals – the author] [Navier, 1833/1878, p. 297]. Expressed simply, Navier considers resolved beams as perforated beams; some bridge-builders were to use this approach, for example, for calculating close-mesh lattice girders [Mehrtens, 1908, pp. 530–531]. Despite these shortcomings, his *Résumé des Leçons* was *the* pioneering work for the entire discipline-formation period of structural theory (1825–1900).

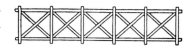

FIGURE 2-53
Lattice girder after Navier
[Mehrtens, 1900, p. 11]

### 2.4.8.5 The trussed framework theories of Culmann and Schwedler

In his famous travelogue, Culmann speaks about the great diversity of timber bridges in North America and how many of them show signs of damage – indeed, how many of them have collapsed – despite their generous use of materials. In those bridges, Culmann found various loadbearing systems superimposed on each other which on their own – assuming correct design – would have fulfilled their tasks admirably. He therefore established a trussed framework theory based on the following simplifications:

1. The system of infill bars between top and bottom chords should be arranged such that all bars always form triangles.
2. The bars should be able to rotate without restraint at their joints (pinned connections).

The pinned trussed framework model had thus seen the light of day in theory of structures! With the help of the equilibrium conditions, Culmann was able to calculate the forces in the members of any statically determinate trussed framework with sufficient accuracy.

Almost concurrently, a paper by Schwedler appeared in the journal *Zeitschrift für Bauwesen* in which he developed the principles of a trussed framework theory. The following remark by Schwedler regarding the nature of structural analysis theories is worthy of note: "The above remarks have only been made in order to indicate how a theory based on certain assumptions cannot be applied to building works before one has checked whether all assumptions also apply to the building works. On the contrary, it will be discovered that the theory for every structure, depending on the material, its elasticity, the cross-sections of the parts, the connection details and many other matters will have to be rectified if mistakes are to be avoided. The theory provides only a general scheme according to which the stability of the structure is to be established, but it remains for the builder to flesh out this scheme with his ideas in each particular case" [Schwedler, 1851, p. 167]. The understanding of theory in the establishment phase of the theory of structures (1850–75) – like to all intents and purposes the self-conception of structural theory – is thus precisely defined.

Besides Culmann and Schwedler, both Zhuravsky and Whipple – in 1847 in his book *A Work on Bridge Building* (Fig. 2-54) – proposed trussed framework theories [Stüssi, 1964, p. 18].

## 2.5 The industrialisation of steel bridge-building between 1850 and 1900

The steelwork industry emerged in Great Britain, Germany, France and the USA in the second half of the 19th century. After the American Civil War (1861 – 65), the USA experienced an unprecedented industrial ascent and prepared to overtake that "workshop of the world", England. Attentive visitors to the 1876 World Exposition in Philadelphia could begin to suspect what by the time of the World Exposition in Chicago in 1893 had become clear: the USA was now the world's greatest industrial power. During the final quarter of the 19th century, the steelwork branch in the USA evolved into a huge industry with Philadelphia at its hub. Germany, too, witnessed a considerable expansion on the industrial front after the founding of the Second Reich in 1871. And in the 1890s the German steelwork industry also adopted the linear and continuous type of factory production that had been born in the USA. Like the electrical industry, the huge steelwork industry of the USA and Germany was around 1900 a first-class, science-based, high-technology sector. As part of these developments, the intimate interplay between steel bridge-building and theory of structures was especially typical of Germany.

### 2.5.1 Germany and Great Britain

The building of the large bridge over the River Weichsel at Dirschau is an important example of how large bridge projects were carried out by the state authorities in the Germany of the 1840s and 1850s. This state control started with the composition of the planning committee, the members of which all came from the building authority and local government. Larger states such as Prussia, Bavaria and Saxony were able to control all phases of the building process in terms of organisational and technical aspects. And the workforce, mostly recruited from the surplus of agricultural labour, also required adequate training, especially for iron construction projects.

**FIGURE 2-54**
Cover of Whipple's *A Work on Bridge Building* [Whipple, 1847]

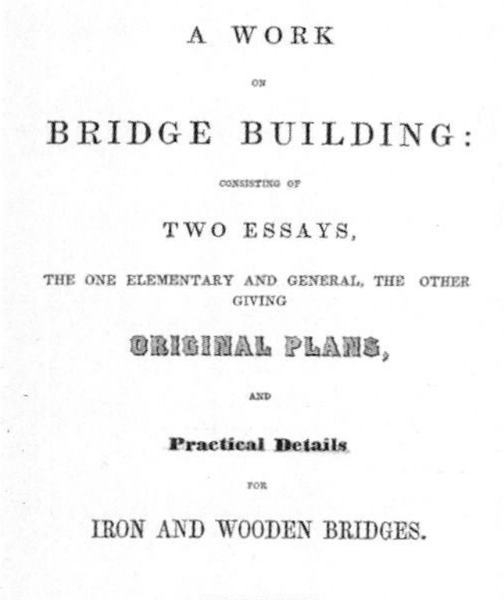

A WORK
ON
BRIDGE BUILDING:
CONSISTING OF
TWO ESSAYS,
THE ONE ELEMENTARY AND GENERAL, THE OTHER GIVING
ORIGINAL PLANS,
AND
Practical Details
FOR
IRON AND WOODEN BRIDGES.
BY S. WHIPPLE, C. E.
UTICA, N. Y.
H. H. CURTISS, PRINTER, DEVEREUX BLOCK.
1847.

To contrast with this, Culmann explained the execution of bridge works in England: "All are witness to the surplus of the most proficient manual labourers at the disposal of the English engineers. The English surpass all other peoples in this respect; the inexhaustible financial resources, the experienced workers and their inexpensive materials are a substitute for their lack of education in theory." He continues thus: "The English engineer knows that this and that construction is flawed not because they do not comply with the simplest laws of statics, but rather because they had not held here and there" [Culmann, 1852, p. 208]. For example, on Stephenson's Britannia Bridge, generous finances were made available for preliminary tests and the mathematician Hodgkinson, "who was certainly fully acquainted with the theory of the strength of materials", was called in to advise; but Hodgkinson "did not focus quickly enough on the specific purpose" [Culmann, 1852, p. 176]. Culmann's opinion must be explained in three respects:

*Firstly:* The loadbearing structure of the Britannia Bridge was the type of innovation that occurs only once a century, and such a complex innovation cannot be simply calculated; the design could only be safeguarded through tests.

*Secondly:* The Britannia Bridge's completely new type of loadbearing structure concealed many theory of structures issues that could not be solved ad hoc, but instead became objects of scientific activity at a later date, e. g. the buckling problem of the large wrought-iron panels.

*Thirdly:* The erection of the Britannia Bridge was accompanied by the structural calculations of Edwin Clark plus deformation measurements.

This might be enough to explain the differences between British and German bridge-building and theory of structures at that time. The effectiveness in practical terms was, for example, that during the building of the Weichsel Bridge, a Swiss engineer, Schinz, calculated this bridge according to the trussed framework theories of Schwedler and Culmann without having to resort to elaborate tests comparable with those carried out by Fairbairn for the Britannia Bridge.

Of course, the building of the Britannia Bridge and other bridges gave scientists like Airy (an astronomer) the chance to carry out research into areas like elastic theory [Raack, 1977], but they did not find their way into the design activities of engineers. The above examples can be summarised in the form of the following thesis:

Whereas the use of structural analysis theories for public-sector bridge projects was already part of the design process in the 1850s, private-sector bridge projects relied on tests; the scientists of the time provided commentaries and assessments, but did not intervene. Public-sector bridge projects were only common in the German states during the Industrial Revolution (1845–65), and even during this period, the consolidation of capitalist production conditions in Germany pointed to increasing privatisation in the building of large bridges. At the end of these developments, the tasks of the building authorities had been reduced to

- preparing a preliminary design,
- selecting a design from the competitors' submissions, and
- supervising all works and deliveries to the foundry and factory as well as the building site.

The preparation of designs was carried out by an "excellent staff of civil servants with thorough theoretical and practical training which was difficult for an individual person to match. This is proved by the more recent tenders for bridges in which the contracts are regularly awarded to the larger companies" [Mehrtens, 1900, p. 93]. Mehrtens gave an impressive account of German bridge-building in the 19th century in his book *A Hundred Years of German Bridge Building*, which was published on the occasion of the World Exposition in Paris in 1900. He was commissioned to write this commemorative volume by six leading steelwork companies, and the publication appeared in English, French and German editions. It exudes the spirit of scientifically founded bridge-building and clearly expresses the convergence of the end of the discipline-formation

**FIGURE 2-55**
Trussed arch of the Worms road bridge over the Rhine (1897–1900) during assembly in the Gustavsburg Works of M.A.N. AG [Mehrtens, 1900, p. 107]

period of structural theory, in the form of classical theory of structures (1875–1900), plus the engineering and scientific development of the bridge-building industry into a major industry (Fig. 2-55).

### 2.5.2 France

At the 1900 World Exposition in Paris, Eiffel played his trump card in the form of two magnificently bound large folio volumes entitled *La tour de trois cents metres* [Eiffel, 1900]. The 500 numbered copies destined for important personalities were simultaneously balance, gift and advertisement of Eiffel's work, as Bertrand Lemoine notes in the reprint of these monumental tomes containing his commentaries (Lemoine in [Eiffel, 2006, p. 9]). Maurice Koechlin calculated the internal forces of the Eiffel Tower, inaugurated in 1889, using graphical statics, and three splendid plates are included. Some 40 engineers and craftsmen produced 700 general arrangement drawings plus 3,600 working drawings (Lemoine in: [Eiffel, 2006, p. 10]). It took less than 200 workers 21 months to erect the tower following finely tuned preparations on an excellently organised building site, and there was only one death during the work. Lemoine attributes Eiffel's success as an engineering entrepreneur – especially in the field of iron bridges – to the convergence of several factors: "Those include technical innovation, primarily in construction techniques, the mastery of industrial production methods, high quality demands, the mobilisation of talents and financial resources thanks to Eiffel's charisma, the excellent organisation of production and management, negotiating skills and the unceasing upkeep of a network of supporters who could be activated at the right moment. Eiffel developed his company like he created his myth: through his technical achievements, his distinctive business acumen and his advertising skills. This enabled him to focus the successes of his company on himself and his name" (Lemoine in [Eiffel, 2006, p. 10]). The 326 engineers, foremen and workers involved in the conception, design and construction of the tower are all listed by name in his book. But that was not enough for him: above the four main arches of the Eiffel Tower, the great man himself instructed that the surnames of 72 natural and engineering scientists, engineers and entrepreneurs be preserved on 72 cast-iron plates as homage to the triumph of scientifically based engineering. The names include those whose creative activities had a profound influence on bridge-building, theory of structures and applied mechanics: Marc Seguin (1786–1875), Henri Tresca (1814–85), Jean-Victor Poncelet (1788–1867), Jacques Antoine Charles Bresse (1822–83), Joseph Louis Lagrange (1736–1813), Eugène Flachat (1802–73), Claude-Louis-Marie-Henri Navier (1785–1836), Augustin-Louis Cauchy (1789–1857), Gaspard de Prony (1755–1839), Louis Vicat (1786–1861), Charles Augustin Coulomb (1736–1806), Louis Poinsot (1777–1859), Siméon Denis Poisson (1781–1840), Gaspard Monge (1746–1818), Antoine-Rémi Polonceau (1778–1847), Benoît-Pierre-Emile Clapeyron (1799–1864), Jean Baptiste Joseph Fourier (1768–1830) and Gabriel Lamé (1795–1870). Despite this array of the great names of French science, the growing optimism in general and bridge-building in particular had already suffered – after the collapse of the railway bridge over the Firth of Tay in Scotland on 28 December 1879 – a second, bitter setback: the railway bridge over the River Birs in Mönchenstein near Basel, built by Eiffel's company in 1875, collapsed on 14 June 1891 with the loss of 73 lives. The tragedy led to the publication of the first Swiss standard for loadbearing structures one year later: *Berechnung und Prüfung der eisernen Brücken- und Dachkonstruktionen auf den schweizerischen Eisenbahnen* (calculation and checking of iron bridges and roof constructions for railways in Switzerland) [Schweizerischer Ingenieur- & Architekten-Verein, 1994, pp. 13–22].

Although the company Atelier de construction d'Eiffel founded by Eiffel in 1866 was a leading French steelwork company with great successes on the international bridge-building market as well, it never became a large group. Following several mergers, Eiffel Construction Métallique was finally incorporated into the Eiffage Group in 1992. With a production capacity of 40 000 t of steelwork per year and about 1000 employees, Eiffel Construction Métallique is today one of Europe's leading steelwork companies with such spectacular bridge structures to its name as the viaduct over the Tarn Valley at Millau (completed in December 2004): two towers of this cable-stayed bridge are more than 300 m high and are thus taller than the Eiffel Tower [Virlogeux, 2006, p. 85].

### 2.5.3 United States of America

Culmann's famous technical travelogue of the USA (1849/50) begins with the words: "I had never imagined such a difference between the outward appearances of the New and Old Worlds. Everything looks different, the ships, machines, towns and villages with their roads and houses appear before us in highly diverse forms and only the people have remained the same" (cited in [Maurer, 1998, p. 312]). And, we should add, chiefly bigger: 55 years later, Müller-Breslau's pupil, Hans Reissner, published his extensive series of articles on North American iron workshops [Reissner, 1905/06], which is the fruit of his one year of employment in the workshop

**FIGURE 2-56**
Performance data of the four largest plants of the American Bridge Co. from 1 Nov 1901 to 31 Oct 1902 [Reissner, 1905, p. 629]

|  | steel construction in t/year | |
|---|---|---|
|  | production capacity | production from 1.11.1901 to 31.10.1902 |
| Pencoyd *(near Philadelphia)* | 84 000 | 75 500 |
| Lassig *(Chicago)* | 54 000 | 35 631 |
| Keystone *(Pittsburgh)* | 48 000 | 30 875 |
| Edge Moor *(Delaware)* | 30 000 | 28 981 |

offices and a further 14-week-long study trip paid for by Berlin Technical University. Reissner starts his report with the following sentence: "The major building activities of the American railways, state and local government and industry in the United States during the past decade has created such a wealth of remarkable structures and methods of production in the steel construction sector that it appears timely to assess this great amount of material critically and place the knowledge at the disposal of German industry" [Reissner, 1905, p. 593]. He met leading representatives of the US steelwork industry such as C. C. Schneider, P. L. Wölfel and J. Christie from the American Bridge Co. In April 1900, no less than 26 bridge-building companies merged under the umbrella of the American Bridge Co., which itself was absorbed by the United States Steel Corporation in 1902 and hence became part of the Steel Trust. This colossal steelwork company had capital assets amounting to 70 million US dollars. Fig. 2-56 provides an insight into the capacities of the four largest plants.

The steel output of the four largest German steelwork companies for the business year 1898 is listed below by way of comparison [Mehrtens, 1900; pp. 96–108]:

– Gutehoffnungshütte AG, Sterkrade: approx. 18 000 t
– M. A. N. AG, Nürnberg/Gustavsburg: 17 015 t
– Union AG, Dortmund: approx. 15 000 t
– Harkort AG, Duisburg: 13 702 t

With a total output of 63 717 t per year, the four largest German steelwork companies together did not come even close to the production figures of Pencoyd Iron Works in Pencoyd near Philadelphia. Together, the 26 plants of the American Bridge Co. produced between 1 November 1901 and 31 October 1902 approx. 465 000 t of steelwork for bridges and buildings with a production capacity capable of about 576 000 t. In the first years of the 20th century, the American Bridge Co. built the Ambridge Works with an annual production capacity of max. 240 000 t on the Ohio river near Pittsburgh. As well as providing 5 000 employees with their daily bread, from 1910 onwards the town became known as Ambridge, the contracted form of the company name. The Ambridge plant closed its gates for ever in 1983 and thus ended the story of what had once been the world's largest steelmaking operation.

In order to circumvent the antitrust laws passed in 1890, the American Bridge Co. was split into the three companies American Bridge Co. of New York, American Bridge Co. and Empire Bridge Co. The first of these three companies was responsible for the commercial management of the overall group; the latter two companies were divided geographically into three districts, each with the following departments:

1. Operating Department
2. Purchasing Department
3. Auditing Department
4. Engineering Department (divided into structural and mechanical divisions)
5. Erecting Department
6. Traffic Department.

On the district level, the Structural Division was responsible for designs, cost estimates and works standards, whereas the works level looked after the production of fabrication drawings and erection concepts for difficult projects, in each case divided into bridges and structures. The Mechanical Division was responsible for the design and installation of fabrication plant (Fig. 2-57) and materials-handling facilities, the design of new and redesign of old plants, plus strength tests and scientific/technical testing; this division also undertook the monitoring of factory operations and work organisation. The preparation of structural calculations was therefore carried out in the Structural Division with the help of the Mechanical Division.

Pencoyd Iron Works near Philadelphia, established in 1852, was regarded as the best-known model of American bridge-building practice and established vertical group organisation: "They stretch from a ridge to the River Schuylkill in a suburb of Philadelphia and their development of the open-hearth furnace, the rolling shop, the storage yard, the machine and template shops, the design offices, the main bridges shop, the eye-bar shop, testing facilities, forging and accessories shops illustrate the origins of modern iron constructions in an organic process" [Reissner, 1905, p. 649].

FIGURE 2-57
Riveting machine, Pencoyd Iron Works [Reissner, 1906, p. 57]

The German steelwork industry, too, was also integrated into vertical organisations of mining groups such as Gutehoffnungshütte AG, Harkort AG, Union AG and Krupp AG. However, in Germany, mechanical engineering groups such as M. A. N. AG, Demag, Maschinenbauanstalt Humboldt AG and Maschinenfabrik Esslingen AG played an equal role in steelwork production. Nevertheless, it must be said that German steelwork production was mainly in the hands of midsize companies, e.g. Steffens & Nölle AG (Berlin), Hein, Lehmann & Co. (Düsseldorf), Brückenbau Flender AG (Benrath), Eisenbauanstalt der Aktiengesellschaft für Verzinkerei & Eisenkonstruktion (formerly Jakob Hilgers) (Rheinbrohl), August Klönne (Dortmund), etc. The methods of working of the most important steelwork operations in Germany at the start of the 20th century has been described by Elbern [Elbern, 1920].

And on the technical side, there were significant differences between American and German bridge-building production. Whereas Germany used Thomas (= basic Bessemer) steel, in the USA Siemens-Martin (= open-hearth) steel was generally preferred. The assembly shops

FIGURE 2-58
Eye-bar shop, Pencoyd Iron Works
[Reissner, 1906, p. 67]

(see Fig. 2-55) in which larger parts of structures were joined together, common in Germany, were absent from US steelwork fabrication.

However, the most important difference was that in Europe all the joints between bridge members were riveted, whereas in the USA they used bolts in conjunction with eye-bars (Fig. 2-58) to create pinned joints.

The use of the bolted pinned joint in American bridge-building was not just due to the omission of the pre-assembly stage at the works (in which parts were joined to form the largest possible subassemblies), but more to reduce significantly the cost of labour for riveting. Bridges with bolted pinned joints were the reason for Reinhold Krohn's stay in the USA between 1884 and 1886. He rose to become senior engineer at Pencoyd Iron Works, and in his capacity as director of Gutehoffnungshütte AG, he adapted the new American industrial methods of fabrication to German bridge-building in 1892/93 for the new three-bay fabrication works for the bridge-building works in Sterkrade [Cywiński & Kurrer, 2005, p. 56]. Under Krohn's leadership, the bridge-building works in Sterkrade thus became the leading bridge-building operation in Germany with the highest export figures. Notwithstanding, the Sterkrade works still continued to produce steel frame bridges with riveted joints.

The difference between the construction languages of American and European bridge-building had considerable consequences for the formation of theories during the consolidation period of structural analysis (1900–50): whereas in the USA the forces in the members of trussed bridges could be determined on the basis of the pinned trussed framework model, i.e. according to the trussed framework theories of Culmann, Schwedler or Whipple, European bridge-builders had to contend with the secondary stresses of riveted joints, which led to the development of the displacement method in the first quarter of the 20th century (see section 2.8). An extract from the list of projects carried out by the American Bridge Co. reads like a history of steelwork construction in the 20th century: Woolworth Building, New York (1913), Hell Gate Bridge, New York (1916), Chrysler Building, New York (1931), Empire State Building, New York (1932), San Francisco Bay Bridge (1936), Verrazano Narrows Bridge,

New York (1964), 25th of April Bridge, Lisbon (1966), New River Gorge Bridge, West Virginia (1977), Boeing 747 Assembly Building, Everett, Washington (1974), Louisiana Superdome (1974), Sears Tower, Chicago (1974), Sunshine Skyway Bridge, Tampa Bay, Florida (1986). The American Bridge Co. was hived off from the Steel Trust in 1987 and since then has operated successfully on the international bridge-building market as an independent company from its headquarters in Caraopolis, Pennsylvania.

Factory organisation in American industry became a model for German industry in the final quarter of the 19th century. Examples of this are the Berlin-based engineering company Ludwig Loewe & Co. A.-G. [Spur & Fischer, 2000, p. 55], the aforementioned bridge-building works in Sterkrade and the Deutsche Edison-Gesellschaft für angewandte Elektrizität in Berlin, founded by Emil Rathenau in 1883, which later became the Allgemeine Elektrizitäts-Gesellschaft (AEG). Around 1900, the US-led linear and continuous manufacturing processes had replaced older methods of production throughout most of the German steelwork industry: "Like in every factory, the iron construction shops also follow the basic principle of arranging the individual departments of the factory in such a way that the progress of the work takes place constantly in one direction and the parts from which the finished construction is to be assembled plus the works on those parts come together at the right place" [Reissner, 1905, p. 644]. And vice versa: it was during this period that American structural engineers became acquainted with the classical theory of structures that had originated mainly in Germany. One early example is George Fillmore Swain (1857–1931), who in 1883 introduced Mohr's theory of statically indeterminate frameworks in the *Journal of the Franklin Institute* [Swain, 1883].

## 2.6 Influence lines

In the building of structures, structural analysis deals primarily with the question: What are the magnitudes of the internal forces $M(x)$, $Q(x)$ and $N(x)$ at any point $x$ as a result of static loads (e.g. self-weight)? In statically determinate systems, this question is answered according to the internal force distributions (diagrams of $M$, $Q$ and $N$) solely with the help of equilibrium conditions (Fig. 2-59/left). However, the building of bridges and industrial structures involves a second important issue: What are the magnitudes of the internal forces $M_{j,m} = M_j(\chi)$, $N_{j,m} = N_j(\chi)$ and $Q_{j,m} = Q_j(\chi)$ at section $j$ as a result of the moving load $P = 1$ (e.g. a railway train, or an overhead moving crane in an industrial shed)? The treatment of and answers to these questions led to the concept of influence lines for force variables. In the case of statically determinate systems, influence lines can also be drawn with the help of the equilibrium conditions, the principle of virtual displacements and Land's theorem. If the force $P = 1$ or the moment $M = 1$ move along the bar and act at the point m with the abscissa $\chi$, then the influence line for a force variable at point $j$ is characterised by that line obtained when the magnitude of the force variable at $j$ is carried at point $m$ (Fig. 2-59/right).

| **Internal force distributions** | **Influence lines** |
|---|---|

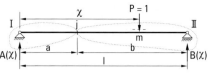

$A = \dfrac{b}{l} = \text{const.}; \; B = \dfrac{a}{l} = \text{const.}$   $\quad\boxed{A(\chi) = \dfrac{1}{l}(l - \chi)}; \quad \boxed{B(\chi) = \dfrac{1}{l}\cdot \chi}\quad \rightarrow$ Influence lines for A and B

Section I for $0 \leq x \leq a$:   $\qquad$ Section I for $0 \leq \chi \leq a$:

$\sum M_{\otimes} = 0 = \widehat{M(x)} - A \cdot x$   $\qquad$ $\sum M_j = 0 = \widehat{M_j(\chi)} - A(\chi)\cdot a + 1 \cdot (a - \chi)$

$\boxed{M(x) = \dfrac{b}{l}\cdot x}$   $\qquad$ $\boxed{M_j(\chi) = \dfrac{\chi}{l}\, b = M_{j,m}}$

$\dfrac{dM(x)}{dx} = \boxed{Q(x) = \dfrac{b}{l} = A}$   $\qquad$ $\sum V = 0 = \uparrow A(\chi) - 1 - Q_j(\chi)$

$\qquad\qquad\qquad\qquad\qquad\qquad\qquad\quad$ $\boxed{Q_j(\chi) = -\dfrac{\chi}{l} = Q_{j,m}}$

Section II for $a \leq x \leq l$:   $\qquad$ Section II for $a \leq \chi \leq l$:

$\sum M_{\otimes} = 0 = \widehat{M(x)} - B \cdot (l - x)$   $\qquad$ $\sum M_j = 0 = \widehat{M_j(\chi)} - B(\chi)\cdot b + 1 \cdot (\chi - a)$

$\boxed{M(x) = \dfrac{a}{l}(l - x)}$   $\qquad$ $\boxed{M_j(\chi) = \dfrac{a}{l}\left(\dfrac{b}{a}\chi - \dfrac{1}{a}\chi + l\right) = M_{j,m}}$

$\dfrac{dM(x)}{dx} = \boxed{Q(x) = -\dfrac{a}{l} = -B}$   $\qquad$ $\sum V = 0 = \uparrow Q_j(\chi) - 1 + B(\chi)$

$\qquad\qquad\qquad\qquad\qquad\qquad\qquad\quad$ $\boxed{Q_j(\chi) = -\dfrac{1}{l}(\chi - l) = Q_{j,m}}$

$Q(x)$ diagram:   $\qquad\qquad\qquad\qquad$ $Q_j(\chi) = Q_{j,m}$ diagram:

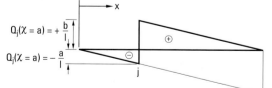

$M(x)$ diagram:   $\qquad\qquad\qquad\qquad$ $M_j(\chi) = M_{j,m}$ diagram:

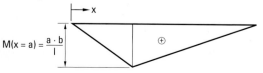

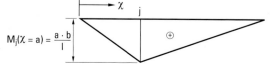

## 2.6.1 Railway trains and bridge-building

In his *Résumé des Leçons*, Navier includes a parabolic, symmetrical, two-pin arch with one degree of static indeterminacy [Navier, 1833/1878, pp. 361–363] for which he derives the influence line for horizontal thrust $H(\chi)$ for the case of a vertical moving load $P$ (Fig. 2-60):

$$H(\chi) = (5/64) \cdot P \cdot (5 \cdot l^4 - 6 \cdot l^2 \cdot \chi^2 + \chi^4)/(l^3 \cdot f) \quad (2\text{-}16)$$

The chapter in which he deals with this problem is entitled *Von den Brücken, welche von Bögen getragen werden* (bridges supported by arches) [Navier, 1833/1878, p. 360].

Just like the physicists and engineers of that time could solve only certain statically indeterminate problems, the derivation of the influence line for horizontal thrust $H(\chi)$ may be just an isolated case, although the calculation of the horizontal thrust in an elastic arch caused problems into the 1850s.

As the railway networks started to spread across all parts of Europe and North America during the Industrial Revolution, and thus placed higher demands on the building of large bridges in qualitative and quantitative terms, so the empirical knowledge about bridges was transformed into scientifically founded bridge theories. One chief problem was how to find the position of the load of a moving railway train that caused the maximum internal forces in the bridge members. At first, it was assumed that all parts of the bridge members (in bending) would be subjected to the greatest stresses when the maximum possible load $max\ q = g + p$ ($g$ = self-weight, $p$ = imposed load) was applied over their full length (Fig. 2-61).

In this approach, the axle loads of the railway train are "spread" into a uniformly distributed load $p$ (Fig. 2-61). For many small- to medium-span $l$ railway bridges, modelling the effect of the moving railway train by means of a vertical uniform load $p$ distributed over the entire span supplies realistic bridge cross-sections. But this means such bridges are designed according to the maximum span moment

$$max\ M = max\ q \cdot l^2/8 \quad (2\text{-}17)$$

and the resulting bridge cross-section is used for the entire span, which means that this method leads to uneconomic bridge cross-sections for larger bridges.

Therefore, starting in the 1850s, bridge-builders such as Schwedler, Bänsch, Laissle and Schübler investigated models of moving loads; using the railway train represented by a partial uniformly distributed load $p$, Laissle and Schübler moved this across the bridge from left to right (Fig. 2-62).

Parameter $\lambda$ is that part of the total length of the train that is just within the bridge span. Once the locomotive reaches the right-hand support, the entire bridge span is subjected to the imposed load $p$ ($\lambda = l$), resulting in the loading case shown in Fig. 2-61. After Laissle and Schübler derived equations for the bending moment $M$ and the shear force $Q$ for a

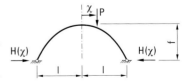

FIGURE 2-60
Calculating the influence line for horizontal thrust H(χ)

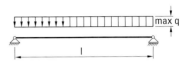

FIGURE 2-61
Maximum uniformly distributed load on a simply supported beam

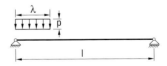

FIGURE 2-62
A railway train load p of length L ≥ l is moved across the span

page 100:
FIGURE 2-59
Internal force distributions and influence lines for a simply supported beam

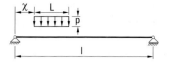

FIGURE 2-63
A railway train load p with length L ≤ l crossing the span of a bridge

given λ (internal force distribution), they moved λ and then determined that parameter λ for each beam section that maximises the magnitude of the internal forces at that point. From these envelopes for $M$ and $Q$, Laissle and Schübler determined the distribution of material over the length of the bridge. They thus calculated a table for the various $g/p$ ratios and appended the corresponding envelopes that can be used directly for a graphical determination of the change in cross-section over the length of the bridge [Laissle & Schübler, 1869, pp. 90 – 92].

Winkler had already suggested an approach in 1860 for large bridges whose spans $l$ exceed the length of a railway train $L$ (Fig. 2-63). He moved the railway train with length $L$ over the bridge span and answered the question of which loading position $\chi$ leads to the maximum bending moment or shear force at a certain point of the beam cross-section.

### 2.6.2 Evolution of the influence line concept

As railway bridge structures resolved into individual members started to appear together with trussed framework theory, the immediate question was how to find the dimensions of members for the most unfavourable positions of railway train loads because the dimensioning leeway between safety and economy is considerably narrower than for masonry arch railway bridges. The influence line concept was added to the world of engineering science models in the form of Winkler's elastic arch theory – even though he had not yet developed this concept for the general case and the term "influence line" was hidden behind more specialised concepts such as "abutment pressure line", "abutment pressure envelope" and "stress curves".

Winkler explained the theory of elastic arches and the concept of the influence line in detail in a presentation on the calculation of arch bridges before the Bohemian Society of Architects & Engineers in Prague in December 1867. Concerning the influence line concept, he writes: "The first bridge in which the most dangerous loading was determined rationally, is, as far as I know, the bridge over the Rhine at Koblenz [completed in 1864 – the author]. However, the assumptions made in this case are not exactly correct and, furthermore, the approach is a very complicated one. For the past three years I have been using a simple, semi-graphical method in my lectures on bridge-building in order to determine the most dangerous loading case exactly and the corresponding cross-sectional dimensions" [Winkler, 1868/1869, p. 6]. So it would seem that Winkler was already in possession of the influence line concept for elastic arches subjected to vertical moving loads in 1864.

Winkler also specifies the evaluation of influence lines for distributed loads. This will be illustrated using the example of the simply supported beam shown in Fig. 2-59 for an imposed load $p$ distributed over the full length of the beam. The influence line of the bending moment $M_j(\chi)$ at point $j$ is made up of two straight lines with the value

$$M = a \cdot b / l \qquad (2\text{-}18)$$

at point $j$.

The bending moment $M(x = a)$ at point $j$ is obtained from the product integral

$$M(x = a) = \int M_j(\chi) \cdot p(\chi) \cdot d\chi \qquad (2\text{-}19)$$

where

$$M(x = a) = (a \cdot b/2) \cdot p \qquad (2\text{-}20)$$

An equilibrium approach would show that eq. 2-20 equals the value of the internal force distribution for the bending moment at point $a$. Eq. 2-19 applies for the general case: the ordinate of an internal force at point $j$ is equal to the product of the influence ordinate for this internal force with the load at point $\chi$.

In 1868 Mohr published a paper in which he used influence lines extensively [Mohr, 1868]. However, whereas Winkler finds influence lines for force variables, Mohr uses influence lines for deformation variables as well. One year prior to that, Fränkel had taken up Navier's eq. 2-16 and, using the statically determinate support reaction $V(\chi)$, had calculated the angle of slope

$$\alpha(\chi) = V(\chi)/H(\chi) \qquad (2\text{-}21)$$

of the abutment pressure for the two-pin arch shown in Fig. 2-60; eq. 2-16 should be used for $H(\chi)$ in eq. 2-21 [Fränkel, 1867]. The intersection of the lines of action of the abutment pressure with those of the moving load is the geometrical position of the abutment pressure enclosing line which Winkler had specified in the same year (see Fig. 6-26).

During the 1870s, the influence line concept was applied to diverse practical engineering tasks. For example, in 1873 Weyrauch calculated the influence line for the support moment of a continuous beam and called it just that, *Influenzlinie* [Weyrauch, 1873], and Fränkel published the first comprehensive work on influence lines in 1876: he calculated the influence lines for force variables in statically determinate trussed systems [Fränkel, 1876]. The influence line concept and its application to specific structural systems reached a certain conclusion at the end of the establishment phase of structural theory (1850–75). However, the general influence lines theory actually belongs to the classical phase of structural theory (1875–1900). Land and Müller-Breslau contributed decisively to this in 1886/87. Fig. 2-64 is intended to show how elegantly the kinematic theory answers the question of the determination of the influence line for force variables.

When designing foundations, we have to ask the following questions (Fig. 2-65):
– What is the magnitude of the ground bearing pressure underneath the foundation?
– What are the magnitudes of the internal forces in the foundation itself?

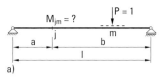
a)
True force condition
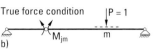
b)
Virtual displacement condition
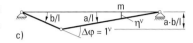
c)

General work theorem:
$W_a^v + W_i^v = 0$
where $W_i^v = 0$ it follows that:
$W_a^v = 0 = P \cdot \eta^v - M_{j,m} \cdot 1^v$
$\boxed{M_{j,m} = \eta^v}$

The projection of the virtual displacement condition onto the direction of the load P is the influence line $M_{j,m}$ (Land's theorem).

**FIGURE 2-64**
Influence line for the bending moment at point j as a result of a vertical moving load determined according to kinematic theory

## 2.7 The beam on elastic supports

**FIGURE 2-65**
a) Rigid foundation,
b) Non-rigid foundation

**FIGURE 2-66**
Bearing pressure distribution for a rigid foundation

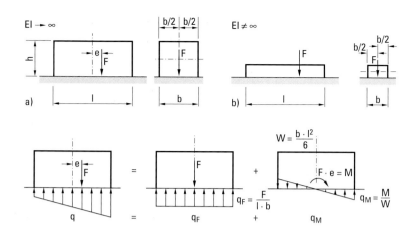

Generally, both questions can only be answered together because the stress distribution in the foundation is dependent on the ground bearing pressure $q$ – and vice versa: the problem is statically indeterminate.

In order to calculate simple foundations, we can assume a ground bearing pressure $q$ bounded by a straight line; with an eccentrically applied force $F$, this results in a trapezoidal bearing pressure distribution which can be determined with the simple equations of the practical bending theory (Fig. 2-66). The internal forces in the foundation itself are then calculated from the equilibrium conditions for the corresponding section: the problem is statically determinate.

### The Winkler bedding

#### 2.7.1

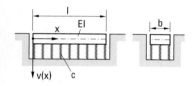

**FIGURE 2-67**
Beam on elastic supports

In the case of slab-type foundations, this approach leads to uneconomic dimensions, and the bearing pressure distribution underneath non-rigid foundations can no longer be determined using the equilibrium conditions alone. Fig. 2-67 shows the structural model of the foundation on elastic supports, which is nothing more than a continuous beam with an infinite number of elastic supports. Such a beam has an infinite degree of static indeterminacy.

In 1867 Winkler implemented Hooke's law for modelling the bedding of railway tracks using

$$q(x, v, b) = c(b, v) \cdot v(x) \qquad \left[\frac{kN}{m^2}\right] = \left[\frac{kN}{m^3}\right] \cdot [m] \qquad (2\text{-}22)$$

Eq. 2-22 would later be called the Winkler bedding (Fig. 2-68). Here, $c$ has the units $[kN/m^3]$ and is known as the bedding constant; in terms of physics it is that ground bearing pressure $q$ that occurs with a settlement $v$ of 1 $[m]$.

**FIGURE 2-68**
Graphic representation of the Winkler bedding

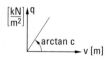

(schematic representation)

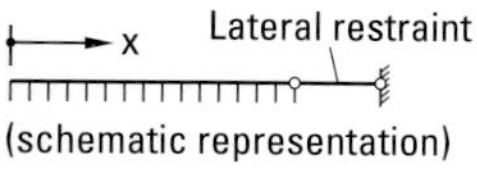

**FIGURE 2-69**
Basic structural system for the beam on elastic supports

The theory of the beam on elastic supports is based on further assumptions:
- beam is rigid in shear, i.e. $G \cdot A_Q \to \infty$,
- individual spans are mutually independent, i.e. there are no friction forces acting in the soil,
- apart from the elastic bed, the surrounding soil is rigid and therefore has no effect on the beam, and

- the system is kinematic in the horizontal direction and therefore a lateral restraint must be provided (Fig. 2-69).

## 2.7.2 The theory of the permanent way

In historical terms, the beginnings of the theory of the beam on elastic supports emerged during the transition from the establishment phase of structural theory (1850–75) to the classical phase (1875–1900) in the form of the structural investigation of the permanent way. Even Max Maria von Weber, the well-known poet and railway engineer, had realised back in the 1860s that displacements in the permanent way are reflected in the internal forces in the longitudinal sleepers and the rails; it was in 1869 that he summarised the state of knowledge regarding the permanent way, which he supplemented with numerous tests, in his monograph entitled *Die Stabilität des Gefüges der Eisenbahn-Gleise* (stability of permanent way structures) [Weber, 1869]. Owing to the inadequacy of the facilities available at that time, corresponding tests supplied results that were unusable in practice.

In that same decade, Winkler calculated the stresses in rails and longitudinal sleepers due to the elastic yielding of the track bed in his book entitled *Die Lehre von der Elastizität und Festigkeit* (theory of elasticity and strength) [Winkler, 1867] [Knothe, 2003/1]. In doing so, he presumed proportionality between ground bearing pressure $q$ and settlement $v$ for the first time (eq. 2-22 and Fig. 2-68); this assumption appeared in all subsequent writings dealing with permanent way theory. As Zimmermann wrote in 1888: "Whether and to what extent this assumption really applies has not yet been determined. But in any case it is permissible for small deformations" [Zimmermann, 1888, p. 1]. Furthermore, Winkler postulates a regular, wave-shaped and constant form for the deflection curve; according to Zimmermann, this presumes a beam of infinite length subjected to equally spaced identical loads (Fig. 2-70).

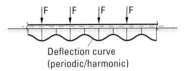

Deflection curve (periodic/harmonic)

**FIGURE 2-70**
Deflection curve for a beam on elastic supports

This simplification may well be approximately valid for railway rolling stock with regularly spaced axles and similar weights, but it certainly does not apply to locomotives! One prerequisite necessary for overcoming this limitation in the beginnings of the evolution of the theory of the beam on elastic supports were tests on the track bed which the German Reichsbahn had been carrying out since 1877 but were not published until later [Häntzschel, 1889]. It was these tests that provided the first insights into the properties of the track bed – the track bed constants $c$ were confirmed in 1899 by Aleksander Wasiutynski (1859–1944) in his dissertation [Wasiutynski, 1899].

Schwartzkopff describes the doubts and uncertainties that permanent way theory had to cope with in the 1870s in his 1882 publication *Der eiserne Oberbau* (the iron permanent way):
- uncertainties in bending theory in general,
- idealised assumptions such as hypothesis eq. 2-22, which inadequately reflects the reality, and
- the neglect of practical building aspects.

From this, Schwartzkopff concludes that a permanent way theory can be used only "for comparing the quality of various arrangements" (cited in [Zimmermann, 1888, p. 6]).

Things did not change until the 1880s when the work of Schwedler and Zimmermann (both leading engineering civil servants in the Prussian Railway Authority) in particular resulted in a permanent way theory that could be used with reasonable effort.

For instance, Schwedler realised the importance of solving the infinite length of the beam on elastic supports with a point load (the solution of which was already known to Winkler), and used it to calculate the permanent way for loads of any magnitude acting at any spacing. Working independently, Zimmermann discovered the same solution around 1882; however, he refrained from publishing this "because he thought it first necessary to investigate the influence of the lifting of the longitudinal sleeper from the track bed, which had not been considered in this solution" [Zimmermann, 1888, p. 7]. Zimmermann carried out extensive successive approximations in order to master this problem, but the difficulty was that the structural system changed with every step in the iteration. Without mentioning Winkler's name, Schwedler [Schwedler, 1882] presented hypothesis eq. 2-22 and its experimental safeguards carried out by Zimmermann and Häntzschel at the Institution of Civil Engineers in London in the same year [Knothe, 2003/2, p. 189]. Schwedler's presentation in London "helped to validate and establish the Winkler model, in particular owing to the agreement between theory and measurements" [Knothe, 2003/2, p. 189].

Zimmermann's book *Die Berechnung des Eisenbahnoberbaues* (calculation of the permanent way) [Zimmermann, 1888] eliminated the misconceptions and contradictions that had existed up to that time. From now on, a coherent and relatively simple permanent way theory was available. Zimmermann developed a strict method of calculating the infinite bar with point loads at any position. Another important innovation was the widespread and systematic application of influence lines. Timoshenko developed the first dynamic track bed model in 1915; it has been frequently quoted, but was not used in permanent way engineering even though it would have been adequate for the needs of railway engineering in the first half of the 20th century [Knothe, 2001, p. 2]. Timoshenko also shows that hypothesis eq. 2-22 can be used for transverse as well as longitudinal sleepers [Knothe, 2001, p. 2]. Nevertheless, railway engineers were still referring to Zimmermann's monograph until well into the 1960s; it formed an important document in the formation of structural analysis theories at the transition from the classical phase of structural theory (1875–1900) to the accumulation phase (1900–25). And "Zimmermann's book," as Klaus Knothe summarises from the historical perspective of railway engineering research, "has contributed considerably to spreading Winkler's ideas in the 20th century and extending them to dynamic problems at the end of the century" [Knothe, 2003/4, p. 450].

### 2.7.3 From permanent way theory to the theory of the beam on elastic supports

During the invention phase of structural theory (1925–50), permanent way theory gradually exceeded its tightly defined purpose. The development of foundation engineering theory in the 1920s brought with it the first signs of a specific theory of elastic supports for beam and slab structures, a development that was completed in the wake of the establishment of reinforced concrete for foundations. For instance, it was in 1921 that the Japanese structural engineer Keiichi Hayashi published a monograph on the use of the theory of the beam on elastic supports for problems in foundation engineering [Hayashi, 1921/1]. But was not until the innovation phase of structural theory (1950–75) that the integration of the theory of the beam on elastic supports into the force and displacement methods was completed.

In order to show how equilibrium considerations can be carried out on the infinitesimal element, the differential equation of the beam on elastic supports is derived below (Fig. 2-71).

As is common in continuum mechanics problems since Euler, theory of structures also counts the method of sections among its indispensable principles for modelling an infinitesimal element.

From the equilibrium of all vertical forces at the infinitesimal element

$$\Sigma V = 0 = \downarrow dQ + p(x) \cdot dx - q(x) \cdot b \cdot dx$$

$$dQ = [b \cdot q(x) - p(x)] \cdot dx$$

and the relationship between shear force $Q$ and bending moment $M$

$$\frac{dQ}{dx} = \frac{d^2M}{dx^2} = b \cdot q(x) - p(x)$$

and the differential equation of the deflection curve

$$\frac{d^2M}{dx^2} = -E \cdot I \cdot v^{IV}$$

it follows that

$$E \cdot I \cdot v^{IV} + b \cdot q(x) = p(x)$$

Entering the Winkler bedding (eq. 2-22) into the latter equation results in the differential equation for the beam on elastic supports:

$$v^{IV} + 4 \cdot \frac{\lambda^4}{l^4} \cdot v = \frac{p(x)}{E \cdot I} \qquad (2\text{-}23)$$

with the parameters $\lambda = l \cdot \sqrt[4]{\dfrac{k}{4 E \cdot I}}$ in [1/m] and $k = c \cdot b$ in [kN/m$^2$].

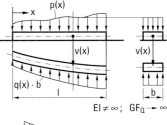

**FIGURE 2-71**
The derivation of the differential equation for the beam on elastic supports

The differential equation for the beam on elastic supports (eq. 2-23) is closely related to the differential equation 2-26 of the bar subjected to compression and shear according to second-order theory. The differential equation of the beam on elastic supports is a linear differential equation with constant coefficients because the bending stiffness $E \cdot I$ and $k = c \cdot b$ do not depend on the load. Therefore, the force and deformation variables are all superimposed, which eases the integration of elastically supported bar elements into the algorithm of the force and displacement method.

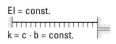

**FIGURE 2-72**
Basic system for the displacement method

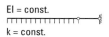

**FIGURE 2-73**
Basic system for the force method

### Displacement method

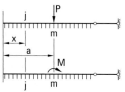

Internal force distributions: $v, \varphi, Q, M$
for $a$ = const. and $x$ = variable

Influence lines: $v_{j,m}, \varphi_{j,m}, Q_{j,m}, M_{j,m}$
for $x$ = const. and $a$ = variable

**FIGURE 2-74**
Basic loading cases for determining internal force distributions and influence lines

**FIGURE 2-75**
Secondary stresses problem in trussed frameworks

Differential equation 2-23 has been solved for various standard loading cases and boundary conditions for the simply supported beam on elastic supports and can be found in diverse manuals (Figs. 2-72 and 2-73). Furthermore, the internal force distributions and influence lines of the basic bar have been tabulated (Fig. 2-74); Pohlmann provided working structural engineers with outstanding aids as early as 1956 in his paper entitled *Balken auf elastischer Unterlage als Teil einer Konstruktion* (beams on elastic supports as part of a construction) [Pohlmann, 1956].

Besides the applications for the permanent way and foundation engineering, the theory of the beam on elastic supports has also served as a basis for engineering models in other areas of civil and structural engineering. For example, Lehmann used the theory of the beam on elastic supports integrated into the force method to model the loadbearing behaviour of a drawn, prestressed ring flange which serves as the connection between the segments of towers for wind energy converters [Lehmann, 2000]. This list can be extended.

## 2.8

In historico-logical terms, the displacement method goes back to the analysis of secondary stresses problems in trussed frameworks. More than 125 years ago, Heinrich Manderla presented a comprehensive solution to this theory of structures problem as his answer to a competition [Manderla, 1880]. From Fig. 2-75 we can see what is meant by a secondary stresses problem. Secondary stresses are stresses that ensue when the connections between the bars are by way of riveted gusset plates (European form of construction) instead of articulated (American form of construction): bending stresses at the joints are superimposed on the axial stresses calculated from the pinned trussed framework model (Fig. 2-75/second from left).

The resulting systems with their many degrees of static indeterminacy formed an important object in the classical phase of structural theory. Numerous structural engineers tried to use successive approximations to quantify the secondary stresses. Manderla was the first to achieve this systematically by entering the deformations as unknowns into the structural analysis and not – as was usual at the time – by working with unknown force variables. However, he concealed the nature of the displacement method with the analysis of the stability problem. The fear of undersizing trussed frameworks with riveted joints when using the pinned trussed framework model drove almost all structural engineers of any importance

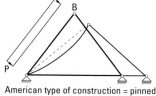
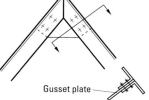
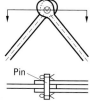

European type of construction = riveted or bolted (which creates secondary stresses due to bending)

American type of construction = pinned

Gusset plate

Pin

to develop models for at least estimating the secondary stresses. It immediately became clear that it is more beneficial to enter deformations, e.g. angles of rotation at the nodes, as unknown quantities. Mohr did this in a very clear way [Mohr, 1892/93] and his contribution represents the concentrated output of the classical phase of structural theory to the theory of secondary stresses. It was not until the subsequent accumulation phase (1900–25) that the displacement method developed closely alongside the emerging theory of frameworks, which received its consequential form from Ostenfeld in the early 1920s. He saw the displacement method as the counterpart to the force method: "It shall now be shown that the 'force method', in which a stress, a member force, a moment, etc. is entered, and the 'displacement method', where a deformation is introduced as an unknown variable, are dualistic, analogous methods existing exactly side by side" [Ostenfeld, 1921, p. 275]. Hence, the accumulation phase of structural theory (1900–25) is crowned by the displacement method.

### 2.8.1 Analysis of a triangular frame

The essence of the displacement method can be shown by using the example of a closed triangular frame that is statically determinately supported externally, but has three degrees of static indeterminacy internally (Fig. 2-76a). At the end of the structural investigation, it will be shown that the triangular frame cannot be analysed with the methods of simple trussed framework theory.

The loadbearing structure consists of cold-formed rectangular hollow sections, 250 × 150 × 8 mm, made from grade S 235 steel to DIN 59411/EN 10210 which are welded together at the joints and have section properties of $I_x = 0.488 \cdot 10^{-4}$ m$^4$ and $A = 0.592 \cdot 10^{-2}$ m$^2$ [Bertram, 1994, p. 169]. Using the elastic modulus $E = 21 \cdot 10^7$ kN/m$^2$ for steel, the bending stiffness $E \cdot I_x = E \cdot I = 102\,606$ kNm$^2$, which can be simplified to $E \cdot I = 100\,000$ kNm$^2$. The uniformly distributed load $p = 50$ kN/m acts on bar $i-k$ (Fig. 2-76a).

The system shown in Fig. 2-76a has three degrees of geometric indeterminacy because the angles of rotation at the nodes $\xi_1$, $\xi_2$ and $\xi_3$ are obviously unknown. The next step is to prevent these node rotations by introducing three rotational restraints which act like rigid, fixed supports; the system is now geometrically determinate (Fig. 2-76b). It should be noted that the strain stiffness of the bars $E \cdot A$ is about twice as high as the bending stiffness $E \cdot I = 100\,000$ kNm$^2$ and therefore the system can be regarded as braced. The duality between this and the force method becomes clear here: in the latter method the restraints are released one by one in order to turn a statically indeterminate system into a statically determinate basic system, e.g. into a pinned system (Fig. 2-75/second from left), but in the displacement method the calculations are based on a geometrically determinate system (Fig. 2-77).

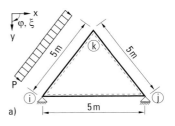
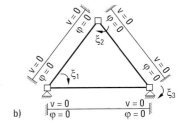

**FIGURE 2-76**
Triangular frame: a) structural system, b) geometrically determinate system

#### 2.8.1.1 Bar end moments

From the 0-state (Fig. 2-77a) we get the bar end moments according to table 2.4 [Duddeck & Ahrens, 1994, p. 288]:

$$M_{ik}^{(0)} = -\frac{p \cdot l^2}{12} = -\frac{50 \cdot 5^2}{12} = -104.17\,\text{kNm}$$

$$M_{ki}^{(0)} = -M_{ik}^{(0)} = +104.17\,\text{kNm}$$

The bar end moments for the unit displacement states $\xi_1 = 1$ (Fig. 2-77b), $\xi_2 = 1$ (Fig. 2-77c) and $\xi_3 = 1$ (Fig. 2-77d) can be taken from table 6.1 [Duddeck & Ahrens, 1994, p. 318]:

$\xi_1 = 1$-state:

$$M_{ik}^{(1)} = 4 \cdot \frac{E \cdot I}{5} \cdot \varphi_i = 4 \cdot \frac{100\,000}{5} \cdot 1 = 80\,000\,\text{kNm}$$

$$M_{ki}^{(1)} = 2 \cdot \frac{E \cdot I}{5} \cdot \varphi_i = 2 \cdot \frac{100\,000}{5} \cdot 1 = 40\,000\,\text{kNm}$$

$$M_{ij}^{(1)} = M_{ik}^{(1)} = 80\,000\,\text{kNm}$$

$$M_{ji}^{(1)} = M_{ki}^{(1)} = 40\,000\,\text{kNm}$$

$\xi_2 = 1$-state:

$$M_{ik}^{(2)} = 2 \cdot \frac{E \cdot I}{5} \cdot \varphi_k = 2 \cdot \frac{100\,000}{5} \cdot 1 = 40\,000\,\text{kNm}$$

$$M_{ki}^{(2)} = 4 \cdot \frac{E \cdot I}{5} \cdot \varphi_k = 4 \cdot \frac{100\,000}{5} \cdot 1 = 80\,000\,\text{kNm}$$

$$M_{kj}^{(2)} = 4 \cdot \frac{E \cdot I}{l} \cdot \varphi_k = 80\,000\,\text{kNm}$$

$$M_{jk}^{(2)} = 40\,000\,\text{kNm}$$

$\xi_3 = 1$-state:

$$M_{ij}^{(3)} = 2 \cdot \frac{E \cdot I}{5} \cdot \varphi_j = 2 \cdot \frac{100\,000}{5} \cdot 1 = 40\,000\,\text{kNm}$$

$$M_{ji}^{(3)} = 4 \cdot \frac{E \cdot I}{5} \cdot \varphi_j = 80\,000\,\text{kNm}$$

$$M_{kj}^{(3)} = 2 \cdot \frac{E \cdot I}{5} \cdot \varphi_j = 40\,000\,\text{kNm}$$

$$M_{jk}^{(3)} = 4 \cdot \frac{E \cdot I}{5} \cdot \varphi_j = 4 \cdot \frac{100\,000}{5} \cdot 1 = 80\,000\,\text{kNm}$$

$$M_{jk}^{(3)} = 80\,000\,\text{kNm} = M_{ji}^{(3)}$$

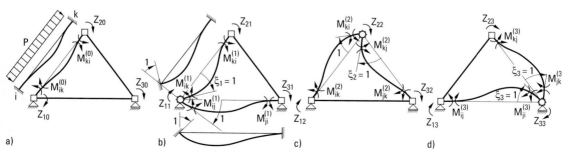

**FIGURE 2-77**
a) 0-state, b) $\xi_1 = 1$-state,
c) $\xi_2 = 1$-state, d) $\xi_3 = 1$-state

### 2.8.1.2 Restraint forces

Whereas in the force method displacement jumps ensue in the statically determinate basic system and hence the continuity conditions are infringed, in the displacement method jumps occur in the force variables in the geometrically determinate system: equilibrium (in this case the sum of the moments at the rotationally restrained nodes) is infringed. Therefore, in order to restore equilibrium, the restraint forces $Z$ (in this case external moments in the direction of the node rotation) shown for the states in Fig. 2-77 must act at nodes $i$, $j$ and $k$. From the node equilibrium at nodes $i$, $j$ and $k$ for the 0-state (Fig. 2-78), the restraint forces $Z_{10}$, $Z_{20}$ and $Z_{30}$ can be determined as follows:

$\Sigma M_i = 0 = Z_{10} - M_{ik}^{(0)} \rightarrow Z_{10} = M_{ik}^{(0)} = -104.17\,\text{kNm}$

$\Sigma M_j = 0 = Z_{20} - M_{ki}^{(0)} \rightarrow Z_{20} = M_{ki}^{(0)} = +104.17\,\text{kNm}$

$Z_{30} = 0$

**FIGURE 2-78**
Equilibrium at nodes i and k in the 0-state

Likewise, equilibrium must also be satisfied at nodes $i$, $j$ and $k$ in the $\xi_1 = 1$-state:

$\Sigma M_i = 0 = Z_{11} - M_{ik}^{(1)} - M_{ij}^{(1)} \rightarrow Z_{11} = 160\,000\,\text{kNm}$

$\Sigma M_k = 0 = Z_{21} - M_{ki}^{(1)} \rightarrow Z_{21} = 40\,000\,\text{kNm} = Z_{12}$

$\Sigma M_j = 0 = Z_{31} - M_{ji}^{(1)} \rightarrow Z_{31} = 40\,000\,\text{kNm} = Z_{13}$

The restraint forces $Z_{i2}$ from the equilibrium at nodes $j$ and $k$ in the $\xi_2 = 1$-state are as follows:

$\Sigma M_k = 0 = Z_{22} - M_{ki}^{(2)} - M_{kj}^{(2)} \rightarrow Z_{22} = 160\,000\,\text{kNm}$

$\Sigma M_j = 0 = Z_{32} - M_{jk}^{(2)} \rightarrow Z_{32} = 40\,000\,\text{kNm} = Z_{23}$

Finally, the restraint force $Z_{33}$ from the equilibrium at node $j$ in the $\xi_3 = 1$-state:

$\Sigma M_j = 0 = Z_{33} - M_{ji}^{(3)} - M_{jk}^{(3)} \rightarrow Z_{33} = 160\,000\,\text{kNm}$

*The solution: the restraint forces must be eliminated*
Whereas in the force method the elasticity equations are continuity expressions for deformation variables, i.e. all displacement jumps must be eliminated, in the displacement method the elasticity equations are continuity expressions for force variables (equilibrium expressions), i.e. the sum of all restraint forces at the respective nodes must be equal to zero:

Restraint forces at node $i$ equal to zero:

$Z_{11} \cdot \xi_1 + Z_{12} \cdot \xi_2 + Z_{13} \cdot \xi_3 + Z_{10} = 0$

Restraint forces at node $k$ equal to zero:

$Z_{21} \cdot \xi_1 + Z_{22} \cdot \xi_2 + Z_{23} \cdot \xi_3 + Z_{20} = 0$

Restraint forces at node $j$ equal to zero:

$$Z_{31} \cdot \xi_1 + Z_{32} \cdot \xi_2 + Z_{33} \cdot \xi_3 + Z_{30} = 0$$

These three equations form a set of equations with the unknowns $\xi_1$, $\xi_2$ and $\xi_3$, which takes on the following form when expressed as a matrix:

$$(Z_{ik}) \cdot (\xi_k) = -(Z_{i0}) \qquad (2\text{-}24)$$

Ludwig Mann called the set of equations (eq. 2-24) "elasticity equations of the second order". In the language of structural matrix analysis, matrix $(Z_{ik})$ is called the stiffness matrix. Entering all the values of the restraint forces into eq. 2-24 results in

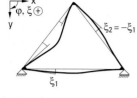

**FIGURE 2-79**
Deformation figure

$$\begin{bmatrix} 160\,000 & 40\,000 & 40\,000 \\ 40\,000 & 160\,000 & 40\,000 \\ 40\,000 & 40\,000 & 160\,000 \end{bmatrix} \cdot \begin{bmatrix} \xi_1 \\ \xi_2 \\ \xi_3 \end{bmatrix} = \begin{bmatrix} +104.17 \\ -104.17 \\ 0.00 \end{bmatrix}$$

with the solution vector

$$\begin{bmatrix} \xi_1 \\ \xi_2 \\ \xi_3 \end{bmatrix} = \begin{bmatrix} 0.000866 \\ -0.000866 \\ 0.00 \end{bmatrix} = \begin{bmatrix} 0.05 \\ -0.05 \\ 0.00 \end{bmatrix} \text{ in } [°]$$

The deformation figure resulting from the solution vector is shown qualitatively in Fig. 2-79.

**Superposition means combining the state variables linearly with the solution**

### 2.8.1.3

As the displacement method is completely linear (statics, geometry, materials), the superposition theorem applies:

$$C_j = C_{j,0}^{(0)} + \sum_{i=1}^{n} \xi_i \cdot C_j^{(i)} \qquad (2\text{-}25)$$

The superposition theorem takes on the form

$$M_{ik} = M_{ik}^{(0)} + \sum_{j=1}^{3} \xi_j \cdot M_{ik}^{(j)}$$

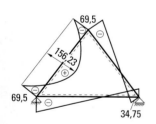

**FIGURE 2-80**
Bending moment diagram

for the bending moment at node $i$ of bar $i$–$k$. The final bar end moments are then as follows:

$M_{ik} = -104.17 + 80\,000 \cdot 0.000866 + 40\,000 \cdot (-0{,}000866) \cong -69.50$ kNm
$M_{ki} = -104.17 + (-40\,000) \cdot 0.000866 + (-80\,000) \cdot (-0.000866) \cong -69.50$ kNm
$M_{kj} = 80\,000 \cdot (-0.000866) \cong -69.50$ kNm
$M_{jk} = (-40\,000) \cdot (-0.000866) \cong +34.75$ kNm
$M_{ji} = (+40\,000) \cdot (0.000866) \cong +34.75$ kNm
$M_{ij} = (-80\,000) \cdot (0.000866) \cong -69.50$ kNm

Fig. 2-80 shows the bending moment diagram, which represents the solution to the structural engineering task per se. This step is followed by the detailed design and/or the analysis of the stresses.

**Comparing the displacement method and trussed framework theory for frame-type systems**

### 2.8.2

To conclude this section, the deformation method will be compared with simple trussed framework theory.

The bending stresses at node $k$ are calculated from

**FIGURE 2-81**
East Liverpool Bridge (built in 1895) over the Ohio downstream of Pittsburgh, USA, is a suspension bridge stiffened by a lattice girder [Mehrtens, 1908, p. 503]

$\sigma_{b,k} = M_{kj}/W_x$, where $I_x/(h/2) = 0.488 \cdot 10^{-4}/(0.250/2) = 3.904 \cdot 10^{-4} \text{ m}^3$

which results in

$\sigma_{b,k} = 69.5/(3.904 \cdot 10^{-4}) = 17.8 \cdot 10^4 \text{ kN/m}^2$

which corresponds to 178 N/mm². This value lies below the yield stress for grade S 235 steel ($\sigma_{yield} = 240$ N/mm²). On the other hand, the stresses at the same point calculated using the pinned trussed framework model (Fig. 2-75/second from left) are as follows:

$\sigma_k = N_{kj}/A = 28.87/(0.592 \cdot 10^{-2}) = 48.7 \cdot 10^2 \text{ kN/m}^2$, or 4.88 N/mm²

The bending stresses at mid-span of bar $i–k$ are as follows:

$\sigma_{b,i-k} = 156.25/(3.904 \cdot 10^{-4}) = 40.02 \cdot 10^4 \text{ kN/m}^2$, or 400.20 N/mm²

These are much higher than the yield stress for grade S 235 steel ($\sigma_{yield} = 240$ N/mm²). From the stress analysis it follows that the secondary stresses at the nodes considerably exceed the primary stresses obtained from the pinned trussed framework model. On the other hand, the bending stress at mid-span calculated using the pinned trussed framework model is more than twice the maximum node stresses given by the displacement method. This means that the pinned trussed framework model does not represent a realistic model for calculating frame constructions. So this is also the logical side of the historical development of the displacement method.

## 2.9 Second-order theory

Navier's suspension bridge theory comes at the start of the discipline-formation period of structural theory, Josef Melan's more exact suspension bridge theory at the end of this period. Fig. 2-81 shows a suspension bridge stiffened by a lattice girder. In such a bridge, the question is: Which forces does the stiffening girder have to accommodate?

### 2.9.1 Josef Melan's contribution

It was in 1883 that Wilhelm Ritter published the first theory for calculating suspension bridges taking into account the elongation of the cables [Ritter, 1883]. Five years later, Melan investigated statically indeterminate bar polygons stiffened by lattice girders using Castigliano's second theorem. As usual, he applied the internal forces to the undeformed system (equilibrium of the undeformed system), i.e. according to first-order theory. Afterwards, he published his *Genauere Theorie des durch den geraden Stab versteiften Stabpolygons* (more exact theory of the bar polygon stiffened by the straight bar) [Melan, 1888/2, pp. 38–42] in which he derives the differential equation of this loadbearing structure using the deformed system (second-order theory) and calculates the bending moments of the stiffening beam from this. He justifies his approach as follows: "The theory developed in the above paragraphs [calculation of bar polygons stiffened by straight beams according to first-order theory – the author] produces satisfactory results only for those systems in which only very small elastic deformations occur. In the systems treated here, the reliability of this approximate method of calculation [first-order theory – the author] therefore depends on the degree of perfection to which the stiffening beam achieves its task of rectifying the flexibility of the system. If this beam is only small, i.e. it has only a small second moment of area (as is the case with older designs of such semi-stiffened suspension bridges), then the static deformations of the bar polygon can no longer be left unconsidered and a more accurate method of calculation should be applied" [Melan, 1888/2, pp. 38–39]. Melan's method of calculation was first used in a practical situation by Leon Solomon Moiseiff for the design of the Manhattan Bridge in New York (1901–09), which has a span of 448 m, [Mehlhorn & Hoshino, 2007, p. 67]. Melan expresses in clear terms the realisation that there is no longer proportionality between loading and deformation, and therefore the superposition theorem cannot be applied. So the dominance of linearity in theory of structures had already been broken by the end of the discipline-formation period. In his Melan obituary, Fritsche acknowledges his achievements in this field: "He was the first person to set up this 'deflection theory' [second-order theory – the author] in a form that, for less stiff structures such as suspension bridges, has not been completely replaced by newer investigations, and he re-analysed a number of large American steel bridges for his friend Lindenthal in New York" [Fritsche, 1941, p. 89].

### 2.9.2
**Suspension bridges become stiffer**

Leon Solomon Moiseiff and David Bernard Steinman in particular continued to develop second-order theory for suspension bridges with the building of the very large suspension bridges in the USA in the 1920s and 1930s. Steinman published his translation of Melan's *Theorie der eisernen Bogenbrücken und der Hängebrücken* [Melan, 1888/2] in 1913 under the title of *Theory of Arches and Suspension Bridges* [Melan, 1913]; he edited a considerably expanded version in 1922 and gave it the title *Suspension Bridges, Their Design, Construction and Erection* [Steinman, 1922], and published it as a separate book entitled *A Practical Treatise on Suspension Bridges* in 1929 [Steinman, 1929]. As one of the owners of consulting engineers Robinson & Steinman (New York), which specialised in the design of long-span suspension bridges, Steinman was responsible for the design of numerous bridges of this type. He presented his theory of continuous suspension bridges according to second-order theory in the forum of the first congress of the International Association for Bridge & Structural Engineering (IABSE) in Paris in 1932 [Steinman, 1934]. Using the example of a suspension bridge with a main span of 240 m, Steinman found that the bending moments in the continuous stiffening girder calculated according to second-order theory could be reduced by 45 % on average when compared to a calculation employing first-order theory [Steinman, 1934, p. 451]. In Germany it was Kuo-hao Lie and Kurt Klöppel in particular who developed the calculation of suspension bridges further [Lie, 1941; Klöppel & Lie, 1941]; these publications were inspired by designs for a suspension bridge over the River Elbe in Hamburg during the latter half of the 1930s [Berg, 2001]. Further German contributions to the theory of suspension bridges according to second-order theory were provided by Hertwig, Krabbe, Hoening, Unold, Cichocki, Müller, Tschauner, Hartmann, Theimer, Müller, Selberg and Bornscheuer [Bornscheuer, 1948]; most of these contributions referred to H. Neukirch's suspension bridge theory. This may be enough to understand the economic aspects behind the application of second-order theory for specific steel structures loaded in tension, e.g. suspension bridges, which essentially consists of establishing the lower bending moments in the stiffening girder resulting from second-order theory, and using these as the basis for the design. Suspension bridges with main spans of several hundred metres were now possible thanks to second-order theory.

### 2.9.3 Arch bridges become more flexible

But Melan was also aware that in the theory of the elastic arch "the deformation of the arch caused by the load is not taken into account when determining the moments" [Melan, 1888/2, p. 100]. Nevertheless, he pleads for basing the calculation of arch bridges on first-order theory because these "are generally constructed considerably stiffer, i.e. with a cross-section having a larger second moment of area, than suspension bridges ..." [Melan, 1888/2, p. 100].

Significant research activities did not start before the construction of long-span arch bridges in the 1930s. And so it was in 1934 that Bernhard Fritz managed to achieve a synthesis between second-order theory and elastic arch theory – in the first half of the consolidation period of structural theory [Fritz, 1934]. He was thus responding to the construction of long-span, slender arch bridges, which started to appear in the late 1920s. Whereas the internal forces calculated for suspension bridges according to second-order theory are less than those determined using first-order theory, the situation is reversed for arch bridges: "The relationships are totally different in arch bridges. Considering the influence of the system deformation increases the stresses, which means that ignoring the deformations results in more favourable stresses and the structure has a lower factor of safety in practice than in theory." [Fritz, 1934, pp. 1–2]. Fritz established second-order theory calculations for three-pin, two-pin and fixed-end arches, and compared the results of the calculations with tests on models. Besides the bending stiffness, he also took the strain stiffness of the arch into account.

## The differential equation for laterally loaded struts and ties

### 2.9.4

The opposite effect of the influence of second-order theory on structural systems in tension and compression, of which Melan was aware [Melan, 1888/2, p. 42], will be shown below using the example of the simply supported beam illustrated in Fig. 2-61, but instead of the uniformly distributed load $q_{max}$, the load $q$ will be applied, and in addition the external horizontal force $S$ at the roller-bearing support. The latter action makes itself felt in the simply supported beam as the axial force $N$. If $S$ acts as a compressive force, the differential equation for vertical displacement $v$ is

$$v^{IV} + \frac{\varepsilon^2}{l^2} \cdot v^{II} = \frac{q}{E \cdot I} \tag{2-26}$$

where

$$\varepsilon = l \cdot \sqrt{\frac{|N|}{E \cdot I}}$$

which is the bar factor. If there is no axial force, eq. 2-26 is changed to the differential equation for the deflection of the laterally loaded beam. The particular solution of eq. 2-26 is

$$v_p = \frac{q}{E \cdot I} \cdot \left(\frac{l}{\varepsilon}\right)^2 \cdot \frac{x^2}{2} \tag{2-27}$$

If the direction of action of $S$ is reversed, then an axial tension acts in the simply supported beam, which changes the sign of the second term in differential equation 2-26:

$$v^{IV} - \frac{\varepsilon^2}{l^2} \cdot v^{II} = \frac{q}{E \cdot I} \tag{2-28}$$

Differential equation 2-28 then has the following particular solution for the laterally loaded simply supported beam in tension:

$$v_p = -\frac{q}{E \cdot I} \cdot \left(\frac{l}{\varepsilon}\right)^2 \cdot \frac{x^2}{2} \tag{2-29}$$

Comparing eq. 2-27 and eq. 2-29 reveals that according to second-order theory, the deflection $v$ of the simply supported beam increases with a compressive stress, but decreases in the case of a tensile stress: compressive forces result in a more flexible, tensile forces a stiffer structural system.

## The integration of second-order theory into the displacement method

### 2.9.5

Ernst Chwalla and Friedrich Jokisch published a systematic extension of the displacement method for second-order theory in 1941. In their publication they analyse the stability problem of storey frames and integrate the minimum load increase factor $v_{krit}$ (for calculating the system buckling) – introduced into aircraft design by Teichmann and Thalau in 1933, which results from the non-trivial solution of the homogeneous elasticity equations according to second-order theory – into the equations of the displacement method [Chwalla & Jokisch, 1941]. The superiority of the displacement method compared to the force method would soon become evident, also when analysing stresses according to second-order theory, which is characterised by the solution of the non-homogeneous elasticity equations according to second-order theory. Teichmann [Teichmann, 1958] and Chwalla [Chwalla, 1959] published a coherent presentation of the solution of the stress and stability problems using the displacement and force methods according to second-order theory: standardising the equations of the force and displacement methods according to first- and second-order theory completes the theory of elastic bar frameworks in the innovation phase of structural theory (1950–75). Teichmann's contributions to *Übersicht über die Berechnungsverfahren für Theorie II. Ordnung* (overview of calculation methods for second-order theory) by Klöppel and Friemann [Klöppel & Friemann, 1964] were ignored almost certainly because of the apparently laborious use of operative symbols. It is interesting to note that the authors of second-order elastic theory worked exclusively on the basis of the displacement method. They therefore anticipated the break with symmetry in the dualist nature of theory of structures on the level of second-order theory still evident in Teichmann's work. The obvious superiority of the displacement method over the force method became evident in the diffusion phase of theory of structures as the automation of structural calculations with computers advanced to become part of the structural engineer's everyday workload.

### 2.9.6 Why do we need fictitious forces?

The fictitious force $P_{Fi} = N \cdot \psi$ introduced by Teichmann in the computational algorithm of second-order theory takes into account the influence of the member chord rotation $\psi$ (Figs 2-82 and 2-83). Hence, the deformation influences on the equilibrium condition of bar frames was split into a component from first-order theory, a member chord rotation component (node displacement) and the component that results from the bending of the individual bar with fixed nodes and axial force. The latter two components are explained in Fig. 2-83 for a cantilever beam.

The advantage of fictitious forces is that the equilibrium conditions can be assumed at the bar with fixed nodes (Figs. 2-83 and 2-84). For example, for the moment $M_1(x)$ (Fig. 2-84):

$$\sum M_x = 0 = M_1(x) - (P_{Fi} + W) \cdot (h - x)$$

$$M_1(x) = N \cdot \psi \cdot (h - x) + W \cdot (h - x) \quad (2\text{-}30)$$

**FIGURE 2-82**
Equilibrium of a cantilever beam in the deformed system

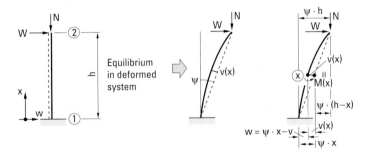

**FIGURE 2-83**
Fictitious forces

The first term in eq. 2-30 is the member chord rotation component and the second term is the first-order theory component. The moment component from the deflection curve is given by Fig. 2-85:

$$\sum M_{(x)} = 0 = M_2(x) - N \cdot v(x) \tag{2-31}$$

$M_2(x) = N \cdot v$ is nothing more than the deflection component $v$ according to second-order theory for fixed nodes.

The sum of $M_1(x)$ and $M_2(x)$ is the total moment $M^{II}$ according to second-order theory:

$$M^{II} = M_1(x) + M_2(x)$$

$$M^{II} = W \cdot (h - x) + N \cdot \psi \cdot (h - x) + N \cdot v \tag{2-32}$$

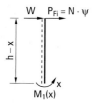

**FIGURE 2-84**
Equilibrium with fictitious forces

This shows that determining the moment $M^{II}$ according to second-order theory with the help of the fictitious force $P_{Fi} = N \cdot \psi$ agrees with the moment that follows from considering the equilibrium directly in the deformed system (Fig. 2-82).

*Sample calculation*

The displacement method according to second-order theory is illustrated below using a simple example (Fig. 2-86a). Fig. 2-86b shows the geometrically determinate system with the unknown node displacement $\xi_1$. The forces in the bars are calculated using the stabilised pinned system (Fig. 2-86c):

$$N_{12} = F = -100 \text{ kN}$$

which results in the bar factor

$$\varepsilon = l \cdot \sqrt{\frac{|N_{12}|}{E \cdot I}} = 3 \cdot \sqrt{\frac{|100|}{400}} = 1.5$$

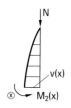

**FIGURE 2-85**
Equilibrium in the deformed system (without member chord rotation)

The bar end moment $M_{12,0}$ in the 0-state (Fig. 2-87a) can be calculated with the equation

$$M_{12}^{(0)} = -\frac{q \cdot l^2}{2 \cdot \alpha} \tag{2-33}$$

from table 7.3 [Duddeck & Ahrens, 1994, p. 328], and the stiffness coefficient

$$\alpha = \frac{\varepsilon \cdot \sin \varepsilon - \varepsilon^2 \cdot \cos \varepsilon}{2 \cdot (1 - \cos \varepsilon) - \varepsilon \cdot \sin \varepsilon} \tag{2-34}$$

from table 7.1 [Duddeck & Ahrens, 1994, p. 327]:

$$\alpha = \frac{1.5 \cdot \sin 1.5 - 1.5^2 \cdot \cos 1.5}{2 \cdot (1 - \cos 1.5) - 1.5 \cdot \sin 1.5} = 3.7 \rightarrow M_{12}^{(0)} = -\frac{5 \cdot 3^2}{2 \cdot 3.7} = -6.08 \text{ kNm}$$

Using first-order theory ($\varepsilon = 0$), the bar end moment $M_{12,0}$ would be $-5.63$ kNm, i.e. 8% less than the moment calculated with second-order theory. The absolute difference $6.08 - 5.63 = 0.45$ is nothing more than the deformation influence for fixed nodes according to second-order theory.

The $\xi_1 = 1$-state is shown in Fig. 2-87b. Taking the applied displacement 1, the angle of rotation of the member $\psi_{12} = 1/3$ [1/m] and the fictitious forces are as follows:

$$P_{Fi}^{(1)} = N_{12} \cdot \psi_{12} = 100 \cdot \frac{1}{3} = 33.33 \, \frac{\text{kN}}{\text{m}}$$

The fictitious forces should be applied as a couple to the ends of the bar in such a way that the couple assists the rotation of the member chord (clockwise rotation in this example, see Fig. 2-87b). The bar end moment for unit displacement $\xi_1 = 1$ (Fig. 2-87b) can be calculated with the equation

$$M_{ik}^{(1)} = -\gamma \cdot \frac{E \cdot I}{l} \cdot \psi \tag{2-35}$$

from table 7.2 [Duddeck & Ahrens, 1994, p. 328], and the stiffness coefficient

$$\gamma = \frac{\varepsilon^2 \cdot \sin \varepsilon}{\sin \varepsilon - \varepsilon \cdot \cos \varepsilon} \tag{2-36}$$

from table 7.1 [Duddeck & Ahrens, 1994, p. 327]:

$$\gamma = \frac{1.5^2 \cdot \sin 1.5}{\sin 1.5 - 1.5 \cdot \cos 1.5} = 2.5 \rightarrow M_{12}^{(1)} = -2.5 \cdot \frac{400}{3} \cdot \frac{1}{3} = -111.1 \text{ kN}$$

Calculation of the restraint forces $Z_{10}$ and $Z_{11}$ is carried out with the help of the principle of virtual displacements applied to the simple kinematically determinate system (Fig. 2-88).

The principle of virtual displacements states that the sum of the virtual external work and internal virtual work is zero. As the latter component vanishes in simple kinematic systems (Fig. 2-88), the virtual external work is equated to zero here; the principle of virtual displacements is equivalent to the equilibrium conditions. In this example the virtual displacement state (Fig. 2-88a) carries out the virtual external work on the true force state (0-state) (Fig. 2-88b):

$$Z_{10} \cdot 1^v + M_{12}^{(0)} \cdot \left(\frac{1}{3}\right)^v + \int_{x=0}^{x=l=3} p \cdot v^v(x) \cdot dx = 0$$

$$\rightarrow Z_{10} = -\left[(-6.08) \cdot \frac{1}{3} + \frac{1}{2} \cdot 3 \cdot 5 \cdot 5\right] = -5.47 \text{ kN}$$

In the same way, the virtual displacement state (Fig. 2-88a) carries out the virtual external work on the true force state $\xi_1 = 1$:

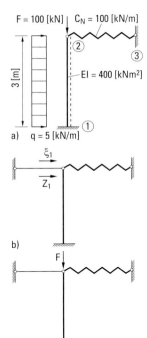

FIGURE 2-86
a) Structural system, b) geometrically determinate system, c) stabilised pinned system

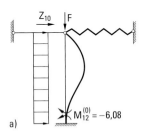

FIGURE 2-87
a) 0-state, b) $\xi_1 = 1$-state

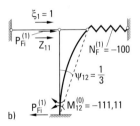

119

$$Z_{11} \cdot 1^v + M_{12}^{(1)} \cdot \left(\frac{1}{3}\right)^v + P_{Fi}^{(1)} \cdot 1^v + N_F^{(1)} \cdot 1^v = 0$$

$$\rightarrow Z_{11} = -\left[(-111.11) \cdot \frac{1}{3} + 33.33 - 100\right] = 103.71 \ \frac{\text{kN}}{\text{m}}$$

As the restraint forces are eliminated (Fig. 2-86a), the following must be true:

$$Z_{11} \cdot \xi_1 + Z_{10} = 0 \qquad (2\text{-}37)$$

$$\rightarrow \xi_1 = -Z_{10}/Z_{11} = 5.47 / 103.71 = 0.053 \text{ m}$$

A calculation according to first-order displacement theory would give $\xi_1 = -Z_{10}/Z_{11} = 5.63 / 144.44 = 0.039$ m. Compared to the result of first-order theory, the node displacement according to second-order theory increases by 35.9 %; the system becomes more flexible.

In 1958 Teichmann specified a computational algorithm for calculating elastic bar frameworks according to second-order displacement theory [Teichmann, 1958]:

1. Restrain a given structural system by introducing geometric unknowns $\xi_1 \dots \xi_i \dots \xi_m$, i. e. make it geometrically determinate (Fig. 2-86b).
2. Calculate or estimate the member forces $N_{ik,j}$ (= axial force in the $j$th iteration step) in the stabilised pinned system (Fig. 2-86c).
3. Calculate the stiffness coefficient $\alpha_{ik}$ or $\gamma_{ik}$ for struts (see eq. 2-34 and eq. 2-36); the bar factor for ties is generally taken to be $\varepsilon = 0$.
4. Calculate the $m$ unit displacement states $\xi_1 \dots \xi_i \dots \xi_m$ of the system with $m$ degrees of geometric indeterminacy and the 0-state (Fig. 2-87):
   – For joint rotation (Fig. 2-87a): Calculate the bar end moments $M_{ik}$ and take $\varepsilon$ into account (calculate spring forces if necessary).
   – For member chord rotation (Fig. 2-87b): Calculate the bar end moments $M_{ik}$ and take $\varepsilon$ into account; calculate the member forces $N_{ik}$ for bars where $E \cdot A \neq \infty$; calculate the fictitious forces $P_{Fi,ik} = N_{ik} \cdot \psi_{ik}$.
5. Calculate the $m \cdot m$ elements of the stiffness matrix $(Z_{ik})$ with the help of the principle of virtual displacements or the equilibrium conditions.
6. Calculate the $m$ element of the load vector $(Z_{i0})$ with the help of the principle of virtual displacements or the equilibrium conditions:

6a. If $(Z_{i0}) \neq (0)$ and failure of the system *is not* to be determined, this is a *stress problem*:
Solve the set of equations $(Z_{ik}) \cdot (\xi_k) = -(Z_{i0})$ (eq. 2-24) and calculate all the internal and deformation variables with the help of the superposition theorem (eq. 2-25), e. g. the axial force diagram $N_{ik,j+1}$.
If $N_{ik,j} \neq N_{ik,j+1}$, repeat calculation steps 2 to 6a until $N_{ik,j} = N_{ik,j+1}$.

6b. If $(Z_{i0}) = (0)$ or $(Z_{i0}) \neq (0)$ and failure of the system *is* to be determined, this is a *stability problem*:
Is the denominator of the stiffness matrix $\Delta(Z_{ik}) = 0$? If this is not the case: increase all loads equally by the factor $v$ and repeat calculation

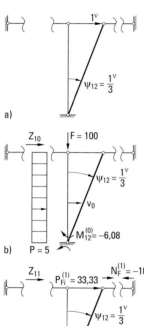

**FIGURE 2-88**
Calculation of the $Z_{ik}$ members with the help of the principle of virtual displacements: a) virtual displacement state, b) 0-state carries out work on the virtual displacement state, c) $\xi_1 = 1$-state carries out work on the virtual displacement state

steps 2 to 6 until the denominator $\Delta(\underline{Z}_{ik}) = 0$, i.e. the critical load increase factor $v_{krit}$ or rather the critical load $P_{crit}$ is reached (check buckling of individual bars).

The advantage of the displacement method is to be found in the uniform analysis of the stress and stability problems. At the same time, a serious disadvantage is evident in the solution of the stability problem: the calculation of the denominator of the stiffness matrix $\Delta(\underline{Z}_{ik})$ is very complex for those structural systems whose geometric indeterminacy is equal to or greater than four ($m \geq 4$). Nevertheless, Teichmann inserted the final piece in the jigsaw of the elastic theory of bar frameworks by presenting the second-order displacement theory in the formal language of classical theory of structures at the start of the integration period (1950 to date).

## 2.10 Ultimate load method

It was not too long ago that structural engineers working with steel always praised the "ingenuity", the "self-help" features of the material, when the results of the calculations using models based on elastic theory did not reflect adequately the true loadbearing behaviour. They were thus praising their plastic, ductile material steel, which had reserves of strength beyond the elastic limit: the metaphysical excess which has become a metaphor. The praises delivered with assuredness nevertheless expressed uncertainty about the quantifiability of the "ingenuity", the "self-help" features of the material, by theory of structures in its accumulation phase (1900–25). Although the dominance of linearity was broken in the invention phase of structural theory (1925–50) through the development of second-order theory on the side of the relationship between loading and internal force conditions, during that same period, scientists from the theory of structures and applied mechanics disciplines exposed the linear-elastic stress-strain relationship (Hooke's law) to the criticism of the reality. This opened up two breaches in the linear trinity (statics, materials, geometry) that characterises classical theory of structures. Non-linear stress-displacement relationships (geometry), on the other hand, developed at the end of the innovation phase (1950–75) into a scientific object of theory of structures in order to calculate with computers the lightweight shell structures that were then appearing.

### 2.10.1 First approaches

The first approach to the experimental measurement of the loads on steel beams was carried out in 1914 by the Hungarian engineer Gábor v. Kazinczy (1889–1964). He investigated beams fixed at both ends with clear spans $l = l_1 = 5.60$ m and $l = l_2 = 6.00$ m consisting of a steel I-beam section encased in concrete [Kazinczy, 1914] (Fig. 2-89). Fig. 2-90 shows Kazinczy's structural model as depicted in a publication by Heyman, carrying a uniformly distributed load $w$ [Heyman, 1998]. Kazinczy concluded from this series of tests that there must be three plastic sections (plastic hinges) in a beam fixed at both ends (two degrees of static indeterminacy) in order to reach the failure mode. The beam must therefore be designed not according to the elastic solution, i.e. based on the bending moment $w \cdot l_2/12$ (Fig. 2-90b), but rather on the value $w \cdot l_2/16$ (Fig. 2-90c).

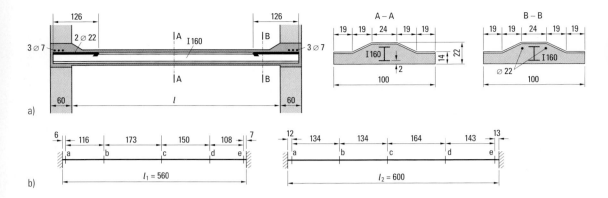

**FIGURE 2-89**
Kazinczy's test beams: structural systems with points a, b, c, d and e for measuring the deflection (in cm) [Kaliszky, 1988, p. 78]

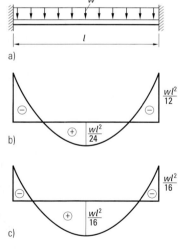

**FIGURE 2-90**
a) Fixed-end beam; b) elastic, and c) plastic solution [Heyman, 1998, p. 128]

In 1917 Prof. N. C. Kist from Delft looked into the ultimate load problem (see [Gebbeken, 1988, pp. 34–36]), and in 1920 he proposed designing constructions made from mild steel according to the ideal-elastic and ideal-plastic material law [Kist, 1920, p. 428]. Today, this idealised material law still forms the basis for the structural steelwork design procedures based on ultimate load theory. The terms "ultimate load theory" and "plastic hinge theory" are used in the following account as synonyms, although the former is the generic term. The same is true for the corresponding hybrid constructions "ultimate load method" and "plastic hinge method".

The realisation that the economic design of steel structures could no longer be assured just on the basis of elastic theory became clear towards the end of the 1920s in Germany, Austria and Czechoslovakia, later in Great Britain and the Soviet Union. In Germany, Prof. Martin Grüning (Hannover TH) started the discussion on the theoretical level, Prof. Maier-Leibnitz (Stuttgart TH) on the experimental level. Grüning expanded Kazinczy's findings about the plastic limit state of the fixed-end beam to systems with $n$ degrees of static indeterminacy [Grüning, 1926].

Maier-Leibnitz (see [Kurrer, 2005]) shed light on the "ingenuity" of the material in 1928 with his test report on the true load-carrying capacity of simply supported and continuous beams made from mild steel grade St 37 (corresponds to today's S 235) and timber: "Many design engineers working in steel construction, mindful of the doubts expressed by Mohr and in other theory of structures textbooks, shun the use of the continuous beam and give preference to single-span and pinned beams, although experienced design engineers have known for a long time that the self-help of ductile mild steel is limited in these types of beam … The author … has carried out the tests … described below in order to clarify whether the maximum value of the support moment calculated is critical for the load-carrying capacity of continuous beams and to what extent yielding supports have an effect on the load-carrying capacity, and also to show – at least in the special case – how to recognise the constructional self-help of the building material, which up to now was used only intuitively, according to its nature, and how one can intentionally use the cross-sectional dimensions of the beam" [Maier-Leibnitz, 1928, p. 11].

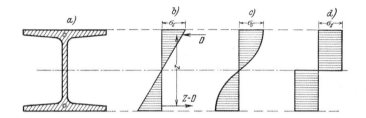

**FIGURE 2-91**
Stress distribution after Maier-Leibnitz: a) cross-section, b) elastic limit state, c) partially plastic, and d) fully plastic cross-section [Maier-Leibnitz, 1929, p. 314]

Maier-Leibnitz summarises his bending tests on beams with two equal spans by saying that a small amount of yielding at the supports has no effect on the load-carrying capacity and the critical design moment is not the support moment, but rather the span moment, the magnitude of which is equal to 75 % of that of the support moment. Otto Mohr had already quantified the unfavourable effect of yielding supports on the internal forces of continuous beams back in 1860 [Mohr, 1860] and hence backed up with figures the concerns about using continuous beams. But Mohr based his work on elastic theory. The work of Maier-Leibnitz resulted in the editors of the journal *Die Bautechnik* receiving many letters from prominent structural engineers, which were published in issue No. 20 in that same year: from Grüning and Kulka, Bohny, Metzler, Beyer, Jagschitz, Gaber, Krabbe, Bernhard and Lienau. Finally, in 1929, Maier-Leibnitz interpreted his ultimate load tests with fixed-end and simply supported I-beams of grade St 37 steel against the background of the ideal elastic/ideal plastic material law proposed by Kist and specified the stress distribution in the fully plastic cross-section (Fig. 2-91).

## 2.10.2 Foundation of the ultimate load method

Papers dealing with the plasticity when calculating statically indeterminate steel structures now appeared in rapid succession in journals such as *Die Bautechnik*, *Der Bauingenieur*, *Der Stahlbau* and *Zeitschrift für Angewandte Mathematik und Mechanik (ZAMM)* (journal of applied mathematics and mechanics). The most important work was carried out by Prof. Josef Fritsche (Prague TH) and Karl Girkmann (Waagner-Biro company, Vienna). While Fritsche was investigating statically determinate simply supported beams, the fixed-end single-span beam and specific continuous beams in 1930 [Fritsche, 1930], Girkmann had by 1931/32 specified a plastic design method for loadbearing frames backed up by tests [Girkmann, 1931 & 1932].

### 2.10.2.1 Josef Fritsche

Fritsche was to the first person to devise equations for the bending moment $M_{pl}$ of fully plastic rectangular and I-beam cross-sections in the case of pure bending [Fritsche, 1930, pp. 852–855]. After he had exemplified the expanded differential equation for the beam bending for the calculation of the elasto-plastic deformation of a simply supported beam, he turned to the analysis of statically indeterminate systems. He now solved the same differential equation for the case of the fixed-end beam with a central point load; in the fully plastic state, plastic hinges occur simultaneously at the points of fixity and at mid-span (Fig. 2-92).

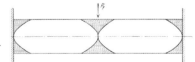

**FIGURE 2-92**
Plastic hinges of a fixed-end beam after Fritsche [Fritsche, 1930, p. 873]

Fritsche uses the same method to determine the ultimate load of two- and three-span beams. Fritsche writes: "As the examples up to now show, calculating the load-carrying capacity is quite simple and avoids the – often tedious – calculation of the statically indeterminate variables, although knowledge of the purely elastic solution, albeit in the general form only, is very helpful. It is therefore within the range of possibilities for the steel loadbearing structures of buildings, where beams made from plain rolled sections are really quite common, to carry out the design according to the load-carrying capacity determined in the way described instead of according to permissible stresses" [Fritsche, 1930, p. 890]. He verified his method of elastic-plastic analysis for steel beam structures by re-analysing the tests of E. Meyer, H. Maier-Leibnitz and J. H. Schaim. Fritsche comes to the conclusion "that the inconstancy in the yield stress value of steel is a considerable obstacle to the exact computational examination of the same; but it does show that the errors, expressed as a percentage of the load-carrying capacity, are never larger than the percentage fluctuations in the value of the yield stress. Further test programmes to check the above method of calculating the load-carrying capacity of steel beams would of course be highly desirable, although in order to determine the reliability of the basis for the calculations exactly, the aim should be to use a material with a yield stress as constant as possible" [Fritsche, 1930, p. 893]. Fritsche clearly recognised that the yield stress of mild steel represents *the* critical material parameter for the ultimate load method.

### 2.10.2.2 Karl Girkmann

Concerning his motives for investigating the elasto-plastic behaviour of steel frames, Girkmann writes: "In my treatise on the design of frames based on an ideal-plastic steel [Girkmann, 1931 – the author], I have looked at the strength case 'bending with axial force', pursued the processes during the course of the loading in plane, frame-like bar constructions and thereupon attempted to evaluate the 'self-help' effect of the steel when designing such structures in order to achieve more economic dimensions for such structures. Apart from the savings in weight that can be achieved, the use of this method makes it possible to reduce the maximum moments, to even out the differences in the thicknesses of the cross-sections required and hence to simplify the construction details and reduce their cost. If the deflections occurring under service loads do not have to be specially calculated, the calculation of frame-like bar frameworks in buildings as statically indeterminate frames could be dispensed with, which would result in a considerable simplification of the design work for storey frames with a high degree of static indeterminacy" [Girkmann, 1932, p. 121].

Girkmann develops a method of successive load increases with interaction, which is today the basis of almost all computer-assisted methods of calculation [Rothert & Gebbeken, 1988, pp. 23–27]. He verified his method for statically indeterminate frameworks by measuring the strains on a test frame with two pin supports. Fig. 2-93a shows the test frame

with tie (top) and cross-beam (bottom) which has the following parameters:
- material: standard mild steel with limit of proportionality 1.8 t/cm², yield stress of 2.58 or 2.67 t/cm² and tensile strength of 4.32 or 4.27 t/cm²
- cross-beam length (system dimension): 1500 mm
- cross-beam cross-section: 2 No. 80 mm channels
- leg length (system dimension): 595 mm
- leg cross-section: 2 No. 80 mm channels
- tie cross-section: 2 No. 30 × 45 × 4 mm angles
- all rolled sections riveted together with gusset plates to form a structural connection at the rigid frame corners, at the bottom of the legs and at the leg-tie connections

a)

Fig. 2-93b shows the two-pin frame under maximum load (measuring instruments already removed). After relieving the load, a distinct curvature can be seen in the middle of the cross-beam (Fig. 2-93c), which indicates the hinge-like effect of this area (plastic hinge). Fig. 2-94 shows the stress diagrams for the centre of the cross-beam. The critical stress condition according to Fig. 2-94d is characterised by the negligible elastic zone; Girkmann derives the plasticity condition for axial force plus bending moment from the equilibrium conditions for this situation. If the load is increased further, the neutral axis merely shifts from $n$ (Fig. 2-94d) to $n'$ (Fig. 2-94e).

b)

Whereas the test frame can be analysed as a system with one degree of static indeterminacy up to the loading case shown by Fig. 2-94d using the elasto-plastic deformations, as the load increases further to Fig. 2-94e it acts as a three-pin frame. Finally, at the critical load $P = P^T = 10.11$ t, plastic hinges form at points $O$, $E$ and $F$. (Fig. 2-95). According to Girkmann, the load $P = P^T = 10.11$ t represents the ultimate load of the test frame determined according to his method of successive load increases (Fig. 2-93). At this ultimate load, hinges can form at any number of points – which in this case actually happens in the legs at points $G$ and $H$ (see Fig. 2-95).

c)

**FIGURE 2-93**
a) Test frame, b) test frame under maximum load, and c) after relieving the load
[Girkmann, 1932, p. 124/25]

Rothert and Gebbeken were able to verify credibly that Girkmann had presented considerable information about the ultimate load method in his paper in the journal *Der Stahlbau*, which many years later had to be laboriously (re-)established through research [Rothert & Gebbeken, 1988, pp. 26 – 27]:

**FIGURE 2-94**
Cross-section and stress distribution in the middle of the frame: a) elastic limit state, b) and c) partially plastic cross-section, d) and e) fully plastic cross-section
[Girkmann, 1932, p. 125]

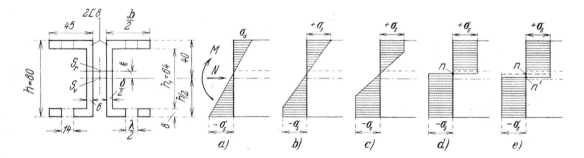

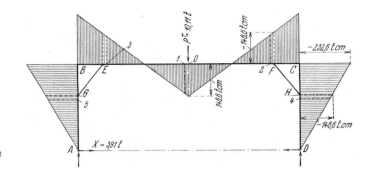

**FIGURE 2-95**
Calculated ultimate load of the test frame
[Girkmann, 1932, p. 126]

1. Migration of the plastic hinges – eccentric plastic hinge: "A cross-section with a critical stress distribution acts like a hinge whose position changes constantly with respect to the axis of the bar" [Girkmann, 1932, p. 122].
2. Demand for positive dissipation work: "The effect of such hinges is always limited because only those mutual rotations … are possible in which these are moved towards each other in the parts of the cross-section in compression, and moved apart in the parts in tension. The material resists opposing rotations (which happens when relieving the loads), and the hinge effect in such cases … is cancelled out again" [Girkmann, 1932, p. 122].
3. Second-order theory: "The design method shown can be applied to loadbearing structures of any form carrying any loads provided the additional bending effects that occur during the deformation of the structures are always negligible" [Girkmann, 1932, p. 123].
4. Critical internal forces: "When drawing the bending moment diagrams, it should be noted that the actual end cross-sections of the bars do not lie in the system nodes …, but rather, depending on the design of the frame corners, at various distances from the nodes" [Girkmann, 1932, p. 123].
5. Local failure – rotation capacity: "The permissible loading on which the design is based can be achieved only if premature folding of parts of the cross-section … of the bars is avoided" [Girkmann, 1932, p. 123].
6. Proof of stability: "The ultimate stability of the loadbearing structure corresponds to attaining the kinematic chain. In addition, plastic local buckling of individual bars must be avoided. Compressive forces should be increased in the ratio of the safety factors for bending and buckling, with the buckling coefficient multiplied and then entered into the plasticity condition" [Rothert & Gebbeken, 1988, p. 27; see also Girkmann, 1932, p. 123].
7. Safety concept and serviceability: "Elastic deformations [should be encouraged] under service loads alone … If an increase in the elastic limit can be reckoned with here …, then the stresses at the extreme fibres may increase up to yield stress and all we need to avoid is that the yield stress already penetrates into the cross-sections under service loads. This condition is … always met if the permissible loads are

specified based on a factor of safety of a least two" [Girkmann, 1932, p. 123].

8. Safety in the case of recurrent loading: "Based on the investigations by Fritsche …, the same degree of safety should be required as for a one-off load" [Rothert & Gebbeken, 1988, p. 27; see also Girkmann, 1932, pp. 123/124].

9. Design of connections: "Connections between members, corner connections, anchorages, etc. should also be designed based on moments and forces resulting from the underlying moments in such a way that failure prior to reaching the required permissible load is prevented" [Girkmann, 1932, p. 123].

Therefore, a comprehensively formulated plastic design method for steel structures was already available in 1932.

### 2.10.2.3 Other authors

The year 1932 was when Felix Kann investigated the distribution of internal forces in continuous beams in the elasto-plastic range; like Girkmann, he assumes the correct stress distribution of partially plastic cross-sections of doubly symmetric steel sections [Kann, 1932, p. 106]. One year before that, Kazinczy wrote on this subject in the journal *Der Stahlbau*, including the following, remarkable statement: "The design engineer was certainly also familiar with the stress-strain curve in the past, but was afraid of the mathematical difficulties involved in its use, although in some cases calculations based on Hooke's law are more complicated. And now we see that the calculation is not more involved, but rather simpler. Up until now, the true strain curve was used only by Kármán for the buckling problem, and the trials of Rŏs and Brunner have confirmed this splendidly. So we can resolutely use the plastic theory in other problems as well" [Kazinczy 1931/1, p. 59]. This is what Kazinczy did – as will be shown later – but unfortunately without causing any impact.

## 2.10.3 The paradox of the plastic hinge method

Gábor von Kazinczy drew attention to a contradiction in the plastic hinge method as early as 1931 [Rothert & Gebbeken, 1988, p. 29; Kazinczy 1931/2], with which Fritz Stüssi and Curt Fritz Kollbrunner caused the engineering world to hold its breath in 1935 [Stüssi & Kollbrunner, 1935]: the paradox of the plastic hinge method (Fig. 2-96).

The load $P$ of the continuous beam is increased from zero to the elastic limit state so that the extreme fibres begin to become plastic at the point of the maximum bending moment; as the load increases further, so plastic hinges form. The limit state of load-carrying capacity (plastic limit state) of the continuous beam is then reached when plastic hinges form at points ①, ② and ③. These three plastic hinges change the system with two degrees of static indeterminacy into a simple kinematically determinate system. The load associated with this series of plastic hinges (kinematic chain) is known as the ultimate load $T$ or the plastic limit load $T$ of the continuous beam.

By contrast, a plastic hinge can only form at point ② in the simply supported beam (Fig. 2-96/left). If we compare the ultimate loads of the

**Problem:**

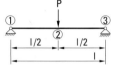
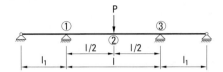

**Material law:**

 elastic
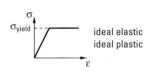 ideal elastic ideal plastic

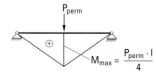
$$M_{max} = \frac{P_{perm} \cdot l}{4}$$

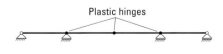 Plastic hinges

This state is reached in 3 steps:

Point ②:

1) ideal elastic  $\sigma \leq \sigma_{yield}$

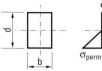

plasticity begins
$M_{elast}$

**Design:**

$$M_{max} = M_{perm} = \sigma_{perm} \cdot \frac{b \cdot d^2}{6}$$
$$\phantom{M_{max} = M_{perm} = \sigma_{perm} \cdot}\ W$$

2) partially plastic

$(\ M_{elast} \leq M \leq M_{pl}\ )$

The load is now increased to form a plastic hinge at ②.

$$M_{pl②} = \sigma_{yield} \cdot \frac{b \cdot d^2}{4}$$

3) fully plastic (plastic hinge)

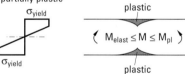

$$M_{pl①} = M_{pl②} = M_{pl③} = \sigma_{yield} \cdot \frac{b \cdot d^2}{4}$$

$$M_{pl} = \frac{P_{pl} \cdot l}{4}$$

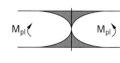

**FIGURE 2-96**
The paradox of the plastic hinge method

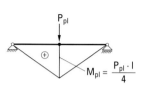

$$P_{pl} = \frac{4 M_{pl}}{l} = T$$

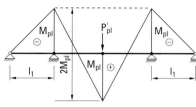

$$2 M_{pl} = \frac{P'_{pl} \cdot l}{4} \qquad P'_{pl} = \frac{8 M_{pl}}{l} = T$$

$2 P_{pl} = P'_{pl}$

Corresponds to ultimate load T

For $l_1 \to \infty$, the above continuous beam is converted into a beam on 2 supports (see top left)

$T = P_{pl} \Leftarrow$ conflict $\Rightarrow T' = 2 P_{pl}$

two systems shown in Fig. 2-96, then the ultimate load of the continuous beam is twice that of the simply supported beam. Stüssi and Kollbrunner now consider two limit states. If $l_1 \to 0$, the continuous beam shown in Fig. 2-96/right changes to a fixed-end beam with span $l$ and ultimate load $T = 8 \cdot M_{pl}/l$; this ultimate load matches that of the continuous beam. But if we take this to the limit $l_1 \to \infty$, the continuous beam is transformed into a simply supported beam with span $l$ and ultimate load $T = 4 \cdot M_{pl}/l$; this ultimate load is only half that of the continuous beam. And that is the paradox of plastic hinge theory. Stüssi and Kollbrunner sum up as follows: "The ultimate load method consequently specifies excessive values for the load-carrying capacity, when compared to simply supported beams, and therefore refrains from using certain internal loadbearing reserves (strain-hardening zone). Compared to this, statically indeterminate loadbearing structures designed according to elastic theory exhibit an excessive factor of safety" [Stüssi & Kollbrunner, 1935, p. 267]. The authors therefore plead for a return to elastic theory because the ultimate load method lies on the unsafe side in the continuous beam in their investigations. Their motto is: "Safety – protection" [Polónyi, 1995].

Rothert and Gebbeken have discovered that Kazinczy recognised clearly the paradox of the plastic hinge method as early as 1931 and even drew attention to the fact that the deflection $f_2$ of the continuous beam when taken to the limit $l_1 \to \infty$ "reaches an unacceptable magnitude", as Kazinczy expressed it [Rothert & Gebbeken, 1988, p. 29].

Although Maier-Leibnitz re-analysed the Stüssi beam (Fig. 2-96/right) in a test one year later [Maier-Leibnitz, 1936] and Fritsche considered the paradox of plastic hinge theory from the viewpoint of plastic theory [Fritsche, 1936], they did not manage a credible explanation. And in the broad discussion surrounding the ultimate load method at the "Berlin Olympics for structural engineers" [Kurrer, 2005, p. 630] in 1936 – the second congress of the International Association for Bridge & Structural Engineering (IABSE) – Maier-Leibnitz, as a protagonist of the plastic theorists [Maier-Leibnitz, 1938; Kazinczy, 1938], was unable to gain ground over the elastic theorists [Stüssi, 1938] in structural steelwork. The scientific game between the plastic and elastic theorists ended in a draw on that occasion [Karner, 1938; Karner & Ritter, 1938], but this would soon change in structural steelwork in favour of the elastic theorists. "After the discussions and meetings at the IABSE Berlin Congress in 1936," wrote Stüssi the elastic theorist in retrospect, "it suddenly became very quiet on this front [ultimate load method as the basis for designing steel structures – the author] and the structural steelwork specialists responsible agreed more and more that the introduction of the ultimate load method would mean a drop in quality for this form of construction" [Stüssi, 1962, p. 53]. It was the insatiable appetite of the Third Reich's rearmament policy after 1936 that capped the use of steel for buildings in Germany; industrial buildings with continuous purlins for the roof construction, a primary proving ground for the ultimate load method, were hit particularly hard. Basically, other forms of construction were preferred to steel and that curbed the interest in the further development of the ultimate load method. After 1936, Maier-Leibnitz, unpopular with the Nazis, had to remain silent on the ultimate load method as well, and Germany therefore lost its leading role in the further development of the plastic method of calculation for structural steelwork (ultimate load method) to Great Britain, the USA and the Soviet Union.

It was not until 1952 that Symonds and Neal cleared up the paradox of ultimate load theory [Symonds & Neal, 1952]. They investigated the deflections, the relative rotations and the strains at the plastic hinge and discovered that, in particular, the rotation of the plastic hinge became unacceptably large above a certain length of end span $l_1$; at the limit $l_1 \rightarrow \infty$, the deflection at point ② $f_2$ (Fig. 2-96/right) also tends to infinity, which means that the calculations lose their meaning and cannot be regarded as a limit state for the simply supported beam [Reckling, 1967, p. 152].

"He who faces the paradox, faces reality" [Dürrenmatt, 1980] is the 20th thesis of the Swiss author Friedrich Dürrenmatt (1921–90) in the appendix of his grotesque comedy *Die Physiker*.

One could almost regard it as a scientific reparation for Maier-Leibnitz as he was awarded an honorary doctorate by Darmstadt Technical University upon the recommendation of the Faculty of Civil Engineering on 13 June 1953. The citation reads: "In recognition of his great achievements in the researching of scientific principles of structural steelwork and theory of structures, especially plastic theory for applications involving statically indeterminate steel constructions." (cited in [Kurrer, 2005, p. 630]). The bad conscience of the German steelwork theory establishment knew very well that it had failed to some extent during the Third Reich and Maier-Leibnitz' greatest scientific achievement was his pioneering work on the ultimate load method which, however, could not expose its originality during the years of the Third Reich.

## 2.10.4 The acceptance of the ultimate load method

If we inquire today about why structural steelwork occupies a leading position in the British and American construction industries, then it becomes clear that one important element was the early and widespread introduction of the ultimate load method for designing steel structures in the UK and the USA, work which is inextricably linked with the names John Fleetwood Baker and William Prager. These two together managed to found the Anglo-American school of ultimate load theory in the 1950s.

### 2.10.4.1 Sir John Fleetwood Baker

It was in 1929 that the British Steelwork Association set up the Steel Structures Research Committee (SSRC), under the leadership of John F. Baker, to review the design, calculation and construction of steel structures with the aim of restructuring the British steel construction industry. Baker's work achieved the triumvirate interaction of steel industry, steelwork theories and steel building codes on a level that permitted a new view of the loadbearing behaviour of steel structures at an early date. The tests he instigated plus the critical assessment of the design of steel frame structures based on elastic theory and its translation into a code of practice were described in extensive reports published by the SSRC in 1931, 1934 and 1936 (see [Heyman, 1998, p. 136]). At the second congress of the International Association for Bridge & Structural Engineering (IABSE) in Berlin in 1936, Baker's contribution *A new method for the design of steel building frames* [Baker, 1936] represented a review of his activities as Technical Officer of the SSRC on the level of design theory. It was at the congress that he learned about Maier-Leibnitz' ultimate load tests, which occasioned him to initiate further test programmes on the plastic behaviour of steel construction. The outcome of this was that by 1948 British Standard 449 *The use of structural steel in building* permitted the design of steel structures according to the ultimate load method. Baker published a trial-and-error method for determining the bending moment distribution in steel frames at the plastic limit state in 1949 [Baker, 1949]; he did not refer to the work of Girkmann from the years 1931/32 (see [Gebbeken, 1988, pp. 49 – 50]).

### 2.10.4.2 Excursion: a sample calculation

Fig. 2-97 shows the calculation of a propped cantilever (one end fixed, one end simply supported) according to the static (Fig. 2-97/left) and kinematic (Fig. 2-97/right) methods. Symonds and Neal devised the kinematic method for frames [Neal & Symonds, 1950/51].

Method I (Fig. 2-97/left) assumes a plausible, permissible force condition for the given load; it is therefore called the static method. The next step here is the calculation of the load increase factor ν taking into account the plasticity condition $M = M_{pl}$ at the plastic hinges and $M < M_{pl}$ in the other zones. Finally, the displacement condition is inspected for geometric compatibility (kinematic check); at the plastic limit state there must be a compatible (simple kinematically determinate) series of plastic hinges.

The kinematic method (method II, Fig. 2-97/right) begins by assuming plausible, permissible series of plastic hinges. Using the principle of virtual displacements, i.e. equality between external virtual and negative internal virtual work, the second step is to calculate the load increase factors. As both the external and the internal virtual work must both be always minimal, the critical series of plastic hinges is the one given by the smallest load increase factor $v_{min}$. Finally, a check must be carried out to ensure that the plasticity condition $M \leq M_{pl}$ is satisfied.

### 2.10.4.3 The Anglo-American school of ultimate load theory

Jacques Heyman completed his engineering studies in 1944 and joined Baker's team at the University of Cambridge in 1946; he quickly took charge of important areas of work and was awarded a doctorate in 1949. Shortly after that he travelled to the USA in order to work with Wilhelm (William) Prager. Prof. Prager had fled Germany immediately after the National Socialists came to power and in 1941 had joined the renowned Department of Mathematics at Brown University, where he carried out research into plasticity. Baker and Prager set up an exchange programme between Brown and Cambridge. This cooperation would soon prove to be decisive. Prager had shown that the three expressions concerning the

$P = 200$ kN
$M_{pl} = 500$ kNm

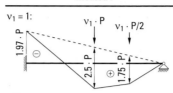

## Method I

$v_1 = 1$:

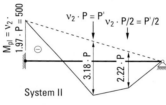

$M^{(1)}_① = -1.97P = 394 < M_{pl}$
$M^{(1)}_② = (2.5 - 1.97/2) P = 1.515P = 303 < M_{pl}$
$M^{(1)}_③ = (1.75 - 1.97/4) P = 1.258P = 255 < M_{pl}$

⟹ When $v_1 = 1$, the beam does not plasticise

⟶ $v_2 = \dfrac{500}{394} = 1.27$

When $v_2 = 1.27$, point ① plasticises

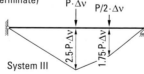

System II

⟹ The structural system has changed because $M^{(1)}_① = M_{pl}$ can be increased no further

$M^{(0)}_② = 1.515P \cdot 1.27 = 1.92P = 394 < M_{pl}$
$M^{(0)}_③ = 1.258P \cdot 1.27 = 1.6P = 320 < M_{pl}$

"New" structural system (statically determinate)

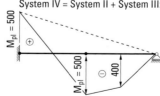

System III

System IV = System II + System III:

⟹ The 2nd plastic hinge forms at point ②:

$394 + 2.5 \cdot P \cdot \Delta v = 500 = M_{pl}$
⟶ $\Delta v = 0.23$
⟶ $v = 1.27 + 0.23 = 1.5$

## Method II

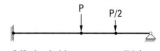

$z = 3$ (3 plastic hinges are possible)
⟹ $u = z - n = 3 - 1 = 2$
⟹ $u_1, u_2$ and the combination $u_3 = u_1 + 2u_2$

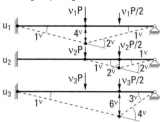

Work equation: $W_a^v = -W_i^v$

$W_a^v \quad = \quad -W_i^v$

$u_1$: $v_1 (P \cdot 4^v + P/2 \cdot 2^v) = M_{pl} \cdot (1^v + 2^v)$
⟶ $v_1 = 1.5 = v_{min}$

$u_2$: $v_2 (P \cdot 0^v + P/2 \cdot 2^v) = M_{pl} \cdot (1^v + 2^v)$
⟶ $v_2 = 7.5$

$u_3$: $v_3 (P \cdot 4^v + P/2 \cdot 6^v) = M_{pl} \cdot (1^v + 4^v)$
⟶ $v_3 = 1.785$

$v_{min} = 1.5$ ⟶ critical factor at which the ultimate load is reached

⟹ series $u_1$ is critical

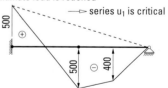

Plasticity check: $M \leq M_{pl}$ is satisfied!

Signs

The internal moments $M_i = M_{pl}$ always act such that they try to eliminate the change in angle $\Delta\varphi$.

$W_i = -M \cdot \Delta\varphi$ ⟶ $-W_i = M \cdot \Delta\varphi$

Negative internal work is always positive.

The ultimate load of the system is therefore

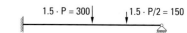

**FIGURE 2-97**
Calculation of a propped cantilever using the static method (left) and the kinematic method (right)

mechanics of solid bodies (equilibrium and elasticity equations plus material properties) in classical elastic theory could be united in one single equation, whereas the picture was completely different in plastic theory (see [Calladine, 1992]). This new approach was to prove fundamental for the later, more rigorous formulation of plastic theory for steel frames. Exchange programmes for scientific employees with a doctorate generally lasted one year, but Jacques Heyman remained at Brown University for three and by his return in 1952 had acquired wide-ranging fundamental theoretical knowledge. Together with J. F. Baker and M. R. Horne, he published the first book on plastic theory for structural steelwork in 1956: *The steel skeleton. Vol. 2: Plastic behaviour and design* [Baker et al., 1956]. This book brought together all the work of the teams in Cambridge from the previous 10 years and was the first to mention the fundamental theorems of ultimate load theory and describe practical applications.

These fundamental theorems had been verified by the Soviet engineer A. A. Gvozdev in 1936. However, although they were published in 1938, there were made available in Russian only and then only through the Moscow Academy of Sciences, whose publications were little known in the West [Gvozdev, 1938/1960]; they therefore remained essentially inaccessible to the international scientific community. The theorems of Prager's team were rediscovered, so to speak, in the early 1950s (Fig. 2-98). The application of the theorems to the analysis of steel structures enabled the consequential use of calculations according to plastic theory, which had been carried out in the UK since 1948 (a supplementary clause had been added to the relevant British Standard).

Theory of structures founded on plastic theory experienced its consolidation phase between 1960 and 1970. Specific studies were replaced by the publication of manuals, which played a crucial role in the widespread application of the theory. The authors of these manuals were prominent engineers and scientists who had been considerably involved in this process, including Beedle, Neal and Horne. Heyman and Baker also published a manual in this series in 1966, which became a prototype for many subsequent contributions [Heyman & Baker, 1966].

Theory of structures founded on plastic theory was initially devised for structural steelwork. Later it was realised that it could be applied to

> Theorem I.—The safety factor against collapse is the largest statically admissible multiplier.
>
> Theorem II.—The work that the collapse loads do on the displacements of their points of application must equal the work that the limit moments in the yield hinges do on the relative rotations of the parts connected by the hinges.
>
> Theorem III.—The safety factor against collapse is the smallest kinematically sufficient multiplier.
>
> Theorem IV.—Collapse cannot occur under the loads obtained by multiplying the given loads by a factor which is smaller than a statically admissible multiplier.
>
> Theorem V.—If a beam or frame is strengthened (that is, if its cross sections are changed in such a manner that the limit moment is increased for, at least, one cross section and decreased for none), the safety factor for a given system of loads cannot decrease as a result of this strengthening.

**FIGURE 2-98**
The theorems of ultimate load design after Greenberg and Prager [Greenberg & Prager, 1951/52]

reinforced concrete structures as well. It also became clear that plastic theory can be used for any constructions with a ductile behaviour, provided there are no stability problems. Jacques Heyman presented this fact, which some engineers had suspected since the beginning of the 20th century, clearly and understandably. He was the first to recognise that the fundamental theorems produced a new paradigm that could be used for all constructions erected using conventional materials. This was obvious for reinforced concrete in some circumstances. For example, Kazinczy [Kazinczy, 1933] and Gvozdev [Gvozdev, 1936/1960] had already looked into the application of ultimate load analysis to reinforced concrete structures back in the 1930s. However, for materials such as timber, not to mention masonry, the application was less obvious. Heyman realised, however, that the theorems of ultimate load theory could be suitably adapted for heterogeneous materials such as natural stone or clay bricks (see chapter 4).

**Controversies surrounding the ultimate load method**

### 2.10.4.4

It was at the second Swiss Steel Construction Conference in 1956 that Fritz Stüssi, who in 1937 had been appointed professor of theory of structures, structural engineering and bridge-building in steel and timber at Zurich ETH, repeated his criticism based on the paradox of plastic hinge theory, and referred to authorities such as Leonardo da Vinci, Navier and others [Rothert & Gebbeken, 1988, p. 34]. Ignoring this, Bruno Thürlimann presented his plastic hinge method (based on tests) in the journal *Schweizerische Bauzeitung* [Thürlimann, 1961]. These tests had been carried out and evaluated at Lehigh University, Bethlehem, a leading centre of steel construction research in the USA; it was there that Thürlimann gained his doctorate and worked as a professor from 1953 to 1960 before he was appointed professor of theory of structures, structural engineering and bridge-building in stone, concrete and prestressed concrete at Zurich ETH in 1960; this professorship was renamed "Theory of Structures & Design" after Stüssi's transfer to emeritus status in 1973. In his comprehensive, clearly structured contribution, Thürlimann investigates structural systems in which he reveals the uncertainties of the elastic method of calculation (paradox of elastic theory); furthermore, he analyses the "Stüssi beam". Thürlimann specifies a general version of the kinematic approach to the plastic hinge method based on the principle of virtual displacements (see also Fig. 2-97/right). Stüssi's reaction to this was allergic! In the first place, he pointed out the paradox of plastic hinge theory, and secondly, he disputed the moment redistribution through the formation of plastic hinges and the admissibility of the use of the principle of virtual displacements which he had discovered: "A statically indeterminate structure remains statically indeterminate also if the limit of proportionality or the yield stress of the material is exceeded in particular cross-sections. This means that besides the equilibrium conditions, the deformation conditions also remain valid even in the post-elastic loading range. The inadequacy of the ultimate load method is based on the fact that it treats this fundamental fact wrongly and upon closer inspection its 'simplicity' is revealed as unacceptable primitiveness. … If, however, favouring the ultimate load method is intended to placate those people who cannot master, and given normal talents cannot learn, the normal methods of calculating statically indeterminate structures, then the introduction of such a 'theory of structures for idiots' should certainly be rejected" [Stüssi, 1962, p. 57].

For Stüssi, an advocate of elastic theory, the idea of disrupting the elastic continuum by plastic hinges and hence negating the deformation conditions, from which – as we know – the elasticity equations for determining the statically indeterminate variables in elastic loadbearing structures are derived, is undoubtedly alien. He insists on the exclusivity of the true value of the selection mechanism of deformation conditions – part of his very nature – with which the true equilibrium state in statically indeterminate systems can be selected from the infinite number of possible equilibrium conditions. On the other hand, plastic hinge theory is based on probable equilibrium conditions whose true value is checked by, for example, the principle of virtual displacements or – even worse for Stüssi – methods of trial and error! In the dispute between Stüssi and Thürlimann, the age-old difference between the geometric view of statics (searching for the true equilibrium conditions) and the kinematic view of statics (physics-based establishment of failure mechanisms from the many possible equilibrium conditions) surfaces again, a difference (see Fig. 2-8) that is founded philosophically in the logical and ontological status of the difference between possibility and reality first recognised by Aristotle.

In his comments, Thürlimann, following Symonds and Neal, demystifies the paradox of plastic hinge theory [Thürlimann, 1962]. However, Stüssi will not let the subject rest and assesses the ultimate load method as an unsatisfactory approximation technique, but drops his claim that it leads to unsafe calculation results [Gebbeken, 1988, p. 66]. Nevertheless, Stüssi's criticism of the ultimate load method encouraged the development of a theory of structures founded on plastic theory during the 1960s. For instance, Charles Massonnet took up the dispute surrounding the plastic hinge method in 1963 and used a triangular diagram to define the areas of application for elastic, plastic and viscoelastic theories [Massonnet, 1963]. Massonnet was also the one who promoted the European recommendations for the plastic design of structural steelwork steel structures in the 1970s [Massonnet, 1976].

Udo Vogel's habilitation thesis *Die Traglastberechnung stählerner Rahmentragwerke nach der Plastizitätstheorie II. Ordnung* (ultimate load calculations for steel frames according to second-order plastic theory) [Vogel, 1965], completed in 1964 and published in 1965, established the plastic hinge method according to second-order theory. In this work, Vogel determines the critical load factor using an iteration method he developed himself. It is similar to the method proposed by Alfred Teichmann for solving the stability problem according to second-order displacement theory: non-linear theories lead to non-linear sets of equations, which as a rule can only be solved iteratively. Iteration methods in turn favour the use of computers for structural calculations and, vice versa, the computer has become a means of conceiving structural analysis theories in the non-linear sphere. This interaction was set to become a distinguishing feature of the diffusion phase of structural theory (1975 to date).

Bill Addis has described the intrusion of plastic theory into structural design based on elastic theory as a "paradigm articulation" in the meaning of Thomas S. Kuhn [Kuhn, 1962/1979; Addis, 1990, p. 97 ff]. Whereas this "paradigm articulation" had already been completed in the UK by 1956 and was soon followed by other countries, the ultimate load theory in structural steelwork in the Federal Republic of Germany did not advance to become an integral component in design practice until after the publication of the new series of structural standards (DIN 18 800) between 1981 and 1990 – admittedly with the systematic transition from the concept of permissible stresses to the concept of partial safety factors. Therefore, the authors and publishers of the commentaries to DIN 18 800 parts 1 to 4 wrote in their foreword: "Today, the essentially trouble-free calculation of bar frameworks according to elastic and plastic theory, i. e. also with the planned use of the plastic reserves of the steel, and this according to second-order theory as well, can be called progress." [Lindner et al., 1993, p. I]. The demystification of the "ingenuity" of the material halted all references to the "ingenuity" of the material by practising structural engineers.

## **Structural law – Static law – Formation law** 2.11

"Architecture is frozen music." This phrase coined by the philosopher Arthur Schopenhauer (1788–1860) is often quoted in the relevant literature. If it also applies to structural engineering, the designs of structural engineers can be regarded as structural compositions. This relationship is stated more precisely as the structural law, static law and formation law, which can be consolidated to form the composition law. How beauty and law appear in the composition law as the synthesis of structural law, static law and formation law will be illustrated using the example of spatial frameworks.

## **The five Platonic bodies** 2.11.1

Did mathematical law exist historically before us and outside us? Will it survive us? Or is it purely the work of man? Like technology, art and science.

Plato (427–348 BC) divided human cognition into sensibility, intellect and reason. However, sensations, perceptions and notions can never reach beyond the remit of subjective thinking and can never create knowledge. According to Plato, external objects can only be reached through cognition-obscuring sensibility. The light of mathematical law does not shine in external objects, but in ourselves – after all, it is already intrinsic to the immortal and frequently revived soul that has seen all things on Earth and in Hell. As Plato said: "Searching and learning … invariably involves recollection" [Seidel, 1980, p. 220].

Mathematical intellect, cleansed of sensibility, deals with numbers and numerical relationships, as expressed in geometry, for example, which leads to true knowledge. Even computer-generated right-angled triangles (Fig. 2-99a), from which Plato (in *Timaios*) assembled the regular polyhedra named after him [Plato, 1994, pp. 55–57], would be only imperfect images of an eternal, quiescent idea in which only the geometric triangle theorems can claim to be real. Of particular interest is Plato's implicit mathematical structural law. Since each triangle consists of right-angled triangles, he selects two triangles from each set: "Of the two triangles, the isosceles triangle comes in one form only [see Fig. 2-99a/right – the author], whereas the scalene triangle can have countless forms. From these countless forms we now have to select the most beautiful … From … the many triangles we regard the most beautiful as that from which the equilateral triangle was formed [see Fig. 2-99a/left – the author] …" [Plato, 1994, pp. 55–56]. From the "most beautiful", i.e. the scalene triangle (Fig. 2-99a/left), Plato constructed

- the tetrahedron, consisting of 24 basic triangles, defining the element fire,
- the octahedron, consisting of 48 basic triangles, defining the element air, and
- the icosahedron, consisting of 120 basic triangles, defining the element water.

The hexahedron, representing the element earth, is formed from 24 right-angled isosceles triangles (Fig. 2-99a/right). This beauty of the elements is somewhat disrupted by the dodecahedron (Fig. 2-99b), which Plato was unable to reduce to basic triangles: "So God used it for the world as a whole" [Plato, 1994, p. 57].

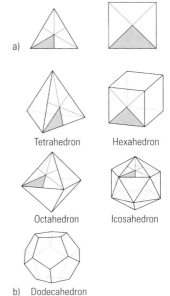

**FIGURE 2-99**
Plato's implicit mathematical structural law; a) basic triangles, b) the five Platonic bodies [Falter, 1999, p. 49]

In Plato's explanation of the transitions between the four elements, the significance of his polyhedron idea already begins to fade and soon becomes lost in the darkness of his description of nature in *Timaios*. Only his polyhedron idea prevailed, as indicated by Johannes Kepler's (1571–1630) representation of the world order in his *Mysterium Cosmographicum*, published in 1596 (Fig. 2-100). Kepler interpreted the relative distances of the planets from the Sun through nested Platonic bodies. Some 2000 years after Plato's death, Leonhard Euler's (1707–83) polyhedron theorem

$$e + f = k + 2 \qquad (2\text{-}38)$$

where $e$ = number of vertices, $f$ = number of surfaces and $k$ = number of edges, brought Plato's non-conforming dodecahedron down from the heavenly spheres to the Procrustean bed of mathematical law. While desperately trying to find proof of the existence of God, the religious Euler thus expelled God from the paradise of the Platonic bodies.

**FIGURE 2-100**
Kepler's representation of the world order dating from 1596 [Wußing, 1983, p. 240]

### 2.11.2 Beauty and law

In his work *Politeia*, Plato placed art after science and measured it based on beauty as the good aspect benefiting the state, and not on the skilfulness of the artist or the imitation of the existing. In contrast, his great disciple Aristotle (384–322 BC), who later taught Alexander the Great, stood up for art. After all, *tekhnē* is the quintessence of all human skills of accomplishment: through work, craftsmanship and skilfulness [Friemert, 1990, pp. 919–920]. Aristotle therefore moved action in technology, art

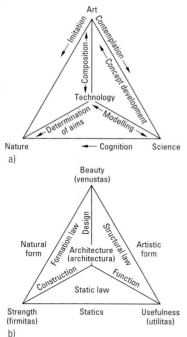

FIGURE 2-101
a) The tetrahedron of the four realms, and b) the tetrahedron of beauty and law in architecture

and science into the focal point of philosophy for the first time. Since the drawing is the language of the engineer, Plato's tetrahedron is used to illustrate important terms. Fig. 2-101a uses a tetrahedron to illustrate the interaction of the four realms of nature, technology, art and science:

– technology and art are linked through composition,
– technology and science are linked through modelling,
– art imitates nature,
– science comprehends, and
– perceives nature.

Nature, science and art form the basic triangle, the foundation of modern technology. Admittedly, in practice the forces between the four realms have different magnitudes, which leads to different edge lengths and the regular tetrahedron is therefore distorted affinely into an irregular tetrahedron.

The situation is similar for the tetrahedron of beauty and law in architecture, which illustrates the balance between strength, beauty, usefulness and architecture (Fig. 2-101b):

– Architecture lends weight to beauty in the design.
– Architecture realises usefulness through function.
– Architecture achieves strength through construction.

Whereas

– beauty is linked with usefulness through artistic form,
– usefulness is linked with strength via statics, and
– strength is linked with beauty through natural form.

Here, too, the symbolic use of Plato's tetrahedron reflects practice in an idealised way only.

The building aesthetics considerations of Vitruvius' *Ten Books on Architecture* are based on Plato's art theory, interpreting the human as a sign of the divine. His building aesthetics is dimensional aesthetics, reflecting the eternal order [Vitruvius, 1981, pp. 37–43].

According to Vitruvius, architecture is divided into the building of structures, clocks and machines. Furthermore, structures may be public buildings and facilities or private buildings. Public buildings serve purposes such as defence, worship and general utility. "These structures," writes Vitruvius, "must be built taking into account strength, usefulness and beauty. Strength is taken into account through foundations extending down to firm subsoil and by the careful selection of building materials (whatever they may be) without excessive frugality. Usefulness is taken into account through the proper arrangement of rooms without obstructing their utilisation, and with the orientation of each room reflecting its purpose. Beauty is taken into account by giving the structure a pleasant and agreeable appearance with proper symmetrical relationships between the components" [Vitruvius, 1981, p. 45].

Vitruvius' dimensional aesthetics was secularised with the discovery of personality by Alberti and others during the Renaissance: the architect no longer realises the divine order, but designs buildings based on empirical and theoretical knowledge which satisfy human needs successfully and respectfully. Characteristic of this period are the geometric proportioning rules, embodying the experience of architects and builders, for coordinating the dimensions of the components with the overall appearance of the structure (Fig 2-102). Such rules were later developed into geometry-based design rules in the form of the geometric composition theory, the authority of which was successfully contested by structural theory, which emerged during the 19th century.

Since the Renaissance, the world has become a sensuous place, with the artist, architect and engineer becoming its active creators and therefore each a subject of his profession. According to Jacob and Wilhelm Grimm's German dictionary, since the end of the 17th century beauty "is no longer associated with desiring, but rather with stimulating and fulfilling that desire, and the gracefulness: beauty of life …; referred to as sensuous beauty …" [J. Grimm, W. Grimm, 1854, pp. 409–410]. So much for beauty. What about law?

In a philosophical sense, law expresses the necessity of a sequence of events. Laws of nature, for example, are objective relationships describing the inevitability of events and the repetitiveness of processes, whereas interpretations of laws describe scientific insights relating to these regular relationships. The latter include the static law describing the equilibrium of forces in space, the formation law of trussed framework theory, and the structural laws for spatial frameworks proposed by Mengeringhausen in 1940.

There is an objective link between static law, formation law and structural law for spatial frameworks (Fig. 2-101b):
– Static law is bounded by the vertices of architecture, usefulness and strength, and by the edges of function, statics and construction.
– Formation law is bounded by the vertices of architecture, strength and beauty, and by the edges of construction, natural form and design.
– Structural law is bounded by the vertices of architecture, beauty and usefulness, and by the edges of design, artistic form and function.

Strength, statics, usefulness, artistic form, beauty and natural form make up the basic triangle, the foundation of architecture. Static law interacts with formation law via construction. Function links static law with structural law. And the latter is coupled with formation law via design. In spatial

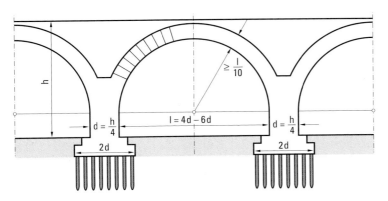

**FIG 2-102**
Alberti's geometric proportioning rules for arch bridges [Straub, 1992, p. 129]

frameworks, static law, formation law and structural law also form a superior unit in the shape of the composition law for spatial frameworks, as recognised and implemented in practice by Mengeringhausen.

### Structural law 2.11.2.1

Plato had already introduced an implicit mathematical structural law by assembling the tetrahedron, hexahedron, octahedron and icosahedron from two basic triangles (see Fig. 2-99). In 1940 Mengeringhausen formulated eight structural laws for spatial frameworks, based on the aforementioned polyhedra. In terms of method, his first structural law is reminiscent of Plato: "*Loadbearing spatial structures* (spatial frameworks) are ideally composed from equilateral and (or) right-angled isosceles *triangles* such that *regular multiples* (polyhedra) are created in the form of tetrahedra, cubes, octahedra and truncated octahedra or cubes or parts thereof" [Mengeringhausen, 1983, p. 114].

Fig. 2-103a illustrates Mengeringhausen's first structural law. The elementary tetrahedron with an edge length of 1/m (Fig. 2-103a/left) can be stacked to fill completely a tetrahedron-shaped, limited space (Fig. 2-103a/right). Normalised to 1, this space is characterised mathematically by four sets of coordinate planes that intersect in straight lines and points in such a way that each point is defined by four coordinate numbers, the sum of which must always be 1. A spatial framework is formed if the straight lines are transformed into real linear members and the vertices into joints. In 1971 such a spatial framework (from the MERO company) formed the basis of the Siemens pavilion at the German Industry Exhibition in São Paulo, Brazil (Fig. 2-104), although in this case the spatial framework created by Mengeringhausen's first structural law (Fig. 2-103a/right) was not infilled but left as three double-layer tetrahedron frameworks.

### Static law 2.11.2.2

In its most elementary form, the static law for spatial frameworks manifests itself as a free equilibrium system in a closed tetrahedron framework which is subject to tensile forces $F$ in the four axes of symmetry and "responds" with six internal tensile forces $S$ in the linear members (Fig. 2-103b). Static law is based on nothing other than different types of equilibrium statements as described in sections 2.2.1, 2.2.2 and 2.2.5.

**FIGURE 2-103**
Elements of the composition law, illustrated using the example of the tetrahedron framework;
a) Mengeringhausen's first structural law,
b) static law of equilibrium, and c) first formation law for spatial frameworks

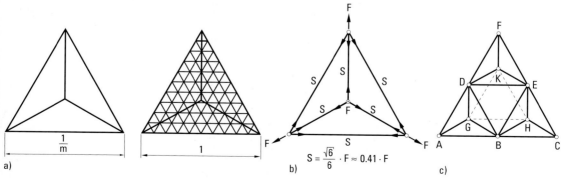

FIGURE 2-104
Siemens exhibition pavilion in the shape of a tetrahedron at the German Industry Exhibition, São Paulo, Brazil, 1971 [Mengeringhausen, 1975, p. 37]

### 2.11.2.3 Formation law

The formation law was illustrated for plane frames in section 2.2.9 using the example of the historico-logical development of the Polonceau truss.

In composite spatial frameworks such as that in São Paulo, an equilibrium system according to Fig. 2-103b is unlikely to come about in practice, as a swift look at the simplest formation law for spatial frameworks reveals (Fig. 2-103c). The law states that, based on the stably supported ball joints of basic triangle $ABD$ formed by three straight members, joint $G$ can be connected stably, provided $G$ does not lie in a plane formed by basic triangle $ABD$. The same procedure can be applied to basic triangles $BEC$ and $DEF$, resulting in three tetrahedron frameworks with vertices $G$, $H$ and $K$. If additional connections are introduced between high points $G$, $H$ and $K$, an octahedron filling the space between the three tetrahedra is created. The complete spatial framework thus created from three tetrahedra has three degrees of static indeterminacy and can be analysed only by solving the elasticity equations. This game with the formation law can be continued until a spatial framework as shown in Fig. 2-104 emerges. Ultimately, there will be $n$ redundant members, requiring $n$ elasticity equations with $n$ unknown member forces to be solved. Since $n$ will be hopelessly large, calculation of the $n$ statically indeterminate member forces using the manual means of classic structural theory is practically impossible. The reason for the unsynchronised development of the theory and practice of spatial frameworks with a high degree of static indeterminacy is thus obvious.

In the construction of spatial frameworks, beauty and law express themselves in the composition law, consisting of the formation law, structural law and static law. The historical and logical development of the cognition of the static law and the formation law for spatial frameworks is discussed in chapter 8.

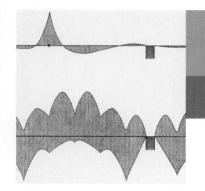

# CHAPTER 3

# The first fundamental engineering science disciplines: theory of structures and applied mechanics

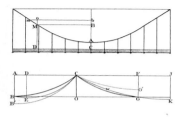

Is theory of structures a specific developmental form of applied mechanics? Or can theory of structures and applied mechanics claim independence on the level of scientific theory and epistemology? And the eternal question: What is the nature of engineering? The author has been searching for answers to these questions since the early 1980s – probing the philosophy of engineering and looking for works dealing with the *specifica* of the engineering sciences. Approaches from the tradition of system theory and Marxist thinking, which since the late 1990s have been contributing to the emerging theory of the engineering sciences, presented one opportunity. With the help of five case studies from the history of applied mechanics, theory of structures and the theory of bridge-building, the author has tried to focus the results of his philosophically flavoured studies into a concrete form. If theory of structures is understood as a fundamental engineering science discipline, then the discourse between the philosophy and history of structural theory is essential.

Since the appearance of their first disjointed elements in the 18th century, the scientific character of engineering sciences has been measured by the degree of their advancement, initially in mathematics and later in theoretical mechanics. The final building block in the development of mechanics, which had stretched over 1500 years, was laid in 1687 by Isaac Newton with the publication of his work *Mathematical Principles of Natural Philosophy*, in which he divorced mechanics completely from the natural sciences and established it deductively based on three axioms:
1. Newton's law of inertia: Every body continues in its state of rest or of uniform motion in a straight line unless it is acted upon by some external impressed force.
2. Newton's law of force: The rate of change of momentum of a body is proportional to the impressed force and takes place in the direction of that force.
3. Newton's law of reaction: To every action there is an equal and opposite reaction, i.e. when two bodies interact, the force exerted by the first body on the second body is equal and opposite to the force exerted by the second body on the first.

Newtonian mechanics represented more than just the final full stop at the end of more than 1500 years of research into the mechanical forms of motion of matter. It marked the close of the scientific revolution that had been initiated by Copernicus with his heliocentric picture of the world. The axiomatically organised system of Newtonian mechanics meant that mathematical principles could now be applied to describe all those technical artefacts that function primarily according to the laws of the mechanical motion of matter. The contradiction between the mastery of the principles and the complexity of the forms of such technical artefacts permeates the scientific works of all the mathematicians of the 18th century. Only rarely were they successful in dealing with the questions arising out of manufacturing operations. The arch, beam and earth pressure theories represent such exceptions – theories that formed vital cornerstones of construction engineering. It is against this background and the systematic thinking that arose out of the cataloguing of technical knowledge that we now view the emergence of theory of structures and applied mechanics in the early 19th century as the prototypical engineering sciences. By the end of the 19th century, the encyclopaedic cataloguing in the system of the classic engineering sciences had come to an end, and theory of structures and applied mechanics had acquired the status of fundamental engineering science disciplines.

## 3.1 What is engineering science?

It was at the start of the classical phase of the system of classic engineering sciences (1875–1900) that Ernst Kapp (1808–96) founded modern engineering philosophy with his monograph *Grundlinien einer Philosophie der Technik* (principles of a philosophy of technology) [Kapp, 1877]. That prompted the systematic philosophising about engineering, which in historico-logical terms preceded the philosophising about engineering sciences. Kapp understands engineering as "organ projection", as

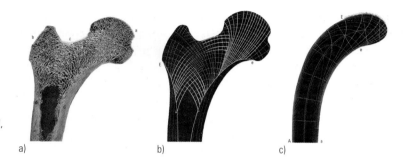

**FIGURE 3-1**
Organ projection after Kapp:
a) human right thigh bone [Kapp, 1877, p. 110], b) associated tensile and compressive stress trajectories [Kapp, 1877, p. 115], c) organ projection in the form of a crane after Karl Culmann [Kapp, 1877, p. 112]

a disembodied organ. For instance, the hammer emulates the fist, the emerging telegraphy the nervous system, and the telescope, unwittingly, the internal structure of the eye. An example used again and again since Kapp for analogies between nature and engineering is Culmann's crane which, following Kapp's thinking, is a projection of the human thigh bone. Fig. 3-1 shows a longitudinal section (viewed from the front) through the top end of a human right thigh bone (Fig. 3-1a), the tensile and compressive stress trajectories in the spongiosa (Fig. 3-1b) (see [Gerhardt et al., 2003]) and Kapp's organ projection in the form of a crane after Karl Culmann (Fig. 3-1c). But is Culmann's crane, or the framework of the Eiffel Tower, not the incarnation of a specific engineering science subdiscipline, i. e. graphical statics?

**First approximation**    **3.1.1**

It was only with the consolidation of the system of classical engineering sciences after 1900 that attempts were made to emphasize explicitly the gnosiological independence of the engineering sciences within the realm of science.

At the start of the 20th century, a not insignificant number of engineers and engineering scientists, e. g. Franz Reuleaux (1829–1905), Max Eyth (1836–1906), Alois Riedler (1850–1936) and Peter Klimentyevich Engelmeyer (1855–1942), were investigating general engineering issues. For instance, in his paper on the Russian engineer Engelmeyer, Hans-Joachim Braun draws attention to the fact that Engelmeyer had been highly influential in providing engineering with an independent foundation, divorcing it from the natural sciences [Braun, 1975, p. 310]. Engelmeyer understands engineering as "the art of bringing to life natural phenomena systematically and based on the acknowledged natural interactions between objects". We should point out here that engineering is an art and that Engelmeyer understands art to be any objectivistic activity "through which an idea precedes an action as a goal … The goal presupposes the action teleologically" [Braun, 1975, p. 310]. But it is not only his emphasis on the intentionality of engineering – i. e. the recognition of the purpose-means nature of engineering – that enables Engelmeyer's theory of engineering to stand out among those of his contemporaries. Rather, he also examined the problems of the theory of engineering sciences, the relationship between natural and engineering sciences, and the theme of engineering, or engineers, and society; in particular, the consequences for the practical side arising out of the theory of engineering sciences are still relevant today [Braun, 1975, p. 309], as the compendium *Erkennen und Gestalten* (cognition and design) [Banse et al., 2006] reveals. Engelmeyer's influence on the philosophical discourse on engineering and engineering science in Russia has been shown in a study by Vitaly G. Gorokhov [Gorokhov, 2001].

The rudiments of a theory of engineering sciences developed by Eberhard Zschimmer (1873–1940) played a significant role in German engineering philosophy up until World War 2. In a publication commemorating the 100th anniversary of Karlsruhe Technical University, Zschimmer, as engineering philosopher and engineering scientist, summarised his thoughts on the "epistemology of engineering science" [Zschimmer, 1925].

After he initially draws attention to the research desideratum of such an epistemology, Zschimmer develops the concept of "engineering science" philosophically from the more comprehensive "cultural science", on the same lines as the analogy between the structure of the world of culture as the realisation of ideas and the world of nature as the realisation of natural laws, as advocated by the south-west German school of the new Kantians Wilhelm Windelband (1848–1915) and Heinrich Rickert (1863–1936). The parallel between the world of nature and the world of culture produces a "similar structure of research methodology" for the natural and cultural sciences. So, like natural history corresponds to cultural history, natural theory and cultural theory, natural elucidation and cultural elucidation are similarly related. Accordingly, the epistemological dimension of engineering science research can be broken down into three functions:

a) the graphic systematic presentation, which exists in space and time in the accomplishments of engineering creativity – embracing both past and present;
b) the abstract systematic knowledge of those rules and principles the engineering creators (inventors) obeyed when organising natural processes;
c) the explanation of the tangible creations of engineering (e.g. machines) through abstract theory [Zschimmer, 1925, p. 536].

Intrinsic to the history of engineering is therefore an epistemological function of engineering science research because we can learn the "true categories of engineering or engineering science" [Zschimmer, 1925, p. 539]. "It is the exact knowledge of the natural processes and inventive goals organised in the technical methods (inventions) plus the systematic summation of the observations, rules and principles from practice and the laboratory to form one non-contradictory whole" [Zschimmer, 1925, p. 542] that turns the lessons of engineering into a science. Like any cultural science, engineering science is nothing more than "the creative spirit's knowledge of itself", i.e. "a tool of the creator of engineering" [Zschimmer, 1925, p. 542]. Zschimmer's objective-idealistic, philosophical view of the engineering sciences was in line with the established engineering interpretation of the protagonists of the engineering science organisations in the Germany of the 1920s under the topos *Technik und*

*Kultur* (engineering and culture) as a spiritual reaction to the conservative criticism of engineering, but remained an exception until well into the 1960s. Although the history of engineering had already achieved notable status by the time of the Weimar Republic, e. g. through the works of Conrad Matschoß (1871–1942) and the Karlsruhe historian Franz Schnabel (1887–1966), Zschimmer's epistemological function definition of engineering history as a component in the engineering science research process is original, even if it is expressed in a Hegelian and new Kantian dialect. It is original because although other engineering scientists such as August Hertwig (1872–1955) were writing on issues of engineering history and, occasionally, engineering science history, they did not recognise, let alone illustrate, the informative value for engineering science research methodology. In 1958 the Darmstadt professor of theory of structures and structural steelwork Kurt Klöppel (1901–85) emphasized the definitive structure of the engineering science research process and distanced this from that of physics: the laboratory of the engineering scientist is "not just – and sometimes never – his testing room at the institute, but rather practical operations, where the engineered products and hence also the scientific prognoses must prove themselves" [Klöppel, 1958, p. 15]. However, an interdisciplinary discussion on the specific of the engineering sciences could not be initiated for the following reasons:

1. Liberal arts studies at technical universities in the Federal Republic of Germany after 1945 were characterised entirely by the teaching of neo-humanistic cultural traditions mixed with traditional engineering philosophy.
2. The institutionalisation of engineering history in the Federal Republic of Germany did not take place until the 1970s. And in the German Democratic Republic the evolution of the history of engineering sciences first appeared towards the end of the 1970s, signalled by the 1st International Conference on Philosophical and Historical Questions in Engineering Sciences, which took place in 1978 on the occasion of the 150th anniversary of Dresden Technical University [Striebing et al., 1978].
3. The principal metaphysical feature in the philosophising about engineering in the Federal Republic of Germany and the associated dualising of engineering apologetic and anti-technical critique of civilisation excluded a differentiated contemplation of the scientific theory and epistemological problems in engineering sciences.

### 3.1.2

**Raising the status of engineering sciences through philosophical discourse**

The publication of Simon Moser's (1901–88) *Kritik der Metaphysik der Technik* (critique of the metaphysics of engineering) [Moser, 1958] in 1958, which at the time went unheeded by philosophers in this field, heralded a paradigm change in Federal German engineering philosophy, which acquired a way of thinking devoted to critical rationalism and in the 1970s advanced to become the dominating form of philosophising about engineering. Moser managed to involve engineering scientists in the discourse on engineering at the University of Karlsruhe (formerly Karlsruhe

Technical University). As a systematist of mechanical process engineering and head of the largest institute devoted to this discipline in the German-speaking world, Hans Rumpf (Fig. 3-2) was an obvious candidate to achieve the separation of the engineering sciences from the natural sciences in theory of science terms. In the introduction to his detailed research programme for a science of comminution founded on physics, he writes: "Every engineering science faces a dual task. Firstly, we have to describe the existing state of the art and its development in a systematic arrangement. All the insights into the principles and methods observed to be gained from such an analysis have to be presented ... and the ensuing instructions and rules for the solution of technical tasks have to be devised. Such a compendium of knowledge is generally an indispensable foundation, not only for the sensible processing of concrete engineering assignments, but also for the purposeful consolidation of a quantitative science founded on natural sciences. The development of this is the second function, and actually the scientific task per se. We have to employ theoretical and experimental methods to discover which natural science processes determine the technical process, which functional relationships exist between the critical influencing variables, and which laws and which behaviour result from this for the optimum solution of technical assignments" [Rumpf, 1966, p. 422].

FIGURE 3-2
Hans Rumpf (1911–1976) [archives of the Institute of Process Engineering, University of Karlsruhe]

Whereas Zschimmer in addition to systemising engineering practice and analysing the state of the art also integrates their historic identity into the engineering science research process, Rumpf sees the knowledge of causal relationships modelled in technical entities and methods as the principal content of the engineering sciences, under the aspect of the transformation of natural causality into technical finality. As Rumpf groups together the physically measurable features of original Nature and the physically measurable technical features into a field of physically measurable phenomena, he cannot divorce the engineering sciences from the natural sciences according to the type of artefact. He therefore distinguishes between three types of scientific question:

- *Question 1:* Which findings can be established about a matter or a process on the basis of an observation or a measurement?
- *Question 2:* Which theory can be propounded about a causal relationship between variables?
- *Question 3:* How is something carried out? [Rumpf, 1973, p. 92]

Like in the natural sciences questions 1 and 3 depend on question 2, in the engineering sciences questions 1 and 2 are dominated by question 3. "Therefore, phenomena complexes are prescribed as a preferred task for engineering science, or it [engineering science] simply develops from the knowledge of individual phenomena and the principles of the feasibility of technically fruitful phenomena complexes as an object of their research" [Rumpf, 1973, p. 97]. Just like the natural sciences, the fundamental engineering science disciplines, e.g. theory of structures, applied mechanics and materials science, must investigate individual phenomena; resolving the phenomena complexes into the relevant individual phenomena is, in the fundamental engineering science disciplines, "a methodical step when answering question 3" [Rumpf, 1973, p. 97]. Further distinguishing criteria nominated by Rumpf are the systematic methodology, on which the classical methods of construction are based, and the specific structure of the statement. According to Rumpf, the engineering sciences certainly have to consider the following statements: descriptive statements regarding individual observations; statistical statements covering observations; inductive-statistical statements; deductive-nomological statements; deductive-statistical statements; prognoses; imperative technical statements [Rumpf, 1973, pp. 101, 102].

The complex phenomenon "engineering", assembled from heterogeneous elements, appears to engineering philosophy, with its theory of science orientation, in the form of sets of scientific statements. By grouping together the physically measurable features into one field of physically measurable phenomena covered by both the natural and the engineering sciences, Rumpf has transformed the theory of science search for the *differencia specifica* between nature and engineering into a search for the *differencia specifica* between the natural and engineering sciences.

A selection of articles on engineering philosophy and engineering sociology from the estate of Hans Rumpf was published by Hans Lenk, Simon Moser and Klaus Schönert in 1981 [Lenk et al., 1981].

### 3.1.2.1 The contribution of system theory

If the reconstruction of the philosophical reflection on engineering manifests itself in the theory of engineering as a methodical reflection on engineering science knowledge, where the social quality of the technical appears as an organisation of experience with the subject of the knowledge, then the system theory of engineering is expanded in that it infers the function of the individual building element from the function of the whole. Contrasting with the theory of engineering science, owing more to the forms of thinking of the analytical theory of science, the advancement of knowledge in the system theory of engineering (developed by Günter Ropohl in particular) consists of having differentiated the term *Realtechnik* (real engineering), coined by Friedrich von Gottl-Ottlilienfeld (1868–1958), into the artefacts themselves, their manufacture by human beings and their use within the scope of purposeful actions, and to have subjected technical systems to a comprehensive theoretical model analysis under ontic, genetic and ultimate aspects. With the help of knowledge from mathematical-cybernetics system theory, which Ropohl worked into an exact model theory, a theoretical integration potential had been found to enable an interdisciplinary engineering concept and to "embrace the artefact calculated for use as well as the contexts of its origins and uses" [Ropohl, 1979, p. 314]. The disciplinary fragmentation of engineering sciences into small and miniature packages was abolished by Ropohl and moulded into a homogeneous science – *Allgemeine Technologie* (general technology) (Fig. 3-3) –, the object of which is the principles of technical feasibility and technical artefacts. Using this, he acknowledged critically the existing approaches to *Allgemeine Technologie*: cameralistic technology, scientific and social philosophy of engineering, special sociologies (e.g. industrial sociology), human resources studies, history of engineering, theories of technical progress, technical prognostic, system engineering, construction theory, traditional philosophy of engineering and the Marxist theory of engineering [Ropohl, 1979, p. 207].

According to Ropohl, engineering comprises the artefact itself (natural dimension), its construction by human beings (human dimension), and its use within the framework of purposeful actions (social dimension). These three dimensions of engineering correspond to different perspectives of knowledge: the natural dimension (ecological, of the natural and engineering sciences), the human dimension (anthropological, physiological, psychological and aesthetic), and the social dimension (economic, sociological, political and historical) [Ropohl, 1979, p. 32]. The engineering science knowledge perspective appears as a parameter of one of the three aforementioned independent variables of engineering. Consequently, the technical aspects cannot be adequately grasped by engineering and social philosophy, nor by the natural, engineering and social sciences. Rather, they have to be reflected via the synthesis into a general technology accomplished on the basis of general system theory.

Ropohl likes to differentiate between *Ingenieurwissenschaften* (engineering sciences) as the "conventional expression of these disciplines" [Ropohl, 1979, p. 34], and *Technikwissenschaften* (technical sciences), which he reserves for his system theory concept of *Allgemeine Technologie*, the object of which is knowledge of the three-dimensional world of engineering. The specific of the formation of engineering science theory consists, according to Ropohl, of the following [Ropohl, 1979, p. 34]:

– prognosis of the behaviour of planned engineering artefacts,
– prognosis of the structure of engineering artefacts for desired effects,
– theoretical analysis of existing engineering artefacts,
– practicological reformulation and concretisation of known natural science theories and hypotheses,
– recognition, separation and determination of the laws of nature effective in an engineering artefact,
– development of engineering science theories for specific engineering artefacts for which there is no natural science theory yet available, and
– the search for various engineering realisation options for one and the same natural science effect.

FIGURE 3-3
The 2nd edition of Ropohl's *Allgemeine Technologie* [Ropohl, 1999/1]

#### 3.1.2.2 The contribution of Marxism

As Gerhard Banse intimated in 1976, in the German Democratic Republic the treatment of philosophical issues in engineering sciences remained for a long time the province of engineering and social science disciplines, but without the representatives of these branches of science taking any particular notice of each other [Banse, 1976]. It was not until the 1960s that the Marxist dispute surrounding engineering and engineering sciences unfolded within the scope of the social sciences discussion about the nature of the scientific-technical revolution – a term that had been introduced into the Marxist debate on science and technology in highly developed industrial societies by the British natural scientist and researcher John Desmond Bernal (1901–71). Associated with this was the new definition of the productive resources and science concept, the criticism of the engineering philosophy renewed through theories of industrial society, and the clarification of the social status of natural scientists and engineers. With few exceptions, this dispute remained confined to the USSR and GDR.

In connection with his deliberations surrounding the term engineering, Kurt Teßmann specifies the object of engineering sciences as "the targeted, complex interaction of various forms of motion and manifestation of matter according to social needs and based on the status and development tendencies of the productive resources" [Teßmann, 1965, p. 132]. Picking up this thread, Lothar Striebing adds that the object of engineering sciences "results from the design of technical systems created artificially by man" [Striebing, 1966, p. 803]. According to Striebing, engineering sciences examine "the specific complex laws functioning in technical systems with the inclusion and systematic exploitation of the knowledge of the objective laws of nature" [Striebing, 1966, p. 804]. Owing to their universality, the definitions linked to Teßmann's specification of the engineering concept did not become widely accepted, a fate that also awaited those who favoured this definition of the object of engineering sciences. In the German Democratic Republic, the definition of engineering and engineering science formulated back in the 1960s by Johannes Müller had become entrenched: "Engineering is the totality of objects and processes that human beings set up and constantly reproduce at a particular stage of their evolution owing to the given objective opportunity, and in such combinations, dimensions and forms that the properties of these objects or processes work for mankind's purposes under certain conditions. The object of engineering sciences is therefore the engineering in the aforementioned sense" [Müller, 1967, p. 350]. The content of engineering sciences is the ordered systematic presentations of the principles of technical means, elements, systems, operations and methods developed previously, the systemised results of experiments and measurements carried out on technical entities and methods, and their theoretical generalisations as well as the research into the thinking processes behind technical development work.

Müller's definition of the content is obviously related to the aforementioned three functions of the epistemological dimension of engineering science research according to Zschimmer. The specific of engineering sciences worked out by Müller was adopted in the several editions of a German dictionary of the philosophical issues of the natural sciences. In this book, Gerhard Banse defines the engineering sciences as "sciences whose object is engineering" and whose purpose is "the analysis of technical systems and the theoretical synthesis of new technical objects" [Banse, 1978, p. 904].

Using Müller's ideas, Banse developed this definition into four features: engineering sciences exhibit an intentional, constructive, operational and integrative orientation. The methodological analysis and philosophical discussion of the process of discovery in the engineering sciences was concluded for the time being in the book by Gerhard Banse and Helge Wendt *Erkenntnismethoden in den Technikwissenschaften* (knowledge methods in engineering sciences) [Banse & Wendt, 1986], which also included contributions from engineering scientists.

In the German Democratic Republic from the late 1970s onwards, the history of engineering sciences evolved into a historical discipline [Buchheim & Sonnemann, 1986], and together with the inclusion of Soviet works on the specific of engineering sciences [Volossevich et al., 1980], their origins and development [Ivanov & Chechev, 1982], the historical characteristic of the materialistic analysis of the engineering sciences really took shape. Whereas even in the early 1980s, knowledge about the nature of engineering sciences was essentially still governed by the historical aspects in the dialogue with Soviet science and engineering historians [Albert et al., 1982; Buchheim, 1984], historical case studies on the genesis of individual engineering science disciplines appeared in quick succession. These works acknowledged by the international community of science and engineering historians [Sonnemann & Krug, 1987; Blumtritt, 1988] enable a philosophically accentuated history of the engineering sciences.

### 3.1.2.3 The theory of engineering sciences

The theory of engineering sciences currently evolving brings together contributions from system theory, general technology, Marxism, history of engineering, engineering philosophy and engineering sociology. Gerhard Banse, Armin Grunwald, Wolfgang König and Günter Ropohl therefore published the compendium *Erkennen und Gestalten. Eine Theorie der Technikwissenschaften* (cognition and design – a theory of engineering sciences) [Banse et al., 2006] in which they draw up an interim balance for the genesis of a theory of engineering sciences. The authors nominate two reasons why they feel it is necessary to consider the development of a theory of engineering sciences [Banse et al., 2006, p. 343]:

- The engineering sciences operating with their multidisciplinary approach exploit numerous individual scientific disciplines for the solution of their tasks, whereas the known theory of science is restricted to single disciplines.
- The knowledge generated by the engineering sciences is basically a means for successful design; the practice-oriented form of the knowledge of engineering sciences is a theory about the right skills, whereas the known theory of science concentrates in the first place on the right knowledge.

In the engineering sciences, cognition and design are interwoven via the knowledge and the products (Fig. 3-4): "engineering sciences are not there to serve theory, but instead practice" [Banse et al., 2006, p. 343]. This also manifests itself in fundamental engineering science disciplines such as applied mechanics and theory of structures because in the end their success is measured in terms of practical actions. Wolfgang König splits the engineering sciences into product-, function- and occupation-oriented engineering sciences [König, 2006/2, p. 41]. Whereas the product-oriented

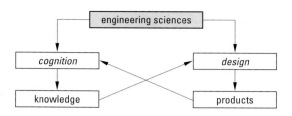

FIGURE 3-4
Interplay between cognition and design
[Banse et al., 2006, p. 344]

engineering sciences have evolved from engineering practice and relate to major engineering fields such as construction, mechanical engineering, mining, metallurgy, electrical engineering and computer science, the function-oriented engineering sciences are to a certain extent positioned transverse to these. "Their themes are the functional aspects that are relevant for all or at least a large number of engineering fields and hence also the product-oriented engineering sciences" [König, 2006/2, p. 41]. Therefore, in König's view, applied mechanics, applied thermodynamics, materials research and theory of structures could be allocated to the function-oriented engineering sciences. On the other hand, the occupation-oriented engineering sciences integrate the knowledge from the product- and function-oriented engineering sciences and present this in the form of courses of study; in addition, they investigate systematically the requirements in the respective occupations and generate specific knowledge databases and methical procedures for these [König, 2006/2, p. 42]. König lists design, production, environmental technology, quality assurance and industrial engineering among the occupation-oriented engineering sciences [König, 2006/2, p. 42]. Nonetheless, the three categories of product-, function- and occupation-oriented engineering sciences represent an ideal breakdown [König, 2006/2, p. 42]. This is shown by elastic theory, which in the last quarter of the 19th century evolved to become the theoretical basis of strength of materials and classical theory of structures and influenced both of these fundamental engineering science disciplines until well into the 20th century.

Among the selected case studies in the *Erkennen und Gestalten* compendium is one that deals with the practice and science of civil engineering. "Construction theory," notes Peter Jan Pahl, "is based on the observation that for some areas of building, rules can be established that enable forecasts of the behaviour of structures and components as well as the sequence of building processes" [Pahl, 2006, p. 283]. In doing so, construction theory in recent decades has not restricted itself to the physical behaviour of structures and natural systems. Instead, the computer-assisted use of structural mechanics (especially graph theory) for logical tasks that form part of the planning, organisation and management of construction projects plus the use of structures has opened up an important new area that will occupy research and practice for a long time and change the face of civil engineering [Pahl, 2006, pp. 283–284]. In their monograph *Mathematical Foundations of Computational Engineering* [Pahl & Damrath, 2000], Peter Jan Pahl and Rudolf Damrath (1942–2003) demonstrate how such a structural mechanics could function.

The compendium on the theory of engineering sciences closes with the following sentence: "Theory without practice is weak, but practice without theory is blind" [Banse et al., 2006, p. 348]. Therefore, in the next section, after defining engineering, we will investigate the origins of engineering sciences and in section 3.2 use concrete examples from applied mechanics, theory of structures and bridge-building to explain this.

### 3.1.3 Engineering and engineering sciences

Engineering is the object of engineering sciences. It is "the socially organised renunciation of the metabolic process taking place between mankind and nature in the form of the functional modelling of the external, artificial organ system of human activities" [Kurrer, 1990/1, p. 535]. The first and indefeasible fundamental quality of engineering is the reversal of the purpose-means relationship (Fig. 3-5b): whereas in primates the stick fashioned into a tool may be used to obtain a particular fruit and may have served its purpose once the fruit is eaten (Fig. 3-5a), in hominids it serves as a universal means for obtaining fruits. In other words, the means is there, so to speak, before the purpose allotted to it by society, and the purpose becomes concrete in the expediency of the means, i.e. in the artefact. Hence, engineering is the instrument for the purpose of satisfying needs. Engineering is insofar a constituent moment in mankind's self-creation as, with the reversal of the purpose-means relationship, the knowledge of the natural context develops in embryonic form as an inversion of the transformation of natural causality into technical finality. For only with the potential existence of the means before the allotted purpose, due to the reversal of the purpose-means relationship, will the effective result of the reason-consequence relationship become the desired consequence of the causal relationship and hence, as the engineered effect, also become the object of the knowledge of the natural causal relationship.

The engineering sciences that evolved during the Industrial Revolution represent a specific developmental form of the relationship of the transformation of natural causality into technical finality and the inversion of this transformation. This definition is developed in four steps below in a historico-logical sense:

*Step 1* (Fig. 3-5d): Constructional and technological modelling of causal relationships in technical entities and methods (classical engineering form).

*Step 2* (Fig. 3-5e): Knowledge of the causal relationship realised in the technical model (emergence of the first fundamental engineering science disciplines).

*Step 3* (Figs 3-5e and 3-5f): From the coexistence of the knowledge of the causal relationship realised in the technical model, with the constructional or technological modelling of causal relationships in technical entities and methods, to the cooperation (emergence of the system of classical engineering sciences).

*Step 4* (Fig. 3-5g): The space-time integration of the knowledge of the causal complex existing as the object in the technical model with the constructional and technological modelling of the causal complex in the

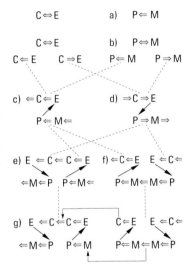

**FIGURE 3-5**
Development of the reversal of the purpose-means relationship (historico-logical structural scheme): a) means-purpose relationship for the use of tools in subhuman hominids, b) reversal of the means-purpose relationship in hominids as an elementary form of engineering, c) classical natural sciences (emerged from the scientific revolution), d) classical engineering form (emerged from the Industrial Revolution), e) natural science-based engineering, f) engineering-based natural science, g) integration of natural science-based engineering with the engineering-based natural sciences

engineering system (automation and the emergence of non-classical engineering science disciplines).

*Explanation of step 1:* If the purpose-means relationship in the technical model is idealised as a relationship between function and structure, where function represents the purpose and structure the means of its idealised realisation, then the inversion of the transformation of natural causality into technical finality represents a conversion of the function of the technical model into an effect of the causal relationship in the technical model, and the expression of the structure of the technical model appears as an instigator of the causal relationship. The result of the reason-consequence relationship, as the technically generated effect, becomes the object of the knowledge of the causal relationship existing as the object in the technical model.

This mathematical natural science bias is especially evident in the characteristics of the first fundamental engineering science disciplines (applied mechanics and thermodynamics) that began to emerge at the École Polytechnique, founded in 1794 in Paris. Nicolas Léonard Sadi Carnot (Fig. 3-6) carried out a thermodynamic analysis of Watt's steam engine during the 1820s. He concludes that the function of converting the chemical energy embodied in the fossil fuel into mechanical energy is the expression of the purpose materialised in the steam engine entity comprehended as an effect of an ideal transformation process of a theoretical heat engine model, which consists of the optimum exploitation of the thermal energy and, in the entropy, exhibits the cause to be identified. Starting with the social need for knowledge of the relationship between the coal consumed and the mechanical energy generated for the economic operation of steam engines, Carnot's analysis of the Nature imagined in the technical model of the Carnot machine achieved fundamental knowledge about a universal natural relationship, which was later superseded by the second principle of thermodynamics.

*Explanation of step 2:* The historico-logical condition for the knowledge of the causal relationship existing as the object in the technical model, i.e. the aforementioned inversion of the transformation of the natural causal relationship into a purpose-means relationship, was not only the mathematical and natural science disciplines evolving during the scientific revolution, but also the transformation of the natural causality into technical finality.

If the purpose-means relationship in engineering is objectified as the relationship between function and structure, where function represents the purpose and structure the means to achieve that function, then the transformation of the cause-effect relationship into a purpose-means relationship presents itself as a conversion of the effect into a function of the technical system and a conversion of the expression of the cause in the structure of the technical system. This transformation is characterised by the fact that it embodies a reversal of the causal effect into the ultimate reason; the result of the causal reason-consequence relationship becomes the aim of the desired function of the technical system and hence, as the

**FIGURE 3-6**
The 17-year-old Nicolas Léonard Sadi Carnot (1796–1832) in the uniform of the École Polytechnique
[Krug & Meinicke, 1989, p. 116]

socially desirable purpose, the cause of the existence of a technical system [Krämer, 1982, pp. 35 – 36].

Jacob Ferdinand Redtenbacher (Fig. 3-7) – one of the founding fathers of mechanical engineering – expressed this in 1852 with the classical engineering form evolving during the Industrial Revolution: "For man, natural forces appear to be foes, but this semblance only lasts until he has become more familiar with the actions of these forces; once this is the case, he discovers in their spirit the means whereby the forces can be organised so that while obeying the original nature in their being they can nevertheless cause such changes to substances, and thus make such changes usable and useful for his purposes. This is where engineering now appears in its full importance and seriousness, for its duty is to harness and master the natural forces in such a way that they supply the means to accomplish the multifarious purposes of mankind to greater degrees, at the most suitable places and with the most capable and best characteristics without being unfaithful to their intrinsic nature in doing so" [Redtenbacher, 1852, p. 190].

FIGURE 3-7
Jacob Ferdinand Redtenbacher (1809 – 1863) (archives of the University of Karlsruhe)

*Explanation of step 3:* Whereas the classical engineering form represented by the practising engineer and the fundamental engineering science disciplines represented by the theory-based academic symbolised two sides of the same coin as the system of classical engineering sciences began to emerge, their cooperation formed the critical moment in the formation of the system of classical engineering sciences and their consolidation in the ensemble of branches of science. This was because this cooperation marked the end of the duality of the classical engineering form and the fundamental engineering science disciplines in the form of the dispute over principles in the engineering sciences [Braun, 1977] which was becoming significant to engineers. The knowledge of the causal relationship existing as the object in the technical model is the starting point for the constructional and technological modelling in technical entities and methods. Only in this way was it possible for the interaction of the analytic with the synthetic, the deductive with the inductive, the theoretical with the practical, the abstract with the concrete, the receptive with the anticipative and the universal with the specific to become practically effective in the system of classical engineering sciences.

In terms of both the empirical and the theoretically accentuated knowledge of the causal relationship existing as the object in the technical model, the orientation towards the relationship's purposeful objectification in technical entities and methods now became the knowledge ideal of the engineering scientist. Carl von Bach (Fig. 3-8) – one of the pioneers in the experimental work on materials in the final three decades of the 19th century – notes in the preface to his work *Elasticität und Festigkeit* (elasticity and strength) published in 1889/90: "May this work, too, no longer be a step in a new direction through exposing the significance of the knowledge of the actual behaviour of the materials and by clarifying the inadequacy of utilising solely the laws of proportionality between strains and stresses to build the whole entity of elasticity and strength on the basis of mathematics. Instead, it will show the need for the design engineer to realise again and again the conditions for the individual equations he uses on the basis of experience, where such is available, i. e. when he stipulates the dimensions in full knowledge of the true relationships and does not wish to remain in the mould of standard forms. It will also demonstrate the need for deliberation to assess the relationships obtained from mathematics with regard to the degree of their accuracy, insofar as this is at all possible with the current level of knowledge, and that wherever the latter and the deliberation – including selection of and training in new methods – are inadequate, then our first aim should be to try to direct the question at Nature – for promoting engineering and hence contributing to industry" [Bach, 1889/90, pp. VII – VIII].

**FIGURE 3-8**
Carl von Bach (1847–1931) [Bach, 1926]

*Explanation of step 4:* Automation is the essential developmental form in which the hegemonic form of engineering unfolds in the scientific-technical revolution. The transformation of the natural cause-effect relationship into an engineering purpose-means relationship converges in this in such a way that the means reacts directly to that cause that was the original source of the natural causal relationship ultimately restructured in the engineering. Automatic machines are such means because the segment of the universally interactive automated and self-organising management of the matter modelled functionally in such machines reflects itself in the means. It appears autonomous, so to speak, and hence perpetuates the cycle of transformation and retransformation in a self-contained space-time sphere. The system of non-classical engineering sciences is fashioned in this engineering form, where the knowledge of the cause-effect relationships aggregated into a causal complex in the technical model – especially in automatic machines – is integrated in space-time terms with its constructional and technological modelling in the technical system. As in the socio-historical process of the scientific-technical revolution the inversion of the transformation of the natural cause-effect relationship tends towards its retransformation in a technical purpose-means relationship, i. e. the scientific knowledge of the technically formulated natural relationship, automatic machines are ascribed the mechanisation of the formalisable side of the scientific cognition process. The analysis of the causal complexes existing as the object in such technical models for the purpose of their synthesis in the form of technical systems, as is characteristic of the non-classical engineering sciences, corresponds in the modern natural sciences to the ending of natural causality in technical finality. This happens in such a way that the purpose of the knowledge of causal relationships is impressed on the means.

In their monograph *Mathematical Foundations of Computational Engineering* (Fig. 3-9), Peter Jan Pahl and Rudolf Damrath span the whole spectrum of logic, set theory, algebraic, ordinal and topological structures, counting systems, group theory, graph theory, tensor analysis and stochastic. They have created the first manual of engineering mathematics in which the mathematical principles of the non-classical engineering sciences that evolved with the computer from the mid-20th century onwards are explained systematically and comprehensively. The upheaval in the mathematical foundations of engineering sciences induced by the computer are expressed by the authors in their preface as follows:

"Before computers were introduced into engineering, numerical solutions of the mathematical formulations of engineering problems involving irregular geometry, varying material properties, multiple influences and complex production processes were difficult to determine. Nowadays, computers amplify human mental capacities by a factor of $10^9$ with respect to speed of calculation, storage capacity and speed of communication; this has created entirely new possibilities for solving mathematically formulated physical problems. New fields of science, such as computational mechanics, and widely applied new computational methods, such as the finite element method, have emerged.

"While computers were being introduced, the character of engineering changed profoundly. While the key to competitiveness once lay in using better materials, developing new methods of construction and designing new engineering systems, success now depends just as much on organization and management. The reasons for these changes include a holistic view of the market, the product, the economy and society, the importance of organization and management in global competition as well as the increased complexity of technology, the environment and the interactions among those participating in planning and production.

"Given the new character of engineering, the traditional mathematical foundations no longer suffice. Branches of mathematics which are highly developed but have so far been of little importance to engineers now prove to be important tools in a computer-oriented treatment of engineering problems" [Pahl & Damrath, 2001, p. V].

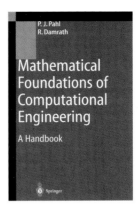

FIGURE 3-9
The first manual of non-classical engineering mathematics [Pahl & Damrath, 2001]

## 3.2 Revoking the encyclopaedic in the system of classical engineering sciences: five case studies from applied mechanics and theory of structures

The encyclopaedia is "archive and anticipation" [Sandkühler, 1990, p. 768]. This is the conclusion reached by Sandkühler based on Diderot's definition in his "Encyclopédie" article in the *Encyclopédie*; it is not simply the summation of catalogued knowledge, but always also a normative programme. Five case studies from applied mechanics and theory of structures will be used to illustrate the metamorphosis of anticipation in Sandkühler's sense as a constituent of classical engineering sciences:

– Gerstner's *Handbuch der Mechanik* (manual of mechanics) (see section 3.2.2),
– Weisbach's *Lehrbuch der Ingenieur- und Maschinen-Mechanik* (Principles of the Mechanics of Machinery and Engineering) (see section 3.2.3),
– Rankine's *Manual of Applied Mechanics und Manual of Civil Engineering* (see section 3.2.4),
– Föppl's *Vorlesungen über technische Mechanik* (lectures on applied mechanics) (see section 3.2.5), and
– *Handbuch der Ingenieurwissenschaften* (manual of engineering sciences) (see section 3.2.6).

At the start we have Gerstner's contribution to theory of structures and applied mechanics. Totally in keeping with Diderot's definition, Gerstner, influenced by the Josephinian enlightenment, sees building not as a beautiful art but rather as a mechanical one. In his *Einleitung in die statische Baukunst* (introduction to structural engineering) [Gerstner, 1789], he uses the example of the masonry arch to place the programmatic claim of theory of structures in its rightful historical position with respect to the traditional proportion theory of architecture. And in the *Handbuch der Mechanik*, Gerstner honoured this claim in a generalised sense to a certain extent and in doing so created *the* outstanding German-language compendium on the scientific basis of construction and mechanical engineering for the transition from workshop to factory. Both works can therefore be read as prolegomena of the classical engineering sciences. Nonetheless, Gerstner is alone in realising the relationship between science-based engineering (Fig. 3-5e) and engineering-based science (Fig. 3-5f), which he clearly identified.

At the end we have Föppl's *Vorlesungen über technische Mechanik* and the *Handbuch des Brückenbaus* (manual of bridge-building) by Schäffer, Sonne and Landsberg. However, the historico-logical development from the coexistence of the knowledge of causal relationships realised in technical models and the constructional and technological modelling of causal relationships in technical entities and methods to the cooperation between these two aspects did not appear in print until the second edition of the five-volume *Handbuch des Brückenbaus* (which appeared within the scope of the *Handbuch der Ingenieurwissenschaften* and Föppl's *Vorlesungen*). The *Handbuch der Ingenieurwissenschaften* is "archive and anticipation" with practical intentions, organised according to construction theory principles, with the classical theory of structures at its heart. Anticipation has now changed on a theoretical level into a *tekhnē* (art, craft) of the *epistēmē* (epistemic).

In terms of its presentation, Gerstner's *Handbuch* inserted the final piece of the jigsaw for the preparatory period of applied mechanics (1575–1825). At the same time, in terms of content it contains isolated transitions to the discipline-formation period of applied mechanics (1825–1900), which in terms of form and content is concluded by Föppl's *Vorlesungen* in the classical style. Nonetheless, Föppl's *Vorlesungen* already demonstrate the theory and presentation styles of applied mechanics that were to characterise the consolidation period (1900–50). However, Weisbach's *Lehrbuch der Ingenieur- und Maschinen-Mechanik* and Rankine's *Manual of Applied Mechanics* form the golden section of the discipline-formation period of applied mechanics. Weisbach's *Lehrbuch* is encyclopaedic in character, but in its fifth edition prepared by Gustav Herrmann it reaches the limits of an encyclopaedia of applied mechanics. Not until Föppl applied calculus to applied mechanics was it possible to structure the vastly expanded quantity of data and overcome these limits.

**On the topicality of the encyclopaedic**

### 3.2.1

As the purpose of the reversal of the purpose-means relationship constituting modern engineering is no longer evident, we lose the core of the encyclopaedic. Whereas in the use of tools by subhuman hominids the purpose for which the tool was produced ends when the job is finished (Fig. 3-5a), this role is reversed in hominids, for whom the purpose of the tool becomes the means of the action (Fig. 3-5b). The epistemic of the hominids is thus set in an elementary way. On the other hand, on the level of engineering science knowledge, the purpose is today neither evident in the engineering-based natural sciences nor in the natural science-based engineering, but the purpose is always incorporated through means and causes (Fig. 3-5g). Consequently, the crux of the social discourse on anticipation as the target of the knowledge is at risk. The encyclopaedia, too, as "archive and anticipation", mutates into a monumental lexicon, into nothing more than an archive. The tendency – evident since the last century – to blur the distinction between encyclopaedia and lexicon is thus complete because the encyclopaedia could no longer be read systematically and consecutively owing to the wealth of knowledge inside; it merely provides the respective desired knowledge at its appropriate alphabetical point just like a lexicon. Therefore, the holistic view gave way to a linear sequence of knowledge, the knowledge about the individual, the positive knowledge, the currently available knowledge. But epistemic means "to be knowledgeable of the world, to have knowledge of things, to possess skills to master those things, to perceive generally through theory the multiplicity of phenomena, to monitor the reference to reality through the self-assuredness of the reflection, in the end to be able to act knowledgeably" [Sandkühler, 1988, p. 205].

So what is it that holds the construction theories together at their very innermost level? First of all, two observations on this which are then substantiated by five case studies from the field of applied mechanics and theory of structures:

*First observation:* The inner signature of the schism of construction theories is the use of calculus in mathematics which appeared around 1600, the purpose of which is "to express a problem with the help of a man-made language such that the steps to the solution of the problem can be designed as a step-by-step rearrangement of symbolic expressions, whereby the rules of this successive rearrangement refer exclusively to the syntactic design of the symbolisms, and not to that for which the symbols stand" [Krämer, 1991/3, p. 1]. The history of applied mechanics and theory of structures can be understood as the history of the creation of model worlds that become increasingly more removed from construction and are dominated more and more by the non-interpretative use of symbols. In the first half of the 19th century, the engineer achieved this by algebraicising his texts (Gerstner's *Handbuch*), around the middle of the 19th century by geometricising the design, calculation and construction of his real artefacts (Weisbach's *Lehrbuch* initiated development in this area), in the second half of the 19th century by placing more emphasis on the analytical side of modelling in applied mechanics and theory of structures (Rankine's *A Manual of Applied Mechanics* and *A Manual of Civil Engineering* set standards here), and around the turn of the 20th century by the extensive use of calculus for his intellectual artefacts (vector calculus in Föppl's *Vorlesungen*).

Today, the engineer's perspective is increasingly governed by that of the symbol worker, a fact that is verified impressively by the evolution of computational mechanics (Fig. 3-10) in recent years. On the other hand, the architect, for example, is in the end still reliant on the semantic burden of his products.

*Second observation:* Nonetheless, the history of fundamental engineering science disciplines generally, and applied mechanics and theory of structures particularly, cannot be entirely resolved into the genesis of the practical use of symbols, even on the logical side. The disciplines expanded with the Industrial Revolution and form a specific developmental form of the relationship of the transformation of natural causality into technical finality as well as the inversion of this transformation. The first step can be seen in the constructional and technological modelling of causal relationships in technical entities and methods (Fig. 3-5d). Gerstner's critical analysis of suspension bridges in his *Handbuch* can be regarded as an example of the constitution of the classical engineering form. The second step came with the creation of the first fundamental engineering science disciplines, i.e. applied mechanics and theory of structures, which are characterised by the knowledge of the causal relationship realised in the technical model (Fig. 3-5e); Weisbach's *Lehrbuch*, his manual *Der Ingenieur* (the engineer) and *Civilingenieur* (civil engineer), the journal on which he had a great influence, represent the grid lines of this development. In the third step, the system of classical engineering sciences takes on its form. The logical nucleus is marked by the transition from the coexistence of the first two steps to their cooperation (Figs. 3-5e and 3-5f). Rankine's *Manual of Applied Mechanics* and *Manual of Civil Engineering*, Föppl's *Vorlesungen* and the *Handbuch der Ingenieurwissenschaften* are the publications that express this third step. Whereas Rankine returns again and again in his *Manuals* to the relationship between natural science-based engineering and engineering-based natural science, and uses this to solve problems, Föppl's *Vorlesungen* is regarded as a major attempt to introduce calculus into applied mechanics, and the *Handbuch der Ingenieurwissenschaften* is seen as an encyclopaedia

**FIGURE 3-10**
*Encyclopedia of Computational Mechanics*:
a) volume 1 [Stein et al., 2004/1],
b) volume 2 [Stein et al., 2004/2] and
c) volume 3 [Stein et al., 2004/3]

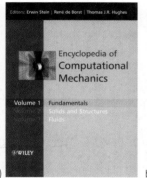
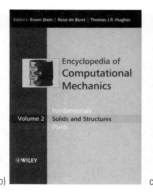

of classical civil engineering theory. In the fourth step the space-time integration of the knowledge of the causal complex existing as the object in the technical model with the constructional and technological modelling of the causal complex in the technical system is accomplished in the form of automation and the evolution of non-classical engineering science disciplines (Figs. 3-5g and 3-9). Examples of this are the design of materials through materials research and computational mechanics (Fig. 3-10).

### 3.2.2 Franz Joseph Ritter von Gerstner's contribution to the mathematisation of construction theories

Franz Joseph Ritter von Gerstner's three-volume *Handbuch der Mechanik* (Fig. 3-11), which was published by his son Franz Anton Ritter von Gerstner (1793–1840), was the first comprehensive book on applied mechanics in the German language. It marked the culmination of the technical-scientific life's work of Franz Joseph Ritter von Gerstner and as such had a lasting influence on this fundamental engineering science discipline during the constitution phase of applied mechanics in the German-speaking countries from 1830 to 1850 alongside the classical French contributions of Messrs. Navier, Poncelet, Coriolis, etc., which, however, did not appear until a decade later. But more besides: what father and son Gerstner published under the modest title of *Handbuch der Mechanik* became *the* outstanding German-language compendium on engineering-based science (Fig. 3-5e) in the first half of the 19th century. In this book, contemporary construction and mechanical engineering, at the transition from workshop manufacture to the technology of the Industrial Revolution, were given such scientific treatment for engineering buffs that, looking back from the present day, we can assess it as a watershed in the evolution of the system of classical engineering sciences.

#### 3.2.2.1 Gerstner's definition of the object of applied mechanics

Gerstner was the first German-speaking engineering scientist to define the object of applied mechanics and introduce this terminology into the German language: "The object of applied mechanics (mécanique industrielle or mécanique appliquée aux arts) is the performance of all those works necessary for creating the products of commercial and artistic energy and are presented according to the necessities of life to be satisfied. All human endeavours are achieved partly by hand, partly by using instruments and machines. The teaching of mechanics has to specify laws and rules for both types of work according to which the workers are to behave, set up their instruments and machines purposefully, check them and use them. A force is necessary to execute any work ... Therefore, we have in mechanics merely to observe exactly the laws of nature, examine with due care the properties of the objects to be processed and thereby find the most appropriate means to accomplish our works" [Gerstner, 1833, p. 3]. Among those forces, Gerstner lists the muscle power of human beings, cohesion, gravity and "many more besides which are put to the service of man and which have been used to carry out his work" [Gerstner, 1833, p. 4]. In this context he mentions the muscle power of animals, water and wind power, springs, steam and the explosive force of gunpowder.

After Gerstner has named a number of customary energy sources involved in work, he explores the moments of the simple working process as therein lie the object of the work and the means (tools, machines) but excluding the enlivening working forces. The latter appears merely in the form of a physical force that is used purposefully in the working process together with other technically harnessed forces. Finally, Gerstner includes in his observation of the working process the "determination of the relationships of the components of the machine … in order to achieve the best possible exploitation of the force and consequently also the largest possible quantity of worked products in a certain time, e. g. in one day … If several simple devices are combined for one common purpose and thereby the work be more ordered, the arbitrariness of the hand eliminated, or several purposes achieved simultaneously, we call this a machine" [Gerstner, 1833, pp. 4 – 5]. The calculation of the power requirement and its minimisation by machines built by adding together machine elements or simple machines at the same time specify the "relationships of machine components" [Gerstner, 1833, p. 5], and according to Gerstner is one of the most important aspects of applied mechanics.

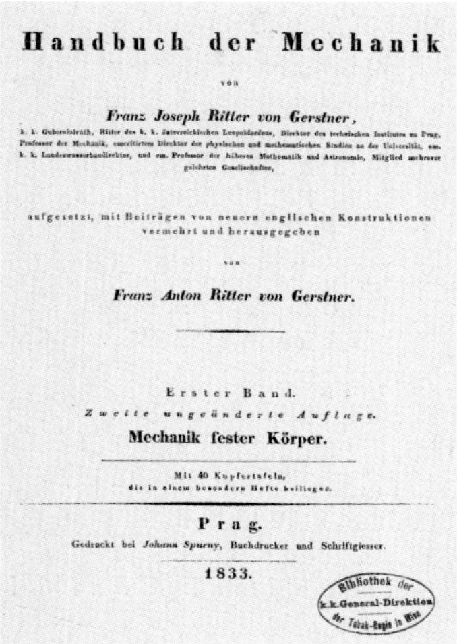
Handbuch der Mechanik

von

Franz Joseph Ritter von Gerstner,

k. k. Gubernialrath, Ritter des k. k. österreichischen Leopoldordens, Direktor des technischen Institutes zu Prag, Professor der Mechanik, emeritirtem Direktor der physischen und mathematischen Studien an der Universität, em. k. k. Landeswasserbaudirektor, und em. Professor der höheren Mathematik und Astronomie, Mitglied mehrerer gelehrten Gesellschaften,

aufgesetzt, mit Beiträgen von neuern englischen Konstruktionen vermehrt und herausgegeben

von

Franz Anton Ritter von Gerstner.

Erster Band.

Zweite ungeänderte Auflage.

Mechanik fester Körper.

Mit 40 Kupfertafeln, die in einem besondern Hefte beiliegen.

Prag.

Gedruckt bei Johann Spurny, Buchdrucker und Schriftgiesser.

1833.



**FIGURE 3-11**
Cover of the 2nd edition of volume I of Gerstner's *Handbuch der Mechanik* (1833)

Owing to his observations regarding the material side of the working process, his mechanising and economising with the help of applied mechanics, Gerstner arrives at a precise definition of the object of applied mechanics (Fig. 3-12): applied mechanics can be broken down into dynamics, "which deals with the forces of human beings, animals, water, air, steam, etc. in terms of their magnitude and the laws of their effectiveness", and "the theory of resistance, wherein the magnitude of the resistances occurring and the laws are discussed according to which such resistances counteract the forces applied in all works" [Gerstner, 1833, p. 6]. Gerstner divides the latter further into statics and "true mechanics," or rather, "the theory of motion, in which the magnitude of the movement, or the work carried out, is calculated as the final purpose of the force application" [Gerstner, 1833, p. 7]. Here, statics has the task of analysing the conditions for the equilibrium between force and resistance at rest or under uniform motion.

Although the word energy had not yet become one of the fundamental terms of natural science in Gerstner's lifetime, his breakdown of applied mechanics is implicitly energy-based because he places the energy (dynamics) on one side of the equation, the analysis of the use of the energy employed in the mechanised working process on the other. Accordingly, he divides the content of the first volume of his *Handbuch der Mechanik* into the analysis of animal and human energy sources, their use through simple machines (lever, wedge, screw, wheel on axle, jack, balance and capstan), i. e. dynamics, and the "theory of resistance", represented by strength of materials, structural design and the theory of friction resistances.

Whereas the purpose behind Gerstner's dynamics is to exemplify the knowledge of the causal relationship realised in the technical model or artefact, his "theory of resistance" is determined more by the constructional

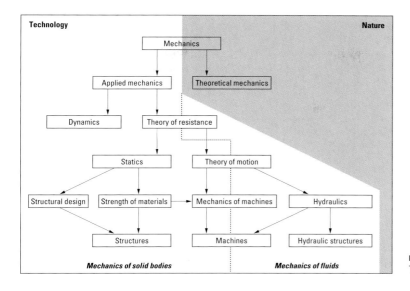

FIGURE 3-12
The structure of mechanics after Gerstner

and technological modelling of causal relationships in technical entities and methods, in particular in their useful application to freight wagons, roads and railways. This dualising into engineering-based natural science (Fig. 3-5f) and natural science-based engineering (Fig. 3-5e) permeates the second volume of his *Handbuch der Mechanik* [Gerstner, 1832], which contains engineering hydromechanics and its application to water supply systems plus river and canal works, and the design of water-mills. Contrastingly, the third volume [Gerstner, 1834] is dedicated to the scientific establishment of machines for building, mining and metallurgy. The emphasis in Gerstner's mechanical engineering is on engineering based on natural science. The mechanisation of natural science in mechanical engineering did not assert itself until the Industrial Revolution in Central Europe (1830 – 60), i.e. about a decade after the mechanisation of natural science in building.

Nonetheless, the interlocking of the knowledge of the causal relationship realised in the technical model or artefact, which characterises classical engineering sciences, with the transformation of the acknowledged natural cause-effect relationship into a technical purpose-means relationship (Fig. 3-5e), while emphasising unambiguously the time-saving principle, is formulated by Gerstner into the credo of applied mechanics: "The object and scope of the investigations are hereby quite clearly prescribed for applied mechanics, and we can therefore specify their purpose more accurately if we say that their primary task is to find the most reliable and most appropriate means by way of which the physical force and skill of mankind is assisted, how this force is used when carrying out works necessary for human needs, and how the works can be realised either with the greatest savings in time and effort, or at the lowest cost" [Gerstner, 1833, p. 7].

Therefore, in Gerstner's work the acquisition of knowledge by means of the artefact transformed into the means of knowledge (Fig. 3-5f) takes second place to a construction and mechanical engineering founded on natural science (Fig. 3-5e). The encyclopaedic aspect of Gerstner's *Handbuch der Mechanik* is hence more an archive of the knowledge of contemporary construction and mechanical engineering founded on natural science than anticipation conveyed via differential and integral calculus.

### The theory and practice of suspension bridges in the *Handbuch der Mechanik*

#### 3.2.2.2

Nevertheless, Gerstner was able to model individual real artefacts of construction engineering in such a way that he could express these in the language of differential and integral calculus and develop methods of structural analysis. Thomas Telford's (1757–1834) suspension bridge (opened in 1826) over the Menai Strait in Wales was just one such real artefact. Gerstner developed the system of hinges illustrated in Fig. 8 of Fig. 3-13 into a mechanical model of the bridge (Fig. 3-13, Fig. 10) by taking this to the limit: "It would be too long-winded to wish to calculate the stress in each chain-link [= funicular force – the author] and the angular position ($\alpha$, $\beta$, …) of the same; in this respect we consider the chain links as infinitesimal, which transforms the line of the positions into a curve" [Gerstner, 1833, p. 474]. Fig. 10 in Fig. 3-13 illustrates such a mechanical model of a suspension bridge with span $l$ and the given central sag $h$. Gerstner placed the origin of the $xy$ system of coordinates and the arc coordinates $s$ at point $A$, with the direction of the $y$-axis to the left and the $x$-axis downwards.

**FIGURE 3-13**
On the theory of suspension bridges after Gerstner [Gerstner, 1832, plate 2]

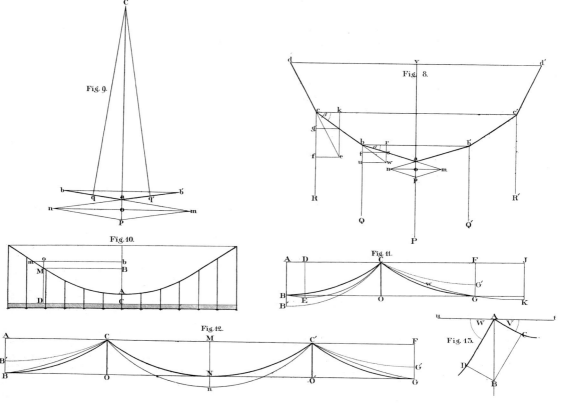

After Gerstner has used integration to develop transcendental equations for the y-axis and the x-axis depending on the angle of the tangent to the funicular curve in an ascending power series, he compares the form parameters obtained theoretically for the funicular curve ($s$ = arc length of chain, $r$ = radius of curvature of chain at $A$) with the values determined by Telford from a 1:4 scale model of the bridge. In conclusion, Gerstner specifies the stress equation as follows:

$$\sigma_{ten,\,exist} = H_q/f \leq \sigma_{ten,\,fail}/\nu \tag{3-1}$$

where

$$H_q = (G \cdot F + g \cdot f + p) \cdot r \tag{3-2}$$

i.e. the horizontal tension due to the loads uniformly distributed over the entire span: the dead load of the deck $G \cdot F$, the dead load of the chain $g \cdot f$ ($f$ = chain cross-section at $A$) and the assumed imposed load

$$p = 1.0606 \cdot (G \cdot F + g \cdot f) \tag{3-3}$$

again using Telford's figures. Using the tensile strengths $\sigma_{ten,\,fail}$ determined in tests, Gerstner calculates a factor of safety against tensile failure $\nu$ (which was a little over 3) for die Menai Strait bridge. The factor of safety of Hammersmith Bridge in London lies in the same order of magnitude. Gerstner had thus formulated a suspension bridge theory based on infinitesimal calculus.

Gerstner summarises the theory of structures foundation of the static-constructional design of suspension bridges in five steps using the example of a bridge over the Moldau near Prague ($l$ = 142.25 m, $h$ = 10.75 m):

1. Determine the spans, especially the main span $l$.
2. Determine the central sag $h$ in conformity with English suspension bridges already built, and calculate the chain lengths using

$$s = y \cdot \{l + (2x^2/3y^2) + 0.8 \cdot [(2/9) \cdot \mu - 0.5] \cdot (x/y)^4\} \tag{3-4}$$

where

$$\mu = g \cdot f/(G \cdot F + g \cdot f) \tag{3-5}$$

3. Determine the radius of curvature $r$ at $A$ and the length of the hangers using

$$y^2 = 2 \cdot r \cdot x - (2/3) \cdot \mu \cdot x^2 + (4/45) \cdot \mu^2 \cdot (x^3/r) \tag{3-6}$$

4. Determine the chain cross-section $f$ at point $A$ using equations 3-1 and 3-2.
5. Calculate the deformation of the suspension bridge:
- elastic deformation for the loading cases dead load, and dead + imposed loads
- elastic deformation for the loading case dead + asymmetric imposed loads
- deformation for the loading case uniform temperature change.

When designing the chain cross-section $f$ at point $A$ (Fig. 10 in Fig. 3-13) using equation 3-2, Gerstner assumes $\sigma_{ten,fail}$ = 463.2 N/mm² and $v$ = 3, i. e.

$\sigma_{ten,perm} = \sigma_{ten,fail} / v = 463.2 / 3 = 154.4$ N/mm²

This value is well below the yield stress for conventional grade S 235 mild steel ($f_{y,k}$ = 240 N/mm²) according to Eurocode EC 3. The value $\sigma_{ten,fail}$ = 463.2 N/mm² corresponds to the mean of 17 tensile tests which Telford and Brown carried out with wrought-iron samples of different lengths with round and square cross-sections [Gerstner, 1833, p. 256].

Gerstner's suspension bridge theory is genuine engineering science because he was able to link Telford's suspension bridge engineering based on natural science (Fig. 3-5e) with the engineering-based natural science (Fig. 3-5f), i. e. the mechanically generated catenary. Gerstner's suspension bridge theory was joined in the early 1820s by the monographs of Marc Seguin (1786–1875), Guillaume Henri Dufour (1787–1875) and Claude Louis Marie Henri Navier (1785–1836) [Wagner & Egermann, 1987; Peters, 1987]. Therefore, the design, calculation and building of suspension bridges constitutes an early example of the cooperative relationship between natural-science based engineering (Fig. 3-5e) and engineering-based natural science (Fig. 3-5f), which would not characterise structural engineering until the last quarter of the 19th century.

### 3.2.3 Weisbach's encyclopaedia of applied mechanics

The *Lehrbuch der Ingenieur- und Maschinen-Mechanik* (Weisbach, 1845–87) forms the heart of Weisbach's encyclopaedia of applied mechanics, and it was this that made an impression on, indeed popularised, applied mechanics in the middle of the discipline-formation period (1850–75). The *Lehrbuch* was translated into English (*Principles of the Mechanics of Machinery and Engineering*) [Weisbach, 1847/48], Russian, Swedish, Italian, French, Spanish and Polish. In its original German, it was the most influential book on applied mechanics in the 19th century, containing numerous original chapters, in particular those on hydraulics and strength of materials. Weisbach's *Lehrbuch* was supplemented by his manual entitled *Der Ingenieur* and the journal *Der Civilingenieur*.

#### 3.2.3.1 The *Lehrbuch*

"My chief aim in writing this work," writes Weisbach in the introduction to volume I of his *Lehrbuch* (Fig. 3-14a), "was the attainment of the greatest simplicity in enunciation and proof; and with this to give the demonstration of all problems, important in their practical application, by the lower mathematics only" [Weisbach, 1845, p. V]. As a teacher of and author for engineers, he saw it as his duty "to render well-grounded study of science easy by simplicity of explanation, by the use of only the best known and easiest auxiliary sciences, and by eschewing everything that is unnecessary" [Weisbach, 1845, pp. V–VI]. The "bestseller" nature of Weisbach's *Lehrbuch* can be attributed to the faithful implementation of those three principles, which introduced new groups of readers to the literature of applied mechanics.

 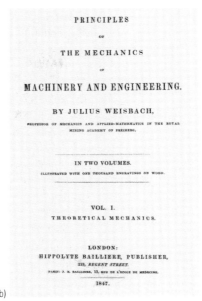

FIGURE 3-14
Cover of a) 1st German edition of Weisbach's *Lehrbuch* and b) its English translation

In writing his book, Weisbach was at pains to "preserve the right medium between *generalizing* and *individualizing*" [Weisbach, 1845, p. VI]. In doing so, he preferred induction to deduction: "It is also undeniable, that in treating a general case, the knowledge which might be gained by the treatment of a specific case, is frequently lost, and that it is not unfrequently easier to deduce the compound from the simple, than to eliminate the special from the general" [Weisbach, 1845, p. VI]. Weisbach was therefore the first person to apply the principle of induction – philosophically founded for all sciences by the British philosopher and science historian William Whewell (1794–1866) in his 1840 publication *The Philosophy of the Inductive Sciences* – to applied mechanics. This principle would later play a great role in English science and would help to ensure that Weisbach's *Lehrbuch* would enjoy several English-language editions.

According to Weisbach, applied mechanics is not mechanical engineering, but rather should be understood "merely as an introduction to or preparatory science for this" [Weisbach, 1845, p. VI], and in this respect its relationship with mechanical engineering is like that between descriptive geometry and engineering drawing.

Weisbach split his *Lehrbuch* into one volume entitled "Theoretical Mechanics" and one entitled "Applied Mechanics". He explained his reason for doing so thus: "… this work is to furnish instructions on all mechanical relations, in architecture and the science of machines … In order to form a complete opinion of a building or machine, the most various doctrines of mechanics … must be taken into consideration; the material for the study of the mechanics of a building or machine must, therefore, be collected from all parts of mechanics. Now, as it is much more useful practically to be able to study the doctrines relative to every individual machine in connection, than to have to collect them from all departments of mechanical science, the utility of the adopted division seems to be beyond all doubt" [Weisbach, 1845, p. VII].

Weisbach divides "Theoretical Mechanics" into
- phoronomy; or the pure mathematical science of motion (= kinematics),
- mechanics in the physical science of motion in general (= kinetics of the material point),
- statics of rigid bodies (= statics of rigid and elastic bodies, strength of materials),
- dynamics of rigid bodies,
- statics of fluid bodies (= hydrostatics and aerostatics), and
- dynamics of fluid bodies (= hydrodynamics and aerodynamics),

and "Applied Mechanics" into
- application of mechanics in buildings (= theory of structures), and
- application of mechanics to machinery (= analysis of prime movers).

In terms of number of pages (German editions)
- the 1st edition of volume I [Weisbach, 1845] has 535 pages, and volume II [Weisbach, 1846] 618 pages,
- the 2nd edition of volume I [Weisbach, 1850] has 696 pages, and volume II [Weisbach, 1851] 704 pages (theory of structures: 118 pages; prime movers: 586 pages), and
- the final, 5th edition of volume I [Weisbach, 1875] has 1312 pages, and volume II [Weisbach, 1882 & 1887] 1870 pages (theory of structures: 614 pages; prime movers: 1256 pages).

The 1st edition of volume III on the mechanics of production and intermediate machines (published in 1860) managed an impressive 1360 pages by itself!

The sheer size of the book enables us to gain some insight into the qualitative development of theory structures and the mechanics of machines. The establishment phase of theory of structures spanned from 1850 to 1875 and formed the "middle" of the discipline-formation period (1825–1900), which was characterised by the emergence of trussed framework theory; this is analogous with how mechanical engineering experienced the "mechanisation" of thermodynamics in the form of the development of a theory of steam engines. Both discipline-formation processes coincided with the Industrial Revolution in Germany. So Weisbach's *Lehrbuch* forms not only the historical but also the logical "middle point" in the evolution of the discipline of applied mechanics. One factor behind the success of his *Lehrbuch* was the invention of the engineering manual by Weisbach, which was set to boost the establishment of practical engineering on the basis of applied mechanics.

**The invention of the engineering manual**

**3.2.3.2**

The publication of the manual *Der Ingenieur* (the engineer) [Weisbach, 1848/2] in 1848 meant that Weisbach had provided a companion volume to his *Lehrbuch* which contained "a compact and orderly compilation of carefully selected rules, formulas and tables based on the most reliable theories and facts gained through experience, and intended for applications in engineering, practical geometry and mechanics, machines, architecture and technical matters in general" [Weisbach, 1848/2, p. VI].

*Der Ingenieur* is divided into three parts:

*1st part:* arithmetic (147 pages)
- tables: e.g. roots and logarithms
- rules and formulas: e.g. arithmetic operations, roots, logarithms, equations, series;

*2nd part:* geometry (205 pages)
- tables: measures, trigonometry, circles
- rules and formulas for theoretical geometry: planimetry, stereometry
- rules and formulas for practical geometry: geodesy;

*3rd part:* mechanics (255 pages)
- formulas, rules and tables for theoretical mechanics: weight tables, formulas, rules and tables for general mechanics, statics, dynamics, hydraulics
- formulas, rules and tables for practical mechanics: statics of building structures, mechanics of prime movers, heat theory and steam engines, intermediate machines, production machines.

Fig. 3-15 shows the composite strength concept first introduced by Weisbach. The cantilever $AB$ of length $l$ fixed at the support at an acute angle $\alpha$ undergoes bending due to the shear force $P \cdot \sin\alpha$ and tension due to the axial force $P \cdot \cos\alpha$. Weisbach was the first person to specify a stress equation for this case of combined actions [Weisbach, 1848/2, p. 425]:

$$\sigma = \frac{P \cdot \cos\alpha}{F} + \frac{e}{W} P \cdot \sin\alpha \cdot l \qquad (3\text{-}7)$$

where
- $F$   cross-sectional area
- $W$   second moment of area
- $e$   distance of extreme fibres from neutral axis, where $W/e$ corresponds to the section modulus of the cross-section.

On the following two pages, Weisbach derives further stress equations for bending plus axial force and bending plus torsion.

*Der Ingenieur* thus met the demands of practising engineers for a book containing the dynamic developments in knowledge in the emerging classical engineering sciences of the second half of the 19th century but structured and prepared in a way they could use for their daily business. Weisbach's engineering manual would be followed by others, one example of which is *Des Ingenieur's Taschenbuch. Hütte* (the engineer's pocket-book), published by Ernst & Korn (today Ernst & Sohn) in 1857, which had originally been published by the members of the iron and steel works society *Die Hütte* founded in 1846 at the Royal Academy of Industry in Berlin; it became the most successful engineer's pocket-book in the German language. This was followed in England in 1866 by Rankine's *Useful Rules and Tables* [Rankine, 1866/1], the fourth and last part of his *Manuals*. Some years later, John C. Trautwine (1810–83) conceived the pocket-book for civil engineers together with the New York-based publishing house John Wiley & Son, which appeared in 1872 under the title of *Civil*

**FIGURE 3-15**
Bending plus axial force after Weisbach [Weisbach, 1848/2, p. 425]

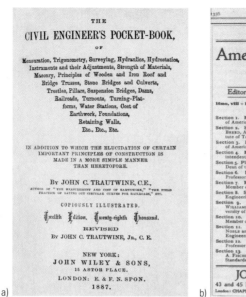
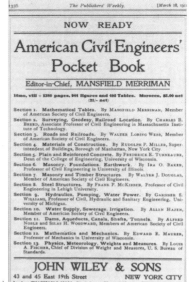

**FIGURE 3-16**
a) The 12th edition of Trautwine's *Pocket-Book* and b) advertisement by the publisher John Wiley & Sons for Merriman's *Pocket-Book* in *The Publisher's Weekly*

*Engineer's Pocket-Book* [Trautwine, 1872] and went through an impressive 17 editions by 1894. From the 1870s onwards, the former civil engineer William Halsted Wiley (1842–1925) had put the publishing house (which traded under the name John Wiley & Sons after 1875) among the leading publishers of technical and scientific books and journals in the USA, and this fact contributed to the success of Trautwine's *Pocket-Book* (Fig. 3-16a). William Halsted Wiley replaced Trautwine's book in 1911 by one written by Mansfield Merriman (1848–1925): *The American Civil Engineer's Pocket-Book* [Merriman, 1911] (Fig. 3-16b). By 1930 Merriman's *Pocket-Book* had seen five editions. Quite rightly, the Rensselaer Polytechnic Institute honoured its former civil engineering student William Halsted Wiley with the title "Giant of Scientific and Technical Publishing"!

Nevertheless, we should not forget that the several editions of Weisbach's engineering manual founded this literary genre.

**The journal** **3.2.3.3**

Georg Zöllner suspects that the observations and impressions gained by Weisbach on his trip to Paris in 1839 inspired him to found an engineering journal together with his colleagues from Freiberg Bornemann, Brückmann and Röting [Zöllner, 1956, p. 30]. The planning work got underway in 1846 and came to fruition in 1848 in the form of the journal *Der Ingenieur. Zeitschrift für das gesammte Ingenieurwesen* (the engineer – journal for all forms of engineering). During the first year of publication, Weisbach published his very significant paper *Die Theorie der zusammengesetzten Festigkeit* (theory of composite strength) [Weisbach 1848/1]. The journal was renamed *Der Civilingenieur. Zeitschrift für das Ingenieurwesen* (the civil engineer – journal of engineering) in 1853 and publication continued with Weisbach's pupil Gustav Anton Zeuner (1828–1907) as chief editor, becoming the mouthpiece of the Sächsischer Ingenieur- & Architekten-Verein. Weisbach published 29 out of his total of 59 journal articles in *Der Civilingenieur*, including in 1863 and 1868 his exemplary teaching methods for applied mechanics.

### 3.2.3.4 Strength of materials in Weisbach's *Lehrbuch*

Strength of materials appears as part of the section "Statics of Rigid Bodies" [Weisbach, 1855, pp. 146–468] in the chapter "Elasticity and Rigidity" [Weisbach, 1855, pp. 305–468] in volume I. Elasticity in the broader sense of the word is defined by Weisbach as the ability of a body that undergoes deformation when acted upon by forces to restore itself fully after removing those forces. He regards elasticity in the more precise meaning of the word as "the resistances which a body opposes to change of form", and on the other hand, strength is "the resistance which a body opposes to a separation of its parts" [Weisbach, 1855, p. 306]. Weisbach therefore clearly differentiates in a terminological sense between the serviceability and ultimate states. Depending on how external forces act on bodies and change these in spatial relationships, we can divide the elasticity and strength of bodies into

    I. simple, and
    II. composite,
and the first of these further into
    1. absolute or tensile,
    2. reactive or compressive,
    3. relative or bending, and
    4. torsional or rotational elasticity and strength [Weisbach, 1855, p. 306].

Weisbach's revolutionary contribution to strength of materials is the introduction of the composite elasticity and strength comprising the simple actions 1 to 4 which occur most frequently in practical engineering.

Weisbach deals with the following strength of materials topics: simple actions, material parameters, beam theory (cantilever and simply supported beams – including statically indeterminate examples, beams with customary cross-sections, beams of equal strength), torsion theory, composite strength, buckling theory and beams in tension according to second-order theory (cantilevers). Weisbach, with his numerous experimental findings prepared conveniently in tables, rapidly changed the face of strength of materials, and introduced remarkable innovations.

### Bending strength

Weisbach was the first person to use the graphical presentation of bending moments when dealing with the various support and loading cases (Fig. 3-17).

Weisbach's occasional errors are revealed by the bending moment diagram for the single-span beam fixed at both ends and subjected to a uniformly distributed load (Fig. 3-17b). Although Weisbach's equations for the fixed-end and span moments are correct, the curve in the drawing is not. He used implicitly the fact that the curvatures $1/R(x)$ and bending moments $M(x)$ at points $D$ and $E$ must be equal to zero, which had already been specified in Eytelwein's linearised differential equation for the elastic curve $y(x)$

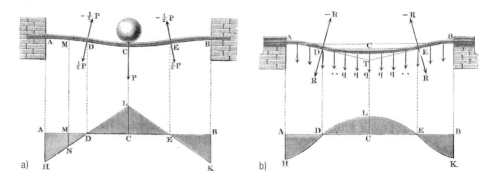

**FIGURE 3-17**
a) Bending moment diagram for a single-span beam fixed at both ends and subjected to a central point load P, and b) a uniformly distributed load q [Weisbach, 1855, pp. 410/411]

$$E \cdot I \cdot \frac{d^2 y(x)}{dx^2} = -M(x) = \frac{1}{R(x)} \qquad (3\text{-}8)$$

However, this does not infer that the points $D$ and $E$ on the bending moment diagram $M(x)$ are also points of contraflexure – as Weisbach shows in his drawing. We can see that this is not correct by considering the double integration of the differential equation

$$\frac{d^2 M(x)}{dx^2} = q(x) \qquad (3\text{-}9)$$

Here, $q(x) = q = $ const. results in a bending moment diagram $M(x)$ in the shape of a quadratic parabola and not a 4th order polynomial with points of contraflexure at $D$ and $E$. The reason for this incorrect drawing is that Weisbach used the means of elementary mathematics, which do not specify the curve of the function for $M(x)$, and he was unaware of the differential equation 3-9. Nonetheless, he derived $M(x)$ correctly for statically indeterminate single-span beams subjected to a point load (Fig. 3-17a).

### Tensile strength

In contrast to Gerstner, Weisbach specifies force-deformation diagrams for the most diverse materials both qualitatively (Fig. 3-18a) and quantitatively (Fig. 3-18b). In Fig. 3-18 the X-axis represents the force $F$ and the Y-axis the elongation or compressive strain $\Delta l / l$. Interesting in this respect is that Weisbach interprets the area enclosed by $AON$ or $AO_1N_1$ as deformation energy in the meaning of Clapeyron (Fig. 3-18a) and thus for the first time popularised the energy principle in strength of materials. Fig. 3-18b shows as an example the characteristic force-deformation dia-

**FIGURE 3-18**
Force-deformation diagrams after Weisbach: a) schematic [Weisbach, 1855, p. 313], b) to scale for timber, cast iron and wrought iron [Weisbach, 1855, p. 331]

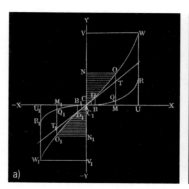
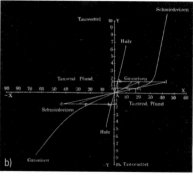

gram for wrought iron with the limit of proportionality and the yield zone. Weisbach specifies the elastic range of stress (where Hooke's law is obeyed) for all materials and focuses on the elastic modulus in his observations. He is thus able to list in a table the elastic moduli, the elastic limits and the tensile and compressive strengths for diverse materials [Weisbach, 1855, pp. 335–337]. In volume II, Weisbach uses applied mechanics to solve earth pressure, masonry arch and timber and iron construction problems [Weisbach, 1846, pp. 5–118]. Weisbach concentrates on mechanical engineering and hydraulics even more than Gerstner.

### 3.2.4 Rankine's *Manuals*, or the harmony between theory and practice

Rankine's *Manuals* covering civil and mechanical engineering influenced applied mechanics and thermodynamics until well into their consolidation period (1900–50). In 1864 he divided his *Manuals* into four independent parts (Fig. 3-19):

- Applied Mechanics [Rankine, 1858]
- The Steam Engine and Other Prime Movers [Rankine, 1859]
- Civil Engineering [Rankine, 1862]
- Useful Rules and Tables [Rankine, 1866/1].

This structuring of classical engineering sciences enabled Rankine to complete the divorce between civil and mechanical engineering and therefore went far beyond Weisbach's concept. The distinction in the publicised representation of engineering is also remarkable in formal terms. Whereas Weisbach employed the engineering manual and the engineering journal for the first time alongside the textbook type of publication, Rankine remained faithful to the engineering manual and presented the unity of classical engineering sciences formerly in the shape of his *Manuals* (Fig. 3-19). He therefore set the standard for the "instruction manual" type of publication in British engineering literature.

In 1858 Rankine introduced the term "applied mechanics" into the English language in the 1st edition of his monograph *A Manual of Applied Mechanics*: "The branch to which the term *'Applied Mechanics'* has been restricted by custom, consists of those consequences of the law of mechanics which relate to works of human art. A treatise on applied mechanics must commence by setting forth those first principles which are common to all branches of mechanics; but it must contain only such consequences of those principles as are applicable to purposes of art" [Rankine, 1858, p. 13]. A total of 21 English editions of this book appeared, the last of them in the 1920s. However, even this remarkable feat was exceeded by Rankine's *A Manual of Civil Engineering*, which was first published in 1862 [Rankine, 1862] and had already reached the 24th edition by 1914! Both *Manuals* were translated into several languages.

**FIGURE 3-19**
Advertisement for Rankine's *Manuals* by the Charles Griffin publishing house dating from 1864

#### 3.2.4.1 Rankine's *Manual of Applied Mechanics*

Rankine's *A Manual of Applied Mechanics* was published within the scope of the 2nd edition of the *Encyclopaedia Metropolitania* (1849–58) (Fig. 3-20a). The 1st edition of the *Encyclopaedia Metropolitania* had been published in 28 volumes and 59 parts between 1817 and 1845 and was based

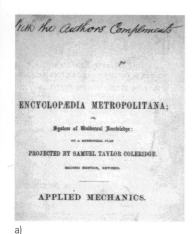

**FIGURE 3-20**
a) *Applied Mechanics* as published in the *Encyclopaedia Metropolitania* with Rankine's handwritten dedication, and b) the title sheet thereto

on suggestions by England's leading advocate of romantic literature at that time, Samuel Taylor Coleridge (1772–1834). It is divided into

I. Pure Sciences
II. Mixed and Applied Sciences
III. History and Biography
IV. Miscellaneous.

Unlike rival works, the *Encyclopaedia Metropolitania* was in the end unable to assert itself because its portrayal of technical, natural science and engineering science knowledge was on the whole poor and gave precedence to the humanities. Coleridge's *Encyclopaedia Metropolitania* therefore represents a counter-project to Diderot's *Encyclopédie*, which was dedicated to enlightenment and whose core was the applied arts, i. e. engineering. Even Rankine's *A Manual of Applied Mechanics* (Fig. 3-20b) appeared like foreign matter in the body of the *Encyclopaedia Metropolitania*, dedicated as it was to the harmony between theory and practice in mechanics. This is evident in Rankine's introduction "Preliminary Dissertation on the Harmony of Theory and Practice in Mechanics" [Rankine, 1858, pp. 1–11], where he develops very clearly the independence of applied mechanics as a fundamental engineering science discipline. The introduction corresponds to the written edition of his inaugural lecture (1855) at the University of Glasgow's Chair of Civil Engineering and Mechanics (established in 1840).

In his lucid historico-critical time-line analysis, Rankine traces the relationship between theory and practice in mechanics from the time of Aristotle right up to the middle of the 19th century. He comes to the conclusion that theory and practice harmonise in the form of applied mechanics: "Theoretical and Practical Mechanics are in harmony with each other, and depend on the same first principles, and that they differ only in the purposes to which those principles are applied, it now remains to be considered, in what manner that difference affects the mode of instruction to be followed in communicating those branches of science" [Rankine, 1858, p. 8]. Between the practical and theoretical knowledge of mechanics, Rankine sees engineering science knowledge as an independent type of knowledge: "Mechanical knowledge may obviously be distinguished into three kinds: purely scientific knowledge, – purely practical knowledge – and that intermediate knowledge which relates to the application of scientific principles to practical purposes, and which arises from understanding the harmony of theory and practice" [Rankine, 1858, p. 8].

The independence of engineering science knowledge was Rankine's *raison d'être* for the Chair of Civil Engineering and Mechanics at the University of Glasgow: "The third and intermediate kind of instruction, which connects the first two, and for the promotion of which this Chair was established, relates to the application of scientific principles to practical purposes" [Rankine, 1858, p. 8]. The engineering science knowledge taught to the students enables them, when designing structures and machines, to

- "compute the theoretical limit of the strength or stability of a structure, or the efficiency of a machine of a particular kind …,
- ascertain how far an actual structure or machine fails to attain that limit …,
- discover the causes of such shortcomings …,
- devise improvements for obviating such causes …, and …
- judge how far an established practical rule is founded on reason, how far on mere custom, and how far on error" [Rankine, 1858, p. 9].

Accordingly, Rankine divided mechanics into "Pure Mechanics" – consisting of kinematics, dynamics and statics – and "Applied Mechanics", which is concerned with machines and structures (Fig. 3-21).

Applied mechanics transforms kinematics into machine kinematics, dynamics into machine dynamics, statics into machine statics on the one hand and theory of structures on the other. If in machine kinematics we are to concentrate only on the analysis of machine motion, then machine dynamics and machine statics take into account the forces acting on the machine. Examining the equilibrium of the forces in building structures, on the other hand, is the task of theory of structures. This is why Rankine divided his *Manual of Applied Mechanics* into six parts:

I. Principles of Statics
II. Theory of Structures
III. Principles of Cinematics, or the Comparison of Motions
IV. Theory of Mechanisms
V. Principles of Dynamics
VI. Theory of Machines.

Rankine further divided each part into chapters and these in turn into sections. For example, Part II (Theory of Structures) consists of a brief chapter on definitions and general principles (2 pages), a chapter on the stability of building constructions (139 pages) and a chapter on the strength and stiffness of building constructions (109 pages). Out of a total of 630 pages, Part II (Theory of Structures) accounts for 250. This part contains pioneering contributions on masonry arch, earth pressure and beam theories. For example, Rankine defines the fundamental principles stress $\sigma$ and strain $\varepsilon$ [Rankine, 1858, p. 270] for the first time in English-language engineering literature. Furthermore, he quantifies the influence of shear stresses $\tau$ on the deflection of elastic beams, and for a beam on two supports subjected to a uniformly distributed load, Rankine estab-

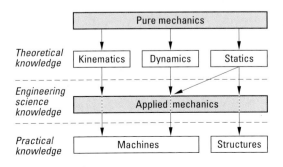

FIGURE 3-21
The knowledge system of mechanics after Rankine

lished that in practice the shear stresses have a negligible influence on deflection [Rankine, 1858, p. 342].

Rankine's readily understandable description of the energy principle of elastic theory formulated by George Green (1793–1841) represented a major leap forward in establishing the energy-based imperative in theory of structures during the establishment phase of structural theory (1850–75). His breakdown of applied mechanics into machine mechanics, machine dynamics and theory of structures speeded up the historico-logical process of development in the aforementioned engineering science subdisciplines.

### 3.2.4.2 Rankine's *Manual of Civil Engineering*

The *Manual of Civil Engineering* published in 1862 is the prototype of the civil engineering handbook, a publication which even today is an important aid for the civil engineer. Modern examples are the *Standard Handbook for Civil Engineers* [Ricketts et al., 2003] and *Handbuch für Bauingenieure* [Zilch et al., 2002], to name just two. Such civil engineering manuals present the whole gamut of civil engineering knowledge relevant to practical issues, mostly arranged in subdisciplines such as structural steelwork, reinforced concrete, highways and hydraulic engineering. The differentiation of civil engineering theory in its classical phase (1875–1900) also forced the authors of civil engineering manuals to divide up their work. Whereas Rankine was still personally responsible for every sentence, every illustration and every table in his *Manual of Civil Engineering*, the emergence of subdisciplines made it necessary to appoint different authors under the auspices of one editor or a group of editors. A typical example of this is the *Handbuch des Brückenbaus* (manual of bridge-building), which is described in section 3.2.6.

Rankine divided his *Manual of Civil Engineering* (Fig. 3-22) into three parts:

I. Engineering Geodesy, or Field-work
II. Materials and Structures
III. Combined Structures.

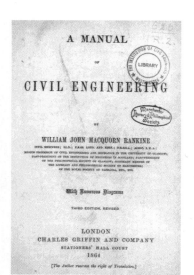

FIGURE 3-22
The 3rd edition of Rankine's *Manual of Civil Engineering*

In his preface, Rankine explains the reason for subdividing the book into three: "This work is divided into three parts. The first relates to those branches of the operations of engineering which depend on geometrical principles alone … The second part relates to the properties of the *materials* used in engineering works, such as earth, stone, timber, and iron, and the art of forming them into *structures* of different kinds, such as excavations, embankments, bridges, & c. The third part, under the head of *combines structures*, sets forth the principles according to which the structures described in the second part are combined into extensive works of engineering, such as Roads, Railways, River Improvements, Water-Works, Canals, Sea Defences, Harbours, & c" [Rankine, 1862, p. V].

Out of a total of 783 pages, Part II, covering statics, strength of materials and constructional disciplines, accounts for 487 pages – roughly 62 % of the whole book. Part II deals with the following subjects: principles of stability and strength of loadbearing structures, earthworks, masonry, timber and metal structures, tunnels and foundations. This list covers the most important areas of civil engineering, which in the final 25 years of the 19th century developed into independent subdisciplines in classical civil engineering theory.

### 3.2.5 Föppl's *Vorlesungen über technische Mechanik*

The hitherto most influential textbook on applied mechanics in Germany was written by August Föppl (1854–1924) during his time in Munich (1894–1924). The six-volume *Vorlesungen über technische Mechanik* (lectures on applied mechanics) (1897–1910) appeared in numerous editions and was translated into several languages [Föppl, 1897–1910], and by 1925 more than 100 000 individual copies had been sold. This comprehensive work was complemented in 1920 by the two-volume *Drang und Zwang* (pressure and restraint), which he compiled together with his son Ludwig Föppl (1887–1976) [A. Föppl & L. Föppl, 1920]. In 1917 he succeeded in extending the St. Venant torsion theory, to which Constantin Weber (1885–1976) would add more later. Föppl's most important pupil was Ludwig Prandtl (1875–1953), who gained his doctorate with Föppl in 1899 with a dissertation on the lateral buckling of beams with slender rectangular cross-sections. Any history of the teaching of applied mechanics in German universities would have to analyse, in particular, the textbooks of Franz Joseph Ritter von Gerstner, Julius Weisbach, August Föppl, Otto Mohr and István Szabó. But one thing is clear: only the mechanics books of Julius Weisbach and August Föppl managed to attain the level of "mass communication media". No wonder August Föppl established the most influential school of applied mechanics in the early 20th century, the teachings of which had a lasting effect on the, at the time, young Stepan P. Timoshenko (1878–1972): "The best textbooks dealing with mechanics and strength of materials," writes Timoshenko in his *Memoirs*, "were at that time those of August Föppl" [Timoshenko, 2006, p. 90].

#### 3.2.5.1 The origin and goal of mechanics

The first volume of Föppl's *Vorlesungen* (Fig. 3-23) contains an introductory chapter on the origin and goal of mechanics [Föppl, 1898, pp. 1–12]. This introduction is not only a self-assured scientific and epistemological review of mechanics, but also a programme for applied mechanics which was set to dominate academic activity in this field in the first half of the 20th century.

Föppl's introduction begins with the terse remark: "Mechanics is part of physics" [Föppl, 1898, p. 1]. In the second sentence he comments that the lessons of mechanics, like all other natural sciences, are, in the end, based on experience. And by the third sentence Föppl has already demarcated mechanics from experimental physics: "Our task is not, however, to demonstrate here in detail all the observations that can be obtained from planned experimentation and which were used originally for the consolidation of mechanics" [Föppl, 1898, p. 1]. That is the task of experimental physics.

Using the example of the practical thoughts and actions of a factory manager, Föppl continued to develop the difference between mechanics and experimental physics into "the degree of difference between the theories of science and the theoretical views that the true practitioner prepares consciously or subconsciously, that he tries to summarise the pool of knowledge, for which we are grateful for the work of the researchers of all ages, in the most sensible order. The more the material as a whole piles up through the unceasing work, all the more necessary will it be to look for guiding concepts to improve the overview and save us the effort of recording many individual facts separately in our minds" [Föppl, 1898, p. 3]. Here, Föppl warns us that "the direct comparison of theoretical views and the facts of reality is unduly neglected" [Föppl, 1898, p. 3] and calls for "scholars of science to practise the highest form of study" [Föppl, 1898, p. 4]. Thereupon, Föppl formulates two epistemological questions:
- How is it at all possible to predict the course of a natural event through logical deduction?
- From where does the bond come "that ties the laws of our thoughts with the laws of the reality outside us so closely that both lead to the same results?" [Föppl, 1898, p. 4]

Föppl's answers must have surprised his contemporaries. It is not the human spirit that has coerced nature or conquered its secrets for itself, but rather "constant occurrences according to fixed rules in nature have – I hesitate to say conquered, but it is so – steered, urged and compelled it (the human spirit) until it was capable of accepting a picture of the outside world" [Föppl, 1898, p. 7].

Where does Föppl see the difference between mechanics and applied mechanics? Firstly, "to use their lessons advantageously in engineering ... The more important reason for treating applied mechanics separately as a special branch of science is that the generally applicable lessons of mechanics are in no way adequate for solving strictly and accurately all the questions that can arise in the realm of mechanics" [Föppl, 1898, p. 11]. If the practising engineer is confronted by some restrictive or helpful phenomenon, he must "prepare a theoretical opinion of this as good as he can" [Föppl, 1898, pp. 11–12]. The engineer therefore initially seeks solutions that "in applied mechanics for the present are grouped under the heading of approximation theories" [Föppl, 1898, p. 12], and are incorporated into the body of mechanics afterwards – if this is at all possible. "However, he who first establishes the right theory of a process, may – if an intervention be at all possible – lead as he wishes, and that is why science is the most potent weapon at the disposal of man and peoples" [Föppl, 1898, p. 12]. That is the closing sentence of Föppl's introduction.

How did Timoshenko, in his *Memoirs*, describe Föppl's *Festigkeitslehre* (strength of materials) [Föppl, 1897], which appeared in 1897 as the third volume of his *Vorlesungen*? "Föppl provided a theoretical introduction to the subject, and I liked that" [Timoshenko, 2006, p. 86]. So Föppl's view of mechanics inspired the scientific thinking of the most important representative of applied mechanics in the 20th century.

**FIGURE 3-23**
The first volume of Föppl's *Vorlesungen*

### 3.2.5.2 The structure of the *Vorlesungen*

Föppl's *Vorlesungen* reflects the level of applied mechanics reached in German universities, but at the same time sets new standards for university teaching practices in this fundamental engineering science discipline. The book was initially intended for students attending Föppl's lectures at Munich Technical University and practising engineers with a university education "who have not totally forgotten how to differentiate ... Admittedly, we repeatedly hear the claim that engineers in practice know how to apply only elementary mathematics; however, expressed in this way, I cannot really believe this claim. In my opinion, elementary mathematics is harder to learn and also harder to remember than those simple segments of differential and integral calculus normally required for applications" [Föppl, 1897, p. VI]. This is based on the fact that it was already obvious to Föppl that operating in the language of calculus is simpler than operating with the symbols of algebra. Whereas the latter uses symbols that stand for numbers (second historico-logical stage of the practical use of symbols), calculus represents a formalised language where symbols have only an intrasymbolic meaning – i.e. are on the level of the non-interpretative use of symbols (third historico-logical stage of the practical use of symbols). Was it a surprise, then, when Föppl used the linguistic power of vector calculus in applied mechanics and electrical engineering for the first time? Nonetheless, he always considers analytical developments "as merely a means of perceiving the internal relationships between facts" [Föppl, 1897, p. VII]. Föppl had thus composed a counterpoint to Weisbach's *Lehrbuch*. To enhance the usability of his six-volume *Vorlesungen*, Föppl appended thoroughly-worked-out assignments to each section.

Föppl's *Vorlesungen* is divided into

- Introduction to mechanics (volume I) [Föppl, 1898]: introduction, point mechanics, rigid body mechanics, theory of centre of gravity, energy transformations, friction, elasticity and strength, hydromechanics
- Graphical statics (volume II) [Föppl, 1900]: addition and resolution of plane forces, funicular polygons, forces in space, plane frames, space frames, elastic theory of frameworks, masonry arches and continuous beams
- Strength of materials (volume III) [Föppl, 1897]: general stress condition, elastic material behaviour, bending theory of straight beams, deformation work, bending theory of curved beams, elastically bedded beams, plates, vessels, torsional strength, buckling strength, principles of mathematical elastic theory
- Dynamics (volume IV) [Föppl, 1899]: point dynamics, dynamics of rigid bodies and scatter diagrams, relative motion, dynamics of combined systems, hydrodynamics,
- The most important lessons of higher elastic theory (volume V) [Föppl, 1907]: stress condition and risk of failure plus moments of masses, plate and slab theory, torsion of prismatic bars, rotationally symmetrically loaded cylindrical shells and thermal stresses,

principles of deformation work and residual stresses, various applications
- The most important lessons of higher dynamics (volume VI) [Föppl, 1910]: relative motion, equations of motion for mechanical systems with several degrees of freedom, gyroscopes, various applications, hydrodynamics.

The first four volumes constitute the entire content of Föppl's lectures that he delivered to students of civil and mechanical engineering at Munich Technical University between the first and fourth semesters; the latter two volumes "were intended to satisfy further needs" [Föppl, 1910, p. III]. Föppl's *Vorlesungen* constituted not only the applied mechanics curriculum in German universities in the 20th century, but also shaped the content and form of applied mechanics during its consolidation period (1900 – 50).

### 3.2.6 The *Handbuch der Ingenieurwissenschaften* as an encyclopaedia of classical civil engineering theory

Like the steam locomotive, the bridge carrying the locomotive is also an energy-based machine (see section 6.1.3). Both the steam locomotive and the iron railway bridge are symbols of the Industrial Revolution. It is therefore no surprise to discover that after 1850 the civil engineer was first and foremost a railway engineer responsible for the design, calculation and construction, but also the operation and technical supervision of the railway network. The encyclopaedia of classical civil engineering theory is essentially an encyclopaedia of modern transport dominated by railways.

The *Handbuch der Ingenieurwissenschaften* (manual of engineering sciences) compiled in the 1880s covers the routing of railway lines, bridge-building, tunnelling, railway stations and other associated buildings, and railway operations engineering among its many chapters. The material is divided among four volumes which are in turn split into several parts. The 2nd edition of the *Handbuch des Brückenbaus* (manual of bridge-building) (Fig. 3-24) by Schäffer, Sonne and Landsberg, for instance, became volume II of the *Handbuch der Ingenieurwissenschaften* and appeared between 1886 and 1890 in five parts covering the following areas:

*Part 1* [Schäffer & Sonne, 1886]:
    I. Bridges generally (T. Schäffer & E. Sonne)
    II. Stone bridges (F. Heinzerling)
    III. Building and maintenance of stone bridges (G. Mehrtens)
    IV. Timber bridges (F. Heinzerling)
    V. Aqueducts and canal bridges (E. Sonne)
    VI. Artistic forms in bridge-building (R. Baumeister)

*Part 2* [Schäffer et al., 1890]:
    VII. Iron bridges generally (J. E. Brik & T. Landsberg)
    VIII. The bridge deck (F. Steiner)
    IX. Theory of iron beam bridges (F. Steiner)
    X. Construction of iron beam bridges (F. Steiner)

*Part 3* [Schäffer & Sonne, 1888/1]:
    XI. Moving bridges (W. Fränkel)

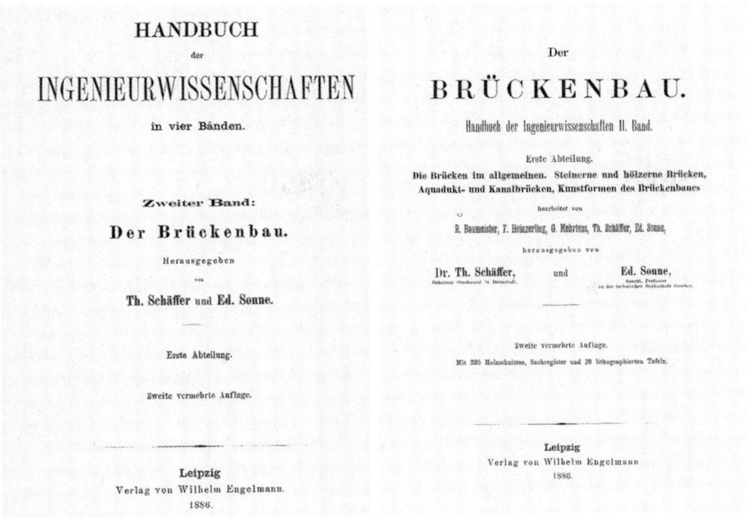
HANDBUCH
der
INGENIEURWISSENSCHAFTEN
in vier Bänden.

Zweiter Band:
Der Brückenbau.
Herausgegeben
von
Th. Schäffer und Ed. Sonne.

Erste Abteilung.
Zweite vermehrte Auflage.

Leipzig
Verlag von Wilhelm Engelmann.
1886.

Der
BRÜCKENBAU.
Handbuch der Ingenieurwissenschaften II. Band.

Erste Abteilung.
Die Brücken im allgemeinen. Steinerne und hölzerne Brücken, Aquadukt- und Kanalbrücken, Kunstformen des Brückenbaues
bearbeitet von
R. Baumeister, F. Heinzerling, G. Mehrtens, Th. Schäffer, Ed. Sonne,
herausgegeben von
Dr. Th. Schäffer, und Ed. Sonne,

Zweite vermehrte Auflage.
Mit 226 Holzschnitten, Sachregister und 28 lithographierten Tafeln.

Leipzig
Verlag von Wilhelm Engelmann
1886.

**FIGURE 3-24**
The *Handbuch des Brückenbaus* (1886)

*Part 4* [Schäffer & Sonne 1888/2]:
XII. Theory of iron arch bridges and suspension bridges (J. Melan)
XIII. Construction of suspension bridges (J. Melan)
XIV. Construction of iron arch bridges (T. Schäffer & J. Melan)

*Part 5* [Schäffer & Sonne, 1889]:
XV. The iron bridge pier (F. Heinzerling)
XVI. Building and maintenance of iron bridges (W. Hinrichs).

This encyclopaedia of bridge-building with its 16 chapters denoted with Roman numerals is not only a record of the evolution of the respective structures on an international scale; far more than that, the chapters link the historical with the logical developments in the analysis using the current art and epistemic of the bridge-builder. All the chapters with their many wood engravings are uniform in structure and there is an extensive international bibliography so that the reader can delve deeper into each subject. The formal connections between the chapters of each part are realised through cross-references, an index and an atlas with lithographic plates. The tree-like arrangement of the engineering knowledge in which art and epistemic are networked cooperatively in many ways corresponds in its formal layout to the scientific theory of enlightenment pursued in Diderot's *Encyclopédie*.

The link between the five parts of the encyclopaedia is conveyed by presenting classical theory of structures as a theory of bridge-building. Nonetheless, the authors create reminders of the schism of architecture – albeit in the prevailing positive language of the close of the 19th century: "The relationships between bridge-building and architecture manifest themselves in particular in stone bridges, for obvious reasons, whereas it is primarily the iron bridges that create the bond between bridge-building and the fundamental sciences of engineering, especially mechanics. In particular, since the widespread use of wrought iron and steel has rendered possible structures of amazing size and boldness, it has become ever more necessary to provide a sound basis for the forms and dimensions of iron bridges by way of calculation. This has led to the formation of a special branch of science – known as engineering mechanics or the theory of constructions [theory of structures – the author] – which has contributed significantly to the mathematical sciences just as much as the engineering sciences" [Schäffer & Sonne, 1886, pp. 1–2]. In the opinion of the authors, the theory of iron constructions is a constituent part of classical structural theory, whereas the masonry arch and earth pressure theories form more of an appendix to classical structural theory. And architecture, as a beautiful art, compared to theory of structures as a mechanical one, is assigned only the role of a niche player in bridge-building.

On the other hand, graphical analysis brings "significant simplifications" to construction theory, the authors continue [Schäffer & Sonne, 1886, p. 2]; they refer here to the rationalisation of engineering work in design and construction activities. Nevertheless, the authors could not decide how to present succinctly the engineering science foundations of bridge-building by way of classical structural theory in the encyclopaedia. They write: "The theoretical foundation of the constructions is now so indispensable in many cases and so intrinsic to the design of the same that it would have been highly desirable to base the following discussion on bridge-building on a coherent treatment of the theory of construction. As, however, a corresponding expansion of our manual continuing the established division of work would have been difficult to reconcile and therefore is not intended, we had no choice but to insert the scientific foundation necessary for the constructions at suitable points in this work" [Schäffer & Sonne, 1886, p. 2].

The fact that the engineering science foundation for bridge-building is to be found in those areas where they originated (Fig. 3-5f) and where they have to prove themselves in the future (Fig. 3-5e) shows that the objective reality of the development of the cooperation between engineering sciences and science-based engineering had already overtaken the thinking of the authors. The perspective of the structural engineer's work is then successful only if he operates in the two-dimensional field of engineering-based natural science and natural science-based engineering, and consequently the relationship between the engineering science object and the engineering artefact is identified and functionalised for the engineering work.

**Iron beam bridges**  **3.2.6.1**

The expansion of knowledge in the classical phase of structural theory (1875–1900) and its summation in the classical theory of structures prompted the authors of the *Handbuch des Brückenbaus* to dedicate a chapter to the theory of iron beam bridges. Both this chapter and the one on the construction of iron bridges were written by Friedrich Steiner. Steiner, formerly assistant to Emil Winkler and thereafter professor at the German Technical University in Prague, based the theory of statically indeterminate beam systems on the energy doctrine. The mathematical background to classical structural theory is clearly evident in the development of the theory: linear algebra. With the help of this, Steiner provided solutions to sets of equations as occur in statically indeterminate systems with several degrees of indeterminacy [Schäffer et al., 1890, pp. 246-250], so that performing the statically indeterminate calculation is mere art, but the use of symbols is schematic and non-interpretative. Admittedly, the symbols before and after performing the statically indeterminate calculation must have a physical basis and interpretation.

The three steps become clear through the example of the "highly comprehensive and difficult calculation for the Kaiser Franz Joseph Bridge in Prague" [Schäffer et al., 1890, p. 309], which Steiner carried out in 1883 and 1884 for the Prague city authorities. The loadbearing structure consists of three continuous wrought-iron plate girders supported by steel chains whose forces stemming from the change of direction were carried down to the foundations as compressive forces via cast-iron pylons (Fig. 3-25).

The system has seven degrees of static indeterminacy. In concrete terms, the three steps, consisting of the allocation of physical variables to the symbols for the static system (semantic burden) and their formation into a set of equations (semantic relief), the non-interpretative transformation of the symbols and their physical re-interpretation (semantic burden), are as follows:

*First step:* Present the funicular forces as a linear function of the static indeterminates. Set up the seven elasticity equations using the principle of minimum deformation energy with the help of influence lines for the deflection due to a vertical travelling load according to type, location and direction of the corresponding static indeterminates on the statically determinate basic system.

*Second step:* Solve the set of equations.

*Third step:* Calculate the internal forces in the statically indeterminate system (e.g. funicular forces) for various loading cases.

Steiner does not neglect to mention the pleasing agreement between his calculated values and the data obtained experimentally on the bridge itself (force and deflection measurements, also test loads). Nevertheless, shortly after the bridge was handed over in 1888, additional measures to guarantee its stability had to be considered.

The operative and non-interpretative treatment of symbols in the three steps of the statically indeterminate calculation would reach new heights with the δ-symbols introduced by Heinrich Müller-Breslau and spread internationally by the Berlin school of structural theory, in which calculus would be introduced for the first and third steps, too (see section 6.5).

### 3.2.6.2 Iron arch and suspension bridges

The 1st edition of the part covering iron and suspension bridges was completely reworked for the 2nd edition by a new author: Joseph Melan (1853–1941). Like Steiner, Melan had also worked as assistant to Winkler. In 1886 he had been appointed professor at Brünn Technical University and after Steiner's death in 1902 he became his successor and remained true to the German Technical University in Prague (the university founded in 1806 by Franz Joseph Ritter von Gerstner) until his transfer

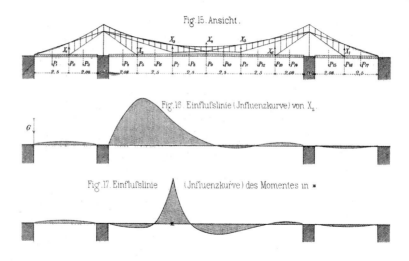

**FIGURE 3-25**
Diagrams for the structural analysis of the Kaiser Franz Joseph Bridge in Prague after Steiner
[Schäffer et al., 1890, plate V]

to emeritus status. It was here that Melan progressed to become one of the leading bridge-builders working in the German language. He earned lasting praise for providing bridge-building with a scientific foundation. The fourth part of the *Handbuch des Brückenbaus*, in which he was instrumental, can be regarded as the testing point of this work [Schäffer & Sonne, 1888/2]. He anticipated the style of theory of the consolidation phase of scientific bridge-building after 1900.

As an example, the initial assumptions of second-order theory (equilibrium of the deformed system) can serve in the analysis of the polygon of bars stiffened by a straight beam. Melan recognised that the influence of the deformation reduces the bending moment in the beam of a suspension structure (Fig. 3-26a). Thus Melan had for the first time opened up a breach in classical structural theory, the logical core of which is characterised by the trinity of the linearity of material behaviour, deformed state and force condition. As the relationship between the load and the internal forces is not linear, the law of superposition loses its validity. Melan's conclusion: "The method of influence lines cannot therefore be applied; rather, the horizontal force and the moments and the transverse forces [= shear forces – the author], which cause the internal stresses, must be calculated separately for each loading case [Schäffer & Sonne, 1888/2, p. 42].

Melan's perception was not unique and this is revealed in the paragraph on the "more exact theory of arched beams taking into account the deformation caused by the load" [Schäffer & Sonne, 1888/2, pp. 100–101]. Here, too, he observes that the elastic arch theory of classical structural theory is merely an approximation theory which only supplies satisfactory values for the internal forces and horizontal thrust for the subsequent arches of adequate stiffness. For very shallow arches with comparatively low stiffness, he specifies an iteration method in order to determine the additional bending moment due to the influence of the deformation. Such arches did not become a subject in theory of structures until the appearance of long-span solid-web arch bridges in the 1930s [Fritz, 1934].

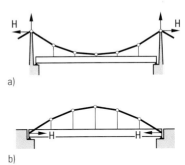

**FIGURE 3-26**
A polygon of bars stiffened by a straight beam: a) stiffened suspension structure, b) arch structure

The cooperation between engineering based on natural science (Fig. 3-5e) and natural science based on engineering (Fig. 3-5f) would become a necessary prerequisite, especially in the building of long-span arch and suspension bridges. The break-up of the linear structure of theory of structures initiated by Melan prompted a development that these days, in the form of non-linear structural mechanics, increasingly determines the everyday workloads of structural engineers.

# CHAPTER 4

# From masonry arch to elastic arch

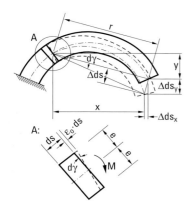

The masonry arch is still one of the mysteries of architecture. Anybody who looks into the history of theory of structures quickly encounters this puzzle, the solution to which has occupied countless numbers of scientists and engineers right up to the present day. Since completing his diploma at the theory of structures department of Berlin TU in 1981, the author can be counted as belonging to that group. These studies introduced him to Jacques Heyman's work on the history of theory, which the latter interprets in his ultimate load theory masonry arch model. A lecture given at the faculty of civil engineering at the University of Stuttgart, which had been instigated by Prof. Ekkehard Ramm, resulted in a work summarising the development of masonry arch theories since Leonardo da Vinci – and forms the crux of this chapter. Section 4.1.1 was written by Andreas Kahlow and section 4.1.2 by Holger Falter; I would like to take this opportunity of thanking both of them for their kind permission to reproduce their work in this book. The excellent researches of Antonio Becchi, Federico Foce and Santiago Huerta contributed to the success of sections 4.2.1, 4.3.1, 4.3.7 and 4.6; friendships grew out of our many years of cooperation in the fields of history of science and history of construction engineering. The author's dream of a historical theory of structures within the framework of a historical engineering science took shape through the works of the aforementioned researchers.

Jakob Grimm (1785–1863) and Wilhelm Grimm (1786–1859) describe the German noun *Bogen* (= bow, curve, arch) as "… that which is curved, is becoming curved, is rising in a curve" [Grimm, 1860, p. 91], the roots of which lie in the German verb *biegen* (= to bend). A bow (i.e. arch, from *arcus*, the Latin word for arc, bow) in the structural sense is consequently a concave loadbearing structure whose load-carrying mechanism is achieved by way of rigid building materials such as timber, steel and reinforced concrete. When loading such a curved loadbearing structure, a not inconsiderable part of the external work is converted into internal bending work. Therefore, in German the verb *biegen* not only constitutes the etymological foundation for the noun *Bogen*, but also characterises the curved loadbearing structure from the point of view of the load-carrying mechanism in a very visual and impressive way.

The genesis of the German noun *Gewölbe* (= vault, from *voluta*, the Latin word for roll, turn) is much more complex. Its roots are to be found in Roman stone buildings, as opposed to timber buildings, and in particular the Roman camera, i.e. initially the arched or vaulted ceiling or chamber: "Actually only the word for the curved ceiling, … 'camera' gradually became the term for the whole room below the ceiling. And it is this shift in meaning, which is repeated similarly in '*Gewölbe*', that leads to the majority of uses for which the latter is regarded as characteristic" [Grimm, 1973, p. 6646].

It was the German building terminology of the 18th century that adopted the word *Gewölbe* in its two-dimensional meaning, whereupon the three-dimensional sense was quickly forgotten. The reason for this may well have been the masonry arch theories that began to surface in the century of Enlightenment, which started the transitions from loadbearing structure to loadbearing system as a masonry arch model abstracted from the point of view of the loadbearing function – and therefore permitted a quantitative assessment of the load-carrying mechanism in the arch. The beam theory that began with Galileo acted as complement to this terminological refinement. In Zedler's *Universal-Lexikon* dating from 1735, for example, *Gewölbe* is defined totally in the two-dimensional sense, "a curved stone ceiling" [Zedler, 1735, p. 1393], and is differentiated from the suspended timber floor subjected to bending. In 1857 Ersch and Gruber expanded the definition on the basis of the two-dimensional term by mentioning, in addition to dressed stones and bricks, rubble stone material (with mortar joints) as a building material for vaults and arches [Ersch & Gruber, 1857, p. 129]. This became apparent in the material homogenisation of the masonry arch structure that began around 1850 in France, which in the plain and reinforced concrete structures of the final decades of the 19th century paved the way – in the construction engineering sense – for the transition from the theories linked with the materials of the loadbearing structure masonry arch to the elastic masonry arch theories of Saavedra [Saavedra, 1860], Rankine [Rankine, 1862], Perrodil [Perrodil, 1872, 1876, 1879, 1880 & 1882], Castigliano [Castigliano, 1879], Winkler [Winkler, 1879/1880] and others, and from there

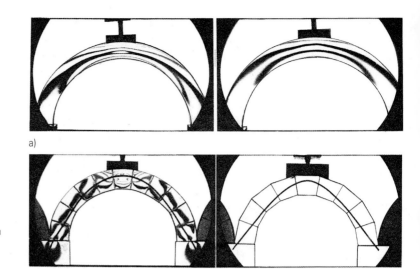

a)

b)

**FIGURE 4-1**
Photoelastic experiment carried out on a model subjected to a central point load: isochromatic lines of a) monolithic arch model, and b) masonry arch model [Heinrich 1979, pp. 37/38]

**FIGURE 4-2**
Historico-logical developments:
a) corbelled arch, b) three-hinge system, and c) from lintel to masonry arch [Heinrich, 1979, pp. 24/25]

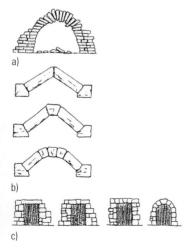

a)

b)

c)

to elastic arch theory. The logical nucleus of this historical process is the transition from the loadbearing system to the structural system of the elastic arch, a sort of concave elastic bar fixed at the abutments. Notwithstanding, the German term *Gewölbe* is still used erroneously for the designations of a number of modern arch structures, e. g. *Stahlgewölbe* and *Stahlbetongewölbe* [Badr, 1962, p. 43] (steel and reinforced concrete arches respectively). This contradicts the view that such loadbearing structures work not only in compression, but also in bending as linear-elastic, concave continua. The photoelastic experiments of Bert Heinrich proved the conceptual difference between *Bogen* and *Gewölbe*. Whereas the parallel isochromatic lines in the homogeneous arch indicate high bending stresses (Fig. 4-1a), the loadbearing quality of the (inhomogeneous) masonry arch is characterised purely by the propagation of compression in the direction of the thrust line (Fig. 4-1b).

Summing up, the following definition is proposed: a concave loadbearing structure is a masonry arch when the provision of the loadbearing function is realised through rigid building materials joined together with negligible tensile strength. Weber has refined this definition and proposed one based on the two-dimensional concept of differential geometry [Weber, 1999, pp. 30–37].

The invention of the masonry arch is, like that of the wheel, impossible to date. In the Berlin Museum of Prehistory & Ancient History, visitors can admire a Mesopotamian burial chamber more than 5000 years old which is in the form of a barrel vault with a span of a little over 1 m. "False and true arches as used over canals and crypts," writes Ernst Heinrich, "could well date from about the same period even if the one is known to us from the Uruk age, the other from the Mesilim. Both remain ... in use until the time of the Seleucids" [Heinrich, 1957–71, p. 339]. There are without doubt various historico-logical chains of development that culminate in the masonry arch. It is not difficult to imagine that during the construction of a false or corbelled arch the upper stones may have fallen inwards and wedged themselves into an arch shape (Fig. 4-2a), or one or more wedges could have been inserted between two mutually supportive stone slabs to enable the use of shorter slabs (Fig. 4-2b). The same technical motive to reduce the length of a beam and hence increase the bending strength may have encouraged ancient builders to switch from the lintel to the arch (Fig. 4-2c).

More than 2000 years certainly passed before the Etruscans' masonry arch with specially cut joints appeared. But the span of time from the first masonry arch theories of the late 17th century to the elastic arch theory is less than 200 years. And the analysis of masonry arches based on the ultimate load method did not appear on the scene until the 1960s.

## 4.1 The geometrical thinking behind the theory of masonry arch bridges

Whereas the large bridges of the late Renaissance demonstrated innovations primarily through the use of geometry, the application of the methods of statics in construction remained the province of the Baroque.

More precise variation in possible design geometries, the centering, the foundations and the construction sequence, etc. was now feasible through the use of drawings, ever-better dimensional accuracy and precision in the designs. Using the examples of the Ponte S. Trinità in Florence and the Fleisch Bridge in Nuremberg it will be shown how these new design approaches gradually became accepted for bridge-building.

During the first decades of the 18th century, bridge-building progressed via the intermediate stages of the first attempts to quantify this subject (La Hire, Couplet, Bélidor) to become the number one object of masonry arch theory. The idea of the thrust line became, indirectly, the hub of all deliberations: conceptual designs concerning the functional mechanism of bridges and intensive communication between experts advanced the formation of bridge-building theories.

### 4.1.1 The Ponte S. Trinità in Florence

The end of the 16th century marked the start of a new evolutionary era in the building of masonry arch bridges. The Renaissance initially took the structures and forms of construction of the Romans as its models. Due to its rise/span ratio of 1 : 2, the semicircular arch permits only very restricted functionality and is therefore unsuitable for urban structures in particular. This functional disadvantage gave rise to new arch forms, which were considerably shallower than the Roman arch.

Besides longer spans, the rise/span ratio also increased. The classical ratio was around 1 : 3, but this increased during the late Middle Ages to 1 : 6.5 with the Ponte Vecchio (5 m rise, 32 m span) by Taddeo Gaddi (1300 – 66). However, a new approach to design was the main aspect that signalled the leap in quality of the Renaissance compared to ancient times. The circle or the circular segment as the ideal form for the bridge arch was no longer matter-of-course; the three-centred arch (basket arch), the ellipse and the inverted catenary became the new forms. As the ramps to bridge structures had to be kept as shallow as possible, especially in urban environments, the new forms also served practical requirements.

Cosimo I instructed Bartolomeo Ammannati (1511–92) to build the Ponte S. Trinità (built between 3 April and 15 September 1569) (Fig. 4-3). Although this structure is attributed to Ammannati, a letter from Giorgio Vasari (1511–74) to Cosimo I reveals that Vasari had discussed the problem of the design of the bridge with a very old Michelangelo (1475–1564), who may indeed have had an influence on the design [Gizdulich, 1957, p. 74].

The reasoning is based entirely on art history findings: the expression of the opposing forces, which illustrate simultaneously the strain and the control of the strain, is a typical feature of Michelangelo's work – both his sculptures and his architecture. For Michelangelo, the outward form was not only a decorative finish, but also an expression of the inner substance of the sculpture or the structure. Michelangelo emphasizes the idea of vivacity and dynamic within a sculpture or a structure (Fig. 4-4).

If we follow this line of reasoning, we can attribute a certain desire on the part of Michelangelo to reveal the equilibrium present within a load-bearing structure, to demonstrate this artistically as a conflict between opposing forces and to awaken the impression that, once released, these forces would lead to movement. Art historians have had all sorts of discussions about the relationship between the arches of the sarcophagus embellishments at the Medici Chapel in Florence and the arches of the Ponte S. Trinità. The architects Enrico Falleni and Pero Bargellini drew attention to the similarity between the forms in 1957 and 1963 respectively [Paoletti, 1987, p. 137].

The three-centred arch (basket arch) and the elliptical arch, which appear for the first time in the architecture of the 16th century, should be regarded as the means behind this compositional intention. This is why the publications of Dürer (1525) and Serlio (1545) specified ways of constructing ellipses; their construction with string was described by Bachot in 1587 [Heinrich, 1983, pp. 110–111]. The year 1537 saw the publication of the first printed edition of *Conica* by that ancient mathematician Apollonius of Perga (c. 262–190 BC), which deals with conic sections (hyperbola, parabola, ellipse) [Wußing & Arnold, 1983, p. 44].

**FIGURE 4-3**
Ponte S. Trinità, photo taken prior to the bridge's destruction in World War 2 (photo: Gizdulich Archive)

A new, non-circular form was tried out on the Ponte S. Trinità for the first time. Starting at the centre of the respective span, the radius of the arch decreases towards the abutments. As the centre span measures 32 m and was therefore longer than the two side spans (29 m), the arches, had they been circular segments like on the Ponte Vecchio, would have intersected with the piers at different angles, and that would have resulted in an unsatisfactory aesthetic. The ever-decreasing radii, on the other hand, harmonise the disparity. Furthermore, the increasing curvature of the arches at the piers emphasizes their strength and stability.

It is interesting to note that the modified form causes the observer to see the loadbearing behaviour of bridges in a completely new light. It is not the geometrically tranquil image of a circle or circular segment that is behind the stability of a bridge, but rather a laboriously achieved equilibrium which counteracts the actions and gives the bridge its load-carrying ability. We can surmise the high compressive forces in the centre of the arch and understand the neutralisation of the forces acting on the sides by the load of the piers and the attachment to the banks.

FIGURE 4-4
Interior of the Medici Chapel in the New Vestry (1520–34), Florence (Michelangelo Buonarotti); tomb for Giuliano de' Medici with two reclining figures symbolising day and night (photo: Kahlow Archive)

Unfortunately, the bridge was maliciously destroyed by German troops in the night of 3/4 August 1944. The discussion surrounding the rebuilding of the bridge led soon after the war to the question of how the bridge had been originally designed. The variants up for discussion were a parabolic form, a three-centred arch and a catenary. The architect Riccardo Gizdulich, who was very committed to rebuilding the bridge in its old form using as many of the original stones as possible, developed a photogrammetric method with which old, but very good quality, photographs [Viviani, 1957, p. 22] of the undamaged condition could be evaluated (see Fig. 4-3). The controversy with the engineer E. Brizzi, who assumed a parabola [Brizzi, 1951], reinforced the efforts to research Ammannati's design principles.

In the history of this bridge, one version had been principally assumed up to the time of rebuilding, a version that had prevailed since 1808: naming a contemporary description as his source, which came from Alphonso Parigi, Ferroni assumed a three-centred design [Ferroni, 1808]. Some decades previously, the famous French bridge-builder Jean-Rodolphe Perronet (1708–94) had still been assuming an ellipse [Paoletti, 1987, p. 122].

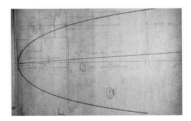

FIGURE 4-5
The application of the catenary for finding the form of the Ponte S. Trinità; the shape of the bridge was determined by a suspended chain turned through 90° (photo: Gizdulich Archive)

Gizdulich tried to analyse the form-finding process very exactly and, based on history of architecture reasons and a very precise evaluation of the photographs, ruled out a three-centred circular arch and a parabola, and discovered the design to be based on a catenary. Experiments with suspended chains then enabled Gizdulich to reconstruct the old form (Fig. 4-5).

Owing to the agreement between the measurements of the old bridge based on the photographs and Gizdulich's reconstruction, we can assume that his hypothesis was confirmed by a "large-scale experiment", i.e. the rebuilding of the bridge (Fig. 4-6).

**FIGURE 4-6**
Basic structure during rebuilding of the Ponte S. Trinità, Florence, in 1957; concrete was used only at the abutments (photo: Gizdulich Archive)

### Galileo and Guidobaldo del Monte

#### 4.1.1.1

Whether – as Gizdulich tried to prove – Michelangelo himself, during the last years of his life, had a direct influence on the building of the Ponte S. Trinità, or whether the arch form can be attributed to Ammannati, is undoubtedly interesting from the history of art viewpoint. But from the history of science standpoint it would be more important to know to what extent suspended chain experiments were carried out and how these experiments were connected with some kind of theory of structures concept centred around the idea of equilibrium of forces and the inversion of the suspended rope or chain to form the line of thrust for an arch.

The use of the catenary for the Ponte S. Trinità, the first bridge for which such an approach can be verified, is not evidence of an approach in the meaning of today's theory of structures. Using a hanging chain or rope for finding the form and not taking the next, seemingly logical, step, i.e. inverting it to form the line of thrust, seems totally illogical to us today. Would it not have been so easy to attach weights to the hanging chain at certain points to simulate the self-weight of the bridge and then to derive the line of thrust from the resulting catenary? What actually happened was obviously something totally different: the catenary used for the form-finding is the shape of the hanging chain turned through 90° (!) with respect to the base line (parallel to the surface of the water). In order to smooth out the kink in the middle of the span, desirable from the aesthetic viewpoint but excessive, the curve was subsequently turned through a few more degrees (see Fig. 4-5).

Why was the chain not understood as the model for the inverted line of thrust? Could Ammannati or Michelangelo recognise the play of the forces perhaps just as well in a catenary turned through 90° to the arch of the bridge? Then the weight would have been considered as an analogy to the thrust of the arch – at the place where this has to be directed into the subsoil there is, on the chain turned through 90°, the lowest point at which the left part of the chain is connected to the right part and only a horizontal force is effective here.

Interestingly, the chain idea plays a central role in Simon Stevin's (1548–1620) book *Beghinselen des Waterwichts* (principles on the weight of water) which dates from 1586 – just a few years after the building of the Ponte S. Trinità; the chain is used to explain the equilibrium of an inclined plane. Both here and elsewhere, Stevin presumes the impossibility of the perpetuum mobile and he explains the stable state of the chain, also when cutting off the lower, suspended part, the impossibility of a movement from the state of rest. From his deliberations he also derives the parallelogram of forces and was the first person to draw a catenary with weights attached (Fig. 4-7).

He proceeds likewise in his explanation of the impossibility of an autonomous circular movement of sinking and rising water particles; he therefore proves the necessity of hydrostatic pressure increasing with the depth of the water. Critical here is the rising and sinking of weights, whether they be water particles or chain elements. Galileo (1564–1642) also employed this line of reasoning by comparing the upward and downward swings of a pendulum with the rolling of beads up and down inclined planes [Galileo, 1638/1964, pp. 122–123].

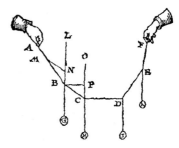

**FIGURE 4-7**
Simon Stevin's *De Beghinselen des Waterwichts* (1586): equilibrium of forces in the funicular polygon with weights attached [Stevin, 1634, p. 505]

What is lost on one side is added to the other. Galileo, in the introduction to his 1593 essay *Le mecaniche* [Galileo, 1987, pp. 68–70], mentions the fact that this again and again becomes a guiding principle in the emerging branch of physics. Balancing formed the path from a miraculous multiplication of forces through skilful mechanics to a natural explanation.

At this point we shall investigate the significance of equilibrium experiments and their theoretical foundation in conservation principles.

As is well known, Galileo had discovered the law of gravitation shortly after 1600 through the abstract separation of falling bodies and straight-line uniform motion, and had described the path of a projectile, neglecting the resistance of the air, as a parabola.

It had been hitherto assumed that this development had been completed in numerous stages either by 1604, or perhaps not until 1609, and the knowledge had been assembled from various sources. However, newer investigations show that Galileo as early as 1592, in joint tests with Guidobaldo del Monte (1545–1607), was on the brink of discovering what he published in his famous *Dialogue* in 1638, i.e. the connection between the law of gravitation and the parabola of a projectile resulting from the superimposition of motions [Renn, et al., 2000, p. 303]. At the end of his *Dialogue* of 1638, Galileo concludes that the parabola of a projectile is related not only to the form of the catenary, but also to its very structure. His reasoning is as follows: whereas the trajectory of a projectile is governed by two forces, one that drives it forward, acting horizontally, and one that pulls the weight downwards, a rope is also deformed by a horizontal, pulling force and a force due to the weight. Both are very similar processes. The corresponding diagrams remind us immediately of the relationship between catenary and line of thrust (Fig. 4-8).

Galileo and Del Monte generated the parabolic path of a projectile experimentally by rolling a bead down an inclined mirror. Renn, Damerow and Rieger have proved that Galileo regards the methods specified in the *Dialogue* for constructing a parabola, i.e. by representing it as the locus of a rolling bead or generating it from a hanging chain, as equivalent, and that he really did carry out the corresponding experiments [Renn, et al., 2000, pp. 313–315]. The known problem, which is linked to the question of the true trajectory of a cannonball fired from a gun, had already appeared in Tacola's *Nova Scienzia*. Whereas he still makes a distinction between enforced and natural motion, totalling in keeping with Aristotle, and accordingly assumes a vertical downward motion at the end of the trajectory, Del Monte was already talking about the symmetry of the projectile's trajectory and setting up the relationship to the catenary [Renn, et al., 2000, p. 318].

The construction of the catenary is left to experimentation, to Nature herself. Galileo tried to prove the agreement between the catenary and the abstract, geometrically founded parabola construction by analysing both curves exactly. The speculative nature of this equivalence must certainly have been clear to Galileo. In 1637 René Descartes (1596–1650) made a very clear distinction between geometric and mechanical curves in his *Geometrie*; and the catenary belongs to the latter category [Rühlmann, 1885, p. 77]. The identification of the trajectory of a projectile with the form of a chain or a rope loaded by its self-weight only is highly interesting. This is where geometry gives way to physics, and the hope arises that Nature herself will reveal her laws, if we were only smart enough to ask her through experimentation.

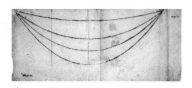

a)

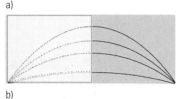

b)

**FIGURE 4-8**
a) Diagram of a traced hanging chain (ms. Gal. 72, f. 41/42v), b) Galileo inverts the chain and constructs a curve from it, point by point, which he looks upon as the parabolic path of a projectile (ms. Gal. 72, f. 113r) after [Renn et al., 2000, p. 305]

**Hypotheses**

#### 4.1.1.2

Just how old is the idea of the catenary as the inversion of the line of thrust for an ideal masonry arch? Surveys of the Sassanids' Palace (531–79) show indications of this, but there is no direct proof [Trautz, 1998, p. 97]. But the use of the catenary in the form-finding exercise for the Ponte S. Trinità does appear to be conclusive [Benvenuto, 1991, p. 328]. The fact that special significance was attached to the natural sag of a chain or rope could explain why Galileo was attracted to this curve as the parabola of a projectile.

The first written works describing the static functional mechanism of a masonry arch appeared in 1621 (Baldi – see section 4.3.1) and 1667 (Fabri – see section 4.3.2). The latter bases his observations on a roof structure in which the rafters are supported on the walls. Del Monte's notes of Galileo's projectile trajectory trials dating from around 1590 also show a very similar, three-part roof structure immediately prior to the treatment of the analogy between projectile trajectory and catenary. Was perhaps even in Del Monte's time the masonry arch seen as a roof-type supporting structure, and was the analogy with a hanging chain known to architects? If this is the case, and the Ponte S. Trinità and many polygon-type supporting structures, like the famous masonry arch bridge of Palladio, indicate that this could well be so, then considering the projectile trajectory as a catenary would represent a logical next step after discovering the analogy between line of thrust and catenary.

### 4.1.2 Establishing the new thinking in bridge-building practice using the example of Nuremberg's Fleisch Bridge

In 1587 the Senate of Venice passed a resolution to invite tenders for a stone bridge over the Grand Canal, to replace the existing bridge in the Rialto quarter of the city. A design submitted by Andrea Palladio (1508–80) shows a grandiose three-arch bridge totally in keeping with Roman architecture. However, it would seem that the new way of thinking had already become established because the Senate instead appointed Antonio da Ponte (c. 1512–97) to carry out the work. The desire for a bridge without piers in the river, which hinder river traffic, and the desire for a level crossing inevitably led to a single, shallow, segmental arch. The geometry of the 28.38 m span arch roughly corresponds to that of a quarter-circle. Da Ponte's design is recorded in several drawings (Fig. 4-9).

The tender for the building of the Rialto Bridge resulted in much more than just the building of the bridge; it also influenced further bridge-building projects in neighbouring countries. Examples of this technology transfer in the late Renaissance are the Fleisch Bridge in Nuremberg (1598) and the Barfüßer Bridge in Augsburg (1610), built according to Palladio's concept for the Rialto Bridge. Elias Holl (1573–1646), a master-builder from Augsburg, had visited Venice in 1600/1601.

Owing to the close trading ties between Nuremberg and Venice, the bridge plans of the Venetians were known in Nuremberg. The conditions for the bridges of both cities were similar: the marshy subsoil and the problem of not being able to drain the surface waters, which had to be considered in the design. A model of the Rialto Bridge as designed by Da Ponte, which is still in the possession of the descendants of Wolf-Jakob Stromer (1561–1614), senior master-builder of the City of Nuremberg, proves that the Nuremberg planners knew how the problems had been overcome in Venice. Stromer was in charge of the Nuremberg Building Department and he represented the interests of the client. Nuremberg, too, drew up a tender which was sent to the stonemasons and carpenters of the city, but also to master-builders elsewhere. The unique documentation for this structure permit us an insight into the planning work. In her dissertation (completed in 2005), Christiane Kaiser investigates the Fleisch Bridge in Nuremberg using modern research techniques, drawing on historical documents, photographs and surveys, a historical statics-constructional analysis and tests on 1:10 scale models [Kaiser, 2005/vol. 1]. A second volume presents the design and working drawings [Kaiser, 2005/vol. 2]. And some of the printed records that form part of this historic monument, showing how important this structure is to structural engineers and the general public, have been brought together in a third volume [Kaiser, 2005/vol. 3]. According to Kaiser, Wolf-Jakob Stromer was not the designer of the Fleisch Bridge. That honour goes to master-carpenter Peter Carl for the foundations and the centering, and master-stonemason Jakob Wolff for the bridge arch and abutments [Kaiser, 2005/vol. 1, p. 256].

**FIGURE 4-9**
The Rialto Bridge in Venice as designed by Antonio da Ponte; model built by Christian von Maltzahn and Yorck Podszus (Potsdam Polytechnic) (photo: Volker Döring)

**Designs for the building of the Fleisch Bridge**

**4.1.2.1**

Jakob Wolff proposed twin arches in one of the two designs he submitted (Fig. 4-10). Each arch in this design had a span of 43 feet (= 13.063 m, 1 Nuremberg foot = 0.304 m) and a rise of 12 feet (= 3.646 m), a rise/span ratio of 1 : 3.58. Although Bamberg-based Jakob Wolff was in the end responsible for the masonry works of the bridge, neither of his two design proposals bears any resemblance to the actual bridge as built.

In both of his designs the rise/span ratio is much smaller than the 1 : 6.2 of the bridge finally built. In addition, the design for the twin-arch bridge reveals how complex it is to join two banks at different heights; one bank would have a steeper approach ramp, impairing the aesthetics of the bridge.

Another design, by David Bella, shows an almost semicircular bridge with a span of 50 feet (= 15.190 m) and a rise of 21 feet (= 6.380 m) (Fig. 4-11). The rise/span ratio of 1 : 2.38 is only a little larger than that of a semicircle. The problems of such a high arch are obvious. Bella's submission also included his thoughts on the foundations; he proposed timber piles plus masonry with horizontal bed joints. Firstly, the rise could not be so large because of the approach ramps, secondly, a maximum rise/span ratio of approx. 1 : 5 was considered feasible [Borrmann, 1992, p. 80], and thirdly, the majority of master-builders preferred a single arch in order to avoid the problems of the difficult central pier foundation and the risk of it being undermined.

As the Ponte S. Trinità had been completed only a short time before, we can surmise a certain Florentine influence for the unsigned design for a three-centred arch. The span is about 95 feet (= 28.861 m) and the approach ramps are correspondingly shallow, but the size of the rise is not mentioned.

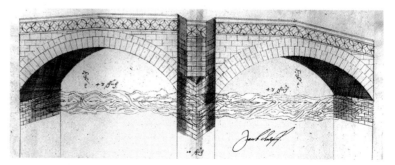

**FIGURE 4-10**
Design by Jakob Wolff
[Pechstein, 1595, p. 81]

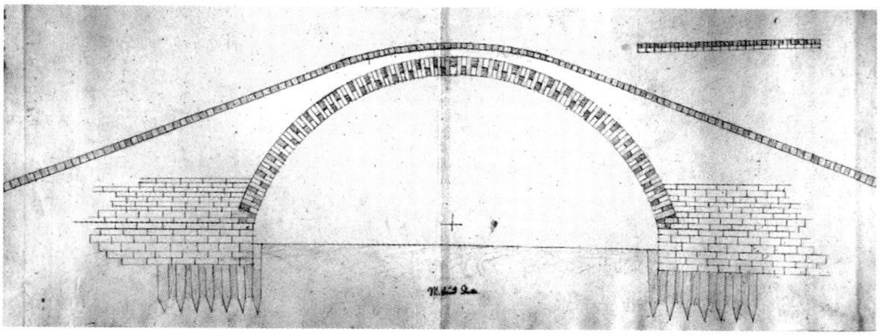

**FIGURE 4-11**
Design by David Bella
[Pechstein, 1595, p. 84]

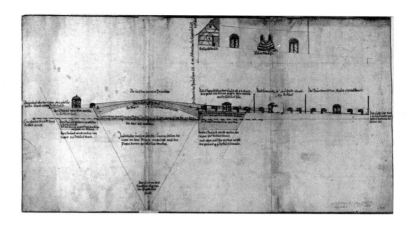

**FIGURE 4-12**
Sketch showing the effects of various arch geometries on the approaches to the Fleisch Bridge
(ink drawing by Wolf-Jakob Stromer, 1596; City of Nuremberg drawings collection)

The bridge marking the end of this design history is a shallow, single arch over the Pegnitz. This initiated a discussion on the technical and functional aspects regarding the rise of the planned bridge. Fig. 4-12 shows the Fleisch Bridge with its planned rise of 14 feet (= 4.253 m) plus the embankments and approach ramps required for the banks with their different heights; disadvantages for local residents were expected. The bridge was finally built in 1598 with a rise of 12 feet (= 3.646 m) [Borrmann, 1992, p. 171].

### 4.1.2.2 Designs and considerations concerning the centering

Various designs for the arch centering have been preserved, some from the pen of Wolf-Jakob Stromer in his *Baumeisterbuch* and others in the "Nicolai Collection" belonging to the City of Stuttgart archives.

All four designs show joints dividing up the upper timber ribs on which the timber boarding (laggings) is laid (Fig. 4-13). The loads are carried by transverse timbers and beams to the piles standing in the river. We do not know exactly which design was actually used for the construction.

Design A (Fig. 4-13a) shows 14 ribs supported on radial struts. The struts are supported on transverse beams which are in turn supported on a grid of 14 × 14 piles driven into the riverbed. Apart from the ribs, none of the timber members is loaded in bending. However, the large number of piles in the river is a serious disadvantage. Although the load per pile is substantially reduced, the work involved in erecting the centering is considerable. The significant obstruction to the flow of the river during the building works is also problematic.

In design B (Fig. 4-13b) the number of piles in the river has been considerably reduced. Now only every second rib is supported directly by five piles in the direction of the span. Beams and struts transfer the loads to the piles.

Design C (Fig. 4-13c) shows even fewer piles standing in the river and only three loadbearing ribs. The purlins between the ribs are additionally braced at mid-span. It remains unclear as to whether the first and last loadbearing ribs are only a suggestion, or whether it really was the intention to use this much-reduced loadbearing system.

Design D (Fig. 4-13d) could well represent the method actually used. Besides the arch centering itself, the entire building operations, tools and even the workers and their working platforms are shown. Compared to the other designs, the centering has been simplified yet again. Again, this drawing indicates that several parallel loadbearing ribs were erected across the width of the bridge. The wedges between the centering and the piles in the river are interesting; removing these gradually would allow the centering to be lowered step by step in a controlled fashion. All designs show the attention paid to the actual building of the bridge.

### 4.1.2.3 The loadbearing behaviour of the Fleisch Bridge

The arch itself consists of dressed sandstone blocks and, in some places, thick mortar joints. Similar structures from this period lead us to assume that the mortar used was a lime mortar with only a very slight hydraulic effect (see [Schäfer & Hilsdorf, 1991; Kraus et al., undated]). Characteristic are the low elastic modulus and the high creep component in the overall deformation in the first months. Various assumptions regarding the material behaviour coupled with an investigation of the loadbearing behaviour, the stages of the work and the loads on the centering can be found in Holger Falter's work [Falter, 1999].

The method of erection had to guarantee that the compressive strength of the masonry was not exceeded at any stage of the work and that major cracks or serious, permanent deformation did not occur. Furthermore, the supporting centering and the piles in the river had to be able to carry the loads (Fig. 4-14a).

At first, the keystone remains unloaded. Not until the earth fill is in place does the change in load cause deformation of the arch and hence the activation of the arching effect. The stiffness of the arch is critical here. If the stiffness of the arch is low in relation to the centering, the loads are carried mainly by the centering and the temporary piles, which is why the load on the arch increases only marginally. Only a stiff arch can contribute significantly to the load-carrying effect. If the centering is not supported on piles in the river, it deforms so severely that the arch itself is loaded immediately.

The centering now has to be lowered. The wedges visible in Fig. 4-13d (design D) point to the fact that this procedure was critical. Lowering the centering slightly means that the arch itself no longer rests on the centering over the full span and must help carry the loads. However, the associated deformations result in the full span of the arch dropping onto the centering again. Lowering the centering in many tiny steps prevents the critical plastic strain being exceeded, stops the mortar being squeezed out of the joints and enables the arch to take on its final equilibrium and deformation states gradually (Fig. 4-14d).

Striking the centering step by step has another advantage: a considerable reduction in the length of time needed for the mortar to achieve its final stiffness.

In a masonry arch with no mortar joints, or one where the mortar in the joints is very stiff and not capable of plastic deformation, the ten-

**FIGURE 4-13**
Designs A to D for the arch centering to the Fleisch Bridge (*Baumeisterbuch* of Wolf-Jakob Stromer and "Nicolai Collection", City of Stuttgart archives)

a)

b)

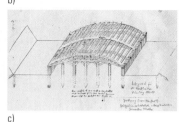
c)

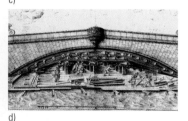
d)

sile stresses on the top side of the impost are much higher. Gaping joints are the result, which no longer close owing to the lack of plasticity in the masonry. Furthermore, loading the foundations suddenly by striking the centering in one operation would cause immediate displacement of the supports. In this loading case a masonry arch with exclusively elastic material behaviour also cannot adapt to the stresses through plastic deformations. Displacement of the supports results in high tensile stresses on the top side of the impost and the soffit at the crown, which causes gaping joints at these points. The result is a statically determinate three-pin arch. The hinges caused by the cracks are not able to transfer the forces and the outcome is either spalling at the edges of the stones or slippage of the entire arch.

Masonry arches with a plastic material behaviour have proved the better choice. Although here, too, cracks on the top side of the impost are inevitable, the remaining overcompressed area in the crack cross-section remains large enough to carry the loads. Indeed, viscoplastic deformations allow the cracks to close again partly.

Kaiser's static-constructional analysis of the Fleisch Bridge using various structural models reveals that the imposts were oversized for dead and imposed loads. She demonstrates that increasing the width of the arch profile towards the imposts was unnecessary from the calculations point of view "because the formation of cracks in the extrados near the imposts relieves the load on part of the arch and redistributes loads to the intrados" [Kaiser, 2005/vol. 1, p. 257]. On the other hand, Kaiser's tests on 1 : 10 scale models confirm that the builders of the Fleisch Bridge were right to assume that with correspondingly high asymmetric loads, the provision of radial joints throughout the arch and the overlying masonry was better than building up the arch with horizontal courses of masonry.

The static-constructional knowledge of Alberti and other masterbuilders of the Renaissance in the field of masonry arch construction resulted in numerous consequences: the constructional measures the builders chose to bring about the desired behaviour of the bridge show that the mechanical behaviour of structures could actually be seen and, interpreted correctly, could influence the design of a masonry arch. This approach was based on the wealth of experience gained by builders. However, the disadvantage of this approach is that purely empirical knowledge is difficult to pass on and wide application is therefore ruled out. Not until the appearance of the first masonry arch theories in the late 17th and early 18th centuries did a slow process of the separation of objectivisable knowledge from the possessors of that knowledge start to take hold. The outcome was that towards the end of the 18th century a construction engineering system of knowledge was available which was not only easily reproducible, but brought with it numerous tactics to expand that reproduction.

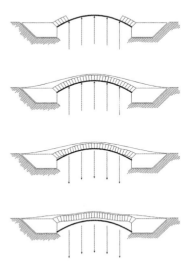

FIGURE 4-14
Probable phases during the building of the Fleisch Bridge: a) masonry arch prior to fitting the keystone, b) arch completed and fill in place, c) step-by-step lowering of the centering, d) centering completely struck [Falter, 1999, p. 151]

## 4.2 From the wedge to the masonry arch – or: the addition theorem of wedge theory

Roman masonry arch theory was dominated by the small-format one-and-a-half-foot clay brick and pozzolanic mortar, which was later superseded by *opus caementicium*. This mortar, made from building lime, sand

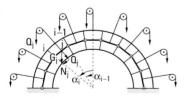

**FIGURE 4-15**
Leonardo da Vinci's wedge model

**FIGURE 4-16**
Trial arrangements for measuring the horizontal thrust of masonry arches after Leonardo da Vinci [Heinrich, 1979, p. 100]

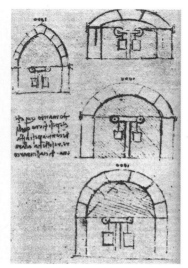

and pozzolanic earths, "is used in walls and – since the time of the first Emperors – in vaults, too, built up in horizontal layers alternating with the *caementa*, irregular blocks of different natural stones or clay bricks, although the *caementa* is normally larger in vaults than in walls. The fact that domes – with few exceptions – are built up in this way horizontally and no longer radially shows that they were already perceived constructively as a homogeneous structure and no longer as an assembled component" [Rasch, 1989, p. 20].

On the other hand, the Italian master-builders of the Renaissance used dressed stones, especially the relatively easy-to-work Lias limestone and gneiss found in parallel strata at the foot of the Alps and quarried there. This dressed-stone technique certainly contributed to Leonardo da Vinci dividing his arch into wedge-shaped, discrete elements matching the dressed stones (Fig. 4-15), and to considering each single stone of the arch as a wedge. He thus founded a wedge theory for masonry arches which, from the early 18th century onwards, was to play a significant role in establishing a theory of masonry arches for more than 100 years.

So what makes up the addition theorem of wedge theory? The addition of loadbearing elements to form the vault or arch, emulating the arch construction technique of the bricklayer and stonemason, is the moment that literature discovered production. As one of the five simple machines, the mechanical capability of the wedge had been known since ancient times, also from the theoretical viewpoint. Playing with combinations of these simple machines, putting them together to form more complex machines, was the daily bread of the engineers of the Renaissance. It is therefore no surprise to see Leonardo da Vinci building an arch as a machine made up of wedges, ropes and pulleys. Consequently, he is adhering to the kinematic view of statics [Kurrer, 1985/1], which goes back to the time of Aristotle, in which the equilibrium effect of the five simple machines and their synthesis to form more complex machines is perceived indirectly through a geometrical disruption of the state of equilibrium (see section 2.2.7). Leonardo da Vinci's loadbearing system synthesis is therefore additive. Although it satisfies the equilibrium conditions for the loadbearing element, it does not form an equilibrium configuration when considered as a whole because the transition conditions for the equilibrium at all the joints in the masonry arch are infringed [Kurrer, 1991/1].

Leonardo da Vinci was fully aware of the fact that masonry arches generate a horizontal thrust effect at the springings. "Do the experiment," he implores, and specifies a simple measuring arrangement with which the horizontal thrust can be determined with ropes, pulleys and weights (Fig. 4-16). More than 300 years would pass before Ardant would use this measuring arrangement to determine the horizontal thrust of long-span timber arches [Kurrer & Kahlow, 1998] – without any knowledge of the thought experiments of the "giants in power of thought" [Engels, 1962, p. 312].

Nevertheless, the additive activity of the stonemason when building masonry arches, as illustrated by Leonardo da Vinci in the loadbearing system model using the addition theorem, determined the historico-logical development of wedge theory.

The first mathematical formulations concerning the mechanics of masonry arches date from the years 1690 to 1720. The writings of Hooke, Stirling, Gregory, Bernoulli and La Hire established the foundations of masonry arch theory and departed from the empirical terrain on which the practical rules had evolved hitherto. Those rules had originated in the traditions of the Middle Ages and had been handed down by Gil de Hontañon, Martínez de Aranda and Derand.

After the publication of Galileo's *Dialogue* in 1638, masonry arch theory emerged alongside beam theory and within a few years had advanced to become an indispensable element of the new thinking in the fields of architecture and applied mechanics. The traditional view of history here generally draws on two famous documents by Philippe de La Hire (1640–1718): *Proposition 125* from the *Traité de Mécanique* [La Hire, 1695] and the report *Sur la construction des voûtes dans les edifices* [La Hire, 1712/1731]. However, Antonio Becchi has been able to show that these contributions had been preceded by other important work that La Hire had carried out for the Académie Royale d'Architecture de Paris, where he was active as member and professor from 1687 to 1718 [Becchi & Foce, 2002, p. 31; Becchi, 2002 & 2003]. The masonry arch problem is treated here with a clear interest for the *règles de l'art*, for which a scientific explanation was being sought to identify the constructional intuition in the planning and building of masonry arches (Fig. 4-17). This aspect, ignored by La Hire in his preceding works for the Académie des Sciences [La Hire, 1695 & 1712/1731], clarifies the historico-logical context and permits a new interpretation closely related to the themes discussed in the Académie d'Architecture: La Hire's masonry arch theory was not merely a purely academic exercise in classical mechanics, but rather drew its sustenance from the need for a scientific legitimation of masonry arch design – hence the founding of the *règles de l'art* through classical mechanics.

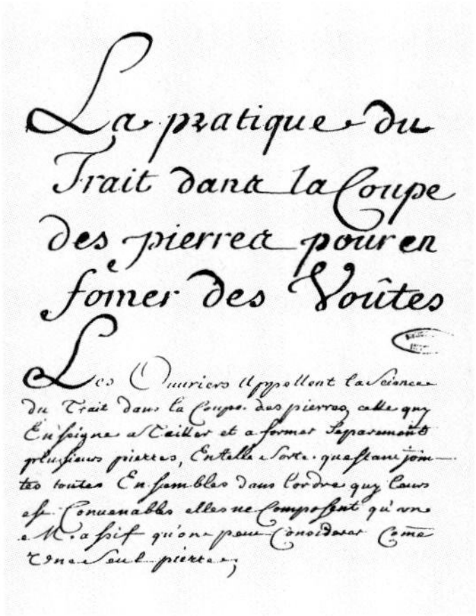

### 4.2.1 Between mechanics and architecture: masonry arch theory at the Académie Royale d'Architecture de Paris (1687–1718)

**FIGURE 4-17**
Cover of La Hire's *Traité de la coupe des pierres* [La Hire, 1687–90]

### 4.2.2 La Hire and Bélidor

We shall now look at the academic side of La Hire's masonry arch theory as described in his *Traité de Mécanique* [La Hire, 1730] presented to the Académie des Sciences de Paris, which had been founded by Colbert in 1666. Like Leonardo da Vinci, La Hire considers the masonry arch as a machine made up of wedges. After La Hire has asked the question of how large the weights $M$, $N$ and $O$ need to be so that the resulting funicular polygon forms a tangent with a quarter-circle arch, (Fig. 4-18a), he switches to modelling the semicircular arch under self-weight (Fig. 4-18b). By inverting the funicular polygon to form the line of thrust he has specifically formulated the second prime task of thrust line theory: How heavy does the voussoir (= wedge-shaped stone) have to be so that the semicircular arch remains in unstable equilibrium? However, La Hire's inversion presumes no friction in the joints between the voussoirs, a premise

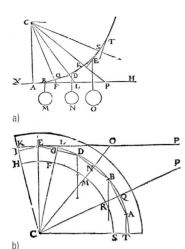

**FIGURE 4-18**
La Hire's first masonry arch model a) derived from the funicular polygon, and b) after inversion to form the wedge model [La Hire, 1730]

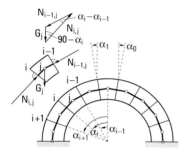

**FIGURE 4-19**
Bélidor's wedge model

that is infeasible in practice. That is the price he has to pay for his loadbearing system synthesis, free from mechanical contradictions. Whereas Leonardo da Vinci's voussoir at the impost has a weight equal to zero, in La Hire's version it has to be excessive. It was very probably Varignon – the first to recognise the ordered relationship between polygon of forces and funicular polygon in his *Nouvelle Mécanique ou Statique* of 1725 [Varignon, 1725] – who inspired, indirectly, La Hire's frictionless wedge theory of the semicircular arch; for it was in 1687 that Varignon obtained membership of Colbert's Academy with his preliminary studies leading to this work, which was published posthumously. And La Hire was possibly influenced by the theory of the frictionless wedge postulated by Borelli.

Bélidor, the engineering officer and teacher of mathematics and physics at the La Fère Artillery School and later Director of the Paris Arsenal as well as Inspector-General of the engineering troops, based his work mainly on the wedge theory of La Hire (Fig. 4-19). In his engineering handbook published in 1729, *La sciences des ingénieurs*, Bélidor departs from the academic questioning of La Hire by prescribing the weight of the voussoir for equilibrium, but retaining the concept of the frictionless joint. Bélidor therefore infringes the transition conditions for the normal forces in the masonry joints, and therefore the equilibrium of the entire arch. His addition of the frictionless wedges is wrong because he finds the second prime task of thrust line theory, i.e. the question of the loading function for a given arch centre-of-gravity axis, to be impractical. Bélidor is aware of this: "Like the voussoirs left and right of the keystone can resist that, i.e. the abutments trying to drive them forcefully apart, so one calls the total force of all these voussoirs together the pressure. But this does not express itself completely in the way I have described" [Bélidor, 1757, p. 4]. Bélidor now fills the contradiction between the arch and his wedge model, between loadbearing structure and loadbearing system, with mortar, and writes: "It is clear that all the voussoirs from which an arch is built cannot support themselves when they are not bonded together with cement or mortar because the upper voussoirs press with a greater force on those below than those below can counteract" [Bélidor, 1757, p. 4]. And this is exactly where Bélidor infringes the transition conditions. "It is therefore agreed," he continues, "that those which needed the least force so that the upper ones are free to fall down, which would eliminate the entire order of the voussoirs and as a result the arch itself would collapse. And one can easily see that if the voussoirs without the assistance of some material bonding them together should remain in equilibrium" [Bélidor, 1757, p. 4]. And here Bélidor refers to La Hire's masonry arch theory and comes to the conclusion that the weight of the voussoirs "must steadily increase from the keystone to the abutments … As now a masonry arch cannot be maintained without cement, so one does not have to consider the true forces of the voussoirs, but rather only their endeavours to work" [Bélidor, 1757, p. 4].

So that's Bélidor. He knows the difference between loadbearing structure and loadbearing system, between as-built reality and model, and

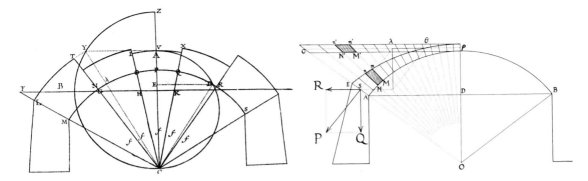

FIGURE 4-20
Couplet's determination of the position and size of the voussoirs according to La Hire [Benvenuto, 1991/2, p. 341]

therefore he also knows the difference between the normal forces in the masonry arch and the masonry arch model. It is clear to him that his analysis of the loadbearing system element is correct, but the addition of wedges to form the arch model, the loadbearing system synthesis, is incorrect. But for Bélidor it does not depend on the knowledge of the true force condition; for him it is enough to have an insight into the loadbearing quality backed up by figures.

In 1729 Bélidor's fellow countryman Couplet specified a way of determining the size of the voussoirs for the segmental arch, based on the premises of La Hire's wedge theory, such that the loadbearing system as a whole is in equilibrium (Fig. 4-20).

### 4.2.3 Epigones

The evolutionary path of the history of masonry arch theory based on the mechanical model of the wedge continued through the adaptation of that model to the physical reality but without overcoming the addition theorem. Once the basis for the classical theory of dry friction was established, to which famous mechanics researchers of the 18th century such as Ammontons (1699/1702), Euler (1750/1769), Musschenbroek (1762) and Coulomb (1781/1785) contributed, it became possible to analyse the masonry arch in the sense of a machine composed of wedges influenced by friction. Despite the introduction of friction into the arch model of wedge theory, La Hire's theory continued to be refined and formed the starting point for the further development of a wedge theory taking friction into account. But this is exactly where the machine concept of the pre-Industrial Revolution fails. Although the loadbearing system element of the wedge affected by friction could be investigated from the statics viewpoint, the difficulties of the loadbearing system synthesis multiplied because only the shear forces of the limiting states of the upward and downward sliding of the wedges were ascertainable, and therefore the kinematically feasible configurations of the arch model expanded beyond all imagination. Whereas classical friction research was able to lend powerful momentum to the formation of a scientific machine theory in France and Germany in the early 19th century, it could not add anything to masonry arch theory. The wedge theory therefore quickly entered a period of stagnation and by the 1840s was of historical interest only.

## From the analysis of masonry arch collapse mechanisms to voussoir rotation theory

### 4.3

A stable masonry arch is a problematic loadbearing structure when in service. Only at the price of failure does it relinquish the forces acting within itself, and indeed in a very simple way: in the form of collapse mechanisms. With a little fantasy, students of building can imagine many such collapse or failure mechanisms in the structural engineering laboratory. Thinking is fun! A number of French and Italian scientists of the 18th century sharpened their intellect by way of such thought experiments, possibly at the expense of frightening builders in every country, who with every crack in their masonry arches lost sleep anyway. For example, the Alsatian fortifications master Frézier, who included the failure tests of Danyzy, carried out on small mortar models and masonry arches [Danyzy, 1732/1778], in his book *La théorie et la pratique de la coupe des pierres* [Frézier, 1737–39] (Fig. 4-21).

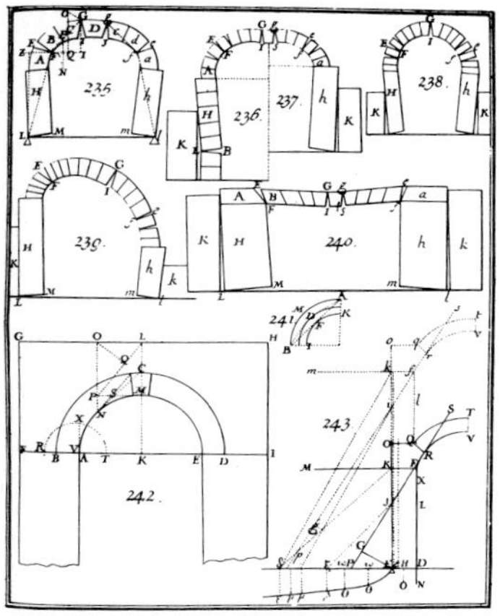

In order to shed light on the load-carrying action in masonry arches at rest, the contemporaries of the century of Enlightenment had to destroy them first in many ways in their heads – like the Enlightenment philosophers did with the *ancien régime*. On account of its massiveness, a very stable masonry arch crushes the knowledge of the play of forces active within it because it expresses the very opposite of motion. As the masterbuilders built very stable masonry arches, they concealed their structural analysis, unless, of course, they risked something, perhaps only in terms of an imaginary movement … Their enemy was any spreading of the abutments because that resulted in settlement at the crown; and excessive settlement at the crown could lead to collapse of the arch (Fig. 24/25 in Fig. 4-22).

#### Baldi

### 4.3.1

Bernardino Baldi (1553–1617) tackled the masonry arch problem as part of a detailed and original commentary to Aristotle's *Meccaniche* [Baldi, 1621]. Antonio Becchi has edited this major contribution from the orien-

**FIGURE 4-21**
Masonry arch collapse mechanisms after Danyzy and Frézier [Frézier, 1737–39]

**FIGURE 4-22**
Collapse mechanisms of semicircular arches after Schulz [Schulz, 1808]

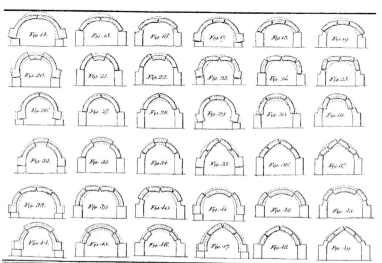

tation phase of structural theory (1575–1700) and added a commentary from the history of science perspective [Becchi, 2004 & 2005]. Baldi's *Exercitationes* is not mentioned by La Hire or other authors who looked into the masonry arch problem during the course of the 18th century, e. g. Danyzy, Frézier, Coulomb.

Aristotle's problem is well known: "Why are pieces of timber weaker the longer they are, and why do they bend more easily when raised, even if the short piece is, for instance, two cubits and light, while the long piece of a hundred cubits is thick?" (cited in [Becchi, 2002]). This problem constitutes the principal theme on the second day of Galileo's *Dialogue* [Galileo, 1638], although masonry arch theory is not explicitly mentioned in that publication.

Once he has tackled Aristotle's problem directly, Baldi expands his analysis to cover other topics: the column carrying a vertical load, the strength of floor joists, roof structures and masonry arches. This latter topic accounts for one half of the commentary to *Quaestio XVI*. We shall examine only the masonry arch collapse mechanisms according to Baldi [Baldi, 1621, pp. 112–114].

Baldi claims that a semicircular arch fails when the abutments are displaced by the amount $ES$ to the left and $ET$ to the right (Fig. 4-23a). The parts of the arch attached to the abutments $FQVH$ and $GRXP$ remain stable and together make up two-thirds of the total arch because Baldi divides the arch into three equal rigid bodies. The stability of the rigid bodies $FQVH$ and $GRXP$ had been presumed by Baldi in an earlier essay [Baldi, 1621, p. 109], which states that the centre of gravity of the rigid body $FQVH$ or $GRXP$ always lies on a line running perpendicular to the horizontal plane and passing through $F$ or $G$. This means that the line of action of the weight of rigid body $FQVH$ or $GRXP$ always passes through $F$ or $G$; the stability of the individual rigid bodies $FQVH$ and $GRXP$ is thus just guaranteed. Baldi's assumption that the perpendicular centre-of-gravity lines of rigid bodies $FQVH$ and $GRXP$ pass through $F$ and $G$ respectively applies only when the thickness of the arch $d$ is selected such that it is equal to about 30% of the external radius $R$. In the case of significantly thicker arches (Fig. 4-23b), Baldi's assumption deviates even further from the exact solution, whereas the deviation for thinner arches lies well below 10%. According to the rule of Alberti, the thickness of a semicircular arch should be at least

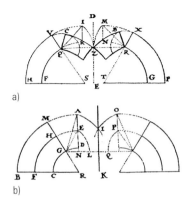

FIGURE 4-23
a) Collapse mechanisms for arch thicknesses d common in practice, which lie in the range given in eq. 4-2 [Baldi, 1621, p. 112],
b) according to Baldi, a very thick arch where $d \geq 0.4 \cdot R$ exhibits greater stability [Baldi, 1621, p. 113]

$$d_{u,Alberti} = 0.167 \cdot R \qquad (4\text{-}1)$$

(see Fig. 2-102). For an arch thickness $d$, Baldi's assumption for the range

$$0.167 \cdot R \leq d < 0.4 \cdot R \qquad (4\text{-}2)$$

deviates from the exact solution by less than 10%. The range given in eq. 4-2 for the arch thickness $d$ essentially covers the thicknesses of semicircular arches customary in practice at that time. So much to the high quality of Baldi's assumption concerning the perpendicular centre-of-gravity line for the rigid bodies $FQVH$ and $GRXP$.

Turning to the two central rigid bodies $QKIV$ and $RNMX$, according to Baldi these rotate about the points $Q$ and $R$ respectively. This rotation can be prevented if the distance $QR$ is no larger than the sum of the distances $QI$ and $RM$. Baldi's collapse mechanism agrees with that given by Schulz in Fig. 39 of Fig. 4-22. Baldi uses this collapse mechanism as the basis for assuming the greater stability of thick arches according to Fig. 4-23b from the kinematic viewpoint. Hence, Baldi can be allocated to the tradition of the kinematic view of statics (see section 2.2.7) founded by Aristotle.

From our modern viewpoint, the physics content of Baldi's "reasoning" appears very naive, although he does emphasize three important aspects of the problem which crop up again and again in subsequent discussions concerning the statics of masonry arches:

1. The division of the arch into three equal parts permits the definition of two stable rigid bodies attached to the abutments and one central part with the collapse mechanism.
2. The central part does not form one piece, but instead breaks in the middle along a line passing through the crown.
3. The two rigid bodies created in this manner rotate about the intrados (points $Q$ and $R$ in Fig. 4-23a).

Baldi's "reasoning" leads us to suspect the motives that might have led to Derand's empirical rules for determining the thickness of abutments for arches in 1643 [Derand, 1643; Becchi, 2004, p. 97]. Baldi's line of reasoning seems to exhibit a tendency to quantify in a unique way the masonry arch analysis in particular and structural calculations in general.

### 4.3.2 Fabri

In his *Physica* published in 1669, the Jesuit pastor Honoré Fabri (1607–88) specified a structural masonry arch model (Fig. 4-24). Fabri reduces the line of thrust of the semicircular arch to the straight lines $CA$ and $CD$, which intersect the extrados $AICKD$ at points $A$, $C$ and $D$ and act like hinges at those points. In the modern modelling language of structural analysis we would describe this as a three-pin system loaded at $C$ by the dead load of the arch. As half the dead load of the arch is proportional to the distance $R = CB$ and the horizontal thrust is proportional to the distance $R = BD$, it follows for the semicircular arch that the magnitude of the horizontal thrust is equal to half the dead load of the arch. Incidentally, triangle $ACD$ can be regarded as an inverted funicular polygon with the total dead load of the arch applied as a point load at $C$. Johann Esaias Silberschlag (1721–91) also uses a three-pin system to calculate the horizontal thrust of segmental arches in 1773 [Silberschlag, 1773, p. 257]. Using this masonry arch model, the values obtained for the horizontal thrust of segmental arches are about 5 % higher than those calculated with elastic theory; this approach is therefore on the safe side.

In Fabri's masonry arch model (Fig. 4-24) we can see a geometrical construction for determining the thickness of the abutments and the arch $ND = d_1$: the line of thrust $CD$ is a tangent to the semicircle $APN$ inscribed in the figure $AICD$ and it intersects line $AD$ at $N$. Drawing a line

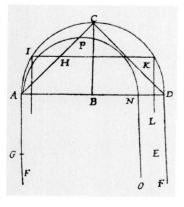

**FIGURE 4-24**
Fabri's masonry arch model
[Benvenuto, 1991/2, p. 319]

perpendicular to tangent $CD$ divides line $AD$ into the portions $MA$ and $MD$ ($M$ = centre of semicircle $APN$). The unknown radius $MA = AN/2$ of semicircle $APN$ is obtained from the geometrical condition that distance $MD = 2 \cdot R - MA$ is at the same time the diagonal of a square with side length $a = |MA|$. Translating Fabri's geometrical construction into the language of algebra results in the following thickness for the abutments and arch $d_1$

$$d_1 = ND = 2 \cdot R \cdot (3 - 2 \cdot \sqrt{2}) = 0.343 \cdot R \qquad (4\text{-}3)$$

From this it follows that the radius $r$ of the intrados of the semicircular arch is

$$r = R - d_1 = BN = R \cdot (4 \cdot \sqrt{2} - 5) = 0.657 \cdot R \qquad (4\text{-}4)$$

As the line of thrust $CD$ lies completely within the profile of the arch, the arch is stable. If tangent $CD$ were less than 45° (angle of rupture $\alpha = 45°$), the thickness of the arch would be

$$d_{o,Fabri} = 0.5 \cdot R \cdot (2 - \sqrt{2}) = 0.293 \cdot R \qquad (4\text{-}5)$$

The arch thickness $d_{o,Fabri}$ represents an upper bound; all masonry arches with thickness $d > d_{o,Fabri}$ are always stable. And conversely, masonry arches of thickness $d \leq d_{o,Fabri}$ may be unstable.

### 4.3.3 La Hire

The year 1712 saw La Hire propose a masonry arch theory for calculating the thickness of the abutments, which could be called an ultimate load theory (Fig. 4-25). However, he is unable to shake off the shackles of the frictionless wedge model. La Hire groups together the crown portion to form the wedge $LMFN$ and the abutment with the first quarter of the arch to form the rigid body $HSILMB$. The masonry abutment rotates about point $H$ and at the same time the wedge of weight $2G$ at the crown slides on the plane of rupture $ML$; but La Hire leaves the position of the plane of rupture undefined [La Hire 1712/1731]. Bélidor was the first to set the angle of rupture at 45° [Bélidor, 1729]. If we take this value for the angle of rupture, then force $D$ is the projection of the normal force $F = G \cdot \sqrt{2}$ acting at right-angles to the plane of rupture $ML$ on a line perpendicular to the lever $LH$ (Fig. 4-25). The thickness of the abutment

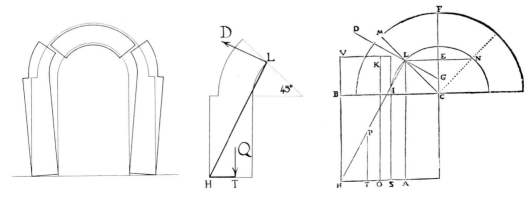

FIGURE 4-25
La Hire's mechanism for calculating the thickness of the abutments [Benvenuto, 1991/2, p. 333]

necessary follows from the lever principle for the cranked lever $LHT$. Despite the modelling of several voussoirs bonded with mortar to form a rigid body, La Hire is still rooted in the ancient thinking of the five simple machines, he does indeed abstract the arch with wedge and cranked lever.

Bélidor was the first (in 1725) to use La Hire's theory to set up a table for calculating the abutment thicknesses for gunpowder magazines [Bélidor, 1725]. Four years later, he published his book *La science des Ingénieurs* [Bélidor, 1729] in which he modifies La Hire's approach so that it can be used directly by any engineer. In addition to assuming a 45° angle of rupture, he shifts the intersection point of the arch thrust from the intrados to the centre of the arch: using the normal force $F = G \cdot \sqrt{2}$ applied to the centre of gravity of the plane of rupture $ML$, it is possible to work out the moment equation about the centre of rotation $H$ for calculating the abutment thickness.

This approach enabled Bélidor to eliminate the uncertainty in La Hire's theory and create a simple method for calculating the abutment thicknesses for barrel vaults and any combinations of barrel vaults. Perronet modified the position of the plane of rupture for arches with a three-centred intrados [Perronet & Chezy, 1810]: he positioned it where the curvature changes and at that point applied the arch thrust tangentially to the intrados. The theory of La Hire and Bélidor was accepted throughout continental Europe almost without question during the rest of the 18th century and was still to be found in some books in the 19th century, e.g. that of Sganzin [Sganzin, 1840–44].

The masonry arch theory of La Hire and Bélidor is a wedge theory with frictionless planes of rupture, but the upper portion of the arch behaves like a wedge that spreads the two abutments perpendicular to the planes of rupture; the collapse mechanism matches that specified by Schulz in Fig. 20 of Fig. 4-22 (see also Fig. 4-25/left). The theory results in abutment thicknesses that coincide well with the rules of proportion of the old builders; a factor of safety was not applied.

### Couplet   4.3.4

Using a different approach to that of Fabri, Pierre Couplet did not base his second *Mémoire* (1730) on an assumed line of thrust, but rather on observed collapse mechanisms [Heyman, 1972/1, p. 172]. Couplet specifies the minimum arch thickness as follows (Fig. 4-26):

$$d_{u,Couplet} = 0.101 \cdot R \qquad (4\text{-}6)$$

All masonry arches with thickness $d < d_{u,Couplet}$ are always unstable, and for arches of thickness $d_{u,Couplet} \leq d \leq d_{o,Fabri}$ the collapse mechanism according to Fig. 24 in Fig. 4-22 can occur. The parts of the arch behaving like rigid bodies then rotate about the points on the extrados at the crown and impost joints, and also about those points on the intrados where the joints have an angle less than the angle of rupture $\alpha$ ($\alpha = 45°$ in Fig. 24 of Fig. 4-22).

Such points can be regarded as hinges from the statics viewpoint through which the line of thrust must pass in the unstable equilibrium condition. In the given case the position of the line of thrust would be overdefined because just three hinges are sufficient to determine their position unequivocally (stable equilibrium condition, statically determinate), as is the case in Fabri's masonry arch model with hinges at $A$, $C$ and $D$ (see Fig. 4-24).

Couplet derives the two limiting values $d_{u,Couplet}$ and $d_{o,Fabri}$ for an angle of rupture $\alpha = 45°$. To determine the angle of rupture $\alpha$ (measured from the vertical) of a masonry arch of minimum thickness, Heyman specifies a transcendental equation which results in $\alpha = 58.9°$ [Heyman, 1972/1, p. 173]. Heyman [Heyman, 1982, p. 55] gives the minimum value for the arch thickness as

$$d_{u,Heyman} = 0.106 \cdot R \qquad (4\text{-}7)$$

The sizing according to Fabri, i.e. according to eq. 4-3 and 4-4, leads to excessively thick arches – approx. 300% higher than the minimum arch thickness given by eq. 4-7. On the other hand, the arch thickness according to Alberti's empirical equation, i.e. according to eq. 4-1, is a little under 66% higher than $d_{u,Heyman}$. In 1835 Petit specified the minimum arch thickness as

$$d_{u,Petit} = 0.1078 \cdot R \qquad (4\text{-}8)$$

This value given by Petit [Petit, 1835] is slightly greater than that of Milankovitch:

$$d_{u,Milankovitch} = 0.1075 \cdot R \qquad (4\text{-}9)$$

Milankovitch calculates an angle of rupture $\alpha = 54.5°$ [Milankovitch, 1907]. Both values have been confirmed by Ochsendorf [Ochsendorf, 2002, p. 75]. The minimum arch thickness according to eq. 4-6 as specified by Couplet in 1730 lies only 6% below the exact value given by eq. 4-9. The intention behind these examples is to show that the significance of the masonry arch models of Couplet and others should be given due regard today.

The tests carried out by Danyzy in 1732 on small mortar models or masonry arches [Danyzy, 1732/1778] verify the correctness of Couplet's approach. They show convincingly that sliding is impossible and that hinges form between the voussoirs. Finally, in 1800, Boistard set up a more comprehensive series of tests on arches spanning 2.60 m which were regarded as decisive [Boistard, 1810].

Couplet analyses the semicircular arch in a way that, since the mid-1960s, Jacques Heyman has regarded this as the historico-logical starting point of ultimate load theory approaches to masonry arch statics [Heyman, 1982]. In contrast to La Hire, Couplet departs from the premises of wedge theory:

- he assumes infinite friction in the joints of the masonry arch,
- he presumes an infinite compressive strength for the material of the arch, and
- he rules out any tensile stresses in the arch.

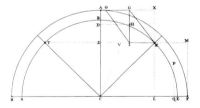

FIGURE 4-26
Determining the minimum thickness of semicircular arches subjected to dead loads after Couplet (1730) [Heyman, 1982, p. 53]

Heyman uses these three basic assumptions in Couplet's masonry arch theory in order to re-formulate masonry arch theory within the framework of ultimate load theory and hence investigate numerous historical arches (see section 4.6). Based on this, Sinopoli, Corradi, Foce and Aita have defined mathematically the upper and lower bounds for stable masonry arches from rupture kinematics (see, for example, [Sinopoli et al., 1997 & 1998; Sinopoli, 2002; Foce & Aita, 2003; Foce, 2005]).

### 4.3.5
**Bridge-building – empiricism still reigns**

Initially, builders dared only apply empirical rules to the bolder bridge arches. And some paid a high price for this learning curve! For example, the Welsh master-bricklayer William Edwards had the misfortune to see his bridge in Pontypridd collapse no less than three times [Ruddock, 1979, pp. 46-53]. Firstly, two years after completing the bridge over the River Taff, floodwaters washed away the pier in the river, which convinced Edwards that he should span the river with just one arch. During the rebuilding work, the centering collapsed into the water taking the bridge with it. And at the third attempt the crown was pushed upwards because the masonry infill between the first quarter-point of the arch and the impost was too heavy and the crown portion too light. In principle, the collapse mechanism may well have been that illustrated by Fig. 37 in Fig. 4-22. Despite these setbacks, Edwards tried for a fourth time in 1756: he specified three cylindrical openings on both sides and increased the height of the masonry at the crown (Fig. 4-27). After 10 years of practical building trials so to speak, Pontypridd Bridge, with a span of 43 m, was the boldest arch bridge in Great Britain.

Masonry arch collapse mechanisms can still be seen today. For example, Fig. 4-28 shows an arch bridge spanning a mountain stream in northern Italy which despite the assistance of an additional trestle has in the meantime collapsed. The asymmetric collapse mechanism caused by asymmetric loading is clearly visible. A hinge is forming at the extrados in the vicinity of the additional trestle – an ideal plastic hinge according to Heyman [Heyman, 1972/1]; but such a hinge is forming at the intrados in the other half of the arch. Plastic hinges form in the intrados at the impost joint adjacent to the additional trestle and in the extrados at the impost joint on the other side. A fixed-end arch system with four plastic hinges is, however, kinematic and failure is inevitable: the arch sub-

**FIGURE 4-27**
View of Pontypridd Bridge showing the three cylindrical openings on both sides [Ruddock, 1979, p. 49]

sides in the vicinity of the additional trestle and the other half of the arch rises.

Investigations into the failure mechanisms of masonry arches by way of experiments and real arches did not involve analysing the wedge as a loadbearing system element, but rather the arch, or a model thereof, as a whole (Fig. 4-22). The assembly and abstraction of the two-component system voussoir-mortar to form a rigid body created a machine model in the form of collapse mechanisms which, against the backdrop of the rigid body mechanics emerging in the 18th century, finally broke away from the traditional thinking of the ancients, i.e. the five simple machines and their combinations. By separating construction engineering and mechanical engineering, in theory and in practice, the ultimate load theory approaches of the 18th century were superseded by voussoir rotation theory because what was now interesting was not "what could be", the potential, but rather "what is", the actual – no longer the kinematic view of statics, but rather the geometric view (see section 2.2.7).

This brought to the forefront of theory formation the question of the exact position of the line of thrust in the service condition of masonry bridge arches.

FIGURE 4-28
Bridge in northern Italy shortly before its collapse (photo: Guilio Mirabella Roberti)

### 4.3.6 Coulomb's voussoir rotation theory

The French engineering officer and civil engineer Charles Auguste Coulomb presented his *Essai sur une application des maximis règles et de minimis à quelques problèmes de statique relatifs à l'architecture* to the French Academy of Sciences in 1773. Following a benevolent report by Academy members Bossut and De Borde, the paper (nearly 40 pages long) was published in the *Mémoires* of the Academy in 1776 [Coulomb, 1776]. This work not only completed the beam statics that had been initiated by Galileo's beam failure problem (see section 5.3), but also explored new solutions in earth pressure and masonry arch theories. His paper can be regarded as a review of theory of structures and strength of materials for civil engineers in the 18th century. Like he calculates the maximum active earth pressure and the minimum passive earth pressure, so Coulomb determines the maximum value of the arch thrust $H$ with the help of the extreme value calculations of differential calculus. In doing so, he leaves open the position of application of the horizontal thrust $H$ at the crown of the arch. He defines the horizontal thrust $H$ by analysing the four limiting cases (Fig. 4-29), whose limiting values $H_{1,max}$, $H_{2,min}$, $H_{3,max}$ and $H_{4,min}$ he determines depending on the position of the joint $Mm$ from the corresponding equilibrium conditions at the rigid arch body $aMmG$. After that, he calculates the angle of rupture $\varphi_1$ for downward sliding and then the horizontal thrust $H_{1,max}$ (case 1) etc. from the required maximum or minimum conditions. In contrast to Couplet, Coulomb's collapse mechanisms are not kinematic, but rather statically determinate because he models the arch impost plus abutment as one rigid body subjected to a small horizontal displacement. As Coulomb regards cases 1 and 2 as irrelevant in practice (because these cases occur only with very thick arches), he first uses cases 3 and 4 to formulate a voussoir rotation

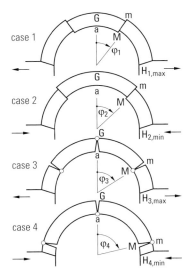

FIGURE 4-29
On the voussoir rotation theory by Coulomb: limit values of horizontal thrust for sliding downwards (case 1) and upwards (case 2); limit values of horizontal thrust for voussoir rotation about M (case 3) and m (case 4)

theory for masonry arches: the horizontal thrust $H$ in the service condition is a maximum at $H_{4,min}$ and a minimum at $H_{3,max}$; its magnitude and also its point of application at the crown of the arch must for the time being remain undefined.

Coulomb's treatise remained forgotten for nearly 50 years. It was not until 1820 that the French engineer Audoy rediscovered Coulomb's voussoir rotation theory [Audoy, 1820], whereupon he set up equations for the most common masonry arch profiles and evaluated them. Upon applying the calculated arch thrusts to the abutments, he discovered that the abutment thicknesses obtained were much too small and therefore a factor of safety had to be introduced. Thereupon, Audoy proposed multiplying the horizontal thrust at the crown joint by a certain numerical value (factor of safety) in order to obtain safe abutment thicknesses.

**Monasterio's *Nueva Teórica***     **4.3.7**

The Spanish civil engineer Joaquín Monasterio helped the kinematic view of statics, as applied to masonry arch theory, to gain important new ground in the first decade of the 19th century with his unpublished manuscript *Nueva teórica sobre el empuje de bovedas*. In this work, Monasterio cites Coulomb's *Mémoire* [Coulomb, 1776] but goes way beyond his masonry arch theory. For example, he investigates for the first time the collapse mechanisms of asymmetric masonry arch profiles of varying thickness (Fig. 4-30). Monasterio designates the translation of a rigid body with $t$ and the rotation with $r$; he represents a collapse mechanism symbolically as a permutation of $t$ and $r$ [Monasterio, undated, pp. 4–6]. By way of example, Fig. 3 in Fig. 4-30 shows three rigid bodies; while the left-hand and central rigid bodies rotate, the right-hand rigid body displaces on the rupture plane $M''N''$ – so the symbolic representation would be $rrt$. Monasterio specifies the permutation $trr$ for Fig. 4 in Fig. 4-30.

Monasterio's manuscript was discovered by Santiago Huerta in 1991. It consists of cover, 90 pages of text and two plates, and has been analysed in detail by Huerta and Federico Foce [Huerta & Foce, 2003]. Monasterio divides his work into an introduction and four chapters:

- In his introduction (pp. 3–11), Monasterio stands up for the scientific methods of analysing masonry arches with the help of mechanics and develops his symbols for characterising the collapse mechanisms using permutations of $t$ and $r$.
- In the first chapter (pp. 12–33), Monasterio investigates collapse mechanisms in which the rigid bodies undergo translation only, e.g. Fig. 1 in Fig. 4-30 (permutation $tt$).
- In the second chapter (pp. 34–48), Monasterio dedicates himself to analysing collapse mechanisms in which the rigid bodies undergo rotation only, e.g. Fig. 2 in Fig. 4-30 (permutation $rrr$).
- In the third chapter (pp. 49–66), Monasterio covers collapse mechanisms involving translation and rotation of the rigid bodies, e.g. Fig. 4 in Fig. 4-30 (permutation $trr$).
- And in the fourth chapter (pp. 67–90), Monasterio analyses the failure modes of masonry arch abutments.

Monasterio's *Nueva teórica* rounds off the masonry arch theories based on the mathematical analysis of collapse mechanisms. Unfortunately, his masonry arch theory was never published, and so the pioneering *Nueva teórica* remained an invisible milestone in the initial phase of theory of structures (1775–1825).

Looked at in terms of the history of theory, the ultimate load theory approaches of Couplet, Mascheroni [Benvenuto, 1991, p. 412] and Monasterio were superseded by the generalised voussoir rotation theory of Coulomb. The analysis of kinematic collapse mechanisms (Fig. 4-22) is now less interesting than assumptions regarding the position of that line within the arch profile that results from the geometrical position of the intersection points of the ensuing internal force in the masonry joints: the line of thrust.

## 4.4 The line of thrust theory

The mechanics of the chain or rope formed one of the objects on which classical mechanics and differential and integral calculus were tried out with considerable success. In order to fill the space on one page of his 1675 publication *Helioscopes and some other instruments* [Hooke, 1675], the experiment specialist of the Royal Society, Robert Hooke, embedded

**FIGURE 4-30**
Collapse mechanisms after Monasterio [Monasterio, undated, plate 1]

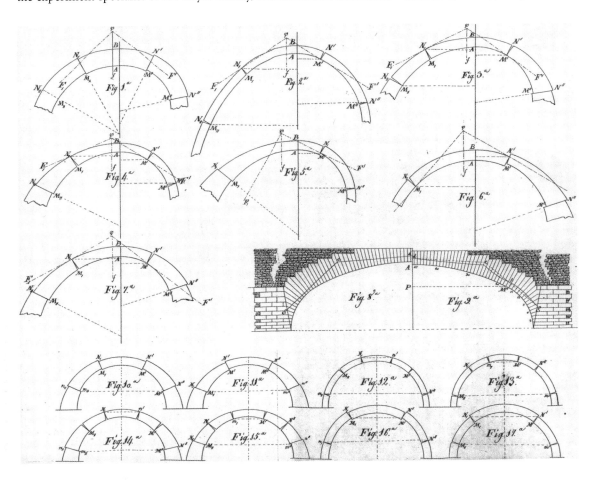

the form problem of masonry arch statics – i.e. the first prime task of thrust line theory – in a Latin anagram: "The true mathematical and mechanical form of all manner of arches for building, with the true butment necessary to each of them. A problem which no architectonick writer hath ever yet attempted, much less performed. abcccddeeeeefggiiiiiiiillm mmmnnnnnooprrsssttttttuuuuuuuvx" [Hooke, 1675, p. 31].

The solution to the puzzle is "ut pendet continuum flexile, sic stabit contiguum rigidum inversum", which translated into English means: "As hangs the flexible line, so but inverted will stand the rigid arch."

Hooke had advised the architect of St. Paul's Cathedral in London, Christopher Wren, on the basis of this insight into the geometry of the catenary arch. According to Hambly, he even suggested to Wren that he should determine the form of the dome of St. Paul's by using a chain model loaded with the corresponding weights at the respective centres of gravity [Hambly, 1987]. The evaluation of a Wren drawing by Allen and Peach has revealed that in an intermediate design the dome and its system of piers had indeed been shaped in part according to an inverted catenary [Hamilton, 1933/1934]. And finally, Heyman has verified that Hooke regarded the cubic parabola as the best form for the dome, and really did discover that precisely this curve appears in one of the preliminary designs for the dome [Heyman, 1998/2].

However, it was left to the English mathematician David Gregory (1659–1708) to describe the shape of the catenary mathematically, with the help of Newton's laborious method of fluxions, in a paper published in 1697 [Gregory, 1697]. He repeated Hooke's words and added a significant remark: "… none but the catenaria is the figure of a true legitimate arch … And when an arch of any other figure is supported, it is because in its thickness some catenaria is included" [Gregory, 1697]. Although Hooke specified the direction for solving the form problem of masonry arch statics (first prime task of thrust line theory), Gregory's supplementary remark anticipates the third prime task of thrust line theory: a masonry arch whose centre-of-gravity axis is not formed according to the line of thrust is then and only then stable when the line of thrust resulting from the given loading case lies completely within the profile of the arch. Unfortunately, Gregory's supplementary remark went unheeded. Probably without being aware of the solution to the Hooke anagram first published in 1701, Gregory understands the catenary as a very thin arch consisting of rigid, infinitesimal spheres with smooth surfaces. The form of the thin catenary arch was determined in 1704 by Jakob Bernoulli using the new differential and integral calculus [Radelet-de Grave, 1995]; in his derivation he employs the principle of virtual work.

After Hooke and Gregory, British mathematicians and engineers searched for a long time to find solutions to the first prime task of thrust line theory, i.e. they determined the associated line of thrust for a given loading case. For example, William Emerson (1701–82) [Emerson, 1754] and Charles Hutton (1737–1823) [Hutton, 1772 & 1812] specified catenary arches for particular loading cases. However, the masonry arch of

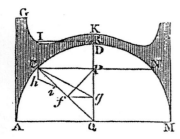

FIGURE 4-31
Determining the loading case for a given intrados (second prime task of thrust line theory) after Hutton [Hutton, 1812]

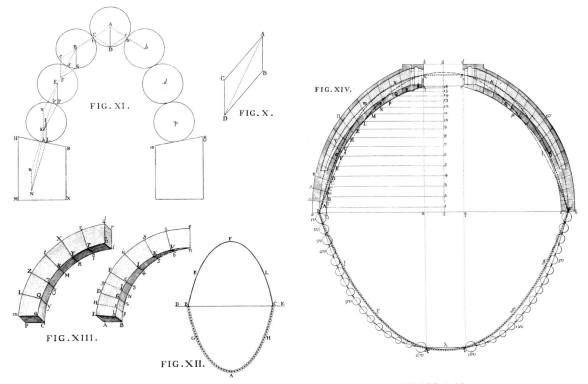

**FIGURE 4-32**
Poleni's masonry arch model
[Szabó, 1996, plate IX]

**FIGURE 4-33**
Poleni's investigation of the dome to
St. Peter's in Rome [Szabó, 1996, plate X]

constant thickness carrying merely its own weight is a purely theoretical problem. The course of the intrados and extrados define the loads on a real masonry arch. There were two fundamental problems: determining the geometry of the extrados for a given intrados (second prime task of thrust line theory) (Fig. 4-31) and determining the geometry of the intrados for a given extrados (first prime task of thrust line theory). The solution can be found by calculation, but in 1801 John Robison (1739–1805) proposed determining the course of the curve through experimentation [Robison, 1801].

In terms of physics, the masonry arch is interpreted here as an equilibrium configuration of voussoirs in the meaning of the "academic" masonry arch theory of La Hire, whose joints were always perpendicular to the intrados (see Figs. 4-18 and 4-20). This approach led to a certain, invariable form for transferring the compressive forces in the arch. The theory supplied no information about the thickness of the arch and could not explain common phenomena such as crack formation in arches.

Gregory's concept was first applied by the Italian mathematician and engineer Poleni in 1748 to the structural analysis of the dome to St. Peter's in Rome. Fig. 4-32 shows the catenary and Poleni's concept of the masonry arch model based on the work of Gregory and Stirling.

He idealises the voussoirs as geometrically and mechanically identical frictionless spheres that form an unstable equilibrium figure. The abstraction from the loadbearing structure to the loadbearing system was carried out by Poleni as follows: he first reduces the dome to spherical segments with identical geometry bounded by the meridians; this simplification is permissible because hoop tension forces occur in the bottom part of the dome which lead to cracks in the direction of the meridian.

After Poleni has divided two such plane, mutually supportive spherical segments into pieces, he represents the weights of the arch pieces, including the load from the lantern at the crown, by way of interlinked spheres of proportional weight. From the number of twin-parameter sets of thrust lines statically possible, he determines the one that passes through the centres of gravity of the impost and crown joints while remaining completely within the profile of the arch: the dome is stable (Fig. 4-33). Investigations into the load-carrying action of historical solid domes can be found in the writings of [Thode, 1975], [Stocker, 1987], [Ramm & Reitinger, 1992], [Falter, 1994], [Trautz & Tomlow, 1994], [Pesciullesi & Rapallini, 1995], [Heinle & Schlaich, 1996], [Trautz, 1998 & 2001], [Falter, 1999], [Mainstone, 2003], [Huerta, 2004] and others.

The line of thrust theory was first made possible through recognition of the problem of needing to satisfy the transitional conditions for the equilibrium between the neighbouring voussoirs from the viewpoint of the loadbearing system synthesis. Neither the structural analysis of the individual wedges as loadbearing elements and their addition to form the arch model, nor the self-fulfilling transitional conditions in the chain model resulting from differential and integral calculus or the method of fluxions, seemingly behind the backs of their users, could demonstrate any real progress after 1750. Whereas the analysis of the loadbearing system of the masonry arch remained restricted to the wedge loadbearing system element in all the wedge theory approaches and the loadbearing system synthesis was understood only in the form of addition, the approaches based on the chain model suffered from a lack of analysis of their individual elements; like in the ultimate load theory approaches and in the voussoir rotation theory of rigid bodies, it was only the chain as a whole that was interesting. Without the merger of the loadbearing system synthesis with the loadbearing system analysis, this line of masonry arch statics theory was destined to stagnate.

**Gerstner**     **4.4.1**

Franz Joseph Ritter von Gerstner brought about this merger in 1789 and 1831. In 1789 he constructed an unstable equilibrium figure from a bar subjected to self-weight only according to its structural analysis (Fig. 4-34), and then completed the transition to the catenary arch by infinitesimalising the bars (Fig. 12 in Fig. 4-35). In the first volume of his popular three-volume work *Handbuch der Mechanik*, Gerstner introduces the line of thrust concept and formulates the three prime tasks of thrust line theory [Gerstner, 1831, p. 406]:

*1st prime task:* Determine the line of thrust for a given loading case. Jakob Bernoulli had already solved this task for the specific case of the free-standing arch.

*2nd prime task:* Determine the loading case for a given arch centre-of-gravity axis such that said axis coincides with the line of thrust. Solutions to this task had already been specified using wedge theory for various specific cases, e.g. by La Hire, Bossut, Coulomb, etc.

*3rd prime task:* Take into account the line of thrust for a given loading case and masonry arch centre-of-gravity axis. This prime task was first formulated by Gerstner. In the historico-logical sense, it was set to merge with voussoir rotation theory to create the elastic theory of arches.

The formulation of these three prime tasks meant that Gerstner had laid the foundation for thrust line theory [Kurrer, 1991/2, pp. 27–31].

In French literature, the line of thrust concept is attributed to Méry [Méry, 1840] and in English literature to Moseley [Moseley, 1835]. Both Moseley and Méry combined thrust lines with the formation of failure mechanisms. Méry compared the results of the ultimate loading tests carried out by Boistard in 1800 [Boistard, 1810] and interpreted them in the context of the new theory by drawing lines of thrust for the failure modes of the models, and also calculating the minimum thickness of the corresponding masonry arch [Méry, 1840]. His concept resulted in a merger between the voussoir rotation and thrust line theories, and the work formed the springboard for further investigations. In Great Britain Barlow proposed a graphical method for drawing the line of thrust and devised a series of ingenious experiments to demonstrate that such a line "exists in practice" [Barlow, 1846]. The treatise by Snell dating from 1846 is also noteworthy [Snell, 1846].

But the first person to present a complete masonry arch theory on the basis of the line of thrust concept was Thomas Young, who in 1817 wrote an article on bridges for the *Encyclopaedia Britannica* [Young, 1817 & 1824]. Young's masonry arch theory was discovered for the history of theory of structures by Huerta [Huerta, 2005] and has already been described in detail in section 2.4.8.2. Unfortunately, Young's article had no influence on masonry arch theory because he wrote it in his typically laconic and incomprehensible style.

Nevertheless, Gerstner earned his place in history by bringing together the various approaches to line of thrust theory in his formulation of the three prime tasks and by introducing the genuine engineering science concept of the line of thrust.

The contribution of thrust line theory to masonry arch statics was not only in finding the most purposeful shapes for masonry arches, as Emil Winkler has said. The application of the catenary arch had to overcome several practical obstacles. The cutting of stones and building of centering were considerably more difficult than for the customary arch forms such as the semicircular, segmental and three-centred profiles; moreover, many architects viewed the circle and semicircle as the most pleasing arch forms. The line of thrust theory adapted to this situation by forcing the

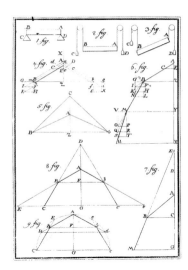

**FIGURE 4-34**
Synthesis of the loadbearing system element to form the loadbearing system after Gerstner [Gerstner, 1789]

**FIGURE 4-35**
Gerstner's infinitesimalising to form the catenary arch [Gerstner, 1789]

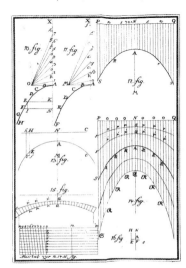

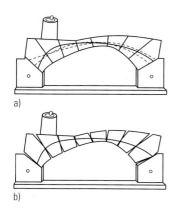

**FIGURE 4-36**
a) Which is the true line of thrust?,
b) Model by F. Jenkin for illustrating the true line of thrust

### The search for the true line of thrust

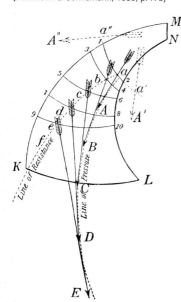

**FIGURE 4-37**
Moseley's distinction between line of thrust (line of resistance) and the direction of the pressure (line of pressure) in masonry arches [Buchheim & Sonnemann, 1990, p. 176]

first and second prime tasks more and more into the background and favouring the third prime task. Gerstner realised, more intuitively than anything else, that a masonry arch becomes more stable as the number of statically possible lines of thrust increases. He therefore had to stipulate two geometrical parameters (ordinate, angle) of the line of thrust in order to solve the third prime task solely with the help of the equilibrium conditions.

Just like Coulomb completed the change from the ultimate load theory approaches to voussoir rotation theory, there was also a change from the kinematic view to the geometric view of statics, which is interested in real rather than potential static conditions, i.e. in the state of forces in the service condition. In the line of thrust theory the interest shifted to the question – intrinsic to the geometric view of statics – of the true line of thrust in the masonry arch (Fig. 4-36) from the infinite number of statically possible equilibrium conditions. This change from the kinematic view of statics is the logical nucleus of the convergence of voussoir rotation theory with line of thrust theory.

#### 4.4.2

Henry Moseley's contributions to masonry arch statics, their adoption by the railway engineer Hermann Scheffler and his extensive detailed rearrangement to form a masonry arch theory, which had a practical value for civil engineers that should not be underestimated, gave the line of thrust theory a significant evolutionary boost. By formulating the thrust line problem, Gerstner had indeed pushed open the door to the solution of the third prime task; however, the knowledge that in the service condition of the masonry arch one and only one reality can prevail among an infinite number of equilibrium conditions, i.e. the fact that a true line of thrust exists, is due to those approaches to line of thrust theory that tried to solve the statically indeterminate masonry arch problem with the help of principles.

Just one such principle was formulated by Moseley in 1833 [Moseley, 1833] – generally for the statics of rigid bodies, but ready for immediate application in masonry arch statics [Moseley, 1843]. The principal states that of all the statically possible force systems or equilibrium conditions, the one that prevails is the one in which the resistance is minimal. Moseley distinguishes the line of resistance (which is constructed by joining together the intersections of the resultants with the joints), from the line of pressure (which is constructed from the directions of the resultants in the joints) (Fig. 4-37). The line of resistance is identical with the line of thrust.

The line of thrust and the line of pressure then always coincide when the intersection of the two resultants acting on the voussoir forms an obtuse angle. What form does the principle now assume for the specific case of the symmetrical, free arch (Fig. 4-38)? It is the minimal line of thrust that passes through points $G$ and $M$, i.e. case 3 according to Coulomb's voussoir rotation theory (see Fig. 4-29). So Moseley claims that the true horizontal thrust $H_{exist}$ passes through points $M$ and $G$ because it is a minimum in this case. From the standpoint of Coulomb's voussoir rotation theory this is wrong. Moseley does not consider the other limiting case of the maximum line of thrust. Basically, Moseley has turned the masonry arch problem into a statically determine one by introducing three hinges. It is this procedure that has characterised the merger of the voussoir rotation and line of thrust theories since 1840.

The masonry arch statics of Moseley therefore represents an incomplete but original translation of voussoir rotation theory based on the concepts of line of thrust theory. The eminent heuristic nature of this reformulation is proved by Scheffler's 1857 monograph. He recognises the problem and writes: "This results in an idiosyncratic difficulty because under the conditions assumed hitherto, an infinite number of force systems is possible in a stable arch without being able to declare which of them really occurs in nature" [Scheffler, 1857, p. 203].

According to Scheffler, the minimal line of thrust is the true line of thrust only for masonry arches with a rigid mass of voussoirs. However, as the material of the voussoirs is not rigid, but rather elastic, the true line of thrust lies between the minimum and maximum lines (Fig. 4-38), i.e. like with Coulomb the true horizontal thrust $H_{exist}$ must have a value between $H_{min}$ and $H_{max}$. For symmetric masonry arches with symmetric loading, Scheffler classifies the position of the minimal line of thrust geometrically and statically (Fig. 4-39). Although for Scheffler the consideration of the elasticity of the voussoir material is theoretically crucial for selecting the true line of thrust from the infinite number of statically possible lines of thrust, his knowledge did not yet have an effect on his practical calculations because, in the end, he still adhered to the rigid body model in practice.

Scheffler's masonry arch theory influenced numerous authors in Europe and North America, but not all of them regarded the theory as correct. For example, Culmann [Culmann, 1864/1866] tried to fix the position of the line of thrust and for this purpose set up his "principle of minimum loading", which states that of all the possible lines, the actual or true line is that one with the smallest deviation from the axis of the arch.

The loadbearing system analysis of the voussoir rotation and line of thrust theories is pure statics and it did not manage to integrate the real material behaviour into the masonry arch model. This would be left to elastic theory. A masonry arch theory that was still dedicated to the tradition of voussoir rotation theory but already incorporated elements of elastic theory was published by Alfred Durand-Claye [Durand-Claye, 1867]. Federico Foce and Danila Aita have investigated this in detail [Foce & Aita, 2003].

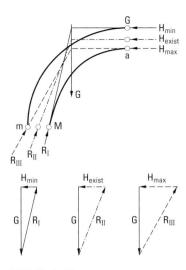

FIGURE 4-38
Significant equilibrium conditions after Moseley and Scheffler

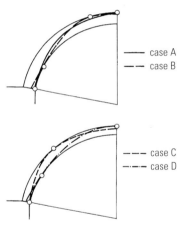

FIGURE 4-39
Minimal lines of thrust for a circular masonry arch after Scheffler

## 4.5 The breakthrough for elastic theory

Navier was the one who started to establish the paradigm of linear elastic theory in strength of materials [Picon, 1995], a view that would assume a dominating position in strength of materials for civil and mechanical engineers well into the second half of the 20th century, and together with tensile and compressive tests on building and other materials would form the basis of traditional strength analyses. The theory of the elastic arch gradually became established for the analysis of masonry arches in the discipline-formation period of structural theory (1825–1900) (see, for example, [Mairle, 1933/1935; Hertwig, 1934/2; Timoshenko, 1953/1; Charlton, 1982; Kurrer, 1995]).

## The dualism of masonry arch and elastic arch theory under Navier

### 4.5.1

In his masonry arch theory, Navier adheres to the voussoir rotation theory of Coulomb (Fig. 4-40); in addition, he permits horizontal loads and assumes a triangular distribution for the compressive stress in the joints under consideration [Navier, 1826]. He devised equations for masonry arches of any shape; however, there is seldom an exact solution for Navier's integrals. Thereupon, some engineers such as Petit [Petit, 1835] and Michon [Michon, 1848] computed new tables for masonry arch thrusts and abutment thicknesses. Poncelet developed a graphical method in 1835; it was rather laborious but did promise a certain time-saving [Poncelet, 1835].

Navier's most important contribution to masonry arch theory was the introduction of stress analysis. Independently of this, he developed the elastic arch theory in the chapter on the theory of timber and iron constructions. For example, Navier exploits symmetry to investigate a parabolic two-pin arch with one degree of static indeterminacy subjected to a point load in the centre of the arch, and specifies the stress analysis for this arch. A comparison with the *Statik der hölzernen Bogen-Brücken* (statics of timber arch bridges) [Späth, 1811] written by the Munich-based professor of mathematics Johann Leonhard Späth, who was totally committed to the tradition of the semi-empirical proportion rules of beam theory, allows the great progress of Navier's elastic arch theory to shine through.

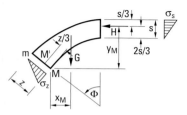

FIGURE 4-40
On the stress analysis in masonry arches after Navier

Although Navier introduced stress analysis into both masonry arch theory and elastic arch theory, it would seem that he did not arrive at the idea of modelling the masonry arch as an elastic arch. The indirect determination of the elastic moduli [Kahlow, 1990] of timber and iron structures by comparing deflection measurements with those of the linearised differential equation for the deflection, which Navier introduced so successfully for the constitution of his practical beam theory (see section 5.6.3.2), was not possible for the masonry arches of that time for the following reasons:

– The force-deformation behaviour of masonry arch materials in the service condition was not researched experimentally until the final third of the 19th century, in particular by Johann Bauschinger (1834–93) [Bauschinger, 1884].

– The small compressions under service conditions could not be quantified reliably with the testing apparatus available at that time.

– Deformation measurements on the generally oversized masonry arch structures could not be meaningful because the effects of arch settlement, dimensional stability, etc. were in the same order of magnitude as the deformations under service conditions.

## 4.5.2 Two steps forwards, one back

That there were undoubtedly alternatives to elastic theory for analysing the stress distributions in the joints of masonry arches is shown by Bandhauer in his book *Bogenlinie des Gleichgewichts oder der Gewölbe und Kettenlinien* (arch profile of equilibrium or the masonry arch and catenaries) [Bandhauer, 1831]. It is not Bandhauer's determination of the centre-of-gravity coordinates of the catenary arch for voussoirs of equal weight with a constant stress that is new, but rather his approach to the stress distribution in the joints. In such a catenary arch with a factor of safety of 1 "the form of the average compression course may not be more than a hair's-breadth different from that given in the calculation [rectangular stress distribution across the joint – the author] without failure occurring. On the other hand, in arches 2, 10 and 20 times stronger [i.e. factors of safety of 2, 10 and 20 respectively – the author], the average compression course can deviate by 1/2, 9/10, 19/20 times half the thickness from the centre before the same case, namely failure, occurs, which happens with a strength of just 1 [i.e. a factor of safety of 1 – the author] and the smallest (math.) deviation. It is this and only this condition that we have to thank for the stability of all our free-standing masonry arches designed according to the catenary [catenary arch – the author]" [Bandhauer, 1831, pp. 141–142]. Bandhauer's approach translated into the language of the stress concept (Fig. 4-41) results in the compressive stress

$$\sigma_B(v) = \frac{N}{b \cdot d} \cdot \frac{1}{(1-v)} \qquad (4\text{-}10)$$

for a masonry arch cross-section of thickness $d$, width $b$ and an eccentricity $e$ for the normal force $N$, with the parameter

$$v = \frac{2 \cdot e}{d} \qquad (4\text{-}11)$$

If the first factor in eq. 4-10 is replaced by the average compressive stress $\sigma_m$ corresponding to eq. 2-14, then eq. 4-10 can be converted into the following dimensionless form:

$$\frac{\sigma_B(v)}{\sigma_m} = \frac{1}{(1-v)} \qquad (4\text{-}12)$$

Eq. 4-12 should here be called Bandhauer's hyperbolic function of compressive stress distribution in the masonry arch cross-section. In the same way, the extreme fibre stress $\sigma = \sigma(v)$ resulting from the compressive stress distribution according to Young (see Fig. 2-51), i.e. eq. 2-13, can be presented as follows:

$$\frac{\sigma(v)}{\sigma_m} = (3 \cdot v + 1) \qquad (4\text{-}13)$$

Fig. 4-41 shows the compressive stress diagrams of Bandhauer and Young/Navier (see also Fig. 4-40) for the case of normal force $N$ applied to the boundary of the middle-third of the masonry arch cross-section, i.e. $v = 1/3$. Furthermore, eq. 4-12 and 4-13 are also illustrated. Bandhauer's hyperbolic function eq. 4-12 has its origin at $v = 1$, where the compressive stress in the masonry arch material is infinite. This is one of the three conditions of the masonry arch theories of Couplet (see section 4.3.4) and

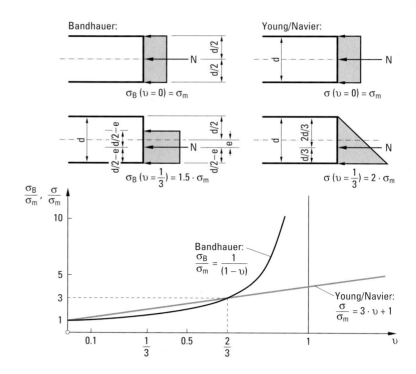

**FIGURE 4-41**
Stresses at the extreme fibres in the masonry arch cross-section according to Bandhauer and Young/Navier

Heyman (see section 4.6), according to which the compressive strength of the arch material is infinite (rigid-plastic material behaviour).

When $v = 0$, the functional value is 1 for eq. 4-12 and 4-13, i.e. the average compressive stress $\sigma_m$. Bandhauer's hyperbolic function eq. 4-12 assumes the functional value 3 for $v = 2/3$. The resulting stress diagram according to Bandhauer exactly matches the equation customarily used internationally these days (e.g. according to EC 6 [Jäger et al., 2006, pp. 379–384] and DIN 1053-100 [Graubner & Jäger, 2007, pp. 22–26]) for calculating the load-carrying capacity of masonry $N = N_T$ according to the equation

$$\sigma_B\left(v = \frac{2}{3}\right) = \beta_R = \frac{N_T}{b \cdot d} \cdot \frac{1}{\left(1 - \frac{2}{3}\right)} = 3 \cdot \frac{N_T}{b \cdot d} \tag{4-14}$$

or eq. 4-14 solved for $N_T$

$$N_T = \frac{1}{3} \cdot b \cdot d \cdot \beta_R \tag{4-15}$$

In eq. 4-14 and 4-15, $\beta_R$ stands for the characteristic compressive strength of the masonry being used in accordance with the relevant standards, e.g. EC 6 or DIN 1053-100. With his hyperbolic function for the compressive stress distribution in the arch cross-section, Bandhauer thus anticipated a considerable part of our modern understanding of masonry in the sense of ultimate load theory. Perhaps it was Bandhauer's tragic destiny and the justification of his masonry arch theory [Bandhauer, 1831] (see section 2.4.1.3) that led to his pioneering approach to the compressive stress distribution in the arch cross-section at the limit state being forgotten.

### 4.5.3 From Poncelet to Winkler

The fact that Navier's masonry arch and elastic arch theories were adopted at the same time is shown by the experiments carried out by Ardant on timber arch structures [Ardant, 1847] and the elastic analysis of an iron two-pin arch by Bresse [Bresse, 1848] plus the latter's monograph on the theory of the elastic arch [Bresse, 1854]. It was Saavedra [Saavedra, 1860] who first took up the idea of using elastic theory for masonry arches in 1860, at the suggestion of Poncelet [Poncelet, 1852]; however, his approach was not suitable for the practical design of masonry arch structures. Two years prior to this, Rankine had coupled the third prime task of thrust line theory with elastic theory; he postulated the theorem that the stability of a masonry arch is guaranteed "when a line of thrust due to a given loading case can be drawn through the middle-third of all arch cross-sections" [Rankine, 1858], i. e. passes through the kern cross-section of the arch joints. He therefore recognised the connection between the middle-third rule, anticipated by Young (see section 2.4.8.2) and worked out by Navier (see also Fig. 4-40), and the line of thrust concept. Later, the so-called line of thrust method, with which masonry arches with small and medium spans were analysed structurally as late as the 1960s, was formed in Germany by combining the masonry arch theories of Coulomb, Scheffler and Rankine [Schreyer et al., 1967, pp. 143–147].

With the boom in the building of wrought-iron arch bridges in France and Germany during the 1860s, the adoption of elastic arch theory by practising civil and structural engineers initiated the calculation of such structures. For instance, Hermann Sternberg (1825–85) used basic elements of this theory to design a bridge (completed in 1864) over the Rhine for the Koblenz-Lahnstein railway line. The Koblenz Bridge was the first structure to have trusses between curved concentric chords, and in addition was hinged at the supports – the structural system is therefore a two-pin arch. It became not only the reference structure for the wrought-iron arch bridges built over the Rhine in the following years, but indeed advanced to become a preferred object when discussing the application of elastic arch theory in construction engineering [Kurrer & Kahlow, 1998].

Sternberg's criticism of the determination of the critical loading case for calculating the forces in the arch chords inspired Emil Winkler to devise the influence line concept for three-pin, two-pin and fixed arches in 1867/1868 (see section 6.3.1.2). Despite the successes in the ongoing development of the theory of the elastic arch, Winkler left the question of its applicability to masonry arches unanswered.

The year 1868 saw Schwedler publish a design concept based on beam theory for jack arches spanning between rolled sections [Schwedler, 1868]. Up until the introduction of reinforced concrete slabs, such jack arches were the most common form of suspended floor construction for heavily loaded floors. Schwedler splits the imposed load $q$ on one half of the arch into a symmetrical component $q/2$ and an antisymmetrical component $q/2$ (Fig. 4-42). If we take the width of the masonry arch as being 1, then the equation for the compressive stress according to Schwedler takes on the following form

$$\sigma_{exist} = \frac{Z_0 \cdot r}{c} + \cdot \frac{\frac{q \cdot l^2}{64}}{\frac{c^2}{6}} \leq \sigma_{perm} \tag{4-16}$$

where

$$Z_0 = c + e + \frac{q}{2} \tag{4-17}$$

$r$ is the radius of curvature at the crown of the arch, and the horizontal thrust is

$$H = Z_0 \tag{4-18}$$

for the symmetrical loading case. The fixity moment $M$ for the antisymmetrical loading case appears in eq. 4-16 in the form of $q \cdot l^2 / 64$. The minimum thickness of the keystone $c_{min}$ can be calculated from the compressive stress equation 4-16 after solving the quadratic inequality.

Design engineers will have often used this method of modelling the jack arch, based partly on elastic theory. In 1868 Schwedler was promoted to First Secretary in the Prussian Ministry of Trade, Commerce & Public Works; from this high-ranking administrative post he was to exert a greater influence on the statics-constructional side of Prussian building than any other person over the next 20 years.

With Bauschinger's discovery of the non-linear stress-strain diagrams for the building materials commonly used in masonry arches, the deformation measurements of Claus Köpcke (1831–1911) carried out on existing masonry arches and the homogenisation of the masonry arch material, which had already been seen in French masonry bridge construction around 1850 [Mehrtens, 1885], the ice had been broken in favour of a masonry arch theory based on elastic theory. The loadbearing system analysis of wrought-iron, timber arches and masonry arches could now take place on the more abstract level of the structural system. This abstraction process was reflected in the German language by a shift in meaning for the terms *Gewölbe* and *Bogen* – with the latter becoming favoured.

In 1879 Weyrauch laid the foundation for a relatively autonomous further development of elastic arch theory by drawing up an arch analysis theory based on elastic theory [Weyrauch, 1879]. His radical departure from the loadbearing system analysis of the masonry arch and the elastic arch helped practising engineers become aware of the fact that a critical discussion surrounding the assumptions of structural masonry arch models was lacking. This was the hour of Emil Winkler's historico-critical work on the statics of masonry arches.

Winkler's presentation entitled *Lage der Stützlinie im Gewölbe* (the position of the line of thrust in a masonry arch) given at the Berlin Society of Architects on 17 March 1879 and again on 12 January 1880 was later published in the journal *Deutsche Bauzeitung* [Winkler, 1879/1880]. And 1879 was also the year in which Castigliano used elastic theory to calculate the masonry arch of the Ponte Mosca in Turin – assuming masonry

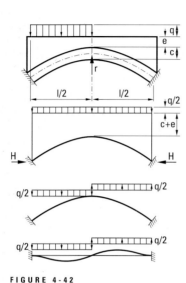

FIGURE 4-42
Schwedler's modelling of a shallow barrel vault as a fixed elastic beam in order to assess the stresses

with zero tensile strength [Castigliano, 1879] (see also [Perino, 1995]). Finally, Fernand Perrodil overcame the dualism between elastic arch theory and masonry arch theory in his monograph *Résistance des voûtes et arcs métalliques employés dans la construction des ponts* [Perrodil, 1879]. Around 1880, i. e. at the start of the classical phase of theory of structures (1875–1900), the synthesis of elastic arch and masonry arch theories, with the paradigm of elastic theory prevailing, was in the air. Winkler's presentation went beyond the other aforementioned contributions because he reflected the engineering science modelling and demonstrated the limits of the structural analysis model.

In his introduction, Winkler describes, in three steps, the object of a masonry theory founded on engineering science principles:

In the first step Winkler defines the line of thrust as the geometrical position of the point at which the resultants intersect the masonry joints. His second step is to state clearly that the crux of the object of masonry arch theory is to determine the position and not the form of the line of thrust. If the form of the line of thrust were to result simply from satisfying the equilibrium conditions, its position would require additional mechanical-mathematical assumptions. As the quality of such assumptions determines the applicability of a masonry arch theory for practical engineering purposes, the assumptions also form a logical set of tools for the critical assessment of historical masonry arch theories. Finally, in his third step, Winkler differentiates the *normal state* of the masonry arch from the *disrupted state*. Winkler treats the problem of assigning defined material properties and boundary conditions to real masonry arches. The masonry arch is subjected to disruptions: incompletely cured mortar, temperature changes, yielding centering during construction and, first and foremost, sinking abutments after striking the centering, which lead to visible cracks and considerable changes to the course of the line of thrust. Thereupon, Winkler proposes calculating the masonry arch for a certain ideal case: mortar evenly and completely cured in all joints, fully rigid centering, constant temperature and infinitely rigid abutments, i. e. the masonry arch has fully restrained points of fixity at both springings. Winkler designates this condition of the masonry arch as the normal state. The result of calculating the masonry arch in this normal state with the aid of elastic theory would supply the *correct line of thrust*.

After determining the object of a masonry arch theory founded on engineering science, Winkler turns to the structural analysis of the masonry arch in the normal state. Like Poncelet [Poncelet, 1852], he classifies and assesses the value of the engineering science knowledge and the practical engineering usefulness of masonry arch theories in a brilliant history-of-theory analysis.

Winkler then adapts his elastic arch theory, which had been tried out successfully on timber and iron arches, to the analysis of masonry arches. Fig. 4-43 illustrates the masonry arch model for which Winkler derives the three elasticity conditions in descriptive form, from which the position of the line of thrust in a masonry arch with three degrees of static indeterminacy can be determined. As the released impost joints may neither rotate ($\Delta\varphi = 0$) nor move horizontally ($\Delta u = 0$) or vertically ($\Delta v = 0$), the three elasticity conditions

$$\Delta\varphi = 0 = \int d\gamma = \int \frac{1}{E \cdot I_z} \cdot M(s) \cdot ds \qquad (4\text{-}19)$$

$$\Delta u = 0 = \int \Delta ds_x = \int \frac{1}{E \cdot I_z} \cdot M(s) \cdot y \cdot ds \qquad (4\text{-}20)$$

$$\Delta v = 0 = \int \Delta ds_y = \int \frac{1}{E \cdot I_z} \cdot M(s) \cdot x \cdot ds \qquad (4\text{-}21)$$

must be satisfied. $M$ is the bending moment in the masonry arch cross-section with the arc coordinate $s$, $E$ is the elastic modulus, and $I_z$ is the second moment of area of the arch cross-section about the $z$-axis. Only the influence of bending elasticity is taken into account in the above elasticity conditions.

Winkler also formulates a theorem equivalent to the three elasticity conditions, which is now named after him and which states that "for a constant thickness, the line of thrust close to the right one [is] the one for which the sum of the squares of the deviations from the centre-of-gravity axis is a minimum" [Winkler, 1879/1880, p. 128]:

$$I = \int [z(s)]^2 \cdot ds = Minimum \qquad (4\text{-}22)$$

Besides the function for the masonry arch centre-of-gravity axis, the required thrust line function, in the meaning of the smallest Gaussian error sum of squares is entered into the difference function $z(s)$ together with its three generally unknown position parameters $a$, $b$ and $c$. The necessary condition for eq. 4-22 is

$$\frac{\partial I}{\partial a} = 0, \quad \frac{\partial I}{\partial b} = 0, \quad \frac{\partial I}{\partial c} = 0 \qquad (4\text{-}23)$$

The vertical support reaction $X_1$, the horizontal thrust $X_2$ and the fixity moment $X_3$ at the right-hand arch impost (Fig. 4-43) can be used for the three position parameters $a$, $b$ and $c$ of the line of thrust, for example. Winkler's theorem is thus a special case of Menabrea's principle valid for elastic systems, or rather the principle of the minimum deformation complementary energy [Kurrer, 1987, p. 6]

$$\Pi^* = \int \frac{1}{E \cdot I_z} \cdot [M(X_1, X_2, X_3, s)]^2 \cdot ds = Minimum \qquad (4\text{-}24)$$

with the necessary conditions

$$\frac{\partial \Pi^*}{\partial X_1} = 0, \quad \frac{\partial \Pi^*}{\partial X_2} = 0, \quad \frac{\partial \Pi^*}{\partial X_3} = 0 \qquad (4\text{-}25)$$

In the integrand of eq. 4-24, $M$ is the bending moment depending on the static unknowns $X_1$, $X_2$ and $X_3$ plus the arc coordinate $s$. From the triple sequence of statically possible lines of thrust, only that line of thrust that causes the deformation complementary energy $\Pi^*$ to be a minimum becomes established in the masonry arch; the necessary conditions of eq. 4-25 can be easily converted into the three elasticity equations 4-19, 4-20 and 4-21.

Although Winkler did not consider the structural engineering implications when formulating his theorem, he concluded the search for the

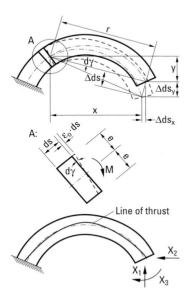

FIGURE 4-43
Winkler's determination of the position of the line of thrust in a masonry arch using elastic theory

true line of thrust in the masonry arch with the help of elastic theory. His theorem therefore superseded the structural modelling concepts of the thrust line and voussoir rotation theories.

In contrast to Winkler, Jean Résal had by 1887 classified the old methods of masonry arch theory not in four, but rather in three categories [Résal & Degrand, 1887]:

- *Méthodes des courbes des pression hypothétiques*: Méry [Méry, 1840], Dupuit [Dupuit, 1870], Scheffler [Scheffler, 1857]
- *Recherche du profil théorique des voûtes le plus advantageux au point de vue de la stabilité*: Villarceau [Villarceau, 1844, 1853, 1854], Denfert-Rochereau [Denfert-Rochereau, 1859], Saint-Guilhem [Saint-Guilhem, 1859]
- *Méthodes des aires de stabilité*: Durand-Claye [Durand-Claye, 1867].

Résal placed the new method for the structural analysis of masonry arches, which for him was identical with the theory of the elastic arch, ahead of the old methods.

### 4.5.4 A step back

Winkler's ground-breaking presentation inspired Engesser [Engesser, 1880], Föppl [Föppl, 1881] and Müller-Breslau [Müller-Breslau, 1883/1] to carry out their work on the elastic theory of masonry arches, which became a major factor in integrating masonry arch statics into the classical theory of structures taking shape in the 1880s. Nevertheless, there was no respite in the criticism of the elastic theory of masonry arches and its successive integration into the emerging general theory of plane elastic trusses as the engineering science answer to the supremacy of iron construction. For example, in 1882 the Bavarian Railways engineer Heinrich Haase declared "the recent attempt to apply the theory of the elastic arched beam to masonry arches [to be] a failed attempt to create a remedy for the obvious shortcomings of previous theories because it is only apt to make masonry arch theory more laborious, more complicated and more unacceptable to practice than before" [Haase, 1882 – 85, p. 90]. Taking up a position opposing that of Winkler, Haase pleads for "a theory appropriate to stone constructions which with corresponding modification is totally suitable for use on iron constructions as well" [Haase, 1882 – 1885, p. 77]. Haase actually carried out this remarkable about-turn by modifying the masonry arch theory of Hagen [Hagen, 1862] and then using it to re-analyse the Maria Pia Bridge – the 160 m-span trussed iron arch designed by Théophile Seyrig (1843 – 1923), the partner of Gustave Eiffel (1832 – 1923) – over the River Douro near Porto in Portugal [Trautz, 2002, p. 106]. His hope that his "important results of protracted and thorough studies" [Haase, 1882 – 85, p. 82] would be enthusiastically accepted by practising engineers proved to be an illusion. His lengthy article in the Viennese journal *Allgemeine Bauzeitung* went unnoticed.

### 4.5.5 The masonry arch is nothing, the elastic arch is everything – the triumph of elastic arch theory over masonry arch theory

The Masonry Arch Committee of the Austrian Society of Engineers & Architects carried out extensive tests on 23 m-span masonry arches made from rubble stones, clay bricks, plain concrete and reinforced concrete

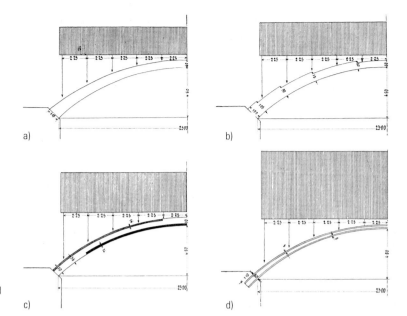

**FIGURE 4-44**
Comparison of the ultimate loads on trial arches made from a) rubble stone, b) clay bricks, c) plain concrete, and d) reinforced concrete [Spitzer, 1908, p. 354]

in the early 1890s and discovered that the deformations of the centre-of-gravity axis of every trial arch in the service condition increased more or less proportionately with the load [Spitzer, 1908]. Further load increases resulted in greater deformations, cracks and non-proportional displacements, until the arch finally collapsed. Fig. 4-44 shows the ratio of the measured ultimate loads on the rubble stone arch (Fig. 4-44a), the clay brick arch (Fig. 4-44b), the plain concrete arch (Fig. 4-44c) and the reinforced concrete arch (Fig. 4-44d); this ratio is 1 : 1.2 : 1.5 : 2.5. In terms of the ultimate load, the reinforced concrete arch therefore proved to be far superior to the masonry arch.

Furthermore, masonry arch materials and their elastic moduli were tested; but on the other hand, elastic arch theory permitted the elastic modulus to be determined from the measured displacements. The difference was enormous: the arches were far more elastic than had been predicted by the material tests, as Paul Séjourné (1851–1939) pointed out [Séjourné, 1913–16, vol. 3, p. 374].

The report on the masonry arch tests carried out by the Austrian Society of Engineers & Architects written by Brik in 1895 sums up the situation as follows: "The application of elastic theory enables the calculation of bridge arches without the help of arbitrary assumptions. However, this application can only be valid when the prerequisites of the theory are fulfilled through the construction of the structure" [Spitzer, 1908, p. 352]. Reinforced concrete fulfils these prerequisites.

Emil Mörsch, that old master of reinforced concrete construction in Germany, published a clear structural analysis method based on the theory of the elastic arch in 1906 in the journal *Schweizerische Bauzeitung* [Mörsch, 1906/1]. Using this method, every engineer was now able to design and analyse arches of plain and reinforced concrete, which permitted the direct determination of the influence lines for the statically indeterminate and kern point moments (Fig. 4-45). The goal of evaluating these influence lines for the varying trains of loads (e.g. railway trains) is to calculate the kern point moments in the relevant arch cross-sections for the most unfavourable loading case. Thus, the metamorphosis of the masonry arch to the elastic arch was essentially complete.

The method for the rational calculation of arch bridges of masonry plus plain and reinforced concrete published by Mörsch in 1906 appeared

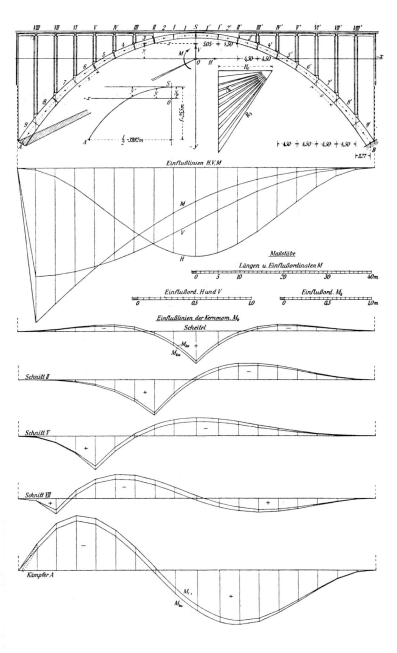

**FIGURE 4-45**
Influence lines of static unknowns H, V, M as well as the upper and lower kern point moments for the Gmündertobel Bridge, Switzerland, after Mörsch [Mörsch, 1947, p. 188]

in 31 editions of the influential concrete yearbook *Beton-Kalender* between 1906 and 1952, and in fact a reprint of the 1952 edition on arched bridges appeared in the year 2000 edition of *Beton-Kalender* [Mörsch, 2000]. Numerous masonry and monolithic arch bridges were still being designed and analysed according to this method as late as the early years of the integration period of theory of structures (from 1950 to date). The American consulting engineer Joseph W. Balet also analysed masonry, steel and reinforced concrete arches all in the same way using elastic arch theory [Balet, 1908].

The shift in the meaning of the word *Gewölbe* in the German language followed in the footsteps of the rise in popularity of reinforced concrete. Marginalised in theory of structures and superseded by the term *Bogen*, the *Gewölbe* retained its importance for the building industry only in the historical sense.

After the completion of the Pont Antoinette near Vielmur-sur-Agoût in 1884 (a railway bridge with a masonry arch spanning 47.53 m), the building of large masonry arch bridges underwent a renaissance closely linked with the name of the French civil engineer Paul Séjourné which extended into the first half of the consolidation period of theory of structures (1900 – 50). Like Mörsch, Séjourné knew how to use influence line theory in a creative and elegant way for the structural analysis of large masonry arch bridges. One example of this is his structural calculations for the Pont Antoinette. However, whereas Mörsch had plain and reinforced concrete arch bridges in mind, Séjourné's interest lay in masonry arch bridges. His epic six-volume work *Grandes Voûtes* [Séjourné, 1913 – 16] constitutes a literary monument to masonry arch bridges. But the age of the *Grandes Voûtes* was already drawing to a close by the end of the second decade of the 20th century, pushed aside by reinforced concrete bridges.

Alfred John Sutton Pippard (1891 – 1969), Eric Tranter and Letitia Chitty invited discussion on extensive tests carried out in 1936 on a model masonry arch at Imperial College, London, on behalf of the Building Research Board [Pippard et al., 1936/1937]. The model masonry arch with span $l = 121.92$ cm and rise $f = 30.48$ cm consisted of 15 machined steel model "voussoirs" of thickness $d = 7.62$ cm and depth $t = 3.81$ cm. Fig. 4-46 shows the model masonry arch together with the testing apparatus.

The research focused on the static behaviour of the masonry arch upon movement of the imposts. Fixed, three-pin and two-pin arches were investigated. The variable to be measured was the horizontal thrust $H$, which was also determined analytically and shown graphically in the form of influence lines. In doing so, the authors calculated the statically indeterminate systems according to the necessary condition 4-25 derived directly from the principle of Menabrea. The agreement between tests and calculations based on elastic arch theory are very good, which is not very surprising for a model masonry arch consisting of steel blocks, which, despite the joints, form a quasi linear-elastic continuum in single curvature. Another important finding of this research was that displacement of the two abutments causes the masonry arch to behave statically like an elastic three-pin arch – corresponding to cases 3 and 4 of Coulomb's voussoir rotation theory (see Fig. 4-29). At the end of their paper, the authors recommend that practising engineers should analyse masonry arches as elastic three-pin arches [Pippard et al., 1936/1937, p. 303]. In Germany, for example, the line of thrust theory (see section 4.5.3) follows on from the model of the three-pin arch [Schreyer et al., 1967, p. 143].

In the 1961 James Forrest Lecture, Pippard stood up in a very elegant way for the paradigm of elastic theory in structural analysis in general and the theory of the elastic arch for analysing masonry arches in particular [Pippard, 1961, pp. 145–149]. Nevertheless, even a cantankerous character like Pippard realised that plastic methods of design (see section 2.10) were justified too: "The behaviour of structures in the plastic region has many able exponents and the success in design achieved by this approach speaks for itself, but I would say again what I have said elsewhere, that the elastic and plastic approaches to design must, for the best results, be considered as complementary and not conflicting. … Elastic and plastic theory both have much to offer. They are concerned with different but equally important aspects, not of 'the fantastic behaviour of things' referred to in my opening remarks, but of that orderly and predictable behaviour upon which the engineer must rely if his designs are to be anything but guesswork" [Pippard, 1961, p. 153]. Many things characterise the innovation phase of theory of structures (1950–75), one of them being the fact that elastic theory lost ground to plastic theory as a fundamental approach in structural analysis, and that plastic methods of design became more and more significant in the everyday workloads of structural engineers – also and particularly in the analysis of masonry arch loadbearing systems.

**FIGURE 4-46**
Model masonry arch with testing apparatus [Pippard et al., 1936/1937, p. 284]

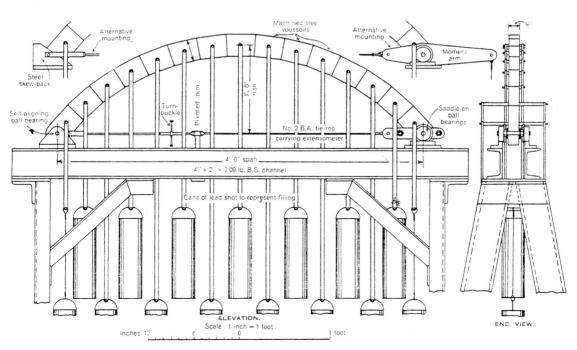

Doubts about the use of elastic arch theory for analysing masonry arches had already been voiced by George Fillmore Swain (1857–1931) as early as 1927 (see [Foce, 2005]). This influential professor of civil engineering at Havard University dedicates one chapter of the third volume of his work *Structural Engineering* to masonry arches and points out the difference between masonry arches and elastic arches: "Since the stone arch is an elastic arch, differing only in degree from a monolithic concrete arch, it is impossible to distinguish sharply between the two. Before the student of structural engineering begins the study of elastic arches, it is desirable that he should study carefully this chapter on the stone arch, notwithstanding the fact that stone voussoir arches are now seldom built" [Swain, 1927, p. 400]. As a student of Emil Winkler, he was fully familiar with Winkler's elastic theory for masonry arches, especially the difference between the normal and disrupted state of the arch (see section 4.5.3), which in the end indicates the difference between the loadbearing structure as a material form of the masonry arch and the structural system as the static model of the masonry arch, hence the model assumptions that deviate considerably from the disrupted state of the masonry, e.g. unintentional displacement of the abutments, or the influence of the change in the mortar consistency over time on the internal forces in the arch: "… The elastic theory seems to be firmly intrenched in American engineering literature. Perhaps some who use it do not realize its defects and assumptions, and like it because it is complex and mathematical. It seems to be a curious characteristic of the human mind that it so often prefers complexity to simplicity, and mistakes obscurity for profundity … The writer believes in elastic methods, if they are necessary; not if they are unnecessary and if a simpler method is just as good" [Swain, 1927, p. 400]. The descent from the concrete to the abstract, or rather the idealised assumptions of elastic theory, on the other hand, lead to a more complex method of structural analysis. When it comes to the structural analysis of masonry arches, Swain prefers methods that are simpler when compared with elastic arch theory, which Federico Foce sees not only as a historicological product of pre-elastic masonry arch theories, but also as a premonition of an ultimate load theory for masonry arches: "… Swain's peroration in favour of a 'weighted' use of the elastic methods for the analysis of the masonry arch has the value of a methodological choice whose last consequences lead to the structural philosophy of limit analysis that, after Heyman's lesson, is nowadays considered as the basis for the study of the stone skeleton" [Foce, 2005, p. 140].

## 4.6
**Ultimate load theory for masonry arches**

Ultimate load theory (see section 2.10), originally formulated for structural steelwork, can be applied to masonry structures provided the masonry material complies with certain conditions. Drucker was the first person to suggest using ultimate load analysis for investigating the equilibrium and failure of voussoir arches. Following Drucker's lead (see also [Drucker, 1953]), Kooharian published the first modern work on this subject in 1952 [Kooharian, 1952], which was followed one year later by Onat and Prager's input [Onat & Prager, 1953]. In his introduction to plastic theory published in 1959, Prager describes the material conditions the voussoirs have to satisfy so that ultimate load theory can be rigorously applied and the corresponding yield surfaces drawn [Prager, 1959]. Another milestone was Heyman's publication of 1966 in which he explains and discusses, rigorously and universally, for the first time the applicability of ultimate load theory for any masonry loadbearing structures and not just voussoir arches [Heyman, 1966]. In the following years, Heyman wrote copiously about the application of ultimate load theory to various masonry structures such as plane arches, domes, fan vaults, groined vaults, towers and spires, plus ways of assessing the safety of such structures or loadbearing systems totally in masonry. Heyman's contributions are so fundamental that it is difficult to imagine today's state of the art without his *œuvre*. Heyman summarises the theory of masonry loadbearing structures in his monograph [Heyman, 1995/1]. The most important aspects of this theory are presented below in order to highlight the way in which current masonry arch problems can be solved with the failure theory of the 18th and early 19th centuries that Heyman introduced into ultimate load theory.

In order to be able to present the theory of masonry loadbearing structures in the context of ultimate load theory, the masonry material must satisfy three conditions, which Heyman called the "principles of ultimate load analysis of masonry constructions":

1. The compressive strength of the masonry is infinite.
2. The tensile strength of the masonry is zero.
3. Adjacent masonry units cannot slide on one another.

These three conditions agree with Couplet's assumptions mentioned in section 4.3.4. The first statement is suspect because obviously no material has infinite strength. However, even in the largest masonry constructions, the actual stresses are one or two orders of magnitude below the compressive strength of the material itself. This is therefore a reasonable assumption that can be checked at the end of the analysis. Boothby presents other stress-strain principles apart from Heyman's rigid-plastic material law for masonry [Boothby, 2001]. The second statement also lies on the safe side because the mortar between the masonry units does indeed exhibit a certain adhesion. The third statement is connected with the high coefficient of friction of masonry ($\mu$ = 0.6 to 0.7, which corresponds to an angle of friction of 35° to 39°).

When the masonry material satisfies these conditions, the component of the resultant of the effective stresses acting perpendicular to the cross-sectional area must be a compressive force $N$ for each cross-section whose intersection point lies within the cross-section. If the compressive force $N$ acts at the edge of the cross-section, a hinge forms (Fig. 4-47a). This leads to a yield surface bounded by two straight lines [Prager, 1959; Heyman, 1966] (Fig. 4-47b). The moment $M$ is nothing more than the product of the normal force $N$ and the eccentricity e, i.e. $M = N \cdot e$; here, the eccentricity must satisfy the condition $-h \leq e \leq +h$ (Fig. 4-47).

**FIGURE 4-47**
a) Formation of a hinge between two voussoirs [Heyman, 1982, p. 31], b) moment-normal force interaction diagram with yield surface in rigid, unilateral masonry [Heyman, 1982, p. 32]

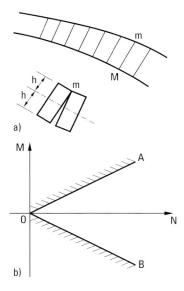

For a masonry arch of thickness $2 \cdot h = d$ (Fig. 4-47a), the equation for the straight line $0A$ (Fig. 4-47b) is

$$M_{0A}(N) = +h \cdot N = +\frac{d}{2} \cdot N \qquad (4\text{-}26)$$

and, correspondingly, for the straight line $0B$

$$M_{0B}(N) = -h \cdot N = -\frac{d}{2} \cdot N \qquad (4\text{-}27)$$

For pairs of values of $M$ and $N$ that lie within the area $A0B$ bounded by the straight line equations 4-26 and 4-27, i.e. also not on the straight lines $0A$ and $0B$, the normal force $N$ acts within the cross-section of the masonry arch. If this condition is satisfied for all the masonry cross-sections, then the line of thrust lies completely within the arch profile: the masonry arch is stable in the kinematic sense. Hinges then form when $N$ acts on the edge of the cross-section and therefore the pairs of $M$ and $N$ values lie on the straight lines given by eq. 4-26 or 4-27. If the pairs of values $M$ and $N$ are above or below the straight lines $0A$ and $0B$, then the normal force $N$ acts outside the arch cross-section: the masonry arch is unstable in the kinematic sense.

## Of cracks and the true line of thrust in the masonry arch

### 4.6.1

In sections 4.4.2 and 4.4.3 it was shown that knowledge of the position of the true line of thrust was of fundamental importance for the engineers of the 19th century, and that the use of an elastic analysis for plane masonry arches was supported by the fact that the position of the line of thrust could be calculated from the elasticity equations 4-19, 4-20 and 4-21. This problem will now be discussed in the context of ultimate load theory.

To do this, we shall consider a masonry arch supported on centering (Fig. 4-48b). After the centering is struck, the masonry arch begins to press against its abutments. Real abutments are not rigid and inevitably yield by a certain amount. This spreading increases the span and the masonry arch has to adapt to the change in geometry. The question is how does a masonry arch – built from the aforementioned rigid, unilateral material – manage this? The answer can be seen in many bridges and also in tests on masonry arch models: cracks form to permit the necessary movement. A downward crack appears at the keystone and two upward ones at the abutments.

The masonry arch develops three hinges, and that makes it statically determinate, i.e. the position of the line of thrust is fixed by the three hinges (Fig. 4-48a). The movement may be asymmetrical – it could be that the right-hand abutment yields not only horizontally, but vertically as well. Every possible movement corresponds to a certain pattern of cracks, and the masonry arch responds to changes to the boundary conditions by opening and closing the cracks. This can also be seen on models. Even simple, "plane" cardboard models supply very good results [Huerta, 2004] (Fig. 4-48b). Cracks are therefore not dangerous. It is precisely the load-bearing structure's ability to form cracks that enables it to react to changing boundary conditions. And this ability is a function of the material properties: infinite compressive strength, zero tensile strength, no sliding.

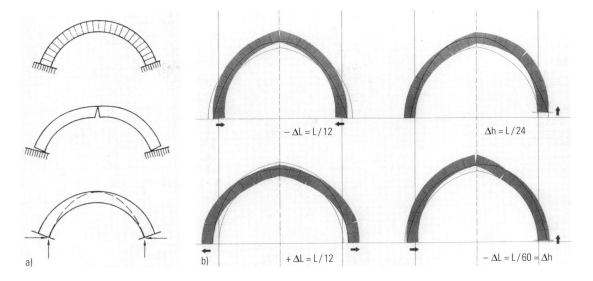

**FIGURE 4-48**
a) Crack formation in a masonry arch after striking the centering [Heyman, 1995/1, p. 15], b) various crack patterns due to movement at the abutments [Huerta, 2004, p. 78]

The local distribution of the cracks defines the position of the line of thrust unequivocally. If the distribution of the cracks in the arch profile changes, then the position of the line of thrust also changes and hence the internal forces condition too (Fig. 4-48a). The movements in the model are very large, but even small movements invisible to the naked eye have the same effect. Unfortunately, it is not possible to know or predict this type of disruption, and so establishing the true line of thrust in the masonry arch is impossible, i. e. to know the particular equilibrium condition of the arch at any one moment. However, there are two extreme positions of the line of thrust that correspond to the minimum and maximum horizontal thrust (see Fig. 4-38).

The cracks behave like hinges and it is precisely the material properties specified in the previous section that enable this hinge formation. The research work of Jagfeld and Barthel, backed up by experiments, confirms this concept of the hinge formation in historical masonry structures [Jagfeld & Barthel, 2004].

### 4.6.2 Masonry arch failures

As the material has infinite compressive strength, collapse must occur after the formation of a kinematically permissible failure mechanism (see Fig. 4-21). If the line of thrust touches the edge of the arch profile, a "hinge" forms and rotation is possible. Three hinges add up to a statically determinate arch, but one more hinge changes the arch into a kinematically permissible hinge mechanism (Fig. 4-49). An increase in the load beyond the amount necessary for the formation of a hinge mechanism therefore leads to failure of the entire arch without the material being crushed. In a stable masonry arch, this can happen if additional loads are applied that deform the line of thrust beyond a certain point. Again, the catenary model illustrates this limit state for the equilibrium of masonry arches (Fig. 4-49).

In his dissertation, John A. Ochsendorf investigates the failure of masonry arches due to displacement at the abutments and presents his computer program "Arch Spread" with which parameters such as the position of the joint of rupture for circular arches of constant thickness can be calculated [Ochsendorf, 2002, p. 84 ff.] (see also [Ochsendorf, 2006]).

### The maximum load principles of the ultimate load theory for masonry arches

#### 4.6.3

If a line of thrust can be drawn within the profile of the arch, then there is a least one possibility that the arch can resist the given loading. But does this also mean that the arch really is stable? Or might the arch find a way of collapsing? Could a small, unforeseen movement lead to cracks that finally result in failure?

The solution to this problem was first possible in the 20th century through adapting ultimate load theory to masonry structures. Therefore, the factor of safety theorem states that the loadbearing structure will not collapse and can be considered as "safe" when an equilibrium condition can be found that does not infringe the hinge condition; such an equilibrium condition is statically permissible. This is the lower bound of the ultimate load. The upper bound of the ultimate load is given by the kinematic maximum load principle, which results from a permissible, inevitable kinematic sequence and can be quantified using the principle of virtual displacements; such a kinematic sequence is kinematically permissible (see Fig. 4-49c).

In the case of a masonry arch, every line of thrust drawn for a given loading satisfies the equilibrium conditions. The requirements placed on the material have already been mentioned above. The most important condition is that the material must resist compressive stresses, i.e. the stress resultants in every cross-section must lie within the arch profile. So if the line of thrust runs completely within the profile of the arch (hinge condition), then this is sufficient evidence for the fact that the arch is stable and will not collapse under the given loading.

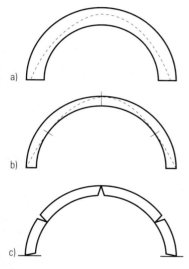

FIGURE 4-49
Failure of a semicircular arch subjected to a point load [Heyman, 1995/1, p. 19]

The factor of safety theorem does not make any statements about the boundary conditions. Cracks form in the masonry arch as a reaction to movements at the abutments, as Fig. 4-48b shows. A new equilibrium condition is set up, i.e. the position of the line of thrust changes, but it always remains within the profile of the arch and never forms a sufficient number of hinges to convert the arch into a failure mechanism. The factor of safety theorem of ultimate load theory therefore offers one solution for the problem of determining a characteristic line of thrust. It is not possible to know the true line of thrust, but this is also not important because the safety of the loadbearing structure can be calculated without having to make assumptions regarding its actual status.

### The safety of masonry arches

#### 4.6.4

The safety of masonry arches can be calculated with the two ultimate load theorems. Heyman proposes a geometrical factor of safety. This is the result of comparing the geometry of the true masonry arch with that of an arch of minimum thickness that would just carry the given loads. From Fig. 4-50a we can see that the arch is safe according to the factor of safety theorem because there is one possible line of thrust that lies completely within the profile of the arch. If we now reduce the thickness of the arch, we will finally reach a certain value that can accommodate only one single line of thrust within the profile (Fig. 4-50b). This line of thrust touches the edge of the arch profile at five places (for reasons of symmetry). This means there are five hinges; and according to the kinematic ultimate load theorem the arch is in unstable equilibrium and would fail after only a minimal further increase in load (upper bound of ultimate load).

The factor of safety of the original masonry arch can now be quantified by comparing the thickness of the arch (Fig. 4-50a) with the minimum thickness of this limiting arch (Figs 4-50b and 4-50c). If the real arch has twice the thickness of the limiting arch, then the geometrical factor of safety is 2. In the case of a bridge, the limiting arch has to be determined for the most unfavourable loading case [Heyman, 1969]. After investigating existing arches and bridges, Heyman recommends a value of 2 for the most unfavourable loading case [Heyman, 1982].

The determination of the exact value of the geometrical factor of safety sometimes involves elaborate calculations. However, a lower bound is very easy to determine. In order to show, for example, that the geometrical factor of safety for a certain masonry arch and a given loading is ≥ 2, it is sufficient to draw a line of thrust within the middle-half of the arch profile. For a geometrical factor of 3 or more, the line of thrust must lie within the middle-third, i. e. within the kern of the arch cross-section; this is nothing other than the middle-third rule. The ultimate load theorems furnish strong confirmation of the intuitive insights of masonry arch theorists such as Gregory [Gregory, 1697] in the 17th century, Couplet [Couplet, 1729/1731 & 1730/1732] in the 18th, Rankine [Rankine, 1858] in the 19th, and Swain [Swain, 1927] in the 20th.

Of course, determining the line of thrust according to elastic theory is a verification of the safety of the masonry arch provided the ensuing line of thrust runs completely within the arch profile. The only difference is that the advocates of elastic theory believe they have found the "true" line of thrust, whereas the advocates of plastic theory know that the line of thrust calculated with the help of elastic theory is only one of an infinite number of lines of thrust running within the arch profile that are in equilibrium with the loading.

One important result of ultimate load analysis applied to masonry arches is that it enables the equilibrium approach for the analysis of the loadbearing system, i. e. simple statics in the sense of the theory of statically determinate systems. Heyman was the first to point out the equilibrium approach explicitly and its importance for assessing structural stability; that was in 1967 [Heyman, 1967]. He mentioned it again more generally in 1969 [Baker & Heyman, 1969] and later emphasized the extraordinary significance of this corollary in many of his publications.

It is not the task of the structural engineer to determine the true equilibrium condition for a particular loadbearing structure, but rather sensible equilibrium conditions. And all the great engineers and architects

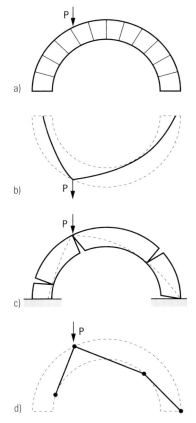

FIGURE 4-50
a) Stable semicircular masonry arch,
b) semicircular masonry arch of minimum thickness showing line of thrust,
c) and d) associated kinematically permissible hinge mechanism
[Heyman, 1995/1, p. 21]

**Analysis of a masonry arch railway bridge**

really have followed this approach. The equilibrium approach was implicit in the rules of proportion applied by the builders of ancient times. And, explicitly, it is to be found in the works of Maillart, Torroja, Nervi, Candela and Gaudí, to name but a few of the great engineers and architects of the last century.

**4.6.5**

In his dissertation completed in 1999 at Munich TU, Wilmar Weber examines the masonry arch single-track railway bridge at Schauenstein in Upper Franconia [Weber, 1999]. The three-centred arch built in 1919 has a clear span of $l_j = 14$ m and a clear rise of $f_j \approx 4.20$ m. The thickness at the crown is $d_S = 0.75$ m and this increases towards the springings. The arch consists of tamped concrete containing plate-like, angular aggregates with a facing of dressed granite stones from the Fichtel Mountains, which serve as permanent formwork. Fig. 4-51 shows the collapse mechanism of this railway bridge at Schauenstein for the static ultimate load $F^*$.

Strictly speaking, this railway bridge is not a true masonry arch in the sense of the proposed definition of this type of structure given at the start of chapter 4 because the arch is not built from jointed masonry. According to Weber's findings, the material parameters of the tamped concrete exhibit distinct anisotropy in the radial and tangential directions to the extent that modelling the bridge as an elastic arch is also inappropriate. The loadbearing behaviour of the bridge is akin to that of a masonry arch, and so only masonry arch theories can be considered for analysing the loadbearing system. For simplicity therefore, the bridge will be discussed below in the sense of a masonry arch.

Based on the maximum load principles of ultimate load theory, Weber develops a hybrid method for determining the upper bound to the load-carrying capacity of single-span masonry arch railway bridges. To validate his method, static failure load tests on a single-track masonry arch railway bridge were carried out for the first time in 1989. To do this, the point load $F$ was applied at one of the quarter-points of the masonry arch railway bridge at Schauenstein and increased up to a total compressive force of

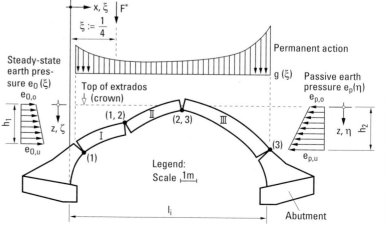

**FIGURE 4-51**
Failure mechanism of a single-span arch bridge with vertical and horizontal actions plus fully restrained abutments after Weber [Weber, 1999, p. 90]

$F^* = 3405$ kN; at this point the failure mechanism shown in Fig. 4-51 became established, which in principle agrees with that of Heymann (see Fig. 4-49c). Taking this kinematic sequence from the test, Weber calculates the upper bound of the maximum load-carrying capacity as $F^* = 3308$ kN according to the kinematic principle of ultimate load theory; he bases this work on the principle of virtual displacements and takes into account, in particular, horizontal actions such as earth pressure. The difference between the calculated maximum load-carrying capacity and the capacity as measured is only 4.1 %. The equations derived by Weber for calculating the horizontal and vertical components of measured, large displacement fields within the scope of his kinematic examination of the railway bridge at Schauenstein represent a new approach. Weber sees a need for research into the derivation of a measure of "voussoir rotation resistance", which plays a fundamental role in the formation of the hinges. Jagfeld and Barthel have reported on tests on the formation of hinges in masonry arches [Jagfeld & Barthel, 2004]. In the meantime, research projects looking into masonry subjected to an eccentric compressive loading have been completed and incorporated in Eurocode EC 6 and diverse National Application Documents (NAD). For example, DIN 1053-100 (August 2006 edition) assumes rigid-plastic deformation behaviour for masonry subjected to an eccentric load, i.e. a rectangular stress distribution [Graubner & Jäger, 2007]. According to DIN 1053-100, for a masonry cross-section of thickness $d$ and width $b$, the permissible normal force $N_T$ acts at an eccentricity of $e = d/3$. The distance from $N_T$ to the nearest extreme fibre is therefore $d/6$, i.e. the rectangular stress distribution has a length of $d/3$; taking $\beta_R$ as the characteristic compressive strength, the permissible normal force is $N_T = (d/3) \cdot b \cdot \beta_R$ [Jäger et al., 2006, p. 161]. This value corresponds exactly with that obtained from Bandhauer's hyperbolic function (see eq. 4-14). It should be noted that through the transition to the rectangular stress distribution in DIN 1053-100, the desired increase in the load-carrying capacity has been taken into account in EC 6 and the hinge-formation phenomenon is not relevant. Nevertheless, the analysis of the cross-sectional capacity of masonry arches is more realistic. Back in 2003, Alfred Pauser used this as his basis for developing an interaction diagram for verifying adequate (= safe) loading capacity in his report on the load-carrying capacity of existing circular masonry arches which he carried out for Austrian Railways [Pauser, 2003, p. 33 ff.].

Weber's dissertation also includes a table listing the bridge surveys of European railway operators, most of which date from the early 1990s. The numbers of masonry arch railway bridges belonging to the former State Railways are as follows:
– France, 13 167 (Nov. 1992)
– United Kingdom, approx. 13 000 (Feb. 1993)
– Germany, 9146 (1993)
– Czechoslovakia, 3213 (Jan. 1992)
– Spain, 3205 (Jun. 1992).

These are all bridges still in use for rail traffic. The percentages in terms of the total railway bridge stock for the above railway networks is considerable: France 47%, United Kingdom 50%, Germany 29%, Czechoslovakia 34%, Spain 50%. From this it is clear that there is a great need to be able to assess masonry arch railway bridges realistically in case a further increase in loads due to changing railway operations is envisaged. And methods of analysis based on ultimate load theory could play a major role here. An overview and discussion of the literature published to date on the calculation of masonry arch bridges can be found in the publication by Proske, Lieberwirth and van Gelde 2006 [Proske et al., 2006].

The introduction of ultimate load theory approaches into structural methods of analysis from the 1950s onwards enabled the paradigm of elastic theory to be discussed by Heyman and others in the sense of the structural analysis of historical masonry arches as well [Heyman, 1980/1 & 1982]. Heyman has interpreted the kinematic failure approaches in the masonry arch statics of the 18th century as ultimate load theory, and has developed and applied simple methods to analyse historical masonry arches [Heyman, 1972/2, 1972/3 & 1980/2]. The heuristic character of a history of theory can be demonstrated by way of such theory formation processes. Becchi, Corradi and Foce have attempted to gain an insight into the history of masonry arch theory for a better understanding of the load-carrying ability of historical masonry arches, especially with a view to refurbishment and restoration work [Becchi et al., 1994/2; Sinopoli et al., 1997]. A systematic use of historical knowledge in the engineering science cognition process presumes a prolegomenon of a historical engineering science. But the history has not even begun here!

The edition of the journal *Meccanica* issued by Christopher R. Calladine on the occasion of Heyman's transfer to emeritus status is dedicated to the structural mechanics analysis of masonry constructions. It includes pioneering contributions from Livesley [Livesley, 1992], Di Pasquale [Di Pasquale, 1992], Como [Como, 1992] and Augusti and Sinopoli [Augusti & Sinopoli, 1992], all of which show an allegiance to Heyman's ultimate load theory for masonry constructions.

Ochsendorf has developed Heyman's masonry arch ultimate load theory further and used this to investigate masonry arch abutment systems, e.g. buttresses, in particular [Ochsendorf, 2002]. Huerta included a splendid description of Heyman's masonry arch ultimate load theory in the historical context within the scope of his lengthy monograph *Arcos, bóvedas y cúpulas. Geometría y equilibrio en el cálculo tradicional de estructuras de fábrica* [Huerta, 2004]. Huerta also arranged for Spanish editions of Heyman's books (e.g. [Heyman, 1995/2 & 1999]).

Boothby, too, is convinced of the benefits of Heyman's historically based ultimate load theory for the analysis of historical masonry arches: "The rigid-plastic analysis is a compelling treatment of the stability of masonry arches and vaults, and has advanced modern understanding of ancient construction techniques considerably. It is particularly useful in its appeal to physical intuition ..." At the same time, however, he draws attention to its fundamental assumptions, which would limit its application noticeably: "… The assumptions made in invoking rigid-plastic analysis limit the ability to address apparent phenomena in the behaviour of actual arch and vault structures. Although the assumption of rigidity is fundamentally justified for the units in ashlar masonry, the mortars used in ancient masonry construction are deformable and have been found to exhibit significant deformation over time … Neither the equilibrium method nor the mechanism method lend themselves comfortably to application to the analysis of structures with complex three-dimensional geometry, such as cathedral vaults or skew bridges" [Boothby, 2001, p. 250]. Owing to these restrictions placed on the ultimate load theory of masonry arches, Boothby sees two successful development paths for research: firstly, in the direction of practical plastic theory, and secondly, in the direction of developing computer tools for assessing complex geometries and material laws.

## 4.7 The finite element method

The first elastic analysis of a Gothic building can certainly be attributed to Robert Mark. In the 1960s, he used the photoelastic technique to determine experimentally the stress states of a number of French cathedrals. As this method can only measure the stress states of elastic plates, Mark investigated plate-form models of the cross-sections of Gothic cathedrals. Criticism of this approach has been voiced by Klaus Pieper, who says that such measurements on masonry structures can hardly be expected to agree adequately with the reality [Pieper, 1983]. Notwithstanding, elastic analysis – primarily influenced by Mark's work [Mark, 1982] – were carried out in the 1970s and 1980s with the help of the finite element method (FEM). The problem here is that the loadbearing structure is modelled as a continuum with known elastic properties and precise, known boundary conditions. As this does not apply to masonry buildings, the FEM analysis of masonry structures according to elastic theory is a purely "academic" exercise. As recently as 2001, Herrbruck, Groß and Wapenhans criticised the linear-elastic analysis of masonry arch bridges in the form permitted by the DIN 1053 and EC 6 editions current at that time, calling it "destructive calculation", and therefore pleaded for a non-linear analysis. "However, non-linear analysis is not a miracle cure. But an excellent alternative in many apparently hopeless cases" [Herrbruck et al., 2001]. Christiane Kaiser, too, in her 2005 dissertation on the Fleisch Bridge in Nuremberg (1596–98), comes to the conclusion that modelling as a fixed-end elastic arch achieves an acceptable result [Kaiser, 2005, p. 207].

Some FEM programs permit the simulation of a unilateral material without any tensile stresses. There are various possibilities here, but as a rule an iterative process is involved. Following an initial elastic analysis, those nodes at which tension occurs are "broken", and discontinuity lines form. The new structure is analysed anew and after a number of iterations the process converges on a solution in which only compressive forces occur. This is, of course, far better than the normal linear-elastic analysis.

In 2000 Schlegel and Rautenstrauch presented an elasto-plastic calculation model for the three-dimensional investigation of masonry structures, verified by tests, in which the masonry is described with the help of a spread equivalent continuum [Schlegel & Rautenstrauch, 2000]. Using this as a basis, it was able to re-analyse the Göltzsch Viaduct (78 m high and 574 m long, built between 1846 and 1851, see section 2.4.4) [Schlegel et al., 2003; Schlegel, 2004]. Fig. 4-52 shows the three-dimensional FEM model consisting of 46 504 elements and 63 336 nodes with a total of 190 008 degrees of freedom.

Nine material parameters were necessary to describe the strength of the masonry bond. The stability of the railway bridge and the admissibility of the new direction of travel was able to be verified in compliance with the applicable standards. The authors point out "that linear calculations cannot map, for example, the activation of the arch effect under dead loads. … Likewise, load redistributions, which are evident due to thermal effects in particular, could lead to noticeable non-linear effects which could be regarded as highly problematic. Another necessity for the realistic assessment of the stresses in the structure is the three-dimensional modelling. Only with a 3D model is it possible to consider the eccentricity and unfavourable superimposition of various actions and to guarantee the full activation of the reserves of strength in the masonry construction" [Schlegel et al., 2003, pp. 22 – 23].

The question that crops up almost involuntarily is which FEM models should be used in order to take account of the discontinuity and irregularity so intrinsic to the loadbearing behaviour of masonry? Of course, the loading history must also be considered – for the final condition also depends on the chronological sequence of loads and movements to which

**FIGURE 4-52**
FEM model of the Göltzsch Viaduct
[Schlegel et al., 2003, p. 17]

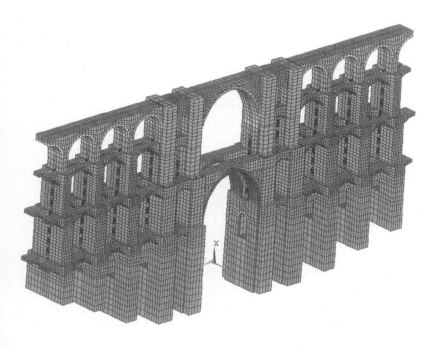

the loadbearing structure has been subjected. Therefore, in 2005 Schlegel, Konietzky and Rautenstrauch proposed modelling the structural mechanics of masonry within the scope of discontinuum mechanics with the help of the distinct element method (DEM) [Schlegel et al., 2005].

However, programs can only be implemented successfully when their users are able to interpret the results and, if necessary, adapt the analysis.

Barthel had written a splendid review of Gothic groined vaults back in 1993/94 [Barthel, 1993 & 1994], which takes into account the main aspects of the loadbearing system analysis of such vaults and contains a historical overview of the methods employed. In addition, the book [Barthel 1993/1] includes plenty of information about measurements of and damage to buildings that is otherwise difficult to obtain from the relevant literature. Barthel presents the results of a non-linear FEM analysis that permits predictions about the cracks caused by certain types of yielding at the abutments for various types of masonry arch (Fig. 4-53). One of his findings is that even a minor displacement of 0.5 to 2 mm for a span of 10 m (i.e. 1/20 000 to 1/5000 of the span!) leads to cracks and a drastic change in the internal stress condition and the value of the horizontal thrust. The latter drops by up to 64 % depending on the form of the masonry arch (see table in Fig. 4-53). It is obvious that a tool with such a degree of uncertainty for such tiny displacements (which cannot even be measured on masonry structures!) is unusable. Barthel's study indicates that we should refrain from using linear-elastic FEM analyses of masonry arches and other masonry loadbearing structures.

Non-linear FEM analyses of three-dimensional masonry arches are necessary of course, like the elastic analyses of two-dimensional masonry arches at the end of the 19th century. However, they are not sufficient to establish fully the loadbearing behaviour. To do that, we need to observe the loadbearing systems and perform experiments, and last but not least we need knowledge about the constructional and scientific history in order to do justice to the nature of the masonry structures that have been handed down to us. For example, Anne Coste was able to show conclusively in the early 1990s, using the example of Beauvais Cathedral, how FEM analyses can be integrated into the strategy for preserving historically important structures [Coste, 1995]. Nevertheless, when using non-linear FEM analyses we must always remember that the solution to the set of equations for material laws plus equilibrium and compatibility conditions reacts very sensitively to changes in the boundary conditions and the internal relationships, e.g. crack growth. The mechanical behaviour of historical masonry structures has its own history; knowledge of that history is indispensable for an adequate understanding of the loadbearing behaviour of such structural systems.

## 4.8 On the epistemological status of masonry arch theories

The historical epistemology project, in particular as drawn up by Gaston Bachelard, studies the deformations and corrections to the scientific concepts that have actually led to a cognition process in an experimental phenomenon-technical framework acknowledged by the scientific com-

| | s/b = 1/1 | s/b = 3/2 | s/b = 2/1 | s/b | v | $V/(g\frac{s}{2}\frac{b}{2})$ | $H/(g\frac{s}{2}\frac{b}{2})$ | $h/f_k$ |
|---|---|---|---|---|---|---|---|---|
| Groin vault | | | | 1/1 | 0.0 | 1.07 | 1.03 | 0.49 |
| | | | | 1/1 | 2.0 | 1.07 | 0.77 | 0.34 |
| Rib vault | | | | 3/2 | 0.0 | 1.19 | 0.83 | 0.44 |
| | | | | 3/2 | 0.1 | 1.19 | 0.61 | 0.29 |
| | | | | 1/1 | 0.0 | 1.17 | 0.73 | 0.45 |
| | | | | 2/1 | 0.0 | 1.24 | 0.91 | 0.44 |
| Rib vault with cambered cells | | | | 3/2 | 0.0 | 1.41 | 0.76 | 0.29 |
| | | | | 3/2 | 0.5 | 1.41 | 0.73 | 0.28 |
| Domical rib vault | | | | 3/2 | 0.0 | 1.11 | 0.89 | 0.46 |
| | | | | 3/2 | 2.0 | 1.11 | 0.64 | 0.27 |
| | | | | 1/1 | 0.0 | 1.13 | 0.78 | 0.44 |
| | | | | 2/1 | 0.0 | 1.12 | 0.94 | 0.48 |
| Sail vault | | | | 3/2 | 0.0 | 1.14 | 0.83 | 0.34 |
| | | | | 3/2 | 2.0 | 1.14 | 0.68 | 0.27 |
| | | | | 1/1 | 0.0 | 1.19 | 0.70 | 0.33 |
| | | | | 2/1 | 0.0 | 0.82 | 0.89 | 0.39 |

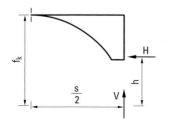

- s  Major span
- b  Minor span
- $f_k$  Rise at intersection of vaults
- g  Weight per unit area
- v  Support displacement in mm on calculated system
- V  Vertical force of one quarter
- H  Horizontal force of one quarter
- h  Height of horizontal force above impost

**FIGURE 4-53**
Analysis of groined vaults after Barthel [Barthel, 1993/1]: important types of groined vault and results of linear and non-linear FEM analyses

munity [Michaux, 1990, p. 758]. According to Michaux, the historical epistemology relieves itself from its philosophical foundations, from being a knowledge theory in the sense of a rational study of the nature, aims and means of the knowledge, and tends to merge with the history of science; according to that it should have a regional character [Michaux, 1990, p. 758]. Michaux sees the reason for this in the concept of epistemological recourse contained in the historical epistemology, which is acknowledged as the standard for distinguishing between confirmed history and obsolete history, and this distinction is such that we can detect it in the actual and provisional status of a science. Accordingly, the elastic theorist Emil Winkler should have read the history of masonry arch theory against the backdrop of practical elastic theory, and, on the other hand, Jacques Heyman, praised for the development of ultimate load theory, should have reconstructed the history of masonry arch theory from the point of view of the plastic theorist. The concept of epistemological recourse appears to apply to both, but they stand at the paradigm change in theory of structures towards practical elastic theory and then turn away from that in favour of plastic theory because they tested the paradigm change on masonry arches and therefore understand the history of masonry arch theory from the viewpoint of the new paradigm. But they transcend the epistemological recourse insofar as their reconstruction of the history of masonry arch theory does not simply reproduce the cognition horizon of current science on a particular object, but reproduce it in expanded form. The epistemological recourse denies itself by being concrete in the first place – historical knowledge has been transformed into engineering science knowledge and turned the latter into history. This is the very essence of a historical engineering science. A historical engineering science cannot be initiated without a philosophical inflection to the history of engineering sciences. The following is therefore an attempt to demonstrate the epistemological status of masonry arch theories in the form of a thesis.

### 4.8.1 Wedge theory

On the production level, wedge theory corresponds to "heterogeneous manufacture" in which the product is the result of "the mere mechanical fitting together of partial products made independently" [Marx, 1979, p. 362], i.e. various trades grouped together under one command. It is characterised by
- splitting the production into a mass of heterogeneous processes,
- the minimal use of "common means of production",
- the outward relationship of the finished product to its various elements, and (in a very close sense)
- the arbitrary combination of "detail labourers" in the same workshop [Marx, 1979, p. 362f.].

The wedge, representing the voussoir, is understood as a simple machine, and several wedges are assembled to form the masonry arch model just like building a masonry arch itself. The loadbearing system synthesis prevails. The disparity of the individual loadbearing system elements appears in the model in a geometry of masses form. The synthesis achievement of wedge theory is ascribed to the outward relationship of the masonry arch model to its loadbearing system elements because the transitional conditions of the equilibrium in the modelling work are ignored. The combination of the individual wedges to form the masonry model is therefore also arbitrary, as Bélidor's example shows us (see section 4.2.2). Nevertheless, the existence of heterogeneous manufacturing forms the basis of the first attempts to create a scientific foundation for engineering thinking from the Renaissance to the 18th century. Wedge theory outlasts its usefulness with heterogeneous manufacture.

### 4.8.2 Collapse mechanism analysis and voussoir rotation theory

In the organic manufacture as the consummate form of manufacturing, the manual work is divided into various suboperations; the product "owes its completed shape to a series of connected processes and manipulations"

[Marx, 1979, p. 362]. On the cognition level, it corresponds to the masonry arch theories that are based on the analysis of collapse mechanisms and point to ultimate load theory, including their transformation into voussoir rotation theory. Organic manufacture transforms the chronological sequence of production processes into a spatial sequence. This means that the workshop-type division of the work is not only simplified by the "qualitatively different parts of the social collective labourer", but rather by "the qualitative sub-division of the social labour-process, a quantitative rule and proportionality for that process" [Marx, 1979, p. 366f.].

Whereas the modelling work of wedge theory reflects the loadbearing system elements in the emulation of the real arch-building process on the chronological axis for the masonry arch model, the collapse mechanism approaches and the voussoir rotation theory break down the real masonry arch into separate rigid bodies according to crack patterns and determine the specific equilibrium conditions from their kinematic behaviour. The qualitative division of the masonry arch into rigid bodies bounded by cracks allows the kinematic proportions between the rigid bodies to be found and expressed in rules for calculating limiting equilibrium conditions. The loadbearing system analysis prevails. The modelling work corresponding to the peculiarity of production in organic manufacture is consummated in Coulomb's voussoir rotation theory. Compared with wedge theory, voussoir rotation theory has gained emancipation from emulating the real arch-building process precisely because it is based on the true crack patterns in the masonry arch and the masonry arch has been split into rigid bodies for the model. The fact that voussoir rotation theory reached its height in France and Italy with Coulomb and Mascheroni was due to the strong influence of organic manufacture on pre-Industrial Revolution production as the consummate form of manufacture. This is proved not only by the ultimate load theory approaches worked out historico-logically from the loadbearing system analyses of the 18th century by Heyman, Benvenuto, Sinopoli, Corradi, Foce and Huerta, but also by the incomparable analysis of the technical-organisational basis of the manufacturing period in the *Encyclopédie* published by Diderot.

**Line of thrust theory and elastic theory for masonry arches**

### 4.8.3

The heart of the Industrial Revolution in the late 18th century was marked by the substitution of hand tools by machine tools. Whereas it was the workers themselves who created the upheaval in the manner of production in the manufacturing period, it was the tools that revolutionised the manner of production in the large industries that emerged with the Industrial Revolution. "The implements of labour, in the form of machinery, necessitate the substitution of natural forces for human force, and the conscious application of science, instead of rule of thumb" [Marx, 1979, p. 407]. It was not until later that the triad of the fully developed machine was formed with the prime mover, the transmission mechanism and the machine tool as its constituent parts. The subjective principle of division typical of manufacturing disappears in this because the total process, seen objectively, is analysed in its constituent phases and the various subprocesses are linked via the technical application of natural sciences. "In Manufacture the isolation of each detail process is a condition imposed by the nature of division of labour, but in the fully developed factory the continuity of those processes is, on the contrary, imperative" [Marx, 1979, p. 401].

Just like the technical basis for large industries was born in the manufacturing period, every age brings forth attempts to model the masonry arch as an inverted catenary. The continuity principle intrinsic to differential and integral calculus first appears in masonry arch models and then, from the mid-19th century onwards, in real masonry arches; however, it cannot unfold its theoretical potency because it is still too closely related to the real masonry arch, e.g. when Poleni models the voussoirs by means of weights. This does indeed determine the shape of the line of thrust, but not its position. Establishing the position of the line of thrust remains subjective and is based on the rules of voussoir rotation theory. The continuity principle in masonry arch theory gradually became established with the method of sections of mechanics [Kahlow, 1995/1996, p. 67], the thrust line concept introduced by Gerstner and the discovery of a material constant – the elastic modulus discovered by Young in England, the birthplace of the Industrial Revolution.

The calculation of the line of thrust initially took place subjectively against the background of voussoir rotation theory, was then objectivised by means of the principles of statics, and finally the method of sections brought about full emancipation from the material jointing of the masonry arch. The modelling work had itself become emancipated not only from the arch-building process, but also from the real masonry arch. And with the establishment of elastic theory in masonry arch theory, the relic consisting of the assumption of certain position parameters for the line of thrust also disappeared. The line of thrust now becomes objective according to the principles of elastostatics. The masonry arch, on the cognition level, has been transformed into a one-dimensional curved elastic continuum, into an elastic arch. Like the fully developed machine, so the structural system of the elastic arch also has a triadic organisation: the prime mover appears as the transformation of external work into deformation energy in a sense of the energy conservation principle, and the transmission mechanism is made up of the line of thrust obtained from Menabrea's principle in the sense of economic load-carrying. In the end, the model of the loadbearing system as a whole, abstracted to the structural system, conveys the relationship between loadbearing function and loadbearing system in the sense of a machine tool: the elastic arch works in bending. Such structural systems can be objective, looked at separately, and with the help of the method of sections can be broken down into and analysed as their constituent basic elements (linear, planar and spatial elements or loadbearing system elements). On the level of the model, the rise from the abstract to the concrete in joining these subprocesses through the practical application of elastic theory is complete. From now on, loadbearing system analysis and loadbearing system synthesis form a unity at

**Ultimate load theory for masonry arches as an object in the historical theory of structures**

the abstraction stage of the structural system in the modelling work of the structural engineer.

**4.8.4**

The rediscovery of the pre-elastic masonry arch theories began during the 1930s as engineers were asked to provide definitive statements about the abilities of old masonry arch bridges to carry heavier loads. Then, starting in the 1960s, the masonry arch theories of the 18th century were discovered, modernised and the state of knowledge concerning ultimate load theory considerably expanded by Heyman from the history of science and history of construction engineering perspectives for the purpose of investigating the structural analysis behind historical structures. Giving the statics of masonry arches a history was a necessary condition. Heyman identified implicitly for the first time, using the example of his ultimate load theory of masonry arches, that knowledge of the historical development of theory of structures can bring about significant progress in the formation of structural analysis theories. This recognition from theory of structures and the philosophical interpretation of engineering sciences presented in section 3.12 can be summarised in five theses on historical engineering science:

I. Knowledge of the history of engineering science can be turned into a productive energy for the current engineering science cognition process.

II. The ongoing development of the productive energy of history of engineering science knowledge is the purpose of historical engineering science.

III. Establishing the history of engineering sciences is a necessary condition for unfolding the productive energy of history of engineering science knowledge for historical engineering science.

IV. The ensuing theory of engineering sciences (see section 3.1.2.3) is a sufficient condition for unfolding the productive energy of history of engineering science knowledge for historical engineering science.

V. Historical theory of structures is a subdiscipline of the historical engineering science.

The productive energy of history of engineering science knowledge is a secondary productive energy. It only becomes a primary productive energy when – as in the case of historical structural analysis – it becomes the means of loadbearing structure analysis in the sense of the culturally sensible and economically justifiable preservation of the loadbearing elements of historical structures.

**The finite element analysis of masonry arches**

**4.8.5**

Automation is

– firstly, the social manifestation of the "substitution and re-application of manual and formalisable mental work through intensive changes to all productive energies for computer-integrated production …," and

– secondly, "a social process for transferring manual and formalisable mental work to technical and database systems and system documentation, or for modifying human work through interactive collaboration with such systems for the purpose of processing energy, materials and information" [Böhm & Dorn, 1988, pp. 11–12].

The possible transformation of production processes into "industrial natural processes" [Marx, 1981, p. 581] through automation, the core of which lies in the mechanisation of formalisable logical operations of human activities, becomes an object in the form of that universal symbolic machine the computer, also computer networks and finally the Internet. The finite element method (FEM) that evolved in the late 1950s is the negation of the Leibniz continuity principle, which still dominated elastic theory and hence also the theory of the elastic arch. In principle it is possible to solve any practical problem numerically by means of FEM. For example, according to standard FEM in the form devised by Turner, Clough, Martin and Topp, the "direct stiffness method" [Turner et al., 1956], the masonry arch is handled as follows:

1. The arch is broken down into one-, two- or three-dimensional elements – finite elements.
2. Approximation functions for the displacements are formulated for the finite elements.
3. The deformation state is determined for this displacement condition.
4. The stress, or rather internal force, condition and hence the forces at the extreme fibres in the finite element are calculated from this deformation state via the material law.
5. The distributed forces at the extreme fibres are converted into statically equivalent node forces and these are placed in equilibrium with the forces from the neighbouring finite elements.

Such a process of intellectual engineering is generally transformed by the software developer into an algorithm, implemented in a program and installed on the computer. All the user has to do is to translate the computational model into the input data for the software, and then let the computer do the calculations and produce the drawings, and finally interpret the results. Thus, structural calculations have themselves developed into the object of automation.

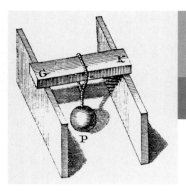

# CHAPTER 5

# The beginnings of a theory of structures

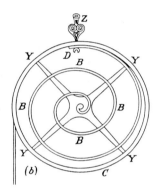

Like all roads in the Roman Empire led to Rome, so we can trace strength of materials back to Galileo's *Discorsi* of 1638. Historically, statics and strength of materials had to be found in theory of structures. The author's interest in Galileo stems not only from Bertolt Brecht's famous play *The Life of Galileo*, but also from an in-depth study of the philosophical history of mechanics writings of Pierre Duhem, Eduard Jan Dijksterhuis, Michael Wolff, Gideon Freudenthal and Wolfgang Lefèvre. In 1980, while still a student, the author purchased a copy of Franz Joseph Ritter von Gerstner's scientific life's work, the three-volume *Handbuch der Mechanik* (manual of mechanics) with its magnificent copperplate engravings. An intensive study of this publication and the work of Johann Albert Eytelwein led to the author's view that these two personalities rounded off the preparatory period of structural theory, but in the end were unable to formulate the programme of structural theory. That was to be left to Navier, who in 1826 fused together statics and strength of materials to form theory of structures. And that's where the history of theory of structures, in its narrower sense, really begins. The results have been published by the author in the yearbook *Humanismus und Technik* edited by Prof. Rudolf Trostel.

The process of the scientific revolution in the 17th century was characterised by the fact that the emergence of the modern natural sciences shaped by Galileo, Descartes and Newton resulted in the natural sciences leaving the production sphere on the social scale and progressing to a separate sphere of social activities.

The natural sciences approach to analysing simple technical artefacts is evident in Galileo's important work *Dialogue Concerning Two New Sciences* (1638) due to the fact that he embraces both nature and engineering mathematically, i. e. describes them as a world of idealised objects. Galileo ignores disturbing influences in the formulation of the law of falling bodies and realises concrete technical artefacts as idealised theoretical models (tensile test, bending failure problem). Galileo's questioning of the difference of geometrical and static similarity for objects in nature and engineering forms the very heart of his strength of materials investigations; its origin lies in his idealisation of objective reality through mathematics, which for him is essentially another theory of proportions.

Whereas mechanics before Galileo and Newton was able to express theoretically simple problems of engineering, like the five simple machines (lever, wedge, screw, pulley, wheel on axle), only in isolated instances, the scientific system of theoretical mechanics evolving dynamically in the 18th century was now able to express all those technical artefacts whose physical behaviour is determined principally by the laws of mechanics. The contradiction between the scientific understanding and complexity of the technical artefacts primarily analysed by mechanics (beam, arch, earth pressure on retaining walls, etc.) is evident not only in all the scientific works of the most important mathematicians and mechanics theorists of the 18th century, but also in the gulf between those and the advocates of the practical arts – the hands-on mechanics, the mill-builders, the mining machinery builders, the instrument-makers and the engineers. In the scientific system of mechanics, theoretical mechanics dominated until its social impact was directed at the creation of a comprehensive scientific conception of the world for the rising middle classes.

It was not until the Industrial Revolution took hold in Great Britain around 1760 – i. e. the transition from workshop to factory – did the conscious link between natural science knowledge activities and engineering practice become the necessary historical development condition for the productive potency of society. As criticism of the static theory of proportion in strength of materials investigations became loud during this period, so a scientific basis for this branch of knowledge began to emerge, simultaneously with the mechanisation of building, finally crystallising in the first three decades of the 19th century as the fusion of these two processes in the form of a theory of strength of materials and a theory of structures. What was now needed was neither a geometrical theory of proportion, which we find in the master-builders of the Renaissance, nor a static proportions theory, as Gerstner was still using to size his beams, but rather the unity of strength test and theoretical modelling of load-bearing structures in construction engineering. The discipline-formation

**What is the theory of strength of materials?**

period of structural theory began with the formulation of Navier's theory of structures programme in 1826.

**5.1**

The resistance of a solid body to its mechanical separation by external mechanical-physical actions is a principal property of solid bodies. We call this property "strength".

For instance, in a tensile test a steel wire of cross-section $A = 1$ mm² opposes the external tensile force $F$ by means of an equal internal tensile resistance (Figs 5-1a and 5-1b). If the applied tensile force increases beyond the yield point $F_{yield}$ until failure $F_B$ of the steel wire, the internal tensile resistance $F_B$ is overcome. We then say that the tensile strength of the steel wire has been reached (Figs 5-1c and 5-1d). Fig. 5-1d illustrates the associated stress-strain diagram for the steel wire (mild steel grade S 235 to Eurocode EC 3, i.e. with a minimum tensile strength of 360 N/mm²). Force $F$ increases linearly with the deformation $\Delta l$ until it reaches the yield point at $F_{yield} = 240$ N. In the yield zone, the steel wire continues to extend while the force $F_{yield}$ remains constant at 240 N. Afterwards, it enters the plastic phase until it fails (ruptures) at $F_B = 360$ N. For calculations according to plastic hinge theory, the stress-strain diagram according to Fig. 5-1d is simplified to Fig. 5-1e: we then speak of ideal elastic (range: $0 \leq F \leq F_{yield} = 240$ N) and ideal plastic (range: $F = F_{yield} = 240$ N) material behaviour. Tensile strength is today described in terms of stress, i.e. in the magnitude of a force per unit surface area (e.g. N/mm²). Consequently, the tensile test can reveal the internal tensile resistance at every stage, e.g. by way of a calibrated force scale (Fig. 5-1d).

This rendition of the invisible, this exposure of the internal tensile resistance, is particularly evident at the moment of failure, when the specimen splits into two parts (Fig. 5-1c). The method of sections of mechanics (see Fig. 5-1b) is in its simplest form the imaginary emulation of the rupture process in the tensile test. If the engineer wishes to calculate the internal forces of the members of a truss at a joint, he cuts them apart at the joints in a thought experiment and thus "exposes" the internal member forces at the joint under consideration. He determines their magni-

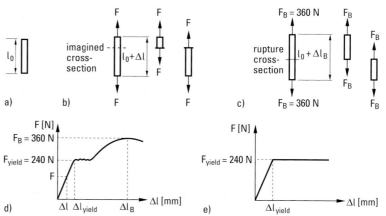

**FIGURE 5-1**
Schematic diagram of a tensile test with associated stress-strain diagram for mild steel grade S235: a) specimen of length $l_0$ and cross-sectional area $A = 1$ mm², b) stress-strain condition in the elastic range of stress, and c) at failure; d) stress-strain diagram, e) simplified stress-strain diagram for ideal elastic and ideal plastic material behaviour.

tude and direction (tension or compression) in such a way that the joint remains in a state of equilibrium. Without the thought experiments of Leonardo da Vinci (1452–1519) concerning the tensile strength of wires, without those of Galileo Galilei (1564–1642) concerning the tensile strength of copper wires, ropes and marble pillars, without the numerous tensile tests of the natural scientists and engineers of the 18th and early 19th century on the one hand and the kinematic analysis of simple machines (especially the pulley) on the other, the method of sections, which Lagrange's work *Méchanique analytique* (1788) had already explained in theory, would not have become the fundamental method of mechanics in the 19th century. The tensile test therefore marks the beginnings of a theory of strength of materials, in the historical and the logical sense. Although modern strength of materials theories take account of bending, compression, shear and torsional strength as well as tensile strength, Galileo's tensile tests in the form of thought experiments formed the germ cell of a theory of strength of materials. Only very few textbooks on strength of materials do not begin by describing the tensile test (Fig. 5-1); not only the method of sections, but also the codified relationship between the two principal mechanical variables can be verified empirically in a particularly simple and convincing way by measuring the tensile force $F$ and the elongation $\Delta l$. By contrast, identifying the effects due to bending, shear and torsion are much more complicated. After the formulation of the bending problem by Galileo in 1638, almost two centuries passed before Navier's practical bending theory (1826) enabled engineers to understand reliably the bending strength of beam-type construction elements. Fig. 5-2 (Fig. 1 to Fig. 9) gives the reader an impression of the strength problems scientists were already analysing in the early 18th century. Besides the loading due to tension (Fig. 1 and Fig. 8), we have beams subjected to bending, such as the cantilever beam (Fig. 2) plus the simply supported and the fixed-end beams (Figs 3 and 4 respectively) on two supports; there is also a column fixed at its base loaded in compression (Fig. 5). Even the strength of prismatic bodies to resist compression – at first sight just as easy to analyse as the strength to resist tension – turned out to be a difficult mechanical problem in the case of slender struts (buckling strength), which was not solved satisfactorily for steelwork much before the end of the 19th century. The tensile test and its mechanical interpretation is therefore at the historico-logical heart of the emerging theory of strength of materials.

According to Dimitrov, strength of materials is the foundation of all engineering sciences. It aims to "provide an adequate factor of safety against the unserviceability of the construction" [Dimitrov, 1971, p. 237]. That historian of civil engineering, Hans Straub, regards strength of materials as "the branch of applied mechanics that supplies the basis for design theory", with the task of "specifying which external forces a solid body … may resist" [Straub, 1992, p. 389]. Whereas Straub places elastic theory firmly in the strength of materials camp, Istvan Szabó distinguishes the aforementioned scientific disciplines according to their objectives: the aim

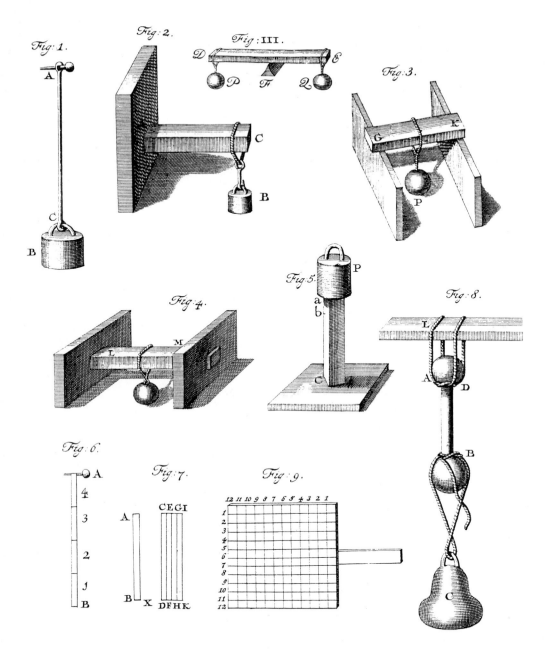

**FIGURE 5-2**
Some strength problems of the early 18th century

of elastic theory is to determine the deformation or displacement condition for a body with a given form subjected to an applied load, "whereas strength of materials regards the load on a body as known when the (internal) stresses, for which we prescribe permissible limits depending on the material, are calculated in addition to the displacement" [Szabó, 1984, p. 84]. On the other hand, the "Old Master" of applied mechanics, August Föppl, places the examination of the displacement condition and the associated stress condition in the focal point of the object of strength of materials, which can therefore be understood as the "mechanics of internal forces" [Föppl, 1919, p. 3]. "Strength of materials," write Herbert Mang and Günter Hofstetter, "is a customary abbreviation of the engineering science discipline of the applied mechanics of deformable solid bodies. The mechanics of deformable bodies is a discipline belonging to continuum mechanics … The main task of strength of materials is to calculate stresses and deformations, primarily in engineering constructions" [Mang & Hofstetter, 2004, p. 1].

The purpose of strength of materials is to portray quantitatively and qualitatively the resistance of solid bodies to their mechanical separation by mechanical-physical actions with the help of experiments and theoretical models, and to prepare these in the form of an engineering science knowledge system in such a way that they can be used as a resource in engineering activities. Consequently, strength of materials is based, on the one hand, on the practical experiences of materials testing plus the science of building and materials, but, on the other, also the theoretical models of the applied mechanics of deformable solid bodies.

## 5.2 On the state of development of structural design and strength of materials in the Renaissance

When we look at the domes of the late ancients, the delicate loadbearing systems of the Gothic period and the long-span masonry arch bridges of the Renaissance, it is not unusual to ask the question of whether their builders did not perhaps have some knowledge of theory of structures on which to base their bold designs. As A. Hertwig, that aficionado of the history of building and building theory, wrote in 1934: "If we study the Hagia Sophia [in Istanbul – the author] built in 537 AD or the Pantheon [built in 27 BC in Rome – the author] from a structural viewpoint, then we discover a cautious exploitation of the various material strengths which could not have been achieved just by using the simple rules for the design of individual elements. The builder of the Hagia Sophia, Anthemios, was revered by his contemporaries as a mathematician and mechanical engineer. That can mean nothing other than he prepared structural calculations for his structures. The knowledge of mechanics at that time would have been wholly adequate for this purpose. For with the help of Archimedes' (287 – 212 BC) principle concerning the equilibrium of forces in a lever system it is certainly possible to investigate the limit states of equilibrium in masonry arches and pillars by considering all the parts as rigid bodies" [Hertwig, 1934/2, p. 90]. On the other hand, based on careful history of building and history of science studies, R. J. Mainstone comes to the following conclusion: "No quantitative application of statical theory is recorded before the time of Wren" [Mainstone, 1968, p. 306]. The architect and engineer Christopher Wren (1632 – 1723), friend of Isaac Newton (1643 – 1727), was born in the year that Galileo's *Dialogue* was published.

However, were not individual principles of structural theory and strength of materials recognised and integrated in qualitative form into the engineering knowledge surrounding buildings and machines? In a paper, S. Fleckner presents his comparative structural investigations of the great Gothic cathedrals and formulates the thesis that their characteristic buttresses are dimensioned on the basis of structural calculations

[Fleckner, 2003, p. 13]; however, he considers only the circumstantial evidence without naming any positive contemporary sources. Nonetheless, the above question cannot be rejected for the period of the Renaissance. In the drawings of Leonardo da Vinci (1452–1519), it is possible to find numerous examples of the principles of structural theory and strength of materials which eclipse the resources of outstanding contributors from the Hellenistic phase of ancient science such as Archimedes, Heron (around 150 BC) and Ktesibios (around 250 BC), and are closely linked with Leonardo's engineering thinking. Also famed as the painter of "Mona Lisa" and "The Last Supper", this engineer provides the first written evidence of a strength experiment (Fig. 5-3) in his notebooks from around 1500 well known to historians of culture, science and technology under the title of *Codex Atlanticus*.

Leonardo describes his test thus: "Experiment concerning the load that wires of different length can carry. Perform the experiment to find out how much weight an iron wire can hold. You should proceed as follows in this experiment: Hang an iron wire 2 cubits long [1 Milanese cubit = approx. 600 mm – the author] from a place that holds it firmly. Then hang a basket or similar on the wire in which there is a small hole with which to fill to fill a basketful with fine sand from a funnel. When the iron wire can no longer carry the load, it breaks. ... Note the magnitude of the weight as the wire broke and also note at which point the wire breaks. Perform the experiment again and again in order to discover whether the wire always breaks at the same point. Then halve the wire and observe whether it can now carry more weight. Then shorten it to one-quarter of the initial length and gradually, using different lengths, you will discover the weight and the place at which the wire always breaks. You can carry out this test with any material – wood, stone, etc. Set up a general rule for each material" [Krankenhagen & Laube, 1983, p. 31]. When we read the description of the experiment, it is easy to gain the impression that Leonardo had not recognised the cause of the different tensile failure forces for the different specimen lengths. However, this impression is totally refuted when we study the relevant passages in his notebooks rediscovered in 1965 in the Madrid National Library, which is why they are called *Codex Madrid I and II*. Leonardo recognised firstly that "it is possible for a vertical, suspended rope to break due to its own weight", and secondly that "the rope breaks where it has to carry the greatest weight, i. e. at the top where it is connected to its support" [Leonardo da Vinci, 1974/2, p. 204]. Leonardo had therefore not only anticipated qualitatively the notion of the breaking length, but had also indirectly answered the question regarding the rupture cross-section in the wire strength test he so carefully described: the wire in the test, too, must theoretically break at the point of its suspension because this is where the total self-weight of the iron wire of length $l$ is added to the weight of the basket and the sand (Fig. 5-3). If we reduce the length of the iron wire in the test, the shortened wire can carry an additional weight of sand corresponding to the weight of the "missing" length of wire. And vice versa: lengthening the wire in the test results in

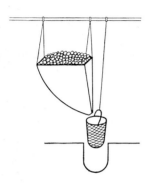

**FIGURE 5-3**
Leonardo da Vinci's test setup for determining the tensile strength of a wire

an equivalent reduction in the weight of sand that can be carried at the moment of failure; when we reach the breaking length, the iron wire parts under its own weight. But the *Codex Madrid*, with their sensational findings for the history of science and technology, also contain statics-constructional knowledge which, had it been available in the Renaissance, would have helped Galileo's strength experiments enormously.

Probably inspired by his designs for a giant crossbow, Leonardo analysed the relationship between external load and deformation on pre-bent and straight elastic bars. The intuitive knowledge of the codified relationship between pretensioning force and deformation is a prerequisite for giant crossbows that cannot be pretensioned manually because the deformation energy stored in the elastic bar after releasing the pretension is converted almost exclusively into the kinetic energy of the projectile; as the load increases, so the elastic deformation also increases – and hence the stored deformation energy. Fig. 5-4a shows an elastic bar loaded in the middle successively with the weights $G$, $2G$, $3G$, $4G$ and $5G$. Leonardo now wanted to know "the curvature of the [elastic] bar, i.e. by how much it differs, larger or smaller, from the other weights. I believe that the test with twice the weight in each case will show that the curvatures behave similarly" [Leonardo da Vinci, 1974/1, p. 364].

Although the notion of the curvature $\kappa$ of a curve as the inverse of the radius of curvature was not expressed mathematically until the turn of the 18th century in the investigations into elastic lines, Leonardo had recognised the material law for specific elastic lines: from the proportionality between external load $G$ and internal bending moment $M$ on the one hand and the proportionality between $M$ and the curvature $\kappa$ of the elastic line on the other, it follows that there should be proportionality between external load $G$ and curvature $\kappa$ as asserted by Leonardo. Contrasting with this is Leonardo's mistaken assertion that the bending deflection $f$ of two beams subjected to the same central load $G$ is identical when the longer beam is four times the length and the cross-section twice as wide and twice as deep (Fig. 5-4b). If the bending deflection of the short beam

$$f = \frac{G \cdot l^3}{48 \cdot E \cdot I} \qquad (5\text{-}1)$$

where span = $l$, elastic modulus of the material = $E$ and second moment of area $I = b \cdot h^3/12$ ($b$ = width of cross-section, $h$ = depth of cross-section), then if instead of $l$, $l' = 4l$ and instead of $I$, $I' = [2 \cdot b \cdot (2 \cdot h)^3]/12$ are used for the longer beam,

$$f' = \frac{G \cdot l^3}{48 \cdot E \cdot I} = 4 \cdot f \qquad (5\text{-}2)$$

i.e. the longer beam exhibits four times the bending deflection of the shorter one.

Leonardo's assertion would be right if the bending deflection increased by only the square of the span, i.e. the bending line was a quadratic and not a cubic parabola. However, Leonardo's answer to the following problem is correct (Fig. 5-4c): "I shall take three bars of the same thickness, one of which is twice as long as the others. And each shall be subjected to

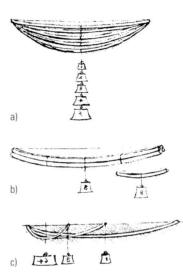

**FIGURE 5-4**
a) Deformations of elastic bars of constant cross-section and equal length subjected to various point loads at the middle of the bar, b) deformations of elastic bars of different cross-sections and lengths subjected to an identical point load in the middle of the bar, c) deformations of elastic bars of constant cross-section but different lengths subjected to various point loads in the middle of the bar.

a load in the middle such that the curvatures exhibit the same deflection" [Leonardo da Vinci, 1974/1, p. 364]. According to Leonardo, the proportion of the loads to the spans must be $G : 8G : 64G = 1 : 0.5 : 0.25$.

The bending deflection $f$ of the longest beam of span $l$ under load $G$ can be calculated using eq. 5-1. The load must equal $8G$ to produce the same bending deflection $f$ when the span is halved, which is easy to establish by entering the values into eq. 5-1. If we use the span $l/4$ instead of $l$ in eq. 5-1, the load must increase to 64 times the value of the original load for the beam of span $l$ in order to achieve the same bending deflection.

As the treatment of construction engineering issues in Leonardo's notebooks is secondary to his qualitative analysis of machine elements and does not form a coherent system of construction engineering knowledge, the following is merely a summary of some potential solutions that today would fall within the remit of theory of structures and strength of materials:

- Leonardo recognised the principle of resolving a force into two components, but without determining those components quantitatively [Wagner & Egermann, 1987, p. 179].
- He used the notion of the static moment (force multiplied by lever arm) for the first time on inclined forces [Straub, 1992, p. 91].
- Leonardo anticipated the linear strain distribution (beam cross-sections remain plane) of elastic beams with a rectangular cross-section as assumed by Jakob Bernoulli in 1694 [Kurrer, 1985/1, p. 3].
- In the static analysis of masonry arches, he developed a wedge theory in which he satisfies the moment equilibrium of each voussoir but neglects the displacement (translation) equilibrium despite being aware of the parallelogram of forces; further, he specified possible collapse mechanisms of asymmetrically and symmetrically loaded masonry arches [Zammattio, 1974, p. 210].
- Leonardo sketched out a method in which the horizontal thrust of diverse arch forms and couple roofs could be analysed experimentally and, as a result, how thick the abutments would have to be [Zammattio, 1974, p. 211].
- The load-carrying capacities of concentrically loaded fixed-end columns behave inversely proportional to their height; whereas with a concentrically loaded column there are no flexural deformations, the fibres of eccentrically loaded columns are extended on the side opposite to the load but compressed on the loaded side.

In fact, the French science historian Pierre Duhem claims that Leonardo's pupil Francesco Melzi passed on the notebooks, either as originals or copies, to Cardano, Benedetti, Guidobaldo del Monte and Bernardino Baldi, and via these scientists could have influenced Galileo [Straub, 1992, p. 94]. But we can also assume that Leonardo's "chaotic collection of notes in which ingenious ideas and everyday extracts from acknowledged works are intermingled" [Dijksterhuis, 1956, p. 283] may not have been able to accelerate scientific and engineering progress precisely because of the lack of an orderly presentation and their coincidental distribution. Mind you, these two remarks were expressed prior to the rediscovery of the *Codex Madrid*! The fact that at least some of the aforementioned structural theory findings of Leonardo were not taken up in the subsequent two centuries is illustrated by using Philippe de la Hire's masonry arch theory (first published in 1695) as an example. Here, in contrast to the masonry arch theory of Leonardo, the translation equilibrium of the frictionless-jointed voussiors of a semicircular masonry arch is assumed, i.e. the theory requires the magnitude of the external forces required to keep the – expressed in modern terms – system of hinges (with multiple kinematic indeterminacy) just in (unstable) equilibrium. Therefore, Leonardo's (building) design theory findings remain as erratic uplands in the emerging new scientific landscape between which paths of scientific knowledge pass. The grandiose individual (building) design theory findings of Leonardo da Vinci were not recognised by subsequent generations of scientists and engineers in the age of the scientific revolution.

In 1586 the Dutch mathematician Simon Stevin (1548–1620) carried out the "wreath of spheres" experiment to prove the law of the inclined plane in his work *Beghinselen der Weeghconst* (The Elements of the Art of Weighing). As professor of mathematics and senior waterways engineer in the Netherlands, Stevin had noticed the difference between theoretical and applied mechanics but had underestimated its extent: "Once he had determined theoretically the ideal equilibrium condition between force and load in a tool, then he was of the opinion that a very minor increase in the force could now set the load in motion. Nonetheless, owing to the many applications that made use of his theoretical investigations (in weighing and lever tools, in windmills, the horse's bridle and in military science), he advanced both *Weeghconst* (the art of weighing = theoretical statics) and *Weghattet* (the practice of weighing = practical statics)." [Dijksterhuis, 1956, p. 365]

The relationship with engineering, as we experience in Stevin's scientific work on statics, remained more of an exception in the late Renaissance. And conversely, we cannot speak of progress in engineering steered by theory. The technical problems of construction, mining and mechanical engineering were often too complex to be adequately catered for scientifically in the laws of nature effective in engineering entities and methods – indeed, even to be anticipated theoretically. And that is why the Fleisch Bridge in Nuremberg, completed in 1598, with its spectacular rise-to-span ratio of 4 m to 27 m ($f:l = 1:6.75$) and highly complex subsoil conditions cannot possibly be the result of an analysis based on a structural theory [Falter et al., 2001]. Assuming elastostatic behaviour in the uncracked condition, Karl Krauß discovered in his re-analysis that the Fleisch Bridge should exhibit cracks on the underside of the crown of the arch. "Despite several detailed investigations of the masonry, I found neither cracks nor signs of repairs" [Krauß, 1985, p. 220]. During his examination of the archive material, Krauß noticed the short construction period (about two months to construct the masonry arch of the bridge) and he turned his attention to the problem of the lime mortar, which would have been still soft upon completion. Another structural check assuming plastic deformation of the fresh mortar in the joints revealed that the crack zone at the crown calculated previously had now disappeared. "Sensitised by this probing of the mortar's influence, I read Alberti's work [*On the Art of Building in Ten Books*, Florence, 1485 – the author] again and discovered in chapter 14 of the third book a method of striking the centering of masonry arches, the significance of which had not been realised: 'And apart from that it is good in centred masonry arches to relieve the support a little (forthwith), where they are completed by the uppermost voussoirs (by which the centering is supported so to speak) so that the freshly built voussoirs do not float between their bed and the lime mortar, but rather assume a balanced, steady position of complete equilibrium with one another. If this happens during the drying, however, the masonry would not be compressed together and hold, as is necessary, but would leave cracks upon settling. The work should therefore be carried out as follows: the centering should not be simply removed, but instead gradually loosened from day to day so that the fresh masonry does not follow if you take the centering away prematurely. After a few days, however, depending on the size of the arch, loosen the centering a little more. And then continue until the voussoirs fit to the arch and to one another and the masonry stiffens'" [Krauß, 1985, p. 220]. The structural re-analysis by Krauß taking into account the plastic deformation of the joints in the masonry arch and the passage from Alberti's work quoted by him is an impressive demonstration of how the builders of the Fleisch Bridge were able to "influence the static behaviour of the masonry by controlling the progress of the work" [Krauß, 1985, p. 221] even without structural calculations. In her dissertation *Die Fleischbrücke in Nürnberg 1596 – 98* (2005), which is based on a critical evaluation of the extensive archive material available, Christiane Kaiser arrives at the conclusion that merely qualitative statics-constructional deliberations were represented graphically in the designs for this structure [Kaiser, 2005].

Until well into the discipline-formation period of structural theory in the 19th century, the relevance and practicability of the experience of builders accumulated at the interface between the construction process, structural design and static behaviour remained far superior to that gained in theoretical trials. It was not until the conclusion of the consolidation phase of structural theory (in the 1950s), as civil and structural engineers had at their disposal the knowledge of materials in a scientific form and even industrialised methods of building had access to a scientific footing (and therefore the interaction between progress on site and statics-constructional scientific attained relevance), that we can speak for the first time of the supremacy of the experience at that interface being overtaken by scientifically founded knowledge.

## 5.3 Galileo's *Dialogue*

Long chapters have often been devoted to the *Dialogue* (Fig. 5-5) and its embedment in the process of forming the new natural sciences in monographic summaries on the history of the natural sciences. With only a few exceptions, these contributions to the history of science concentrate their analysis on the dynamic of Galileo as one of the "two new sciences" (Galileo), whereas the individual problems of Galileo's strength tests are mentioned only in passing in the bibliography on the history of technology and engineering sciences, like in M. Rühlmann [Rühlmann, 1885], S. P. Timoshenko [Timoshenko, 1953], F. Klemm [Klemm, 1979], T. Hänseroth [Hänseroth, 1980], E. Werner [Werner, 1980], K. Mauersberger [Mauersberger, 1983], Hänseroth and Mauersberger [Hänseroth & Mauersberger, 1987], H. Straub [Straub, 1992], and P. D. Napolitani [Napolitani, 1995]. Although historians of engineering sciences have already achieved significant successes in unravelling the early evolutionary stages of structural mechanics [Hänseroth, 1980] and applied mechanics [Mauersberger, 1983], a detailed analysis of Galileo's strength experiments still represents a gap in the research.

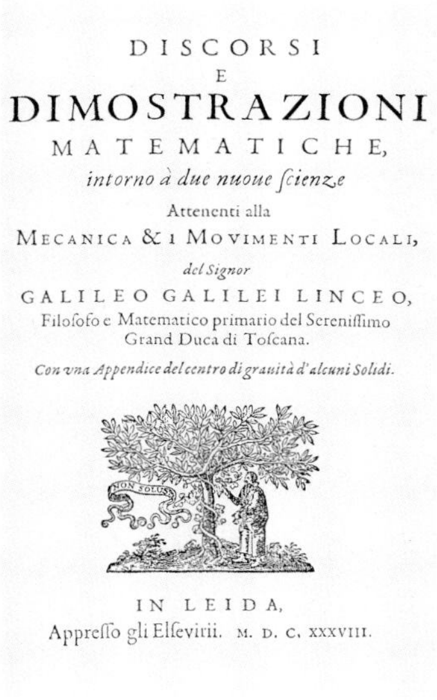
DISCORSI
E
DIMOSTRAZIONI
MATEMATICHE,
*intorno à due nuoue scienze*
Attenenti alla
MECANICA & I MOVIMENTI LOCALI,
*del Signor*
GALILEO GALILEI LINCEO,
Filosofo e Matematico primario del Serenissimo
Grand Duca di Toscana.
*Con vna Appendice del centro di grauità d'alcuni Solidi.*

IN LEIDA,
Appresso gli Elsevirii. M. D. C. XXXVIII.

**FIGURE 5-5**
Cover of Galileo's important work *Dialogue Concerning Two New Sciences* (1638)

Galileo's *Dialogue* is a discussion between Salviati (the voice of Galileo), Sagredo (an intelligent layman) and Simplicio (representing Aristotelian philosophy) about two new sciences, and unfolds over a period of six days:

*First day:* Tensile strength of marble columns, ropes and copper wires; reference to cohesion; mathematical considerations; free fall in a vacuum and in a medium; pendulum motion, etc.
*Second day:* Consideration of the ultimate strength of beams with different forms, loadings and support conditions under similarity mechanics aspects.
*Third day:* Law of falling bodies
*Fourth day:* The motion of projectiles
*Fifth day:* Theory of proportion
*Sixth day:* Force of percussion

Only the first two days are of immediate interest for the history of applied mechanics or the theory of strength of materials.

### 5.3.1 First day

Through the dialogue between Salviati and Sagredo, Galileo explains to the reader how engineering can play a great role as an object of natural science knowledge, i. e. that the analysis of the transformation of the engineering purpose-means relationship manifests itself in the form of the knowledge of the cause-effect relationship of the technically formulated nature. Salviati begins the dialogue: "The constant activity which you Venetians display in your famous arsenal suggests to the studious mind a large field for investigation, especially that part of the work which involves mechanics; for in this department all types of instruments and machines are constantly being constructed by many artisans, among whom there must be some who, partly by inherited experience and partly by their own observations, have become highly expert and clever in explanation." Sagredo replies: "You are quite right. Indeed, I myself, being curious by nature, frequently visit this place for the mere pleasure of observing the work of those who, on account of their superiority over other artisans, we call 'first rank men'. Conference with them has often helped me in the investigation of certain effects including not only those which are striking, but also those which are recondite and almost incredible. At times also I have been put to confusion and driven to despair of ever explaining something for which I could not account, but which my senses told me to be true …" [Galileo, 1638/1964, p. 3].

Such a causal relationship, which Galileo refers to again and again in examples on the first and second days of his *Dialogue*, represents the difference between the geometric and static similarity of objects in nature and engineering. "Therefore, Sagredo," says Salviati, "you would do well to change the opinion which you, and perhaps also many other students of mechanics, have entertained concerning the ability of machines and structures to resist external disturbances, thinking that when they are built of the same material and maintain the same ratio between parts, they are able equally, or rather proportionally, to resist or yield to such external disturbances and blows. For we can demonstrate by geometry that the large machine is not proportionately stronger than the small. Finally, we may say that, for every machine and structure, whether artificial or natural, there is set a necessary limit beyond which neither art nor nature can pass; it is here understood, of course, that the material is the same and the proportion preserved" [Galileo, 1638/1964, p. 5].

As will be shown below, Galileo's strength of materials investigations combine the question of the ultimate strength of simple loadbearing structures with the construction of transfer principles for such loadbearing structures. The latter is even today the task of the similarity mechanics founded by Galileo, which consists of deriving the mechanical behaviour of the large-scale construction by means of the mechanical findings acquired through experimentation on the model and the transfer principles. As Klaus-Peter Meinicke was able to show within the scope of his case study on the historical development of similarity theory, "Galileo's statements on similarity are a qualitative intervention in the scientific explanation of problems of scale transfer" [Meinicke, 1988, p. 15].

Salviati tries by way of qualitative examples to convince the others, Sagredo and Simplicio, that the geometric similarity may not be identified with the static. A cantilevering wooden stick that only just supports itself must break if it is enlarged; if the dimensions of the wooden stick are reduced, it would have reserves of strength. Galileo identifies here a very singular aspect of the collapse mechanism of loadbearing structures. But Sagredo and Simplicio still do not seem to have grasped the point; Salviati has to make things much clearer for them, and asks: "Who does not know that a horse falling from a height of three or four cubits will break his bones, while a dog falling from the same height or a cat from a height of eight or ten cubits will suffer no injury? Equally harmless would be the fall of a grasshopper from a tower or the fall of an ant from the distance of the moon" [Galileo, 1638/1964, p. 5]. Galileo quickly abandons such plausibility considerations appealing to "commonsense" which certainly contradict the premises of Galileo's similarity mechanics-based strength of materials considerations for the same material and geometric proportions. He describes how a marble obelisk supported at its ends on two timber baulks (statically determinate beam on two supports with span $l_0 = 2l$) fails exactly over the timber baulk added later in the middle (statically indeterminate beam on three supports with spans $l_1 = l_2 = l$) because one of the end supports has rotted away, i.e. the static system has been changed from a beam on three supports with one degree of static indeterminacy with spans $l_1 = l_2 = l$ into a statically determinate system with one span $l_1 = l$ and a cantilevering length $l_2 = l$. According to Galileo, yielding of one support in the original loadbearing system (statically determinate beam on two supports with span $l_0 = 2l$) would not have had any consequences, whereas for the beam on three supports with one degree of static indeterminacy with spans $l_1 = l_2 = l$ the force condition with an increasingly yielding support would be redistributed until the support reaction of the end support affected becomes zero, i.e. the statically indeterminate system has changed back to a statically determinate system, and the ensuing force condition causes the obelisk to fail over the (originally) central support. Galileo has thus identified the nature of the yielding support loading case for the simplest statically determinate system, but has not answered the question about the relationship between static and geometric similarity. It is therefore not surprising when even Sagredo, the intelligent layman, is not quite satisfied with Galileo's résumé that this would not have happened with a smaller but geometrically similar marble obelisk. Sagredo expresses his confusion: "… and I am the more puzzled because, on the contrary, I have noticed in other cases that the strength and resistance against breaking increase in a larger ratio than the amount of material. Thus, for instance, if two nails be driven into a wall, the one which is twice as big as the other will support not only twice as much weight as the other, but three or four times as much." To which Salviati replies: "Indeed you will not be far wrong if you say eight times as much …" [Galileo, 1638/1964, p. 7]. Thus, Galileo hints for the first time at the failure mechanism of the fixed-end beam. Sagredo: "Will you not then, Salviati, remove these difficulties and clear away these obscurities if possible? …" [Galileo, 1638/1964, p. 7].

Galileo begins his explanation with the tensile test (Fig. 5-6), "for this is the fundamental fact, involving the first and simple principle which we must take for granted as well known" [Galileo, 1638/1964, p. 7]. Galileo proposes a weight $C$ of sufficient size that it breaks the cylinder of timber or other material where it is fixed at point $A$. Even non-fibrous materials such as stone or metal exhibit an ultimate strength. A distribution of the tensile resistance opposing the weight $C$ over the area of the cross-section does not take place here. After Galileo attempts to explain qualitatively the tensile strength of a hemp rope whose fibres do not match the length of the test specimen by discussing the friction in the rope, he digresses from the explanation of the tensile resistance of non-fibrous materials and spends many pages discussing the "aversion of nature for empty space" [Galileo, 1638/1964, p. 11] and the cohesive force of the particles of such a body – these two together supposedly constitute said body's tensile resistance.

FIGURE 5-6
Galileo's tensile test as a thought experiment

In this context, he answers the question regarding the breaking length of a copper wire quantitatively: "Take for instance a copper wire of any length and thickness; fix the upper end and to the other end attach a greater and greater load until finally the wire breaks; let the maximum load be, say, 50 pounds. Then it is clear that if 50 pounds of copper, in addition to the weight of the wire itself which may be, say, 1/8 ounce, is drawn out into wire of this same size we shall have the greatest length of this kind of wire which can sustain its own weight [breaking length – the author]. Suppose the wire which breaks to be one cubit in length and 1/8 ounce in weight; then since it supports 50 pounds in addition to its own weight, i. e. 4800 eighths of an ounce, it follows that all copper wires, independent of size, can sustain themselves up to a length of 4801 cubits and no more …" [Galileo, 1638/1964, p. 17]. This is one of the two places in the *Dialogue* where we could say that Galileo assumes a constant distribution of stress over the cross-section through the proportionality between tensile resistance and cross-sectional area.

Without having satisfied Sagredo's wish to discuss the failure problem of the fixed-end beam on the first day, Galileo instead devotes the major part of his discussion to a comprehensive explanation of mathematical questions and problems. On the first day of the *Dialogue* the reader gains the impression that Galileo's intention is to describe each question which will then be answered in detail the next day.

## Second day

### 5.3.2

Salviati: "Resuming the thread of our discourse, whatever the nature of this resistance which solids offer to large tractive forces, there can at least be no doubt of its existence; and though this resistance is very great in the case of a direct pull, it is found, as a rule, to be less in the case of bending forces … It is this second type of resistance which we must consider,

**FIGURE 5-7**
Galileo's bending failure problem

seeking to discover in what proportion it is found in prisms and cylinders of the same material, whether alike or unlike in shape, length, and thickness" [Galileo, 1638/1964, p. 94]. The failure problem of the cantilever beam (beam fixed at one end) (Fig. 5-7) forms the true crux of Galileo's statements on the second day of his *Dialogue*.

After he has presented the lever principle and has distinguished clearly the force condition due to self-weight from one of "an immaterial body devoid of weight" [Galileo, 1638/1964, p. 96], he explains the collapse mechanism of the cantilever beam in three steps (Figs 5-8a to 5-8c):

a) The beam fails at $B$, which makes $B$ the point of support and rotation of the kinematic collapse mechanism. Whereas the weight $E_B$ acting at $C$ exhibits the lever arm $\overline{BC} = l$, the tensile resistance $W_Z$ at the fixed end acts at the lever arm $\overline{AB}/2 = h/2$, so that the beam can be idealised mechanically by the angle lever $ABC$ (Fig. 5-8a).

b) From the tensile test (see Fig. 5-6), Galileo finds that the force at failure $T_B$ is identical with the tensile resistance at the point of fixity $W_Z$ (Fig. 5-8b).

c) Since $W_Z = T_B$, by applying the lever principle to the angle lever $ABC$, it follows that (Fig. 5-8c)

$$\frac{T_B}{E_B} = \frac{l}{\left(\dfrac{h}{2}\right)} \tag{5-3}$$

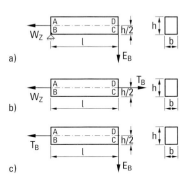

**FIGURE 5-8**
Schematic representation of Galileo's bending failure problem

As Galileo expressly notes, the steps a) to c) also apply when considering the self-weight of the prismatic cantilever beam. Like Leonardo, Galileo does not consider translation equilibrium at all; he should have applied the support reactions equivalent to $T_B$ and $E_B$ at $B$.

It has often been asked why Galileo applied the failure force $W_Z = T_B$ at the centre of gravity of the symmetrical beam cross-section and in doing so ignored the equilibrium conditions in the horizontal and vertical directions. This question can only be answered satisfactorily if we fully realise how crucial the tensile test was to his thought experiment. As the scientists and engineers prior to Newton (1643–1727) could only understand applied forces essentially in the form of weights, it is hardly surprising that Leonardo and Galileo thought of their test specimens being suspended from a fixed point. Acted upon by the force of gravity, the longitudinal axis of the specimen plus the suspended test weight aligned itself with the centre of the Earth. However, as in Galileo's tensile test (Fig. 5-6) the test weight is obviously introduced concentrically into the cylindrical cross-section via a hook at $B$ and must always remain aligned with the centre of the Earth, the line of action coincides with the axis of the cylinder and hence also that of the tensile resistance at $A$. If we now analyse Galileo's cantilever beam, the axis of which is perpendicular to the direction of the force of gravity but is loaded at the end of the cantilever with a weight $E_B$, then the tensile resistance $W_Z = T_B$ must act in the axis of the centre of gravity of the beam cross-section because $T_B$ was determined previously from a tensile test (Fig. 5-8b). It is important to realise that Galileo can imagine the force $T_B$ in his tensile test only in the form of

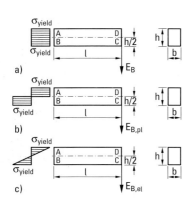

FIGURE 5-9
Comparison of the failure or ultimate loads:
a) based on Galileo's stress theorem,
b) with a fully plastic cross-section AB according to plastic hinge theory (see Fig. 5-1e), and c) when yielding starts at points A and B according to elastic theory.

a weight that causes the concentric ultimate tensile force $W_Z$ at the fixing point of the cross-section (compare Fig. 5-8b with Fig. 5-6). He imagines his test specimen loaded in tension and with its axis aligned with the direction of the centre of gravity now placed in a horizontal position, and instead of $T_B$ applies the weight $E_B$ perpendicular to the beam's centre-of-gravity axis, which results in the tensile force $W_Z = T_B$ at half the depth of the fixed-end cross-section $\overline{AB}$ (Figs. 5-8a and 5-8c). Galileo can determine the relationship between the ultimate tensile force $W_Z = T_B$ and the weight $E_B$ using the lever principle applied to the angle lever $ABC$.

Knowledge of the support reactions at $B$ (Fig. 5-8a) is wholly unnecessary for solving this task. Whereas Leonardo's tensile test merely served to determine the tensile strength of various materials by way of experimentation, Galileo discovered in his beam problem the relationship between the tensile test and the static effect of the angle lever in the form of a proportion (see eq. 5-3). If we assume a constant stress distribution in Galileo's tensile test (Galileo never did this explicitly; from our modern viewpoint we can always assume a constant stress distribution when all the fibres parallel to the axis of the bar undergo the same change in length), i.e.

$$T_B = W_Z = \sigma_{yield} \cdot (b \cdot h) \tag{5-4}$$

(where $\sigma_{yield}$ = yield stress of material, see Fig. 5-1e), then the failure load for the beam fixed at $\overline{AB}$ with length $l$, width $b$ and depth $h$ (Fig. 5-9a) is

$$E_B = \frac{\sigma_{yield}}{l} \cdot \frac{b \cdot h^2}{2} \tag{5-5}$$

As Galileo calculates the bending failure with eq. 5-3, from the modern viewpoint the comparison with the ultimate load

$$E_{B,pl} = \frac{\sigma_{yield}}{l} \cdot \frac{b \cdot h^2}{4} \tag{5-6}$$

for materials with distinct yield points and a fully plastic cross-section $\overline{AB}$ according to plastic hinge theory (Fig. 5-9b) would seem to apply, and not, as many authors still assume today, with

$$E_{B,el} = \frac{\sigma_{yield}}{l} \cdot \frac{b \cdot h^2}{6} \tag{5-7}$$

as the elastic ultimate load calculated from elastic theory at the start of yielding of the extreme top and bottom fibres at cross-section $\overline{AB}$ (Fig. 5-9c). (Many authors who have analysed Galileo's collapse mechanism for the cantilever beam compare eq. 5-7, which applies in the elastic range only – where the true stress $\sigma = \sigma_{yield}$ should be assumed instead of $\sigma_{yield}$ – with eq. 5-5; but they are therefore comparing the serviceability limit state with the ultimate limit state.)

The ratio $E_B : E_{B,pl}$ is 2 : 1. This means that Galileo assumes the ultimate load to be twice the value of the maximum load according to plastic hinge theory. However, Galileo is interested in the ratios between the loads, and therefore statements about ultimate load ratios regarding "prisms and cylinders of the same material, whether alike or unlike in shape, length, and thickness" [Galileo, 1638/1964, p. 94]. What is still to be shown is that Galileo's ultimate load ratios are correct.

Firstly, Galileo calculates the ultimate ratio of a beam placed on edge (beam depth $h$, beam width $b$) to that of a beam placed flat (beam depth $b$, beam width $h$) of length $l$ (Fig. 5-10). According to eq. 5-3, the following is valid for the beam placed on edge

$$\frac{T_B}{T} = \frac{l}{\left(\frac{h}{2}\right)} \qquad (5\text{-}8)$$

and the following for the beam placed flat

$$\frac{T_B}{X} = \frac{l}{\left(\frac{b}{2}\right)} \qquad (5\text{-}9)$$

Dividing eq. 5-9 by eq. 5-8 produces the following ultimate load ratio

$$\frac{T}{X} = \frac{h}{b} \qquad (5\text{-}10)$$

After Galileo has proved that for the self-weight loading case the fixed-end moment of a prismatic cantilever beam is proportional to the square of its length, he goes on to analyse cantilever beams with a solid circular cross-section (Figs. 5-11a to 5-11c). Galileo begins by considering the tensile test for bars of diameters $d_1$ and $D_1$ with a solid circular cross-section (Fig. 5-11a). In doing so, he assumes that the ultimate tensile forces $T^2_{B,d}$ and $T^2_{B,D}$ are proportional to $d_1^2$ and $D_1^2$ respectively because "the [tensile] strength of the cylinder [with diameter $D_1$] is greater than that [with diameter $d_1$] in the same proportion in which the area of the circle [= cross-sectional area – the author] [with diameter $D_1$] exceeds that of circle [with diameter $d_1$]; because it is precisely in this ratio that the number of fibres binding the parts of the solid together in the one cylinder exceeds that in the other cylinder" [Galileo, 1638/1964, p. 100]. After equating the ultimate tensile forces $T^2_{B,d}$ and $T^2_{B,D}$ with the tensile resistances at the point of fixity of the two beams (Fig. 5-11b), eq. 5-3 takes on the following form:

$$\frac{T^2_{B,d}}{E^2_{B,d}} = \frac{l_1}{\left(\frac{d_1}{2}\right)} \qquad (5\text{-}11)$$

$$\frac{T^2_{B,D}}{E^2_{B,D}} = \frac{l_1}{\left(\frac{D_1}{2}\right)} \qquad (5\text{-}12)$$

Dividing eq. 5-12 by eq. 5-11 while taking into account the ultimate tensile forces $T^2_{B,d}$ and $T^2_{B,D}$ with the square of the diameter $d_1$ or $D_1$ produces an ultimate load ratio of

$$\frac{E^2_{B,d}}{E^2_{B,D}} = \frac{d_1^3}{D_1^3} \qquad (5\text{-}13)$$

In the third step, Galileo varies the length of the cantilever beam as well as the diameter (Fig. 5-11c). From the proportionality of the ultimate tensile forces to the squares of the diameters

$$T^3_{B,d} \sim d_2^2 \quad \text{or} \quad T^3_{B,D} \sim D_2^2 \qquad (5\text{-}14)$$

plus the moment equilibrium with respect to point $B$ (see Fig. 5-8)

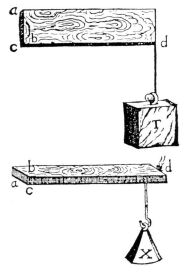

FIGURE 5-10
Galileo's consideration of the ultimate load ratio of a beam on edge (top) to one placed flat (bottom), (depth h or b, and width b or h depend on the respective orientation of the beam).

267

$$T_{B,d}^3 \cdot \frac{d_2}{2} = E_{B,d}^3 \cdot l_2 \quad \text{or}$$

$$T_{B,D}^3 \cdot \frac{D_2}{2} = E_{B,D}^3 \cdot L_2 \tag{5-15}$$

we get the following proportional relationships

$$E_{B,d}^3 \cdot l_2 \sim d_2^3 \quad \text{or} \quad E_{B,D}^3 \cdot L_2 \sim D_2^3 \tag{5-16}$$

and therefore the ultimate load ratio becomes

$$\frac{E_{B,d}^3}{E_{B,D}^3} = \frac{d_2^3}{D_2^3} \cdot \frac{L_2}{l_2} \tag{5-17}$$

For geometrical similarity in particular, i.e. when

$$\frac{d_2}{D_2} = \frac{l_2}{L_2} = k = const. \tag{5-18}$$

is satisfied, eq. 5-17, taking into account the moment equilibrium of eq. 5-15, becomes

$$\frac{E_{B,d}^3}{E_{B,D}^3} \cdot \frac{l_2}{L_2} = \frac{M_{B,d}^3}{M_{B,D}^3} = k^3 = \left[ \frac{T_{B,d}^3}{T_{B,D}^3} \right]^{\frac{3}{2}} \tag{5-19}$$

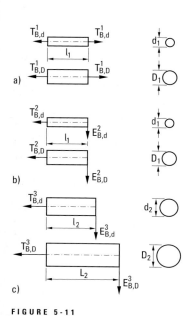

**FIGURE 5-11**
a) Tensile tests on bars of equal length but different diameters, plus the ultimate load ratios of b) cantilever beams of equal length but different diameters, and c) cantilever beams with unequal lengths and different diameters.

In response to eq. 5-19, Simplicio is astonished: "This proposition strikes me as both new and surprising: at first glance it is very different from anything which I myself should have guessed: for since these figures are similar in all other respects, I should have certainly thought that the forces and the resistances of these cylinders would have borne to each other the same ratio" [Galileo, 1638/1964, p. 104]. Salviati consoles him with the fact that he, too, at some stage also thought the resistances of similar cylinders would be similar, "but a certain casual observation showed me that similar solids do not exhibit a strength which is proportional to their size, the larger ones being less fitted to undergo rough usage ..." [Galileo, 1638/1964, p. 104].

And Galileo went further. He proves the assertion of the first day that among geometrically similar prismatic cantilever beams there is only one "which under the stress of its own weight lies just on the limit between breaking and not breaking: so that every larger one is unable to carry the load of its own weight and breaks; while every smaller one is able to withstand some additional force tending to break it" [Galileo, 1638/1964, p. 105].

After Galileo has also solved the task of calculating – for a cantilever beam of given length with a solid circular cross-section – the diameter at which, subjected to its own weight only, the ultimate limit state is reached, he can sum up: "From what has already been demonstrated, you can plainly see the impossibility of increasing the size of structures to vast dimensions either in art or in nature; likewise the impossibility of building ships, palaces, or temples of enormous size in such a way that their oars, yards, beams, iron-bolts, and, in short, all their other parts will hold together; nor can nature produce trees of extraordinary size because the branches would break down under their own weight ..." [Galileo, 1638/1964, p. 108]. The conclusion of Galileo's similarity mechanics-based strength considerations regarding the cantilever beam is – analogous to the breaking length of the bar in tension – the answer to the question of the maximum length of a cantilever beam loaded by its own weight only.

At the start of his examination of the symmetrical lever and the beam on two supports (Fig. 5-12), Galileo verifies that such beams may be twice as long as a cantilever beam. From that it follows that the fixed-end moment of the cantilever beam is equivalent to the maximum span moment of the beam on two supports and also the support moment of the symmetrical lever. Afterwards, he specifies the proportions of the ultimate limit states of the symmetrical with respect to the asymmetrical lever: "Another rather interesting problem may be solved as a consequence of this theorem, namely, given the maximum weight which a cylinder or prism can support at its middle-point where the resistance is a minimum, and given also a larger weight, find that point in the cylinder for which this larger weight is the maximum load that can be supported" [Galileo, 1638/1964, p. 115].

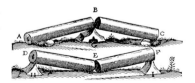

FIGURE 5-12
Failure mechanisms of the symmetrical lever and the beam on two supports (after Galileo)

The problem is illustrated in Figs 5-13a and b. (Without mentioning this explicitly, Galileo releases the beam on two supports and introduces a support at each point of load application $E_B$ or $E'_B$; he has thus reduced the problem to the statically equivalent systems of the symmetrical or asymmetrical lever.) From the equivalence of the ultimate moments

$$M_{B,E} = 0.25 \cdot l \cdot E_B = M_{B,E'} = E'_B \cdot \frac{a \cdot b}{l} \quad (5\text{-}20)$$

it follows that the ratio of the ultimate loads is

$$\frac{E'_B}{E_B} = \frac{l^2}{4 \cdot a \cdot b} \quad (5\text{-}21)$$

Galileo now turns his attention once again to the cantilever beam. He poses the question of which longitudinal form such a beam must have when loaded with a point load at the end of the cantilever so that the ultimate limit state is reached at every cross-section. Galileo proves that the longitudinal form must be that of a quadratic parabola (Fig. 5-14), and immediately identifies the engineering advantages: "It is thus seen how one can diminish the weight of a beam by as much as thirty-three per cent without diminishing its strength; a fact of no small utility in the construction of large vessels, and especially in supporting the decks, since in such structures lightness is of prime importance" [Galileo, 1638/1964, p. 118]. He provides the practical builder with methods for constructing parabolas – theoretically correct, but less than practicable, and theoretically incorrect, but a good approximation.

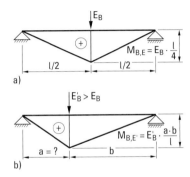

FIGURE 5-13
Bending moment diagram for a beam on two supports at the ultimate limit state for a) central maximum point load, and b) eccentric maximum point load

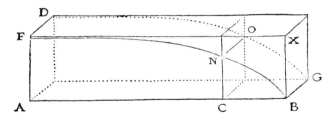

FIGURE 5-14
A beam fixed at AFD with a point load at the end of the cantilever BG must have the longitudinal form of a quadratic parabola (FNBGOD) if the ultimate limit state is to be reached at every cross-section (after Galileo).

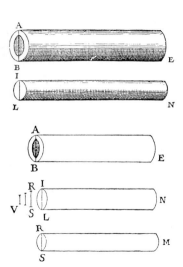

**FIGURE 5-15**
Relationship between the ultimate loads of cantilever beams with solid and hollow circular cross-sections fixed at the left-hand end (after Galileo)

## Developments in the strength of materials up to 1750

At the end of the second day of the *Dialogue*, Galileo also investigates cantilever beams with annular cross-sections "for the purpose of greatly increasing strength without adding to weight" [Galileo, 1638/1964, p. 123] compared to cantilever beams with a solid circular cross-section.

He discovers that the relationship between the ultimate loads is the same as that between the diameters $\overline{AB}$ and $\overline{IL}$ (Fig. 5-15). Using this proportion and the eq. 5-13 valid for the cantilever beam with solid circular cross-section (diameters $\overline{IL}$ and $\overline{RS}$), Galileo deduces that the ratio between the ultimate loads at $E$ and $M$ must be the same as for the product of $\overline{IL}^2 \cdot \overline{AB}$ to $\overline{RS}^3$.

This deduction ends the second day of Galileo's *Dialogue* and hence also his similarity mechanics-based theory of the failure of simple beam-type loadbearing systems. It almost sounds like Galileo is commanding future generations of scientists to carry out further research when he says (through Salviati): "Hitherto we have demonstrated numerous conclusions pertaining to the resistance which solids offer to fracture. As a starting point for this science, we assumed that the resistance offered by the solid to a straight-away pull was known; from this base one might proceed to the discovery of many other results and their demonstrations; of these results the number to be found in nature is infinite" [Galileo, 1638/1964, p. 123].

### 5.4

In almost all the history of science works dealing with the development of beam theory from Galileo to Navier (1785–1836), the bending failure problem of Galileo (Fig. 5-7) and its proposed solution is interpreted as though Galileo was already aware of the notion of stress. However, this notion, crucial to strength of materials, was not generally defined until 1823 – by the civil engineer and mathematician A. L. Cauchy (1789–1857) after he had employed the limit state notion of d'Alembert for the theoretical foundation of differential and integral calculus two years before. Stress, too, is a limiting value because the divided difference $\Delta P / \Delta F$ ($\Delta P$ is the internal force that acts on a finite area $\Delta F$) translates to the differential quotient $dP/dF$ for ever smaller areas $\Delta F$, i.e. $\Delta F$ tries to attain zero. The prerequisite for the establishment of this notion was that the solid body under consideration should not consist of a finite number of indivisible elements of finite size such as atoms or molecules, but rather that the material be distributed throughout the solid body (continuum hypothesis). The continuum hypothesis, too, did not advance to become the generally accepted structural model of the body in the emerging fundamental engineering science discipline of strength of materials until long after Euler's work on hydromechanics (1749) and Cauchy's founding of continuum mechanics in the 1820s [Truesdell, 1968, pp. 123–124]. As could be shown, Galileo's embryonic strength of materials theory was limited to the knowledge that the ultimate tensile forces from tensile tests were related to the geometrically similar cantilever beams abstracted to angle levers at failure. It was only for this reason that Galileo could ignore the relationship between force and deformation conditions conveyed by the material law in his tensile test and beam problem.

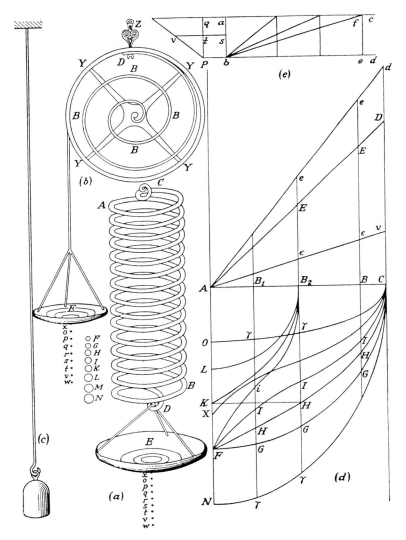

**FIGURE 5-16**
Hooke's spring trials from his *Lectures de potentia restitutiva, or of spring, explaining the power of springing bodies* (London, 1678)

Exactly 40 years after the publication of Galileo's *Dialogue* (i.e. in 1678), the highly inventive experimenter of the London-based Royal Society, Robert Hooke (1635–1703), published his law of deformation for elastic springs, which was based on his extensive trials with watch springs: *ut tensio sic vis*, i.e. the power of any spring is in the same proportion as the tension thereof (Fig. 5-16). "The same will be found," writes Hooke, "if trial be made, with a piece of dry wood that will bend and return, if one end thereof be fixed in a horizontal posture, and to the other end be hanged weights to make it bend downwards ... From all which it is very evident that the Rule or Law of Nature in every springing body is, that the force or power thereof to restore it self to its natural position is always proportionate to the distance or space it is removed therefrom ..." [Szabó, 1996, p. 356].

Hooke's knowledge of the material law

$$F = c_N \cdot \Delta l \tag{5-22}$$

(in words: the force $F$ is equal to the spring constant $c_N$ multiplied by the associated extension $\Delta l$, see Fig. 5-1) was arrived at neither in a qualitative (Leonardo) nor quantitative way through a thought experiment; Hooke's law is the result of real experiments carried out under defined practical testing conditions. Although Hooke worked out his law not on a natural object but rather with the help of a technical artefact (watch springs), he calls it a Law of Nature. This generalisation shows that Hooke's intention was obviously to discover in a technical artefact an objective, universally necessary, essentially natural relationship which can be repeated under the same conditions. Hooke's method therefore embodies the very germ of the engineering science experiment, which in the age of the Enlightenment and the Industrial Revolution was increasingly gaining emancipation from natural science experimentation and was to become the basis of a theory of strength of materials in the early 19th century. For only by measuring the force on the one hand and the associated extension of the test specimen in the experiment on the other is it possible to determine the material-related spring constants $c_N$ of metal, timber, silk and other springy substances. But Hooke had not only opened the door to research into strength via experiments; he definitely inspired the first elastic theory analysis of the beam problem (by Jakob Bernoulli in 1694), without which the theory tradition of the elastic line (elastica), created essentially by mathematicians in the 18th century, would have had no foundation. But at the same time there emerged a deep rift in the structural and strength of materials theories of the 18th century between the theory of elastica and Galileo's failure principle, which had been extended to the analysis of masonry arches and retaining walls primarily by French civil engineers. The failure load of a masonry arch or the earth pressure exerted by a body of soil on a retaining wall therefore formed the focal point of the theoretical interests of civil engineers, whereas the internal forces at the serviceability limit state could not yet be readily ascertained. It was the Dutch physicist Christiaan Huygens (1629 – 95) who explained – in a letter to Leibniz (1646 – 1718) dated 20 April 1691 – that Hooke's law reflected the material behaviour of springy bodies correctly only for small deformations [Stiegler, 1969, p. 111]. However, owing to the heavyweight forms of construction that prevailed until well into the 19th century, small deformations were difficult to observe; the serviceability limit state of structures remained largely an unknown quantity. Contrastingly, by observing cracks or assuming ruptured joints in masonry, e.g. stone arches, the ultimate load could be calculated with the equilibrium conditions. Accordingly, research into the strength of building and other materials began with the failure load in strength tests; it was not until the close of the 18th century that the deformation of test specimens could be measured thanks to the further development of testing apparatus. Galileo's *Dialogue* found its way via the Minorite monk Marin Mersenne (1588 – 1648) into scientifically aware France and from there to England; Mersenne, one of the most prolific authors of scientific letters with a natural philosophy, natural science and engineering content, had good connections to the leading scientists of Europe. The meetings proposed and organised by him and attended by French scientists and philosophers such as Gassendi (1592–1655), Descartes (1596–1650) and Pascal (1623–69) continued to take place at regular intervals in the house of Habert de Montemor after Mersenne's death, and became public in 1666 in the form of Colbert's Academy of Sciences. As one of the first members, Edme Mariotte (1620–84) delved intensively into the bending failure problem of Galileo, the reason being the building of the water main to the Palace of Versailles. Besides tensile tests on specimens of different materials, he was the first person to carry out bending failure tests on beams of timber and glass under various support conditions. Mariotte came to the conclusion that the fibres of beams in bending deform prior to rupture. According to him, the fibres of a fixed-end beam (Fig. 5-17) extend in the zone $AI$, but are compressed by the same amount in the zone $ID$; the distribution of the strains over the beam cross-section is linear. He placed the neutral fibres $IF$ subjected to neither tension nor compression correctly at half the depth of the beam cross-section symmetrical in two directions.

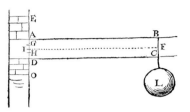

FIGURE 5-17
Fixed-end beam from Mariotte's *Traité du mouvement des eaux et des autres fluids* (published posthumously in 1686)

Like Galileo, Mariotte placed the tensile resistance $T_B$ from the tensile test in a relationship with the failure load $E_B$ from the bending test (see Figs. 5-8a to 5-8c). Mariotte's trials did not agree with the theoretical value given by Galileo

$$\frac{T_B}{E_B} = \frac{l}{\left(\frac{h}{2}\right)} \qquad (5\text{-}23)$$

nor with his value derived from assuming a linear distribution of stress over the beam cross-section

$$\frac{T_B}{E_B} = \frac{l}{\left(\frac{h}{3}\right)} \qquad (5\text{-}24)$$

Instead, some of his test results tended to be in the direction of

$$\frac{T_B}{E_B} = \frac{l}{\left(\frac{h}{4}\right)} \qquad (5\text{-}25)$$

According to plastic hinge theory, the same ratio of tensile resistance to failure load will result for the cantilever beam of rectangular cross-section (compare Fig. 5-9b with Fig. 5-17) using eq. 5-4 and eq. 5-6, i.e.

$$\frac{T_B}{E_{B,1}} = \frac{\sigma_{yield} \cdot b \cdot h}{\left(\frac{\sigma_{yield}}{l}\right) \cdot \left(\frac{b \cdot h^2}{4}\right)} = \frac{l}{\left(\frac{h}{4}\right)} \qquad (5\text{-}26)$$

Therefore, Mariotte must have achieved full plastification of the fixed-end cross-section $AD$ (Fig. 5-17) in the tensile and bending tests whose results were in the vicinity of the ratio $l/(0.25 h)$!

The most significant progress in the field of strength of materials experiments in the first half of the 18th century was achieved by the

**FIGURE 5-18**
Musschenbroek's apparatus for tensile tests

Dutchman Petrus von Musschenbroek (1692–1761), professor of physics at the universities of Utrecht and Leiden. In his book (written in Latin) *Physicae experimentales et geometricae* [Musschenbroek, 1729], he summarises problems of strength, test methods for use in strength research, and strength tests. Musschenbroek was the first person to build a testing apparatus for tensile tests (Fig. 5-18). He was helped by his brother Johann, who as a *mechanicus* in Leiden had also helped Wilhelm Jacob S'Gravesande to develop the testing apparatus described in his book *Physices elementa mathematica, experimentis confirmata* (1720–21) [Ruske, 1971, p. 14]. Musschenbroek's tensile testing apparatus was based on the principle of the beam scale. On the longer lever, the weight C generates the tensile force acting at the very short lever at A. The tensile force is transferred via a hook F and shackle O to the test specimen. The lever arm is moved into the horizontal position by means of a threaded bar E. In contrast to the simple tensile test, this testing apparatus allowed a continuous range of tensile forces to be applied to the test specimen, which enabled a considerable improvement in establishing the accuracy of the tensile forces at failure. It is therefore no surprise to learn that Musschenbroek's testing apparatus became the prototype for later materials testing machines.

Musschenbroek's other original contribution to experiments for research into materials is his buckling test apparatus (Fig. 5-19). By placing weights symmetrically on a platform guided at its four corners, he was able to transfer load concentrically into a bar until it lost its stability at the buckling load and deflected sideways. Years ahead of Euler, Musschenbroek used this testing apparatus to establish that the buckling load $P_{crit}$ is inversely proportional to the square of the length of the compression member. Whereas Musschenbroek's predecessors were only able to establish the tensile strength of a few building and other materials (mostly timber) with considerable scatter owing to the low level of mechanisation of strength experiments, Musschenbroek was now in a position to tabu-

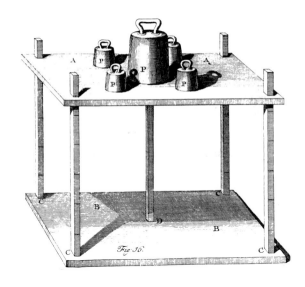

**FIGURE 5-19**
Musschenbroek's apparatus for buckling tests

late systematically the tensile strength of the most diverse materials in comprehensive test series. For example, he tested specimens with square and round cross-sections (60–70 mm side length or diameter, 160 mm long) made from lime, alder, pine, oak, beech and other species of wood. According to Winkler, Musschenbroek established a mean tensile strength of 81 N/mm² for softwoods, and values of 128 N/mm² and 143 N/mm² for oak and beech respectively [Winkler, 1871, p. 32]. Furthermore, he determined the tensile strength of copper, brass, lead, tin, silver, gold, wrought iron, cast iron and steel. Musschenbroek thus quantified the tensile strengths of the most important building materials.

His bending tests (Fig. 5-20), on the other hand, remained essentially confined to timber beam models with a rectangular cross-section. Musschenbroek's strength tests were quickly adopted in engineering publications, which started to appear after 1750, especially in France. One hundred years later, F. J. Ritter v. Gerstner was still referring to Musschenbroek's tensile and bending tests in his lectures. He calculated the mean value of the proportionality factor $m$ from Musschenbroek's bending tests on cantilever beams of length $l$ and cross-section $b \times h$ ($b$ = beam width, $h$ = beam depth) for various species of wood [Gerstner, 1833, p. 297]. He obtained the relationship

$$E_B = m \cdot \frac{b \cdot h^2}{l} \tag{5-27}$$

for the bending failure force $E_B$ (see Figs. 5-8a to 5-8c) applied at the end of the cantilever. The proportionality factor $m$ depends on the material and the cross-section. If we assume an ideal elastic and ideal plastic material behaviour according to Figs. 5-1e and 5-9a to 5-9c, where $\sigma_{yield}$ is the yield stress, then a successive evaluation of eq. 5-5 to eq. 5-7 results in the following values of $m$ for building materials with a distinct yield point:

$$m = \frac{\sigma_{yield}}{2} \tag{5-28}$$

(constant stress distribution over the fixed-end cross-section),

$$m = \frac{\sigma_{yield}}{4} \qquad (5\text{-}29)$$

(fully plastic fixed-end cross-section according to plastic hinge theory), and

$$m = \frac{\sigma_{yield}}{6} \qquad (5\text{-}30)$$

(elastic limit state according to practical bending theory). Eq. 5-23 had a different significance for Gerstner. As he knew the ultimate bending force $E_B$, the proportionality factor $m$ and the dimensions $b$, $h$ and $l$ of the test specimen from Musschenbroek's bending tests on cantilever beams, he could calculate the ultimate bending force $E'_B$ of a cantilever beam made from the same material but with the dimensions $b'$, $h'$ and $l'$; and vice versa: for a given load

$$P = \frac{E'_B}{\nu} \qquad (5\text{-}31)$$

($\nu$ = factor of safety against bending failure) and the dimensions $b'$ and $l'$, he could calculate the depth $h'$ required for the cantilever beam. Musschenbroek's tests enabled engineers to design beams by way of a comparative calculation.

It was exactly this point that was criticised by the natural science researcher Buffon (1707 – 88). He was dedicated to the French early enlightenment and therefore carried out bending tests on oak members with the dimensions commonly used on building sites (square timbers with side lengths of 100 to 200 mm and cantilevers of 2.1 to 8.5 m). Besides the force at failure, which in the case of oak differs only marginally from Musschenbroek's figure when using eq. 5-23, Buffon was the first person to measure the deflection at the end of the cantilever.

At the end of the 18th century, the French civil engineer P. S. Girard summarised the history of the theory of strength of materials since Galileo in his book *Traité analytique de la résistance des solides* [Girard, 1798].

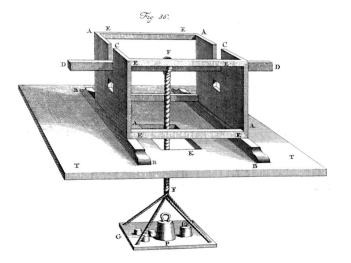

**FIGURE 5-20**
Musschenbroek's apparatus for bending tests

This book was the first monograph devoted exclusively to this subject and expressed the situation in the middle of the initial phase (1775–1825) of the evolution of the discipline of strength of materials. In terms of bending theory, Girard was a disciple of the Euler tradition. Although Girard tried to simplify its derivation, the bending theory was not limited to small deformations, which meant that its equations became complex mathematical formulations and were therefore unsuitable for practical engineering applications. His theory developments starting in 1787 were accompanied by comprehensive strength tests on timber beams in Le Havre. For example, he carried out buckling tests on full-size timber columns in order to check Euler's buckling theory. However, the findings of his experiments deviate considerably from ideal elastic material behaviour [Timoshenko, 1953, p. 58]. From the viewpoint of the historical development of elastic theory, Todhunter and Pearson have the following to say about Girard's monograph: "The whole book forms at once a most characteristic picture of the state of mathematical knowledge on the subject of elasticity at the time and marks the arrival of an epoch when science was free itself from the tendency to introduce theologico-metaphysical theory in the place of the physical axiom deduced from the results of organised experience" [Todhunter & Pearson, 1886, p. 77]. Like the thinking of the Enlightenment drove metaphysics out of strength of materials, too, so it gradually bid farewell to the theory of proportions. The process of the Industrial Revolution which started in the 1760s and continued into the early 1860s was to lead Europe to a new type of science: the fundamental engineering science disciplines of applied mechanics, theory of structures and applied thermodynamics.

## 5.5 Civil engineering at the close of the 18th century

The last three decades of the 18th century marked a heroic period for technology in Great Britain: Hargreaves invented the spinning jenny and Watt revolutionised Newcomen's steam engine such that it became a prime mover not only in mining, but in the production of machine tools and the textile industry as well. The Scottish country gentleman MacAdam developed the road wearing course named after him and the principles that enabled Great Britain to be criss-crossed by roads with a tough, even finish that could cope with the immense increase in the volume of traffic that was soon to come. Brindley – a mill-builder by profession – built the first English canal between Worsley and Manchester, which would later be used for transporting huge quantities of cheap coal to the manufacturing towns of central England.

As the Industrial Revolution began to take hold in Great Britain around 1780, so infrastructure engineering in the form of roads, canals and bridges became coveted objects for financial investors. More than just a few of them bought shares in the canal-building companies that were springing up everywhere in England at the end of the 18th century, speculating on an increase in the demand for transport services exceeding the volume of transport at that time. Once Abraham Darby II had succeeded in replacing charcoal with coke for smelting pig iron in 1750,

Henry Cort built the first furnace (puddling furnace) fired with coal in 1785. At last large quantities of usable wrought iron for building purposes could be produced. And after Parnell had been granted a patent for the rolling of wrought iron in 1788, the material and technological basis for the coming century of iron had been established. Great Britain advanced to become not only the world's most prominent manufacturing centre, but also the world's most prominent building site, where self-taught engineers, e. g. Thomas Telford, were active in the building of stone and iron bridges, canals and harbours, plus other civil engineering activities. No wonder Schinkel, Navier, Franz Anton von Gerstner and many other continental Europeans who enter the history of engineering at a later date spent some time visiting the factories and building sites of this industrious island!

On the continental mainland, the transition from technological or constructional-technical invention and innovation to diffusion into building activities was much slower because the mercantile characteristic was intrinsic to the promotion of trade and industry in the absolutist states, which eliminated the natural vigour of such diffusion processes but hampered their dynamic. After a delay of several decades, the institutionalisation of engineering science education emerging in the last 30 or so years of the 18th century in the more important absolutist states developed opposing forces as an element of the engineering science intervention institution, which likewise was intended to help throw off the chains of such an institution and evolve into the place where the system of classical engineering sciences would form. The fact that theory of structures in France became the prototype for classical engineering sciences and acted like a paradigm for the constitution of other engineering sciences in the 19th century was due to the following reasons:

*Firstly:* Civil engineering played a key role in the creation of the material and technical basis for manufacturing operations. Without the creation of the general infrastructure a society requires, e. g. roads, bridges and canals, a significant exchange of goods across the continent would have been impossible. Whereas in Britain the building of such works was mainly in private hands, especially after the onset of the Industrial Revolution around 1770, and the state refrained from all forms of intervention, France, as an absolutist state, trained its engineers initially in military academies and later at schools of civil engineering. The trained engineers had knowledge of mathematics and mechanics and they were employed as civil servants on state-run building sites operated, however, by companies with a manufacturing structure. Based on such practical experience, Bélidor published the first manual for engineers in 1729, and so beam, masonry arch and earth pressure theories took on the status of prescribed engineering science principles.

Although it can be shown that mathematics and mechanics were applied to construction engineering issues in Great Britain even as early as the 18th century, e. g. the masonry arch problem [Ruddock, 1979], they remain isolated examples because this theoretical knowledge was not generated, handed-down and further developed in educational establishments for the building industry. Furthermore, in contrast to their continental European colleagues, English authors did not make use of Leibniz' differential calculus until much later in the 19th century, and instead used Newton's very cumbersome method of fluxions.

*Secondly:* When France founded the École Polytechnique in Paris amid revolutionary upheaval in 1794, it created the first technical university at which leading mathematicians and natural scientists could employ descriptive geometry, differential calculus, mechanics and chemistry to prepare students for studies at the Écoles d'Applications for civil engineering, mining, military engineering, etc. For the first time, mathematics and the natural sciences were institutionalised for engineering, and given the task of improving French industry by Monge, the inventor of descriptive geometry and the minister responsible for the French Navy. This created the necessary organisational conditions for fusing together the theoretical knowledge already extant for certain technical artefacts of the building industry into a fundamental discipline of the system of classical engineering sciences such as applied mechanics and construction engineering. But it also rendered possible the formation of new engineering science disciplines with a mathematical and natural science foundation.

### 5.5.1 Franz Joseph Ritter von Gerstner

Franz Joseph was born on 23 February 1756 in Komotau in Bohemia, the son of a harness-maker, and was later elevated to the aristocracy through the granting of the title Ritter von Gerstner (Fig. 5-21). Together with the important Prussian civil engineer J. A. Eytelwein (1764–1848), he is one of the German-speaking engineers who – in addition to the French polytechnicians – helped considerably to place construction engineering in Germany and Austria on a scientific footing by 1850.

Besides studying the classics at grammar school, the young Gerstner became acquainted with some of the manual trades located in his hometown, in particular baking, brewing, soap-making, tanning, carpentry, joinery, bricklaying, blacksmithery and locksmithery. He attended the University of Prague from 1773 to 1779, where he studied philosophy, theology, Greek and Hebrew, and attended lectures on elementary mathematics given by Wydra, astronomy by Steppling and higher mathematics by Tessanek. Alongside his studies he had to earn part of his living by playing the organ and teaching mathematics and physics privately. Despite his speech impediment, he undertook two public doctor examinations on astronomy and Newton's *Principia*, which he passed with flying colours. Thereafter, he worked until 1781 as a surveyor on a royal commission set up by Emperor Joseph II in the course of abolishing serfdom [Bolzano, 1837, p. 7], which involved considerable surveying work.

Inspired by the enlightened Emperor's 1781 decision to turn the great Viennese poorhouse into a universal hospital [Wyklicky, 1984, p. 7], Gerstner studied medicine in Vienna. He nevertheless retained a serious interest in mathematics and astronomy and worked at Vienna's observatory under Hell, who advised Gerstner to give up medicine and turn to astronomy and mathematics. As a junior civil servant at the Prague

**FIGURE 5-21**
Franz Joseph Ritter von Gerstner (1756–1832), wood engraving by J. Passini, 1833

**FIGURE 5-22**
Cover and frontispiece of Gerstner's *Einleitung in die statische Baukunst*, 1789

observatory, his publications brought him acclaim from his contemporaries.

In 1788/89 Gerstner gave lectures on higher mathematics on behalf of Tessanek at the University of Prague. On the occasion of a public examination of his students in July 1789, he published his paper *Einleitung in die statische Baukunst* (introduction to structural engineering) (Fig. 5-22). In this paper he showed not only the usefulness of higher mathematics for the building industry using the example of the masonry arch problem, but also initiated the concept of structural engineering from the science of history standpoint. The success of his students in the public examination, the distribution of his publication *during* the examination and the recommendation of the acclaimed Paris-based astronomer de La Lande to the Prince of Kaunitz, the Austrian State Chancellor, earned him on 4 December 1789 the appointment as ordinary professor of higher mathematics at the University of Prague. "From now on … his main concern was to raise the status of the study of higher mathematics, hitherto largely ignored in Bohemia, and to attract a larger number of students; to this end, he considered it necessary to take into account the needs of commerce and industry in his lectures and not just concentrate on the objects of higher analysis and astronomy (as had been customary up to that time), and therefore also include higher mathematics and hydrodynamics and applications for everyday commerce and mechanical engineering in his lessons" [Bolzano, 1837, p. 12].

Gerstner, who right from his student days had recognised society's need for a theoretical treatment of engineering knowledge, now started testing the analytical power of mathematics and mechanics on technical artefacts. As a result, the number of students attending his lectures rose from three or four to 70 to 80! Within a short time, Gerstner was a much-sought-after engineering adviser to the Habsburg monarchy: "It was rare for civil servants from private and state organisations *not* to consult Gerstner when they encountered difficulties during their engineering business, primarily connected with iron ore mining, blast-furnaces, hammer-works and the construction of the most diverse types of machinery" [Bolzano, 1837, p. 13].

The shift in Gerstner's scientific interests in the final decade of the 18th century – from astronomy to engineering – is understandable when we consider the fact that in contrast to many other German states at the time, the Bohemian aristocracy had earned a considerable reputation for its industrialisation of the region and had turned Bohemia into the No. 1 export state of the Habsburg monarchy [Manegold, 1970, p. 35].

Just one year after the founding of the École Polytechnique in Paris, Emperor Franz II set up a Royal Commission for the Reform of Public Education Establishments [Gerstner, 1932, p. 256], with Duke Rottenhan, the owner of Bohemia's calico and mining industry, in the chair; Gerstner was appointed to report on the situation regarding natural sciences, agriculture, mathematics and technology in the schools and universities. The curriculum prepared by Gerstner drew the commission's attention to the great accomplishments of British industry, "which threatened to displace the industry of the entire continent from the market-places of major international trade", and at the same time contradicted the widely held view "that for the continental states it would be most convenient to concentrate on agriculture and mining alone, and to leave the processing of the materials thereby obtained to the machines of England" [Bolzano, 1837, p. 16].

Gerstner's proposals for reforming education were warmly received in Vienna and also by the Bohemian Parliament. This culminated in the opening of the Polytechnic Institute in Prague on 3 November 1806, hardly three months after the Austrian ruler Franz had given up the title of Holy Roman Emperor following massive threats by Napoleon and at the same time decreed the break-up of the Holy Roman Empire of the German Nation [Zöllner, 1970, p. 337]. Like Monge had founded the École Polytechnique on the basis of improving French industry, Gerstner saw the purpose of the Polytechnic Institute financed by the Bohemian Parliament as "raising the status of the Fatherland's industry through scientific teaching" [Jelinek, 1856, p. 36].

Besides those interested in industry, future agricultural and waterways engineers also attended the lectures on mechanics and hydraulics at the Polytechnic Institute headed by Gerstner. All the professors at the Polytechnic Institute undertook a pledge "... to teach and advise all those who seek explanation of an object concerning his industry, in particular from the subjects of chemistry, mechanics and building" [Bolzano, 1837, p. 19]. It goes without saying that it was Gerstner who "attracted the greatest number of persons seeking advice" [Bolzano, 1837, p. 20] and in addition he was consulted by state departments on waterways matters. Gerstner therefore provided reports on projects in the Elbe and Moldau (now Vltava) rivers; in an extensive report he spoke out against the planned building of the Tabor Bridge near Vienna and instead proposed building a bridge near Nußdorf. His advice was taken.

The publication of his theory of water impacts in straight channels (1790), his theory of waves and the theory of dyke profiles (1804) derived therefrom, and his mechanical theory of overshot waterwheels (1809) represented considerable contributions to the "advancement ... of hydromechanical theories in Germany" [Hänseroth, 1987, p. 91]. These, together with the pertinent works of Woltmann, Eytelwein and Hagen, were later to become an important source for applied hydromechanics.

Gerstner, appointed Director of Bohemian Waterways by the Emperor in 1811, had already realised back in 1807 that it was more economic to abandon plans for a canal between the Moldau and Danube rivers, which had existed since the 14th century, and instead proposed building an "iron roadway" between Linz on the Danube and Joachimsmühle on the Moldau [Knauer, 1983, p. 12].

Gerstner was the first engineer in continental Europe to recognise the immense technical and scientific significance of Britain's renewal and would thus become the initiator of Austrian railways. His pioneering ideas were published in 1813 under the title of *Zwei Abhandlungen über*

*Frachtwägen und Straßen, und über die Frage, ob und in welchen Fällen der Bau schiffbarer Kanäle, Eisenwege oder gemachten Straßen vorzuziehen sey* (two treatises on goods wagons and roads and on the question of whether and in which cases the building of navigable canals, iron roadways or made-up roads should be preferred) [Gerstner, 1813].

This paper was translated into French in 1827 and Hungarian in 1828. On 7 September 1824 the son of F. J. v. Gerstner – Franz Anton von Gerstner – was granted his request to build a horse-drawn railway between Linz and Budweis by Emperor Franz I. The first part of the line, from Budweis to Pramhöf, was built under the supervision of F. A. v. Gerstner. Fig. 5-23 shows the first (left) and last (right) sheets of the six-page principal balance of accounts for the period 1 February to 31 December 1825 for this railway project; the accounts from 28 January 1826 form part of F. A. v. Gerstner's *Bericht über Linz–Budweiser Bahn 1825* (report on the Linz–Buweis railway of 1825) and are kept in the archives of the Technical Museum in Vienna [Gerstner, F. A. v., 1826]. According to F. A. v. Gerstner, the costs incurred for the accounting period amounted to a total of 121 313 gulden C. M. (C. M. = convention coin), which he carefully broke down into

| I | Cost of land: | 2 470 gulden C. M. | 2.04 % |
|---|---|---|---|
| II | Stone: | 13 054 gulden C. M. | 10.76 % |
| III | Timber: | 23 382 gulden C. M. | 19.27 % |
| IV | Iron: | 9 588 gulden C. M. | 7.90 % |
| V | Building work in 1825: | 24 288 gulden C. M. | 20.02 % |
| VI | Rolling stock: | 15 003 gulden C. M. | 12.37 % |
| VII | Buildings: | 1 647 gulden C. M. | 1.36 % |
| VIII | Tools: | 7 450 gulden C. M. | 6.15 % |
| IX | Horses: | 3 500 gulden C. M. | 2.88 % |
| X | Labour: | 20 931 gulden C. M. | 17.25 % |

Goods trains ran for the first time between Trojern (near Wullowitz) and Budweis in 1827. On 2 April 1829, after just under four years of building work, scheduled rail operations began between Budweis and Pramhöf; the first scheduled passenger trains started running five years later. This was the first railway line on the continental mainland and so F. A. v. Gerstner became the first international railway engineer.

For health reasons, F. J. v. Gerstner asked to be released from his position as professor of higher mathematics and director of physics and mathematics studies at the University of Prague in 1822. However, he continued as director of waterways and as director and professor of mechanics at the Prague Polytechnic Institute. Gerstner's final dedication to a comprehensive scientific footing for engineering was the underlying cause, and the ban on reading, writing and drawing "on overcast days and by candlelight" imposed by his doctors occasioned him to resign his post as director of waterways; his application "to grant his earlier request for personnel" [Gerstner, 1932, pp. 258–259] to help him with the publication of his lectures on mechanics and hydraulics was refused, which meant that his son undertook the editing of these lectures after his return

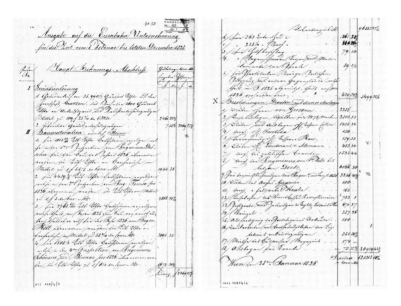

**FIGURE 5-23**
First (left) and last (right) sheets of F. A. v. Gerstner's accounts for the year 1825 for the Linz–Budweis railway line, the first in continental Europe

from England in 1829. The work took five years and Franz Joseph Ritter von Gerstner would only live to see the publication of the first volume of his *Handbuch der Mechanik* (manual of mechanics). Just a few weeks after his honourable discharge from state service, he died on 25 June 1832 on the country estate of his son-in-law near Gitschin.

Together with Eytelwein's multi-volume *Handbuch* [Eytelwein, 1801 & 1808], the *Handbuch der Mechanik* [Gerstner, 1832, 1833 & 1834] can be regarded as *the* outstanding German-language compendium of science-based engineering of the early 19th century. It also presents the building and machine technologies prevalent at the transition from workshop manufacture to the production techniques of the Industrial Revolution in a scientific way such that we can regard it as a watershed in the evolutionary process of the system of classical engineering sciences.

## 5.5.2 Introduction to structural engineering

After Gerstner notes in his *Einleitung in die statische Baukunst* that mechanical engineering had been able to achieve more progress in the 18th century with the help of higher mathematics than construction engineering, he immediately turns to the development of masonry arch theory as "the most difficult part of higher construction engineering" [Gerstner, 1789, p. 5]. Although Leibniz and the Bernoullis "declared the catenary as the most advantageous for masonry arches …, the builders of such arches did not, or knew not how to, take advantage of this because they appeared to see the focus of their work as the voussoirs and not the centering". Furthermore, the inverted catenary was not in harmony with the "good taste of the ancients", who preferred semicircular arches, which is why Bélidor, very probably because of the simpler stone-cutting, analysed semicircular arches, and in doing so made use of "arbitrary hypotheses", and "for safety reasons … advised increasing the thickness" of the abutment as calculated [Gerstner, 1789, p. 2].

Despite the fact that French civil engineering literature had permeated German construction engineering since the 1740s [Hänseroth, 1987, p. 93], in particular the publication of Bélidor's *Science des ingénieurs* (1729) translated by the rationalistic Enlightenment philosopher Christian Wolff, the use of even simple structural calculations was not widespread among builders in Germany. The reference to Bélidor's design equations based on statics was certainly a mark of authority [Hänseroth, 1987, p. 93], but was way behind the hands-on experience of the builders themselves. When we also consider the fact that even experienced builders were in real life faced with damage to, even the collapse of, buildings and structures, the importance of Gerstner's recognition of the conflict between theory and practice and his proposed solution for establishing a theory of structures can hardly be overestimated. As Gerstner writes: "Correspondingly, our public buildings ... are either built according to the prototypes of the ancients or according to practical judgements by experienced builders, but without being able to vouch for their whole long-term safety (at least when seen on average); there is still no large masonry arch constructed by mankind – when we look at the works of the most famous architects – in which sooner or later some cracks, due to whatever reasons, appear; and in this case the dome of St Peter's in Rome can serve as a single example for all others. Only a thorough mechanics derived from the nature of building itself can control this weakness ... It is a pity that the greatest exponents of this art, starting with Vitruvius, saw the mastery of building only in the sense of good intercolumniation and arch forms, were carried away by wonder, which we ourselves cannot deny in the ruins of magnificent old buildings, were destined to mere imitation, and thought they were in possession of the best rules of building. Since then, strength, which is the essential element of great structures, has been so superseded by the love of beauty that the question of whether their magnificence be appropriate to their lifespan was almost the last one" [Gerstner, 1789, p. 4]. Only someone who is fully aware of the historical process of separating the useful from the beautiful arts in building, has decided in favour of the useful in building and with the help of mathematics proposes mechanics corresponding to the nature of building can write in this way. Gerstner realised the first step towards a methodical preliminary outline of structural theory by taking masonry arch theory as an example.

**Gerstner's analysis and synthesis of loadbearing systems**

#### 5.5.2.1

Gerstner evolved his masonry arch theory in *Einleitung in die statische Baukunst* in four logical development stages.

*First development stage*

Gerstner assumes a prismatic "beam or stone" [Gerstner, 1789, p. 7] supported on two columns which has to carry a uniformly distributed self-weight $g$ and results in the support reactions $0.5 \cdot g \cdot l_{12}$ (Fig. 5-24a); he makes the latter easier to understand for the reader by replacing each column by a permanent pulley whose ropes are loaded with $G_1 = G_2 = 0.5 \cdot g \cdot l_{12}$ (Fig. 5-24b). The forces in the ropes also remain equal if such a released beam is placed at an angle and the ropes are guided parallel (Fig. 5-24c).

*Second development stage*

Gerstner leans the inclined bar 1-2 (Fig. 5-24c) "on a smooth wall" [Gerstner, 1789, p. 7] in such a way that the static system shown in Fig. 5-25a results; he replaces the uniformly distributed load $g$ due to the self-weight of the bar 1–2 by the point loads at the nodes $G_1 = G_2 = 0.5 \cdot g \cdot l_{12}$. As the load $G_1$ at node 1 cannot be accommodated by the lateral support, it must be neutralised by the horizontal thrust

$$H_1 = (0.5 \cdot g \cdot l_{12}) / \tan\alpha_{12} \quad (5\text{-}32)$$

and the force in the bar

$$S_{12} = (0.5 \cdot g \cdot l_{12}) / \sin\alpha_{12} \quad (5\text{-}33)$$

to produce equilibrium. When $S_{12} = S_{21}$, it follows that for the equilibrium conditions at node 2

$$H_2 = H_1 \quad (5\text{-}34)$$

and

$$V_2 = 2 \cdot G_1 = 2 \cdot G_2 = g \cdot l_{12} \quad (5\text{-}35)$$

Gerstner now extends bar 1–2 symmetrically to form a three-pin system 2–1–2' (Fig. 5-25b) with the support reactions

$$H_2 = H_{2'} = (0.5 \cdot g \cdot l_{12}) / \tan\alpha_{12} \quad (5\text{-}36)$$

and

$$V_2 = V_{2'} = 2 \cdot G_2 = 2 \cdot G_{2'} = g \cdot l_{12} \quad (5\text{-}37)$$

Using eq. 5-36 and eq. 5-37, "it is easy to calculate the pressure [= support reaction as vector sum of $H_2 + V_2$ – the author] with which common roofs rest on their abutments" [Gerstner, 1789, p. 8]. But not only that: "the question of whether the higher German or the lower foreign roofs place more load on their abutments can also be easily deduced from this" [Gerstner, 1789, p. 9] (Fig. 5-25b). So if in the equation for the support reaction derived from eq. 5-36 and eq. 5-37

$$A_2 = A_{2'} = \sqrt{V_2^2 + H_2^2} = \sqrt{(g \cdot l_{12})^2 + (0.5 \cdot g \cdot l_{12})^2 / \tan^2\alpha_{12}} \quad (5\text{-}38)$$

$l_{12}$ is taken to be $\sqrt{s_2 + h_2}$ and $\tan^2\alpha_{12}$ to be $(h/s)^2$, then it follows that

$$A_2(h) = A_{2'}(h) = g \cdot \sqrt{s^2 + h^2 + s^4/(4h^2) + s^2/4} \quad (5\text{-}39)$$

Using the constant $s$, the calculation of the necessary condition for the existence of an extreme value

$$dA_2(h)/dh = 0 \quad (5\text{-}40)$$

means that the roof pitch is

$$h = (\sqrt{2}/2) \cdot s \quad \text{or} \quad \alpha_{12} = 35.26° \quad (5\text{-}41)$$

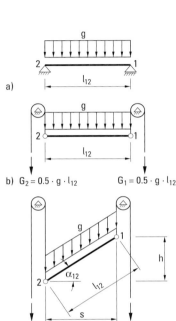

**FIGURE 5-24**
Prismatic beam on two columns subjected to self-weight only:
a) horizontal, b) released, and c) inclined

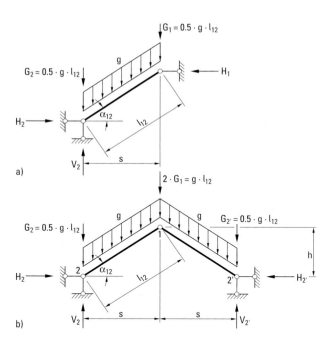

**FIGURE 5-25**
Support reactions of a) the inclined pin-jointed bar, and b) the symmetrical three-pin system

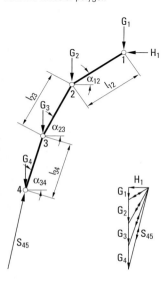

**FIGURE 5-26**
Inverted funicular polygon

From the second derivation of eq. 5-39 for $h$, it is easy to see that the support reaction $A_2(h)$ or $A_{2'}(h)$ is a minimum for $h = (\sqrt{2}/2) \cdot s$. From this it follows that the "higher German roofs" place "more load" on their abutments than the "lower foreign roofs" [Gerstner, 1789, p. 9].

*Third development stage*
After Gerstner has shown the relevance of the three-pin system for construction engineering using the calculation of the support reactions of couple roofs as an example, he removes all lateral supports (Fig. 5-25a) and sets up a system with $(i + 2)$ degrees of kinematic indeterminacy at bar 1–2 using identical bars 2–3, 3–4, ..., $j-(j + 1)$, ..., $i-(i + 1)$. Fig. 5-25 illustrates such a system for $i = 3$ bars. The kinematically indeterminate system subjected to the node forces $G_1 = G_{i+1} = 0.5 \cdot l_{12}$ and $G_2 = G_3 = G_4 = ... = G_j = G_{j+1} ... = G_i = g \cdot l_{12}$ is then – and only then – in unstable equilibrium when the condition

$$tan\, \alpha_{j,j+1} = (1/H_1) \cdot \sum_{k=1}^{j} G_k \tag{5-42}$$

(where $H_1$ is the horizontal thrust quantified with eq. 5-32, Fig. 5-26) is satisfied for the direction of all bars, i.e. $1 \leq j \leq i$. From the equilibrium conditions at node $j$, we can also calculate the force in the bar

$$S_{j,j+1} = (1/sin\, \alpha_{j,j+1}) \cdot \sum_{k=1}^{j} G_k \tag{5-43}$$

A kinematically indeterminate hinged system set up by eq. 5-42 is known as an inverted funicular polygon. Again, Gerstner discusses the practical uses of his observations for the statics-constructional analysis of mansard roofs and comes to the following conclusion: "The two-pitched roofs therefore have the advantage that besides creating more interior space and requiring shorter timbers for their building, the lateral compression

286

[= horizontal thrust $H_1$ – the author], which is always the most dangerous in any building, is smaller" [Gerstner, 1789, p. 12]. This is the carpenter talking!

*Fourth development stage*

"What was said about the constitution of rafters up to now also applies to voussoirs if they are to remain in equilibrium" [Gerstner, 1789, p. 13]. This means that eq. 5-42 and eq. 5-43 also apply to the calculation of the inverted funicular polygon of a stone arch with constant and varying depth provided the nodes 1, 2, 3 ..., $j$, ($j$ + 1), ..., ($i$ + 1) represent the centres of gravity of the voussoirs and their self-weight is represented by the statically equivalent point loads $G_1$, $G_2$, $G_3$, ..., $G_j$, $G_{j+1}$, ..., $G_{i+1}$. Gerstner now completes the transition from inverted funicular polygon to line of thrust: "The smaller the voussoirs are assumed to be, the closer the resulting polygon approximates a regular curve" [Gerstner, 1789, p. 14]. He achieves this by replacing eq. 5-42 by the differential equation

$$dy/dx = (1/H_1) \cdot g \cdot s_1 \qquad (5\text{-}44)$$

where

$$V(x) = \int_0^x g \cdot ds_1 = g \cdot s_1 \qquad (5\text{-}45)$$

i.e. the weight of the arch from the crown to the point under consideration $x$ (Fig. 5-27a). For an arch of constant depth, i.e. $g$ = const. and after introducing the shortcut $m = H_1/g$ = const., eq. 5-44 takes on the following form

$$m \cdot (dy/dx) = s_1 \qquad (5\text{-}46)$$

After Gerstner squares eq. 5-46, adds $s_i^2 dy^2$ to both sides and considers the geometrical relationship, he can convert the expression

$$ds_1^2 = dx^2 + dy^2 \qquad (5\text{-}47)$$

into

$$s_1 = \sqrt{y^2 - m^2} \qquad (5\text{-}48)$$

Using eq. 5-48, he substitutes the arc length $s_1$ into eq. 5-46 and after integrating arrives at the equation for the inverted catenary (line of thrust)

$$y = 0.5 \cdot m \cdot (e^{x/m} + e^{-x/m}) = m \cdot \cosh(x/m) \qquad (5\text{-}49)$$

Gerstner tabulated eq. 5-49 for $m = 10$. He derives the curve of the soffit (intrados) for arches with a homogeneous masonry infill (Fig. 5-27b)

$$y = 0.5 \cdot b \cdot (e^{x/m'} + e^{-x/m'}) = m \cdot \cosh(x/m) \qquad (5\text{-}50)$$

using the parameter $m'^2$ as the quotient of the horizontal thrust $H$ and the specific weight of the masonry infill $y$. If $m' = b$, the curve of the soffit is an inverted catenary according to eq. 5-49. Therefore, as the ratio $b : m'$ is constant for every point $x$, all curves can be very easily constructed according to eq. 5-50, also in the case of $b \neq m'$ (Fig. 5-28a). Gerstner had thus solved the second prime task of thrust line theory.

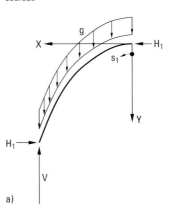
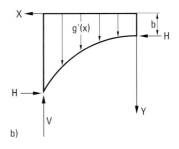

**FIGURE 5-27**
Analysing an arch a) with constant depth, and b) with an infill of horizontal masonry courses

Gerstner summarises the results of the four development stages in the form of eight rules for engineers and builders:

I. "For masonry arches with a single depth throughout that carry nothing other than their own weight, the standard catenary applies." (Fig. 5-27a)

II. "Masonry arches with an infill of horizontal masonry courses must be built according to a curve from the set of catenaries, the uppermost horizontal line of which is always the abscissa." (Fig. 5-27b)

III. "… the shallower the masonry arch at its crown, the flatter is the soffit s – r – r – a, and … the steeper the arch, the more curved is the soffit, for a uniform width." (Fig. 5-28a)

IV. "The voussoirs should be long enough to reach from one curve s' – r' – r' to the other s – r – r." (Fig. 5-28b)

V. "Their side faces r – r' should be perpendicular to all curves that pass through their depth. To do this it would be necessary that these side faces themselves be cut to follow a curved line, but their length is seldom such that this curvature is noticeably different from a straight line." (Fig. 5-28b)

VI. "The length of the voussoirs depends on their quality and the load to be carried. These circumstances must govern the choice the builder has to make from the infinite number of soffit lines drawn." (Fig. 5-28a)

VII. "The flatter the soffit lines are, the less they deviate from an arc. It just depends on the builder specifying the radius that deviates least from the associated curve in each case."

VIII. "The soffit has a horizontal tangent at the crown. … the Gothic arches are not built like nature. However, there are cases where even our rules call for a kink in the soffit, namely at those places where a particular load is to be applied to the arch; but as this case contradicts the apparent strength and contradicts good taste in architecture, it would be superfluous to provide a special calculation for this." [Gerstner, 1789, pp. 18 – 19]

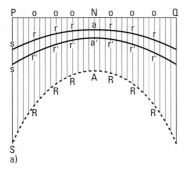

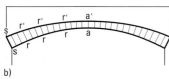

FIGURE 5-28
Constructing the inverted catenary: a) for masonry arches with an infill of horizontal masonry courses; b) voussoirs for an arch with varying depth plus an infill of horizontal masonry courses

**Gerstner's method of structural design**

### 5.5.2.2

A study of the four logical development steps with the help of Büttner and Hampe's systematic analysis and synthesis of loadbearing systems reveals that Gerstner had already identified the formation of a loadbearing system consisting of loadbearing system elements as the crux of the design of loadbearing structures, and had worked out a method for this.

The "loadbearing structure" is the material reality, it is the tangible timber beam or the tangible stone arch, whereas a "loadbearing system" is an abstract model of the loadbearing structure taking into account the loadbearing function [Büttner & Hampe, 1977, p. 10].

After Gerstner, in his first development stage, assumes a beam on two supports and presents its structural equivalence with the same, but inclined beam using the method of sections illustrated with two simple pulleys, he analyses, in his second development stage, the pin-jointed bar as the loadbearing system element under various boundary conditions without especially mentioning the modelling of the loadbearing structure that has been implicitly undertaken. This levelling of material and modelled reality comes to light in the second development stage through Gerstner's imagined fluctuation between loadbearing structure (e.g. couple and mansard roofs) and the loadbearing system made up of pin-jointed bars (e.g. three-pin system). The addition of any number of identical loadbearing system elements, carried out in the third development stage, to form a kinematically indeterminate articulated system made up of pin-jointed bars results in a system in equilibrium when, and only when, the articulated system assumes the form of an inverted funicular polygon. A loadbearing system formed in this way is theoretically possible, but cannot be built as a loadbearing structure. In order to achieve a stable loadbearing structure, Gerstner, in his fourth development stage, considers the inverted funicular polygon as a model of a stone arch. The infinitising of the inverted funicular polygon to form the line of thrust enables differential and integral calculus to be used to determine the shape of the stone arch. Gerstner therefore calculates the upper and lower intrados lines for the masonry arch with an infill of horizontal masonry courses, a type frequently used in practice. Such an arch is therefore in stable equilibrium because any number of lines of thrust can be drawn between the two intrados lines for the corresponding masonry infill, which allowed Gerstner to point out the structural indeterminacy of the arch problem.

**5.5.2.3** *Einleitung in die statische Baukunst* **as a textbook for analysis**

The final third of Gerstner's *Einleitung in die statische Baukunst* contains 130 theorems and exercises for the analysis of higher equations. The 119 theorems, which read more like the titles of lectures, are divided up as follows: solution of algebraic equations (1 to 17), logarithmic calculation (18 to 26), trigonometry (27 to 33), infinite series (34 to 40), analytical geometry (41 to 81), differential calculus (82 to 109), and integral calculus (110 to 119). The order and content correspond to the two- to three-semester courses in mathematics that form the foundation courses in civil/structural engineering at German polytechnics (converted from state building schools in the 1970s). If we exclude matrix calculus, then the current bachelor courses of studies in this field differ only marginally from the aforementioned content. It is probable that Gerstner's theorems formed the basis of his lectures in higher mathematics that he gave in 1788/89 at the University of Prague while standing in for his mentor Tessanek. He was therefore certainly the first author working in the German language who provided paradigms for successfully dealing with constructional and building technology problems in building with the help of analysis.

**5.5.3** **Four comments on the significance of Gerstner's** *Einleitung in die statische Baukunst* **for theory of structures**

*Firstly:* The nucleus of the classical type of engineering that established itself during the Industrial Revolution in Great Britain is the transformation of the natural cause-effect relationship into an engineering purpose-means relationship. The condition necessary for this type of engineering, which prevailed into the second half of the 20th century, was the scientific knowledge of the natural cause-effect relationship realised in engineering entities and methods as a purpose-means or function-structure relationship. By modelling the stone arch – in terms of its loadbearing function – as a loadbearing system consisting of pin-jointed bars, Gerstner discovered the causal relationship between loadbearing function and catenary realised in the catenary arch.

*Secondly:* Gerstner, with his loadbearing system analysis and synthesis, had not only taken the first step towards realising his call for a "mechanics derived from the nature of building itself", but for the first time had formulated the discovery of the natural laws realised in the loadbearing function/loadbearing system relationship as a prime task of the engineer. And that with a view to creating a knowledge system for theory of structures. His *Einleitung in die statische Baukunst* placed the first milestone in the transition from the initial phase of the theory of structures to the first stage of its discipline-formation period. Through his subsequent contributions to the theory of structures, he had a considerable influence on that transition in the German-speaking countries.

*Thirdly:* According to the first structure formation law, loadbearing systems are created by adding equivalent loadbearing system elements. But on the other hand, the formation of a loadbearing system is carried out according to the second structure formation law by combining different loadbearing system elements [Büttner & Hampe, 1984, p. 16]. The first structure formation law for loadbearing systems, a specific form of engineering science modelling which is an effective design aid in modern structural engineering, can indeed be seen in Gerstner's four-stage masonry arch theory. However, the masonry arch theory reaches its limits where Gerstner does not distinguish sufficiently between the loadbearing structure, as a constituent of the material reality of the finished structure, and the loadbearing system, as an abstract model of the loadbearing structure looked at from the loadbearing function viewpoint. The upshot of this is that his *Einleitung* is sometimes written from the viewpoint of the builder on site and sometimes from the viewpoint of the design engineer or the practical mathematician.

*Fourthly:* Gerstner was able to omit the treatment of the strength problem because the quantification of the horizontal thrust is in the first place dependent on the form of the masonry arch and, moreover, the great reserves of strength turn the question of strength into a secondary issue. Consequently, the link between statics and strength of materials that characterises theory of structures was not put in place – an aspect that was left for Eytelwein to start and Navier to finish.

## 5.6 The formation of a theory of structures: Eytelwein and Navier

The beginnings of engineering science theories in the fields of masonry arch statics, beam statics, earth pressure statics, strength of materials and the mathematical analysis of the elastic line which were beginning to appear in the "preparatory period of the engineering sciences" [Buchheim, 1984] were first moulded into a theory of structures by Navier in 1826 [Navier, 1826], in the "formation period of the engineering sciences"

[Buchheim, 1984]. The new relationship between science and production in general and mechanics and engineering practice in particular which had been brought about by the Industrial Revolution was seen by the French polytechnicians as a redefinition of the relationship between empirical findings and theory. In his introduction to the second edition of his *Résumé des Leçons* on the mechanics of building, Navier writes: "The majority of design engineers determine the dimensions of parts of structures or machines according to the prevailing customs and the designs of works already completed; only rarely do they consider the compression that those parts have to withstand and the resistance with which those parts oppose said compression. This may have only few disadvantages as long as the works to be built are similar to those others built at other times and they remain within conventional limits in terms of their dimensions and loads. But one should not use the same method if the circumstances require one to exceed those limits or if it is a whole new type of structure of which there is as yet no experience. This book is intended to specify the conditions for the erection of such edifices that are carried out under the supervision of engineers; it will also specify the degree of resistance for the individual parts … We believe it is necessary to outline briefly the principles that we have used as our basis for the treatment of the most important questions" [Navier, 1833/1878, pp. XV – XVI]. So the father of modern structural theory was formulating not only the methods, problems and aims, but also a programme that would turn out to be the theoretical foundation of construction-oriented engineering sciences over the course of the 19th century. Fig. 5-29 shows the cover of the Brussels edition of Navier's *Résumé des Leçons*, which also contains a biography of Navier written by Gaspard de Prony (1755 – 1839) [Prony, 1839, pp. x – lj].

In contrast to this, Eytelwein regarded statics as a branch of mathematics. Nevertheless, Eytelwein was able to deal with problems of construction and mechanical engineering in terms of mechanics. One of the most remarkable things to come out of this was his solution to the continuous beam problem.

The analysis of the continuous beam (the first statically indeterminate problem calculated in structural engineering) by Navier and Eytelwein will be used as an example to compare the different levels that theory of structures had attained in France and Germany in the early 19th century.

FIGURE 5-29
Cover of Navier's *Résumé des Leçons* [Navier, 1839]

## 5.6.1 Navier

Claude-Louis-Marie-Henri Navier, born on 10 February 1785 in Dijon, the son of a highly respected lawyer, experienced at an early age the wide chasm between the customary handing-down of constructional-technical knowledge through empirical rules and the engineering sciences starting to take shape at the École Polytechnique on the basis of natural sciences and mathematics [McKeon, 1971]. After the death of his father, Navier grew up under the guardianship of his uncle Emil Gauthey, who at that time occupied the highest official post in French civil engineering and whose main task was to look after the condition of the canals, roads and bridges that had been built during the mercantile economic policies of the *ancien régime*. Gauthey's manuscripts on bridge-building, edited by Navier in 1813, exude the whole spirit of the 18th century. This must have represented a challenge to the graduates of the École Polytechnique and the École des Ponts et Chaussées because in the end these establishments attracted the best mathematicians and natural scientists in the world to an extent that has never been seen again. As an engineer in the Bridges & Highways Department, Navier was involved in the building of important stone bridges and experienced at first hand that in practice engineers were still "determining the dimensions of parts of structures … according to the prevailing customs and the designs of works already completed" [Navier, 1833/1878, p. XV]. For him, that was reason enough to republish the engineering manuals of the French military engineer Bélidor (originally published between 1729 and 1737) together with a critical commentary.

Although French civil engineers benefited from the very best scientific training in the fundamentals of their art, they were still using the natural building materials stone and timber for their bridges. The situation was different across the English Channel, where inventive manual tradesmen and empirically oriented engineers were at work. New foundry processes permitted the manufacture of cast iron with its high compressive strength, and puddling iron with its high tensile strength. In the hands of British builders and engineers, these became the materials of a new world of constantly changing technical artefacts in which there appeared to be no natural constraints: steam engines as the "agents" of the factories, mobile steam engines as universal means of transport, prime movers of all kinds, cast-iron arch bridges and wrought-iron chain suspension bridges. Since 1820 Britain had become a favourite destination of highly educated continental European tourists armed with pencil and paper. Navier, too, travelled several times to England on official business. In 1823 he published a structural theory of chain suspension bridges alongside a description of Britain's chain suspension bridges [Navier, 1823], which remained the theoretical guideline for engineers for some 50 years. This excellent *Rapport* earned him membership of the Paris Academy of Sciences and the job of planning a suspension bridge spanning 160 m over the Seine. Navier's design included in the aforementioned *Rapport* was used with only minor alterations and financed through a public company. Shortly before its completion (September 1826), cracks appeared in the masonry anchorage for the chains, which then worsened due to unfavourable weather conditions. The Council of Paris, which had been against the project from the very outset because the bridge would impair the view of the Dôme des Invalides, demanded the immediate demolition of the bridge without even considering any possible repair measures. The public company ordered the bridge demolished and then appointed the engineer Vèrges to design a suspension bridge along the lines of Hammersmith Bridge in London. In order to start earning a return on their investment through bridge tolls, the public company erected an oversized suspension bridge not far from Navier's bridge which, even down to the construction details, was a copy of its far bolder predecessor in London. Antoine Picon has sounded out the influence of Navier on the building of suspension bridges in France [Picon, 1988] and Eda Kranakis includes Navier in her comparative study of the engineering cultures of France and North America in the 19th century [Kranakis, 1996].

It was in that unlucky year of 1826 that Navier published his *Résumé des Leçons* on the mechanics of building – the lectures he held at the École des Ponts et Chaussées, the top address for French civil engineers [Navier, 1826]. It was in this publication that Navier brought together the mathematical-mechanical analysis of the elastic line, carried out by Jakob Bernoulli, Leonhard Euler and others, and the primarily engineering-based beam statics, and replaced these two traditional 18th-century theories with his practical bending theory [Navier, 1833/1878, pp. 33 – 80].

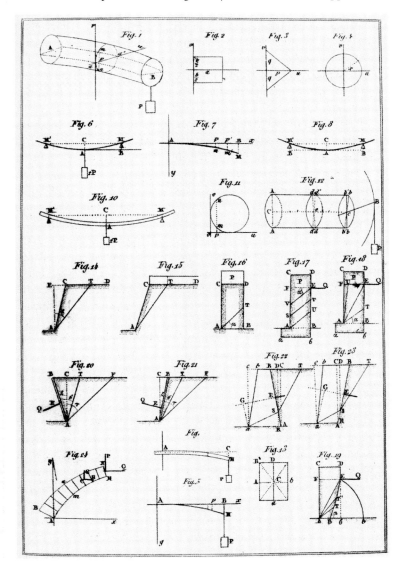

**FIGURE 5-30**
Navier's drawings of structural theory problems in beam, earth pressure and masonry arch theories [Navier, 1833/1838]

Using this genuine engineering science theory it was possible to provide a general solution for the equilibrium and deformation questions of components loaded in bending both for statically determinate and statically indeterminate systems on the empirical basis provided by strength tests. Furthermore, this classic work of modern structural theory contains helpful approaches for solving important problems in civil engineering (Fig. 5-30): stability of retaining walls subjected to earth pressure [Navier, 1833/1878, pp. 98–137], statics of masonry arches [Navier, 1833/1878, pp. 137–75], buckling of columns [Navier, 1833/1878, pp. 190–209], elastic arched beams [Navier, 1833/1878, pp. 217–246], elastic slabs [Navier, 1833/1878, pp. 327–330], and membranes [Navier, 1833/1878, pp. 331–340].

On the problem of the loss of stability of columns due to lateral buckling, "a series of severe setbacks and accidents," writes Fritz Stüssi, "could have been avoided if the knowledge of Navier had not been forgotten for a long period, but instead had been kept alive in everyday building practice" [Stüssi, 1940]. Nonetheless, theory of structures began to shake off its passive role in relation to the practice of building more and more just a few years after Navier's death on 21 August 1836. Characteristic of this development is:

– firstly, the establishment of its independence within the scope of the constructional side of engineering activities thanks to new constructional-technical demands in the field of railways;
– secondly, the dominance of its anticipatory nature in the further development of framework types of construction (Pauli truss, Schwedler truss, etc.);
– thirdly, the standardisation of the entire stock of theory of structures knowledge on the basis of elastic theory and the principles of mechanics (principle of virtual forces, principle of virtual displacements, energy principles);
– fourthly, the faithful modelling of loadbearing structures (masonry arch, beam, trussed framework) as elastic bar systems – therefore the formation of the notion of the structural system –, which thus rendered possible for the first time the replacement of the masonry arch, beam and trussed framework theories by the general theory of trussed frameworks that characterises classical theory of structures.

### 5.6.2 Eytelwein

Until the founding of the Regional Building Department in 1770, public building works in Prussia were controlled by building engineers, inspectors and directors on the level of the individual province authorities within the scope of the War and Land Boards. The civil servants in the Prussian public building authority, which was just starting to become a uniform entity in the final few decades of the 18th century, were retired soldiers, officers and master artisans. The public building system, in which waterways played a leading role, was mainly in the hands of practical men, "who pursued the momentary purpose too zealously without the necessary circumspect for the ensuing consequences" [Encke, 1849, p. XIX].

Not until 1773 did those involved in building have to sit an examination (prepared by the Regional Building Department) to prove their knowledge of mechanics, civil and waterways engineering, geometry, drawing and surveying. Owing to the lack of training and education establishments, the examination candidates were forced to acquire this elementary knowledge for themselves. Of the new members of the Regional Building Department, only the consistorial counsellor and waterways builder J. E. Silberschlag had published anything on building. Even J. A. Eytelwein, born in Frankfurt on 31 December 1764, the son of a merchant [Scholl, 1990, p. 47], and a recruit from the Tempelhof Artillery Academy, had to acquire the construction engineering and mathematics-mechanics knowledge he required for the surveying examination in his spare time between his activities as a bombardier. He experienced the lack of German literature on applied mechanics at first hand and therefore the 28-year-old Eytelwein published a textbook with the title *Aufgaben größtentheils aus der angewandten Mathematik zur Übung der Analysis* (exercises mainly from applied mechanics for practising analysis). On the strength of this first publication, he was promoted one year later from dyke-building inspector to Privy Counsellor at the Regional Building Department because the most senior building authority at that time had no examiner familiar with mathematics-mechanics subjects. Together with David Gilly, Eytelwein published the *Sammlung nützlicher Aufsätze und Nachrichten, die Baukunst betreffend* (collection of useful papers and bulletins on architecture) between 1797 and 1806. This first German-language journal for the building industry became a vehicle for leading civil servants to publish articles on waterways, buildings for agriculture and prestigious public buildings. The use of mathematical and natural science findings is evident in only a few contributions. In contrast to the education of engineers at the École Polytechnique founded in 1794 and the École des Ponts et Chaussées established by Trudaine in 1747, the curriculum of the Building Academy set up by Eytelwein, Riedel and Friedrich Gilly in 1799 looked very modest in terms of its form and content. Nevertheless, the theoretical content of the curriculum of the Building Academy was too much for the Prussian king! In 1802 he reminded the three directors Eytelwein, Gilly and Riedel "never to forget that practical civil servants of building had to be trained and not professors in the academy" [Teut-Nedeljkov, 1979, p. 75]. The three directors responded as follows: "We cannot do without our scientific collegiate because these alone can produce fundamental judgments and statements for the master-builders" [Teut-Nedeljkov, 1979, p. 75].

Eytelwein's three-volume work *Handbuch der Statik fester Körper mit vorzüglicher Rücksicht auf ihre Anwendung in der Architektur* (manual of the statics of solid bodies with careful respect to their use in architecture) [Eytelwein, 1808, vol. I] appeared in 1808. In volume I Eytelwein analyses the simple machines with the help of statics and investigates machines assembled from these machines on the basis of the laws of friction. Volume II begins with a chapter on the statics of tensioned ropes taking into account the friction and stiffness of the ropes; the next four chapters are devoted to the statics of beams, Eytelwein's masonry arch theory and his own strength tests as well as those of Parent, Musschenbroek, Buffon, Bélidor and Girard carried out on timber beams. The "theory of those transcendental curved lines that occur primarily in structural investigations" is the content of the third volume; here, Eytelwein discusses the theory of the elastic line and other topics, and calculates the support reactions of continuous beams with one and two degrees of static indeterminacy.

During the era of the Stein-Hardenberg reforms (1807–15), Eytelwein rose to become director of the Regional Building Department and was nominated lecturing counsellor at the Ministry of Trade & Commerce. Like many leading engineers in 19th-century Germany, Eytelwein equated the theoretical basis of construction engineering knowledge with its mathematical basis. For instance, in the preface to his *Handbuch der Statik fester Körper*, he writes: "Among those parts of applied mathematics that are indispensable to the builder as a helpful science, statics of solid bodies occupies first place" [Eytelwein, 1808, vol. I, p. III]. The development of civil engineering in Germany with the Industrial Revolution after 1840 is indebted to Eytelwein, who died on 18 August 1848, for providing a scientific footing for the knowledge of building acquired in the late 18th and early 19th centuries. Also for preparing and extending the significant beginnings in the engineering science theories of the 18th century for German builders – for waterways, beam and masonry arch statics, strength of materials and the analysis of the elastic line.

### 5.6.3
**The analysis of the continuous beam according to Eytelwein and Navier**

The first statically indeterminate problem was solved by Euler in 1774 [Oravas & McLean, 1966]: he calculated the support reactions of a four-legged table idealised as a rigid body subjected to a vertical load. In order to solve this problem with one degree of static indeterminacy, Euler assumes that all four table legs are supported on linear-elastic springs because only by considering the proper relationship between force and displacement is it possible to quantify the support reactions. Such a relationship based on laws and used by Euler is Hooke's law in its simplest version as expressed by its discoverer in 1678: *ut tensio sic vis* (the power of any spring is in the same proportion as the tension thereof).

The hegemony of rigid body mechanics in the stock of structural theories (masonry arch and beam theory) of the 18th century, primarily promoted by Euler, plus the experiments designed to back up a universal approach to linear-elastic loadbearing behaviour in complex loadbearing systems (e.g. timber beams), which had not yet been completed, resulted in mathematicians and engineers initially considering statically indeterminately supported loadbearing systems as rigid bodies. One characteristic of static indeterminacy had been recognised by the engineer A. A. Vène as early as 1818 [Oravas & McLean, 1966]: there is an infinite number of force conditions that satisfy all the equilibrium conditions. The historico-logical development therefore extended to the design of a selection mechanism that sorted out the true force condition from the wealth of possible force conditions. Totally in keeping with the natural theory style of thinking of the 18th century, mathematicians and engineers saw such a selection mechanism in the setting-up of extremal principles. It is therefore not surprising that in the first half of the 19th century A. A. Vène, H. G. Moseley, H. Scheffler and others tried to circumvent the material law in their analyses of statically indeterminate problems by way of special hypotheses and extremal statements about the force condition in rigid bodies [Kurrer, 1991/2]. On the other hand, Eytelwein (1808) and Navier (1826) investigated a statically indeterminate problem relevant to construction engineering taking into account the elasticity of the building material – the continuous beam. Both engineers therefore laid the foundation for the theory of continuous beams, which became part of everyday engineering after 1850, and besides trussed framework and masonry arch statics played an important role in the constitution of the general theory of trusses during the discipline-formation period of structural theory up to 1900.

### 5.6.3.1 The continuous beam in Eytelwein's *Statik fester Körper*

Euler was able to calculate easily the support reactions by assuming a linear-elastic, statically indeterminate support condition for the rigid body. However, his assumption was a blatant contradiction of the true mechanical behaviour of the four-legged table. Eytelwein focused his attention on this contradiction as he reconstructed the Eulerian solution [Eytelwein, 1808, vol. III, pp. 65–69] and also drew attention to the conflict with experience for the specific case of solving the rigidly supported statically determinate three-legged rigid body on the two-span beam with one degree of static indeterminacy (Eytelwein allows the third support in the line connecting the remaining supports to drop and obtains indeterminate expressions in the form of $\frac{0}{0}$ for the support reactions): "If a straight, rigid line is supported at three points and one attempts to find the compression in each column caused by a load suspended from the rigid line spanning between said columns, then all triangles = 0, i.e. the compression at the individual point is indeterminate. The magnitude of the compression at the individual columns can be determined very easily with the help of the assumption of Euler [linear-elastic support – the author], but one realises very quickly that this assumption must lead to results that are untenable for a straight line because then the column further from the load must experience a greater compression than the one nearer the load" [Eytelwein, 1808, vol. II, pp. 69/70].

Eytelwein takes the decisive step by considering the continuous beam itself no longer as a rigid body, but rather as "flexible and elastic" [Eytelwein, 1808, vol. II, p. 70]: "If now the long building components are also of such a nature that they bend under their own weight and even long stones, but especially beams, are not regarded as rigid bodies, then the assumption of a completely rigid line is even less acceptable because a long, loaded body spanning between two supports can only press on more distant columns in one uniform horizontal plane provided its own weight bends it downwards onto the more distant supports … It is therefore necessary that a line supported at several points is attributed a certain, albeit perhaps only extremely small, flexibility and elasticity" [Eytelwein, 1808, vol. II, p. 70].

Eytelwein's insight into the nature of statically indeterminate problems based on his constructional-technical knowledge about the need of building practice for an understanding of the load-carrying behaviour of continuous beams, e.g. continuous rafters in collar roofs [Eytelwein, 1808, vol. II, pp. 94–101], and his secure mastery of the analysis, enabled him to calculate the support reactions of two- and three-span beams. To do this, he makes use of the classical, original contributions of Jakob Bernoulli (1694/95 and 1705) and Leonhard Euler (1728, 1744, 1770 and 1775) concerning the theory of the elastic line [Eytelwein, 1808, vol. III, pp. 185/186]. Eytelwein models a beam as an "elastic rod" and assumes a "prismatic form and equal elasticity throughout" [Eytelwein, 1808, vol. III, p. 129]: "If one fixes an elastic rod at one or more points, it will take on a curvature due to forces applied to it or even due to its own weight, with some parts of the rod being stretched, others pressed together. Between these there must be parts or fibres that are neither extended nor compressed, and if we draw a line through these, then one can assess the curvature of the rod. This line is called the elastic line (*curva elastica*) and at this point it is admissible to consider the elastic line of the rod to be without depth …" [Eytelwein, 1808, vol. III, p. 129]. After Eytelwein linearises the equation of the elastic line according to practical building requirements for "low flexibility"

$$E^2 \frac{d^2y}{dx^2} = -M(x) \tag{5-51}$$

(in our modern notation $E^2$ corresponds to the bending stiffness $E \cdot I$), he derives equations for the deflection and the radius of curvature at specific points on a cantilever beam and a simply supported beam subjected to point and uniformly distributed loads. Eytelwein notes that the value "$E^2$ can be determined from a test of each material of the rod" [Eytelwein, 1808, vol. III, p. 143] when the deflection $f = y(x = l/2)$ of a simply supported beam subjected to a central point load $P$ is measured and entered into the equation

$$E^2 = E \cdot I = \frac{P \cdot l^2}{48 \cdot f} \tag{5-52}$$

The deformation observations of the two simplest statically determinate systems are followed by certainly the first investigation of a two-span

**FIGURE 5-31**
Two-span beam with one degree of static indeterminacy after Eytelwein

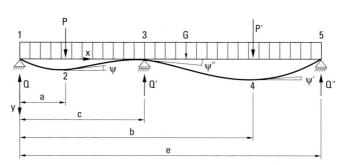

beam with one degree of static indeterminacy (Fig. 5-31) [Eytelwein, 1808, vol. III, pp. 148–156].

From the requirement for equilibrium of all vertical forces (Fig. 5-31), Eytelwein uses

$$Q + Q' + Q'' = P + P' + e \cdot G \qquad (5\text{-}53)$$

to calculate the first condition equation for determining the support reactions $Q + Q' + Q''$. He then divides the two-span beam into four continuity zones (Fig. 5-31):

zone 1–2: $0 \leq x0 \leq a \quad \rightarrow \quad y(x)$
zone 2–3: $a \leq a + x' \leq c \quad \rightarrow \quad y'(x')$
zone 3–4: $c \leq c + x'' \leq b \quad \rightarrow \quad y''(x'')$
zone 4–5: $b \leq b + x''' \leq e \quad \rightarrow \quad y'''(x''')$

Fig. 5-32 shows the structural substitute system with which Eytelwein determines the bending moment $M(x)$ in zone 1–2 as a static moment

$$M(x) = (c-x)\cdot Q' + (e-x)\cdot Q'' - (a-x)\cdot P - (b-x)\cdot P' - \frac{1}{2}\cdot(e-x)^2\cdot G \qquad (5\text{-}54)$$

at the cantilever $x$–5. For $x = 0$, eq. 5-54 is converted into the second condition equation for determining the support reactions

$$M(x=0) = 0 = c\cdot Q' + e\cdot Q'' - a\cdot P - b\cdot P' - \frac{1}{2}\cdot e^2\cdot G \qquad (5\text{-}55)$$

With the help of eq. 5-53 and eq. 5-55, the support reactions $Q'$ and $Q''$ are eliminated in the moment equation eq. 5-54 for zone 1–2

$$M(x) = x\cdot\left(Q - \frac{1}{2}\cdot x\cdot G\right) \qquad (5\text{-}56)$$

The resulting equation is entered into the differential equation for the elastica, eq. 5-51

$$E^2\cdot\frac{d^2y}{dx^2} = -x\cdot\left(Q - \frac{1}{2}\cdot x\cdot G\right) \qquad (5\text{-}57)$$

and integrated for $x$

$$E^2\cdot\frac{dy}{dx} = \frac{1}{6}\cdot x^3\cdot G - \frac{1}{2}\cdot x^2\cdot Q + K' \qquad (5\text{-}58)$$

Eytelwein determines the integration constant from $\dfrac{dy(x=a)}{dx} = \tan\Psi$

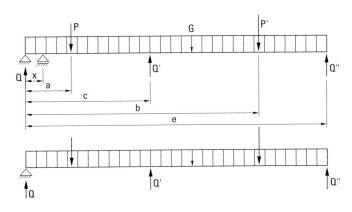

**FIGURE 5-32**
Eytelwein's substitute structural system for determining the static moment M(x) in zone 1–2

$$K' = -\frac{1}{6} \cdot a^3 \cdot G + \frac{1}{2} \cdot a^2 \cdot Q + E^2 \cdot \tan\Psi \tag{5-59}$$

Eq. 5-59 entered into eq. 5-58 and integrated taking into account the boundary condition $y(0) = 0$ results in the $E^2$-fold bending line

$$E^2 \cdot y = -\frac{1}{6} \cdot \left(a^3 - \frac{1}{4}x^3\right) \cdot x \cdot G + \frac{1}{2} \cdot \left(a^2 - \frac{1}{3}x^2\right) \cdot x \cdot Q + E^2 \cdot \tan\Psi \tag{5-60}$$

in zone 1–2. Eytelwein worked out similar equations for the bending line in zones 2–3, 3–4 and 4–5 by first obtaining the moment equations $M(x')$, $M(x'')$ and $M(x''')$ from observations of corresponding substitute structural systems (see Fig. 5-32) and then repeating the mathematical procedure expressed in eq. 5-55 to eq. 5-60. Like the equation for the bending line (eq. 5-60), the three derived equations contain the bending angles $\Psi$, $\Psi'$ and $\Psi''$ (Fig. 5-31). The elimination of these deformation variables leads to the third condition equation for determining the support reactions $Q$, $Q'$ and $Q''$:

$$(a+2c)\cdot(e-c)\cdot(c-a)^2\cdot P + c\cdot(3e-2c-b)\cdot(b-c)^2\cdot P' + \frac{3}{4}\cdot \\ c\cdot e\cdot(e-c)\cdot(e^2-3c\cdot e+3c^2)\cdot G = 2c^3\cdot(e-c)\cdot Q + 2c\cdot(e-c)^3\cdot Q'' \tag{5-61}$$

The elasticity equation eq. 5-61 and the equilibrium conditions eq. 5-53 and eq. 5-55 represent a set of equations for $Q$, $Q'$ and $Q''$, with the following solution

$$\left.\begin{array}{l} Q = \dfrac{A\cdot C\cdot(2c\cdot e - a\cdot c - a^2)\cdot P + c\cdot B\cdot(e-b)\cdot(b+c-2e)\cdot P' + \frac{1}{4}\cdot c\cdot e\cdot C\cdot(c^2+3c\cdot e-e^2)\cdot G}{2c^2\cdot e\cdot(e-c)} \\[2mm] Q' = \dfrac{C\cdot a\cdot(2c\cdot e - a^2 - c^2)\cdot P + c\cdot(e-b)\cdot(2b\cdot e - b^2 - c^2)\cdot P' + \frac{1}{4}\cdot c\cdot e\cdot C\cdot(e^2+c\cdot e-c^2)\cdot G}{2c^2\cdot(e-c)^2} \\[2mm] Q'' = \dfrac{B\cdot c\cdot(3b\cdot e - b\cdot c - c\cdot e - b^2)\cdot P' - a\cdot A\cdot C\cdot(c+a)\cdot P + \frac{1}{4}\cdot c\cdot e\cdot C\cdot(3e^2 - 5c\cdot e + c^2)\cdot G}{2c\cdot e\cdot(e-c)^2} \end{array}\right\} \tag{5-62}$$

where $A = c - a$, $B = b - c$ and $C = e - c$ as parameters. Eytelwein notes with relief "that the perpendicular compression on the individual supports is not dependent on the elasticity or flexibility of the beam, or that this compression remains unchanged but the elasticity of the beam may be large or small" [Eytelwein, 1808, vol. III, p. 153].

The static indeterminacy calculation presented above was, admittedly, of only minimal use in the everyday workloads of German structural or civil engineers in the early 19th century. Even the generalisation to form a theory of continuous beams was limited by the fact that the tedious, unclear, excessive mathematics overloaded the key problem of static indeterminacy. The reasons for this lay in the following shortcomings in Eytelwein's approach:

*Firstly:* In order to model the beam-in-bending problem just with the notions of the theory of elastica and the limited task of quantifying the support reactions $Q$, $Q'$ and $Q''$ for a constant bending stiffness $E^2$, it is not necessary to know the variable $E^2$ nor is it necessary to distinguish between the bending moment resulting from the distribution of stress over the beam cross-section from the (external) static moment.

*Secondly:* Eytelwein determines the moments $M(x)$, $M(x')$, $M(x'')$ and $M(x''')$ not by cutting through the given system at $x$, $x'$, $x''$ and $x'''$, but rather by considering the corresponding substitute structural system; Eytelwein can thus reduce this equilibrium task to determining the fixed-end moments of the cantilever beam already investigated by Galileo.

*Thirdly:* Eytelwein had not yet recognised the characteristic unity between equilibrium and deformation tasks in the analysis of static indeterminacy problems because he counted statics as a branch of applied mathematics and understood strength of materials as a branch of physics [Eytelwein, 1808, vol. I, pp. V/VI].

### 5.6.3.2 The continuous beam in Navier's *Résumé des Leçons*

Although Eytelwein adopted Coulomb's work on beam statics (1773) in the chapter on the strength of materials in volume II of his *Statik fester Körper* [Eytelwein, 1808, vol. II] and in volume III successfully applied the theory of elastica to the calculation of specific continuous beams [Eytelwein, 1808, vol. III], it remained to Navier to eliminate the dualism of beam statics and the theory of elastica. The reason for the new quality in engineering science theory formation achieved with Navier's practical bending theory can be found in the industrial manufacture of iron that was causing an upheaval throughout structural and civil engineering (pressure on economic design of cross-sections, new loadbearing systems such as chain suspension bridges, etc.) and the elastic theory that became popular in France between 1820 and 1830 [Szabó, 1987]. Only after the evaluation of numerous strength tests and an engineering-type projection of the elastic theory laid down by Cauchy, Navier, Lamé, Poisson, Clapeyron and others based on the scientific object of the construction-oriented engineer was it possible, in principle, to dimension building components and provide a general solution for static indeterminacy problems. In the preface to his *Résumé des Leçons*, Navier emphasizes the experimental basis of theory of structures and gives precedence to the quantitative ascertainment of the serviceability state (deformations plus force or stress conditions under service loads) over the ultimate state widely considered in the 18th century: "The elasticity [its magnitude, the elastic modulus $E$, is defined by Navier – the author] and the cohesion [compressive or tensile strength – the author] must be determined for the various materials by way of tests; we have tried to gather together all such results that appear to be applicable. If the elasticity is known, then one is in the position of being able to determine the magnitude by which a part of the construction will shorten, lengthen or bend under a given load. If the cohesion is known, one can determine the maximum weight a body can still carry. But this is not enough for the design engineer because it is not his job to know which weight will crush a body, but rather to find out the weight a body can carry without the deformation it suffers increasing over time" [Navier, 1833/1878, p. XVI]. On the engineering use of the laws and hypotheses of elastic theory for the dimensioning of building components, Navier notes that "the rules that were given earlier for the equilibrium of elastic bodies … could be used only in very few cases; now, however, for the solution of the new, aforementioned questions, one has the means to consider the size of the main parts of a timber construction with the same ease and accuracy as the size of the chains for a suspension bridge" [Navier, 1833/1878, p. XVIII].

In the first section of *Résumé des Leçons* [Navier, 1833/1878, pp. 1 – 97], Navier presents numerous strength tests for customary building materials (tensile, compressive and buckling strength tests plus bending tests on timber, stone, mortar, cast iron and wrought iron) and summarises the figures in clear tables. The first section also contains Navier's practical bending theory [Navier, 1833/1878, pp. 33 – 80] in which he assumes a linear differential equation for the bending line of the elastic beam and integrates this for the statically determinate beam on two supports and the cantilever [Navier, 1833/1878, pp. 38 – 43]. Without any particular commentary, he superposes the deflections due to the self-weight and point load loading cases [Navier, 1833/1878, p. 42]. He now compares these theoretical deflections with the results of corresponding deflection measurements in the elastic range. Afterwards, Navier combines the practical bending theory with the deflection tests in the elastic range and determines the elastic modulus $E$ from the relationships for quantifying the deflections [Navier, 1833/1878, pp. 43 – 55]. What Eytelwein mentions only in passing with eq. 5-52 [Eytelwein, 1808, vol. III, p. 143], Navier places at the centre of his observations. The new type of relationship between strength test and bending theory – i.e. between empirical findings and theory – that is expressed in the practical bending theory enables for the first time a complete representation of the true force and deformation condition of elastic beam-type loadbearing systems that is adequate for practical engineering purposes. An approach to calculating the support reactions of the statically indeterminate continuous beam that goes beyond Eytelwein's approach restricted to specific cases is now conceivable. It is not yet necessary to distinguish between statically determinate and statically indeterminate bending structures because Navier assumes the linear differential equation of the bending line and then integrates this taking into account the given boundary conditions. Like Eytelwein, Navier considers knowledge of the elastic material behaviour of the loadbearing system to be a necessary condition for calculating the statically indeterminate continuous beam: "If a rigid bar loaded with weights is supported on more than two supports, the loads on each individual support are indeterminate within certain limits. The limits can always be determined by means of the principles of statics. However, if one assumes the bar to be elastic, the indeterminacy is eliminated completely. We shall investigate here only the simplest questions of this kind" [Navier, 1833/1878, p. 187].

In a chapter devoted to the equilibrium of a bar supported at three or more points [Navier, 1833/1878, pp. 187 – 189], Navier analyses the

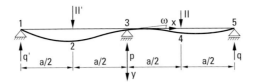

FIGURE 5-33
Two-span beam with one degree of static indeterminacy after Navier

continuous beam shown in Fig. 5-33. Navier "requires the form of the bar after bending and the reaction at each point of support" [Navier, 1833/1878, p. 187]. To do this, he uses

$$II + II' = p + q' + q \tag{5-63}$$

to satisfy the equilibrium condition "sum of all vertical forces = 0", and

$$II - II' = 2(q - q') \tag{5-64}$$

the equilibrium condition "sum of all static moments about point 3 = 0".

Like Eytelwein, Navier integrates the differential equation of the bending line eq. 5-51 section by section; Navier calls the bending strength $E \cdot I$ the elastic moment $\varepsilon$ because he derives a general equation for the $E$-fold second moment of area about the main axis of bending [Navier, 1833/1878, p. 36] and applies this to common forms of cross-section (rectangle, circle, annulus). Eq. 5-51 for section $0 \leq x \leq = \frac{a}{2}$ takes on the form

$$\varepsilon \cdot \frac{d^2y}{dx^2} = II \cdot \left(\frac{a}{2} - x\right) - q \cdot (a - x) \tag{5-65}$$

Eq. 5-65 integrated twice for $x$ and taking into account the conditions $d(y = 0)/dx = \tan \omega$ and $y(x = 0)$ supplies the $\varepsilon$-fold bending line

$$\varepsilon \cdot y = II \cdot \left(\frac{a \cdot x^2}{4} - \frac{x^3}{6}\right) - q \cdot \left(\frac{a \cdot x^2}{2} - \frac{x^3}{6}\right) + x \cdot \varepsilon \cdot \tan \omega \tag{5-66}$$

Navier obtains a corresponding equation for the $\varepsilon$-fold bending line from

$$\varepsilon \cdot \frac{d^2y}{dx^2} = - q \cdot (a - x) \tag{5-67}$$

by including the transition conditions at point 4 for the section $\frac{a}{2} \leq x \leq a$:

$$\varepsilon \cdot y = - q \cdot \left(\frac{a \cdot x^2}{2} - \frac{x^3}{6}\right) + \left(II \cdot \frac{a^2}{8} + \varepsilon \cdot \tan \omega\right) \cdot x + II \cdot \frac{a^3}{48} \tag{5-68}$$

"The equations for the sections $AN'$ [section: $-\frac{a}{2} \leq x \leq 0$ – the author], $N'M'$ [section: $-a \leq x \leq -\frac{a}{2}$ – the author] are obtained from those preceding [eq. 5-66 and eq. 5-68 – the author] if one substitutes $II'$ for $II$, $q'$ for $q$ and reverses the sign of tan" [Navier, 1833/1878, p. 188]. The $\varepsilon$-fold bending line takes on the following form in the section $-\frac{a}{2} \leq x \leq 0$

$$\varepsilon \cdot y = II' \cdot \left(\frac{a \cdot x^2}{4} - \frac{x^3}{6}\right) - q' \cdot \left(\frac{a \cdot x^2}{2} - \frac{x^3}{6}\right) - x \cdot \varepsilon \cdot \tan \omega \tag{5-69}$$

and the following form in the section $-a \leq x \leq -\frac{a}{2}$

$$\varepsilon \cdot y = - q' \cdot \left(\frac{a \cdot x^2}{2} - \frac{x^3}{6}\right) + \left(II' \cdot \frac{a^2}{8} + \varepsilon \cdot \tan \omega\right) \cdot x - II' \cdot \frac{a^3}{48} \tag{5-70}$$

Navier then enters the boundary condition $y(x = a) = 0$ into eq. 5-68 and $\varepsilon \cdot y(x = -a) = 0$ into eq. 5-70 and obtains

$$\varepsilon \cdot y(x=a) = 0 = -q \cdot \frac{a^2}{3} + II \cdot \frac{5a^2}{48} + \varepsilon \cdot \tan\omega \qquad (5\text{-}71)$$

plus

$$\varepsilon \cdot y(x=-a) = 0 = -q' \cdot \frac{a^2}{3} + II' \cdot \frac{5a^2}{48} - \varepsilon \cdot \tan\omega \qquad (5\text{-}72)$$

Eq. 5-63, eq. 5-64, eq. 5-71 and eq. 5-72 form a set of equations with the following solutions:

$$\left.\begin{aligned} \tan\omega &= \frac{a^2 \cdot (II - II')}{32\,\varepsilon} \\ p &= \frac{22\,(II + II')}{32} \\ q &= \frac{13\,II - 3\,II'}{32} \\ q' &= \frac{-3\,II + 13\,II'}{32} \end{aligned}\right\} \qquad (5\text{-}73)$$

"Substituting these values [$\tan\omega$, $p$, $q$ and $q'$ – the author] into the above equations [eq. 5-66, eq. 5-68, eq. 5-71 and eq. 5-72 – the author], results in the form of the bar" [Navier, 1833/1878, p. 188].

A comparison of the methods of calculating continuous beams used by Navier and Eytelwein reveals that Navier goes way beyond Eytelwein in the following points despite his more specific example (two-span beam with equal spans, central point loads and ignoring the self-weight loading case):

- The elimination of the dualism between beam statics and the theory of elastica in the practical bending theory plus – from the point of view of elastic theory – the consequential interpretation of strength tests permit the quantitative representation of the deformation condition and hence contain a general approach for solving the continuous beam problem.
- When calculating the internal moments $M(x)$ for the differential equation of the bending line, Navier manages without a substitute structural system, and instead applies the method of sections.
- Navier calculates the bending moments at points 2, 3 and 4 of the continuous beam [Navier, 1833/1878, p. 189] based on the maximum value of his stress analysis [Navier, 1833/1878, pp. 94–97].

In 1843, H. Moseley presented Navier's beam theory for British engineers in his book *The Mechanical Principles of Engineering and Architecture* [Moseley, 1843]. In particular, Moseley describes Navier's derivation of the equations for the support reactions $p$, $q$ and $q'$ (see eq. 5-73). Moseley refers to measurements of the angle of slope at point 1 of a wrought-iron two-span beam according to Fig. 5-33 with a square cross-section (side length of cross-section $a$ = 12.6 mm) which Hatcher carried out at King's College, London [Charlton, 1982, p. 19]. Thanks to Moseley's work, the theory of the continuous beam, albeit backed up by experiments, was first applied successfully in the building of the Britannia Bridge (1846–50) by W. Pole and E. Clark, [Clark, 1850]. In Germany it was Prof. C. M. Rühlmann (1811–96) from the Polytechnic School in Hannover, who since 1845 had been using Navier's *Résumé des Leçons* in his lectures, who encouraged his pupil G. Westphal to translate *Résumé des Leçons* into German [Navier, 1833/1878], the first German edition of which was published in Hannover in 1851. The adoption of Navier's *Résumé des Leçons* by British and German civil and structural engineers rounded off the constitution phase of structural theory (1825–50).

In the subsequent establishment phase of structural theory (1850–75), the theory of continuous beams – encouraged by the boom in the building of wrought-iron bridges after 1850 – underwent further development by Bertot in 1855, Lamarle [Lamarle, 1855], Rebhann [Rebhann, 1856], Köpcke [Köpcke, 1856], Scheffler [Scheffler, 1857, 1858, 1860], Clapeyron [Clapeyron, 1857], Molinos and Pronnier [Molinos & Pronnier, 1857], Belanger [Belanger, 1858, 1862], Bresse [Bresse, 1859, 1865], Heppel [Heppel, 1860, 1870], Mohr [Mohr, 1860, 1868], Winkler [Winkler, 1862, 1872/1], and Weyrauch [Weyrauch, 1873]. The final piece in the jigsaw of continuous beam theory in the establishment phase of theory of structures was inserted by Weyrauch in 1873 with his monograph *Allgemeine Theorie und Berechnung der kontinuierlichen und einfachen Träger* (general theory and calculation of continuous and simply supported beams) [Weyrauch, 1873]. This work by Weyrauch was not only the first comprehensive presentation of the theory of continuous beams in a book, but also represents a milestone in the history of the development of the influence lines concept, which during the classical phase of structural theory (1875–1900) was expanded into a theory widely used in bridge-building. The theory of continuous beams based on the concept of influence lines enabled engineers to solve a group of static indeterminacy problems completely, universally, rationally, clearly and elegantly for the first time.

# CHAPTER 6

# The discipline-formation period of theory of structures

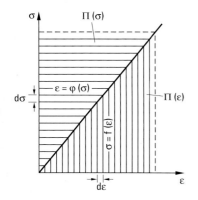

Between 1978 and 1990, Prof. Rolf Sonnemann of Dresden TU was the central figure in a large group of scientists who created a new type of history of science discipline in the shape of the history of the engineering sciences. The papers in the periodical *Dresdener Beiträge zur Geschichte der Technikwissenschaften* (Dresden's contributions to the history of the engineering sciences) and also the letters of this scientific community, in which they express their thoughts on models of the origins of the engineering science disciplines, first came to the attention of the author in 1984. That provided the inspiration for a thorough study of the historical development of structural theory. Particularly helpful in this respect were the works of Thomas Hänseroth on the history of structural mechanics, and those of Klaus Mauersberger on the history of applied mechanics, but also the publications of Gisela Buchheim and Martin Guntau concerning the problem of dividing the engineering sciences into historical periods. Prof. Gleb Mikhailov assisted the author with documents on Kirpichev relevant to section 6.5.2. Hubert Laitko's clear exposition on scientific schools helped the author to understand better the Dresden school of applied mechanics headed by Mohr and the Berlin school of structural theory headed by Müller-Breslau. Taking this as a basis, the author was able to continue formulating the evolution of structural analysis. The journal *Bautechnik* under its chief editor Doris Greiner-Mai was always on hand to publish the results.

According to Guntau, the formation of a scientific discipline is characterised by the development of a summarising and explanatory theoretical conception [Guntau, 1978, p. 16]. The individual coexistent pieces of knowledge that had accumulated in the early history of theory of structures were now joined together by one fundamental theory. To do this, it was necessary to separate the individual pieces of structural theory knowledge from their historical genetic relationship, i.e. the logical had to shed the historical. This objectivisation in the sense of the classical scientific ideal was achieved by Navier in 1826 in his *Résumé des Leçons* in the form of his practical bending theory (see section 5.6.1). Of the three great strides forward taken during the discipline-formation period of structural theory (1825–1900), Navier made the first and greatest. The second stride was that of Culmann, with the expansion of his trussed framework theory (1851) to form graphical statics (1864/66) – an attempt to translate the entire statics repertoire into graphical solutions, which gave engineers like Eiffel and Koechlin a rational means of analysing statically determinate systems quickly. Nevertheless, the first two strides were still characterised by pluralism in the theories of masonry arch statics (see section 4.5). This did not change until the third and last stride, which saw Maxwell, Castigliano, Winkler, Mohr and Müller-Breslau apply elastic theory to trusses and create a general theory of elastic trusses – also embracing masonry arch statics. The dispute surrounding the foundation of a classical theory of structures (1883–89), which arose again shortly after 1900, heralded the end of the discipline-formation period of structural theory and the start of the consolidation period. The force method gave classical theory of structures a fundamental concept for statically indeterminate systems.

Based on the proposals of Guntau [Guntau, 1978 & 1982], Ivanov et al. [Ivanov et al., 1980] and Buchheim [Buchheim, 1984] for dividing engineering sciences into periods, the discipline-formation period of structural theory (1820–1900), which succeeded the preparatory period of structural theory, will be divided into three phases:

*Constitution phase (1825–50)*: The creation of a "system of structural mechanics knowledge" [Hänseroth, 1985], or rather the formulation and first step towards realising a theory of structures programme by Navier (1826), including the adoption and expansion of his practical bending theory.

*Establishment phase (1850–75)*: The development of trussed framework theories in the late 1840s and their replacement by graphical statics, in particular by Culmann, Rankine, Maxwell and Cremona.

*Classical phase (1875–1900)*: The formation and completion of classical theory of structures by Maxwell, Winkler, Mohr, Castigliano, Müller-Breslau, Weyrauch, Kirpichev and others. The classical phase closed the discipline-formation period of structural theory, which was succeeded by the consolidation period (1900–50).

**Clapeyron's contribution to the formation of classical engineering sciences**

## 6.1

In the field of science, Clapeyron made a name for himself with two fundamental contributions: firstly, the elastic theory that bears his name and the mathematical formulation of Sadi Carnot's pioneering work on thermodynamics, and secondly, the formulation of the theorem of three moments. Clapeyron's work should always be seen in conjunction with his commitment to the expansion of the French railway network. The way he united science and industry made him one of the most important supporters of the Industrial Revolution in France. Seen from the history of science viewpoint, Clapeyron stands at the transition between the constitution and establishment phases of theory of structures.

*Les Polytechniciens*: the fascinating revolutionary élan in post-revolution France

### 6.1.1

After its reorganisation on 10 July 1793, The Committee of Public Safety, the central organ of the revolutionary French government under the leadership of Lazare Carnot (1753–1823), began to establish the Revolutionary Army. The national needs of the new republic – the military engineering ones resulting from the French Revolutionary Wars and also the civil engineering ones – were obvious reasons for setting up the École Polytechnique, which represents the most important creation of the French Revolution in the teaching of natural and engineering sciences [Klemm, 1977, p. 25].

Although the plans for setting up the École Polytechnique had already been drafted before the fall of Maximilien Robespierre on 27 July 1794 and a corresponding commission had been inaugurated, teaching at the École Polytechnique did not begin until May 1795. The figures of those early days of the École Polytechnique were Lazare Carnot and Gaspard Monge (1746–1818). The central role played by Monge's descriptive geometry at the École Polytechnique can be seen from the original curriculum (Fig. 6-1).

Descriptive geometry occupied half the teaching time, with physics, chemistry, freehand drawing and mathematics sharing the other half. After three years of study, the graduates of the École Polytechnique departed as military engineering officers, artillery officers and civil engineers. It was not until some years later that the curriculum at the École Polytechnique was changed to a two-year mathematics/natural sciences foundation course. That was followed – depending on the degree of success in the final examination – by further training at the various (meanwhile reactivated) Écoles d'applications of the *ancien regime*; in order of merit the first three were the École des Ponts et Chaussées in Paris (for

FIGURE 6-1
Original curriculum for the École Polytechnique after Monge

|  | 1st year | 2nd year | 3rd year |
|---|---|---|---|
| Mathematics (applications) | Analytical geometry of spaces | Mechanics of solid and fluid bodies | Machine theory |
| Descriptive geometry (applications) | Stereometry and stereotomy | Construction and maintenance of engineered structures | Building of fortifications |

the historical development of this, the oldest civil engineering school in the world, see [Chatzis, 1997]), the École des Mines in Paris, and the École d'application d'Artillerie et de Génie militaire in Metz, which was the result of the amalgamation of Mézière's military engineering school and Châlon's artillery school.

The 18th Brumaire (9 November 1799), when Napoleon assumed power as First Consul, signalled the immediate start of the militarisation of the École Polytechnique; it was removed from the portfolio of the Interior Ministry and placed under the control of the War Ministry. Napoleon recruited students from the school at ever more frequent intervals to meet his incessant need for artillery and engineering officers. Nonetheless, the École Polytechnique experienced its heroic age during the Napoleonic years. It retained its military character more or less throughout the 19th century.

The Egyptian Expedition was the first ordeal for the military engineering corps, which was made up of polytechnicians. Without them, Napoleon's march would have been unthinkable, starting with the crossing of the St. Bernhard to building bridges over the Beresina in the Russian campaign. Franz Schnabel describes many more feats of military engineering heroism during the time of Napoleon (see also [Bradley, 1981, pp. 11–12]). Therefore, Napoleon's Waterloo was also the Waterloo of the polytechnicians: Gaspard Monge was discharged, Lazare Carnot was sent into exile and the majority of the students of the École Polytechnique, who had been born only a few years before 1800 and had already inhaled the republican spirit, were thrown into conflict with the restored Bourbon monarchy. One of those students was Benoît-Pierre-Emile Clapeyron, born in Paris on 26 January 1799 (Fig. 6-2 bottom) and his slightly older friend Gabriel Lamé (1795–1870) (Fig. 6-2 top).

Lamé's turbulent graduation party so enraged the authorities that they dissolved the École Polytechnique by way of a royal decree on 13 April 1816; 250 polytechnicians were forced to quit the school. But a year later the King allowed the École Polytechnique to reopen, but this time under strict clerical control and political supervision. Clapeyron was therefore able to continue his studies at the École Polytechnique, where secret societies were increasingly influential, in particular the philosophies of the social and science theorist Claude-Henri de Saint-Simon (1760–1825) [Bradley, 1981, p. 294].

**FIGURE 6-2**
Two friends: Gabriel Lamé (top) and Benoît-Pierre-Emile Clapeyron (bottom) (Collection École Nationale des Ponts et Chaussées)

The ideas of Saint-Simon regarding scientific teaching had their roots in the work of the French encyclopaedists of the 18th century (see section 3.2). Based on materialistic philosophy, Saint-Simon drew up a classification of sciences whose social addressee was the Third Estate, i.e. that body whose political triumph had begun in 1789 with the seizing of the Bastille and whose social claim to power was later given a European context by Napoleon. According to Saint-Simon, human cognition passes through three stadia: polytheism, deism and physicism. He called physicism the research into reality and placed this in opposition to deism and polytheism. The progress in the sectors of the natural and engineering sciences were, so to speak, proportional to the waning of interest in the belief in God. Human thinking, according to Saint-Simon, passes from polytheism to deism and then finally to physicism, thus liberating itself from religious prejudices and placing itself firmly on a scientific basis [Kedrow, 1975, pp. 110–111].

Clapeyron and Lamé, together with numerous other polytechnicians, searched for ways to implement this scientific theory programme in practice. Their search took them via the École des Mines in Paris (1820) to St. Petersburg, into the political heart of the European restoration.

## 6.1.2 Clapeyron and Lamé in St. Petersburg (1820–1831)

Tsar Alexander I, who reigned from 1801 to 1825, had understood better than any other monarch the history lesson of the French Revolution: only scientific knowledge and its application to military engineering and commerce could in the long-term secure Russia's influence on the political developments in Europe after the Congress of Vienna of 1815.

The Tsar offered attractive jobs to numerous unruly polytechnicians in Russian exile, which resulted in a considerable science and technology transfer from France to Russia during the 1820s [Gouzévitch, 1993]. Lamé and Clapeyron taught at the Institute of Engineers of Ways of Communication (Institut du Corps des Ingénieurs des Voies de Communication) which had been founded in St. Petersburg in 1809/10 and which would give rise to the civil engineering school in 1832. The course of study lasting six years trained civil engineers and, to a lesser extent, also military engineers. During the first two years of study the emphasis was on elementary mathematics, which corresponded to the material of the third and fourth years at the École Polytechnique, and the last two years focused on applications just like they had been taught at the École des Ponts et Chaussées. Three polytechnicians lectured in design theory alone (Antoine Raucourt, Guillaume Ferradin-Gazan, André Henri), while Lamé taught differential and integral calculus plus physics and astronomy, and Clapeyron concentrated on pure and applied mechanics, chemistry and design theory [Gouzévitch, 1993, p. 351].

The preferred subjects of the two friends had already became evident at this early stage. Their lectures appeared in lithographed form, some in print. They published numerous articles on mathematics, applied mechanics, suspension bridges, cement, steamships and structural problems on the building of the St. Isaac Cathedral in the following journals: *Annales des Mines*, *Annales de Chimie et de Physique*, *Journal für reine und angewandte Mathematik*, *Journal du Génie Civil, des Sciences et des Arts*, *Journal de l'École Polytechnique* and *Journal des Voies de Communication*. This latter publication was the in-house journal of the Institute of Engineers of Ways of Communication and was published from 1826 to 1836 in both Russian and French editions. These articles were closely connected with the involvement of Clapeyron and Lamé as consulting engineers on prestige projects in Russia: St. Isaac Cathedral in St. Petersburg, suspension bridges [Fedorov, 2000], the Schlüsselburger locks and the Alexander Column in St. Petersburg [Gouzévitch, 1993, p. 351].

Examples of this close interweaving of their activities as consulting engineers and as scientists can be seen in their publications on masonry arch theory [Lamé & Clapeyron, 1823] and on the funicular polygon [Lamé & Clapeyron, 1828]. The former paper was praised by Prony in a detailed *Rapport* [Prony, 1823] and was directly related to the building of the St. Isaac Cathedral [Lamé & Clapeyron, 1823, p. 789]. Clapeyron and Lamé calculated the position of the joint of rupture for symmetrical masonry arches formed by two arcs with different radii (Fig. 6-3 top) with the help of the equilibrium conditions. They showed that the calculations for joints of rupture assumed to be vertical could be simplified and specified a simple graphical method for doing this (Fig. 6-3 bottom).

Clapeyron and Lamé's work on the funicular polygon and the polygon of forces [Lamé & Clapeyron, 1828] enabled them to apply Varignon's two fundamental theorems published in 1725 for the first time in construction engineering. They used the funicular polygon and the polygon of forces (Fig. 6-4) to analyse the dome of the St. Isaac Cathedral. Rühlmann [Rühlmann, 1885, p. 474] contains original quotations which can be translated as follows: "Therefore, successive lines, proportional and parallel to the forces and acting at the corners of a funicular polygon, form a polygon of forces, open or closed, which possesses the property that lines extending from the ends and corners of said polygon and meeting at one point, or rather pole, $A$ (Fig. 6-4), are proportional and parallel to the tensile forces that act in the various sides of the associated funicular polygon. This principle is universally applicable, regardless of whether the funicular polygon is horizontal or inclined." If all the forces in the funicular polygon act perpendicular, "any lines $Ae$, $Ag$, $Ah$, $Ak$, $Al$, $Am$ are drawn through any point $A$ parallel to the sides $EG$, $GH$, $HK$, $KL$, $LM$ and $MN$ and cut by a vertical $em$ line; the lines $eg$, $gh$, $hk$, $kl$, $lm$ are proportional to the forces required and the lengths of these lines, between pole $A$ and the section line, represent the tensile and compressive stresses that act in the sides parallel to them if such forces act at the corners $G$, $H$, $K$, $L$, $M$ of the given funicular polygon." Clapeyron and Lamé continue: "It appears to us that the theory of polygons of forces and poles of forces throws new light on the theory of funicular polygons which enables them to be used in many more applications than many other theorems whose verification is much more complex."

Even at this early stage we can see the projective relationship between the funicular polygon and the polygon of forces, which Culmann discovered 40 years later and which became a crucial element in graphical statics (see section 6.2). According to Jaques Heyman [Heyman, 1972/1, p. 185], the masonry arch theory of Clapeyron and Lamé is therefore much more than just the conclusion of the phase of rediscovery of Coulomb's masonry arch theory dating from 1776.

Senatsplatz in the direct vicinity of the St. Isaac Cathedral building site was the scene of the bloody defeat of the Decembrist uprising of 14 December 1825 which had been directed at Nikolaus I, the successor to Tsar Alexander I, who had died in November 1825. Allegedly, workers on

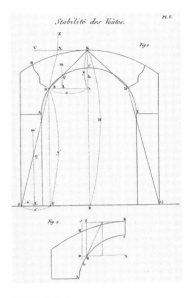

**FIGURE 6-3**
The masonry arch theory of Clapeyron and Lamé [Lamé & Clapeyron, 1823, plate V]

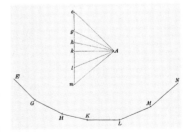

**FIGURE 6-4**
Polygon of forces and funicular polygon after Clapeyron and Lamé [Rühlmann, 1885, p. 473]

the site threw stones and baulks of timber at Nikolaus I and his followers as the 3000 rebelling soldiers were besieged by the 10 000-strong tsarist forces. The foreign engineers and scientists in Russia were aware of the seriousness of the social situation, in particular the worsening conditions in the education sector under Nikolaus I, who tried to fill vacant educational posts with Russian teachers although their technical shortcomings were obvious. Nevertheless, Clapeyron and Lamé were pleased by the goodwill of their new ruler. Clapeyron wrote in a letter to his mother in the summer of 1829 that he and his friend Lamé were to receive 2000 roubles each from Nikolaus I as an expression of his gratitude, and could reckon with military promotion [Bradley, 1981, p. 299]. Lamé's house in St. Petersburg gradually evolved into the meeting place for French scientists and engineers plus their friends. There they could discuss politics undisturbed, in particular talk about the ideas of Saint-Simon. Whether and how his development into a critical-utopian socialist, achieved around 1820, was adopted into the intellectuality of this group would be worth a separate study as utopian socialism together with classical German philosophy and the value of labour theory of Adam Smith and David Ricardo form the sources of Marxism.

Clapeyron met Alexander von Humboldt in St. Petersburg and described the friendly encounter in glowing tones in a letter to his mother in 1829: "Mr. Von Humboldt, a Prussian by birth and a Frenchman at heart, divides his inclinations between his country of residence and his country of birth; I have a thousand reasons for assuming that I, as a Frenchman and student of the École Polytechnique plus the recommendation of my person, would be more than enough … to pay me a thousand compliments" [Bradley, 1981, p. 300]. In the first two decades, Alexander von Humboldt appraised the colossal amount of material from his great research study trip to Paris. The universal natural researcher had numerous contacts among the polytechnicians; he was a very close friend of Lazare Carnot, the man of the early days of the École Polytechnique. Bradley suspects that Clapeyron and Alexander von Humboldt spoke about the Carnots, in particular the younger son of Lazare Carnot, Sadi Carnot, the clever and rebellious polytechnician, the founder of applied thermodynamics.

Bétancourt, too, the engineer who had fled to Russia in 1808 to escape the Spanish Inquisition, was a friend of Clapeyron and Lamé. He not only organised the Institute of Engineers of Ways of Communication in St. Petersburg (see section 2.3.5), but from 1819 onwards was also responsible for public civil engineering works in Russia. One of the projects within his remit was the building of the St. Isaac Cathedral. However, following irregularities in public building projects, Bétancourt was forced to retire in 1822 and appointed Clapeyron and Lamé to carry on the project. The list of names is easily continued: Pierre Dominique Bazaine, who succeeded Bétancourt and after returning to his homeland in 1832 became inspector-general of road- and bridge-building, and, finally, Eugène Flachat, who was to play a leading role in the early days of the building of

French railways. Together, Flachat, Clapeyron and Lamé wrote numerous political articles on the improvement of commerce and the transport infrastructure in France.

Clapeyron was to work on the realisation of these ideas in the spirit of Saint-Simon from his return to France after the July revolution of 1830 until his death.

In their voluntary exile in Russia, Clapeyron and Lamé had at least one ideal of the French Revolution they could enjoy: the brotherhood of their friendships. A rare piece of luck under social conditions that were the opponents of freedom and most certainly equality.

### 6.1.3 Clapeyron's formulation of the energy doctrine of classical engineering sciences

The two most influential scientific works of Clapeyron are most probably his *Abhandlung über die bewegende Kraft der Wärme* (treatise on the motive power of heat) [Clapeyron, 1926] and the theorem that bears his name in elastic theory. Naming this latter theorem after Clapeyron can be attributed to Lamé, who in 1852 had presented the theorem of Clapeyron for the general case of the spatial elastic continuum in the first monograph on elastic theory [Lamé, 1852, pp. 80 – 92]. Both these works of Clapeyron had a decisive influence on the fundamental engineering science disciplines of theory of structures and applied thermodynamics, which had been taking shape since 1820.

Although the main principles of thermodynamics that had been formulated prior to 1850 by Sadi Carnot (1796 – 1832), James Prescott Joule (1818 – 89), Robert Mayer (1814 – 79), Hermann Helmholtz (1821 – 94), Rudolf Clausius (1822 – 88) and William Thomson (1824 – 1907) were counted among the most important discoveries of the 19th century by many prominent contemporaries, it was not until many decades later that they had an effect on the theoretical basis of the discipline-formation period of structural theory and applied thermodynamics, which had stretched across almost the whole century (1825 – 1900). Their establishment was the outcome of the development of the steam engine, not as an invention for a particular purpose, but rather as "an agent universally applicable in Mechanical Industry" [Marx, 1979, p. 398]. The steam locomotive hauled the Industrial Revolution to the farthest flung corners of the continent.

The establishment of the energy doctrine in theory of structures – a topos introduced by the founder of physical chemistry, Wilhelm Ostwald (1853 – 1932), around 1900 – is nothing more than the projection of the real steam engine onto the engineering science model of the trussed framework by James Clerk Maxwell (1831 – 79) [Maxwell, 1864/2]: the energy-based machine model of trussed framework theory (see section 6.3.2). This becomes clear when we compare the pressure-volume diagram of a heat engine (Fig. 6-5a) with the force-displacement diagram of a linear-elastic trussed framework structure (Fig. 6-5b). In both cases the circulatory integral, i.e. the area, specifies the energy of the respective technical artefact replaced in the form of mechanical work. Clapeyron formulated the mathematical principles behind both diagrams.

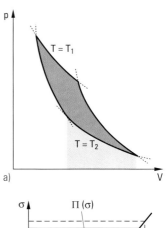
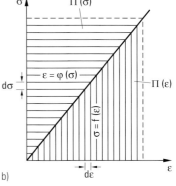

**FIGURE 6-5**
a) Pressure-volume diagram after Clapeyron, and b) stress-strain diagram of a one-dimensional Hookean body

Clapeyron's student companion Sadi Carnot at the École Polytechnique stands out thanks to one single completed treatise (1824): *Betrachtungen über die bewegende Kraft des Feuers und die zur Entwicklung dieser Kraft geeigneten Maschine* (observations on the motive power of fire and the machine suitable for developing this power) [Carnot, 1824]. Carnot financed this epic publication himself. Klaus Krug and Klaus-Peter Meinicke have called it the "birth certificate" of applied thermodynamics [Krug & Meinicke, 1989, p. 117]. Without using one single equation, Carnot proves that the amount of work performed by the steam engine is proportional to the heat transferred from the boiler – the heat reservoir at a higher temperature – to the condenser – the reservoir at a lower temperature – and the highest degree of efficiency theoretically possible for a heat machine must be less than 1 because some heat is always lost without being used. Contrasting with this, the trussed framework portrayed by Maxwell as an energy machine based on the paradigms of elasticity and the conservation of energy principle (theorem of Clapeyron), therefore excluding energy dissipation, has a degree of efficiency of 1.

Sadi Carnot was not able to enjoy the establishment of his fundamental knowledge. Together with almost all his manuscripts and his possessions, he was buried by his friends and fellow students in 1832. One of those was Clapeyron, who made known the works of his friend to the world of science in a very original way in 1834, 1837 and 1843. Clapeyron's service involved translating Carnot's manuscripts into a mathematical framework and illustrating the essential results in a pressure-volume diagram (Fig. 6-5a). His work was therefore of a formal nature, but we can say categorically that Clapeyron thus laid the mathematical foundations of thermodynamics. Without Clapeyron's intervention, Carnot's work would probably have gone unnoticed because only a few of his contemporaries would have been able to understand it. It was Clapeyron who first made it comprehensible and hence useful [Clapeyron, 1926, p. 41]. Pioneers of thermodynamics such as Clausius and Thomson were therefore not able to get hold of Carnot's original work [Carnot, 1824]; instead, it came to their notice via the final publication of Clapeyron's reworking in *Poggendorfs Annalen* (Poggendorf's chronicles) in 1843. It is fascinating to see with what tenacity Clapeyron proved his duty to his friend Carnot: he published the work in a French scientific journal in 1834 [Clapeyron, 1834], an English one in 1837 [Clapeyron, 1837] and a German one in 1843 [Clapeyron, 1843]. And with the third edition, Clapeyron finally gave his friend Carnot a lasting monument in science.

**Bridge-building and the theorem of three moments**

### 6.1.4

After his return from Russia in 1830, Clapeyron rose to become the leading railway engineer in France. As part of a study of railway engineering, for which students of Saint-Simon were able to secure a budget of 500 000 francs, Clapeyron conceived the Paris – St. Germain railway line. As, however, the funding was slow in coming, Clapeyron took up a post as professor at the École des Mineurs in St. Etienne. By 1835 he was able to realise his plans, which gave France her first railway line. During the

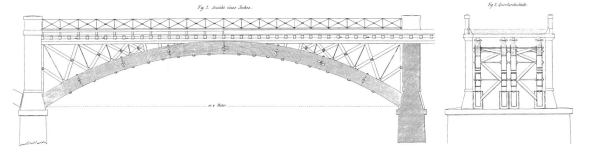

**FIGURE 6-6**
Railway bridge on the Paris–St. Germain line [n. n., 1849/1, sht. 267]

building of the Versailles Railway (right bank), he was responsible for the design of the locomotives. However, as Robert Stephenson did not want to supply the locomotives for this railway, which had a gradient of 1 : 200 over a length of 18 km, Clapeyron appointed the Manchester-based company Scharp & Roberts to build the locomotives according to his drawings [Rühlmann, 1885, p. 453]. Therefore, Clapeyron also became the spiritual father of French locomotive-building.

He was occupied with the planning and management of the Northern Railway from 1837 to 1845, and he remained a consulting engineer for this company until his death. In 1842 he presented an important *Mémoire* on slide valves for steam engines to the Parisian Academy of Sciences [Clapeyron, 1842]. This was certainly the outward reason for being appointed to the post of professor of steam engine-building at the time-honoured École des Ponts et Chaussées in 1844. Clapeyron also played a great part in the plan for an atmospheric railway between Paris and St. Germain. This railway line (completed in 1846) included a viaduct over the Seine with six timber arches each spanning 31.8 m (radius = 27.78 m, rise = 5.0 m, cross-section $h \times b = 1.2 \times 0.45$ m) [n. n., 1849/1, p. 173].

The reporter writes as follows about the arch bridge shown in Fig. 6-6: "According to the assessment of the independent expert, this bridge combines lightness of construction with long-lasting strength, and together those things will make this work one of the best in the art of bridge-building" [n. n., 1849/1, p. 175]. But the era of iron bridge-building had already begun in Great Britain and in that respect, too, Clapeyron would become a pioneer for his country. After 1852 he was involved with the building of the Southern Railway from Bordeaux to Cette and Bayonne. He exerted a major influence on the design of the large railway bridges over the Seine at Asnières, over the Garonne, the Lot and the Tarn in the course of this railway. He was helped with the bridge over the Seine by the German engineer Karl von Etzel (1812–65), who had already worked with Clapeyron during the building of the railway to St. Germain. Thanks to his apprenticeship years with Clapeyron, Karl von Etzel quickly became the leading railway engineer in the German-speaking countries; he later took charge of many railway projects, including the building of the Brenner Railway, the completion of which he did not live to see. Clapeyron developed the theorem of three moments that bears his name during the building of the bridges for the Southern Railway.

Clapeyron designed the bridge over the Seine at Asnières as a five-span continuous beam (Fig. 6-7). The loadbearing bridge cross-section consisted of two 2.25 m deep rectangular hollow boxes made from plates and angles riveted together (Fig. 6-8).

The results of his engineering activities for the bridges over the Seine at Asnières, over the Garonne, the Lot and the Tarn were published in 1857 in his *Mémoire* on the calculation of continuous beams with unequal spans [Clapeyron, 1857].

Clapeyron begins his pioneering *Mémoire* with the following words: "The high capital outlay for railway lines has given the science of theory of structures enormous impetus; a few years ago, engineers were confronted with difficult practical problems, and often proved incapable of solving these" [Clapeyron, 1857, p. 1076]. Just one such difficult problem for which a solution had still to be found was noticed by Clapeyron in the analysis of Robert Stephenson's continuous four-span Britannia Bridge over the Menai Strait. Here, like everywhere, practice would race ahead of theory. "It is [therefore] necessary to take account of the facts of the latest developments and establish rules for these cases where our predecessors let themselves be guided merely by vague intuitions" [Clapeyron, 1857, p. 1076]. And it is such rules that concern Clapeyron when dealing with the continuous beam – appraising his experiences in the building of continuous beam bridges. His theorem of three moments for continuous beams with a constant bending stiffness was obtained from the continuity and equilibrium conditions [Clapeyron, 1857, p. 1077]. The four three-moment equations for the continuous beam shown in Fig. 6-7 take on the following form:

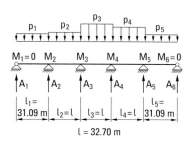

FIGURE 6-7
Structural system after Clapeyron for the railway bridge over the Seine at Asnières

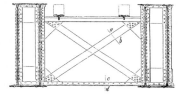

FIGURE 6-8
Section through the railway bridge at Asnières [n. n., 1853, p. 377]

$$\begin{aligned} l_1 \cdot M_1 + 2(l_1 + l_2) \cdot M_2 + l_2 \cdot M_3 &= \tfrac{1}{4}(p_1 \cdot l_1^3 + p_2 \cdot l_2^3) \\ l_2 \cdot M_2 + 2(l_2 + l_3) \cdot M_3 + l_3 \cdot M_4 &= \tfrac{1}{4}(p_2 \cdot l_2^3 + p_3 \cdot l_3^3) \\ l_3 \cdot M_3 + 2(l_3 + l_4) \cdot M_4 + l_4 \cdot M_5 &= \tfrac{1}{4}(p_3 \cdot l_3^3 + p_4 \cdot l_4^3) \\ l_4 \cdot M_4 + 2(l_4 + l_5) \cdot M_5 + l_5 \cdot M_6 &= \tfrac{1}{4}(p_4 \cdot l_4^3 + p_5 \cdot l_5^3) \end{aligned} \quad (6\text{-}1)$$

If $M_1 = M_6 = 0$, the matrix equation for calculating the unknown support moments $M_2$, $M_3$, $M_4$ and $M_5$ is as follows:

$$\begin{bmatrix} 2(l_1+l_2) & l_2 & 0 & 0 \\ l_2 & 2(l_2+l_3) & l_3 & 0 \\ 0 & l_3 & 2(l_3+l_4) & l_4 \\ 0 & 0 & l_4 & 2(l_4+l_5) \end{bmatrix} \cdot \begin{bmatrix} M_2 \\ M_3 \\ M_4 \\ M_5 \end{bmatrix} = \frac{1}{4} \begin{bmatrix} p_1 \cdot l_1^3 + p_2 \cdot l_2^3 \\ p_2 \cdot l_2^3 + p_3 \cdot l_3^3 \\ p_3 \cdot l_3^3 + p_4 \cdot l_4^3 \\ p_4 \cdot l_4^3 + p_5 \cdot l_5^3 \end{bmatrix} \quad (6\text{-}2)$$

By introducing the support moments as static indeterminates, Clapeyron has taken the most rational calculation route to analysing continuous beams: the matrix of coefficients for the set of equations 6-2 is a band matrix with one element to the left and one to the right of the main diagonals; it was from this that Bertot derived the name "three-moment equation" in 1855. The set of equations 6-2 can thus be solved through successive solving of the set of equations with two unknowns. Clapeyron specifies very clearly the set of equations for a seven-span continuous beam structurally equivalent to the set of equations 6-2 [Clapeyron, 1857, p. 1080].

It should be noted that Henri Bertot had already published the three-moment equation for a continuous beam on $n$ supports in 1855. It is possible that Bertot had already received information on the preliminary work to Clapeyron's *Mémoire* [Benvenuto, 1991, p. 485]. Nevertheless, the theorem of three moments carries Clapeyron's name, and he used it for his continuous beam theory as early as 1848 when building the bridge over the Seine at Asnières. Clapeyron's *Mémoire* on continuous beam theory and the 1858 *Mémoire* on his theorem [Clapeyron, 1858], which makes use of Lamé's book on elastic theory [Lamé, 1852] (elastic theory with two material constants), earned him membership of the Paris Academy of Sciences in 1858, where he succeeded the famous mathematician Augustin-Louis Cauchy (1789–1857).

In the course of his professional career, Clapeyron always managed to maintain close ties between science and practical engineering. Clapeyron not only made fundamental contributions to the foundation of classical engineering sciences, but can also be regarded as an important liberal brain behind the Industrial Revolution in France. He was the personification of the joint civil and mechanical engineer, the practical engineer and the academic. When he died in Paris on 28 January 1864, besides his professional work it was more than anything else his engaging qualities, of which he had many, that assured him lasting fame.

## 6.2 From graphical statics to graphical analysis

The first two chapters of Culmann's principal work *Die graphische Statik* appeared in the year that Clapeyron died, and the other six chapters were published in 1866 [Maurer, 1998, p. 151] by the Zurich-based publishing house Meyer & Zeller [Culmann, 1864/1866]. This work laid the foundation stone for graphical statics and graphical analysis. Whereas Culmann formulated explicitly the programme of theory for graphical statics and was engaged in implementing this in polytechnic curricula and everyday engineering, his monograph still contains the core of the programme of theory for graphical analysis. The historico-logical evolution of graphical analysis in the classical phase of the discipline-formation period of classical structural theory (1875–1900) would negate the emerging graphical statics with its claim to a mathematical foundation through projective geometry. Like the machine kinematics of Reuleaux, the programme of theory for graphical statics is the attempt to save the classical scientific ideal for the emerging engineering sciences in order to cancel out the technicising of mathematical and natural science theory fragments it had brought about in the form of the mapping of incomplete sets of symbols. It was therefore not graphical statics that became established in engineering practice, but rather the graphical analysis Culmann had so vehemently opposed because only the latter was in the position of being able to compress the conceptual and design activities into a graphic form, and therefore rationalise the engineer's workload. It was the tool-like character of graphical analysis, already part of Culmann's graphical integration machine (reverting to rope statics from beam statics) and the Cremona diagram, that allowed this to become a modern rules of proportion theory for engineers in the analysis and synthesis of loadbearing systems in the classical phase of theory of structures. As equations were developed for the theory of statically indeterminate systems and their linear-algebraic structure was recognised at the start of the consolidation period of structural theory after 1900, so graphical analysis ossified in formulas on the practical side and ceased to exist as a subdiscipline of theory of structures. Concurrently with this, the rationalisation movement in engineering practice concentrated on the calculus integral to structural engineering methods.

### 6.2.1 The founding of graphical statics by Culmann

Graphical statics started life with the publication of Culmann's monograph in 1864/1866. Culmann had been lecturing in graphical statics at Zurich Polytechnic (which later became the Swiss Federal Institute of Technology Zurich) since 1860. According to the claim of its creator, it is the attempt "to use the newer geometry to solve those tasks of engineering suitable for geometric treatment" [Culmann, 1866, p. VI]. By "newer geometry" Culmann means the projective geometry based on Jean-Victor Poncelet's (1788–1867) paper *Traité des propriétés projectives des figures* [Poncelet, 1822] dating from 1822, a form of geometry that examines the projective properties of figures. For example, central projection is the projection of figures (triangle $A'B'C'$ in Fig. 6-9) in plane $\beta$ from the centre of projection 0 onto a plane $\alpha$ other than plane $\beta$. In order to study his graphical statics, Culmann presumes knowledge of the elements of projective geometry at the very least, as can be found in the work by Karl Georg Christian von Staudt (1798–1867) dating from 1847 [Staudt, 1847]. But what is the relationship between this projective geometry freed from the metric, i.e. from dimensions and numbers (which is interesting only for the properties of the reciprocal positions of geometrical figures), and graphical statics, where the latter is precisely concerned with the measurement of geometric variables?

The answer has been given in the form of an example, and indeed by means of Culmann's mathematical evidence of the projective relationship between the funicular polygon and the polygon of forces plus their origi-

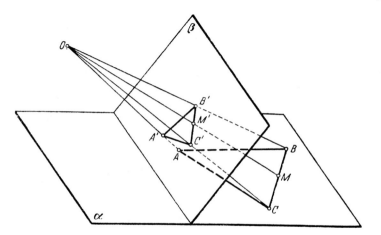

FIGURE 6-9
Central projection of a triangle A′B′C′ from the centre of projection 0 onto plane α after Efimow [Efimow, 1970, p. 226]

nal engineering science interpretation and introduction as an important tool for analysing plane, statically determinate trusses.

Pierre Varignon (1654–1722) introduced the funicular polygon and the polygon of forces in his work *Nouvelle Mécanique ou Statique* published posthumously in 1725. An inelastic, dimensionless rope with a defined length is suspended from points $A$ and $B$ and loaded with the weights $K$, $L$, $M$ and $N$ (Fig. 6-10). The resulting equilibrium position $ACDPQB$ of the rope is called the funicular polygon; it is defined by the polygon of forces $SEFGHRI$. The polygon of forces is a succession of triangles of forces with which the equilibrium at nodes $C$, $D$, $P$ and $Q$ is satisfied successively, e.g. the triangle of forces $SEF$ satisfies the equilibrium at node $C$. Varignon also specifies the construction of a funicular polygon for forces in any direction. Apart from Poncelet, who used the funicular polygon to determine the centre of gravity in his lectures at the artillery and military engineering school in Metz, the application of the funicular polygon for determining the equilibrium positions of ties and struts, e.g. in suspension bridges and masonry arch structures, remained limited. This development from Varignon to Poncelet is traced by Konstantinos Chatzis in the first part of his profound history-of-science analysis of the adoption of graphical statics in France [Chatzis, 2004].

So what progress did Culmann achieve? For the special case of a system of forces in equilibrium with one point of application, Culmann discovered the "structural relationship between funicular polygon and polygon of forces through a planar correlation of the projective geometry" [Scholz, 1989, p. 174]. We start with a central system of forces with the point of application 0 (Fig. 6-11). Using the associated polygon of forces after assuming point 0' as the pole, we can draw the closed funicular polygon $A_1A_2A_3A_4$ belonging to this pole. If now the four straight lines 0'0, 0'1, 0'2 and 0'3 are considered as the lines of action of four forces in equilibrium applied at point 0', the funicular polygon $A_1A_2A_3A_4A_1$ can now be regarded as a closed polygon of forces with pole 0 because it is parallel with the lines of action 0'0, 0'1, 0'2 and 0'3; the funicular polygon belonging to this polygon of forces is then the former polygon of forces. The funicular polygon and the polygon of forces are interchangeable insofar as it is irrelevant which of the polygons is taken to be the system of the lines of action besides the associated funicular polygon. Culmann calls such figures reciprocal. Working independently, James Clerk Maxwell had already proved in 1864 that the only case in which there are two reciprocal figures for non-central systems of forces is when one figure can be considered as a projection of a polyhedron; the other figure then also appears as a projection of a polyhedron. In Fig. 6-11 the two figures can be interpreted as the projections of four-sided pyramids with their apexes at 0 and 0'. The knowledge of this mathematical relationship, also known as the duality of funicular polygon and polygon of forces, could be used by Culmann only for determining the loading function for elliptical, parabolic and hyperbolic masonry arch forms. The heuristic function for the formation of an engineering science theory of graphical statics

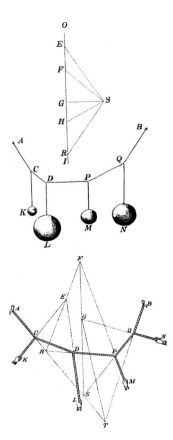

FIGURE 6-10
Funicular polygon and polygon of forces after Varignon [Varignon, 1725]

FIGURE 6-11
On the duality of the funicular polygon and polygon of forces for plane force systems after Culmann

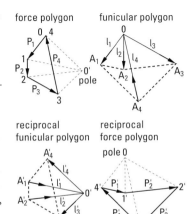

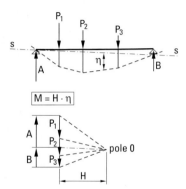

**FIGURE 6-12**
Culmann's solution to the beam problem with the help of the funicular polygon

founded on projective geometry therefore remained an illustrative fringe concept.

Even though the less successful attempt at a geometrical foundation occupies considerable space in Culmann's *Graphische Statik*, its original core is the solution of the internal forces problem in statically determinate plane beam structures with the help of the funicular polygon (Fig. 6-12). The foundation stone of graphical statics lies in the reduction of beam statics to the much simpler rope statics conveyed by the funicular polygon and polygon of forces as the graphical means at the disposal of the structural engineer. Fritz Stüssi (1901–81) quite rightly referred to the funicular polygon used to solve the beam problem as an "integration machine" because we obtain the bending moment in the beam, obtained mathematically by integrating the loading function twice, as a product of the pole distance $H$ in the polygon of forces and the ordinate $\eta$ of the funicular polygon bounded by Culmann's closing line $s$.

### 6.2.2 Rankine, Maxwell, Cremona and Bow

Rankine used the funicular polygon for investigating statically determinate trussed frameworks as early as 1858. Calladine has compared the Wiegmann-Polonceau truss (see section 2.2.9) analysed graphically by Rankine in 1858 with his more elegant solution dating from 1872 (Figs. 6-13a, c, d) [Calladine, 2006, p. 3]. After defining the pole 0, Rankine satisfies the equilibrium for node 1 using the triangle of forces, i.e. he determines the member forces $S_{15}$ and $S_{12}$ graphically. As the relationship $S_{ij} = S_{ji}$ is valid for all member forces, Rankine can move on to node 2 with the member force $S_{12} = S_{21}$ now known, draw the triangle of forces and work out the member forces $S_{23}$ and $S_{25}$. Member forces $S_{34}$ and $S_{35}$ are then calculated from the triangle of forces for node 3 with the help of the known member force $S_{23}$. The last unknown member force $S_{45}$ is obtained from the triangle of forces for node 4, into which the member force $S_{34}$ determined previously and the support reaction $W/2$ are entered. Rankine has thus quantified all the member forces of the trussed framework shown in Fig. 6-13a. He obtained his solution from the additive combination of the four triangles of forces of the bottom chord to form the polygon of forces (Fig. 6-13c) – the associated funicular polygon (or system of hinges) is shown in Fig. 6-13b: from the given geometry of the funicular polygon (or system of hinges) 1–2–3–4 and the two support reactions $W/2$ at nodes 1 and 4, the external loads $S_{15}$, $S_{25}$, $S_{35}$ and $S_{45}$ as well as the member forces $S_{12}$, $S_{23}$ and $S_{34}$ can be determined such that equilibrium prevails. The solution of this task corresponds to the solution of the second prime task of thrust line theory (see section 4.4.1). Rankine's polygon of forces does not apply to node 5.

It was Maxwell who first gradually developed a theory of reciprocal diagrams [Maxwell, 1864/1, 1867 & 1870]. Cremona followed up this work and combined the theory with Culmann's theory formation programme for graphical statics based on projective geometry [Scholz, 1989, pp. 193–199]. Therefore, Cremona [Cremona, 1872] generalised the Maxwell theory of reciprocal diagrams for the case of plane frameworks whose

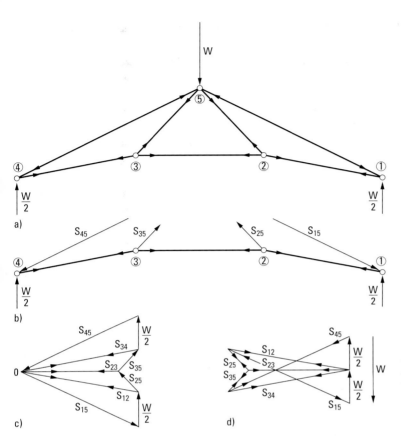

**FIGURE 6-13**
a) Statically determinate trussed framework [Rankine, 1858, Fig. 76], b) bottom chord as funicular polygon, c) Rankine's solution of 1858 [Rankine, 1858, Fig. 76*] and d) 1872 [Rankine, 1872, Fig. 101]

external forces represent a non-central system of forces [Scholz, 1989, p. 194], e.g. Rankine's trussed framework exercise (Fig. 6-13a). Fig. 6-13d is nothing more than the Cremona diagram for the structural system shown in Fig. 6-13a – both diagrams are in accordance with Cremona's [Cremona, 1872] dual, or rather reciprocal, diagrams. Using the example of Rankine's trussed framework exercise, the construction instructions for the Cremona diagram (see, for example, [Schreyer et al., 1967/1, p. 254]) will be explained below (Figs 6-13a, b):

1. Scale drawing of the structural system
2. Determination of support reactions
3. Scale drawing of the addition of the external forces in the clockwise direction to form a closed polygon of forces
4. Determination of the member forces at one node with just two unknown member forces (node 1 in this example). Enter the direction of the first **unknown** member force proceeding in the clockwise direction at the end point of the last **known** force in the **separated node diagram** in the triangle of forces. The second unknown member force **must** close the triangle of forces (node equilibrium). This is why a line parallel to this has to be drawn through the starting point of the force diagram. Its intersection with the line parallel to the first unknown gives us the magnitude of the two member forces required. Their signs

are established by moving around the triangle of forces once more, always in the same direction. Arrows pointing towards the node stand for compressive forces, those pointing away are tensile forces. In this example all members are in tension apart from the upper chord.

5. Those nodes at which there are no more than two unknown member forces are now handled as described in point 4 above (in this example nodes 2, 3, 4 and 5).

6. Final check: the last triangle of forces, too, must also be closed (in this example the triangle of forces for node 5).

The Cremona diagram can therefore be drawn according to a formula without knowledge of projective geometry – and it would soon develop into the incarnation of graphical analysis. This operative side is matched by the causal side, which exists in Cremona's generalisation of the dual, or rather reciprocal, diagrams of Maxwell [Cremona, 1872].

One first milestone in the rationalisation of engineering work through the practical application in the sense of graphical analysis was reached with Robert Henry Bow's (1827–1909) monograph *The Economics of Construction in relation to Framed Structures* [Bow, 1873]. In that work he divides 136 different types of trussed frameworks into four different classes and specifies the dual polygons of forces. This catalogue made the formula-type knowledge of the Cremona diagram obsolete.

### 6.2.3
**Differences between graphical statics and graphical analysis**

Whereas knowledge of projective geometry was crucial to the derivation of the Cremona diagram by its creator, this was no longer necessary for deriving the analogy between rope and beam statics (see Fig. 6-12). But then why the graphic representation of this analogy and not its interpretation by the differential equations of the rope and the beam, with their identical mathematical formulation? And: What are the specific differences between this graphic representation and that implied by projective geometry?

In Culmann's graphic representation of beam statics by means of the funicular polygon, which together with the Cremona diagram forms the heart of graphical analysis, the operational side locked in the means-purpose relationship prevails. Relying on dimensions and figures, it rationalises and mechanises an important aspect of the engineer's work process by condensing the information into the form of graphical methods. Graphical analysis is hence the process exposing the causal relationships modelled as functions in the structural system for the purpose of analysing and synthesising loadbearing systems; it is therefore an intermediary between the work of the engineer and the object of his work, which in this case is present in the form of idealised structural models. Graphical analysis represents an important intellectual aid for the engineer, which in the last three decades of the 19th century called for a marriage between classical fundamental engineering science disciplines and practical structural engineering. Therefore, Culmann's integration machine could only be a graphical integration machine. The tool-like character of Culmann's graphical analysis becomes even clearer in his introductory section on graphical calculations, which follows on from the work of Barthélemy-Édouard Cousinéry [Cousinéry, 1839]. He translates arithmetical into graphical operations and specifies applications as they occur in engineering practice. For example, he determines the relationship between cut and fill in earthworks for a railway line using graphical means (Fig. 6-14). The transport costs for the volume of soil depend directly on the quality of the graphical longitudinal profile. The rationalisation of the work of the railway engineer with the help of graphical methods was therefore one element in the economising of the building process.

Graphical analysis superseded the rules of proportion for the conception and design of buildings until well into the consolidation period of classical fundamental engineering science disciplines. Indeed, we can even claim that the formation of graphical analysis represented a modernised rules of proportion for civil and structural engineers. At a certain stage of the loadbearing system evolution, which began in the second decade of the 20th century with the spread of frame-type loadbearing systems brought about by reinforced concrete construction (see section 9.4.2.1), graphical analysis quickly lost ground to specific numerical computational methods based on linear algebra. This was because it had exhausted its rationalisation potential, primarily in the analysis and synthesis of statically determinate beam and trussed framework systems.

**FIGURE 6-14**
Graphical longitudinal profile for minimising the transport costs of soil during the building of railway lines after Culmann [Culmann, 1875, plate 6]

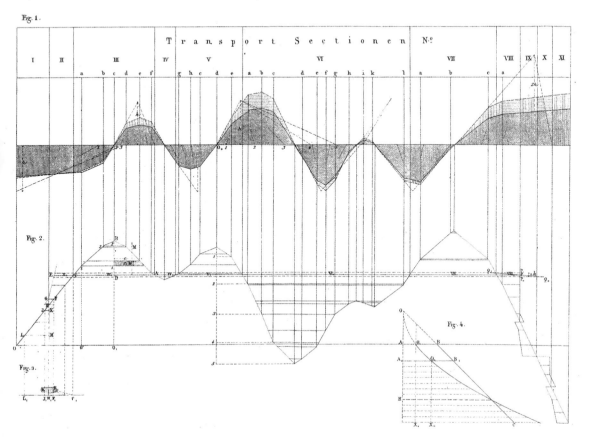

In contrast to graphical analysis, the graphic representation of the mathematical basis of statics, which will be called graphical statics here and which was brought into play in the formation of engineering science theories by projective geometry, can be understood neither as a rationalisation movement in engineering practice nor as the foundation of modern rules of proportion for building; it was weighed down too much by the concepts of projective geometry which were still unfolding. Concepts such as pencil of lines, points at infinity, harmonic and involute figures hindered the adoption of graphical statics by engineers. As theory developments in projective geometry, indeed geometry as a whole, in the 19th century were gradually steering towards a geometry without dimensions and numbers, its projection onto structural relationships was blocked. The dead weight of its system of concepts was therefore only capable of promoting theory developments in graphical statics in the early years, acting like a brake to the rationalisation of engineering practice, but also the reflection of conception and design activities. Graphical statics therefore became separated from its applications in the form of graphical analysis soon after the publication of Culmann's *Graphische Statik*. One indication of this is the second edition dating from 1875, which is based more on mathematics and in which Culmann promises the reader that the applications will be expanded in a second volume [Culmann, 1875, p. XIV]. Culmann never managed that because his scientific thinking was too focused on the ideal of an axiomatically organised foundation to graphical statics.

### 6.2.4 The breakthrough for graphical analysis

Nevertheless, Culmann's theory programme was like two souls in one breast: graphical statics and graphical analysis. Just a few years after the publication of the first edition of Culmann's *Graphische Statik*, engineers opted for graphical analysis. Bauschinger, professor of applied mechanics and graphical statics at the Polytechnic School in Munich writes in the preface to his 1871 book *Elemente der Graphischen Statik* (elements of graphical statics): "Graphical statics is so crucial to the study of engineering sciences and the practising engineer that we hope to see it spread widely, and this will certainly be the case to some extent. Perhaps my book can help proliferate this knowledge further because it is not necessary to be familiar with the so-called newer geometry. I did not plan it that way, it happened by itself" [Bauschinger, 1871, preface]. I shall not explore the discussion about graphical statics and the graphical statics "stripped of its spirit" [Culmann, 1875, p. VI], i.e. graphical analysis, which was kindled by Culmann's preface to the second edition of his *Graphische Statik*, because this has already been analysed splendidly by Erhard Scholz [Scholz, 1989] from the history of mathematics viewpoint.

If in the discipline-formation period of structural theory graphical statics represents the final piece in the jigsaw of the establishment phase (1850–75), which began with trussed framework theory, then the emancipation of graphical analysis from graphical statics and its cognitive expansion forms an important moment in the classical phase of structural

**FIGURE 6-15**
Cover of Ritter's *Anwendungen der graphischen Statik* [Ritter, 1888]

theory (1875–1900). Even the student of and successor to Culmann in Zurich, Wilhelm Ritter (1847–1906), did not remain true to graphical statics in his four-volume work *Anwendungen der graphischen Statik. Nach Professor Dr. C. Culmann* (applications of graphical statics after Prof. Dr. C. Culmann) (Fig. 6-15), published between 1888 and 1906, and his principle formulated in the preface after its founding via projective geometry. He wrote a book about graphical analysis in which this principle appeared merely in the form of a "manner of speaking" [Scholz, 1989]. During the classical phase, graphical analysis became generalised as an intellectual tool for design and structural engineers. In 1882 Ludwig Tetmajer (1850–1905) paid tribute to the influence of Culmann's graphical analysis with the following words: "The magnificent arch bridges built in Switzerland since 1876 were all calculated according to Culmann's theory, and bridge-building establishments such as Holzmann & Benkieser in Frankfurt, Eiffel in Paris and others employed their own personnel to apply the results of Culmann's research in their design offices" (cited in [Stüssi, 1951, p. 2]). For a quarter of a century graphical analysis grew to become a modern set of rules of proportion for structural and bridge engineering in which conception, modelling, statics, dimensioning and design were once again moulded into a graphical formulation.

### 6.2.4.1 Graphical analysis of masonry vaults and domes

Graphical analysis was able to achieve noteworthy successes in the investigation of Gothic churches (see, for example, [Huerta & Kurrer, 2008]) in the structural engineering studies for architects, which became a separate entity during the classical phase of structural theory. During the 1870s, graphical analysis methods for masonry arches in combination with the strip method rendered possible the loadbearing system analysis of any masonry vault or dome. In 1879 Wittmann [Wittmann, 1879] published what was probably the first correct graphical analysis of three-dimensional masonry constructions such as rib vaults and domes. Planat approached this subject in 1887 [Planat, 1887]; in doing so he did not use the polygon of forces or the funicular polygon, but instead preferred resolving the forces directly on the same drawing. Barthel explains in detail the various methods of loadbearing system analysis for rib vaults [Barthel, 1993/1].

The most complete presentation of structural investigations into stone vaults is that by Karl Mohrmann in the third edition of Ungewitter's *Lehrbuch der gotischen Konstruktionen* (textbook of Gothic constructions) in 1890 [Ungewitter, 1890], in which Mohrmann deals in particular with the rib vaults of neo-Gothic churches. The first two chapters on masonry vaults and buttresses, which Mohrmann rewrote completely for the third edition, contain the best presentation of the construction and structural analysis of Gothic masonry structures. Mohrmann first describes the application of the strip method for determining the resulting compressive forces in a Gothic vault: the vault segments are divided into strips, the form of which depends on the shape of the vault (Fig. 6-16a). The resulting compressive forces due to the vault strips then act as loads on the ribs of the vault (Fig. 6-16b), are transferred from these to the buttresses as

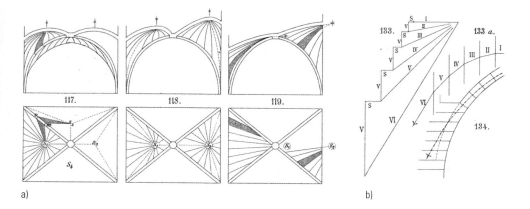

**FIGURE 6-16**
Mohrmann's explanation of the strip method: a) possible division of the vault segment into strips, b) line of thrust in the rib of a rib vault [Ungewitter, 1890]

support reactions, and are carried by the buttresses down to the foundations – together with other actions, e. g. the self-weight of the buttress.

The method is obvious, but can involve tedious calculations and graphical analysis operations, especially if the division of the vault results in many differently shaped strips. Mohrmann therefore proposed a more rational method: he calculated the horizontal thrust (per square metre of plan area of the vault segment of the rib vault generating a load) for different materials, loads and rise ratios from the global equilibrium for the half-vault of the individual vault span, and compiled these in a table together with other interesting parameters. The table can still be used today advantageously for the structural re-analysis of rib vaults. Heyman includes a simplified version in his 1995 book *The Stone Skeleton. Structural Engineering of Masonry Architecture* [Heyman, 1995/1].

The loadbearing system analysis of domes turns out to be considerably simpler because of the rotational symmetry. The dome is divided along its meridians into equal segments in such a way that each pair of opposing segments forms an arch of varying width. The hoop forces in the upper part of the dome are also sometimes considered. In this way, Wittmann developed a method for constructing the line of thrust [Wittmann, 1879]; he was followed by other German authors such as Körner [Körner, 1901] plus Lauenstein and Bastine [Lauenstein & Bastine, 1913]. In 1878 Eddy published a graphical method for determining the hoop forces (both tension and compression) for the membrane stress condition of domes [Eddy, 1878 & 1880], which Föppl applied to masonry domes in 1881 [Föppl, 1881] (Fig. 6-17). The method permitted the calculation of rotationally symmetrical domes of any form. This approach was made known in North America by Dunn [Dunn, 1904], and Rafael Guastavino Jr. was still making full use of this in the first decades of the 20th century in order to design thin domes of clay bricks [Huerta, 2003]. The hoop forces were normally neglected so the theory of plane masonry arches could be applied (see, for example, [Lévy, 1886 – 88]).

In 1928 Dischinger specified a general method for determining the membrane stress condition of any thin reinforced concrete shell [Dischinger, 1928/1], but without referring to the relevant work by Eddy

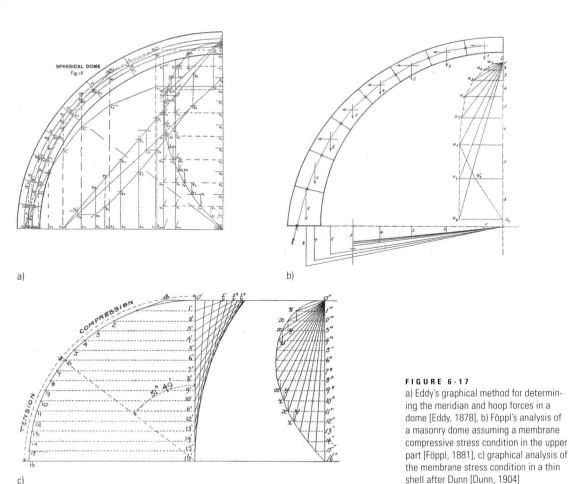

FIGURE 6-17
a) Eddy's graphical method for determining the meridian and hoop forces in a dome [Eddy, 1878], b) Föppl's analysis of a masonry dome assuming a membrane compressive stress condition in the upper part [Föppl, 1881], c) graphical analysis of the membrane stress condition in a thin shell after Dunn [Dunn, 1904]

and Föppl. Therefore, Dischinger's publication was the first monograph on shell theory and not Flügge's 1934 book entitled *Statik und Dynamik der Schalen* (statics and dynamics of shells) [Flügge, 1934]. The shell structures of Dyckerhoff & Widmann designed by Dischinger, Rüsch and Finsterwalder, lent the development of a structural shell theory a decisive impulse in reinforced concrete construction during the invention phase of structural theory (1925 – 50).

### 6.2.4.2 Graphical analysis in engineered structures

On the cognitive side, too, graphical analysis underwent an enormous expansion during the classical phase of structural theory (1875 –1900). Noteworthy here is the kinematic trussed framework theory with which it was possible to determine the influence of travelling loads (as occur in bridge-building) on the force condition in the truss members under investigation (Fig. 6-18). The interpretation of the deflection curve as the beam loaded with the associated moment function in the sense of a funicular polygon as discovered by Otto Mohr in 1868 (Mohr's analogy) enabled displacements and hence also statically indeterminate problems, e.g. the continuous beam, to be solved with the help of graphical analysis

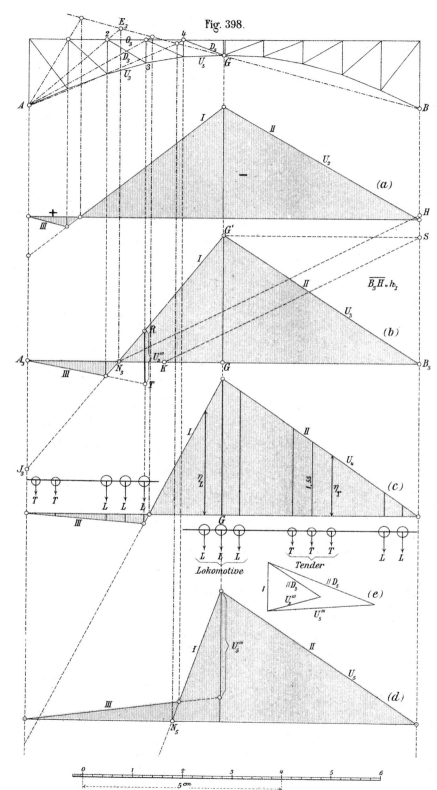

FIGURE 6-18
Müller-Breslau's graphical determination of the influence of travelling loads on member forces in statically determinate trussed frameworks with the help of kinematic methods [Müller-Breslau, 1887/1, plate 6, Fig. 398]

as well. (The deflection curve is based mathematically on the same differential equation, just like the relationship discovered by Culmann between bending moment diagram and loading function in the sense of a funicular polygon.) Graphical analysis concentrated on these two applications.

Culmann, too, uses Mohr's analogy extensively in the second edition of his *Graphische Statik* in the presentation of the elastic beam theory [Culmann, 1875, pp. 627–644]. He criticised the omission of graphical statics from the curriculum: "In the French schools graphical statics is not taught anywhere; the awkward *conseils d'études* only permit innovations to appear slowly in France" [Culmann, 1875, p. IX]. An independent development in France did not appear until 1874 in the form of Maurice Lévy's (1838–1910) book *La statique graphique et ses applications aux constructions* [Lévy, 1874], which in the classical phase of structural theory generated copious writings on graphical analysis (see [Chatzis, 2004]). The situation was similar in the United States, where Augustus Jay Du Bois (1849–1915) [Du Bois, 1875] and Henry Turner Eddy (1844–1921) [Eddy, 1877 & 1878] were the main protagonists.

Müller-Breslau's *Graphische Statik der Baukonstruktionen*, published in several volumes starting in 1887, gives an impression of the powerful influence graphical analysis had on theory developments in classical theory of structures. Although this major work of classical theory of structures covers much more than just graphical analysis, Müller-Breslau retained this title for the subsequent volumes as well. Despite the energy principle basis of classical theory of structures (see section 6.3.2), he regards graphical analysis as important: "A wide field is still open to graphical methods: the plotting of the displacement diagrams and the use of these line figures for deriving the influence lines and influence figures gives the clearest answer to all the questions that arise when investigating a given trussed framework" [Müller-Breslau, 1903/1, p. VI]. The method of the displacement diagram reproduced by Müller-Breslau (Fig. 6-19) can be attributed to Victor-Joseph Williot [Williot, 1877 & 1878]; Williot's displacement diagram became popular during the 1880s in France, Germany and other countries as well [Chatzis, 2004, p. 22]. Graphical analysis had thus been put fully into practice in terms of determining both the force condition and the displacement condition, devoid of those mathematical foundation attempts through projective geometry; even such phrases as, for example, "newer geometry" and "projective geometry" are absent from Müller-Breslau's classic work of classical theory of structures.

The star of graphical analysis began to set shortly after 1900; no significant further advances were made. In the preface to the third edition of *Graphische Statik*, which appeared as the second volume of Föppl's highly popular *Vorlesungen über Technische Mechanik* (lectures in applied mechanics) in 1911, August Föppl notes: "As just a few years after the publication of the first edition it became clear that a second edition of this volume was necessary, I suggested to the publishers that they print enough copies to cover demand for a longer period. I had assumed that significant progress in graphical statics, which had to be discussed in this textbook,

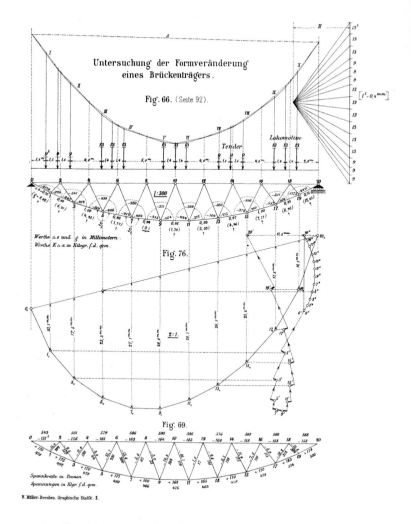

**FIGURE 6-19**
Displacement diagram for a three-pin truss after Müller-Breslau
[Müller-Breslau, 1903/1, plate 2]

was hardly to be expected in the near future. And I was not wrong. Although the sales of the greatly enlarged second edition were spread over nearly nine years, in my current work on the third edition I feel it is hardly necessary to make any significant alterations" [Föppl, 1911, p. IV].

Graphical analysis soon stagnated on the operational side in the shape of formulas, ceased to exist as a subdiscipline of theory of structures and wasted away into an archaeological relic of its three resplendent decades before the turn of the 20th century. And with it we lost an extremely compact form of rendering visually the play of forces in the analysis and synthesis of loadbearing systems in the conception and design activities of the civil and structural engineer.

**The classical phase of theory of structures**

## 6.3

I shall pay tribute to the engineering science work of Emil Winkler (1835–88) starting with the relationship between construction engineering and theory of structures during the Industrial Revolution in Germany. Using the example of a history-of-science analysis of masonry arch theories and his solution to the masonry arch problem based on elastic theory, it will be shown how the founding process through elastic theory, with its repercussions, was so characteristic of the last phase of the creation of the theory of structures discipline.

### 6.3.1 Winkler's contribution

Emil Winkler (Fig. 6-20), born on 18 April 1835 in Falkenberg near Torgau (Saxony), the son of a forester, was one of those outstanding structural engineers whose engineering science work prepared the way for the consummation of classical theory of structures [Knothe, 2000]. After attending primary school and grammar school, he was apprenticed to a bricklayer and attended the building trades school in Holzminden in order to prepare for the subsequent four-year course of study in building at Dresden Polytechnic. Until 1868, Johann Andreas Schubert, the designer of the first German steam locomotive and the Göltzsch Viaduct, and the "creative spirit behind Saxony's Industrial Revolution" [Brendel, 1958], was on the teaching staff there and taught all the subjects of civil engineering. It was from him that the young Winkler learned of the difficulties that engineers had to face when analysing masonry arches; Schubert himself had added to the multitude of theoretical approaches to masonry arch statics, but this, the unanswered "puzzle of architecture" [Hänseroth, 1982], still needed a solution after 150 years. Even the usefulness of Navier's practical bending theory published in 1826 for everyday engineering was still being questioned in the trade literature in 1855 [Werner, 1974, p. 16].

FIGURE 6-20
Emil Winkler [Stark, 1906]

Winkler had published an important paper on the deformation and strength of curved bodies in the journal *Zivilingenieur* as early as 1858 [Winkler, 1858]. This contribution already shows that he was one of the few engineers who, independently, could make use of the basic equations of elastic theory for solving problems in theory of structures. Three years later he gained his Dr. phil. at the University of Leipzig with a dissertation on the pressure inside masses of earth. Starting with Cauchy's equilibrium conditions for the two-dimensional continuum and taking into account the boundary conditions for friction, Winkler was able to describe the earth pressure problem with the help of differential equations. His dissertation was published in 1872 under the title of *Neue Theorie des Erddruckes* (new theory of earth pressure) [Winkler, 1872/3]. Winkler first had to earn a living in the Saxony Waterways Department and later in the Standard Calibration Commission in Dresden. At the same time, Winkler's interest in theory encouraged him to take a post as field survey assistant to Prof. Nagel, then as a private lecturer in strength of materials and finally as a teacher of technical drawing at the Trade & Industry School in Dresden. It was not until 1863 that he gained a position as teacher at the grammar school for boys in Dresden-Friedrichstadt (Freemason's school) – a preparatory school for Dresden Polytechnic – with an annual salary of 400 thaler, which just sufficed for a secure material existence. From 1861 onwards, Winkler had also been working part-time as assistant to Prof. Schubert: "This is how he managed to get a foot in the door of Dresden Polytechnic, where he worked as a teacher of engineering sciences in general, but in particular was active in that subject he later enriched so amazingly, i.e. the calculation of bridges" [n.n., Zentralblatt der Bauverwaltung, 1888/1, p. 387]. And this is where Winkler's odyssey ended; his interest in theory of structures finally became his profession.

Just after his 30th birthday in 1865, Winkler was appointed professor of civil engineering at Prague Polytechnic, which had been founded in 1806 and after the École Polytechnique was the oldest such school. It was here that the polytechnic's founder, Franz Joseph Ritter von Gerstner, had developed mechanics into the main pillar of polytechnic education (see sections 2.3.5 and 3.2.2). The fact that a "mechanics derived from the nature of building itself", as Gerstner had called for as early as 1789 in his *Einleitung in die statische Baukunst* (introduction to structural engineering), did not become an independent subject until 1852 at Vienna Polytechnic at the instigation of Georg Ritter Rebhann von Aspernbruck (1824–92) and not until 1864 at Prague Polytechnic at the behest of Karl von Ott (1835–1904), points to an uneven development in theory of structures in the polytechnic schools of continental Europe during the discipline-formation period. Despite this, the decisive momentum in the ongoing development of theory of structures between 1855 and 1875 did not come from its academic institutionalisation, but rather arose gradually from the relatively unspecific cognitive structure of civil engineering – and that was not until the 1870s. With the exception of structural mechanics, Winkler covered the whole gamut of civil engineering in his lectures. That did not stop him from continuing his study of mathematical elastic theory which he had started in Dresden and which later became a useful tool for civil engineers through his 1867 book *Die Lehre von der Elasticität und Festigkeit* (theory of elasticity and strength) [Winkler, 1867]. The verdict of the author of a Winkler obituary was that "Winkler's elastic theory is to the civil engineer what Grashof's recently published [in 1866 – the author] strength of materials is to the mechanical engineer" [n.n., Zentralblatt der Bauverwaltung, 1888/1, p. 388]. As mathematical elastic theory was transformed into a practical elastic theory by Grashof and Winkler, elastic theory became the strength theory paradigms in the core disciplines of classical engineering sciences.

Winkler's successes in establishing elastic theory in bridge-building were certainly the reason for his taking up a post as professor of railway and bridge engineering at Vienna Polytechnic in 1868. Without doubt, this very specialised professorship enabled him to develop further the "theoretical-constructional approach" [Melan, 1888/1, p. 186] in bridge-building. It was in the early 1870s that he began to publish his lectures on bridge-building, a project that was intended to encompass all aspects of bridge-building in five parts and provide bridges with an engineering science foundation: theory of bridge beams (4 booklets), bridges in general plus stone bridges, timber bridges (4 booklets), iron bridges (6 booklets), construction of bridges. Unfortunately, this engineering science encyclopaedia of bridge-building remained merely an outline. In spite of this, the booklets that were released for publication (theory of bridge beams Nos I and II, timber bridges No. I, iron bridges Nos II and IV) provided the most comprehensive insight into bridge-building at that time. Bridge-building was a part of structural engineering crucial to the foundation of a classical theory of structures, and owing to the enormous expansion in this field during the 1880s, even an extraordinarily competent scientific personality like Emil Winkler could not complete such a work alone; for engineering science activities were also subject to the increasing division of work that was affecting the whole of society. So it was not until 1890 that an encyclopaedia of bridge-building on a scientific footing appeared – the fruits of a joint project involving highly specialised engineers (see section 3.2.6).

Winkler, well-versed in all fields of civil engineering, was paid another honour – admittedly, suiting his inclinations – as in 1877 he was appointed professor of theory of structures and bridge-building at the Berlin Building Academy (which was amalgamated with the Trade & Industry Academy to form Berlin-Charlottenburg TH in 1879). Whereas Winkler had not lectured in structural mechanics in Prague or Vienna, but had had to cover all aspects of railways instead, the fruitful interaction between the special disciplines of bridge-building and theory of structures – an expression that can be attributed to Winkler – could be realised in one person. It was this early institutional bond between bridge-building and theory of structures (Fig. 6-21) that helped the Berlin school of structural theory to gain international recognition after 1888 through Heinrich Müller-Breslau (see section 6.5).

According to Winkler, theory of structures (Fig. 6-21) could be broken down into four components: strength of materials into its fundamentals and its application to structural theory, statics of bar systems into its general interpretation and its application to structural theory, the theory of earth pressure and the statics of stone constructions, in particular revetments and masonry arches; this classification of the cognitive make-up of theory of structures was customary up to the middle of the consolidation phase of structural theory, i. e. the end of the 1920s. Besides his teaching and research activities plus several presentations held at the Berlin Architects Society, Winkler also served as chairman of the civil engineering department in 1880/81 and 1885/86, and he was also elected dean of Berlin-Charlottenburg TH for 1881/82. "The time I have been allotted is short; I must work fast if I wish to reach the goal" [n. n., Zentralblatt der Bauverwaltung, 1888/1, p. 387] were the words he said to his wife on many occasions, playing down his serious physical handicaps. He was just 53 when he died on 27 August 1888 as the result of a stroke. If Winkler's goal was to give the constructional disciplines of civil engineering a theoretical foundation, then he had achieved that goal.

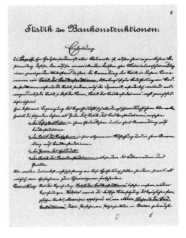

FIGURE 6-21
First page of the printed manuscript of Winkler's lecture notes *Vorträge über Statik der Baukonstruktionen* [Winkler, 1883]

### 6.3.1.1 The elastic theory foundation to theory of structures

The elastic theory foundation to theory of structures is among the three main threads in Winkler's scientific work, the other two being the ongoing development of structural analysis itself and bridge-building. Winkler's most important publication on the elastic theory foundation to theory of structures was his *Lehre von der Elasticität und Festigkeit* (theory of elasticity and strength) published in 1867. In his introduction he writes: "The publication of this book was primarily occasioned by the wish to have a solid basis for the numerous studies in the fields of elastic theory and strength of materials, which are necessary in my lectures on civil engineering. As by far the largest part of this theory can be applied, I therefore resolved to compile the work in the form of a general textbook on elastic theory and strength of materials for engineers" [Winkler, 1867, p. III]. It is interesting to compare this work with Carl Bach's 1889/90 publication *Elasticität und Festigkeit* (elasticity and strength). Whereas Bach, a pioneer of materials research in Germany, assumes "that it is not sufficient to establish the whole fabric of elasticity and strength on a mathematical basis focused solely on the principle of proportionality between stress and strain, and that it appears far more necessary for the design engineer ... to visualise again and again the prerequisites of the individual equations he uses in the light of his experience" [Bach, 1889/90, p. VII], Winkler has to derive the real loadbearing behaviour of loadbearing systems essentially from elastic theory because strength of materials experiments were still at an embryonic stage of development. Winkler overcomes the need for strength of materials in elastic theory by determining existing stresses in loadbearing systems on the basis of elastic theory and compares these with the help of permissible stresses, which multiplied by a safety factor results in the corresponding material strength determined empirically. Bach, on the other hand, sees elastic theory as part of strength of materials experiments by defining concepts like strength, limit of proportionality and elastic limit based on "knowledge of the actual behaviour of the materials". Winkler subsumes chapter 5 (strength of materials), containing fewer than seven pages, in the first section, entitled *Allgemeine Theorie der Elasticität* (general theory of elasticity). Although he was an enthusiastic supporter of Wöhler's theory, his strength of materials theory amounts to the comparison of the calculation results founded on elastic theory provided with a safety factor.

Although the subtitle to Winkler's *Lehre von der Elasticität und Festigkeit* denotes the target readership as not just civil engineers, but also mechanical engineers and architects (Fig. 6-22), his book is primarily aimed at the former.

The book is divided into two parts, and in the first part Winkler deduces a comprehensive bending theory from the general theory of elasticity, which in principle could be used to calculate all the bar-like bending structures customary at that time. He regards it as "reasonable to divide the book into two parts, the first part of which contains the most important theories for building, especially bridge-building. It includes general elastic theory, normal elasticity (tensile and compressive elasticity), shear elasticity and bending elasticity of straight and curved bars. As these parts may be sufficient for gaining an understanding of structural design, this part can be seen as complete in itself" [Winkler, 1867, p. III].

FIGURE 6-22

Cover of Winkler's *Lehre von der Elasticität und Festigkeit*

The second part of Winkler's book was supposed to include the "torsional elasticity of straight bars, the elasticity of curved bars, rotational bodies plus flat and curved plates, the dynamics of elasticity and, finally, a history and bibliography of elastic theory" [Winkler, 1867, pp. III – IV]. Unfortunately, this part did not make it to publication. This might have been chiefly due to Winkler's lack of time, but might also have been due to the fact that it was not until reinforced concrete construction became established after 1900 that engineers in everyday practice needed to carry out structural analyses of one- or two-dimensional elastic continuua in double curvature. His promise to write a history of elastic theory was only partly realised [Winkler, 1871].

But back to the first part of *Lehre von der Elasticität und Festigkeit*. After Winkler has introduced "normal elasticity" (strain stiffness $E \cdot A \neq \infty$, $G \cdot A_Q \to \infty$, $E \cdot I \to \infty$), "shear elasticity" (shear stiffness $G \cdot A_Q \neq \infty$, $E \cdot A \to \infty$, $E \cdot I \to \infty$) and "bending elasticity" (bending stiffness $E \cdot I \neq \infty$, $E \cdot A \to \infty$, $G \cdot A_Q \to \infty$), he considers straight and curved bars in bending in succession subjected to certain practical boundary conditions and loading cases: straight bars in bending ($E \cdot I \to \infty$) under vertical load (cantilever beam, beam simply supported at both ends, propped cantilever with one end fixed, one end simply supported, beam fixed at both ends, continuous beam), straight bars in bending with or without vertical load and subjected to concentric or eccentric tension and compression (four Euler buckling cases, simple problems of second-order theory), straight beams in bending on elastic supports under vertical load, catenaries and inverted catenaries (lines of thrust) under vertical load and, finally, the arched beam with infinite bending and strain stiffness ($E \cdot I \neq \infty$, $E \cdot A \neq \infty$, $G \cdot A_Q \to \infty$) without hinges, or with two or three hinges, under a travelling vertical load and the concentric thermal loading case $T_s$. Winkler derives the basic equations between the deformations and the internal forces for elastic arched beams for the general case of significant curvature of the arch axis. Furthermore, he presents the exact bending theory according to St. Venant, determines the shear stress diagrams for important cross-sectional forms and compares them with those obtained using Navier's bending theory. Winkler points out possible applications in practical engineering by way of many derivations. His *Lehre von der Elasticität und Festigkeit* is therefore a compendium of structural engineering based on elastic theory. Besides the general theory of the plane elastic arch, it contains new approaches and solutions which were not introduced into engineering practice until many years later, e.g. the equation for calculating the shear stresses in the cross-section of beams, the calculation of influence lines for internal forces under travelling vertical loads on a beam simply supported at both ends, plus two-pin, three-pin and fixed arches, the analysis of the beam on elastic supports, the solution of specific tasks according to second-order theory (stress problem) and, finally, the extension of beam theory to $\mathcal{I}$-shaped cross-sections.

Winkler also adopted the findings of leading structural engineers in a creative way. For example, he continued adapting the theory of the kern

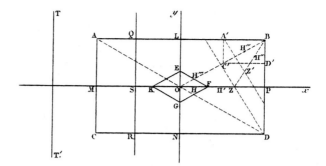

**FIGURE 6-23**
Kern of a rectangular cross-section after Bresse [Bresse, 1854, plate I]

which Bresse had developed in 1854 within the scope of his elastic arch theory [Bresse, 1854]. This is at the same time the foundation of an elegant theory of bending about two axes. His adaptation consisted of introducing kern lines for characterising the longitudinal profile of elastic arch structures. Fig. 6-23 shows the construction of the rhombus-shaped kern $EFGK$ for a rectangular cross-section after Bresse with width $b = NL$ and depth $d = PM$. If the axial compressive force $N$ acting at a right-angle to the $x$-$y$-plane (i.e. acting in the $z$-direction) is applied in the kern $EFGK$, then the rectangular cross-section is subjected to compression only – otherwise tensile stresses also occur. For an axial compressive force $N$ acting at the centre of gravity 0, the stress over the entire cross-section is constant, i.e. stress $\sigma_z$ is $\sigma_z = \sigma_m = N/(b \cdot d)$. If, for example, $N$ lies at point $E$ (i.e. at $y = d/3$), the stresses are reduced to zero along edge $CD$ ($\sigma_z (y = -b/2) = 0$), are linear with respect to the $y$-axis and take on the value $\sigma_z (y = +b/2) = 2 \cdot \sigma_m$ along edge $AB$: the distribution of the compressive stresses is triangular with respect to the $y$-axis (see Fig. 4-41). The reasoning is similar for the x-direction. If the point of application of $N$ is neither on the $x$- nor on the $y$-axis, the plane of the stress distribution of $x$ **and** $y$ exhibits a linear relationship, i.e. $\sigma_z (x, y) = c + d \cdot x + e \cdot y$ and the cross-section is therefore subjected to bending about two axes. For the special case of bending of an elastic arch about one axis, Winkler defines the upper or lower kern line as the geometrical point of the upper or lower kern point of all arch cross-sections and enters both kern lines into the longitudinal profile of the elastic arch. If the point of application of $N$ always lies between the upper and lower kern lines for every arch cross-section, then only compressive stresses are present throughout the entire elastic arch. For the special case of a rectangular cross-section, this is nothing other than the middle-third rule (see section 4.5.3).

In his book of 1953, Timoshenko assesses Winkler's *Lehre von der Elasticität und Festigkeit* as follows: "Winkler's book is written in a somewhat terse style and it is not easy to read. However, it is possibly the most complete book on strength of materials written in the German language and is still of use to engineers. In later books in this subject, we shall see a tendency to separate strength of materials from the mathematical theory of elasticity and to present it in a more elementary form than that employed by Winkler" [Timoshenko, 1953, p. 155].

The transformation of the mathematical elastic theory into Winkler's practical elastic theory was the historico-logical condition for the elastic-theory foundation of structural theory, and therefore the implementation of the paradigms of elastic theory in the emerging classical theory of structures. As the content of Winkler's practical elastic theory was overtaken by classical theory of structures and applied mechanics, it lost its germ-like disciplinary independence.

### 6.3.1.2 The theory of the elastic arch as a foundation for bridge-building

After Winkler had generalised Bresse's theory of the elastic arch [Bresse, 1854] in 1858 with a view to using it in mechanical engineering for significantly curved arched beams [Winkler, 1858], construction engineering problems became prominent in his creative output in connection with his professional activities as a civil engineering scientist at Dresden Polytechnic, and later at Prague Polytechnic.

Although Winkler in developing the basic equations for the theory of significantly curved elastic arched beams in his *Lehre von der Elasticität und Festigkeit* was aiming to use them in the first instance on iron and timber arch constructions, he had no objections to using these basic equations for the structural analysis of masonry arches, provided the line of thrust was within the kern of all voussoir joints, i.e. no tensile stresses could develop in those joints [Winkler, 1867, p. 277]. In connection with shallow elastic arched beams in particular, the relationship between axial force $P$ and bending moment $M$ on the one hand, and the element deformations $\Delta ds$ and $\Delta d\varphi$ related to the arch element of length $ds$ on the other, takes on the form

$$\frac{\Delta ds}{ds} = \frac{P}{E \cdot A} \quad (6\text{-}3)$$

$$\frac{\Delta d\varphi}{ds} = \frac{M}{E \cdot I} \quad (6\text{-}4)$$

where $E \cdot A$ denotes the strain stiffness and $E \cdot I$ the bending stiffness of the elastic arch. Taking into account eq. 6-3 and eq. 6-4, Winkler's displacement equations for the rotation $\Delta\varphi$, vertical displacement $\Delta y$ and horizontal displacement $\Delta x$ of any point along the arch axis are converted into the basic equations

$$\Delta\varphi = \int \frac{M}{E \cdot I} \cdot ds \quad (6\text{-}5)$$

$$\Delta y = \iint \frac{M}{E \cdot I} \cdot ds \cdot dx + \int \frac{P}{E \cdot A} \cdot dy \quad (6\text{-}6)$$

$$\Delta x = \iint \frac{M}{E \cdot I} \cdot ds \cdot dy + \int \frac{P}{E \cdot A} \cdot dx \quad (6\text{-}7)$$

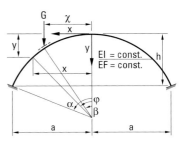

FIGURE 6-24
Structural system for a fixed symmetrical arch of constant bending and strain stiffness subjected to a travelling vertical load G

These basic equations are used with mathematical geniality by Winkler in the chapter on arches without hinges but with fixed abutments [Winkler, 1867, pp. 328–358] (Fig. 6-24). The arch with three degrees of static indeterminacy is subjected to a travelling vertical load $G$, which in the case of a bridge could be a moving railway or road vehicle. The task now is to determine the internal forces at point $\{x,y\}$ due to the travelling vertical load $G$. If we plot the ordinates of bending moment $M$ and axial force $P$

at point {x,y} due to the travelling vertical load G at χ, we obtain the influence line of bending moment M and axial force P at section {x,y} for the travelling vertical load G. In order to find the cross-section at risk, we determine the corresponding influence lines for several sections {x,y} and evaluate them for a given imposed load in such a way that the internal force of the section under consideration is a maximum; the geometrical position of these maxima for all {x,y} points finally produces the maximum line of the force variable concerned, which is used as the basis for sizing the cross-sections at {x,y}.

Winkler solves this classical task of bridge-building statics for the first time, not only for the fixed arch (Fig. 6-24), but also for the statically determinate three-pin arch and the two-pin arch with its one degree of static indeterminacy. He derives his solution in seven steps [Winkler, 1867 & 1868/69]:

1. Determine the internal forces M, P and Q at {x,y} on the statically determinate basic system due to the travelling vertical load G (influence lines $M(\chi)$, $P(\chi)$ and $Q(\chi)$ for the statically determinate basic system).
2. Calculate the static indeterminates H, V and $M_0$ due to the travelling vertical load G from basic equations 6-5, 6-6 and 6-7 taking into account the boundary and transitional conditions (influence lines for static indeterminates $H(\chi)$, $V(\chi)$ and $M_0(\chi)$).
3. Construct the abutment pressure line and abutment pressure envelope (influence lines for abutment pressure $D(\chi)$ and $D'(\chi)$).
4. Determine the resulting internal forces for various sections {x,y}, or in the case of resolved arched beams the forces in the upper and lower chords $S_o$ and $S_u$, due to the travelling vertical load G (influence lines for $S_o(\chi)$ and $S_u(\chi)$).
5. Evaluate the influence lines $S_o(\chi)$ and $S_u(\chi)$ for various sections {x,y} under imposed load p and dead load g (Winkler obtains the maximum line for the upper and lower chord forces by evaluating the most unfavourable loading position p).
6. Take into account the loading case for concentric thermal load $T_s$.
7. Determine the chord cross-sections required for the loading cases based on permissible axial stresses.

Steps 1 to 3 will be discussed below using the example of an arch with a circular axis of radius r and a constant cross-section.

*Step 1:* Winkler introduces the arch cut through at the crown joint as a statically determinate basic system (Fig. 6-25a). He applies the statically indeterminate internal forces $M_0$, H and V to this cantilever beam at the crown. Fig. 6-25b shows the equilibrium system $M_0$, H, V and G with the internal forces M, P and Q defined by Winkler as positive. From the equilibrium conditions in the portion $\chi \leq x \leq a$ it follows that

$$M = M_0 - H \cdot r \cdot (1 - \cos\varphi) + (G - V) \cdot r \cdot \sin\varphi - G \cdot r \cdot \sin\beta \tag{6-8}$$

$$P = -H \cdot \cos\varphi - (G - V) \cdot \sin\varphi \tag{6-9}$$

and in the portion $0 \leq x \leq \chi$

$$M' = M_0 - H \cdot r \cdot (1 - \cos\varphi) - V \cdot r \cdot \sin\varphi \qquad (6\text{-}10)$$

$$P' = -H \cdot \cos\varphi + V \sin\varphi \qquad (6\text{-}11)$$

It can be seen that the static indeterminates $M_0$, $H$ and $V$ plus the internal forces $M$ and $P$, or $M'$ and $P'$, depend on the position $\chi$ of vertical load $G$, and therefore represent the influence lines for the statically determinate basic system (Fig. 6-25a).

*Step 2:* From basic equation 6-5, taking into account eq. 6-8 and eq. 6-10 as well as $ds = r \cdot d\varphi$, the transitional condition $\Delta\varphi(\varphi = \beta) = \Delta\varphi'(\varphi = \beta)$ produces the relationship

$$A - A' = G \cdot r^2 \cdot (\cos\beta + \beta \cdot \sin\beta) \qquad (6\text{-}12)$$

Apart from the integration constants $A$ and $A'$, from the two boundary conditions (both ends fixed) of basic equation 6-5

$$\Delta\varphi\,(\varphi = \alpha) = 0$$

and $\qquad(6\text{-}13)$

$$\Delta\varphi'(\varphi = -\alpha) = 0$$

it follows that the first condition equation for the static indeterminates is

$$2M_0 \cdot \alpha - 2H \cdot r \cdot (\alpha - \sin\alpha) = G \cdot r \cdot [(\cos\alpha - \cos\beta + (\alpha - \beta) \cdot \sin\beta)] \quad (6\text{-}14)$$

Similarly, basic equations 6-6 and 6-7, where $ds = r \cdot d\varphi$, $dx = r \cdot \cos\varphi \cdot d\varphi$ and $dy = r \cdot \sin\varphi \cdot d\varphi$, taking into account the corresponding boundary and transitional conditions, where

$$2M_0 \cdot \sin\alpha - H \cdot r \cdot (2\sin\alpha - \sin\alpha \cdot \cos\alpha - \alpha) = -0.5\,G \cdot r \cdot (\sin\alpha - \sin\beta)^2$$

and $\qquad(6\text{-}15)$

$$V \cdot r \cdot (\alpha - \sin\alpha \cdot \cos\alpha)$$
$$= -0.5\,G \cdot r \cdot (\alpha - \beta - \sin\alpha \cdot \cos\alpha - \sin\beta \cdot \cos\beta + 2\cos\alpha \cdot \sin\beta) \qquad (6\text{-}16)$$

result in the second and third condition equations for the static indeterminates $M_0$, $H$ and $V$. The solution to the set of eq. 6-14 to 6-16 is

$$V = G \cdot \frac{\alpha - \beta - \sin\alpha \cdot \cos\alpha - \sin\beta \cdot \cos\beta + 2\cos\alpha \cdot \sin\beta}{2(\alpha - \sin\alpha \cdot \cos\alpha)} \qquad (6\text{-}17)$$

$$H = G \cdot \frac{2\sin\alpha \cdot (\cos\beta - \cos\alpha + \beta \cdot \sin\beta) - \alpha \cdot (\sin^2\alpha + \sin^2\beta)}{2\,[\alpha \cdot (\alpha + \sin\alpha \cdot \cos\alpha) - 2\sin^2\alpha]} \qquad (6\text{-}18)$$

$$M_0 = \frac{H \cdot r}{2} \cdot \left(2 - \cos\alpha - \frac{\alpha}{\sin\alpha}\right) - \frac{G \cdot r}{4} \cdot \frac{(\sin\alpha - \sin\beta)^2}{\sin\alpha} \qquad (6\text{-}19)$$

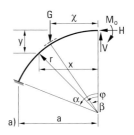
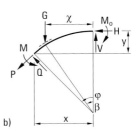

**FIGURE 6-25**
Statically determinate basic system:
a) for the circular arch subjected to a travelling vertical load, and b) the portion of the structure for determining the internal forces M, P and Q

Eq. 6-17 to 6-19 are the influence lines for the static indeterminates $V$, $H$ and $M$ as a result of the travelling vertical load $G$ at $\chi = r \cdot \sin\beta$.

*Step 3:* With the help of the static indeterminates $V(\chi)$, $H(\chi)$ and $M_0(\chi)$, Winkler introduces the abutment pressure line $\eta(\chi)$ and the abutment pressure envelope $\{v(\chi), w(\chi)\}$ (Fig. 6-26/Fig. 1). If the travelling vertical load G at point $\chi$ is placed to intersect the abutment pressure line $\eta(\chi)$ (curve JK in Fig. 6-26/Fig. 1) and if we lay the two tangents through this intersection on the abutment pressure envelope (the lower curve in Fig. 6-26/Fig. 1), then abutment pressure $D(\chi)$ (at abutment A) and abutment pressure $D'(\chi)$ (at abutment B) are determined in terms of direction (by the two aforementioned tangents to the abutment pressure envelope) and in terms of magnitude (by the equilibrium of forces $D(\chi)$, $D'(\chi)$ and G). Winkler derives general equations for the abutment pressure line

$$\eta(\chi) = \frac{V \cdot \chi - M_0}{H} \qquad (6\text{-}20)$$

and the abutment pressure envelope

$$\left. \begin{array}{l} v(\chi) = \dfrac{H \cdot \dfrac{dM_0}{d\chi} - M_0 \cdot \dfrac{dH}{d\chi}}{V \cdot \dfrac{dH}{d\chi} - H \cdot \dfrac{dV}{d\chi}} \\[2em] w(\chi) = \dfrac{V \cdot \dfrac{dM_0}{d\chi} - M_0 \cdot \dfrac{dV}{d\chi}}{V \cdot \dfrac{dH}{d\chi} - H \cdot \dfrac{dV}{d\chi}} \end{array} \right\} \qquad (6\text{-}21)$$

for the fixed arch. Here, $\eta(\chi)$ is plotted upwards from the horizontal through the centroid of the crown, $w(\chi)$ downwards, and $v(\chi)$ from the axis of symmetry to the right (Fig. 6-26/Fig. 1). If eq. 6-20 and 6-21 are expressed for the specific case of the circular arch by inserting the influence line of the static indeterminates $V(\chi)$, $H(\chi)$ and $M_0(\chi)$, i.e. by eq. 6-17 to 6-19, then indirectly we obtain the influence line of the abutment pressures $D(\chi)$ and $D'(\chi)$ due to a travelling vertical load G (Fig. 6-26/Fig. 1). With the help of his two influence lines, Winkler constructs the influence lines for the upper and lower chord forces due to a travelling vertical load G for various arch sections $\{x,y\}$. The influence lines (Fig. 6-26/Fig. 2), which Winkler calls "stress curves" [Winkler, 1868/69, p. 9], are evaluated by him for the imposed load $p$ in such a way that the chord forces in the cross-sections considered take on extreme values; the extreme values of all sections $\{x,y\}$ obtained in this way produce the lines shown in Fig. 6-26/Fig. 3.

Winkler was also the one who established in quantitative terms the influence of the important loading case concentric thermal load $T_s$ for statically indeterminate arch structures [Winkler, 1868/69]. Winkler's theory, which he demonstrated using the example of a fixed circular arch, solved not only the central problem of bridge-building for an important class of arch structures, but also gave classical theory of structures a basic concept in the shape of the influence line.

**The beginnings of the force method**

### 6.3.2

The formulation of the theory of statically indeterminate trusses in the form of the force method stands at the focus of the classical phase of the discipline-formation period of structural theory.

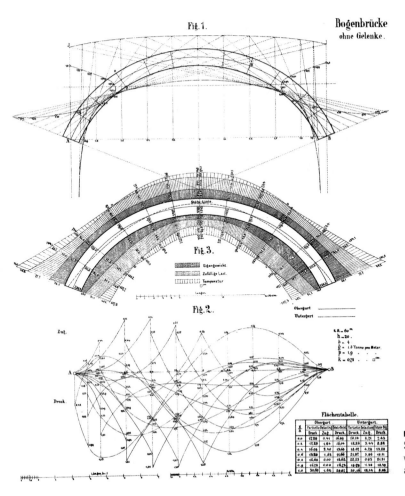

**FIGURE 6-26**
Semi-graphical analysis of a fixed arch with three degrees of static indeterminacy after Winkler [Winkler, 1868/1869, plate 4]

### 6.3.2.1 Contributions to the theory of statically indeterminate trussed frameworks

Whereas during the 18th century constructional-technical progress in construction engineering concentrated on masonry arches, the focus in the course of the 19th century turned to resolved methods of construction, initially in timber, then in cast iron, wrought iron and composite systems (loadbearing structures made from more than one material), and later exclusively in wrought iron. The theoretical interest of engineers also changed after 1850 and gave more attention to the structural analysis of trussed framework structures, although the areas of beam and masonry arch statics were not altogether neglected. In trussed framework construction, the transition in structural modelling from the loadbearing system to the structural system took place considerably faster than in solid construction (i.e. masonry and concrete) because the geometrical and physical properties of trussed frameworks were the logical indication of the abstraction to the elastic pinned truss. For instance, in 1849 Warren introduced pinned joints in the one-piece truss with diagonal struts only [Mehrtens, 1900, p. 15]; two years later, Schwedler defined the elastic pinned truss [Schwedler, 1851, p. 168] and used it as the basic model for a trussed framework theory. British and North American engineers continued to make use of the pin-jointed framework for several decades, whereas their colleagues in continental Europe preferred the rigid (riveted) joint in their frameworks and in constructional terms moved further and further away from the basic model of trussed framework theory. So while engineers in continental Europe constantly had to deal with the contradiction between trussed framework practice and structural modelling, the basic model in British and North American construction practice was confirmed with every pin-jointed framework and could therefore divorce itself from its original range of applications, i.e. could become an object of reflection in mathematics and physics. It was left to the theoretical physicist James Clerk Maxwell to conceive a general theory of statically indeterminate trussed frameworks in 1864 based on the energy concept of the mathematical elastic theory, which at that time was still inaccessible to civil and structural engineers [Maxwell, 1864/2].

## Maxwell

After Maxwell has introduced the criterion for statically determinate plane frames $s = 3k - 6$ ($s = 2k - 3$) and statically indeterminate plane frames $s > 3k - 6$ ($s > 2k - 3$) ($s$ = number of members, $k$ = number of joints), he investigates the node displacement $\delta_F$ for a given force condition in a statically determinate trussed framework (Fig. 6-27a). In order to solve the deformation question, he applies the force $1^v$ to the same framework in the direction of the required node displacement $\delta_F$ and can calculate the associated member force $S_i^v$ (Fig. 6-27b). Maxwell calculates the displacement condition $\Delta l_i^v$ associated with this virtual force condition $\delta_{F,i}^v$ as follows: apart from the elastic bar $i$, all bars are assumed to be rigid. Maxwell now considers this framework as a machine in which the force $1^v$ overcomes an elastic resistance $S_i^v$ (Fig. 6-27b). Clapeyron's theorem for this infinitely slow loading process up to the final value $1^v$ then supplies the condition

$$\frac{1}{2} 1^v \cdot \delta_{F,i}^v = \frac{1}{2} S_i^v \cdot \Delta l_i^v \quad \text{or} \tag{6-22}$$

$$\delta_{F,i}^v = S_i^v \cdot \Delta l_i^v \tag{6-23}$$

Eq. 6-23 is valid for any small values $\Delta l_i^v$. If $\delta_{F,i}^v$ – as shown in Fig. 6-27c – for the true displacement condition, where

$$\Delta l_i^v = \Delta l_i = \frac{S_i \cdot l_i}{E \cdot A_i} \tag{6-24}$$

is selected for $\Delta l_i^v$, then – considering all $s$ elastic bars of the framework – the required total displacement becomes

$$\delta_F = \sum_{i=1}^{s} \delta_{F,i} = \sum_{i=1}^{s} S_i^v \cdot \frac{S_i \cdot l_i}{E \cdot A_i} \tag{6-25}$$

Eq. 6-25 corresponds to the solution of the deformation task for elastic pin-jointed frameworks on the basis of the principle of virtual forces. Maxwell employs eq. 6-25 to obtain the relationship $\delta_{ik} = \delta_{ki}$ [Maxwell,

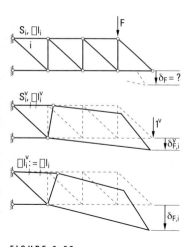

**FIGURE 6-27**
On Maxwell's interpretation of the general work theorem for linear-elastic frameworks in the version of the principle of virtual forces as the sum of the external and internal virtual work: $W_a^v + W_i^v = 0$

1864/2, p. 297], which in 1886 Müller-Breslau designated Maxwell's theorem as a tribute to this marvellous physicist.

Maxwell's idiosyncratic derivation of eq. 6-25 from Clapeyron's theorem, reduced to just a few lines and without diagrams, would play a great role in the dispute between Mohr and Müller-Breslau concerning the theoretical basis of classical theory of structures which raged during the 1880s.

After solving the deformation issue on the statically determinate trussed framework, Maxwell turns to the analysis of statically indeterminate trussed frameworks. To do this, he selects a statically determinate primary system with $s - n$ bars and notes the superposition equation of the member force $S_j^{(n)}$ in bar $j$ for the system with $n$ degrees of static indeterminacy

$$S_j^{(n)} = S_{j,0}^{(0)} + \sum_{k=1}^{n} S_{j,k}^{(0)} \cdot X_k \tag{6-26}$$

He then determines the member forces $S_{j,0}^{(0)}$ for the statically determinate primary system as a result of the 0-state and the member forces $S_{j,k}^{(0)}$ due to the $X_k = 1$-state. Using superposition eq. 6-26, Maxwell obtains the change in length

$$\Delta l_j^{(n)} = S_j^{(n)} \cdot \frac{l_j}{E \cdot A_j} = \left( S_{j,0}^{(0)} + \sum_{k=1}^{n} S_{j,k}^{(0)} \cdot X_k \right) \cdot \frac{l_j}{E \cdot A_j} \tag{6-27}$$

of bar $j$ for the system with $n$ degrees of static indeterminacy. Applying work eq. 6-25 to the changes in length $\Delta l_1^{(n)}, ..., \Delta l_n^{(n)}$ of the redundant bars in the statically indeterminate system and to the $n$ force conditions $X_k = 1$ in the statically determinate basic system and taking into account eq. 6-27, we get $n$ elasticity equations in the form

$$\Delta l_i^{(n)} = \sum_{j=1}^{s-n} \left[ \left( S_{j,0}^{(0)} + \sum_{\substack{k=1 \\ i \neq k}}^{n} S_{j,k}^{(0)} \cdot X_k \right) \frac{l_j}{E \cdot A_j} \right] \cdot S_{j,i}^{(0)} \quad \text{where } i = 1...n \tag{6-28}$$

for calculating all statically indeterminate member forces $X_k$. Eq. 6-28 can be rearranged to produce

$$0 = \underbrace{\left( \sum_{j=1}^{s-n} S_{j,i}^{(0)} \cdot S_{j,0}^{(0)} \cdot \frac{l_j}{E \cdot A_j} \right)}_{\equiv \delta_{i0}} + \sum_{k=1}^{n} \underbrace{\left( \sum_{j=1}^{s-n} S_{j,i}^{(0)} \cdot S_{j,k}^{(0)} \cdot \frac{l_j}{E \cdot A_j} \right)}_{\equiv \delta_{ik}} \cdot X_k \quad \text{where } i = 1...n \tag{6-29}$$

Maxwell calculates all coefficients in the set of equations 6-29 for the statically determinate primary system taking into account his reciprocal theorem

$$\delta_{ik} = \sum_{j=1}^{s-n} S_{j,i}^{(0)} \cdot S_{j,k}^{(0)} \cdot \frac{l_j}{E \cdot A_j} = \sum_{j=1}^{s-n} S_{j,k}^{(0)} \cdot \frac{l_j}{E \cdot A_j} \cdot S_{j,i}^{(0)} = \delta_{ki} \tag{6-30}$$

Using eq. 6-25, eq. 6-26 and eq. 6-29, Maxwell's reciprocal theorem (eq. 6-30) and the definition of the statically indeterminate trussed framework, Maxwell had created the principles of the force method for calculating statically indeterminate pin-jointed frameworks. Although he developed the method for calculating the static indeterminate $X_k$ and the displacements in the statically indeterminate system very clearly in seven steps [Maxwell, 1864/2, pp. 297 – 298], his classic work was not adopted by civil and structural engineers until many years later.

**Mohr**

In a presentation to the Bohemian Architects & Engineers Society in December 1867, Winkler introduced a semi-graphical method for calculating a solid-web two-pin arch and a fixed arch based on elastic theory, which he used to determine the member forces of an externally statically indeterminate two-pin arch resolved into a lattice structure (crossing diagonals) [Winkler, 1868/1869]. Such lattice structures – hybrid forms between solid-web beams and trussed frameworks – with their high degree of internal static indeterminacy were in widespread use in German bridge-building up until the 1870s and led to uncertainties in their modelling as a structural system that reflected adequately the loadbearing behaviour of the lattice structure.

It was the rigorous practical realisation of the resolution of the solid-web beam or arched beam into the trussed beam or trussed arch clearly designed according to simple trussed framework theory that created the springboard for the further development of simple trussed framework theory into a general theory of statically indeterminate frameworks. It was left to Mohr to recognise clearly this contradiction between constructional-technical progress in frame construction and the structural analysis of such loadbearing systems with beam or arch theory plus simple trussed framework theory, and supersede this theoretically in 1874/75 with a general theory of statically indeterminate frameworks [Mohr, 1874/1, 1874/2 & 1875]. "Arched beams of larger dimensions generally do not have solid webs, but rather framework sides, for reasons of construction and economy. If I am not mistaken, an exact and simple method of calculating such arched girders without a hinge at the crown [arched truss with one external degree of static indeterminacy – the author] has not yet been published. Those treatises known to me that deal with this subject make the daring assumption, without further reasoning, that when investigating the elastic deformations of the girder, which are essential for determining the horizontal thrust, only the changes in length of the chord elements need be considered, and that the influence of the infilling members may be neglected. Despite this simplification, it has never proved possible to portray the results in a simple way convenient for application. The purpose of the following short observation is to develop a method of calculation without any such preconditions. The observation is confined to beams with a hinge at the crown, but extending this to beams without impost hinges would admittedly not have caused any particular problems" [Mohr, 1874/1, p. 233].

After Mohr has expressed the change in length $\Delta l_j^{(1)}$ of the truss bars due to the 0-state in the statically indeterminate two-pin trussed girder as a function of the unknown horizontal thrust $X_1$, he imagines the arch arranged in such a way "that the supports can move in the horizontal direction and that the changes in length $[\Delta l_j^{(1)}$ – the author] of the individual parts of the construction do not occur simultaneously but rather successively" [Mohr, 1874/1, p. 229]. He applies the true changes in length $\Delta l_j^{(1)}$ to this system bar for bar in succession for the corresponding kinematic system and determines their contribution to the change in the span $\Delta s_j^{(1)}$. Mohr applies the work

$$X_1 \cdot \Delta s_j^{(1)} + S_{j,1}^{(0)} \cdot X_1 \cdot \Delta l_j^{(1)} = 0 \tag{6-31}$$

to this true displacement condition according to the "principle of virtual velocities" [Mohr, 1874/1, p. 232] for the $X_1$-state, where $S_{j,1}^{(0)}$ is the member force in truss member $j$ as a result of force condition $X_1$ in the statically determinate system (restraint removed at abutment). The sum of the span changes $\Delta s_j^{(1)}$ for all bars must obey the condition

$$\sum_j \Delta s_j^{(1)} = 0 \tag{6-32}$$

because the abutments may not spread apart in the true displacement condition. Taking into account eq. 6-31, eq. 6-32 is transformed into

$$\sum_j S_{j,1}^{(0)} \cdot \Delta l_j^{(1)} = 0 \tag{6-33}$$

Taking

$$\Delta l_j^{(1)} = S_j^{(1)} \cdot \frac{l_j}{E \cdot A_j} \tag{6-34}$$

and the superposition equation

$$S_j^{(1)} = S_{j,0}^{(0)} + X_1 \cdot S_{j,1}^{(0)} \tag{6-35}$$

eq. 6-33 can be written as follows:

$$0 = \underbrace{\sum_j S_{j,1}^{(0)} \cdot S_{j,0}^{(0)} \cdot \frac{l_j}{E \cdot A_j}}_{\equiv \delta_{10}} + \underbrace{\left(\sum_j S_{j,1}^{(0)} \cdot S_{j,1}^{(0)} \cdot \frac{l_j}{E \cdot A_j}\right)}_{\equiv \delta_{11}} \cdot X_1 \tag{6-36}$$

Comparing Maxwell's general elasticity eq. 6-29 with eq. 6-36 discovered by Mohr shows that both equations agree for statically indeterminate frameworks with $k = 1$.

Like Maxwell, Mohr solves the deformation problem of the statically determinate trussed framework: he applies a load $P$ in the direction of the required displacement variable $\delta_F$ (Fig. 6-28a), cuts through any truss member $j$ (Fig. 6-28b) and allows the member force $S_{j,p} = u \cdot P$ to do work on the true change in length $\Delta l_j$ of the member (Fig. 6-28c). "In this condition one can consider the trussed framework as a simple machine be-

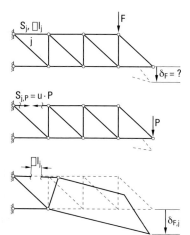

FIGURE 6-28
On Mohr's interpretation of the general work theorem in the version of the principle of virtual forces as a sum of the external virtual work: $W_a^v = 0$

cause it forms a movable connection of solid bodies which by virtue of their geometrical relationship move along prescribed paths and by means of this one can overcome the two resistances $u \cdot P$ through the driving forces $P$" [Mohr, 1874/2, p. 512]. In a similar way, he notes the work equations for the remaining bars so that he obtains $s$ work equations for $s$ truss members; with the summation of all $s$ truss members, the required displacement variable $\delta_F$ – just like with Maxwell – is determined by work eq. 6-25.

In the following "contributions to the theory of the trussed framework" [Mohr, 1874/2, pp. 509–526 & 1875, pp. 17–38], Mohr turns his theory of trussed arches with one degree of static indeterminacy into a general theory of trussed frameworks with $n$ degrees of static indeterminacy:

1. Definition of the statically determinate plane framework $s = 2k - (2a + b)$ and the framework with $n$ degrees of static indeterminacy $s > 2k - (2a + b)$ or $n = s - 2k + (2a + b)$ ($s$ = number of members, $k$ = number of joints, $a$ = number of immovable supports, $b$ = number of movable supports) [Mohr, 1874/2, pp. 510, 511]; in contrast to Maxwell, Mohr includes the nature of the supports in his definition.
2. Selection of a statically determinate primary system [Mohr, 1874/2, p. 514].
3. Mohr obtains $n$ condition equations (see eq. 6-32) from the $n$-fold application of work eq. 6-25 to the true displacement condition $\Delta l_i^{(n)}$ of the system with $n$ degrees of static indeterminacy and the force condition $X_i$ on the statically determinate basic system (see eq. 6-31) [Mohr, 1874/2, p. 514].
4. With the introduction of the deformation principle (eq. 6-34) – possibly also taking into account a temperature term – and the superposition principle for member forces (eq. 6-35) into the $n$ condition equations derived with the help of the principle of virtual forces, Mohr sets up $n$ elasticity equations for calculating the $n$ statically indeterminate force variables $X_i$ [Mohr, 1874/2, pp. 517, 518].

Using the principle of virtual forces, Mohr proves his theorem $\delta_{ik} = \delta_{ki}$ concerning the reciprocity of displacements for trussed frameworks [Mohr, 1875, p. 28].

Mohr drafted his theory of statically indeterminate trussed frameworks without any knowledge of Maxwell's work [Maxwell, 1864/2], simply to satisfy the need of civil engineers for a realistic representation of the force and displacement conditions and hence an economic and reliable dimensioning of iron trussed frameworks. Therefore, Mohr developed his theoretical concept using a readily comprehensible, everyday loadbearing system in a critical appreciation of its previous structural analyses and only after that did he set up the general theory of statically indeterminate trussed frameworks. Maxwell, on the other hand, developed his theory strictly deductively and without diagrams by obtaining the elasticity equation for statically indeterminate trussed frameworks from Clapeyron's theorem. Only at the end of his treatise does he explain the theory using the example of the Warren truss so popular in Great Britain and the USA.

## 6.3.2.2 From the trussed framework theory to the general theory of trusses

Until the introduction of rigid loadbearing systems (e.g. frames) through the reinforced concrete construction that emerged after 1900, the trussed framework, as the most advanced constructional-technical artefact of construction engineering, governed the object of engineering science theory formation in theory of structures. Around 1880, theory of structures was still broken down into three areas corresponding to the three main areas of structural engineering (solid, timber and iron construction) whose internal logical relationship had only been partly worked out theoretically: masonry arch statics, beam statics (including continuous beam theory) and the general theory of statically determinate and indeterminate trussed frameworks. At that time, the engineering science penetration of the trussed framework on the basis of elastic theory was more advanced than that of the masonry arch and the solid-web beam. This was because

- the changeover in the modelling of the loadbearing system to the structural system could be completed by engineers simply and clearly, but at the same time this modelling embodied a mathematical-physical development potential (graphical statics as one area of applied mathematics),
- the laws of elastic theory could be reduced to a single elasticity principle with an easily calculable material constant (the elastic modulus) owing to the modelling of truss members as bars with pinned ends and the use of an industrially manufactured building material (iron) for these, and
- the theoretical foundation of trussed framework theory using the principle of virtual displacements was based on the traditional thinking of the kinematic view of statics and the theoretical model of the simple kinematic series of hinges influenced by the emerging machine kinematics of Reuleaux (Fig. 6-28b), and could be carried out in a very simple manner (see section 6.3.3).

There were no end of suggestions for mapping the loadbearing behaviour of iron solid-web beams theoretically with the help of the general trussed framework theory; no less a person than Mohr expressed his opinion on this in 1885 in a comparative analysis of the theoretical reasoning behind trussed framework theory using the principle of virtual displacements, Clapeyron's theorem and Castigliano's theorems: "An exact determination of the deformations of beams with solid webs, in particular plate girders, presents insurmountable difficulties ... anyhow, the calculation of the stresses and deformations based on the customary bending theory of a homogeneous beam does not produce even an approximate depiction of the reality. It is therefore meaningless to derive equations as incorrect as they are long in the way described. Without doubt, one will achieve a more accurate result in a shorter way when one converts the plate girder into a trussed framework for the purpose of calculating the deformations and the statically indeterminate support reactions" [Mohr, 1885, p. 306].

However, the formation of structural analysis theories did not progress towards – as Mohr was hoping – the modelling of beam-like loadbearing systems within the scope of trussed framework theory. Instead, the structural concept of Müller-Breslau and others, which was based on trussed framework theory, became separated from the original object and was developed further into a general theory of trusses with the help of the energy expressions of elastic theory.

At the heart of the classical phase (1875–1900) of turning theory of structures into a fundamental engineering science discipline of civil engineering was the wrestling concerning its theoretical foundation on which the force method created by Müller-Breslau and his students arose in the form we know today.

Müller-Breslau considerably expanded his journal papers on the theory of statically indeterminate trusses published between 1882 and 1885 in his 1886 book on strength of materials and the theory of structures based on the principle of virtual displacements and deformation work (Fig. 6-29). In this monograph the characteristic constructional-technical loadbearing system analysis problems resulting from everyday building underwent a treatment based on the uniform theoretical basis of the principle of virtual displacements (in the form of the principle of virtual forces), Castigliano's second theorem (based on the energy principle) and Menabrea's principle (see section 6.4.1). This work, which enjoyed a total of five editions (1886, 1893, 1904, 1913, 1924), not only concluded the discipline-formation period of structural theory, which had been initiated by Navier's *Résumé des Leçons* [Navier, 1826], but also enabled classical theory of structures to supersede statics and strength of materials. The heart of this synthesis was the rigorous formulation of the force method for trusses in its current structure and form.

In the first edition, Müller-Breslau introduces the concept of the statically determinate primary system and the unit force conditions $X_k = 1$ acting on this system. Pursuing Mohr's idea but expanding this to cover bending structures, Müller-Breslau allows these unit force conditions to do work on the true displacement condition (i.e. on the given system with $n$ degrees of static indeterminacy) and in this way arrives at $n$ elasticity equations; he arrives at the same equations via Menabrea's principle and the superposition equations for internal forces. Almost every statically indeterminate task is solved by Müller-Breslau using Menabrea's principle. However, when determining the influence lines for the static indeterminates for travelling load $P_m$, he assumes the principle of virtual forces directly and with the help of Maxwell's theorem $\delta_{mn} = \delta_{nm}$, which Müller-Breslau had generalised for rotations, expresses the first version of the force method [Müller-Breslau, 1886, pp. 138–140].

Directly after publication of *Die neueren Methoden...*, Mohr published a strongly worded polemic in the journal *Zivilingenieur* opposing Müller-Breslau's concept of "idealised deformation work", which he had extended to cover the special loading cases of thermal effects and displacement of

**The newer methods of strength of materials**

**FIGURE 6-29**
Cover and dedication of the second edition of Müller-Breslau's *Die neueren Methoden der Festigkeitslehre* …

the supports; one of his criticisms was that "a whole series of other designations, e.g. principle of work, principal of virtual work, principal of virtual displacements, was being used for the principle of virtual velocities ... by the advocates of the newer methods" [Mohr, 1886, p. 398]. Müller-Breslau replied to this correct objection in 1892 by distinguishing symbolically the actual displacement condition and the causal force condition from the theoretical force condition [Müller-Breslau, 1892, pp. 9–11]. The principle of virtual forces was therefore granted an existence for the first time independent of that of the principle of virtual displacements on the level of small displacements, not in name but in the equation apparatus of classical theory of structures.

**The force method**

As the principle of virtual forces began to take shape, Müller-Breslau achieved a consistent formulation of the force method for statically indeterminate trussed frameworks [Müller-Breslau, 1892, pp. 35, 36] – and one year later for trusses as well (Fig. 6-30):

"A particularly clear determination of the statically indeterminate variables, which will now be designated $X_a$, $X_b$, $X_c$, ..., is achieved by considering $X$ initially as one of the loads acting on the statically determinate primary system and expressing the displacements $\delta_a$, $\delta_b$, $\delta_c$, ... of the loads $X_a$, $X_b$, $X_c$, ... as follows:

$$\delta_a = \sum P_m \delta_{am} - X_a \delta_{aa} - X_b \delta_{ab} - X_c \delta_{ac} - \ldots + \delta_{at} + \delta_{aw}$$

$$\delta_b = \sum P_m \delta_{bm} - X_a \delta_{ba} - X_b \delta_{bb} - X_c \delta_{bc} - \ldots + \delta_{bt} + \delta_{bw}$$

$$\delta_c = \sum P_m \delta_{cm} - X_a \delta_{ca} - X_b \delta_{cb} - X_c \delta_{cc} - \ldots + \delta_{ct} + \delta_{cw}$$

where:
$\delta_{am}$ is the influence of the cause $P_m = 1$ on the displacement $\delta_a$
$\delta_{aa}$ is the influence of the cause $X_a = -1$ on the displacement $\delta_a$
$\delta_{ab}$ is the influence of the cause $X_b = -1$ on the displacement $\delta_a$
$\delta_{at}$ is the same for the influence of thermal effects
$\delta_{aw}$ is the same for the influence of displacements of the support points for
the primary system, and the other $\delta$ values with double suffixes can be interpreted similarly.
However, as according to Maxwell's theorem the letters of the double suffixes can be swapped, then $\delta_{am} = \delta_{ma}$, $\delta_{bm} = \delta_{mb}$, ... and the coefficients of the loads $P_m$ are fully determined by the displacement conditions corresponding to the causes $X_a = -1$, $X_b = -1$, ... . [The elasticity equations] lead to the unknowns $X$ after the values $\delta_a$, $\delta_b$, $\delta_c$, ... are subjected to certain conditions. If, for example, $a$ is the point of application of a support reaction $X_a$, and if this abutment is rigid, then $\delta_a = 0$" [Müller-Breslau, 1893, pp. 188–189].

Müller-Breslau determines the deflection curve $\delta_{ma}$, $\delta_{mb}$, $\delta_{mc}$, ... and the coefficient $\delta_{ik}$ for the system matrix with the help of Mohr's analogy; it was not until the third edition (1904) that he specified the equation obtained directly from the principle of virtual forces [Müller-Breslau, 1904, p. 205]

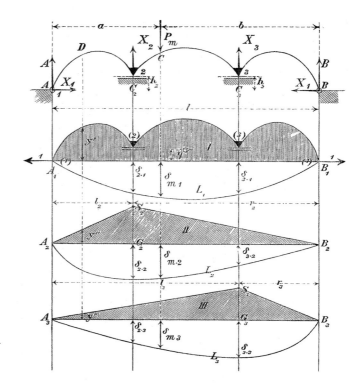

**FIGURE 6-30**
Statically indeterminate calculation according to the force method [Müller-Breslau, 1893, p. 193]

$$\delta_{ik} = \int \frac{N_i \cdot N_k}{E \cdot A} \cdot dx + \int \frac{M_i \cdot M_k}{E \cdot I} \cdot dx + \sum_{j=1}^{s} \frac{S_i \cdot S_k}{E \cdot A} \cdot l_j \qquad (6\text{-}37)$$

for calculating the elements of the system matrix. The force method was therefore the first main method for calculating statically indeterminate trusses to experience a classical characteristic.

### 6.3.3

**Loadbearing structure as kinematic machine**

The quantitative determination of the influence of travelling loads on the internal force variables of certain cross-sections in bridge structures – in particular iron trussed framework bridges for railways – advanced during the classical phase (1875–1900) of the discipline-formation period of structural theory to become the theory of influence lines. This is where the kinematic machine concept of Franz Reuleaux (1829–1905) explained in his *Theoretische Kinematik*, which was incorporated in Mohr's trussed framework theory (1874/1875), gained ground during the 1880s with respect to Maxwell's trussed framework theory, based on Clapeyron's energy-based machine concept (1864), in the form of the kinematic trussed framework and beam theory of Robert Land. Therefore, the graphical examination, also in the quantification of the influence of travelling loads on force and displacement variables in statically determinate trusses, could quickly attain the rank of a highly efficient intellectual aid for civil and structural engineers.

**Trussed framework as machine**

#### 6.3.3.1

Winkler was the main force behind turning the mathematics- and natural science-oriented elastic theory into a practical elastic theory, thus creating the conditions for the elastic theory foundation underpinning the entire theory of structures. Winkler was the first German-speaking engineering scientist who helped Castigliano's main work, *Théorie de l'Équilibre des Systèmes Élastiques et ses Applications* [Castigliano, 1879] become known in the German-speaking world. In that work, Carlo Alberto Pio Castigliano (1847–84) uses the energy principle to synthesise statics and elastic theory to form an energy-based theory of structures (see section 6.4.2). The steam locomotive carried the industrial revolution to the farthest-flung corners of Europe. The emergence of the energy doctrine in theory of structures (see section 6.1.3) was nothing more than the projection of the physical steam engine onto the engineering science model of the trussed framework by James Clerk Maxwell (1831–79) in 1864 – the energy-based machine model of trussed framework theory (Fig. 6-27).

Maxwell considers the trussed framework as a machine with a degree of efficiency of 1, where the force 1 overcomes an elastic resistance $S_i$ (Fig. 6-27b); he writes: "… the frame may be regarded as a machine whose efficiency is perfect" [Maxwell, 1864/2, pp. 295, 296]. The trussed framework modelled as an energy-based machine converts the external work $W_a$ according to the energy conservation principle without losses into the deformation energy $\Pi$.

Otto Mohr (1835–1918), on the other hand, developed a kinematic machine model of the trussed framework in 1874 (Fig. 6-28). The difference between the energy-based and the kinematic machine models of the trussed framework is merely that Maxwell starts with the internal virtual work and the energy conservation law, and Mohr with the external virtual work and the general work theorem.

The solution to the basic task of trussed framework theory can also be found within the scope of the energy-based machine model of trussed framework theory according to Castigliano's second theorem: the required displacement variable $\delta_F$ results from the first partial derivation of the total deformation energy of the trussed framework for $P$ expressed as a function of the external force $P = 1$ (Fig. 6-27b).

The dispute between Heinrich Müller-Breslau and Otto Mohr, which raged from 1883 to 1886, concerned the question of whether the theorems of Castigliano and Maxwell on the one side and Mohr's work theorem on the other were equivalent for the foundation of classical theory of structures (see section 6.4.2).

### 6.3.3.2 The theoretical kinematics of Reuleaux and the Dresden school of kinematics

Franz Reuleaux' *Theoretische Kinematik*, published in 1875, did the same for mechanical engineering as Karl Culmann's *Graphische Statik* did for theory of structures. Both are indebted to the method ideology of classical natural sciences insofar as they achieve their theorems via deductive means. However, whereas Culmann used the language of drawing, Reuleaux used the language of symbols – a kinematic sign language – with the intention of rationalising the kinematic analysis and synthesis of machines. Both Culmann's graphical methods and Reuleaux' kinematic sign language claim to be both a *tekhnē* of the engineer and an *epistēmē* of the engineering scientist. Whereas the graphical statics of Culmann were aimed at a continuation of their basis, the objective of Reuleaux is to concentrate on a machine design heuristic. The machine concept developed by Reuleaux advanced to become the *epistēmē* of the theory of elastic trusses.

Reuleaux illustrates his machine concept using a four-bar linkage mounted on a sturdy pedestal (Fig. 6-31). He calls the cylindrical pin $b$ and the sleeve $a$ that surrounds this a pair of elements. All pairs of elements $ab$, $cd$, $ef$ and $gh$ are joined by bars and together form an inevitably closed kinematic series (mechanism, transmission).

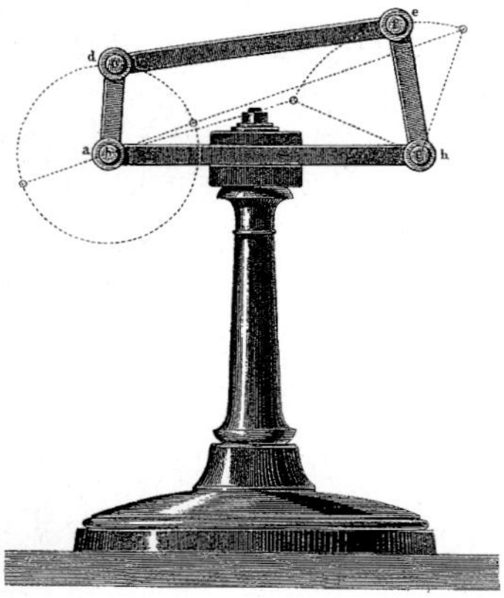

If, for example, the position of the bar between the pairs of elements $ab$ and $cd$ changes, then this movement inevitably leads to certain changes in the positions of the two other bars; the mechanism is kinematic to one degree: "A kinematic transmission or machine," writes Reuleaux, "is set in motion when a mechanical force able to change the position of one of its movable elements is applied to said element. In doing so, the force performs mechanical work, which takes the form of certain movements; the whole is thus a machine" [Reuleaux, 1875, pp. 53–54]. The cause of this movement, be it now brought about by the piston in a steam engine or a hand-operated crank handle, is not the object of Reuleaux' *Theoretische Kinematik*; he is concerned with the transmission of the movement which lies between the source of the force causing the movement and the location of the effect of the tool on the work object. His kinematic machine concept reduces the "developed machinery" [Marx, 1979, p. 393] consisting of prime mover, transmission mechanism and machine tool to its middle part: the transmission mechanism. His kinematic doctrine could therefore not establish itself in machine science. Some years later, Reuleaux' radical abstraction from the source of the force causing the movement became the prerequisite for an alternative to the theory of elastic trusses based on the energy concept – an alternative incorporating a model world in which only external mechanical work exists and the energy concept remained excluded. Reuleaux' *Theoretische Kinematik* had already been published beforehand from 1871 onwards in the proceedings of the Society for the Promotion of Commercial Enterprise in Prussia.

Otto Mohr switched from Stuttgart Polytechnic in 1873 to succeed Claus Köpcke (1831–1911) in the chair for railways, waterways and graphical analysis at Dresden Polytechnic. Why did Mohr include the kinematic machine model in his trussed framework theory of 1874/75? Three observations on this point:

*First observation:* The initially classification-oriented mechanical engineering, which had been freeing itself from the shadows of descriptive geometry since the first decades of the 19th century and which André-Marie Ampère (1775–1836) had raised to the level of a scientific discipline and called it kinematics in 1830, did not find an adequate object until after 1850 as developed machinery appeared. The transmission mechanism had to integrate the machine worlds of prime movers and machine tools, which had already developed relatively independently,

**FIGURE 6-31**
Sturdily mounted four-bar linkage
[Reuleaux, 1875, p. 51]

to form the technical heart of large industry – the developed machinery. Reuleaux' kinematic doctrine is nothing other than the view through the window of the transmission mechanism to the prime mover on the one side and to the machine tool on the other. The trussed framework bridge is also a transmission mechanism. Its bars and joints change their positions owing to the loads acting and transfer them in the form of support forces into the ground. Looked at from this transmission viewpoint, the trussed framework can be regarded as a machine tool whose working parts act on the ground; the trussed framework can also be seen through the window of the transmission mechanism as a prime mover that performs only external work. Both perspectives are indivisible and form the heart of the kinematic doctrine of Mohr's trussed framework theory, which was first developed by Robert Land (1857–99) to create the kinematic theory of trusses. As the loaded trussed framework bridge can be interpreted as developed machinery, it embodies yet another cognition perspective: that of the prime mover or – totally alien to Mohr – the energy doctrine of structural theory. The dispute between Mohr and Müller-Breslau about the foundation of theory of structures thus turned out to be one element in the argument over methods [Braun, 1977] in the fundamental engineering science disciplines in the last quarter of the 19th century.

*Second observation:* In 1874 Trajan Rittershaus (1843–99) became professor of pure and applied kinematics and subjects related to mechanical engineering at Dresden Polytechnic. Rittershaus had worked previously as a private lecturer at the Trade & Industry Academy in Berlin and at the same time as an assistant to Reuleaux [Sonnemann, 1978, p. 72]. In the year after his professorial appointment, he published a paper with the title *Zur heutigen Schule der Kinematik* (on the current school of kinematics) in the journal *Zivilingenieur*, an article that agreed with the principle but disagreed with certain details of Reuleaux' *Theoretische Kinematik*; however, he did not touch the latter's machine concept [Rittershaus, 1875]. But the paper did include a proof of the inevitability of the movement of the four-bar linkage (see Fig. 6-31) he felt was missing in Reuleaux' account. Land was later to rationalise this proof graphically with the help of the pole diagram within the scope of his trussed framework theory (Fig. 6-32).

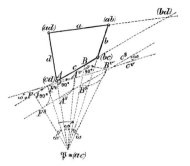

**FIGURE 6-32**
Pole diagram of the four-bar linkage after Land [Land, 1887/1, p. 367]

Proofs of the inevitability of the movement of simple and compound kinematic series soon became the object of numerous publications – especially in the journal *Zivilingenieur*, where both Rittershaus and Mohr were active. Beginning in 1879, Mohr published numerous articles on the kinematics and kinetics of transmissions. The construction of criteria for the movement of kinematic series was able to become the model for the construction of stability criteria for trussed frameworks in the early 1880s. That initiated the prelude to the kinematic trussed framework theory which Land soon extended to become the kinematic theory of trusses.

*Third observation:* The year 1872 saw Ludwig Burmester (1840–1927) appointed to the third mathematics chair of the mathematics/natural science department of Dresden Polytechnic, where he had previously worked as a private lecturer with a permit to teach mathematics and descriptive geometry. Burmester also taught projective geometry from 1875 onwards, and from 1879 onwards he lectured in kinematics [Sonnemann, 1978, p. 81]. Soon afterwards, his paper *Über die momentane Bewegung ebener kinematischer Ketten* (on the momentary movement of plane kinematic series) [Burmester, 1880] appeared in the journal *Zivilingenieur*. This article develops the pole plan from the velocity diagram (already specified by Aronhold) for determining the relative movement of transmission elements with respect to each other. It was this paper that encouraged Martin Grübler in 1883 to publish *Allgemeine Eigenschaften der zwangsläufigen ebenen kinematischen Ketten* (general properties of inevitable plane kinematic series) [Grübler, 1883], also in *Zivilingenieur*; Grübler makes use of Reuleaux' concepts and introduces the term hinge for the pair of elements (cylindrical pin/sleeve). Kinematics was thus positioned alongside trussed framework theory as a neighbouring discipline. Who would open the door? Who would amalgamate kinematics and Mohr's kinematic machine model of trussed framework theory?

Dresden Polytechnic had the triumvirate of kinematics in the shape of Mohr, Rittershaus and Burmester – three men who developed and promoted this discipline from three different perspectives: theory of structures, mechanical engineering and mathematics. It was there that Robert Land took his final examinations in the engineering department (1878 and 1880) and the teaching department (1883). His examiners were none other than Mohr, Fränkel and Burmester.

## 6.3.3.3 Kinematic or energy doctrine in theory of structures?

Whereas the mechanical engineer must fulfil the criteria for the movability of transmissions in practice, the civil or structural engineer would be well advised to ensure the stability of loadbearing systems. Nevertheless, the civil or structural engineer can play the movement game in the model world of structural systems by converting a statically determinate trussed framework into a kinematic series by releasing a bar. After Mohr, Robert Land and Müller-Breslau were masters of this game.

In January 1887 Land published a paper on the reciprocity of elastic deformations as a basis for a general description of influence lines for all types of beam plus a general theory of beams [Land, 1887/2] in the journal *Wochenblatt für Baukunde*. In his six-page work, he formulates the entire basis of classical theory of structures based on one principle. Müller-Breslau, on the other hand, had founded classical theory of structures on the energy doctrine in numerous publications step by step and, from 1886 onwards, had drawn up pragmatic kinematic solutions for the rationalisation of engineering work when determining influence lines in the sense of graphical analysis. Land's brilliant fundamental work was forgotten after 1900, overshadowed by Mohr's 1906 publication *Abhandlungen aus dem Gebiete der Technischen Mechanik* (treatises from the field of applied mechanics) and Müller-Breslau's multi-volume *Graphische Statik der Baukonstruktionen* (graphical statics of structural theory), which was published after 1887 and prevailed until the 1930s. In that work, Müller-Breslau transforms graphical statics rigorously into graphical analysis, finally places classical theory of structures on an energy foundation and makes extensive use of his δ-symbols, which in the 1930s were to become the seed of computer development through Konrad Zuse (1910–95).

In his fundamental work on classical theory of structures, Land derives the principle of interaction from the general work theorem [Land, 1887/2] without knowledge of Betti's publication dating from 1872. Although in his method he follows the trussed framework theory of Mohr, he generalises this into a theory of linear-elastic plane bars. Müller-Breslau had already had the idea of generalising Maxwell's theorem; he had shown its validity for moments and rotations plus force couples and displacement jumps, but the knowledge of the dual nature of theory of structures remained hidden from him. Müller-Breslau, always a pragmatist, had already left the strict theory formation style of the classical scientific ideal. That was not the case with Land.

The general work theorem conceals two souls in one breast: the principle of virtual forces for calculating displacement variables, and the principle of virtual displacements for calculating force variables. Mohr had paid little attention to the latter and Müller-Breslau had instrumentalised it in the eclectic sense. Land includes the principle of virtual displacements in the derivation of his influence lines theorem; he expressed this in the formulation of his kinematic beam theory.

According to Land's theorem, for example, the influence line of the bending moment $M_{j,m}$ is identical to the deflection curve $v_{m,\Delta\alpha=1}$, which comes about because the bending moment at point $j$ is released by a hinge and here the angular difference $\Delta\alpha = 1$ is applied (Figs 2-64 and 6-33). Land had therefore returned the calculation of influence lines for force variables to the solution of the deformation problem.

Land derives his theorem in general terms by speaking of the static causes and effects which he then specifies as the influence line of the bending moment, shear force, axial force, support reaction, etc. Only the principle of virtual forces becomes visible in such a generalised influence lines principle for force variables (solution of the deformation problem), whereas the principle of virtual displacements remains concealed. This is the reason why the principle of virtual displacements was not able to unfold its anticipatory strengths until 50 years later, in the displacement variable method, with which the dual nature of theory of structures was revealed (see section 10.2). So, in determining the influence lines for statically indeterminate systems, Land has to rely on the force method that Müller-Breslau had developed back in 1886 on the basis of Maxwell's theorem (which he had generalised himself), Castigliano's theorems and the principle of virtual forces.

Land's theorem can also be formulated with the help of the internal virtual work $W_i^v$. So, for example, the influence line $M_{j,m}$ of the structural system shown in Fig. 6-33a can also be determined such that the rotation

### Land's theorem

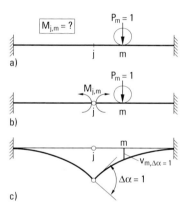

**FIGURE 6-33**
Land's theorem for determining the influence line $M_{j,m}$

jump $\Delta\alpha = 1$ can be applied to the hingeless system (Fig. 6-33a) as a kink at $j$ instead of on the system with hinge at $j$ (Fig. 6-33c). The projection of the resulting deflection curve with kink of magnitude $\Delta\alpha = 1$ at $j$ in the direction of the travelling load $P_m$ is nothing other than the required influence line $M_{j,m}$. However, Land did not pursue the route via internal virtual work at the inconstant point $j$ – he always used the corresponding hinges here, e. g. moment hinge, shear force hinge. Applied kinks, vertical and horizontal displacement jumps were certainly foreign to him because such singularities seemed at first glance to contradict the continuum hypothesis of elastic theory that had finally taken hold in the 1880s.

## Kinematic theory of statically determinate beams

Whereas Land did not initially have the idea of applying his influence line theorem to statically indeterminate systems, he did extend it to create a theory of influence lines, to create the kinematic theory of statically determinate beams [Land, 1888].

A brief description of the method can be found in the aforementioned fundamental work he handed over to Müller-Breslau. Land now responded to two publications of Müller-Breslau, which had appeared in the *Schweizerische Bauzeitung* on 14 May 1887 [Müller-Breslau, 1887/2] and 26 November 1887 [Müller-Breslau, 1887/3], with a paper entitled *Kinematische Theorie der statisch bestimmten Träger* (kinematic theory of statically determinate beams), published on 24 December 1887 in the journal *Schweizerische Bauzeitung* [Land, 1887/3], and shortly afterwards with a very detailed paper with the same title in the January 1888 issue of the *Zeitschrift des Österreichischen Ingenieur- & Architekten-Vereins* [Land, 1888]. In his first paper, Müller-Breslau specifies a method for determining the pole diagram of a kinematic trussed framework based on the work of Burmester [Burmester, 1880], and calculates the member forces according to the principle of virtual displacements; the follow-up article contains the resulting method for determining the influence lines of force variables for statically determinate beams, which to a large extent is the basis of his *Graphische Statik der Baukonstruktionen. Band I* (graphical statics of structural theory, vol. I) published in the same year [Müller-Breslau, 1887/1].

Müller-Breslau was not afraid of expressing the basis of classical theory of structures by way of his own method pluralism, i. e. drawing on energy considerations (Maxwell, Castigliano, principle of virtual forces) for the theory of statically indeterminate systems, but making use of pragmatic kinematic methods for the theory of statically determinate beams. By contrast, Land concentrates on the kinematic theory of statically determinate beams and in doing so ignores the theory of statically indeterminate systems. Land's work represents the kinematic doctrine of classical theory of structures because he uses the concepts of Reuleaux, and therefore introduces pairs of elements (Fig. 6-34) into a theory of structures. He therefore understands the beam – in the condition of the kinematic series for the purpose of calculating influence lines for force variables – as a kinematic machine.

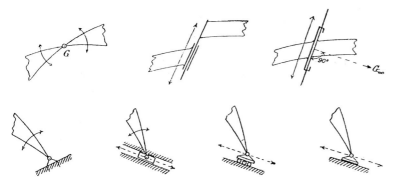

FIGURE 6-34
Pairs of elements after Land
[Land, 1888, p. 12]

In Land's kinematic theory of statically determinate beams, it becomes clear that the principle of virtual displacements is the dual principle in theory of structures alongside the principle of virtual forces. In this respect, Land was way ahead of his influential rival, Müller-Breslau. Concerning the principle of virtual work, he writes: "This principle, originally introduced into the theory of **trussed frameworks** by Mohr through the fundamental work 'Beitrag zur Theorie des Fachwerks' [contribution to the theory of trussed frameworks] [*Zeitschrift des Architekten- & Ingenieur-Vereins zu Hannover*, 1874, p. 512], is used here in order to determine the **kinematic** relationships between $\Delta l$ and displacements $\delta$ of the nodes when the member stresses $S$ are found beforehand via the laws of **statics** [solving the deformation problem with the help of the principle of virtual forces – the author]. In the present treatise the same principle is used in a **reversed** and essentially more general way in order to derive **static** relationships through **kinematic** principles [solving the equilibrium problem with the help of the principle of virtual displacements – the author]" [Land, 1888, p. 17]. Land had thus clearly recognised the dual nature of theory of structures on the level of the theory of statically determinate systems.

His elegant graphical method can be demonstrated splendidly using the example of determining the influence lines for the force variables of a three-pin arch (Fig. 6-35). To determine the influence line of the horizontal thrust $H$ due to a travelling load $P$, the restraint corresponding to the horizontal thrust $H$ must be released; the unit displacement jump 1 is now applied to the resulting kinematic series in the direction of the released restraint. The projection of the displacement figure of the load chord, normally determined using the pole diagram, in the direction of travelling load $P$ is then the required influence line of the horizontal thrust (Fig. 6-35, top). The principle of virtual displacements serves only to determine the dimension of the influence line. The determination of the influence line for the bending moment $M$ at point $Z$ is performed similarly if a moment hinge is introduced here and the angular difference 1 is applied (Fig. 6-35, centre). The influence line of the shear force $V$ at point $Z$ follows from releasing the shear force restraint at $Z$ and applying the vertical displacement jump 1, and in a similar way the influence line of the axial

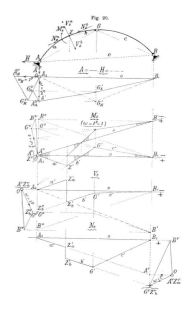

FIGURE 6-35
Graphical determination of the influence lines for the force variables of a three-pin arch [Land, 1888, p. 172]

force *N* at point *Z* follows from releasing the axial force restraint at *Z* and applying the horizontal displacement jump 1 (Fig. 6-35, bottom).

## The Pyrrhic victory of the energy doctrine in theory of structures

### 6.3.3.4

Unfortunately, Land failed to create a method for determining influence lines for the force variables of statically indeterminate systems that could be applied rationally to civil engineering practice, although his influence line principles did establish the foundation for this. The expansion of the displacement method – to complement the force method – would not profit from that until many decades later.

Like Reuleaux' *Theoretische Kinematik* had already lost the battle for the foundation of mechanical engineering as early as 1890, Land was defeated around the same time in his battle against the energy-based foundation of classical theory of structures. The energy doctrine celebrated a successful comeback in the classical fundamental engineering science disciplines of theoretical mechanical engineering and theory of structures.

The hesitant adoption of the anticipatory elements of Land's contributions to the theory of influence lines in the 1930s was connected with the further expansion of the displacement method during the consolidation period of structural theory. And this signalled structural theory's departure from the theory formation style of the Berlin school of structural theory. This history-of-theory evolution process, which culminated in the clear recognition of the dual nature of theory of structures in the matrix-based formulation of the force and displacement methods during the 1950s, flowed into the great torrent of structural mechanics (see section 10.4.3).

## Theory of structures at the transition from the discipline-formation to the consolidation period

## 6.4

While Mohr was disputing the validity of Castigliano's theorems for the foundation of classical theory of structures, Müller-Breslau was completing classical theory of structures by successively expanding the deformation energy expression for elastic trussed frameworks. The energy doctrine was therefore already dominating theory and practice around 1900. Nevertheless, during the first decade of the 20th century, Weingarten and Mehrtens tried to break the dominance of Castigliano's theorems. The accompanying debate was reminiscent of the old dispute between Mohr and Müller-Breslau, which had been considerably affected by priority issues. But by 1910 the debate had been concluded by Weyrauch in favour of the Castigliano theorems in the classical theory of structures.

## Castigliano

### 6.4.1

In his admirable Castigliano (Fig. 6-36) obituary of 1884, Emil Winkler (1835–88), the founder of the Berlin school of structural theory, wrote as follows: "Theory of structures was to a certain extent founded by Italians such as Galileo, Marchetti, Fabri, Grandi, etc; but said theory has made important progress in recent times as a result of the needs brought about by the introduction of railways, and the Italians are again playing a prominent role in this progress. The more recent works of Allievi, Biadego, Canevazzi, Ceradini, Clericetti, Cremona, Favaro, Favero, Figari, Guidi, Jung, Modigliani, Saviotti and Sayno, to name but a few; outstanding among these works are those of Castigliano. If we Germans also wish to claim that our efforts in these areas are also worthy of note, then we must admit that we have learned much from our Italian colleagues and that, regretfully, language barriers still prevent a faster dissemination of their theories" [Winkler, 1884, p. 570].

Together with the papers of Maxwell and Mohr, Castigliano's main work *Théorie de l'Équilibre des Systèmes Élastiques et ses Applications* (Fig. 6-37), which was published in 1879 and did not appear in English until 1919 (*Elastic Stresses in Structures*), constitutes the foundations of classical theory of structures. In this work, Castigliano synthesises statics and elastic theory to form a theory of structures based on the energy principle. It can therefore be said that Castigliano assisted in the birth of the greatest discovery of the 19th century: the law of energy conservation for structural mechanics. His contribution to this synthesis was not only the key to the historico-logical development of structural theory in the final third of the 19th century, but at the same time also the object of dispute in this fundamental engineering science discipline in the 1880s and the first decade of the 20th century – intensifying the theory and pushing it beyond the classical disciplinary boundaries.

Carlo Alberto Pio Castigliano was born on 8 November 1847 in Asti, Italy, the son of Michele Castigliano and Orsola Maria Cerrato. He grew up in deprived conditions and lost his father while still a child. The family's conditions did not even improve after his mother remarried. Notwithstanding, Castigliano's mother and stepfather managed to ensure that he attended school, and as his stepfather lay ill, his mother earned the family's upkeep by selling fruit. Following his excellent results at the primary school in Asti, prosperous citizens provided financial assistance so that he could attend the newly established Istituto Industriale, Sezione di Meccanica e Costruzioni. And from December 1865 to July 1866 he was a student at the Reale Istituto Industriale e Professionale in Turin; however, a shortage of funds stopped him from gaining an engineering diploma. But with the help of a government scholarship, he obtained a distinction in the Diploma di Professore di meccanica examination after just three months. After that, Castigliano worked for three years as a teacher in Turin. During this period he used his meagre salary to help support his needy relatives, but by giving private lessons and translating scientific books, primarily from German and French into Italian, he still managed to save enough to attend university (Reale Università degli Studi Turin). Despite these adversities, Castigliano managed to achieve something no other student of this university had ever achieved before: he shaved two years off the three-year course of study and graduated (with excellent results) after one year with a *diploma di licenza in mathematiche pure*. This was followed by civil engineering studies at the Reale Scuola d'Applicazione degli Ingegneri in Turin. After just two years of study he was awarded a diploma in civil engineering on 30 September 1873 – a course that normally took five years! His diploma dissertation bore the title *Intorno ai sistemi elastici* (on elastic systems). In this work he analyses systems with internal static indetermi-

FIGURE 6-36
Carlo Alberto Pio Castigliano (1847–84) [Oravas, 1966/2]

FIGURE 6-37
Cover of Castigliano's main work [Castigliano, 1879]

nacy but with statically determinate supports on the basis of the theorem of minimum deformation work (deformation energy) $\Pi$, which was soon to be revealed as the

$$\Pi = minimum \qquad (6\text{-}38)$$

principle discovered by Luigi Federigo Menabrea (1809–96) back in 1857. This slight of Menabrea by Castigliano forced the former to publish a *Mémoire* in 1875 at the Accademia dei Lincei in which he claimed his priority. Castigliano reacted within a few months with his 150-page essay entitled *Nuova teoria intorno all'equilibrio dei sistemi elasticiti* (new theory of the equilibrium of elastic systems), which went way beyond Menabrea and formed the heart of his main work, which was to appear in 1879.

Castigliano's main work is founded on three statements concerning deformation energy:

"If we express the internal work of a body or elastic structure as a function of the relative displacements of the points of application of the external forces, we shall obtain an expression whose differential coefficients with regard to these displacements give the values of the corresponding forces," [Castigliano, 1886, p. 42]:

$$\partial\Pi(\ldots, \delta_k, \ldots)/\partial\delta_k = F_k \qquad (6\text{-}39)$$

*Castigliano's first theorem* had already been applied to physical problems by George Green (1793–1841). The inversion of this theorem is a genuine creation of Castigliano and had been formulated by him for the first time in 1873 in his diploma dissertation on elastic systems.

*Castigliano's second theorem* is expressed thus: "If we express the internal work of a body or elastic structure as a function of the external forces, the differential coefficient of this expression, with regard to one of the forces, gives the relative displacement of its point of application" [Castigliano, 1886, p. 42]:

$$\partial\Pi(\ldots, F_j, \ldots)/\partial F_j = \delta_j \qquad (6\text{-}40)$$

*Castigliano's third theorem* can be deduced from eq. 6-40: "The stresses which occur between the molecular couples of a body or structure after strain are such as to render the internal work to a minimum, regard being had to the equations which express equilibrium between these stresses around each molecule" [Castigliano, 1886, p. 47]. This theorem corresponds to the principle of Menabrea (eq. 6-38), to which Castigliano refers explicitly in his introduction, but adds that he gave a rigorous proof of eq. 6-38 in his 1873 diploma dissertation [Castigliano, 1886, p. V].

In the introduction to his main work, Castigliano formulates the claim "that the present book, comprising the complete theory of elastic stresses in structures, … is wholly based on the theorems of the differential coefficients of internal work" [Castigliano, 1886, p. V]. He has thus introduced the energy principle into theory of structures.

In just a few years, Castigliano rose from building site manager to chief of the design office in the Northern Italy Railways Company. He therefore

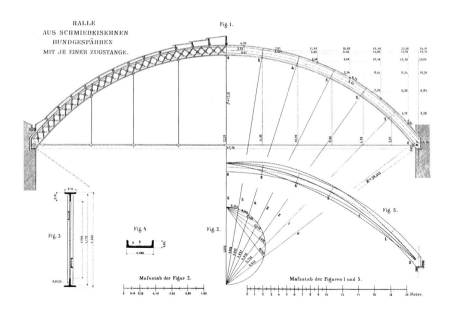

FIGURE 6-38
Arched roof truss with single tie-rod
[Castigliano, 1886, plate III]

saw his goal as "not only to expound a theory, but further to show its advantages of brevity and simplicity in practical applications" [Castigliano, 1886, p. V]. He therefore dedicated the second part of his main work to applications, analysing timber beams trussed with wrought-iron bars, wrought-iron trusses and arch bridges, plus arches made from clay bricks and dressed stones. Fig. 6-38 shows, for example, a wrought-iron arched truss with one degree of static indeterminacy for which Castigliano calculates the internal forces due to the loading cases dead load, dead load + random one-sided loading and constant temperature change with the help of eq. 6-38 and eq. 6-39 in the form

$$\partial \Pi(X_1)/\partial X_1 = 0 \qquad (6\text{-}41)$$

In eq. 6-41, $X_1$ is the statically indeterminate axial force acting in the tie bar. Castigliano calculates thrust lines for the first two loading cases, which are plotted in Fig. 5 of Fig. 6-38. A loadbearing structure of this kind was used for the roof over the railway station in Genoa; however, Castigliano altered the depth of the arched truss as well as the cross-sectional area and the shape of the sections, arguing that "a considerable saving in material might have been effected, and a sufficient factor of safety still obtained" [Castigliano, 1886, p. 354]. Accordingly, the legitimation of theory of structures for practical building, in addition to proving the stability of loadbearing structures, was also their economic design. This union between necessary and sufficient legitimation of structural analysis plays a leading role in the self-image of the structural engineer even today.

Castigliano's output goes well beyond his *Elastic Stresses in Structures*. For example, he wrote about an instrument he had devised for measuring strains, and which he actually used in practice on the Fella a Pontedimura Bridge. After his death, a book on the theory of bending and torsion springs was published posthumously in 1884. In his position as a member of the advisory council to the pension fund of the Alta-Italia railway company (Strade Ferrate Alta Italia), he investigated the reorganisation of this fund based on investment mathematics studies. His mastery of shorthand enabled him to write more effectively. Nonetheless, his *Manuale practico per gli ingegneri* (practical manual for engineers), which would have run to several volumes, remained only an outline. After a long and painful illness, Castigliano died on 25 October at the age of just 37.

### 6.4.2 The foundation of classical theory of structures

It was no less a figure than Emil Winkler who drew his German colleagues' attention to Castigliano's main work *Théorie de l'Équilibre des Systèmes Élastiques et ses Applications*. In 1882, unaware of the relevant work of Menabrea and Castigliano, Wilhelm Fränkel (1841–95) derived the principle of Menabrea for plane frames with $n$ degrees of static indeterminacy and the linear-elastic continuum from the principle of virtual forces [Fränkel, 1882]. Fränkel, too, hoped that his principle would permit "a uniform understanding of a whole series of structural engineering problems" [Fränkel, 1882, p. 63]. In that same year, Matthias Koenen (1848–1924) took up the general work theorem used by Mohr so successfully for trussed framework theory and used it to calculate displacements in statically determinate simply supported beams and the support reactions of a continuous beam in the form of the principle of virtual forces [Koenen, 1882]; in doing so, he formulated for the first time the work equation in the form of the known product integral:

$$1 \cdot \delta_j = \int_{(l)} \frac{M_1 \cdot M_j}{E \cdot I} \cdot dx \qquad (6\text{-}42)$$

The evaluation of the product integral according to eq. 6-42 is still carried out today with the help of integral tables (Fig. 6-39). These integral tables were drawn up by the engineers Richard Schadek von Degenburg and Karl Demel [Schadek v. Degenburg & Demel, 1915], who worked for the Vienna-based Ignaz Gridl bridge-building company (since 1934 part of the Waagner-Biró company). "The method [integral tables – the author]," the authors note in their preface, "supplies the solution without intermediate calculations for simple relationships and, on the other hand, its ease of use enables the solution of otherwise difficult tasks. Its clarity also enables a reliable prior assessment of the aptness of a method of calculation to be employed" [Schadek v. Degenburg & Demel, 1915, p. V]. For example, the integral tables can be used very simply to calculate the elements of the system matrix according to eq. 6-37. The integral tables brought about a significant rationalisation in the calculation of deformations in general and the force method in particular.

But back to Koenen, Castigliano and Müller-Breslau. Inspired by Koenen's publication and Castigliano's main work, Heinrich Müller-Breslau (1851–1925) lectured on Mohr's method extended to beam-type structures and the principle of Menabrea using the example of some statically indeterminate systems frequently found in practice [Müller-

**Werte der Integrale**  $\int M_i M_k \, dx = l \cdot$ (Tafelwert)

| Nr. | $\overleftrightarrow{l}$ | $k \boxed{\phantom{xxx}} k$ | $\boxed{\phantom{xxx}} k$ | $k_1 \boxed{\phantom{xxx}} k_2$ | $k \boxed{\phantom{xxx}}{-k}$ | $k \boxed{\phantom{xxx}}{-\frac{k}{2}}$ | $\boxed{\leftarrow \alpha l \rightarrow \beta l \rightarrow}$ | $\int j^2 \, dx$ |
|---|---|---|---|---|---|---|---|---|
| 1 | $j \boxed{\phantom{xxx}} j$ | $jk$ | $\frac{1}{2}jk$ | $\frac{1}{2}j(k_1+k_2)$ | $0$ | $\frac{1}{4}jk$ | $\frac{1}{2}jk$ | $j^2$ |
| 2 | $\boxed{\phantom{xxx}} j$ | $\frac{1}{2}jk$ | $\frac{1}{3}jk$ | $\frac{1}{6}j(k_1+2k_2)$ | $-\frac{1}{6}jk$ | $0$ | $\frac{1}{6}jk(1+\alpha)$ | $\frac{1}{3}j^2$ |
| 3 | $j \boxed{\phantom{xxx}}$ | $\frac{1}{2}jk$ | $\frac{1}{6}jk$ | $\frac{1}{6}j(2k_1+k_2)$ | $\frac{1}{6}jk$ | $\frac{1}{4}jk$ | $\frac{1}{6}jk(1+\beta)$ | $\frac{1}{3}j^2$ |
| 4 | $j_1 \boxed{\phantom{xxx}} j_2$ | $\frac{1}{2}k(j_1+j_2)$ | $\frac{1}{6}k(j_1+2j_2)$ | $\frac{1}{6}[j_1(2k_1+k_2) + j_2(k_1+2k_2)]$ | $\frac{1}{6}k(j_1-j_2)$ | $\frac{1}{4}j_1k$ | $\frac{1}{6}k[j_1(1+\beta) + j_2(1+\alpha)]$ | $\frac{1}{3}(j_1^2 + j_1 j_2 + j_2^2)$ |
| 5 | $j \boxed{\phantom{xxx}}{-j}$ | $0$ | $-\frac{1}{6}jk$ | $\frac{1}{6}j(k_1-k_2)$ | $\frac{1}{3}jk$ | $\frac{1}{4}jk$ | $\frac{1}{6}jk(1-2\alpha)$ | $\frac{1}{3}j^2$ |
| 6 | $j \boxed{\phantom{xxx}}{-\frac{j}{2}}$ | $\frac{1}{4}jk$ | $0$ | $\frac{1}{4}jk_1$ | $\frac{1}{4}jk$ | $\frac{1}{4}jk$ | $\frac{1}{4}jk\beta$ | $\frac{1}{4}j^2$ |
| 7 | $-\frac{j}{2}\boxed{\phantom{xxx}} j$ | $\frac{1}{4}jk$ | $\frac{1}{4}jk$ | $\frac{1}{4}jk_2$ | $-\frac{1}{4}jk$ | $-\frac{1}{8}jk$ | $\frac{1}{4}jk\alpha$ | $\frac{1}{4}j^2$ |
| 8 | $\boxed{\phantom{xxx}}$ (j) | $\frac{1}{2}jk$ | $\frac{1}{4}jk$ | $\frac{1}{4}j(k_1+k_2)$ | $0$ | $\frac{1}{8}jk$ | $\frac{jk}{12\beta}(3-4\alpha^2)$ | $\frac{1}{3}j^2$ |
| 9 | $\leftarrow \gamma l \rightarrow \delta l \rightarrow$ $j$ | $\frac{1}{2}jk$ | $\frac{1}{6}jk(1+\gamma)$ | $\frac{1}{6}j[k_1(1+\delta) + k_2(1+\gamma)]$ | $\frac{1}{6}jk(1-2\gamma)$ | $\frac{1}{4}jk\delta$ | $\frac{jk}{6\beta\gamma}(2\gamma-\gamma^2-\alpha^2)$ $\gamma \geq \alpha$ | $\frac{1}{3}j^2$ |
| 10 | $j \boxed{\phantom{xxx}}$ Quadr. Parabel | $\frac{2}{3}jk$ | $\frac{1}{3}jk$ | $\frac{1}{3}j(k_1+k_2)$ | $0$ | $\frac{1}{6}jk$ | $\frac{1}{3}jk(1+\alpha\beta)$ | $\frac{8}{15}j^2$ |
| 11 | $j \boxed{\phantom{xxx}} j$ Quadr. Parabel | $\frac{1}{3}jk$ | $\frac{1}{6}jk$ | $\frac{1}{6}j(k_1+k_2)$ | $0$ | $\frac{1}{12}jk$ | $\frac{1}{6}jk(1-2\alpha\beta)$ | $\frac{1}{5}j^2$ |

**FIGURE 6-39**
Integrals table
[Duddeck & Ahrens, 1998, p. 374]

Breslau, 1883/2]. In his introduction, Müller-Breslau remarks that "Mohr's method [general work theorem in the form of the principle of virtual forces – the author] … is equivalent to the 'principle of least deformation work' [principle of Menabrea – the author] established recently by Castigliano" [Müller-Breslau, 1883/2, p. 87]. It was this remark that initiated the controversy between Mohr and Müller-Breslau regarding the foundation of classical theory of structures. The dispute encouraged Müller-Breslau to complete classical theory of structures in the form of his 1886 monograph *Die neueren Methoden der Festigkeitslehre* (the newer methods of strength of materials) [Müller-Breslau, 1886], the second edition of which (1893) he dedicated to Castigliano, as well as the first part of his *Graphische Statik der Baukonstruktionen* (graphical statics of structural theory) published in 1887 [Müller-Breslau, 1887/1].

The dispute about the foundation of classical theory of structures had already been decided in favour of Müller-Breslau at the beginning of the 1890s because the theorems of Castigliano were regarded by the majority of structural engineers with a scientific outlook and by mathematicians [Klein, 1889] as equivalent to the general work theorem introduced by Mohr into trussed framework theory in the form of the principle of virtual forces. In addition, Friedrich Engesser (1848–1931) was able to show in 1889 that thermal load cases and non-linear-elastic trusses could also be analysed with these theorems [Engesser, 1889]. One essential discovery of Engesser was distinguishing deformation energy $\Pi$ from deformation complementary energy $\Pi^*$ ($\Pi^*$ must always be used in the theorem of Castigliano). This conceptual differentiation is not necessary with linear-elastic systems because $\Pi = \Pi^*$ applies (see Fig. 6-5b).

Mohr's criticism was aimed at halting Müller-Breslau's expansion of the concept of deformation energy [Müller-Breslau, 1884]. The latter specified the following formula for the total deformation energy of a trussed framework with $m$ bars, $l$ applied displacements and $q$ support displacements:

$$\Pi_{FW} = \sum_{i=1}^{m} \frac{1}{2} \cdot \left( \frac{S_i^2 \cdot l_i}{E \cdot A_i} \right) + \sum_{i=1}^{m} \alpha_i \cdot T_i \cdot l_i + \sum_{p=1}^{l} \frac{1}{2} \cdot \left( \frac{R_p^2 \cdot s_p}{E \cdot A_p} \right) + \sum_{k=1}^{q} C_k \cdot \delta c_k \quad (6\text{-}43)$$

where
- the first sum represents the energy component from the elastic deformations of the truss bars as a result of the member forces $S_i$,
- the second sum represents the energy component from the temperature differences $T_i$,
- the third sum represents the energy component from $m$ applied displacements simulated by the elastic deformations of $m$ fictitious bars, and
- the fourth sum represents the energy component from $q$ support displacements.

In the contest between the general work theorem in Mohr's version (principle of virtual forces) and the theorem of Castigliano for the foundation of trussed framework theory, eq. 6-43 from Müller-Breslau ensured the equivalence of the two methods. Nevertheless, Mohr's method was more rational; Müller-Breslau had to extend the trussed framework model by $l$ fictitious bars in order to accommodate applied displacements. Mohr spoke out against eq. 6-43 in 1886, calling it, ironically, the "elasticity of deformation work" [Mohr, 1886], seeing four different types of deformation work within, namely
- deformation work No. 1: the first sum in eq. 6-43,
- deformation work No. 2: the first plus second sums in eq. 6-43, designated by Müller-Breslau as "ideal deformation work" [Müller-Breslau, 1884],
- deformation work No. 3: the first plus third sums in eq. 6-43, and
- deformation work No. 4: the first plus fourth sums in eq. 6-43.

Mohr regarded this as confusing the concepts. He was suspicious of the addition of the energy terms. This is where the contrast between the kinematic and the energy doctrines in theory of structures comes to light (see section 6.3.3.1). Added to this contrast is a completely different way of abstracting the loadbearing system to the structural system. Whereas in the end Mohr can imagine the solid-web beam (Fig. 6-40a) only as a trussed framework with a great number of bars (Fig. 6-40c), Müller-Breslau and the other advocates of the "newer methods of strength of materials" work with the practical bending theory (Fig. 6-40b). For Mohr, the linear-elastic trussed framework theory consequently becomes the model for the theory of linear-elastic trusses, as the main object of classical theory of structures.

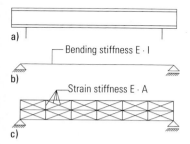

FIGURE 6-40
Beam models: a) solid-web beam, b) bar in bending, c) internally statically indeterminate trussed framework

Therefore, according to Mohr, internal forces must ensue in the externally statically determinate solid-web beam in the case of temperature changes because it contains an internally statically indeterminate trussed framework. On the other hand, the solid-web beam modelled as a simply supported beam in bending could convert the temperature loading cases $T$ (constant temperature change over the beam cross-section) and $\Delta T$ (linear temperature change over the beam cross-section) into deformations without restraint. The expression for the deformation energy in trusses based on Castigliano and extended by Müller-Breslau then takes on the following form:

$$\Pi_B = \frac{1}{2}\int_{(l)} \frac{M^2}{E \cdot I} \cdot dx + \frac{1}{2}\int_{(l)} \frac{Q^2}{G \cdot A_Q} \cdot dx + \frac{1}{2}\int_{(l)} \frac{N^2}{E \cdot A} \cdot dx + \qquad (6\text{-}44)$$

$$+ \frac{1}{2}\int_{(l)} \frac{M_T^2}{G \cdot I_T} \cdot dx + \int_{(l)} \varepsilon \cdot \Delta T \cdot \frac{M}{h} \cdot dx + \int \varepsilon \cdot T \cdot N \cdot dx$$

In his book *Die neueren Methoden der Festigkeitslehre*, Müller-Breslau specifies, as he calls it, "the ideal deformation work" for composite structural systems as follows:

$$\Pi_{FW+B} = \Pi_{FW} + \Pi_B \qquad (6\text{-}45)$$

For the application of eq. 6-45 to concrete cases, Müller-Breslau ignores the corresponding terms in eq. 6-43 and eq. 6-44; he of course realises that eq. 6-44, too, can be extended by energy terms from applied displacements. If now the superposition equations for internal variables $M$, $Q$, $N$, etc. are used for an elastic truss with $n$ degrees of static indeterminacy and we differentiate according to the static indeterminates $X_j$ ($j = 1 \ldots n$) according to eq. 6-41, then we obtain $n$ elasticity equations for determining the $n$ static indeterminates. Müller-Breslau had thus integrated the theorems of Castigliano into the form of the force method he had developed (see section 6.3.2.2).

### 6.4.3 The dispute about the fundamentals of classical theory of structures is resumed

The genesis of the controversy surrounding the foundation of classical theory of structures can be broken down into two clearly separate time periods. The first of those was the conclusion of the classical phase of the discipline-formation period of structural theory in the 1880s (see section 6.4.2); the other was the first decade of the 20th century, as the body of classical theory of structures was showing the first signs of consolidation. Some 15 years lay between these two, a time-span in which the two main opponents, Mohr and Müller-Breslau, did not publish anything. Whereas the first phase of this controversy (1883–89) was concerned with the foundation of classical theory of structures, the second phase was also characterised by a priority dispute.

#### 6.4.3.1 The cause

The formal structure of the second phase of the controversy is much more complex than that of the first phase. It started with two separate disputes: that between Müller-Breslau and Georg Christoph Mehrtens (1843–1917)

(Fig. 6-41) on the one hand (1901–03), and that between Müller-Breslau and Mohr on the other (1902–03). This was to become entwined with Mehrtens' comprehensively formulated priority claim in favour of Mohr in 1905, which occasioned Müller-Breslau to formulate a comprehensive reply in 1906 and to refer back to the object of the dispute. During this period, the professor of mathematics Julius Weingarten (1836–1910) wrote three papers between 1901 and 1904 criticising the deformation energy concept and in this context hence the validity of the theorems of Castigliano. August Hertwig (1872–1955) summarised the debates in 1906 [Hertwig, 1906] and triggered a polemic between Weingarten and Johann Jakob Weyrauch (1845–1917) concerning the deformation energy concept, which continued for many years. The reason behind this extensive paper was Mehrtens' three-volume work *Vorlesungen über Statik der Baukonstruktion und Festigkeitslehre* (lectures on theory of structures and strength of materials) completed in 1905 [Mehrtens, 1903–05] plus Mohr's *Abhandlungen aus dem Gebiet der technischen Mechanik* (treatises from the field of applied mechanics) dating from 1906 [Mohr, 1906].

In three chapters Hertwig deals critically with
- priority issues concerning the kinematic theory of trusses, the theory of influence lines and the equivalent member method,
- the differences in the theory of statically indeterminate systems by Maxwell, Mohr and Castigliano, and
- Mehrtens' presentation of the theory of the elastic arch and the theory of secondary stresses in trussed frameworks.

Finally, he expresses the hope of "contributing to clarifying the aforementioned priority disputes and promoting the universal acceptance of the Castigliano-Engesser theorems" [Hertwig, 1906, p. 516].

**FIGURE 6-41**
Georg Christoph Mehrtens (1843–1917)
(university archives, Dresden TU)

**The dispute between the deputies**

#### 6.4.3.2

In terms of published scientific information, the dispute surrounded the different presentation of the fundamentals of classical theory of structures in
- Müller-Breslau's three-volume main work *Die graphische Statik der Baukonstruktionen* [Müller-Breslau, 1887/1 & 1892] (Fig. 6-42), and
- Mehrtens' three-volume *Vorlesungen über Statik der Baukonstruktionen und Festigkeitslehre* [Mehrtens, 1903–05] plus Mohr's *Abhandlungen aus dem Gebiet der technischen Mechanik* [Mohr, 1906] (Fig. 6-43).

The works by Müller-Breslau and Mehrtens were the first rival theory of structures textbooks in the classical phase (1875–1900). For the first time in the civil engineering literature, Mehrtens' goes beyond the customary listing of the sources used by referring to these in the text in such a way that the reader can himself deduce the sources used by the author. Contrasting with this, Müller-Breslau and Mohr merely added commented bibliographies to their monographs. Mohr's book essentially amounts to a summary of his published papers in 12 treatises, some of which go beyond the object of classical theory of structures. One of the questions dealt with in the aforementioned monographs is whether the contributions of Maxwell and Castigliano, based on the deformation energy concept,

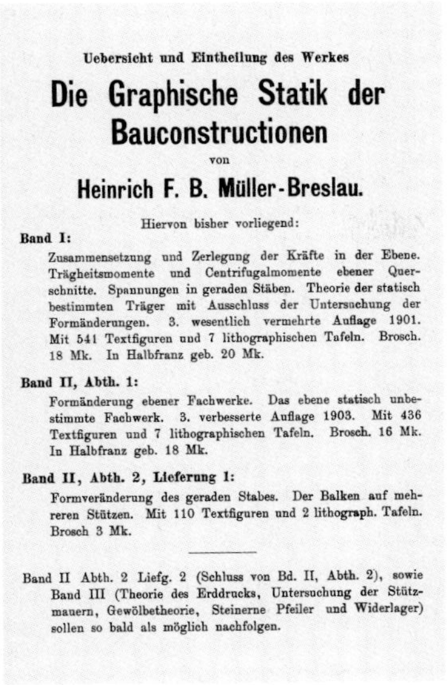

Uebersicht und Eintheilung des Werkes

**Die Graphische Statik der Bauconstructionen**

von

**Heinrich F. B. Müller-Breslau.**

Hiervon bisher vorliegend:

**Band I:**
Zusammensetzung und Zerlegung der Kräfte in der Ebene. Trägheitsmomente und Centrifugalmomente ebener Querschnitte. Spannungen in geraden Stäben. Theorie der statisch bestimmten Träger mit Ausschluss der Untersuchung der Formänderungen. 3. wesentlich vermehrte Auflage 1901. Mit 541 Textfiguren und 7 lithographischen Tafeln. Brosch. 18 Mk. In Halbfranz geb. 20 Mk.

**Band II, Abth. 1:**
Formänderung ebener Fachwerke. Das ebene statisch unbestimmte Fachwerk. 3. verbesserte Auflage 1903. Mit 436 Textfiguren und 7 lithographischen Tafeln. Brosch. 16 Mk. In Halbfranz geb. 18 Mk.

**Band II, Abth. 2, Lieferung 1:**
Formveränderung des geraden Stabes. Der Balken auf mehreren Stützen. Mit 110 Textfiguren und 2 lithograph. Tafeln. Brosch 3 Mk.

Band II Abth. 2 Liefg. 2 (Schluss von Bd. II, Abth. 2), sowie Band III (Theorie des Erddrucks, Untersuchung der Stützmauern, Gewölbetheorie, Steinerne Pfeiler und Widerlager) sollen so bald als möglich nachfolgen.

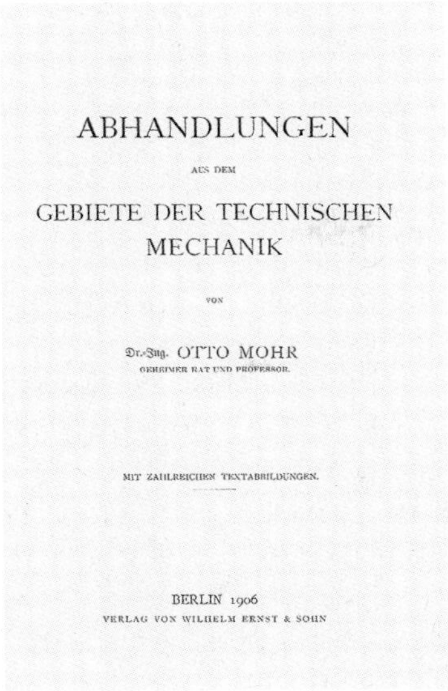

ABHANDLUNGEN

AUS DEM

GEBIETE DER TECHNISCHEN MECHANIK

VON

Dr.-Ing. OTTO MOHR

GEHEIMER RAT UND PROFESSOR.

MIT ZAHLREICHEN TEXTABBILDUNGEN.

BERLIN 1906

VERLAG VON WILHELM ERNST & SOHN

**FIGURE 6-42**
Publisher's advertisement for Müller-Breslau's *Graphische Statik der Baukonstruktionen* [Müller-Breslau, 1903]

**FIGURE 6-43**
Cover of Mohr's *Abhandlungen aus dem Gebiete der Technischen Mechanik*

permit a development of the theory of statically indeterminate systems that is equivalent to the general work theorem introduced into theory of structures by Mohr. Whereas this was accepted by Müller-Breslau, the main representative of the Berlin school of theory of structures, and his former scientific assistant Hertwig, who in 1902 was appointed professor of theory of structures and iron construction at Aachen TH and in 1924 became Müller-Breslau's successor at Berlin TH, plus other representatives of the energy doctrine in theory of structures, e. g. Weyrauch, this was vehemently opposed by Mohr and Mehrtens. Since 1895 Mehrtens had been giving lectures in theory of structures and bridge-building at Dresden TH, and he was also responsible for strength of materials after Mohr's departure in 1900.

### 6.4.3.3 The dispute surrounding the validity of the theorems of Castigliano

Mohr and Mehrtens were indirectly supported by a series of articles (1901 – 05) by Weingarten [Weingarten, 1901, 1902 & 1905]. The dispute between Müller-Breslau and Mohr regarding the validity of the theorems of Castigliano as a foundation for classical theory of structures shifted after 1906 to a dispute between their "seconds": Hertwig versus Mehrtens and Weingarten; Weyrauch versus Weingarten.

**Weingarten versus Föppl**

Weingarten (Fig. 6-44) used the perspective of the criticism of the technicising of mechanics and its development into applied mechanics as the starting point for criticising the theorems of Castigliano developed by August Föppl (Fig. 6-45) in his *Vorlesungen über Technische Mechanik* (lectures in applied mechanics) [Föppl, 1897]. In essence, Weingarten's criticism is directed against the concept of deformation work (deformation energy), which he specifies with

$$\Pi = \frac{1}{2} \sum_i R_i \cdot r_i \qquad (6\text{-}46)$$

Here, $r_i$ is the projection of the displacement difference between the start and end positions of the *ith* point of application onto the direction of force $R_i$; $\Pi$ is determined by the end position of the structural system only, but not by the individual terms of eq. 6-46. The latter is not taken into account in the second theorem of Castigliano (eq. 6-40). "This theorem has therefore no precise content" [Weingarten, 1901, p. 344]. The partial differential quotients $\partial \Pi / \partial R_i$ have certain values, whereas the $r_i$ values to be shown have an infinite number of solutions depending on the initial position of the deformed body [Weingarten, 1901, p. 344]. Further, as all $R_i$ values must satisfy the equilibrium conditions, a certain $R_i$ cannot be changed without "changing some or all of the others at the same time. To speak of a differential of deformation work, formed by changing just one force, i.e. to speak of a partial differential of deformation work, is nonsense" [Weingarten, 1901, p. 344].

According to Weingarten, $\Pi$ as a function of the external forces, which remain independent of each other due to the equilibrium conditions, can therefore be attributed various forms depending on the selection of these forces. "The theorem of Castigliano is therefore without content and ambiguous even in the case of precisely defined content. This means that all the conclusions drawn from this alone are actually invalid" [Weingarten, 1901, p. 345]. Weingarten develops his criticism by using a calculation of Föppl (Fig. 6-41a) as an example. After Föppl has derived the support reactions

FIGURE 6-44
Julius Weingarten (1836–1910)
[Knobloch, 2000, p. 395]

$$B = \frac{q \cdot (a+b)}{2} - \frac{b}{a+b} \cdot Z \qquad (6\text{-}47)$$

and

$$C = \frac{q \cdot (a+b)}{2} - \frac{a}{a+b} \cdot Z \qquad (6\text{-}48)$$

from the equilibrium conditions as a function of the required support reaction $Z$ and has presented the two bending moment diagrams $M_I$ and $M_{II}$ depending on $B$ and $C$, he enters the functions found in this way into the equation for the deformation energy

$$\Pi = \frac{1}{2} \int_0^a \frac{M_I^2}{E \cdot I} \cdot dx + \frac{1}{2} \int_a^b \frac{M_{II}^2}{E \cdot I} \cdot dx. \qquad (6\text{-}49)$$

FIGURE 6-45
August Föppl (1854–1924)
[Föppl et al., 1924]

According to the principle of Menabrea, eq. 6-38 or eq. 6-41, the following relationship must apply:

$$\frac{\partial \Pi}{\partial Z} = \frac{1}{2 E \cdot I} \cdot \left[ \left( \frac{2 B \cdot a^3}{3} - \frac{q \cdot a^4}{4} \right) \cdot \frac{\partial B}{\partial Z} + \left( \frac{2 C \cdot b^3}{3} - \frac{q \cdot b^4}{4} \right) \cdot \frac{\partial C}{\partial Z} \right] = 0 \qquad (6\text{-}50)$$

Here,

$$\frac{\partial B}{\partial Z} = -\frac{b}{a+b} \quad \text{and} \qquad (6\text{-}51)$$

$$\frac{\partial C}{\partial Z} = -\frac{a}{a+b} \qquad (6\text{-}52)$$

are the partial derivations of eq. 6-47 and eq. 6-48. The relationships of eq. 6-47, eq. 6-48, eq. 6-51 and eq. 6-52 entered into eq. 6-50 produce the static indeterminate

$$Z = q \cdot \frac{(a^3 + 4a^2 \cdot b + 4a \cdot b^2 + b^3)}{8a \cdot b} \qquad (6\text{-}53)$$

So that is Föppl's calculation for one degree of static indeterminacy. Weingarten complains now that eq. 6-51 and eq. 6-52 do not vanish, are not equal to zero, as the immovability of the two supports demands. "It is easy to see that these displacements correspond to the deformation of the beam from an initial position, which is found by rotating the end position about the central support" (Fig. 6-46b) [Weingarten, 1901, p. 346]. Weingarten now continues to reason that, similar to eq. 6-51 and eq. 6-52, the expressions

$$\frac{\partial Z}{\partial B} \neq 0 \quad \text{or} \quad \frac{\partial A}{\partial B} \neq 0 \qquad (6\text{-}54)$$

would result for $\Pi$ as a function of $B$, and notes that any other presentation of $\Pi$ corresponds to a different manner of displacement of all points. Only the elimination of $B$ and $C$ and the retention of the central support reaction $Z$ alone would lead to a displacement figure in which the support points remain undisplaced. "Castigliano's theorem is therefore indeterminate, and has an infinite number of solutions depending on the choice of the function for $A$ [= $\Pi$ – the author], and is in no way suitable for applications without prior investigations" [Weingarten, 1901, p. 346].

In his response, Föppl draws attention to the fact "that the 'external forces'," to which Castigliano refers in his second theorem, "include those which one – in contrast to those with conditional support reactions – is able to change at will. In doing so, one may, if it seems appropriate, also make use of the trick of considering individual support reactions as 'external forces' or 'loads' provided the remaining support reactions are adequate for establishing equilibrium for any selection of external forces" [Föppl, 1901/1, p. 354]. Based on these explanations, Föppl continues: "It will be impossible to raise any objections to my presentation of Castigliano's method and everything connected with it" [Föppl, 1901/1, p. 354]. Föppl's reasoning is correct, understanding the support reaction $Z$ implicitly as statically indeterminate (Fig. 6-46c), i. e. as an external force for which he derives condition equation 6-50 from the condition of the immovability of the central support in order to quantify this. The fact that eq. 6-51 and eq. 6-52 resulting from the chain rule of differential calculus do not vanish is due to a different physical significance to that assumed by Weingarten. With the help of the principle of virtual forces according to eq. 6-42, Weingarten's rigid body displacement turns out to be a bending moment diagram of the virtual unit force state (Fig. 6-46d); the quantities of eq. 6-51 and eq. 6-52 are nothing other than the shear forces due to the unit force state (Fig. 6-46e). Of course, Weingarten also allows $B$ to be entered as a static indeterminate into the expression for $\Pi$ and the static indeterminate to be calculated by setting the partial derivation for $B$ to zero. This means that $\Pi$ depends on the choice of the static indeterminates, but the second theorem of Castigliano is certainly not indeterminate if $\Pi$ is presented as a function of that force variable for which the displacement variable is to be determined according to the

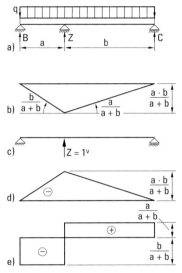

FIGURE 6-46
Föppl's sample calculation: a) continuous beam on three supports, b) Weingarten's rigid body displacement, c) virtual unit force state $Z = 1^v$, d) bending moment diagram of virtual unit force state $Z = 1^v$, e) shear force diagram of virtual unit force state $Z = 1^v$

nature, position and effect of this force variable. Although in the case of a system with one degree of static indeterminacy $\Pi$ generally represents an infinite number of equilibrium conditions, the definition of the static indeterminates means that the static indeterminate is unequivocally calculable from $\Pi$ according to the elasticity condition expressed by the second theorem of Castigliano (eq. 6-40) or the principle of Menabrea (eq. 6-41). We may look for Weingarten's understanding of the "indeterminacy of Castigliano's rule" in the imprecise concept of the "external force". If we pursue the method of sections, then in Föppl's sample calculation it is not only the static indeterminate $Z$ that is an external force, but the support reactions $B$ and $C$ as well. Nevertheless, the decision as to which of the three variables is entered into the expression for $\Pi$ as a static indeterminate cannot be formalised in the mathematical sense; it is rather a question of the appropriateness that the analyst has to answer first in order to translate his choice into a concrete structural model. In the case of Weingarten, this essential relationship between engineering science model and mathematical-physical law was lost in favour of the latter.

In another work, Weingarten argues that the theorems of Castigliano concern a "principle of elastic theory" [Weingarten, 1902, p. 233]; rather, that they are "tied to the limitation of the disappearance of the sum of the elastic work of the support reactions in relation to the displacements of the supports" [Weingarten, 1902, p. 233]. Again in this reasoning, characteristic is the lack of a relationship between Weingarten's mathematical-physical deductions and the engineering science model.

Weingarten investigated Castigliano's proof of eq. 6-40 in 1905 and again came to the conclusion that the second theorem of Castigliano, in the way he expressed it, would lack any exact sense. Nevertheless, he converts it "into a correct principle with exact meaning" and notes that "Castigliano himself had used it in this meaning in later developments of his work but without referring to it" [Weingarten, 1905, p. 187]. According to Weingarten, the second theorem of Castigliano should have read as follows: "If we express the internal work of a body or elastic structure by means of the equilibrium equations for a rigid body as a function of independent external forces, then the differential coefficient of this function, with regard to one of the forces, gives the relative displacement of its point of application in that displacement system that occurs when those points of the body, whose applied forces are eliminated, are prevented from taking on relative displacements" [Weingarten, 1905, p. 188]. Weingarten had thus provided the mathematical completion of the foundation of classical theory of structures through the second theorem of Castigliano and silently revised his objections to Föppl's sample calculation. In the same way, he acknowledged the first theorem of Castigliano (eq. 6-39), as the conditions for the applicability of both theorems are the same. But Weingarten continued to reject the principle of Menabrea. He asked those authors who spoke of Castigliano's 'principle of least work' to "identify the deformation, which deformations, lead to a minimum" [Weingarten, 1905, p. 192]. Weingarten's disapproval of the concept here is essentially one of criticising the language used without acknowledging the successful progress of the deformation energy concept in engineering practice.

## The debate between Hertwig and Weingarten

Hertwig took up the aforementioned publications of Weingarten and criticised his definition of the deformation energy (eq. 6-46) in which the rigid body displacement is mixed with the elastic deformation. The upshot of this was that the displacements are dependent on the initial position and therefore can no longer be determined through differentiation. Hertwig returns to Weingarten's more precise definition of the second theorem of Castigliano and can quite rightly conclude that "even Weingarten in his most recent work acknowledges that the theorems of Castigliano are correct to the extent that the tasks of statics can be solved without problems, like Castigliano himself did" [Hertwig, 1906, p. 506].

In his reply, Weingarten says that the more precise definition of the second theorem of Castigliano is "detailed proof of the claims made in 1901" [Weingarten, 1907/1, p. 109]. To this, Hertwig responded with a remark that "no engineer has ever performed the task of calculating the deflection of a beam supported on three rigid supports as wrongly as Mr. Weingarten did so in 1901 in his criticism of Föppl's book on mechanics" [Hertwig, 1907/1, p. 375] (see Fig. 6-46b). In the second part of his criticism of Weingarten, Hertwig illustrates the concept of deformation energy using the model of an internally and externally statically indeterminate system and in particular points out that in the principle of Menabrea deformation energy always means the deformation energy of the corresponding statically determinate basic systems in the theory of statically indeterminate systems (see Fig. 6-46c). And that ended the debate between Hertwig and Weingarten.

## The polemic between Hertwig and Mehrtens

Mehrtens' book contains the first comprehensive historico-logical presentation of classical theory of structures, and is backed up by a comprehensive list of references [Mehrtens, 1903–05]. In this book, Mehrtens develops the first phase of the dispute between Mohr and Müller-Breslau surrounding the foundation of classical theory of structures and interprets the sources in favour of Mohr. Hertwig's paper dating from 1906 [Hertwig, 1906] can be regarded as an independent, contrasting model of the history of classical theory of structures. However, Mehrtens' reply [Mehrtens, 1907] is surprisingly weak. In the first part he complains that Hertwig is a nobody in theory of structures and appears merely as Müller-Breslau's lawyer! And concerning the validity of the theorems of Castigliano for classical theory of structures disputed by Mohr and Mehrtens, Mehrtens merely refers to Weingarten's criticism. Mohr's open brief to the editors of the journal *Zeitschrift für Architektur- und Ingenieurwesen*, which begins "Dear Sir!", forms the second part of Mehrtens' reply. Here, Mohr formulates priority claims in nine points concerning the evolution of classical theory of structures. This strongly worded open letter had been triggered by the 12-page supplement to the Müller-Breslau book *Erddruck auf Stützmauern* (earth pressure on retaining walls) published in 1906 [Müller-

Breslau, 1906]. That supplement was in turn Müller-Breslau's reaction to the monographs of Mehrtens [Mehrtens, 1903–05] and Mohr [Mohr, 1906]; it forms the logical heart of Hertwig's paper dating from 1906 [Hertwig, 1906]. In the second point of his open letter, Mohr reiterates his old objections to the theorems of Castigliano. Mehrtens completes his reply with a third part in which he defends the independence of his work generally against the plagiarism accusation of Hertwig and questions the latter's authority as a critic.

Hertwig's reply is correspondingly general [Hertwig, 1907/2]. It remains to note that in the polemic between Hertwig and Mehrtens, the evaluation of the theorems of Castigliano for the foundation of classical theory of structures play almost no role; it is merely a dispute about priorities.

### The polemic between Weingarten and Weyrauch

Silenced by Hertwig, Weingarten immediately resurfaced with a publication that renewed his criticism of the theorems of Castigliano by taking into account thermal effects on elastic bodies [Weingarten, 1907/2]. In essence, Weingarten aimed his comments at Müller-Breslau's consideration of the terms for thermal effects in the expression for deformation energy, which the latter had called "ideal deformation work" in [Müller-Breslau, 1884]; Weingarten had thus opened up the same wound as Mohr had done more than two decades previously. But it was Weyrauch who replied [Weyrauch, 1908], not Müller-Breslau. According to Martin Grüning (1869–1932) [Grüning, 1912], in his reply Weyrauch (Fig. 6-47) developed the expression for deformation energy for the case of non-isothermic deformations (in the case of non-isothermic state transitions the temperature of every body element is variable) by introducing the stress components as a function of the distortion components and the temperature change into the principle of virtual displacements. This expression can assume different values for one and the same final condition of the deformed body.

FIGURE 6-47
Johann Jakob Weyrauch (1845–1917)
(archives of the University of Stuttgart)

Weyrauch understands deformation energy to be the work done by external forces in overcoming internal forces, even in the case of non-isothermic deformations. Weingarten criticised this "elasticity of deformation work" (Mohr) by claiming "that in the case of the occurrence of thermal effects, the consumed heat converted into work is understood as external work" [Grüning, 1912, p. 436]. The deformation energy of an elastic body is therefore that quantity of work that has to be applied to the body by an external force in order to transform it from its natural position into the given deformed equilibrium position, presuming isothermic deformations [Weingarten, 1909/1, p. 518]. According to Weingarten, the deformation energy is not identical with the work of the internal forces actually occurring and does not disappear when the stress components become zero. Weyrauch therefore regrets "that Mr. Weingarten has himself added a new species to the numerous types of deformation work" [Weyrauch, 1908, p. 101]. But Weingarten took the scientific feud to the mathematics-physics circle around Felix Klein in Göttingen. In 1909 he believed he had delivered his opponent a knockout blow with the following final remark: "We do not believe that engineers share the opinion that the expansion of an unloaded trussed framework caused by heat takes place without energy, i.e. with a deformation work equal to zero" [Weingarten, 1909/2, p. 67]. Weyrauch responded to this a few months later: "No one has claimed that the expansion takes place without energy; Prof. Weingarten had simply neglected certain energies, especially thermal energy. The statement 'i.e. with a deformation work equal to zero' relates to Weingarten's fictitious deformation work. If we understand deformation work to be the work done in overcoming the internal forces, then the heat rise in an unloaded trussed framework is connected with deformation work only if member forces are induced by the heat. This generally applies to statically indeterminate trussed frameworks, whereas Prof. Dr. Weingarten will obtain a deformation work even if no forces are applied that do work or could require work" [Weyrauch, 1909, pp. 245–246].

The overwhelming majority of engineers, mathematicians and physicists agreed with Weyrauch's concept definition. Weingarten therefore became isolated before his death in 1910 and turned himself into the last scientific victim of the anti-mathematics movement among German engineers that was already declining in the 1890s.

The acknowledgement – at last – of the theorems of Castigliano as a means supporting classical theory of structures – or, expressed differently, the long overdue establishment of the energy doctrine in theory of structures around 1910 – closed the first phase of the consolidation period of this fundamental engineering science discipline. The other occurrences related to this theme are written on a different page of science history: the beginnings of the creation of the foundations of modern structural mechanics on the basis of functional analysis by the Göttingen-based school around Felix Klein [Prange, 1999].

### 6.4.4 The validity of Castigliano's theorems

Friedel Hartmann was able to stake out the validity of the theorems of Castigliano (eq. 6-39 and eq. 6-40) in 1982 within the scope of his habilitation thesis at the University of Dortmund [Hartmann, 1982, 1983/1 & 1983/2]. His thesis appeared in 1985 in the form of a monograph with the title *The Mathematical Foundation of Structural Mechanics* [Hartmann, 1985]. Based on the embedding theorem of Sergei Lvovich Sobolev (1908–89), he specified the inequality

$$m - i > n/2 \qquad (6\text{-}55)$$

in which $m$ is the order of the energy, $i$ is the index of singularity, and $n$ is the dimension of the continuum. The theorems of Castigliano are valid only if the inequality of eq. 6-55 is satisfied [Hartmann, 1985, p. 181]. Therefore, the differential equation of the deflection curve of the order $2m = 4$ applies for a beam structure ($n = 1$), i.e. the order of the energy is $m = 2$. If point loads are applied to the beam, the index of singularity is $i = 0$. If we enter these three values into eq. 6-55, the result is $2 - 0 > 1/2$, i.e. the inequality of eq. 6-55 is satisfied and Castigliano's theorems apply.

The inequality is also satisfied by planar structures ($n = 2$) and solid-body structures ($n = 3$) if point loads act on the structure. If, on the other hand, individual moments act ($i = 1$), then the theorems of Castigliano apply for beams only, but not for planar or solid-body structures. In the case of applied rotation jumps $\Delta\alpha = 1$ ($i = 2$) and vertical displacement jumps $\Delta v = 1$ ($i = 3$), the theorems of Castigliano cannot be used for beam, planar or solid-body structures. The latter is evident in the case of the applied kink $\Delta\alpha = 1$ at point $j$ for the structural system shown in Fig. 6-33a: here, deformation energy $\Pi$ at point $j$ would be infinitely large, meaning that here the derivation according to the theorems of Castigliano (eq. 6-39 and eq. 6-40) does not exist.

## 6.5 Lord Rayleigh's *The Theory of Sound* and Kirpichev's foundation of classical theory of structures

Without doubt, Rayleigh's two-volume work *The Theory of Sound* [Rayleigh, 1877 & 1878] belongs to the library of physics classics and is the first book on the shelves devoted to acoustics. In his review of the first volume of this work for the magazine *Nature*, Hermann Helmholtz writes: "The author will earn the undying gratitude of all those studying physics and mathematics when he continues the work in the same way in which he has begun the first volume … Through his apt systematic arrangement of the whole for the most difficult problems of acoustics, the author has made it possible to study this subject with far greater ease than was previously the case" (cited in [Rayleigh, 1879, German translator's foreword]). At the suggestion of the "Chancellor of German Physics", Hermann Helmholtz, Rayleigh's two-volume work was immediately translated into German [Rayleigh, 1879 & 1880]. The second, improved and expanded edition in English appeared in 1894 (Fig. 6-48) and 1896 [Rayleigh, 1894 & 1896]; it was this edition that was reprinted unaltered by Dover Publications in 1945.

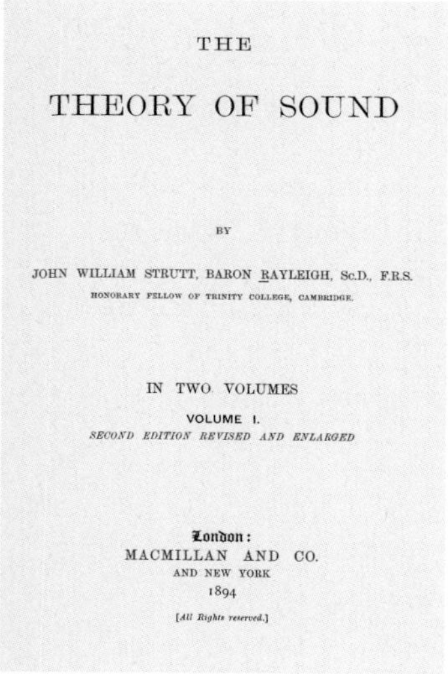
THE

THEORY OF SOUND

BY

JOHN WILLIAM STRUTT, BARON RAYLEIGH, Sc.D., F.R.S.

HONORARY FELLOW OF TRINITY COLLEGE, CAMBRIDGE.

IN TWO VOLUMES

VOLUME I.

SECOND EDITION REVISED AND ENLARGED

London:

MACMILLAN AND CO.

AND NEW YORK

1894

[All Rights reserved.]

**FIGURE 6-48**
Cover of the second edition of the first volume of Lord Rayleigh's *The Theory of Sound* [Rayleigh, 1894]

But what does the classic of acoustics have to do with the classical theory of structures? A glance at the first volume of Rayleigh's *The Theory of Sound* provides a clue. After he deals with the vibrations of systems in general in the first four chapters, Rayleigh analyses the vibrations of ropes, bars, membranes and plates; in the second edition he includes chapters on the vibrations of shells [Rayleigh, 1894, pp. 395 – 432] and electrical vibrations [Rayleigh, 1894, pp. 433 – 474]. His second volume is dedicated to the aerodynamic problems of acoustics.

The energy concept stands at the focal point of Rayleigh's acoustics. For example, in the third chapter, *Vibrating systems in general* [Rayleigh, 1894, pp. 91 – 169], he starts with potential energy $\Pi$ and kinetic energy $T$ and generalises the reciprocity theorem of vibrating elastic systems formulated by Betti in 1872 with the help of the concept of generalised forces $F_k$ and the corresponding generalised coordinates (displacements) $\delta_k$ and the D'Alembert principle: "If a harmonic force of given amplitude and period acts upon a system at the point $P$, the resulting displacement at a second point $Q$ will be the same both in amplitude and phase as it would be at the point $P$ were the force to act at $Q$" (cited in [Timoshenko, 1953, p. 321]). In his book *Redundant Quantities in Theory of Structures*

[Kirpichev, 1903 & 1934], Kirpichev congenially takes up Rayleigh's mathematical formulation of the vibration problem in the language of the energy principle and the Lagrange formalism of the generalised coordinates and forces on the level of structural analysis.

## 6.5.1 Rayleigh coefficient and Ritz coefficient

Rayleigh is the originator of the principle [Rayleigh, 1894, pp. 109–112] that in an enclosed system in the sense of thermodynamics in which the energy conservation law takes on the form

$$\Pi = T \tag{6-56}$$

the first natural frequency $\omega_1$ of a vibrating system can be calculated from

$$\omega_1^2 = \frac{\Pi}{\left(\dfrac{T}{\omega_1^2}\right)} \tag{6-57}$$

Eq. 6-57 is known as the Rayleigh coefficient. For the vibrating and simply supported beam of length $l$ on two supports with constant mass per unit of length $\mu$ and constant bending stiffness $E \cdot I$, with deformation energy

$$\Pi = \frac{E \cdot I}{2} \cdot \int_0^l \left|\frac{d^2 \overline{w}(x)}{dx^2}\right|^2 \cdot dx \tag{6-58}$$

and kinetic energy

$$T = \omega_1^2 \cdot \frac{\mu}{2} \cdot \int_0^l [\overline{w}(x)]^2 \cdot dx \tag{6-59}$$

the natural frequency obtained from the energy conservation law (eq. 6-56) is

$$\omega_1^2 = \frac{\dfrac{E \cdot I}{2} \cdot \displaystyle\int_0^l \left|\dfrac{d^2 \overline{w}(x)}{dx^2}\right|^2 \cdot dx}{\dfrac{\mu}{2} \displaystyle\int_0^l [\overline{w}(x)]^2 \cdot dx} \tag{6-60}$$

Eq. 6-60 is the specific expression of the Rayleigh coefficient (eq. 6-57) as results from comparing the coefficients of the denominators in eq. 6-60 and eq. 6-57. If we enter the static deflection curve $w(x)$ for the vibration deflection curve $\overline{w}(x)$ into eq. 6-58, eq. 6-59 and eq. 6-60, then for eq. 6-60 we get the useful approximation

$$\omega_1^2 \approx \frac{E \cdot I \cdot \displaystyle\int_0^l \left|\dfrac{d^2 w(x)}{dx^2}\right|^2 \cdot dx}{\mu \cdot \displaystyle\int_0^l [w(x)]^2 \cdot dx} \tag{6-61}$$

So Rayleigh succeeded in calculating the first natural frequency of vibrating systems from the energy conservation law (eq. 6-56) without solving the corresponding vibration differential equation. Later, Walter Ritz developed Rayleigh's method further within the scope of the calculus of variations to form a general approximation method [Ritz, 1909] – the Rayleigh-Ritz method. With this method for the direct solution of variational problems, Ritz provided not only mathematical physics, but also applied mechanics and theory of structures with an approximation method that could be used to solve elegantly problems of elasto-statics and elasto-kinetics, e.g. calculating the buckling loads of bars with varying stiffness, during the consolidation period of structural theory. For example, calculating the buckling load $F_{crit}$ of a bar leads to a coefficient whose formula is almost identical with eq. 6-61 and takes on the form

$$F_{crit} \approx \frac{E \cdot \int_0^l I(x) \cdot \left[\frac{d^2 w_1(x)}{dx^2}\right]^2 \cdot dx}{\int_0^l \left[\frac{dw_1(x)}{dx}\right]^2 \cdot dx} \tag{6-62}$$

for the fixed-end bar of length $l$ and varying second moment of area $I(x)$. Eq. 6-62 represents a special form of the Ritz coefficient (for the derivation please refer to Peter Zimmermann in [Straub, 1992, p. 342]). The only difference between eq. 6-61 and eq. 6-62 is that in the derivation of eq. 6-62, the energy conservation law is

$$\Pi = W_a \tag{6-63}$$

where $W_a$ is the external work,

$$W_a = \frac{F_{crit}}{2} \cdot \int_0^l \left[\frac{dw_1(x)}{dx}\right]^2 \cdot dx \tag{6-64}$$

is the applied force $F_{crit}$, and $w_1(x)$ is an approximation function for the deflection curve of the fixed-end bar, which must satisfy the geometric boundary conditions

$$w_1(x=0) = \frac{dw_1(x=0)}{dx} = 0 \tag{6-65}$$

at least. Peter Zimmermann specifies the following comparison function for the cantilever beam with constant second moment of area $I$, whose unsupported end deflects vertically by $f$ (see [Straub, 1992, p. 348]):

$$w_1(x) = \frac{f}{2 \cdot l^3} \cdot x^2 \cdot (3 \cdot l - x) \tag{6-66}$$

The comparison function 6-66 satisfies not only the geometric boundary conditions (eq. 6-65), but also the dynamic boundary condition

$$\frac{d^2 w_1(x=l)}{dx^2} = 0 \tag{6-67}$$

In terms of physics, eq. 6-67 signifies that the bending moment at the unsupported end of the cantilever bar becomes zero. If eq. 6-66 is entered into the Ritz coefficient (eq. 6-62), the critical load according to Zimmermann (see [Straub, 1992, p. 348]) is approximately

$$F_{crit} \approx 2.5 \cdot \frac{E \cdot I}{l^2} \tag{6-68}$$

Euler had already specified the exact solution in 1744 [Euler, 1744]:

$$F_{crit} = \left(\frac{\pi}{\beta}\right)^2 \cdot \frac{E \cdot I}{l^2} = \left(\frac{\pi}{2}\right)^2 \cdot \frac{E \cdot I}{l^2} = 2.4674 \cdot \frac{E \cdot I}{l^2} = F_{crit,1} \tag{6-69}$$

The buckling formula (eq. 6-69) for $F_{crit,1}$ is known in engineering literature as the 1st Euler case ($\beta = 2$). For the other three Euler cases, parameter $\beta$ in eq. 6-69 takes on the following values:
- bar pinned at both ends (2nd Euler case): $\beta = 1.00$ in eq. 6-69 → $F_{crit,2}$
- bar pinned at one end, fixed at the other (3rd Euler case): $\beta = 0.699$ in eq. 6-69 → $F_{crit,3}$
- bar fixed at both ends (4th Euler case): $\beta = 0.500$ in eq. 6-69 → $F_{crit,4}$

The critical loads of the four Euler cases are thus

$$F_{crit,1} : F_{crit,2} : F_{crit,3} : F_{crit,4} = 0.25 : 1 : 2.0467 : 4 \qquad (6\text{-}70)$$

Both the Rayleigh coefficient and the Ritz coefficient cleared the way for the common energy foundation for elasto-kinetics on the one hand and elasto-statics on the other.

### 6.5.2 Kirpichev's congenial adaptation

Kirpichev (Fig. 6-49) concluded the discipline-formation period of structural theory in Russia with his book *Redundant Quantities in Theory of Structures* [Kirpichev, 1903] – a mere 140 pages. It explains the entire theory of statically indeterminate trussed frameworks in an extraordinarily clear style. Therefore, he and Müller-Breslau can be regarded as having rounded off the classical theory of structures.

Like Rayleigh, Kirpichev based his work on potential energy $\Pi$ (deformation energy) and introduced the concept of generalised forces $F_k$ and the corresponding generalised coordinates (displacements) $\delta_k$. For a system with $n$ degrees of freedom and in the specific case of the time-related independence of $\Pi$ (conservative system), Kirpichev specifies the Lagrange equation as follows:

$$\sum_{k=1}^{n}\left(F_k - \frac{\partial \Pi}{\partial \delta_k}\right) \cdot d\delta_k = 0 \qquad (6\text{-}71)$$

**FIGURE 6-49**
Viktor Lvovich Kirpichev (1845–1913)
[Timoshenko, 2006, p. 89]

Kirpichev obtains the Lagrange equation 6-71 specific to the statics case from the principle of virtual displacements with subsequent comparison of coefficients. As $d\delta_k \neq 0$, the term within the brackets in eq. 6-71 must vanish, i. e.

$$\frac{\partial \Pi}{\partial \delta_k} = F_k \qquad (6\text{-}72)$$

must apply. Eq. 6-72 is nothing more than the first theorem of Castigliano (see eq. 6-39). Fig. 6-50 shows an excerpt from page 27 of Kirpichev's book with eq. 6-71 and eq. 6-72, where
- $U$ stands for deformation energy $\Pi$
- the Greek capital letters $\Phi$, $\Psi$, $\Theta$, ... stand for the generalised forces $F_1$, $F_2$, $F_3$, ...
- the Greek lower-case letters $\varphi$, $\psi$, $\theta$, ... stand for the generalised displacements $\delta_1$, $\delta_2$, $\delta_3$, ...

In the upper equation in Fig. 6-50, which corresponds to eq. 6-71, there is a mistake in the middle expression: factor $\partial \psi$ should read $d\psi$.

The generalised forces and coordinates will be briefly explained using the example of the elastic arch structure with three degrees of static

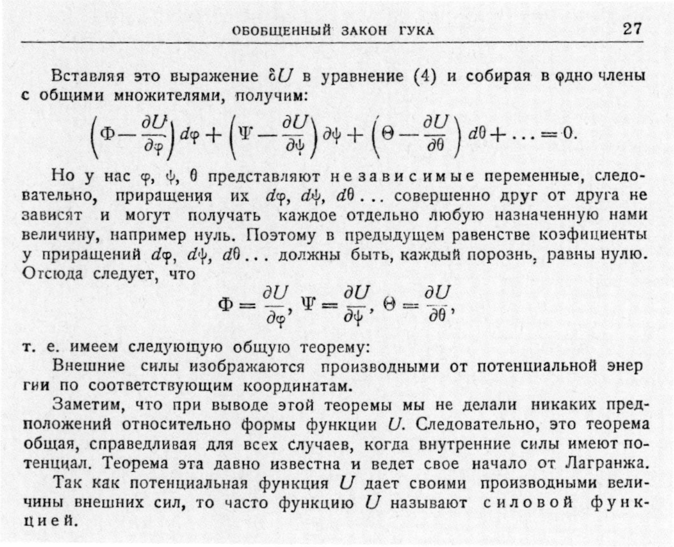


Вставляя это выражение $\delta U$ в уравнение (4) и собирая в одно члены с общими множителями, получим:

$$\left(\Phi-\frac{\partial U}{\partial\varphi}\right)d\varphi+\left(\Psi-\frac{\partial U}{\partial\psi}\right)\partial\psi+\left(\Theta-\frac{\partial U}{\partial\theta}\right)d\theta+\ldots=0.$$

Но у нас $\varphi$, $\psi$, $\theta$ представляют независимые переменные, следовательно, приращения их $d\varphi$, $d\psi$, $d\theta\ldots$ совершенно друг от друга не зависят и могут получать каждое отдельно любую назначенную нами величину, например нуль. Поэтому в предыдущем равенстве коэфициенты у приращений $d\varphi$, $d\psi$, $d\theta\ldots$ должны быть, каждый порознь, равны нулю. Отсюда следует, что

$$\Phi=\frac{\partial U}{\partial\varphi},\ \Psi=\frac{\partial U}{\partial\psi},\ \Theta=\frac{\partial U}{\partial\theta},$$

т. е. имеем следующую общую теорему:

Внешние силы изображаются производными от потенциальной энергии по соответствующим координатам.

Заметим, что при выводе этой теоремы мы не делали никаких предположений относительно формы функции $U$. Следовательно, это теорема общая, справедливая для всех случаев, когда внутренние силы имеют потенциал. Теорема эта давно известна и ведет свое начало от Лагранжа.

Так как потенциальная функция $U$ дает своими производными величины внешних сил, то часто функцию $U$ называют силовой функцией.

**FIGURE 6-50**
Lagrange equation and first theorem of Castigliano in Kirpichev's version [Kirpichev, 1903, p. 27]

indeterminacy shown in Fig. 4-45. The generalised forces correspond to the static indeterminates $H$, $V$ and $M$, all of which are applied at the elastic centre 0 of the structural system. The horizontal displacement $w_x$, the vertical displacement $w_y$ and the rotation $\varphi_z$ of the elastic centre are the associated generalised displacements. From eq. 6-72 we now obtain three equations for the static indeterminates $H$, $V$ and $M$:

$$\frac{\partial\Pi\,(w_x,w_y,\varphi_z)}{\partial w_x}=H;\quad \frac{\partial\Pi\,(w_x,w_y,\varphi_z)}{\partial w_y}=V;\quad \frac{\partial\Pi\,(w_x,w_y,\varphi_z)}{\partial\varphi_z}=M \quad (6\text{-}73)$$

According to Kirpichev, generalised forces are not only forces in the narrower sense, e.g. $H$ and $V$, but rather force variables which also include moments $M$, for instance. Generalised displacements are allocated to these, which are not only displacements in the narrower sense, e.g. $w_x$ and $w_y$, but rather displacement variables which also include rotations $\varphi_z$, for instance. For Kirpichev, all those are special cases of the generalised forces (= force variables) and generalised displacements (= displacement variables). This would lead to the displacement method in the first half of the consolidation period of structural theory (1900 – 50).

Running parallel with the displacement method was the force method, which did not pursue the route via deformation energy $\Pi$, but rather deformation complementary energy $\Pi^*$ ($\Pi = \Pi^*$ always applies in the case of linear-elastic material behaviour) and the principle of virtual forces. The equations corresponding to eq. 6-73 are as follows:

$$\frac{\partial\Pi^*(H,V,M)}{\partial H}=w_x;\quad \frac{\partial\Pi^*(H,V,M)}{\partial V}=w_y;\quad \frac{\partial\Pi^*(H,V,M)}{\partial M}=\varphi_z \quad (6\text{-}74)$$

Eq. 6-74 corresponds to the second theorem of Castigliano (see eq. 6-40). As the displacement variables $w_x$, $w_y$ and $\varphi_z$ of the elastic centre 0 must disappear in the statically indeterminate system (elasticity conditions), the statically indeterminate variables $H$, $V$ and $M$ can be calculated from this equation. Kirpichev develops this way of calculating statically indeterminate trusses using the second theorem of Castigliano and the principle of Menabrea in chapters 6 and 7 of his book. Again, he bases his deductions on the concept of generalised forces and displacements. In doing so, he assumes equality between deformation energy $\Pi$ and deformation complementary energy $\Pi^*$, designating both with the letter $U$. The description with generalised forces and displacements plus the restriction to linear-elastic material behaviour means that Kirpichev's presentation is extremely clear and focused.

Besides Castigliano's theorems, Kirpichev also derives the reciprocity theorem. The fabric of his theory of statically indeterminate trusses, rigorously founded on the energy principle, the concepts of generalised forces and generalised displacements and the Lagrange equation, is very appealing owing to its universality, which means Kirpichev can present it in a concise form, the clarity and elegance of which were to remain unsurpassed for many years. For example, the section on influence lines becomes particularly transparent thanks to his use of the reciprocity theorem [Kirpichev, 1903, p. 57], which in its most general form had of course been verified by Rayleigh. Kirpichev achieves a general and consistent formulation of the theory of statically indeterminate, linear-elastic trusses that is obligated to neither the energy nor the kinematic doctrine in theory of structures, and therefore leaves open the doors to both the force method and the displacement method. The common base beneath the dual make-up of theory of structures, so clearly presented in Kirpichev's book, anticipates the path to be taken by the ongoing founding of theory of structures in the first half of the 20th century.

Kirpichev was heavily influenced by Rayleigh's *The Theory of Sound* and in his introduction recommends this work to those of his readers concerned with structural theory. It is therefore Kirpichev we have to thank for making Rayleigh's methods known, first in Russia and then in other countries. One famous student of Kirpichev was Timoshenko, who wrote of him: "His advice was of great benefit to me. It helped me to define the direction of my later activities" [Timoshenko, 2006, p. 90]. Nonetheless, the influence of Kirpichev's congenial adaptation of the methods in Rayleigh's *The Theory of Sound* to suit the theory of statically indeterminate systems lagged behind that of the Berlin school of structural theory, which was to dominate the theory formation processes in the first half of the consolidation period of structural theory.

## 6.6 The Berlin school of structural theory

The theory of structures founded by Emil Winkler at Berlin-Charlottenburg TH took on the features of an engineering science school due to the dispute between Otto Mohr and Winkler's successor Heinrich Müller-Breslau: the Berlin school of structural theory. Its rise to become a school of world renown can only be understood in conjunction with the technical and scientific progress in Germany in general during the reign of Wilhelm II and the development of Berlin from the seat of the Prussian royal court to a world metropolis. Müller-Breslau not only concluded the discipline-formation period of structural theory (1825–1900) by rounding off this discipline, but also, together with his students, had a great influence on the consolidation period of structural theory (1900–50). After the death of Müller-Breslau in 1925, August Hertwig carried on the work of his teacher. As the Nazi Party prepared to conquer Berlin, the big city, the decline of the Berlin school of structural theory was settled too. When Hans Reissner departed, they lost their most prominent advocate. Nevertheless, Hertwig managed to pass on the Berlin school of structural theory at Berlin-Charlottenburg TH, even though it no longer had a direct, substantial influence in the German-speaking countries. After the collapse of Nazi Germany, Alfred Teichmann, in lone scientific work at Berlin TU, integrated second-order theory into member analysis as a whole in the sense of Müller-Breslau. His scientific seclusion and the suppression of his research activities in the Third Reich is symbolic for the isolation of this city that began with the division of Berlin in 1948. As Teichmann was transferred to emeritus status, the Berlin school of structural theory also went into retirement. The international status of theory of structures, in the meantime superseded by structural mechanics, did not manage to gain a footing in teaching and research at Berlin TU until the student-induced social reforms took place. The newer representatives of theory of structures adopted calculus bit by bit in structural analysis, which had been introduced by Müller-Breslau, and its consummation in the form of structural matrix analysis by John Argyris.

**The notion of the scientific school**

### 6.6.1

During the 1970s scientific schools played a notable role as an object of scientific research in the USSR and GDR; that is clear from the two compendiums published simultaneously in German [Mikulinskij et al., 1977 & 1979] and Russian in 1977 and 1979 by the Institute of History of Natural Science & Technology at the USSR Academy of Sciences and the Institute of Theory, History & Organisation at the GDR Academy of Sciences. Besides general statements, the compendiums contain numerous case studies regarding school formation in medicine and the natural sciences; but the field of engineering sciences was ignored. This author asked a prominent scientist of the latter institute, Prof. Dr. Hubert Laitko, about current literature on the subject of scientific schools; the result was disappointing. The computer-assisted literature researches carried out by Dr. Regine Zott revealed that the term "school" features in the title of only a few papers, and then only with a secondary significance [Laitko, 1995]. So the only works of reference remain the aforementioned compendiums. In those publications, Laitko develops definitory components for the term "school": "A scientific school is the finite collection of interrelated research activities carried out by various individuals, at least partly, and there exists a chronological relationship between their elements, at least in part. Because of the relationship, the various research activities are elements of one and the same collection, which makes those carrying out the work members of one of the same social group" [Laitko, 1977, p. 267]. The designation "scientific school" can therefore be applied to certain collections of research activities, certain collections of individuals and – most thoroughly – for the collection formed by assigning both quantities to each other [Laitko, 1977, p. 268]. Laitko develops his definition in four hypothetical formulations [Laitko, 1977, pp. 275–278]:

1. An idea constituting a school is a paradigm in a competitive situation, the realisation of which in a programme of research goals is linked with a social interest related to the situation.
2. The objective possibility of the school formation exists in gnosiological terms when a social need for knowledge can be assigned several non-equivalent scientific problem formulations or several non-equivalent potential solutions can be assigned to one problem (competing schools).
3. From the social point of view, a scientific school directly represents a problem-related formation of a group of scientific actors for implementing the paradigm, i. e. the idea constituting the school; this competitive situation forms elements that determine the style of scientific thinking of the respective group.
4. The common interest between the various research activities belonging to a scientific school is the relationship to one and the same paradigm, which is realised through communicative relationships between the scientists.

In the fourth point, Laitko distinguishes between schools with a centric structure and direct communication, schools with a centric structure and indirect communication, schools with no centric structure but direct communication, and schools with no centric structure and only indirect communication.

Laitko's four hypothetical formulations concerning the definition of a scientific school can be seen in practice in the development of theory of structures at Berlin TH from 1880 to 1970.

## 6.6.2 The completion of classical theory of structures by Heinrich Müller-Breslau

It was in the 1880s that an engineering scientist entered the world of iron construction and structural theory who would have an influence on the scientific basis of structural engineering for more than half a century: Heinrich Müller-Breslau (Fig. 6-51). After completing his secondary education and taking part in the Franco-Prussian war of 1870/71, he turned to studies in civil engineering at the Berlin Building Academy. In addition, he attended the mathematics lectures of Karl Weierstraß (1815–97) and Elwin Bruno Christoffel (1829–1900) [Hees, 1991, p. 327]. Christoffel had already completed the decisive step to the use of calculus in tensor calculation in 1869, and this would later become the mathematical basis for the general theory of relativity and the theory of shell structures [Herbert, 1991, p. 136]. It is possible that this is also one of the sources of Müller-Breslau's δ-symbols, which he devised some 20 years later and opened up the perspective for the formal use of symbols on the level of calculus in the theory of statically indeterminate systems.

As a student, Müller-Breslau coached his fellow students in the dreaded second state examination set by Schwedler. He opened a consulting engineering practice in Berlin in 1875. His activities as a coach and consulting engineer quickly led to his first publications on the subject of theory of structures. Particularly noteworthy are his contributions to the sections on elasticity, strength and structural mechanics starting with the 11th edition (1877) of the most influential German engineering pocketbook, *Hütte*. In 1883 Müller-Breslau, who had never taken a final examination, was appointed lecturer and two years later professor at Hannover TH [Lorenz, 1997, p. 298]. It was there that Müller-Breslau gave lectures in structural theory for mechanical engineers, the principles of structural engineering for architects and mechanical engineers, and the theory of statically indeterminate bridge beams. It was through his contributions to the theory of statically indeterminate systems that Müller-Breslau rose, between 1883 and 1888, to become one of the leading authorities on theory of structures.

**FIGURE 6-51**
Heinrich Müller-Breslau (1851–1925) at the age of 60 [Hertwig & Reissner, 1912]

He was just 32 years old when he was criticised in 1883 by the famous old master of applied mechanics Otto Mohr (Fig. 6-52). That began the dispute about the foundation of classical theory of structures, created a competitive situation and led to the formation of schools: the Berlin school of structural theory on the one side and the Dresden school of applied mechanics on the other. The object of the dispute was Müller-Breslau's conviction of the equivalence between the "method of Mohr" (Müller-Breslau), which he and Matthias Koenen (1849–1924) had extended to cover trusses, and the theorems of Castigliano. The crux of Mohr's objection to the principles character of the theorems of Castigliano was the difference in the theoretical modelling of the loadbearing system. Whereas Mohr advanced the linear-elastic trussed framework theory and thus the kinetic machine model of trussed framework theory (see section 6.3.3) to become the basic model of the complete theory of trusses and the main object of classical theory of structures, Müller-Breslau based his work on the simplifications of the practical bending theory, which Navier had essentially prepared, extending the energy-based machine model of trussed framework theory to all trusses. The energy (see Fig. 6-27) and the kinematic (see Fig. 6-28) doctrines of classical theory of structures represent the two ideas constituting the aforementioned schools, the paradigms in a competitive situation, the realisation of which in the respective programmes of research goals were worked on by Müller-Breslau and Mohr, goading each other in the process.

Laitko's second proposed criterion for school formation – the gnosiological – also fits in with the argument between Mohr and Müller-Breslau. Mohr's criticism of Müller-Breslau consists precisely of the fact that the theory of statically indeterminate systems permits several non-equivalent solutions, i.e. besides those of Mohr, those of Maxwell and Castigliano as well. Therefore, up until 1886, Müller-Breslau was working feverishly on expanding the applications for the theorems of Maxwell and Castigliano from trussed framework systems to trussed systems, and summarised his

**FIGURE 6-52**
Otto Mohr (1835–1918) at the age of 80 with the work equation for trussed frameworks in the background (university archives, Dresden TU)

knowledge on the theory of statically indeterminate trusses in his book of 1886 *Die neueren Methoden der Festigkeitslehre und der Statik der Baukonstruktionen* (see Fig. 6-29). This book already contains the formulation of the force method expressed in terms of his δ-symbols. Müller-Breslau expressed scrupulously the tendency inherent in the δ-symbols towards the symbols of calculus for the theory of statically indeterminate trusses in the second edition of his book, which appeared in 1893 (see section 6.3.2).

Müller-Breslau discovered the mathematical structure of the theory of statically indeterminate systems with the elasticity equations of the force method in the language of the δ-symbols: the theory of linear sets of equations. Mohr's criticism of the *Die neueren Methoden der Festigkeitslehre…* published under the polemic title of *Über die Elastizität der Deformationsarbeit* (on the elasticity of deformation work) in the journal *Zivilingenieur* was intended to destroy Müller-Breslau's scientific reputation: "Of the various paths that always lead to the goals, the newer methods contain, remarkably, only one, albeit to be thoroughly recommended: the whole theory of the trussed framework and at the end the theorems of Castigliano, too, are derived from the principle of virtual velocities. Accordingly, those theorems are treated as appendices to the same investigations, whose foundation they should form according to the view of their author … The newer methods have not added anything other than the unproven claim that I have overlooked something" [Mohr, 1886, p. 400]. In the dispute about the formation of classical theory of structures, the social aspect of school formation, which expresses itself directly as a problem-related formation of a group of scientific actors for implementing the idea constituting the school, shaped the elements determining the style of scientific thinking at the Berlin school of structural theory and the Dresden school of applied mechanics as listed in Fig. 6-53. By the end of the 1880s the dispute about the foundation of classical theory of structures had been decided in favour of Müller-Breslau because the theorems of Castigliano were regarded by the majority of structural engineers with a scientific outlook and by mathematicians [Klein, 1889] as equivalent to the general work theorem introduced by Mohr into trussed framework theory in the form of the principle of virtual forces [Klein, 1889].

So it was not Mohr, but rather Müller-Breslau who completed the classical theory of structures; and what's more, classical theory structures would soon appear as a scientific creation of the Berlin school of structural theory.

### 6.6.3

As the 37-year-old Müller-Breslau was appointed professor of theory of structures and bridge-building at Berlin-Charlottenburg TH in late 1888 after the death of Emil Winkler, his establishment of the energy doctrine in theory of structures during the dispute with Mohr had intensified and broadened the idea constituting the school (gnosiological aspect and paradigm character of school formation) to such an extent that the social and communicative aspect of school formation was able to develop in a particularly ideal way within the first few years of his professorship.

| Berlin school of structural theory | Dresden school of applied mechanics |
|---|---|
| Principal advocate: Müller-Breslau | Principal advocate: Mohr |
| receptive | reflective |
| extensive reproductive | transcendental |
| method pluralism | single method |
| eclectic | original |
| applications-oriented | fundamentals-oriented |
| inductive | deductive |
| synthetic | analytic |
| method-oriented | problem-oriented |
| syntactic | semantic |
| international | national |

**FIGURE 6-53**
The elements determining the style of scientific thinking at the Berlin school of structural theory and the Dresden school of applied mechanics

**Classical theory of structures takes hold of engineering design**

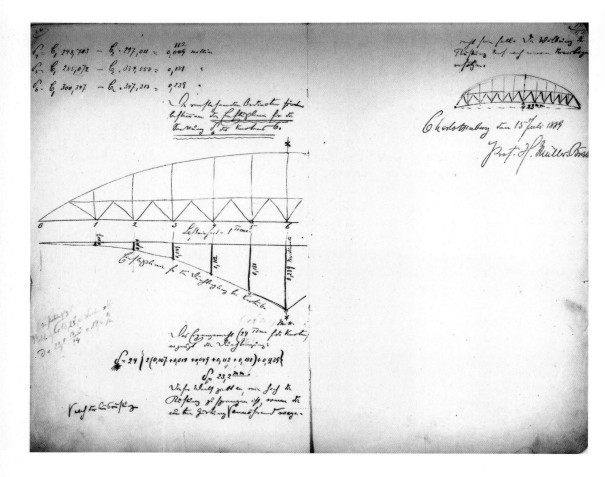

**FIGURE 6-54**
The final sheet of the structural calculations by Müller-Breslau for the bridge over the River Ihme in Hannover, completed in 1889 and first replaced in 1950

Nevertheless, Müller-Breslau's completion of the classical theory of structures was not in the form of the establishment of the energy doctrine in this fundamental building science discipline. Its classicism consisted of the creation of a uniform theory of statically indeterminate systems in the shape of the force method, which would dominate the disciplinary discourse for decades as well as practical structural calculations in steelwork and, after 1900, reinforced concrete, which revolutionised the whole of construction engineering; Müller-Breslau would also carry out the first successful trial applications in airship and aircraft construction. The δ-symbols introduced by Müller-Breslau for the force method not only inserted the final piece of the jigsaw in the discipline-formation period of structural theory that Navier had started in the 1820s, but also laid the first piece of the next jigsaw, i. e. the use of calculus for its mathematical form, which was to culminate in computer mechanics in the 1950s.

Like Leibniz' differential and integral calculus pushed forward analysis through the practical use of symbols and condemned Newton's method of fluxions to oblivion, Müller-Breslau's δ-symbols invaded the entire theory and practice of structural analysis, and pushed aside other formulations,

e. g. the Dresden school around Mohr. They therefore transcended the skill of theory formation and structural calculations, which for Culmann, Schwedler and Winkler were still an integral part of the engineer's personality, to become a method. This objectivisation of structural calculations generated the hegemony of the structural model over design calculations in civil and structural engineering. Civil and structural engineers created real constructions from the world of the idealised models of classical theory of structures.

By way of his Berlin consultancy practice, Müller-Breslau carried out creative and highly diversified design and assessment activities in the field of iron construction (Fig. 6-54). He thus became the prototype of the scientifically active consulting engineer and at the same time the engineering scientist with practical experience. The Berlin school of structural theory and its principal proponent Müller-Breslau therefore represents the dwindling influence of the civil servants trained at the Building Academy and the rise of the consulting engineer.

Just how the structural model influenced design in engineering is shown by the "Kaisersteg" footbridge over the River Spree at Oberschöneweide, completed in 1898. Müller-Breslau used the bending moment diagram as the basis for designing a three-span continuous beam, the upper chord of which he designed as a hinge at mid-span [Müller-Breslau, 1900]. Another impressive example of the supremacy of the structural model over design calculations in engineering is the Dome of Berlin Cathedral (Fig. 6-55) which was built in 1897/98 according to the drawings of Müller-Breslau. This structure designed by the architect Julius Carl Raschdorff (1823–1914) is today regarded as a symbol of the eclecticism of the reign of Wilhelm II. That matches the eclecticism of Müller-Breslau in the realm of structural analysis theory formation; he was more interested in knowledge about how to do something rather than about its foundation. The lack of style in the Berlin school of structural theory was an essential component in its success.

FIGURE 6-55
The Dome of Berlin Cathedral under construction
[Müller-Breslau, 1898, p. 1208]

The literary monument of the eclecticism of the Berlin School of classical theory of structures, like theory of structures itself, can be found in Müller-Breslau's multi-volume work *Die graphische Statik der Baukonstruktionen* (Fig. 6-42), which first appeared in 1887 and was translated into many languages. Müller-Breslau extended the object of theory of structures with each new volume, covering all the structural analysis methods and techniques appropriate for the practice of structural engineering. The work grew with theory of structures' claim to hegemony in civil and structural engineering like Wilhelminian Germany claimed a place in the sun among the nations of Europe; it grew with the claim to supremacy of the Berlin school of structural theory in this discipline like Prussia claimed first place among the German federal states. Müller-Breslau's *Graphische Statik der Baukonstruktionen* therefore became the leading source of advice for practical engineering at home and abroad, and theory of structures the king among the disciplines of civil and structural engineering, with Müller-Breslau as its king-maker.

On the other hand, Mohr's *Abhandlungen aus dem Gebiete der Technischen Mechanik* (Fig. 6-43), the first edition of which appeared in 1906, did manage three editions, but was never translated into any other language and was not significantly expanded, despite its very helpful, modern content.

It is thanks to Müller-Breslau, Alois Riedler (1850–1936), professor of mechanical engineering and dean of Berlin-Charlottenburg TH during that memorable year in which this establishment celebrated its 100th anniversary, and Adolf Slaby (1849–1913), professor of theoretical mechanical and electrical engineering and the imperial engineering adviser, that in 1899 Wilhelm II granted the technical universities the right to award doctorates despite the protests of the other universities.

The work of this triumvirate, all born shortly after the failed revolution in Germany, inserted the final piece into the jigsaw of the engineers' battle for academic emancipation; the right to award doctorates and the associated university constitution of the technical universities represents one element in the historical compromise between the nobility and the middle classes so characteristic of Prussian Germany. It is therefore the realisation of the demands of the rebels of 1848 regarding the abolition of the directorial constitution of higher technical schools, like that demanded by, for example, the students of the Berlin Building Academy at that time.

The granting of the right to award doctorates brought with it social recognition for civil, structural, mechanical and electrical engineering – the key scientific disciplines of the technical universities with which Wilhelminian Germany wished to safeguard its claim to authority in the rivalry among the imperialist nations. Adolf Slaby personifies the rise of science in electrical engineering, Alois Riedler played a similar role in mechanical engineering, characterised primarily through large-scale laboratory tests, and Müller-Breslau did the same for classical theory of structures, so closely related to iron construction. At the turn of the century in Germany, the aforementioned disciplines were backed up by major electrical and mechanical engineering industries as well as the large steel construction companies integrated into the vertical business structures of the steel-producing industry.

Therefore, honours were heaped on Müller-Breslau:
- 1889 full member of the Prussian Building Academy
- 1895/96 and 1910/11 dean of Berlin-Charlottenburg TH
- 1901 full member of the Prussian Academy of Sciences, Berlin
- 1902 gold medal for services to building
- 1902 honorary member of the American Academy of Arts & Sciences, Boston
- 1903 Dr.-Ing. E.h. awarded by Darmstadt TH
- 1908 member of the Swedish Academy of Sciences, Stockholm
- 1910 honorary member of the Institute of Engineers of Ways of Communication, St. Petersburg
- 1913 lifetime representative of the German technical universities at the Prussian parliament

- 1921 Dr.-Ing. E. h. awarded by Berlin-Charlottenburg TH
- 1921 honorary member of Breslau TH
- 1921 honorary fellow of Karlsruhe TH
- 1923 honorary fellow of Berlin-Charlottenburg TH.

When Heinrich Müller-Breslau died in 1925 (Fig. 6-56), Karl Bernhard (1859–1937), one of his outstanding students, wrote the following lines in the obituary: "One of the greatest from the realm of engineering has passed away. Heinrich Franz Bernhard Müller-Breslau died on 23 April after a long illness shortly before his 74th birthday ... His creative works were masterly and his teachings have been disseminated throughout the world by a large circle of students and admirers. For it is certainly not exaggerated to say that it is due to his influence that the faculty of civil engineering at Berlin University exerted a great attraction for many years. It therefore gave rise to a Müller-Breslau school that produced the majority of leading engineers in practice and outstanding teachers now working at universities in Germany and elsewhere" [Bernhard, 1925, p. 261].

On the eve of World War 1, the Berlin school of structural theory had established a dominating position for itself in the disciplinary fabric of civil engineering; among the community of civil and structural engineers, which this school helped to internationalise to a large extent, the Berlin school of structural theory was esteemed throughout the world.

FIGURE 6-56
Tombstone of the Müller-Breslau family at the Berlin-Wilmersdorf crematorium (as of March 2002)

### 6.6.4 Müller-Breslau's students

Müller-Breslau taught at Berlin-Charlottenburg TH from 1888 to 1923. After 1900, the social and communicative aspects gained more and more ground over the paradigmatic and gnosiological aspects in the development of the Berlin school of structural theory. In terms of its communication relationships, during this period the Berlin school of structural theory qualifies as an engineering science school with centric structure and direct communication, i. e. a classical school with individual authorship of the constituent idea and a student community grouped around the central person [Laitko, 1977, p. 277]. Due to the national and international dissemination of the constituent idea of the Berlin school of structural theory in particular, the school gradually became an engineering science school with a centric structure but indirect communication after 1910; and with that the consolidation period of structural theory (1900–50) really got underway.

Two important students of Müller-Breslau were August Hertwig (1872–1955) and Hans Reissner (1874–1967), who in 1912 published a commemorative document celebrating their teacher's 60th birthday (Fig. 6-57).

In their preface, Hertwig and Reissner write on behalf of their collaborators: "Highly esteemed teacher and friend! Portraying a picture of your life's work, which still grows from year to year, in this publication is not our task. It is portrayed in the history of structural theory and steel construction. It is not possible to reflect that picture in the works of your students and friends. That army of your students is working outside in practice, not with words and documents, but with actions and works in

FIGURE 6-57
Cover of the commemorative publication for Müller-Breslau [Hertwig & Reissner, 1912]

stone and iron. They are the best witnesses to your spirit. Only those who proclaim your life's work in the work of the young have collaborated on this commemorative document. Some have set up the field with you side by side, the majority follow in your footsteps. May we be honoured with seeing you at our head for a long time to come" [Hertwig & Reissner, 1912, p. I].

The preface reveals not only that the communication of the centric Berlin school of structural theory was gradually taking on an indirect character, but that its advocates were combining science-based practice with practice-based science. One of those students who was achieving great things through "actions and works", primarily in steel, but who did not contribute to the commemorative publication, was Karl Bernhard (Fig. 6-58). Werner Lorenz reviews examples of his impressive steel structures for industry and transport infrastructure (e. g. AEG turbine hall in Moabit, 1909, locomotive works of Linke-Hofmann-Lauchhammer AG in Breslau, 1916–18, diesel engines factory in Glasgow, 1913, German prime-mover machinery hall at the Brussels World Exposition, 1910) and pays tribute to Bernhard's contributions to engineering aesthetics. He comes to the following conclusion: "Bernhard's radical view of the autonomous structural engineer solely responsible for the aesthetics, coupled with great dedication and sound professional competence [is] a model with which Berlin's steel construction can attain recognition and acknowledgement beyond Germany's borders" [Lorenz, 1997, p. 307]. In addition to his achievements in engineering practice, Müller-Breslau encouraged Bernhard to become a private lecturer for steel construction, structural engineering and bridge-building at Berlin-Charlottenburg TH, where he worked from 1898 to 1930 [Baer, 1929, p. 794].

**FIGURE 6-58**
Karl Bernhard [Baer, 1929, p. 794]

The contributions of Karl Bernhard, who died in 1937, were not honoured in any relevant journals, certainly for racist reasons. Hans Reissner (Fig. 6-59), a student of Müller-Breslau who had been appointed to the chair of mechanics at Berlin-Charlottenburg TH in 1913, had to flee Nazi Germany in 1938 in order to save his life. He settled in the USA and carried on his pioneering fundamental research into aircraft construction there [Szabó, 1959, p. 82]. The Berlin school of structural theory therefore lost two important representatives little more than a decade after Müller-Breslau's death. Nevertheless, August Hertwig, Müller-Breslau's successor at Berlin-Charlottenburg TH, managed to uphold the position of the Berlin school of structural theory and maintain its centric structure until the collapse of Hitler's Germany.

**FIGURE 6-59**
Hans Reissner (university archives, Aachen RWTH)

### 6.6.4.1
**August Hertwig**

August Hertwig's (Fig. 6-60) life is the incarnation of the historical compromise between the nobility and the industrial middle class in Prussia from the point of view of the self-endured economic decline to the academic middle class. His memoirs, written in 1947, remain as a typewritten manuscript [Hertwig, 1947]. They read like the account of an engineering personality between the Scylla and Charybdis of two world wars. Born in Mühlhausen just after the founding of the Second Reich as the son of a factory owner, he experienced at first hand the practical failure of the liberal ideas of the 1848 revolution with the rise of Prussian Germany even before he started attending Berlin-Charlottenburg TH in 1890. The closure of his father's business not only fired his professional ambitions, but also heightened his awareness of the cultural inheritance of the German middle classes. Hertwig was already developing into a scientific personality with a broad range of knowledge during his time as professor at Aachen RWTH (1902 – 24), with political and social activities in the sense of fulfilling a cultural mission that went way beyond the civil engineering profession.

In the 1906 dispute between August Hertwig, a student of Müller-Breslau, and Georg Christoph Mehrtens (1843 – 1917), a student of Mohr, concerning the history of theory of structures, the idea was not only to secure the priority claims of the two mentors. The crux of that quarrel is to be found in the historico-logical configuration of a uniform disciplinary self-image of classical theory of structures. It was precisely the eclecticism and the pluralism of methods in Müller-Breslau's structural theory (see Fig. 6-53) that motivated Mohr, and later Mehrtens, to object. So Hertwig and Mehrtens were the first to approach a history of structural analysis based on comprehensive, dependable sources which could be placed in the first decade of the previous century in the context of the history of engineering science being written by Conrad Matschoß. Notwithstanding, only Hertwig managed to exploit the historical reconstruction of theory of structures for the progress of engineering science knowledge. Was it the engineering science and method pluralism character of Müller-Breslau's structural theory that forced him to track down its internal logical bond? Mehrtens, on the other hand, could quite happily fall back on Mohr's original and single-method style of theory formation (see Fig. 6-53). Whereas Mehrtens, Mohr's successor, battled on behalf of Mohr in his three-volume work *Vorlesungen über Statik der Baukonstruktionen und Festigkeitslehre* completed in 1905 with a long chapter on the history of theory of structures, in that same year Hertwig, who in 1902 at the age of 30 had been appointed professor of structural analysis at Aachen RWTH, became intensively involved in the mathematical structure of trussed framework theory. Hertwig's first scientific publication came as a result of his activities as a structural engineer working on the palm houses for the botanic gardens in Berlin (Fig. 6-61) and a detailed study of the book by Arthur Schoenfließ (1853 – 1928) entitled *Kristallsysteme und Kristallstruktur* (crystalline systems and structures) (1891). He continued this work later by means of further studies looking into the relationship between the symmetry properties of planar and spatial frames and determinant theory. Hertwig thus uncovered the mathematical structure of an important part of classical theory of structures: the theory of sets of linear equations. In theory of structures, Hertwig was therefore the first person to follow rigorously the path of the practical use of symbols mapped out by Müller-Breslau's δ-symbols right up to calculus, which determined the scientific work of Hertwig from that moment on.

**FIGURE 6-60**
August Hertwig at 70 (university archives, Berlin TU)

Hertwig's logical side was soon joined by a historical side. The starting point for this was his second publication of 1906, which explored the historical development of the principles of structural analysis. It was Hertwig's direct reaction to the historical reconstruction of theory of structures in Mehrtens' *Vorlesungen*; and more besides: it contains the clear recognition of the equivalence between the theorems of Maxwell, Mohr and Castigliano, and their summary through Müller-Breslau's δ-symbols on the level of the classical, i.e. linear, theory of structures, which consists of the trinity of linearity in material behaviour (Hooke's law), force condition (statics) and displacement condition (geometry). Therefore, the union between the application of calculus and the history of theory of structures is evident right from the outset of Hertwig's scientific life's work.

Hertwig succeeded Müller-Breslau in the chair of structural steelwork and theory of structures at Berlin-Charlottenburg TH in 1924. On 15 June 1925 he gave a memorable speech at an event in memory of the late Müller-Breslau held at Berlin-Charlottenburg TH. He summarised his teacher's contributions to theory of structures and steel construction; pointing out the line of tradition in the Berlin school of structural theory, he immediately took up the work of his teacher. The baton had been handed over. Kurt Klöppel (1901–85) arranged for Hertwig's speech to be reprinted in the journal *Der Stahlbau* (where Klöppel was chief editor) in 1951 on the occasion of the 100th anniversary of Müller-Breslau's birth for the structural steelwork conference held on 10/11 May 1951 in Karlsruhe [Hertwig, 1951].

During his time in Berlin, Hertwig's creative work became an enormous driving force, whether as academic teacher, independent expert, specialist author or co-designer of joint engineering science activities:

- 1926 membership of the Building Academy; editor of the report on the major Niederfinow ship lift project
- first editor of the journal *Der Stahlbau* founded in 1928 upon the suggestion of Gottfried Schaper, head of bridge-building for German Railways and a leading German steel bridge-builder of the interwar years
- 1934–37 chairman of the German Aircraft Committee
- corresponding member of the German Aviation Academy
- 1935 chairman of the German Building Association
- chairman of the VDI history of engineering study group
- chairman of the German Soil Mechanics Association set up in 1928 with financial assistance from the Transport Ministry, German Railways and the Education Ministry
- initiator of the Berlin-Charlottenburg TH civil engineering laboratory completed in 1933
- chairman of the German Structural Steelwork Committee (Schaper's successor).

Hertwig's work is therefore a focus for the disciplinary differentiation in civil and structural engineering: the establishment of foundation dynam-

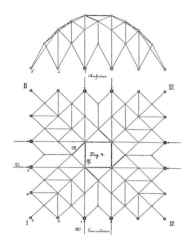

**FIGURE 6-61**
Structural system of the dome over the corner glasshouses of the botanic gardens in Berlin-Dahlem [Hertwig, 1905, p. 212/Fig. 4]

ics, the intensification and expansion of structural steelwork, the bringing to fruition of knowledge of structural engineering in aircraft construction, the organisation of laboratory research for civil and structural engineering, the history of civil and structural engineering as part of the cultural inheritance, and further development of theory of structures.

Despite this diversity, Hertwig was drawn again and again to theory of structures, his specialist subject. In a larger work dating from 1933, he developed clearly the dual nature of the entire field of structural analysis. And that enabled him to create the foundation for comprehensive formalisation in theory of structures.

In 1942, at the age of 70, "Vater Hertwig", as he was lovingly called by his students, was honoured by the German Structural Steelwork Society with a research publication dedicated to his name with papers on theory of structures, elastic theory, stability theory and soil mechanics [DStV, 1943]. The list of authors reads like a "Who's Who" of the civil and structural engineering profession in Germany at the time – supplemented by a few scientists from Austria, Norway and Czechoslovakia (occupied countries at that time). This is the last document recording the centric character of the Berlin School of structural theory. The school was still firmly rooted in Berlin-Charlottenburg TH. Nevertheless, it had been leading a dual existence for many years, sometimes compelled by outside forces, like Hans Reissner's enforced flight from Germany. Towards the end of the war, the Berlin School of structural theory and August Hertwig had become quieter.

In 1941 his only child Rolf became a victim of euthanasia. In his memoirs August Hertwig notes: "As during the war years the order was given to free all hospitals of terminally ill patients, Rolf, too, became a victim of this decree. In 1941 we received a letter telling us that he had been transferred from Eberswalde and 14 days later the notification of his death from Bernburg an der Saale" [Hertwig, 1947, p. 188]. What was Hertwig thinking as he read that letter informing him of his son's death? What was he thinking as on his 70th birthday in 1942 he was awarded the Goethe Medal for Art & Science by the Fuehrer of the German people [DStV, 1943, p. V] ?

#### 6.6.4.2 August Hertwig's successors

Siegmund Müller (1870–1946), who lectured in structural steelwork, industrial buildings and structural engineering for about 35 years at Berlin-Charlottenburg TH, retired in 1936, one year before Hertwig. The successor to Hertwig and Müller was Ferdinand Schleicher (1900–57), who continued to expand the civil engineering laboratory; Karl Pohl (1881–1947), a scientific assistant to Müller-Breslau and Hertwig, took over structural engineering studies for architects.

Berlin-Charlottenburg TH gained a first-class civil engineering scientist in the shape of Ferdinand Schleicher, who enriched not only theory of structures, but almost all facets of structural engineering. From 1935 until his death in 1957, Schleicher (Fig. 6-62) was in charge of the journal *Der Bauingenieur*. Under his editorship, *Der Bauingenieur* was able to assert and increase its lead over the journal *Die Bautechnik*, which focused more on construction management. The *Taschenbuch für Bauingenieure* (pocket-book for civil engineers) [Schleicher, 1943], which was edited by Schleicher and appeared in 1943, was a publication in which the entire scientific fabric of civil engineering was given a precise summary by first-class authors; the second edition of this unique publication for the civil engineering industry appeared in 1955, once again edited by Schleicher [Schleicher, 1955]. In 2002 the *Schleicher*, as the Taschenbuch für Bauingenieure was soon christened in professional circles, formed the basic publishing idea behind the *Handbuch für Bauingenieure* (civil engineering manual), edited by Konrad Zilch, Claus Jürgen Diederichs and Rolf Katzenbach [Zilch et al., 2002].

Schleicher joined the National Socialist German Workers' Party on 1 May 1933 because his "political disorientation" led him "to take propaganda to be the reality and believe in 'ideal goals'" [Schleicher, 1949, p. 10]. "I never had any function in the Nazi Party or any of its organs," Schleicher added to his CV of 15 June 1949, plus the general remark: "Neither did I play an active part in National Socialism in any other way" [Schleicher, 1949, p. 10]. According to his own accounts, Schleicher repeatedly had disputes with party officials. Also as editor of the journal *Der Bauingenieur*, he found it difficult to ignore "political effusiveness" etc. "I think it is worth mentioning," Schleicher wrote, "that there was never a mention of the 'Fuehrer's birthday' etc. in the *Bauingenieur*. As the *Bauingenieur* was subject to the same stipulations of the Propaganda Ministry as, for example, *Bautechnik* or *Zentralblatt der Bauverwaltung*, one can only appreciate the difficulties by comparing the *Bauingenieur* with these other journals. It is only possible to sell journals these days by censoring certain passages, indeed many issues were banned completely, but this has *not* been the case with the *Bauingenieur* – with one exception: as Reichsminister TODT was killed, tributes had to be included, but I removed all trace of praise from those in the *Bauingenieur* to the extent that the editors received a warning" [Schleicher, 1949, pp. 10/11]. But although Schleicher was able to resume his post as professor at Berlin-Charlottenburg TH on 15 July 1945, by the end of that year the military government had decreed that all persons with a political involvement should be removed from the university, without exception. Thus, Schleicher's university career in Berlin was at an end, too.

From then on until 1952, Hertwig covered the whole field of structural steelwork, despite his great age. Pohl took over as professor of theory of structures, and after his death in 1948 it was Alfred Teichmann's (Fig. 6-63) turn to be responsible for structural theory, under the title of "theory of structures and mechanical engineering".

We know little about the private life of Alfred Teichmann (1902–71), who gained his doctorate at Berlin-Charlottenburg TH in 1931 with a dissertation on the flexural buckling of aircraft spars. Klaus Knothe has provided a comprehensive tribute to his scientific activities in aircraft construction and theory of structures [Knothe, 2005]. As a student of Hans

**FIGURE 6-62**
Ferdinand Schleicher (federal archives, Berlin)

Reissner, Teichmann quickly found aircraft construction to be a worthwhile field of activity. In 1937 he became a departmental head at the Institute of Strength of Materials at the German Aviation Testing Authority (DVL) in Berlin-Adlershof; in that same year he was appointed associate professor in the service of the Reich as a tribute to his scientific work. With the help of the "Z3", the first computer in the world which had been developed by Konrad Zuse and completed in 1941, flutter calculations were carried out on behalf of the DVL. By the end of 1941, Teichmann was emphasizing the significance of electronic computing: "The basic requirement for a desirable arrangement of flutter calculations is the use of automatic [= electronic – the author] computing systems. We can expect that devices of this kind will be widely used in the future" [Teichmann, 1942, p. 35].

After 1945 Teichmann was not in the forefront of scientific communication and had little contact with others in his profession. He passed on industrial commissions to his senior engineer Gerhard Pohlmann, who joined him in 1950. The perfectionistic, conscientious professor of structural theory dedicated himself totally to teaching at Berlin TU. According to the reports of former students, he delivered his lectures on structural theory with a rare pithiness. But he held back in research. Only once did he supervise a doctorate. For Teichmann, the fabric of theory of structures was complete. Notwithstanding, he did publish a three-volume work entitled *Statik der Baukonstruktionen* (theory of structures) towards the end of the 1950s. In that work, he integrated second-order theory (determination of internal forces in the deformed structural system) into a consistently formulated member analysis – the final contribution of the Berlin school of structural theory. Due to his introverted way of working, Teichmann reorganised the theory of structures of the Berlin school into a

**FIGURE 6-63**
Alfred Teichmann (university archives, Berlin TU)

**FIGURE 6-64**
Table for solving sets of linear equations after Teichmann/Pohlmann
[Teichmann & Pohlmann, 1952, sht. 1]

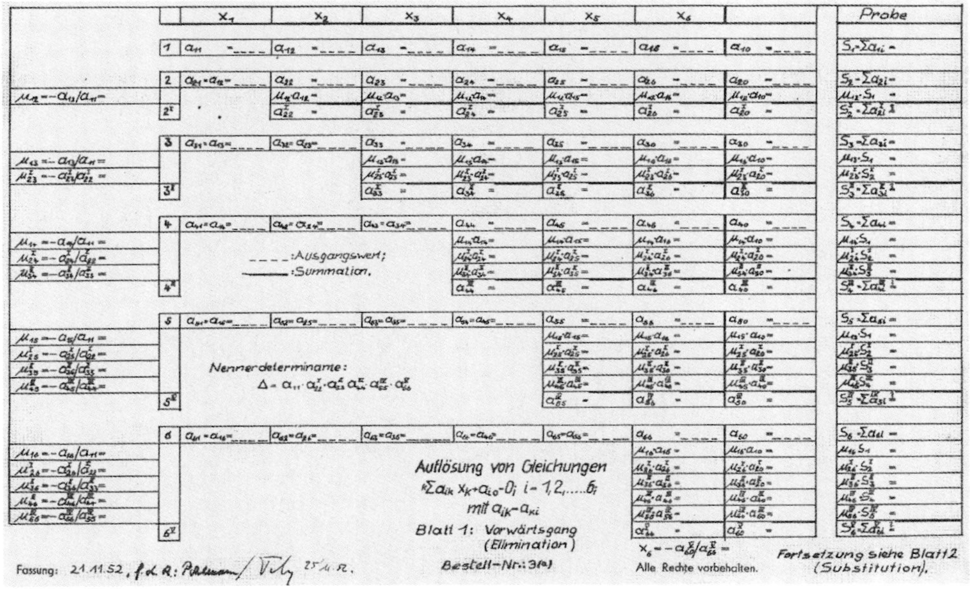

theory of structures that was a means to an end in itself. Unfortunately, his *Statik der Baukonstruktionen* suffered from excessive equations expressed in the tendency to index structural engineering variables. This is one reason why Teichmann failed to formalise theory of structures. Nevertheless, Teichmann advanced the resolution of structural calculations into tabular calculations and right up to programmability (Fig. 6-64). He pointed out as early as 1958 that the iteration method for solving large sets of equations would lose ground once computers started to be used [Teichmann, 1958, p. 99]. The fact that he did not actively support the process of establishing the computer in theory of structures was due to the suppression of his scientific activities prior to 1945, in which he called for the use of Zuse's computer in the course of armaments research. The other reason is his conservative insistence on the customary mathematical methods of solution in structural theory. Therefore, Teichmann, as the last representative of the Berlin school of structural theory, failed where John Argyris, who gained his diploma in civil engineering in Munich in 1936 and furthered his knowledge of mechanics, mathematics and aerodynamics at Berlin-Charlottenburg TH in 1940, succeeded in England between 1954 and 1957: complete formalisation of structural analysis by means of matrix algebra right up to structural matrix analysis and hence the step to computer-based structural mechanics.

And so the Berlin school of structural theory in its historico-logical sense ended in London by being superseded by computer mechanics.

# BOOK RECOMMENDATION

**Nixdorf, S.**
**Stadium Atlas**
Technical Recommendations
for Grandstands in Modern Stadia
2008. 352 pages with 700 figures
350 in color. Hardcover.
€ 149,–*/sFr 235,–
ISBN: 978-3-433-01851-4

## Stadium Atlas

**Unique technical guide of planning stadia combining the requirements of different decision authorities**

This Stadium ATLAS is a building-type planning guide for the construction of spectator stands in modern sports and event complexes. In this handbook, the principles of building regulations and the guidelines of important sports associations are analyzed and inter-related in order to clarify dependencies and enable critical conclusions on the respective regulations. The Stadium ATLAS aims to illustrate the constructional and geometrical effects of certain specifications and to facilitate decision-making for planners and clients regarding important parameters of stadium design.

*In EU countries the local VAT is effective for books and journals. Postage will be charged.
Our standard terms and delivery conditions apply.
Prices are subject to change without notice.

**Ernst & Sohn**
Verlag für Architektur
und technische Wissenschaften GmbH & Co. KG

www.ernst-und-sohn.de

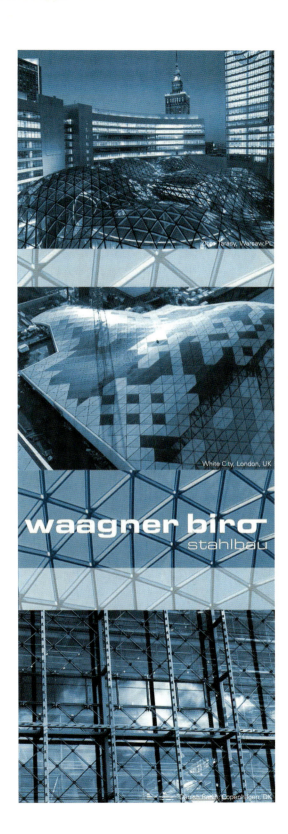

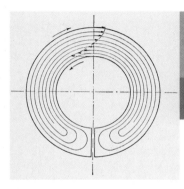

CHAPTER 7

# From construction with iron to modern structural steelwork

The author got to know Edoardo Benvenuto in 1992. It was Benvenuto who compiled a comprehensive work on the history of structural mechanics as early as 1981, and in 1993, together with Patricia Radelet-de Grave, established a group dedicated to the theme of *Between Mechanics and Architecture*. Benvenuto invited the author to contribute, and this began a period of intensive discussions with Benvenuto and his collaborators – Antonio Becchi, Massimo Corradi and Federico Foce – concerning the relationship between construction theory formation and the history of design. After the author became chief editor of the journal *Stahlbau*, published by Ernst & Sohn, in 1996, he was able to combine his passion for the history of science with professional interests by tracing the history of steelwork science. And so, for example, he published an abridged history of steelwork science in *Stahlbau* on the occasion of the 100th anniversary of the German Steelwork Association (DStV) in 2004. The most important results of this work have been incorporated in sections 7.4 and 7.5; the author benefited here from Holger Eggemann's research into the history of composite steel-concrete construction.

Iron construction took the Industrial Revolution into the building industry. It advanced the programme of theory of structures outlined by Navier in 1826, which more than 100 years ago characterised the structural analysis theories of Rankine, Maxwell, Culmann, Winkler, Castigliano, Mohr and Müller-Breslau, indeed the planning work of the structural engineer. Construction with iron had a permanent influence on the self-image of civil and structural engineers and the social status of their work.

By using Bessemer steel for the construction of the large bridges over the River Weichsel at Dirschau, Marienburg and Fordon (1889–93), Mehrtens helped the technological innovation of the iron industry to achieve a breakthrough in the building industry as well: steel took the place of wrought iron, and construction in iron made way for construction in steel. It was also Mehrtens who, commissioned by the six most important bridge-building companies in Germany, published the book entitled *A Hundred Years of German Bridge Building* on the occasion of the World Exposition in Paris in 1900. This book describes the engineering science theories plus the design, fabrication and erection of iron and steel bridges as well as the associated engineering science theories in a harmonised historico-logical unity, a unity that attracted international acclaim and generated exports for Germany's steel industry up until 1914.

From 1910 until the years of hyperinflation, the journal *Der Eisenbau – Constructions en Fer – Steel Constructions* was the first German-language trade journal for structural steelwork with an international outlook. Besides showing the technical state of the art by means of the exemplary steel structures of those times, in contrast to the up-and-coming reinforced concrete construction, it concentrated on developments in theory of structures during the consolidation period of this fundamental engineering science discipline.

It was not until 1928, at the instigation of Gottwalt Schaper, head of the Bridges & Structural Engineering Department in the Transport Ministry and for many years chairman of the German Committee for Structural Steelwork (DASt), that a new publication forum for structural steelwork was set up – by Wilhelm Ernst & Sohn – in the form of the journal *Der Stahlbau*. Under the editorship of August Hertwig, Müller-Breslau's successor at the chair of theory of structures & structural steelwork at Berlin TU, the new publication was modelled on the journal *Der Eisenbau* and – up until 1914 – its successful promotion of international collaboration among steelwork fabricators and contractors [Der Stahlbau, 1928, p. 1], and placed particular emphasis on the balance between detail and overall picture in terms of content and form. The aim of this concept was to
- demonstrate the load-carrying capacity and quality of newly developed structural steels for economic use in building,
- describe exemplary steel projects,
- identify the special features and qualities of structural steelwork,
- provide information for clients and contractors,
- promote the science and art of construction with steel, and
- investigate the economic relationships in structural steelwork.

Iron construction evolved into the independent discipline of steel construction within the activities of the structural engineer. Just how laborious the theory formation process was is shown by the integration of Saint-Venant's torsion theory into the scientific foundation of structural steelwork, alongside the establishment of new subdisciplines such as crane-building. This also applies to the paradigm change from elastic to plastic design methods in structural steelwork, which slowly became established during the invention phase of structural theory (1925 – 50) in the area of the buckling theory of steel stanchions. By the end of this development phase there existed a stability theory that owed its genesis to steel and lightweight metal construction, and structural steelwork and theory of structures had merged to form the steelwork science of Kurt Klöppel and his school, which was set to enrich significantly the innovation phase of structural theory (1950 – 75). The development of structural analysis theories during this period was therefore inspired by the construction of steel shell, composite steel-concrete and lightweight structures. The trend towards system-based structural steelwork, evident since the late 1970s, has loosened the close ties to statics-constructional thinking not only in structural calculations, but also in the formation of structural theories, and has established technological thinking in its own right. In the current diffusion phase of structural theory (1975 to date), synthesis activities have been successful only in exceptional cases. One such exception is the scientific life's work of Christian Petersen: his three monographs on theory of structures [Petersen, 1980], structural steelwork [Petersen, 1988] and building dynamics [Petersen, 1996] have created an as yet unsurpassed encyclopaedic summary and engineering science foundation for these three bodies of knowledge and cognition in structural engineering.

## 7.1

**Torsion theory in iron construction and theory of structures from 1850 to 1900**

Whereas the foundations of torsion theory had been laid right at the start of the establishment phase of structural theory (1850 – 75), classical theory of structures had only managed to integrate these into its fabric to a very limited extent.

### 7.1.1

**Saint-Venant's torsion theory**

Adhémar Jean-Claude Barré de Saint-Venant (1797 – 1886) had already developed the principles of his torsion theory by 1847. Its application by construction and mechanical engineers in the 19th century was initially associated with a simplification.

According to Isaac Todhunter (1820 – 84) and Karl Pearson (1857 – 1936), the Saint-Venant torsion theory was first seen in 1847 in the form of three *Mémoires*

- *Mémoire sur l'équilibre des corps solides, dans les limites de leur élasticité, et sur les conditions de leur résistance, quand les déplacements éprouvés par leurs points ne sont pas très-petits* (memoir on the equilibrium – against torsion – of solid bodies at the elastic limits plus their resistances when the displacements experienced by their points are not very small) [Saint-Venant, 1847/1],

- *Mémoire sur la torsion des prismes et sur la forme affectée par leurs sections transversales primitivement planes* (memoir on the torsion of prismatic bars plus the form of their cross-sectional surfaces subjected to torsion which were originally plane) [Saint-Venant, 1847/2], and
- *Suite au Mémoire sur la torsion des prismes* (supplement to the memoir on the torsion of prismatic bars) [Saint-Venant, 1847/3].

"These mémoires," they write, "are epoch-making in the history of the theory of elasticity; they mark the transition of Saint-Venant from Cauchy's theory of torsion to that which we must call after Saint-Venant himself. The later great mémoir on Torsion is really only an expansion of these three papers" [Todhunter & Pearson, 1886, p. 868].

What Todhunter and Pearson mean by the "later great mémoir on Torsion" is the *Mémoire sur la Torsion des Prismes, avec des considérations sur leur flexion, ainsi que sur l'équilibre intérieur des solides élastiques en général, et des formules pratique pour le calcul de leur résistance à divers efforts s'exerçant simultanément* (memoir on the torsion of prismatic bars, including considerations of their bending, and on the internal equilibrium – against torsion – of elastic solid bodies in general, and practical formulas for the calculation of their equilibrium with respect to diverse actions taking place simultaneously) [Saint-Venant, 1855], presented to the Paris Academy of Sciences on 13 June 1853 and used in lectures by Cauchy, Poncelet, Piobert and Lamé. It is regarded in the pertinent literature as the foundation of the Saint-Venant torsion theory, e. g. in Love's *A treatise on the mathematical theory of elasticity* [Love, 1892/93] and in Timoshenko's *History of strength of materials* [Timoshenko, 1953].

What reasons do Todhunter and Pearson give for this premature birth of the Saint-Venant torsion theory?

*Firstly:* Although in his first *Mémoire* [Saint-Venant, 1847/1] Saint-Venant develops only the basic equations, they contain the realisation that the cross-sectional surface of a prismatic bar with a non-circular cross-section subjected to torsion is no longer plane, but rather curved, i.e. warped. Saint-Venant mentions this in conjunction with a number of tests on models. He writes: "I have checked this theory experimentally in a different way. I subjected two 20 cm long prismatic rubber bars to torsion: one square with a side length of 3 cm, the other with a rectangular cross-section measuring $4 \times 2$ cm. The lines drawn on the side faces prior to applying the torsion would have remained straight and perpendicular to the axis if there had been no warping of the cross-sectional surfaces; they would have merely assumed a lateral inclination to the axis of the rectangular prismatic bar and would have remained straight if only the first bending had taken place. Instead, these lines were bent into an S-shape such that the ends of these lines remained perpendicular to the edges, which certainly proves the aforementioned warping" [Saint-Venant, 1847/2, p. 263].

*Secondly:* The second *Mémoire* [Saint-Venant, 1847/2] contains the solution of the torsion problem of the prismatic bar with a rectangular cross-section (Fig. 7-1). Saint-Venant was able to confirm the resulting torsion moment through experiments carried out by Duleau and Savart; furthermore, he notes that Cauchy's results are valid for very slender rectangular cross-sections only. The solution shapes the one pursued in the *Mémoire* of 1853 [Saint-Venant, 1855].

*Thirdly:* In his third *Mémoire* [Saint-Venant, 1847/3], Saint-Venant finally derives the warpage function $u$ of the prismatic bar with an elliptical cross-section due to a torsion moment. This solution, too, is reproduced in his *Mémoire* dating from 1853 [Saint-Venant, 1855].

Through Todhunter and Pearson we can therefore establish that Saint-Venant had already formulated his torsion theory in the three *Mémoires* of 1847. This does not apply to his semi-inverse method ("méthode mixte ou semi-inverse"), which is formulated explicitly and expounded comprehensively in his torsion theory *Mémoire* of 1853 [Saint-Venant, 1855, pp. 232–236].

Saint-Venant's real success in torsion and bending theory is his semi-inverse method. Whereas in the inverse method the displacements are based on a known solution from which the distortions and stresses are determined with the help of boundary conditions, in the semi-inverse method only some of the unknowns are given; the missing defining pieces are searched for such that the differential equations and the boundary conditions are satisfied. The semi-inverse method is explained below using Saint-Venant's torsion theory as an example.

The task involves a prismatic bar with a solid cross-section subjected to a torsion moment $M_t$ along its axis (Fig. 7-2). Subjected to torsion, the point $P(y;z)$ rotates about the axis of the bar through the angle of twist

$$\vartheta_x = \frac{x}{l} \cdot \vartheta$$

in such a way that point $P$ now has the coordinates

$$y + v = r \cdot \cos(\beta + \vartheta_x)$$

$$z + w = r \cdot \sin(\beta + \vartheta_x)$$

As $\vartheta_x$ is small,

$$\cos \vartheta_x = 1 \quad \text{and}$$

$$\sin \vartheta_x = \vartheta_x$$

can be used such that in the end the displacements of point $P$ in the $z$-$y$-plane are

$$v = -\frac{\vartheta}{l} \cdot x \cdot z \quad \text{and} \tag{7-1}$$

$$w = \frac{\vartheta}{l} \cdot x \cdot y \tag{7-2}$$

Saint-Venant now assumes that all cross-sections of the bar are displaced (warped) in the same way in the direction of the bar axis, i.e. remain congruent to one another; for the displacements independent of $x$ Saint-Venant sets

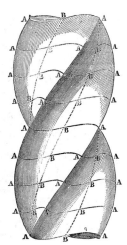
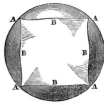

**FIGURE 7-1**
Twisted prismatic bar with a square cross-section after Saint-Venant [Pearson, 1889, p. 27]

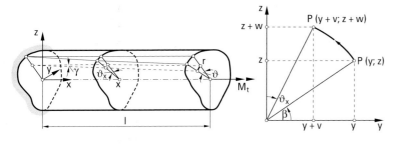

FIGURE 7-2
Prismatic solid bar subjected to a torsion moment $M_t$

$$u(y,z) = \frac{\vartheta}{l} \cdot \varphi(y,z) \qquad (7\text{-}3)$$

where $\varphi = \varphi(y,z)$, a warpage function still to be determined. As in contrast to warping torsion the normal stresses in Saint-Venant torsion are

$$\sigma_x = \sigma_y = \sigma_z = 0$$

then neglecting the force of gravity, the equilibrium conditions are as follows:

$$\frac{\partial \tau_{yx}}{\partial y} + \frac{\partial \tau_{zx}}{\partial z} = 0 \qquad (7\text{-}4)$$

$$\left.\begin{array}{l} \dfrac{\partial \tau_{xy}}{\partial x} + \dfrac{\partial \tau_{zy}}{\partial z} = 0 \\[1em] \dfrac{\partial \tau_{xz}}{\partial x} + \dfrac{\partial \tau_{yz}}{\partial y} = 0 \end{array}\right\} \qquad (7\text{-}5)$$

If we now enter eq. 7-1, eq. 7-2 and eq. 7-3 into the deformation equations

$$\left.\begin{array}{l} \tau_{xy} = G \cdot \left( \dfrac{\partial v}{\partial x} + \dfrac{\partial u}{\partial y} \right) \\[1em] \tau_{xz} = G \cdot \left( \dfrac{\partial u}{\partial z} + \dfrac{\partial w}{\partial x} \right) \\[1em] \tau_{yz} = G \cdot \left( \dfrac{\partial v}{\partial z} + \dfrac{\partial w}{\partial y} \right) \end{array}\right\} \qquad (7\text{-}6)$$

it follows that

$$\left.\begin{array}{l} \tau_{xy} = G \cdot \dfrac{\vartheta}{l} \cdot \left( \dfrac{\partial \varphi}{\partial y} - z \right) \\[1em] \tau_{xz} = G \cdot \dfrac{\vartheta}{l} \cdot \left( \dfrac{\partial \varphi}{\partial z} + y \right) \\[1em] \tau_{yz} = 0 \end{array}\right\} \qquad (7\text{-}7)$$

Eq. 7-7 satisfies the Laplace differential equation for $\varphi$

$$\frac{\partial^2 \varphi}{\partial y^2} + \frac{\partial^2 \varphi}{\partial z^2} = \Delta \varphi = 0 \qquad (7\text{-}8)$$

That means that the general stress and deformation equations are exhausted and it merely remains to satisfy the boundary condition for the shear stress vector $\tau$. As $\tau$ must form a tangent with the boundary curve (Fig. 7-3), it follows that

$$tan\,\psi = \frac{dz}{dy} = \frac{\tau_{xz}}{\tau_{xy}}$$

from which it follows that

$$\tau_{xz} \cdot dy - \tau_{xy} \cdot dz = 0$$

or, using eq. 7-7

$$\left(\frac{\partial \varphi}{\partial z} + y\right) \cdot dy - \left(\frac{\partial \varphi}{\partial y} - z\right) \cdot dz = 0 \tag{7-9}$$

follows on the boundary curve.

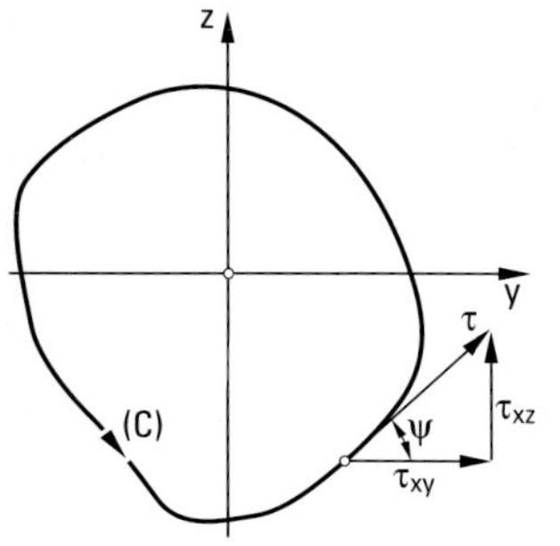

Saint-Venant now proceeds in such a way that he first determines the warpage function $\varphi(y, z)$ that satisfies the boundary value problem given by eq. 7-8 and eq. 7-9. From the condition that the shear stresses at every cross-section must produce a moment in equilibrium with the external torsion moment $M_t$

$$M_t = \int_F (\tau_{xz} \cdot y - \tau_{xy} \cdot z) \cdot dF \tag{7-10}$$

Saint-Venant determines the angle of rotation $\vartheta$ taking into account eq. 7-7 and in the end the shear stresses $\tau_{xy}$ und $\tau_{xz}$ by entering $\varphi(y, z)$ and $\vartheta$ in eq. 7-7. The critical variable in the Saint-Venant torsion theory in the version by August Föppl [Föppl, 1917/1] is the determination of the proportionality factor in eq. 7-3:

$$D = \frac{\vartheta}{l} = \frac{M_t}{G \cdot I_t} \tag{7-11}$$

where $D$ is the torsion, $G \cdot I_t$ is the torsional stiffness of the bar cross-section, $G$ is the shear modulus, and $I_t$ is the torsion constant.

The innovation in Saint-Venant's semi-inverse method is the combination of the practical elasticity theory and the theoretical interpretation of tests. Constant throughout the historico-logical development of torsion problems in mechanical and structural engineering is the determination of the torsion $D$ and the torsion constant $I_t$ through theory and tests. Saint-Venant's *Mémoires* dating from 1847 and 1853 overcame the one-sidedness of the merely empirical and merely theoretical approaches. Saint-Venant had clearly asserted the distinguishing feature of classical engineering sciences.

**FIGURE 7-3**
On the boundary condition for the shear stress vector

### 7.1.2 The torsion problem in Weisbach's *Principles*

One very early application of the special results of the Saint-Venant torsion theory can be found in the third edition of Julius Weisbach's *Principles of the Mechanics of Machinery and Engineering* dating from 1855. Weisbach is concerned with the sizing of timber as well as cast- and wrought-iron shafts as an important machine element in water-power systems. Fig. 7-4 shows a timber shaft with square and circular cross-sections subjected to torsion. After Weisbach describes the older torsion theory going back to Coulomb, Navier and others, he uses Saint-Venant to determine their limits. He writes: "The given theory gives us torsion moments somewhat different from the reality because in its development it was assumed that the end surfaces of the prism subjected to torsion remain plane while the torsion acts, whereas in reality these surfaces are warped.

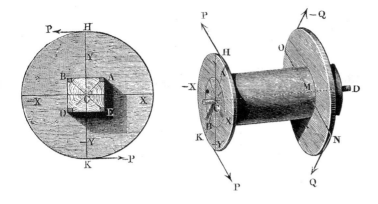

**FIGURE 7-4**
Shaft with square and circular cross-sections after Weisbach [Weisbach, 1855, p. 428]

According to the investigations of Saint Venant," and here Weisbach refers to the *Mémoire* dating from 1853 [Saint-Venant, 1855], "for a square shaft" [Weisbach, 1855, p. 429]

$$M_t = 0.0526 \cdot \frac{\vartheta}{l} \cdot E \cdot h^4 \tag{7-12}$$

Using different notation to that of Weisbach, $h$ is the side length of the square and $E$ is the elastic modulus, for which he specifies the empirical relationship with the shear modulus $G$ as

$$G = 0.3756 \cdot E \tag{7-13}$$

Using eq. 7-13, the torsion moment can be expressed as follows:

$$M_t = 0.140 \cdot \frac{\vartheta}{l} \cdot G \cdot h^4 \tag{7-14}$$

The exact solution is

$$M_t = 0.141 \cdot \frac{\vartheta}{l} \cdot G \cdot h^4 \tag{7-15}$$

Comparing the coefficients of eq. 7-11 and eq. 7-14 leads to the torsion stiffness

$$G \cdot I_t = G \cdot 0.140 \cdot h^4 \tag{7-16}$$

Weisbach calls the product $G \cdot 0.14$ or $0.3756 \cdot E \cdot 0.14 = 0.0526 \cdot E$, converted from radian measure to angular measure with the factor $\vartheta = (\pi/180) \cdot \vartheta° = 0.017453 \cdot \vartheta°$, the rotation coefficient $D$ (not be confused with torsion $D$, see eq. 7-11); he specifies $D$ for a square cross-section as

$$D = 0.0526 \cdot 0.017453 \cdot E = 0.000918 \cdot E \tag{7-17}$$

so eq. 7-16 can be written as follows:

$$G \cdot I_t = D \cdot h^4 \tag{7-18}$$

Weisbach produced a table (Fig. 7-5) for practical applications in which he specifies the rotation coefficient for shafts with a square cross-section $D$ and a circular cross-section $D_1$ using the elastic moduli of timber, cast iron, brass, basic pig iron and wrought iron. And Weisbach's interest in torsion theory ends there.

430   Dritter Abschnitt. Sechstes Capitel.

Torsion.  $$Pa = \frac{3\pi}{16}\frac{\alpha Er^4}{l} = 0{,}59\,\frac{\alpha Er^4}{l}.$$

Bei Körpern, deren Querschnittsdimensionen sehr von einander abweichen, fallen die Abweichungen größer aus. Z. B. für ein Parallelepiped, dessen Höhe $h$ von seiner Breite $b$ vielfach übertroffen wird, ist nach Saint-Venant und Cauchy:

$$Pa = \frac{2}{0{,}841}\frac{\alpha Eb^3h}{16l} = 0{,}149\,\frac{\alpha Eb^3h}{l}.$$

Legt man die letzten Formeln von Wertheim zu Grunde und verwandelt man die Torsionsbogen $\alpha$ in Winkel, setzt also:

$$\alpha = \frac{\pi\alpha^0}{180} = 0{,}017453\cdot\alpha^0,$$

so erhält man für den quadratischen Schaft:

$$Pa = 0{,}0009180\,\frac{\alpha^0 Eb^4}{l},$$

und für die kreisrunde Welle:

$$Pa = 0{,}010297\,\frac{\alpha^0 Er^4}{l}.$$

Setzt man nun:

$$Pa = \frac{\alpha^0 b^4}{l}D \quad\text{und}\quad Pa = \frac{\alpha^0 r^4}{l}\cdot D_1,$$

so hat man für die Wellen mit quadratischem Querschnitte:
$$D = 0{,}0009180\,E,$$
und für die Wellen mit kreisförmigem Querschnitte:
$$D_1 = 0{,}010297\,E,$$

und nimmt man nun für die bei Wellen gewöhnlich angewendeten Stoffe die Elasticitätsmodel $E$ aus der Tabelle in §. 200, so läßt sich folgende Tabelle der Drehungscoefficienten $D$ und $D_1$ zusammensetzen.

| Materie der Wellen. | Quadratischer Querschnitt. | | Kreisförmiger Querschnitt. | |
|---|---|---|---|---|
| | Französisches Maaß. | Preußisches Maaß. | Französisches Maaß. | Preußisches Maaß. |
| Holz . . . . . . . | 92 | 1400 | 1030 | 15000 |
| Stahl u. Schmiedeeisen . . . . . . | 1836 | 27500 | 20600 | 310000 |
| Gußeisen . . . . . | 872 | 13000 | 10000 | 150000 |
| Messing . . . . . | 1500 | 22000 | 17000 | 250000 |

**FIGURE 7-5**
Rotation coefficients for square D (see eq. 7-17 and eq. 7-18) and circular cross-sections $D_1$ after Weisbach [Weisbach, 1855, p. 430]

Summing up, we should note that:

*Firstly:* As Weisbach was writing more for mechanical engineers than civil and structural engineers, his torsion theory deals only with the design of shafts.

*Secondly:* From that it follows that Weisbach's work is based on results. He is not interested in providing a foundation for torsion theory, but rather in compiling practical formulas for the design of shafts subjected to torsion. Therefore, Saint-Venant's semi-inverse method is uninteresting for Weisbach.

*Thirdly:* The practical character of Weisbach's applied mechanics is revealed in the development of the proportionality factor $D$ for determining the torsion stiffness $G\cdot I_t$ according to eq. 7-18, and its numerical evaluation for practical cases in mechanical engineering. Nevertheless, he did not manage to express the concept of the torsion constant $I_t$: Weisbach's proportionality factor $D$ did not yet have the status of an engineering science concept.

*Fourthly:* Weisbach's applied mechanics unites the practice-based scientific mechanics and the mechanical and construction engineering given a scientific footing through mechanics on the level of a ready-made formula. For this reason, Weisbach's applied mechanics advanced to become the most important compendium of applied mechanics for civil, structural and mechanical engineers in the period 1850–75.

### 7.1.3 Bach's torsion tests

The foundation of the practical elastic theory and strength of materials through experimentation achieved a new quality in Carl Bach's book *Elasticität und Festigkeit* (elasticity and strength), which was published in 1889/90. In his book, Bach assumes that "knowledge of the actual behaviour of the materials is of paramount importance" [Bach, 1889/90, p. III]; for it is no longer sufficient "to base the whole mathematical fabric of elasticity and strength solely on the law of proportionality between stress and strain". Bach continues: "[For the design engineer] specifying the dimensions who is fully aware of the true relationships and does not wish to follow traditions, [it appears necessary], again and again, to update the premises of the individual equations he uses in the light of experience, if such is available, and to assess the relationships achieved through mathematical derivation with regard to their degree of accuracy. Insofar as our current level of knowledge renders this possible, and that wherever the latter and the deliberation (including the consultation and evolution of new methods) are inadequate we should first try to direct the question at nature – through experimentation – in order to contribute to promoting technology and hence industry" [Bach, 1889/90, pp. VII–VIII].

Concerning the adoption of Saint-Venant's torsion theory by engineers, Bach notes that "despite its scientific thoroughness it has found its way into the technical literature only in isolated instances" [Bach, 1889/1890, p. VI]. According to him, the reason for this is "the lack of the relative simplicity of the calculations leading to the solution" [Bach, 1889/90, pp. VI–VII]. This is why Bach, in his book, carries out the "individual developments singly as far as possible" and limits "the mathematical apparatus required for this by taking into account tests wherever possible" [Bach, 1889/90, p. VI]. "The ever-more urgent need," Bach writes, "to be able to determine [the torsion in prismatic bars with a non-circular cross-section] with more dependability than was possible hitherto called for a detailed treatment of the associated tasks … In doing so, it became necessary to face up to the deformations that up until now had been totally neglected when assessing the material stresses and strains" [Bach, 1889/90, p. VI].

In this last sentence Bach is alluding to the influence of the resistance to warping of cast-iron bars with square and rectangular cross-sections, resulting in fracture due to torsion (Fig. 7-6).

**FIGURE 7-6**
Torsion-induced fracture of cast-iron bars with square (left) and rectangular (right) cross-sections after Bach [Bach, 1889/90, plate VIII]

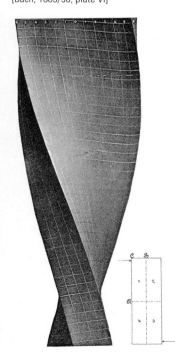

**FIGURE 7-7**
Deformation of a bar made of hard lead with a rectangular cross-section subjected to torsion after Bach [Bach, 1889/90, plate VI]

Bach devotes 50 pages to the torsion problem in his 376-page book *Elasticität und Festigkeit*. He introduces every Saint-Venant torsion case phenomenologically and subsequently interprets it theoretically. He begins with the simplest case of the bar with a circular cross-section and then proceeds to bars with elliptical and rectangular cross-sections. For example, Fig. 7-7 shows the deformation of a twisted bar made of hard lead with a cross-section measuring 60 × 20 mm. Bach used photography to a great extent, something that Saint-Venant was unable to do.

After Bach has described precisely the phenomena revealed by the torsion test, he interprets these with the mathematics of mechanics and the help of a simple apparatus. At the end of this procedure he presents the engineer with the simple proof equation

$$M_{t,exist} \leq \varphi \cdot \tau_{perm} \cdot \frac{I_{min}}{b} \tag{7-19}$$

where

$\varphi$ is a proportionality factor with the following values:
– 2 for a full circle and an annulus where $b = d/2$
– 2 for a full ellipse and an elliptical annulus
– 8/3 for a rectangle

$\tau_{perm}$ is the maximum shear stress of the corresponding cross-section
$I_{min}$ is the smallest second moment of area of the cross-section
$b$ is the radius of the circle, the smallest semi-axis of the ellipse, and the smaller side of the rectangle.

Using Bach's equation, it is now easy for the engineer to check the shear stress for the Saint-Venant torsion in the given bar cross-sections. He also provided similar equations for twisted bars with cross-sections in the shape of an equilateral triangle and a regular hexagon. Finally, Bach establishes that "the equations given in this and the preceding paragraphs for calculating bars subjected to rotation ... [were] devised under the implicit assumption that the warping of the cross-section can take place unhindered" [Bach, 1889/90, p. 162]. Bach therefore clearly distinguishes between Saint-Venant torsion and warping torsion.

Following this phenomenologically oriented introduction to the Saint-Venant torsion problem, Bach discusses his own and Bauschinger's torsion tests on bars with the cross-sections common at that time. Like Bauschinger, he based his work on a modern version of eq. 7-11 discovered by Saint-Venant. Fig. 7-8 shows the results of Bach's tests in the form of a table.

For the twisted bar with a square cross-section, for example, we can infer the equation for the angle of rotation related to the unit length

$$\vartheta = 3.56 \cdot M_d \cdot \frac{b^2 + h^2}{b^3 \cdot h^3} \cdot \beta \qquad (7\text{-}20)$$

When $M_d = M_t$, $b = h$ and $\beta = 1/G$, the torsion moment $M_t$ is

$$M_t = 0.14 \cdot \vartheta \cdot G \cdot h^4 \qquad (7\text{-}21)$$

This result matches the solution already given by Weisbach (see eq. 7-14).

| No. | Querschnittsform | Drehungsmoment $M_d$ | Drehungswinkel $\vartheta$ | $K_d : K_z$ für Gusseisen |
|---|---|---|---|---|
| 1 | 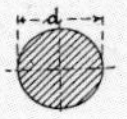 (circle, $d$) | $\frac{\pi}{16} k_d d^3$ | $\frac{32}{\pi} \frac{M_d}{d^4} \beta$ | reichlich 1 |
| 2 | 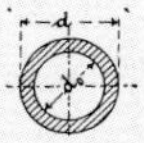 (hollow circle, $d$, $d_o$) | $\frac{\pi}{16} k_d \frac{d^4 - d_o^4}{d}$ | $\frac{32}{\pi} \frac{M_d}{d^4 - d_o^4} \beta$ | „ 0,8¹) |
| 3 |  (ellipse, $a > b$) | $\frac{\pi}{2} k_d a b^2$ | $\frac{1}{\pi} M_d \frac{a^2 + b^2}{a^3 b^3} \beta$ | 1 bis 1,25²) |
| 4 | 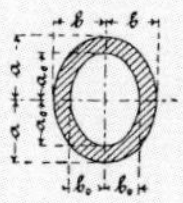 (hollow ellipse, $a > b$, $a_o:a = b_o:b = m$) | $\frac{\pi}{2} k_d \frac{a b^3 - a_o b_o^3}{b}$ | $\frac{1}{\pi} M_d \frac{a^2 + b^2}{a^3 b^3 (1-m^4)} \beta$ | 0,8 bis 1³) |
| 5 | 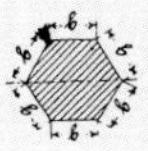 (hexagon) | $\frac{1}{1,09} k_d b^3$ | $0{,}967 \frac{M_d}{b^4} \beta$ | — |
| 6 | 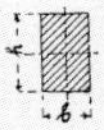 (rectangle, $h > b$) | $\frac{2}{9} k_d b^2 h$ | für $h:b = 1:1$   $3{,}56 M_d \frac{b^2 + h^2}{b^3 h^3} \beta$, <br> für $h:b = 2:1$   $3{,}50 M_d \frac{b^2 + h^2}{b^3 h^3} \beta$, <br> für $h:b = 4:1$   $3{,}35 M_d \frac{b^2 + h^2}{b^3 h^3} \beta$, <br> für $h:b = 8:1$   $3{,}21 M_d \frac{b^2 + h^2}{b^3 h^3} \beta$. | 1,4 bis 1,6²) |
| 7 |  (triangle) | $\frac{1}{20} k_d b^3$ | $46{,}2 \frac{M_d}{b^4} \beta$ | — |
| 8 | 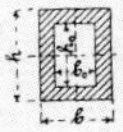 (hollow rectangle, $h > b$, $h_o:h = b_o:b$) | $\frac{2}{9} k_d \frac{b^2 h - b_o^2 h_o}{b}$ | — | 1 bis 1,25³) |

**FIGURE 7-8**
Summary of the torsion tests after Bach [Bach, 1889/90, pp. 185–187]

Some remarks on Bach's contribution to the torsion problem:

*Firstly:* In contrast to Weisbach, Bach – professor of mechanical engineering at Stuttgart TH – was also considering the needs of civil and structural engineers. Notwithstanding, Bach is seen as one of the leading representatives of materials testing in Germany, which was beginning to emerge in the last quarter of the 19th century and which made inroads into both mechanical and construction engineering after 1890.

*Secondly:* Bach's equation for the shear stresses in twisted bars (eq. 7-19) supplemented the stress equation concept set up by Navier. Therefore, Bach's work to a certain extent concludes the development of the proof equations for the engineer.

*Thirdly:* By extending his torsion tests to the cross-sections common in structural steelwork at that time (Fig. 7-8), Bach had given engineers working with steel a means of estimating quantitatively the influence of torsion on the stress state in a loadbearing system.

### 7.1.4 The adoption of torsion theory in classical theory of structures

The evolutionary process of turning theory of structures into a fundamental engineering science discipline for structural engineering was concluded around 1900 in the form of the classical theory of structures. But structural steelwork was the incarnation of the classical theory of structures and both were dominated by the trussed framework until the 1920s.

In the four editions of his book *Die neueren Methoden der Festigkeitslehre und der Statik der Baukonstruktionen* (newer methods of strength of materials and theory of structures), Heinrich Müller-Breslau included only one paragraph on the torsion of prismatic bars.

In the first edition [Müller-Breslau, 1886], he derives the differential equation of Saint-Venant torsion for a bar with a circular cross-section:

$$\frac{d\vartheta}{dx} = \frac{M_t}{G \cdot I_p} \qquad (7\text{-}22)$$

where $I_p$ is the polar moment of inertia, which is only identical with the torsion constant $I_t$ in the case of rotationally symmetric cross-sections. According to a publication by Saint-Venant dating from 1879, the value

$$I_t = \frac{A^4}{\kappa \cdot I_p} \qquad (7\text{-}23)$$

should be used instead of $I_p$ for non-rotationally symmetric cross-sections, where $A$ is the cross-sectional area and $\kappa$ is a factor for which a value of 40 is sufficiently accurate [Müller-Breslau, 1886, p. 161]. Müller-Breslau's original contribution to the torsion problem now consists of having added the torsion term to the virtual internal work (see eq. 6-37)

$$\int_{(l)} M_t(x) \cdot \frac{d\vartheta}{dx} \cdot dx = \int_{(l)} M_t(x) \cdot \frac{M_t(x)}{G \cdot I_t} \cdot dx \qquad (7\text{-}24)$$

After he has transformed the entire virtual work into an expression for the deformation energy $\Pi$, he applies the principle of Menabrea together with the second theorem of Castigliano to set up $n$ elasticity equations for the truss with $n$ degrees of static indeterminacy. The two examples calculated by Müller-Breslau are curved beams subjected to bending and torsion. He calculates only the internal forces for these statically indeterminate systems and neglects to calculate the stresses in the cross-section necessary for analysing the stresses. Classical theory of structures is a means to an end in itself!

This paragraph was incorporated unchanged into the second [Müller-Breslau, 1893] and third [Müller-Breslau 1904] editions of the *Neueren Methoden* ... In the fourth edition of this standard work of classical theory of structures, Müller-Breslau added a remarkable footnote to this paragraph [Müller-Breslau, 1913, p. 254] and included an extra exercise [Müller-Breslau, 1913, pp. 262–268]. The footnote reads: "At this point we must refer to the source and merely remark that for the important L-, T-, U- and I-shaped-shaped cross-sections it cannot yet be said that a practical solution to the torsion problem has already been found. Fritz Kötter made an important contribution in his treatise on the torsion of an angle in the proceedings of the Royal Prussian Academy of Sciences, 1908, XXXIX" [Müller-Breslau, 1913, p. 254]. Looked at from the point of view of classical theory of structures, we must agree with Müller-Breslau, as Bach's cross-section values for torsion over the entire cross-section demonstrate. In order to be used in work equation 7-24, we need the functional relationship between $\vartheta$, $M_t$ and the torsional stiffness $G \cdot I_t$ (eq. 7-22) or the corresponding $\kappa$-value from eq. 7-23 – and the Bach table (Fig. 7-8) contains no details. Owing to the symmetry of the system and the loading conditions, the elasticity equations are independent of $I_p$ and $I_t$ (eq. 7-23) in Müller-Breslau's sample calculation (Fig. 7-9). This ring structure, which is supported by vertical hangers at eight points and carries a uniformly distributed load, represents the structure carrying the lantern at the crown of the dome of Berlin Cathedral. Therefore, Müller-Breslau was not placed in the embarrassing situation of having to make assumptions regarding the torsional stiffness of the curved I-beam. In another leading work of classical theory of structures, Georg Christoph Mehrtens' *Statik und Festigkeitslehre* (theory of structures and strength of materials), the torsion problem is dealt with on just a few pages because "in engineered structures," Mehrtens explains to the reader, "the shear stresses caused by rotations at first play only a secondary role. As a rule, the member loads lie in one plane of forces and if twisting moments have to be taken into account in exceptional circumstances, the shear stresses to be calculated only assume the importance of secondary stresses" [Mehrtens, 1909, pp. 321–322]. Mehrtens, the leading steel bridge engineer from 1890 to 1910, lagged behind his scientific rival Müller-Breslau in this respect; although he also quotes the aforementioned work by Kötter, he regards his calculations as complicated [Mehrtens, 1909, p. 322].

The contemporary literature on structural steelwork also contains only isolated references to the torsion problem in common steel sections. Luigi Vianello (1862–1907) devotes less than three pages to this problem in his book *Der Eisenbau* (steel construction) [Vianello, 1905], a popular manual for bridge-builders and steel designers. Here, the reader finds the moment

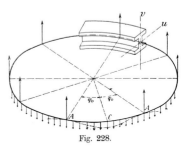

Fig. 228.

**FIGURE 7-9**
Ring structure subjected to torsion and bending at the top of the dome of Berlin Cathedral after Müller-Breslau [Müller-Breslau, 1913, p. 265]

of resistance against torsion $W_t$ for common cross-sections. Bach's equation for the shear stresses based on Saint-Venant's torsion (eq. 7-19) then takes on the following form:

$$M_{t,exist} = \tau_{exist} \cdot W_t \leq \tau_{perm} \cdot \varphi \cdot \frac{I_{min}}{b} \quad (7\text{-}25)$$

where

$$W_t = \cdot \varphi \cdot \frac{I_{min}}{b} \quad (7\text{-}26)$$

The proof of the shear stresses based on Saint-Venant's torsion is therefore identical with the form of the bending stresses equation:

$$M_{b,exist} = \sigma_{exist} \cdot W_b \leq \sigma_{perm} \cdot W_b \quad (7\text{-}27)$$

This method is very advantageous for designers and checkers. Vianello provides valuable constructional advice for the torsion-resistant support of T-beams and channels plus laced stanchions with a triangular cross-section and I-/L-sections or channels at the corners of the triangle.

The book *Die Statik des Eisenbaus* (theory of structural steelwork) published by W. L. Andrée in 1917 [Andrée, 1917] contains not a single mention of the torsion problem, although the book covers the whole range of steel applications interesting to the structural engineer: buildings, sheds, crane rails, airship hangars, slipway structures, conveyor structures for mining, cooling towers and bridges.

Even in the journal *Der Eisenbau (Steel Constructions, Constructions en Fer)*, with its international outlook, the effects of torsion moments were confined almost exclusively to isolated instances in papers on spatial structures. The contributions of Ludwig Mann, *Über die zyklische Symmetrie in der Statik mit Anwendungen auf das räumliche Fachwerk* (on the cyclic symmetry in theory of structures with applications for spatial structures) [Mann, 1911], and Henri Marcus, *Beitrag zur Theorie der Rippenkuppel* (contribution to the theory of ribbed domes) [Marcus, 1912], are just two examples. In both these papers, the terms of the virtual internal work due to torsion (eq. 7–24) or the deformation energy due to torsion moments is considered. However, these contributions are aimed at exploiting the topological properties for simplifying the elasticity equations or the stiffness matrix.

How can we explain this miserable situation regarding the application of the Saint-Venant torsion theory to structural steelwork between 1890 and 1920? The answer is given below in the form of six theses:

*First thesis:* Up until 1920 the trussed framework, which converted the external load actions into tensile and compressive forces, prevailed in structural steelwork; solid-web beams made from steel plates and carrying the loads via bending were only relevant in the building of bridges with small to medium spans. The preferred use of equal angle sections and flats enabled the construction of simple riveted joints practically free from torsion.

*Second thesis:* The bar-like rolled products of the steel foundries were predestined for the orthogonal construction of loadbearing elements throughout the structure. This is closely allied to the nature of steelwork drawings developing on the basis of orthogonal projection. The relationship between the geometrical basis of the method of drawing and the technical artefact drawn was more obvious in this discipline than any other.

*Third thesis:* The analysis and synthesis of the entire loadbearing system of the structure in plane sub-loadbearing systems is readily possible with the orthogonal construction principle of structural steelwork. The way the loads are carried is directly and geometrically obvious, whether based on graphical analysis constructions, steelwork drawings or the loadbearing system of the structure as built.

*Fourth thesis:* Seen in the light of the first three theses, it becomes clear that the plane, linear-elastic trussed framework theory forms the heart of classical theory of structures – regardless of the integration of beam theory and the masonry arch theory gradually developing into elastic arch theory.

*Fifth thesis:* The good disciplinary organisation of classical theory of structures disciplines its subjects insofar as these newer developments in related disciplines such as theoretical and even applied mechanics are almost overlooked. This blindness was particularly evident in the area of torsion theory; fans of the Old Testament would say it weighed like a curse on classical theory of structures!

*Six thesis:* As spatial loadbearing structures with sub-loadbearing systems curved in three dimensions started to appear, the torsion problem, in the form of torsional stiffness in the energy integral, gained a foothold in theory of structures. However, determining the torsion constant $I_t$ was still the scientific object of materials testing and mechanics. Not until 1917–21 did August Föppl succeed in backing up the concept of the torsion constant $I_t$ through experiments.

## 7.2 Crane-building at the focus of mechanical and electrical engineering, structural steelwork and theory of structures

Between 1890 and 1920, crane-building developed out of mechanical and electrical engineering and structural steelwork into a special technical field whose scientific self-image was in the first place derived from theory of structures.

The name Rudolph Bredt (1842–1900) is familiar to engineers. His two equations – Bredt's first and second equations – were both published in 1896 and made a decisive contribution to publicising the Saint-Venant torsion theory. However, Bredt wrote other pioneering works on applied mechanics.

But his greatest achievement was the development of the Ludwig Stuckenholz company, founded in 1830 in Wetter a. d. Ruhr, to become the first crane-building company in Germany and one of the parent companies of Demag (now Demag Cranes & Components GmbH). His ingenious crane designs were widely acclaimed at home and abroad and his "proven constructions" were regrettably also copied illegally.

Therefore, Bredt, as entrepreneur, as designer and as practical engineering scientist, created the foundations of crane-building.

**Rudolph Bredt – the familiar stranger**

### 7.2.1

If you ask students of structural and mechanical engineering "Who was Bredt?" directly after passing their examinations in applied mechanics, they will immediately be able to mention the Bredt equations. And if they have passed with distinction, they should be able to quote both the first Bredt equation

$$\tau = \frac{M_t}{2A_m \cdot t} \quad \text{or} \quad \max \tau = \frac{M_t}{2A_m \cdot \min t} \tag{7-28}$$

and the second Bredt equation

$$I_t = \frac{4 A_m^2}{\oint \frac{ds}{t}} \tag{7-29}$$

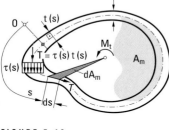

**FIGURE 7-10**
Torsion of thin-wall hollow sections after Bredt

(see Fig. 7-10), where
$M_t$ torsion moment
$\tau$ shear stress averaged over the wall thickness of the thin-wall hollow section
$A_m$ area enclosed by the centre-line of the wall of the thin-wall hollow section
$t$ thickness of the cross-section at the point under consideration
$s$ arc coordinate of the centre-line of the wall of the thin-wall hollow section
$I_t$ torsion constant

The circulatory integral in the denominator of eq. 7-29 takes on the following value for a constant wall thickness $t$ of the thin-wall hollow section

$$\oint \frac{ds}{t} = \frac{U_m}{t} \tag{7-30}$$

so eq. 7-29 can be rewritten as

$$I_t = \frac{4 A_m^2 t}{U_m} \tag{7-31}$$

where $U_m$ is the circumferential length of the centre-line of the wall of the thin-wall hollow section.

Taking the torsion

$$D = \frac{d\vartheta}{dx} = \vartheta' = \frac{M_t}{G \cdot I_t} \tag{7-32}$$

and the above equations by Bredt, the torsion theory of thin-wall hollow sections would be complete. This theory fits nicely into the fabric of the Saint-Venant torsion theory. And thus the name of Bredt is familiar to every structural and mechanical engineer.

**The Ludwig Stuckenholz company in Wetter a. d. Ruhr**

### 7.2.2

In the oil painting *Das Eisenwalzwerk (Moderne Cyklopen)* (The iron rolling mill – a modern Cyclops) (1872–75) (Fig. 7-11) by Adolph Menzel, a manually operated iron crane with a complicated gear mechanism in the top right corner of the canvas overlooks the hectic scenery of the iron rolling mill. The viewer suspects that this "modern Cyclops" has already done its work, and will soon do some more, according to the golden rule of mechanics – the less force you need, the more distance you need – or

**FIGURE 7-11**
*Das Eisenwalzwerk* by Adolph Menzel (1815–1905)

rather, according to the principle of virtual displacements. The upper half of Menzel's monumental work is dominated by the transmission linkages, the centre by the machinery of great industry described by Karl Marx in 1867 in his principal work *Capital*. The iron crane still stands alongside the trinity of prime mover, transmission and machine tool. The workers are still handling the workpiece on the roll stand in the centre of the picture. But Menzel's composition anticipates the germ of industrialisation: the tool as the vanishing point of the Industrial Revolution. Steam, the "agent [in the sense of motive power – the author] universally applicable in Mechanical Industry" as Marx liked to call it, had not yet reached the iron crane. But it would come with an iron will to celebrate frenetically its – albeit Pyrrhus – victory over human muscle power and water power before the close of the 19th century. As early as the 1890s, electricity was getting ready to lend a linear technical and organisational framework to industrial production.

### 7.2.2.1 Bredt's rise to become the master of crane-building

Rudolph Bredt (Fig. 7-12), born in Barmen on 17 April 1842, was the son of an evangelist couple. His father, Emil Bredt (1808–74), married Adelheid Heilenbeck (1813–84) and thus gained a foothold in the management of the Heilenbeck strips and strands factory in Barmen and through the church also took on a number of voluntary offices in the local community.

Rudolph Bredt inhaled the spirit of the Protestant ethic of the industrial middle class so splendidly analysed by Max Weber in 1904/05, in particular the Calvinistic idea of the necessity of preserving religious beliefs in worldly activities [Weber, 1993, p. 81], which formed the source of the scheduled organisation of one's own life, its methodical rationalisation. The resulting ascetic character of the religiousness and the onward march of the Protestant industrial middle class towards education in practical matters, as the antithesis of the neo-humanistic classics, would leave a deep and lasting impression on Rudolph Bredt. He passed his university entrance examination at the secondary school in Barmen in 1859 and enrolled in the second mathematics class at Karlsruhe Polytechnic in 1860–61. It was there that he must have heard Alfred Clebsch (1833–72), who as the teacher of Felix Klein (1849–1925) was to have a lasting effect on applied mathematics and mechanics. For instance, in 1862 Clebsch completed the first German-language monograph on elastic theory: *Theorie der Elasticität fester Körper* (theory of elastic rigid bodies) [Clebsch, 1862], which was translated (with a commentary) into French in 1883 by no less a person than Saint-Venant. Bredt studied mechanical engineering at the same establishment from 1861 to 1863, attending lectures given by that authority in theoretical mechanical engineering Ferdinand Redtenbacher (1809–63). But Bredt had already left by the time Redtenbacher's successor, Franz Grashof (1826–93), co-founder of the Association of German Engineers (VDI), arrived because thanks to his excellent references, including one provided by Clebsch, he was able to continue his studies at Zurich Polytechnic without having to pass the entry examination. He left there with good results in mechanical engineering and machine design (Franz Reuleaux) and excellent results in theoretical mechanical engineering (Gustav Zeuner). These engineering personalities, but especially Reuleaux, Grashof, Zeuner and later Emil Winkler, were working on the adaptation and expansion of mathematics and the natural sciences – following on from the technicisation of the production process so closely linked with the Industrial Revolution – for the purpose of modelling technical artefact classes: machine kinematics evolved from kinematics (Reuleaux), applied mechanics from mechanics (Grashof), applied thermodynamics from thermodynamics (Zeuner) and the practical elastic theory from elastic theory (Winkler). Rudolph Bredt was to remain true to this style of theory formulation during the discipline-formation period of the classical engineering sciences.

**FIGURE 7-12**
Rudolph Bredt (1842–1900)
(Mannesmann archives, Düsseldorf)

As a newly graduated mechanical engineer, Rudolph Bredt learned about the world of practical mechanical engineering from F. Wöhlert in Berlin (1864–67), probably Waltjen & Co. (the predecessor of AG Weser) in Bremen, and from John Ramsbottom's (1814–97) locomotive and machine factory in Crewe (England). It was in England that he carried out an intensive study of the latest lifting equipment, which back in Germany was still struggling to get off the ground. Furnished with considerable experience, Bredt returned to Germany in 1867 and took over the machine factory of Ludwig Stuckenholz in Wetter a. d. Ruhr together with Gustav Stuckenholz and Wilhelm Vermeulen, the son and son-in-law of the founder respectively. Very soon, Rudolph Bredt had turned Ludwig Stuckenholz' successful steam boiler and machine factory with iron foundry into the first factory for lifting equipment. The Hagen Chamber of Trade reported in 1871 that in this machine factory "the building of mechanically operated lifting machines, and preferably overhead travelling cranes with rope and shaft drives, began just a few years ago" and owing to the demand for such lifting devices "the factory could specialise exclusively in the building of such devices" (cited in [Rennert, 1999, p. 62]). In England, the "workshop of the world", Bredt was introduced to the overhead travelling crane with rope drive developed by Ramsbottom in Crewe in 1861. Fig. 7-13 shows the crane installation erected by Ramsbottom in 1861 for the repair workshops of the London & North Western Railway at Crewe.

The crane structure is a composite construction consisting of timber members with wrought-iron trussing. The steam-powered driving pulley is located at *A* (Fig. 10 in Fig. 7-13). The transmission from the driving pulley to the crane itself consists of a 16 mm dia. endless cotton rope. Bredt improved and simplified Ramsbottom's innovation (Fig. 7-14). This form of crane quickly became popular in Germany and had the following advantages over its predecessor:

- Positioning the drive mechanism at one end of the crane beam gave the crane operator a better overview of the factory floor and load movements.
- The simplified transmission obviated the need for the transverse rope beneath the shaft.
- Bredt accepted that omitting the worm gear would lead to a lower rope speed, but that meant he could reduce the energy input and increase the service life of the rope transmission.

Thus, in Germany as well, steam had reached the crane as the Industrial Revolution's "agent universally applicable in Mechanical Industry" (Marx). The crane was integrated into the trinity of steam engine, transmission and machine tool. The mechanical overhead travelling crane would become the leading type of factory crane prior to the introduction of electricity. In 1873 an overhead travelling crane with rope drive and 25 t lifting capacity designed by Bredt moved proudly above the exhibits in the pavilion of the German Reich at the World Exposition in Vienna (Fig. 7-15). As early as 1877, the British journal *Engineering* published a detailed description of this Bredt crane [n. n., 1877, p. 87]. And so the 35-year-old Rudolph Bredt advanced to become the master from Germany in the field of crane-building, and Bredt's products certainly did not fall into the category of "Made in Germany = cheap and poor", the verdict of Reuleaux at that time. Nevertheless, during the 1870s, cranes powered by human muscle power still remained the norm and overhead travelling cranes with mechanical transmission the exception. Bredt therefore designed overhead travelling cranes with a crank drive. But the replacement of human power as a source of energy in major industry was unstoppable, and in 1887 Bredt built the first electrically driven crane. He recommended such cranes only for large factories where the boiler and machine

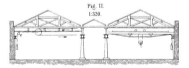
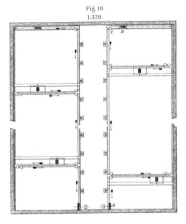

**FIGURE 7-13**
Ramsbottom's overhead travelling crane
[Ernst, 1883/2, plate 32]

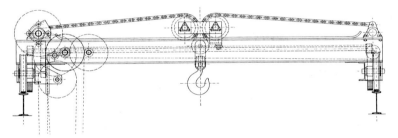

**FIGURE 7-14**
Bredt's overhead travelling crane
[Bredt, 1894/95, p. 13]

station were centralised and therefore efficient working was possible as soon as an electrical distribution network was installed.

In that same year, the Ludwig Stuckenholz company supplied the world's largest crane, a slewing jib crane with 150 t lifting capacity (Fig. 7-16). Bredt's crane designs were described again and again in the British journal *Engineering* [n. n., 1881, pp. 192–194; 1884/1, pp. 166–167; 1884/2, pp. 308, 315], but alas were also copied by rival companies and marketed in their catalogues and advertisements as, for example, "overhead travelling cranes in proven construction". For Bredt this was a reason to publish his catalogue *Krahn-Typen der Firma Ludwig Stuckenholz* [Bredt, 1894/95] shortly after the World Exposition in Chicago (1893); another reason was "to provide a large number of my customers with an overview of my designs", in particular "those that are used frequently and can be regarded as types" [Bredt, 1894/95, p. 7]. It should be mentioned that Bredt's slewing jib crane was awarded a medal in Chicago in 1893; the accompanying certificate reads as follows: "The photographs on display show large port cranes lifting heavy items. These cranes are used extensively, are of a standard form of construction and place the work of the factory, as is generally acknowledged, on a high level of engineering" (cited in [Bredt, 1894/95, p. 82]).

**FIGURE 7-15**
Bredt's overhead travelling crane for the World Exposition in Vienna [Bredt, 1894/95, p. 19]

**FIGURE 7-16**
Bredt's slewing jib crane for the Port of Hamburg [Bredt, 1894/1895, p. 75]

| 7.2.2.2 | **Crane types of the Ludwig Stuckenholz company** |

In his catalogue *Krahn-Typen der Firma Ludwig Stuckenholz* (Fig. 7-17), the master of German crane-building not only demonstrated his ingenious technical accomplishments, he also created a classic of German-language crane-building literature, which had repercussions far beyond the boundaries of the German Reich. It is a symphony of crane-building conducted by Bredt according to music written by his customers. The catalogue is divided into four chapters:

I   Cranes for machine factories, foundries and boiler forges
   1. Overhead travelling cranes
   2. Slewing jib cranes
   3. Travelling jib cranes
   4. Cranes for special purposes
II  Cranes for factory yards, railways and stone quarries
III Cranes for steelworks and rolling mills
   1. Hydraulic cranes
   2. Travelling slewing jib cranes

    3. Travelling gantry cranes
    4. Ingot cranes for charging furnaces
 IV Cranes for ports and harbours
    1. Cranes for bulk goods
    2. Cranes for heavy loads
    3. Cranes with very high lifting capacity

**Cranes for machine factories, foundries and boiler forges**

Whereas slewing jib cranes (see Fig. 7-11) prevailed in the early days of industrially driven mechanical engineering operations, by the end of the Industrial Revolution (after 1870) overhead travelling cranes (see Figs 7-13 and 7-14) had become standard. In particular, the economic transport of heavy items in foundries, assembly shops and boiler forges was "extremely important for rapid fabrication" [Bredt, 1894/95, p. 9]. After Bredt has explained the advantages of overhead travelling cranes operating immediately below the roof trusses, he analyses the lifting gear, the most important element in the design:

– welded chains which wind onto a windlass
– chains which engage with a sprocket (Fig. 7-18a)
– flat link articulated chains (Fig. 7-18b)
– ropes.

The advances in the production as well as the mechanical and technological quality of steel wire ropes in the early 1880s were exploited systematically by Bredt in his crane designs. He regarded the advantages as obvious: economic efficiency, easy replacement, high tensile strength and completely smooth operation, which permitted high speeds and "eases the exact positioning for all accurate operations" [Bredt, 1894/95, p. 12]. Bredt also pointed out the fatigue in ropes subjected to bending in the vicinity of the windlass – an engineering science problem which would later attract the attention of some outstanding researchers. Nevertheless, this problem could not prevent the wire rope triumphing as the preferred lifting gear for cranes. "As a result of these considerations [concerning the pros and cons of the rope as a lifting gear for cranes – the author]," Bredt comments, "I make considerable use of wire ropes" [Bredt, 1894/95, p. 12].

His electrically operated overhead travelling crane of 1887 enabled Bredt to achieve a comprehensive synthesis of structural steelwork, mechanical engineering and electrical engineering, which from then on would determine the design of cranes for industry. Fig. 7-19 shows a Bredt

**FIGURE 7-17**
Cover of Bredt's cranes catalogue
[Bredt, 1894/95]

**FIGURE 7-18**
Lifting gear; a) chain and sprocket,
b) flat link articulated chain and sprocket
[Bredt, 1894/95, p. 10]

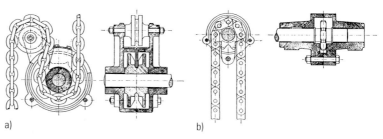

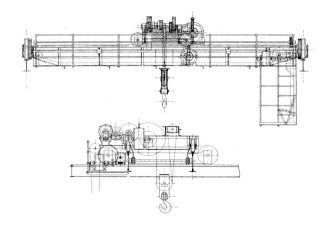

FIGURE 7-19
Overhead travelling crane
with three reversible electric motors
[Bredt, 1894/95, p. 23]

overhead travelling crane with reversible electric motors for each of the three crane movements and a rope system as the lifting gear. This type of crane conquered the factories in the period of high industrialisation after 1890 – in mechanical engineering, shipbuilding, mining and foundries, structural steelwork, electrical engineering and, later, the automotive industry.

In total, by 1892 Bredt's crane-building works had supplied more than 400 large overhead travelling cranes, most of which had mechanical transmission or electric motors, adding up to a total lifting capacity of 70 000 t!

The travelling jib crane invented by Ramsbottom in the early 1860s and used by him in the large wheel- and axle-turning shop of his locomotive plant in Crewe was a technical compromise between the slewing jib crane (see Fig. 7-11) and the overhead travelling crane. This crane ran on the floor on just a single rail and was secured laterally by one or two rails fitted below the roof trusses (Fig. 7-20). Travelling jib cranes were preferred for foundries and the ancillary bays of metalworking industries. However, the need to accommodate larger horizontal forces at the eaves level of the shed structure was a disadvantage. This peculiar crane type disappeared as the design of the overhead travelling crane was refined

FIGURE 7-20
Travelling jib crane
with two reversible electric motors
[Bredt, 1894/95, p. 31]

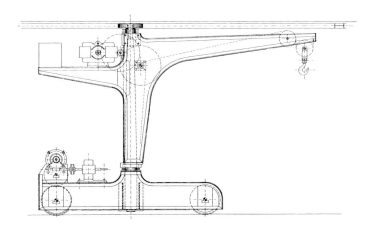

and the concept of the steel shed was developed further – especially by Maier-Leibnitz – in the first half of the 20th century. Bredt closes this section with reports on two special cranes that he developed for use in the casting pits of pipe foundries and for serving riveting machines in boiler forges.

**Cranes for factory yards, railways and stone quarries**

In this chapter Bredt describes a travelling gantry crane and several travelling slewing jib cranes. For example, his company supplied travelling slewing jib cranes for use on major building sites. Bredt designed several travelling slewing jib cranes for the construction of the 170 m-span trussed arch of the railway bridge over the Wupper Valley at Müngsten (1894 – 97) [Schierk, 1994]. Fig. 7-21 shows his design for a travelling slewing jib crane with electric drive, 7 t lifting capacity and 70 m lifting height. The Müngsten Bridge was erected as a balanced cantilever with the help of Bredt's slewing jib cranes. To avoid placing too much load on the bridge beams, these cranes had movable counterweights which enabled the load to be distributed evenly. The cranes had reversible electric motors for lifting, slewing and travelling, but the positioning of the counterweight was carried out manually [Bredt, 1894/95, p. 38]. Bredt's travelling slewing jib cranes represented a significant innovation in the building of steel bridges and helped ensure that the entire balanced cantilever of the Müngsten Bridge, with a total of 5100 t of steel, was completed in just seven months.

**Cranes for steelworks and rolling mills**

Hydraulic cranes were widely used in steelworks and rolling mills until well into the 20th century. It was Henry Bessemer (1813 – 98) who "with remarkable ingenuity [solved] the mechanical part of this task" [Bredt, 1894/95, p. 41] at the same time as inventing the blasting process for steel production (1855), which enabled the production of about 200 times the same amount as the puddling process in the same time. Although the use of water pressure had been known since 1846 thanks to the work of William George Armstrong, Bredt acknowledged that it was Bessemer

**FIGURE 7-21**
Brect's travelling slewing jib crane for the Müngsten Bridge
[Bredt, 1894/95, p. 40]

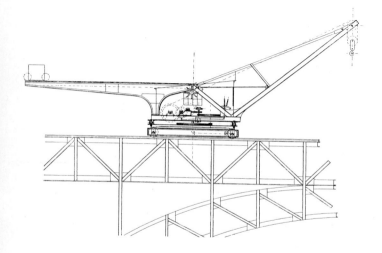

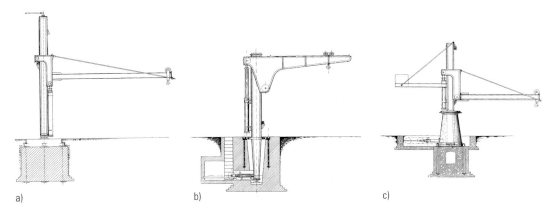

**FIGURE 7-22**
a) Free-standing hydraulic crane slewing about a solid column [Bredt, 1894/95, p. 42], b) with a Fairbairn foundation [Bredt, 1894/95, p. 43], and c) with a conical steel plate pedestal [Bredt, 1894/95, p. 44]

who first tried out a special arrangement for the specific purpose of lifting loads in the steel industry. Nevertheless, Bessemer's hydraulic cranes suffered teething troubles, which occasioned Bredt to separate the hydraulic cylinder from the crane structure as early as 1876; but he was not able to implement his idea until 1881. Three years before that, Wellmann had been granted a US patent for the same concept, which was used for an unusual application within a very short time. Bredt designed several variations of the hydraulic crane

- with a forged crane column (Fig. 7-22a),
- with a Fairbairn foundation beneath the floor (Fig. 7-22b), and
- with a conical steel plate pedestal above floor level (Fig. 7-22c).

Fig. 7-22 shows the separation between hydraulic cylinder and crane column; Bredt sold many of these cranes as well.

Bredt designed steam-driven locomotive cranes with slewing jibs for operating in factory yards and for the transport from steelworks to rolling mill. In addition, he introduced travelling gantry cranes for replacing the rolls in rolling mills and describes an ingot crane for transporting steel ingots and placing them in the furnace upstream of the rolling process, which he supplied to Thyssen & Co. in Mühlheim a. d. Ruhr.

## Cranes for ports and harbours

Whereas in the past cargo had been loaded from sea-going vessels into lighters or barges for transport to the warehouses, as the Industrial Revolution really took hold, port logistics underwent fundamental changes in the final three decades of the 19th century. The time-is-money principle became intrinsic to port operations; sea-going vessels were unloaded directly by cranes on the quayside and the goods placed in warehouses and storage depots for later transport by road, rail or canal.

Bredt's craneworks supplied numerous steam-powered quayside cranes for bulk goods such as ores, stone and coal plus large and very large cargos. In doing so, he exploited the whole range of his crane-building skills masterly and worked through the specific conditions of port operations at that time ingeniously. The erection of a heavy-duty crane in the Port of Hamburg in 1887 (see Fig. 7-16) marks the climax of his accomplishments in this field.

## Overhead travelling cranes within the factory organisation during the period of maximum industrialisation

Bredt always derived the crane types described in the individual chapters from the standpoint of serviceability, or rather their usefulness for his customers. Therefore, at the start of the section on overhead travelling cranes, he emphasizes their advantages over slewing jib cranes [Bredt, 1894/95, p. 9]:

- they serve an area of any size
- they serve the entire production area
- they require no floor area.

Bredt's information on overhead travelling cranes covers not only chapter I, but in fact occupies just over one-quarter of the entire catalogue; only

**FIGURE 7-23**
The bridge fabrication plant of Gesellschaft Union in Dortmund, built in 1898/99 [Mehrtens, 1900, p. 110]

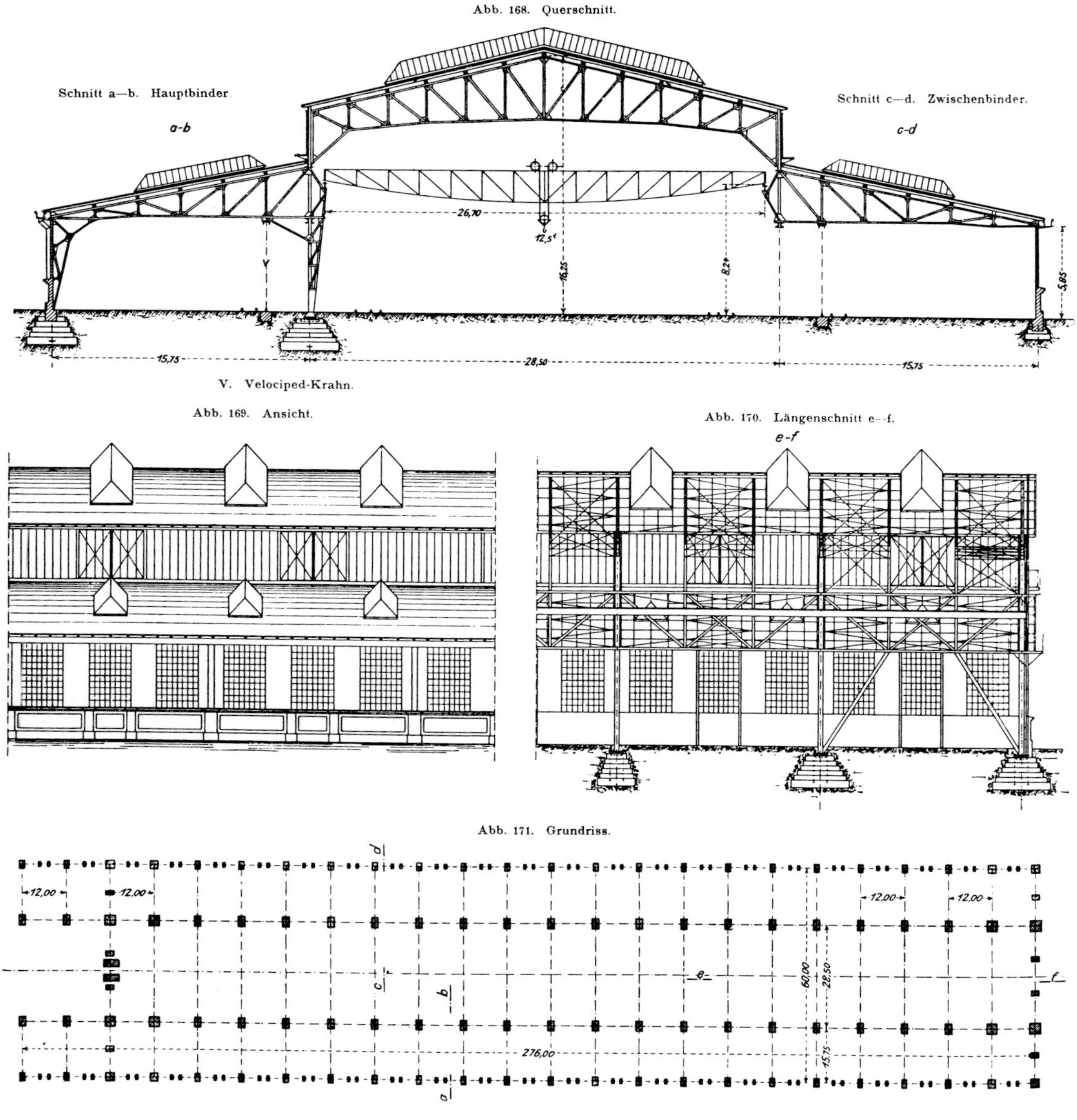

the chapter on quayside cranes is longer – and then only marginally. The chapter on cranes for steelworks and rolling mills takes up almost one-quarter of the catalogue. From this we can see that Bredt dedicates half of his catalogue to the mechanisation of vertical and horizontal transportation in the plants of large industries.

Overhead travelling cranes with an electric drive formed an important element in the new linear organisation of industrial manufacture during the 1890s, the phase of maximum industrialisation – presenting a totally different picture to the one painted by Menzel (see Fig. 7-11).

One example is Georg Mehrtens' commemorative volume *A Hundred Years of German Bridge Building* [Mehrtens, 1900] published by the six largest German steel bridge-building companies on the occasion of the World Exposition held in Paris in 1900. The book includes detailed descriptions of those companies' fabrication plants, all of course equipped with electrically driven overhead travelling cranes, e.g. the enormous three-bay shed (floor area 16 500 m$^2$) erected by Gesellschaft Union in Dortmund in 1898/99 for the fabrication of steel bridges (Fig. 7-23). Even today, overhead travelling cranes preside over the manufacturing facilities of large industries; the main difference is that solid-web beams are preferred these days and not trussed girders. As will be described in detail below, Rudolf Bredt laid the theoretical foundations for the structural analysis of solid-web beams with hollow sections subjected to torsion loads.

To finish this section, Kornelia Rennert's thoughts on the value of Bredt's crane catalogue: "This catalogue reflects not only the achievements of Bredt and his company; it also provides an overview of the development of lifting equipment in Germany up to the mid-1890s, and represents – as the first-ever crane catalogue published in Germany – a milestone in the history of advertising and marketing in the engineering industry. It was not until after the turn of the 20th century that the publication of such catalogues became the norm – also in various languages with a view to enticing an international clientele. In terms of format, scope and layout with respect to groups within the production programme, many later catalogues show amazing similarities with Rudolph Bredt's publication" [Rennert, 1999, p. 64].

### 7.2.3 Bredt's scientific-technical publications

In his scientific-technical publications, Bredt analyses the most important elements of the crane designs he had developed: crane hooks, struts (e.g. crane columns), pressure cylinders (e.g. for hydraulic cranes), foundation anchors (e.g. for anchoring stationary cranes), bending theory of curved members (e.g. for crane hooks), elastic theory and torsion theory. In addition, he published a detailed description of a machine for testing the tensile, compressive, bending and shear strength of iron and steel. And in the year before he died he expressed his opinion on the issue of engineers' training.

#### 7.2.3.1 Bredt's testing machine

It was in 1867 that Bredt designed a machine for testing the elasticity and strength of iron and steel. The machine was used in the Ludwig

Stuckenholz company primarily for testing steel plates for locomotive boilers. However, the company also built large numbers of the machine for sale to ironworks, railways and other users. Bredt says the motive behind this was the needs of buyers and factory owners "to have a way of classifying the various iron and steel grades as precisely as possible so that the prices of their products would no longer be governed by meaningless names such as fine-grain, charcoal or low moors grade (which were more or less dependent on the whims of the individual factories), but instead by other important physical properties, primarily the strength coefficients" [Bredt, 1878, p. 138].

Bredt's testing machine could be used to determine the tensile, compressive, bending and shear strength. It is interesting to note that Bredt investigated the causes of the considerable discrepancies in strength values when using different types of specimens and testing machines, and suggested that "through a railway authority, the Weights & Measures Office or a strength-testing institute connected with the Trade & Industry Academy it would be desirable, by means of longer series of tests, to establish how the results are affected by the shape of the specimens and the method of clamping them as well as the duration of the test and the influence of the hydraulic operation" in order to achieve general testing standards that would guarantee the reliability of strength tests [Bredt, 1878, p. 140]. Bredt therefore became an early spokesman for materials testing in Germany, which – starting with the analysis of failures of axles and tyres in the railway industry by August Wöhler (1819–1914) – culminated in the establishment of the first materials testing institution in the form of the "Mechanical-Technical Testing Authority for Testing the Strength of Iron, Other Metals & Materials" in 1870/71 at the Berlin Trade & Industry Academy (now Berlin TU) and one decade later led to the classification of iron and steel.

#### 7.2.3.2 The principle of separating the functions in crane-building

Bredt describes the hydraulic cranes (see Fig. 7-22) he designed for steelworks in his publications of 1883 [Bredt, 1883] and 1885 [Bredt, 1885/1]. He emphasizes the fact that separating the crane column from the pressure cylinder and ram avoids subjecting the latter to bending moments. This intentional use of the principle of separating the functions resulted not only in a clear constructional division of cranes into lifting gear on the one hand and the machine elements of the drive and the force transfer on the other, but also simplified the mechanical analysis of the crane components. For instance, a cylinder subjected to internal pressure only is considerably easier to analyse than a compression cylinder that is subjected to bending moments as well. The compression cylinder therefore becomes an object of strength of materials, whereas the crane structure – in the case of statically determinate systems – falls under the remit of structural analysis and only needs to take strength of materials into account when sizing the components. Bredt thus created the basis for the clear division of work between mechanical engineering and structural steelwork in crane-building, which after 1900 led to the rigorous scientific footing for crane-building – and therefore contributed to the international success of the German crane industry.

### 7.2.3.3 Crane hooks

The crane hook (Fig. 7-24) was just one such machine element investigated by Bredt [Bredt, 1885/2].

He criticised the view, published in the *ZVDI* in that same year, that the cross-sectional area of a crane hook is a minimum when the values of the maximum tensile stresses are equal to the maximum compressive stresses in the hook, based on Grashof's version of the theory of curved elastic bars [Grashof, 1878]. He specifies the values $P/\sigma_{i,A}$ ($\sigma_{i,A}$ inner extreme fibre stress at $A$) for the ratios $h_A/a = 1$, 2 and 3 ($h_A = e_1 + e_2$ = depth of cross-section at $A$) – in each case for rectangular, triangular and trapezoidal cross-sections. He compares these values with those from the crane hook modelled as an eccentrically loaded straight bar – a model described by Adolf Ernst (1845–1907) in his book [Ernst, 1883/1, pp. 32–36]. Bredt discovered considerable discrepancies in the $h_A/a$ ratios common in practice and realised that the crane hook designed by Ernst with a ratio of $h_A/a > 2.5$ was hopelessly undersized.

Whereas Ernst had needed several pages to reach results of doubtful relevance in practice, Bredt's treatise on crane hooks – so valuable to crane designers – covered just 1½ pages.

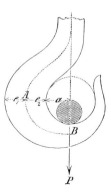

**FIGURE 7-24**
Crane hook [Grashof, 1878, p. 289]

### 7.2.3.4 Struts

Bredt's publications on struts [Bredt, 1886 & 1894] are considerably longer. In the 1886 paper he solves the stress problem according to second-order theory for the cantilever bar and simply supported member taking into account imperfections relevant in practice. He of course includes a table in which he compares his results with the empirical buckling equations [Bredt, 1886, p. 625]. After a detailed analysis of Euler's buckling theory, Bredt concludes that this "does not guarantee the long-term safety of members in compression for loads causing a direct risk of buckling because experience shows that disruptive bending moments and deflections occur even under low loads" [Bredt, 1894, p. 878]. The further development of the design theory for struts would prove that Bredt was right.

### 7.2.3.5 Foundation anchors

Bredt's paper on foundations [Bredt, 1887] ended a debate between Richard Schneider [Schneider, 1886] and Adolf Ernst [Ernst, 1886] as well as one between the Stuttgart professor of applied mechanics Edmund Autenrieth (1842–1910) [Autenrieth, 1887/1 & 1887/2] and Bredt himself with an extremely simple and mechanically sound model. He showed that for an anchor plate fixed with $n$ anchors in the foundation (on which a crane column stands, for example) such that the bending moment $M$ and the axial force $N$ act on the anchor plate, the maximum tensile and compressive stresses at the anchor points can be calculated with the equation for bending plus axial force:

$$\left.\begin{array}{l} |max\,\sigma_Z| \\ |max\,\sigma_D| \end{array}\right\} = \left| \pm \frac{M}{W_A} - \frac{N}{n \cdot A_s} \right| \qquad (7\text{-}33)$$

where $W_A$ is the moment of resistance of the anchor cross-sections related to the axis of the crane column, and $A_s$ is the net cross-sectional area of one anchor. This mechanical model for calculating foundations, developed on hardly more than half a page, would be accepted fully by Ernst a few years later [Ernst, 1899, pp. 470–473].

### Pressure cylinders

#### 7.2.3.6

In another paper, Bredt criticises the design formulas for pressure cylinders [Bredt, 1893] (Fig. 7-25).

For pressure cylinders with a high internal pressure $p_i$ and a large wall thickness $t$, the use of the equation given by Emil Winkler in 1860

$$t = r_a - r_i = r_i \cdot \left[ -1 + \sqrt{\frac{m \cdot \sigma_{perm} + p_i \cdot (m-2)}{m \cdot \sigma_{perm} - p_i \cdot (m+1)}} \right] \qquad (7\text{-}34)$$

had become standard practice in design offices [Bach, 1889/90, pp. 341–342], whereas the boiler formula

$$t = r_i \cdot \frac{p_i}{\sigma_{perm}} \cong r \cdot \frac{p_i}{\sigma_{perm}} \qquad (7\text{-}35)$$

had to be used for pressure cylinders with a low internal pressure $p_i$ and a small wall thickness $t$. The elastic constant $m$ is 10/3, the permissible stresses $\sigma_{perm}$ act as a tangential tensile stress $\sigma_t$. Bredt proved that eq. 7-35 did not represent an approximation of eq. 7-34 because when $t \ll r_i$, $p_i \ll \sigma_{perm}$ and $m = 10/3$, eq. 7-34 becomes

$$t = 0.85 \cdot r \cdot \frac{p_i}{\sigma_{perm}} \qquad (7\text{-}36)$$

"With these considerable differences [between eq. 7-35 and eq. 7-36] it would seem appropriate to clarify the basic premises" [Bredt, 1893, p. 903]. It is remarkable that Bredt derived the basic equations of elastic theory from the corpuscular model and not from the continuum model. Bredt regarded the elementary tetrahedron, whose corners are occupied by atoms, as a space frame and specified the stress-strain relationships. His modelling concept can therefore be assigned to the rari-constant concept of elastic theory. Admittedly, the works of Woldemar Voigt (1850–1919) had assured victory for the multi-constant theorists by 1889 [Love, 1907, pp. 17–18].

Bredt finally specifies general design equations for pressure cylinders which also take into account the longitudinal stresses $\sigma_x$ and the external pressure $p_a$; from that he deduces specific cases. However, Winkler had already specified the formula for pressure cylinders subjected to $p_i$ and taking into account $\sigma_x$ back in 1860.

### Curved bars

#### 7.2.3.7

In a long paper [Bredt, 1895] on the theory of curved elastic bars, Bredt develops equations based on physics for designing crane hooks, rings and chain links, i.e. those construction elements forming part of the crane engineer's daily workload. The clarity, simplicity and convenience of his equations made them very popular.

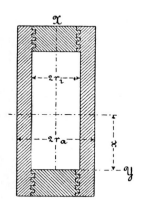

**FIGURE 7-25**
Pressure cylinder [Bach, 1889/90, p. 333]

### 7.2.3.8 Elastic theory

Bredt's last published work in the field of applied mechanics concerns non-linear-elastic material behaviour [Bredt, 1898]. He recommends the parabolic stress-strain relationship: "Several equations will probably assert themselves for specific cases in practice, the linear equation perhaps for cast iron, the hyperbolic one for stone; but the parabolic can be recommended for all cases" [Bredt, 1898, p. 699]. His recommendation would have an effect in practice at least for the theoretical determination of the stress-strain ratio in normal-weight concrete according to EC 2 part 1 [Zilch & Rogge, 1999, pp. 408, 409].

He left one final manuscript on elasticity unfinished. His widow supposedly donated it to Karlsruhe Technical University, but it has since disappeared (see [Rennert, 1999, p. 73]).

### 7.2.3.9 The teaching of engineering

VDI member Bredt became briefly involved in the discussion about the training of engineers from the viewpoint of the scientifically educated mechanical engineer [Bredt, 1899]. The discussion had started with the presentations of Adolf Ernst and Alois Riedler at the 35th AGM of the VDI in 1894 and reached its climax in the controversy between Reuleaux and Riedler.

"Engineers trained ready for practice," Bredt writes, "cannot be educated in universities because the creative work of the engineer calls for more experience and skill than science, and benefits can be gleaned from science only after an extended period of practical experience is added to natural talent" [Bredt, 1899, p. 662]. According to Bredt, the engineer trained at a polytechnic is just as useful to the design office of an engineering company as the engineer trained at university. Nevertheless, he did not wish to restrict the education of engineers to just the necessary minimum. But he did criticise the fact that all too often "the activities of the engineer are primarily seen as the use of algebraic formulas" [Bredt, 1899, p. 663] and pleaded for a two-year period of training in the design office of an engineering company as practical preparation for studying engineering at university; only in that way was the student of engineering protected against overestimating the theory.

Furthermore, he warned against overestimating the direct use of specialist lectures for practical engineering, and described his experiences with the lectures of Redtenbacher, Zeuner and Reuleaux, arguing for dropping specialist studies in favour of more fundamental subjects such as mathematics, physics, mechanics, strength of materials and graphical statics, which would enable "shortcomings in specialist sciences ... to be overcome easily with the help of the extensive literature already extant as soon as the general scientific foundation is available" [Bredt, 1899, p. 663].

One no less important source of knowledge was the engineering science test, which should not be limited to just the strength test, but instead should be "extended in the physical direction" [Bredt, 1899, p. 663]. "It is enlightening," Bredt continues, "to see the imperfections of all research work through the many discrepancies between the results of tests and the mathematical derivation" [Bredt, 1899, p. 663].

Bredt's plea for freedom in teaching and learning culminated in his recommendation to allow the students to deviate from the standard curricula of the specialist faculties of the universities: "In my opinion the final examination is worthless for the mechanical engineer; practice places little value on a good leaving certificate. The examination restricts the freedom of the student, prolongs the period of study and places an unnecessary burden on the student. The examination provides a measure of the knowledge, but not of the skill, and it is the latter that is crucial for the mechanical engineer" [Bredt, 1899, p. 663]. Of course, Bredt replies to the objection that without examinations students might not attend the course of study with the necessary eagerness. He refers to his experiences as a student under Redtenbacher in Karlsruhe. As an active Protestant, Bredt counted on the self-discipline of the Protestant ethic of the industrial middle class as analysed by Max Weber [Weber, 1993], which had a permanent influence on Bredt's scientific-technical thoughts and actions. Looked at in this way, Bredt's final publication on the issue of engineers' education can be seen today as the legacy of an introverted engineer and manager of a midsize industrial company standing on the threshold of the "The Age of Extremes" [Hobsbawm, 1994].

**Torsion theory**

### 7.2.3.10

Rudolph Bredt's greatest contribution to applied mechanics was achieved in 1896 in the field of torsion theory [Bredt, 1896/1], which he introduces as follows: "In the theoretical studies with which the engineer is concerned it is beneficial to turn the task into a clear geometrical view wherever possible. Spatial relationships are always familiar to the engineer and simply correctly drawn figures will ease the understanding. The one-sided algebraic treatment preferred hitherto is fully understood by only a few; it can even happen that in such calculations a mistake goes unnoticed by the mathematicians themselves. This albeit rare case can be seen in the elasticity of torsion; in the abstract treatment of this task, which is carried out in various textbooks in similar ways, the necessary geometrical relationship has been ignored and that is why the correct solution has not been found" [Bredt, 1896/1, p. 785]. Bredt did indeed offer a promising solution to the Saint-Venant torsion problem, which he demonstrated fully for thin-wall hollow sections. Based on a geometrical consideration of the twisted bar, he derives the equation

$$\oint \tau \cdot ds = 2 A_m \cdot \vartheta' \cdot G \qquad (7\text{-}37)$$

Expressed in words: "The sum of the tangential shear forces per unit area on a closed curve within the cross-section subjected to twisting is equal to twice the product of rotation, shear modulus and the area enclosed by the forces" [Bredt, 1896/1, p. 787]. This principle applies not only to thin-wall hollow sections. If we integrate eq. 7-28 via $ds$

$$\oint \tau \cdot ds = \oint \frac{M_t \cdot ds}{2 A_m \cdot t} = \frac{M_t}{2 A_m} \oint \frac{ds}{t}$$

and take into account eq. 7-32 in the form

$$M_t = \vartheta' \cdot G \cdot I_t$$

as well as eq. 7-29 in the form

$$\oint \frac{ds}{t} = \frac{4 A_m^2}{I_t}$$

then we arrive directly at eq. 7-37, which August Föppl called the "first Bredt equation".

Bredt specifies his second equation for thin-wall hollow sections ($t$ = const) with

$$\vartheta' = \frac{M_t \cdot U_m}{4 A_m^2 \cdot t \cdot G} \tag{7-38}$$

[Bredt, 1896/1, p. 815]. If we enter eq. 7-32 into eq. 7-38 and solve for $I_t$, the result is the second Bredt equation (eq. 7-31) for the specific case of a constant wall thickness $t$. Bredt of course also specified an equation for non-constant wall thicknesses, which Föppl christened the "second Bredt equation". But Bredt's torsion theory went unnoticed – with one exception: August Föppl. Föppl allowed his letter concerning Bredt's paper [Bredt, 1896/1] to be published [Föppl, 1896] – to which Bredt replied [Bredt 1896/2]; the pair also exchanged letters on the torsion problem. In his reply to Föppl's letter, Bredt makes it clear that his criticism was generally directed "at the prevailing algebraic method, which regards the help of geometry as unnecessary or indeed as a nuisance" [Bredt, 1896/2, p. 943]. He still held theorists such as Clebsch, Saint-Venant and Grashof in high esteem.

In the case of torsion theory, Grashof had overlooked errors in the section on the elasticity of torsion in his widely read book [Grashof, 1878] which, concealed by the academic style of the derivations, could lead the reader to believe that the statements were reliable. Bredt experienced this, too: "It was not until I tried to draw a picture of the deformed end surface did I notice that one very important condition was missing!" [Bredt, 1896/2, p. 944]. The high level of Bredt's understanding can be seen in Fig. 7-26, in which he has drawn the shear flow in a twisted slit pipe cross-section. He could therefore run any business systematically because he was fully aware of the geometrical properties of the shear flow.

The uncommon power of the Bredt view impressed August Föppl even years afterwards: "This work [Bredt's torsion theory – the author] is very curious; it is certainly a very rare example of how a shrewd mind – despite little schooling in theory and without any detailed knowledge of the previous work in an admittedly difficult field – can identify the truth easily and indeed look beyond the frontiers that others had reached before him, but without his knowing that." Bredt's torsion theory "would reach some considerable way beyond the results of Saint-Venant" [Föppl, 1917/1, p. 18].

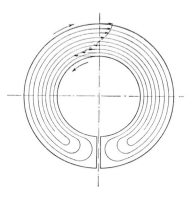

**FIGURE 7-26**
Shear flow in a slit pipe cross-section [Bredt, 1896/1, p. 814]

### 7.2.4 The engineering industry adopts classical theory of structures

Under Bredt's management, Ludwig Stuckenholz, the first crane-building company in Germany, employed a workforce of 250–300 in its best years. Wolfgang Reuter (1866–1947) started work in the design office of Ludwig

Stuckenholz in 1888 as a young engineer under the direct leadership of Bredt. By the time Reuter was 30, Bredt had made him a partner of the firm and in 1899 he became the sole proprietor of Ludwig Stuckenholz. His mentor, Rudolph Bredt, retired in that same year and died on 18 May 1900. Just six years later Reuter's company merged with its neighbour, Märkische Maschinenbau-Anstalt, to form Märkische Maschinenbau-Anstalt, Ludwig Stuckenholz A. G. This was a difficult decision for Reuter, to switch from independent factory owner to manager and hence an employee of a public company [Matschoss, 1919, p. 211]. In 1910 came the hour of the Deutsche Maschinenfabrik in Duisburg – known worldwide by its telegram abbreviation "Demag" (now Demag Cranes & Components GmbH) – in the form of the merger between the company headed by Reuter, Duisburger Maschinenbau-A. G. (formerly Bechem & Keetman) and Benrather Maschinenfabrik A. G. Wolfgang Reuter, as managing director, moulded Demag into the top address for heavy engineering in Germany. In the cranes sector, Demag became the technical and economic leader – Benrather Maschinenfabrik had already been announcing in its advertising since 1906 that it was the largest crane factory in Europe. The Cyclops of crane technology, the hammerhead crane Demag built for the Blohm & Voß shipyard in Hamburg in 1913, broke all records at that time (Fig. 7-27):

- 250 t lifting capacity at a radius of 34.5 m
- 110 t lifting capacity at max. radius of 53 m
- 55.4 m long boom able to be raised to 104 m above water level
- travelling slewing jib crane on boom with lifting capacity of 20 t at 10 m radius and 10 t at 18 m
- working area covered by crane hook has a diameter of 147 m and an area of 17 000 m$^2$

Crane designs like the hammerhead crane could no longer be conceived and calculated with the methods of structural analysis and strength of materials that had been developed by Bredt and were in widespread use in

**FIGURE 7-27**
Hammerhead crane built by Demag in 1913 [Bachmann et al., 1997, p. 75]

mechanical engineering. Therefore, an engineer employed by Duisburger Maschinenbau A.-G., W. Ludwig Andrée (1877–1920), published the first large compendium of crane design in 1908: *Die Statik des Kranbaues* (structural theory of crane-building) [Andrée, 1908] enjoyed several editions up until Demag's 100th anniversary celebrations in 1919. Andrée made use of the entire set of tools in classical theory of structures, as completed by Heinrich Müller-Breslau, in particular the influence line theory which hitherto had been confined to bridge-building. As a Demag engineer, Andrée worked through the structural theory of the entire Demag product range in the cranes sector plus conveying and lifting technology by means of examples: overhead travelling cranes, crane rails, transporter cranes, tower cranes, slewing jib cranes, portal frames, slipway frames, suspended platforms, cableways, floating cranes, dockyard cranes, grabs, lifting gear, pithead gear, skew bridges, swing bridges, lifting bridges [Andrée, 1913, pp. VII–X]. So, in the form of Demag, the heavy engineering industry put classical theory of structures to work.

Many of these examples were drawn from Andrée's work as a crane engineer at Demag. This book was followed in 1919 by *Die Statik der*

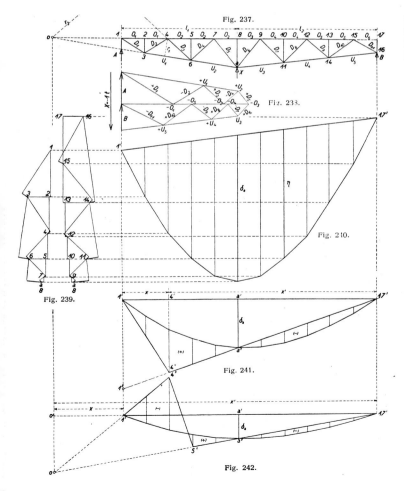

**FIGURE 7-28**
Semi-graphical analysis of a crane rail beam with one degree of static indeterminacy [Andrée, 1917, p. 186]

*Schwerlastkrane* (structural theory of heavy-duty cranes) [Andrée, 1919/2]. In that same year, Conrad Matschoss (1871–1942), VDI director and the master of the German history of engineering, published his magnificent Demag-sponsored book *Ein Jahrhundert Deutscher Maschinenbau …* (a century of German engineering) [Matschoss, 1919, p. II].

The self-assured crane engineer Andrée was also active in the field of structural steelwork in general as well as structural analysis. The year 1917 saw the appearance of *Die Statik des Eisenbaues* (theory of structural steelwork) [Andrée, 1917], which became a standard work of reference with more 100 sample calculations for sheds, cranes (Fig. 7-28), airship hangars, slipway frames, pithead gear, cooling towers and bridges. Two years later he gathered together individual papers on the analysis of symmetrical statically indeterminate systems in his book *Das B-U-Verfahren* (load redistribution method) [Andrée, 1919/1]. The load redistribution method enabled Andrée to take a big step towards the rationalisation of structural calculations [Kurrer, 2000/2]. One year before his death, Andrée became editor of *Der Eisenbau*, the international journal for the theory and practice of structural steelwork.

This outline of the professional career of W. Ludwig Andrée contrasted with that of Rudolph Bredt is intended to demonstrate the following:

1. The new relationship between science and industry was reflected in the employment of the entire classical theory of structures and electrical engineering for the everyday work of the crane engineer in the world of big industry.
2. The three-dimensional form of crane structures made an elaborate three-dimensional structural analysis unavoidable, and therefore the rationalisation of structural calculations.
3. The internationalisation of the large companies in the crane-building sector fostered the internationalisation of classical theory of structures, the product of the Berlin school of structural theory [Kurrer, 2000/1].
4. Crane-building went beyond the boundaries of mechanical engineering to form a synthesis with theory of structures and structural steelwork.
5. The influence of crane-building (or in a more general sense materials handling and lifting technology) still visible today in mechanical and electrical engineering on the one hand and structural steelwork on the other became established in the second decade of the 20th century.
6. Bredt united the profession of crane-builder in one person: as ingenious designer, as mechanical engineer with scientific ambitions, and as entrepreneur; he thus guaranteed the quality of his products. Demag and the other large crane manufacturers, e. g. MAN, Krupp-Gruson, guaranteed the quality of their products with their names. Andrée new how to work fully through the scientific foundation of his engineering activities on product quality to convey this effectively.
7. Bredt's modest way of linking his achievements in the field of crane-building with the name of his company was a way of expressing the notion that the engineering achievements of the 20th century would

no longer carry the names of their creators, but would instead appear as trademarks. He therefore ensured that his name would not be forgotten by writing scientific papers. Andrée had no other choice but to make his name through science.

Rudolph Bredt paved the way for major industry to take responsibility for crane-building in the 20th century. W. Ludwig Andrée placed crane-building on the foundation of classical theory of structures by integrating it with structural steelwork.

## 7.3 Torsion theory in the consolidation period of structural theory (1900–1950)

Following on from Rudolph Bredt's contribution of 1896 [Bredt, 1896/1], August Föppl (1853–1924) was able to achieve a great step towards a practical torsion theory in 1917. It was in 1909 that Bach published the results of bending tests on channel sections from which it could be seen that it was not sufficient to refrain from examining such cross-sections with the customary moments of resistance for bending about the main axes of the section [Bach, 1909]; one year later he reported on further tests in this direction [Bach, 1910]. These "cracks" in the fabric of the classical practical bending theory, detected by tests, led in the 1920s and 1930s to extensive discussions and to the discovery of the shear centre plus the formation of the theory of warping torsion, which gradually became a parallel theory to the Saint-Venant torsion theory.

Up until 1950, torsion problems were met with only rarely in structural steelwork applications. This situation did not change until the trussed framework lost its dominating position in structural steelwork and, after World War 2, lightweight construction was no longer the exclusive province of aircraft designers.

Around 1950, the Saint-Venant torsion theory was also noticed by the structural steelwork community as it started to absorb the knowledge from lightweight construction that had been accumulating since the 1930s about cellular hollow sections subjected to torsion in steel bridges. Nevertheless, the main focus of research into structural steelwork was soon devoted to warping torsion – an object of research in lightweight construction which had been explored successfully in the building of aircraft even before World War 2.

### 7.3.1 The introduction of an engineering science concept: the torsion constant

August Föppl achieved a significant expansion of the Saint-Venant torsion theory in 1917 [Föppl, 1917/1]. At a meeting of the mathematics/physics class of the Royal Bavarian Academy of Sciences on 13 January 1917, he presented for the first time the general equation (eq. 7-11) for the torsion $D$ and introduced the designations "torsional stiffness" for the product $G \cdot I_t$ as well as "torsion constant" for $I_t$ [Föppl, 1917/1, p. 6]. Although exact equations had been derived for $I_t$ for simpler cross-sectional forms such as ellipse, rectangle, triangle, sector and a number of others, $I_t$ for the common rolled sections were still being calculated with the Saint-Venant approximation (eq. 7-23) using $\kappa = 40$. In this context, Föppl referred to the 22nd edition (1915) of the very popular work of reference *Hütte, des Ingenieurs Taschenbuch* (engineer's pocket-book). Admittedly, this book does not state that Saint-Venant had also referred to exceptions with respect to eq. 7-23 and that "the figures given for assessing the accuracy are valid only for cross-sectional forms that are in no way similar to ⊥-sections" [Föppl, 1917/1, p. 13].

In order to fight effectively against Saint-Venant's eq. 7-23, Föppl stated that enginners must therefore use the weapons of Saint-Venant himself, i.e. they must verify through theoretical means that the theory can lead to totally incorrect results within the limits of validity currently assumed [Föppl, 1917/1, p. 14]. Starting with the hydrodynamics allegory of W. Thomson (Lord Kelvin) and P. G. Tait, the soap-bubble (membrane) allegory of Ludwig Prandtl and, in particular, Bredt's publication of 1896 (see section 7.2.3.10), Föppl achieved this proof for thin-wall open sections whose cross-section can be approximated adequately by means of $n$ rectangles (Fig. 7-29):

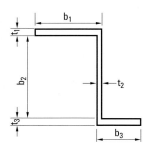

**FIGURE 7-29**
An open section assembled from narrow rectangles

$$I_t = \frac{1}{3} \sum_{i=1}^{n} t_i^3 \cdot b_i \tag{7-39}$$

where $t_i$ is the thickness and $b_i$ the width of the $ith$ rectangle. As eq. 7-39 applies to $b_i/t_i \to \infty$ only, Föppl introduced a correction factor $\eta$ into eq. 7-39:

$$I_t = \frac{\eta}{3} \sum_{i=1}^{n} t_i^3 \cdot b_i \tag{7-40}$$

In the next step, Föppl compares eq. 7-23 and eq. 7-39 for common cross-sections. Eq. 7-23 is unusable in the case of bars with a cruciform cross-section and equal angles as well as the wide-flange ⊥-sections (width of flange = depth of section) which started to spread after 1900. On the other hand, eq. 7-23 supplies acceptable values for older standard sections (⊥-sections where width of flange = ½ depth of section).

Finally, Föppl develops a proof format for the torsion problem that corresponds in formal terms fully with the practical bending theory of Navier. For thin-wall open sections assembled from $n$ rectangles, Föppl determines the maximum shear stress as

$$\tau_{max} = \frac{M_t \cdot t_{max}}{I_t} = \frac{M_t}{W_t} \tag{7-41}$$

where

$$W_t = \frac{I_t}{t_{max}} \tag{7-42}$$

In a further meeting of the mathematics/physics class of the Royal Bavarian Academy of Sciences, Föppl verified his version of the Saint-Venant torsion theory extended by eq. 7-39 to eq. 7-42 by way of comprehensive test results; Föppl specified correction factor $\eta$ for common rolled sections [Föppl, 1922].

Although Föppl published his fundamental work [Föppl, 1917/1] as a summary in the journal *Zeitschrift des Vereines deutscher Ingenieure* in 1917 [Föppl, 1917/2], and this was referred to later that year in the journal review in the journal *Der Eisenbau*, and A. Müllenhoff writing in the latter journal in 1922 presented Föppl's test report [Föppl, 1922] in detail [Müllenhoff, 1922], Föppl's extension of the Saint-Venant torsion theory at first could not gain a foothold in structural steelwork design offices. Even a discussion of this within the scope of his monograph *Drang und Zwang* (pressure and restraint) (1920) plus later editions of *Hütte* (23rd edition in 1920, 24th edition in 1923, 25th edition in 1925) did not bring any significant progress. And Föppl's request back in 1917 to include the torsion constant $I_t$ in the tables of sections [Föppl, 1917/1, pp. 7–8] was not put into practice until more than 25 years later.

So it was left to Bornscheuer in 1952 [Bornscheuer 1952/2] and in 1961 [Bornscheuer, 1961], within the scope of his version of the theory of warping torsion [Bornscheuer, 1952/1], to publish tables of torsion parameters – including the torsion constant $I_t$ – for rolled sections, which were immediately incorporated into the appropriate German DIN standards.

W. Wagner, R. Sauer and F. Gruttmann have recently used the finite element method (FEM) to show that the DIN table values of torsion parameters in some cases deviate from the exact values by more than 10 % [Wagner et al., 1999].

The reason for August Föppl's extension of the torsion theory not being incorporated into practical structural steelwork design until much later was the divergence in the disciplinary development of the engineering science disciplines involved from the 1920s onwards: applied mechanics on the one side and theory of structures – still closely entwined with structural steelwork – on the other.

### 7.3.2 The discovery of the shear centre

Fig. 7-30a shows a thin-wall channel section subjected to transverse bending, perhaps due to the common case of a point load applied in the centre of a simply supported beam.

The bending moment $M_y$ generates the axial forces $\sigma_x$ shown in Fig. 7-30b. The shear force $Q_z$ applied at the centroid S causes the shear stresses $\tau_Q$ (Fig. 7-30c), which in the top flange add up to

$$H_\tau = \int_{s_1=0}^{s_1=b_1} \tau_Q(s_1) \cdot t_1 \cdot ds_1 \qquad (7\text{-}43)$$

Force $H_\tau$ in the bottom flange is of the same magnitude but of opposite sign (Fig. 7-30c):

$$-H_\tau = \int_{s_3=0}^{s_3=b_1} \tau_Q(s_3) \cdot t_1 \cdot ds_3 \qquad (7\text{-}44)$$

The shear force $Q_z$ can also be determined from the equivalence of the stresses:

$$Q_z = \int_{s_2=0}^{s_2=b_2} \tau_Q(s_2) \cdot t_2 \cdot ds_2 \qquad (7\text{-}45)$$

The flange forces $H_\tau$ according to eq. 7-43 and eq. 7-44 result in a couple with the lever arm $h - t_1$ (Fig. 7-30c). As most of the shear force $Q_z$ passes through the centre of the web, the moment equilibrium condition

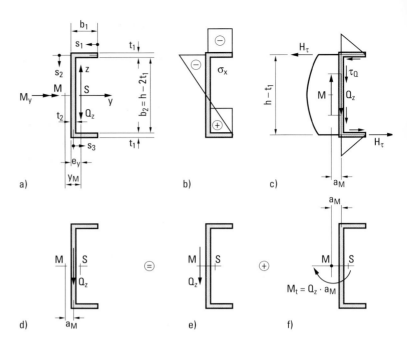

**FIGURE 7-30**
On the determination of the position of the shear centre of a thin-wall channel section

$$Q_z \cdot a_M = H_\tau \cdot (h - t_1) \tag{7-46}$$

must be satisfied (Fig. 7-30c). Eq. 7-46 can be used to determine $a_M$, from which it follows that

$$y_M = a_M + e_y \tag{7-47}$$

(Fig. 7-30a). Point $M$ is known as the shear centre. If the line of action of the shear force $Q_z$ passes through $M$, pure bending is present (Fig. 7-30e). If, on the other hand, the shear force $Q_z$ acts in the centre of the web (Fig. 7-30d), then a torsion moment of magnitude

$$M_t = Q_z \cdot a_M \tag{7-48}$$

is added to the pure bending (Fig. 7-30f). The system of forces shown in Fig. 7-30d is therefore made up of pure bending (Fig. 7-30e) plus torsion (Fig. 7-30f). So much to the discovery of the shear centre on the logical side; the history of this discovery is outlined below.

**Julius Bach**    **7.3.2.1**

It was in 1909 that Bach published the results of tests he had carried out on steel beams comprising channel cross-sections [Bach, 1909]. The structural system was a simply supported beam with a point load acting at mid-span; the point load was transferred through the centroid $S$ (see Fig. 7-30a) and through the centre of the web (see Fig. 7-30d). Based on the deflections measured, he determined the true moment of resistance from the equation for deflection (see eq. 5-1) and came to the following conclusions:
- The resistance decreases as the flange width $b_1$ increases.
- The resistance when transferring the load via the centre

of the web is greater than that when transferring the load via the centroid S.

In both cases the true moment of resistance determined from the deflection measurements is lower than the value found through equations. Using the theoretical moment of resistance, the bending stresses would be too low, i.e. channel sections designed for bending would have an inadequate safety margin, which is why Bach regarded the design as inadmissible. The discrepancies would have to be calculated according to the method proposed by Saint-Venant. In a second test report, Bach took up the theme of the bending resistance of channel sections once more and published the results of tests on ⊥-beams [Bach, 1910]. With respect to the channel sections, Bach describes the deformation figure of a test beam in which the plane of loading coincides with the centre of the web (see Fig. 7-30d) very precisely: the top flange bends in the negative $y$-direction due to $-H_\tau$ and the bottom flange in the positive $y$-direction due to $+H_\tau$ (see Fig. 7-30c); the web deforms, the channel section takes on an S-form and the member cross-sections warp.

### 7.3.2.2 Louis Potterat

One decade later, Louis Potterat, a professor at Zurich ETH, referred to the test reports of Bach and triggered a debate about the "Archimedes point" of the practical bending theory in the journal *Schweizerische Bauzeitung*. Potterat was interested in getting the practitioners to check the assumptions of the strength of materials theorists. For example, he complained that strength of materials had accepted assumptions from the mechanics of rigid bodies, e.g. equivalent force systems: "When it comes to assuming the displacement of a couple parallel to its plane, this is just as inadmissible in strength of materials as replacing a force system by its resultant" [Potterat, 1920/1, p. 142]. As an example of this, he quotes the shift in the plane of the forces from the centre of the web to the centroid of the channel section, which according to Bach's tests "brings with it a change in the load on the extreme fibres amounting to 25 %" [Potterat, 1920/1, p. 142]. Unfortunately, Potterat ignored the moment due to the offset. Equivalent force systems are also admissible in strength of materials just like the resolution of the force system shown in Fig. 7-30d into pure bending (Fig. 7-30e) and torsion (Fig. 7-30).

### 7.3.2.3 Adolf Eggenschwyler

In his letter, Adolf Eggenschwyler attributes the discrepancy between Bach's tests and the calculation according to the practical bending theory to the "incorrect test setup and questionable assumptions made when assessing the readings … The results cannot even be verified because the stresses had merely been concluded from the deflections and nowhere does it state where the marks observed were located, which means that it is not possible to establish to what extent their movements were also influenced by the twisting of the beam, lateral buckling of the compression flange, outward bending of and vertical compression loads on the web, etc." [Eggenschwyler, 1920/1, p. 207]. The reason why the calculation according to the practical bending theory did not cover the phenomenon brought to light by Bach's tests is to be found in the theory on which the evaluation of the tests was based. Eggenschwyler supplies the right interpretation by rigorously separating bending and torsion. Without supplying equations, he specifies the correct method for determining the position of the shear centre $M$. In particular, he clearly explains that the force system shown in Fig. 7-30e can be dealt with solely with the stress equations of the practical bending theory. However, as soon as the plane of the applied forces, e. g. shear force $Q_z$, is displaced by the amount $a_M$ from point $M$, the moment due to the offset $M_t = Q_z \cdot a_M$, which twists the channel, must be taken into account (Fig. 7-30); Eggenschwyler explains this point as well [Eggenschwyler, 1920/1, p. 207]. Potterat does not mention this problem in his commentary [Potterat, 1920/2]; he is mainly interested in criticising the Bernoulli hypothesis that sections remain plane, which Potterat believed applies only to beams with a symmetrical cross-section and whose axis of symmetry is identical with the plane of the applied forces. It would soon be shown that Potterat's criticism of the Bernoulli hypothesis was unfounded. In his letter following Potterat's reply [Potterat, 1920/2], Eggenschwyler fleshes out his statements about the channel section and frankly confesses to an error in his letter to the journal *Der Eisenbau*: "At that time I still shared the generally accepted view that a beam does not rotate when subjected to bending if the loads acting upon it pass through the axis of the centroid, and I realised only later that the loads must pass through another axis, which one could call the bending axis and only coincides with the axis of the centroid in the case of point-symmetric cross-sections, but not in channels and other asymmetric cross-sections or those with only one axis of symmetry" [Eggenschwyler, 1920/2]. Eggenschwyler corrected his error in a letter to *Der Eisenbau*, but it was not published by the time of the controversy with Potterat. So Eggenschwyler provided a concrete answer to the question of the axis of the shear centre, which he called the "bending axis".

**Robert Maillart**  7.3.2.4

Robert Maillart (Fig. 7-31), a practising civil and structural engineer, entered the debate surrounding the shear centre on 30 April 1921. In several brief contributions he managed to develop the theory of the shear centre.

Maillart reported in detail on Bach's bending tests on channel beams (see section 7.3.2.1), the results of which Bach incorporated into the eighth edition of his book *Elastizität und Festigkeit* (elasticity and strength) [Bach, 1920]. Maillart carried out a linear interpolation of the discrepancies between Bach's test results and the results of calculations using the practical bending theory and claimed that the excess discovered by Bach must disappear when the shear force $Q_z$ transmitted through the centre of the web in the test is shifted to the shear centre $M$ (Fig. 7-30e). Maillart also verified that in a doubly symmetric I-section with an eccentrically applied shear force $Q_z$ the result will be a different distribution of bending stresses to that calculated using the practical bending theory. He therefore disproved Bach's claim that the asymmetry of the channel section would exclude the use of the practical bending theory.

In contrast to Bach, Maillart recognises the importance of the shear force $Q_z$. He makes clear distinctions between the force systems shown in Figs 7-30d, 7-30e and 7-30f and also handles the equivalence relationships for the shear stresses (see eq. 7-43 to 7-45) correctly. Like Eggenschwyler [Eggenschwyler, 1920/1] (to whom he refers), Maillart finds the position of the shear centre $M$ (see also eq. 7-46). In an analogy to the position of the resultants of the compressive or tensile stresses, i.e. the centre of compression or tension, Maillart calls the point at which the resultant of the shear stresses must act the "shear centre" [Maillart, 1921/1, p. 196]. Maillart had thus clearly formulated the equivalent relationship of the shear stresses and laid the foundation for the theory of the shear centre. Based on this, he recalculated the stress distribution given by Bach and came to a satisfactory answer. Eggenschwyler's harsh criticism of Bach's tests was therefore essentially unjustified. In the September issue of the journal *Der Eisenbau*, Eggenschwyler published a paper dealing with the rotation stresses of thin-wall symmetrical channel-shaped sections [Eggenschwyler, 1921] in which he determines the position of the shear centre of channel-shaped sections without mentioning the shear centre term introduced by Maillart shortly before.

In another publication, Maillart does not hold back in his criticism of acknowledged authorities such as Hermann Zimmermann and Wilhelm Ritter. In the end he calls for tests to confirm his hypotheses drawn from the theory of the shear centre:

*Hypothesis 1:* All cross-sectional forms obey the practical bending theory provided only bending moments are present.

*Hypothesis 2a:* "If a shear force (transverse bending) is added to the bending moment, then a certain normal stress diagram results … for any cross-sectional form … if the plane of the shear force parallel to the axis of the member is also parallel to the line connecting the centre of tension and centre of compression of the diagram and passes through the shear centre of the section" [Maillart, 1921/2, p. 19].

*Hypothesis 2b:* If this shear force plane is shifted away from the shear centre but remains parallel with it, then any cross-sectional form will exhibit partial outward bending and rotation proportional to the extent of the parallel displacement.

In 1922 Maillart complained that tests had still not been carried out and the authority of Bach had in many instances led to asymmetric sections not being fully utilised [Maillart, 1922, p. 254]. At the same time, Maillart also referred to Eggenschwyler's paper published in the journal *Der Bauingenieur* in 1922 concerning the opposite bending of the flanges due to the couple $+H_\tau$ and $-H_\tau$ (Fig. 7-30c), which results in axial stresses. Eggenschwyler was of the opinion that the phenomenon of axial stresses caused by opposite bending of the flanges could not be covered by the Saint-Venant torsion theory. It was left to Constantin Weber [Weber, 1926], Herbert Wagner [Wagner, 1929] and Robert Kappus [Kappus, 1937] to explain this phenomenon from the viewpoint of the theory of warping torsion for which they had created the basis.

**FIGURE 7-31**
Robert Maillart (1872–1940)
[Billington, 2003, p. 47]

Finally, on 8 March 1924, Maillart was able to report on bending tests on channel beams, carried out in the meantime by Mirko Roš and François Louis Schüle, in which the plane of the shear force passes sometimes through the shear centre $M$ and sometimes through the centroid $S$. These experiments confirmed Maillart's hypotheses 1 and 2a [Maillart 1924/1, p. 109]. And so the shear centre acquired the status of an engineering science concept.

#### 7.3.2.5
**Rearguard actions in the debate surrounding the shear centre**

The debate about the shear centre could have been wound up if Maillart had not criticised the book by professors Föppl and Föppl *Grundzüge der Festigkeitslehre* (principles of strength of materials) which appeared in 1923 [A. Föppl & O. Föppl, 1923] and which still clung to the idea that if the plane of the loads passes through the centroid of the section, then only bending stresses are present [Maillart 1924/1, p. 109]. Maillart, ever the practical engineer, made it quite clear to the two engineering scientists that the discrepancy between theory and tests could not be explained by the shortcomings of and corrections to the latter, "but instead can only be eliminated by rectifying the theory" [Maillart 1924/1, p. 110]. Thereupon, Arthur Rohn, professor of bridge-building and theory of structures at Zurich ETH, stood up for the representatives of science. He claimed that the shear centre could not be generally regarded as a fixed point of the cross-section like, for example, the centroid [Rohn, 1924, p. 129] – to which Maillart replied with further observations [Maillart 1924/2]. Stickforth was the first person to show that the shear centre depends on Poisson's ratio, i.e. is definitely not a variable of the cross-section [Stickforth, 1986]. This was followed by letters from Eggenschwyler [Eggenschwyler, 1924] and Maillart [Maillart, 1924/3] which, however, could not advance the theory of the shear centre very much in the direction of the theory of warping torsion. Thereupon, the editors of the journal *Schweizerische Bauzeitung* ended the debate about the shear centre on 21 June 1924 [n. n., 1924]. The historico-logical termination to the first development phase of the theory of the shear centre is the dissertation by Constantin Weber, which he completed in 1924 and which appeared in an abridged version in that same year in the *Zeitschrift für angewandte Mathematik und Mechanik* [Weber, 1924]. Based on a historical and critical analysis of the literature concerning the shear centre, Andreaus and Ruta have assessed this concept from the viewpoint of modern continuum mechanics [Andreaus & Ruta, 1998].

#### 7.3.3
**Torsion theory in structural steelwork from 1925 to 1950**

The further development of torsion theory shifted more and more from the journals of civil and structural engineering to the more theoretically oriented journals such as the *Zeitschrift für angewandte Mathematik und Mechanik* (journal of applied mathematics and mechanics). Nevertheless, there were also exceptions, like the paper by E. G. Stelling on rotationally rigid, trihedral bridge beams with examples of more recent structures in Hamburg's elevated railway [Stelling, 1929], which appeared in the second year of publication of the journal *Der Stahlbau*.

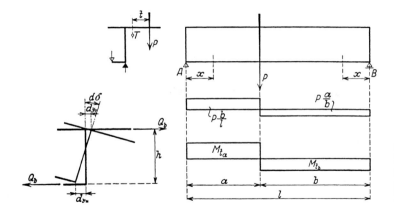

FIGURE 7-32
Bridge beam for Hamburg's elevated railway – asymmetric cross-section subjected to torsion after Stelling [Stelling, 1929, p. 80]

Stelling adapted the differential equation for the rotation of a channel section (specified by Ludwig Föppl in 1925) to suit the trihedral bridge beam (Fig. 7-32). Solving the differential equation supplies the internal forces for the beam restrained against torsion at supports $A$ and $B$ via the rotation $\delta = \vartheta$. However, he concludes from the results of the calculations for the bridge beam as built that "the calculation of the rotational stresses according to knowledge of the position and direction of points of bending [shear centre – the author] and bending loads for the bending situation parallel to flange and web can be solved with the simple principles of moment and shear force theory for beams in bending" [Stelling, 1929, p. 81].

Stelling was thus the first to investigate stresses due to warping torsion in the journal *Der Stahlbau*; he could only do this because he adopted the pioneering work of Constantin Weber [Weber, 1924] in particular. Stelling therefore exceeded the area of validity of Saint-Venant's torsion theory, one prerequisite of which was no restraint to twisting along the longitudinal axis of the member. But this excursion was to remain the exception in structural steelwork literature until 1950. The paper by Karl Schaechterle on the general principles of strength calculations [Schaechterle, 1929], which appeared in the same journal, shows that steelwork engineers had other scientific problems to solve. In that paper, the paradigm change from elastic theory to plastic theory was conspicuous. The analysis of torsion problems in structural steelwork are absent in Schaechterle's contribution, although Arpad L. Nádai had already provided a graphic interpretation of the elastic-plastic torsion of bars in 1923 [Nádai, 1923].

In 1930 Rudolf Bernhard published a programme of measurements for establishing the torsional stiffness of trussed framework twin-track railway bridges loaded on one side that had been carried out by staff of the State Railways Authority in Berlin [Bernhard, 1930]. However, the theoretical interpretation remained wholly in the model world of classical theory of structures. By comparing theory and experiments, Bernhard obtained simple empirical equations for determining the torsional stiffness, which enabled the calculation of spatial frameworks with multiple degrees of static indeterminacy to be avoided; torsion theory was not a theme for Bernhard.

Likewise, torsion loads on spatial frameworks were resolved into tension and compression forces in the case of the loadbearing systems common in aircraft construction up until the early 1930s, for the structures required for the rapidly expanding radio networks, especially short-wave antennas, and for the pylons carrying electricity cables.

For many years, the annual *Stahlbau-Kalender*, which had been published since 1935 by the German Steelwork Association (DStV), had included two pages of equations for the section values for the torsion of selected steel sections. Not until 1942 did this publication include a reference to the paper published in the journal *Zeitschrift für Mathematik und Physik* in 1910 by Timoshenko [Timoshenko, 1910] concerning the problem of torsion in a partially warped I-section beam [DStV, 1943, pp. 71–75].

As early as 1938, Prof. O. Eiselin at the University of Gdańsk drew attention to the beneficial structural effect and the economic benefits of hollow sections in steel bridges [Eiselin, 1938]. In the case of thin-wall hollow sections with $n$ cells, Karl Marguerre (1905–79) managed to solve this problem with $n$ degrees of static indeterminacy in 1940 by means of the elegant equations of the force method [Marguerre, 1940, pp. 321–322]. In that same year, Fritz Leonhardt (1909–99) pleaded for the adaptation of the methods of lightweight construction systematically developed in aircraft design – especially the use of hollow sections – to suit structural steelwork [Leonhardt, 1940]. One year later, Reinitzhuber once again pointed out to bridge engineers the favourable effects of hollow sections and referred to the aforementioned works of Eiselin, Marguerre and Leonhardt plus the relevant publications by Heinrich Hertel (1931) and Hans Ebner (1937) from aviation research [Reinitzhuber, 1941]. Furthermore, he presented a simple way of calculating the shear flow in a three-cell hollow cross-section. But not until 1949 did Leonhardt once again list the advantages of hollow forms of construction, this time in a paper describing the new road bridge over the Rhine from Cologne to Deutz in the journal

**FIGURE 7-33**
Section through the Cologne-Deutz road bridge [Leonhardt, 1950, p. 3]

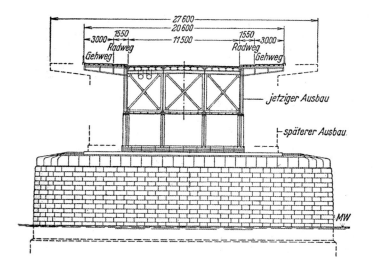

*Die Bautechnik*. In this bridge constructed as a three-cell hollow box (Fig. 7-33), the shear stresses due to the torsion load resulting from a one-sided imposed load were quantified with the help of Saint-Venant's torsion theory and Bredt's theorem [Leonhardt, 1950, pp. 20–21].

The work of Marguerre [Marguerre, 1940] marked the historical conclusion of the Saint-Venant torsion theory in structural steelwork. After 1950, the discussions surrounding the torsion problem in structural steelwork were outside the assumptions of the Saint-Venant torsion theory. Bornscheuer opened up a new chapter in the history of torsion theory in structural steelwork in 1952 with his systematic depiction of the bending and twisting process taking into account, in particular, warping torsion [Bornscheuer, 1952/1].

### 7.3.4 Summary

*Firstly:* In the consolidation period of structural theory (1900–50), too, structural steelwork theory followed in the footsteps of classical theory of structures, as was pointed out in the third to sixth theses of section 7.1.5.

*Secondly:* Structural steelwork in the 1930s and 1940s was characterised by the appearance of steels with higher strengths and a new method of jointing – welding.

*Thirdly:* The use and ongoing development of the research into the torsion of thin-wall sections carried out during the 1920s and 1930s was only of benefit to aircraft construction – part of Hitler's rearmament programme. The benefits of lightweight construction went into military aircraft and could not be exploited by structural steelwork; up until 1945, lightweight construction was a secret science essentially reserved for the military.

*Fourthly:* The end of the war put an end to this problem immediately. Lightweight construction took on non-military forms, and this science transfer via structural steelwork also touched on torsion theory.

## 7.4 Searching for the true buckling theory in steel construction

Euler's buckling theory in extended form was confirmed brilliantly during the 1920s in buckling tests carried out by the Committee for Tests in Steel Construction, which were largely funded by the German Steelwork Association (DStV) and accompanied scientifically by Hermann Zimmermann (Fig. 7-34a).

A practical theory of steel struts that also covered satisfactorily the inelastic range began to develop in Austria and Czechoslovakia. Further work was carried out in the 1930s by Ernst Chwalla (Fig. 7-34b), Friedrich Hartmann (Fig. 7-34c) and Karl Jäger (Fig. 7-34d). With the *Anschluss* of Austria and the "liquidation of the remainder of Czechoslovakia" (Hitler), this research work shaped the development of DIN 4114.

### 7.4.1 The buckling tests of the DStV

Right from the very first annual report of the DStV (1906), the board was of the opinion that the DStV "had a duty to promote the science from which German bridge-building had grown" [DStV, 1954, p. 42]. Prior to this, meetings had taken place between the DStV chairman at that time, Leonhard Seifert, also a director of the Harkort company in Duisburg,

a) Hermann Zimmermann   b) Ernst Chwalla   c) Friedrich Hartmann   d) Karl Jäger

**FIGURE 7-34**
The most important figures in the development of buckling theory for steel columns: a) Hermann Zimmermann (1845–1935) [Schaper, 1935/1, p. 225], b) Ernst Chwalla (1901–1960) [Beer, 1960, p. 223], c) Friedrich Hartmann (1876–1945) [Vienna TU archives], d) Karl Jäger (1903–1975) [picture archive of the Austrian National library, Vienna]

and Hermann Zimmermann, the most senior engineering civil servant at the Prussian Ministry of Public Works, the aim of which was "to bring about clarification of important theoretical and constructional problems in structural steelwork by way of tests, involving as required appointed representatives from other authorities and bodies responsible" (cited in [DStV, 1929, p. 50]). At the suggestion of the DStV, a neutral commission was set up, chaired by Zimmermann and comprising representatives from relevant Prussian authorities, the State Naval Department, the Materials Testing Institute (MPA Berlin, which had been founded in 1904), the Jubilee Foundation of German Industry, the Association of German Engineers (VDI), the Association of German Steelworks and the DStV; and on 11 January 1908 that became the Committee for Tests in Steel Construction (since 1935 known as the German Committee for Structural Steelwork, DASt).

The first and most important task of the DASt was to establish a plan of work. After the tests in structural steelwork prior to 1900 had been reviewed and evaluated, the DASt realised that they were of only limited validity because the wrought iron used in those days had in the meantime been superseded by mild steel. In order to avoid repetitions and questions "that could be answered in a simple way and at relatively little cost and which would help clarify and simplify further tests" [Kögler, 1915, p. 6], the DASt decided on preliminary tests.

Franz Kögler (1882–1939) regarded the most important and most extensive part of the plan of work as the tests to determine the buckling strength of single and compound struts [Kögler, 1915, pp. 6–7] because an adequate buckling theory and design principle for struts was still lacking.

In 1910 the DStV set aside 100 000 marks (which amounted to approx. 10 % of the surplus the DStV had generated in that year) for the series of tests; by 30 June 1910 the accumulated reserves for this purpose had reached an impressive 350 991 marks [V. d. B. u. E.-F., 1910, p. 15]. We can be sure that the above-average funding provided by the DStV in 1910 for test purposes was influenced by the collapse of the large gasometer at Großer Grasbrook in Hamburg on 7 December 1909, which claimed 20 lives and resulted in injuries to 50 others [Förster, 1911, pp. 178–179]. The two experts in charge of producing a report on the tragedy – Krohn for the City of Hamburg and Müller-Breslau for the two steelwork companies – agreed that an undersized resolved strut had caused the collapse. This failure poured oil on the fire of the discussion surrounding the buckling strength of resolved struts; the failure of such a member in the bottom chord of the bridge over the St. Lawrence River at Quebec in Canada in 1907 had shown that the limits of the self-confident "faster – further – higher" credo had been far exceeded – on that occasion 74 building workers had died. The message for steel bridge-building, regarded as a high-tech discipline at that time, was: concentrate forces to solve the buckling problem.

### 7.4.1.1 The world's largest testing machine

According to Leonhard Seifert, the DStV had already ordered a 3000 t testing machine from Haniel & Lueg in Düsseldorf in February 1910 for carrying out tests with very large specimens and high loads. The machine was ready for use in 1912, set up together with its own hydraulic system in a specially constructed building measuring 30 × 13 m on the premises of the MPA (Fig. 7-35).

The technical specification was impressive, and represented a quantitative and qualitative leap in the development of materials testing:
- max. compression: 3000 t
- max. tension: 1500 t
- hydraulic pressure for compressive tests: 400 at
- hydraulic pressure for tensile tests: 200 at
- max. length of test specimens: 15 m
- weight: 350 t
- footprint (L × W): 28 × 4.50 m
- total cost: 250 000 marks
- year of manufacture: 1911/12
- owner: DStV

Following the satisfactory acceptance tests, which were not reported in detail until 1920 (see [Rudeloff, 1920]), full-size copies of the failed struts of

**FIGURE 7-35**
The DStV testing machine first used in 1912 [V. d. B. u. E.-F., 1912, p. 34]

the Hamburg gasometer were fabricated (Fig. 7-36/top) and tested in the new machine. The tests resulted in average value of 84.63 t for the buckling load of the struts shown in Fig. 7-36/bottom [Kögler, 1915, p. 50]. Müller-Breslau, in his report into the collapse of the Hamburg gasometer commissioned by the two steelwork companies, had established a load-carrying capacity of 88 t and calculated that an eccentricity of loading or member axis out of true by just 13.3 mm would have been enough to trigger failure of the member.

So at the start of the second decade of the 20th century the "perfection of technology" (1946) of a Friedrich Georg Jüngers (1898–1977) proved to be a feuilletonistic illusion of the conservative revolutionaries. Technology is always imperfect, as the struts of the Hamburg gasometer failure demonstrated. Engineers experience the imperfection of technology in their everyday work. Nevertheless, the high precision intrinsic to building with steel, theory of structures modelling and the skill of structural calculation misled users into placing all this above the constructional and technological possibilities of practical steelwork. One impressive example of this is the search for the true buckling theory.

**The perfect buckling theory on the basis of elastic theory**

### 7.4.1.2

The series of tests published by Kögler in 1915 (see [Kögler, 1915]) met with objections. In a letter to DASt chairman Max Carstanjen (1856–1934) (director of the MAN Gustavsburg works) written in October 1915 by DStV employee Hermann Fischmann but never sent, Fischmann notes that "a great amount of useful work has already been done," but complains that since the DASt was founded "it has not got one step nearer to answering the burning question of the most efficient form of struts, which had

**FIGURE 7-36**
The struts tested [Kögler, 1915, p. 42]

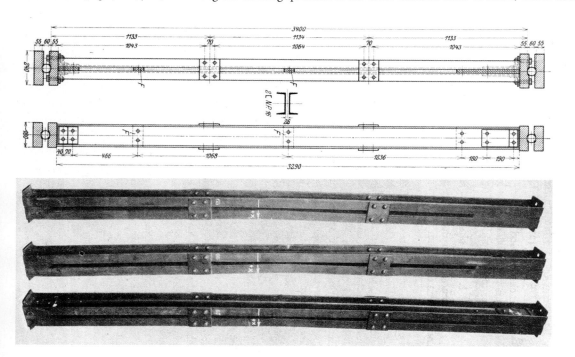

been one of the most important points of the programme" [Fischmann, 1915, pp. 1 – 2]. In particular, Fischmann criticises the fact that hitherto only a few buckling tests had been carried out on the 3000 t testing machine, the series of tests had been changed more than once, Kögler's programme was too costly and was tantamount to a repetition of the buckling tests of Ludwig von Tetmajer (1850 – 1905) [Fischmann, 1915, pp. 10 – 14].

As an alternative, he proposed using tests to confirm the theoretical buckling theorems of Leonhard Euler (1707 – 83), Friedrich Engesser (1848 – 1931), Theodore von Kármán (1881 – 1963) and Heinrich Müller-Breslau (1851 – 1925), and concentrating on practical tests with the sections and resolved members common in structural steelwork [Fischmann, 1915, pp. 14 – 16]. Fischmann was due to take over as director of the DStV on 1 January 1918 and preside until 31 December 1923, but the outcome was different. What happened was that the programme of work for "theoretical tests", proposed in 1920 by Schaper, Müller-Breslau and Rein, was adopted, and this was successfully accompanied by a large number of theoretical papers by Zimmermann (see also [Nowak, 1981, pp. 220 – 231]). What was at stake here was the search for the true buckling theory on the basis of elastic theory, the assumptions of which had to be reproduced as accurately as possible for the tests. Characteristic of this are Zimmermann's instructions for calculating the error adjustment for buckling tests – suggested by the MPA and sent to the DASt on 22 June 1922 [Zimmermann, 1922/1]; they were published later that same year (see [Zimmermann, 1922/2; 1922/3]). The structural and geometrical imperfections of the test specimen could therefore be compensated for through a correction of the load application during the test and an ideal buckling load achieved. Zimmermann's test regime devised theoretically enabled the MPA to verify experimentally that perfect test specimens subjected to compression obey the Euler curve and exhibit a yield plateau in the lower slenderness range: the Euler curve and the yield point describe struts adequately (Fig. 7-37).

**FIGURE 7-37**
Buckling stress lines and results of the DASt buckling tests carried out at the MPA in Berlin: a) for grade St 37 steel, and b) for grade St 48 steel [Eberhard et al., 1958, p. 21]

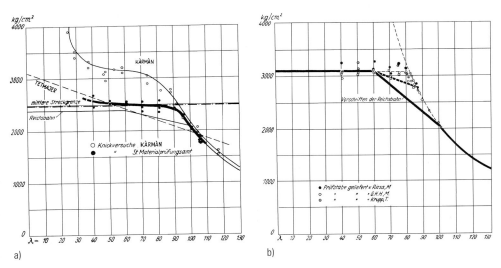

FIGURE 7-38
Cover of Hermann Zimmermann's
*Lehre vom Knicken auf neuer Grundlage*
[Zimmermann, 1930]

The 85-year-old Zimmermann crowned his 30 or so publications on buckling theory in 1930 with his *Lehre vom Knicken auf neuer Grundlage* (theory of buckling on a new basis) (Fig. 7-38) [Zimmermann, 1930]. So the shrewd mathematician and old master of structural steelwork with a sense for the aesthetics of symbolic forms in theory of structures remained the perfect elastic theorist in buckling theory as well, which was confirmed by the DASt buckling tests of the 1920s which owed much to his methods. Buckling in the inelastic range remained a taboo subject in buckling theory for a few more years. So it was not Zoroaster who spoke (Friedrich Nietzsche), but rather Zimmermann who in 1930 had the last word on buckling theory based solely on elastic theory. During the 1930s, a paradigm change from elastic theory to plastic theory in the development of theory of structures became evident here as well – one of the main threads of development during the invention phase of structural theory (1925 – 50).

But the results of the buckling tests, unsurpassed in their execution, came too late to have an influence on practical steelwork design because they were not published until 1930 [Rein, 1930] – and German State Railways had to act faster.

## German State Railways and the joint technical-scientific work in structural steelwork

### 7.4.2

Only after the establishment of the Ministry of Transport on 9 January 1920 and the amalgamation of the railways of the federal states into German State Railways in 1924, independent in terms of policies, administration and finances, did a uniform set of standards come into force for steelwork in railways (see [Werner & Seidel, 1992, pp. 59 – 75]). Gottwalt Schaper (1873 – 1942), as the head of the Bridges & Structural Engineering Department at the Ministry, was in charge of the most influential post in structural engineering, which would make its mark on joint technical-scientific work in German steel construction during the inter-war years [Siebke, 1990, pp. 114 – 121]. This work embraced the introduction of uniform codes of practice throughout the country, steels with higher strengths and welding.

## Standardising the codes of practice for buckling in structural steelwork

### 7.4.2.1

The codes of practice published by German State Railways were mainly shaped by Schaper; the design of steel struts was covered by the
- Code of Practice for Steel Structures. Principles for the design and calculation of steel railway bridges of May 1922 [Deutsche Reichsbahn, 1922]
- Code of Practice for Steel Structures. Principles for the calculation of steel railway bridges dated 25 February 1925 [Deutsche Reichsbahn-Gesellschaft, 1925]

These codes of practice published in 1922 and 1925 explore completely new paths in the area of strut design. The evaluation of various accidents had shown that struts with slenderness ratios $\lambda$ between 60 and 100 according to the old standards were often too heavily loaded, which is why the second of the above codes was based on the Tetmajer straight line (see Fig. 7-37a).

The introduction of the proof format for the $\omega$-method (identical with the stress proof)

$$\omega \cdot (N/A) \leq \sigma_{perm} \qquad (7\text{-}49)$$

where the compressive stress $N/A$ multiplied by $\omega$ is compared with the permissible ideal compressive stress $\sigma_{perm}$, provided the steelwork engineer with a simple design formula for concentrically loaded members, without having to consider possible safety factors or elastic/inelastic zones. The aforementioned codes of practice also made provision for checking the ideal compressive stresses for eccentrically loaded members:

$$\omega \cdot (N/A) + (M/W) \leq \sigma_{perm} \qquad (7\text{-}50)$$

Bernd Nowak suspects that the $\omega$-method can be attributed to an Austrian buckling standard dating from 1907 [Nowak, 1981, p. 200]. But as early as 1924 Friedrich Bleich (1878–1950) criticised that the formalisation, rationalisation and standardisation of the design of struts "completely masks the nature of the buckling problem" [Bleich, 1924, p. 115].

### 7.4.2.2 The founding of the German Committee for Structural Steelwork (DASt)

On 3 December 1935, the Committee for Tests in Steel Construction was reformed as the German Committee for Structural Steelwork (DASt).

Steel construction was therefore following the prevailing standardising nationalisation form in Germany in the sense of the basic pattern of technical-scientific joint activities – an organisational invention of reinforced concrete construction that first saw the light of day in 1907 as the German Committee for Reinforced Concrete (DAfStb, see section 9.2.3): "As in recent times it has become more and more necessary to extend the remit of the committee beyond the actual tests and research activities through the production of relevant guidelines, principles and standards for structural steelwork, it has been decided, with the consent of the authorities involved, to place the committee on a wider footing. From now it will be known as the 'German Committee for Structural Steelwork'" [Schaper, 1935/2, p. 762]. The powers of the Specialist Standards Committee for Structural Steelwork (part of the German Standards Committee) were transferred to the DASt.

The 28 DASt members were spread fairly evenly over the areas of administration, industry and science. The establishment of the DASt in 1935 completed the integration of the six areas of action: science- and administration-related association policies, science and industry policies, plus management- and industry-type science in structural steelwork (see Fig. 9-9).

### 7.4.3 Excursion: the Olympic Games for structural engineering

The "golden twenties" were also the years in which international political, cultural and technical-scientific organisations became established. For example, structural engineers exchanged their research findings for the first time on an international level in Zurich in 1926, at the International Bridges & Structural Engineering Congress. The second congress was held two years later in Vienna. These congresses led to the formation of the

International Association for Bridge & Structural Engineering (IABSE), which was founded on 29 October 1929 in Zurich. The first IABSE congresses took place in 1932 (Paris) and 1936 (Berlin).

But even the peaceful Olympic contests surrounding the technical-scientific ideas in structural engineering could not escape the long shadows of the Spanish Civil War, the war in Abyssinia and the vigorous, incessant rearmament programme of Hitler's Germany. Like the perfectly staged 1936 Berlin Olympics (1–16 August) was exploited in the propaganda of the Third Reich to enhance its international reputation, so the second International Congress for Bridges & Structural Engineering which took place in Berlin later that year (1–11 October) was also (ab)used for political purposes. Under the patronage of Hitler's government, Hitler's "new Germany" courted the international structural engineering community by way of festivities, excursions and grand receptions in Berlin, Dresden, Bayreuth and Munich. In particular, the political leaders played their trump card in the form of the engineering works accompanying the national network of motorways, ensured that the congress ran perfectly smoothly and created the illusion of a peaceful Third Reich through the "thoughtful besieging" (Heinrich Böll) of the 1500 participants from more than 40 countries.

The technical-scientific balance of the second IABSE congress is impressive. Out of a total of 10 sessions, four were devoted to steelwork, three to reinforced concrete and one to foundations; the other two sessions were used for a miscellany of presentations on other subjects. Merely the titles of the steelwork sessions will be mentioned here:

- The significance of the toughness of steel for the calculation and dimensioning of steel structures, especially statically indeterminate constructions
- Practical issues concerning welded steel structures
- Theory and testing of details of steel structures for riveted and welded constructions
- The use of steel in bridges, buildings and hydraulic engineering.

It becomes clear from the preliminary report of the congress that welding and the problems of plastic theory (fundamentals, ultimate load method) represented a focal point of interest in structural steelwork at that time. For example, two giant strides in development that triggered upheavals in steelwork and the formation of steelwork science appeared simultaneously at the congress:

- welding, which became a technical revolution in structural steelwork, and
- the scientific revolution in structural steelwork heralded by the partial paradigm change from elastic theory to plastic theory, the "paradigm articulation" of Thomas S. Kuhn [Kuhn, 1962 & 1979].

This paradigm change could be seen in the steelwork research of the late 1920s in the shape of the first approaches to the ultimate load method and non-elastic buckling theory.

The second IABSE congress was accompanied by publications both before and after the event, which set qualitative and quantitative standards in structural engineering:

- The preliminary report (Fig. 7-39) [IABSE, 1936/1] published to accompany the congress itself contains nearly 1600 pages and covers all those contributions that were not presented, but instead were grouped together for each session by general writers (including Klöppel) to form an introduction to the discussions.
- The final report (approx. 1000 pages) [IABSE, 1938] published in 1938, which reflects the expert discussions in all sessions.

Both the preliminary and the final report were published separately in German, English and French, i. e. a total of approx. 3 × 2600 = 7800 printed pages – a record that has yet to be beaten by any large scientific congress!

The reports published by Wilhelm Ernst & Sohn were joined by vol. 4 of the IABSE *Treatises* [IABSE, 1936/2] (650 pages), which was handed out to every participant of the Berlin Congress together with the preliminary report. This mammoth publishing task comprising approx. 3250 printed pages was financed by the German organising committee of the Berlin Olympics for structural engineering, which was led by Fritz Todt, the inspector-general of German roads, and included many high functionaries of the Third Reich. Owing to the immense consumption of steel by the rearmament machinery of the Third Reich, Todt's agenda for 1936 already included saving steel in the construction industry, which he soon had reissued as a decree and would reinforce the trend towards lightweight steelwork construction.

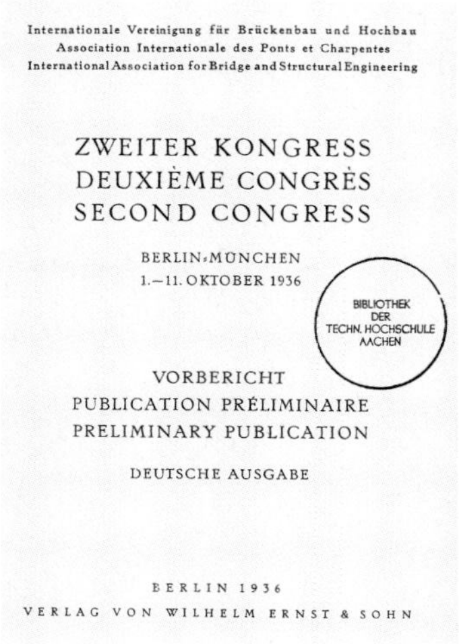
Internationale Vereinigung für Brückenbau und Hochbau
Association Internationale des Ponts et Charpentes
International Association for Bridge and Structural Engineering

ZWEITER KONGRESS
DEUXIÈME CONGRÈS
SECOND CONGRESS

BERLIN-MÜNCHEN
1.–11. OKTOBER 1936

VORBERICHT
PUBLICATION PRÉLIMINAIRE
PRELIMINARY PUBLICATION

DEUTSCHE AUSGABE

BERLIN 1936
VERLAG VON WILHELM ERNST & SOHN



**FIGURE 7-39**
Cover of the preliminary report on the second IABSE congress [IABSE, 1936/1]

### 7.4.4 A paradigm change in buckling theory

Roš (Zurich EMPA), Brunner (Zurich EMPA), Chwalla (Brno TH), Hartmann (Vienna TH), Jäger (Vienna TH) and Fritsche (Prague TH) made important contributions to the strut in the inelastic range. Scientists were already arguing about the further development of buckling theory at the first international congresses for bridges and structural engineering in Zurich (1926), Vienna (1928) and Paris (1932). For example, Fillunger (Vienna TH), Roš and others proved that their Warsaw colleague Broszko had made a mistake in his calculations, which triggered a fierce controversy between Hartmann and Broszko in the *Zeitschrift des Österreichischen Ingenieur- und Architekten-Vereins*. Broszko threw down the gauntlet to his opponents by claiming that the buckling theory of Engesser and Kármán was founded "on an evidently unsound principle" (cited in [Nowak, 1981, p. 237]). To prove this, he carried out buckling tests in 1932 commissioned by the DStV (see [Rein, 1930; Broszko, 1932]). But Broszko made the same mistakes as Wilhelm Rein, who misinterpreted the buckling tests theoretically (see [Rein, 1930]).

Chwalla, Fritsche, Hartmann and Jäger had determined the status of research regarding the buckling of struts since 1934. "The significance of Chwalla's work," Nowak remarks, "lies in the theory that takes good account of the reality, so all other authors should measure their results against his" [Nowak, 1981, p. 243]. Jäger took heed: Kist had already proposed an ideal-elastic and ideal-plastic material law for steel generally back in 1920, and hence had had a decisive influence on the development of the ultimate load method in the 1920s and 1930s (see section 2.10.1), and so Jäger used the same principle for steel struts in 1934. In this respect he was inspired by the writings of Fritsche and Girkmann on the ultimate load method for continuous beams and frames dating from 1930 and 1931 (see section 2.10.2). Jäger could therefore formulate a purely analytical buckling theory that embraced both the elastic and inelastic ranges in the same way. In his book of 1937, *Festigkeit von Druckstäben aus Stahl* (strength of steel struts) [Jäger, 1937], Jäger summarised the research findings of the Viennese school of buckling theory and inserted the final piece in the jigsaw of buckling theory extended to the inelastic range. Jäger offered the steelwork engineer a large number of ready-made formulas for the most common sections (Fig. 7-40) plus charts for sizing struts, which 15 years later would be included in the German stability standard (DIN 4114). The ultimate load method had thus entered buckling theory for the first time – a theory that matched the maxim of Wilhelm Ostwald (1853–1932), i.e. that there is nothing more practical than a good theory!

### 7.4.5 The standardisation of the new buckling theory in the German stability standard DIN 4114

After the *Anschluss* of Austria in March 1938 and the "liquidation of the remainder of Czechoslovakia" (Hitler) one year later, the once Austrian Buckling Committee, chaired by Hartmann, could now exert a decisive influence on the relevant steelwork standards in the "Greater Germany". As early as January 1939, Austrian representatives took part in the standardisation meetings for the first time and presented a detailed proposal that would shape the first draft of DIN E 4114 of 1 November 1939. The diverse drafts of and proposals for DIN 4114 went through a "considerable metamorphosis" [Klöppel, 1952/3, p. 85], as Klöppel verified in retrospect at the DStV's Munich steelwork conference in 1952. The editorial com-

| Querschnitt | Formel für die kritische Schlankheit | Gültigkeitsbereich | Beiwerte $\mu_1$ | Beiwerte $\mu_2$ | Anmerkung |
|---|---|---|---|---|---|
| 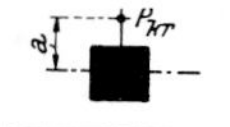 | $\lambda^2 = \frac{\pi^2 E}{\sigma_{kr}}\left[1 - \mu_1 \frac{m\,\sigma_{kr}}{(\sigma_s - \sigma_{kr})}\right]\left[1 - \mu_2 \frac{m\,\sigma_{kr}}{(\sigma_s - \sigma_{kr})}\right]$ | unbeschränkt $0 \leqq \sigma_{kr} \leqq \sigma_s$ | 0,5 | 0,5 | Es bedeutet: $L$ ....... Stablänge; $F$ ....... Querschnittsfläche; $W_{1,2}$ ..... Widerstandsmoment des Biegedruck- bzw. des Biegezugrandes; $i$ ........ Trägheitsradius; $\lambda = \frac{L}{i}$ ... Schlankheit; $a$ ........ Exzentrizität; $m = \frac{aF}{W_1}$ Exzentrizitätsmaß; $\sigma_s$ ....... Fließgrenze; $E$ ....... Elastizitätsmodul; $\sigma_{kr}$ ...... Kritische Spannung; $P_{kr} = F\sigma_{kr}$ Tragkraft |
| 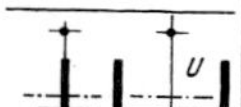 | | | 0,5 | 0,5 | |
| 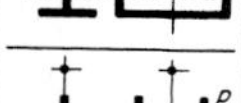 | | | 0,4 | 0,4 | |
| 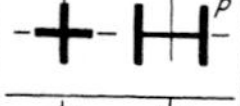 | | | 0,9 | 0,1 | |
| 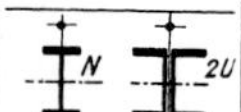 | | | 0,9 | 0,1 | |
| 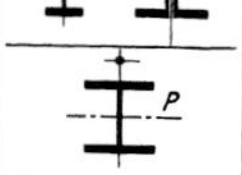 | | $\frac{\sigma_{kr}}{\sigma_s} \geqq \frac{W_1 - W_2}{W_1 + W_2}$ | 0,8 | 0,2 | |
| | $\lambda^2 = \frac{\pi^2 E}{\sigma_{kr}}\left[1 - \mu_1 \frac{W_1 m\,\sigma_{kr}}{W_2(\sigma_s + \sigma_{kr})}\right]\left[1 - \mu_2 \frac{W_1 m\,\sigma_{kr}}{W_2(\sigma_s + \sigma_{kr})}\right]$ | $\frac{\sigma_{kr}}{\sigma_s} \leqq \frac{W_1 - W_2}{W_1 + W_2}$ | | | |

**FIGURE 7-40**
Approximation formulas for the critical slenderness ratios of steel members carrying eccentric compression loads after Jäger [Jäger, 1937, p. 215]

mittee responsible for the drafts of DIN 4114, which appeared in quick succession, was made up of professors Willy Gehler (Dresden TH), Kurt Klöppel (Darmstadt TH) and Ernst Chwalla (Brno TH). By autumn 1943, the final draft of DIN 4114 (Fig. 7-41) to appear before the end of the war (edited by Chwalla) was out for comment; this version included torsional-flexural buckling, the first time this theme had appeared in a code of practice for steelwork [Chwalla, 1943, pp. 8/20 – 21]. Chwalla's draft anticipated the widely acclaimed 1952 edition of DIN 4114 (stability cases) in several respects:

- uniform treatment of the stability cases buckling, lateral buckling and local buckling,
- selective integration of findings from aircraft construction (in the sense of lightweight construction) for structural steelwork,
- direct transfer of current results from steel research into a technical code of practice,
- division of the draft into an applications part (Sheet 1: Specifications) and a background part (Sheet 2: Guidelines), which was intended to introduce users to the steelwork findings behind the provisions.

Nevertheless, the successes of steelwork stability theory and its projection piece by piece into the rapid succession of drafts for DIN 4114 cannot be explained by internal science factors alone.

The appendices of the *Stahlbau-Kalender* editions of the war years contained numerous orders, directives, regulations and decrees for restricting building with steel. For this purpose, the bureaucrats of the Third Reich invented a new class of engineer in 1940, the *Sparingenieur* (economising engineer), whose job was "to investigate, monitor and implement all options for saving steel in the planning and construction of buildings" [n. n., 1941/2, p. 576]. The preface to the 1941 edition of the *Stahlbau-Kalender*, entitled "The tasks of German steelwork during the war" was used by the DStV not just to exchange empty pleasantries, but instead to report on how steel-builders were and should "meet the demands of the German armed forces and the armaments industry" [n. n., 1941/2, p. III]: wide-span aircraft hangars as a prototype for an economic method of construction with steel, the rebuilding of destroyed infrastructure and essential facilities in war-torn regions to secure supplies for the troops, the repair of production plants in the "annexed regions" in order to use them for wartime production. "In all cases, the minimum use of the material was maximum priority" [n. n., 1941/2, p. III].

This "maximum priority" also led to the most successful work of the Buckling Committee under the leadership of Willy Gehler, an active Nazi Party member (see [Hänseroth, 1991]). For instance, Gehler, in the preface to Chwalla's DIN 4114 draft (1943), criticised the size of the Euler safety factor for buckling at that time, saying it was too high at $v_{Ki} = 3.5$, and even suggested that the $v_{Ki} = 3.0$ figure in the previous draft of 1942 could also be reduced to $v_{Ki} = 2.5$. Gehler founded his criticism on more recent research findings and the fact that "an Euler safety factor for buckling of 2.5 had already been proven in practice in the former

FIGURE 7-41
Cover of the final draft of DIN 4114 [Ernst & Sohn archives]

Austria (OeNORM B 1002, 2nd edition, June 1930)" (cited in [Chwalla, 1943, p. 2]). Chwalla, too, regarded the reduction in the Euler safety factor for buckling to $v_{Ki} = 2.5$ in his draft as a feature that distinguished it from the previous draft [Chwalla, 1943, p. 2]. He therefore points out in his guidelines that when comparing the maximum applied load $P$ with the load-carrying capacity $P_{Kr}$ according to Engesser

$$P \leq P_{Kr}/v_{Kr} \qquad (7\text{-}51)$$

the proof

$$P \leq P_{Ki}/v_{Ki} \qquad (7\text{-}52)$$

is also required because $v_{Ki}$ is considerably greater than the safety factor for load-carrying capacity $v_{Kr}$, and therefore in the case of slender struts eq. 7-52 can lead to a lower permissible load than eq. 7-51 [Chwalla, 1943, p. 12]. If Chwalla's draft had come into force, then he would have been able to increase the permissible load by a factor of $3.5/2.5 = 1.4$ over the old provision regarding slender struts!

Chwalla's DIN 4114 draft dating from October 1943 [Chwalla, 1943] was never published. One of the reasons for this was that the majority of the stocks of the publishing house Wilhelm Ernst & Sohn in Berlin fell victim to Allied bombing in November 1943, and so the 1944 edition of the *Stahlbau-Kalender* (published in late 1943) was the last coherent publication: "Like everything in modern Germany," as the Steelwork Study Group of the National Socialist Federation of German Engineering, the new publisher of the *Stahlbau-Kalender*, explained to its readers in the preface, "the *Stahlbau-Kalender* is also fully committed to total war" [n. n., 1943/2, p. IV]. The claim of this study group to be "preserving the tradition of the joint technical and scientific work of the German Steelwork Association," which included the publication and promotion of the *Stahlbau-Kalender* [n. n., 1943/2, p. IV], was punished by the reality of the total war lies. The abolition of the DStV and its replacement by the Steelwork Study Group of the National Socialist Federation of German Engineering also signified the end of one cornerstone of structural steelwork literature in Germany.

It was not until many years after the war that the German stability standard DIN 4114 could be completed. Kurt Klöppel introduced DIN 4114 at the DStV's Munich steelwork conference [DStV, 1952, pp. 84–143], and this standard soon became the model for the stability codes of other countries. So Klöppel's introductory speech can be regarded as a classical document of steelwork science in the innovation phase of structural theory (1950–75).

## 7.5 Steelwork and steelwork science from 1950 to 1975

Like the formation of reinforced concrete construction as a scientific discipline in the fabric of structural engineering can only be understood in conjunction with the technical revolution in the building industry that was triggered by reinforced concrete construction, an analysis of the genesis of steelwork science in the late 1930s cannot be considered without recognising the fact that welding represented a technical revolution in structural steelwork. Both of these structural engineering fields could only develop into separate scientific disciplines because they were embraced completely by the trinity of administration, industry and science. Nevertheless, there are differences to reinforced concrete construction: welding technology was only one prerequisite for the emergence of steelwork science; the main condition was the development of stability theory, the transition from member to continuum analysis, composite construction theory and lightweight steel construction. Kurt Klöppel (Fig. 7-42), who in 1929 took charge of the Technical-Scientific Department of the DStV, was elected managing director of the DASt in 1935 and appointed professor at Darmstadt TH in 1938, recognised this fact; he became chief editor of the journal *Der Stahlbau* one year later, a post he held until 1981.

His understanding of the fundamentals of structural steelwork was clearly emphasized in his presentation *Rückblick und Ausblick auf die Entwicklung der wissenschaftlichen Grundlagen des Stahlbaues* (review of and outlook for the development of the scientific basis of structural steelwork) [Klöppel, 1948, pp. 48–72] at the first DStV steelwork conference after the war (Hannover, 1947). Stability theory (buckling, local buckling, lateral buckling, torsional-flexural buckling), the transition from member to continuum analysis (beam grid, orthotropic plate), welding and materials science played the key roles. For example, structural steelwork was not just the application of stability theory, but itself "has become a mainstay of one branch of natural science research" [Klöppel, 1948, p. 51]. Klöppel thus formulated the programme for steelwork science, which he honed in 1951 within the scope of the definition of the tasks and aims of the journal *Der Stahlbau* [Klöppel, 1951/1, p. 1]:

FIGURE 7-42
Kurt Klöppel at the age of 52
(Darmstadt TU archives)

- The scientific effect of structural steelwork is even greater than its economic significance; it fulfils important tasks in the development of the scientific basis of structural engineering. Besides its role as a classical theme for the civil and structural engineer, structural steelwork creates a bridge to mechanical engineering. "Therefore, for the care of its foundations and the promotion of its areas of application, we want to consider the concept of structural steelwork in a very broad context" [Klöppel, 1951/1, p. 1].
- The theoretical basis of structural steelwork is much more important now than it was in the past.
- The development of the practical basis of structural steelwork must keep pace with its theoretical basis.
- The interaction between materials science and welding technology so typical of structural steelwork has also helped steel production in recent years.
- The main aim of lightweight steel construction is the economic use of steel. "An exchange of experiences in the coming years is especially important in this comparatively new field of application for steelwork." [Klöppel, 1951/1, p. 2]
- Research must close a number of gaps in the foundations of structural

steelwork. *Der Stahlbau* was to report on the ongoing tests supervised by the DASt.

As will be shown in the following sections using the examples of the orthotropic plate, composite steel-concrete construction and lightweight steel construction, steelwork science lent theory of structures important momentum during the innovation phase of structural theory (1950–75).

### 7.5.1 From the truss to the plane frame: the orthotropic bridge deck

It was pointed out in section 2.4.5 that the box construction of the Britannia Bridge was inspired by shipbuilding. And in historico-logical terms, the orthotropic bridge deck can also be attributed to the construction principles employed in the building of steel ships: the system of longitudinal frames introduced by Joseph William Isherwood (1870–1937) integrated the steel plate with the longitudinal ribs and the transverse stiffeners, which improved considerably the longitudinal strength of ships and reduced the weight of steel in tankers by 15–20% [Lehmann, 1999, p. 207]. From the second decade of the 20th century onwards, the Isherwood system, which in structural terms acts like an orthotropic plate, became the norm, initially for ocean-going ships for bulk goods, then for tankers (Fig. 7-43) and finally for all ships. During the inter-war years, Georg Schnadel (1891–1980), the chair of shipbuilding engineering and ship elements at Berlin TH, was the international leader of this engineering science discipline. Together with Fritz Horn (1880–1972), Hermann Föttinger (1877–1945), Hans Reissner (1874–1967) and Moritz Weber (1871–1951), Schnadel made up a quintet at Berlin TH that set standards on the international shipbuilding science scene. For example, the American shipbuilding engineer Henry H. Schade (1900–92) carried out research into the problems of orthotropic plates, the effective width of flanges and plate buckling under Prof. Schnadel [Lehmann, 1999, p. 410], studies which found their way into classical works on the structural design of ships [Schade, 1938, 1940 & 1953].

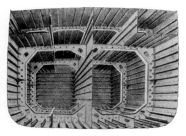

**FIGURE 7-43**
Main frame of a tanker built using the Isherwood system
[Lehmann, 1999, p. 207]

In the steel road bridges common up to the end of the 1930s, the structural elements such as reinforced concrete bridge decks, longitudinal stiffeners, main girders and cross-girders were considered to act separately from the structural design viewpoint; this led to a high consumption of steel and gave reinforced concrete an advantage over steel in the building of road bridges. This situation, the gradual replacement of riveting by welding in steel bridges during the 1930s and the rearmament programme of the Third Reich, which imposed savings measures on the use of steel in the civil sector, forced the introduction of new statics-constructional solutions in the building of steel road bridges. As early as 1934, Karl Schaechterle (1879–1971) had reported on new forms of bridge deck construction for steel road bridges taking special account of new developments in the USA and Germany [Schaechterle, 1934]. The main aim was to secure the economy of the use of steel for road bridges by reducing the weight of the bridge deck – also in bridges made exclusively from steel – to enable the rapid expansion of the motorway network of the Third Reich. One way of reducing the weight was "through using bridge deck plates made from beam grids with a covering of flat plates (cellular decks) which carry a load comprised of point loads without any special base course and sub-base and act as a slab" [Schaechterle, 1934, p. 479]. The first steel bridge deck was erected in 1936 by the MAN company for the Kirchheim u. Teck motorway bridge in the course of the Stuttgart-Ulm motorway [Schaechterle & Leonhardt, 1938, p. 306]. Some 20 years later, Walter Pelikan and Maria Esslinger noted that this deck already included all the essential, constructional features that had come to characterise the orthotropic plates of the MAN company after many years of building experience [Pelikan & Esslinger, 1957, p. 294].

Starting in the mid-1930s, publications on lightweight steel road decks began to appear, e. g. those of Schaper [Schaper, 1935/3], Schaechterle and Leonhardt [Schaechterle & Leonhardt, 1936, 1938] and Otto Graf [Graf, 1937, 1938]. For example, Graf, director of the Materials Testing Institute in Stuttgart, reported on new types of American bridge decks [Graf, 1937] and the investigations into lightweight steel road decks instigated by Schaechterle on behalf of the State Motorways Authority and later the DASt [Graf, 1938]. More or less at the same time, Leonhardt, a close colleague of Schaechterle, completed his dissertation on the calculation of beam grids supported on two sides sponsored by the State Motorways Authority and the DStV [Leonhardt, 1937/1938], a considerably expanded version of which (with co-author Wolfhart Andrä, 1914–96) appeared later [Leonhardt & Andrä, 1950]. One year prior to that, Hellmut Homberg (1909–90) had written on the same subject [Homberg, 1949]. Homberg accused Leonhardt and Andrä of plagiarism and sued both of them, but without success (see [Stiglat, 2004, p. 223]). He advanced the theory of beam grids (see [Homberg, 1951, 1952]) and in doing so used the beam grid as his starting point and later assigned it the qualities of an orthotropic plate (see, for example, [Homberg & Weinmeister, 1956; Homberg & Trenks, 1962]). Nevertheless, the statics-constructional evolution of the lightweight steel road deck into the orthotropic plate in the late 1940s induced the transition from the theory of the beam grid to the theory of the orthotropic plate in the 1950s. Therefore, the transition from member to continuum analysis had been partially completed in steel bridge-building on the theoretical side.

#### 7.5.1.1 The patent

Patent No. 847014 with the title *Straßenbrücke mit Flachblech* (road bridges with flat plates) was published in 1948. The patent was registered in the name of the MAN company and the inventor's name was given as Wilhelm Cornelius (1915–96), a MAN employee. The inventive value of this first patent dealing with an orthotropic road deck plate (Fig. 7-44) was that
- the road deck plate is part of the main girder, cross-girder and continuous longitudinal stiffeners because it forms a common top flange for said structural elements;
- the spacing of the cross-girders is less than one-third of that of the main girders.

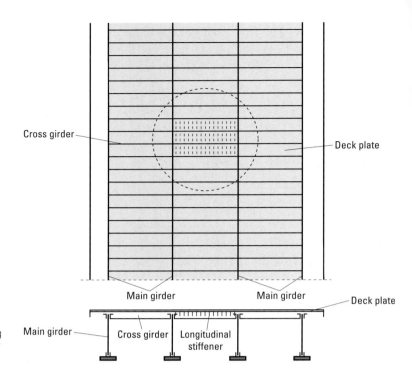

**FIGURE 7-44**
Basic concept of the orthotropic road deck plate for road bridges according to the MAN patent of 1948 (drawing after [Minten et al. 2007])

In contrast to this, the spacing of the cross-girders in conventional steel bridges was 0.5 to 0.8 times that of the main girders. Thanks to the close spacing of the cross-girders according to the MAN patent, the entire structure could be modelled as an orthotropic plate. The stresses in the longitudinal and transverse members could be determined from the continuum mechanics system of the orthotropic plate. The benefits of the transition from the one- to the two-dimensional loadbearing structure in structural steelwork was emphasized by Klöppel: "The inclusion of the road deck and road deck beams – previously provided merely for distributing the load transversely – in the loadbearing bridge cross-section is a special feature of this progress, which is primarily founded on theory. This benefits both the composite construction with the beam grid effect and the true steel road deck plate, which is now calculated as an orthogonally anisotropic plate (i. e. no longer according to member analysis) and makes the reinforced concrete road deck superfluous, which leads to a considerable saving in self-weight" [Klöppel, 1951/1, p. 2].

As welding became firmly established in structural steelwork, the language of design changed fundamentally. Fig. 7-45 shows sections through the old and new Cologne-Mülheim suspension bridges. Whereas the old, riveted bridge erected in 1927–29 and destroyed in 1944 required 12 900 t of steel for the main span alone, the partially welded, new bridge opened for traffic on 8 September 1951 required only 5800 t [Schüßler & Pelikan, 1951, p. 141]. The reason for this 55 % reduction in weight was the use of welding, steel grade St 52 and the orthotropic road deck. One reason why the MAN company was involved with this bridge was that for the first time they incorporated the concept of the orthotropic plate into their bid, the structural calculations for which were based on the orthotropic plate theory in the version by Cornelius.

### 7.5.1.2 Structural steelwork borrows from reinforced concrete: Huber's plate theory

Six months after the new Cologne-Mülheim suspension bridge was opened, Klöppel student Cornelius revealed the theoretical basis behind his recipe for success [Cornelius, 1952]. Cornelius consciously completed the transition from member to continuum analysis. He justified this by saying that in the past steel structures had been products of mechanical engineering and hence represented multi-part trusses whose loadbearing elements were carefully separated from one another in both constructional and theoretical terms. Owing to their origins, such loadbearing structures were considered as machines (see section 6.3.3), and proof of their safety had been carried out almost exclusively with the means of member analysis or kinematics derived from mechanics. "But the character of permanent structures such as bridges is such that they cannot be machines because if we imagine for a moment bridges were to 'grow' in nature, then it would be unthinkable for such 'bridges' to have hinges, joints, springs, etc. Therefore, it is a totally natural development to design the individual elements of permanent loadbearing structures ever more coherently so that over the course of time they lose the features of a machine and approach closer and closer to the organically grown structure, the structure from one mould. In structural terms, this means a considerable increase in the degree of static indeterminacy, right up to complete continuity. As we know, this increases the calculation work very rapidly within the scope of member analysis if the structural engineer does not wish to accept compromises in the loadbearing reserves actually available – at the cost of economy. At this stage it can be advantageous to select a loadbearing system based on continuum analysis instead of member analysis" [Cornelius, 1952, p. 21]. Cornelius' reasoning is reminiscent of the holistic tone of that advocate of organicism William Emerson Ritter (1856–1944) in general and the harmony between man-made structure and nature propagated by the practitioners and theorists of organic building in particular. It is important that Cornelius recognised the genesis of the loadbearing systems as, so to speak, an organic development from discontinuum to continuum,

**FIGURE 7-45**
Sections through the old (left) and new (right) Cologne-Mülheim suspension bridges (dimensions in mm)
[Kurrer, 2006, p. 250]

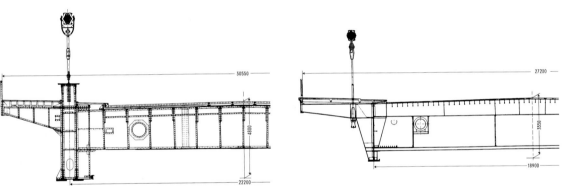

which he also observed in reinforced concrete construction. Such a change in the modelling of loadbearing structures, from member to continuum analysis, was not new because "even the progress in reinforced concrete construction replaced, for example, ... the previous structural design of lattice structures by the structural design of shells and folded plates" [Cornelius, 1952, p. 21]. Cornelius' work was based on the plate theory developed by Maksymilian Tytus Huber (1872–1950) for reinforced concrete construction. He solved the Huber differential equation for orthotropic plates for various types of plates such as a steel plate with a group of rolled sections (see Fig. 7-45) and a beam grid in conjunction with a concrete slab, i.e. he specified integral functions for the deformations and internal forces and tabulated the constants for the integral functions for common types of loading.

Since the second decade of the 20th century, the calculation of reinforced concrete slabs had been based on a simple structural model, essentially based on beam theory. In the method attributed to Franz Grashof, for example, a rectangular slab is divided into two orthogonal strips and the respective deformations and internal forces in the slab strips in the in $x$- and $y$-direction calculated at the points of intersection based on the condition of the equality of the deflections [Grashof, 1878, p. 234]. In this method, the torsion in the slab is neglected. On the other hand, tests on reinforced concrete slabs with the same amount of reinforcement in the $x$- and $y$-direction confirmed the validity of Kirchhoff's plate theory for homogeneous and isotropic slabs [Kirchhoff, 1850]. However, it could not be applied directly to reinforced concrete slabs purely for the reason that the bending stiffness of a reinforced concrete slab, depending on the reinforcement, can assume very different values in different directions [Huber, 1914, p. 557]. This is why Huber, in 1914, developed the general theory of reinforced concrete slabs reinforced in both directions and derived the differential equation for their deflection $w(x,y)$ [Huber, 1914, p. 561]. Later, he published an appendix in Polish [Huber, 1921] and German [Huber, 1923, 1924, 1925, 1926], which he again summarised in his Polish monograph [Huber, 1929].

Huber published a theory simpler and better than the one published in 1914 in a series in the journal *Der Bauingenieur* [Huber, 1923, 1924, 1925, 1926]. The findings presented in the series of articles had been available to Huber since 1918 [Huber, 1923, p. 354] and three years later had appeared in his Polish publication [Huber, 1921].

After Huber has talked about fundamental but also critical points in the theoretical foundation of tests in reinforced concrete construction, he derives the differential equation of deflection $w(x,y)$

$$K_x \cdot \frac{\partial^4 w}{\partial x^4} + 2 \cdot H \cdot \frac{\partial^4 w}{\partial x^2 \partial y^2} + K_y \cdot \frac{\partial^4 w}{\partial y^4} = p(x,y) \tag{7-53}$$

due to load $p(x,y)$ with the help of the energy principle [Huber, 1923, pp. 356–360]. Applied to orthotropic road decks on steel bridges, the Huber differential equation contains the plate bending stiffness transverse to the axis of the bridge (bending stiffness of road deck plate) $K_x$, the plate bending stiffness in the direction of the bridge axis (bending stiffness of longitudinal stiffeners) $K_y$ and the effective torsional stiffness

$$H = \frac{1}{2} \cdot (4 \cdot C + v_y \cdot K_x + v_x \cdot K_y) \qquad (7\text{-}54)$$

for thin, homogeneous-elastic but orthogonally anisotropic plates. Of course, Huber's theory applies to all thin, homogeneous-elastic and orthogonally anisotropic plates such as steel or reinforced concrete. In eq. 7-54

$2 \cdot C$ is the pure torsional stiffness
$v_x$ is the lateral strain due to normal stress in the $x$-direction
$v_y$ is the lateral strain due to normal stress in the $y$-direction.

In the isotropic case, the plate bending stiffnesses or lateral strains in the two directions are equal, i.e. $K_x = K_y = K$ and $v_x = v_y = v$. The pure torsional stiffness in this special case is

$$2 \cdot C = K \cdot (1 - v) \qquad (7\text{-}55)$$

and entered into eq. 7-54 gives the value $H = K$, which means that Huber's differential equation 7-53 is converted into Kirchhoff's differential equation for plates [Kirchhoff, 1850]

$$\frac{\partial^4 w}{\partial x^4} + 2 \cdot \frac{\partial^4 w}{\partial x^2 \partial y^2} + \frac{\partial^4 w}{\partial y^4} = \frac{p(x,y)}{K} \qquad (7\text{-}56)$$

which with the Laplace operator $\Delta$ takes on the form

$$\Delta \Delta w = \frac{p(x,y)}{K} \qquad (7\text{-}57)$$

Huber applied his theory to
- rectangular plates with unequal bending stiffnesses ($K_x \neq K_y$) simply supported on all sides and subjected to a sinusoidal and constant line load $p(x,y)$ [Huber, 1924],
- very long rectangular plates simply supported along their longitudinal edges and subjected to point loads etc. [Huber, 1925/1],
- rectangular plates with unequal bending stiffnesses ($K_x \neq K_y$) simply supported on all sides and subjected to a uniform load of one strip of plate plus a point load [Huber, 1926] and a comparison with the results of the calculations stemming from the plate trials of the German Reinforced Concrete Committee [Bach & Graf, 1915].

It was in 1925 that Huber first shortened the adjective "orthogonally anisotropic" to "orthotropic" [Huber, 1925/2, p. 878]. In the summary to the subsequent paper, he then speaks of orthotropic plate [Huber, 1926, p. 121] and a few sentences further on he explains: "The aforementioned adjective 'orthotropic' is simply a shortened form of 'orthogonally anisotropic'" [Huber, 1926, p. 121].

Differential equation 7-53, which Huber derived in the journal *Der Bauingenieur* [Huber, 1823, pp. 356–360] was used by Cornelius in his version of orthotropic plate theory [Cornelius, 1952, p. 21]. Therefore, structural steelwork borrowed from reinforced concrete and during the 1950s and 1960s encouraged a far-reaching development of the theory of the orthotropic plate, driven by technical progress in steel bridge-building and aircraft construction.

## The theory dynamic in steelwork science in the 1950s and 1960s

### 7.5.1.3

The publication of Cornelius' work was followed by numerous further contributions to the theory of orthotropic plates in the journal *Der Stahlbau* (see also [Weitz, 1975]) – the papers of Mader, Giencke, Klöppel and Schardt to name but a few. For example, Mader [Mader, 1957] and Giencke [Giencke, 1958] dealt with the discontinuity of the cross-girders and considered the orthotropic steel bridge deck as a composite system consisting of Huber's continuum and the discontinuous cross-girders below. In a later paper, Giencke analysed the hollow-rib plate (Fig. 7-46) – a variation of the orthotropic plate whose success first came in the mid-1960s as the Krupp company took on a series of large bridges simultaneously and was forced to rely on large-scale production with maximum standardisation. At the same time, the steel industry switched the production of lightweight sheet piling sections from hot- to cold-rolling, which rendered possible the standardisation of deep trapezoidal profiles with transverse beams at spacings of up to 5 m. This technical progress led to the orthotropic bridge deck plate so typical these days: "Automatic welding and assembly plants for welding hollow ribs to deck plates rendered possible good-quality weld seams with good penetration for the typical solution, and the arrangement of the close-tolerance longitudinal rib penetrations through cut-outs in the cross-girder webs with adequate room for compensating for the tolerances of the trapezoidal profiles plus the design of the longitudinal rib splices ensured details not susceptible to fatigue" [Minten, et al., 2007]. The forerunners of the hollow-rib plate perfected in the 1960s had been produced many years before. The first two examples of hollow-rib plates in use were the road bridges over the River Weser at Porta Westfalica [Dörnen, 1955] and the suspension bridge between Duisburg-Ruhrort and Homberg (Friedrich Ebert Bridge) [Sievers & Görtz, 1955], all completed in 1954.

Klöppel and Schardt achieved a graphic synthesis of the Huber [Huber, 1923] and Pflüger continuum theory [Pflüger, 1947] of anisotropic shell structures with the help of matrix calculations [Klöppel & Schardt, 1960]. Their work progressed to become not only the foundation of a continuum theory for orthotropic plates, but also a way of analysing the lattice domes that would appear later in the 1960s (see section 8.3). The consequential matrix formulation can be regarded as equally important because it considerably simplified the transformation into algorithms for computer programs. Giencke was the driving force behind this development. In 1967 he managed to formulate a finite method for calculating orthotropic plates and slabs [Giencke, 1967]; three years later, Giencke, together with J. Petersen, published a finite method for calculating shear-flexible orthotropic plates, which at that time were being used more and more in building for sandwich constructions.

The growth in the importance of numerical methods of structural analysis tailored to the computer resulted in a major upheaval in the range of trade journals on offer to the engineering industry. For example, in 1969 the pioneers of the finite element method, O. C. Zienkiewicz and

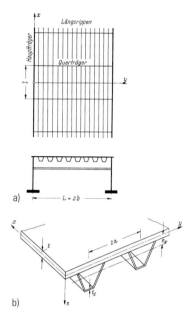

**FIGURE 7-46**
Hollow-rib plate: a) section through orthotropic bridge deck, b) detail of hollow-rib plate [Giencke, 1960, p. 1]

R. H. Gallagher, founded the *International Journal for Numerical Methods in Engineering*, the first journal of its kind (see [Zienkiewicz et al., 1994]). Therefore, starting in the early 1970s, the focus in the reporting on theory in the journal *Der Stahlbau* shifted from the principles of the structural theory of structural steelwork to engineering models. However, on the disciplinary level, composite steel-concrete construction and lightweight construction became subdisciplines of structural steelwork.

### 7.5.2 The rise of composite steel-concrete construction

Composite steel-concrete construction is as old as reinforced concrete construction. In historico-logical terms, it is based on the steel construction that had already developed considerably by the last decades of the 19th century, which was then combined with the emerging concrete construction in that same period. One outstanding example of this is the system invented by Joseph Melan in 1892 in which a steel construction is combined with concrete, without any shear connectors, to form a structurally effective cross-section; this system was able to secure a significant market share in the building of arch bridges. Although a composite column appeared in the form of the Emperger column not long after 1910, it became clear in the middle of the invention phase of structural theory (1925–50) that the bond – made up of friction and adhesion – between rolled steel sections and concrete was not adequate to guarantee a composite effect in beams; shear connectors would have to be welded to the rolled sections. Maier-Leibnitz reported on tests on such composite beams and also proposed a new design approach [Maier-Leibnitz, 1941]. So, like with the orthotropic plate, in composite steel-concrete construction welding was a necessary prerequisite for the technical and scientific developments. By the time of the transition from the invention to the innovation phase of structural theory (1950–75), a theory of composite steel-concrete beams, based on Dischinger's ageing theory, had been formulated, and this underwent further development in the late 1960s on the basis of Trost's theory of viscoelastic bodies [Wapenhans, 1992, p. 14]. In the diffusion phase of structural theory (1975 to date), the first 15 years were characterised by the introduction of the ultimate load method and the establishment of the welded shear stud as an economic shear connector. Besides journals such as *Stahlbau* (founded 1928), *Construction Métallique* (founded 1964) and *Journal of Constructional Steel Research* (founded 1981), which had always included articles on composite steel-concrete construction, 1983 saw the appearance of a further journal – *Composite Structures* – covering composite construction in general, and since 2001 there has also been a journal dedicated to composite steel-concrete structures: *Steel & Composite Structures*.

### 7.5.2.1 Composite columns

Shortly after 1900, Fritz von Emperger began a series of tests on concrete columns reinforced with iron sections and was the first person to formulate the addition principle for such columns [Eggemann, 2003/2, p. 789]:

$$N_{fail} = A_b \cdot \sigma_b + A_s \cdot \sigma_s \tag{7-58}$$

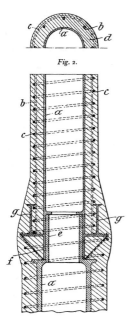

**FIGURE 7-47**
Patented Emperger column
[Eggemann, 2003/2, p. 790]

where $N_{fail}$ is the ultimate load of the column, $A_b$ is cross-sectional area of the concrete, $\sigma_b$ is the compressive strength of the concrete, $A_s$ is the cross-sectional area of the mild steel, and $\sigma_s$ is the yield stress of the mild steel. Emperger was granted a patent for his "hollow cast-iron column with a casing of reinforced concrete" (Fig. 7-47) in 1911. According to Holger Eggemann, this column can be regarded as a composite column in the modern sense because the heavy spiral reinforcement provides for good composite action between the cast-iron core and the concrete casing right up to failure, and the full compressive strength of the cast iron can be utilised. Fritz von Emperger expanded his addition equation 7-58 for this composite column by adding the terms for the cross-sectional area of the cast iron $A_g$ and the compressive strength of the cast iron $\sigma_g$:

$$N_{fail} = A_b \cdot \sigma_b + A_s \cdot \sigma_s + A_g \cdot \sigma_g \tag{7-59}$$

Addition equation 7-59 still appears today in the simplified calculation of the axial strength of composite steel-concrete columns according to Eurocode 4 [Eggemann, 2003/2, p. 792].

At the international building trade fair held in Leipzig in 1913, Emperger presented another concept of this composite construction: the recently completed Schwarzenberg Bridge in Leipzig was an arch bridge constructed from encased cast iron, which Emperger described in detail in an accompanying publication [Emperger, 1913]. In contrast to the related Melan arch bridges, which were first used in the USA, then later in Spain and even today are very popular in Italy and Japan [Eggemann & Kurrer, 2006, p. 914]. Fritz von Emperger's arch bridge of encased cast iron did not become widespread.

However, the Emperger column enjoyed considerable success. Using a modified version of addition equation 7-59, which eased its use considerably, it was considered as a composite column in the American reinforced concrete code of 1920 [Eggemann, 2003/1, p. 41, & 2006, p. 1029]. One highlight in this development was the 16-storey McGraw-Hill building in Chicago, which was built in 1929 and following its demolition in 1998 was rebuilt on the same site with its original facade at the request of the building preservation authority responsible. In addition to the USA, the Emperger column also saw wide use in Czechoslovakia and Austria – but not in Germany. "According to Kleinlogel, the reason for this was that there were no codes of practice in Germany and official trials were refused because of Emperger's patent. Fritz von Emperger relinquished his patent rights in 1928" [Eggemann, 2003/3, p. 702].

Kuno Boll and Udo Vogel reported on the design of concrete-cased steel columns, which were not unlike the Emperger column, in 1969 [Boll & Vogel, 1969]. They proposed concrete-cased steel columns, with concrete grade B 450 and steel grade St 37, for the Spiegel Magazine building, but official approval could not be obtained quickly enough and so conventional columns had to be used instead. Some 30 years later, researchers from Innsbruck published several papers on the 50-storey Millennium Tower in Vienna (completed in 1999), the external columns of which for the lowest standard floor consist of circular composite sections with an outside diameter $D = 406.4$ mm and a solid steel core $d = 200$ mm (Fig. 7-48). The shear transfer within the composite section is in this case achieved with shot-fired fixings [Angerer et al., 1999], the first time this system had been used. In this arrangement, composite column and composite floor slab act together to form a composite frame [Huber & Rubin, 1999; Huber & Obholzer, 1999; Michl, 1999; Taus, 1999]. Two storeys were completed every week. The Millennium Tower is an impressive example of the close interaction between science and practice in composite steel-concrete construction – and more besides: it bears witness to the science-based intelligent combination of reinforced concrete, steel and composite steel-concrete construction in the form of the "Innsbruck composite construction technology" [Tschemmernegg, 1999].

### 7.5.2.2  Composite beams

The history of the development of composite beams can be traced back to investigations into encased rolled sections at the beginning of the 20th century. Like with conventional reinforcement in reinforced concrete construction, a bond – made up of friction and adhesion – between the steel and the concrete had always been presumed in the case of rigid reinforcement (= concrete-encased rolled sections). For example, at the IABSE congress in Paris, Cambournac reported on tests on encased rolled beam sections that had been carried out in 1927. He stated that such construction systems may be designed according to reinforced concrete theory, but called for the inclusion of transverse bars passing through the web of the steel beam [Cambournac, 1932]. Gehler, too, was in 1931 still assuming that composite steel-concrete beams in bridges could be calculated according to the $n$-method of reinforced concrete theory (see section 9.2), common in Germany at that time, without taking into account any shear connectors. However, Koenen had already pointed out in 1905 that the bond of round bars must be considered differently to that of rigid rolled beam sections encased in concrete [Koenen, 1905]. But from 1932 onwards, the new German reinforced concrete standard, DIN 1045, stipulated that the composite effect of rolled beam sections in concrete with a depth representing a considerable part of the overall beam depth may not be taken into account. The history of the development of composite steel-concrete construction thus took a new turning [Wapenhans, 1992, p. 8].

It was in 1937 that Günther Grüning [Grüning, 1937] published his important research findings concerning composite steel-concrete construction, supplemented four years later by Hermann Maier-Leibnitz' work [Maier-Leibnitz, 1941]. Inspired by on-site tests by I. G. Farbenindustrie AG within the scope of an industrial building project in Ludwigshafen in 1937, Hermann Maier-Leibnitz proposed a solution using composite beams, which were tested for their load-carrying capacity in 1940 at the Institute of Building at the Materials Testing Institute of Stuttgart TH (Fig. 7-49a). The possibility of designing the composite beam based on ultimate load theory was addressed, i.e. a non-$n$ design method (Fig. 7-49b), which did not become part of composite construction until the 1970s.

**FIGURE 7-48**
Integration of the composite columns and the composite beams with shear studs into the floor slab [Huber & Obholzer, 1999, p. 629]

Like the ultimate load method in structural steelwork, Maier-Leibnitz was not able to further his work on composite construction theory because no funding was available to him from the tests budget [Kurrer, 2005, p. 631]. Even the savings in steel that composite beams would bring, backed up by figures, did not help. The continuation of the work by way of "typical applications and the question of the economy of composite beams" [Maier-Leibnitz, 1941, p. 270] did not materialise, unfortunately.

The work of Grüning and Maier-Leibnitz was incorporated into the new edition of DIN 1045 in 1943: "Rolled beams and plate girders in concrete whose web depth constitutes a considerable part of the overall beam depth may not be calculated as reinforced concrete beams but instead should be designed so that they can carry all the loads without taking account of the load-carrying capacity of the concrete unless special measures to secure the composite action are taken. Only if this is the case can a composite effect between steel beam and overlying reinforced concrete slab [see Fig. 7-48a – the author] be acknowledged" (cited in [Wapenhans, 1992, p. 11]).

Accordingly, the quantification of the shear connectors, the development of a structural theory for the composite action, design methods and fire protection have formed the cornerstones of composite steel-concrete construction since the late 1940s. The driving force behind composite construction in the post-war years was savings in steel. For example, in his presentation entitled *Welche Möglichkeiten bietet die Verbundbauweise dem Stahlbau?* (What are the opportunities for composite construction in steelwork?) given at the 1949 steelwork conference in Braunschweig, Bernhard Fritz compared the load-carrying capacity of a composite construction with that of a pure steel construction. He came to the conclusion that the use of composite construction resulted in potential savings of between 15 and 55 % over the pure steel solution [DStV, 1950, p. 75]. This explains the great number of contributions to composite action theory in the years 1949–50, which Wilfried Wapenhans has rightly described as "a sudden hurricane of immense power" [Wapenhans, 1992, p. 13]. The 8 December 1949 and 20/21 April 1950 saw conferences on composite steel-concrete construction taking place in Hannover, the results of which

**FIGURE 7-49**
a) Theoretical stresses in a test beam [Maier-Leibnitz, 1941, p. 265], and b) stresses after full plasticisation of the steel cross-section [Maier-Leibnitz, 1941, p. 268]

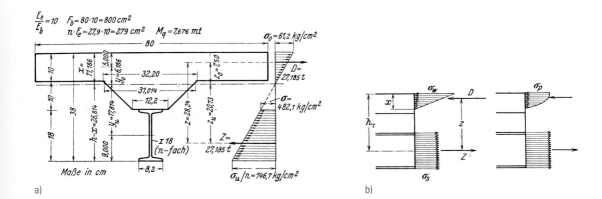

were summarised in two special editions of the journal *Der Bauingenieur* [n. n., 1950/1; n. n., 1950/2]. Franz Dischinger's theory of the creep and shrinkage behaviour of concrete (ageing theory) [Dischinger, 1937, 1939], which links the elastic-plastic deformation behaviour of concrete with the elastic theory in the form of a differential equation, formed the nucleus of composite steel-concrete construction theory until well into the 1960s. H. Fröhlich was the first to apply ageing theory to the analysis of composite beams [Fröhlich, 1949, 1950]. He was inspired by a monograph by Mörsch, who used Dischinger's ageing theory for the structural analysis of prestressed concrete beams [Mörsch, 1943, p. 51]. Shortly afterwards, Bernhard Fritz managed to ascertain the creep and shrinkage of concrete by introducing the "ideal elastic modulus" and derive simple, understandable formulas for designing statically determinate [Fritz, 1950/1] and statically indeterminate continuous [Fritz, 1950/2] composite steel-concrete beams. In 1951 Klöppel formulated a general theory of statically indeterminate composite structures in the formal version of the force method [Klöppel, 1951/2]. W. Danilecki analysed composite beams and columns from the viewpoint of the ultimate load method, but unfortunately did not investigate the shear connectors [Danilecki, 1954]. Konrad Sattler, professor of steelwork at Berlin TU from 1951 to 1962, wrote the first monograph on the theory of composite steel-concrete construction (Fig. 7-50). The book is a compilation of the methods developed by himself and others for any type of composite and prestressed concrete constructions with different types of support and loading. H. J. Sontag specified an approximate solution for composite steel-concrete beams in which the second moment of area of the concrete slab is small compared to that of the steel section [Sontag, 1951], which K. Kunert made more precise by using a limiting parameter [Kunert, 1955].

A second edition of Sattler's monograph was published in 1959, this time in two volumes [Sattler, 1959], and quickly became the standard work of reference for the quasi-rigid composite action. However, in his attempt to achieve a perfect presentation, Sattler tended to overemphasize the formalisation, which resulted in a multiplicity of subscripts and superscripts and considerably restricted the legibility and applicability of his composite construction theory. In this respect, Sattler was very similar to his colleague at the same faculty of the Berlin TU, Alfred Teichmann, professor of theory of structures (see section 6.6.4.2). Fritz Leonhardt criticised this overuse of formalisation in the theory of composite construction, calling it "Sattlerism".

For more than 10 years, Sattler's brochure *Ein allgemeines Berechnungsverfahren für Tragwerke mit elastischem Verbund* (general method of calculation for structures with elastic composite action) [Sattler, 1955] remained the main work of reference for elastic composite action.

Another book on composite construction theory was published by Fritz [Fritz, 1961], which was for many years a reliable source of information for bridge-building engineers. On the whole, the complex creep and shrinkage calculations and the analysis work connected with design-

FIGURE 7-50
Cover of the first monograph on the theory of composite steel-concrete construction [Sattler, 1953]

ing composite steel-concrete beams may well have been the reason why, even in the 1960s, composite construction theory still had the status of an esoteric doctrine – because it could not be applied without extensive special knowledge. Heinrich Trost was the first to break down the barrier with his theory of the viscoelastic body for determining the stress redistributions as a result of creep and shrinkage in the concrete [Trost, 1966, 1967/1, 1967/2]. His approach replaced Dischinger's differential concrete stress-strain relationship with simple algebraic equations, which were based on rheological model concepts and noticeably simplified the theory of composite construction [Trost, 1968].

But the breakthrough for composite steel-concrete construction really came with the welding of the shear studs and the introduction of the ultimate load method. The first ideas about welding the shear studs appeared in the 1920s in the USA [Sattler, 1962], which underwent development in the UK in the 1940s and started to be widely used internationally during the 1950s [Wapenhans, 1992, p. 39]. In Europe, Sattler was the first to use the American Nelson shear studs as shear connectors and carried out extensive tests. And so starting in the early 1960s stud welding enabled shear connectors to be economically welded to the steel sections of composite beams [Muess et al., 2004, p. 791].

The research work carried out at the Bochum School of Structural Steelwork by Karlheinz Roik made a considerable contribution to the rise of composite steel-concrete construction in Germany from the 1970s onwards. It was Roik and his colleagues who in 1974 formulated the *Richtlinien für die Bemessung und Ausführung von Stahlverbundträgern* (guidelines for the design and construction of composite beams) ([n. n., 1974; Roik et al., 1975]), which permitted the use of the ultimate load method and paved the way for supplanting the tedious composite action theory based on elastic theory: the complex calculations for the creep and shrinkage behaviour of concrete were superfluous when calculating the ultimate limit state for composite steel-concrete constructions. During the 1980s, Roik and his colleagues published numerous papers on their research into composite columns, and in 1984 DIN 18806 (composite columns) appeared. The latest results from fire protection research, the revised composite beam guidelines from 1981 and DIN 18806 led to a notable upswing in composite construction in buildings from that time onwards: the painting shops of the Opel plant in Rüsselsheim, the IWF/IPK engineering institutes building in Berlin (Fig. 7-51), the main post office in Saarbrücken, the Siemens plant for safety and signalling engineering in Berlin-Treptow, an extension to the German Science Museum in Berlin, the Commerzbank headquarters in Frankfurt am Main, Düsseldorf's "Stadttor" project, the Millennium Tower in Vienna, the Bonn Post Tower, the "Münchener Tor" high-rise project in Munich and Berlin's new main railway station.

These days, steel bridges without composite members would be inconceivable. In her paper *Perspektiven im Verbundbrückenbau* (outlook for composite bridge construction), Ulrike Kuhlmann reminds engineers to

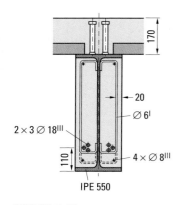

FIGURE 7-51
One of the first applications of a composite steel-concrete floor beam – at the IWF/IPK engineering institutes building in Berlin (dimensions in mm) [Kurrer, 2006, p. 253]

pay more attention to criteria such as economy, ease of maintenance and ease of construction; she comes to the conclusion that solutions that do justice to the material are possible "when those involved stop regarding themselves merely as steel builders or concrete builders, and start thinking of themselves first and foremost as structural engineers. This is an important element in engineering education, which we in Stuttgart are dealing with through a new teaching concept. Composite bridges need engineers with an interdisciplinary view right now" [Kuhlmann, 1996, p. 336]. In the meantime, thinking and acting in more than one material has gone beyond composite steel-concrete construction and has become one of the main developmental threads in structural engineering.

### 7.5.3 Lightweight steel construction

Lightweight construction and steel construction have long been related. One good example of this is the beneficial exchange between steel bridge-building and aircraft construction, e. g. the orthotropic bridge deck and the integral plate (Fig. 7-52). The success story of the orthotropic plate and its engineering science analysis is also the success story of lightweight steel construction and has already been described in section 7.5.1. Another example is sandwich construction – widely used in the building industry, particularly in steel buildings, since 1960 and the subject of ongoing development in the shipbuilding and aircraft industries as well. Closely related to this is the continuous roll-forming method for trapezoidal steel profiles, which enabled the mass production of sheets for roofs, walls and floors. These developments led to issues of lightweight steel construction forming the focal point of the technical and scientific presentations at the 1964 steelwork conference in Aachen. Klöppel's and Jungbluth's demands for optimising the stiffness of thin-wall, cold-formed steel components by way of specific cold-working and shaping was investigated in more depth by P. Stein on the statics-constructional side [DStV, 1964, pp. 10–22] and O. Jungbluth on the technological side [DStV, 1964, pp. 23–24]. Jungbluth's goal here was to capture an ever more important segment of the construction market, in the shape of the industrial and residential sector, which he wanted to exploit better for structural steelwork through the large-scale production of prefabricated steel parts on automated production lines.

K. Stamm and H. Witte summarised the level of knowledge relating to the theory and practice of sandwich construction at the end of the innova-

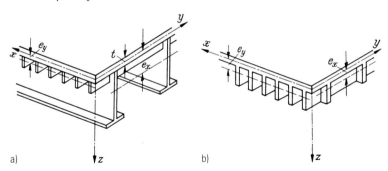

**FIGURE 7-52**
Plate cross-sections around 1960;
a) flat steel plate (bridge-building),
b) integral plate (aircraft construction)
[Giencke, 1961, p. 35]

tion phase of structural theory (1950–75) in their monograph [Stamm & Witte, 1974]. Developments in this area of lightweight construction have been and still are reported regularly in journals like *Stahlbau*, *Construction Métallique* and *Journal of Constructional Steel Research*. And since 1983 there has been an international forum for presenting scientific progress in lightweight construction in general and lightweight steel construction in particular in the shape of the journal *Thin-Walled Structures*.

The development of three-dimensional loadbearing structures made from steel and glass has seen much progress at the hands of Herbert Klimke, Jörg Schlaich, Hans Schober, Werner Sobek and other engineers since the 1990s. The roof structures to Berlin's new main railway station, the Sony Center, the British Museum, the Eden Project and Milan's new trade fair (Fig. 7-53) reflect the whole spectrum of the ingenious artistry and science of the conception, design, calculation and construction of spatial loadbearing structures in a balanced relationship between detail and whole. For example, the freely formed surface of the roof structure to Milan's new trade fair were generated using the NURBS (non-uniform rational B-splines) method. Thanks to the CAD programs available today with integral NURBS functions, it is becoming less and less necessary for users to understand the mathematical principles behind the NURBS method. The design of a truss on an arbitrary, non-optimised irregular surface is very complex [Stephan et al. 2004, p. 566]:

– The structural behaviour and the ensuing dimensioning of the member cross-sections are virtually impossible to predict and rarely identical throughout the global loadbearing system; in the case of single-layer grids in particular, the stresses vary between pure tension or compression and primarily bending.
– The local geometric parameters of the members can vary enormously; even the parameters at one node can vary considerably, which has to be taken into account in the type of connection selected.

Architects increasingly prefer irregular surfaces with neither structural nor geometrical optimisation. The reason for this according to Stephan, Sánchez-Alvarez and Knebel is the availability of CAD programs with NURBS functions and the preference for artistic design without having to take full account of technical limitations. Accordingly, the only sensible way of achieving such grids is to provide "nodes with maximum design flexibility in order to be able to handle both the varying structural behaviour and the changing geometrical parameters with just a few basic node details" [Stephan et al. 2004, p. 566]. This example of interaction between constructional node design and grid of members throws a spotlight on the new relationship between detail and overview, between local and global structural behaviour, between construction and design, and between the part and the whole. Modern structural calculations are more than just experimental geometrical mechanics: the latest software solutions help both architects and engineers in the design and development of interesting loadbearing systems for irregular surfaces. However, they in no way replace the knowledge of the engineer, but instead challenge it

FIGURE 7-53
View of one "Vulcano" area in the east of the approx. 1300 m long irregularly formed roof to the central atrium of Milan's new trade fair building [Kurrer, 2006, p. 254]

in a very particular way so that an irregular form can be turned into an efficient loadbearing structure. Besides the pure form-finding and structural calculations, the complex geometries throw up many 'new types' of design-, fabrication- and erection-related problems which have to be solved adequately when translated into a real structure" [Schober et al., 2004, pp. 550, 551].

## 7.6 Eccentric orbits – the disappearance of the centre

Modern structural steelwork is system-based steel construction and on the academic side is no longer solely concerned with theory of structures like it was in the middle of the innovation phase of structural theory (1950–75); structural steelwork is moving in eccentric orbits. As system-based steel construction, structural steelwork goes way beyond the classical activities of steelwork science theory formation and testing, conception, calculation and design, fabrication and erection; it now has to embrace uses and reuses, preservation and maintenance of the existing building stock, plus the recycling and disposal of structures.

Even the ancient Greek astronomers sought to save heavenly phenomena by constructing eccentric motions of the heavenly bodies in that they shifted the centre of their trajectories from that of the Earth (Fig. 7-54). In poetry, Friedrich Hölderlin (1770–1843) depicts the image of the eccentric orbit: "We all pass along an eccentric orbit and there is no other path possible from childhood to death" (cited in [Müller-Seidel, 1993]). The eccentric orbit therefore reflects not only superhuman nature, but also the flawed human nature as the nature of humankind in all its facets.

The topos of the "disappearance of the centre" invented by Hans Sedlmayr (1896–1984) in 1948 for the history of art is a multi-dimensional metaphor for the development phase of structural steelwork since the end of the 1960s, which will be dealt with here merely in thesis form:

1. Economy of labour took over from economy of material, which had prevailed in structural steelwork up to this point: rationalisation of operations in structural steelwork became the key issue (see [Scheer, 1992]).

2. Since the structural crisis in the mining industry and the 1966/67 economic crisis in Germany, the big groups departed step-by-step from the steelwork sector: the steel construction sector changed wholly to an industry dominated by small and midsize businesses.

3. Industrial steelwork research was reorganised. A new form of application-oriented joint research appeared in Germany alongside the structural steelwork research activities traditionally undertaken by the DASt (see [Bossenmayer, 2004]): as industrial steelwork research could no longer take place in a steel industry organised into large groups, the steel industry founded a research organisation which has since become FOSTA, a non-profit-making body which today is the most important R&D address for industrial steelwork research in Germany.

4. Due to modern materials research and the use of computers, elastic theory lost its universal role in the foundation of theory of structures, steelwork science and design theory: breaking the dominance of the linear in these disciplines in general and the long onward march of the ultimate load method through the institutions in particular.

5. Replacement of the deterministic by the semi-probabilistic safety concept in structural steelwork (see [Petersen, 1977; Siebke, 1977]): the DIN 18 800 standard (see [Lindner et al., 1998]) as the reformation and standardisation of the diverse codes of practice for structural steelwork as well as a model for the whole of structural engineering.

6. Discovery of architects as a target group for the DStV: the award of the German Steelwork Prize since 1972 and the emergence of architectural steelwork (see [Schmiedel, 1994]).

7. The statics-constructional lost ground to the technological in structural steelwork: rationalisation through automation in steel fabrication and engineering work with the help of NC machines, CAD and computer mechanics with a view to coupling the information from these areas in product models for optimising project management (see [Haller et al., 2004]).

8. The design office is no longer the focus of the steel fabricator: important tasks of the design office such as loadbearing structure conception, structural calculations and detailed design are increasingly carried out by consulting engineers.

9. Pure steelwork takes second place to combinations of materials: steel mixes well – or the unstoppable rise of composite construction – steel and glass, steel and membrane, façade construction.

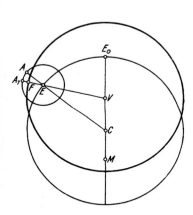

FIGURE 7-54
Eccentric orbits – the disappearance of the centre: a metaphoric model taken from Greek astronomy
[Dijksterhuis, 1956, p. 67]

10. The building envelope imposes its product aesthetics character on a new scale: together with façade construction it develops into an autonomous sector of metalworking in the building industry.
11. In many sub-areas of structural steelwork, e.g. industrial buildings, system-based structural steelwork is currently the most prominent tendency technically and economically: "from steel erector to general contractor" [Goldbeck, 2004].
12. Building within the existing stock and upgrading are becoming increasingly important in structural steelwork as significant elements of a form of construction dedicated to sustainability and sparing of resources: the discovery of the history of construction engineering as a productive force in engineering work is still learning to walk, and making progress here is one task that structural engineers will face in the future.

# CHAPTER 8

# Member analysis conquers the third dimension: the spatial framework

On 24 Oktober 2003, at the invitation of Dr.-Ing. Herbert Klimke, the former head of the engineering office of the MERO company in Würzburg, the author gave a presentation on "space frames from Föppl to Mengeringhausen" at the festive colloquium to mark the 100th anniversary of the birth of Max Mengeringhausen. The event took place in Dessau, in the auditorium of the main Bauhaus building designed by Walter Gropius. There were further presentations by professors Werner Sobek, Volkwin Marg and Werner Nachtigall. In preparing his presentation, the author studied the lives and works of Johann Wilhelm Schwedler, August Föppl, Hermann Zimmermann, Alexander Graham Bell, Vladimir Grigorievich Shukhov, Walther Bauersfeld, Franz Dischinger, Richard Buckminster Fuller and Mengeringhausen himself in order to discover the relationship between the invention of space frames and the development of structural theory. This historico-logical path of knowledge led to a new appreciation of the conception, calculation, design, fabrication and erection of space frames.

## Continually Developing
## Three-Dimensional Structures, Associated Claddings, Structural Glazing, Point-Supported Glass, and More

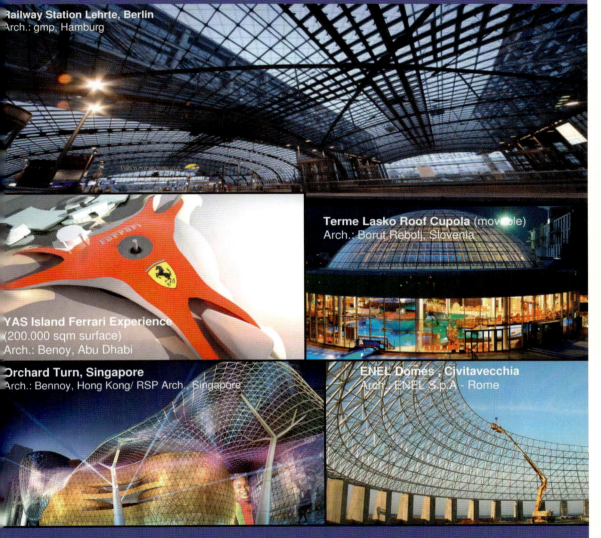

Railway Station Lehrte, Berlin
Arch.: gmp, Hamburg

YAS Island Ferrari Experience
(200.000 sqm surface)
Arch.: Benoy, Abu Dhabi

Terme Lasko Roof Cupola (movable)
Arch.: Borut Rebolj, Slovenia

Orchard Turn, Singapore
Arch.: Bennoy, Hong Kong/ RSP Arch., Singapore

ENEL Domes, Civitavecchia
Arch.: ENEL S.p.A - Rome

As project solution providers, our teams of experts consistently adopt new technologies in design, material, production and project management to meet increasingly sophisticated client requirements.

Continually building on the initial MERO-TSK invention of the classic spherical node and tube space frame system, our international organization leads the way as a specialist in custom three-dimensional structural systems and integrated cladding.

MERO-TSK International GmbH & Co. KG • Max-Mengeringhausen-Str. 7 • D-97084 Würzburg • Germany
Phone: +49 (0) 931 6670-0 • Fax +49 (0) 931 6670-409 • info@mero-tsk.de • www.mero-tsk.de

**COMPENDIUM FOR ALL CIVIL ENGINEERS**

## Beton-Kalender
(Concrete Structural Design)

Ed.: K. Bergmeister, J.-D. Wörner
**Beton-Kalender 2008**
Hydraulic engineering, seismic design

2007. published in German. 1160 pages with 745 figures. Hardcover.
€ 165.–*
Series price:
€ 145.–*
ISBN: 978-3-433-01839-2

Annual focuses:
2003 – high-rise and multi-storey buildings
2004 – bridges and multi-storey car parks
2005 – Precast elements and tunnel structures
2006 – Tower structures, industrial buildings
2007 – Traffic structures, shell structures

## Stahlbau-Kalender
(Steel Construction Design)

Ed. U. Kuhlmann
**Stahlbau-Kalender 2008**
Dynamics, Bridges

2008. published in German approx. 900 pages, approx. 500 figures. Hardcover.
approx. € 135,–*
Series price: € 115,–*
ISBN 978-3-433-01872-9

Annual focuses:
2004 – Slender structures
2005 – Joints
2006 – Durability
2007 – Materials

## Bauphysik-Kalender
(Building Physics Year book)

Ed.: N. Fouad
**Bauphysik-Kalender 2008**
Sealings

2008. published in German 697 pages, 474 figures, 194 tab. Hardcover.
Approx. € 135,–*
Series price: € 115,–*
ISBN 978-3-433-01873-1

Annual focuses:
2003 – Mould infestation
2004 – Algea on façades – nondestructive testing
2005 – Substainable construction, waterproofing of structures
2006 – Fire safety
2007 – Energy performance of buildings

## Mauerwerk-Kalender
(Masonry Year book)

Ed.: W. Jäger
**Mauerwerk-Kalender 2008**

2007. published in German. 856 pages. 464 fig. 235 tab. Hardcover.
€ 135.–*
Series price:
€ 115.–*
ISBN: 978-3-433-01871-2

The yearbook in its 34. volume with a wellbalanced ratio between up-to-date and revised contributions gives consideration to the versatility of masonry as a structural carrying element, wall construction material with buildingphysical and aesthetical functions, as the medium of innovation in prefabricated structures and for energy-saving buildings. All new developments which require approval will be introduced with the up-to-dateness of a yearbook.

\* Prices are valid in Germany, exclusively, and subject to alterations. Prices incl. VAT. Books excl. shipping. Journals incl. shipping.

**Ernst & Sohn**
Verlag für Architektur und technische Wissenschaften GmbH & Co. KG

www.ernst-und-sohn.de

For order and customer service:

**Verlag Wiley-VCH**
Boschstraße 12
69469 Weinheim
Deutschland

Telefon: +49(0) 6201 / 606-400
Telefax: +49(0) 6201 / 606-184
E-Mail: service@wiley-vch.de

In 1866 Schwedler pioneered the development of a structural theory of spatial frameworks, although a satisfactory spatial framework theory did not emerge until the classical phase of structural theory (1875–1900). Particularly noteworthy is August Föppl's work, especially his first monograph on spatial frameworks [Föppl, 1892]. During the consolidation period of structural theory (1900–50), the further development of spatial frameworks was characterised by constructional-technical innovations by Alexander Graham Bell, Vladimir Grigorievich Shukhov, Walther Bauersfeld and Franz Dischinger, Richard Buckminster Fuller, and Max Mengeringhausen. The latter two engineering personalities had a significant influence on spatial framework theory during the innovation phase of structural theory (1950–75). In the late 1970s, MERO, the company founded by Mengeringhausen (today known as MERO-TSK International), achieved a breakthrough with the consistent application of computer techniques to the design, calculation, construction and production of spatial frameworks.

## 8.1 Development of the theory of spatial frameworks

The development of trussed framework theory, which started in 1850, was limited to plane systems until the last decade of the 19th century. During that final decade, classic structural theory also took on its final form as a coherent theory of statically determinate and indeterminate plane elastic trusses. Spatial structural systems for factory buildings, railway stations and bridges employed orthogonal structures, so resolving into plane systems was generally sufficient. In addition, since the beginning of the 19th century engineering thinking in three dimensions had been based on orthogonal projection in the shape of technical drawings, a technique that dominated descriptive geometry. Early spatial structural analysis approaches such as the forward-looking two-volume *Lehrbuch der Statik* (theory of structures textbook) [Möbius, 1837] by August Ferdinand Möbius (1790–1868) or Otto Mohr's (1835–1918) article on the composition of forces in space [Mohr, 1876] attracted little interest from practising structural engineers.

Even for dome structures, load transfer in three dimensions was initially not considered, and the radially arranged trusses were analysed using a plane frame approach. One such dome, over the gasometer at the Imperial Continental Gas Association in Berlin (Hellweg No. 8), collapsed in 1860 while being erected. The engineer responsible for that project, Johann Wilhelm Schwedler (1823–94), improved the design for the dome structure that was rebuilt one year later, although he still used the conventional approach. In 1863 he designed another dome structure for the same client, covering the gasometer at Holzmarktstraße 28, Berlin, and became the first engineer to make the transition to a dome working in three dimensions, which became known as the "Schwedler dome" in the technical literature. Three years later he described five further Schwedler domes in the journal *Zeitschrift für Bauwesen*, providing not only the theory behind them, but also a simplified structural calculation technique: "Existing dome theory and construction practice took account of radial resistances only … However, in considering the dome equilibrium it is necessary to dispense with elastic member theory and instead use thin elastic plates in double curvature as a basis" [Schwedler, 1866, p. 8].

Schwedler developed the membrane theory for axially symmetric shells under load and calculated the membrane forces acting on the meridians and parallels. Fig. 8-1 shows the roof of the municipal gasworks at Fichtestraße, Berlin-Kreuzberg, built in 1875 in the form of a Schwedler dome. The iron loadbearing dome structure with a diameter of 54.9 m and a rise of 12.2 m still exists today. The spatial system is statically indeterminate to a high degree and cannot be calculated in practice using classic member analysis. Schwedler therefore had to "blur" his dome to form a two-dimensionally curved elastic continuum and then resolve it – an approach for which August Föppl (1854–1924) developed a method to cover other shell-type trussed frameworks [Föppl, 1892], and which only became obsolete at the end of the 1960s when computers started to be used for analysing spatial frameworks.

During the classical phase (1875–1900) and the consolidation period (1900–50) of structural theory, spatial framework theory developed in four stages:

1. Static-constructional development of spatial frameworks that could be calculated in practice using classic structural theory. One prominent example was the original, statically determinate dome over the plenary chamber of the Reichstag building in Berlin [Zimmermann, 1901/1], invented and calculated in 1889 by Hermann Zimmermann (1845–1935) and known as the "Zimmermann dome".
2. Development of the theory of spatial frameworks and design of shell-type spatial frameworks by August Föppl in 1892, following on from the Schwedler dome [Föppl, 1982].
3. Integration of spatial framework theory into classic structural theory by
   – Heinrich Müller-Breslau (1851–1925) [Müller-Breslau, 1891/92, 1898, 1899/1, 1899/2, 1902, 1903/2, 1903/3, 1903/4],
   – Otto Mohr (1835–1918) [Mohr, 1876, 1902/1, 1902/2, 1903],
   – Lebrecht Henneberg (1850–1933) [Henneberg, 1886, 1894, 1902, 1903, 1911], and
   – Wilhelm Schlink (1875–1968) [Schlink, 1904, 1907/1, 1907/2].
4. Expansion of spatial framework theory during the consolidation period of structural theory (1900–50) by Mayor [Mayor, 1910, 1926], von Mises [v. Mises, 1917], Prager [Prager, 1926, 1927] and Sauer [Sauer, 1940].

**The original dome to the Reichstag (German parliament building)**

### 8.1.1

Zimmermann's spatial framework invention helped architect Paul Wallot out of a serious predicament in 1889/90 and was probably the reason for Zimmermann's promotion 18 months later to the top building authority post in the Prussian Ministry of Public Works. The structural insight, i.e. the formation law and the static law of the "Zimmermann dome", comprises the following aspects (Fig. 8-2):

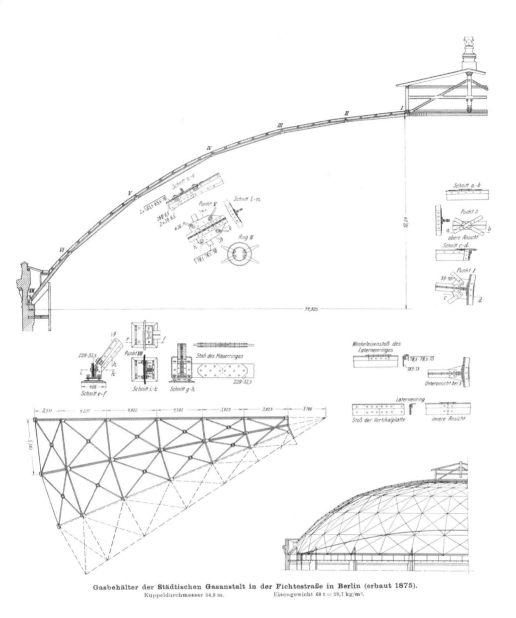

Gasbehälter der Städtischen Gasanstalt in der Fichtestraße in Berlin (erbaut 1875).
Kuppeldurchmesser 54,9 m.    Eisengewicht 68 t = 28,7 kg/m².

**FIGURE 8-1**
Schwedler dome over the gasometer of the municipal gasworks at Fichtestraße, Berlin-Kreuzberg, built in 1875 [Hertwig, 1930/2, plate 7]

- The vertical forces $A_I$ to $A_{IV}$ and $B_I$ to $B_{IV}$ are transferred via the four supports at the corners of the drum masonry rectangular on plan and measuring 38.74 × 34.73 m (Fig. 8-2a).
- The horizontal forces are controlled via an ingenious support system in such a way that the drum masonry can develop its structural plate effect, i. e. only shear forces $T$ act in the plane of the wall: Zimmermann's spatial framework thus resolved the issue of orthogonal horizontal thrust, which builders had dreaded for more than 2000 years.
- The spatial framework has four members in its upper ring, 12 in the lower, 12 in the wall members, eight vertical and four horizontal support reactions acting parallel to the drum masonry, i. e. a total of

477

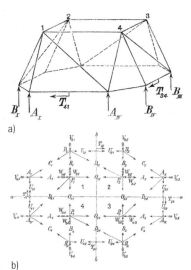

a)

b)

**FIGURE 8-2**
a) Isometric view of the structural system of the Zimmermann dome [Zimmermann, 1901/1, p. 6], b) plan on support and member forces [Zimmermann, 1901/1, p. 9]

40 unknown force variables (Fig. 8-2b). On the other side of the equation, for each of the 12 joints there are three equilibrium conditions, i.e. a total of 36 equilibrium conditions. Through a special arrangement of the four supports at $a$ and $b$, four further equilibrium conditions can be set up so that the 40 unknown member forces of the spatial framework can be determined based solely on the 40 equilibrium conditions, i.e. the Zimmermann dome is statically determinate!

Just like Schwedler, his predecessor in the Prussian Ministry of Public Works, Zimmermann still mastered the design language of the mechanical engineer and the structural engineer: the Zimmermann dome is an ingenious "structural machine".

Zimmermann did not reveal his elegant structural calculations until 11 years after the construction of his dome above the plenary chamber of the Reichstag [Zimmermann, 1901/1, pp. 1–46]. At the same time, he expanded his concept from spatial frameworks with four corners in the upper ring to spatial frameworks with any number of corners [Zimmermann, 1901/1, pp. 46–67] and also derived further spatial framework types [Zimmermann, 1901/1, pp. 67–93]. Through the structuring of the elimination process for the 40 equilibrium equations, he succeeded in demonstrating the relationship with the topology of this class of spatial framework with a rare clarity: static law and formation law merge on a mathematical level by means of defined spatial structural forms. Zimmermann thus lends his structural calculations their own intrinsic aesthetic value. How must Zimmermann have felt two years before his death when Göring's fatuous auxiliaries let the pinnacle of his creative skills go up in flames for political reasons to aid the Nazi cause?

### 8.1.2 Foundation of the theory of spatial frameworks by August Föppl

August Föppl's book *Das Fachwerk im Raume* (spatial frameworks) (Fig. 8-3), published in 1892, was the fruit of his deductive research into the structural law, static law and formation law of spatial frameworks in the shape of the composition law for spatial frameworks, which he had started in 1880 [Föppl, 1880]. Almost 50 years later, Mengeringhausen used this work, which remained unsurpassed until the 1960s, as the basis for his work on spatial frameworks. Anyone who has a chance to read Föppl's monograph on spatial frameworks today and who is reasonably familiar with the work of Mengeringhausen, his successors at MERO-TSK International and other engineering and architectural practices will no longer be surprised by the affinity between classic and modern spatial frameworks. The cover of Föppl's book shows a roof reminiscent of a barrel vault (Fig. 8-3), which he called a "trussed shell". In 1890 the Imperial Patent Office in Berlin denied it any inventive value, claiming that common gasometer guide frames also represented trussed shells [Föppl, 1892, p. 56].

In 1883 Föppl suggested a new domed roof system (the lattice dome), but the calculation method for this was not published until 5 May 1888 in the journal *Schweizerische Bauzeitung* [Föppl, 1888]; after a five-year break he thus resumed his research on the theory of spatial frameworks. Föppl was prompted by Hacker's work [Hacker, 1888] published shortly before, in which he systematically assembled the Schwedler dome from basic triangles and determined the member forces without using membrane theory (even for asymmetric loads) solely with the aid of the node equilibrium conditions of the trussed framework. On 28 March 1891 Föppl reported on the first application of the lattice dome for the roof of the central market hall in Leipzig [Föppl, 1891/1]. This lattice dome over an irregular pentagonal plan spans approx. 20 m, is 6.80 m high and transfers the loads via five 4.40 m high plane frames into five non-sway pinned supports. The structural system consists of 42 members with 42 unknown member forces, 14 free nodes, each with three equilibrium conditions, i.e. a total of $14 \times 3 = 42$ joint equilibrium conditions. The Leipzig lattice dome is thus statically determinate and the 42 unknown member forces can be calculated graphically with force diagrams.

After his rejection by the Patent Office, Föppl published his ideas on trussed shells in the journal *Schweizerische Bauzeitung* on 18 April 1891 [Föppl, 1891/2]. Föppl divided his paper into 13 paragraphs, in all probability to coincide with his patent claims. He interpreted the Platonic bodies as the simplest forms of trussed shells: "Among the regular bodies, the tetrahedron, the octahedron and the icosahedron represent trussed shells. Whereas calculation of the member stresses in the latter two is not always easy, it is always possible without using elastic theory. The hexahedron and the dodecahedron can be converted to trussed shells by introducing one diagonal (for the former) or two diagonals (for the latter) on each side face" [Föppl, 1891/2, p. 95]. In the next paragraph Föppl provides a definition of a statically determinate trussed shell and proof of the formation law with the aid of Euler's polyhedron theorem: "Any enclosed single-shell trussed framework without web members, whose side faces are formed by triangles in different planes, is generally a trussed shell" [Föppl, 1891/2, p. 96]. In conclusion, Föppl points out the benefits of the new trussed shell system compared with the traditional truss system: "The system offers substantial material savings and significant aesthetic benefits" [Föppl, 1891/2, p. 96].

Just two months after Föppl's paper was published, the collapse of the trussed railway bridge over the River Birs at Mönchenstein (today: Münchenstein) near Basel on 14 June 1891 shocked the public; 73 people died and more than 100 were injured. Throughout the second half of 1891 and the whole of 1892, *Schweizerische Bauzeitung* published articles on this bridge disaster almost every week.

One month after the disaster, Föppl was the first specialist from abroad to join in the discussion on the causes of the collapse: "The bridge collapsed, because – as a spatial trussed framework – it was unstable." He concluded that it was unjustified "to limit framework theory almost exclusively to plane trussed frameworks". According to Föppl: "Textbooks mention spatial trussed frameworks only in passing, if at all, and in the past the teaching of spatial framework theory had been largely ignored. This neglect has now led to the Mönchenstein disaster that has shaken the world" [Föppl, 1891/3, p. 15]. In his case analysis contained in the first

FIGURE 8-3
Cover of the first monograph on spatial frameworks [Föppl, 1892]

volume of his work *Versagen von Bauwerken. Ursachen, Lehren* (structural failures – causes and lessons), published in 2000, Joachim Scheer considers Föppl's analysis of the cause to be important [Scheer, 2000, p. 113]. Föppl's views on the bridge collapse were only reported in passing in the discussion on the causes of the failure, a fact that prompted him to write a manuscript on spatial framework theory. He wrote this manuscript between 21 June 1891 and 22 September 1891, although due to a printers' strike it was not published until early 1892 – the aforementioned *Das Fachwerk im Raume* (see Fig. 8-3).

In this book Föppl readdresses his trussed framework definition of 1880 [Föppl, 1880, p. 3]. He defines a trussed framework as "a system composed of material points and certain connecting lines combined in such a way that no movement of the system components relative to each other is possible without changing the length of the connecting lines" [Föppl, 1892, p. 2]. As an example of a customary trussed framework definition, Föppl quotes that of Wilhelm Ritter (1847–1906) [Föppl, 1892, p. 3]: "A trussed framework is a rigid structure composed of straight members and is intended to carry loads" [Ritter, 1890, p. 1]. Whereas Ritter uses the structure as the basis for his definition, Föppl abstracts the trussed framework at the level of a mathematical-physical model, thus laying the foundation for spatial framework theory. Having made a distinction between "free frameworks" and "supported frameworks" in the form of trussed girders, he develops mathematical stability criteria for free spatial frameworks. In 1880 he had already formulated criteria for plane frameworks [Föppl, 1880, pp. 7–11] and provided proof in 1887 [Föppl, 1887]. Föppl's enumeration condition for a free spatial framework with $s$ members and $k$ nodes

$$s \geq 3k - 6 \tag{8-1}$$

is necessary for stability or rigidity, but not sufficient (Fig. 8-4).

According to Föppl, a free spatial framework is only stable, i.e. rigid or not kinematic, if the Jacobian determinant of the order $s = 3k - 6$ is not zero for the implicit functions of the square of the distance of nodes

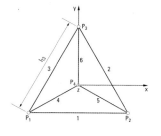

**FIGURE 8-4**
Föppl's stability criterion illustrated using the example of a free tetrahedron framework

$s$ = number of members
$s = 6$

$k$ = number of joints
$k = 4$

Necessary stability criterion:
$s \geq 3 \cdot k - 6$
$6 \geq 3 \cdot 4 - 6 = 6$

$f_1 = (x_1 - x_2)^2 + (y_1 - y_2)^2 + (z_1 - z_2)^2 - l_{12} = 0 = f_1(x_1, x_2, y_1, y_2, z_1, z_2)$
$\vdots$
$f_r = (x_i - x_k)^2 + (y_i - y_k)^2 + (z_i - z_k)^2 - l_{ik} = 0 = f_r(x_i, x_k, y_i, y_k, z_i, z_k)$
$\vdots$
$f_6 = (x_3 - x_4)^2 + (y_3 - y_4)^2 + (z_3 - z_4)^2 - l_{34} = 0 = f_6(x_3, x_4, y_3, y_4, z_3, z_4)$

Jacobian functional determinant of order $3k - 6 = 6$:

$$\Delta = \begin{vmatrix} \frac{\partial f_1}{\partial x_1} & \frac{\partial f_1}{\partial x_2} & \frac{\partial f_1}{\partial y_1} & \frac{\partial f_1}{\partial y_2} & \frac{\partial f_1}{\partial z_1} & \frac{\partial f_1}{\partial z_2} \\ \vdots & & & & & \\ \frac{\partial f_r}{\partial x_i} & \frac{\partial f_r}{\partial x_k} & \frac{\partial f_r}{\partial y_i} & \frac{\partial f_r}{\partial y_k} & \frac{\partial f_r}{\partial z_i} & \frac{\partial f_r}{\partial z_k} \\ \vdots & & & & & \\ \frac{\partial f_6}{\partial x_3} & \frac{\partial f_6}{\partial x_4} & \frac{\partial f_6}{\partial y_3} & \frac{\partial f_6}{\partial y_4} & \frac{\partial f_6}{\partial z_3} & \frac{\partial f_6}{\partial z_4} \end{vmatrix}$$

Sufficient stability criterion:

$\Delta \neq 0$ ⟶ stable (not kinematic)
Example: Tetrahedron

$\Delta = 0$ ⟶ unstable (kinematic)
Example: If $P_4$ falls within the triangular plane formed by $P_1 P_2 P_3$ the necessary stability criterion is met, but not the sufficient criterion.

$f_r(x_i, x_k, y_i, y_k, z_i, z_k)$ connected by members [Föppl, 1892, pp. 6–11]. This sufficient criterion for the stability of free spatial frameworks is "rather useless in practice", as Föppl comments [Föppl 1892, p. 9], although he uses it as a basis for deriving a very important theorem on statically determinate trussed frameworks, for which inequality (8-1) is transformed into

$$s = 3k - 6 \qquad (8\text{-}2)$$

"A stable trussed framework containing only the necessary number of members [i. e. satisfying eq. (8-2) – the author] is also statically determinate and vice versa, i. e. it is stable if it is statically determinate for all applied loads" [Föppl, 1892, p. 30]. Föppl uses this theorem as a basis for a coherent theory of statically determinate spatial frameworks. He went on to formulate an outline for the theory of statically indeterminate spatial frameworks.

In the second part of his book, Föppl composes numerous structural forms of spatial frameworks such as trussed shells, lattice domes, spatial bridge systems and truss systems braced in three dimensions – without using equations. He also develops a new approach for analysing familiar systems such as the Schwedler dome, thus systematically tapping the third dimension for structural theory. His work is a rare example of the great heuristic potential of theoretical thinking in the theory of structures, the horizon of which was not reached until comparatively recently. Nevertheless, his work was not adopted in structural theory. From the historico-logical point of view, Föppl advanced as far as the implicit mathematical structural law of Plato, and merged it with his static law and formation law to form the composition law for spatial frameworks. In Föppl's work beauty and law thus appear in a virtually corporeal-like, aesthetic appreciation of feasible artistic forms of spatial frameworks by means of the mathematical-physical cognition of their composition law.

### 8.1.3 Integration of spatial framework theory into classic structural theory

As a consulting engineer, Müller-Breslau undertook a statics-constructional analysis of the Zimmermann dome above the plenary chamber of the Reichstag in Berlin, which had been completed in 1890. Two years earlier he had succeeded Emil Winkler (1835–88) in the Chair of Structural Design & Bridge-Building at Berlin TH. In this capacity, Müller-Breslau expanded his structural theories, which he had summarised in the form of monographs [Müller-Breslau, 1886, 1887/1], into a theory of linear-elastic frames. He thus rounded off the discipline-formation period of structural theory which stretched from 1825 to 1900. The development of spatial framework theory by Müller-Breslau, using the methodology of plane framework theory, represents a moment of consummation in the development towards classic structural theory [Müller-Breslau, 1891/92]. It differs significantly from Föppl's work [Föppl, 1892] since Müller-Breslau moved within the disciplinary framework of classic structural theory, whose symbols were his invention. Müller-Breslau did not work deductively; instead, he used examples to develop his spatial framework theory, i. e. he worked inductively. The inductive method continued to shape structural theory until well into the second half of the 20th century.

Having explained spatial dynamics using statically determinate domes and a bridge girder as examples, he goes on to suggest a general technique for calculating the member forces of statically determinate spatial frameworks, i. e. his substitute member method [Müller-Breslau, 1891/92, pp. 439–440]. The substitute member method even leads to success in cases where forces cannot be resolved directly in nodes with three to six members: "By removing members and adding the same number of new members, referred to as substitute members, the trussed framework can be transformed into a very simple structure, possibly a structure with tension forces that can be determined by repeatedly solving the task of resolving a given force in three directions. The tension forces of the members removed are applied to new trussed frameworks as external forces, referred to as $Z_a$, $Z_b$, $Z_c$, ..., $Z_n$. The tension forces of the new trussed framework are then presented as a function of the given loads $P$ and the initially unknown forces $Z$. They appear in the form $S = S_0 + S_a \cdot Z_a + S_b \cdot Z_b + S_c \cdot Z_c + ... + S_n \cdot Z_n$, where $S_0$ represents the value of $S$ when all forces $Z$ are set to zero, i. e. when only loads $P$ act on the new trussed framework. $S_a$ represents the value of $S$ for the case when all loads $P$ and forces $Z_b$, $Z_c$, ..., $Z_n$ are zero, whereas the two forces $Z_a$ take on a value of one. This load state is referred to as $Z_a = 1$; $S_b$, $S_c$, ..., $S_n$ can be interpreted as the tension forces for states $Z_b = 1$, $Z_c = 1$, ..., $Z_n = 1$. $S_a$, $S_b$, $S_c$ ... are independent of the loads $P$, whereas the tension forces $S_0$ have to be calculated for each load case to be examined. Setting the tension forces in the substitute members to zero results in the same number of linear equations as there are forces $Z$ present, which means the latter can be calculated, provided the denominator determinant is not equal to zero. Otherwise the trussed framework is unusable, despite the fact that the equation $s = 3k$ is satisfied" [Müller-Breslau, 1891/92, p. 439].

Müller-Breslau's substitute member method, which he had already alluded to in 1887 through two examples [Müller-Breslau, 1887, pp. 207–208 & 213–214], corresponds structurally to his force method for analysing statically indeterminate systems. The substitute member method and the force method are based on equivalent formulations (Fig. 8-5).

The substitute member method can be used to calculate the member forces of complex statically determinate systems using simpler substitute systems, and to verify the stability of the initial systems. For the latter, Müller-Breslau provided a sufficient stability criterion [Müller-Breslau, 1891/92, p. 439] that can be verified directly and much more easily than with Föppl's technique. This demonstrates the superiority of structural techniques based on linear algebra compared with the pure mathematical technique in the form of the Jacobian determinant. Müller-Breslau's substitute member method is particularly suitable for analysing complicated statically determinate spatial frameworks. He also drew attention to the technique Henneberg suggested in 1886 [Müller-Breslau, 1891/92, p. 440], where a statically determinate free plane or free spatial framework

| Force method | Substitute member method |
|---|---|
| System with $n$ degrees of static indeterminacy (initial system) | Statically determinate system (initial system) |
| Transformation into a statically determinate basic system through release of $n$ ties | Transformation into a statically determinate substitute system through replacement of $n$ members |
| Calculation of the displacement steps $\delta_{i0}$ and $\delta_{ik}$ in the statically determinate basic system:<br><br>– $\delta_{i0}$: Displacement step at point $i$ due to initial state (given load)<br>– $\delta_{ik}$: Displacement step at point $i$ due to the force parameter $X_k = 1$ | Calculation of the member forces $S_{i0}$ and $S_{ij}$ in the substitute system:<br><br>– $S_{i0}$: Member force at point $i$ in the member inserted at this point due to the initial state (given load)<br>– $S_{ij}$: Member force at point $i$ in the member inserted at this point due to the member force $Z_j = 1$ resulting from the removal of the member at point $j$ |
| Compliance with the $n$ elasticity conditions of the statically indeterminate system (continuity statements for deformation variables):<br><br>$[\delta_{ik}] \cdot [X_k] + [\delta_{i0}] = [0]$ | Compliance with the $n$ equilibrium conditions of the statically determinate initial system (continuity statements for force variables):<br><br>$[S_{ij}] \cdot [Z_j] + [S_{i0}] = [0]$ |
| Calculation of the $n$ released statically indeterminates $X_k$ of the initial system from $n$ elasticity equations | Calculation of the $n$ removed member forces $Z_j$ of the initial system from $n$ equilibrium conditions |
| | Sufficient stability criterion:<br><br>$\det [S_{ij}] \neq 0$ ⟶ initial system is not kinematic<br>$\det [S_{ij}] = 0$ ⟶ initial system is kinematic |

FIGURE 8-5
Form equivalence of substitute member method and force method according to Müller-Breslau

with $k$ nodes is transformed into a substitute system with $k-1$ nodes and finally into a triangle [Henneberg, 1886, pp. 213–222 & 228–235] or tetrahedron [Henneberg, 1886, pp. 247–268]. However, Henneberg's technique was limited to free trussed frameworks, whereas Müller-Breslau's substitute member method is also suitable for supported systems. Henneberg had therefore provided the methodological basis for Müller-Breslau's substitute member method, but remained within the realm of the mechanics of rigid bodies because Henneberg was more interested in reducing complex framework systems to simpler systems, i.e. in the mathematical-physical derivation of the formation law for trussed frameworks. Müller-Breslau's substitute member method, on the other hand, was aimed at the unambiguous, rational structural analysis of complex trussed frameworks.

Nevertheless, his contribution to spatial framework theory was not confined to the substitute member method. Having analysed the Schwedler dome, he extended the kinematic theory of statically determinate plane trussed frameworks to spatial frameworks. By removing a trussed framework member $i$, Müller-Breslau converts the statically determinate spatial framework into a kinematic sequence. Based on the principle of virtual displacements, he generates the influence line of the member force in the trussed framework member $i$ as a projection of the displacement figure of the kinematic sequence – resulting from an applied displacement step of magnitude 1 – in the direction of the moving load. Using the structural calculations for a regular, octagonal truncated pyramid with two degrees of static indeterminacy as an example, Müller-Breslau finally convinced readers of the practicability of the force method for analysing statically indeterminate spatial frameworks. He thus succeeded in calculating spatial frameworks using the classic structural theory techniques that had been developed for linear-elastic frames.

Zimmermann's publication in 1901 of the structural calculations for the dome of the Reichstag building in Berlin [Zimmermann, 1901/1, 1901/2] and the associated, extraordinarily laborious calculations by Zschetzsche [Zschetzsche, 1901] was followed by a response from Föppl in the same year, without a graphical analysis [Föppl, 1901/3]. Müller-Breslau intervened with a reminder of his contributions on spatial framework theory published in 1891/92, providing a new explanation based on the example of the Reichstag dome [Müller-Breslau, 1902, pp. 49 – 51 & 61 – 63]. During the course of Müller-Breslau's series of articles, Otto Mohr published his calculation of the member forces in the Reichstag dome and the Schwedler dome, derived directly from the principle of virtual displacements [Mohr, 1902/1]. When Müller-Breslau continued his series of articles with kinematic spatial framework theory [Müller-Breslau, 1902, pp. 429 – 439 & 501 – 503] Mohr accused him of having copied the technique he had just published [Mohr, 1902/2, p. 634]. This marked the start of the dispute between Mohr [Mohr, 1903] and Müller-Breslau [Müller-Breslau, 1903/2, 1903/3, 1903/4] concerning the foundation of spatial framework theory. Ultimately, Mohr accused Müller-Breslau of plagiarism regarding spatial framework theory – first of all relating to Henneberg in the context of the substitute member method and then relating to himself in the context of kinematic trussed framework theory [Mohr, 1903]. "If such simple things," Mohr wrote, referring to the principle of virtual work he had introduced into trussed framework theory, "are worthy of scientific merit at all, it is not the solution itself that deserves praise, but the formulation of the task, i.e. the realisation that it can be used for further conclusions" [Mohr, 1903, p. 238]. And this is precisely where Mohr differed from Müller-Breslau. Whereas Mohr's research was based on problems and principles, Müller-Breslau's approach was based on methods and applications – methodologies whose scientific relevance Mohr vehemently disputed. The methodology problem, which is also virulent in other engineering sciences [Braun, 1977], thus took on a tangible form in the dispute between Mohr and Müller-Breslau. During the first few years of the 20th century it grew into a repeat of the 1880s dispute about the foundation of classic structural theory, with claims to the fundamental theorems of structural theory at stake [Müller-Breslau, 1903/4] [Mohr, 1903], and no longer spatial framework theory.

Henneberg incorporated spatial framework theory into the system of theories dealing with the mechanics of rigid bodies as part of a mathematically formulated graphical statics [Henneberg, 1894, 1902, 1903, 1911]. Graphical statics reached its practical limits with the analysis of spatial frameworks and statically indeterminate systems.

Schlink took up Föppl's trussed shell idea and developed multiple trussed shells enclosing one or more voids [Schlink, 1907/2]. The second edition of his work *Technische Statik. Ein Lehrbuch zur Einführung in das technische Denken* (applied statics – an introduction to engineering thinking) appeared in 1946 [Schlink & Dietz, 1946]. In this book, Schlink describes the corresponding plane projection technique developed by Mayor [Mayor, 1910, 1925], von Mises [v. Mises, 1917], Prager [Prager, 1926, 1927] and Sauer [Sauer, 1940], which restores spatial force problems to planar ones [Schlink & Dietz, 1946, pp. 305–314].

This level of knowledge of spatial framework theory remained unsurpassed until the second half of the 20th century. Practical applications included structural calculations for domes, cranes and towers, overburden conveyor bridges plus bridge and aircraft construction.

During its consolidation period (1900–50), the theory of structures thus lost the unity of static law, formation law and structural law in the shape of the composition law for spatial frameworks – that had been alluded to so longingly by Föppl in 1892 – because the primary focus was on the static law.

## 8.2 Spatial frameworks in an era of technical reproducibility

Schwedler domes (Fig. 8-6a), lattice domes (Fig. 8-6b), Zimmermann domes (Fig. 8-6c) and Schlink domes (Fig. 8-6d) consisted of rolled steel sections in different lengths with riveted joints. Despite the fact that industrially produced steel sections were used, each dome was unique.

Only the introduction of welding plus the standardisation of joints and members paved the way for the large-scale production of spatial frame-

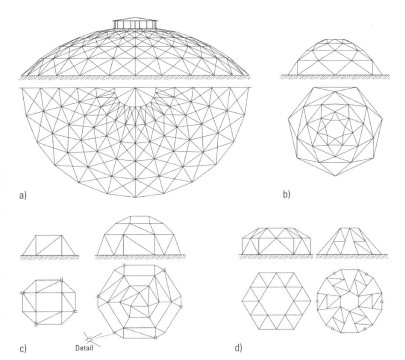

**FIGURE 8-6**
Dome types: a) Schwedler dome, b) lattice dome, c) Zimmermann domes, d) Schlink domes

works. Nevertheless, a conflict between production and assembly of components on the one hand and structural calculations on the other began to emerge during the period of great industrialisation between 1890 and 1914, which became apparent during the inter-war years. Whereas the scientific development of structural theory tended towards a procedural, syntactic and operative use of symbols (therefore creating a special intellectual technology), practical structural calculations continued to favour a hands-on approach, although that, too, was affected by the rationalisation movement that began to spread after World War 1. As part of this rationalisation movement, the structural engineer Konrad Zuse (1910–95) started to automate structural calculations. Between 1936 and 1941 he developed the first functioning computer, although computers only started to be used for the calculation of spatial frameworks towards the end of the 1960s. On the other hand, inventors such as Alexander Graham Bell (1847–1922), design engineers such as Vladimir Grigorievich Shukhov (1853–1939), physicists such as Walther Bauersfeld (1879–1959), civil engineers such as Franz Dischinger (1887–1953), designers such as Richard Buckminster Fuller (1895–1983), and inventors and entrepreneurial engineers such as Max Mengeringhausen (1903–88) had already taken the first steps towards large-scale production of spatial frameworks decades before.

### Alexander Graham Bell

#### 8.2.1

The first prefabricated spatial framework was designed by Alexander Graham Bell in early 1907 (Fig. 8-7/top). It consisted of relatively small tetrahedra. Bell, well known as the inventor of the telephone, was also engaged in research into aircraft, ships, medicine, electrical engineering, biology and engineering sciences. For example, Bell built a glider model consisting of a spatial framework covered with canvas (Fig. 8-7/bottom).

**FIGURE 8-7**
First prefabricated spatial framework of tetrahedra (top) [Makowski, 1963, p. 36] and glider model (bottom) [Makowski, 1963, p. 37], both designed by Bell

Later he developed gliders capable of carrying people. Bell standardised member and joint elements, commissioned the production of standard tetrahedra, and assembled them to form complex spatial frameworks according to the first formation law. In Bell's assembly technique, formation law and structural law for spatial frameworks form a materialised technological unit determined by function, construction, natural form and artistic form. Procedures on construction sites thus started to tend towards standardised forms and developed into the scientific subject of construction site management that started to emerge during the comprehensive rationalisation movement of the 1920s.

### 8.2.2 Vladimir Grigorievich Shukhov

"Engineering is thankless," Shukhov told his grandson, "because you have to possess knowledge in order to understand its beauty" [F. V. Shukhov, 1990, p. 21]. Shukhov invented, designed, calculated and built spatial structures of breathtaking beauty. In 1894 he applied for a patent for his lattice roofs. These tension structures consist of a grid of steel strips and angle sections. Shukhov's lattice shells are similar to the trussed shells of Föppl, although Shukhov abandoned the trussed framework principle. But the lattice roofs and shells are made from identical parts riveted or bolted together at the joints. In his patent applications, Shukhov highlighted the benefits of his spatial structural systems [Graefe 1990, p. 28]:
- Substantial weight reduction compared with ordinary roof structures.
- The components of lattice roofs are subject to tension only, those in lattice shells to compression only.
- High load-carrying capacity of the lattice surfaces, even for point loads.
- Simplified production and assembly through uniform structural elements.

As chief engineer of the Moscow-based company Bari, Shukhov used his new roof designs for eight halls at the 1896 pan-Russian exhibition at Nizhni Novgorod, where he also exhibited a tower structure in the shape of a hyperboloid. This structural system became widespread for the construction of water towers, radio masts and electricity pylons. Fig. 8-8 illustrates the telescopic assembly principle for a such tower structure in the shape of the 120 m high, five-section NiGRES electricity pylon for the River Oka crossing. The hyperbolic sections were assembled inside the structure and raised using five wooden crane trestles.

Shukhov's cost-effective, graceful, engineered structures accompanied the industrialisation of Russia right up until the time of the first Soviet five-year plan. According to Herbert Ricken, Shukhov, like Eiffel, Maillart and Freyssinet, "employed a precise mathematical approximation and an obvious feeling for the beauty of the structure and the spaces thus formed to develop bold, economic structures and realise them in a cost-effective way through efficient labour management" [Ricken, 2003, p. 546].

### 8.2.3 Walther Bauersfeld and Franz Dischinger

At the suggestion of Oskar von Millers, Walther Bauersfeld developed a projection device in 1922 for the Deutsches Museum in Munich to display

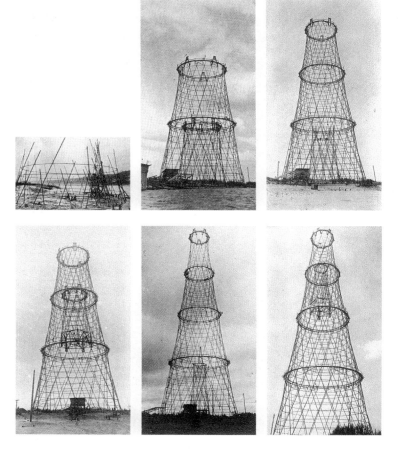

**FIGURE 8-8**
Assembly of the 120 m high, five-section NiGRES electricity pylon by Shukhov for the River Oka crossing, 1927–29 [Petropavlovskaja, 1990, p. 99]

**FIGURE 8-9**
Planetarium dome at Zeiss in Jena before applying the sprayed concrete [Schmidt, 2005, p. 87]

the movements of celestial bodies. It required a hemispherical shell as a projection surface. Bauersfeld constructed a hemispherical spatial framework (derived from the icosahedron) with 3840 steel members and 51 different member lengths. In order to obtain a smooth projection surface for the artificial night sky, Franz Dischinger, at the time chief engineer of the contractors Dyckerhoff & Widmann, recommended sprayed concrete (Fig. 8-9). Wire mesh was used as the substrate, and a 3 × 3 m curved timber panel attached to the spatial framework was used as formwork [Schmidt, 2005, p. 87]. The hemispherical shell with a thickness of 30 mm built on the roof of the Zeiss factory building in Jena in 1922 had a diameter of 16 m.

The first "Zeiss-Dywidag System" element had thus seen the light of day. In the 1920s and 1930s the Zeiss-Dywidag system shaped the theory and practice of reinforced concrete shells. The spatial framework of the Zeiss-Dywidag shell, acting structurally like rigid concrete reinforcement, is based on the reinforced concrete system invented in 1892 by Joseph Melan, which was widely used during the first decades of the 20th century, particularly for arch bridges in the USA and Spain [Eggemann & Kurrer, 2006]. In the Melan system the centering is replaced by the trussed steel arch, in the Zeiss-Dywidag shell it is replaced by the spatial framework. In addition to structural and constructional aspects, the technological benefits for the on-site process therefore also become significant. Nevertheless, the limited degree of rationalisation of the manufacturing technology for this type of spatial framework and the high material costs for this precision construction prevented its further development as rigid reinforcement and permanent formwork. Shortly afterwards, spatial frameworks were developed that served as reusable formwork for shell structures with non-prestressed (i.e. conventional) reinforcement.

### 8.2.4 Richard Buckminster Fuller

In 1954 Richard Buckminster Fuller was awarded a US patent for his geodesic dome design. The principle was based on projecting an icosahedron inscribed in a sphere onto the surface of the sphere (Fig. 8-10). The result is 20 equilateral spherical triangles, each of which can be subdivided into six further triangles through their three right bisectors, which lie on great circles. The frequency of the first subdivision is $v = 2$ because the sides of the basic triangle are split into two identical sections. For large-span domes, further subdivisions with $v = 4$, $v = 8$, $v = 16$, etc. are required because otherwise the slenderness ratio of the members would become excessive. Nevertheless, even for higher frequencies, the number of different

**FIGURE 8-10**
Geodesic domes after
Richard Buckminster Fuller

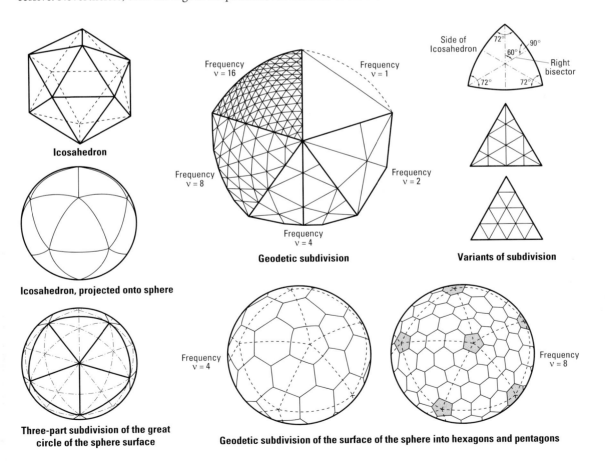

**FIGURE 8-11**
a) Dome in Baton Rouge, Louisiana, built in 1959 [Makowski, 1963, p. 133],
b) close-up of dome structure [Makowski, 1963, p. 134]

member lengths is relatively small. Whereas the surface of the sphere can be divided into a maximum of 20 equilateral spherical triangles, complete division into hexagons is no longer possible; at least 12 pentagons have to be used. This basic fact of sphere geometry is illustrated by a football as a true image of a "Fuller dome".

Fig. 8-11a shows the geodesic steel dome in Baton Rouge, Louisiana, with a span of 115 m, completed in 1959. The double-layer structure with a layer spacing of 1.20 m is characterised by the structural (i. e. load-transferring) connection between the roof covering and the hexagonal skeleton framework, forming a sort of folded-plate-cum-shell structure. The dome consists of 321 hexagonal welded steel panels to which the outer hexagonal layer of steel tubes is connected via diagonal ties and vertical corner members (Fig. 8-11b). In the late 1950s, Fuller's geodesic domes were used to house radar equipment (radomes) for the US early warning system running the 5000 km between the Arctic Circle and latitude 60° N [Marks, 1960, p. 462]. The highlight in the construction of geodesic domes was the three-quarter dome (76 m diameter) forming the US pavilion at EXPO 1967 in Montreal. His idea of the tensegrity structure represents another highlight (see [Calladine, 1978; Schlaich, 2003]).

Fuller's geodesic dome system marks the entry of non-Euclidean geometry, in the shape of sphere geometries, into the construction of spatial frameworks. At this level he investigated the interaction between the formation law and the structural law for spatial frameworks. In this way, Fuller created industrially produced "peculiar webs of space and time" [Benjamin, 1989, p. 355]. Not until Mengeringhausen's composition law for spatial frameworks and the associated work by Helmut Eberlein, Helmut Emde and Herbert Klimke would the spatial framework be freed from the space-time concept linked to the sphere.

### Max Mengeringhausen

#### 8.2.5

Inspired by August Föppl's pioneering publication *Das Fachwerk im Raume* [Föppl, 1892], Walter Porstmann's fundamental work on standardisation theory [Porstmann, 1917] and a critical analysis of Ernst Neufert's building design theory [Neufert, 1936], Mengeringhausen formulated eight structural laws for spatial frameworks in 1940 [Mengeringhausen, 1983, pp. 114–115], which formed the basis for the development of the MERO system between then and 1942 [Mengeringhausen, 1942]. The acronym MERO stands for "MEngeringhausen ROhrbauweise" (Mengeringhausen tubular construction system). On 12 March 1943 Mengeringhausen was awarded a German patent for his "combination of tubular members and joint-forming connection pieces, particularly for demountable framework structures" (Fig. 8-12). Mengeringhausen had formulated his first patent claim in plain language: "Combination of tubular members and joint-forming connection pieces, particularly for demountable framework structures, characterised in that a tubular member (2) with a threaded stud (5) having only one thread (6) carries a bushing (9) with a longitudinal slot (12) in which a coupling pin (13) attached to the threaded stud (5) engages such that upon turning the bushing, e. g. by

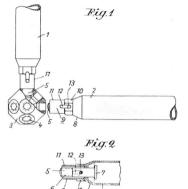

**FIGURE 8-12**
Drawing from the MERO patent dated 12 March 1943 [Mengeringhausen, 1943]

means of a spanner, the threaded stud (5) can be turned in both directions while the joint (3) and the tubular member (2) remain stationary for the purpose of establishing or releasing the connection" [Mengeringhausen, 1943, p. 2].

Mengeringhausen formulated a total of five patent claims. In that same year Mengeringhausen founded the MERO company in Berlin. On 12 March 1953, i.e. exactly 10 years after his "Reichspatent", Mengeringhausen's ingenious invention was protected by a patent for the Federal Republic of Germany. This exceptional structural steelwork invention revolutionised the design, technology and architectural aspects of spatial frameworks.

The development had commenced in the early 1940s in the context of orders from the German Aviation Ministry, under the auspices of Ernst Udet (1896–1941), for the airborne transport of construction kits for gantry cranes, temporary bridges, antenna systems and mobile shed structures stored in manageable crates. Nevertheless, Mengeringhausen immediately recognised the universality of MERO spatial frameworks. In 1944 he wrote: "The MERO system can be used for all kinds of equipment, frames, buildings and structures. The main benefits are: simple and fast assembly of just a few components; the option of quick dismantling and low weight, therefore ideal conditions for transportation in small or large units by road, rail, sea or air, even people or animals; versatile application and unlimited adaptability of the design for any task; easy replacement of damaged parts, straightforward replenishment of supplies and minimum storage of spare parts" [Mengeringhausen, 1944, p. 2]. Mengeringhausen offered steel tubes in the sizes 30 × 1 mm or 60 × 1.5 mm and in six different lengths: 0.5 m, $0.5 \times \sqrt{2}$ m, 1.0 m, $\sqrt{2}$ m, 2.0 m and $2.0 \times \sqrt{2}$ m (measured between joint centres). The lengths of the members form a geometric series with a natural growth factor of $\sqrt{2}$ (Fig. 8-13). Mengeringhausen specified bolts with M12 and M20 threads only. He only moved away from this standardisation gradually with the advent of computers for the design of spatial frameworks in the late 1960s.

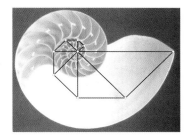

**FIGURE 8-13**
Geometric series of member lengths with the growth factor $\sqrt{2}$ and the natural model for his geometric series – the ammonite shell
[Mengeringhausen, 1983, p. 25]

## 8.3 Dialectic synthesis of individual structural composition and large-scale production

In the late 1950s Mengeringhausen succeeded where the progressive Bauhaus masters – in particular Walter Gropius (1883–1969), whom he admired – had failed in structural steelwork practice, i.e. the synthesis of individual structural composition and large-scale production. A precondition for this dialectic synthesis was his recognition and systematic application of the composition law for spatial frameworks in the form of the higher unity of static law plus formation law and structural law.

### 8.3.1 The MERO system and the composition law for spatial frameworks

The MERO technique represents the practical implementation of the eight structural laws for spatial frameworks formulated by Mengeringhausen in 1940 [Mengeringhausen, 1983, pp. 114–115]:

1. Regular spatial frameworks are composed from equilateral and/or right-angled triangles to create Platonic bodies or shapes derived from them (see Fig. 2-99).

**FIGURE 8-14**
Pavilion for the *Town of the future* at Interbau, Berlin, 1957, with exposed spatial framework as an architectural feature [Mengeringhausen, 1975, p. 129]

2. Owing to their regular structure, spatial frameworks are statically ideal; uniform joints and a limited number of different members enable industrial series production.
3. The member lengths in a spatial framework form a geometric series with a natural growth factor of $\sqrt{2}$ (see Fig. 8-13).
4. Series of similar polyhedra can be built using $n$ different member lengths from the geometric series of natural growth.
5. For the similar polyhedra, the sizes of the surfaces form a geometric series with the factor 2, and the volumes form a geometric series with the factor $2 \cdot \sqrt{2}$.
6. All the elementary bodies mentioned above, their derivatives and the associated composite spatial frameworks can be built using a single universal joint and members from the geometric series of natural growth.
7. The universal joint is a polyhedron with 26 surfaces from the series of semi-regular Archimedean bodies whose 18 squares are equidistant from the centre of the body and exhibit concentric holes.
8. Regular spatial frameworks can be constructed from a universal joint with 18 connections in the form of the standard MERO joint.

The breakthrough for Mengeringhausen's MERO framework for long-span roof structures came in 1957 at Interbau fair in Berlin, where in collaboration with the architect Professor Karl Otto he created a spatial framework grid consisting of truncated octahedra and tetrahedra (Fig. 8-14). The rectangular spatial framework roof covered an area of 52 × 100 m and was constructed exclusively from standard MERO joints and a single type of standard member, with a system dimension of 2 m, plus bolts with M20 threads.

The structural law corresponds to the densest hexagonal sphere packing. The tetrahedra are inserted from above between the upright semi-octahedra to create a rectangular framework (Fig. 8-15). Such plate-like spatial frameworks are statically indeterminate to a high degree and cannot be calculated using member analyses. In the early 1970s engineers therefore had to make do with the finite difference method applied to plate theory [Lederer, 1972]. The basic idea of this discretisation of the spatial elastic continuum through the spatial framework had already been explored by Ernst Gustav Kirsch (1841–1901) [Kirsch, 1868] and Hrennikoff [Hrennikoff, 1941] and was to benefit the development of the finite element method in the 1950s, although that method would not become significant for the structural analysis of spatial frameworks until the 1970s. The relationship between static law and structural law of spatial frameworks thus remained superficial for the time being.

In 1962 Mengeringhausen published a brochure entitled *Komposition im raum* (spatial composition) [Mengeringhausen, 1962], on the occasion of the Debau fair in Essen, in which he made an attempt to find a systematic relationship between his structural laws, the formation law and the static law of spatial frameworks. However, it was not until 1966 that Mengeringhausen presented his *Kompositionslehre räumlicher Stab-Fachwerke* (composition theory for spatial member frameworks) [Mengeringhausen, 1967] at the "International Conference on Space Structures" in London. He coded the elementary units of spatial frameworks by referring to the symbols of crystallography and chemistry in order to classify spatial frameworks composed of such bodies using structural equations. The generalisation of his structural laws to form the composition law for spatial frameworks was based on the *Discourses on symmetric polyhedra* published in 1849 by Auguste Bravais [Bravais, 1892], the co-founder of crystallography. Fritz Kesselring's book *Technische Kompositionslehre* (technical composition theory), published in 1954 [Kesselring, 1954], may well have inspired Mengeringhausen, too. In his dissertation on the geometric derivation of the double-layer frame grids from cubic grids (completed in 1970), Helmut Eberlein significantly expanded Mengeringhausen's classification system for spatial frameworks using the terminology of crystallography [Eberlein, 1970].

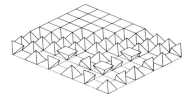

FIGURE 8-15
Composition of a plate-type, rectangular spatial framework consisting of semi-octahedrons (1/2O) and tetrahedra (T): 1/2O + T [Mengeringhausen, 1983, p. 72]

At the end of the 1960s Mengeringhausen finally approved member lengths forming a geometric series with a natural growth factor of $\sqrt{2}$ (see Fig. 8-13) and overcame his fixation with M12 and M20 bolt threads, thus enabling spatial trussed frameworks to be represented as affine deformations through adaptation to given structure dimensions and to be realised in practice. Between 1969 and 1971 Joachim Scheer and his former colleague Uwe Ullrich wrote numerous reports on the application of such construction elements, which were not covered by the 1963 building authority approval for the MERO system, in anticipation of regulations for the new technical approval being discussed at the time, which was finally granted in 1971 [Scheer, 2003]. In 1978 the Institute of Steel Construction at Braunschweig TU, where Joachim Scheer had been director since 1971, developed a design theory for the MERO system which was based on the ultimate load technique and had been verified through experiments.

**FIGURE 8-16**
Grandstand roof, Split, Croatia
[Kurrer, 2004/1, p. 621]

**Spatial frameworks and computers**

### 8.3.2

Towards the end of the 1960s, the use of computers led to a revolution in both the theoretical and the practical composition of spatial frameworks. It started with the construction of the German concert dome at EXPO 1970 in Osaka and ended in 1979 with the construction of the shell roof over the grandstand of the sports stadium in Split (then Yugoslavia, now Croatia) with its span of 200 m (Fig. 8-16). It would have been impossible to calculate manually the 839 different joints (out of a total of 3460) and the 1143 different members (12 382 in total).

The craftsmen of structural calculations who left impressive, intellectually created artistic forms, developed into *Geistwerker* ("mind workers", a term coined by Mengeringhausen) such as Helmut Emde and Herbert Klimke, who advanced the systemic integration of design, calculation, construction and production of derived spatial frameworks through computers. Helmut Emde expanded the geometry from flat to curved spatial frameworks and generated computer-aided framework topologies, thus compiling derived spatial framework geometries by computer [Emde, 1977]. Having become director of the MERO computer centre in 1974, Herbert Klimke succeeded in breaking the linear dominance in the structural analysis of spatial frameworks by utilising the insights gained while researching his dissertation and by modifying the Structural Analysis Program (SAP) finite element software system in such a way that non-linear second-order theory effects and ideal-elastic, ideal-plastic material behaviour could be considered [Klimke, 1976]. For non-standardised grids, Jaime Sanchez [Sanchez, 1980], Martin Ruh and Herbert Klimke [Ruh, Klimke, 1981] described the topological relationships of spatial frameworks in an integer grid and the metric through a coordinate transformation [Klimke, 1983, pp. 258–259], thus faithfully continuing Helmut Emde's work.

C. R. Calladine, A. Affan and S. Pellegrino were able to show that certain kinematic states exist for double-layer space grids with many degrees of static indeterminacy if the number of effective members $s$ according to eq. 8-1 is significantly larger than $3 \cdot k - 6$ ([Affan & Calladine, 1986 & 1989; Pellegrino & Calladine, 1986]). The condition of eq. 8-1, although necessary, is not sufficient.

The introduction of the NC production of members and joints and the automatic generation of location drawings, etc. marked the MERO company's completion of the systemic link between design, calculation, construction and production of derived spatial frameworks through the computer and set the pace in the construction industry, particularly for structural steelwork. Mengeringhausen thus concluded the dialectic synthesis of individual structural composition and large-scale production of spatial frameworks. The composition law for spatial frameworks, comprehensively described in his books [Mengeringhausen, 1975, 1983] (see Fig. 2-101b) can be interpreted as follows:

- Owing to the large number of members $s$ and nodes $k$, the formation law for spatial frameworks ($s = 3k - 6$) can no longer be calculated manually.
- In computer analysis the formation law for spatial frameworks is represented as a pure mathematical transformation.
- In computer-aided design the formation law merges with the structural law such that the difference between geometry and statics disappears.
- The static law no longer appears in the form of force diagrams or elasticity conditions, but instead becomes an inextricable component in the trinity of material, translation and equilibrium laws for finite member elements.
- And finally, the formation law, structural law and static law of spatial frameworks only evolved into the higher unity of the composition law via the linking of design, calculation, construction and production through the computer as the symbolic machine.

Therefore, the composition law not only manifests itself as the opportunity for beauty in the finished spatial framework, but is also part of the formation process leading to it, which is enlivened through human activity. In this formation process we can also discover aspects that stimulate and satisfy desire, grace or sensory beauty. The philosophical discipline of aesthetics reflects this.

Only the holy place, the genius loci, the presiding god or spirit of a place, is able to endow beauty through aura, these "peculiar webs of space and time" [Benjamin, 1989, p. 355]. Without doubt, the man-made Garden of Eden in Cornwall (Fig. 8-17) has such an aura as the project developers, architects and engineers have returned a little piece of the lost Paradise to the descendants of Adam and Eve, to let them inhale the distant light of history.

FIGURE 8-17
View of the roofscape of the Eden Project in Cornwall, England [Knebel et al., 2001]

# CHAPTER 9

# Reinforced concrete's influence on theory of structures

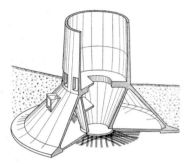

The author is grateful for the receptiveness of Klaus Stiglat, formerly chief editor of the journal *Beton- und Stahlbetonbau*, who published the history of the first design theories for reinforced concrete. The history of the modern building industry begins with the establishment of reinforced concrete construction in the first decade of the 20th century, which also gave rise to the first modern codes of practice in Switzerland, Germany, France, Belgium and Austria. The question arose as to the relationships within the science-administration-industry triad. The author has pursued this line via the interdisciplinary aspect of the history of structural analysis – at an event organised by Werner Lorenz at Cottbus TU and in a series of conferences on the history of reinforced concrete inaugurated by Hartwig Schmidt at Aachen RWTH. And the 100th anniversary of the yearbook *Beton-Kalender* in 2005 encouraged the author to investigate the development of the new structural engineering language that emerged in the 1920s with reinforced concrete shell structures, and to acknowledge prestressed concrete construction and the holistic concept of the truss models of Jörg Schlaich and Kurt Schäfer within the scope of a history of reinforced concrete construction.

The successful introduction of cement and rubble-stone vaults in French industrial building after the 1840s marked the beginning of the end of vaults made from dressed stones and clay bricks. The cement and rubble-stone vault gradually evolved to become the concrete vault whose granular microstructure was bonded together with Portland cement (from 1852 onwards produced industrially in Germany as well) to form a structural concrete continuum. Like wrought iron carried the resolution of the loadbearing system elements too far as it displaced timber and cast iron and rendered visible the play of forces with the help of trussed framework theory, concrete abolished the joint and turned the vault into a curving one-dimensional continuum.

But in bridge-building, concrete did not start to play a role until the late 1880s. And what in elastic theory was only a question of the sign, plus or minus, was for concrete, which had shaken the old division of work on the building site, a break with symmetry with serious consequences: although concrete can accommodate high compressive stresses, even very moderate tensile stresses result in serious cracks. For example, an asymmetric load distribution over the axis of the arch can cause tension cracks in the curving concrete continuum, which suddenly develops "joints" and is partially converted into a stone arch with gaping joints. Cracks are the horrors of elastic theory! And even though masonry work prevailed over concrete during the Bismarck era – using standard-format clay bricks, introduced in Germany in 1872 – and despite the supplanting of the handmade brick by the machine-made brick fired in the Hoffmann rotary kiln, it could not stop the onward march of concrete, whether for foundations, hydraulic structures or fortifications. In order to alter the building site for solid construction (i.e. masonry and concrete) to suit industrial aspects, and therefore place conception, design and construction on a construction theory footing, concrete – just like wrought iron – had to become so universal from the structural-constructional side that the ensuing loadbearing systems could resist both tension and compression. Iron superseded masonry in bridge-building in the second half of the 19th century, and concrete could only survive into the 20th century by joining forces with its greatest rival. For the first time, the civil and structural engineer was asked to turn the invisible, the symbiotic characteristic of every composite material, into the visible – with far-reaching consequences:

1. The mutual cooperation of construction sciences with constructional-technical developments – specifically, the collaboration between reinforced concrete research and reinforced concrete practice that arose soon after 1900 and helped both sides.
2. The upheaval in construction due to the emergence of the building industry.
3. The creation of the science-industry-administration triad as a new incorporative form of scientific-technical cooperation [Kurrer, 1997/1].
4. The evolution of the design diversity possible with this composite building material as a prerequisite for the modern movement in architecture.

5. The totalising and leap in quality in the conception and design process owing to the monolithic character of this composite building material.
6. The progress in theory of structures induced by this composite building material.

## 9.1 The first design methods in reinforced concrete construction

As reinforced concrete construction became established in the final third of the 19th century, solid construction underwent a technical upheaval which not only led to the modern building industry, but also to a closer relationship between progress in construction engineering and the emerging civil and structural engineering disciplines. Using the example of the *Monier-Broschüre* [Wayss, 1887] published in 1887 by Gustav Adolf Wayss (1851–1917) and Matthias Koenen (1849–1924), the cooperation between technical trials and engineering science theory formation on the one hand and building practice on the other – characteristic of the later development of reinforced concrete construction – will become clear. Comparative calculations with various historical calculation theories for strips of slab taken from the *Monier-Broschüre* illustrate the efficiency of the first design method, drawn up by Koenen.

### 9.1.1 The beginnings of reinforced concrete construction

The history of structural engineering in the final decades of the 19th century is characterised by two processes that rebuilt the very foundations of building: on the one hand, as the general theory of trusses appeared in the 1890s, theory of structures attained the rank of a fundamental engineering science discipline in civil engineering; and on the other, the first trials involving a combination of steel and concrete led to a new composite building material with completely new material qualities and a technical revolution in solid construction – in the wake of which the modern construction industry emerged. Whereas classical theory of structures formed in an intensive interaction with the resolved method of construction, especially trussed frameworks, the creation of a practical reinforced concrete theory by Matthias Koenen, Armand Considère, Paul Christophe, Emil Mörsch and others was the result of materials research and tests. This relationship between the constructional-technical and materials science-based examination of the new form of construction so characteristic of the formation of a reinforced concrete theory hinted at a higher level of relationship between the empirical and the theoretical in solid construction even before 1900. As was shown in section 4.5.3, the multiplicity of masonry arch theories had already been made obsolete – on the theoretical side – by Winkler's use of elastic theory around 1880; however, the calculation of solid arches on this basis did not become part of everyday civil engineering practice until after the Berlin trials by Wayss in 1886 and the comprehensive loading tests on tamped and reinforced concrete arches carried out by the Austrian Engineers & Architects Association in Purkersdorf (1892). Building with plain and reinforced concrete therefore gave elastic theory and research trials their rights in solid construction.

The early history of reinforced concrete construction begins with Lambot's reinforced concrete boat and Coignet's plan for a concrete house at the Paris World Exposition of 1855 and ends with Hennebique's monolithic reinforced concrete frame at the Paris World Exposition of 1900, which at the same time announced the arrival of the century of reinforced concrete construction with a triumphant fanfare (for more about the early history of reinforced concrete, see [Huberti, 1964; Jürges, 2000; Newby, 2001; Iori, 2001; Bosc et al., 2001; Simonnet, 2005]). The first simple design method for reinforced concrete slabs appeared in the middle of the classical phase of structural theory (1875–1900); this was joined by others in the 1890s and shortly before 1900 these were combined by Paul Christophe [Christiophe, 1899] to create the first reinforced concrete theory backed up by tests.

"The dominance of stone in building seems to be nearing its end; cement, concrete and steel are destined to take its place" (cited in [Huberti, 1964, p. 16]). It was with these words that the French concrete building contractor François Coignet (1814–88) tried – in vain – to convince the management of the Paris World Exposition of 1855 to let him build a concrete house for display at the exhibition. In that same year he took out a British patent for a suspended floor slab supported on four sides. The year 1861 saw the publication of his large book *Bétons agglomérés appliqués à l'art de construire* [Coignet, 1861] in which he formulates the constructional-technical knowledge that the steel bars and the concrete act as a composite material and the steel bars carry the tensile forces in the cross-section after the tensile strength of the concrete has been exceeded. The American Thaddeus Hyatt (1816–1901) considerably extended the knowledge of the loadbearing quality of reinforced concrete with his US patent of 1878: "Concrete made with cement is combined with strips and hoops of iron to form slabs, beams or arches in such a way that the iron is used only on the tension side" [Huberti, 1964, p. 42]. Although Hyatt recognised the division of work between concrete and steel according to compressive and tensile forces and hence placed reinforced concrete construction on a rational footing, it was left to the Berlin civil engineer Matthias Koenen to publish the first design method for reinforced concrete slabs in bending in the *Zentralblatt der Bauverwaltung* [Koenen, 1886, p. 462] and in the *Monier-Broschüre* [Wayss, 1887, pp. 27–28] in 1887.

The fact that a gardener from Paris, Joseph Monier (1823–1906), despite the pioneering work of Joseph Louis Lambot (1814–87) and François Coignet, was on 16 July 1867 granted a French patent for his *Système de caisses-bassins mobiles en fer et ciment applicables à l'horticulture* (Fig. 9-1), i. e. for the production of "movable tubs and containers for horticulture" made from cement mortar with embedded iron, still remains an unsightly blemish on construction engineering history.

By 1868 Monier had taken out two additional French patents for the production of pipes and stationary containers. Encouraged by the successful building of water tanks, he had numerous applications of the new form of construction protected by patents in France: the construction of flat slabs (1869), arch bridges (1873), stairs (1875), railway sleepers (1877),

**FIGURE 9-1**
Monier's patent of 1867
a) text and b) drawings
[Bosc et al., 2001, pp. 76 & 77]

arches, beams and an improperly thought-out T-beam (1878), suspended slabs with rolled sections and flat reinforced panels (1880), and arched reinforced panels (1881). Monier quite rightly called himself a "Rocailleur en Ciment" [Bosc et al., 2001], a sculptor in cement, because his idea of using the embedded steel as a means of shaping dominated his constructional thinking. His understanding of the loadbearing ability of reinforced concrete members was far inferior to that of Coignet and Hyatt; however, his contribution was "to create all the conditions that prepared the way for the later triumphs of reinforced concrete construction through an incredible strength of will and a practical view" [Förster, 1908, p. 17].

### 9.1.2 From the German Monier patent to the *Monier-Broschüre*

Monier's invention was patented in Germany in 1880 in the class for the clayware and stoneware industry (Fig. 9-2). The patent claim covers a "method for producing articles of all kinds by casting the walls of the article in cement to match the ribs of iron and in particular for the production of railway sleepers according to this method" [Monier, 1880]. It was under this very general claim that the Monier system first gained a footing in southern Germany. Just four years later, Monier was able to sign the first German license agreements with the Martenstein & Josseaux (Offenbach) and Freytag & Heidschuch (Neustadt a. d. Haardt, now Neustadt a. d. Weinstraße) companies. Whereas Martenstein & Josseaux used the German Monier patent only within a radius of 30 km of Frankfurt am Main, Conrad Freytag (1846–1921) had acquired the licence for the whole of southern Germany with an option for the rest of the country. And it was in 1855 that the small Freytag & Heidschuch Company transferred this latter option to the civil engineer and concrete contractor Gustav Adolf Wayss free of charge in the interest of spreading the Monier system; Wayss purchased a licence for the rest of Germany from Monier and in that same year moved his business from the Ruhr region to Berlin [Ramm, 2007].

Berlin offered Wayss the best-possible technical-scientific and organisational conditions for introducing the Monier system and then tapping the national building market. This was where the Berlin school of structural theory with its international eminence emerged from the research and teaching at Charlottenburg TH by Emil Winkler and later Heinrich Müller-Breslau (see section 6.6). Müller-Breslau, Koenen and others exchanged their latest ideas on the foundation of classical theory of structures in the Berlin Architects Society and in the *Wochenblatt für Architekten und Ingenieure*, and also confirmed their work in everyday engineering. They founded the first private civil engineering consultancies and hence were the personification of the successful practising civil engineering scientist whose engineering science work was driven by the collective engineering experience and whose engineering practice was permeated by engineering science findings. The construction theory infrastructure of Berlin that met Wayss was rounded off by the Testing Institute for Building Materials founded in 1875 whose task it was "to carry out tests in relation to strength and other properties of the build-

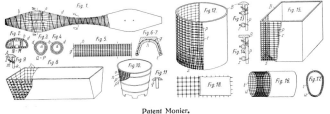

FIGURE 9-2
The German Monier patent
[Wayss & Freytag AG, 1925, annex 5]

ing materials based on commissions from authorities and private companies and to conduct tests in the general interest of science and the public" (cited in [Ruske, 1971, p. 76]). One of the most important clients of the Testing Institute was the Royal Police Presidium, which was responsible for supervising building work in Berlin; the scientifically trained civil servants could only be convinced of the worth of a new form of construction through loading, fire and materials tests plus a scientifically based method of design.

Koenen, the government's chief building officer entrusted with overseeing the construction of the Reichstag, knew this only too well when Wayss visited him in his on-site office in late 1885 in order to praise the stability of reinforced concrete walls and their safety in fire [Ramm, 1998, p. 353]. Koenen expressed his doubts regarding the corrosion resistance and the steel-concrete bond plus the risk of cracks due to be different thermal expansion coefficients of the two materials. To check the first two objections, he recommended that Wayss consult Johann

Bauschinger (1834–93) in Munich, who was head of Germany's leading materials testing institute. After a series of successful preliminary trials with the Monier system heightened Koenen's interest, he was also able to clear up the remaining question of the linear thermal expansion coefficients of steel and concrete. In the 1863 volume of the leading French civil engineering journal *Journal des Ponts et des Chaussées*, he found values for concrete between $1.37 \times 10^{-5}$ and $1.48 \times 10^{-5}$ m/°C in Bouniceau's tables; the linear thermal expansion coefficient of steel is, however, $1.45 \times 10^{-5}$ m/°C. "From this moment on," Koenen writes in his memoirs, "I was resolved to give the matter my full attention because it was completely clear to me that I now had the basic conditions for a new type of construction before me; for besides the concrete body in compression I saw not only the possibility of ties and walls, but primarily also a chance of building elements to resist bending; all one had to do was to increase the insufficient tensile strength of the concrete considerably by placing a suitable number of iron bars in the tension zone which through the known resistance to sliding on the concrete during deflection and the formation of the moment of resistance would have to be called upon to contribute" [Koenen, 1921, p. 348]. To calculate the internal moment of a simply supported reinforced concrete slab, Koenen embeds the steel in the soffit of the slab with 5 mm concrete cover. He assumes that the triangular compressive stress block of the concrete ends on the centre-line of the slab and determines the steel cross-section in the cracked state (Fig. 9-3). But before publishing this first scientifically founded design theory for reinforced concrete construction, he suggested to Wayss that he build reinforced and unreinforced test slabs ($b = 60$ cm; $l = 100$ cm; $d = 5$ cm; $h = d - d_1 = 4.5$ cm) and calculated the steel cross-section – probably based on the loading assumptions for the floor slabs in the Reichstag – to be $A_s = 3$ cm$^2$. The ultimate load of the Monier slab was six times that of the unreinforced slab! With small deflections of the reinforced slab, Koenen observed that there was a linear relationship between loading and measured deflection, and therefore Navier's beam theory could also be used by engineers for reinforced concrete.

Koenen was now sure about his design method and was able to summarise the basic conditions for reinforced concrete construction as follows: "When calculating the depth of the Monier cement slabs with embedded iron which are subjected to bending, an approximate method is obtained when one applies the internal forces that produce the resistance couple …, and in so doing the tensile stresses of the cement mortar are ignored … The desired displacement of the iron bars within the slab due to the tensile force is prevented by the significant adhesion between cement and iron" [Koenen, 1886, p. 462].

After Koenen has also set up design equations for arches, pipes and water tanks, he goes on to develop a scientifically founded programme of tests for Monier constructions with prescribed dimensions and steel areas. The tests were carried out during 1886 under the critical gaze of the civil servants of the Berlin "building police" and other specialists. Together

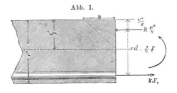

**FIGURE 9-3**
Publication of Koenen's design theory in the *Monier-Broschüre* [Wayss, 1887, p. 27]

with Koenen's design theory they formed the heart of the first engineering science work in reinforced concrete literature: the *Monier-Broschüre* [Wayss, 1887].

### 9.1.3 The *Monier-Broschüre*

Wayss had 10 000 copies of the *Monier-Broschüre* (Fig. 9-4) printed in Berlin and Vienna in 1887 and sent them to building authorities, the "better-known private architects" and civil engineers. Although older publications on the early history of reinforced concrete assess the contributions of Wayss and Koenen to the genesis of the *Monier-Broschüre* differently, all authors regard the *Monier-Broschüre* as a classical work of reinforced concrete. According to Förster, the reason for this is that "the *Monier-Broschüre* is the first publication to deal with the applications of reinforced concrete construction in a comprehensive way that describes the experiences and trials with this form of construction hitherto, but primarily contains for the first time a theory of the new type of construction worked out by the government's chief building officer M. Koenen" [Förster, 1908, p. 19]. Here, Wayss, as an energetic canvasser in matters of reinforced concrete, was thinking more along the lines of soundly based advertising copy

- against attempts to contest his rights to the German Monier patent and in that context to misunderstand reinforced concrete as merely a grid of iron bars surrounded with cement without any structural-constructional function (presentation of the new type of constructional-technical quality offered by the Monier system);
- to convince building authorities, architects and civil/structural engineers of the scientifically based design and the stability of the Monier system (presentation of the engineering science basis of the Monier system by way of calculations and test results);
- to convince clients of the efficiency and the outstanding serviceability properties of the new form of construction (presentation of the areas of application of the Monier system).

#### 9.1.3.1 The new type of structural-constructional quality offered by the Monier system

Wayss, the licence-holder of the Monier patent for northern Germany, had not been in Berlin very long before he had to defend the broad claim of the Monier patent against patent-law objections. The objection of the Berlin masonry master Rabitz, for example, was that the embedded steel in Monier constructions was a copy of his patented plasterwork backing made from wire. Although an injunction against the Monier patent granted by a Berlin Court was revoked by the Royal Court of Appeal at the end of 1886, which enabled Wayss to build floors and walls in reinforced concrete again unhindered, the patent claim of the German Monier patent remained an easy victim of legal challenges as long as the structural-constructional principle of reinforced concrete construction had not been demonstrated. However, even more serious were the doubts expressed by the building community regarding the new form of construction, which concerned the protection against corrosion, the adhesive bond between the steel and the concrete, and the coefficients of linear thermal expansion. In order to answer these questions, the smart building contractor Wayss made use of civil engineering science, one of the first in his profession to do so: "Thanks to the inventor [Monier – the author], these doubts could be refuted by way of tests over a period exceeding 20 years – doubts which even today plague many technicians because the experiences gained with Monier constructions have not yet been studied scientifically" [Wayss, 1887, p. 2]. When investigating these doubts, which had in no way had been refuted by Monier, the signature of Koenen is clearly evident in the *Monier-Broschüre*. In the book, all objections to the Monier system are refuted one by one with the same reasoning [Wayss, 1887, pp. 4 – 8], which had been worked out in essence by Koenen beforehand [Koenen, 1921].

**FIGURE 9-4a**
*Monier-Broschüre*: cover of the Berlin edition

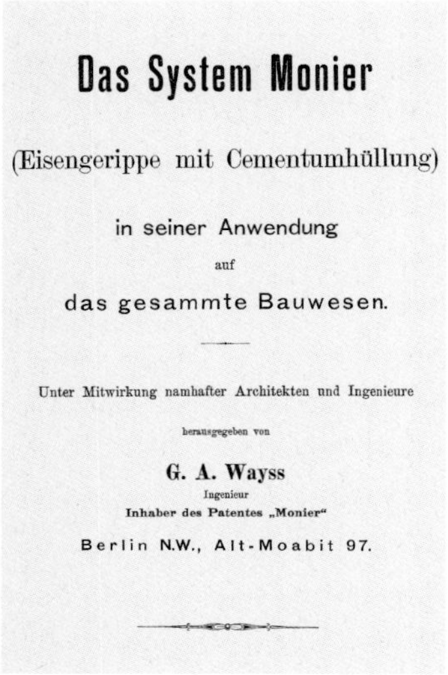
Das System Monier

(Eisengerippe mit Cementumhüllung)

in seiner Anwendung

auf

das gesammte Bauwesen.

Unter Mitwirkung namhafter Architekten und Ingenieure

herausgegeben von

G. A. Wayss

Ingenieur

Inhaber des Patentes „Monier"

Berlin N.W., Alt-Moabit 97.

The *Monier-Broschüre* therefore contains extracts from two independent reports on the aforementioned patent disputes commissioned by Wayss. In the report by Admiralty Counsellor Vogeler, concerning the question of the significance of reinforced concrete construction it merely says that the "embedded iron bars carry the tensile or compressive stresses and the enclosing hardened cement prevents the lateral buckling of the bars under load" [Wayss, 1887, p. 3]. Whereas Vogeler muddles rather than explains the structural division of work between steel and concrete in the reinforced concrete cross-section, this is expressed considerably more precisely in the second report by Prof. Fritz Wolff (Charlottenburg TH): "Every element of the floors ... and walls is essentially loadbearing in the Monier form of construction. Such floors and walls are made up of elements, each one of which represents a beam constructed of cement and an iron bar embedded in this in such a way that the great compressive strength of the cement and the splendid tensile strength of the iron is exploited rationally ... It simply depends on placing the iron bar exactly at the point in the beam cross-section where the tensile stress occurs. The thickness of the iron bar depends on the magnitude of the tensile stress expected" [Wayss, 1887, p. 3]. The other remarkable thing about this report is the suggestion of the existence of an objective relationship between theory of structures and reinforced concrete construction. This admission is initially limited to the theoretical examination of individual loadbearing elements such as the planar simply supported slab spanning in one direction, arches, etc. Owing to the lack of knowledge about concrete as a material and about its interaction with the steel, the abstraction, even on this level of the simplest loadbearing elements, to the structural system could only be incomplete.

### 9.1.3.2 The applications of the Monier system

The *Monier-Broschüre* starts with a list of the most diverse applications of the Monier system in structural engineering, civil engineering, mining, shipbuilding, agriculture, horticulture and industrial buildings. What is conspicuous is the multitude of container systems named in civil engineering and the foodstuffs, paper, textiles and chemicals industries, with their high requirements regarding imperviousness, thermal stability and resistance to acids. The reasons for this are as follows:

*Firstly:* Monier had already successfully built water and gas tanks from reinforced concrete in the 1870s, which meant that later on Wayss had merely to continue this work and impress upon aspiring industries the economies and better serviceability qualities of this new type of construction.

*Secondly:* The thin-wall container systems could not only be derived from Monier's original patent of 1867, but could also be built without knowledge of the structural-constructional principle of the new type of construction; this was especially true for the circular water and gas tanks in which Monier placed the embedded steel – purely by chance correctly – in the centre of the wall.

*Thirdly:* It was precisely the specific properties of reinforced concrete containers such as imperviousness, thermal stability and acid resistance and not the characteristic loadbearing quality of reinforced concrete components loaded in bending that first earned respect for the new method of building from open-minded clients.

In the *Monier-Broschüre* the fire resistance and imperviousness are highlighted in the applications of the Monier system for buildings

FIGURE 9-4b
*Monier-Broschüre*: cover of the Viennese edition

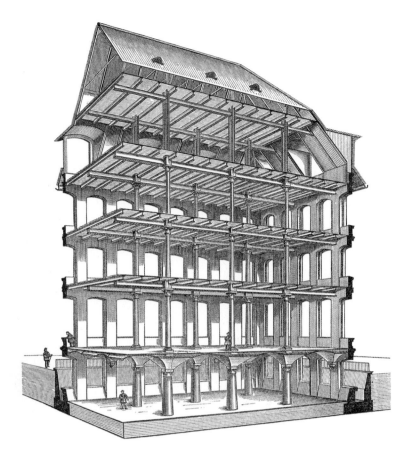

**FIGURE 9-5**
Warehouse with Monier floors and roof
[Wayss, 1887, p. 75]

(Fig. 9-5). Alongside many components of secondary structural importance there are "flat and vaulted suspended floors to suit any span and load …, thin, fire-resistant, self-supporting walls [and] columns" [Wayss, 1887]. Of course, the authors do not forget to mention in detail the advantages of the Monier system over classical forms of solid construction: durability, high load-carrying capacity for a low self-weight, less space required, savings in abutments and anchorages for shallow arches and domes, faster construction on site, economy and, finally, hygiene benefits. And so several pages are devoted to comparing the costs of brick jack arches with and without steel beams and Monier jack arches with a span of 4.50 m.

All the applications of the Monier system are explained in understandable terms for builders and potential clients in the second section [Wayss, 1887, pp. 73–128]. Although this section demonstrates with some imagination the technical and economic possibilities of the new form of construction, it is conspicuous that the examples of construction work and design proposals do not yet form a uniform whole; for the Monier system was just one of the numerous technical systems competing for attention in the building industry. Apart from that, Wayss and Koenen were vaguely aware of the internal development tendency of the Monier system towards the monolithic form of construction for factory buildings only: "The union of various building components made from iron ribs encased in cement to form a building in which otherwise only clay bricks for the enclosing walls, cast iron for the internal columns and rolled iron for the frames of machines are used ... is illustrated most simply in the design for a factory building" [Wayss, 1887, p. 107].

Under such constructional-technical conditions, the structural analysis of reinforced concrete construction had to remain in the domain of elementary theory of structures because the synthesis of the individual loadbearing elements to form the complete loadbearing structure for a building in solid construction was not yet an object of structural theory formation.

### 9.1.3.3 Laying the engineering science foundation of the Monier system

"In order to be able at last to cease making construction using Monier's system more expensive through the unnecessary use of materials, without on the other hand endangering its solidity, it was necessary to depart from the purely empirical path for determining the cross-section of various artefacts" [Wayss, 1887, p. 35].

The theoretical nucleus of the *Monier-Broschüre* is the design method based on Koenen's work. From the assumptions that the concrete resists only the compressive stresses distributed linearly over the cross-section, and the reinforcement only the tensile stresses, the depth of the compression zone is

$$x = 0.5\, d \tag{9-1}$$

and the lever arm of the internal forces takes on the value

$$z = 0.75\, d \tag{9-2}$$

which allows Koenen to derive the design equation for a slab of width $b = 100$ cm, unknown thickness $d$, reinforced in one direction (steel cross-section $A_s$) and subjected to the bending moment $M$ [Koenen, 1886, p. 462]. If we convert these equations according to the known concrete compressive stress at the edge of the compression zone $\sigma_{b,exist}$ and the tensile stress present in the steel $\sigma_{s,exist}$, then we obtain the stress equation for a beam with bottom reinforcement subjected purely to bending:

$$\sigma_{b,exist} = (16/3) \cdot [M/(d^2 \cdot b)] \leq \sigma_{b,perm} \tag{9-3}$$

$$\sigma_{s,exist} = (0.25 d \cdot b) \cdot (\sigma_{s,exist}/A_s) \leq \sigma_{s,perm} \tag{9-4}$$

Koenen had thus integrated his simplified design indirectly into the stress analysis scheme known to civil and structural engineers since the time of Navier. The Koenen method was later interpreted by Völker in such a way that the neutral axis was positioned at half the depth of the effective cross-section [Völker, 1908, p. 232]; in that case the factor 16/3 had to be replaced by 24/5 in eq. 9-3, and in eq. 9-3 and eq. 9-4 the slab thickness $d$ by the distance $h$ from the centre of the tension reinforcement to the edge in compression.

Koenen derived design equations for reinforced parabolic and circular jack arches subjected to a complete and a one-sided uniformly distributed load, for domical vaults, for cylindrical pipes subjected to internal and external pressure, and for free-standing cylindrical water tanks (see Fig. 9-32). He always expresses the thickness $d$ of the reinforced concrete loadbearing member and the steel cross-section $A_s$ by way of the permissible concrete compressive stress $\sigma_{b,perm}$ and the permissible steel tensile stress $\sigma_{s,perm}$. The reinforced concrete loadbearing members widespread 100 years ago were therefore accessible via a calculation from elementary theory of structures.

The *Monier-Broschüre* contains tables for slab thicknesses $d$ for simply supported strip of slab of width $b = 1.00$ m with reinforcement on one side and subjected to a uniformly distributed load $p$ [Wayss, 1887, pp. 68–72]. Koenen had evaluated his eq. 9-3 rearranged for $d$ to obtain $l$ and $p$; the steel cross-section $A_s$ is not specified – certainly due to patent-law considerations. Re-analyses using the slab thicknesses given in the tables have revealed that for a maximum bending moment of $M = 0.125 \cdot p \cdot l^2$, eq. 9-3 always supplies a constant concrete compressive stress $\sigma_b = 6.2$ MN/m$^2$. This value is not only much higher than the permissible concrete compressive stress of 2 MN/m$^2$ assumed by Koenen with a "safety factor exceeding 10" [Koenen, 1886, p. 462], but also higher than the $\sigma_{b,perm} = 4$ MN/m$^2$ which had been prescribed in 1904 for concrete compressive stresses due to bending by the Confederation of German Architects & Engineers Associations and the German Concrete Association (DBV) in their *Vorläufigen Leitsätzen für die Vorbereitung, Ausführung und Prüfung von Eisenbetonbauten* (provisional guidelines for the preparation, construction and checking of reinforced concrete structures). Besides the resulting considerably lower ultimate loads in the tests, the aim of the reduced slab thickness $d$ was to convince potential clients of the technical and economic superiority of the Monier slab compared to competing conventional suspended floor systems.

In order to demonstrate the efficiency of eq. 9-3 and eq. 9-4 and to compare this with newer design methods, Fig. 9-6 shows the steel cross-section $A_s$ according to the $k_h$-method [Grasser, 1987]. The values for $p$ and $\delta = d$ have been taken from the *Monier-Broschüre*. The table is based on concrete grade B 25 and steel grade BSt 420/500 according to DIN 1045.

**FIGURE 9-6**
Steel cross-section $A_s$ for a strip of slab b x l = 1.00 x 3.00 m from the *Monier-Broschüre* according to the $k_h$-method

| $p$ [kN/m] | $M$ [kNm] | $\delta = d$ [cm] | $h$ [cm] | $A_s$ [cm$^2$] |
|---|---|---|---|---|
| 2.0 | 2.25 | 4.4 | 3.1 | 3.53 |
| 4.0 | 4.50 | 6.2 | 4.9 | 4.32 |
| 6.0 | 6.75 | 7.6 | 6.3 | 5.01 |
| 8.0 | 9.00 | 8.8 | 7.5 | 5.58 |
| 10.0 | 11.25 | 9.9 | 8.6 | 6.06 |
| 12.0 | 13.50 | 10.8 | 9.5 | 6.58 |

A stress analysis is given for the strip of slab reinforced with the steel cross-section $A_s$ according to DIN 1045 from the *Monier-Broschüre* – in line with the historical design equations for pure bending. The concrete and steel stresses calculated are related to the permissible stresses for bending

$$\nu_b = \sigma_{b,exist} / \sigma_{b,perm} \tag{9-5}$$

and

$$\nu_s = \sigma_{s,exist} / \sigma_{s,perm} \tag{9-6}$$

Fig. 9-7 is based on permissible stresses of 4 MN/m² and 100 MN/m² respectively after Mörsch [Mörsch, 1906/2]. The other curves drawn in Fig. 9-7 – Koenen II, Koenen III, Ritter and Hennebique – are based on the following design methods:

- *Koenen II* corresponds to the design method proposed by Koenen in 1897 [Koenen, 1906, preface], which was absorbed into the official reinforced concrete codes of practice as the *n*-method (in this example $n = E_s / E_b = 15$).
- *Koenen III*, in contrast to Koenen I, assumes that the neutral axis is positioned at half the depth of the effective cross-section [Völker, 1908, p. 232].
- *Ritter* assumes a parabolic distribution of compressive stress [Ritter, 1899].
- *Hennebique*, like Koenen I, Koenen II and Koenen III, assumes a linear distribution of compressive stress, but infringes the equilibrium condition of the internal forces $D = Z$ in the reinforced concrete cross-section considered [Mörsch, 1901].

Further, Fig. 9-7 includes the values $\nu_b = 2.08$ and $\nu_s = 2.40$ as dotted lines; these result from the $k_h$-method when $\sigma_{b,exist}$ is replaced by $\beta_R / 2.1$ (B 25) in eq. 9-5 and $\sigma_{s,exist}$ by $\beta_s / 1.75$ (BSt 420/500) in eq. 9-6. In comparison to Koenen II (the *n*-method valid in the Federal Republic of Germany until 1971) the values for the concrete compressive stresses in Koenen I, Ritter and Koenen III are too low (Fig. 9-7a). The permissible stress 240 MN/m² from Koenen II, Koenen III and Ritter is exceeded by less than 10 % for all *M*- and *d*-values from Fig. 9-6 (Fig. 9-7b). Koenen I supplies reliable steel tensile stresses for $M > 8$ kNm, i.e. slab thickness $d > 8$ cm, which deviate only marginally from those obtained using Koenen II, i.e. the values calculated according to the *n*-method (Fig. 9-7b).

The basis of Koenen's design method was backed up in 1886 by numerous loading and fire tests; together with fire tests, tests on the adhesive bond between steel and concrete plus impact tests, the Berlin loading tests of 23 February 1886 represent the most comprehensive series of tests and thoroughly convinced the profession of the value of the new form of construction. Wayss arranged for a one-sided uniformly distributed load to be added in stages to reinforced concrete arches (reinforcement in one and two directions) and plain concrete arches with various spans and for

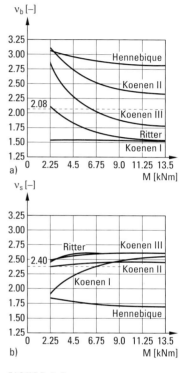

**FIGURE 9-7**
A comparison of historical design theories

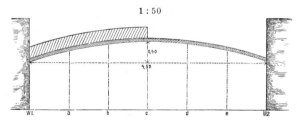

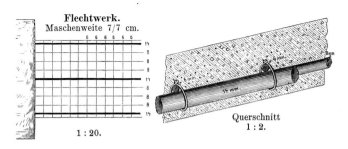

**FIGURE 9-8**
First test specimen of the Berlin loading tests of 23 February 1886 [Wayss, 1887, p. 37]

the associated vertical displacements to be measured at the sixth-points (Fig. 9-8). Furthermore, such loading tests were carried out on continuous strips of slab, strips of slab on two supports, pipes, an ultra-thin free-standing wall and an elliptical arch for a staircase.

The relationship between technical testing and engineering science theory formation was staked out by the publication of the *Monier-Broschüre* and civil and structural engineers only became aware of this very gradually. The recognition of this objective relationship and its practical implementation by Koenen, Mörsch and other engineers became not only a quality feature of the development of reinforced concrete construction, but also an important condition for the industrialisation [Becker, 1930] and scientific basis of solid construction.

## 9.2 Reinforced concrete revolutionises the building industry

The division of work between steel and concrete for the loadbearing action of reinforced concrete components is a metaphor for the industrialisation of the building industry. Both the production of steel and the production of cement for making concrete had already been organised on an industrial scale in the final decades of the 19th century. What this signified from the point of view of the steel and cement industries was that,
– in terms of science, the trend towards giving industrial practice a scientific footing;
– in terms of administration, the trend towards administration in industry's trade association policies (Fig. 9-9a).

For example, the purpose behind the Association of German Portland Cement Manufacturers, founded in 1877, was "to clarify all the technical and scientific issues important for the cement industry in a cooperative venture" [Becker, 1930, p. 9].

Together with the German Association of Brick, Clayware, Lime & Cement Manufacturers, founded in 1876, which was working on drawing up a uniform testing method and preparing specifications for the requirements to be placed on the quality of the cement, and in conjunction with architectural associations, the Berlin building market and the brickmaking industry, the first Prussian standards for testing Portland cement were published in 1878, which immediately became mandatory for all government buildings. The mouthpieces of cement research were the *Notizblatt des Deutschen Vereins für Fabrikation von Ziegeln, Tonwaren, Kalk und Zement*, founded in 1865, and the *Tonindustrie-Zeitung*, founded in 1876 by the editors of the *Notizblatt*, Hermann Seger and Julius Arons [Becker, 1930, p. 84]. The weekly *Tonindustrie-Zeitung* covered the sector of science-related trade association policies in the triad of industry, administration and science from the viewpoint of the clay industry (Fig. 9-9a). All articles in this journal concerned principally cement and cement mortar. By contrast, papers on concrete as a building material were conspicuous through their absence even in the first years of the 20th century, despite the fact that the German Concrete Association, founded in 1898, had selected this journal as its mouthpiece.

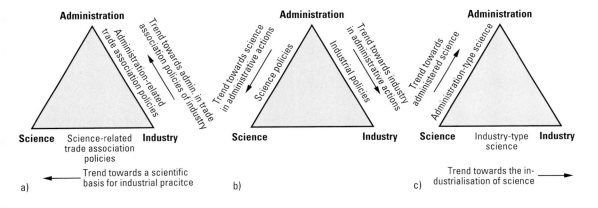

FIGURE 9-9
The triad of industry, administration and science from the point of view of a) industry, b) administration, and c) science

The model of the industry-administration-science triad [Kurrer, 1997/1] renders possible an understanding of the logical dimension in the historical development from the perspective of the ironmaking industry. It is worth considering for a moment the foundation of the Association of German Iron Foundries (VdEh) in 1860, and the journal *Zeitschrift für Eisenforschung* published by this technical-scientific body, which is still published today under the name *Stahl und Eisen*, and their participation in the discussion surrounding the standardisation of rolled iron sections. As the cement and steel industries were given a scientific basis and the administration-related trade association policies of these sectors of industry started to take shape, this created the foundation for the expansion in materials testing which had started to make headway shortly after the founding of the Second Reich. It was only now that the widespread tradition of promoting the trade and industry of the German federal states, in place since 1820, was abandoned. From now on, the relevant state authorities pursued science policies on the one hand and industrial policies on the other. The players in the industry-administration-science triad from the standpoint of administration (Fig. 9-9b) were, for example, the specialist ministries with their building departments, e. g. the Prussian Ministry of Public Works. Administration activities moving in the direction of science manifested themselves in the substantial role of the state in the foundation of materials testing institutes, e. g. the Royal Testing Institute for Building Materials founded in 1875 as part of the Berlin Trade & Industry Academy. On the other hand, the building departments operated industrial policies with a multi-facetted system of building codes and thus characterised the trend towards the building industry in administrative actions. The *Zentralblatt der Bauverwaltung*, founded in 1881 at the instigation of the publisher Wilhelm Ernst and published by Ernst & Korn (Wilhelm Ernst & Sohn after 1891) on behalf of the Prussian Ministry of Public Works, was the expression of this trend in trade journal terms [Wilhelm Ernst & Sohn, 1967].

With the fulfilment of demands by the building authorities for scientifically based methods of analysis for reinforced concrete members, theory of structures changed into an administration form, i. e. the trend towards an administered science is unmistakable. The publication of Koenen's design method for Monier slabs in the *Zentralblatt der Bauverwaltung* was the launch pad for this development. Nevertheless, the industry-administration-science triad was not able to unfold from the scientific perspective (Fig. 9-9c) until the industrialisation of the building industry – induced by reinforced concrete construction – after 1900. Reinforced concrete was therefore not only a technical revolution, but also the nucleus of the second Industrial Revolution in building, which led to the emergence of the modern building industry.

## 9.2.1
**The fate of the Monier system**

Shortly after the publication of the *Monier-Broschüre*, Koenen resigned from the civil service and took a job as engineering manager with the Berlin-based building contractor G. A. Wayss. He thus integrated two views of the industry-administration-science triad into his career: that of industry (Fig. 9-9a) and that of administration (Fig. 9-9b). However, the trend towards providing a scientific basis for the industrial practice of the triad from the industrialist's perspective (Fig. 9-9a) and the trend towards the industrialisation of the science of the triad from the scientist's perspective (Fig. 9-9c) were subjective and confined to the individual fields of application of the new composite material.

In January 1887 the licensee of the Monier patent in Austria, Rudolf Schuster, ensured through the brochure entitled *Bauten und Konstruktionen aus Zement und Eisen nach dem patentierten System J. Monier* (structures and constructions made from cement and iron according to the system patented by J. Monier) that the tests described in the *Monier-Broschüre* would gain more attention. Furthermore, Schuster described his own projects using the Monier method of construction plus older French structures built by Monier (see [Förster, 1921, p. 25]).

The theoretical part of Schuster's *Broschüre* was revised in 1902 by the government's chief building officer Emil Mörsch with the help of Anton Spitzer (1856–1922), the director of G. A. Wayss & Co. in Vienna.

Wayss purchased the Austrian Monier patent from Schuster, founded a company in Vienna and employed Schuster as its director. (This may have been the reason for the Viennese edition of the *Monier-Broschüre*.) After Wayss had converted his Berlin company into a limited partnership with the name G. A. Wayss & Co., it then became the public company with the name A.-G. für Monierbauten (later Beton- & Monierbau AG), and Wayss remained its sole director for a further two years before he was succeeded by Koenen, who was to lead the company for more than 20 years. While Wayss was still a director, A.-G. für Monierbauten published *Die Monierbauwerke, Album* (an album of Monier structures), which by 1895 needed a second edition (see [Förster, 1921, p. 27]). In 1892 Wayss reached an agreement with Conrad Freytag, the owner of Freytag & Heidschuch, which held the Monier licence for southern Germany, and founded the company Wayss & Freytag Neustadt a. d. Haardt. During the heyday of reinforced concrete construction in Germany, this company, with its engineering director Emil Mörsch, would play a leading role in the science and technology of reinforced concrete [Förster, 1921, p. 28].

Max Förster points out that the military was interested in the Monier form of construction from an early date, a fact that also found its way into the literature [Förster, 1921, p. 34]. Despite this, before 1900 there are only occasional papers on reinforced concrete construction in German engineering journals – apart from Koenen's design method for Monier slabs published in the *Zentralblatt der Bauverwaltung* in 1886 [Koenen, 1886, p. 462]. Worth mentioning here are the papers of the Austrian researchers Paul Neumann (1890), Josef Melan (1890), Max von Thullie (1896, 1897) and Joseph Anton Spitzer (1896, 1898), which appeared in the journals of the Austrian Engineers & Architects Association, or Ritter's critical appraisal of the Hennebique design concept published in 1899 in the *Schweizerische Bauzeitung* [Ritter, 1899].

On the materials testing side too, reinforced concrete became an object of engineering science research – beginning with Bauschinger's investigations of 1887. The research work of Bauschinger, Föppl, Bach and Schüle make up the industry-administration-science triad from the viewpoint of materials testing science (Fig. 9-9c). Writings on reinforced concrete that satisfied technical-scientific criteria first appeared with the integration of the three triads in reinforced concrete construction. Nevertheless, the dawn of a technical revolution initiated by reinforced concrete did appear around 1900, which in historico-logical terms would bring about the second Industrial Revolution in the building industry.

### 9.2.2
### The end of the system period: steel reinforcement + concrete = reinforced concrete

After the expiry of the German Monier patent in 1894, reinforced concrete construction experienced its system period. In 1903 Emil Mörsch remarked: "Reinforced floor constructions have reached such a diversity that the number of systems cannot be listed; there are almost 300 of them and nearly every week sees a new one, which in most cases does not represent any improvement" (cited in [Becker, 1930, p. 18]). Only a few patented systems and registered designs found their way into everyday building: Melan's concrete arch reinforced with steel sections, the ancestor of composite steel-concrete construction [Wapenhans, 1992], the truss-like, factory-prefabricated Visintini beam, Möller's T-beam, the Monier system and Hennebique's T-beam system.

**FIGURE 9-10**
Overview of design methods for reinforced concrete beams [Völker, 1908, p. 232]

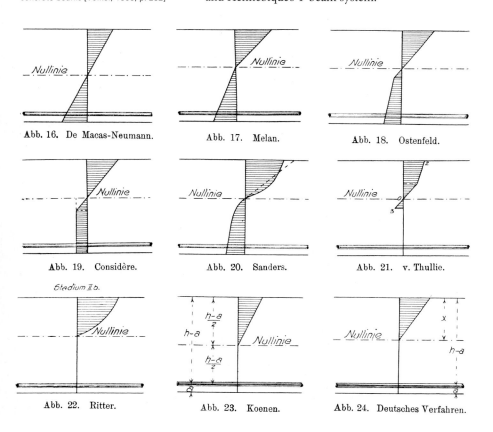

514

And on the design side too, various approaches for determining the stress distribution in the cross-section of a reinforced concrete beam co-existed for a few years after 1900, e. g.
- Paul Neumann [Neumann, 1890]
- Josef Melan [Melan, 1890]
- Edmond Coignet and Napoleon de Tedesco [Coignet & Tedesco, 1894]
- Maximilian Ritter von Thullie [Thullie, 1896, 1897, 1902]
- Asgar Ostenfeld [Ostenfeld, 1898]
- Adrian Sanders [Sanders, 1898]
- Josef Anton Spitzer [Spitzer, 1896, 1898]
- Armand Considère [Considère, 1898/99]
- Wilhelm Ritter [Ritter, 1899]
- Paul Christophe [Christophe, 1899, 1902, 1905]
- Matthias Koenen [Koenen, 1902]
- Emil Mörsch (see [Wayss, 1902; Mörsch, 1903]).

Fig. 9-10 shows the most important assumptions concerning the stress distribution for designing reinforced concrete beams and slabs over the period 1886 to 1904.

Völker's overview does not include the design concept of the Belgian engineer Paul Christophe. The "Deutsches Verfahren" (German method, see Abb. 24 in Fig. 9-10), which was introduced first in Prussia in 1904 and later the other German federal states, is attributed to him. Christophe's method of 1899 marks the beginning of the end of the system period in reinforced concrete construction on the design theory side, a period that was terminated on both the theoretical and practical levels by Wayss & Freytag and its engineering director Emil Mörsch in the first decade of the 20th century.

### 9.2.2.1 The Napoleon of reinforced concrete: François Hennebique

François Hennebique (1842–1921) was trained as a stonemason (for his biography see [Delhumeau, 1999]) and did not think much of the composite construction consisting of reinforced slabs, iron beams and cast-iron columns (see Fig. 9-5). During the 1890s he first joined the beam to the slab to form the T-beam and finally pushed the loadbearing structure synthesis to the limit with his monolithic reinforced concrete frame (Fig. 9-11). Although factory buildings were dominated by structural steelwork, the monolithic reinforced concrete frame proved to be a serious competitor for multi-storey industrial buildings and a little later for industrial sheds as well – initially in the textiles industry and the building of warehouses, then in the paper-processing industry and finally also in the automotive industry, which had started to emerge shortly after 1900. Like wrought iron supplanted cast iron and timber in trusses after 1850, so the new composite material could not tolerate the coexistence of other building materials. From his Paris headquarters on the rue Danton, Hennebique built up an international network of agents and licensees; and he separated design from construction (see also [Gubler, 2004, p. 30]). For a licence fee, which could amount to 10 % or even more of the building

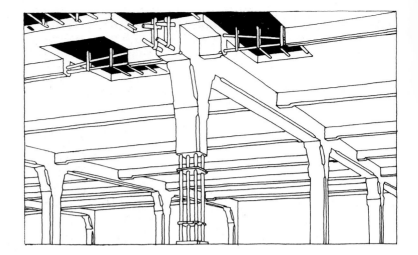

**FIGURE 9-11**
Monolithic reinforced concrete frame after Hennebique
[Christophe, 1902, p. 106]

sum, he could supply sets of drawings to the licensed contractors. He thus helped to spread reinforced concrete construction internationally. His company journal *Le Béton Armé*, founded in 1898, ensured that Hennebique not only maintained communications between his licensees, but also created a powerful piece of technical marketing for his system.

Like structural steelwork was able to celebrate its triumph at the Paris World Exposition of 1889 with the Eiffel Tower and the "Galerie des Machines", the reinforced concrete of François Hennebique was able to attract the attention of millions of visitors in the shadow of the icon of the World Exposition and the landmark of Paris some 11 years later.

The social theorist Walter Benjamin has called World Expositions "places of pilgrimage for the goods fetish" [Benjamin, 1983, p. 50]. In the self-portrayal of middle-class society by way of this pompous department store, in a world reduced to goods organised like an encyclopaedia, the fetish character of the goods takes on religious forms.

The architects knew very well how to aestheticise the "goods fetish" at the Paris World Exposition of 1900: they concealed Hennebique's monolithic reinforced concrete frame behind the historical "clothing" of the "Palais du Costume", the "Palais des Lettres, Sciences et Art" and the Belgian exhibition hall – an imitation of the mediaeval town hall of Oudenaarde in Belgium. And Hennebique? He, too, was hiding something. Hennebique protected the design principles of his reinforced concrete system as though they were the apple of his eye; he treated them like a professional secret and made sure that they never left the drawing offices (Fig. 9-12) of his Paris headquarters [Huberti, 1964, p. 120].

The Austrian Engineers & Architects Association employed Fritz von Emperger to report from the World Exposition in Paris about the exhibits in reinforced concrete [Emperger, 1901/1, 1901/2, 1902/2]; supplemented by three other reports [Emperger, 1902/1, 1902/3, 1902/4], they formed the nucleus for a new journal *Beton und Eisen* (now *Beton- und Stahlbetonbau*), which from 1905, the fourth year of its publication, was

FIGURE 9-12
View of one of the drawing offices at Hennebique's headquarters in 1912 [Simonnet, 2005, p. 67]

published by Wilhelm Ernst & Sohn. It can be regarded as the first technical-scientific journal for reinforced concrete.

After Hennebique had secured a decent market share in France and Belgium, he set up branch offices in all the other countries of Europe. Between 1892 and 1899, a total of 3000 structures were built using his reinforced concrete system [Christophe, 1902, p. 6], and a further 2500 were added in the next two years [Christophe, 1905, p. 5]!

So what constituted the content of the French Revolution in solid construction?

- In contrast to composite construction, the entire loadbearing system of the structure was, so to speak, monolithic, turned into one single body of goods.
- The practical value of this body of goods is functionalised for the purpose of its exchange value.
- The separation of monopolistic planning and pluralistic execution appeared as an element of this functionalisation.
- The contradiction between monopolistic planning and pluralistic execution terminated the system period in reinforced concrete construction and established a building industry based on a standard, unpublished loadbearing system design.

Like Napoleon established the products of the French revolution throughout continental Europe and in doing so swept aside the many duodecimo princedoms, Hennebique marketed his reinforced concrete system throughout Europe as *the* form of reinforced concrete construction, sweeping aside other reinforced concrete systems as he went. So we can call François Hennebique the Napoleon of reinforced concrete. Nevertheless, the history of a fully evolved middle-class society requires not only a political, but also a scientific and an industrial revolution. And reinforced concrete construction was no exception: in the historical process of its establishment as a universal form of construction, the scientific and industrial components appear alongside the administrative.

**The founding father of rationalism in reinforced concrete: Paul Christophe**

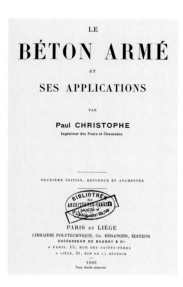

FIGURE 9-13
Cover of Paul Christophe's monograph *Le béton armé et ses applications* [Christophe, 1902]

### 9.2.2.2

Analyses of the early design methods for reinforced concrete structures in bending have been carried out by, for example, Alfred Pauser [Pauser, 1994, pp. 50 – 53] and – in great detail – by Thomas Jürges [Jürges, 2000, pp. 31 – 62]. Both authors highlight the contribution of the Belgian building civil servant Paul Christophe, who in 1899 was sent by his government to attend the International Congress on Reinforced Concrete in Paris, which Hennebique had organised on the occasion of the building works for the Paris World Exposition (1900). His careful studies prior to the congress enabled Christophe to publish an extensive report – containing a design method and the latest scientific findings – about the experiences with reinforced concrete construction very soon after the congress. The report appeared as a three-part series in the journal *Annales des Travaux Publics de Belgique* [Christophe, 1899], and covered a total of 306 printed pages. Christophe used this as the basis for his 1902 monograph *Le béton armé et ses applications* (Fig. 9-13), a German translation of which appeared three years later [Christophe, 1905]. In his preface, Christophe writes: "The goodwill with which our audience received our treatises [by Christophe in 1899 – the author], the encouragements given to us in this matter by important technical and scientific authorities and, specifically, by the highest learned body in our country, the Royal Belgian Academy, made it our duty to continue with our task. All those who showed an interest in our initial work will hopefully consider this book as a pledge of our gratitude" [Christophe, 1902, p. III]. Christophe published the first summary of the whole field of reinforced concrete construction from the point of view of administration (Fig. 9-9b) and science (Fig. 9-9c). He always preserves an unbiased and objective view of detail and overview, and in doing so highlights the essentials. Christophe's entire book exudes the spirit of critical rationalism without becoming arrogant; this is especially evident in the analysis of reinforced concrete systems and design concepts.

His 756-page book is divided into five sections:

I. Principles and types of construction [Christophe, 1902, pp. 1 – 71]: historical review, principles, forms of construction
II. Types of application [Christophe, 1902, pp. 73 – 386]: buildings, bridges and jetties, roofs, vaults, retaining walls, revetments, cantilevers, loading platforms using trusses, foundations, sewers, tanks, waterproofing, miscellaneous
III. Construction [Christophe, 1902, pp. 387 – 463]: materials, workmanship
IV. Theory [Christophe, 1902, pp. 465 – 687]: results of tests on plain and reinforced concrete, theoretical studies of reinforced concrete, common methods and practical equations, useful arrangements
V. Advantages and disadvantages of reinforced concrete [Christophe, 1902, pp. 689 – 726]

The fifth section is followed by an extensive bibliography [Christophe, 1902, pp. 727 – 739] and is rounded off by an index of names plus a general index.

At this point only the fourth section concerning the theory of reinforced concrete will be explored. Christophe begins this section as follows: "A purposeful method of calculation must correspond with the reality as closely as possible. The assumptions serving as a basis for this must be drawn from real experiences. Before we write any kind of equation, we must investigate the properties of the two materials used for the production of reinforced concrete" [Christophe, 1902, p. 465]. Such tests on concrete and steel in the building materials laboratory are necessary, but not sufficient; for the essence of reinforced concrete lies in the bond between the steel and the concrete: "The laws and values determined by experience of the various properties of concrete and steel allow us to set up methods of calculation; but these are still dependent on as yet imperfect assumptions if we do not take the trouble to establish through tests how the strains in the steel are transferred to the concrete and the reciprocal effects of the two materials with respect to the stability of the disparate body. Accurate tests on reinforced concrete workpieces are therefore indispensable" [Christophe, 1902, p. 465]. The first large-scale tests on reinforced concrete loadbearing structures, e. g. beams, slabs, T-beams and columns, together with materials tests, permitted the development of a design theory: "Based on tests, the assumptions and equations that agree best with the reality should be sought in para. 3 [results of tests on reinforced concrete specimens – the author]. This paragraph contains the studies to assess the principal scientific methods of treatment proposed up to now and an explanation of the method of calculation we recommend" [Christophe, 1902, pp. 465 & 466]. At the same time, Christophe does not claim that his design method "solves the question of the stability of reinforced concrete exactly, but rather is intended merely for daily use in the calculations" [Christophe, 1902, p. 466].

After Christophe has summarised the key findings of the research work in four hypotheses [Christophe, 1902, pp. 512–517], he pays critical acclaim to the design concepts of Neumann, Melan, Coignet and De Tedesco, Von Thullie, Ostenfeld, Sanders, Ritter, Considère and Haberkalt [Christophe, 1902, pp. 517–529]. This is followed by further research findings concerning shrinkage, for instance. Only now does he formulate the assumptions on which his design method is based [Christophe, 1902, p. 535]:

*Firstly*: The interaction of the concrete and the steel with the restriction that the embedded steel be arranged taking into account adequate consistency of the reinforced concrete (full composite action).

*Secondly*: All sections remain plane (Bernoulli hypothesis).

*Thirdly*: A constant concrete elastic modulus for compressive strains in the serviceability state (Hooke's law for concrete).

*Fourthly*: No tensile stresses in the concrete.

*Fifthly*: Initial stresses, e. g. due to temperature and shrinkage, are neglected.

Christophe published his reinforced concrete theory as early as 1899, as the last and longest part of his three-part series [Christophe, 1899, pp. 961–1124]. It is demonstrated below using the example of his design for a reinforced concrete beam of width $e$ and depth $h$ subjected to a bending moment $M$ (pure bending) and containing reinforcement top and bottom. Fig. 9-14 illustrates the reinforced concrete beam with the neutral axis $F-N$, the triangular concrete compressive stress block $AOA''$ and the linear strain $A'D'$ over the cross-section. The cross-sectional area of the bottom reinforcement in tension is $A_s$ and that of the top reinforcement in compression is $A'_s$.

From equilibrium of the forces in the horizontal direction it follows that

$$\sigma_s \cdot A_s = \frac{1}{2} \cdot \sigma_b \cdot a \cdot e + \sigma'_s \cdot A'_s \tag{9-7}$$

and from the moment equilibrium

$$M = \frac{2}{3} \cdot a \cdot \frac{1}{2} \cdot \sigma_b \cdot a \cdot e + \sigma'_s \cdot A'_s \cdot b_1 + \sigma_s \cdot A_s \cdot b \tag{9-8}$$

where $\sigma_b$ is the concrete compressive stress in the upper extreme fibres (line $AA''$), $\sigma_s$ is the tensile stress in the bottom reinforcement, and $\sigma'_s$ is the compressive stress in the top reinforcement. From the Bernoulli hypothesis and with the help of the intercept theorem we get the relationships

$$\sigma_s = \sigma_b \cdot n \cdot \frac{b}{a} \quad \text{and} \quad \sigma'_s = \sigma_b \cdot n \cdot \frac{b_1}{a} \tag{9-9}$$

where $n$ is the ratio of the elastic modulus of steel $E_s$ to that of concrete $E_b$

$$n = E_s/E_b \tag{9-10}$$

The two equations in 9-9 can be entered into the equilibrium equations 9-7 and 9-8 to give the following

$$\frac{1}{2} \cdot a^2 \cdot e + n \cdot (A'_s \cdot b_1 - A_s \cdot b) = 0 \tag{9-11}$$

and

$$M = \frac{\sigma_b}{a} \cdot \left[ \frac{1}{3} \cdot a^3 \cdot e + n \cdot (A'_s \cdot b_1^2 + A_s \cdot b^2) \right] \tag{9-12}$$

Eq. 9-12 can also be rearranged in the form of the well-known bending stress equation

$$\sigma_b = M \cdot \frac{a}{I} \tag{9-13}$$

where the second moment of the area of the reinforced concrete cross-section is

$$I = \left[ \frac{1}{3} \cdot a^3 \cdot e + n \cdot (A'_s \cdot b_1^2 + A_s \cdot b^2) \right] \tag{9-14}$$

Entering bending stress equation 9-13 into eq. 9-9 produces

$$\sigma_s = n \cdot M \cdot \frac{b}{I} \quad \text{and} \quad \sigma'_s = n \cdot M \cdot \frac{b_1}{I} \tag{9-15}$$

If the geometric relationships

$$b = h' - a \tag{9-16}$$

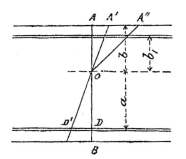
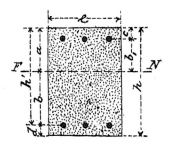

**FIGURE 9-14**
Design of a reinforced concrete beam after Christophe
[Christophe, 1899, pp. 993, 1034]

and

$$b_1 = a - c \qquad (9\text{-}17)$$

are entered into eq. 9-11, then we get the quadratic equation for the position of the neutral fibres $a$

$$\frac{1}{2} \cdot a^2 \cdot e + n \cdot (A'_s + A_s) \cdot a - n \cdot (A'_s \cdot c + A_s \cdot h') = 0 \qquad (9\text{-}18)$$

the solution of which is

$$a = \frac{n \cdot (A'_s + A_s)}{e} + \sqrt{\left[\frac{n^2 \cdot (A'_s + A_s)^2}{e^2}\right] + \frac{2 \cdot n}{e} \cdot (A'_s \cdot c + A_s \cdot h')} \qquad (9\text{-}19)$$

So the position of the neutral fibres depends merely on the steel cross-sections and the ratio $n$ between the elastic moduli of steel and concrete. Using eq. 9-19, eq. 9-16, eq. 9-17 and eq. 9-14, the steel stresses $\sigma_s$ and $\sigma'_s$ for a given bending moment $M$ can be calculated from eq. 9-15 and the concrete compressive stress $\sigma_b$ from 9-13. And vice versa: a permissible bending moment for the given cross-sections when the permissible steel and concrete stresses are entered into the corresponding equations.

In addition, Christophe analyses the shear and principal stresses in reinforced concrete beams with a rectangular cross-section, and T-beams. He specifies an equation for checking the stress in the shear reinforcement of a reinforced concrete beam with a rectangular cross-section. His three-part series of papers and his monograph contain design cases for a

- strut not at risk of buckling,
- tie,
- beam in bending with heavy reinforcement,
- T-beam (Fig. 9-15),
- inverted T-beam, and
- beam in bending with axial force for light, moderate or heavy reinforcement.

Therefore, in 1899 Christophe had already created the first comprehensive reinforced concrete theory for all the design cases relevant to practice, which would soon become established internationally as the standard method. Christophe's standard method is often designated the "German method" (see Abb. 24 in Fig. 9-10), which formed the basis of the reinforced concrete standard DIN 1045 in the Federal Republic of Germany up until 1971. In this context, the "German method" was mostly identi-

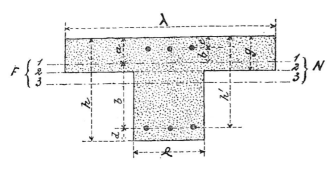

**FIGURE 9-15**
Design of T-beams distinguished according to the position of the neutral fibres after Christophe [Christophe, 1899, p. 1072]

fied with the design method specified in the book published by Wayss & Freytag, *Der Betoneisenbau. Seine Anwendung und Theorie* (reinforced concrete, its application and theory) [Wayss & Freytag AG, 1902], the theoretical part of which was the work of Emil Mörsch. But the "German method" is in reality the work of the Belgian engineer Paul Christophe! We have to thank Thomas Jürges for ascertaining the outstanding role of Christophe in the historical development of reinforced concrete design theory [Jürges, 2000, pp. 56–59]. Christophe himself writes in the German edition of his fundamental work on reinforced concrete that M. Koenen (director of A.-G. für Beton- & Monierbau, Berlin), E. Mörsch (board member of the engineering office of Wayss & Freytag A.-G., Neustadt a. d. Haardt) and the architect E. Turley (Düsseldorf Building Authority) were instrumental in getting his design theory included in the *Vorläufigen Leitsätzen…* (1904) [Christophe, 1905, p. 149] – with success, as Mörsch confirmed in 1906: "In the first edition of the work of Christophe, *Annales des Travaux publics de Belgique*, 1899, the theory today contained in the *Leitsätze* is completely discussed, including the provision that the concrete can withstand no tension" [Mörsch, 1906/2, p. 149]. In this context, Mörsch refers to the publication by Autenrieth in which he uses a graphical method to determine the forces in anchors used to fix plates to flat surfaces, e. g. foundations [Autenrieth, 1887/1] (see also section 7.2.3.5): "The methods there set forth are identical with those herein employed for the calculation of reinforced concrete work, so that the processes indicated for simple bending and for flexure with axial pressure have been carried over, without change, to reinforced concrete work" [Mörsch, 1906/2, p. 149]. Mörsch adapted Autenrieth's graphical method for the design of complex reinforced concrete sections [Mörsch, 1906/2, pp. 116–120]. Despite the existence of this graphical method interesting for the design theory of reinforced concrete, it was neither Autenrieth nor Mörsch, but rather Christophe who in 1899 eliminated the multiplicity of theories for the design of reinforced concrete sections and therefore ended the system period in reinforced concrete construction on the theoretical side. However, the consumption of the industry (Fig. 9-9a), administration (Fig. 9-9b) and science (Fig. 9-9c) triad still required an essential social factor: the building industry. Only when this started to grow, which cannot be understood as a technical revolution without the close interaction with reinforced concrete, did the second Industrial Revolution in the building industry become a reality.

**The consummation of the triad**

#### 9.2.2.3

A carnival joke of the Munich Architects & Engineers Association that did the rounds in the winter of 1902 poked fun at the Hennebique system by calling it the *Hennepeck* (= pecking hen) system; the moving model showed a hen pecking at sand and cement and laying golden eggs! The résumé was that "the patent claim of the clever Frenchman still forces us to buy the individual answers from Paris, but this matter undoubtedly has a great future and German thoroughness and science will bring us the associated theory very soon and allow us to overtake the French" (see

| Year | Turnover (million gold mark) | Total payroll (million gold mark) | Waged employees | Salaried employees |
|---|---|---|---|---|
| 1893 | 0.90 | 0.26 | 300 | 25 |
| 1900 | 2.55 | 0.84 | 1100 | 55 |
| 1905 | 5.50 | 1.70 | 2500 | 95 |
| 1910 | 25.50 | 6.20 | 8000[1] | 400[2] |
| 1915 | 32.80 | 7.60 | 9500[1] | 475[2] |
| 1917 | 51.50 | 10.00 | 12500[1] | 625[2] |

[1] Extrapolated and rounded down from the average payroll of a worker in 1900 and 1905.
[2] Estimated from the ratio of waged workers to salaried workers in 1900.

**FIGURE 9-16**
Principal facts and figures of the Wayss & Freytag company
[Wayss & Freytag AG, 1925, pp. 24, 36, 67]

[Huberti, 1964, p. 126]). Although Belgian thoroughness and science had already provided the associated theory three years before, it was the Germans who brought together science and industry in reinforced concrete construction. The Achilles heel of the Hennebique system lay in separating design and construction, which barred the way to the creation of a standard design method and the standardisation of the theoretical work of the reinforced concrete engineer. Wayss & Freytag in Neustadt a. d. Haardt was showing the way forward here. This company, which with the help of the Pfälzische Bank became a public company in 1900, is an excellent example of the way the industrial realisation of reinforced concrete technology was given a scientific footing. As the negotiations between Hennebique and Wayss & Freytag failed owing to the high licence fees, the building contractor, which had developed into an industrial enterprise, turned independently to the subject of the monolithic reinforced concrete frame. Fig. 9-16 provides an overview of how reinforced concrete construction helped Wayss & Freytag to rise to the level of a large industrial company – an example of what the building industry would become during the second Industrial Revolution in building.

An essential condition for the emergence of a large building industry was the establishment of the engineering offices. For example, the engineering office of Wayss & Freytag quickly became a first-class technical-scientific competence centre for reinforced concrete. In early 1901 Conrad Freytag was able to recruit the Württemberg government's chief building officer Emil Mörsch (just 28 years old at the time) to take charge of the Wayss & Freytag engineering office. It was in that same year that Mörsch published his *Theorie der Betoneisenkonstruktionen* (theory of reinforced concrete) [Mörsch, 1901], which one year later appeared in an expanded edition under the title of *Der Betoneisenbau, seine Anwendung und Theorie* (reinforced concrete, its application and theory) [Wayss & Freytag AG, 1902] (Fig. 9-17). The book summarises the most important test results of the materials testing institutes regarding reinforced concrete, which in the main were the work of the Materials Testing Institute of Stuttgart TH under the leadership of Carl Bach (sponsored by Wayss & Freytag). This not only opened up a breach in the business technique of Hennebique, but also initiated the cooperation between engineer-

**FIGURE 9-17**
Cover of the first German standard work on reinforced concrete
[Wayss & Freytag AG, 1902]

ing science activities and technical developments so typical of reinforced concrete in Germany. The design theory attributed to Christophe was extended by Mörsch in the later editions of his book – especially with respect to shear design ([Mörsch, 1906/2, 1907, 1908; DBV, 1907]).

The newly built five-storey "Hotel zum Bären" in Basel, erected by a Hennebique licensee, collapsed in 1901. Shortcomings in planning, design, structural analysis, workmanship and quality assurance led to this spectacular failure and several deaths. One year later, Fritz von Emperger analysed the causes of the collapse in detail and came to the conclusion that he had uncovered the darkest side of Hennebique's business methods [Emperger, 1902/3]. Thereafter, safety of structures was placed at the top of the agenda of the emerging international community of reinforced concrete engineers.

On 25 May 1902 the Central Committee of the Swiss Engineers & Architects Association (SIA) invited its various association sections to prepare "provisional rules" [SIA, 1994, p. 80] concerning the construction of reinforced concrete structures. Following internal and public discussions, the SIA published the first reinforced concrete standard in August 1903 (Fig. 9-18), which included an explanatory report by Prof. Schüle, the director of the Swiss Federal Materials Testing Laboratory in Zurich. The 16 articles of this standard regulate

- general matters (scope, checking of construction documents),
- principles of structural calculations (loading assumptions, $n = E_s/E_b = 20$, $\sigma_{b,d,perm} = 350$ N/cm$^2$, $\tau_{b,perm} = 40$ N/cm$^2$, $\sigma_{s,z,perm} = 13\,000 - 5 \cdot \sigma_{b,z,perm}$ N/cm$^2$),
- materials (mild steel with the properties prescribed in the decree of 19 August 1892 for the calculation and checking of steel bridges and roof constructions [SIA, 1994, pp. 13–22], min. 300 kg/m$^3$ standardised Portland cement, minimum cube compressive strength after 28 days $\sigma_{b,28,min} = 1600$ N/cm$^2$),
- workmanship (striking times, skilled personnel), and
- inspection and handover of structures (inspections by site manager, records after striking formwork, loading tests).

The standard (excluding the explanatory report) covers eight pages and ends with a provision regarding exceptions: "In order to take into account the newness of this type of construction, deviations from the foregoing standards are permissible provided they are founded on detailed tests and the judgments of competent persons" [SIA, 1994, p. 80]. This internal division of the first reinforced concrete standard would become the model for reinforced concrete standards for many years.

In Germany the standards debate began at the 4th general assembly of the German Concrete Association (DBV) held on 1/2 March 1901 in Berlin and reached its first conclusion with the *Vorläufigen Leitsätze* …, a joint publication of the German Architects & Engineers Association (VDAIV) and the DBV which appeared on 4 June 1904. One highlight in this debate was the presentation by Mörsch entitled *Theorie der Betoneisenkonstruktionen* (reinforced concrete theory) at the 6th general

**FIGURE 9-18**
Cover of the first reinforced concrete standard [SIA, 1994]

assembly of the DBV on 21 February 1903 [DBV, 1903, pp. 105–128]. The presentation outlined his self-contained design theory, backed up by tests and applications, which proved popular among the delegates owing to its clarity, elegance and straightforwardness. In terms of content it corresponded to the design theory already developed by Christophe. But new was the fact that now the results of a series of tests sponsored by a company were incorporated directly into the foundation of the design theory; whether or not Mörsch was already aware of Christophe's design theory at this time is less important here. Therefore, Mörsch's direct development of a design model from test data for the design of reinforced concrete beams in bending will be shown here as an example. Starting with the exponential law of the stress-strain relationship of concrete obtained from Bach's compression tests of 1897

$$\sigma_{b,d}^m = \varepsilon_b \cdot E_b \tag{9-20}$$

Mörsch arrives at an almost linear compressive stress distribution where $m = 1.15$ (from tests) (Fig. 9-19). This is where he criticises existing design models (see Fig. 9-10), the assumptions of which he feels complicate the calculations unnecessarily: "It is predictable that such assumptions [e.g. parabolic compressive stress distribution, consideration of the tensile stress of concrete – the author] produce expressions, the lengths of which are considered by their authors as a particular feature of their accuracy and reliability. But these long equations do not entice the designer. Added to this is the fact that replacing the strain curve by a parabola is less accurate than replacing it with a straight line because for the exponential law $m$ is much closer to 1 than 2 and one must attack the strain curves violently if one wishes to force them into a parabola" [DBV, 1903, p. 116]. The linearisiering of the compressive stress distribution is sufficient for Mörsch's design model (Fig. 9-20) because as he considered tests fundamental to engineering science modelling, the purpose of every structural calculation is to verify a sufficient factor of safety, and the accurate determination of the stress occurring in the construction for any particular loading is less important. Mörsch concluded his presentation with remarks concerning the standards issue: "In the interests of reinforced concrete construction, generally applicable and meaningful codes of practice are desirable because the competition between the individual companies active in the reinforced concrete market easily leads to the permissible limits not being achieved with the minimum amount of steel reinforcement … These codes of practice must also specify the method of calculation plus the permissible stresses and strains" [DBV, 1903, pp. 127–128]. The technical-scientific demand for a new type of standard had thus been expressed by a leading industrial entrepreneur initially for reinforced concrete. Mörsch's reinforced concrete theory, a sort of works standard employed with great success by Wayss & Freytag, would quickly advance to become the standard model for the first German industrial standard.

On 18 June 1903 the board of the Jubilee Foundation of German Industry, an industry-based institution for promoting engineering sciences

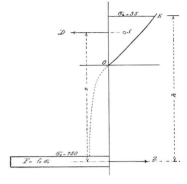

FIGURE 9-19
The compressive stress curve in a reinforced concrete beam determined in tests by Mörsch [DBV, 1903, p. 116]

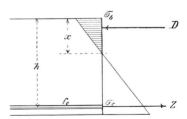

FIGURE 9-20
Christophe's design model in the version by Mörsch for reinforced concrete beams with bottom reinforcement only [DBV, 1903, p. 118]

formed on the occasion of the 100th anniversary of Berlin-Charlottenburg TH, agreed to set up a subcommittee for reinforced concrete chaired by the Stuttgart-based materials researcher Carl Bach; Freytag and Koenen were nominated by the DBV as members of this committee. The subcommittee decided on a research programme, which generalised the triad from the industrial perspective in the trend towards placing industrial practice on a scientific footing in the form of a science-related trade association policy (Fig. 9-9a) and from the science perspective in the trend towards industrialisation of science in the form of an industry-type science (Fig. 9-9c).

After the business technique had been abolished in reinforced concrete construction, many building contractors, in addition to the handful of specialist reinforced concrete companies, took their chance with the construction of reinforced concrete structures: "Accidents and failures were the unavoidable consequence … The provision of uniform codes of practice for the calculation and assessment of structures … was therefore … considered primarily as a survival strategy for the reinforced concrete sector" [Becker, 1930, p. 28] because the building authorities often refused to approve reinforced concrete structures. This was tantamount to a severe restriction of corporate freedom.

Therefore, the VDAIV, as the representative of the civil servants in the building authorities, and the DBV adopted the *Vorläufigen Leitsätze…* in joint consultations on 4 June 1904 and presented these to all the governments of the German federal states with a recommendation to implement them. With only minor amendments, Prussia had already published the *Bestimmungen für die Ausführung von Konstruktionen aus Eisenbeton im Hochbau* (provisions for the construction of reinforced concrete buildings) on 16 April 1904. The dominant theoretical basis was the design concept by Christophe, or rather Mörsch (see Fig. 9-20), who progressed to become professor for theory of structures, bridge-building and reinforced concrete at Zurich ETH in 1904.

The triad from the perspective of administration (Fig. 9-9b) thus expanded in the entire reinforced concrete industry to include the trend towards industry in administrative activities and in the industry perspective to include the trend towards administration in industry's trade association policies (Fig. 9-9a).

Mörsch's real success in reinforced concrete was due to the fact that in his professional activities he could realise the science-industry-administration triad as the integration of the industry (Fig. 9-9a), administration (Fig. 9-9b) and science perspectives (Fig. 9-9c) in an almost ideal way, and therefore the triad was completed not only according to objective, but also subjective aspects. Mörsch thus anticipated the modern engineering scientist of the 20th century, who advanced to become the sponsor of the incorporation processes in the triadic relationship between science, industry and administration.

But that is not all in the thoroughness of the joint technical-scientific work in Germany. In 1904 the board of the DBV turned to the Prussian

Ministry of Public Works with a request for funds for testing. The "Committee for the Execution of Concrete Tests and Tests on the Behaviour of Steel in Masonry" had drafted a schedule for reinforced concrete tests, the execution of which had been given to a separate working group, especially with regard to the updating of the existing Prussian reinforced concrete standards and the preparation of uniform codes of practice for the entire German Reich. Taking part in the consultations of this working group were all the ministries affected, railway authorities, the materials testing institutes of the federal states, the VDAIV, the Association of German Portland Cement Manufacturers and the DBV. In its meeting on 8/9 January 1907 the working group adopted the name of the German Committee for Reinforced Concrete. During the meeting, the Prussian Undersecretary of State Dr. Holle designated reinforced concrete as the "cause of a complete upheaval in the building sector" (cited in [DAfStb, 1982, p. 139]).

The German Committee for Reinforced Concrete (DAfStb) achieved German unification in reinforced concrete construction with its *Bestimmungen für Ausführung von Bauwerken aus Beton* (provisions for the execution of concrete structures) officially adopted by all German federal states in 1916 [DAfStb, 1982, p. 7].

The paradigmatic integration of the six fields of activity of science- and administration-related trade association policies (Fig. 9-9a), science and industry policies (Fig. 9-9b) plus administration- and industry-type science (Fig. 9-9c) in the German Committee for Reinforced Concrete completed the science-industry-administration triad on the objective side as well. It became the incorporative form in this sector and the prerequisite for the outstanding position of reinforced concrete in the disciplinary fabric of structural engineering. But more than that, it anticipated the basic pattern of institutional actions for other fields of joint technical-scientific work, as would be seen in the work of the German Standards Committee (later DIN) founded on 17 May 1917, for instance.

## 9.3 Theory of structures and reinforced concrete

Once reinforced concrete had already become a serious rival to structural steelwork by 1910, it started to equal the role played by structural steelwork in driving forward the formation of theory of structures in the middle of the accumulation phase of structural theory (1900–25). This is especially evident in the theory of elastic frameworks (originally based on structural steelwork), which reinforced concrete placed at the focus of theory of structures and structural calculations in the second decade of the 20th century. Only through the constructional and technological self-examination of reinforced concrete and its discovery by advocates of the classical modern movement in architecture during the 1920s was reinforced concrete able to establish itself through new types of loadbearing systems, the theory of plate and shell structures, during the invention phase of structural theory (1925–50). The theory became established in reinforced concrete with the works of Marcus (1919, 1924, 1925), Nádai (1925), Lewe (1926), Beyer (1927), Dischinger (1928, 1929), Craemer (1929, 1930) and Ehlers (1930), was continued in the shell theory with

Flügge's monograph [Flügge, 1934] and reached an interim conclusion in the general theory of elastic shells developed on the basis of tensor calculus by A. E. Green and W. Zerna ([Zerna, 1949/1; Green & Zerna, 1950/2]). Since the end of World War 2, reinforced concrete construction has evolved into the hegemon of structural engineering. Up until the middle of the innovation phase of structural theory (1950–75), reinforced concrete contributed only moderately to theory of structures developments – through the theory of prestressed concrete – but placed considerably greater demands on structural calculations than ever before. Nevertheless, prestressed concrete was to prove a driving force behind modern materials research in reinforced concrete during the diffusion phase of structural theory (1975 to date).

### 9.3.1 New types of loadbearing structures in reinforced concrete

The loadbearing structure is the part of a structure that takes on the loadbearing functions necessary for securing the functions of the structure: a reinforced concrete beam, a reinforced concrete arch, a reinforced concrete plate, a reinforced concrete slab, a reinforced concrete shell, etc. According to Büttner and Hampe, a loadbearing system is an model of the loadbearing structure abstracted from the point of view of the loadbearing function [Büttner & Hampe, 1977, p. 10], e.g. a simply supported beam, a fixed-end arch, an elastic plate on two supports, an elastic slab simply supported on four sides, etc. A structural system is the loadbearing system refined for the purpose of the quantitative investigation by way of geometrical and material parameters, e.g. a fixed-end arch provided with numerical values for the strain and bending stiffness is a structural system. It is the multi-tiered, detailed analytical breakdown from the engineered artefact via the total structure made up of loadbearing structures to the loadbearing system and then the structural system and the reverse process to the synthesis that forms the essence of the art of structural engineering.

Every chain of structure – loadbearing system/structure – loadbearing system/structural system in engineering activities possesses a historical side in addition to this logical one. In the historical process, the structural engineer gradually discovers the internal logic of the structures he has invented – their "logic of form" (E. Torroja). The evolution of reinforced concrete bears witness to this.

Through loadbearing systems such as plates with out-of-plane loading (e.g. slabs), plates with in-plane loading, folded plates and shells, reinforced concrete opened up the two-dimensional continuum to the structural engineer and hence a close relationship between conception, engineering loadbearing system analyses and syntheses plus applied mathematics and mechanics.

### 9.3.1.1 Reinforced concrete gains emancipation from structural steelwork: the rigid frame

During the first decade of the 20th century, as the system period of reinforced concrete construction came to an end, building materials coexisted in buildings; diverse reinforced concrete systems replaced systems employing established building materials – for suspended floors in particular (see [Voormann, 2005]). Structural modelling of the structure using sim-

FIGURE 9-21
Vierendeel girder bridge in Freudenstadt, Germany [Zipkes, 1906/1, p. 141]

ple theories was adequate for this form of composite construction. Merely the composite action in the beam or slab had to be taken into account quantitatively by way of loading tests and a rough design theory; this had been available since 1887 in the form of the *Monier-Broschüre* (see section 9.1.3). In buildings, loadbearing system analysis concentrated on the loadbearing element. By no stretch of the imagination could engineers speak of a synthesis of the elements into a structural model of the building.

After, initially, Emperger in criticising Brik lent clarity to the determination of the bending strength of reinforced concrete beams [Emperger, 1902/1, 1902/4], there was no longer anything standing in the way of calculating statically indeterminate systems in reinforced concrete. As early as 1906, Kaufmann investigated the two-span reinforced concrete beam with one degree of static indeterminacy [Kaufmann, 1906]; in that same year Frank examined how varying cross-sections influenced the bending moments of continuous beams with the help of an extended formulation of the theorem of three moments [Frank, 1906]. Nevertheless, there were still uncertainties in the abstraction process from loadbearing structure to structural system. For example, Simon Zipkes modelled a Vierendeel girder bridge (Fig. 9-21) not as a rigid frame, but rather as a perforated beam "for the sake of greater safety" [Zipkes, 1906/1, 1906/2]. Zipkes made use of the structural model of the perforated beam because the equations derived by Vierendeel in 1897 "are awkward and [would] produce odd results that are not easy to check" [Zipkes, 1906/2, p. 247].

The analysis of the Vierendeel girder with its many degrees of static indeterminacy was to remain the theme of engineering papers for many years to come. It is interesting that when comparing the triangulated framework system with the Vierendeel girder system, Zipkes highlights the difference between loadbearing structure and loadbearing system quite specifically: in the steel triangulated framework with riveted joints, which is modelled as a pin-jointed framework, the secondary stresses are ignored. As the secondary stresses are often of the same order of magnitude as the axial stresses calculated using the pin-jointed trussed framework model, but a reliable quantification of the secondary stresses calls for time-consuming, tedious calculations, the Vierendeel girder was introduced into structural steelwork. To conclude, Zipkes points out the advantages of the reinforced concrete Vierendeel girder over the steel one. In doing so, he emphasizes the fixity effect of the members of the Vierendeel girder [Zipkes, 1906/2, p. 246]. The structural-constructional relationship between open-frame girder – in particular in the form of the Vierendeel girder – and trussed framework would be debated again and again in the coming years (e.g. [Engesser, 1913/1]). One great disadvantage of the Vierendeel girder is its high internal static indeterminacy, which complicates a reliable structural analysis.

The idea that the structural model of the rigid joint follows directly from the nature of the truss-like reinforced concrete structure quickly became popular. This is the reason for the numerous papers on rigid frames in reinforced concrete in journals such as *Beton und Eisen* and *Armierter Beton* even before 1910. For instance, Charles Abeles derived a number of frame equations with the help of the theorems of Castigliano (see section 6.4.1), already with a view to standardisation [Abeles, 1907]. Furthermore, structural calculations for reinforced concrete rigid frame structures were increasingly appearing (e.g. [Wuczkowski, 1907; Leuprecht, 1907]).

The year 1909 saw the publication of Ejnar Björnstad's book on the calculation of statically indeterminate frames [Björnstad, 1909], one of the first monographs of this type. Although Björnstad worked as an engineer in the steelwork company Beuchelt & Co. in Grünberg (Silesia), his book was also aimed at reinforced concrete engineers. Björnstadt's book followed the one by Karl Schaechterle on the calculation of the elastic arches and frames customary in reinforced concrete construction in which he incorporates his experiences in the design of reinforced concrete structures for the Württemberg State Railways Authority into the theory [Schaechterle, 1910]; however, in contrast to Björnstadt, Schaechterle's book was written exclusively for reinforced concrete engineers. After 1910, numerous monographs on rigid frames appeared. The advertisements of the publishing house Wilhelm Ernst & Sohn from 1914 alone list nine such publications: Strassner (1912), Wuczkowski (1912), Bronneck (1913), Hartmann (1913), Gehler (1913), Engesser (1913/2), Kleinlogel (1914), Schaechterle (1914) and Rueb (1914), of which those by Engesser, Gehler and Kleinlogel are particularly noteworthy. Apart from the book by Schaechterle, all are directed at readers from both structural steelwork and reinforced concrete.

On the nature of the static indeterminacy of reinforced concrete structures, Gehler writes as follows: "In reinforced concrete structures, on the other hand, the individual loadbearing parts, the columns, beams and floor slabs, are produced monolithically, like from one mould, by filling the timber shuttering with concrete mortar. In this form of construction this means that the number of separating joints between the individual loadbearing parts is minimised in reinforced concrete construction, which results in a rigid, i.e. statically indeterminate, connection between said parts. If this circumstance now prevents the full use of the building material as is possible with steelwork, then, on the other hand, the so-called fixing of the ends of the beams has the advantage of a considerable reduction in the positive bending moments … Furthermore, the interconnection of all the loadbearing parts of a reinforced concrete structure guarantees a valuable safety factor for all contingencies and weathering effects during construction which can impair the quality of the product. Therefore, in deliberate contrast to steel structures, the rigid connection of all parts and hence many degrees of static indeterminacy is normally the aim in reinforced concrete structures" [Gehler, 1913, p. 3]. From this quotation we can see that since the middle of the accumulation phase of structural theory (1900–25), reinforced concrete engineers have needed books on rigid frames for the practical side of structural calculations. And indeed, a flood of these appeared on the market, all claiming to ease the work of the reinforced concrete engineer in the calculation of statically indeterminate frames.

The by far most successful rigid frame book was that by Kleinlogel, with its collection of ready-to-use frame equations [Kleinlogel, 1914], the last (17th) edition of which appeared in 1993 [Kleinlogel & Haselbach, 1993]. As he writes in his preface: "I began quite some time ago to calculate the frame cases I met in practice for my own purposes using various methods, and to collect the equations together. The time-savings this brought about during later repetitions and the rapid usability of the equations occasioned me to extend the calculations to more and more frame types and loading cases" [Kleinlogel, 1914]. Kleinlogel's frame equations were of course translated into English, but also Spanish, French, Italian and Greek. Fig. 9-22 shows two loading cases for frame type 41 of the US edition, which contains a total of 114 frame types with hundreds of loading cases.

Just like the trussed framework and its associated theory characterised steel bridge-building and theory of structures during its establishment (1850–75) and classical (1875–1900) phases, the rigid frame evolved out of Hennebique's monolithic reinforced concrete frame to become the prevailing type of loadbearing structure. During the accumulation phase of structural theory (1900-25), this was accompanied by an elastic frame theory which essentially consisted of differentiated applications and new ways of presenting the methods of classical theory of structures and was tailored to the needs of structural calculations in reinforced concrete construction. This strengthened the method-based character of the classical theory of structures of Müller-Breslau even more. One example of this is the semi-graphical fixed-point method attributed to Wilhelm Ritter [Ritter, 1900, pp. 22–43], which Strassner adapted in a masterly way for the investigation of continuous and multi-storey frames in reinforced concrete [Strassner, 1916]. By drawing on the methods already available in the arsenal of resources in graphical analysis, it was possible to rationalise structural calculations significantly during the accumulation phase of structural theory.

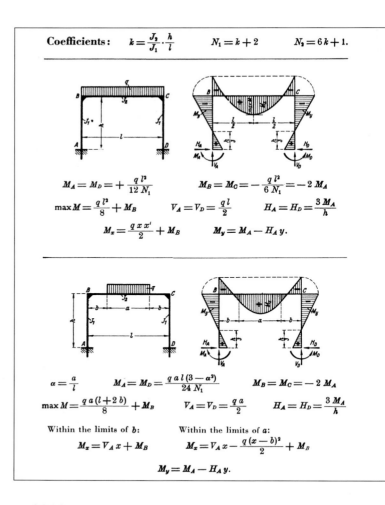

FIGURE 9-22
Frame equations for two loading cases for frame type 41 [Kleinlogel, 1952, p. 146]

**Reinforced concrete takes its first steps into the second dimension: out-of-plane-loaded structures**

### 9.3.1.2

It was reinforced concrete that awakened the theory of out-of-plane-loaded structures from its 100+ years of slumber in the "Palace of Mechanics" – to which the practising civil and structural engineer had no access – in the accumulation phase of structural theory (1900 – 25). By way of Turner's and Maillart's flat slabs and the slab tests in the Materials Testing Institute at Stuttgart TH, reinforced concrete construction was for the first time able to break away completely from traditional methods of building and adapt the knowledge about slabs gained from mathematical elastic theory. And the end of this development was marked by an appropriate structural theory.

**The theory of out-of-plane-loaded structures as an object of mathematical elastic theory**

Todhunter and Pearson (1886, 1893/1, 1893/2), Love (1893/94), Timoshenko (1953) and Szabó (1996) have contributed to the history of the theory of out-of-plane-loaded structures from the viewpoint of elastic theory. The early history of this theory begins in 1764 with Leonhard Euler's work *Tentamen de sono campanarum*, in which he grapples with how a bell produces sounds, reached a climax with Ernst Florens Friedrich

Chladni's 1787 publication entitled *Akustik* and Jakob II Bernoulli's plate theory *Essai théorétique sur les vibrations des plaques élastiques, rectangulaires et libres*, which he presented to the Petersburg Academy in 1788, and was completed in 1813 by Sophie Germain and Joseph Louis Lagrange. The latter two specified the differential equation for plates (eq. 7-56). Seven years later, Navier submitted his *Mémoire sur la flexion des planes élastiques* (see Fig. 1-3) to the Academy of Sciences. In that work he specifies a solution to the differential equation in the form of a double trigonometric series for a rectangular slab simply supported on all sides. Navier's Mémoire on the theory of out-of-plane-loaded structures was first published in 1883 in the appendix to Clebsch's *Theorie der Elasticität fester Körper* [Clebsch, 1862] in the French translation by Saint-Venant and Flamant [Clebsch, 1883, pp. 740–752]. The *Mémoire* initiated the constitution phase of elastic theory in general and the theory of out-of-plane-loaded structures in particular, which was rounded off in 1850 by Gustav Robert Kirchhoff with his theory of thin elastic plates, the first self-contained theory of out-of-plane-loaded structures. The constitution phase of elastic theory, too, was completed around the middle of the 19th century, as Lamé's *Leçons sur la théorie mathematique de l'élasticité des corps solides* [Lamé, 1852] shows.

Kirchhoff derives the differential equation of plate deflection $w(x,y)$ (see eq. 7-56 in section 7.5.1.2) with the help of the principle of virtual displacements [Kirchhoff, 1850, p. 68]. The theory of out-of-plane-loaded structures is the subject of the last of his 30 lectures on mechanics and gets only very brief treatment in the mechanics volume of his four-volume work on mathematical physics [Kirchhoff, 1876, pp. 449–465]. Kirchhoff's differential equation for plates (eq. 7-56) is identical to that given by Siméon Denis Poisson in 1829 [Poisson, 1829], apart from the factor for Poisson's ratio. Whereas Kirchhoff leaves Poisson's ratio open, Poisson assumes a value of 0.5. Sophie Germain, too, arrives at the correct differential equation for plates via two errors which to a certain extent cancel each other out: on the one hand, the Gaussian curvature (product of curvature in $x$- and $y$-direction) is missing in her hypothesis, and on the other, her plate analysis results in incorrect boundary conditions. The amusing thing about Kirchhoff's plate theory, however, is not only its consideration of the Gaussian curvature, but in the consistent formulation of the boundary conditions, i.e. reducing Poisson's three boundary conditions to just two: "Poisson reaches ... the same partial differential equation as Sophie Germain's hypothesis, but using different boundary conditions, and in fact three boundary conditions. I will prove that generally these cannot be satisfied simultaneously. From this it follows that also according to Poisson's theory a plate subjected to out-of-plane loading need not generally have an equilibrium position. However, I shall present the proof only after I have derived the two boundary conditions to be used instead of Poisson's three because the proof naturally follows on from the considerations from which I wish to derive the boundary conditions" [Kirchhoff, 1850, p. 54]. Kirchhoff's sensitive proof became the uncontested mathematical principle that solutions of the biharmonic equation 7-56 can be adapted to two boundary conditions (see also [Szabó, 1996, p. 422]). Here, he makes use of the integral theorems of Gauß and Green which were not available to Germain, Lagrange, Navier and Poisson. Kirchhoff obtains a double integral from this which results in the differential equation for an elastic plate (eq. 7-56) plus a curvilinear integral along the contour line of the midsurface (boundary curve of elastic plate), from which conclusions can be drawn via the boundary conditions.

Although the theory of out-of-plane-loaded structures is presented in detail in the monographs of Rayleigh (1877) and Love (1892/93), it was not adopted by engineers. Exceptions to this are the works of Grashof (1866, 1878), Lavoinne (1872) and Lévy (1899). Like Jakob II Bernoulli in 1788, Grashof proposed a method for calculating the vertical displacements $w(x,y)$ of slabs which later became known in reinforced concrete as the grillage method (see section 7.5.1.2). Lavoinne was the first to supply a solution for slabs on discrete supports like those built for boiler bottoms on stay bolts and flat slabs. Lévy verified that a rectangular slab with Navier-type boundary conditions (two opposite sides of the slab simply supported) can be solved with simple hyperbolic series instead of the double trigonometric series after Navier.

The extension to thick out-of-plane-loaded plates was achieved by Michell (1899) and Dougall (1903/04) with the help of rapidly converging Bessel functions. Within the scope of his lectures in applied mechanics, A. Föppl dealt in detail with the theory of out-of-plane-loaded structures [Föppl, 1907, pp. 97–144], presented the Hertz theory of the elastically supported plate [Föppl, 1907, pp. 112–130], which would later form the basis for the analysis of shallow foundations in reinforced concrete, and developed the equations apparatus for slabs with large vertical displacements $w(x,y)$ [Föppl, 1907, pp. 132–144]. Hadamard (1908) and Happel (1914) introduced integral equation methods into plate theory. As the aforementioned contributions to plate theory made extensive use of the language of mathematical elastic theory, they did not enter the vocabulary of the structural engineer as a genuine structural theory until the findings of tests on slabs became available in the invention phase of structural theory (1925–50).

**Simple engineering models for designing slabs**

Rectangular reinforced concrete slabs of length $l_x$ and width $l_y$ were still being calculated many years into the 20th century as beams with span $l_y$ and a cross-sectional width of 1 m (see section 9.1.3.3). Systematic research into and tests on plates first began in the late 1880s. For example, Bach reports on square and rectangular plates of cast iron and hard lead with an out-of-plane loading being tested to failure [Bach, 1889/90]. Fig. 9-23 shows the equilibrium condition for a square slab simply supported on all four sides subjected to a uniformly distributed load $q$ for the diagonal section.

For reasons of symmetry, the shear forces vanish in the diagonal section so that only the resultant of the moment $M$ acts here; the problem is statically determinate. The resultant of the force of half the slab in the z-direction acts at the centre of gravity S and amounts to

$$Z_S = -\frac{q \cdot l^2}{2} \tag{9-21}$$

Due to the equilibrium of all forces in the z-direction, the support reaction $Z_1$ acting at $x_1 = l/2$ and $y_1 = l$ and the support reaction $Z_2$ acting at $x_2 = 0$ and $y_2 = l/2$ become

$$Z_1 = Z_2 = +\frac{q \cdot l^2}{4} \tag{9-22}$$

To ensure equilibrium of moments in the diagonal section

$$M = Z_1 \cdot h_1 + Z_2 \cdot h_2 + Z_S \cdot h_S \tag{9-23}$$

where the lever arms are

$$h_1 = h_2 = h = \frac{l}{4} \cdot \sqrt{2} \tag{9-24}$$

and

$$h_S = \frac{l}{6} \cdot \sqrt{2} \tag{9-25}$$

so that in the end

$$M = \frac{q \cdot l^2}{24} \cdot l \cdot \sqrt{2} \tag{9-26}$$

results for eq. 9-23. If the resultant moment $M$ in the diagonal section is related to the unit length, then

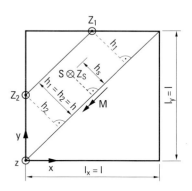

**FIGURE 9-23**
On the derivation of the bending moment in the diagonal section of a square slab after Bach (see [Bach 1890])

$$m = \frac{M}{l \cdot \sqrt{2}} \cdot \frac{q \cdot l^2}{24} \tag{9-27}$$

In the case of homogeneous square slabs, according to Bach's tests eq. 9-27 must be multiplied by 1.12:

$$m_{Bach} = 1.12 \cdot \frac{q \cdot l^2}{24} = \frac{q \cdot l^2}{21.4} \tag{9-28}$$

In 1905 Bosch generalised the Bach slab equation for rectangular reinforced concrete slabs where $l_x < l_y$ and $\lambda = l_y/l_x$ [Bosch, 1905, p. 177]:

$$m_{Bosch} = \frac{1}{3} \cdot \left(\frac{l_x}{2}\right)^2 \cdot \frac{\lambda^2}{1+\lambda^2} \cdot q \tag{9-29}$$

In a letter, Sor criticises the Bosch equation 9-29 [Sor, 1905] and instead recommends an approximation equation by Paul Christophe ([Christophe, 1899, pp. 1029–1034; Christophe, 1902, pp. 618–623]).

Christophe clearly recognises the advantages of the slab over the beam: "If a slab rectangular on plan is supported not just on two sides [i.e. acts as a beam – the author], but rather around its entire perimeter, then its strength increases significantly as its form approaches that of a square. No practical investigations into the strength of reinforced concrete slabs under such conditions have been undertaken hitherto. One can therefore only deal with this case by generalising the method chosen for calculating iron plates" [Christophe, 1902, pp. 620–621]. He therefore follows Grashof's grillage method, which specifies the equation

$$M_x = M_y = \frac{q \cdot l^2}{16} \tag{9-30}$$

for the bending moments acting in the $x$- and $y$-direction on a square slab with uniformly distributed load $q$. Starting with a simply supported rectangular slab where $l_x < l_y$, which Christophe initially models structurally as a beam with span $l_x$ and whose maximum span moment is

$$M_x = \frac{q \cdot l_x^2}{8} \tag{9-31}$$

he considers in a second step the favourable influence of the slab effect by reducing the bending moment (eq. 9-30) as follows:

$$M_x = \frac{l_y^4}{l_x^4 + l_y^4} \cdot \frac{q \cdot l_x^2}{8} \tag{9-32}$$

For a strip of slab where $l_y \to \infty$, eq. 9-32 becomes eq. 9-31 (ideal beam action), whereas eq. 9-32 for a square slab where $l_y = l_x = l$ becomes Grashof's eq. 9-30 (ideal slab action). If $l_y = 2 \cdot l_x$, then according to eq. 9-32 the moment is only reduced by a factor of 0.94 compared to eq. 9-31. Christophe concludes from this that "as soon as the length of the slab is greater than twice the width, one can ignore the influence of the lateral supports and the slab can be calculated like a beam on two supports with span $l_x$" [Christophe, 1902, p. 621].

The moment critical for the $y$-direction amounts to

$$M_y = \frac{l_x^4}{l_x^4 + l_y^4} \cdot \frac{q \cdot l_y^2}{8} \tag{9-33}$$

For a strip of slab where $l_y \to \infty$, the reduction factor in eq. 9-33 approaches zero (ideal beam action), whereas eq. 9-33 for a square slab where $l_y = l_x = l$ becomes Grashof's eq. 9-30 (ideal slab action). If $l_y = 2 \cdot l_x$, then according to eq. 9-33 the moment is only reduced by a factor of 0.06 compared to eq. 9-31. Christophe also specifies equations for slabs fixed on all sides in a similar manner. Eq. 9-32 and eq. 9-33 were included in the *Bestimmungen für die Ausführung von Konstruktionen aus Eisenbeton im Hochbau* of 13 January 1916 published by the German Committee for Reinforced Concrete [Mörsch, 1922, p. 426].

**Innovation dynamic in reinforced concrete: the "mushroom" flat slab**

In 1905 the journal *Engineering News* included a report by the American engineer Claude Allen Porter Turner on a new reinforced concrete floor system which in contrast to Hennebique's floor system (see Fig. 9-11) did not require any downstand beams or secondary beams (Fig. 9-24); Turner was granted a US patent for the system on 11 June 1907. The reinforcement continues in four directions over the top of the columns. The integration of the reinforced concrete columns into the suspended floor slab is achieved partly by an assembly of steel flats which is fixed to the column reinforcement. This accommodates not only part of the slab bending moments, but also prevents the column punching through the slab. Such steel punching protection – in a modern form – still proves useful today in reinforced concrete and composite steel-concrete construction. Turner also adapted the "mushroom" flat slab for concrete bridges [Gasparini & Vermes, 2001]. Turner's flat slab extended substantially the construction vocabulary of structures with out-of-plane loading in reinforced concrete.

The Turner-type flat slab was first used in 1906, for the Johnson Bovey Co. Building in Minneapolis. Further applications followed soon afterwards [Gasparini, 2002, pp. 1247, 1248] and led to the system been referred to in the industry as the "Turner Mushroom System". He proved the stability of his flat slabs by means of loading tests. On 18 February 1909 Turner proudly wrote the following lines in the journal *Engineering News*: "… the writer has been associated with the erection of over 400 acres [1 acre = 4048.86 m² – the author] of floor built without ribs or beams, scattered from Portland, Me., to Portland, Ore., in the United States and from Regina, Saskatchewan, in the north to Melbourne, Australia, in the south. This work has been erected, without an accident to the construction, in temperatures from 24° below zero to 102° in the shade." Just short of a year later, Turner noted "… now that many engineers have become familiar with its design, … the amount constructed and in use is rapidly approaching a thousand acres of floor" (cited in [Gasparini, 2002, p. 1248]). In the case of Turner's "mushroom" flat slab, the three strides of invention, innovation and diffusion took place within just a few years! A similar diffusion dynamic in American reinforced concrete construction had been achieved by Melan's arch bridge system 10 years previously [Eggemann & Kurrer, 2006]. Both the "Turner Mushroom System" and the Melan System

– reduced the cost of formwork considerably,
– prospered well in the technology-based activities of American civil and structural engineers, and
– responded well to the shortage of skilled workers and the relatively high wages in the USA.

The mathematics graduate and director of a large Russian building contractor, Artur Ferdinandovitch Loleit, constructed Russia's first "mush-

**FIGURE 9-24**
Turner's "mushroom" (i.e. column head) flat slab concept dating from 1905 [Gasparini, 2002, p. 1246]

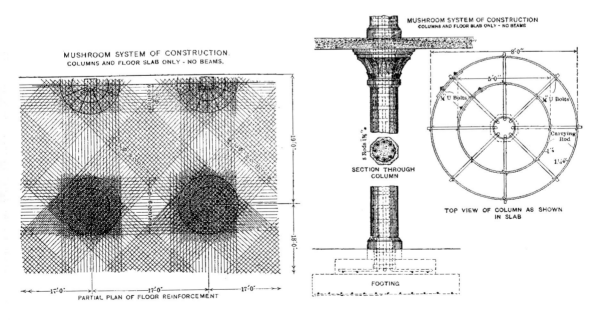

room" flat slab for a textiles factory near Moscow as early as 1907 – certainly unaware of Turner's achievements. In 1913 Loleit gave a presentation on his flat slab at the Forum of the Association of Russian Materials Researchers; the slab had two groups of reinforcing bars and used a method of calculation based on Grashof's grillage method. "Around 1930," writes Kierdorf, "Loleit tried to establish a completely new reinforced concrete theory, which was not accepted by the professional and scientific élite of the time" [Kierdorf, 2006, p. 1804].

Robert Maillart erected two "mushroom" flat slabs for test purposes on the premises of his construction company in Zurich in 1908. The grid of reinforcement was orthogonal and arranged to match the column grid. Maillart's deflection measurements on a flat slab of thickness $d = 8$ cm with $3 \times 3$ bays and column spacings $l_y = l_x = l = 4$ m and the design method he developed on this basis have been analysed by Armand Fürst and Peter Marti [Fürst & Marti, 1997], which they compared with the flat slab elastic theories of Lewe (1920, 1922) and Westergaard and Slater (1921). Maillart placed measuring gauge studs every 25 cm in both directions, applied a load of 10 kN at each of 144 points and read off the deflections. He thus obtained three fields of influence for the vertical deflection for three different boundary conditions [Maillart, 1926]. In 1909 Maillart was granted a Swiss patent for his flat slab system and one year later was able to employ this system in the Giesshübel company warehouse in Zurich [Fürst & Marti, 1997, p. 1102]. Maillart's flat slab (Fig. 9-25) can be called the masterpiece of the "concrete virtuoso" [Marti, 1996] and would become widely used in France, Spain, Russia and the Baltic states.

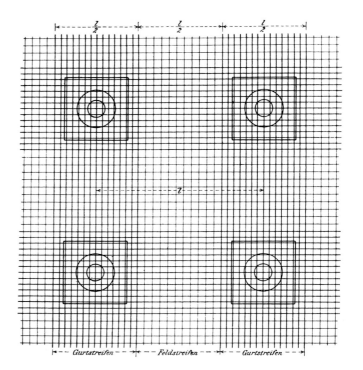

**FIGURE 9-25**
Two-strip system after Maillart
[Mörsch, 1922, p. 381]

The "mushroom" flat slab did not feature in the German reinforced concrete provisions because there was no associated theory and therefore such suspended slabs could not be built: "In Germany, it was initially impossible to build flat slabs [Mayer, 1912] because there was no recognized theory and because any structural engineering calculation was scrutinized by the authorities" [Fürst & Marti, 1997, p. 1102].

Considerable progress was made in understanding the loadbearing behaviour and the structural modelling of reinforced concrete slabs during the second decade of the 20th century. For example, Arturo Danusso published a series of papers on reinforced concrete slabs in the journal *Il Cemento*, which were translated into German by Hugo von Bronneck [Danusso, 1913]. In his papers, Danusso calculates waffle slabs as beam grids without considering the torsional stiffness. He then transfers the structural model of the beam grid in the form of a grillage to the analysis of square, rectangular and triangular reinforced concrete slabs. Danusso's beam grid model is a generalised grillage method à la Grashof, which permitted the calculation of the static indeterminates from the equality of the deflections at the nodes.

As early as 1910, the German Committee for Reinforced Concrete drew up an extensive schedule of tests on simply supported square and rectangular reinforced concrete slabs, which were then carried out in the Materials Testing Institute at Stuttgart TH. Carl von Bach and Otto Graf reported on the results [Bach & Graf, 1915]. Fig. 9-26 shows the pattern of cracks in a 12 cm slab where $l_y = l_x = l = 2$ m and the reinforcement in both the $x$- and $y$-direction consisted of steel bars of 7 mm diameter at a pitch of 10 cm. The condition illustrated in Fig. 9-26 was caused by a total load of 39 t (excluding self-weight) distributed evenly over 16 points in order to simulate a uniformly distributed load $q$. The failure load of the reinforced concrete slab was approx. 40.33 t (excluding self-weight). The lines of fracture that can be seen in Fig. 9-26 were interpreted by the Danish engineer Aage Ingerslev as plastic hinges on which he founded a plastic design method for calculating slabs [Ingerslev, 1921]. This method was extended by Johanson with the help of the general work theorem to create the yield line theory [Johanson, 1932]. Further slab tests to failure carried out by Steuermann on a "mushroom" flat slab in Baku (1936), by Ockleston on a suspended floor system in a warehouse in Johannesburg (1955), and by Guralnick and LaFraugh on a nine-bay point-supported flat slab (1963) verified the yield line theory. Johanson published a two-volume compendium of equations in 1949/50 which was intended to ease the application of the yield line theory in practice [Johanson, 1949/50]. In 1961 Wood tackled the issue of membrane action in reinforced concrete slabs [Wood, 1961], which is the reason why the measured load at failure in full-size tests is always higher than the value calculated by the yield line theory. The monograph *Grenztragfähigkeitstheorie der Platten* (ultimate load theory of slabs) [Jaeger & Sawczuk, 1963], which appeared in the middle of the innovation phase of structural theory (1950–75), provided

**The theory of out-of-plane-loaded plates as an object of materials research and theory of structures**

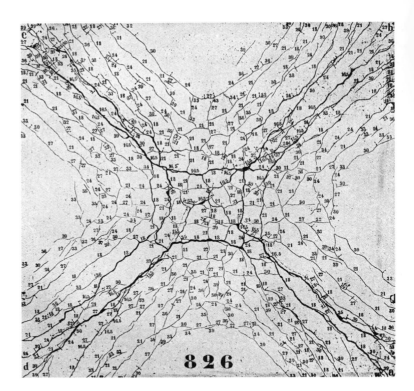

**FIGURE 9-26**
Pattern of cracks in a square reinforced concrete slab [Mörsch, 1922, p. 351]

the structural engineering community with a comprehensive and concentrated summary.

But back to the slab tests of Bach and Graf. Based on 10 tests on reinforced concrete slabs with the same spans $l_x$ and $l_y$ but different thicknesses, amount and direction (45°) of reinforcement, the value of the denominator was not 21.4 as in eq. 9-28, but on average 21.76 [Mörsch, 1922, p. 349]:

$$m_{Bach/Graf} = \frac{q \cdot l^2}{21.76} \qquad (9\text{-}34)$$

The empirical finding expressed by eq. 9-34 corresponds very closely with the denominator determined from the theory of out-of-plane-loaded structures by

– Heinrich Leitz: 20.9 [Leitz, 1914, p. 49] and
– Heinrich Hencky: 21.2 [Hencky, 1913, p. 29].

Max Mayer (1912) and Henry T. Eddy (1913) analysed the "mushroom" flat slab using Grashof's grillage method. Another attempt at a solution for such flat slabs and also other slab problems was given by Karl Hager (1914), who assumed a double Fourier series for the deflection; K. Hruban (1921) and V. Lewe (1920, 1922) then continued his work. In his dissertation completed in 1918 (and published two years later), Nielsen solves the differential equation for plates (eq. 7-56) for the first time with the help of the finite difference method [Nielsen, 1920], which Friedrich Bleich had used as long ago as 1904 for continuous beams and L. F. Richardson had applied to elastic in-plane-loaded plates in 1909 [Richardson, 1909].

The breakthrough to a theory for out-of-plane-loaded structures was achieved by Henri Marcus – at the time director of Huta, Hoch- & Tiefbau-Akt.-Ges., Breslau – with his series of papers on *Die Theorie elastischer Gewebe und ihre Anwendung auf die Berechnung elastischer Platten* (theory of elastic membranes and its application to the calculation of elastic plates) [Marcus, 1919]. If the Kirchhoff differential equation for plates (eq. 7-56 or eq. 7-57) is multiplied by the plate stiffness K, we get the following expression:

$$K \cdot \left( \frac{\partial^4 w}{\partial x^4} + 2 \cdot \frac{\partial^4 w}{\partial x^2 \partial y^2} + \frac{\partial^4 w}{\partial y^4} \right) = p(x,y) \qquad (9\text{-}35)$$

which Marcus rearranges as follows:

$$K \cdot \left( \frac{\partial^2}{\partial x^2} + \frac{\partial^2}{\partial y^2} \right) \cdot \left( \frac{\partial^2 w}{\partial x^2} + \frac{\partial^2 w}{\partial y^2} \right) = K \cdot \nabla^2 \cdot \left( \frac{\partial^2 w}{\partial x^2} + \frac{\partial^2 w}{\partial y^2} \right)$$

$$= K \cdot \nabla^2 \nabla^2 w = p(x,y) \qquad (9\text{-}36)$$

Marcus now enters the sum of the moments

$$K \cdot \left( \frac{\partial^2 w}{\partial x^2} + \frac{\partial^2 w}{\partial y^2} \right) = K \cdot \nabla^2 w = -M \qquad (9\text{-}37)$$

and can rewrite eq. 9-36 as follows

$$\nabla^2 M = -p(x,y) \qquad (9\text{-}38)$$

He now compares eq. 9-38 with the known equation for the elastic bar (see eq. 3-7 in section 3.2.3.4), where the designation for the external load $q(x)$ is replaced by $p(x)$:

$$\frac{d^2}{dx^2} M(x) = -p(x) \qquad (9\text{-}39)$$

Marcus noted that eq. 9-38 and eq. 9-39 exhibit a similar mathematical structure. This becomes clear when the differential equation for the elastic membrane

$$\nabla^2 w = \frac{\partial^2 w}{\partial x^2} + \frac{\partial^2 w}{\partial y^2} = -\frac{p}{S} \qquad (9\text{-}40)$$

($S$ = surface tension, $p$ = compression perpendicular to membrane, $w$ = deformation of membrane in $z$-direction) is compared with eq. 9-38. From this, Marcus obtains the following theorem: "The midsurface of an elastic membrane loaded with an overpressure $p$ and subjected to the surface tension $S = 1$ represents a map of the moment area of the elastic slab" [Marcus, 1919, p. 110]. According to Marcus, the analogy between the linear and the planar goes even further. If the elastic membrane is loaded with the elastic weights $p_i = M/K$, then for the same surface tension $S = 1$ it undergoes a vertical deformation $w_i$ in the $z$-direction which must satisfy the differential equation

$$\nabla^2 w_i = \frac{\partial^2 w_i}{\partial x^2} + \frac{\partial^2 w_i}{\partial y^2} = -p_i = -\frac{M}{K} \qquad (9\text{-}41)$$

"The midsurface of an elastic membrane loaded with the elastic weights $p_i = M/K$ and subjected to the surface tension $S = 1$ represents a map of the elastic area of the slab" [Marcus, 1919, p. 110]. Hence, Marcus had generalised Mohr's analogy for out-of-plane-loaded structures. Mohr had specified the following method for bars: from the form equivalence of dif-

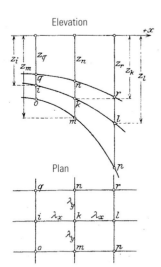

**FIGURE 9-27**
Mesh for the finite difference method for rectangular slabs after Marcus [Marcus, 1919, p. 131]

ferential equation 9-39, which links bending moment $M(x)$ with external loading $p(x)$, and the differential equation for the deflection curve (see eq. 3-8 in section 3.2.3.4)

$$\frac{d^2}{dx^2} w(x) = -\frac{M(x)}{E \cdot I} \qquad (9\text{-}42)$$

which creates the relationship between deflection $w(x)$ and bending moment $M(x)$, it follows that when determining $w(x)$, the bar with the elastic weight $M(x)/(E \cdot I)$ is loaded and the bending moment is calculated for that, which is identical with the deflection $w(x)$ of the bar. The two second-order differential equations 9-39 and 9-42 can be combined for bars in bending in the ordinary fourth-order differential equation

$$\frac{d^4}{dx^4} w(x) = \frac{p(x)}{E \cdot I} \qquad (9\text{-}43)$$

Marcus' generalisation of Mohr's analogy from the bar in bending to the out-of-plane-loaded plate is complete. Just like the ordinary fourth-order differential equation for the bar (eq. 9-43) can be split into two ordinary second-order differential equations (eq. 9-39 and eq. 9-42), so the partial fourth-order differential equation for the plate (eq. 7-56 or eq. 9-35) can also be split into two partial second-order differential equations (eq. 9-40 and eq. 9-41). The funicular curve served Culmann as a graphical analysis model of the analysis of beams for determining the bending moments $M(x)$ (double integration of eq. 9-39, see also Fig. 6-12) and Mohr for determining the deflection curve $w(x)$ (double integration of eq. 9-42). In a similar way, Marcus based his analysis of out-of-plane-loaded structures on the model of the elastic membrane, which he used to calculate the sum of the bending moments $M(x,y)$ (double integration of eq. 9-38) and the bending surface $w(x,y)$ (double integration of eq. 9-41). So according to Marcus, an out-of-plane loading problem could be solved in two steps. He used the finite difference method for the numerical solution (Fig. 9-27).

The two-step solution to out-of-plane loading problems using the finite difference method according to Marcus enabled the number of differential equations to be reduced. Fig. 9-27 shows the mesh for setting up the differential equations for rectangular slabs according to eq. 9-41. Marcus specifies suitable mesh geometries for rectangular, triangular and circular slabs, devises the corresponding differential equations for eq. 9-38 or eq. 9-41 and calculates numerical examples. His series of papers [Marcus, 1919], which he later collected into a book and expanded [Marcus, 1924], gave the engineer a clear algorithm for the numerical analysis of out-of-plane-loaded structures for which there had been no definite solution so far. The German Committee for Reinforced Concrete commissioned Marcus to draft a simplified method for calculating slabs [Marcus, 1925], which was immediately incorporated into the German reinforced concrete standard. Therefore, the practical side of structural calculations had at its disposal a structural approximation method for the analysis of reinforced concrete slabs with which standard cases could be quickly and reliably quantified.

Besides the aforementioned works by Marcus, the following monographs were also extremely important for the formation of a theory of out-of-plane-loaded structures in the accumulation phase of structural theory (1900–25):
- *Die elastischen Platten* (elastic plates) [Nádai, 1925, 1968], and
- *Pilzdecken und andere trägerlose Eisenbetonplatten* ("mushroom" flat slabs and other reinforced concrete slabs without beams) [Lewe, 1926/1].

A reviewer of Nádai's 1925 monograph wrote the following: "Among the hitherto known attempts to build a bridge between the customary training of the engineer in the area of mathematical elastic theory and the current situation of the latter, this ... book must be regarded as one of the best and most successful" [Lewe, 1926/2]. In the brochure by Marcus (1925) and the book by Lewe (1926/1), the practical character of the analysis of slabs is emphasized (Fig. 9-28).

Fig. 9-28 shows the sketch of the loading case for the tabular determination of deflection $w(x,y)$, curvature in the $x$- and $y$-direction (second partial derivative of $w$ according to $x$ or $y$) and the torsion (mixed partial derivative of $w$ according to $x$ and $y$) for a square "mushroom" flat slab with the load applied in alternate bays. Lewe specifies many more out-of-plane loading cases to simplify the structural calculations; he thus founded the genre of the design table, which culminated in the classic tables of Klaus Stiglat and Herbert Wippel [Stiglat & Wippel, 1966]. Contrastingly, the theory of the orthotropic plate developed by Huber between 1914 and 1926 did not benefit reinforced concrete but rather, after 1950, structural steelwork (see section 7.5.1.2). Rudolph Szilard has compiled an up-to-date and comprehensive study of the theory of out-of-plane-loaded structures in all its forms, e.g. Kirchhoff, Huber, Reissner, with numerous examples [Szilard, 2004].

The structural plate theories of Marcus, Nádai and Lewe revealed a new aspect in the relationship between theory and practice in structural analysis which emphasized the mathematics-mechanics foundation but at the same time did not lose sight of the need for a user-friendly presentation of the structural analysis methods. The relationship between numerical mathematics and theory of structures achieved a new quality with the monograph *Die gewöhnlichen und partiellen Differenzengleichungen der Baustatik* (ordinary and partial differential equations in theory of structures) [Bleich & Melan, 1927] by Friedrich Bleich and Ernst Melan. Thus, the theory of out-of-plane-loaded structures forms the historico-logical prerequisite for the development of the theory of plate and shell structures, which represents one main thread in the invention phase of structural theory (1925–50).

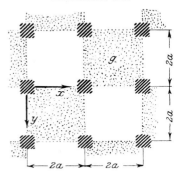

**FIGURE 9-28**
Sketch of the loading case for a design table for square "mushroom" flat slabs with the load applied in alternate bays [Lewe, 1926/1, p. 88]

### 9.3.1.3 The structural-constructional self-discovery of reinforced concrete

Concerning the oil tank in Bertolt Brecht's poem *700 Intellektuelle beten einen Öltank an* (*700 Intellectuals Worship an Oil Tank*) (see [Knopf, 2001, pp. 144–146]) it says:

*Du bist nicht gemacht aus Elfenbein
Und Ebenholz, sondern aus
Eisen.*
…
*Und du verfährst mit uns
Nicht nach Gutdünken noch unerforschlich
Sondern nach Berechnung.*

*Thou art not made of ivory
And ebony, but instead of
Iron.*
…
*And thou goeth with us
Not according to discretion nor the unfathomable
But instead according to calculation.*

Brecht could just as well have directed his stinging criticism of the technicism of the *Neue Sachlichkeit* (new objectivity), which in architecture would make a further impression in the form of the so-called international style, at any reinforced concrete grain silo of the 1920s in mid-western USA. The structural theories on which modern engineered structures are based were certainly part of the reason why the followers of the *Neue Sachlichkeit* appeared awestruck: the theory of plate and shell structures – a closed book to the protagonists of any artistic movement. But also the reinforced concrete silo allows Brecht – contradicting the intellectuals of the *Neue Sachlichkeit* – to say:

*Du Häßlicher,
Du bist der Schönste!
Tue uns Gewalt an,
Du Sachlicher.*

*Thou ugly One,
Thou art the most beauteous!
Do violence unto us,
Thou 'sachlich' One.*

The misunderstood discovery of the aesthetics of pure engineered structures and its adoration by the advocates of the *Neue Sachlichkeit*, its dialectic reversal by Brecht in the form of the Lord's Prayer, has a less-well-known true story for architectural writers which manifests itself in reinforced concrete construction in the historico-logical loadbearing system development: beams, T-beams, continuous beams, frames, plates with in-plane and out-of-plane loading, folded plates and, finally, shells. Besides the out-of-plane-loaded plate, the in-plane-loaded plate and the folded plate are important objects of the theory of plane shell structures inspired by reinforced concrete in the first half of the invention phase of structural theory (1925 – 50). It was this period that witnessed the formation of a structural plate and shell theory.

## In-plane-loaded plates and folded plates

Hermann Craemer and Georg Ehlers were prominent in developing the principle of the plane shell structure. In the introduction to the paper entitled *Scheiben und Faltwerke als neue Konstruktionselemente im Eisenbetonbau* (in-plane-loaded plates and folded plates as new construction elements in reinforced concrete) [Craemer, 1929/1], the author outlines not only the historico-logical loadbearing system development of reinforced concrete from the beam to the shell, but rather sees its economic efficiency in the stress equilibrium brought about by the seamlessness of reinforced concrete structures: "Designing economically means ... exploiting this through the stress equilibrium due to the seamlessness and if necessary consciously bringing this about by way of a suitable arrangement" [Craemer, 1929/1, p. 254]. This continuity principle is invariable, is the logical nucleus of the historical unfolding of the loadbearing systems of reinforced concrete. But that's not enough. Craemer warns us of the new construction language of reinforced concrete, whose grammar he sees implemented in the folded plate: "I shall now deal with an area," Craemer continues in his introduction, "in which due to the too literal translation of the customary design principles of timber and iron construction into the new language of reinforced concrete we have – apart from a few long-standing approaches – overlooked almost all the advantages we could have achieved through the conscious exploitation of the seamlessness" [Craemer, 1929/1, p. 255]. Using the example of the wall-floor junction in large bunkers, Craemer demonstrates how conventional modelling using beam theory throws away the advantages of reinforced concrete. By contrast, in the folded plate, "which consists of several in-plane-loaded plates in different planes whose ends are secured against displacement and which are connected seamlessly, [the fold] completely takes over the role of a downstand beam at that point" [Craemer, 1929/1, p. 270]. In a letter written on behalf of the Dyckerhoff & Widmann company [Dyckerhoff & Widmann, 1929/1 & 1929/2], the folded plate is classed as a special form of the patented Zeiss-Dywidag shell (Fig. 9-29) because the structural action of these prismatic roofs is almost identical to that of the cylindrical ones; the company insisted that its patent rights be respected.

Craemer's objection reached its peak in the statement that shell and folded plate actions would cancel each other out [Craemer, 1929/2, p. 339]. The controversy between Dyckerhoff & Widmann and Craemer concerning the loadbearing quality of the folded plate and a shell continued via a new move (see [Craemer, 1929/3; Dyckerhoff & Widmann, 1929/3]). In 1930 Craemer published a paper on the general theory of folded plates in the journal *Beton und Eisen* [Craemer, 1930]. In the same issue, Ehlers made it known through his paper *Die Spannungsermittlung in Flächentragwerken* (determining the stresses in plate and shell structures) that he had developed the principle of plate and shell structures back in 1924/25 and it had been used for the first time in the building of a boilerhouse bunker for the Märkische power station in Finkenheerd in 1925 [Ehlers, 1930/2, p. 281]. A paper entitled *Ein neues Konstruktionsprinzip* (a new

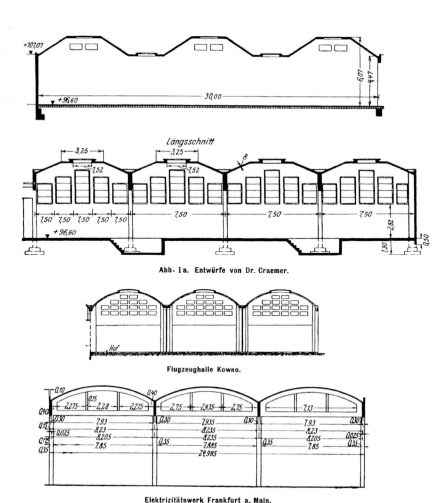

**FIGURE 9-29**
A comparison between folded plates and cylindrical shells [Dyckerhoff & Widmann, 1929/2, p. 338]

construction principle) published shortly before in the journal *Der Bauingenieur* announced Ehlers claim to priority in the matter of the folded plate, which he called a "shell structure" [Ehlers, 1930/1, pp. 125 & 127]. Ehlers presented his idea in 1927 at a conference organised by Wayss & Freytag. Fig. 9-30 shows the difference between the conventional structural modelling of a bunker compartment (left) and considering the load-bearing action as a folded plate (right). In the first case the two in-plane-loaded plates $ABB'A'$ and $EDD'E'$ are modelled as a beam without taking into account how the bunker bottom slabs $BCC'B'$ and $DCC'D'$ contribute to carrying the loads. If we believe that the lateral pressure on the bunker walls and bottom slabs must be resisted by special measures such as the inclusion of beams along the axis $BB'$, transverse rails through the bunker compartment between $BB'$ and $DD'$, Ehlers realised that the resultant force at nodes $B$, $C$ and $D$ in each case must be spread into the bunker walls and bottom slabs (Fig. 9-30/right). In such shell structures, Ehlers says there are two questions to be answered [Ehlers, 1930/1, p. 127]:

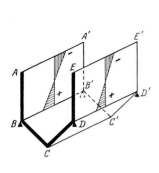
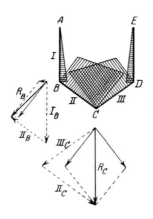

FIGURE 9-30
Comparison of the modelling of a bunker compartment as a beam (left) and as a folded plate (right) [Ehlers, 1930/1, p. 127]

- Which force actions ensue between two or more beams connected along their longitudinal edges?
- Which changes to the stress distribution do these force actions cause in the beams themselves?

According to Ehlers, the reason why the folded plate as a construction principle of reinforced concrete was not recognised earlier can be found in the fact that structural theory formation was closely linked to structural steelwork: "Initially trained in iron construction and its properties, theory of structures up to now has preferably dedicated itself to planar problems and as a result has neglected spatial relationships; only recently has a primarily spatial approach started to break through. However, there is no reason for reinforced concrete to cling to the planar arrangement of the loadbearing structure and to single out such a two-dimensional loadbearing structure and ignore other relationships" [Ehlers, 1930/1, p. 132]. The new language of reinforced concrete called for a new form of expression in structural theory in the shape of the theory of plate and shell structures! As one important pillar of the theory of plate and shell structures, the theory of folded plates continued to develop during the 1930s and 1940s. The first summary of this work was Joachim Born's monograph *Faltwerke. Ihre Theorie und Berechnung* (folded plates – theory and calculation) [Born, 1954], which simplified noticeably the structural analysis of such loadbearing structures.

## The construction of reinforced concrete shells and the school of shell theory

The discovery of the shell as a loadbearing structure is a work of reinforced concrete and started around 1900 with the erection of water and gas tanks plus thin-wall domes. By the end of the accumulation phase of structural theory (1900–25) the Zeiss-Dywidag shell by Bauersfeld and Dischinger (see section 8.2.3) had appeared. This was also the period of the technicisation of shell theory, which had evolved in the tradition of the mathematical elastic theory. And by the middle of the invention phase of structural theory (1925–50) the new language of reinforced concrete had been completed by the shell constructions of Dischinger, Rüsch and Finsterwalder, which found their adequate engineering science expression in structural shell theory.

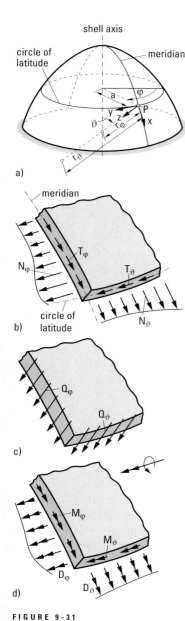

**FIGURE 9-31**
a) Rotational shell with designations, b) element axial forces and element thrust forces, c) element shear forces, d) element bending moments and element torsion moments

In fact, Navier had investigated rotational shells in the membrane stress condition in his *Résumé des Leçons* [Navier, 1826], i.e. shells in which the element axial forces $N_\vartheta$ and $N_\varphi$ plus the element thrust forces $T_\vartheta$ and $T_\varphi$ act at their midsurface, i.e. where the element shear forces $Q_\vartheta$ and $Q_\varphi$, the element bending moments $M_\vartheta$ and $M_\varphi$ plus the element torsion moments $D_\vartheta$ and $D_\varphi$ vanish (Fig. 9-31).

Lamé and Clapeyron calculated the stresses and deformations in a spherical shell subjected to internal or external pressure [Clapeyron & Lamé, 1833], and in 1854 the latter managed a complete solution to the deformation problem of spherical shells subjected to any distributed loads [Lamé, 1854]. Aron was the first person to consider all moments and formulated the bending theory of any curved elastic shell for the static and dynamic cases [Aron, 1874]. Unfortunately, Aron's bending theory for shells was not adopted. It was not until 14 years later that Love, working independently of Aron, set up a bending theory for shells [Love, 1888], which he presented in his famous textbook *A treatise on the mathematical theory of elasticity* [Love, 1892/93]. Rayleigh, too, published separate contributions on shell theory in 1881, 1888 and 1889, and collected these together in a separate chapter in the second edition of the first volume of his *The Theory of Sound* [Rayleigh, 1894]. A generalisation of the bending theory for shells that did not use the Bernoulli hypothesis of plane sections remaining plane was provided by the Cosserat brothers [E. & F. Cosserat, 1909]. And that concluded the constitution phase of mathematical shell theory.

Only by using the symbolic notation of H. Lamb in the second edition of his textbook [Love, 1906 & 1907] did Love manage to bring shell theory to the attention of engineering researchers. Nevertheless, a reviewer of the German edition of Love's book writes [Love, 1907]: "This book is written by an expert for mathematicians and physicists … A study of this book will not benefit the practising engineer because apart from a few cases the golden bridge leading from vague theory to practice is missing. However, the enthusiastic researcher who senses the familiar results of elastic theory on their theoretical foundations will derive much and new inspiration from this work" [Schönhöfer, 1907, p. 296].

Practising engineers initially approached shell theory cautiously via the analysis of the simplest shell form, the single-curvature, fixed cylindrical shell; but the representatives of fundamental engineering science disciplines such as applied mechanics and theory of structures were no different (Fig. 9-32). Using this structural model, engineers attempted to size vessels of steel and later reinforced concrete – the works of E. Winkler (1860), F. Grashof (1878), G. A. Wayss (1887), V. G. Shukhov (1888) (see [Ramm, 1990]), R. Maillart (1903) (see [Schöne, 1999]), C. Runge (1904), Panetti (1906), H. Müller-Breslau (1908), H. Reissner (1908), K. Federhofer (1909 & 1910), T. Pöschl and K. v. Terzaghi (1913), A. and L. Föppl (1920) to name just some. In 1923 V. Lewe summarised the methods for the structural calculation of tanks for fluids in a longer article for the *Handbuch für Eisenbetonbau* (reinforced concrete manual) [Lewe, 1923].

In his *Monier-Broschüre*, G. A. Wayss specifies an equation for determining the wall thickness $t(z)$ of a reinforced concrete water tank [Wayss, 1887, p. 34] which he derived from the boiler formula (eq. 7-35) (Fig. 9-32):

$$t(z) = t = r \cdot \frac{p_i}{\sigma_{perm}} = r \cdot \frac{\gamma \cdot z}{\left[\sigma_{b, perm} + \dfrac{1}{n}(\sigma_{s, perm} - \sigma_{b, perm})\right]} \quad (9\text{-}44)$$

In eq. 9-44

| | |
|---|---|
| $r$ | is the internal radius of the water tank |
| $t(z)$ | is the wall thickness |
| $\gamma \cdot z$ | is the hydrostatic pressure at depth $z$ below the surface of the water |
| $\sigma_{b, perm}$ | is the permissible tensile stress of concrete |
| $\sigma_{s, perm}$ | is the permissible tensile stress of steel, and |
| $n$ | is the ratio of the concrete cross-sectional area $A_b$ to the steel cross-sectional area $A_s$ (i.e. the amount of reinforcement per unit length in the $z$-direction). |

Theoretically, eq. 9-44 should always result in $t(z=0) = 0$ when $z = 0$, but in practical terms a certain wall thickness $t_0$ with a steel cross-section $A_{s0} = t_0/n$ always results. For this reason, Wayss proposes a wall thickness $t_0$ with a steel-cross-section $A_{s0}$ up to a height $z = z_0$, which according to eq. 9-44 would produce the value $t_0$, and only after that would the linear change in wall thickness down to the base of the tank be determined for $z = h$ according to eq. 9-44. For this latter section, Wayss specifies a simple construction according to the intercept theorem (see Fig. 9-32):

$$\frac{t}{t_h} = \frac{z}{h} \quad (9\text{-}45)$$

Design equation 9-44 only takes into account the hoop tension stresses in the $\varphi$-direction (see Fig. 9-31b); the normal stresses in the $\vartheta$-direction (see Fig. 9-31b) are not entered into the boiler formula. Maillart would be the first to consider the bending stresses due to $M_\vartheta$ – in 1903 in the structural analysis of the reinforced concrete gas tank in St. Gallen [Schöne, 1999]. One year later, Runge published an approximation calculation for the cylindrical water tank.

The cylindrical shell with a varying wall thickness (Fig. 9-33) investigated by Reissner led to a fourth-order differential equation with varying coefficients that cannot be solved in itself. Reissner resolved this differential equation using power series and prepared it in the form of tables and charts.

But uncertainties still existed in the structural analysis of tanks and this was expressed in the work of Emil Reich [Reich, 1907]. Reissner criticises not only Reich's complex solution, but also his sample calculation, which results in a value for the wall thickness seven times the radius of the tank! Following Reissner's work, Federhofer proposed a graphical method for determining the stress distribution in cylindrical tank walls with any wall thickness [Federhofer, 1909 & 1910]. The bending theory for cylindrical shells as a practical structural model for reinforced concrete

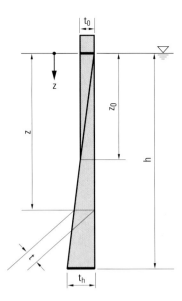

**FIGURE 9-32**
On the design of a reinforced concrete cylindrical water tank with a partially varying wall thickness after Wayss [Wayss, 1887, p. 34]

**FIGURE 9-33**
A cylindrical shell with a varying wall thickness and fixed at the base [Reissner, 1908, p. 150]

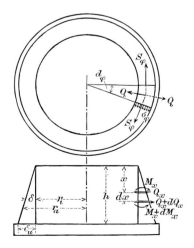

tanks was fully developed by the middle of the second decade of the 20th century.

Another decade was to pass before the structural bending theory for rotational shells could be completed with the work of J. W. Geckeler (1926). The first major step in the direction of a structural membrane theory for rotational shells was taken by J. W. Schwedler (1863 & 1866). He realised that in the structural analyses of domes it was not only the radial stresses $\sigma_\vartheta$ that had to be quantified – as had been the case in the past –, but also the tangential stresses $\sigma_\varphi$. Schwedler derived the equilibrium conditions for a dome-type rotational shell with any geometry (see Fig. 9-31a) and specified them for shallow shell surfaces and spherical surfaces [Schwedler, 1863]. In a further paper, he uses his structural membrane theory for rotational shells to calculate the member forces in the space frame he had invented, the Schwedler dome [Schwedler, 1863] (see section 8.1). As the internal forces in the radial and tangential directions of rotationally symmetric membrane shells can be determined from the equilibrium conditions alone, i.e. this is an internally statically determinate system, graphical analysis was already being used to analyse such loadbearing systems in the 1870s (see section 6.2.4.1). In the *Monier-Broschüre*, Schwedler's membrane theory was used to design dome-type reinforced concrete shells [Wayss, 1887, pp. 31–33]. The reinforcement was laid in the radial and tangential directions. Spherical reinforced concrete shells would be calculated according to this method up until the middle of the accumulation phase of structural theory (1900–25).

Nevertheless, it was clear to reinforced concrete engineers that the membrane stress condition in reinforced concrete shells was "disrupted" by bending stresses at the supports. The internally statically determinate membrane theory had to be supplemented by the internally statically indeterminate bending theory for shells in double curvature: the 10 unknown internal forces (see Fig. 9-31) were matched against only six equilibrium conditions. The following made contributions to the bending theory of rotational shells:
- A. Stodola, with his calculation of conical shells of constant thickness [Stodola, 1910],
- H. Reissner, with his investigation of spherical shells subjected to a rotationally symmetric loading [Reissner, 1912], which was extended to cover any rotational shell by E. Meissner [Meissner, 1913],
- O. Blumenthal, who introduced asymptotic integration, which replaced the integration of poorly converging power series [Blumenthal, 1913],
- E. Schwerin, with his analysis of spherical shells, in particular subjected to asymmetric loads such as wind [Schwerin, 1919], and finally,
- J. Geckeler, with his simplified solution [Geckeler, 1926], "which covers all rotational shells with a meridian and also any varying wall thickness in the same way and is simple enough to be used by practising engineers" [Dischinger, 1928/1, p. 153].

Geckeler obtained the structural bending theory for rotational shells from his theoretical work on the development of the planetarium domes of Walther Bauersfeld (Carl Zeiss company) and Franz Dischinger (Dyckerhoff & Widmann company) (see section 8.2.3). Both companies were able to expand the planetarium dome from a circular base to square and then polygonal forms. This was quickly followed by the structural shell theory of Dischinger, Rüsch and Finsterwalder. This new form of shell construction, given momentum by the building industry in both practical and theoretical terms and quite rightly called the Zeiss-Dywidag shell or the Zeiss-Dywidag system, represents one main developmental thread in the invention phase of structural theory (1925–50).

The invention phase of structural theory began with a bang in shell construction: Dischinger's presentation *Fortschritte im Bau von Massivkuppeln* (progress in the construction of concrete domes) given on 23 February 1925 at the 28th general assembly of the German Concrete Association in Berlin, which was published in the October edition of the journal *Der Bauingenieur* [Dischinger, 1925]. In his presentation, Dischinger embraced the whole historico-logical spectrum from the Roman Pantheon and the Byzantine Hagia Sophia to the domes of the Renaissance and St. Paul's Cathedral and the reinforced concrete ribbed dome of the Breslau Centennial Hall and the Zeiss-Dywidag shells of the early 1920s (see also [Schmidt, 2005]).

The first summary of the theory and practice of building reinforced concrete shells appeared in 1928 in the form of the influential 14-volume *Handbuch für Eisenbetonbau* (reinforced concrete manual) [Dischinger, 1928/1], the "encyclopaedia of reinforced concrete construction" [Kurrer, 1999, p. 48]. One year later, Dischinger published his dissertation *Die Theorie der Vieleckkuppeln und die Zusammenhänge mit einbeschriebenen Rotationsschale* (theory of polygonal domes and the relationships with inscribed rotational shells), supervised by Professors Kurt Beyer and Willy Gehler (Dresden TH), in the journal *Beton und Eisen* [Dischinger, 1929]. This was followed by his series of articles – written together with Hubert Rüsch – on the large market hall in Leipzig [Dischinger & Rüsch, 1929]. In the article, Dischinger and Rüsch report on a new dome construction system comprising Zeiss-Dywidag shells (Fig. 9-34).

Just over two years later, Franz Dischinger and Ulrich Finsterwalder were able to report on the further development of the Zeiss-Dywidag shell: the octagonal dome to the large market hall in Basel and other shell structures built according to the shell concept of Dyckerhoff & Widmann, the barrel-vault shell roofs for quayside sheds in Hamburg and further structures involving cylindrical shells plus the latest research into and tests on shells and the resulting applications [Dischinger & Finsterwalder, 1932]. Although the first reinforced concrete shell roofs had been erected shortly after 1900 [Schöne, 1999], shell construction first really came to fruition in the "new language of reinforced concrete" [Craemer, 1929/1, p. 255] after 1925 with the work of the Dischinger-Finsterwalder-Rüsch threesome. Its structural-constructional grammar underwent consider-

**FIGURE 9-34**
Construction of the dome to the large market hall in Leipzig [Dischinger & Rüsch, 1929, p. 439]

able development with the hyperbolic paraboloid shells (Fig. 9-35) of B. Lafaille (1935) and F. Aimond (1936) in France and later E. Torroja (1948) in Spain, which were realised masterly by F. Candela in shell roofs of breathtaking beauty. Hyperbolic paraboloids are generated geometrically by two opposing boundary curves. Fig. 9-35 shows hyperbolic paraboloid shell roofs bounded by straight lines; the formwork for such reinforced concrete shells is simple to construct because only straight boards and panels are needed in the direction of the generators.

The two-volume work *Die Statik im Eisenbetonbau* (theory of structures in reinforced concrete construction) by Kurt Beyer (1927) was a textbook and manual of theory of structures commissioned by the German Concrete Association, the second edition of which (1933) was later reprinted. In the work, the linear and planar structures of reinforced concrete construction are for the first time given a comprehensive and uniform theory of structures treatment. Not without reason was Beyer's work christened the "Beyer Bible" in Germany. In 1927 Wilhelm Flügge gained his doctorate (supervised by Prof. Beyer) at Dresden TH, and from then until 1930 worked on the development of reinforced concrete shells in the building industry before completing his habilitation thesis in 1932 at the University of Göttingen. Two years later he published the standard work *Statik und Dynamik der Schalen* (statics and dynamics of shells) [Flügge, 1934], the second, revised edition of which appeared in 1957 [Flügge, 1957] and three years later the English translation with the title *Stresses in Shells* [Flügge, 1960]. In contrast to Dischinger's book-type contribution to shells in the *Handbuch für Eisenbetonbau* [Dischinger, 1928/1], Flügge does not restrict himself to reinforced concrete: "Over the course of recent decades, engineering has been successively faced with questions in various fields which belong to the material dealt with in this book, and their happy solution has occasionally lead to epoch-making progress in structural design. Examples that immediately come to mind are tank construction, the shaping of non-rigid airships, various issues in turbine and steam engine construction, long-span roof constructions in reinforced concrete and, recently, the new questions surrounding aircraft construction (monocoque fuselage etc.). Two circles of specialists are involved in solving these questions. Therefore, this book is aimed at two groups of readers: design engineers who are interested in the results of shell theory and require them in a form that allows them to apply them directly to the solution of specific tasks, and researchers in the field of applied mechanics whose duty it is to advance our knowledge in this field and to find new solutions to new, more complex issues. They are interested not only in the existing stock of solutions, but also the methods by way of which that stock was obtained, and the outlook for further development options" [Flügge, 1934, preface]. Shell theory was propelled forward by reinforced concrete well into the 1930s. Nonetheless, it advanced to become an important object of the Göttingen school around Richard von Mises and Ludwig Prandtl. This school was devoted to the mathematics-mechanics principles of civil and structural engineering, mechanical engineering and, later, aircraft construction, and which with the founding of the *Zeitschrift für Angewandte Mathematik und Mechanik* (*Journal of Applied Mathematics and Mechanics*) by Von Mises in 1921 created a new type of journal which immediately inspired the founding of further engineering journals dedicated to fundamental disciplines.

In their *Beitrag zur allgemeinen Schalenbiegetheorie* (article on the general bending theory of shells) [Zerna, 1949], Wolfgang Zerna and A. E. Green [Green & Zerna, 1950/2] inserted the final piece of the shell theory jigsaw in the invention phase of structural theory (1925 – 50); at the same time, they together laid the foundation stone for the impressive ongoing development of shell theory in the integration period of structural theory (1950 to date). This was only made possible through the rigorous use of the symbolic power of tensor analysis, which is ideal for derivatives in curvilinear coordinates. Green and Zerna also formulated the basic equations of elastic theory in the language of tensor analysis ([Green & Zerna, 1950/1; Zerna, 1950; Green & Zerna, 1954]). Fig. 9-36a shows a shell element with the two curvilinear coordinates $\vartheta^1$ and $\vartheta^2$ [Zerna, 1949, p. 155], which are used to derive the equilibrium relationships of the general shell bending theory. For the derivation of the equilibrium relationships of the spatial elastic theory, Zerna assumed a volume element with the curvilinear coordinates $\vartheta^1$, $\vartheta^2$ and $\vartheta^3$ (Fig. 9-36b) [Zerna, 1950, p. 217].

By 1953 Zerna had described the essential findings of shell theory in the journal *Beton- und Stahlbetonbau*; he concludes his article with the following words: "We now have the possibility of setting up the principal equations for all practically conceivable shell forms within the scope of the validity of the theory without having to rely on geometrical graphic considerations. That means we have now created the prerequisites for dealing with many individual problems" [Zerna, 1953, p. 89]. It was this path that the theory of plate and shell structures, as the school of shell theory, would take.

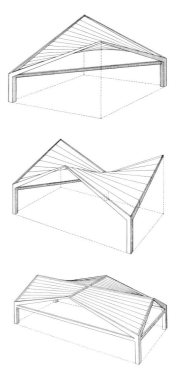

**FIGURE 9-35**
Roof surfaces in the form of hyperbolic paraboloids [Aimond, 1936, p. 9]

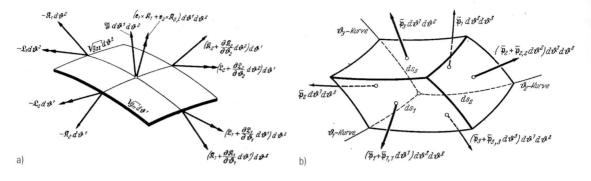

**FIGURE 9-36**
a) Shell element [Zerna, 1949, p. 155] and
b) volume element [Zerna, 1950, p. 217] in the language of tensor analysis

As the constructional language of reinforced concrete shells in the middle of the invention phase of structural theory (1925–50) was completed, so shell theory in the mathematical language of tensor analysis took on its original form through Green and Zerna and started the school of shell theory around the middle of the 20th century. Furthermore, tensor analysis written in index notation was set to give continuum mechanics a completely new mathematical form during the integration period of structural theory (1950 to date), which would meet the programming needs of computers ideally. It is therefore no surprise that Zerna – during the early period of electronic calculation – was one of the initiators of *Computational Civil Engineering*. The brochure *Theory of Shell Structures* [Başar & Krätzig, 2001], written by Yavuz Başar and Wilfried B. Krätzig, both students of Zerna, provides a concise, clear introduction to the state of shell theory as it was in the year 2000. Its masterful handling of the language of tensor analysis makes it well worth reading and it shows how the high standard of non-interpretive symbol usage in structural mechanics was achieved through tensor analysis using the example of the formalisation of shell theory.

### 9.3.2 Prestressed concrete: "Une révolution dans les techniques du béton" (Freyssinet)

It was already evident in the 1930s that prestressed concrete represented a new form of construction, which after 1950 would bring about major changes in concrete bridge-building. For example, Karl Walter Mautner reported on Freyssinet's prestressing technique in 1936 [Mautner, 1936]. Two years prior to that, Dischinger had proposed external prestressing for beam-type bridges in particular. And in 1937 Wayss & Freytag board member Kurt Lenk published a paper on the production of prestressed concrete beams and pipes [Lenk, 1937], and concludes his contribution with the words: "I am convinced that prestressed concrete represents a material that will enjoy widespread use" [Lenk, 1937, p. 169].

And Mautner? Apart from his consultancy contract with Wayss & Freytag, Mautner lost all his positions owing to his Jewish background. He was arrested on 9 November 1938 during the *Reichskristallnacht* ("night of broken glass") pogrom organised by the Nazis and taken to Buchenwald concentration camp with 10 000 others; however, this camp was not built for so many prisoners and therefore many had to be released, one of whom was Mautner. A. Kirkwood-Dodds was able to convince the British government that Mautner's special knowledge of prestressed concrete would be useful for the UK's defences. In the summer of 1939, Kirkwood-Dodds helped Mautner and his wife to flee Germany together with a heavy suitcase with all his prestressed concrete test reports – and that's how prestressed concrete reached the UK! In their book *Freyssinet. Prestressing and Europe 1930–1945* [Grote & Marrey, 2000], Jupp Grote and Bernard Marrey outline the route from Freyssinet's invention via Mautner's contribution to the practical introduction and the misuse of prestressed concrete in World War 2. For example, the authors report on bombproof floor slabs with cyclopic prestressed concrete beams for German U-boat bunkers which were built using enforced labour. Grote and Marrey's book is written in English, French and German and contains the following dedication: "To the memory of those who suffered and gave their lives labouring to realize this technical venture" [Grote & Marrey, 2000, p. 7].

In 1939 Eugène Freyssinet (1879–1962) published a book with the title *Une révolution dans les techniques du béton* [Freyssinet, 1939]. In the book he describes the development of prestressed concrete, in which he played a leading role, as a technical revolution in concrete construction. Indeed, prestressed concrete thoroughly changed concrete construction in the technical-economic and the technical-scientific senses. He initiated numerous new developments such as high-strength concretes and reinforcing steels, raised concrete research (material laws, design theory) and the building industry to a new level (mechanisation, further development of the prefabrication industry, internationalisation), helped concrete bridge-building to become the dominant form in many countries and helped the structural engineer to gain far better social recognition for his work well into the 1980s. However, he also focused on the closeness between prestressed concrete research and prestressed concrete practice in personalities whose works ensured that they would gain acknowledgement in the history of construction engineering and civil engineering theory in the second half of the 20th century: Eugène Freyssinet, Franz Dischinger, Gustave Magnel, Ulrich Finsterwalder, Eduardo Torroja, Hubert Rüsch, Hans Wittfoth, Jean Muller and Fritz Leonhardt.

### 9.3.2.1 Leonhardt's *Prestressed Concrete. Design and Construction*

The first monograph on prestressed concrete construction was published by the Belgian civil engineering professor Gustave Magnel in French [Magnel, 1948/3] and English [Magnel, 1948/2]. Another outstanding example of the closeness between science and industry in prestressed concrete is Fritz Leonhardt's 1955 book *Spannbeton für die Praxis* [Leonhardt, 1955], the second, expanded edition of which appeared in English in 1964 with the title *Prestressed Concrete. Design and Construction* [Leonhardt, 1964]. Leonhardt begins his book in Old Testament style with the "Ten Commandments for the prestressed concrete engineer" [Leonhardt, 1964, p. XI] (Fig. 9-37). Leonhardt divides his "Ten Commandments" into five for the design office and five for the construction site, stressing the internal relationship between science-based design practice and engineered building practice. The first commandment of design is:

**FIGURE 9-37**
"Ten Commandments for the prestressed concrete engineer" by Fritz Leonhardt [Leonhardt, 1964, p. XI]

> **Ten Commandments for the prestressed concrete engineer**
>
> **In the design office:**
>
> 1. Prestressing means compressing the concrete. Compression can take place only where shortening is possible. Make sure that your structure can shorten in the direction of prestressing.
> 2. Any change in tendon direction produces "radial" forces when the tendon is tensioned. Changes in the direction of the centroidal axis of the concrete member are associated with "unbalanced forces", likewise acting transversely to the general direction of the member. Remember to take these forces into account in the calculations and structural design.
> 3. The high permissible compressive stresses must not be fully utilized regardless of circumstances! Choose the cross-sectional dimensions of the concrete, especially at the tendons, in such a way that the member can be properly concreted — otherwise the men on the job will not be able to place and vibrate the stiff concrete correctly that is so essential to prestressed concrete construction.
> 4. Avoid tensile stresses under dead load and do not trust the tensile strength of concrete.
> 5. Provide non-tensioned reinforcement preferably in a direction transverse to the prestressing direction and, more particularly, in those regions of the member where the prestressing forces are transmitted to the concrete.
>
> **On the construction site:**
>
> 6. Prestressing steel is a superior material to ordinary reinforcing steel and is sensitive to rusting, notches, kinks and heat. Treat it with proper care.
>    Position the tendons very accurately, securely and immovably held in the lateral direction, otherwise friction will take its toll.
> 7. Plan your concreting programme in such a way that the concrete can everywhere be properly vibrated and deflections of the scaffolding will not cause cracking of the young concrete. Carry out the concreting with the greatest possible care, as defects in concreting are liable to cause trouble during the tensioning of the tendons!
> 8. Before tensioning, check that the structure can move so as to shorten freely in the direction of tensioning. For distributing the pressure, insert timber or rubber packings between tensioning devices and the hardened concrete against which they may be thrusting. Make it a rule always to cover up high-pressure pipelines.
> 9. Tension the tendons in long members at an early stage, but at first only apply part of the prestress, so as to produce moderate compressive stresses which prevent cracking of the concrete due to shrinkage and temperature.
>    Do not apply the full prestressing force until the concrete has developed sufficient strength. The highest stresses in the concrete usually occur during the tensioning of the tendons.
>    When tensioning, always check the tendon extension and the jacking force. Keep careful records of the tensioning operations!
> 10. Do not start grouting the tendons until you have checked that the ducts are free from obstructions. Perform the grouting strictly in accordance with the relevant directives or specifications.

"Prestressing means compressing the concrete. Compression can take place only where shortening is possible. Make sure that your structure can shorten in the direction of prestressing." In the fifth commandment for the construction site, the last of the "Ten Commandments", the maxim is: "Do not start grouting the tendons until you have checked that the ducts are free from obstructions. Perform the grouting strictly in accordance with the relevant directives or specifications." (A rather simplified translation! The original German goes into much more detail.) As damage to prestressed concrete bridges would later demonstrate, Leonhardt's 10th commandment was all too often ignored.

Leonhardt divides his monograph into 20 chapters. After he has explained the basic principles of prestressed concrete in chapter 1, he deals with the properties of the materials (chapter 2), of which creep and shrinkage of the concrete – which Freyssinet had been researching since 1911 – are especially interesting for prestressed concrete. Chapters 3 to 9 cover the technology of prestressed concrete construction such as anchorages for and splices in prestressing tendons, the prestressing plant and prestressing procedure, degree of prestress, importance of the bond, the longitudinal mobility and sliding resistance of tendons, the pressure-grouting of tendons for subsequent bond and the transfer of prestressing forces. Leonhardt deals with the structural-constructional side of prestressed concrete in chapters 10 to 15: principles for proper construction

## Recommendations on Excavations EAB

The aim of these recommendations is to harmonize and further develop the methods, according to which excavations are prepared, calculated and carried out.

Since 1968, these have been worked out by the TC "Excavations" at the German Geotechnical Society (DGGT) and published since 1980 in four German editions under the name EAB. The recommendations are similar to a set of standards.

They help to simplify analysis of excavation enclosures, to unify load approaches and analysis procedures, to guarantee the stability and serviceability of the excavation structure and its individual components, and to find out an economic design of the excavation structure.

For this new edition, all recommendations have been reworked in accordance with EN 1997-1 (Eurocode 7) and DIN 1054-1. In addition, new recommendations on the use of the modulus of subgrade reaction method and the finite element method (FEM), as well as a new chapter on excavations in soft soils, have been added.

Deutsche Gesellschaft für Geotechnik e.V. (ed.)
**Recommendations on Excavations – EAB**
2nd revised edition
2008. Approx 300 pages. Hardcover.
Approx EUR 69.–

ISBN: 978-3-433-01855-2

€ Prices are valid in Germany, exclusively, and subject to alterations. Prices incl. VAT. Books excl. shipping. Journals incl. shipping. 008358016_my

www.ernst-und-sohn.de

Ernst & Sohn Verlag für Architektur und technische Wissenschaften GmbH & Co. KG
Für Bestellungen und Kundenservice: Verlag Wiley-VCH, Boschstraße 12, D-69469 Weinheim
Tel.: +49(0)6201 606-400, Fax: +49(0)6201 606-184, E-Mail: service@wiley-vch.de

# Geotechnical engineering Handbook

## Editor: Ulrich Smoltczyk

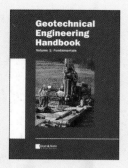

**Volume 1: Fundamentals**
2002.
829 pages, 616 fig.
Hardcover.
€ 179,-*/ sFr 283,-
ISBN 978-3-433-01449-3

This is the English version of the Grundbau-Taschenbuch - a reference book for geotechnical engineering. The first of three volumes contains all information about the basics on the field of geotechnical engineering. The book is written by authors from Germany, Belgium, Sweden, the Czech Republic, Australia, Italy, U.K., and Switzerland.

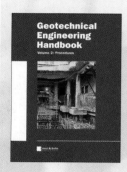

**Volume 2: Procedures**
2002.
679 pages, 558 fig.
Hardcover.
€ 179,-*/ sFr 283,-
ISBN 978-3-433-01450-9

Volume 2 of the Geotechnical Engineering Handbook covers the geotechnical procedures used in manufacturing anchors and piles as well as for improving or underpinning the foundations, securing existing constructions, controlling ground water, excavating rocks and earthworks. It also treats such specialist areas as the use of geotextiles and seeding.

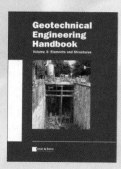

**Volume 3:
Elements and structure**
2002.
646 pages, 500 fig.
Hardcover.
€ 179,-*/ sFr 283,-
ISBN 978-3-433-01451-6

Volume 3 of the Geotechnical Engineering Handbook deals with foundations. It presents spread foundations starting with basic designs right up the necessary proofs. There is comprehensive coverage of the possibilities for stabilizing excavations, together with the relevant area of application, while another section is devoted to the useful application of trench walls. The entire book is an indispensable aid in the planning and execution of all types of foundations found in practice, whether for academics or practitioners.

Ernst & Sohn
Verlag für Architektur und
technische Wissenschaften GmbH & Co. KG

Für Bestellungen und Kundenservice:
Verlag Wiley-VCH
Boschstraße 12
69469 Weinheim
Telefon: +49(0) 6201 / 606-400
Telefax: +49(0) 6201 / 606-184
E-Mail: service@wiley-vch.de

Special Set Price
(three volumes)
€ 499,-* / sFr 788,-
ISBN 3-433-01452-3

www.ernst-und-sohn.de

* €-price is valid in Germany only.
001415066_my  Prices are subject to change without notice.

(chapter 10), calculation of prestressed structures (chapter 11), dealing with the influences of creep and shrinkage of the concrete in the calculations (chapter 12), ultimate strength and safety against failure (chapter 13), the behaviour when subjected to vibration loads (chapter 14) and stability problems of prestressed components (chapter 15). Special areas of prestressing such as prestressed tanks, pipes, carriageways, railway sleepers, masts, piles, shells and ground anchors are left to chapter 16. In the following two chapters Leonhardt talks about fire protection and tests to failure. Advice concerning on-site works, centering and such matters are covered in chapter 19. And Leonhardt rounds off his book with a chronology of prestressed concrete construction, stretching from 1886 to 1953 (chapter 20). His statements regarding the relationship between theory of structures and prestressed concrete in chapter 11 on the calculation of prestressed structures are remarkable: "There is no 'special' structural theory for prestressed concrete. The usual methods can therefore be applied. For this reason we shall only consider, on the basis of such practical experience as is available at the present time, how to proceed with the design and analysis of prestressed concrete structures by means of familiar principles and methods. Indeed, in the case of statically indeterminate prestressed structures the usual methods are even in better agreement with the actual conditions than they are in reinforced concrete construction because in prestressed concrete the entire concrete section remains effective and the sudden change in the modulus of elasticity on passing from the uncracked ['state I' in German technical literature – the author] to the cracked ['state II' in German technical literature – the author] is absent so long as (no excessive) tensile stresses are permitted in the concrete. Only the shrinkage and creep deformations may, in the cases considered in Section 12.4 [where the influence of shrinkage and creep on the internal forces in statically indeterminate systems is examined – the author], cause deviations from the normal pattern of internal forces" [Leonhardt, 1964, p. 333].

In the first place, prestressed concrete complicated the practice of structural calculations (Fig. 9-38).

If the two-span continuous beam shown in Fig. 9-38a is prestressed using a straight tendon below the neutral axis, it deflects upwards in the statically determinate basic system (support restraint B released) by the amount $\delta_B = \delta_{B,V}$ (Fig. 9-38b); the associated bending moment in the statically determinate basic system is

$$M_V^0 = V \cdot y_V \qquad (9\text{-}46)$$

The upward deflection $\delta_B = \delta_{B,V}$ must be cancelled out by the support reaction $B_V$ acting at B, which causes the deflection (Fig. 9-38c); the associated bending moment in the statically determinate basic system takes on the value

$$M_V' = \frac{B_V \cdot l}{2} \qquad (9\text{-}47)$$

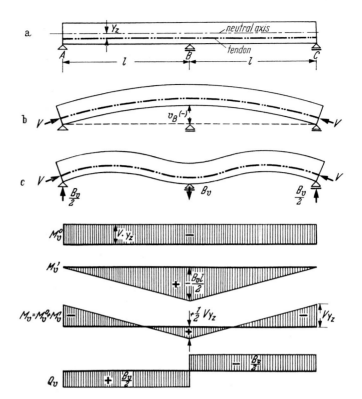

FIGURE 9-38
How statically indeterminate support reactions arise as a consequence of the prestress – illustrated for a two-span continuous beam with straight tendon [Leonhardt, 1964, p. 345]

where the static indeterminate $B_V = X_1$ from the elasticity condition

$$\delta_{B,V} + \delta_{B_V} = 0 = -\frac{M_V^0 \cdot l^2}{2 \cdot E \cdot I} + \frac{M_V' \cdot l^2}{3 \cdot E \cdot I} \tag{9-48}$$

taking into account eq. 9-46 and eq. 9-47 can be calculated as

$$B_V = X_1 = \frac{3 \cdot V \cdot y_V}{l} \tag{9-49}$$

The bending moment at $B$ due to the prestress in the statically indeterminate system is then

$$M_{B,V} = M_V^0 + M_V' = -V \cdot y_V + \frac{B_V \cdot l}{2} =$$
$$= -V \cdot y_V + \frac{3}{2} \cdot V \cdot y_V = \frac{1}{2} \cdot V \cdot y_V \tag{9-50}$$

The equations for analysing prestressed statically indeterminate systems become even more complex when using curved tendons. Added to this is the none-too-simple quantification of the loss of the prestressing force due to creep and shrinkage of the concrete on the basis of Dischinger's ageing theory [Dischinger, 1937, 1939] (see section 7.5.2.2), which complicates the setup of the elasticity equations. Nevertheless, the equations of the force method remain unchanged (see section 6.3.2.2) – a separate theory for prestressed concrete is not necessary. The quality of the approach and equations of the force method as a theoretical core of classical theory of structures celebrated a belated triumph in the prestressed concrete construction of the innovation phase of structural theory (1950–75).

### 9.3.2.2 The first prestressed concrete standard

The world's first standard for prestressed concrete construction (DIN 4227) first saw the light of day in October 1953. Within a few months it had been published in the 1954 edition of the annual *Beton-Kalender* together with the article *Bemessen von Spannbetonbauteilen* (design of prestressed concrete components) written by an authoritative person.

"This edition of the *Beton-Kalender*," it says in the preface to the 1954 edition, "wishes to give itself some credit. We include for the first time a textbook of prestressed concrete design in the shape of a contribution by [Prof. Hubert] Rüsch entitled 'Bemessung von Spannbetonbauteilen'." [n.n., 1954/1, p. III]. Co-author of this article was Rüsch's colleague and successor at Munich TU Herbert Kupfer. As prestressed concrete at that time was developing highly dynamically, the compilers of DIN 4227 took great care to include as a priority the basic ideas regarding design and construction in the text of the standard and as far as possible to avoid including any ready-made solutions for certain types of construction. "The general rule," Rüsch and Kupfer write, "that the designing and contracting engineers must take the responsibility for every structure alone, and therefore are forced to choose the best way themselves in every case, applies to prestressed concrete components to an even greater extent" [Rüsch & Kupfer, 1954, p. 401]. Owing to the engineering responsibility in prestressed concrete construction, but also to improve clarity, the authors of DIN 4227 refrained from giving reasons for the requirements they prescribed in the standard. With classical lucidity, Rüsch and Kupfer distinguish between the intentions in DIN 4227 [n.n., 1954/2], the accompanying explanatory notes [Rüsch, 1953] and their article *Bemessen von Spannbetonbauteilen* [Rüsch & Kupfer, 1954]: whereas DIN 4227 only establishes **what** is to be verified, the explanatory notes contain the answer to the question as to **why**, and the article *Bemessen von Spannbetonbauteilen* covers in detail **how** the calculations are to be carried out to satisfy the requirements of DIN 4227. The triad of administration, science and industry developed in section 9.2 reappears here (see Fig. 9-9): authors of standards (as a rule found in building administration) establish **what** is to be verified, authors from science answer the question as to **why**, and authors from industry (which includes the group of consulting engineers and checking engineers) explain in detail **how** the calculations are to be carried out to satisfy the requirements of the standards. The prestressed concrete "textbook" by Rüsch und Kupfer, updated yearly in the *Beton-Kalender*, the prestressed concrete standard DIN 4227 and the associated explanatory notes ensured that readers always had the latest findings and experiences in prestressed concrete construction at their fingertips in the form of practical knowledge. This was a critical prerequisite for the unstoppable rise of prestressed concrete in Germany.

### 9.3.2.3 The unstoppable rise of prestressed concrete reflected in *Beton- und Stahlbetonbau*

The reconstruction of the war-torn Federal Republic of Germany lent the building industry an exceptional status in the national economy up until the 1970s. In housing and in infrastructure projects, not only was it necessary to replace what had been damaged in the war, but – driven by an expanding population and growing mobility – there were additional building works to be undertaken. So structural engineering for roads, bridges, tunnels and towers, but also in industry, plants and buildings was exposed to a hitherto unknown set of challenges in both quantitative and qualitative terms. Prestressed concrete became the symbol of progress in structural engineering. Bridge-building in the Federal Republic of Germany quickly became dominated by prestressed concrete, which took the place of structural steelwork (see [Pelke, 2007, p. 268]) because the former adopted the constructional language of the latter. The trend towards thin-wall hollow-box beams in prestressed concrete bridge-building is just one example. In 1955, the 50th year of its publication, the bridges listed in the annual index of the journal *Beton- und Stahlbetonbau* were nearly all of the prestressed concrete variety. Just a few years later, dividing bridges according to building material was abandoned because apart from a few exceptions *Beton- und Stahlbetonbau* reported exclusively on prestressed concrete bridges. For example, in 1975 almost 58 % of the 13.5 million m$^2$ of bridge deck on Germany's trunk roads consisted of prestressed concrete [Thul, 1978, p. 1]; according to Hans Wittfoth this share had risen to 65 % of 18.81 million m$^2$ by 1982 [Wittfoht, 1986, p. 35].

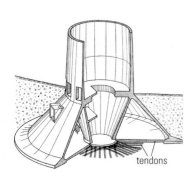

**FIGURE 9-39**
Foundation of Stuttgart TV tower
[Leonhardt, 1956, p. 76]

This expansion in prestressed concrete bridge-building was helped considerably by the incremental launching method, which first appeared around 1960 [Leonhardt & Baur, 1971] and helped industrial methods of fabrication to gain a foothold in the building of concrete bridges. The structural-constructional progress achieved through prestressed concrete in bridge-building was accompanied by technological progress.

The spread of television and VHF radio in the 1950s called for tall towers; Stuttgart TV tower by Fritz Leonhardt played a model role here. In 1956 Leonhardt reported in *Beton- und Stahlbetonbau* [Leonhardt, 1956] and other journals on how the advantages of shell construction and those of prestressed concrete construction had been united in the construction of the tower's foundations (Fig. 9-39). This constructional synthesis would also characterise the structural-constructional progress in reinforced concrete construction over the coming decades.

Reinforced concrete was also able to expand its dominating position in buildings, and the delicate reinforced concrete frame set standards in high-rise buildings. A listing of the authors writing in *Beton- und Stahlbetonbau* shows why this journal evolved into the scientific-technical key voice of structural engineering in the German-speaking countries and also enjoyed high international acclaim:

– leading engineers in the building industry such as
  Hermann Bay, Hans Wittfoth and Ulrich Finsterwalder,
– consulting engineers such as Kuno Boll, Heinrich Bechert,
  Max Herzog and Klaus Stiglat,
– leading engineers from the building authorities such as
  Bernhard Wedler, Hanno Goffin, Heribert Thul and
  Friedrich Standfuß,

- engineering scientists such as Hubert Rüsch, Fritz Leonhardt, Karl Kordina, Wolfgang Zerna and Jörg Schlaich.

Even today, the editors of the journal *Beton- und Stahlbetonbau* led by Konrad Bergmeister, ensure a dynamic equilibrium between authors from the building industry, the independent consultants, the building authorities and the universities.

### 9.3.3 The paradigm change in reinforced concrete design takes place in the Federal Republic of Germany too

After many years of preparatory work, the draft of the new edition of DIN 1045 "Structural use of concrete – design and construction" was published in March 1968; in agreement with the chairman of the German Committee for Reinforced Concrete (DAfStb), Prof. Bernhard Wedler, the editor of the *Beton-Kalender*, Gotthard Franz, included the draft together with the proposals for the revision of DIN 4224 "Aids for the practical application of DIN 1045" in the 1969 edition of the *Beton-Kalender* [n. n., 1969]. One year prior to that, readers of the *Beton-Kalender* had been able to find out about the non-$n$ cross-section design process of the GDR [Aster, 1968]. Many contributed to the discussion and the long overdue conversion of DIN 1045 to the non-$n$ cross-section design process was finally completed in 1971; the A to Z of the process was printed in the *Beton-Kalender* of 1971 [n. n., 1971]. Almost more important than the provisions of DIN 1045 concerning **what** had to be verified, was the detailed explanation of **how** the calculations should be carried out in order to comply with the requirements of DIN 1045. This latter process was described by E. Grasser, K. Kordina and U. Quast in their article entitled *Bemessung der Stahlbetonbauteile* (design of reinforced concrete components) ([Grasser, 1971; Kordina & Quast, 1971]). In the first part, E. Grasser develops the new, non-$n$ design process for reinforced concrete components in bending and bending plus axial force without buckling (Fig. 9-40), and provides the design diagrams and charts necessary for this [Grasser, 1971].

As the changeover from the principles of the design theory for bending – valid since 1904 – marked a fundamental change (see sections 9.2.2.2 and 9.2.2.3), Grasser's contribution contains numerous explanations and equations (background knowledge). Reinforced concrete in that period focused on the cross-section, which becomes clear from the

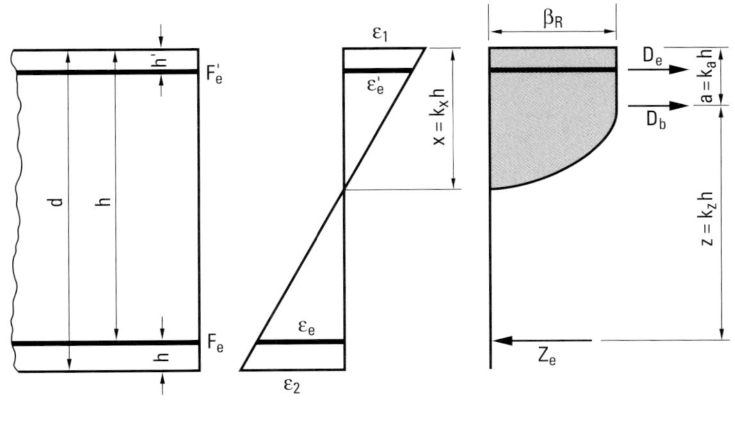

**FIGURE 9-40**
Internal stress and strain condition for the rectangular cross-section of a reinforced concrete beam at the ultimate limit state [Grasser, 1971, p. 518]

cover of the *Beton-Kalender*, which from 1984 to 1999 was decorated by T-beam sections. Not as intense as in Grasser's work is the mixing of practical knowledge with background knowledge in the second part written by K. Kordina and U. Quast and entitled *Bemessung von schlanken Bauteilen – Knicksicherheitsnachweis* (design of slender components – checking buckling) [Kordina & Quast, 1971], although here, too, the upheaval that the ultimate load method brought to the design system is clearly evident. The new design procedures were flanked in the 1972 edition of the *Beton-Kalender* by Heinz Duddeck's article on *Traglasttheorie der Stabtragwerke* (ultimate load theory for trusses) [Duddeck, 1972] and in the *Beton-Kalender* of 1973 by Fritz Leonhardt's article on *Das Bewehren von Stahlbetontragwerken* (reinforcement for reinforced concrete structures) [Leonhardt, 1973].

The aforementioned article by Grasser, Kordina and Quast was published in the *Beton-Kalender* in an updated version every year until 1997 ([Grasser, 1997; Kordina & Quast, 1997]) – a total of 27 editions. Rüsch and Kupfer's *Bemessen von Spannbetonbauteilen*, updated annually, also appeared in the same number of *Beton-Kalender* editions, the last one being that of 1980 [Rüsch & Kupfer, 1980], and after Rüsch's death the work was continued by Kupfer alone. However, the longevity of both chapters was surpassed marginally by E. Mörsch's chapter on arch bridges, which appeared in 31 editions up until 1952 and accompanied the development of concrete arch bridges in their heyday; a reprint of the 1952 version appeared in the *Beton-Kalender* of 2000 [Mörsch, 2000].

So from 1968 onwards *Beton-Kalender* accompanied the changeover from cross-section design based on elastic theory to that based on ultimate load theory.

### 9.3.4
### Revealing the invisible: reinforced concrete design with truss models

Around 1900 the structural modelling of steel loadbearing structures gained sustenance from the modelbuilding reserves of classical theory of structures, the focus of which was the theory of linear-elastic truss systems with trussed framework systems forming the most important aspect of this. Therefore, Kirsch understands the three-dimensional elastic continuum as a spatial trussed framework [Kirsch, 1868] and Mohr the solid-web beam in bending as a planar one (see Fig. 6-40c). Later, the trussed framework model would advance to become one of the historicological crystallisation points of the finite element method in the analysis of elastic plates with out-of-plane loading by Hrennikoff (1940, 1941). But the thinking of the structural engineer in the language of bar-like loadbearing systems in general and trussed framework-like loadbearing systems in particular was at this time not limited to structural steelwork, but indeed during the first half of the accumulation phase of structural theory (1900–25) – due to the enormous scientific legitimation power of classical theory of structures – made inroads into crane-building (see section 7.2.4) and reinforced concrete construction. In contrast to steel loadbearing structures, the structural modelling of reinforced concrete loadbearing structures needed an x-ray view of the internal workings of the loadbearing structure in order to comprehend the mechanical division of work between steel and concrete. Revealing the invisible therefore became a necessary prerequisite not only for the scientific analysis, but also for the constructional synthesis of loadbearing systems in reinforced concrete.

### 9.3.4.1 The trussed framework model of François Hennebique

The evolution from the first trussed framework model of a reinforced concrete beam (Fig. 9-41) to the truss models for consistent design in reinforced concrete construction is at the same time the evolution of the grammar of reinforced concrete construction with the aim of placing the "art of reinforcement" (Fritz Leonhardt) on a rational basis.

The first trussed framework model for a reinforced concrete beam (Fig. 9-41) can be attributed to François Hennebique (Fig. 9-41). According to Hennebique, the bars of flat steel (called links or, formerly, stirrups) placed around the longitudinal round steel bars should resist the shear forces. To calculate the maximum allowable shear force, Hennebique starts with the trussed framework model shown in Fig. 9-41: "Here, we normally assume that the stirrups together with the (longitudinal) bars and the concrete form a sort of trussed framework in which the stirrups represent the ties and the concrete, acting in the direction of the dotted lines, represents the struts. These lines are assumed to be at an angle of 45°, corresponding to the compression curves. Because of this approach, the stirrups are now calculated using the equation

$$Q = 2 \cdot \sigma \cdot b \cdot d \qquad (9\text{-}51)$$

where $Q$ is the shear force and $b$ and $d$ the width and thickness of the flat steel bars respectively. The factor of 2 is based on the fact that each stirrup has two legs. Amazingly, the inventor [= Hennebique – the author] remarks that the factor of 2 is added because one can assume that half of the shear stresses are taken by the round bars" [Ritter, 1899, p. 59, 60]. According to Ritter, eq. 9-51 from Hennebique assumes that the spacing $e$ of the shear links matches the lever arm $z$ of the internal forces because according to the trussed framework model shown in Fig. 9-41, the shear force must take on a value of

$$Q_{Ritter} = 2 \cdot \sigma \cdot b \cdot d \cdot \frac{z}{e} \qquad (9\text{-}52)$$

and eq. 9-51 is only correct for $e = z$. Paul Christophe [Christophe, 1902, pp. 588–589] also came to the same conclusion, and deals with Hennebique's and other design formulas under the heading of *Méthodes empiriques* [Christophe, 1902, pp. 572–597]. Ritter considers the trussed framework model and the ensuing eq. 9-52 for sizing the shear links as a hypothesis that had been neither verified nor discredited by any tests. To increase the shear capacity, Ritter suggests placing the shear reinforcement at the supports at 45° in the direction of the inclined principal tensile stresses.

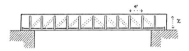

FIGURE 9-41
Hennebique's trussed framework model (after [Ritter, 1899, p. 60])

## The trussed framework model of Emil Mörsch

### 9.3.4.2

As early as 1903, Emil Mörsch, as the manager of the engineering office at Wayss & Freytag, arranged for four tests to be carried out on T-beams in order to establish the influence of the shear stresses in the reinforced concrete beam and the roles of shear links and longitudinal bars bent up at 45°. Mörsch reported on the results in the second edition of *Der Eisenbeton. Seine Theorie und Anwendung* [Mörsch, 1906, pp. 120–135]. The most important finding was the realisation that some of the longitudinal bars in the bottom of the beam have to be bent up through 45° in order to resist the shear forces. Nevertheless, he hoped that further shear tests would be undertaken: "Such shear tests are suitable for setting up general, practical design rules and forcing the system approach out of reinforced concrete construction" [Mörsch, 1906, p. 135]. On 23 February 1907 Mörsch presented the findings of further shear tests on 12 reinforced concrete beams, which had been organised at his request and sponsored by Wayss & Freytag, at the 10th general assembly of the German Concrete Association, ([DBV, 1907, pp. 129–158; Mörsch, 1907]). For the first time, a consistent design method for the quantitative assessment of the shear effect in reinforced concrete beams, backed up by tests, had been established, which Mörsch interpreted in structural theory terms by way of his trussed framework model (Fig. 9-42).

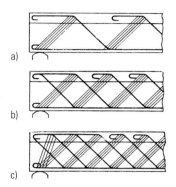

**FIGURE 9-42**
Mörsch's trussed framework model:
a) single strut system, b) double strut system, c) triple strut system
(after [Mörsch, 1922, p. 34])

He writes: "In this arrangement [with bent-up reinforcing bars – the author], the reinforced concrete beam can also be considered as a trussed girder with a system of single or double struts [Mörsch refers to Figs. 9-42a and 9-42b – the author], where the shaded strips of concrete designate the struts. One obtains an equal tensile force in the bent-up bars regardless of whether one resolves the shear force in the direction of the diagonals, similar to the situation with a parallel-chord girder with single or multiple systems, or whether one calculates them from the shear stress $\tau_0$" [DBV, 1907, p. 144; Mörsch, 1907, p. 224]. Mörsch introduced the triple strut system (Fig. 9-42c) somewhat later. The considerably shallower upward bend in the reinforcing bars of Hennebique's system is described by Mörsch as "trussing" reinforcement. Mörsch's presentation was acknowledged with lively applause. After the break, four persons spoke in the discussion, one of whom contributed in detail to the structural modelling of reinforced concrete beams based on trussed framework theory [DBV, 1907, pp. 160–161].

By the end of 1907 Mörsch had presented the schedule of work for shear tests to be carried out on behalf of the German Committee for Reinforced Concrete. which had been founded earlier that year. The tests were performed between 1908 and 1911 in the Materials Testing Institute at Stuttgart TH and reported on in the two publications by Carl Bach and Otto Graf [Bach & Graf, 1911/1 & 1911/2]. The final system of the schedule of work was based on the arrangement principle of the trussed framework model by Mörsch [Bach & Graf, 1911/2, p. 8], to which the model with triple struts had been added in the third edition of his book *Der Eisenbeton ...* [Mörsch, 1908, p. 169]. The "Tests on reinforced concrete beams for determining the shear resistance of various reinforcement arrangements" – the title of the two test reports – verified Mörsch's trussed framework model for the shear design of reinforced concrete beams which he had proposed shortly before. The care with which the tests were planned, carried out, evaluated and described were without equal in reinforced concrete research up to that time, and the series of tests can be regarded as a classic in reinforced concrete research. Fig. 9-43 shows three test beams from the test series of the German Committee for Reinforced Concrete which would prove crucial to the further development of reinforced concrete construction.

The 4 m-span test beams were T-beams (flange thickness 10 cm, flange width 60 cm, beam width 20 cm, beam depth 30 cm + 10 cm = 40 cm, cross-sectional area of bottom reinforcement $A_s$ = 25.3 cm²) loaded to failure with a uniformly distributed load. The maximum loads on the test beams were

- straight reinforcing bars 6 t/m = 24 t (Fig. 9-43a),
- straight reinforcing bars plus shear links 11 t/m = 44 t (Fig. 9-43b), and
- bent-up reinforcing bars 11.4 t/m = 45.6 t (Fig. 9-43c).

The figures next to the cracks in each case represent the load in tonnes at which the crack propagated at that point. From the crack patterns it can be seen that with no or insufficient shear reinforcement (e. g. shear links and bent-up longitudinal bars), premature failure of the beam at the point of maximum shear can occur such that the load-carrying capacity of the beam with respect to the bending moment is not fully utilised. Mörsch concluded from this that when designing a reinforced concrete beam, the task should be "to resist safely the effects of shear forces just as much as the effects of the moments" [Mörsch, 1922, p. 7]. Besides the shear cracks at approx. 45°, the principal stress trajectories are also drawn in Fig. 9-43c.

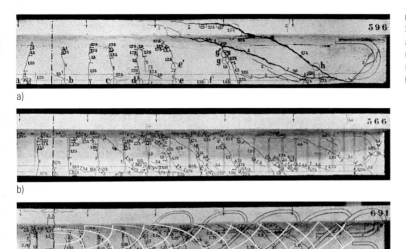

FIGURE 9-43
Stuttgart shear tests from 1908 to 1911: a) straight reinforcing bars, b) straight reinforcing bars with shear links, c) bent-up reinforcing bars showing stress trajectories (after [Mörsch, 1922, p. 5])

Whereas the bent-up reinforcing bars follow the principal stress trajectories, the concrete strips visible between the shear cracks resist the principal compressive stresses. At failure, a system of struts alternating with diagonals in the other direction at 45° is visible, the compression chord for which is formed by the concrete slab and the tension chord by the straight reinforcing bars; this trussed framework model was recorded in the annals of reinforced concrete construction as that of Mörsch (see Fig. 9-42).

#### 9.3.4.3 A picture is worth 1000 words: stress patterns for plane shell structures

When discussing plane shell structures, trajectory diagrams of the principal stresses (stress patterns) represent an important means of rendering visible the hidden flow of forces. Even in the early 1920s, Theophil Wyss was investigating the planar stress condition of riveted steel truss joints to ascertain the secondary stresses through experiments and drew stress patterns of these experiments [Wyss, 1923]. But it was the new language of reinforced concrete in the form of plane shell structures that first placed the stress pattern in its proper light. Together with the analysis of crack patterns from loading tests, such plane stress patterns could be condensed into trussed framework models. Mörsch, for example, understood the web of a reinforced concrete T-beam in the uncracked state (state I) as an elastic in-plane-loaded plate on two supports where the load is applied to its upper chord. He used the in-plane-loaded plate model to obtain a stress pattern and to determine the maximum principal stresses at selected points in order to interpret the shear effect of the corresponding loading test theoretically and to condense it into his trussed framework model. As, theoretically, the reinforcement must follow the principal tensile stresses, the reinforced concrete engineer relied on trajectory diagrams which had to be firmly rooted in his library of ideas and therefore formed the sounding board for the art of reinforcing. Only in that way can the engineer position the reinforcement sensibly and design the reinforced concrete structure economically. For instance, Hermann Bay, who helped Emil Mörsch in his reinforced concrete research, implemented the thinking in terms of stress patterns ingeniously in the new language of reinforced concrete. Bay's research into three-pin arch plates (Fig. 9-44) in the 1930s originated from considerations of how the longitudinal walls between the reinforced concrete arch and the road deck could be used to help carry the

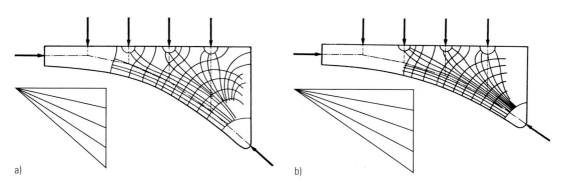

**FIGURE 9-44**
Stress patterns of three-pin arch plates (only one symmetric half shown) with a circular intrados from photoelastic tests;
a) rise f : span l = 1 : 4, b) f : l = 1 : 6
[Bay, 1960, p. 95]

loads instead of being excluded by introducing joints, as was frequently the case.

"With a uniform load," Bay writes, "a system of principal stress lines ensues in the arch plate irrespective of the $f:l$ value which is characterised by the position of the line of thrust within the plate and is essentially determined by the positions of line of thrust and underside of arch" [Bay, 1960, p. 94]. Using the numerical evaluation of the photoelastic measurements and stress-theory considerations, he developed a graphical method for determining the effective cross-sections of arch plates [Bay, 1960, pp. 111–113], which he supplemented with a practical method of calculation [Bay, 1960, pp. 113–121]. Wilhelm Fuchssteiner criticised the application of the theory of the arch plate to three-pin arch bridges of reinforced concrete within, on the one hand, the scope of the dichotomy of practical structural theory he had worked out, the task of which is to investigate a suitable equilibrium condition of the loadbearing system, and on the other hand, the scientific structural theory based on elastic theory [Fuchssteiner, 1954, p. 18]: the arch plate would be separated from the road deck (upper chord) and the arch (bottom chord) and actually investigated for itself alone. The holistic investigation of complex loadbearing systems formed the focus during the innovation phase of structural theory (1950–75) for the emerging subdiscipline of modelling (cf. [Müller, 1971; Hossdorf, 1976]), but quickly lost its importance as computer-based numerical methods gained ground.

Nevertheless, the illustrative power of stress patterns and their condensing to form trussed framework models in reinforced concrete construction led to a differentiated understanding of the internal mechanics of loadbearing systems. And therefore after the middle of the innovation phase of structural theory (1950–75) the trussed framework model enjoyed a comeback in the structural-constructional thinking of the reinforced concrete engineer. The following list of research work is just a small selection of the many proposals that were made (see also [Jürges, 2000, pp. 92–123; Schlaich & Schäfer, 2001; Sigrist, 2005]):

- Hubert Rüsch staked out the boundaries of the trussed framework model for calculating the shear strength [Rüsch, 1964].
- Herbert Kupfer extended Mörsch's trussed framework model and proved with the help of the principle of Menabrea (see section 6.4.1) that the struts must exhibit an angle less than 45° [Kupfer, 1964]; later, Utz Walthelm used the principle of Menabrea for selecting truss models (see, for example, [Walthelm, 1989]).
- Based on extensive shear tests, Fritz Leonhardt concluded that the distribution of the internal forces in reinforced concrete beams was not only due to equilibrium conditions, but that the deformation behaviour of the building materials and the compatibility of the deformations between the stiff concrete compression members and the ductile steel tension members was also relevant; Leonhardt proposed a trussed framework model with inclined compression chords [Leonhardt, 1965].

- Jungwirth was the first person to use the computer to re-analyse a truss geometry determined on the basis of the crack patterns of shear tests [Jungwirth, 1968].
- Cook, Mitchell and Collins considered the idealised force transfer and force redirection in joints as so-called disturbed regions [Cook & Mitchell, 1988; Collins & Mitchell, 1991].

Like reinforced concrete construction around 1900 had still not overcome its system period and several methods of design for bending coexisted, there was a market for trussed framework models in the field of shear measurements.

#### 9.3.4.4 The concept of the truss model: steps towards holistic design in reinforced concrete

Drucker presented the first stress fields for designing reinforced concrete structures based on plastic theory [Drucker, 1961]. In doing so, he used the static limit load principle of ultimate load theory (see section 4.6.3) and introduced the equilibrium approach for describing the flow of forces in reinforced concrete structures; Heyman's masonry arch theory (see section 4.6.4) is based on the same theoretical foundation. According to this, a structure made of plastically deformable materials does not fail when some (any) stress field can be found for the given loading which is in equilibrium and in addition satisfies the hinge condition, i.e. the yield point is not exceeded at any point in the structure. The Zurich school of reinforced concrete construction around Bruno Thürlimann continued this line of development in design theory systematically [Müller, 1978; Thürlimann et al., 1983; Marti, 1985 & 1986; Muttoni et al., 1996].

In 1984 Jörg Schlaich and Kurt Schäfer generalised the trussed framework-type flow-of-forces model to form the concept of the truss model, which formed the theoretical core of their reinforced concrete design article in the *Beton-Kalender* [Schlaich & Schäfer, 1984]; in the following years, updated versions of the article by Schlaich and Schäfer (1989, 1993, 1998, 2001) appeared in the same yearbook. In the concept of the truss model "the stress trajectories of individual stress fields in the loadbearing structure and the associated reinforcement forces are summarised as the struts and ties of the truss model and straightened, or the internal flow of forces is traced ... in some other way and idealised through a corresponding truss model" [Schlaich & Schäfer, 2001, p. 341]. Fig. 9-45 shows how the truss model is based on the linear-elastic stress trajectories.

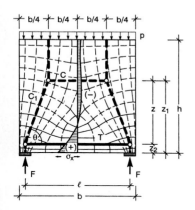

FIGURE 9-45
Truss model of a plate with in-plane loading and span l subjected to a uniformly distributed load p [Schlaich & Schäfer, 2001, p. 348]

Concrete has only a minor plastic deformation capacity. In plastic theory the truss model is selected such that the deformation capacity is not already exceeded locally before the stress field assumed becomes established in the rest of the structure. Even in highly stressed areas, with high concrete grades and heavy reinforcement, the aforementioned prerequisite is automatically satisfied by adapting the truss model and the arrangement of the reinforcement to the flow of forces according to a plastic theory. "By aligning the model on the flow of forces according to elastic theory it is possible to use the same model for the serviceability and ultimate limit states" [Schlaich & Schäfer, 2001, p. 341]. Determining stress fields according to elastic theory is therefore in no way superfluous. Using truss models it has been possible to design those areas rationally where in the past the engineer had to rely on his experience and intuition. Thanks to their practicality, truss models – and not least via their publication in *Beton-Kalender* – became very popular; they were even incorporated into Eurocode 2 without any assistance from their authors [Schlaich & Schäfer, 2001, p. 311].

# CHAPTER 10

# From classical to modern theory of structures

The philosophical writings of Sybille Krämer on the history of the operational use of symbols in mathematics and logic formed the starting point for a rethink of the history of structural theory. The formalisation of theory of structures by Müller-Breslau and Zuse's initial attempts to automate structural calculations were the subject of a discussion between the author and Konrad Zuse, the inventor of the computer, in January 1995. Heartened by this meeting, the author was able to conclude his work on the operative use of symbols in the genesis of the discipline. In the consolidation period of structural theory, the displacement method joined the force method to become the second principal method on the historical stage. Together with matrix analysis, this led initially to structural matrix analysis and later formed the hard core of computer-based structural mechanics. Computer statics would afterwards be superseded by the computational mechanics that was emerging during the 1980s: computer shapes theory. From now on, the technological character of such numerical engineering methods such as the finite element method would enable the sovereignty of the linear to be overthrown on an interdisciplinary scale and would also lead to a new mathematical foundation for the engineering sciences in the form of computational engineering.

The turning point in the genesis of the theory of structures discipline, the change from the classical to the modern, occurred directly after the classical phase of structural theory and was felt way into the first half of the consolidation period. The accumulation phase (1900 – 25) is characterised by
- a stagnation in the work on fundamental theories,
- the inclusion of new objects in theory of structures as new technical artefacts such as reinforced concrete loadbearing structures were embraced,
- the interpretation of disciplinary advances in structural analysis as a rationalisation movement for a specific intellectual engineering, and
- the formalisation of structural calculations increasingly becoming the focus of progress in theory.

As part of all this, the unfolding of the formalisation idea came more and more to the fore, both in the terminology and the organisation of structural analysis.

The arrival of the displacement method completed the dual nature of theory of structures. Around the middle of the 20th century, the displacement method became the foundation for computational mechanics, which together with the matrix theory reformulation of the whole of structural analysis was superseded by structural mechanics. During the innovation phase of structural theory (1950 – 75), new opportunities for loadbearing system analysis and synthesis appeared in the shape of numerical engineering methods tailored to the computer, e.g. the finite element method (FEM). And building microprocessors into computers after the mid-1970s brought about an exponential increase in computing power. Non-linear problems in theory of structures could now be systematically researched and turned into software for users. Ending the dominance of the linear and creating the algorithms for non-linear procedures were thus finalised in the final quarter of the 20th century, and on the disciplinary level this heralded the paradigm change from the scientific to the technological, the final step being taken with the advent of the Internet in the early 1990s.

**10.1 The relationship between text, image and symbol in theory of structures**

As the author of this book worked through the paragraphs on strength of materials in the first volume of Gerstner's *Handbuch der Mechanik* (manual of mechanics) (see section 3.2.2) published in 1831, he found it necessary to supplement his notes with numerous hand-drawn sketches in order to gain a proper picture of the mechanical model ideas, empirical findings and interpretations that Gerstner had expressed in the form of text, mathematical equations and tables of figures. The author added Gerstner's ideas on mechanics – expressed in a mass of text – to the engineering in order to grab hold of them – literally. The author admits that he relied on the pictorial manifestations of Gerstner's ideas in his excerpts because otherwise the writing work involved would have been immense. The copperplate engravings of the appendix made no difference here. This reading experience was not to remain an isolated instance; again and again this proved to be the case when studying the monographs dating from the early evolutionary days of classical fundamental engineering science disciplines. The engineering science model often strutted along in the ponderous trappings of algebraic scholastics. The establishment of lithography in the emerging engineering science literature and the scientific formulation of the knowledge of technical artefacts – initially only in the form of the graphical representation of the real technical artefacts through descriptive geometry and later also as a picture of the model idea of graphical statics – initiated a disarmament programme in favour of the image. This not only supplanted algebraic scholastics, but also made the text subservient to the picture. Even Weisbach's *Lehrbuch der Ingenieur- und Maschinenmechanik* (Principles of the Mechanics of Machinery and Engineering) (see section 3.2.3), published prior to the middle of the 19th century, included numerous wood engravings to break up the mass of text and thus announced – despite its conservative technical form – the decline of the text, the web of words. The appearance of graphical statics hardly two decades later is more than just a failed attempt to rescue the classical scientific ideal for the engineer via an axiom-based geometric formulation of mechanics. By understanding their axiomatisation attempts as a graphical representation, the advocates of this method used lithography and its easily technically reproducible world of images to roll a Trojan horse into the heavily fortified citadel of this discipline; in the final quarter of the 19th century it changed – "stripped of its spirit" [Culmann, 1875, p. VI] – into graphical analysis, into a mere tool of the engineer which survived the turn of the century only in the serial existence of its transfers. "And with reservations," writes Günther Anders, "the expression 'immortal' could even be justified for our products, even the frailest of them. And this is because there is a new variety of immortality: the industrial reincarnation, i. e. the serial existence of the products … Does not every lost or broken piece continue to exist in the image of its model idea? … Has it not become 'eternal' through its replaceability, in other words through reproduction technology?" [Anders, 1985, p. 51]. According to Anders the "Plato effect" here is that the pictures of graphical statics are standardised, intellectual serial products, "which have seen the light of day as imitations or copies of models, blueprints or dies; thanking ideas for their existence" [Anders, 1985, p. 52]. It was with the completion and consolidation of classical theory of structures by the Berlin school of structural theory through its model idea of line diagram analysis and $\delta$-symbols that society's need for an automated intellectual technology of the engineer first appeared, initially on the level of individual sciences.

Anders' thesis, which he developed in 1956 for the purpose of his philosophical observations on radio and television, is given a concrete form in the following using the example of idealised technical artefacts from structural analysis. Fig. 10-1 shows a subset from the world of idealised technical artefacts which characterised the final years of the accumulation phase of structural theory (1900–25). Such idealised technical artefacts are called structural systems: simply supported beam, two-span beam with cantilever, balanced cantilever, continuous beam, rigid frame, frame and arch systems, beam grid, Vierendeel and trussed girders. With the help of these and the idealised technical artefacts of line diagram analysis synthesised from these, engineers created and still create the built environment. Hopefully, the picture of the model idea of line diagram analysis will appear in the mind's eye of the reader.

### 10.1.1 The historical stages in the idea of formalisation

Before we undertake an analysis of the ambivalence of thinking in pictures in engineering activities from the establishment phase (1850–75) to the end of the consolidation period (1900–50) of structural theory, it is important to explore the idea of the practical use of symbols, which Sybille Krämer has impressively revealed via the history of the symbolic machine. The core thinking behind this idea, which gained a place in mathematics with the invention of algebraic notation by Vieta, is the "schematic, non-interpretive handling of written symbols" [Krämer, 1988, p. 176]. Her basic idea consists of "separating the manipulation of expressions from their interpretation", the aim being "to relieve the mind from the tediousness of interpretation" [Krämer, 1988, p. 176]. Formal operations are not only linked with the use of written symbols, they involve the linearisa-

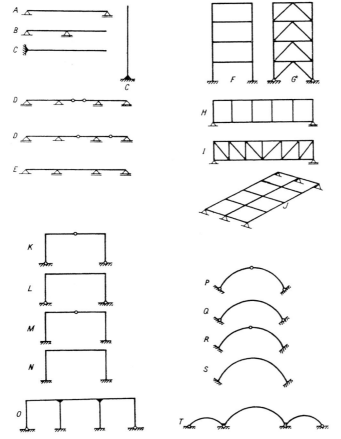

**FIGURE 10-1**
Model pictures of line diagram analysis [Zumpe, 1975, p. 70]

Structures of linear members:
A – Simply supported beam, B – Simply supported beam with cantilever, C – Cantilever beam, D – Balanced cantilever, E – Continuous beam, F – Rigid frame, G – Trussed frame, H – Vierendeel girder, I – Trussed girder, J – Beam grid, K – Three-pin frame, L – Two-pin frame, M – Single-pin frame, N – Fixed-based frame, O – Multi-bay frame, P – Three-pin arch, Q – Two-pin arch, R – Single-pin arch, S – Fixed-end arch, T – Continuous arch

tion of perception; the schematic use of symbols is also inherent to formal operations: "As we form and reform expressions," writes Sybille Krämer, "we must behave as though we were a machine" [Krämer, 1988, p. 178]. Sybille Krämer has proposed three historical stages of formal operations [Krämer, 1988, pp. 181–182]:

*1st Stage:* Symbols represent objects specified prior to and independently of their symbolic representation; they embody an extra-symbolic meaning (e. g. the algebra of the pre-modern age).

*2nd Stage:* Symbols are regarded as variables representing unspecified objects of a specific range of interpretation; these objects can only be symbols, e. g. the letters of Vieta's algebra stand for numbers. Symbols that serve as variables are symbols standing for symbols.

*3rd Stage:* If variables have the functions of formalised theory, i. e. serve as the basic characters of a formalised language, the stage of non-interpretive use of symbols has been reached; such symbols have an intra-symbolic meaning because the rules by which the symbolic expressions are formed and transformed do not contain any reference to the meaning of the symbols. Their range of interpretation is basically unspecified. Various models can be found for this formalised theory (e. g. Boolean algebra, matrix algebra, tensor analysis).

The first two stages of formal operations with symbols can also be used to analyse thinking in pictures in engineering activities, provided we understand graphical characters in the sense of geometry and geometrical mechanics as a specific form of symbols.

**First example: establishment phase of structural theory (1850–1875)**

During the constitution phase of structural theory (1825–50), texts and images were still kept totally separate. For example, Gerstner's *Handbuch der Mechanik* comprised three volumes of text and three volumes of copperplate engravings. It was in 1798 that Alois Senefelder (1771–1834) invented lithography, a method that was initially employed for printing texts and music. In contrast to the copperplate engraver, the lithographer did not have to overcome the resistance of the material; furthermore, lithography, in conjunction with printing presses, was suitable for the mass production of printed works, large print runs. "Lithography," writes Walter Benjamin, "enabled reproduction technology to move up to a whole new level. This very much more precise method, in which the drawing is transferred to a stone instead of being cut in a block of wood or etched into a copper plate, gave graphics its first-ever chance not only to market its wares on a massive scale (as heretofore), but rather in new designs changing by the day. Lithography enabled graphics to illustrate everyday life on a daily basis. It began to keep pace with printing" [Benjamin, 1989, p. 351]. What Benjamin noted on the "works of art in the age of their technical reproducibility" (Benjamin) also applies to engineering science literature. This is why in the history of engineering literature the lithographed lecture manuscripts of Jean Victor Poncelet, who taught geometry and applied mechanics at the École d'Application de l'Artillerie et du Génie in Metz between 1824 and 1838, play such a sig-

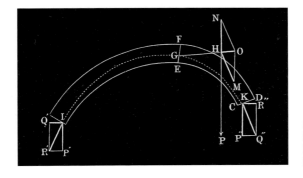

**FIGURE 10-2**
Scheffler's theory of structures model illustrated with the help of wood engraving [Scheffler, 1857, p. 139]

nificant role; they were produced in the school's own lithography shop in print runs of about 500 copies each (see [Kahlow & Kurrer, 1994, p. 100]).

Nevertheless, it was not until the establishment phase of structural theory (1850–75) that text illustrations started to enjoy widespread use in the engineering literature. Prominent examples are Weisbach's *Lehrbuch der Ingenieur- und Maschinenmechanik* (see section 3.2.3) and Rankine's *Manual of Applied Mechanics* (see section 3.2.4). And Hermann Scheffler's book *Theorie der Gewölbe, Futtermauern und eisernen Brücken* (theory of masonry arches, revetments and iron bridges) [Scheffler, 1857] was another that made use of text illustrations (Fig. 10-2); it was accompanied by an appendix with plates, with lithographs.

All the lithographs in the appendix of Scheffler's book are graphical representations of the mechanics of real technical artefacts which the engineer can now refer to in his conception and design work. Fig. 10-3 shows the graphical determination of thrust lines in asymmetric masonry arches. As in this case scale and dimensions are important and especially high demands must be placed on the accuracy of the graphic calculation of the thrust line in the masonry arch profile, a wood engraving is unsuitable as a die. The text illustrations, on the other hand, can be in the form of wood engravings (Fig. 10-2) because the theoretical principles are developed in the text and need to be shown in the form of sketches, not to scale but in the right proportions, e. g. constant arch thickness, the asymmetry of the masonry arch. Such wood engravings had already turned graphical analysis into an intellectual means of rationalisation in engineering activities in the sense of modern rules of proportion for the civil and structural engineer (see section 6.2) because text and text illustrations highlight the operations and methods of engineering work, and model these in anticipatory fashion. In Scheffler's case the text illustration had not yet attained the status of the image of the model idea of line diagram analysis (see Fig. 10-1), i. e. the second stage of formal operations is present here only in vague terms. Although it is already realised in the form of the translation of the algebraic equations of static equilibrium into the graphical representation of the external forces (parallelogram of forces), the mechanical model of the masonry arch, however, had not yet become fully detached from its real technical artefact, was essentially still governed by its extra-symbolic meaning (first stage). Had Scheffler interpreted the centre-of-gravity axis of the masonry arch as an arch and called the masonry material a Hookean body, i.e. searched for the thrust line of an elastic arch, then this model could have been allocated fully to the second stage of formal operations. His adherence to the geometric proportions of the masonry arch and the inconsequential modelling of the material behaviour prevented this. Linearising the masonry arch proportions to form an elastic curved line fixed at the abutments would be the second constituent of the image of the model idea of line diagram analysis. Scheffler could have reduced dozens of text illustrations of masonry arches to a single one; and more besides, this one model illustration could have represented iron and timber arches as well, provided the material of these arches had been abstracted to the Hookean body. Such a compilation did not happen until classical theory of structures formed at the end of the discipline-formation period of structural theory, with Heinrich Müller-Breslau its most famous advocate.

## Second example: graphical statics and graphical analysis between the establishment and classical phases of structural theory

Like Scheffler's monograph, Culmann's *Graphische Statik* (graphical statics) has a dual structure. In the text with its equations and wood engravings, he develops graphical statics on the basis of projective geometry explicitly, and graphical analysis implicitly, whereas the lithographs of the appendix are dominated by the graphical representations of engineering practice. However, the main difference is that Culmann does not simply translate text and equations into graphics, but aims to provide a geometric reasoning for his mix of geometry and statics (graphical statics). At the same time, he develops – in graphical analysis – a model for rationalising engineering activities by geometrising, i.e. treats *epistēmē* and *tekhnē* as parallel components. In the end, the failure of Culmann's theory programme can be found in the fact that graphical analysis did not allow

**FIGURE 10-3**
Graphical calculation of the thrust line in a masonry arch, for which Scheffler employs lithography [Scheffler, 1857, Figure XV]

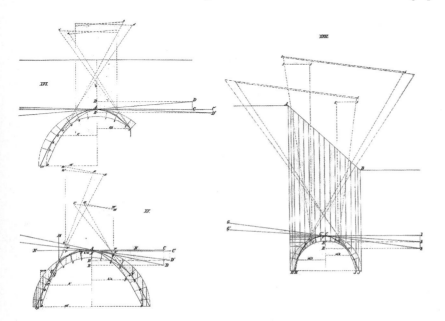

*tekhnē* to reach the status of an *epistēmē*, and graphical statics essentially had to rely on itself as an *epistēmē*.

Although graphical analysis experienced an enormous cognitive expansion during the classical phase of the discipline-formation period of structural theory (1875–1900) and during this time determined the conception and design activities of civil and structural engineers, it quickly lost its significance for structural theory development after 1900 because compared with line diagram analysis based on linear algebra its rationalisation potential had been exhausted and it could make only an insignificant contribution to placing engineering activities on a scientific footing. Graphical analysis, too, reached the second stage of formal operations only on the level of the geometrising of mechanical relationships (polygon of forces and funicular polygon) because its rationalisation potential unfolded precisely in the graphical representation of the unity of conception, calculations and design. In all technical rules of proportion the graphical symbols represent objects, i. e. the first stage of formal operations prevails. On the other hand, graphical statics tends to employ the graphical symbols schematically in the sense of projective geometry and liberate them from the burden of interpretation. Nonetheless, graphical statics was able to achieve the second stage of formal operations to a limited extent only because its graphical symbol language is only suitable for specific statically indeterminate systems.

### Third example: classical phase of structural theory (1875–1900)

Müller-Breslau's *Graphische Statik der Baukonstruktionen* (graphical statics of structural theory) is another publication with a dual structure. "The starting point here," he writes in the second edition of volume II of this work (1903), "forms an analytical basis through the law of virtual displacements [he means the principle of virtual forces – the author] and the principle of the reciprocity of elastic deformations resulting from this, first proved for the simple specific case by Maxwell and extended by the author, which at first sight appears somewhat unsuitable for a textbook on graphical statics. However, those who look into the issue of elastic theory are always forced to perform certain preliminary calculations, and in the light of this situation abandoning such a magnificent tool as we have in the newer analytical theory and replacing this by more awkward aids seems hardly justified. A broad field still remains open to the graphical method" [Müller-Breslau, 1903, pp. V – VI]. In his text with equations and text illustrations, he devises – and this is where he differs from Culmann – the theory of statically indeterminate trussed frameworks based entirely on one single principle: the principle of virtual forces. And he gives precedence to the idea of formal operations with algebraic symbols in the form of the elasticity equations in trussed framework theory. He prepares numerical examples for relevant graphical analysis methods in the text and specifies their construction in lithographed plates. These graphical methods of graphical analysis are now merely *tekhnē* and supplement the rational analysis of trussed frameworks. Graphical analysis is now a set of formulas devoid of any opportunity for assimilating engineering science

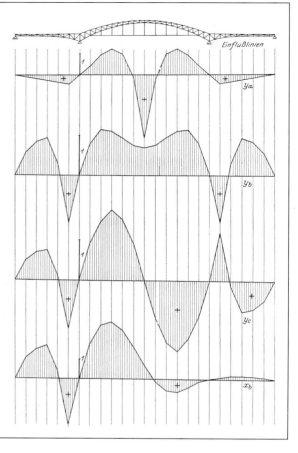

in der Reihenfolge *22* bis *12*. Ist $a_{12}$ der letzte Wert der zweiten Additionsreihe
$$a_{12} = \tfrac{1}{2}\lambda \cdot M w'_{12} + 6c,$$
dann ist $c$ aus der Bedingung
$$\tfrac{1}{2}\lambda M w'_{12} = \tfrac{1}{2}\lambda \delta'_{12c} = -\tfrac{1}{2}L_{12c}\cdot h'_{12}$$
zu bestimmen.
$$c = \tfrac{1}{6}(a_{12} + \tfrac{1}{2}L_{12c}\cdot h'_{12}).$$

Um $\tfrac{1}{2}\lambda \cdot \delta'_{n_1c}$ zu erhalten, ist $(6 - \tfrac{1}{2}n_1)c$ von $a_{n_1}$ abzuziehen.

Aus $\delta'_{14}$ und $\delta'_{12}$ wird weiter $\delta'_{10}$ mit Hilfe von $w'_{12}$ berechnet. Da im Zustand $Y_a = -1$ $w'_{12} = 0$ ist, ergibt sich $\delta''_{10a} = -\delta'_{14a}$. Die Biegungslinie *14, 12, 10* ist eine Gerade. Für die Zustände $Y_b = -1$, $Y_c = -1$ berechnet man die Verschiebung $\delta''_{10}$ gegen die Gerade durch *12* und *14* aus
$$\lambda^2 \delta''_{10} = -\lambda^2 w'_{12}\cdot 2\lambda$$
und addiert $\delta''_{10}$ zu $-(\delta'_{14} - \delta'_{12}) + \delta'_{12}$ Mithin ist
$$\tfrac{1}{2}\lambda \cdot \delta'_{10b} = -\tfrac{1}{2}\lambda \cdot \delta'_{14b} + \lambda \cdot \delta'_{12b} - \lambda^2 \cdot w'_{12b},$$
$$\tfrac{1}{2}\lambda \cdot \delta'_{10c} = -\tfrac{1}{2}\lambda \cdot \delta'_{14c} + \lambda \cdot \delta'_{12c} - \lambda^2 \cdot w'_{12c}.$$

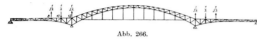

Abb. 266.

In den Knotenpunkten *0* bis *10* verläuft die Biegungslinie aus $Y_a = -1$ geradlinig. Für die Zustände $Y_b = -1$ und $Y_c = -1$ haben die $w'$-Gewichte dieselben Werte, die Ordinaten der Biegungslinien unterscheiden sich nur durch die Verschiedenheit der Ordinate $\delta'_{10}$. Man rechnet zweckmäßig nach folgender Tabelle, in der
$$b_n = \lambda^2 V_n + C, \qquad a_n = \tfrac{1}{2}\lambda M_n + nC$$
ist.

| $n$ | $\lambda^2 w'_m$ | $b_n$ | $a_n$ | $nC_b$ | $nC_c$ | $a_n - nC_b$ | $a_n - nC_c$ |
|---|---|---|---|---|---|---|---|
| 2 | $\lambda^2 w'_2$ | $b_4 + \lambda^2 w'_2$ | $b_2$ | | | | |
| 4 | $\lambda^2 w'_4$ | $b_6 + \lambda^2 w'_4$ | $a_2 + b_4$ | | | | |
| 6 | $\lambda^2 w'_6$ | $b_8 + \lambda^2 w'_6$ | $a_4 + b_6$ | | | | |
| 8 | $\lambda^2 w'_8$ | $\lambda^2 w'_8$ | $a_6 + b_8$ | | | | |
| 10 | 0 | 0 | $a_8$ | | | | |

Nach Ermittlung von $a_{10}$ erhält man
$$C_b = \tfrac{1}{10}(a_{10} - \tfrac{1}{2}\lambda \delta'_{10b}),$$
$$C_c = \tfrac{1}{10}(a_{10} - \tfrac{1}{2}\lambda \delta'_{10c})$$
und weiter die beiden letzten Spalten der Tabelle, welche $\tfrac{1}{2}\lambda \delta'_{mb}$ und $\tfrac{1}{2}\lambda \delta'_{mc}$ angeben.

Man kann auch ein $w'$-Gewicht in den Knoten *10* ansetzen, dessen Wert aus der in Abb. 266 dargestellten Belastung zu berechnen und

**FIGURE 10-4**
Fusing of text and illustrations in Grüning's book using the example of the development of influence line theory [Grüning, 1925, pp. 378/379]

knowledge. In the rigid frame system that started to become popular after 1900 in reinforced concrete construction and later in structural steelwork (see section 9.3.1.1), graphical analysis played a role only in the practice of structural calculations.

Formal operations – with their mathematical consummation in the range of interpretation of linear algebra – by means of the algebraic symbols of structural theory supplanted the graphical methods of graphical analysis apart from a few minor leftovers. A decisive indicator of this is Müller-Breslau's conversion of his formalisation of the elasticity equations from trussed to rigid frameworks in 1908 [Müller-Breslau, 1908]. And this, too, operates under the name of *Graphische Statik der Baukonstruktionen* – and with only two lithographs in the appendix!

### Fourth example: structural analysis in the consolidation period (1900–50)

Martin Grüning supplanted trussed framework theory, rigid framework theory and beam theory in 1925 in his *Statik des ebenen Tragwerkes* (structural analysis of plane loadbearing structures) by referring merely to the idealised technical artefact of line diagram analysis (see Fig. 10-1) and by departing completely from the real object of structural theory. In his text he develops the linear algebra of structural theory still without employing matrix formulation and depicts this in text illustrations (Fig. 10-4); numerical examples solved with the help of the diagrammatic methods of graphical analysis are absent. The latter can be found in compendiums for building trade or state-run building school students. The symbolic graphic representation has completed the metamorphosis to the model world of line diagram analysis (Fig. 10-1), concealing the representation in symbols of linear algebra. This was also in line with the typesetting and printing methods; Grüning's book was typeset with the help of hot-lead composition and printed on flat-bed presses. Electrotype, a method in which the drawings can be easily incorporated in the text, was certainly used for the printing plate. And hence the dual organisation of the engineering science book according to text and appendix of plates was made obsolete by reproduction technology.

Theory of structures during the consolidation period reflected its linear-algebraic foundations and on the disciplinary level attained the second stage of formal operations. The linear-algebraic symbols of structural theory were valid for the graphical symbols of the structural systems, which in turn represented real technical artefacts symbolically. Nevertheless, the $\delta$-symbols introduced by Müller-Breslau and disseminated internationally by the Berlin school of structural theory seemed to pave the way from the second to the third stage of formal operations. In formal terms, the principle of Maxwell and Betti based on the general work theorem

$$\delta_{jm} = \delta_{mj} \tag{10-1}$$

consists of exchanging the indices. If, for example, we require the influence line $\eta_m$ for the displacement $w_j$ caused by the travelling load 1 at $m$ (Fig. 10-5a), i.e.

$$\eta_m = \delta_{jm} \tag{10-2}$$

where with the $\delta$-symbol the first index stands for type, location and effect, the second for the cause of the magnitude, then in eq. 10-1 the indices are exchanged and then interpreted as a deflection curve due to the point load 1 at $j$ (Fig. 10-5b):

$$\delta_{mj} = w_j \tag{10-3}$$

Cause and effect have been formally exchanged and according to the general work theorem changed substantially by converting the influence line according to eq. 10-2 into a specific internal force distribution – the deflection curve according to eq. 10-3. The transformation of the cause-effect relationship into a means-purpose relationship (the *tekhnē*) completed with eq. 10-2 is reflected with eq. 10-1 in the transformation of the purpose-means relationship into an effect-cause relationship (the *epistēmē*) completed with eq. 10-3 (see section 3.1.3). The $\delta$-symbol therefore had the function of a formalised theory as it advanced to become the basic symbol of the not fully formalised language of the Berlin school of structural theory and represented the applicability of the influence and state variables staked out by the school. The influence variables (e.g. influ-

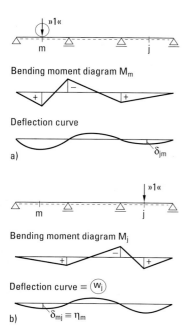

**FIGURE 10-5**
Transformation of the influence line (a) into a deflection curve (b) for a continuous beam

**The structural engineer – a manipulator of symbols?**

ence line $\eta_m$ in eq. 10-2) and the state variables (e.g. the deflection curve $w_j$ in eq. 10-3) are algebraic symbols and belong to the second stage of formal operations. Although symbolic expressions can be formed and transformed with the $\delta$-symbol and are as such intra-symbolic because eq. 10-1 contains no reference to the meaning of the $\delta$-symbol, the symbol transformation is, however, only possible when we assume that the interpretation of the input and output symbols are algebraic symbols of the second stage of formal operations in the sense of classical theory of structures; the range of interpretation of the $\delta$-symbols has thus been defined in principle. The $\delta$-symbol is therefore a pseudo formalised theory of engineering science. Apart from these limitations, *tekhnē* had in principle achieved the status of *epistēmē* in theory of structures.

Based on this concrete historical developmental stage of structural theory (which is best illustrated by the Berlin school of structural theory) and the need to rationalise the extensive calculations needed for systems with multiple degrees of static indeterminacy (resolutely pursued by the Berlin school in particular), the Berlin-based engineer Konrad Zuse resolved the calculation of a bridge frame with nine degrees of static indeterminacy into basic arithmetical operations and prepared these in such a way in a computing plan that the calculations could be carried out without knowledge of theory of structures [Schwarz, 1979, p. 362]. That was in 1934. Zuse generalised this computing plan still based on the model image of line diagram analysis (Fig. 10-1) to become the "computing plan or program method" [Zuse, 1993, p. 168]. This shift also allowed Zuse to supersede the pseudo formalisation proposed by the Berlin school of structural theory, and so he was able to use symbols not only schematically but soon completely freed from any interpretation.

Zuse's computing plan formed the starting point for the world's first program-controlled computing machine, the Zuse Z3 built in 1941. And so the third stage of formal operations had been realised technically. *Tekhnē* had been transformed into *epistēmē* and vice versa. The system of non-classical engineering sciences took shape with automation engineering (see section 3.1.3).

**10.1.2**

We shall once again take up Sybille Krämer's principal notion of the historical route of the idea of formalisation to the symbolic machine; she writes: "A process can be described in formal terms provided it is possible to represent this with the help of synthetic symbols in such a way that the conditions for the typographic, schematic and non-interpretive use of symbols are satisfied. A process that satisfies these conditions can also be performed as an operation of a symbolic machine … Computers are machines that can imitate any symbolic machine" [Krämer, 1988, pp. 2–3].

The world of images of the practising engineer was parcelled up in the computational mechanics absorbed into the non-classical fundamental engineering science discipline of structural mechanics to form model images suitable for discretisation and algorithmisation. At first this allowed the computer to generate endless columns of figures and later practically any reproducible flood of images, a world of the engineer à la Plato, which tended to turn him into a manipulator of symbols, into the objective idealist. Must he model the built environment according to the serial product of the illustrations of his discretisable and algorithmisable model world? Must he crouch all his life ignorant in Plato's cave, perceiving not the original idea, but rather only the shadow on the cave wall to which he directs his attention? Must he recognise his presence in the light of real moments in order to, in the end, forget the difference between essence and semblance, between reality and image?

## 10.2 The development of the displacement method

One hundred years ago, the force method formed the nucleus of classic structural theory. Today, the displacement method is one of the most important mainstays of modern structural mechanics, and played a decisive role during the transition from classic structural theory to modern structural mechanics in the 1950s and 1960s. The internal structure of this method was ideally suited to implementation on computers, and today it is also used as an introduction to the principles of modern structural mechanics. A study of the evolution of the displacement method reveals very clearly the dialectic interaction between the logical and the historical.

Goethe's proposition that "the history of science is science itself" was also a central theme for Edoardo Benvenuto, as indicated by the title of his book *La Scienza delle Costruzioni e il suo sviluppo storico* [Benvenuto, 1981]. Benvenuto was the first person to demonstrate that Alfred Clebsch had already developed the basic concept of the displacement method in his book of 1862 *Theorie der Elasticität fester Körper* (theory of elastic rigid bodies) [Clebsch, 1862].

This section will show that the displacement method also has genuine engineering science origins in the problem of secondary stresses in riveted iron and steel trussed frameworks. The theory of secondary stresses in trussed frameworks developed by Manderla, Winkler, Engesser, Ritter, Landsberg, Müller-Breslau and others in the 1880s was relevant for structural theory because it advanced the basic ideas of second-order theory and introduced displacement variables as unknowns for determining the forces in statically indeterminate trussed frameworks. Otto Mohr's contribution of 1892/93 was not only the final piece in the historico-logical puzzle surrounding the theory of secondary stresses in trussed frameworks, but also the springboard for the development of the displacement method. But it was not until the use of reinforced concrete became widespread after 1900 that a clear, rational and coherent theory of analysis commensurate with the monolithic nature of this form of construction became indispensable for statically indeterminate frames. In 1914 the Danish engineer Alex Bendixsen latched onto Mohr's clear and yet so simple procedure and applied it not only to rigid-jointed frameworks, but also to braced and sway frames. Seven years later, Asger Ostenfeld arranged the conditional equations for displacements in the same form as the elasticity equations for the force method, which were already well known to structural engineers;

he introduced the term "displacement method" and recognised its formal duality with the force method. And it was in 1927 that Ludwig Mann investigated the displacement method against the background of Lagrange's *Mécanique analytique*. Up until the 1950s, reflection on the duality of the force method and the displacement method remained a favourite subject of research into theory of structures. It was only the application of matrix formulation that enabled aircraft construction researchers to integrate the whole of structural theory into computer mechanics and modern structural mechanics. This also marked the beginning of the replacement of the force method by the displacement method.

### 10.2.1 The contribution of the mathematical elastic theory

The mathematical elastic theory was well established by the close of the 19th century. In the first edition of his book *A treatise on the mathematical theory of elasticity* [Love, 1892/93], August Edward Hough Love (1863–1940), provided a complete overview of this scientific discipline. The second edition of the book was translated into German by Aloys Timpe at the suggestion of Felix Klein (1849–1925). Timpe's translation, published in 1907 under the title *Lehrbuch der Elastizität* [Love, 1907], starts with a 38-page historic introduction describing the significant stages in the development of the mathematical elastic theory and culminates in the formulation of its disciplinary identity: "The history of the mathematical theory of Elasticity shows clearly," remarks Love, "that the development of the theory has not been guided exclusively by considerations of its utility for technical Mechanics. Most of the men by whose researches it has been founded and shaped have been more interested in Natural Philosophy than in material progress, in trying to understand the world than in trying to make it more comfortable" [Love, 1907, p. 37]. In particular, Love notes that the solution methods of the mathematical elastic theory form a substantial part of analytical theory, "which is of great importance in pure mathematics" [Love, 1907, p. 37]. Even for the analysis of more technical problems it is suggested that "attention has been directed, for the most part, rather to theoretical than to practical aspects of the questions. To get insight," Love continues, "into what goes on in impact, to bring the theory of the behaviour of thin bars and plates into accord with the general equations – these and such-like aims have been more attractive to most of the men to whom we owe the theory than endeavours to devise means for effecting economies in engineering constructions or to ascertain the conditions in which structures become unsafe" [Love, 1907, p. 38].

Love's men of mathematical elastic theory of the 19th century
- were more interested in causality than in finality,
- moved more in the world of the ideal objects of mathematics than in the world of real objects of engineering,
- saw themselves more as discoverers of the laws of nature than as inventors of technical artefacts,
- regarded their discipline more as a theoretical natural science than as a practice-based discipline of the classic engineering sciences,

- interpreted their discipline more in the context of the contemporary philosophical discourse than in the context of the material needs of industrialisation, and
- saw themselves more as men of academic science than as men of practical engineering.

What did the mathematical elastic theory of the 19th century contribute to the displacement method?

### 10.2.1.1 Elimination of stresses or displacements? That is the question.

As is generally known, the equilibrium conditions, the law of materials and the kinematic relationships produce 15 equations or partial differential equations having 15 unknown scalar functions with three variables, namely

- three displacements,
- six strains, and
- six stresses.

The logical core of elastic theory is characterised by this triadic structure. In principle, elastic theory offers two routes to a solution: elimination of stresses, and elimination of displacements.

If – for the case of complete linearity and homogeneous and isotropic bodies – the stresses and strains are eliminated from the set of equations, a vectorial differential equation remains:

$$\Delta \underline{w} + \frac{1}{(1-2\cdot v)} \cdot grad\,(div\,\underline{w}) = \frac{2\cdot(1+v)}{E} \cdot \underline{k} \qquad (10\text{-}4)$$

These three coupled partial differential equations for the calculation of the displacement vector $\underline{w}$ from the volume forces $\underline{k}$ and the two material constants $E$ (modulus of elasticity) and $v$ (Poisson's ratio) plus the geometric boundary conditions were named after Gabriel Lamé (1795–1870) and Claude Louis Marie Henri Navier (1785–1836). One could subsume approaches leading to the solution of the Lamé-Navier displacement differential equations under the term "displacement method" of mathematical elastic theory.

The second route leads via elimination of displacements and strains – again for the case of complete linearity and homogeneous and isotropic bodies – to the tensorial differential equation named after Eugenio Beltrami (1835–1900) and John Henry Michell (1863–1940):

$$\Delta \underline{S} + \frac{1}{(1+v)} \cdot grad\,(grad\,\sigma) = -[grad\,\underline{k} + grad^T\underline{k} + \frac{v}{(1-v)} \cdot (div\,\underline{k}) \cdot \underline{I}] \quad (10\text{-}5)$$

Taking into account the dynamic boundary conditions, the components of the stress tensor $\underline{S}$ ($\sigma$ = diagonal sum of stress tensor, $\underline{I}$ = unit tensor) can be calculated from these six coupled partial differential equations. Approaches leading to the solution of the Beltrami-Michell stress differential equations could be called the "force method" of mathematical elastic theory.

In the literature, the first route via the Lamé-Navier displacement differential equations and the geometric boundary conditions (specifying the displacements on the surface of the body) is called the first boundary value problem, and the second route via the Beltrami-Michell stress differential equations and the dynamic boundary conditions (specifying the forces on the surface of the body) is called the second boundary value problem of elastic theory [Leipholz, 1968, pp. 116 – 121]. As is so often the case, elastic theory offers yet another, third route: the third boundary value problem arises when forces are specified for one part of the surface of the body, displacements for another part [Leipholz, 1968, p. 121].

Having thus provided a brief description of the logical side of the displacement and force methods in the context of three-dimensional elastic theory, we will now move on to a historical-logical description of the displacement method in the context of elastic truss systems.

### 10.2.1.2 An element from the ideal artefacts of mathematical elastic theory: the elastic truss system

The aim of *Theorie der Elasticität fester Körper* [Clebsch 1862] (Fig. 10-6), published in 1862 by Alfred Clebsch (1833 – 72), who taught mathematics at Karlsruhe Polytechnic, was to provide a textbook of elastic theory "which ... should cover fully the theory and practice of this discipline" [Clebsch, 1862, p. V]. Having developed the basic equations of three-dimensional elastic theory and substantiated them with examples over 189 pages, Clebsch elaborates on the basic equations for bars and thin plates in the 166 pages of the second part. In the 69-page third part, Clebsch applies the equations derived in the second part to elastic truss systems; on pages 409 to 420, we find Clebsch's displacement method, which was later analysed by Edoardo Benvenuto (1940 – 98) [Benvenuto, 1991/2, pp. 492 – 498]. Clebsch describes the basic idea of the displacement method for calculating the displacements of the joints in elastic truss systems as follows: "Here, too, the general principle will be to treat the displacements of the joints initially as known parameters, to determine from them the elastic forces with which the bars react at their joints, and, finally, to establish the equilibrium conditions for the external and elastic forces acting at the joints; these equations will then enable the displacements induced to be calculated" [Clebsch, 1862, p. 413].

Clebsch developed the displacement method analytically, i. e. without any diagrams. Benvenuto can take credit for discovering Clebsch's displacement method for the scientific discipline of the history of structural mechanics and presenting it in a comprehensible manner. Clebsch explained his displacement method using the example of the calculation of the joint displacement in a simple pin-jointed trussed framework (without bending) and generalised it for elastic truss systems. Fig. 10-7 shows the three-pin system investigated by Clebsch with the vertical tie $1-2$ subjected to load $P$ at joint 1. The unknown quantities are the displacements of joints 1 ($w_1$) and 2 ($w_2$) in the $y$-direction. In the first step, Clebsch expresses the forces in bars $a$, $b$ and $c$ as a function of displacements $w_1$ and $w_2$; these displacements are then determined from the equilibrium conditions for joints 1 and 2.

Clebsch plainly recognised that the joint equilibrium conditions form a set of linear equations for determining the joint displacements: "The problem," he writes, "is therefore reduced to solving a set of linear equations, and can thus be regarded as fully solved" [Clebsch, 1862, p. 418].

**FIGURE 10-6**
Cover of the first German monograph on elastic theory

But the displacement method for the analysis of elastic truss systems as formulated by Clebsch was not adopted by engineers – for the following reasons:
- at the time, trussed framework theory was more concerned with member forces than joint displacements;
- it was much easier to investigate statically determinate pin-jointed trussed frameworks using the methods of graphical analysis that had emerged during the 1860s;
- Clebsch's monograph did not meet the clarity criterion for engineering textbooks.

Clebsch's textbook was thus used mainly by men whose cognitive objects represented elements of the ideal artefacts of the mathematical elastic theory. Nevertheless, Clebsch's displacement method anticipated the mathematical methodology of theory formation for a fundamental engineering sciences discipline based on elastic truss systems: theory of structures.

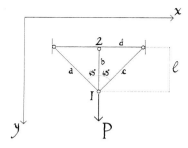

FIGURE 10-7
Analysis of an elastic truss system after Clebsch [Benvenuto, 1991/2, p. 497]

### 10.2.2 From pin-jointed trussed framework to rigid-jointed frame

Trussed framework theory undoubtedly occupies the "golden section" of the discipline-formation period of structural theory. And trussed frameworks dominated the loadbearing structures of civil and structural engineering in the second half of the 19th century. Whereas numerous trussed frameworks had been constructed from timber in the past, the trussed framework construction method only really took off when iron materials began to be used as structural elements. In the 1840s and early 1850s, trussed frameworks were dominated by composite systems made of timber, cast iron and wrought iron. It was with such structures in mind that Culmann introduced the term "trussed framework" in 1851 in the first of his two travelogues [Culmann, 1851] and also developed a trussed framework theory, implicitly assuming frictionless pins at the joints. In the same year, Schwedler managed to differentiate between the material reality of the trussed framework and its structural system – the pin-jointed trussed framework (Fig. 10-8): "Whereas the frames are thought to be of rigid construction, the small resistances caused by the small elastic bending at points *a*, *d*, *c*, etc. are negligible when compared with the resistance of the strut, or, which is the same thing, the individual framework components can be assumed to be capable of rotation at points *a*, *d*, *c*, etc." [Schwedler, 1851, col. 168]. For the first time, Schwedler thus accomplished the abstraction process that typifies structural theory: from the physical loadbearing structure (real trussed framework, e.g. timber frame) via the abstract loadbearing system (trussed framework model or Culmann's trussed framework notion) to the structural system (pin-jointed trussed framework), i.e. the reduced loadbearing system described by way of geometric/material properties for the purposes of quantitative examination. The invention of the structural system in the shape of the pin-jointed trussed framework became the guiding concept for the development of structural theory in the second half of the 19th century.

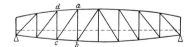

FIGURE 10-8
Schwedler's pin-jointed trussed framework model [Schwedler, 1851, col. 168]

**A real engineering artefact: the iron trussed framework with riveted joints**

### 10.2.2.1

The pin-jointed trussed framework model of trussed framework theory cannot conceal its close relationship with the design language of mechanical engineering. Werner Lorenz investigated the design thinking of August Borsig (1804–54), who was successful both as a mechanical and a structural engineer [Lorenz, 1990, p. 5; 1997, pp. 293–294], and showed through examples that Borsig regarded his buildings as machines [Lorenz, 1990, p. 5]. Structural engineering had not yet quite completed its divorce from the disciplines of mechanical and railway engineering, also shipbuilding, mining and metallurgy – the typical professions of the Industrial Revolution. "The early history of structural engineering," Wieland Ramm notes, "is largely identical with the history of iron construction" [Ramm, 2001, p. 640]. Iron construction, the hub of structural engineering, began to form the basis of an independent design language as early as the late 1850s.

The joining of structural elements in the fashion of the carpenter was superseded by the use of bolts and later rivets. For the bridge over the River Brahe (today Brda) near Czersk, built in 1861 to a design by Schwedler, the truss joints were designed as pins (Fig. 10-9a). Nine years later, another bridge over the Brahe was built near Bromberg (today Bydgoszcz), again designed by Schwedler, this time, however, using riveted joints (Fig. 10-9b). In a paper divided into 100 paragraphs, which was later repeatedly identified as a catechism of iron bridge construction, Schwedler stated that "the material of most iron bridges is rolled wrought iron" [Schwedler, 1865, col. 333]. The disappearance of composite systems from trussed framework construction not only signals the substitution of the carpenter by the metalworker, but also simplifies the theoretical treatment of iron construction practice through structural theory.

The iron trussed frameworks constructed across continental Europe after 1870 gradually became less reminiscent of the mechanical engineering tradition of early iron structures. Whereas in 1872 Emil Winkler

**FIGURE 10-9**
a) Pinned top chord joint of the bridge across the Brahe (now Brda) at Czersk – built in 1861 [Winkler, 1872, p. 135]; b) rigid top chord joint of the bridge across the same river near Bromberg (now Bydgoszcz) – built in 1870 [Winkler, 1872, p. 130]

(1835–88) still provided detailed descriptions of both riveted and bolted truss joints, he recommended giving riveted joints preference over bolted ones [Winkler, 1872, p. 115]. Winkler was aware that the pin-jointed trussed framework model contradicted the as-built reality of the iron trussed framework with riveted joints. During the last decades of the 19th century, this contradiction led to the development of the theory of secondary stresses and – based on that – the displacement method.

### 10.2.2.2 On the theory of secondary stresses

The truss members converging at the riveted joints are not only subjected to axial tension or compression forces, but to bending moments, too (Fig. 10-10); the latter generate bending stresses, which Friedrich Engesser (1848–1931) and Winkler grouped together under the term "secondary stresses" (see [Steinhardt, 1949, p. 13]). Whereas the pin-jointed trussed framework model can only be used to calculate the stresses due to axial forces, the quantification of secondary stresses requires rather cumbersome calculations because the structural system of the trussed framework with rigid joints is highly statically indeterminate. As early as 1872 Winkler stated that, according to his calculations, the secondary stresses can amount to up to 30 % of the primary stresses resulting from the pin-jointed trussed framework model [Winkler, 1872, p. 115]. However, the first writings on the subject did not appear until 1879.

In 1878 Heinrich Manderla (1853-89), a scientific assistant at Munich Technical University, submitted the complete solution for a competition set up by Prof. Johann Gottfried Asimont (1834–98): "What stresses ensue in the members of a trussed girder due to a change in angle at the trussed framework connections caused by the load?" Manderla's solution, which appeared in the 1878/79 annual report of Munich Technical University and was made available to a larger audience as a paper in the journal *Allgemeine Bauzeitung* published in Vienna in 1880 [Manderla, 1880], enabled the calculation of the secondary stresses in simple trussed frameworks with rigid joints on the basis of second-order theory [Kurrer, 1985/2, pp. 327–328]. Forty years later, A. Berry published a similar method without any knowledge of Manderla's work (see [Samuelsson & Zienkiewicz, 2006, p. 151]). After Clebsch first introduced unknown displacement parameters into the mathematical elastic theory for the calculation of truss systems in 1862, Manderla did the same in 1879 for structural theory.

In that same year, Engesser published an approximation method for determining the secondary stresses [Engesser, 1879]. He neglected the bending stiffness of the web members and analysed the top and bottom chords as continuous beams with imaginary supports at the joints. Engesser was well versed in the basic principles and may have been familiar with Clebsch's book on elastic theory, but there is no evidence of Clebsch's displacement method in any of Engesser's writings on the theory of secondary stresses. The same applies to Emil Winkler, a true expert in the literature on basic principles, who made his name through comprehensive contributions to the theory of secondary stresses. Following Manderla's line of thinking, Winkler introduces the difference between end tangents

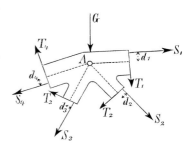

**FIGURE 10-10**
Internal forces at a rigid concentric joint in a trussed framework [Winkler, 1881/1, p. 297]

and member chord angles of rotation at the joint, resulting in *k* linear equations for *k* joints from *k* moment equilibrium conditions [Winkler, 1881/2]. In contrast to Manderla, Winkler employs first-order theory for the member moments. Winkler thus reduces the problem of secondary stresses to an exclusively linear problem, where the superposition principle applies without restrictions. On the other hand, he also considers eccentric truss joints (Fig. 10-11), thus complicating the calculations once more. In one of the many trussed frameworks analysed by Winkler, the increase in stresses compared with those calculated from the pin-jointed trussed framework model is, on average, 14 % for concentric joints (Fig. 10-12b/left) and 20 % for eccentric joints (eccentricity $e = 50$ mm) (Fig. 10-12b/right).

Apart from Manderla, Engesser and Winkler, Wilhelm Ritter (1847 – 1906), Theodor Landsberg (1847 – 1915) and Heinrich Müller-Breslau also worked on the theory of secondary stresses in the 1880s. Like Georg Christoph Mehrtens (1843 – 1917), Schwedler, too, had his own method; unfortunately, he never published it. Using this method, during a six-week period in 1887 Schwedler calculated the secondary stresses in the main girders of the second bridge across the River Vistula at Dirschau and handed the results to Mehrtens for use in the design of the bridge [Mehrtens 1912, p. 226].

The final piece in the theory of secondary stresses jigsaw was inserted by Otto Mohr (1835 – 1918) in 1892/93 in the shape of his work *Die Berechnung der Fachwerke mit starren Knotenverbindungen* (calcu-

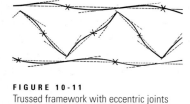

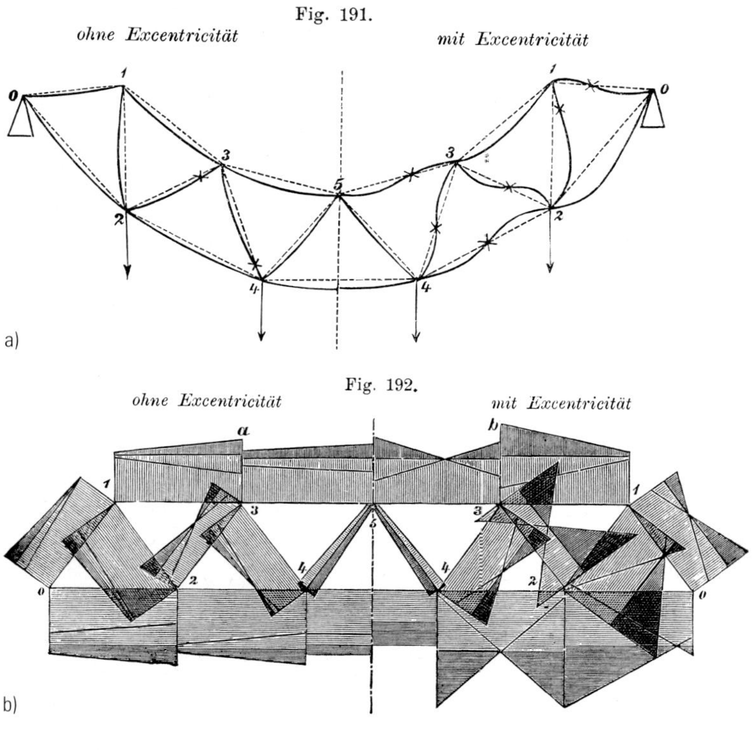

**FIGURE 10-11**
Trussed framework with eccentric joints [Winkler, 1881/1, p. 319]

**FIGURE 10-12**
Analysis of a) deformations, and b) secondary stresses in a trussed framework with rigid joints with and without eccentricity [Winkler, 1881/1, p. 312]

lation of trussed frameworks with rigid connections) [Mohr, 1892/93]. Mohr's crucial contribution consisted of the clear differentiation between the joint angles of rotation $\xi$ and the member angles of rotation $\psi$ for the unambiguous determination of the deformed state of a trussed framework with rigid joints. Mohr developed his procedure using a simple trussed framework with $k = 7$ joints (Fig. 10-13). In the first step, applying the principle of virtual forces, which he introduced into trussed framework theory in 1874/75 independently of James Clerk Maxwell (1831–79), Mohr calculated the member angles of rotation $\psi_{ij}$, $\psi_{ik}$, $\psi_{il}$ and $\psi_{im}$ of the pin-jointed trussed framework under load (Fig. 10-13b). The member end moments are subsequently determined for every joint – e.g. for joint $i$ the following moments:

$$M_{ij} = 2\,(E I_{ij}/l_{ij})\,(2\,\xi_i + \xi_j - 3\,\psi_{ij}) \qquad (10\text{-}6)$$

$$M_{ik} = 2\,(E I_{ik}/l_{ik})\,(2\,\xi_i + \xi_k - 3\,\psi_{ik}) \qquad (10\text{-}7)$$

$$M_{il} = 2\,(E I_{il}/l_{il})\,(2\,\xi_i + \xi_l - 3\,\psi_{il}) \qquad (10\text{-}8)$$

$$M_{im} = 2\,(E I_{im}/l_{im})\,(2\,\xi_i + \xi_m - 3\,\psi_{im}) \qquad (10\text{-}9)$$

The moment equilibrium for joint $i$

$$M_{ij} + M_{ik} + M_{il} + M_{im} = 0 \qquad (10\text{-}10)$$

can then be used to derive an equation with the unknown joint angles of rotation $\xi_i$, $\xi_j$, $\xi_k$, $\xi_l$ and $\xi_m$ and the member angles of rotation $\psi_{ij}$, $\psi_{ik}$, $\psi_{il}$ and $\psi_{im}$ calculated previously.

For the given trussed framework, Mohr notes the moment equilibrium conditions for every joint (eq. 10-10); he thus obtains a set of linear equations for determining the $k = 7$ joint angles of rotation. Having solved the set of linear equations, Mohr calculates the member end moments for every joint.

Mohr solved the problem of secondary stresses with virtually unbeatable clarity and practical elegance, and at the same time laid the foundation for the displacement method. However, Mohr's contribution to the displacement method disappeared from the focus of attention of structural theory for two decades because

- the energy doctrine of structural theory advocated by Müller-Breslau ousted the kinematic doctrine advocated by Mohr;
- the force method resulting from the energy doctrine rounded off the discipline-formation period of structural theory and, for the time being, would not tolerate a second method alongside;
- Mohr published his article in a less-well-known journal;
- Mohr was too preoccupied with trussed framework theory, which certainly prevented him from generalising his work to include elastic truss systems.

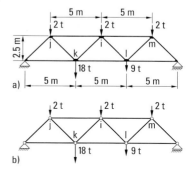

FIGURE 10-13
Sample analysis after Mohr; a) trussed framework with rigid joints, and b) associated pin-jointed trussed framework model

### 10.2.3 From trussed framework to rigid frame

In 1910 Willy Gehler (1876–1953) summarised the development of the theory of secondary stresses in his monograph *Die Entwicklung der Nebenspannungen eiserner Fachwerkbrücken und das praktische Rechenverfahren*

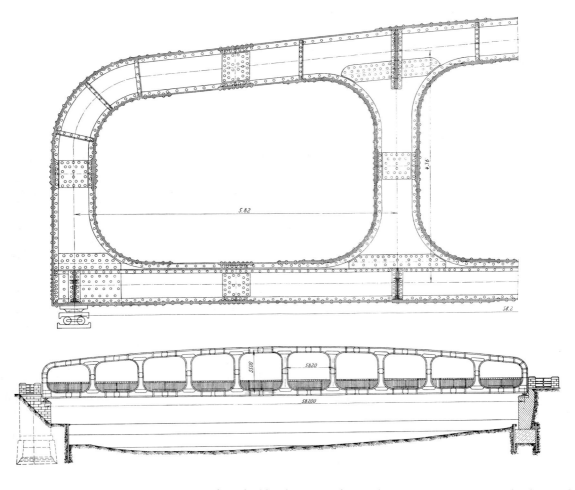

**FIGURE 10-14**
Design for a road bridge in the form of a Vierendeel girder across the River Ems near Salzbergen (not built) [Busse, 1912, p. 215]

*nach Mohr* (development of secondary stresses in iron truss bridges and practical calculation procedure after Mohr) [Gehler, 1910]. He reports on comprehensive measurements carried out on the railway bridge across the River Elster on the Dresden–Elsterwerda line just outside Elsterwerda station. Gehler concludes that Mohr's procedure for determining the secondary stresses provided results "that are totally in compliance with the values observed in reality" [Gehler, 1910, p. 69]. At the time Gehler made his observations, rigid-jointed frameworks were already being used in iron and reinforced concrete construction.

Since 1897, several iron bridges had been built in Belgium according to the Vierendeel system [Vierendeel, 1911]. The Vierendeel girder is a framework without diagonals, i. e. it is not triangulated, but instead comprises only top and bottom chords connected by vertical members (Fig. 10-14). In contrast to a triangle of members, a rectangle of pin-jointed members is kinematic, i. e. a mechanism. It has to be stabilised by additional bars; the four-panel kinematic system shown in Fig. 10-15b therefore requires four additional bars to form a statically determinate, or rather, a stabilised pin-jointed system (Fig. 10-15c).

The loadbearing behaviour of the first Vierendeel girder, a steel bridge spanning 31.50 m in Tervueren which Vierendeel had built at his own cost for the Brussels World Exposition of 1897, was explained in an article written by Albert Lambin and Paul Christophe and published in the journal of the Belgian Ministry of Public Works [Lambin & Christophe, 1898]. Nevertheless, many were still very unsure about the structural analysis of the Vierendeel girder. None other than Otto Mohr considered the reliability of the calculations for trussed frameworks with rigid joints to be better than that for Vierendeel girders. It was for this reason that he advocated lower permissible stresses for the latter, which would make the Vierendeel girder less economic than the trussed girder [Mohr, 1912, p. 96]. In his reply, Vierendeel rejected Mohr's objections and asserted that these applied to trussed frameworks [Vierendeel, 1912]. However, Mohr would be proved right about the reliability of the calculations for Vierendeel girders. He was justified in criticising the simple beam model commonly used to calculate the statically indeterminate Vierendeel girder. After 1910, numerous articles appeared that tried to redress these shortcomings and master the intrinsic static indeterminacy of the Vierendeel girder, among them Mohr's work of 1912 [Mohr, 1912] (see above). Most of the authors used the force method, although this led to confusing, complicated algorithms which were unsuitable for practical structural calculations.

Nevertheless, in a 1912 paper about the historico-logical aspects of the Vierendeel girder in iron construction, Franz Czech expands on numerous arguments in favour of the Vierendeel girder and concludes that both theoreticians and practitioners could no longer ignore it [Czech, 1912, p. 113]. According to the paper, the Vierendeel girder "had proved itself to be the most rational structure" in reinforced concrete construction [Czech, 1912, p. 113].

Rigid joints are an essential feature of reinforced concrete frame construction, which was invented by François Hennebique (1842–1921) and became established after 1900. Its monolithic character precluded modelling as a pin-jointed system a priori. Whereas the trussed framework was the ideal loadbearing structure for traditional steel construction, in reinforced concrete the frame was seen as the appropriate loadbearing structure as early as 1910. The frame and frame analysis thus developed predominantly against the background of reinforced concrete construction as it became widely accepted in structural engineering between 1910 and 1920. During this time, a comprehensive body of frame analysis literature appeared, e. g. the books by Willy Gehler [Gehler, 1913] and Adolf Kleinlogel (1877–1958) [Kleinlogel, 1914]. The focus of structural theory moved from the artefacts of steel construction to the artefacts of reinforced concrete construction.

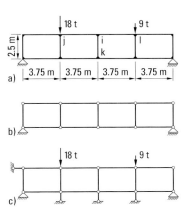

FIGURE 10-15
Vierendeel girder: a) structural system, b) kinematic pin-jointed system, c) stabilised pin-jointed system

### 10.2.4 The displacement method gains emancipation from trussed framework theory

The emancipation of the displacement method from trussed framework theory was the moment of the emancipation of structural theory from the artefacts of steel construction. Although the displacement method incorporated the theory of secondary stresses that originated in steel construc-

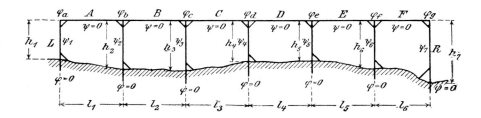

**FIGURE 10-16**
Series of horizontal frames
[Gehler, 1916, p. 103]

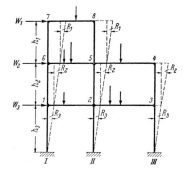

**FIGURE 10-17**
Series of horizontal and vertical frames
[Takabeya, 1967, p. 46]

tion, up until the 1920s it developed principally as reinforced concrete theory.

In the design of industrial buildings, but also for the construction of long bridges, orthogonal frames were strung together horizontally (Fig. 10-16). Vertical and horizontal series of orthogonal frames lead to multi-bay, multi-storey frames, which in the 1920s and 1930s became the structural synonym for high-rise buildings (Fig. 10-17). Again and again, the force method reached its limits in the analysis of such systems because the set of linear equations grew with the number of statically indeterminate variables $n$. Structural theory therefore developed in two directions: the Berlin school of structural theory around Müller-Breslau (see section 6.6), preferring the rationalisation of the force method, e.g. through orthogonalisation techniques, and the engineering scientists influenced by the Dresden school of applied mechanics around Mohr, who favoured the displacement method.

The reduction in the number of linear equations for statically indeterminate systems was the most important driving force behind the ongoing development of the displacement method. Fig. 10-18 shows the number of equations according to the force method $n$ and the displacement method $m$ for trussed girders with rigid joints (Fig. 10-13a), Vierendeel girders (Fig. 10-15a) and orthogonal frames (Fig. 10-17) with $s$ members and $k$ joints. With the exception of the Vierendeel girder, $m$ is invariably smaller than $n$. It is thus apparent that the displacement method is particularly suitable for the structural analysis of series of frames. However, for the Vierendeel girder, the force method is more advantageous because the resulting set of equations will invariably contain two unknowns less than the set of equations for the displacement method. For the case of trussed girders with rigid joints, Mohr had already demonstrated the superiority

**FIGURE 10-18**
Comparison of the number of linear equations according to the force method ($n$ statically indeterminate parameters) and according to the displacement method ($m$ unknown deformation parameters) for trussed girders with rigid joints, Vierendeel girders and orthogonal frames

| Structural system | n | m | n – m | n = m |
|---|---|---|---|---|
| Trussed girder (Fig. 10-13a): k = 7, s = 11 | 3k – 6<br>15 | k<br>7 | 2k – 6<br>8 | 2k = 6 |
| Vierendeel girder (Fig. 10-15a): k = 10, s = 13 | s – 1<br>12 | s + 1<br>14 | –2<br>–2 | n < m |
| Orthogonal frame (Fig. 10-17): k = 8, s = 13 | 3 (s – k)<br>15 | 3k – s<br>11 | 4s – 6k<br>4 | 2s = 3k |

of the displacement method, since the number of unknowns is reduced by $2k - 6$.

Whereas the only unknowns in the set of linear equations for trussed girders with rigid joints are the joint angles of rotation $\xi$, sway frames include further unknowns in the form of member angles of rotation $\psi$ or joint displacements $w$. In the trussed girder with rigid joints, the stabilised pinned system ensues virtually automatically (Fig. 10-13b), but sway frames (Fig. 10-15a) result in kinematic pinned systems (Fig. 10-15) which have to be stabilised by additional bars (Fig. 10-15c). This forms the logical core of the emancipation of the displacement method from its dependence on trussed framework theory.

### 10.2.4.1 Axel Bendixsen

In 1914 the Danish engineer Axel Bendixsen extended Mohr's procedure [Mohr, 1892/93] to braced and sway frames [Bendixsen, 1914]. Bendixsen's approach has two stages (Fig. 10-19):

The first stage yields $\alpha$ loading equations from the analysis of the braced (i.e. non-sway) frame; these are the equations for calculating the joint angles of rotation $\alpha_i'$ for $i = 1, \ldots, 6$ (Fig. 10-19b). Apart from the introduction of the lateral support members, the first stage is identical with Mohr's procedure. Restraint forces $Z_{10}$ and $Z_{20}$ are generated at the horizontal supports (Fig. 10-19b).

In the second stage, Bendixsen formulates the $\alpha$ displacement equations for calculating the joint angles of rotation $\alpha_i''$ for $i = 1, \ldots, 6$ by applying the unknown joint displacements $w_1$ and $w_2$ one by one:

$$\alpha_i'' = a_{i1} w_1 + a_{i2} w_2 \quad \text{for } i = 1, \ldots, 6 \tag{10-11}$$

These equations can also be used to express the joint angles of rotation $\alpha_i''$ due to the member angles of rotation. The second stage was completely new, despite the fact that in formal terms it is similar to the force method and uses unit displacement states instead of unit force states. Joint displacement $w_1 = 1$ results in restraint forces $Z_{11}$ and $Z_{21}$ (Fig. 10-19c), joint displacement $w_2 = 1$ in restraint forces $Z_{22}$ and $Z_{12}$ (Fig. 10-19d). The superposition of both unit displacement states leads to the following restraint forces:

joint 1: $Z_1'' = Z_{11} w_1 + Z_{12} w_2$ (10-12)

joint 2: $Z_2'' = Z_{21} w_1 + Z_{22} w_2$ (10-13)

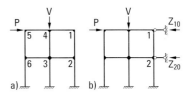
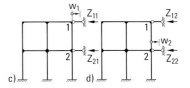

**FIGURE 10-19**
Multi-storey frame: a) in the loaded condition, b) with lateral supports at joints 1 and 2 in the loaded condition, c) with lateral supports and applied displacement $w_1$, or d) $w_2$

Bendixsen then determines the member angles of rotation due to $w_1$ and $w_2$ and inserts these values into the $\alpha$ displacement equations. The resulting joint angles of rotation $\alpha_i''$ and member angles of rotation are used to calculate restraint forces $Z_{11}$, $Z_{12}$, $Z_{22}$ and $Z_{21}$ via the joint equilibrium conditions (only $\Sigma H = 0$ here). The sums of the reaction forces from the $\alpha$ loading equations and the $\alpha$ displacement equations must be zero because in reality joints 1 and 2 are not restrained laterally:

joint 1: $Z_{10} + Z_1'' = 0 = Z_{10} + Z_{11} w_1 + Z_{12} w_2$ (10-14)

joint 2: $Z_{20} + Z_2'' = 0 = Z_{20} + Z_{21} w_1 + Z_{22} w_2$ (10-15)

From these two equations, Bendixsen determines the joint displacements $w_1$ for joint 1 and $w_2$ for joint 2. Through evaluating the appropriate $\alpha$ displacement equations for each joint, the joint displacements yield the joint rotations $\alpha_i''$. The final joint rotations $\xi_i$ are calculated from the sum of $\alpha_i''$ and the appropriate $\alpha$ loading equation $\alpha_i'$:

$$\xi_i = \alpha_i' + \alpha_i'' \quad \text{for } i = 1, \ldots, 6 \tag{10-16}$$

Only two significant technical journals responded with reviews of and commentaries on Bendixen's pioneering contribution to the displacement method. Bendixsen himself used his displacement method for the calculation of ribbed dome structures [Bendixsen, 1915]. Working independently, W. M. Wilson and G. A. Maney published a method in a 1915 edition of the Bulletin of the University of Illinois in Urbana which they called the "slope deflection method" and which was not dissimilar to Bendixsen's method (see [Samuelsson & Zienkiewicz, 2006, p. 151]). A comprehensive description of the slope deflection method can be found in J. A. L. Matheson's works (see [Matheson, 1959, pp. 249–264; Matheson & Francis, 1960, pp. 159–177]).

### 10.2.4.2 Willy Gehler

Gehler, coming from the inner circle of the Dresden school of applied mechanics, published his angle of rotation method in 1916 [Gehler, 1916]. He derives the member end moments $M_{ij}$ as a function of the difference in the joint and member angles of rotation. Having satisfied the joint equilibrium conditions, he obtains a set of equations with $m$ unknown joint angles of rotation $\xi$ and member angles of rotation $\psi$. According to Gehler, one advantage of his displacement method is that "the statics part of the problem is separated from the purely mathematical problem" [Gehler, 1916, p. 104]. Finally, he points out the reduction in the number of unknowns and emphasizes the clear character of the unknown deformation parameters $\xi$ and $\psi$, which can be measured directly on the structure.

### 10.2.4.3 Asger Ostenfeld

Around 1920, Asger Ostenfeld, professor at Copenhagen Technical University, extended Bendixsen's displacement method in parallel with the force method. As early as 1921 he created the dual concept of the displacement method [Ostenfeld, 1921]:

- displacement method versus force method,
- $m$ unknown displacement parameters $\xi_j$ versus $n$ unknown force parameters $X_i$,
- restraint forces $Z_{ij}$ versus displacement jumps $\delta_{ji}$,
- $m$ linear equations for $\xi_j$ versus $n$ linear equations for $X_i$.

Ostenfeld transforms the latter into the following form (using Einstein's summation convention):

$$-Z_{i0} = Z_{ij} \cdot \xi_j \quad \text{for } i, j, \ldots, m \tag{10-17}$$

$$-\delta_{j0} = \delta_{ji} \cdot X_i \quad \text{for } j, i, \ldots, n \tag{10-18}$$

In Ostenfeld's displacement method, the inconsistencies inherent in Bendixsen's approach, i.e.
- application of Mohr's procedure [Mohr, 1892/93],
- separate determination of joint and member angles of rotation, and
- two-stage quantification of joint and member angles of rotation,

are eliminated.

The significant progress in Ostenfeld's displacement method lies in the fact "that it enables previously analysed structures, structural elements so to speak, to be built on" [Ostenfeld, 1921, p. 288]. Ostenfeld achieves this not only through the introduction of restraints to prevent joint displacements (like Bendixsen), but through the introduction of restraints to prevent joint rotations. This enables the complete frame to be subdivided into finite elements. "It can therefore be expected," Ostenfeld concludes, "that this method will allow even rather complicated systems to be processed without difficulty because, unlike with the force method, it is not necessary to start from scratch every time" [Ostenfeld, 1921, pp. 288–289]. Using Ostenfeld's approach, Mohr's equation for the member end moment $M_{ij}$, for example, is broken down into the elementary cases of the bar fixed at both ends (Fig. 10-20):

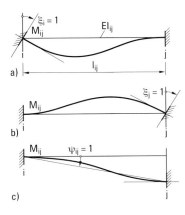

$\xi_i = 1$: $M_{ij} = 4(EI_{ij}/l_{ij})$ (10-19)

$\xi_j = 1$: $M_{ij} = 2(EI_{ij}/l_{ij})$ (10-20)

$\psi_{ij} = 1$: $M_{ij} = -6(EI_{ij}/l_{ij})$ (10-21)

**FIGURE 10-20**
Elementary displacement load cases for the fixed bar: a) $\xi_i = 1$, b) $\xi_j = 1$, c) $\psi_{ij} = 1$

This leads to eq. 10-19 if $\xi_i = 1$, $\xi_j = 0$ and $\psi_{ij} = 0$ are used in Mohr's eq. 10-6; the other two equations are derived similarly. What appeared at first glance to be an unnecessary complication would soon turn out to be an important contribution to the rationalisation of structural calculations.

In his 1926 book *Die Deformationsmethode* (displacement method) [Ostenfeld, 1926], Ostenfeld summarised his articles that had appeared in the engineering journals *Ingeniören* (1920, 1922), *Der Eisenbau* (1921) and *Der Bauingenieur* (1923).

#### 10.2.4.4 Ludwig Mann

One year after Ostenfeld's monograph, Ludwig Mann (1871–1959) published his book *Theorie der Rahmenwerke auf neuer Grundlage* (a new basis for frame theory) [Mann, 1927]. Mann elaborated the displacement method against the background of Joseph Louis Lagrange's *Mécanique analytique*. For example, Mann introduced the notion of base coordinates for the unknown displacement parameters $\xi_j$, which represented a specification of Lagrange's generalised coordinates, where

- the first group of base coordinates represents the joint angles of rotation,
- the second group represents the member chord angles of rotation, and
- the third group represents the independent elongations of the member chords.

For calculating the restraint forces $Z_{ij}$, Mann uses the principle of virtual displacements for the first time. Here, too, he adheres to Lagrange, who had based his complete mechanics on this principle. Finally, Mann calls the set of equations for calculating the displacement parameters "type 2 elasticity equations" and that for calculating the statically indeterminate parameters "type 1 elasticity equations". In 1939 he extended his displacement method to the calculation of spatial truss systems [Mann, 1939]. Mann became confronted with spatial truss systems in the context of calculations for large open-cast lignite mining plant in the former East Germany. For his work in lignite mining, which was vital to the energy sector of the German Democratic Republic (GDR) he was awarded an honorary doctorate by Dresden Technical University in 1957. In his speech held on the occasion of the award ceremony, Mann developed the duality of the force and displacement methods, based on the principle of virtual forces and the principle of virtual displacements, with characteristic clarity [Mann, 1958].

With Mann's logical development of the force method parallel with the displacement method and his reasoning drawn from rational mechanics, the emancipation of the displacement method from trussed framework theory was complete. Historically, this completion constitutes the midpoint in the consolidation period of structural theory (1900–50), which followed the discipline-formation period (1825–1900).

## 10.2.5 The displacement method during the invention phase of structural theory

From 1930 to 1950, the displacement method developed in three directions:

*Firstly*, force and displacement method were compared with each other, but also with different forms of the displacement method. In 1934 Kruck examined the methods of Bendixsen, Ostenfeld and Mann together with Bleich's theorem of four moments [Kruck, 1934]. In 1937 Kruck developed Mann's complex displacement method further to form the "method of base coordinates" and formalised it through tables in such a way that the member parameters could be inserted directly into the summation terms of the basic equations [Schrader, 1969, p. 8].

*Secondly*, further insight was gained into the dual nature of structural theory; relevant contributions were supplied by Pasternak (1922, 1926) and Hertwig (1933). The clear differentiation between the concepts of the principle of virtual forces and those of the principle of virtual displacements, based on the variation principles of elastic theory [Schleusner, 1938], enabled the displacement method to catch up with the force method in terms of its formal structure.

*Thirdly*, the displacement method opened up areas of application that significantly exceeded those of structural theory. Examples are truss dynamics ([Fliegel, 1938] and [Koloušek, 1941]) and the theory of stability, which was heavily influenced by metalworking practice. It was precisely the analysis of components with uncertain stability from aircraft construction, shipbuilding, mechanical engineering and structural steelwork that called for the influence of deformations on the equilibrium condition to be taken into account. The formulation of such a second-order theory in the language of the force method would be very cumbersome, whereas the displacement method posed no such problems, as was shown by Chwalla and Jokisch in 1941 [Chwalla & Jokisch, 1941], and later by Teichmann [Teichmann, 1958]. Soviet scientists extended the displacement method significantly during the invention phase of structural theory (see section 10.3.3).

In historical terms, the emancipation of the displacement method from structural theory, and the laying of its foundations in the shape of the two complementary principles of mechanics, forms the content of the preparatory phase in the development of modern structural mechanics after 1950.

## 10.3 The groundwork for automation in structural calculations

As classical theory of structures was completed by Heinrich Müller-Breslau in the final decade of the 19th century, another scientific discipline appeared on the schedule alongside astronomy and geodesy which required extensive numerical calculations for its practical application.

Private structural engineering practices did not start to appear in the German federal states until after the conclusion of the Industrial Revolution, and they achieved a noticeable significance with the rise of the great iron construction companies in the 1880s and the profession of the consulting engineer (Fig. 10-21) alongside the activities of the structural engineers in the building authorities. The building authorities, which expanded enormously in Germany after the great stock market crash of 1873 due to Bismarck's nationalisation of the private railway companies, objectivised structural engineering practice increasingly through administrative activities legitimised by a construction theory, which led to administration-type engineering activities in the private and public sectors as well as science. The theory and practice of structural calculations formed the crux of administration-type engineering activities because only structural calculations with a scientific basis were accepted as "correct" structural designs in both the physical and the social sense. This synthesis of calculation and building to create calculated building meant that the practical use of symbols in structural engineering became the norm. These new qualitative requirements placed on structural calculations, the quantitative expansion in structural engineering practice and on top of that the demand for quick, reliable structural analyses, brought about by the economising of the structural engineer's working hours, called for adequate aids.

Just one such calculation aid appeared in 1889 with the first edition of a set of tables [Zimmermann, 1889] published by Hermann Zimmermann (1845–1935), who in 1891 was promoted to the post of senior structural engineer in the Prussian Railway Authority. Zimmermann's design tables became an everyday tool for the structural engineer. The 12th edition was published in 1946, but by then, the battle for the rationalisation of structural calculations had theoretically already been lost – Konrad Zuse (1910–95) had built the first successful computer in 1941. But it was not until the full formalisation of structural engineering, which matrix algebra turned into matrix formulation, did Argyris manage to take the decisive practical step to computational mechanics in the years 1954 to 1957.

FIGURE 10-21
One of Ruff engineering practice's own advertisements [Ruff, 1903]

### Remarks on the practical use of symbols in structural analysis

**10.3.1**

Sybille Krämer's concept of the practical use of symbols, interpreted from the historical development of arithmetical, algebraic and logical formalised theory [Krämer, 1988], forms the backdrop to the prehistory of computational mechanics. She has also provided critical accounts of analytical geometry [Krämer, 1989], Leibniz' infinitesimal calculus [Krämer, 1991/2] and the prehistory of informatics [Krämer, 1993] from the perspective of this philosophical theory of the intellect.

Her historical reconstruction of the formalisation idea is based on two systematic theses:

*Thesis 1*: A process can be described in formal terms provided it is possible to represent this with the help of artificial symbols in such a way that the conditions of the typographic, schematic and non-interpretive use of the symbols are fulfilled. A process that satisfies these conditions can also be performed as an operation of a symbolic machine [Krämer, 1988, p. 2].

*Thesis 2*: Every process that can be described in formal terms can be represented as an operation of a symbolic machine and – in principle – be performed by a real machine (e.g. mechanical calculator). Computers are machines that can imitate any symbolic machine [Krämer, 1988, p. 3].

"Before the computer was invented as a real machine," writes Sybille Krämer summing up her methodic prolegomena on the genesis of the practical use of symbols, "we developed the 'computer in ourselves'" [Krämer, 1988, p. 4]. Like Sybille Krämer traces this protracted and laborious history of the practical use of symbols for the aforementioned mathematical disciplines, an attempt will be made here to interpret the relevant sources from that perspective related to the prehistory of computational mechanics.

Is it legitimate to transfer the Krämer concept to non-mathematical disciplines like, for example, the engineering sciences? In order to answer this question, we must first look at how engineering sciences became a branch of science with an independent epistemological status. The historico-logical development of the engineering sciences can be broken down into four steps (see section 3.1.3):

*First step*: Knowledge of the causal relationship realised in the engineering model (e.g. beam model) (emergence of the first fundamental engineering science disciplines, e.g. Navier's beam theory).

*Second step*: Constructional and technological modelling of causal relationships in engineering artefacts and methods (classical engineering type, e.g. trussed framework structures).

*Third step*: The step from the coexistence of the knowledge of the causal relationship realised in the engineering model, with the constructional or technological modelling of causal relationships in engineering artefacts and methods, to the cooperation (emergence of the system of classical engineering sciences, e.g. design theories for structural engineering).

*Fourth step*: The space-time integration of the knowledge of the causal complex objectively extant in the engineering model with the constructional and technological modelling of the causal complex in the engineering system (automation and formation of non-classical engineering science disciplines, e.g. structural matrix analysis/modern structural mechanics).

The historical development of structural analysis really allows a derivation of the tendency towards processes that can be described in formal terms which map a mathematical theory on the engineering science level:

– The theory of sets of linear equations becomes evident in the $\delta$-symbols introduced into the theory of statically indeterminate systems by Müller-Breslau in 1893 [Müller-Breslau, 1893]. By means of the displacement jumps $\delta_{ik}$ ($i = 1, ..., n; k = 1, ..., n$) and $\delta_{i0}$ ($i = 1, ..., n$), the $n$ stactically indeterminate force variables $X_k$ ($k = 1, ..., n$) can be calculated from $n$ elasticity equations of the force method (type 1 elasticity equations) of the system with $n$ degrees of static indeterminacy and $n$ equations (see eq. 10-18).

– Ostenfeld's Z-symbols [Ostenfeld, 1926], which in formal terms follow on from the $\delta$-symbols and lead to type 2 elasticity equations (see eq. 10-17) and from which the $m$ geometrically indeterminate displacement variables $\xi_l$ ($l = 1, ..., m$) can be calculated with the restraint forces $Z_{jl}$ ($j = 1, ..., m; l = 1, ..., m$) and $Z_{j0}$ ($j = 1, ..., m$), also make use of the knowledge of the linear structures of classical structural theory; although they enable the dual construction of the whole of structural theory, at first they do not overstep the bounds laid down by the theory of sets of linear equations.

– In mapping the theory of sets of linear equations on the individual science level in theory of structures, formal operations with symbols comprised formal operations with variables (symbols for symbols), which referred to only a subset of the whole area of the cognitive objects of structural theory (inhomogeneity of the practical use of symbols). In the practice of structural calculations that corresponded to the application of mechanical calculating aids (e.g. the mechanical calculator) – in other words specific symbolic machines.

– Due to the calculation of systems with many degrees of static indeterminacy, the pressure to rationalise structural calculations for buildings and aircraft led to structural analysis iteration methods in the late 1920s and to Konrad Zuse's step-by-step realisation of the computer from the mid-1930s onwards. Although the first functioning computer, the Z 3, was built in 1941 (which could imitate any symbolic machine), users of Zuse's computer saw it in the first place as a technical tool for rational engineering science calculation in the sense of the numerical evaluation of equations. The inhomogeneity in the practical use of symbols in structural analysis, i.e. the coexistence of various formal languages (infinitesimal calculus, algebra, arithmetic), could therefore not be overcome. On the formal level, the formation of structural analysis theories lagged behind the technical developments.

– The full transition to formalised theory in structural analysis was not completed until the matrix algebra reformulation of structural analysis was intentionally switched to the computer (homogenisation of the practical use of symbols in structural analysis purely through matrix algebra). In contrast to the theory of sets of equations, in matrix algebra the variables function as formalised theory, i.e. they are the basic symbols of a formalised language; consequently, the step to the non-interpretive use of symbols has been taken. Such symbols have an intra-symbolic significance because the rules by way of which the symbolic expressions are formed and transformed do not make any reference to the meaning of the symbols. Their range of interpretation is, in principle, not defined [Krämer, 1988, p. 182]. And this is also the reason why the advocates of fundamental engineering science disciplines such as theory of structures first had to cut a path laboriously through the "matrix forest" [Felippa, 2001, p. 1317] before Argyris was able to reformulate structural analysis through matrix algebra in the mid-1950s.

### Rationalisation of structural calculation in the consolidation period of structural theory

**10.3.2**

Around 1910, theory of structures was regarded as that discipline in the system of classical engineering sciences in which mathematicisation had progressed furthest. The fundamentals can be found in numerous contributions to the *Enzyklopädie der mathematischen Wissenschaften* (encyclopaedia of mathematical sciences) edited by Felix Klein (see [Chatzis, 2007]). Its role as the premier discipline in construction theory was undisputed because it essentially legitimised the scientific basis of structural engineering. The legitimation of this scientific base through the mathematicisation of the fundamental engineering science disciplines advanced during the 1890s to become the reasoning component in the movement towards equivalence between Germany's technical universities and its traditional universities. One example of this is Müller-Breslau's nomination as a full member of the Prussian Academy of Sciences in 1900, the members of which up until then had only been drawn from the university-based sciences [König, 1999, p. 392].

However, the mathematicisation of structural theory was only the formal prerequisite for the rationalisation of structural calculation on a theoretical level.

The practical level took shape with the wave of industrialisation in building brought about by iron construction in the final decades of the 19th century. The scope of structural calculations expanded in the design offices of the large structural steelwork companies as the theory of statically indeterminate systems took shape. This development was helped by the evolving model world of line diagram analysis (i.e. all the structural systems of classical theory of structures, see Fig. 10-1), which for the first time placed calculated building quite rightly within the structural engineer's remit, who was now in the position of being able to find or invent new forms of construction from the synthesis of structural subsystems. The engineer working on steel buildings gradually became the structural engineer who understood how to reduce the reality of the structures to the "correct" structural analyses. The rationalisation of structural calculation therefore became not only an essential factor in the rationalisation of structural engineering as a whole, but also the springboard for the development of theories in the consolidation period of structural theory, and therefore the essence of the prehistory of computational mechanics.

#### 10.3.2.1 Orthogonalisation methods

The systematic use of $\delta$-symbols enabled the formation of methods in which type 1 elasticity equations (eq. 10-18) became completely independent, i. e.

$$\delta_{ik} = 0 \quad \text{for } i \neq k \tag{10-22}$$

Müller-Breslau had already specified such a method in 1889 and 1897 and explained it in his 1903 publication *Graphischen Statik* (graphical statics) using the example of a trussed framework system with several degrees of static indeterminacy [Müller-Breslau, 1903]. The motive behind his orthogonalisation method was the general force method's sensitivity to errors, for which eq. 10-22 is generally not satisfied. In symmetrical systems with several degrees of static indeterminacy, however, it started to be used in practical calculations after 1900 together with the concept of the elastic centre of gravity (Fig. 10-22).

Müller-Breslau's method inspired Siegmund Müller to formulate his group loads method for systems with $n$ degrees of static indeterminacy [Müller, 1907]. In mathematical terms, it boils down to splitting the force condition of the system with $n$ degrees of static indeterminacy into $n$ independent linear combinations. The clear recognition of the mathematical structure of the force method and the extensive formal treatment with $\delta$-symbols led Müller to believe he had created a universally applicable orthogonalisation method, the use of which required neither the inventiveness of the calculating engineer nor "kinematic affiliations" [Müller, 1907, p. 25] – an unmistakable sign of the tendency towards the schematic and non-interpretive use of symbols during the consolidation period of structural theory.

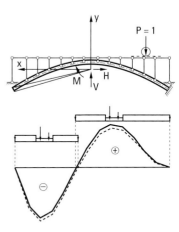

FIGURE 10-22
Determining the influence lines of the upper and lower kern point moments at the arch springings with the help of the elastic pole according to a paper published in 1906 by Emil Mörsch [Kurrer, 1995, p. 346]

#### 10.3.2.2 Specific methods from the theory of sets of linear equations

Hertwig and Pirlet interpreted the Müller-Breslau/Müller method described in section 10.3.2.1 specifically from the standpoint of the theory of sets of equations, in particular determinant theory; based on this they developed further structural analysis solutions ([Hertwig, 1910; Pirlet, 1910]). Hertwig set up four methods for solving the set of equations in 10-18 in the mathematical language of determinant theory [Hertwig, 1910]:
1) General solution through substitution
2) Solution through elimination
3) Solution through substitution and elimination
4) Solution through linear substitution with constant term.

The first method corresponds to that of Müller, whereas that of Müller-Breslau represents a special case of the third method. The fourth method is – in the case of continuous beams – essentially W. Ritter's graphical method of fixed points often used for such systems (see section 9.3.1.1). Hertwig compares methods 1) to 4) from the viewpoint of the calculation work required and their sensitivity to errors. None of the methods offered any decisive advantages. Therefore, at the end of his paper, Hertwig hints at another method that would "reach the objectives most easily and most reliably" [Hertwig, 1910, p. 120]. In a paper published two years later, he develops the coefficients of the inverse matrix belonging to the $\delta_{ik}$-matrix in infinitely convergent series [Hertwig, 1912]. Hertwig clearly recognised the universality of his method: "The method of calculation can of course be used for other tasks which require the solution of linear equations, e. g. in the equivalent member method, in the calculation of electrical networks" [Hertwig, 1912, p. 59]. The chance to employ formalised theory in the engineering science disciplines through formalised engineering methods is evident here.

In his long article on theory of structures for the *Handbuch der physikalischen und technischen Mechanik* (manual of physical and applied mechanics), which appeared nearly 20 years later, Hertwig deals with the entire theory of statically indeterminate systems uniformly from the perspective of the theory of sets of linear equations. Nevertheless, in mathematical terms he essentially remained true to the level of determinant theory, supplemented by iteration methods. Despite his insight into the dual nature of theory of structures, Hertwig did not take the final step to complete use of formalised theory in structural analysis by means of matrix algebra.

### 10.3.2.3 From structural iteration methods to numerical engineering methods

By the early 1920s it was already becoming clear that there was a contradiction between the usefulness of the theory of statically indeterminate systems for practical calculations and construction practice in reinforced concrete, indeed also in structural steelwork. For instance, reinforced concrete structures were increasingly making use of multi-storey frames. Such systems had multiple degrees of static indeterminacy and could only be solved by means of lengthy, extensive calculations, even using the customary techniques based on the force method. To a lesser extent this was also true for the displacement method, which, although there were generally fewer equations to be solved, remained a fringe method in everyday structural calculations because engineers were primarily interested in force variables. The frame tables published in numerous monographs between 1910 and 1930 (see section 9.3.1.1) of course provided inestimable help here; nevertheless, the calculating engineer's handling of such systems with many degrees of structural indeterminacy remained subjective.

The introduction of the rigid frame for providing stability against wind loads, which started in the 1890s in the tall buildings of Chicago, led to the steel frames of the skyscrapers of the 1920s and 1930s in the big cities of the USA. Fig. 10-23 shows the scheme of the loadbearing frame to Penobscot House in Detroit, designed and built by the American Bridge Co. and the Detroit Steel Construction Co. and completed in March 1928. The structural calculations for such high-rise structures were carried out according to R. Fleming's cantilever method and the portal method.

When calculating the wind forces in the frame according to the cantilever method, the entire building is modelled structurally as though it were a vertical cantilever fixed at the base:
- The forces in the columns are therefore proportional to their distance from the neutral axis of the overall cross-section of the building.
- The increase in the flange forces is transferred by the infill members (longitudinal beams, diagonals, end stiffeners, etc.).
- The bending moments in these beams are derived from the said increase in forces and the support moments from the beam moments.

In the portal method, it is assumed that owing to the various types of connections between the beams and the columns, the inner columns are subjected to twice the shear force and twice the bending moment of the outer columns because they are each connected to two beams. This method of calculation was criticised by A. Dürbeck who noted that "in reality ... the distribution into the beams and columns depends on the ratio of the second moments of area, as every frame formula for simple cases clearly shows" [Dürbeck, 1931, p. 238].

In his two-volume monograph *Bridge Engineering* [Waddell, 1916], J. A. L. Waddell turned his attention to the method of successive approximation for calculating the secondary stresses in trussed girders with rigid joints – a problem with many degrees of static indeterminacy (see Fig. 10-13a). K. A. Čališev, Timoshenko's successor at Zagreb Technical University, had been using this same method for calculating frames since 1922 [Čališev, 1922/23]. Using Čališev's structural method it was possible to calculate sway and non-sway systems with many degrees of static indeterminacy iteratively. This work written in Serbo-Croatian [Čališev, 1922/23] first came to the attention of the international community through a German translation with sample calculations [Čališev, 1936], which was presented at the second conference of the International Association for Bridge & Structural Engineering (Berlin, 1936). Čališev's iterative method essentially corresponds to that reached independently by Hardy Cross in 1930. The latter based his work on L. E. Grinter's further development of the method of successive approximation. An iterative method of the same mathematical type had been published in 1929 by N. M. Byernadskii, which he had already used to carry out structural calculations for real structures (see [Rabinovich, 1960, p. 26]).

Cross' iterative method was quickly taken up by the profession. A paper by Cross appeared in 1932 in the *Transactions* of the American Society of Civil Engineers, together with more than 30 articles discussing his work and a concluding evaluation of the whole discussion by Cross himself. Cross introduces his paper as follows: "The purpose of this paper is to explain briefly a method which has been found useful in analyzing frames which are statically indeterminate. The essential idea which the writer wishes to present involves no mathematical relations except the simplest arithmetic" [Cross, 1932/1, p. 1]. Users of Cross' method were relieved of the tediousness of interpretation insofar as they had to adhere directly to the practical use of arithmetic symbols; knowledge about the theory of

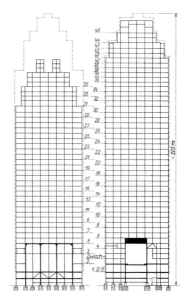

**FIGURE 10-23**
Scheme of the loadbearing frame to Penobscot House in Detroit [Herbst, 1928, p. 231]

statically indeterminate systems was unnecessary. The basic equations for the three static variables required – member end moment, member stiffness, carry-over factor – were not only elementary, but easily found in the structural compendiums of the day. In the calculation of high-rise buildings, the cantilever and portal methods could therefore be replaced by an efficient and realistic iterative method. In the discussion on the paper by Cross [Cross, 1932/1], Grinter applied his extended method of successive approximation to storey frames with unbraced nodes [Grinter, 1932]. One year later, he discussed an algorithm with which the iteration could be speeded up [Grinter, 1933].

Cross' method triggered numerous activities in the 1930s and 1940s, the object of which was to set up structural iteration methods. One example is the method published by Gaspar Kani in 1949 [Kani, 1949] which – in contrast to the original edition of Cross' method – also takes into account the displacement of the nodes. Whereas Cross uses the summation method as the iteration prescription, Kani iterates according to the single-step method (Gauß-Seidel); Kani's method is therefore also superior to that of Cross from the calculation technicalities viewpoint.

The iteration methods according to Cross (Fig. 10-24) and Kani played an important role in the practice of calculations for statically indeterminate structures well into the 1970s. In 1940 Fukuhei Takabeya proposed an iterative method and together with his assistants used it to calculate a 200-storey frame with 13 bays (using the force method that would have meant about 7800 elasticity equations!) in 78 hours [Takabeya, 1967, p. V]. Nevertheless, the race between manual structural calculations and automated structural calculations, between the tortoise and the hare, had long since been decided.

At Imperial College in London, R. V. Southwell and his assistants developed the relaxation method, which he used in 1935 to investigate three-dimensional structures [Southwell, 1935].

Samuelsson and Zienkiewicz summarise Southwell's method as follows: "The 3D truss has three unknown displacement components for

**FIGURE 10-24**
Iteration calculation after Cross for a two-storey frame with 18 degrees of static indeterminacy
[Dernedde & Barbré, 1961, p. 53]

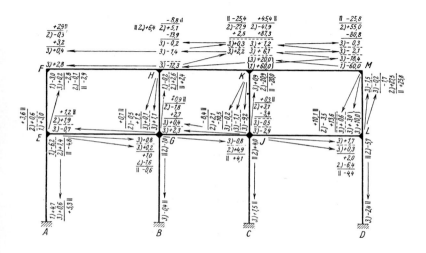

each joint. Southwell starts by locking all joints by imaginary constraints. He then proceeds to unlock or relax the displacement component in the joint with the largest imbalance. He calculates the forces and their components in the connection bars and adds them to the neighbour joint constraints. He then unlocks a new joint component with the largest imbalance and so on. He thus calls the method 'systematic relaxation of restraints'. Slowly, the imbalances go down" [Samuelsson & Zienkiewicz, 2006, pp. 152–153]. The fundamental calculations of Southwell's method for calculating elastic trusses corresponds to that of Čališev. Southwell published his work on the relaxation method in a monograph in 1940 [Southwell, 1940] and solved in an articulate way problems in theory of structures, the theory of electrical networks, stability theory and vibration theory. Southwell therefore created a general numerical engineering method that had repercussions way beyond theory of structures. L. Fox has presented this in the context of the historical development of numerical mathematics in the UK from the first decade of the 20th century up until the 1950s [Fox, 1987, pp. 26–31]. On the differences between the iteration method of Cross and the relaxation method of Southwell, Samuelsson and Zienkiewicz write: "It should be noted that the major difference between the methodologies of Southwell and Cross is that in the former the basic equations are not lost, whereas in the Cross method they are. Thus only the Southwell method is self-checking. Indeed, the reader will recognize that Southwell's relaxation method is simply a form of the well-known Jacobi-Gauss iteration method" [Samuelsson & Zienkiewicz, 2006, p. 153]. Cross' method can be used without knowledge of the theoretical background and represents practical knowledge only. The structural engineer performing his calculations according to Cross or Kani only requires skills in the four fundamental operations of arithmetic, i.e. he works like a calculator. Iteration calculations require only addition and subtraction – carried out mechanically in those days with an "Addiator"; merely the initial values of the said methods require our human calculator to carry out multiplication and division – performed in those days with a slide rule.

The symposium *Numerical Methods of Analysis in Engineering* [Grinter, 1949], organised by Grinter and held at the Illinois Institute of Technology in Chicago in honour of Hardy Cross, marks a provisional climax for the numerical engineering methods tied to manual calculations in general and the structural iteration methods in particular. The symposium, at which leading American, British and Australian engineering scientists such as L. E. Grinter, F. S. Shaw, R. V. Southwell, M. M. Frocht, F. Baron, N. M. Newmark and T. J. Higgins presented the fruits of their research activities, not only inserted the final piece of the historico-logical jigsaw in the field of calculation methods in the consolidation period of structural theory (1900–50), but also pointed the way forward to the integration period of structural theory (1950 to date) – at least in terms of its non-classical theoretical foundations. The contribution of the professor of electrical engineering at the University of Wisconsin, T. J. Higgins, is particularly worthy of mention here: *A Survey of the Approximate Solution of Two-Dimensional Physical Problems by Variational Methods and Finite Difference Procedures* [Higgins, 1949]. This also started the process of the superseding of theory of structures by structural mechanics.

### 10.3.3
**The dual nature of theory of structures**

The creation of the displacement method, and connected with that the recognition of the dual nature of theory of structures, formed the most influential progress in knowledge in this fundamental construction theory discipline during its consolidation period. The general work theorem used successfully by Mohr since 1874/75 for trussed frameworks had a dual make-up: the principle of virtual forces as the foundation of the force method, and the principle of virtual displacements as the foundation of the displacement method (Fig. 10-25). In the theory of statically indeterminate systems and the practice of structural calculations, however, the force method, founded on the principle of virtual forces, quickly adopted the dominant position, regardless of whether directly or indirectly in the form of Castigliano's theorems (Castigliano's 2nd theorem); the main reason for this was the tendency towards the formalisation throughout structural analysis inherent in the $\delta$-symbols. Although Mohr and his student Robert Land (1857 – 99) also acknowledged the fundamental role of the principle of virtual displacements, the first approaches to the displacement method remained on the second stage of formal operations. One reason for this is certainly that formulations with equilibrium conditions were preferred to those of the principle of virtual displacements because engineers were more familiar with these. Ostenfeld therefore used the equilibrium conditions for calculating the restraint forces of the set of equations in 10-17. Nonetheless, the $\delta$-symbols closely linked with the force method anticipated the displacement method in formal terms, and therefore rendered possible Ostenfeld's displacement method in the first place.

The displacement method developed by Ludwig Mann in 1927 on the basis of the Lagrange formalism (see section 10.2.4.4) leads to the set of equations in 10-17 with the displacement variables $\xi_l$ as unknowns and are known as type 2 elasticity equations; for the first time, he used the principle of virtual displacements for calculating the coefficients (restraint forces) in 10-17. In formal terms, the development of the displacement method followed the force method. The dual nature had already become apparent in 1903 in Kirpichev's book *Überzählige in der Baumechanik* (redundancy in structural mechanics) [Kirpichev, 1903], where he develops the theoretical basis of both the force method and the displacement method (see section 6.5.2). A second edition of this outwardly unassuming but in terms of content seminal work was published 21 years after the death of its author [Kirpichev, 1934]. Unfortunately, Kirpichev's book never appeared in English, French or German. It was therefore only of benefit to the formation of structural analysis theories in Russia and the Soviet Union; it was ignored in the West owing to the lack of knowledge of the Russian language. The extremely sparse adoption in the West of the

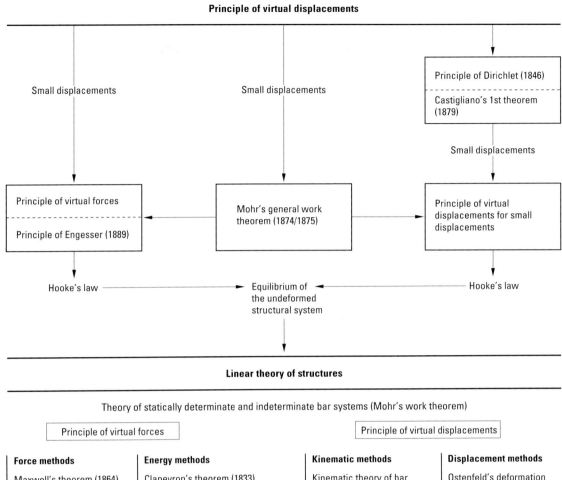

FIGURE 10-25
The dual nature of linear theory of structures

scientific progress achieved in Russia and the Soviet Union would first be partly relieved by the shock of the Sputnik project (1957); one example of this is *Structural Mechanics in the U.S.S.R. 1917–1957* [Rabinovich, 1960], the English title of George Herrmann's translation from the Russian original. In that work Rabinovich also discusses the most important publications on the force method [Rabinovich, 1960, pp. 6–16] and the displacement method [Rabinovich, 1960, pp. 16–19] which had appeared in the Soviet Union over those 40 years. The monograph by A. A. Gvozdev dating from 1927 is outstanding in this respect because it was the first time the displacement method had been presented in full in the Russian engineering literature (see [Rabinovich, 1960, p. 16]). Y. M. Rippenbein (1933) had already created the foundation for a three-dimensional displacement method in 1933 (see [Rabinovich, 1960, p. 17]), which B. N. Gorbunov and Y. V. Krotov developed three years later with the help of the tensor algebra of mechanics in the form of the "motor symbolism" of Richard v. Mises [Mises, 1924] (see [Rabinovich, 1960, p. 18]). The conclusion to this formalisation of the displacement method was undertaken by D. V. Vainberg and V. G. Chudnovskii, whose 1948 monograph presented the three-dimensional displacement method in a tensorial form. Together with the formulation of elastic theory and shell theory in the language of tensor analysis carried out by Zerna and Green (see section 9.3.1.3), the monograph by Vainberg and Chudnovskii reflects the status of the use of calculus at the transition from the consolidation period of structural theory (1900–50) to its integration period (1950 to date) on the theory formation side.

Another chance for the displacement method to match the formal status of formalisation in structural analysis as achieved by the force method appeared in 1938 with Arno Schleusner's work in which he proposes a clear conceptual divorce between the principle of virtual forces and the principle of virtual displacements for small displacements in the light of the calculus of variations [Schleusner, 1938/3]. Despite this, the force method continued to dominate not only the theory of statically indeterminate systems, but also the practical side of structural calculations. The practising structural engineer was therefore less interested in formalisation in structural analysis and more on the development and application of iterative methods.

**First steps in the automation of structural calculations**

### 10.3.4

The starting point for the development of computers in Germany was the numerical analysis of systems with a high degree of static indeterminacy derived from the force method in the form of the $\delta$-symbols.

The engineering student Konrad Zuse was working on a student project – supervised by Karl Pohl (1881–1947), an important student of Müller-Breslau – to calculate a system with nine degrees of static indeterminacy. The year was 1934. In order to formalise the computing process, Zuse initially worked out a scheme for the computing sheets with the aim of being able to enter just the figures (initial values). The sequence of calculations based on fundamental arithmetic operations should then result automatically from the structure of the sheets, hopefully so that adjacent numbers in rows had to be multiplied, adjacent numbers in columns added, and fixed values (equation constants) would be preprinted at the right places [Zuse, 1993, p. 165]. Very soon after joining the structural department of Henschel-Flugzeugwerke AG in 1935, his idea of automating structural calculations was confirmed on a practical level. Zuse's ideas gradually took shape between 1934 and 1936, and step by step he devised an automatic calculating machine [Zuse, 1995].

He decided as early as 1935 to devote his entire energy to the construction of such a machine. He handed in his notice at Henschel and became a computer inventor [Zuse, 1993, p. 31].

In his manuscript entitled *Die Rechenmaschine des Ingenieurs* (the engineer's calculating machine) dating from 30 January 1936 (Fig. 10-26), he speaks of a "computing plan" with which a scheme can be worked out for a calculation of any length "by writing down the successive computing operations according to type and order, and numbering consecutively the values that occur in the course of the computation, or arranging them according to another scheme without first determining their magnitude" [Zuse, 1936, p. 3]. Zuse concludes as follows: "The engineer needs calculating machines that perform these computations automatically by fixing the computing plan on a punched tape which feeds the commands for the individual computations automatically and successively into the machine" [Zuse, 1936, p. 3].

Zuse has therefore turned the basic idea into a machine with which any symbolic machine can be imitated. Based on an analysis of the work of the practising design engineer, Zuse specified the content of the computing plans in a hierarchical system using the example of aircraft construction: total plan, group plans (e.g. statically indeterminate trusses) and individual plans. For example, the group plan for "statically indeterminate truss" contains individual plans such as:

- Development of dimensions and determination of truss components
- Calculation of stiffness
- Member forces in the statically determinate system as a function of the static indeterminates
- Calculation of displacement jumps $\delta_{i0}$ and $\delta_{ik}$
- Solution of the set of equations in 10-18, i.e. calculation of static indeterminates $X_k$
- Calculation of the member forces in the statically indeterminate system
- Checking.

**FIGURE 10-26**
First page of Konrad Zuse's typewritten manuscript *Die Rechenmaschine des Ingenieurs*, 30 January 1936

Zuse divides the variables in the plan into initial values, intermediate values, result values, checking values and constants. "When setting up the total plan, i.e. dividing it into group plans and individual plans, it is necessary to be very disciplined with respect to the initial and result values. These are the threads that weave together the individual plans ... The relationships between the plans must fit together like electric plugs and sockets" [Zuse, 1936, pp. 10-11]. These remarks show that Zuse had already anticipated an essential element in the structure of program systems – the division into routines and subroutines.

But Zuse did not confine his work to just the analysis and synthesis of the intellectual techniques of the design engineer; he also sounded out the possibilities of the machine. These possibilities concerned the rationalisation of everyday engineering calculations (e.g. including tables of steel sections in the machine), the development of new methods to solve technical problems, and expansion into areas that up to then had not been accessible to calculations [Zuse, 1936, p. 12].

The latter two possibilities were seen by Zuse as the internal relationship between computer development and engineering science theory formation. As an example of the development of new methods for solving technical problems, he suggested harnessing the knowledge gained – through mechanics – in the 1920s and 1930s in plastic theory and aerodynamics for aircraft design. This meant that the "thinking of the theorist would be conserved to a certain extent"; the engineer would acquire "the formulas ex works so to speak" and therefore did not need to be aware of the theories on which they were founded [Zuse, 1936, p. 20]. For technical fields in which progress hitherto had only been possible through expensive trials, e.g. engine-building, Zuse saw the chance of opening up calculations through the computer [Zuse, 1936, p. 23].

Although Zuse dedicated himself to the technical realisation of the computer after 1936, the interaction between computer development and engineering science theory formation he was hoping for did not become a reality in Germany until the late 1950s. Even the sponsor of his project after 1940, the aircraft engineer Alfred Teichmann (see section 6.6.4.2) from the German Aviation Testing Authority (DVL) in Berlin-Adlershof (later professor of theory of structures at Berlin TU), still considered the computer to be merely a means of rationalising structural calculations, even after 1950, not recognising that the computer could technically reform the entire theory of structures by means of matrix formulation.

**The diffusion of matrix formulation into the exact natural sciences and fundamental engineering science disciplines**

### 10.3.5

Zuse's plugs-and-sockets metaphor for the relationships between the computing plans was expressed formally on an individual science level in the matrix theory of structural analysis formulated by John H. Argyris [Argyris, 1955 & 1957]; the prerequisite for this was the historico-logical coincidence with the realisation of the computer in the 1940s. As a universal symbolic machine, the computer is in the position of being able to obtain statements on the cognitive objects of any system of notation representing such objects through algorithmic manipulations of the system of notation concerned. Matrix algebra is one such specific system of notation. Structural matrix analysis is therefore computational mechanics at the same time, the artistry of which consists of organising the structural calculations in such a way that they can be carried out almost without having to delve into the theoretical background. The historical timescale from the construction of the matrix concept to the first applications in the exact natural sciences and fundamental engineering science disciplines spans a little over 70 years.

**Matrix formulation in mathematics and theoretical physics**

### 10.3.5.1

Starting with the matrix concept introduced by J. J. Sylvester into the theory of sets of linear equations in 1850, his friend and colleague A. Cayley created matrix formulation in 1858. Cayley realised that symbolic operations with linear transformations could be attributed to a few basic operations with the scheme of coefficients, the matrix of transformations – symbolic operations which Cayley, taking arithmetic as his guide, defined as addition, multiplication and division of matrices.

Symbolic operations with linear transformations therefore become calculations with matrices, which lends clarity to the whole expanded and multifaceted theme of linear relationships, gives it a lightness of form and content, and advances matrix calculations – precisely because of its calculus associations – to become a means of knowledge (*epistēmē*), an art (in the sense of *tekhnē*) of the discovery of new findings in the natural and engineering sciences. By the end of the 19th century, B. Peirce, C. S. Peirce, F. G. Frobenius and C. Hermite in particular had shaped matrix formulation for internal mathematical applications into a self-contained mathematical discipline [Wußing, 1979, p. 255].

But was not until 1926 that matrix formulation would play a decisive role in the formalisation of a non-mathematical discipline – in the quantum mechanics of M. Born, W. Heisenberg and P. Jordan, which was ready to supersede the classical physical conception of our world. There was a good reason why this direction of quantum physics was called "matrix mechanics", there being a need to express that, on the one hand, it was in formal terms a mechanics of matrices, operations involving matrices, and that, on the other hand, its content was different to that of classical mechanics. Like a quarter of a century later matrix formulation became a means to knowledge as structural analysis migrated to structural matrix analysis, it synthesised the *tekhnē* with the *epistēmē* in the historico-logical development of quantum physics in the late 1920s.

#### 10.3.5.2 Tensor and matrix algebra in the fundamental engineering science disciplines

The first ideas for using matrices in structural analysis were expressed by Edward Study (1862–1930) as early as 1903 [Norris & Wilbur, 1960; Corradi, 1984]. These ideas were initially forgotten because his monograph *Geometrie der Dyname* (geometry of the resultant) [Study, 1903] was taken up by the advocates of the fundamental engineering science disciplines only marginally. Study's contribution to screw theory was in keeping with the tradition of the geometrical mechanics of rigid bodies, the methods of which are to a large extent identical with those of projective geometry; both had already lost much of their significance in geometrical research by the end of the 19th century because mathematics was tending towards arithmetisation and axiomatisation and hence "was aimed at overcoming the view through calculus" [Ziegler, 1985, p. 209].

Added to this was the fact that the theory of rigid frames that evolved with reinforced concrete took hardly a decade to become established and only after that did the practice of statically indeterminate calculation express a need for formalisation with respect to structural analysis theory formation; the reinforced concrete designer's need for structural analysis aids was essentially met by the publications on rigid frames that appeared in numerous versions in that period (monographs on the rigid frames customary in practice which contained ready-made formulas) (see section 9.3.1.1). Structural steelwork, too, which in the second decade of the 20th century started to rival reinforced concrete, exerted only little pressure in the direction of formalisation in structural analysis. Nevertheless, at an early stage, theory of structures got to grips with the problems of systems with many degrees of indeterminacy, which plagued the practice of structural calculations in both of these building disciplines; one typical example of this is the debate on Vierendeel girders which continued throughout the second decade of the 20th century in the journals *Der Eisenbau, Beton und Eisen* and *Armierter Beton*. So the symbolic operations with linear transformations in structural analysis did not take place on the level of matrix algebra, but instead in the theory of sets of linear equations.

First of all, the closer relationship between applied mathematics and mechanics since the early 1920s allowed the mathematical foundations of structural analysis to become a distinct object of research – a theme covered by, for example, the mathematician Richard von Mises. Following on from the concept of the "motor" introduced by Study in 1903 [Study, 1903, pp. 51 ff], Mises developed a mathematical aid for mechanics in the shape of the "motor symbolism": "For us, the most important thing to come out of Study's elementary results is that like the vector is determined by a pair of points (start and end points of the directed line segment), the resultant or the motor can be shown pictorially by a pair of straight lines, and that based on this geometrical representation it is possible to explain a 'geometric addition' of motors. I shall now go one step further than Study and – completely in keeping with the analogy to the two product forms of vector calculation [scalar product and vector product – the author] – introduce a scalar and motoric product of two motors which, as is shown, possess both direct and elementary significance in mechanics … It is wrong to use the convenient vector notion for the mechanical concepts of force, velocity, acceleration, etc., but then to refrain from the totally analogous advantages for the higher mathematical functions such as second moment of area, stress condition, deformation, etc." [Mises, 1924/1, p. 156]. Such "higher mathematical functions" are second-order tensors, e.g. the mass inertia tensor and the stress tensor with, generally, nine components. Both the vectors as first-order tensors and also the second-order tensors were summarised by Mises in a tensor algebra of mechanics in symbolic notation: "With the new motor symbolism it is the case that this, too, proves its full benefits in mechanics only when one includes the second-order concepts, i. e. the motor dyad (or the motor tensor, the motor matrix) in our considerations" [Mises, 1924/1, p. 156]. In the second part of his paper [Mises, 1924/2], Mises tests his tensor algebra of mechanics on rigid body mechanics, general dynamics, continuum mechanics, theory of structures, fluid mechanics and the external mechanics of aircraft. Mises takes a little less than seven pages to formulate

- the linear-elastic theory of trussed frameworks (for loads at the nodes),
- the reciprocity theorem (eq. 10-1) in a generalised form as a symmetry condition of the corresponding tensor, and
- the energy principle of theory of structures for linear-elastic trussed frameworks

in the language of tensor algebra [Mises, 1924/2, pp. 199–205].

Fig. 10-27 shows the unloaded bar element for deriving the relationship between the vector of the internal force variables $S$ in the centre of the bar and the difference vector of the displacement variables between cross-sections 2 and 1 $U_2 - U_1$, which is conveyed by the double tensor and the following tensor equation:

$$U_2 - U_1 = K \cdot S \tag{10-23}$$

Here, Mises assumes the origin of the body system of coordinates to be in the centre of the bar. The z-axis coincides with the undeformed bar axis and points towards point 2 (right); the x- and y-axes are the principal axes of the cross-section. The vector of the force variables $S$ in eq. 10-23 contains all six internal forces in the middle of the bar as components. Mises represents all tensors by Gothic capital letters in bold typeface, but these have been replaced in tensor equation 10-23 by Latin capital letters in bold typeface. The component presentation of tensor equation 10-23 corresponds to the following matrix equation:

$$\begin{Bmatrix} \varphi_{x,2} \\ \varphi_{y,2} \\ \varphi_{z,2} \\ u_2 \\ v_2 \\ w_2 \end{Bmatrix} - \begin{Bmatrix} \varphi_{x,1} \\ \varphi_{y,1} \\ \varphi_{z,1} \\ u_1 \\ v_1 \\ w_1 \end{Bmatrix} = \begin{bmatrix} \frac{l}{EJ_{xx}} & 0 & 0 & 0 & 0 & 0 \\ 0 & \frac{l}{EJ_{yy}} & 0 & 0 & 0 & 0 \\ 0 & 0 & \frac{l}{GI_T} & 0 & 0 & 0 \\ 0 & 0 & 0 & \frac{l^3}{12EJ_{yy}} & 0 & 0 \\ 0 & 0 & 0 & 0 & \frac{l^3}{12EJ_{xx}} & 0 \\ 0 & 0 & 0 & 0 & 0 & \frac{l}{EA} \end{bmatrix} \cdot \begin{Bmatrix} S_1 \\ S_2 \\ S_3 \\ S_4 \\ S_5 \\ S_6 \end{Bmatrix} \tag{10-24}$$

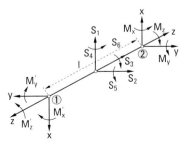

**FIGURE 10-27**
Element of an elastic bar for deriving the relationship between the vector of the force variables and the difference vector of the displacement variables after R. v. Mises [Mises, 1924/2, p. 200]

In eq. 10-23, or rather 10-24, $U_2$ is the state vector of the displacement variables of the right-hand border 2 with the first three components as angle of rotation about x-, y- and z-axes and the last three components as displacement in the x-, y- and z-axes; the same applies similarly to the state vector $U_1$ of the displacement variables of the left-hand border 1. The internal forces are specified in the middle of the bar in the vector of the force variables $S$ – expressed by the internal forces of the right-hand border 2:

$$\begin{Bmatrix} S_1 \\ S_2 \\ S_3 \\ S_4 \\ S_5 \\ S_6 \end{Bmatrix} = \begin{Bmatrix} X \\ Y \\ Z \\ M_x - 0.5 \cdot Y \cdot l \\ M_y - 0.5 \cdot X \cdot l \\ M_z \end{Bmatrix} \tag{10-25}$$

Eq. 10-23 expresses the following physical facts in the language of tensor algebra: the relative displacement of ends 1 and 2 of an unloaded elastic bar is the tensor product of tensor $K$ determined by the bar constants and the vector of the internal force variables $S$ for the system of coordinates selected in the middle of the bar (see Fig. 10-27). Contrastingly, in the language of matrix algebra the 36 components of tensor $K$ can be summarised as a 6 × 6 matrix (the flexibility matrix, see eq. 10-24), whose matrix product with the internal force variables in the middle of the bar (eq. 10-25) results in the relative displacement of ends 1 and 2 of an unloaded elastic bar. The tensor of the bar elasticity $K$ in eq. 10-23, or rather the flexibility matrix in eq. 10-24, is completely symmetrical and only the diagonals are occupied, which means that the system of coordinates selected in the middle of the bar represents a system of principal axes.

Unfortunately, at first only Soviet scientists (see section 10.3.3) and Václav Dašek (1930) from Prague were able to benefit from applying tensor algebra to structural analysis.

Siemens researchers F. Strecker and R. Feldtkeller employed matrix calculation in 1929 for the calculation of quadripoles in electrical engineering theory [Strecker & Feldtkeller, 1929]. In his classic monograph on quadripole theory, Feldtkeller, who was appointed professor of electrical telecommunications technology at Stuttgart TH in 1936 following his work in Siemens' central laboratory in Berlin, systematically used the formal potential of matrix calculation for calculating linear electrical networks [Feldtkeller, 1937]. W. Quade finally provided an overview of the most important applications of matrix calculation to electrical networks and vibrations [Quade, 1940]. Feldtkeller's 1937 monograph helped the quadripole theory to become the showcase of matrix calculation in the fundamental engineering science disciplines. Two years later, the electrical engineer G. Kron, an employee of General Electric, published his book entitled *Tensor Analysis of Networks* [Kron, 1939]. Kron unfortunately mixed tensor and matrix theory. So the introduction of matrix theory into electrical engineering experienced an unlucky start due to a number of less-than-fortunate publications [Zurmühl, 1950, p. 347]. Notwithstanding, Kron was able to cross the boundary between electrical engineering and mechanics. For example, he used the analogy between electrical and mechanical networks (elastic trussed frameworks) known to Maxwell and Kirchhoff for analysing three-dimensional trussed frameworks and formulated them in the language of matrix theory [Kron, 1944]. Kron's work inspired the aircraft engineer B. Langefors, an employee of the Swedish SAAB company, to summarise the force method in matrix form [Langefors, 1952]. Working independently, H. Falkenheiner released two articles in French [Falkenheiner, 1950, 1951], which Alf Samuelsson has compared with that of Langefors (1952): "The papers by Falkenheiner and Langefors are very similar. Both use the principle of deformation minimum according to Menabrea-Castigliano to deduce the matrix of influence coefficient expressing point displacements as functions of point loads. They also both describe a substructure technique. Langefors uses force in hypothetical cuts as redundants while Falkenheiner uses superposition coefficients of equilibrium systems as redundants, The method of Falkenheiner is then more general than that by Langefors" [Samuelsson, 2002, p. 7]. In 1953 Falkenheiner discussed his two articles in the light of the work of Langefors [Falkenheiner, 1953].

### 10.3.5.3 On the integration of matrix formulation into engineering mathematics

One of the historical trails of matrix formulation in structural mechanics leads back to the Aerodynamics Department set up in 1925 by R. A. Frazer at the National Physics Laboratory in Teddington near London. Together with W. J. Duncan, he researched the flutter of aircraft wings and in 1928 published the so-called *Flutter Bible* [Felippa, 2001]. Six years later Duncan and A. R. Collar formulated conservative vibration problems in the language of matrix algebra [Duncan & Collar, 1934], and one year after that date wrote a work on the motion equations of damped vibrations with the help of the powerful mathematic means of matrix algebra [Duncan & Collar, 1935]. Looking back, Collar describes this discovery of matrix algebra for a reformulation of vibration mechanics as follows: "Frazer had studied matrices as a branch of applied mathematics under Grace in Cambridge; and he recognized that the statement of, for example, a ternary flutter problem in terms of matrices was neat and compendious. He was, however, more concerned with formal manipulation and transformation to other coordinates than with numerical results. On the other hand, Duncan and I were in search of numerical results for the vibration characteristics of airscrew blades; and we recognized that we could only advance by breaking the blade into, say, 10 segments and treating it as having 10 degrees of freedom. This approach also was more conveniently formulated in matrix terms, and readily expressed numerically. Then we found that if we put an approximate mode into one side of the equation, we calculated a better approximation on the other; and the matrix iteration procedure was born" [Collar, 1978, p. 17]. The year 1938 saw Frazer, Duncan and Collar publish the first monograph in which areas of structural dynamics such as aeroelasticity were formulated systematically in terms of matrix algebra (Fig. 10-28) [Frazer et al., 1938]; since the end of the consolidation period of structural theory (1900–50), this has become a standard work for engineers who wish to find out something about solving vibration problems using matrices. Fig. 10-28 shows the eigenvalue analysis of a system of bars with the three degrees of freedom $q_1$, $q_2$ and $q_3$, which was investigated with the help of matrices; Fig. 10-28 is taken from the seventh, unaltered reprint of the original edition dating from 1938. The monograph thus remained relevant until the middle of the innovation phase of structural theory (1950–75).

Zurmühl's monograph *Matrizen. Eine Darstellung für Ingenieure* (matrices – an explanation for engineers) of 1950 (Fig. 10-29) represents a milestone in the use of matrix formulation in the German-speaking countries. Dr. Zurmühl recognised that matrix formulation provided linear algebra with a means of expression that – through equations of unsurpassed conciseness and clarity that always concentrate the user's attention on the essentials – could be used to express the linear relationships prevailing in physics and the engineering sciences for operations that were uniform but difficult to present in customary mathematical language (see [Zurmühl, 1950, p. I]). Matrix theory will "become more and more common in engineering mathematics and perhaps soon play a similar role to

Take as generalised coordinates $q_1$, $q_2$, $q_3$ the linear displacements of $B$, $F$, $G$, respectively. Then the displacement of $D$ is $\frac{1}{2}(q_1+q_2)$. In a general static displacement of the system, the elastic moments at $A$, $C$, $E$, $D$, will be $\frac{1}{3}q_1$, $\frac{2}{3}(q_1+q_2)$, $\frac{1}{3}q_2$, $\frac{1}{9}(q_3-q_1-q_2)$, and since the lever arms are all of unit length, the vertical forces at $B$, $D$, $F$, $G$ are also $\frac{1}{3}q_1$, $\frac{2}{3}(q_1+q_2)$, $\frac{1}{3}q_2$, $\frac{1}{9}(q_3-q_1-q_2)$. To find the flexibility matrix, apply unit load at $B$, $F$, $G$ in succession. When unit load is applied at $B$, we have by moments about $AE$,

$$1 = \tfrac{1}{3}q_1 + \tfrac{2}{3}(q_1+q_2) + \tfrac{1}{3}q_2 = q_1+q_2,$$

while by moments about $AB$,

$$\tfrac{2}{3}q_2 + \tfrac{2}{3}(q_1+q_2) = 0, \quad \text{or} \quad q_1 + 2q_2 = 0.$$

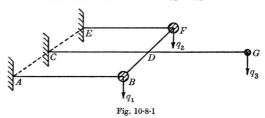

Fig. 10·8·1

Hence $q_1 = 2$, $q_2 = -1$, and since the moment at $D$ is zero, $q_3 = q_1 + q_2 = 1$. The displacements are thus $\{2, -1, 1\}$. Similarly, when unit load is applied at $F$, the displacements are $\{-1, 2, 1\}$. When unit load is applied at $G$, we have by moments about $AE$,

$$2 = \tfrac{1}{3}q_1 + \tfrac{2}{3}(q_1+q_2) + \tfrac{1}{3}q_2 = q_1+q_2,$$

and, since the displacement is symmetrical, $q_1 = q_2 = 1$. Moreover, by moments about $BF$,

$$1 = \tfrac{1}{9}(q_3-q_1-q_2) \quad \text{or} \quad q_3 = 11.$$

Hence in this case the displacements are $\{1, 1, 11\}$. The flexibility matrix is thus

$$\Phi = \begin{bmatrix} 2 & -1 & 1 \\ -1 & 2 & 1 \\ 1 & 1 & 11 \end{bmatrix}.$$

The inertia matrix is evidently

$$A = \begin{bmatrix} 4 & 0 & 0 \\ 0 & 4 & 0 \\ 0 & 0 & 1 \end{bmatrix}.$$

**FIGURE 10-28**
Eigenvalue analysis of a system of bars with three degrees of freedom after Frazer, Duncan & Collar [Frazer et al., 1963, p. 323]

vector theory, which today is indispensable" [Zurmühl, 1950, p. I]. Zurmühl's vision would very soon become reality because during the 1950s his monograph became the standard work of engineering mathematics. The book had been backed up since 1945 by the work of Alwin Walther (1898–1967), who tested numerical methods and procured obscure literature. It was at the Institute for Practical Mathematics (IPM), headed by Walther, at Darmstadt TH that Zurmühl investigated a matrix-based iteration method in the early 1940s, which he tested using the example of the calculations for a three-dimensional trussed framework with many degrees of static indeterminacy (see [Zurmühl, 1950, p. 282]).

Even before World War 2, Walther's IPM was being called a "computations factory", and in 1939 up to 70 female workers equipped with mechanical tabletop calculating machines were performing tasks associated with ballistics, lightweight construction, radiolocation and optics (see [Petzold, 1992, p. 226]). The thinking work of engineering science calculation had thus been placed in a scheme and divorced completely from the engineering work. What could have been more obvious than to automate this calculation work, as Zuse had suggested back in 1936?

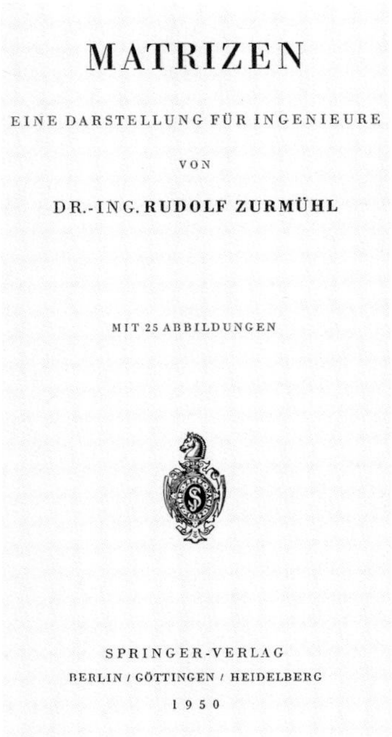
MATRIZEN

EINE DARSTELLUNG FÜR INGENIEURE

VON

DR.-ING. **RUDOLF ZURMÜHL**

MIT 25 ABBILDUNGEN

SPRINGER-VERLAG
BERLIN / GÖTTINGEN / HEIDELBERG
1950

**FIGURE 10-29**
Cover of the first German book on the application of matrices to engineering and the engineering sciences

Plans for a large, powerful, automatic program-controlled computing installation, which was to be assembled from parts for current calculating machines, were therefore discussed as early as 1943 at the IPM, which Walther had made available for research into wartime issues. Spurred on by the message concerning Aiken's large Mark I Automatic Sequence Controlled Calculator (ASCC), the generals of the German armed forces allocated the highest priority to Walther's project, which meant that he could procure the parts he needed to assemble the machine within a very short time. But a few days later the new installation disappeared into the bombed-out ruins of the IPM (see [Petzold, 1992, p. 228]). Through Prof. Herbert Wagner, manager of Special Department F at Henschel-Flugzeugwerke AG, and as such Zuse's superior (Zuse had headed the structural analysis group since 1940), Walther first met Zuse in late 1942 [Zuse, 1993].

Wagner, that pioneer of aviation engineering and ingenious manipulator of numbers, had recognised the universal importance of Zuse's computer and had actively supported the project. Zuse wanted to work towards his doctorate on the theme of the theory of general calculation with Walther, who at that time regarded the computer primarily as a technical tool for rational engineering science calculations, in the sense of the numerical evaluation of formulas. Zuse's doctorate unfortunately remained only an outline. Petzold suspects that it would have proved difficult to carry out such work with Walther, who gave priority to analogue technology (see [Petzold, 1992, p. 197]).

### 10.3.5.4 A structural matrix method: the carry-over method

Nevertheless, Walther, by promoting Zurmühl, had recognised the enormous power of matrix theory for physics and the fundamental engineering science disciplines. And therefore the Darmstadt doctorate project of H. Fuhrke on the determination of beam oscillations with the help of matrices could be completed in the early 1950s [Fuhrke, 1955].

Even more important for structural analysis was the carry-over method for calculating continuous beams with any number of spans created by S. Falk in 1956 [Falk, 1956], which translated the solution of the differential beam equation fully into the language of matrix formulation (Fig. 10-30). The carry-over method only exists through matrix operations and in the case of continuous beams leads to systems with a maximum of two linear equations. The degree of statical or geometrical indeterminacy does not appear in the carry-over method, which belongs to the group of reduction methods; far more significant are the topological properties of the structural system. Consequently, the dual nature of theory of structures – brought about by the force and displacement methods – is insignificant in the carry-over method.

Joachim Scheer was probably the first engineer in the German-speaking countries to investigate in detail the use of program-controlled automatic calculators for structural tasks in conjunction with the carry-over method [Scheer, 1958]. The program presented by Scheer in 1958 was employed for practical tasks, e.g. a number of projects for the engineering practice of Dr. Homberg in Hagen [Scheer, 1998]. Scheer told this author in 1998 that his dissertation on the problem of the overall stability of singly-symmetric I-beams published in the journal *Der Stahlbau* in 1959 had only been rendered possible through the use of the carry-over method and computers in 1957/58 [Scheer, 1998]. Despite this, the influence of the carry-over method, like other reduction methods, remained limited in the theory and practice of structural analysis because matrix theory covered only some of the structural systems. At the same time, Klöppel and Scheer employed matrix theory successfully for preparing the programming of the buckling theory of stiffened rectangular steel plates according to the energy method. With the help of the IBM 704 computer donated to Darmstadt TH by IBM Deutschland in 1958 they were able to calculate the buckling values of stiffened rectangular plates for stan-

**FIGURE 10-30**
Carry-over method after Falk in the representation by Scheer [Scheer, 1958, p. 228]

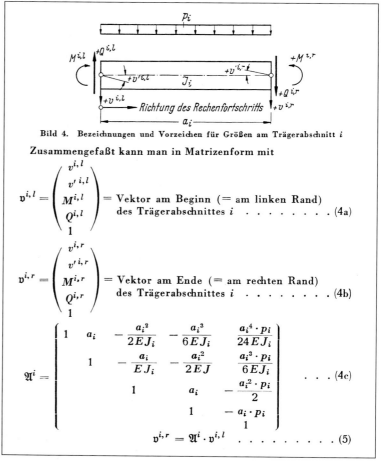

dard cases from the buckling matrix in a relatively short time and publish these as tables of curves for everyday structural steelwork calculations [Klöppel & Scheer, 1960]; a second volume followed eight years later [Klöppel & Möller, 1968]. Such tables for the stress analyses of plate and shell structures calculated with the help of sophisticated research programmes provided important assistance in the production of structural calculations carried out partly by hand and partly with the computer even after the innovation phase of structural theory (1950–75).

The carry-over method was suitable for manual and computerised calculations; this latter point had already been mentioned by S. Falk in 1956 (see [Falk, 1956, p. 231]). The carry-over method could be used to multiply an $m \times r$ matrix (left) by an $r \times n$ matrix (right) in a particularly simple and clear fashion according to the scheme introduced by Falk [Falk, 1951]. The $r \times n$ matrix is positioned to the right above the $m \times r$ matrix such that the extended $n$ columns of the $r \times n$ matrix and the extended $m$ rows of the $m \times r$ matrix overlap to form the result matrix, the $m \times n$ matrix. For example, the element in the $i$th row and the $k$th column of the result matrix is calculated from the sum of the products of the respective elements in the $i$th row of the $m \times r$ matrix and the associated elements in the $k$th column of the $r \times n$ matrix. Fig. 10-31 shows a numerical example of a matrix multiplication according to the Falk scheme. The $m \times r$ matrix ($m = 3$, $r = 2$) is to be multiplied from the right by the $r \times n$ matrix ($r = 2$, $n = 4$). The element in the 3rd line and the 3rd column of the $m \times n$ result matrix ($m = 3$, $n = 4$) then becomes $(6 \times 6) + (1 \times 8) = 36 + 8 = 44$. In the Falk scheme the arithmetisation of the matrix calculation for the purpose of programming is obvious; the suitability of the Falk scheme for manual calculations does not contradict this, but ensures that manual calculations, too, undergo further formalisation, and therefore the practical use of symbols becomes ever more established in the everyday work of the practising structural engineer.

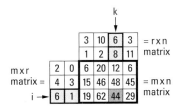

**FIGURE 10-31**
Numerical example of matrix multiplication according to the Falk scheme

## 10.4 "The computer shapes the theory" (Argyris): the historical roots of the finite element method and the development of computational mechanics

The finite element method (FEM), or the weighted residual method, today forms the basis of computational mechanics (CM) (see [Zienkiewicz, 2004, p. 3]). As, in principle, every field problem, e. g. electrodynamic, elasto-mechanical or fluid-mechanical, can be solved numerically with FEM, other CM methods, e. g. boundary element method (BEM), finite differences method (FDM), will not be considered in the following.

The programmable digital computer (simply called the computer in the following) enabled numerical simulation to take its place as an equal alongside the interaction between theory and experimentation that had characterised the natural sciences since the time of Galileo and later the engineering sciences (Fig. 10-32). Today, numerical simulation, theory and experimentation are the three supporting pillars of CM in particular. Theory formation therefore takes place not only on the theory–computer–experiment level, but on the experiment–computer–simulation and the theory–computer–simulation levels as well (see Fig. 10-32). This latter level, which Argyris described splendidly in 1965 in his prophetic essay

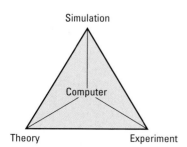

FIGURE 10-32
The tetrahedron of computer, theory, experiment and simulation

*The Computer shapes the theory* [Argyris, 1965], forms the outer framework in which the formation of structural mechanics theories has been mainly taking place since the middle of the innovation phase of structural theory (1950–75).

The historico-logical development of the computer goes hand in hand with the formation of FEM and therefore forms the beginning and the end of the innovation phase of structural theory (1950–75). Talking about the linguistic relationship between the use of the words "computer" and "computations", Zienkiewicz finds the following, telling words: "The development of computational mechanics clearly owes much to the presence of the electronic computer which came on the scene only in the middle of the last century. However, the words 'computer' and 'computations' are much older. In the very first paper on finite differences in the 20th century, Richardson, in 1910, uses the word 'computers' to describe his assistants who were boys from the local high-school, employed to do the numerical calculations at each iteration. It is interesting to note that Richardson paid a price of $N/18$ pence per co-ordinate point calculation in which $N$ was the number of digits used and, as his note says, he did not pay if the computers committed errors" [Zienkiewicz, 2004, p. 3]. Richardson's mention of the human computer hits the bull's-eye as the historical movement of the idea of formalisation (see section 10.1.1) can be understood as the evolution of the "computer in ourselves" [Krämer, 1988, p. 4], in the course of which human beings, step by step, shook off their formalisable mental stocks in the form of "symbolic machines" [Krämer, 1988, p. 4]. Such symbolic machines do not exist in reality, instead only in symbolic form – they exist only on paper, but could be realised technically in principle. A symbolic machine does nothing other "... than transform series of symbols. Their states can be described fully by a succession of system configurations by means of which a certain initial configuration can be transformed into a desired final configuration" [Krämer, 1988, p. 3]. As an example of a symbolic machine, Krämer nominates the execution of written multiplication in the decimal place value system with the mechanical calculating machine as a technical realisation (see [Krämer, 1988, p. 3]). Another example is matrix algebra; the matrix has the function of a formalised theory symbol and is as such the basic symbol of the symbolic machine of matrix algebra. It achieves the level of the non-interpretive use of symbols, the third stage of the formalisation notion. Matrices have an intra-symbolic meaning because the rules of matrix algebra by means of which the matrix expressions are formed and transformed do not have any relationship to the meaning of the matrix. Like every formalised theory, matrix theory, too, is the "production centre for an infinite number of configurations of symbols" [Krämer, 1994, p. 93]. This is one side of formalised theory in the special case of matrix theory. The operative force does not unfold until the formalised theory is interpreted and can be applied in the form of a mental or intellectual technology – that's its other side. Krämer calls interpreted formalised theories "symbolic machines" (see [Krämer, 1994, p. 94]). For example, structural matrix analysis is one interpreted matrix theory, the matrix mechanics of quantum theory another. And the shell theory of Zerna and Green (see section 9.3.1.3) or the gravitation theory of Albert Einstein are interpreted formalised theories, interpreted tensor theories. Finally, we should mention calculus of variations, which constituted the formal basis for the development of the variation methods of modern structural analysis, or rather structural mechanics, and represents the interpreted variation theories. According to Krämer, interpreted formalised theories represent not only certain cognitive objects, but instead also create new objects: interpreted formalised theories or symbolic machines "constitute and structure only those domains that can be considered as interpretation models of formalised theories" [Krämer, 1994, p. 94].

Structural matrix analysis therefore reformed not only the cognitive objects of structural analysis, e.g. the force and displacement methods, but enabled the construction of new kinds of cognitive objects; the carry-over method (see section 10.3.5.4) is just one example. The construction of cognitive objects is, however, a genuine topic of theory formation in mathematics, the exact natural sciences and the fundamental engineering science disciplines. Interpreted formalised theories, or rather symbolic machines, therefore have a major influence on the formation of scientific theories. The artistry of a symbolic machine is "that although its cognitive purpose is related to the assertion status of the symbolic expressions, this purpose is realised by means that relate exclusively to its object status" [Krämer, 1994, p. 97]. With the computer as the technical realisation of the multitude of all symbolic machines, the divorce between symbol interpretation and symbol procedure, between cognitive purpose and technical means has been completed in a universal fashion.

Carlos A. Felippa divides the historico-logical development of modern structural mechanics into two phases (Fig. 10-33). In the first phase (Fig. 10-33a), the matrix methods of structural mechanics are determined by
– manual calculations,
– discrete mathematical models, and
– different matrix formulations of the same object.

This phase defined by Felippa corresponds to the invention phase of structural theory (1925–50), which began with the work of Richard v. Mises on the tensor formulation of mechanics in general and the theory of elastic trussed frameworks in particular (see section 10.3.5.2), and ended with the beginnings of automation in structural calculations as early as the 1940s.

Felippa's second phase (Fig. 10-33b) is when FEM started to take shape, which is determined by
– computers,
– continuum mathematical models, and
– the direct displacement method, or rather the direct stiffness method.

This phase corresponds to the first half of the innovation phase of structural theory (1950–75), which ended with the establishment of the direct stiffness method as the principal method of FEM in the mid-1960s. Here,

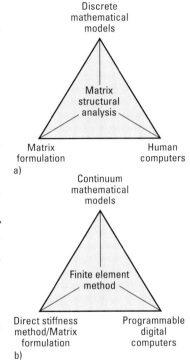

FIGURE 10-33
Two historico-logical development phases of structural mechanics after Felippa: a) matrix methods, and b) FEM (redrawn after [Felippa, 2001, p. 1314])

**Truss models for elastic continua**

## 10.4.1

The formation of trussed framework theory during the establishment phase of structural theory (1850–75) provided a far-reaching theoretical tool for analysing elastic trusses in which tension and compression forces prevail. Every successful theory tends to exceed its original remit and this was no exception: the structural model of the pin-jointed framework on which trussed framework theory was based also served as a model for the dimensioning of reinforced concrete structures (see section 9.3.4) and for elastic continua as well as truss-like structures.

**Kirsch's space truss model**

### 10.4.1.1

It was in 1868 that Gustav Kirsch (1841–1901) managed to derive the basic equations for the homogeneous, isotropic and linear-elastic body from the trussed framework model. Kirsch considers a system of points "connected by elastic struts" in order to investigate the question of "whether it is possible to form an elastic system of points having the character of an isotropic medium when assuming an infinite number of points" [Kirsch, 1868, Sp. 484]. Instead of deriving the equilibrium conditions for the infinitesimal cube continuum, Kirsch assumes a linear-elastic space truss consisting of 12 perimeter bars, four diagonal bars and eight spherical joints $A$ to $H$ (Fig. 10-34).

In doing so, Kirsch assumes that the two Lamé elastic constants

$$\lambda = \frac{E \cdot \nu}{(1 + \nu) \cdot (1 - 2 \cdot \nu)} \qquad (10\text{-}26)$$

and

$$\mu = G = \frac{E}{2 \cdot (1 + \nu)} \qquad (10\text{-}27)$$

are equal. In eq. 10-26 and eq. 10-27, $E$ is the elastic modulus, $G$ or $\mu$ the shear modulus and $\nu$ Poisson's ratio. Assuming eq. 10-26 and eq. 10-27 are equal means that $\nu = 0.25$. According to DIN 18 800-1, an elastic modulus $E$ of 210 000 N/mm² and a shear modulus $G$ of 81 000 N/mm² may be assumed for steel grades S 235, S 275 and S 355 [Eggert & Henke, 2007, p. 15], which results in $\nu = 0.30$. If $E = 210\,000$ N/mm² and $\nu = 0.25$ are entered into eq. 10-27, then $G = 84\,000$ N/mm². Whereas the deviation in Poisson's ratio resulting from Kirsch's assumption that $\lambda = \mu = G$ amounts to about 15.5 %, the shear modulus deviation $G$ is only about 3.7 %. This means that Kirsch's simplification has only a small effect in practical terms. However, for the rigorous mathematical description of the homogeneous, isotropic and linear-elastic continuum, two independent elastic constants $\lambda$ and $\mu = G$ or $E$ and $\nu$ or $E$ and $G$ are required. Consequently, Kirsch's simplification is theoretically incorrect because it assumes that the Lamé elastic constants (eq. 10-26 and eq. 10-27) are equal, i.e. in essence he assumes only one elastic constant. So his simplification belongs to the tradition of rari-constant theories and not to the theoretically cor-

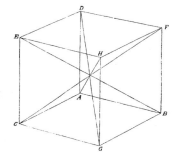

**FIGURE 10-34**
Trussed framework model for the homogeneous, isotropic and linear-elastic continuum after Kirsch [Kirsch, 1868, col. 486]

rect elastic theory with two constants (multi-constant theory). Nevertheless, Kirsch knows very well that the multi-constant theory represents the exact theory but can offer solutions for a small number of cases only and therefore is not taught in the engineering schools. Kirsch hopes, however, that his modelling of the elastic continuum as a pin-jointed framework "will become the basis for investigating simpler cases when solving more difficult tasks (bending of plate-type bodies etc.); for although a solution of the fundamental equations covering all cases is virtually impossible, the discovery of an adequate approximation method for single classes of tasks is certainly not beyond the bounds of probability if only sufficient forces could be united to carry out the search" [Kirsch, 1868, Sp. 638]. Although the modelling of the homogeneous, isotropic and linear-elastic continuum as a pin-jointed framework was taken up again and again by leading engineering scientists such as Otto Mohr (see Fig. 6-40) during the classical phase of structural theory (1875–1900), they did not contribute to specifying a model for plate and shell structures. The reason for this is that the epistemological interest of classical structural analysis was focused on the investigation of linear-elastic trusses. It was not until the consolidation period of structural theory (1900–50) that reinforced concrete turned plate and shell structures into an object of theory of structures.

### 10.4.1.2 Trussed framework models for elastic plates

In 1904 Felix Klein and Karl Wieghardt proposed determining the stress distribution of a thin elastic plate not by way of the self-contained solution of the biharmonic equation derived by J. H. Michell in 1899 [Michell, 1899]

$$\Delta\Delta F(x,y) = \frac{\partial^4 F}{\partial x^4} + 2 \cdot \frac{\partial^4 F}{\partial x^2 \partial y^2} + \frac{\partial^4 F}{\partial y^4} = 0 \qquad (10\text{-}28)$$

for the Airy stress function $F(x,y)$, but instead by a solution based on the equivalence of the stress states of a plate with a close-mesh trussed framework with identical boundaries and loads [Klein & Wieghardt, 1904]. Two years later, Wieghardt investigated this equivalence in more detail and how the member forces of trussed frameworks with a high degree of static indeterminacy could be quantified for the stress distributions for plates

$$\sigma_x = \frac{\partial^2 F}{\partial y^2} \qquad \sigma_y = \frac{\partial^2 F}{\partial x^2} \qquad \tau = -\frac{\partial^2 F}{\partial x \partial y} \qquad (10\text{-}29)$$

obtained from the Airy stress function $F(x,y)$ [Wieghardt, 1906]. He therefore "smears" the highly statically indeterminate trussed framework into a two-dimensional elastic continuum in order to calculate the member forces from the latter by way of the stress state. Schwedler used the same method to determine the member forces of the spatial framework he invented (see section 8.1). Wieghardt was able to prove that as the fineness of the truss mesh increases, so the member forces approach ever closer to the stress state in the plate, i.e. they represent its limiting values. But it was not until 1927 that W. Riedel returned to the inverse operation of discretising the two-dimensional elastic continuum by calculating the stress state of an elastic plate loaded by two completely rigid plungers with the help of an equivalent trussed framework system based on quadratic trussed elements. Fig. 10-35 shows the undeformed (dotted lines) equivalent framework and its deformed final position due to compression loads caused by the upper and lower plungers.

In Riedel's work, the quadratic trussed element represents nothing more than the simplest type of bar cell from which the gridwork (or framework) method for modelling plate and shell structures would develop later, during the innovation phase of structural theory (1950–75). The only difference between this and FEM is that the gridwork method is based on a discontinuous element, the bar cell, whereas all structural mechanics models in FEM are based on continuous elements, the finite elements.

### 10.4.1.3 The origin of the gridwork method

A. P. Hrennikoff gave the gridwork method a decisive impulse with his dissertation *Plane stress and bending of plates by method of articulated framework* [Hrennikoff, 1940] completed at the MIT in 1940. In his dissertation, Hrennikoff calculates the cross-sectional values of the truss members from the condition that the joint displacements of the identically bounded trussed framework model coincide with the corner points of the continuum to be substituted. Hrennikoff carried out such calculations for the following bar cells: cuboid, equilateral triangle, rectangle and square. He investigated, for example, elastic plates with the help of the rectangular bar cell. Furthermore, he modelled the elastic plate as a beam grid. Hrennikoff therefore completed the transition of structural analysis modelling from plane plate and shell structures via trussed frameworks to trusses. He published the essentials of his dissertation one year later in the *Journal of Applied Mechanics*, an annex to the *Transactions of the American Society of Mechanical Engineers* [Hrennikoff, 1941]. Nevertheless, this paper covered more than his dissertation. For example, he proposed employing the gridwork method for calculating cylindrical shells; the decoupled plate and slab effect should be taken care of by way of appropriate square trussed systems (Fig. 10-36). In the highly respected *Journal of the Institution of Civil Engineers*, McHenry published his thoughts on the quantitative treatment of the plane stress condition by the trussed framework model in 1943 [McHenry, 1943].

Later, Hrennikoff used the gridwork method for the linear buckling analysis of rectangular slabs as well [Hrennikoff et al., 1972] and developed trapezoidal bar cells for calculating elastic plates [Hrennikoff & Agrawal, 1975].

The bar cell introduced in 1963 by S. Spierig [Spierig, 1963] marked a new stage in the development of the gridwork method in the middle of the innovation phase of structural theory (1950–75). The reason behind this was to adapt the gridwork method formally to FEM. Fig. 10-37a shows a perforated elastic plate subjected to a constant tensile stress $\sigma_x$. This perforated plate is subdivided into rectangular and triangular bar cells in order to determine the plane stress state (Fig. 10-37b). The bar cell consists of a bar model whose nodes have displacement degrees of freedom. For example, the bar cell illustrated in Fig. 10-37c has two degrees of freedom at

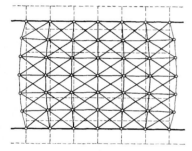

**FIGURE 10-35**
Trussed framework model for the homogeneous, isotropic and linear-elastic plate after Riedel [Riedel, 1927, p. 174]

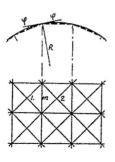

**FIGURE 10-36**
Truss model for a cylindrical shell after Hrennikoff [Hrennikoff, 1941, p. A171]

each of the four nodes, i.e. a total of eight unknown displacements. Correspondingly, in the triangular bar cell shown in Fig. 10-37d, we need assume only six unknown displacements. Starting with the rectangular bar cell, Spierig developed a bar cell in the form of the scalene triangle; this latter is in turn made up of two right-angled triangular elements with the internal bars at angles $\alpha$ and $\beta$ (Fig. 10-37d), which correspond to the internal bars of the rectangular bar cell (Fig. 10-37c). Using the bar cell in the form of the scalene triangle, Spierig was able to model curving boundary forms.

An element stiffness matrix can be set up for every type of bar cell which like with FEM is arranged using index notation to form the total stiffness matrix $K_{ij}$ [Szilard, 1982, pp. 172; 1990, p. 241]. The global displacement state specified unequivocally by the displacement vector $u_j$ is calculated from the matrix equation in index notation

$$p_i = K_{ij} \cdot u_j \quad \text{for } i, j = 1, \ldots, n \qquad (10\text{-}30)$$

where

$n$ number of nodal degrees of freedom of truss model
$K_{ij}$ total stiffness matrix
$p_i$ vector of external nodal loads on truss model

and yields

$$u_i = K_{ij}^{-1} \cdot p_j \quad \text{for } i, j = 1, \ldots, n \qquad (10\text{-}31)$$

The physical expression of the set of equations in 10-30 is equivalent to the set of equations in 10-17 set up by Ostenfeld for plane trusses according to the displacement method because the relationships $p_i = -Z_{i0}$, $K_{ij} = Z_{ij}$ and $u_j = \xi_j$ are valid for $i, j = 1, \ldots, n$. This situation draws attention to the fact that the displacement method represents an important historico-logical source for FEM, also in formal terms. But matrix equation 10-30 goes way beyond eq. 10-17 physically because it can also be used for analysing truss models for flat and curved plate and shell structures. The practical implementation of the generalised gridwork method and its further development to become the computer-aided tool of the design engineer took place first of all in the automotive industry.

### 10.4.1.4 First computer-aided structural analyses in the automotive industry

It was in 1963 that Alfred Zimmer completed the first computer program for structural mechanics analyses in automotive design, which he wrote in assembly language and ran on the new IBM 1620 (see Fig. 10-49) at Daimler-Benz AG (now Daimler AG) in Stuttgart-Untertürkheim. Zimmer had previously been employed as a senior structural engineering assistant at the chair of aviation engineering at Dresden TH. Using this computer program, three-dimensional loadbearing systems like those common in car bodies could be calculated. Later that year, this program confirmed frame calculations for the analysis of the frame floor of the W100 measuring results with such accuracy "that the persons carrying out the measurements assumed that the data had been stolen" [Groth, 2000, p. 338]. The

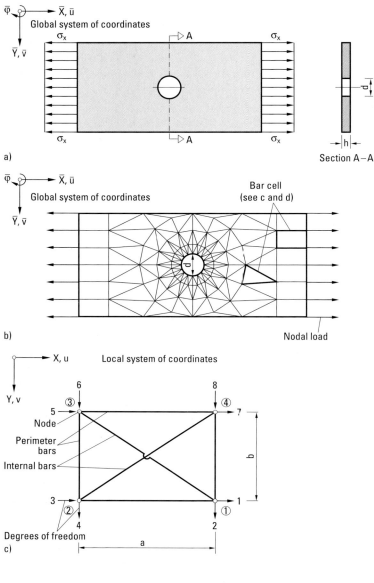

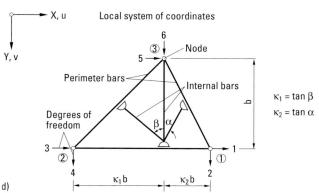

**FIGURE 10-37**
a) Perforated elastic plate subjected to a constant external tensile stress $\sigma_x$, b) bar model according to gridwork method, c) rectangular, and d) triangular bar cell (drawings after [Szilard, 1990, pp. 220, 221])

abbreviation W100 stands for the legendary Mercedes 600 limousine, the preferred means of transport for many heads of state, and which Daimler-Benz AG presented for the first time at the Frankfurt Motor Show in September 1963. Starting with the gridwork method of Hrennikoff, Zimmer and his senior assistant Peter Groth wrote computer programs for analysing structures with a high degree of static indeterminacy, like car bodies. Over the years 1963 to 1968, Werner Dirschmid employed programs to carry out the first bodywork calculations for the DKW 102 and the Audi 100 C1 of Auto-Union GmbH (now Audi AG), which until the end of 1964 was part of Daimler-Benz AG. The W113 sports car (Mercedes-Benz 230 SL) was also reanalysed with the programs in order to obtain more measurement/calculation comparative values. For these applications, the extension of the bar elements to form shell elements acquired great significance, primarily because of the shell- and plate-like configuration of bodywork parts.

Pointing the way forward here was a dissertation by Spierig [Spierig, 1963], which made available the corresponding elements based on Hrennikoff's gridwork method (see section 10.4.1.3). The mathematical models were much too coarse to be able to supply details of stresses and strains, primarily because of the complexity of a car body. This fact did not change until the early 1970s when powerful computers became available and finite elements based on the direct stiffness method could be employed. However, it was still worthwhile being able to assess the flow of forces for conceptual decisions; furthermore, it was also possible to forecast or influence the deformations and hence the stiffness of the loadbearing system. Dirschmid's work provided a method with which it was possible to determine how individual elements of the loadbearing system influence the overall stiffness [Dirschmid, 1969]. It was thus possible to minimise the weight for a given stiffness. This bodywork optimisation procedure gradually evolved into a general standard in vehicle development and it was not until the 1980s that it was superseded by mathematical optimisation methods. The first vehicle to have its weight minimised using this computer method was the Audi 100 [Dirschmid, 2007], which first went into production in 1968 and inaugurated the successful Audi 100 era of Audi AG. Dirschmid was responsible for FEM applications at Audi AG; he took over as head of the Technical Calculations Department in 1989 and from 1998 to 2002 was in charge of the IT Process Chains Department, which advanced the computer-based integration of calculations (CAE = Computer Aided Engineering), conception and design (CAD = Computer Aided Design) plus testing (CAT = Computer Aided Testing) within the scope of product development at Audi AG.

Besides Dirschmid, Zimmer set up another milestone for the computer-viable treatment of the theoretical principles in his 1969 dissertation; in the chapter on large deformations, for example, he describes a computer program for the non-linear analysis of vehicle axles [Zimmer, 1969], which before long would pass the test in practical engineering. While the IBM 1620 was still working with punched cards as an external memory, the IBM 360 with its disk memory and magnetic tape units in conjunction with the FORTRAN programming language instigated a leap in the development of computer-aided structural analysis. Alfred Zimmer's preliminary car development study group at Daimler-Benz AG in Stuttgart-Untertürkheim wrote the first computer program for general structural analysis developed in an industrial company – the Elasto-Statics Element Method (ESEM), details of which were described by Zimmer and his team in 1969 in a series of articles in *IBM-Nachrichten* [Zimmer et al., 1969]. ESEM was based on the gridwork method that had been formally adapted to FEM by Spierig. One year later, Zimmer and Groth, with the assistance of Helmut Faiss from the State Engineering School in Esslingen (now Esslingen University), summarised their knowledge in a monograph entitled *Elementmethode der Elastostatik – Programmierung und Anwendung* (element method of elasto-statics – programming and application) [Zimmer & Groth, 1970]. The calculation model for the structural analysis of the bodywork/frame floor combination of the W115 (Mercedes-Benz 220D) was taken from this book (Fig. 10-38).

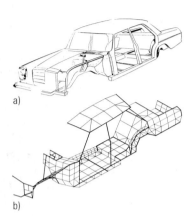

**FIGURE 10-38**
a) Bodywork/frame floor combination of the Mercedes-Benz 220D, and b) associated gridwork model [Zimmer & Groth, 1970, p. 333]

Two loading cases were analysed with the gridwork model (Fig. 10-38b) consisting of 319 nodes, 443 elements and 1684 degrees of freedom (unknowns):

– Torsion loading case: effect of a torsion moment in the plane of the front axle with a supported rear axle plane (twisting of the structure prevented completely in the plane of the rear axle)
– Bending loading case: effect of individual loads in the floor of the passenger compartment with the structure supported in the planes of both axles

The book contains further practical sample applications for computer programs: stress distribution in an annular plate, non-linear analysis of a precambered leaf spring, deformations and stresses in a plate simply supported on four sides and subject to a uniformly distributed line load, determining the member forces of a pylon for electric overhead cables, calculation of a ribbed slab, calculation of a plane storey-height frame with yielding supports, calculating the stresses in a perforated quadratic plate (cf. Fig. 10-37) and the torsion in a prismatic bar assembled from cubic elements (cf. Fig. 10-34). The examples show that the authors regarded their computer programs not only as aids for automotive engineers, but for structural and mechanical engineers, too. In the end, Zimmer and Groth drew up detailed plans for the organisation of programs for mainframe computers [Zimmer & Groth, 1970, pp. 267–292]. For example, they reported on the program-related technicalities of integrating the Calcomp plotter and its upgrading with the first graphic monitor IBM 2250 with light pen. The authors therefore paved the way for the automated input and output of data in the form of the pre-processor or post-processor; today, FEMAP (Finite Element Mapping), for example, guarantees a multi-functional Windows-based engineering analysis environment. Practical implementation began in the autumn of 1969 with the computer-aided calculation accompanying the bodywork and axle development of the Wankel C111 sports car by Zimmer's study group at Daimler-Benz AG [Groth, 2000, p. 339]. The pioneering development of the C111 test vehicle with its Wankel engine was featured in the Daimler-Benz/IBM advertising film *Das Auto, das aus dem Computer kam* (the car that came out of the computer), which was screened at the Cannes Film Festival and won the first prize in the Industrial Films category. This film showed for the first time the use of the plotter (Fig. 10-39a) and the graphic monitor (Fig. 10-39b, c).

Although in the late 1960s NASA – through the MacNeal-Schwendler Corporation (MSC Software) – started employing NASTRAN (NASA Structural Analysis System) as a universal FE program suite in applications beyond the aerospace industry, the American automotive industry was initially slow to react. For example, Ford and General Motors used plotters and graphic screens with light pens for generating production drawings from three-dimensional models, but computer-aided calculations were confined to small parts. This situation changed when IBM showed leading American automotive managers an English version of their film *Das Auto, das aus dem Computer kam* and convinced them of the need for computer-assisted analyses of the overall loadbearing systems in vehicles. So computer-aided structural analyses based on NASTRAN advanced shortly before the beginning of the diffusion phase of structural mechanics (1975 to date) to become the standard tool of the American automotive industry [Groth, 2000, p. 340]. In the Federal Republic of Germany, Esslingen Technical Academy, an institute of the refresher course at Stuttgart University, ensured as early as 1970 that the computer-aided gridwork method of Zimmer, Groth and Faiss was disseminated throughout the metalworking industries with three courses of study on the use of digital computers for the structural and strength problems in practice. The courses taught by the three engineering personalities themselves were announced in leaflets as follows:

"Faster than ever predicted, computer methods are being developed in industry and research that for the first time permit **practical solutions to stiffness problems, stress calculations and optimisations in complex structures**.

"These methods enable not only a reliable insight into the problems that cannot be solved with self-contained mathematical approaches or were only accessible via an expensive roundabout route involving tests. They also form **the basis for solving general dynamic problems, vibrations, stability investigations** and large deformations for large or small strains. These methods are generally conceived for electronic mainframe computers.

"This course of study is intended to familiarise the **practising engineer** with the fundamentals of the **element methods** and show him the way from his view of the problem to the problem definition for the computer. In doing so, the options for using **smaller computer systems** are given special attention. There will be ample opportunity for practical computer tests and detailed discussions.

a)

b)

c)

**FIGURE 10-39**
a) Plotter, b) correcting an incorrect node position on a computer model (outline) for the C111 test vehicle with a light pen, and c) corrected node position
[Zimmer & Groth, 1970, pp. 279 & 281]

"As the range of applications for the method dealt with in the course covers the whole field of technology, this topical theme will attract great interest" [Esslingen Technical Academy, 1970].

The companies that sent delegates to these three courses reads like a "Who's Who" of the metalworking industry in the Federal Republic of Germany – but there were only a few representatives from the construction industry. So those courses on computer-assisted structural analysis in general and the Elasto-Statics Element Method (ESEM) of Zimmer, Groth and Faiss [Zimmer & Groth, 1970] in particular became engraved on the consciousness of design engineers in the West German metalworking industries. The exponential growth in the power of the computer after the mid-1970s, instigated by the introduction of the microprocessor, FE program suites demanding high memory capacity, e. g.

– NASTRAN,
– ASKA (<u>A</u>utomatic <u>S</u>ystem for <u>K</u>inematics <u>A</u>nalysis) developed by Argyris and his assistants in 1965 on which PERMAS (<u>P</u>owerful <u>E</u>fficient <u>R</u>eliable <u>M</u>echanical <u>A</u>nalysis <u>S</u>ystem) by INTES GmbH was based,
– the TPS10 and TPNOLI systems, developed by T-Programm GmbH (founded in 1971), which today have been superseded by the professional FEM system WTP2000 (<u>W</u>ölfel <u>T</u>echnische <u>P</u>rogramme, Höchberg/Würzburg) [Groth, 2002, pp. 2–3],

became interesting for practising design engineers. On the other hand, these technical possibilities powered the formation of structural mechanics theories within the scope of computational mechanics.

### 10.4.2 Modularisation and discretisation of aircraft structures

Whereas the gridwork method was a priori based on the notion of discretisation on the level of the structural mechanics model in the sense of the theory of trusses, aircraft construction followed the path of the modularisation and discretisation of aircraft structures. This development stretches from the lattice girder with rectangular cross-section via the cell tube and the shear field layout right up to the faithful resolution of the cell tube into shear and bar elements. This was the decisive impulse for the formation of FEM from aircraft construction as a consequence of the industrial mass production of military aircraft triggered by World War 2.

#### 10.4.2.1 From lattice girder with rectangular cross-section to cell tube and shear field layout

Herbert Wagner adapted space frame theory to the idiosyncrasies of aircraft construction of those years [Wagner, 1928]: parallel transverse stiffening of the fuselage and the box-like wings, plus X-bracing. At the end of his paper, Wagner introduces guidelines for calculating statically indeterminate trussed frameworks. One year later, Hans Ebner introduced the idea of framework cell discretisation in his dissertation on the calculation of space frames in aircraft [Ebner, 1929/1], which in that same year was published in the *Jahrbuch der Deutschen Versuchsanstalt für Luftfahrt* (yearbook of the German Aviation Testing Authority) [Ebner, 1929/2]. Ebner expanded his dissertation with the help of the finite differences method in 1931 [Ebner, 1931] and published it in extracts with applications for bridge- and tower-building in the journal *Der Stahlbau* [Ebner,

1932]. In those articles he investigates transversely stiffened space frames consisting of four levels (Fig. 10-40a, d) or interrupted (Fig. 10-40b, c) longitudinal walls stiffened by $n + 1$ parallel transverse walls. That leads to $n$ framework cells. Every framework cell consists of four generally trapezoidal plane frames belonging to the longitudinal walls and two parallel rectangular frames (transverse walls).

Ebner called the overall loadbearing system formed by $n$ such framework cells a "cellular truss" – in other words, a three-dimensional lattice girder. Ebner's framework cell mirrors the original form of sectional assembly in aircraft construction and the segment-based erection of the trussed framework tower based on the structural model.

Whereas Föppl's "trussed shell" (see section 8.1.2) has only transverse end walls and is statically determinate internally, the three-dimensional distribution of the member forces of Ebner's lattice girder is achieved by intermediate transverse walls; the cross-sectional geometry of the latter is that of a lattice girder with rectangular cross-section. Every intermediate transverse wall increases the degree of three-dimensional static indeterminacy by one, which means that a lattice girder with $n$ framework cells has $n - 1$ degrees of static indeterminacy internally. When choosing the static indeterminates, Ebner proceeds as follows (Fig. 10-41): "One now imagines the system to be pulled apart at every intermediate transverse wall and the insertion of 12 additional bars corresponding to the four fixed connecting points. Of these, five bars form a new intermediate transverse wall immediately adjacent to the existing one; the other seven represent the connecting bars for the cells, which are now independent. As only six bars are necessary for a statically determinate connection, one of the four longitudinal connecting bars is redundant at every transverse wall" [Ebner, 1932, p. 2]. The longitudinal connecting bars are perpendicular; Ebner introduces a normal force hinge at each longitudinal bar and hence releases the static indeterminates $X_r = 1$ and $X_{r+1} = 1$. The statically indeterminate axial connecting forces form, together with their opposing forces, antisymmetric axial force groups at the three other longitudinal connect-

**FIGURE 10-40**
Space frames after Ebner: a) aircraft frame, b) associated aircraft fuselage, c) bridge, and d) tower [Ebner, 1932, p. 1]

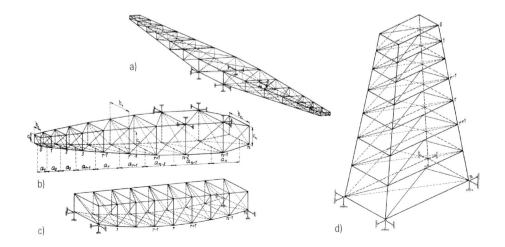

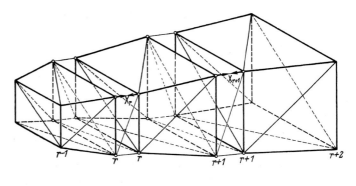

**FIGURE 10-41**
Main system of cells after Ebner: bar cell with the static indeterminates $X_r$ and $X_{r+1}$ for setting up the elasticity equations [Ebner, 1932, p. 2]

ing bars, which are in equilibrium at the individual cells. The influence of each group $X_r = 1$ on the main system of cells now extends to the two neighbouring cells only.

According to the force method, the $r$th elasticity equation for transverse walls rigid in their plane is

$$\delta_{r(r-1)} \cdot X_{r-1} + \delta_{rr} \cdot X_r + \delta_{r(r+1)} \cdot X_{r+1} = -\delta_{r0} \quad (10\text{-}32)$$

Accordingly, the $r$th elasticity equation for transverse walls elastic in their plane is

$$\delta_{r(r-2)} \cdot X_{r-2} + \delta_{r(r-1)} \cdot X_{r-1} + \delta_{rr} \cdot X_r + \delta_{r(r+1)} \cdot X_{r+1} + \delta_{r(r+2)} \cdot X_{r+2} = -\delta_{r0}$$
$$(10\text{-}33)$$

A three-term set of equations can be set up using elasticity equation 10-32, which corresponds to Clapeyron's theorem of three moments (see eq. 6-2) for continuous beams on rigid supports. However, elasticity equation 10-33 yields a five-term set of equations; such sets of equations occur with continuous beams on elastic supports if the support moments are chosen as static indeterminates (theorem of five moments). The coefficient matrix $\delta_{ji}$ of the elasticity equations (see eq. 10-18) exhibits a distinctive band structure in both cases. Ebner's method for calculating statically indeterminate space frames is an outstanding example of how the heuristic possibilities of the operative use of symbols – intrinsic to the force method – can be used to create computational algorithms, not only for enabling the analysis of complicated loadbearing systems, but also for simplifying and rationalising the practical side of structural calculations, even for unusual loadbearing systems. Ebner only managed to achieve this by way of very careful discretisation of the entire space frame for the purpose of analysing the loadbearing system. If the framework cell represents, so to speak, a "macroelement" in the first step of the loadbearing system analysis, the second step consists of resolving the framework cell into individual plane frames. Fig. 10-42 shows such an elementary frame, not as a free body, but instead as an externally statically determinately supported framework.

The historico-logical transition from the three-dimensional lattice girder (10-40a, b) to the cell tube took place in aircraft construction between 1929 and 1938 in four stages:

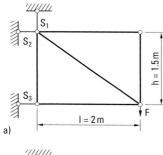

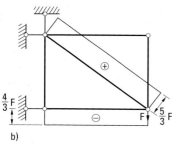

**FIGURE 10-42**
a) Rectangular statically determinate elementary frame carrying a point load F, b) associated force state

- First stage: tension field theory for plane plate girders with very thin webs
- Second stage: solid-web box beams
- Third stage: stiffened shell structures
- Fourth stage: cell tube and shear field theory.

This development in loadbearing systems can only be understood after first contemplating the loadbearing behaviour of the shear field as a load-bearing system element of the cell tube.

Fig. 10-43 represents a statically determinately supported shear field as the simplest form of the cell tube. Four bars form the flanges which are connected via hinges. If the diagonal bar is removed, the result is a system with one degree of kinematic determinacy; in order to brace this system, a shear field (Fig. 10-43) is provided instead of the diagonal bar (Fig. 10-42). In the cell tube, the longitudinal stresses are resisted by the axially elastic flange bars and the shear stresses by the thin shear field of thickness $s$. This division of work saves weight in the construction. The prerequisite for this division of work is the continuous force transfer from the flange bars to the shear field, which in practical terms can be achieved with a continuous weld, closely spaced rivets or adhesive; at the same time, the hinges must be kept free theoretically so that they can satisfy the intended structural function (Fig. 10-43a). The external forces $S_1$, $S_2$ and $F$ cause a triangular tensile stress distribution in the bar, whereas $S_3$ causes a triangular compressive stress distribution. In order to calculate the shear forces $T_1$, $T_2$, $T_3$ and $T_4$, the flanges are theoretically separated from the shear field. From the equilibrium of all vertical forces for the right-hand flange bar of the shear field

$$\Sigma V = 0 = F - T_1 \cdot h \rightarrow T_1 = \frac{F}{h} = \frac{F}{1.5} = \frac{2}{3} \cdot F \qquad (10\text{-}34)$$

we get a shear force distributed constantly over the length $T_1 = s \cdot \tau_1$ in force per unit length (in metres here) and a shear stress $\tau_1$ distributed constantly over the sheet with thickness $s$. For the upper flange bar, the equilibrium of all horizontal forces

$$\Sigma H = 0 = S_2 - T_2 \cdot l = \frac{l}{h} \cdot F - T_2 \cdot l \rightarrow T_2 = \frac{F}{h} = \frac{2}{3} \cdot F \qquad (10\text{-}35)$$

yields a shear force distributed constantly over the length $T_2 = s \cdot \tau_2$ in force per unit length (in metres here) and a shear stress $\tau_2$ distributed constantly over the sheet with thickness $s$.

The shear forces $T_3 = s \cdot \tau_3$ and $T_4 = s \cdot \tau_4$ can be determined directly from the equilibrium conditions at the shear field for the vertical direction

$$\Sigma V = 0 = T_3 - T_1 \rightarrow T_3 = T_1 = \frac{F}{h} = \frac{2}{3} \cdot F \qquad (10\text{-}36)$$

and for the horizontal direction

$$\Sigma H = 0 = T_4 - T_2 \rightarrow T_4 = T_2 = \frac{F}{h} = \frac{2}{3} \cdot F \qquad (10\text{-}37)$$

**Loadbearing behaviour of the shear field**

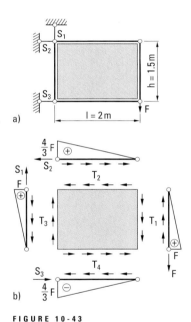

FIGURE 10-43
a) Rectangular statically determinate shear field of thickness s carrying a point load F,
b) associated force state

Equations 10-32 to 10-35 can be summarised as

$$T_1 = s \cdot \tau_1 = T_2 = s \cdot \tau_2 = T_3 = s \cdot \tau_3 = T_4 = s \cdot \tau_4 =$$
$$= T = s \cdot \tau = \frac{F}{h} = \frac{2}{3} \cdot F \qquad (10\text{-}38)$$

i. e. all shear forces and shear stresses in the shear field are constant. Like the elementary framework (Fig. 10-42) can be assembled to form a lattice girder with rectangular cross-section, so the shear field represents the element for constructing complex loadbearing systems.

**First stage: tension field theory for plane plate girders with very thin webs**

In his tension field theory for analysing plane plate girders, Wagner assumes the tension field to be a loadbearing system element [Wagner, 1929/2]. He developed the idea of the tension field from the quadratic elementary framework ($h = l$) with not just one diagonal bar (Fig. 10-42), but rather two, crossing diagonal bars. Next, Wagner replaces the diagonal bars by a plate of solid thin sheet metal and thus creates a tension field, which in structural terms functions exactly like that in Fig. 10-43 with the difference that Wagner assumes inextensible bars instead of axially elastic bars. His pictorial representation of the elementary framework or tension field [Wagner, 1929/2, p. 203], corresponds – apart from the differences mentioned – to Fig. 10-42 or 10-43 respectively.

**Second stage: solid-web box beams**

In 1918 Eggenschwyler investigated for the first time box beams in torsion with a rectangular cross-section and pairs of walls with the same thickness assuming no deviation from the cross-sectional form upon applying the torsion load [Eggenschwyler, 1918]; in doing so he considered the warping normal stresses. An analysis of the box beam with deviation from the cross-sectional form was proposed by Hans Reissner in 1925/26 within the scope of a series of lectures on new problems in aviation engineering at Berlin-Charlottenburg TH [Reissner, 1926, 1927]; he, too, spoke about the case of the box beam with a cross-sectional depth changing linearly over the length, which at that time was being used for aircraft wings. Reissner can be regarded as the primus inter pares of the scientific basis of lightweight construction, which during the invention phase of structural theory (1925 – 50) lent momentum to structural steelwork and the whole of structural mechanics in addition to aircraft construction.

Ebner was the first to achieve a general analysis of the thin-wall box beam in torsion with constrained warping of the cross-section [Ebner, 1933]. In terms of method, Ebner pursued his theory of the box-type space frame with the difference that he replaced the elementary framework by thin sheet metal with no hinges at the nodes. He thus modelled the sheet metal panels firstly as shear-resistant and then as tension-resistant in the meaning of Wagner's tension field theory; the latter modelling is based on the Wagnerian assumption that with very thin sheet metal the shear strength of the panels is negligible because the strain stiffness of the longitudinal bars is very large (rigid bars). Conversely, when modelling the shear-resistant panel, he assumes that no flange bars are necessary on the perimeter. The loadbearing behaviour of the shear field therefore lies between the two models as it assumes elastic flange bars joined by hinges at the nodes. So Ebner did not quite achieve the transition to the box-type cell tube, but closed in on it from the two extreme cases.

Hans Ebner and Hermann Köller published their findings on the stiffened shell structures of aircraft construction in 1937 [Ebner, 1937/1; Ebner, 1937/2; Ebner & Köller, 1937/1; Ebner & Köller, 1937/2]. They thus reconstructed the successful use of shell structures in American aircraft construction from the experimental and theoretical viewpoint. Their main line of reasoning was that the shell should be constructed as thin as possible and the saving in weight should be concentrated in the longitudinal stiffeners, the so-called stringers [Ebner 1937/2, p. 223]. Using this method, Ebner and Köller modelled aircraft fuselages as stiffened cylindrical shells [Ebner 1937/1; Ebner & Köller, 1937/1; Ebner & Köller, 1937/2]. Ebner investigated shell-type aircraft wings [Ebner, 1937/2]. In their publication on longitudinally stiffened cylindrical shells, which are stiffened in the transverse direction by rigid annular frames, Ebner and Köller were already making use of the (curved) shear field: "The longitudinal stiffness of the shell should be concentrated in the flanges, with the contribution of the skin being taken into account by way of an addition to the flange cross-sections. The shear force in the peripheral direction is then also constant between two flanges. The skin therefore serves exclusively for transferring the shear between the longitudinal stiffeners considered collectively at the centres of gravity of the sections. A linear force progression is established in these within each cell owing to the constant shear forces in the longitudinal direction" [Ebner & Köller, 1937/1, p. 244]. The shear field theory was developed by Ebner and Köller using the example of the analysis of the force progression in stiffened cylindrical shells [Ebner & Köller, 1937/2], an idea that Argyris would take up in his reformulation of structural mechanics through matrix theory [Argyris, 1954, p. 354]. Like with a lattice girder with rectangular cross-section, they assemble the cells separated by rigid annular frames – the cylindrical shell modules in the sense of macroelements – to form the entire loadbearing structure of the aircraft fuselage. Here, too, we see in the structural mechanics model the sectional assembly that became standard in aircraft production during the 1930s. In formal terms, the statically indeterminate calculation adheres to the method worked out by Ebner for lattice girders with rectangular cross-section according to the force method. Although the lattice girder structure was soon to disappear from aircraft construction, practical calculations still continued to be based on Ebner's statically indeterminate calculation method for the shell-type aircraft fuselage assembled from sections. And the aircraft wings?

**Third stage: stiffened shell structures**

In 1938 Ebner and Köller introduced the cell tube (Fig. 10-44) made up of many shear fields (Fig. 10-43) for the aircraft wing and formulated a general shear field theory for this [Ebner & Köller, 1938/1] (reprinted in:

**Fourth stage: cell tube and shear field theory**

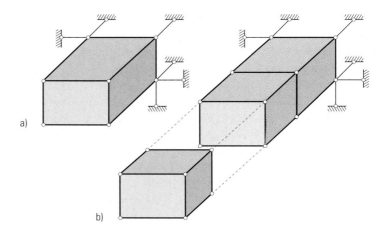

**FIGURE 10-44**
a) Statically determinate cell tube,
b) cell tube assembled from individual boxes (see [Czerwenka & Schnell, 1967, p. 119])

[Ebner & Köller, 1939]). Like Ebner assembled and calculated the elementary framework illustrated in Fig. 10-42 to form a framework cell (Fig. 10-41) and this in turn to form his "cellular truss" (Fig. 10-40), the box-type cell tube can be assembled in two steps (Fig. 10-44):
–   In the first step, the single shear field plus its flanges (see Fig. 10-43) is assembled to form a box (Fig. 10-44a), which is statically determinate.
–   In the second step, further boxes each consisting of eight flange bars and five shear fields are appended to the first box (Fig. 10-44b).

If the cell tube comprises $n$ such elementary boxes, then the complete cell tube has $n-1$ degrees of static indeterminacy internally (Fig. 10-44b). The elementary box is, so to speak, the macroelement of the cell tube.

If the shear fields shown in Fig. 10-44 are additionally stiffened internally by $i$ transverse ribs and $j$ longitudinal ribs, then such a panel is in structural terms an orthogonally stiffened plate with $(i+1) \times (j+1)$ shear fields (= "shear field layout" [Ebner & Köller, 1939, p. 122]) according to Fig. 10-43: "The framework-type construction of the stiffened plates suggests, like with truss members, ignoring the frame action at the nodes for the stiffening [i.e. the introduction of hinges – the author] and regarding axial forces as the most important way in which the stiffening bars work ... The stress state in the shear field model described is represented by the shear forces constant in each field $t = s \cdot \tau$ [$= T = s \cdot \tau$, see eq. 10-38 – the author] acting at the stiffeners. A linear force progression then prevails within the individual stiffening bars between two nodes [see Fig. 10-43b – the author], the rise or fall of which depends on the difference between the shear forces in adjoining panels. The force state in the stiffeners is therefore given by the normal forces at the individual nodes" [Ebner & Köller, 1939, p. 122]. This was the foundation on which Ebner and Köller set up a completely statically indeterminate shear field theory with the help of the force method, which they also extended to shear fields with minimal curvature. One essential prerequisite for their shear field theory is the consideration of the longitudinal elasticity of the stiffening bars. In this context, they criticised a paper by O. S. Heck in which he had postulated a shear field theory with inextensible stiffening bars [Heck, 1937];

neglecting the longitudinal elasticity of the stiffening bars in the static indeterminacy calculation would in the majority of cases lead to unacceptable results [Ebner & Köller, 1939, p. 122]. The experimental and mathematical investigation of a shell wing model loaded in bending [Schapitz et al., 1939] would prove them right. The triumphal march of the cell tube for aircraft wings had begun. The cell tube is an excellent example of the "thin skin – stringer – transverse stiffener" trilogy [Czerwenka, 1983, p. 55], a characteristic feature of lightweight construction which was realised in practice not only in aircraft (see, for example, [Wagner & Kimm, 1940]), but later also in vehicles, railway rolling stock and steel bridges. For example, in the middle of the invention phase of structural mechanics (1925–50), the scientific solutions to problems of lightweight construction in the aircraft industry were primarily worked out in Germany at the German Aviation Testing Authority (DVL) in Berlin-Adlershof, as Nicolas Hoff mentions in a letter to Gerhard Czerwenka [Czerwenka, 1983, p. 57].

The concept of the "shear field layout" that started to appear after the mid-1930s – as the constructional realisation and structural mechanics modelling in the context of the symbols of the force method – therefore designates a method that can be interpreted as a prototype of the dissection of a loadbearing system into the basic elements of the extensible bar and the shear plate in the meaning of the finite element method. Aviation research work carried out in the USA in the second half of the 1940s points to this, and this will be dealt with in the next section.

### 10.4.2.2 High-speed aerodynamics, discretisation of the cell tube and matrix theory

Industrial mass production of military aircraft in World War 2 called for aircraft designs to be broken down into modules and elements in the form of sectional assembly, which on the level of the formation of structural mechanics theory brought with it the discretisation of the part-systems of the aircraft structure in order to master the calculations of these complex loadbearing systems. Research programmes in this direction were first initiated towards the end of World War 2, in the USA and UK especially. In May 1984 Raymond L. Bisplinghoff, in his Lester B. Gardner Lecture at MIT, provided an insight into the history of the elastic theory fundamentals of aircraft construction in the first half of the 20th century [Bisplinghoff, 1997]. Quite rightly, he concentrated on the pioneering contributions of the Aeroelastic & Structures Research Laboratory at MIT, founded in 1946. From the wealth of research work, the contributions to FEM and the systematic use of computers for calculating aircraft structures are particularly worthy of note. All the researchers at the Aeroelastic & Structures Research Laboratory accepted the findings on the matrix formulation of structural mechanics problems (see section 10.3.5.3) laid out in the monograph *Elementary Matrices* [Frazer et al., 1938] by Frazer, Duncan and Collar – especially from the viewpoint of the elastic theory foundations of aircraft construction: "… we recognized that casting aeroelastic problems in terms of matrices was neat, orderly and compendious, and that matrix iteration provided a powerful method for calculating the latent roots and vectors of eigenvalue problems. Thus, each member of the computer group became a specialist in at least one of the matrix operations, and we attempted to cast all of our numerical work into a sort of matrix production line. In parallel with these activities, the laboratory pioneered some of the earliest work on finite-element modelling, which was ideally suited to the digital-computer revolution that was about to take place" [Bisplinghoff, 1997, p. 24]. Bisplinghoff took the decision to build the test model shown in Fig. 10-45 for measuring the influence coefficients. The test model is a box beam with shear fields fixed to its support at an angle (cf. Fig. 10-44). Under the leadership of Prof. Theodore H. H. Pian, experimental studies on the aerodynamic behaviour of aircraft wings were carried out in 1947.

The test model therefore imitates a sweptback aircraft wing attached to the aircraft fuselage. For example, with wings at an angle of 45° (see Fig. 10-45), the critical speed at which the drag rises rapidly can be increased by about 200 km/h: "Sloping the wings backwards increases the critical speed at which this steep increase in drag and other undesirable flow effects begin, and a higher flying speed is possible with the same engine power" [Heinzerling, 2002, p. 4]. The discovery of the aerodynamic effects of angling the wings can be attributed to the German aerodynamics expert Adolf Busemann (1901–86), who presented his findings at the 5th International Volta Conference in Rome [Busemann, 1935]. In 1942 he and Albert Betz (1885–1968) were granted German secret patent No. 732/42 on aircraft with speeds in the proximity of the speed of sound

**FIGURE 10-45**
Test model of an aircraft wing dating from 1947
(see [Bisplinghoff, 1997, p. 32])

[Heinzerling, 2002, p. 6]. Together with the Ohain jet engine factory, there began a feverish development of jet aircraft with sweptback wings within the scope of the German war programme which culminated in series production of the twin-engined Me 262 jet aircraft. The "SS State" [Kogon, 1946] used many prisoners from the German concentration camps for production of the Me 262. Directly after the war, a group of American aircraft experts led by Theodore von Kármán toured the German Aircraft Research Establishment (DFL) in Braunschweig, of which Busemann was the director, and all were amazed by the wealth of research results. On 5 November 1945, George S. Schairer, the leading aerodynamics expert of the Boeing Airplane Company, reported on this visit in a seven-page letter to Ben Cohn, his superior in Seattle at that time: "The Germans have been doing extensive work on high speed aerodynamics. This has led to one very important discovery. Sweepback and sweepforward have a very large effect on critical Mach No." (cited in [Heinzerling, 2002, p. 8]). Independently of developments in Hitler's Germany, Robert T. Jones published his theoretical work on sweptback wings in 1945 in Report No. 863 for the National Advisory Committee for Aeronautics (NACA) and gave it the title *Wing Plan Forms for High-Speed Flight* [Jones, 1945]. Research into sweptback wings formed another starting point for Bisplinghoff's research group at MIT's Aeroelastic & Structures Research Laboratory.

But let us return to the test model shown in Fig. 10-45. Besides the transverse stiffeners, the test model has longitudinal stiffeners (stringers) in order to prevent buckling of the thin shear plates. Bisplinghoff dissected the test model shown in Fig. 10-45 into shear field and bar elements (Fig. 10-46) in order to carry out a structural mechanics analysis of this highly statically indeterminate system. The aim of Hubert I. Flomenhoft's drawing (Fig. 10-46) dating from 1947 was to formulate the equations clearly

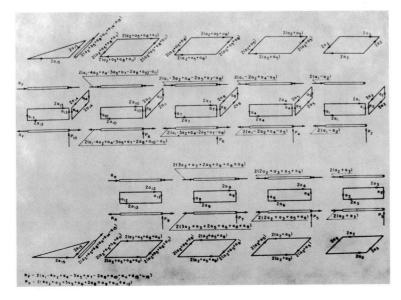

**FIGURE 10-46**
Flomenhoft's 1947 drawing of the discretisation of the test model shown in Fig. 10-45 into shear fields and extensible bars
(see [Bisplinghoff, 1997, p. 33])

in matrix theory obtained via the notion of the principle of minimum deformation energy [Flomenhoft, 2007]. Flomenhoft, Bisplinghoff's assistant at that time, writes that calculating machines were used to invert the matrices [Flomenhoft, 2007]. Fig. 10-46 demonstrates the consequential further development of the shear field layout to form the discretised loadbearing system.

At the 15th Annual Meeting of the Institute of Aeronautical Sciences (IAS), which took place in New York from 28 to 30 January 1947, Samuel Levy from the National Bureau of Standards gave a pioneering talk which was published in the October issue of the *Journal of the Aeronautical Sciences* with the title *Computation of influence coefficients for aircraft structures with discontinuities and sweepback* [Levy, 1947]. In this paper, Levy introduced a system for dissecting the box-type loadbearing systems of aircraft into shear fields and bar elements (cf. Fig. 10-46): "The present report describes a general method of computing influence coefficients for stressed-skin structures with discontinuities characteristic of aircraft design. The method makes use of the stress analysis of the aircraft structure based primarily on equilibrium considerations, together with Castigliano's energy theorem for deriving displacements under load from the elastic energy stored in the structure. In those cases where equilibrium conditions are not sufficient in themselves to specify the stress distribution, use is made of the fact that the actual distribution corresponds to a minimum of the strain energy. Numerical examples are presented for the case of box beams with large cutouts, sweepback, and D-nose sections" [Levy, 1947, p. 547]. The method for discretising the box-type loadbearing systems of aircraft published by Levy would help the progress towards FEM, as the works of Argyris [Argyris, 1955, pp. 83 – 89] or Turner, Clough, Martin and Topp [Turner et al., 1956, p. 810] prove. Without doubt, Levy was inspired to create a method for the discretisation of box-type loadbearing systems by the shear field theory of Ebner and Köller developed using the example of the analysis of the force progression in stiffened cylindrical shells [Ebner & Köller, 1937/2] – he includes the English version of this publication [Ebner & Köller, 1938/2] in his list of references [Levy, 1947, p. 560].

So the unification of the discretisation of the box-type loadbearing systems of aircraft and matrix theory provided significantly more impetus for the creation of the system of non-classical engineering sciences than the findings of high-speed aerodynamics limited to that individual scientific discipline. This instigated the formation of a structural mechanics theory that – in conjunction with high-speed aerodynamics – quickly resulted in the stage of development splendidly described by Carlos A. Felippa in *The delta wing challenge* [Felippa, 2001, p. 1317], and finally culminated in the matrix algebra reformulation of structural mechanics together with the theory of structures it had absorbed in the mid-1950s.

### 10.4.3
**The matrix algebra reformulation of structural mechanics**

The application of formalised theory to structural mechanics could only become standard once the theory and practice of structural mechanics calculations were closely tied to the computer. This meant that the operative use of symbols throughout structural mechanics had to take on the form of a single symbolic machine, which for its part could be easily imitated by the computer as the realisation of a universal symbolic machine.

The means of formalised theory in structural mechanics was matrix algebra. As was mentioned in the previous section, the first step towards the use of matrix algebra was not taken in structural engineering, but rather in aviation engineering, where the analysis of systems with high degrees of static indeterminacy was part of the daily workload of the practising engineer. The introduction of formalised theory into structural mechanics in the first half of its innovation phase (1950 – 75) gave rise to modern structural mechanics, which on a methods level was quickly to develop into the global coordinates system of fundamental engineering science disciplines for individual engineering disciplines such as theory of structures or the theory of ship stability (see, for example, [Paulling, 1963; Lehmann, 2004, p. 91]).

### 10.4.3.1 The founding of modern structural mechanics

As early as 1947, S. Levy mentioned matrix theory in conjunction with the calculation of part-systems of aircraft structures broken down into bar and shear field elements according to the force method [Levy, 1947]. Falkenheiner [Falkenheiner, 1950, 1951, 1953], A. L. Lang and R. L. Bisplinghoff [Lang & Bisplinghoff, 1951], B. Langefors [Langefors, 1952] plus L. B. Wehle and W. Lansing [Wehle & Lansing, 1952] took Levy's publication as their basis and developed the matrix formulation of the force method (flexibility method). P. H. Denke then used this work to introduce the Newton-Raphson method for solving non-linear problems in the flexibility method [Denke, 1956]. The birth certificate of FEM, issued by Turner, Clough, Martin and Topp in 1956, contains the following reference to the flexibility method: "The method is, of course, perfectly general. However, the computational difficulties become severe if the structure is highly redundant, and the method is not particularly well adapted to the use of high-speed computing machines" [Turner et al., 1956, p. 806]. Levy obviously noticed this himself because by 1953 he had published the matrix formulation of the displacement method (stiffness method) [Levy, 1953]. Looking ahead to the use of computers, the creators of FEM acknowledged Levy's work as follows: "In a recent paper Levy has presented a method of analysis for highly redundant structures which is particularly suited to the use of high-speed digital computing machines" [Turner et al., 1956, p. 807].

If in this quote we replace "Levy" by "Argyris", "method" by "general method" and "particularly" by "generally", then we obtain an appraisal of the matrix algebra reformulation of the whole of structural mechanics by Argyris over the years 1954 to 1957.

The series of papers in the journal *Aircraft Engineering* [Argyris, 1954, 1955; Argyris & Kelsey, 1954]) running to more than 80 pages is based on the lectures given by Argyris at Imperial College, London, starting in 1950/51; the series begins as follows: "The increasing complexity of aircraft structures and the many exact or approximate methods available for their analysis demand an integrated view of the whole subject, not only in order to simplify their applications but also to discover some more general truths and methods. There are also other reasons demanding a more comprehensive discussion of the basic theory. We mention only the increasing attention paid to temperature stresses and the realization of the importance of nonlinear effects. When viewed from all these aspects, the idea of presenting a unified analysis appears more than necessary" [Argyris, 1954, p. 347].

By means of an exemplary historico-logical reconstruction of the discipline-formation and consolidation periods of structural and elastic theory, Argyris lays the foundations of structural mechanics, unfolds them in the dual matrix algebra presentation of the force and displacement methods, and illustrates the force of symbolic operations with matrix theory using the example of the complex loadbearing systems of aircraft construction ([Argyris, 1954 & 1955; Argyris & Kelsey, 1954]):

- introduction with historical highlights,
- basic equations of continuum mechanics (equilibrium relationships),
- work and complementary work, deformation energy and deformation complementary energy, principle of virtual displacements, including Castigliano's first theorem, the principle of minimum total potential energy (Dirichlet, Green) and the Rayleigh-Ritz and Bubnov-Galerkin methods (see also [Taylor, 2002]),
- examples for the principle of virtual displacements (two-span beams on non-linear spring supports, statically indeterminate truss, open tubular cross-section subjected to torsion),
- principle of virtual forces, including Castigliano's second theorem and the principle of minimum complementary potential energy (Menabrea),
- examples for the principle of virtual forces,
- applications of the principles of virtual displacements and virtual forces applied to structural mechanics systems under temperature loading plus torsion problems taking into account non-linear stress-strain relationships,
- theory of statically indeterminate systems in matrix algebra formulation: flexibility of system (flexibility matrix), stiffness of system (stiffness matrix), displacement method, dual nature of force and displacement methods, and
- application of the force method in aircraft construction (e. g. shear force distribution in multi-cell wing cross-sections, see also Figs. 10-44, 10-45 and 10-46).

In 1957 Argyris expanded the theory of statically indeterminate systems in matrix algebra formulation to form structural matrix analysis [Argyris, 1957]. Argyris describes his motives for this as follows: "We have known for some years that none of the conventional statical methods are really suitable for determining the stress distribution and flexibility matrices of highly statically indeterminate systems of modern aircraft designs. Similar difficulties occur in other applications of statics. The iterative methods can be useful in certain cases, but are generally too laborious and have not proved worthwhile for the membrane- and shell-type loadbearing structures of aircraft. We can overcome these difficulties by way of the matrix formulation of statics in conjunction with automatic electronic digital computers. Matrix formulation not only enables us to configure the calculations in a much clearer fashion, but is also the ideal form of notation for automatic digital computers. Apart from that, the theoretical derivations of matrix theory are so transparent and elegant that new, practical and valuable relationships, which in conventional notation were impossible or difficult to discern, are now obtained very easily" [Argyris, 1957, p. 174].

Only three principal, simple matrices plus a column matrix for loads are now required for all structural calculations. And structural matrix analysis enables non-linear-elastic and dynamic problems to be handled as well. Argyris maps the dual nature of the whole of theory of structures (principle of virtual forces/force method, principle of virtual displacements/displacement method) in a synoptic, matrix algebra formulation (Fig. 10-47), and provides a dictionary for the practical concepts [Argyris, 1955, p. 90; 1957, p. 175]:

$$
\begin{aligned}
\textbf{Method of forces} &\leftrightarrow \textbf{Method of displacements} \\
\text{Forces} &\leftrightarrow \text{Displacements} \\
\text{Stresses} &\leftrightarrow \text{Strains} \\
\text{External forces} &\leftrightarrow \text{Joint displacements} \\
\text{Flexibility} = \text{Displacement} : \text{Force} &\leftrightarrow \text{Stiffness} = \text{Force} : \text{Displacement} \\
\text{Unit load method} &\leftrightarrow \text{Unit displacement method} \\
\text{Statically determinate system} &\leftrightarrow \text{Kinematically indeterminate system} \\
\text{Statically indeterminate system} &\leftrightarrow \text{Kinematically indeterminate system} \\
\text{Flexibility matrix} &\leftrightarrow \text{Stiffness matrix} \\
\text{Generalised forces} &\leftrightarrow \text{Generalised displacements} \\
\ldots &\leftrightarrow \ldots
\end{aligned}
$$

Argyris has therefore succeeded in giving structural analysis a complete formalised theory; he summarised his pioneering series of papers together with S. Kelsey in the monograph *Energy theorems and structural analysis* [Argyris & Kelsey, 1960]. Looking back, Ray W. Clough quite rightly acknowledges this monograph with the following words: "In my opinion, this monograph ... certainly is the most important work ever written on the theory of structural analysis ..." [Clough, 2004, p. 286].

With the books of E. C. Pestel and A. Leckie [Pestel & Leckie, 1963] as well as R. K. Livesley [Livesley, 1963], elastomechanics and structural matrix analysis came to a provisional conclusion around the middle of the innovation phase of structural theory (1950 – 75). Since then, the whole of theory of structures can be understood as a specific symbolic machine in the form of structural matrix analysis, which can be readily and fully imitated by a universal symbolic machine – the computer. So the structural matrix analysis of Argyris is at the same time computational mechanics; and more besides: owing to its formalised theory nature it is structural

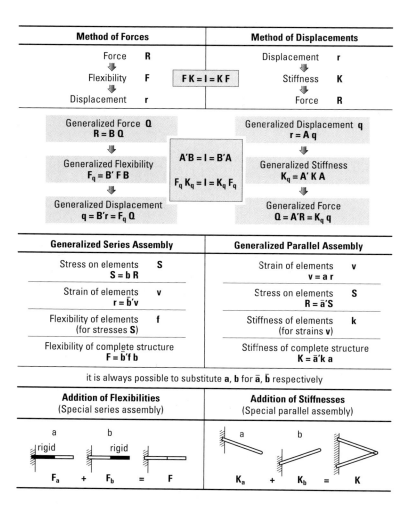

**FIGURE 10-47**
Conceptual rendering of the dual nature of theory of structures in the language of matrix theory (redrawn after [Argyris, 1955, p. 50]

mechanics because the distinction according to fields of application is weak. The tendency towards structural mechanics is also evident in Argyris' early contributions to FEM. For example, Erwin Stein pointed out in 1983 that Argyris had already presented his first works on FEM to the Research Council of the British Royal Aeronautical Society in 1943, 1947 and 1949 in the form of secret memorandums; but these works were regarded as "nonsense" by the Council and not approved for publication [Stein, 1985, p. 10]!

**Early steps in the use of electronic computations in structural analysis**

### 10.4.3.2

Within the scope of research into the use of computers in the engineering sciences at Manchester University, R. K. Livesley solved structural problems with the help of the Manchester University Electronic Computer (MUEC). As early as 1953 and 1954, he published two papers on the computer-aided investigation of rigid frames in the journal *Engineering*. And in 1954 he completed his dissertation at Manchester University on the theme of *The Application of an Electronic Computer to Problems of Structural Analysis and Design*. Shortly afterwards, he published two con-

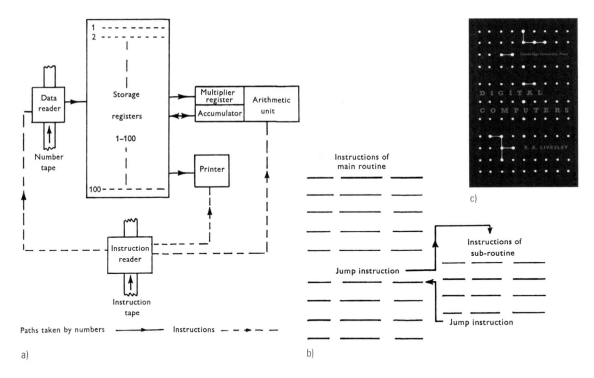

FIGURE 10-48
a) Structure of a simple computer [Livesley, 1957, p. 6], b) invoking a subroutine from the main routine [Livesley, 1957, p. 32], and c) cover of Livesley's book on computers

tributions that were to give decisive momentum to electronic calculation in engineering in general and structural analysis in particular:

*Firstly:* In a paper for the influential journal *The Structural Engineer*, Livesley presented the program that had been implemented on the MUEC in late 1953, with which plane elastic trusses could be calculated according to the displacement method based on second-order theory [Livesley, 1956]. Independently of this, Teichmann published a computational algorithm without using matrix algebra in 1958 (see section 2.9.6). Although in his treatment of iterative methods (see section 10.3.2.3) he pointed out that they would in future lose ground to the computer for the solution of large sets of equations [Teichmann, 1958, p. 99], he did not explore this avenue any further; however, he had in 1941 already pleaded for the use of the computer in flutter calculations (see section 6.6.4.2) for aircraft (for basic information on flutter calculations see [Teichmann, 1939]). The difference between Livesley and Teichmann is that Livesley formulated the displacement method according to second-order theory in the language of matrix algebra and concentrated on finding a solution to the stability problem. The problem prepared mathematically with the help of a formalised theory – in this case matrix algebra – is one that is predestined for the computer and its programming: "The user of an automatic computer must be prepared to adopt new mathematical techniques if he is to get the most out of his machine. These new techniques, like the matrix method described in this paper, are often based on the formal operations of linear algebra, since these operations are easily carried out by standard library routines. In most problems it is best to use the methods of classical mathematics rather than the short-cut techniques of human computers, since the former are logically simpler and therefore lead to easier programming" [Livesley, 1956, p. 11].

*Secondly:* In his book *An Introduction to Digital Computers* (Fig. 10-48c), which appeared in 1957 as the second volume of J. F. Baker's *Cambridge Engineering Series*, Livesley describes not only the structure of the computer (Fig. 10-48a), but also the solution of engineering problems, e. g. with the help of linear algebra or matrix algebra (e. g. elastic trusses, electrical networks), linear programming, ordinary and partial differential equations and non-linear algebra (e. g. second-order theory). Livesley's book made the use of computers popular for the practical calculations of the engineer.

Alwin Walther's presentation on analogue and digital computers and their application possibilities in structural steelwork problems, given at the instigation of Kurt Klöppel at the 1952 German Structural Steelwork Conference in Munich [Walther, 1952, pp. 144 – 197], was the first comprehensive and systematic presentation of the options that the automation of engineering calculations provided at that time. It was certainly the first time that many steelwork engineers had heard of the "calculable reason" (Sybille Krämer) of the *automate spirituel ou formel* anticipated philosophically by Leibniz, which Konrad Zuse first realised technically in 1941 as the program-controlled calculating machine – the computer.

"It was the dawn of a new 'experimental mathematics,'" writes Walther, "in which one could try out in countless variations the consequences of natural science hypotheses or technical or economic plans. So the engineer can pick out, mathematically, the most favourable solution from many possibilities. Expensive physical and technical test series are spared and … replaced by the rapid and inexpensive mathematical calculation of various options" [Walther, 1952, p. 195]. In this context, who can avoid thinking about modern methods of structural and material optimisation, which "computational mechanics" has made possible in recent years? Who doesn't associate Walther's talk of "finitism" or "finite procedures" [Walther, 1952, p. 145] with FEM? The listeners on that day received Walther's enthusiastic concluding remarks – "The view into totally new areas has been opened up. New methods of observation are appearing. We should consider ourselves lucky that we can be the contemporaries of such an amazing expansion of our horizons and intensification of our insights to an extent that would have seemed inconceivable just a short time ago!" – with some astonishment, almost anxiously, as a 1961 review of electronic calculation in structural steelwork [Walther & Barth, 1961, p. 152] records.

In 1961 Prof. Volker Hahn initiated the foundation of the Rechen-institut im Bauwesen (RIB – Computing Institute for Building, now RIB Software AG), the first West German software company for structural engineering: "None of those I knew," Hahn recalls, "were interested in my 'electronic computing office for the building industry'. I then mentioned my ideas to Prof. Leonhardt and was actually able to win him as a supporter of such a plan. He suggested I speak to Prof. Bornscheuer – and that's how the RIB was born" (cited in [Gollert, 2007, p. 19]). As early as 1963, staff at the RIB implemented the first program chains for the electronic calculation of bridges and the design of roads on the IBM 1620 (Fig. 10-49) [Gollert, 2007, p. 19]. One year prior to that, Hans Schröder, in an essay on structural analysis in the first part of the *Beton-Kalender* yearbook [Schröder, 1962], was still talking about traditional methods of calculation without mentioning the use of computers. But the remarkable thing here is that at the end of this essay there is an advertisement for the Zuse KG company founded by Konrad Zuse (Fig. 10-50)! The advertisement reads: "ZUSE has considerable experience in the field of program-controlled computing systems for numerous applications in research, technology and industry. Dr. Konrad Zuse built the world's first program-control computing system (Z 3) in 1941. ZUSE can solve the structural problems of the engineering office's daily workload – with the new program-controlled computing system using transistor technology. Get in touch with ZUSE. Comprehensive programs are available for the calculation of loadbearing structures" [n. n., 1962, A III].

In the second part of the same yearbook, G. Worch concluded his contribution on linear equations in structural analysis with the following words: "Our structural methods should also be inspected to see whether they are suitable for automation. Some methods, e. g. those involving the elastic centre of gravity, are precisely concocted so as to overcome the solution of elasticity equations regarded in the past as too tedious. But other rules apply for the use of automatic computers; we must adapt to these in the future, also in structural analysis" [Worch, 1962, p. 384]. Worch's prophetic words would quickly come true. In 1965 Klöppel and Friemann published a detailed review of the literature on computer-assisted structural analysis in the *VDI-Zeitschrift*, which evolved into computational mechanics after the middle of the innovation phase of structural theory (1950 – 75) [Klöppel & Friemann, 1965]. In that same year, Worch wrote an introduction to electronic calculation in structural analysis in the *Beton-Kalender* [Worch, 1965]. Despite this, the thematic treatment of the automation of structural calculations did not appear in the leading German literature on structural analysis until after 1965 [Rothe, 1967], which first appeared in the form of numerous special programs. The three most important events of the scientific revolution in structural mechanics are the magnificent matrix algebra reformulation of structural mechanics [Argyris, 1955], the ingenious concept of FEM [Turner, Clough, Martin and Topp, 1956] and the direct stiffness method based on matrix theory and FEM (representing a further development of the displacement method) [Turner, 1959]. These broke through the chains of the classical fundamental engineering science disciplines in a radical way and steered theory formation into totally new areas. So the hope expressed by Ostenfeld in 1926 that by dissecting the truss into finite bar segments it would be possible to deal with complicated systems without difficulties, not only for one-dimensional, but also for multi-dimensional continua, had been fulfilled.

**FIGURE 10-49**
The first electronic computer system at the RIB: the IBM 1620 with a capacity of 12.5 KB [Gollert, 2007, p. 19]

**FIGURE 10-50**
The first advertisement for structural calculations with computers in *Beton-Kalender* [n. n., 1962, A III]

## 10.4.4 FEM – formation of a general technology engineering science theory

The term "finite element method" first appeared in 1960 at the second Conference on Electronic Computation of the American Society of Civil Engineering (ASCE) in the contribution by R. W. Clough [Clough, 1960]. Clough's contribution went relatively unnoticed at the time. But at the Symposium on the Use of Computers in Civil Engineering held in Lisbon in 1962, his presentation *Stress analysis of a gravity dam by the finite element method* [Clough, 1962] attracted some interest. In a relatively short time, the term "finite element method" became part of everyday engineering language [Clough, 2004, p. 286], and in the form of the abbreviation "FEM" since the beginning of the 1970s has summarised an intellectual technology of the engineer with a status undoubtedly comparable with that of the infinitesimal calculus created by Leibniz and its application to continua. In a certain way, FEM is in fact the opposite of that: whereas infinitesimal calculus generates the world of symbols of the partial differential equations of continuum physics from the infinitely small continuous elements existing only as an ideal object, finite mathematics starts with the finitely small continuous element – the finite element – in order to solve the partial differential equations of continuum physics approximately. So we are the spectators of the race between the infinitesimal dimensions and – due to the ever-increasing power of the computer – the ever-smaller infinitesimal dimensions.

### 10.4.4.1 On the classical publication of a non-classical method

The tests on aircraft wings with 45° sweepback in order to quantify aeroelastic effects in structural mechanics terms carried out in the early 1950s by the Structural Dynamics Unit of the Boeing Airplane Company in Seattle is the historical crystallisation point of FEM. In these tests, the wings were simulated by a box-type cell tube fixed at an angle of 45° (see also Fig. 10-45). The tests were closely related to the development of long-range bombers for the United States Air Force (USAF), like the B-47 and the B-52. On 29 June 1952 the first B-52 was delivered to a USAF combat squadron, and for several decades formed the backbone of the USAF's Strategic Air Command (SAC). In his review, R. W. Clough (who studied structural engineering) explains how, as part of the Boeing Summer Faculty Program, the head of the Structural Dynamics Unit, M. J. Turner, was inaugurated into the R&D work at Boeing concerning the structural mechanics behaviour of sweptback wings, how he was assisted, and what the results were:

"The job that Jon Turner had for me was the analysis of the vibration properties of a fairly large model of a 'delta' wing structure that had been fabricated in the Boeing shop. This problem was quite different from the analysis of a typical wing structure which could be done using standard beam theory, and I spent the summer of 1952 trying to formulate a mathematical model of the delta wing representing it as an assemblage of typical 1D beam components. The results I was able to obtain by the end of the summer were very disappointing, and I was quite discouraged when I went to say goodbye to my boss, Jon Turner. But he suggested that I come back in Summer 1953. In this new effort to evaluate the vibration properties of a delta wing model, he suggested I should formulate the mathematical model as an assemblage of 2D plate elements interconnected at their corners. With this suggestion, Jon had essentially defined the concept of the finite element method.

"So I began my work in Summer 1953 developing in-plane stiffness matrices for 2D plates with corner connections. I derived these both for rectangular and for triangular plates, but the assembly of triangular plates had great advantages in modelling a delta wing. Moreover, the derivation of the in-plane stiffness of a triangular plate was far simpler than that for a rectangular plate, so very soon I shifted the emphasis of my work to the study of assemblages of triangular plate 'elements', as I called them. With an assemblage of such triangular elements, I was able to get rather good agreement between the results of a mathematical model vibration analysis and those measured with the physical model in the laboratory. Of special interest was the fact that the calculated results converged toward those of the physical model as the mesh of the triangular elements in the mathematical model was refined" [Clough, 2004, p. 285].

Clough notes that he had already formulated the content of the publication regarded as the birth certificate of FEM dating from 1956 [Turner et al., 1956] in a report to Turner in the summer of 1953, and this was presented at the annual meeting of the Institute of Aeronautical Sciences (IAS) in January 1954. In the famous four-man work [Turner et al., 1956], the transition from the discretisation of real structures (see Fig. 10-46) to a mathematical discretisation of the continuum was completed radically in the sense of finite elements (Fig. 10-51).

The mathematical genesis of the element stiffness matrix is explained by the authors using the example of a triangular element of thickness $t$ (Fig. 10-51b) since the triangular element is suitable for modelling plane plate structures of any plan shape. Integrating the strain-displacement relationships

$$\frac{\partial u}{\partial x} = \varepsilon_x = a = \frac{1}{E} \cdot (\sigma_x - \nu \cdot \sigma_y)$$

$$\frac{\partial v}{\partial y} = \varepsilon_y = b = \frac{1}{E} \cdot (\sigma_y - \nu \cdot \sigma_x) \qquad (10\text{-}39)$$

$$\frac{\partial u}{\partial y} + \frac{\partial v}{\partial x} = \gamma_{xy} = c = \frac{1}{E} \cdot (\sigma_x - \nu \cdot \sigma_y)$$

**FIGURE 10-51**
a) Rectangular element, and
b) triangular element (redrawn after [Turner et al., 1956, p. 813])

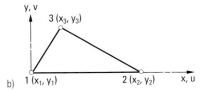

yields the displacements

$$u = a \cdot x + A \cdot y + B$$
$$v = b \cdot y + (c - A) \cdot x + C \tag{10-40}$$

with the Hookean relationships standing for $a$, $b$ and $c$ according to eq. 10-39 plus the integration constants $A$, $B$ and $C$, which are defined by the two translations and the rotation of the triangular element as a rigid body. In the second step, the stresses are expressed by the six node coordinates and node displacements of the triangular element taking into account eq. 10-39 and eq. 10-40:

$$\begin{Bmatrix} \sigma_x \\ \sigma_y \\ \tau_{xy} \end{Bmatrix} = \frac{E}{1-v^2} \cdot \begin{bmatrix} -\frac{1}{x_2} & \frac{v \cdot x_{32}}{x_2 \cdot y_3} & \frac{1}{x_2} & -\frac{v \cdot x_3}{x_2 \cdot y_3} & 0 & \frac{v}{y_3} \\ -\frac{v}{x_2} & \frac{x_{32}}{x_2 \cdot y_3} & \frac{v}{x_2} & -\frac{x_3}{x_2 \cdot y_3} & 0 & \frac{1}{y_3} \\ \frac{\lambda_1 \cdot x_{32}}{x_2 \cdot y_3} & -\frac{\lambda_1}{x_2} & -\frac{\lambda_1 \cdot x_3}{x_2 \cdot y_3} & \frac{\lambda_1}{x_2} & \frac{\lambda_1}{y_3} & 0 \end{bmatrix} \cdot \begin{Bmatrix} u_1 \\ v_1 \\ u_2 \\ v_2 \\ u_3 \\ v_3 \end{Bmatrix} =$$

$$= [\sigma] = [S] \cdot [\delta] \tag{10-41}$$

In matrix equation 10-41, $x_{ij} = x_i - x_j$ and $\lambda_1 = (1-v)/2$. The third step consists of calculating the statically equivalent node forces from the stresses distributed uniformly over the sides of the triangle. This relationship is derived by the authors for the case of shear stresses (Fig. 10-52).

The node loads on the triangular element equivalent to the shear stresses $\tau_{xy}$ follow from Fig. 10-52b:

$$F_{x_1}^{(3)} = -(x_2 - x_3) \cdot (t/2) \cdot \tau_{xy}$$
$$F_{y_1}^{(3)} = -y_3 \cdot (t/2) \cdot \tau_{xy}$$
$$F_{x_2}^{(3)} = -x_3 \cdot (t/2) \cdot \tau_{xy}$$
$$F_{y_2}^{(3)} = +y_3 \cdot (t/2) \cdot \tau_{xy} \tag{10-42}$$
$$F_{x_3}^{(3)} = +x_2 \cdot (t/2) \cdot \tau_{xy}$$
$$F_{y_3}^{(3)} = -0$$

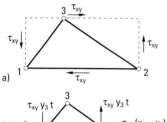
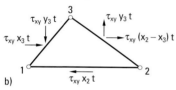

**FIGURE 10-52**
Transferring the a) shear stresses into b) statically equivalent node loads (redrawn after [Turner et al., 1956, p. 815])

The sets of equations for normal stresses $\sigma_x$ and $\sigma_y$ can be derived correspondingly in each case so that the three sets of equations can be summarised in the matrix equation

$$\begin{Bmatrix} F_{x_1} \\ F_{y_1} \\ F_{x_2} \\ F_{y_2} \\ F_{x_3} \\ F_{y_3} \end{Bmatrix} = \frac{t}{2} \cdot \begin{bmatrix} -y_3 & 0 & -(x_2 - x_3) \\ 0 & -(x_2 - x_3) & -y_3 \\ y_3 & 0 & -x_3 \\ 0 & -x_3 & y_3 \\ 0 & 0 & x_3 \\ 0 & x_2 & 0 \end{bmatrix} \cdot \begin{Bmatrix} \sigma_x \\ \sigma_x \\ \tau_{xy} \end{Bmatrix} =$$

$$= [F] = [T] \cdot [\sigma] \tag{10-43}$$

Entering matrix equation 10-41 into eq. 10-43 results in

$$[F] = [T] \cdot [S] \cdot [\sigma] \tag{10-44}$$

The first two matrices on the left-hand side of the matrix equation correspond to the stiffness matrix of the triangular element (element stiffness matrix):

$$[K] = \frac{E \cdot t}{2(1-v^2)} \cdot \begin{Bmatrix} \frac{y_3}{x_2}+\frac{\lambda_1 x_{23}^2}{x_2 y_3} & -\frac{\lambda_2 x_{32}}{x_2} & -\frac{y_3}{x_2}+\frac{\lambda_1 x_3 x_{23}}{x_2 y_3} & \frac{v x_3}{x_2}+\frac{\lambda_1 x_{32}}{x_2} & -\frac{\lambda_1 x_{23}}{y_3} & -v \\ -\frac{\lambda_2 x_{32}}{x_2} & \frac{x_{23}^2}{x_2 y_3}+\frac{\lambda_1 y_3}{x_2} & \frac{v x_{32}}{x_2}+\frac{\lambda_1 x_3}{x_2} & \frac{x_3 x_{23}}{x_2 y_3}-\frac{\lambda_1 y_3}{x_2} & -\lambda_1 & -\frac{x_{23}}{y_3} \\ -\frac{y_3}{x_2}+\frac{\lambda_1 x_3 x_{23}}{x_2 y_3} & \frac{v x_{32}}{x_2}+\frac{\lambda_1 x_3}{x_2} & \frac{y_3}{x_2}+\frac{\lambda_1 x_3^2}{x_2 y_3} & -\frac{\lambda_2 x_3}{x_2} & -\frac{\lambda_1 x_3}{y_3} & v \\ \frac{v x_3}{x_2}+\frac{\lambda_1 x_{32}}{x_2} & \frac{x_3 x_{23}}{x_2 y_3}-\frac{\lambda_1 y_3}{x_2} & -\frac{\lambda_2 x_3}{x_2} & \frac{x_3^2}{x_2 y_3}+\frac{\lambda_1 y_3}{x_2} & \lambda_1 & -\frac{x_3}{y_3} \\ -\frac{\lambda_1 x_{23}}{y_3} & -\lambda_1 & -\frac{\lambda_1 x_3}{y_3} & \lambda_1 & \frac{\lambda_1 x_2}{y_3} & 0 \\ -v & -\frac{x_{23}}{y_3} & v & -\frac{x_3}{y_3} & 0 & \frac{x_2}{y_3} \end{Bmatrix} \quad (10\text{-}45)$$

In eq. 10-45, $\lambda_2 = (1 + v)/2$. Matrix equation 10-44 can therefore be simplified to

$$[F] = [K] \cdot [\delta] \quad (10\text{-}46)$$

Turner, Clough, Martin and Topp summarise FEM according to the displacement method in six steps:

"(1) A complex structure must first be replaced by an equivalent idealized structure consisting of basic structural parts [finite elements – the author] that are connected to each other at selected node points.

(2) Stiffness matrices [see, for example, eq. 10-45 – the author] must be either known or determined for each basic structural unit appearing in the idealized structure.

(3) While all other nodes are held fixed, a given node is displaced in one of the chosen coordinate directions. The forces required to do this and the reactions set up at neighboring nodes are then known from the various individual member stiffness matrices. These forces and reactions determine one column in the overall stiffness matrix. When all components of displacement at all nodes have been considered in this manner, the complete stiffness matrix will have been developed. In the general case, this matrix will be of order $3n \times 3n$, where $n$ equals the number of nodes. The stiffness matrix so developed will be singular.

(4) Desired support conditions can be imposed by striking out columns and corresponding rows, in the stiffness matrix, for which zero displacements have been specified. This reduces the order of the stiffness matrix and renders it nonsingular.

(5) For any given set of external forces at the nodes, matrix calculation applied to the stiffness matrix then yields all components of node displacement plus the external reactions.

(6) Forces in the internal members can be found by applying the appropriate force-deflection relations" [Turner et al., 1956, p. 810].

Whereas engineers are required for steps (1) and (2), the remaining steps (3) to (6) can also be carried out by persons without engineering qualifications; this routine mental work can, however, also be performed by computers, and in future even the generation of the overall stiffness matrix: "It is worth while to notice that once the stiffness matrix has been written, the solution follows by a series of routine matrix calculations. These are rapidly carried out on automatic digital computing equipment. Changes in design are taken care of by properly modifying the stiffness matrix. This cuts analysis time to a minimum, since development of the stiffness matrix is also a routine. In fact, it may also be programmed for the digital computing machine" [Turner et al., 1956, pp. 809–810]. Their prophecy would very soon be fulfilled and later, with the automatic generation of FE meshes, even include step (1).

**The heuristic potential of FEM: the direct stiffness method**

### 10.4.4.2

In 1959 Turner generalised the displacement method (DM) to form the direct stiffness method (DMS) in order to automate step (3), i. e. the construction of the overall stiffness matrix from the element stiffness matrices. At the Aachen Structures & Materials Panel meeting of AGARD (NATO's Advisory Group for Aeronautical Research & Development) on 6 November 1959, Turner included his paper entitled *The direct stiffness method of structural analysis* [Turner, 1959]. Apparently, this contribution was not published, but it must have left a deep impression on the conference delegates for it was quoted again and again in the published documents of the next meeting of the panel (1962) (see [Felippa, 2001, p. 1320]). Turner himself published the first application of his direct stiffness method for non-linear problems in 1960, together with E. H. Dill, H. C. Martin and R. J. Melosh. For this purpose, they presented the direct stiffness method in an incremental formulation [Turner et al., 1960]. Finally, in 1962, Turner, working with H. C. Martin and R. C. Weikel, introduced an extended version of the direct stiffness method presented in 1959, which, however, was not published until 1964 [Turner et al., 1964]. The introduction to the latter reads as follows: "In a paper presented at the 1959 meeting of AGARD Structures & Materials Panel in Aachen, the essential features of a system for numerical analysis of structures, termed the DSM, were described. The characteristic feature of this particular version of DM is the assembly procedure, whereby the stiffness matrix for a composite structure is generated by direct addition of matrices associated with the elements of the structure" (cited in [Felippa, 2001, p. 1320]). Turner develops the direct stiffness method in a few lines:

"For an individual member $e$ the generalized nodal force increments $[\Delta X^e]$ required to maintain a set of nodal displacement increments $[\Delta u]$ are given by a matrix equation

$$[\Delta X^e] = [K^e] \cdot [\Delta u] \qquad (3) \text{ (or (10-47))}$$

in which $[K^e]$ denotes the stiffness matrix of the individual element. Resultant nodal force increments acting on the complete structure are

$$[\Delta X] = \sum[\Delta X^e] = [K] \cdot [\Delta u] \qquad (4)\ (\text{or }(10\text{-}48))$$

wherein $[K]$, the stiffness of the complete structure, is given by the summation

$$[K] = \sum[K^e] \qquad (5)\ (\text{or }(10\text{-}49))$$

which provides the basis for the matrix assembly procedure noted earlier" (cited in [Felippa, 2001, pp. 1320–1321]). In order to satisfy eq. 10-49, the element stiffness matrix $[K^e]$ must be completely transformed into the global system of coordinates. This step is superfluous if eq. 10-49 is replaced by

$$[K] = \sum[L^e]^T \cdot [K^e] \cdot [L^e] \qquad (10\text{-}50)$$

as is usual in the version of the direct stiffness method in common use today [Felippa, 2001, p. 1321]. In eq. 10-50, $[L^e]$ are the Boolean localisation matrices and $[L^e]^T$ their transpositions. Felippa, in his detailed commentary of Turner's pioneering work, highlights further that Turner's direct stiffness method is an incremental formulation and he is specifically looking to solve non-linear problems such as loadbearing systems with large displacements: "[These] problems ... are solved in a sequence of linearized steps. Stiffness matrices are revised at the beginning of each step to account for [changes] in internal loads, temperatures and geometric configurations" (cited in [Felippa, 2001, p. 1321]). The realisation of such calculation procedures as summarised by Turner in the language of matrix theory through eq.10-47 to eq. 10-49 is practically impossible without a computer. Therefore, Turner investigated the implementation of important details of his direct stiffness method on the computer. One example of this is the first mention of user-defined elements (see [Felippa, 2001, p. 1321]). Building the overall stiffness matrix from the element stiffness matrices according to Turner's eq. 10-47 to eq. 10-49 is insensitive with respect to the type of element. For example, the generalised calculation procedures of the direct stiffness method for the two-node beam element when expressed in the language of matrix theory are the same as for the 64-node hexahedron element. Vibration and stability problems, too, can be solved with the formalism of the direct stiffness method.

The direct stiffness method, which Felippa fittingly designates a "paragon of elegance and simplicity" [Felippa, 2001, p. 1321], gave the research and the practising engineer working in the middle of the innovation phase of structural mechanics (1950–75) a powerful computer-viable intellectual technology which later would be ready to expand its validity to the whole range of objects of structural mechanics. But it was initially destined to be restricted to the Boeing Airplane Company (Seattle) and Bell Aerosystems Co. (Buffalo), with M. J. Turner and R. H. Gallagher respectively as its protagonists. At the civil engineering faculty of the University of California at Berkeley it was R. W. Clough and at the University of Washington the aircraft engineer H. C. Martin who taught and further developed the direct stiffness method. After discussions with Clough, O. C. Zienkiewicz, a structural engineering professor from Southwell's

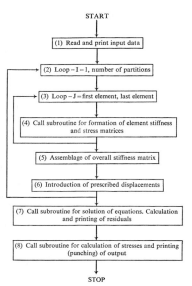

**FIGURE 10-53**
Flow diagram of an FE main program in Fortran IV [Zienkiewicz & Cheung, 1967, p. 229]

**The founding of FEM through variational theorems**

scientific school, and his staff at the University of Wales in Swansea turned their attention to FEM (see, for example, [Zienkiewicz & Cheung, 1964; Zienkiewicz, 1965]). Together with Argyris, Turner and Clough, Zienkiewicz would become one of the champions of FEM and impart decisive impulses at the transition of structural mechanics from its innovation phase (1950 – 75) to its diffusion phase (1975 to date). For example, together with Y. K. Cheung he wrote the first FEM textbook in 1967 and gave it the title *The Finite Element Method in Structural and Continuum Mechanics* [Zienkiewicz & Cheung, 1967] (Fig. 10-53), and two years later founded, together with R. H. Gallagher, the *International Journal for Numerical Methods in Engineering* – a journal that for the first time covered fully the entire field of computer-based numerical engineering methods. The extremely clearly written textbook by Zienkiewicz and Cheung liberated FEM not only from the Procrustean bed of the aerospace industry, but also constituted the prelude to accessing general field problems through FEM.

The force method formulated in matrix notation lost ground to Turner's direct stiffness method because the latter was essentially much better suited to implementation on the computer. There was no lack of contributions during the 1960s aimed at employing the displacement method as a basis for a problem-centric programming language (see [Schrader, 1969]). The direct stiffness method therefore gradually changed into the standard method of modern structural mechanics and, starting in the late 1960s, was the main driving force behind the development of FEM for two decades.

**10.4.5**

The founding of FEM through variational theorems began in 1960 as Clough, with the help of the principle of minimum potential energy, was able to show that with ever smaller element dimensions the approximation of FE calculations converges to the exact mathematical solution [Clough, 1960]. "At this stage both the intuition, which stemmed from the discrete engineering approach, and the purely mathematical reasoning coincided and both approaches were united" [Zienkiewicz, 2004, p. 7]. The variational theorem introduced into FEM by Clough brought answers to two unanswered questions of the direct stiffness method. Firstly, the question of whether the method is mathematically permissible, and, secondly, the question of the universality for any elements, e.g. those with curved boundaries. If, for example, the approximation functions for describing the displacements of the finite elements are entered into a variational theorem, e. g. the principle of the extremum of potential energy, then, mathematically, FEM assumes the form of a specific Ritz method (see, for example, [Knothe & Wessels, 1999, p. 16]). The synthesis of such localised Ritz approaches (in the meaning of the variation formulation) with the matrix formulation (see section 10.4.4) of FEM initiated a breathtaking development in structural mechanics, which in the end would be superseded by computational mechanics in its diffusion phase (1975 to date).

When we speak of the mathematical roots of FEM, then the work of Karl Heinrich Schellbach (1805–92) [Schellbach, 1851] and Richard Courant (1888–1972) [Courant, 1943] are of great significance. Schellbach specified an approximate solution for a minimal area problem with the help of triangular elements within the scope of calculus of variations (see also [Williamson, 1980, pp. 931–933; Oden, 1987, p. 125]). In his analysis of the St. Venant torsion problem, Courant formulates the variational theorem of Dirichlet and Green for triangular elements and in doing so assumes an elementwise linear approximation (see [Oden, 1987, p. 125; Felippa, 1994, pp. 2159–2161]). Frank Williamson jr. began with the method of piecewise linear approximation of the brachistochrone (see Fig. 11-8) after Leibniz in his historical remarks concerning FEM [Williamson, 1980, pp. 930–931]. What Felippa wrote in his introductory commentary to the reprinted edition of Courant's publication [Courant, 1943] (reprinted in [Courant, 1994, pp. 2163–2187]) also applies to the aforementioned works: "Within the present framework of finite elements the influence was insignificant: there is no 'Courant FEM'" [Felippa, 1994, p. 2161]. This is the reason why only those works that were directly involved in the discourse surrounding the founding of structural mechanics will be discussed below.

**10.4.5.1 Variational theorem of Dirichlet and Green**

The variational theorem introduced into FEM by Clough can be attributed to the concept of potential energy $\Pi$ introduced into elastic theory by George Green (1793–1841) [Green, 1839] and the principle of the stability of the elastic equilibrium proven by Peter Gustav Lejeune Dirichlet (1805–59) [Dirichlet, 1846]. According to Dirichlet, the elastic equilibrium – provided the applied forces exhibit a potential – are then stable, and only then, when the total potential energy $\Pi$ of the mechanical system is a minimum, e.g. the centre of gravity is at the lowest point. That is the principle of minimum potential energy. Dirichlet's stability theorem for a system of $n$ rigid bodies was proved very elegantly by Georg Hamel (1877–1954) [Hamel, 1912, pp. 485–487].

**A simple example: the axially loaded elastic extensible bar**

Taking the axially loaded elastic extensible bar shown in Fig. 10-54 with strain stiffness $D(x) = E(x) \cdot A(x)$ – i.e. the product of the varying elastic modulus $E(x)$ and the cross-sectional area $A(x)$ – the variational theorem of Dirichlet and Green takes on the following form:

$$\Pi_{D,G}(x,u,u') = \int_0^l \left\{ \frac{1}{2} \cdot D(x) \cdot [u'(x)]^2 - p(x) \cdot u(x) \right\} \cdot dx = Minimum \quad (10\text{-}51)$$

This variational theorem states that the total potential energy $\Pi_{D,G}$ is a minimum. In eq. 10-51 $u'(x)$ is the first derivation of the displacement $u(x)$ for $x$. Variational theorem 10-51 is a functional – quasi a function of functions. Whereas functions are investigated by conventional analysis (infinitesimal calculus), functionals are the object of functional analysis. As the functional 10-51 is a minimum, the first variation must vanish so that according to the rules of calculus of variations

$$\delta \Pi_{D,G} = \frac{\partial \Pi_{D,G}}{\partial u'(x)} \cdot \delta u'(x) + \frac{\partial \Pi_{D,G}}{\partial u(x)} \cdot \delta u(x) = 0 \tag{10-52}$$

the principle of minimum potential energy (eq. 10-51) can be transformed into

$$\delta \Pi_{D,G} = 0 = \int_0^l D(x) \cdot u' \cdot \delta u' \cdot dx - \int_0^l p(x) \cdot \delta u \cdot dx \tag{10-53}$$

and, following partial integration, finally yields

$$\delta \Pi_{D,G} = 0 = D(l) \cdot u'(l) \cdot \delta u(l) - D(0) \cdot u'(0) \cdot \delta u(0) -$$
$$- \left[ \int_0^l \left\{ [D(x) \cdot u']' + p(x) \right\} \cdot dx \right] \cdot \delta u \tag{10-54}$$

From the first variation of the geometric boundary conditions

$$u(x = 0) = 0; \quad u(x = l) = 0 \tag{10-55}$$

it follows that

$$\delta u(x = 0) = 0; \quad \delta u(x = l) = 0 \tag{10-56}$$

so the first two terms on the right-hand side of eq. 10-54 disappear. As in the end the first variation of the displacement for $0 < x < l$ must be $\delta u \ne 0$, eq. 10-54 can only be satisfied for

$$\left\{ [D(x) \cdot u']' + p(x) \right\} = 0 \tag{10-57}$$

The ordinary differential equation 10-57 is designated the Euler differential equation attributed to the variational theorem of Dirichlet and Green with the boundary conditions according to eq. 10-55 taken from Fig. 10-54.

Eq. 10-53 or eq. 10-54 obtained from the variation of the Dirichlet and Green principle (eq. 10-51) corresponds to the formulation of the principle of virtual displacements for the axially loaded elastic extensible bar. The principle of virtual displacements was presented in section 2.2.2 using the example of the two-arm lever: instead of the Euler differential equation, however, the principle of virtual displacements (eq. 2-3) results in the lever principle on that occasion, i.e. an equilibrium statement. With the help of the relationship

$$N(x) = D(x) \cdot u'(x) \tag{10-58}$$

the Euler differential equation 10-57 for the axially loaded elastic extensible bar can be derived from the equilibrium at the differential element with length $dx$. The variational theorem of Dirichlet and Green can therefore be transformed via the principle of virtual displacements into equilibrium statements in the style of the displacement method (cf. eq. 10-17). The solution of this type 2 elasticity equation leads to displacement variables, the geometrically indeterminate variables.

In a formal analogy with the axially loaded extensible bar, the variational theorem of Dirichlet and Green can be applied, for example, to the bar in bending with transverse load $q(x)$. For the bar in bending with constant bending stiffness $E \cdot I$, the associated Euler differential equation

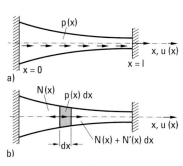

**FIGURE 10-54**
a) Extensible bar subjected to axial loading (redrawn after [Knothe, 1983, p. 3-27]), and b) equilibrium at the differential element (redrawn after [Knothe, 1983, p. 2-15])

would be

$$E \cdot I \cdot \frac{d^4 w(x)}{dx^4} + q(x) = 0 \qquad (10\text{-}59)$$

The ordinary differential equation 10-59 for deflection $w(x)$ corresponds to the linearised differential equations for the elastic curve of Navier and Eytelwein (cf. eq. 3-8 and eq. 3-9). The variational theorem of Dirichlet and Green can also be formulated for two- and three-dimensional continua, e.g. the thin elastic plate; in this case the result would be a Euler differential equation corresponding to Kirchhoff's differential equation for plates (eq. 7-56) – this is how Kirchhoff obtained his equation [Kirchhoff, 1850].

Dirichlet analysed the energy criterion for the stability of the elastic equilibrium for systems with an infinite number of degrees of freedom (Fig. 10-55):

$$\delta^2 \Pi_{D,G} \begin{Bmatrix} < \\ = \\ > \end{Bmatrix} 0 \begin{Bmatrix} \text{unstable} \\ \text{imperfect} \\ \text{stable} \end{Bmatrix} \text{equilibrium} \qquad (10\text{-}60)$$

In the energy criterion (eq. 10-60), $\delta^2 \Pi_{D,G}$ is the second variation of the total potential energy $\Pi_{D,G}$, e.g. functional equation 10-51 must be varied twice for the axially loaded elastic extensible bar.

The idea behind the axially loaded elastic extensible bar is to show by means of an example that the variational theorem of Dirichlet and Green is an interpreted formalised theory in the sense of calculus of variations. The formal force of calculus of variations was exploited by the Göttingen school around Felix Klein for the foundation of elastic theory.

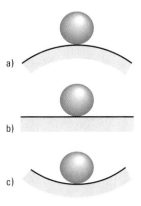

FIGURE 10-55
Pictograms for Dirichlet's energy criterion: a) unstable, b) imperfect, and c) stable equilibrium

**The Göttingen school around Felix Klein**

Influenced by his trip to the USA in 1893, Felix Klein (for a biography see [Tobies, 1981]) specified his thoughts on a synthesis between mathematics, physics and the fundamental engineering science disciplines in his "new agenda for Göttingen". "One essential step for achieving this goal," writes Knothe, "was the founding of the 'Göttinger Vereinigung zur Förderung der angewandten Physik und Mathematik' [Göttingen Society for the Advancement of Applied Physics & Mathematics]" [Knothe, 1999, p. XXI]. Klein was therefore able to attract important mathematicians to Göttingen, e.g. David Hilbert (1862–1943) in 1895, Hermann Minkowski (1864–1909) in 1902, Carl Runge (1856–1927) in 1904 and Edmund Landau (1877–1938) in 1909. It was no less a person than Hilbert who in 1899 succeeded in proving the existence of Dirichlet's variational theorem [Hilbert, 1901]. In 1906 Max Born (1882–1970) completed his dissertation on the stability of the elastic line in two and three dimensions under various boundary conditions [Born, 1906] with Hilbert, and includes a reference to the latter's publication on calculus of variations [Hilbert, 1906]. Erich Trefftz (1888–1937) managed to conclude the theory of elastic equilibrium with his generalisation of the energy criterion (eq. 10-60) based on calculus of variations for arbitrary elastic continua [Trefftz, 1933]. Using this method, Trefftz, Karl Marguerre and Alexander Kromm investigated the supercritical buckling of slab strips ([Marguerre & Trefftz, 1937; Kromm & Marguerre, 1937]). Marguerre wrote an extraordinary lucid portrayal of the application of the energy method [Marguerre, 1938/1 & 1938/3].

One contribution that was to have a profound influence on structural mechanics and the variation formulation of FEM was that of Walther Ritz (1878–1909) [Ritz, 1909]. In this work, Ritz develops the method of solving boundary value problems which bears his name in the theory of partial differential equations; in doing so, he makes use of Hilbert's studies of Dirichlet's variational theorem. At this time, a different variational theorem prevailed in structural analysis together with Menabrea's principle of minimum deformation energy (see eq. 6-39) and Castigliano's second theorem (see eq. 6-40), which was not recognised as such by the advocates of this fundamental engineering science discipline: the principle of the extremum of deformation complementary energy or the variational theorem of Menabrea and Castigliano. This principle, too, leads to a Euler differential equation. In contrast to the variational theorem of Dirichlet and Green, the variational theorem of Menabrea and Castigliano transforms into the principle of virtual forces and the compatibility expressions according to the force method (cf. eq. 10-18). The solution of these type 1 elasticity equations leads to force variables, the statically indeterminate variables. Structural analysis is in the first place interested in force variables in order to design the loadbearing structure in the next step. This is why the force method was for a long time preferred to the displacement method and the underlying variational theorem of Dirichlet and Green. This did not change until the use of Turner's direct stiffness method became widespread.

**The first stage of the synthesis: the canonic variational theorem of Hellinger and Prange**

#### 10.4.5.2

Whereas in the variational theorem of Dirichlet and Green a functional of the displacements $u$ (cf. eq. 10-51), is $v$ and $w$, and, for example, is varied according to eq. 10-54, in the variational theorem of Menabrea and Castigliano the internal forces, or rather the stresses, are entered into the variation. In other words, the variational theorem of Menabrea and Castigliano is a functional of the internal forces (or stresses). The dispute surrounding the foundation of classical theory of structures (see sections 6.4.2 and 6.4.3) was in the first place concerned with the general applicability of various forms of the variational theorem of Menabrea and Castigliano, e. g. the Maxwell-Betti reciprocal equation (see eq. 10-1), the principle of virtual forces (see eq. 6-42) or Castigliano's second theorem (see eq. 6-40). This dispute ended in 1909 with two publications by the antagonists Weingarten [Weingarten, 1909/2] and Weyrauch [Weyrauch, 1909] in the *Nachrichten der königlichen Gesellschaft der Wissenschaften zu Göttingen* (Bulletin of the Göttingen Royal Society of Sciences). "It is highly probable," Knothe writes, "that the controversy was also taken into account by Klein and Hilbert, who in their seminars allowed students to discuss problems from the branches of physics and engineering" [Knothe, 1999, p. XXIX].

It was against this backdrop that Ernst Hellinger (1883–1950) and Georg Prange (1885–1941) took up the challenges of the fundamental dispute from the viewpoint of mathematical physics.

In his overview *Die allgemeinen Ansätze der Mechanik der Kontinua* (general approaches to the mechanics of continua) [Hellinger, 1914] in the *Encyklopädie der mathematischen Wissenschaften* (encyclopaedia of the mathematical sciences) edited by Klein and C. H. Müller, Hellinger, referring to Born [Born, 1906] for the case of three-dimensional continua, shows how the principle of minimum potential energy can first be converted into its canonic form by transferring the canonic transformation of analytical mechanics, whereby displacements and stresses occur as unknown state variables. After that, Hellinger derives the variational theorem of Menabrea and Castigliano by introducing the equilibrium conditions as secondary conditions (see [Knothe, 1999, p. XXIII]). As it says in Wilhelm Busch's tale of *Max and Moritz*: "So much for the opening trick; Worse to follow in a tick" (translation: Percy Reynolds).

The second "trick" was introduced by Prange in his 1915 dissertation [Prange, 1915], completed in Göttingen, and in his habilitation thesis *Das Extremum der Formänderungsarbeit* (the extremum of deformation work) completed in Hannover one year later but unfortunately not published in full until 1999 in a version with a commentary by Klaus Knothe ([Prange, 1916], Prange in: [Knothe, 1999, pp. 1–134]). In his dissertation, Prange prepares mathematically the foundations of elastic theory through calculus of variations by working through the canonic transformation of the Hamilton-Jacobi theory of calculus of variations known from analytical mechanics. That was the first "trick". Both the displacement variables $u$ and also the force variables or stresses $\sigma$ are the unknown varying state variables (Fig. 10-56) in the newly emerging canonic variational problem following the canonic transformation.

**FIGURE 10-56**
Relationships between the variational theorems considered by Prange in 1916 (modified drawing after [Knothe, 1999, p. XXXI])

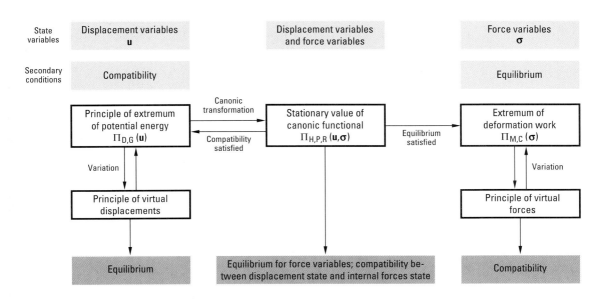

That was the second "trick", which Prange completed in his habilitation thesis.

## Prange's habilitation thesis

Prange begins his habilitation thesis with the following lines: "In the practical elastic theory, especially in theory of structures, 'minimum deformation work' plays a great role, an expression that owing to its vagueness has given rise to many ambiguities because this 'minimum' is interpreted in the most diverse ways such that misunderstandings were inevitable. Sometimes the variables – on which the expression to be made a minimum depends – change; furthermore, the neighbouring states – with which the minimum is to be compared – are seldom specified; the type of variation to be carried out in every individual case is not defined exactly so the same expression can have the most diverse interpretations. This lack of accuracy is exacerbated by the fact that the practical elastic theory was essentially influenced by the study of the trussed framework, for which it was initially expanded. In trussed framework theory, the different interpretations that can occur with the minimum deformation work are really so closely related that one is inclined to say they cannot be strictly separated, and instead are combined in a general interpretation. The trussed framework is the extremum of an ordinary function, specifically a quadratic form, of an infinite number of variables. If these considerations are transferred to the continuous body, then they are replaced by the extremum of the triple integral which is to be treated according to calculus of variations" (Prange in: [Knothe, 1999, p. 3]). Prange manages not only to extract the mathematical sense of the considerations pursued in the dispute about the foundation of theory of structures (see sections 6.4.2 and 6.4.3) in historico-logical terms, but also to discover their similarity to considerations that use analytical mechanics: "The principle of 'minimum deformation work' corresponds to the 'principle of least action', the 'equilibrium conditions' and 'compatibility conditions' correspond to the 'differential equations of motion' and **Castigliano's theorems** are the converse of the description of the integrals of the motion equations through the derivations of the **varying action** introduced by **Hamilton**" [Prange, 1919, col. 83]. Against the background of the historical development of the idea of the practical use of symbols, Prange's habilitation thesis is an agenda for the radical fomalisation of elastic theory while including the whole of structural analysis through calculus of variations in the sense of an interpreted formalised theory. Prange's reformulation of the whole of elastic theory through calculus of variations is revealed in the structure of his habilitation thesis (Prange in: [Knothe, 1999, pp. 7–134]):

*Chapter 1:* The trussed framework
1.1 The trussed framework bar and its deformation.
1.2 The trussed framework, especially its equilibrium.
1.3 The intervention of elasticity. Settling the stress problem for statically indeterminate trussed frameworks.
1.4 Minimum total energy. Canonic transformation. The principle of Menabrea.

1.5 Deformation work as a function of the external forces. Castigliano's theorems. Maxwell's reciprocal theorem.
1.6 The principle of Menabrea in another interpretation.
1.7 The internal stresses of the statically indeterminate trussed framework.
1.8 Temperature stresses. Extension to a non-linear deformation theorem. Engesser's complementary work.
1.9 The historic development.

*Chapter 2:* The continuous elastic body
2.1 The elastic deformation of the continuum. The principle of virtual displacements and the equilibrium conditions.
2.2 The intervention of elasticity.
2.3 Minimum of total energy.
2.4 Canonic transformation. The principle of Menabrea.
2.5 The extremum of deformation work as a functional of the surface displacement and the surface pressures. Castigliano's theorems, Betti's reciprocal theorem.
2.6 A second interpretation of the principle of Menabrea which is valid for the case of multiple coherent bodies.
2.7 The internal stresses.
2.8 The influence of temperature on deformation. Engesser's complementary work.
2.9 Remarks on the historical development.

*Appendix:* Presentation of the extremum of deformation work as a functional of the given surface values.

It becomes clear from this structure that Prange first reformulated trussed framework theory according to calculus of variations and then completed the generalisation for elastic continua. His structure is in this sense strictly periodic, i. e. the sections of chapters 1 and 2 correspond in terms of method; they differ merely in the sense of the two cognition objects treated: the trussed framework and the elastic continuum. According to Prange, the reason for this two-part breakdown is that the investigations into practical elastic theory are mainly carried out for trussed frameworks and their transfer to the continuum takes place in a way "that individual forces are used for these, too. In doing so, no connection is made with the theoretical elastic theory, where individual loads appear only as extreme cases of continuous loading. If, instead, one completes the transfer to the three-dimensional body loaded with continuous forces, then the sense of the considerations leading to minimal demands (principle of least deformation work) becomes clearer. One recognises that in the trussed framework, interpretations coincide that are actually disparate and have to be carefully differentiated in the case of the continuous body. By considering this condition, it is possible to eliminate a number of ambiguities, which were certainly the reasons behind the aforementioned disputes [concerning the foundation of theory of structures, see sections 6.4.2 and 6.4.3 – the author]" [Prange, 1919, col. 83].

The hub of Prange's habilitation thesis is the formulation of the canonic variational theorem for the trussed framework in section 1.4 (Prange in: [Knothe, 1999, pp. 22–30]) and the elastic continuum in section 2.4 (Prange in: [Knothe, 1999, pp. 84–89]). Three years later, Prange specified his canonic variational theorem in detail in the journal *Zeitschrift für Architektur- und Ingenieurwesen* using the straight beam and the arch with different degrees of curvature [Prange, 1919]. It was in the same journal that Hertwig published his paper on the development of some principles in theory of structures in 1906 [Hertwig, 1906], which led to the resumption of the dispute concerning the foundation of classical theory of structures and which inspired Prange to carry out his pioneering work in which he analysed the history of the development of the controversy and finally put an end to it.

Prange's habilitation thesis completed the first use of formalised theory in the whole of structural mechanics on the basis of calculus of variations. The second application of formalised theory in structural mechanics was achieved by Argyris on the basis of matrix algebra (see section 10.4.3). Both have a dual make-up (see Figs. 10-47 and 10-55). It would be left to FEM to fuse the two together in crossing from structural mechanics to computational mechanics at the transition from the innovation to the diffusion phase in the mid-1970s. The sense of Johann Wolfgang v. Goethe's (1749–1832) feeling for Immanuel Kant (1721–1804), spoken to the young Arthur Schopenhauer (1788–1860), applies to both works: "When I read a page of Kant, I feel as if I were stepping into a brilliant room" (translation: H. S. Chamberlain, cited in [Vorländer, 2004, p. 352]). Argyris was read, Prange was largely ignored.

**In the Hades of amnesia**

The representatives of theory of structures at first refrained from entering the "brilliant room" of Prange's canonic variational theorem worked out clearly through the formalisation of mathematic elastic theory, even though Prange published a lengthy paper in an engineering journal on this subject, but in the paper referred to the fact that his habilitation thesis was to be published in book form. However, World War 1 prevented the book from being printed [Prange, 1919, Sp. 83]. An indirect reference can be found in Hertwig's *Lebenserinnerungen* (memoirs), where he writes that Hamel, who gained his doctorate with Hilbert in 1901 and from 1912 to 1918 was professor of mathematics at Aachen RWTH, taught his engineering colleagues Oskar Domke (1874–1945) and August Hertwig (1872–1955) the theory of functions in private tutorials in 1917 and both professorial students worked carefully through the material [Hertwig, 1947, p. 137]. As a deductive and philosophical thinker of theoretical mechanics, Hamel took the *Mécanique analytique* [Lagrange, 1788] as his model; so the central significance of the canonic transformation of William Rowan Hamilton (1805–65) and Karl Gustav Jacob Jacobi (1804–51) – the Hamilton-Jacobi theory – for the axiomatic formulation of mechanics was clear to him, and it may well be that he was aware of the work of Hellinger (1913) and possibly even Prange's dissertation (1915). Domke's scientific work benefited directly from the private tutorials of Hamel, although he had published a paper beforehand on variational theorems in elastic theory besides applications in structural analysis [Domke, 1915]. Later, Schleusner was to write that Domke described the dispute surrounding the foundation of classical theory of structures "clearly and perfectly from the point of view of variational theorems" [Schleusner, 1938/3, p. 185]. Domke's paper appeared in the journal *Zeitschrift für Mathematik und Physik* and was certainly known to Hamel, but not the structural engineering theorists. But it is the Hamilton-Jacobi theory that is decisive, which the structural engineering theorists ignored in the discussion about the variational theorems in the mid-1930s (see section 11.2.11). So in the end the mathematical reasoning behind the duality of the force and displacement methods corresponding to the duality between the variational theorem of Menabrea and Castigliano on the one side and that of Dirichlet and Green on the other remained a closed book to them. They were also cut off from the bridge that Hellinger and Prange had built between the dual variational theorems with the help of their hybrid variational theorem.

### First steps in recollection

In 1942 Schleusner suggested to the Springer publishing house that they should publish Prange's habilitation thesis. Marguerre, Hamel, Grammel and Klotter also highlighted the topical significance of this work. But even Schleusner's attempt to gain approval for publication from the generals of the German armed forces remained unsuccessful. So the publication of Prange's habilitation thesis failed for a second time due to wartime circumstances. It was not until 1999 that this major scientific attempt to find a basis for elastic theory was published in full in the form of Knothe's devoted edition (Prange in: [Knothe, 1999, pp. 7–134]). The first reference to Prange's habilitation thesis can be found at the end of the section on minimal principles of elastic theory in Hamel's *Theoretische Mechanik* (theoretical mechanics) [Hamel, 1949, p. 375]. Prange's variational theorem was first acknowledged on an international level in a paper on historical developments in the principles of elastic theory published in the journal *Applied Mechanics Reviews* [Oravas & McLean, 1966/1] and in the introduction to the new edition of the English translation of Castigliano's *The Theory of Equilibrium of Elastic Systems and its Applications* [Oravas, 1966/2] organised by G. Æ. Oravas. Since then, there have been only individual references to Prange (e. g. [Nemat-Nasser, 1972; Bathe et al., 1977; Gurtin, 1983]). So Prange's foundation for elastic theory had no effect on scientific theory formation. Nevertheless, it is a worthwhile object of research in the history of science, as the publications of Oravas, McLean and Knothe show.

### Eric Reissner's contribution

Parallel lines never cross. Working independently of Hellinger and Prange, Eric Reissner (1913–96) published his famous six-page paper *On a variational theorem in elasticity* in 1950 ([Reissner, 1950; Reissner, 1996, pp. 437–442]). In this paper he develops – without, however, considering the Hamilton-Jacobi theory – a variational theorem corresponding to that of Hellinger and Prange. M. E. Gurtin named this variational theorem after Hellinger, Prange and Reissner [Gurtin, 1983] and the symbol for this in the following is $\Pi_{H,P,R}$. In the spatial case, $\Pi_{H,P,R}$ is a functional of the three components of the displacement vector $\boldsymbol{u}$ and the six components of the stress tensor $\boldsymbol{\sigma}$. The starting point is Reissner's research into the quantification of shear lag in box beams (e.g. [Reissner, 1941]) and his studies of shear-flexible elastic plates, which resulted in the plate theory that bears his name [Reissner, 1944]. These works arose out of the need for a more realistic modelling of aircraft structures, e.g. wings. Reissner solved the aforementioned problems in an elegant way making use of the variational theorem of Menabrea and Castigliano (here designated with $\Pi_{M,C}$), a functional of the six components of the stress tensor $\boldsymbol{\sigma}$. In the light of these experiences and knowledge of the successful use of the variational theorem of Dirichlet and Green $\Pi_{D,G}$ by Kirchhoff in the creation of his plate theory, Reissner once again investigated shear lag in box beams, but this time not with the help of the functional of Menabrea and Castigliano $\Pi_{M,C}$, but instead with that of Dirichlet and Green $\Pi_{D,G}$ [Reissner, 1946]. Looking back, he writes: "Having used the variational theorem for stresses and the variational theorem for displacements, I began to wonder whether this had to be an either-or proposition. The first consequence was a generalization of the variational theorem for stresses, to make this theorem applicable to linear problems of simple harmonic motion [at this point Reissner refers to his brief bulletin [Reissner, 1948] – the author]. The possibility of this generalization depended on the simultaneous introduction of stress and displacement variations, which had to be interdependent to retain the dynamic constraint stipulations. With the concept of independent stress and displacement variations, a natural next step was to think about the possibility of a variational theorem with independent stress and displacement variations" [Reissner, 1996, p. 443]. Reissner generalised his variational theorem for elastic continua with large deflections in 1953 [Reissner, 1953].

In 1949 Reissner was promoted to professor of applied mechanics at MIT, where he had been working since his emigration from Hitler's Germany in 1937, initially as a research assistant in aircraft construction and later at the faculty of mathematics; here, too, he studied in great detail the modelling of the structures important to aircraft construction. At the same time, Prof. Pian was carrying out research into the structural mechanics behaviour of wings at the neighbouring faculty of aircraft design and linked the discretisation with matrix theory (see section 10.4.2.2). Both worked on the same scientific object from different perspectives unaware that these different research directions would give modern structural mechanics an almighty boost 20 years later and from the 1970s onwards would lead to the development of hybrid finite elements based on hybrid variational theorems.

## 10.4.5.3 The second stage of the synthesis: the variational theorem of Fraeijs de Veubeke, Hu and Washizu

Rigid-body mechanics describes the behaviour of elastic continua by means of
- the stress tensor $\sigma$,
- the displacement vector $u$, and
- the strain tensor $\varepsilon$.

Of these three state variables, the stress tensor $\sigma$ and the displacement vector $u$ appear in the variational theorems of
- Menabrea and Castigliano $\Pi_{M,C}(\sigma)$,
- Dirichlet and Green $\Pi_{D,G}(u)$, and
- Hellinger, Prange and Reissner $\Pi_{H,P,R}(\sigma,u)$.

This was the state of knowledge in 1950 with respect to variational theorems. Three scientists quickly posed the question of whether a general variational theorem could be specified in which the strain state could be entered into the functional as well as the stress and displacement states. The answer was found by B. M. Fraeijs de Veubeke (Belgium) [Fraeijs de Veubeke, 1951], H.-C. Hu (China) [Hu, 1955] and K. Washizu (Japan) [Washizu, 1955] working separately. Reissner describes how Washizu visited him during his period of research at MIT between 1953 and 1955 and explained his variational theorem: "... my friend Washizu ... came one day to my office to say that he had a variational theorem with independent variations not only of stresses and displacements, but also of strains, in such a way that not only equilibrium and stress-strain, but also strain displacement relations came out as Euler equations. I first objected that since only stresses and displacements would be encountered in the boundary conditions of problems, it was not natural to consider strain displacement relations in ways other than as defining relations. I was, however, soon persuaded that the 'three-field' theorem which Washizu, and independently Hu, had proposed was a valuable advance which I wished I had thought of myself" [Reissner, 1996, p. 434]. This general variational theorem is therefore named after Hu and Washizu in the literature; it will be symbolised here first of all by $\Pi_{H,W}(\sigma,u,\varepsilon)$.

Baudouin M. Fraeijs de Veubeke had already developed a variational theorem with four varying field or state variables in 1951 in a French publication [Fraeijs de Veubeke, 1951]; he called it "the general variational principle" [Fraeijs de Veubeke, 1965, p. 148]. In his theorem, not only the stresses, displacements and strains are varied independently of each other, but also the surface tractions $t$ ($t_i$ in index notation). Washizu's famous monograph also includes this functional [Washizu, 1968, pp. 31–34], which Felippa named after Fraeijs de Veubeke, Washizu and Hu [Felippa, 2000, 2002] and is here designated with $\Pi_{F,H,W}(\sigma,u,\varepsilon,t)$; $t_i = \sigma_{ij} \cdot n_j$ ($n_j$ = normal vector of the surface, $\sigma_{ij}$ = stress tensor) in the variational theorem of Hu and Washizu $\Pi_{H,W}(\sigma,u,\varepsilon)$, which, accordingly, should also be named after Fraeijs de Veubeke, i.e. here $\Pi_{F,H,W}(\sigma,u,\varepsilon)$.

Fig. 10-57a shows the variation process with the general variational theorem of Fraeijs de Veubeke [Fraeijs de Veubeke, 1965, p. 149], Fig. 10-57b the variation process with the variational theorem of Hellinger,

Prange and Reissner [Fraeijs de Veubeke, 1965, p. 149], and Fig. 10-57c another variational theorem introduced by Fraeijs de Veubeke in which the stresses $\sigma$ and the strains $\varepsilon$ are varied [Fraeijs de Veubeke, 1965, p. 151].

In his development of the general variational theorem for three-dimensional elastic continua, Fraeijs de Veubeke follows the calculus of variations method specified back in 1929 by Kurt Otto Friedrichs (1901 – 82) using the example of one-dimensional elastic continua ([Fraeijs de Veubeke, 1951, 1964] see also [Fraeijs de Veubeke, 1974, p. 783]). Friedrichs writes: "In the numerical solution of variation problems using the Ritz method [see [Ritz, 1909] – the author], it is important to estimate the quality of the approximation of the minimal value. E. Trefftz has specified a method – for Dirichlet's problem and related tasks – for approximating the solution to variation problems in such a way that in doing so the minimum value is approached from below [see [Trefftz, 1926] – the author]. By using both methods it is therefore possible to close in on the minimal value from both sides. The same goal is achieved in the following in a different way and is more general than that of Trefftz. One can, very generally, allocate a maximum problem to a minimum problem where the maximum value is originally equal to the minimum value. The underlying principle is essentially a Legendre transformation. Such a transformation results in, for example, a problem for the conjugated potential function from the variation problem for the solution of the potential equation with given boundary values. Using a Legendre transformation, in elastic theory one can also allocate the principle of 'least deformation work' named after Castigliano, in which variation takes place according to the stresses [i. e. the principle of virtual forces – the author], to the principle of 'virtual displacements'" [Friedrichs, 1929, p. 13]. It is interesting that in the last sentence Friedrichs refers to the principles corresponding to the two variational theorems $\Pi_{D,G}(\boldsymbol{u})$ and $\Pi_{M,C}(\boldsymbol{\sigma})$: the principle of virtual displacements and the principle of virtual forces (see also section 11.2.11). Friedrich's use of the Legendre transformation in the context of his analysis of the two variational theorems $\Pi_{D,G}(\boldsymbol{u})$ and $\Pi_{M,C}(\boldsymbol{\sigma})$ played the decisive role for Fraeijs de Veubeke.

The notes to a lecture on finite methods in design calculations (winter semester 1983/84) by Klaus Knothe at the Aerospace Department of Berlin TU [Knothe, 1983] contain a systematic and very appealing presentation of the seven variational theorems. Three years prior to that, this author was lucky enough to be able to attend this annual lecture, which

**FIGURE 10-57**
a) General variational theorem $\Pi_{H,W}(\sigma,\boldsymbol{u},\varepsilon)$ or $\Pi_{F,H,W}(\sigma,\boldsymbol{u},\varepsilon)$, b) the variational theorem of Hellinger, Prange and Reissner $\Pi_{H,P,R}(\sigma,\boldsymbol{u})$, and c) another variational theorem $\Pi_F(\sigma,\varepsilon)$ introduced by Fraeijs de Veubeke

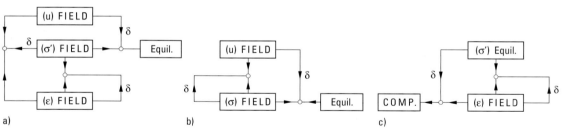

even today can be regarded as a model of research teaching at the university and an example of the successful synthesis of inductive and deductive procedures. Knothe's lecture notes also contain a diagram in which the relationships between the seven variational theorems – with the exception of $\Pi_{F,H,W}(\sigma,u,\varepsilon,t)$ – are brilliantly illustrated (Fig. 10-58).

Knothe highlights the seven variational theorems by means of graphic elements (see Fig. 10-58):
- The vertical shading represents the variation of the field or state variable $\varepsilon$ (strains): $\Pi(\varepsilon)$
- The diagonal shading from top left to bottom right represents the field or state variable $u$ (displacements), i.e. the variational theorem of Dirichlet and Green: $\Pi_{D,G}(u) \equiv \Pi(u)$
- The diagonal shading from bottom left to top right represents the field or state variable $\sigma$ (stresses), i.e. the variational theorem of Menabrea and Castigliano: $\Pi_{M,C}(\sigma) \equiv \Pi(\sigma)$
- The hybrid variational theorems lie in areas where the shading crosses: the variational theorem of Hellinger, Prange and Reissner $\Pi_{H,P,R}(\sigma,u) \equiv \Pi(\sigma,u)$ where the diagonal shading crosses, that of Fraeijs de Veubeke, Washizu and Hu $\Pi_{F,H,W}(\sigma,u,\varepsilon) \equiv \Pi(\sigma,u,\varepsilon)$ where all types of shading cross, that of Fraeijs de Veubeke's additional variational theorem $\Pi_F(\sigma,\varepsilon) \equiv \Pi(\sigma,\varepsilon)$ where the shading from bottom

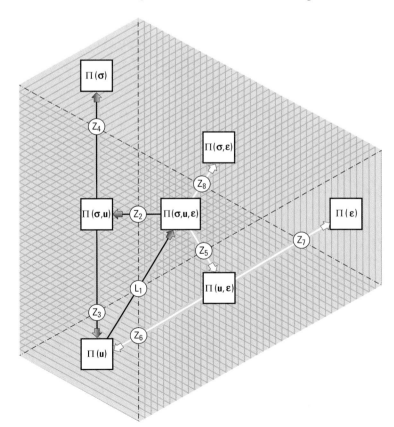

**FIGURE 10-58**
The seven variational theorems in context (modified drawing after [Knothe, 1983, p. 3-31])

left to top right crosses the vertical shading, and the variational theorem $\Pi(u,\varepsilon)$ where the shading from bottom right to top left crosses the vertical shading.

The seven variational theorems can be transformed into one another by the operations $L_i$ and $Z_j$. Symbol $L_i$ stands for the inclusion of secondary conditions and the primary boundary conditions in the variational theorem by means of Lagrange factors. Symbol $Z_j$ stands for the inclusion of the Euler differential equations and the additional boundary conditions in the corresponding variational theorem in the sense of restraint conditions, whereby these conditions become internal secondary conditions and primary boundary conditions.

Knothe explained the seven variational conditions with his graphic representation (Fig. 10-58) using the simplest case of the axially loaded elastic extensible bar (see Fig. 10-54a). How did Karl Culmann express it in the introduction to his graphical statics? "Drawing is the language of the engineer" [Culmann, 1864/1866]. There you have it!

**Variational formulation of FEM**

### 10.4.5.4

In 1965 Fraeijs de Veubeke returned to his general variational theorem $\Pi_{F,H,W}(\sigma,u,\varepsilon,t)$ or $\Pi_{F,H,W}(\sigma,u,\varepsilon)$ that he had presented in 1951, specified it for the simpler variational theorems $\Pi_F(\sigma,\varepsilon)$, $\Pi_{H,P,R}(\sigma,u)$, $\Pi_{M,C}(\sigma)$ and $\Pi_{D,G}(u)$, and gave FEM variational formulations for the latter three theorems [Fraeijs de Veubeke, 1965]. The *International Journal for Numerical Methods in Engineering* included this paper in its *Classic Reprints Series* [Fraeijs de Veubeke, 2001, pp. 290–342] together with a commentary by no less a person than Zienkiewicz [Zienkiewicz, 2001, pp. 287–289]. According to Zienkiewicz, there are five scientific findings of Fraeijs de Veubeke [Fraeijs de Veubeke, 1965 & 2001, pp. 290–342] that gave FEM and its variational formulation decisive impulses:

"1. The introduction of the so-called equilibrating elements based on [the] complementary energy principle [i.e. $\Pi_{M,C}(\sigma)$ – the author].

2. The realization that the standard potential energy formulation [i.e. variational formulation by means of $\Pi_{D,G}(u)$ – the author] and the complementary energy formulation [i.e. variational formulation by means of $\Pi_{M,C}(\sigma)$ – the author] provide bounds to the energy of the system, the first from the lower and the second from the upper limit. The bounding thus is an extremely useful practical measure for assessing the accuracy of any particular solution.

3. The chapter introduces for the first time a quadratic element (here the two-dimensional six-node triangle) which later was to become extremely popular.

4. It introduces mixed formulations which are for the first time discussed in detail.

5. Of particular importance is the introduction of the *limitation principle* for mixed formulation which failed to draw the attention of many later investigators and was only incorporated into texts in the late 1980s" [Zienkiewicz, 2001, p. 288].

Fraeijs de Veubeke's outstanding achievement is that he generalised the unification of calculus of variations and matrix theory, revealed the mathematical basis of the variational theorems and presented this in the form of the variational formulation of FEM.

Only the plane triangular element with nodes 1, 2', 3, 1', 2 und 3' will be investigated here for which Fraeijs de Veubeke derived quadratic approximation functions for displacements of type

$$W = A + B \cdot x + C \cdot y + D \cdot x^2 + E \cdot x \cdot y + F \cdot y^2 \qquad (10\text{-}61)$$

in the $x$-$y$ system of coordinates (Fig. 10-59). For the approximation function $W_{3'}$ in Fig. 10-59a, we get the constants $B$, $C$, $D$, $E$ and $F$ from the condition that $W_{3'}$ at points 1, 2', 3, 1' and 2 must vanish; constant $A$ can be calculated from the condition that $W_{3'}$ at point 3' has a value of 1. Through cyclic permutation of the node numbers, the approximation functions $W_{1'}$ and $W_{2'}$ can be written immediately.

The approximation functions of the apexes $W_3$ (Fig. 10-59b), $W_1$ and $W_2$ are formed similarly. In this way, 12 approximation functions for the displacements can be set up for 12 nodal degrees of freedom, which approximate the displacement field of the triangular element

$$u = \sum_{i=1}^{3} u_i \cdot W_i + \sum_{j=1'}^{3'} u_j \cdot W_j$$

$$v = \sum_{i=1}^{3} v_i \cdot W_i + \sum_{j=1'}^{3'} v_j \cdot W_j \qquad (10\text{-}62)$$

The element stiffness matrix is set up in a similar way to the approach of Turner, Clough, Martin and Topp (see section 10.4.4.1), the difference being that Fraeijs de Veubeke derives the matrix equations consequently from the variational theorem of Dirichlet and Green. Clough had managed this previously [Clough, 1960], but only for linear approximation functions for displacements, which had already been used in 1956 by Turner, Clough, Martin and Topp (see section 10.4.4.1). The further expansion of the direct stiffness method based on the variational theorem of Dirichlet and Green to become the standard method of FEM was car-

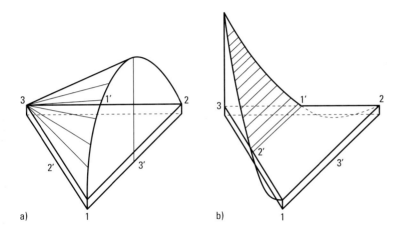

**FIGURE 10-59**
Quadratic approximation functions for a triangular element after Fraeijs de Veubeke: a) local displacement at point 3' only, b) local displacement at point 3 only [Fraeijs de Veubeke, 1965, pp. 162–163]

ried out, in particular, by R. J. Melosh [Melosh, 1963], who was the first person to use FEM for analysing elastic plates while assuming rectangular elements; two years prior to that, he had derived a stiffness matrix for thin elastic plates [Melosh, 1961]. In 1964 Fraeijs de Veubeke – following on from Friedrichs – managed to specify the lower limit of $\Pi_{D,G}(u)$ and the upper limit of $\Pi_{M,C}(\sigma)$ within the framework of the variational formulation of FEM [Fraeijs de Veubeke, 1964]. In that same year, T. H. H. Pian published a paper in the journal of the American Institute of Aeronautics & Astronautics (AIAA) in which he modified the variational theorem of Menabrea and Castigliano for FEM [Pian, 1964]. Pian later disclosed that he had developed the paper against the background of the variational theorem of Hellinger, Prange and Reissner, which he integrated into FEM in 1963 in the final lecture of the MIT course of study on variational and matrix methods in structural mechanics [Pian, 2000, pp. 420–422].

Whereas $\Pi_{D,G}(u)$-based FEM introduced the compatibility of the displacement variables as a secondary condition for all nodes of the structural mechanics model through local approximation functions for displacement variables and in the end leads to equilibrium expressions, $\Pi_{M,C}(\sigma)$-based FEM takes the opposite route: the secondary condition here is the equilibrium of the force variables for all nodes of the structural mechanics model through local approximation functions for force variables, so that the variational theorem of Menabrea and Castigliano in the end leads to compatibility expressions. Pian's great achievement consists of creating a method "for which compatible displacement functions are assumed along the interelement boundary in addition to the assumed equilibrating stress field in each element" [Pian, 1979, p. 891]. The term "hybrid model" was introduced for such elements in 1967 in an MIT course of study on variational and matrix methods in structural mechanics [Pian, 1979, p. 891]. The counterpart to Pian's "hybrid stress model" is the "hybrid displacement model" created by R. E. Jones [Jones, 1964] and later extended by Y. Yamamoto [Yamamoto, 1966]. Jones starts with the variational theorem of Dirichlet and Green and adds an equilibrium expression for the element boundaries to the compatibility of the displacements for every element. The fourth form of the variational formulation of FEM is based on the variational theorem of Hellinger, Prange and Reissner and was designated "mixed method" by T. H. H. Pian and P. Tong: "The fourth method can be derived from Reissner's variational theorem based on an assumed displacement field which is continuous over the entire solid and assumed stress fields for individual elements. This method is called a mixed method" [Pian & Tong, 1969, p. 4]. This path was first explored by L. R. Herrmann in the analysis of incompressible and virtually incompressible elastic bodies [Herrmann, 1965/1] plus elastic plates [Herrmann, 1965/2], and also by C. Prato for shell analysis [Prato, 1968]. The state of development in variational formulation at the end of the 1960s has been summarised by Pian and Tong in their paper *Basis of finite element methods for solid continua* [Pian & Tong, 1969], which was used as the introductory paper to the *International Journal for Numerical Methods in Engineering*, the first journal in which computational mechanics would find a home. This is where J. T. Oden published his pioneering essay *A general method of finite elements* [Oden, 1969] in which he placed FEM on a broader mathematical foundation and hence opened up new perspectives for FEM.

### 10.4.6 Computational mechanics – a broad field

Based on their review of the first 25 years of publication of the *International Journal for Numerical Methods in Engineering* (Fig. 10-60), the editors O. C. Zienkiewicz, R. H. Gallagher and R. W. Lewis predicted the following developments:

"Clearly, fluid mechanics and similar applications will prove most demanding, with boundary layer modelling and incorporation of turbulence being much on the menu. Such dramatic experiments as direct turbulence modelling now applied to very local problems may become practicable in realistic applications. We mentioned that typically a million variables would be required to model inviscid flow around an aircraft today. With inclusion of viscous boundary layers, at least an order of magnitude increase in size of problem will be required. The problems of such modelling are presenting challenges to existing hardware and software today and will certainly have to be addressed.

"To solve the problems of magnitude, further theoretical development will be necessary. Some areas of research, today a province of theoreticians, will doubtless become of practical value. Here we could list:
– *domain decomposition* – a subject in which much interest is already shown, allowing parallel and frequently different methodologies to be linked
– *stochastic computation* – with a probabilistic distribution of loads and material properties (including fuzzy sets)
– *wavelets* – as an interesting possible way to approximation of functions, generated recently in mathematical literature and not yet put to practical application
– *chaos and fractals* – about which much is talked in diverse circles. "Applications to manufacturing processes and reliability assessment are increasing. Many other keywords could be mentioned" [Zienkiewicz et al., 1994, pp. 2155–2156].

In April 1981 R. H. Gallagher, J. T. Oden and O. C. Zienkiewicz convened a meeting with 12 participants at the Georgia Institute of Technology in Atlanta. The goal of this meeting was "to form a group of International Centers of Computational Mechanics". This led to the "formal announcement", the establishment of a "Founding Council" and a "Constitution" for the International Association of Computational Mechanics (IACM), the founding of which was finally completed in 1984. In the constitution, the object of computational mechanics is described as follows: "For the purposes of the Association we define the subject of Computational Mechanics as the development and application of numerical methods and digital computers to the solution of problems posed by

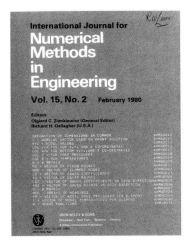

FIGURE 10-60
Cover of the *International Journal for Numerical Methods in Engineering*

Engineering and Applied Science with the objectives of understanding and harnessing the resources of nature.

"While Computational Solid Mechanics (CSM) and Computational Fluid Dynamics (CFD) are at the core of our activity, such subjects as Thermodynamics, Electro Magnetism, Rigid Body Mechanics, Control Systems and some aspects of Particle Physics fall naturally within the scope of the definition. Indeed the provision of a common forum for discussion, education and research information transfer between the diverse disciplines represented is the main 'raison d'etre' of the Association." Of course, these words are also to be found on the homepage of the IACM.

**A humorous plea**     **10.4.7**

The comical side of the history of the development of FEM is revealed by Felippa using the example of diverse stiffness matrices for modelling Timoshenko's theory of the shear-flexible beam [Timoshenko, 1921] – the Timoshenko beam: "So many beam models, so little time. Can they be wrapped into a single package? Yes, by using templates" [Felippa, 2005, p. 18]. The interested reader might find the answer in Felippa's next sentences, but will first let him tell an amusing and entertaining story of the evolution of the species à la stiffness matrices for the Timoshenko beam in a light-hearted way. But his story is not only full of humour, it is essentially a remarkable plea against the "destruction of the past" [Hobsbawm, 1999, p. 17]. This book wishes to agree with that plea.

## JOURNALS FROM ERNST & SOHN

Zeitschrift **Beton und Stahlbetonbau**
(Concrete and Reinforced Concrete Strucutres)

Zeitschrift **Geomechanik und Tunnelbau**
(Geomechanics and Tunnelling)

Zeitschrift **Stahlbau**
(Structural Steelwork)

Zeitschrift **Bauphysik**
(Building Physics)

Zeitschrift **Bautechnik**
(Structural Engineering)

Zeitschrift **Mauerwerk**
(Masonry)

**Free sample copy under www.ernst-und-sohn.de/zeitschriften**

\* Prices are valid in Germany, exclusively, and subject to alterations.
Prices incl. VAT. Books excl. shipping. Journals incl. shipping.

008518026_my

**Ernst & Sohn**
Verlag für Architektur und
technische Wissenschaften GmbH & Co. KG

www.ernst-und-sohn.de

For order and customer service:

**Verlag Wiley-VCH**
Boschstraße 12
69469 Weinheim
Deutschland

Telefon: +49(0) 6201 / 606-400
Telefax: +49(0) 6201 / 606-184
E-Mail: service@wiley-vch.de

# CHAPTER 11

# Twelve scientific controversies in mechanics and theory of structures

Science has always been accompanied by controversies which have had a profound effect on its historico-logical evolution. Thomas S. Kuhn thus attributed controversies an important role in his concept of the scientific revolution (1962). Since the author first became interested in the history of structural theory in the late 1970s, he has therefore followed controversies in the history of the natural and engineering sciences with great interest. When Prof. Kai-Uwe Bletzinger invited the author to give a lecture on the history of theory of structures within the scope of the compulsory curriculum for students at the faculty of civil engineering and surveying at Munich TU, the author was very pleased to accept. Prof. Bletzinger and the author agreed on a presentation about scientific controversies in mechanics and theory of structures, which the author gave on 14 December 2004. The aim of this lecture was to introduce civil engineering students to the history of structural theory using the example of the scientific controversy in a "historical longitudinal section", as it were.

Taking as examples famous but also relatively unknown disputes in mechanics and theory of structures, this chapter will trace the history of developments in theory of structures. During this journey it will become clear that the history of structural analysis, like other authentic stories from the past, has a subjective side in which the facets of the human personality sometimes appear in fascinating but at other times disappointing lights.

## 11.1 The scientific controversy

The scientific controversy is attributed a different standing in the natural, engineering and social sciences and the humanities. Whereas today it is hardly given a second glance in the natural and engineering sciences, it is an important factor in the social sciences and the humanities. The reason for this is that the cognition process in the latter two branches of science genuinely takes place in discursive forms, whereas natural and engineering scientists would appear to have little need of discussions with one another because a clear distinction is still made between cognition subject and cognition object. Nevertheless, there are also engineering science controversies in which the cognition process moves within the field of tension of the omnipresent antagonism of the "unsociable sociability" (Kant) of mankind. Scientific controversies basically come in three different forms:

I. Disputes about the fundamentals of a discipline (foundation)
II. Disputes about disciplinary development (competition)
III. Priority disputes.

As will be shown below, scientific controversy types I, II and III follow the order of the historico-logical evolution of the disciplines of mechanics and theory of structures.

**FIGURE 11-1**
Galilei Galileo [Szabó, 1996, p. 51]

## 11.2 Twelve disputes

Since the emergence of the mechanics of the modern age in the 17th century, the scientific controversy has formed a significant element in the development of the historico-logical evolution of the disciplines of mechanics and theory of structures. In his two dialogue-based main works, Galileo (1564–1642) even uses the fictitious scientific controversy as a formal means of presentation.

### 11.2.1 Galileo's *Dialogo*

In his *Dialogo dei due massimi sistemi del mondo* (Dialogue Concerning the Two Chief World Systems) of 1632, Galileo (Fig. 11-1) allows his three characters, Salviati, Sagredo and Simplicio, to discuss the Ptolemaic planetary system of the ancients and the Copernican planetary system.

In the guise of Salviati (Fig. 11-2/left), Galileo explains to Simplicio (Fig. 11-2/centre), the follower of Aristotelian natural philosophy, the nature of the Copernican planetary system. Galileo checks the latter in the light of the phenomena observed by astronomers (phases of Venus and specific positions of the planets relative to the Earth). Furthermore, Galileo criticises the laws of motion of Aristotle and his epigones and formulates the concept of uniform relative motion.

Sales of the *Dialogo* were banned by church decree soon after its publication and on 1 October 1632 Galileo had to answer to the Inquisition.

On 22 June 1633 Galileo renounced his alleged mistake and resolved in a further oath "in future never to express anything verbally or in writing that could place me under similar suspicions" (cited in [Szabó, 1996, p. 549]). Galileo would break this promise with the publication of his *Discorsi* in 1638.

**Galileo's *Discorsi***     **11.2.2**

Galileo's *Discorsi e Dimostrazioni Matematiche, intorno a due nuove scienze* (Dialogue Concerning Two New Sciences) (see Fig. 5-5) was published by Elsevier [Galileo, 1638] in the Protestant town of Leyden, where Galileo in the discussion between Salviati (who speaks for Galileo), Sagredo (the educated layman) and Simplicio (the representative of the outdated natural philosophy of Aristotle) describes the two new sciences, which are non-Aristotelian dynamics and strength of materials. The crux of Galileo's strength of materials is beam theory (see Fig. 5-7), in the form of rules of proportion founded on mechanics (see section 5.3), which was first solved with Navier's practical bending theory of 1826 (see section 5.6).

Whereas the aim of *Dialogo* is to ensure that the heliocentric system prevails over the Ptolemaic, in his *Discorsi* Galileo undermines the foun-

**FIGURE 11-2**
Cover of Galileo's *Dialogo*
[Szabó, 1996, p. 549]

dations of Aristotelian natural philosophy and at the same time creates two new sciences. Both cases can be classified as type I scientific controversies because they concern the founding of scientific disciplines: physical astronomy, dynamics and strength of materials.

### 11.2.3 The philosophical dispute about the true measure of force

The year 1644 saw René Descartes (1596–1650) formulate the hypothesis that the sum of mass ($m$) times velocity ($v$) is constant throughout the universe, i. e.

$$\sum F = \sum (m \cdot v) = const. \tag{11-1}$$

Leibniz' (1646–1716) (Fig. 11-3) article in the March 1686 issue of the journal *Acta Eruditorum* concerning the "brief description of a remarkable error by Cartesius [= Descartes – the author] and others surrounding a natural law according to which they claim that God always preserves the same amount of motion, and how they misuse this in mechanics" (cited in [Szabó, 1996, p. 62]) initiated a dispute about the true measure of force which raged for half a century.

The dispute surrounding the concept of force (in the meaning of a conservation variable, designated by $F$ here) led to the scientific community being split into two camps: the Cartesians and the Leibnizians. The following are just some of the personalities who became involved one by one:

- Leibniz (1686): statics: $F = m \cdot v = const.$; kinetics: $F = m \cdot v^2 = const.$
- Abbé de Catelan (1686)
- Leibniz (1687)
- Abbé de Catelan (1687)
- Papin (1689)
- Leibniz (1690)
- Daniel Bernoulli (1726):

$$(v^2/2) = \int p \cdot dx \leftrightarrow m \cdot (v^2/2) - m \cdot (v^2/2)_o = \int F \cdot ds$$

FIGURE 11-3
Gottfried Wilhelm Leibniz
[Szabó, 1996, p. 61]

In 1743 D'Alembert (1717–83) (Fig. 11-4) described the dispute surrounding the true measure of force as a "highly unimportant metaphysical discussion …, as an argument not worthy of the involvement of philosophers" (cited in [Szabó, 1996, p. 72]).

Kant (1724–1804) (Fig. 11-5) was not able to contribute anything substantial to the dispute about the true measure of force in his first work (1746) *Gedanken von der wahren Schätzung der lebendigen Kräfte* (*Thoughts on the True Estimation of Living Forces*) [Kant, 1746]. It was not until the 1840s that Joule, Mayer and Helmholtz were able to work out the fundamentals of scientific energetics with the formulation of the conservation of energy law. Nonetheless, the dispute about the true measure of force circled around the philosophical climax of the conservation principle in the sense of the Enlightenment and can therefore be regarded as a type II controversy because it was a competition between hypotheses.

**FIGURE 11-4**
Jean le Rond d'Alembert
[Szabó, 1996, p. 233]

**FIGURE 11-5**
Immanuel Kant
[Szabó, 1996, p. 74]

**FIGURE 11-6**
Pierre-Louis Moreau de
Maupertuis [Szabó, 1996, p. 90]

**FIGURE 11-7**
Johann Samuel König
[Szabó, 1996, p. 95]

## The dispute about the principle of least action

### 11.2.4

In 1740 the Prussian king Friedrich II invited the scientist Maupertuis (1698–1759) (Fig. 11-6) and the philosopher Voltaire to Moyland Palace at Kleve and took Maupertuis with him to Berlin in order to discuss plans for setting up an academy. But Maupertuis was not able to found the academy until 1745, after the end of the Second Silesian War. By 1746 he had declared in the *Mémoires* of the academy a principle valid for bodies in motion and at rest: "If some change takes place in Nature, then the quantity of action necessary for this change is the least possible" (cited in [Szabó, 1996, p. 93]). The "principle of least action" of Maupertuis, who remained president of the academy until 1750, did not go unchallenged.

Johann Samuel König (1712–57) (Fig. 11-7) did not agree with Maupertuis' "principle of least action" and handed Maupertuis a manuscript for publication in the *Mémoires*. In the manuscript, König quotes a letter from Leibniz to Jakob Hermann, dated 16 October 1708, in which Leibniz expresses the principle more precisely, in the meaning of the calculus of variations. The vain Newtonian advocate Maupertuis was infuriated because König thus attributed priority to Leibniz. Maupertuis demanded to see the original of the letter, but all efforts to locate it were unsuccessful. The academy thereupon declared the letter to be a forgery. König defended himself publicly and gained unprecedented support in 1752 through the anonymous satire on Maupertuis entitled *Doktor Akakia*. The author of this publication peppered with jokes and ridicule was none other than Voltaire (see, for example, [Voltaire, 1927])!

In 1696 Johann Bernoulli, writing in the journal *Acta Eruditorum*, asked for help in solving a new problem which was to awaken the interest of mathematicians in the 18th century and that of structural engineers in the first half of the 20th century. The nature of the (specific) variational problem of Johann Bernoulli was as follows (Fig. 11-8):

The object of the exercise is to find the curve that enables a ball starting at *A* to reach the ground in the shortest time. The problem leads to

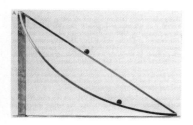

**FIGURE 11-8**
Johann Bernoulli's variational problem
[Szabó, 1996, p. 114]

the extremum of a certain integral, i.e. from the family of possible – quasi varied – curves to find the one whose integral expression forms an extremum. The ball starting at $A$ reaches the ground first via the curve in the model (brachistochrone).

The world had to wait until 1744 for Leonhard Euler to set up the calculus of variations in his work *Methodus inveniendi lineas curvas maximi minimive proprietate gaudentes* ... (method for finding curved lines enjoying properties of maximum or minimum ...) [Euler, 1744].

Consequently, the dispute about the principle of least action is a priority dispute (type III) and at the same time a dispute about disciplinary developments (type II).

## 11.2.5 The dome of St. Peter's in the dispute between theorists and practitioners

In their report dating from 1742, the three mathematicians Jacquier (1711–88), Boscovich (1711–87) and Le Seur (1703–70) calculated the horizontal thrust of the dome of St. Peter's in Rome according to the general work theorem (see section 2.2.3) and the hoop tension resisting the horizontal thrust [Le Seur et al., 1742]. They recommended installing a second tension ring. Fig. 11-9 illustrates the copperplate engraving from the aforementioned report showing the cracks in the dome and the

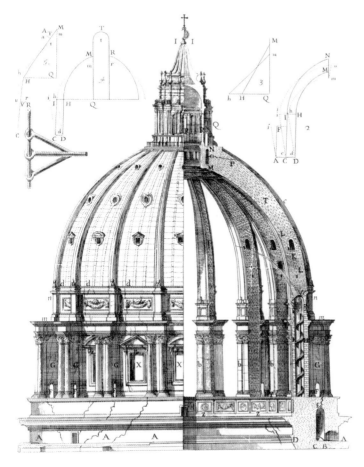

**FIGURE 11-9**
The dome of St. Peter's in Rome
[Le Seur et al., 1742]

kinematic model for calculating the horizontal thrust with the help of the general work theorem. Wilfried Wapenhans (1952–2006) and Jens Richter have translated the Latin text into German [Wapenhans & Richter, 2001] and in doing so have traced "the world's first structural analysis" [Wapenhans & Richter, 2001] in all its details and have evaluated this from the standpoint of modern theory of structures.

The three illustrations in Fig. 11-10 show caricatures by the "Pittore della Camera Apostolica" (artist to the Apostolic Camera), Pier Leone Ghezzi (1674–1755), who recorded the Roman society of that period pictorially in thousands of line drawings and caricatures. The caricatures are accompanied by remarks from Ghezzi, which makes them interesting historical documents. For example, the remark concerning Fig. 11-10b reads as follows: "Father Ruggiero Giuseppe Boscovich from the Society of Jesus is the one without the hat, and the one with the hat is one of his companions, who calls himself Father Melonazi. Father Boscovich is one of the mathematics fathers who together with Fathers Tomaso Le Seur and Francesco Jacquier from the Minorite Order compiled and published the treatise concerning a number of difficulties regarding the damages to and repairs of the dome of St. Peter's. All three are capable mathematicians and said publication appeared on 20 January 1743. I, Cav. Ghezzi, have signed this paper on 15 July 1744 as a reminder …" [Straub, 1960, p. 366]. In the remark to Fig. 11-10a, Ghezzi praises "that very beautiful treatise on St. Peter's dome" [Straub, 1960, p. 365].

In 1748 Giovanni Poleni (1683–1761) criticised the report of the three mathematicians as follows: "If the dome of St. Peter's could be conceived, drawn and built without mathematics and, above all, without the mechanics so esteemed in our days, then it should have been possible to restore it without the help of mathematicians and mathematics … Michelangelo was not trained in mathematics but was nevertheless in a position to build the dome" (cited in [Straub, 1992, p. 159]).

Poleni sliced the dome into 50 hemispherical lunes (orange slices) with identical geometry bounded by the meridians and divided these in turn into segments, the (varying) weight of each being represented by a bead;

**FIGURE 11-10**
a) Francesco Jacquier, b) Ruggero Boscovich and c) Tommaso Le Seur [Straub, 1960, p. 365]

a)     b)     c)

the inversion of the catenary thus formed is the line of thrust, which for reasons of stability must always lie within the profile of the arch. A uniformly loaded inverted catenary (= line of thrust) will intersect the arch intrados twice (see Fig. 4-33).

In a dispute about the structural modelling of St. Peter's dome, the kinematic model developed directly from the theory was competing with the empirically founded model of Poleni. The conflict between the kinematic and the geometric views of statics, which raged fiercely at various times (see section 2.27), is evident here. So the competition between the masonry arch models represents a dispute about disciplinary developments in theory of structures (type II scientific controversy).

### 11.2.6  Discontinuum or continuum?

Navier (1785–1836) understood the linear-elastic solid body as a discontinuum. For example, in 1821 he considered molecules as material points and expressed the attraction and repulsion forces as functions of the displacements of the molecules. His expressions contain triple sums, which he replaces by integrations. He obtains one constant for the isotropic case: the elastic modulus $E$. Using the principle of virtual displacements (see section 2.2.2) according to Lagrange, Navier derives the displacement differential equations obtained beforehand.

In 1822 Cauchy (1789–1857) generalised the concept of hydrostatic pressure attributed to Euler to form the concept of the stress condition, i.e. assumed an elastic continuum. Besides the elastic modulus $E$, his equations based on differential and integral calculus contain one further constant. By 1828 Cauchy had extended his theory to cover the case of crystalline bodies and assumed the molecular hypothesis. That same year saw Poisson criticise Navier's approach of replacing all summations by integrations; nevertheless, he, too, based his ideas on the molecular hypothesis.

George Green (1793–1841) founded elastic theory in 1839 on the principle of conservation of energy [Green, 1839]. Fig. 11-11 illustrates the energy concept in the centre of the constitutive, kinematic and kinetic relationships of continuum mechanics, which in the 19th century were more or less congruent with elastic theory. Green introduces the deformation (strain) energy function $\Pi(\varepsilon_{ij})$. He assumes that this function can be developed according to powers and products of the deformation components and therefore arranges them as a sum of the homogeneous functions of these variables of the first, second and higher orders. The first of these terms may not occur because the potential energy must be a true minimum when the body is undeformed; and as the deformations are all small, only the second term has to be considered. In the simplest case this is the homogeneous quadratic function

$$\Pi(\varepsilon) = \Pi = \frac{1}{2} \cdot E \cdot \varepsilon \cdot \varepsilon \qquad (11\text{-}2)$$

which for the inelastic case takes on the form

$$\Pi(\varepsilon) = \Pi = \frac{1}{2} \cdot f(\varepsilon) \cdot \varepsilon \qquad (11\text{-}3)$$

Eq. 11-3 contains a material law that no longer complies with Hooke's law:

$$f(\varepsilon) = \sigma(\varepsilon) \neq E \cdot \varepsilon \tag{11-4}$$

Using this principle, Green derives the elasticity equations, which in the general case contain 21 constants and in the simplest case just two. Proof of the existence of $\Pi(\varepsilon_{ij})$ based on the first and second fundamental laws of thermodynamics was supplied by Lord Kelvin (1824–1907) in 1855. Ten years prior to that, Stokes had asked the question of whether general elasticity is characterised by 21 (multi-constant theory, continuum) or 15 constants (rari-constant theory, discontinuum). Woldemar Voigt (1850–1919), through experiments between 1887 and 1889, was the first to prove that the multi-constant theory applies.

In the contest between the rari- and multi-constant theories, it was the foundation of general elastic theory that was at stake (type I scientific controversy).

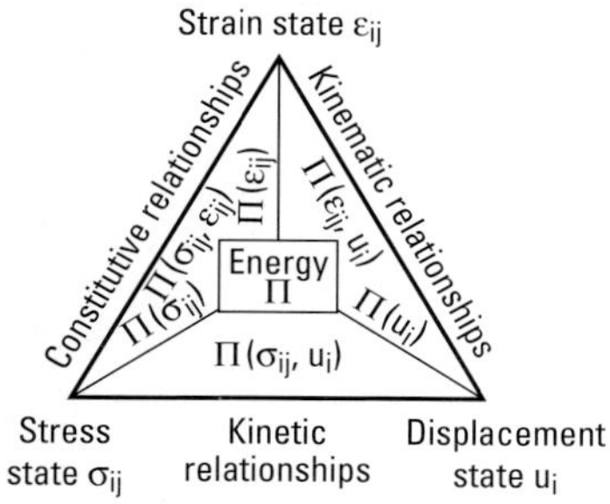

Energy can be expressed as a function of the strain state but also as a function of the stress state (Fig. 11-11). Friedrich Engesser therefore ended the dispute concerning the foundation of the theory of structures (see section 11.2.8) for the time being by introducing the deformation complementary energy [Engesser, 1889/1]

$$\Pi(\sigma) = \Pi^* = \frac{1}{2} \cdot \varphi(\sigma) \cdot \sigma \tag{11-5}$$

The deformation complementary energy (eq. 11-5) is the opposite to the deformation energy (eq. 11-2); eq. 11-5 contains a material law that no longer complies with Hooke's law:

$$\varphi(\sigma) = \varepsilon(\sigma) \neq \frac{\sigma}{E} \tag{11-6}$$

For linear-elastic material behaviour, eq. 11-5 can be simplified to

$$\Pi(\sigma) = \Pi^* = \frac{1}{2} \cdot \frac{\sigma}{E} \cdot \sigma \tag{11-7}$$

The deformation complementary energy (eq. 11-5 or eq. 11-7) corresponds to the area above the line of the stress-strain diagram complementary to the deformation energy (eq. 11-3 or eq. 11-2). This fact is shown in Fig. 6-5b for the linear-elastic case: the law-based relationship given by eq. 11-7 is nothing other than the theorem of Clapeyron (1833) (see eq. 6-22) in the version by Lamé (1852).

### 11.2.7 Graphical statics versus graphical analysis, or the defence of pure theory

Karl Culmann defended his graphical statics, based on projective geometry, against a prescriptive form of graphical statics – graphical analysis. The protagonists of the latter held the view that the further development of graphical statics could also take place without projective geometry. An early, consequential representative of graphical analysis took the stage in the shape of Johann Bauschinger.

In the preface to his *Elemente der graphischen Statik* (elements of graphical statics) (Fig. 11-12) Bauschinger writes: "I am of the opinion that the lack of use of graphical statics by engineers hitherto is mainly due to the fact that a dedicated, systematically drawn-up textbook for this new

FIGURE 11-11
The tetrahedron of energy-based continuum mechanics

science is missing ... For it is certainly the case that graphical statics is so crucial to the study of engineering sciences and the practising engineer that we hope to see it spread widely, and this will certainly be the case to some extent. Perhaps my book can help proliferate this knowledge further because it is not necessary to be familiar with the so-called newer geometry. I did not plan it that way, it happened by itself" [Bauschinger, 1871, p. III]. It was against such opinions that Culmann campaigned. In defence of the pure theory of graphical statics, he wrote in the foreword to the second edition of his *Graphische Statik* (graphical statics) (Fig. 11-13): "Statics is now being stripped of its spirit in numerous major and minor treatises and therefore made more palatable for young engineers who are insufficiently prepared for the study of engineering sciences and thanks to the freedom in studies that is now fashionable enter the highest level of study untrained; and with enormous self-confidence, which is emboldened by **those** professors who want to lecture in a most popular style: the young engineers will work through it. ... only in the **polytechnic** schools should one pursue higher ambitions ... In Prussia the name 'graphical statics' was first changed to 'graphical analysis', but otherwise this achieved very little. The engineers there were lacking in the necessary mathematical, i.e. geometrical, education hitherto. At the Building Academy in Berlin the Building School is not even separated from the Engineering School" [Culmann, 1875, pp. VI, VII, IX].

The conflict between graphical statics and graphical analysis pushed back the boundaries in the disciplinary developments in theory of structures (type II scientific controversy).

**FIGURE 11-12**
Cover of Bauschinger's *Elemente der graphischen Statik*

**11.2.8 Animosity creates two schools: Mohr versus Müller-Breslau**

The controversy between Mohr (see Fig. 6-52) and Müller-Breslau (see Fig. 6-51) became hostile towards the end of the 1880s and led, via several interim steps, to the creation of two scientific schools after 1900 (see Fig. 6-53).

Heinrich Müller-Breslau argued vehemently for Maxwell's interpretation of the general work theorem $A_a + A_i = 0$ (see Fig. 6-27), which he placed on an equal footing with the principle of Menabrea (see eq. 6-38) and the second theorem of Castigliano (see eq. 6-40).

In 1864 Maxwell considered the trussed framework as a machine with a degree of efficiency of 1: in Maxwell's trussed framework modelled as an energy-based machine, the external work $A_a$ is converted into the deformation energy $\Pi$ without losses according to the law of conservation of energy. Contrasting with this, Mohr developed a different interpretation of the general work theorem in 1874 in which only external work occurs, i.e. he assumes $A_a = 0$ (see Fig. 6-28). The difference between the kinematic machine model of the trussed framework by Mohr and the energy-based machine model by Maxwell lies in the fact that Maxwell's approach is based on internal virtual work and the conservation of energy law, and Mohr's, on the other hand, on external virtual work. Nevertheless, both the energy and kinematic machine models are founded on the principle of virtual forces (see section 2.2.4).

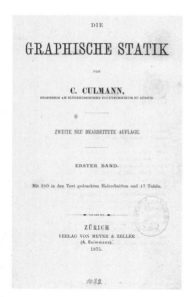

FIGURE 11-13
Cover of Culmann's *Die Graphische Statik*

The dispute between Mohr and Müller-Breslau primarily concerned

- the formulation, adaptation and generalisation of a work and energy concept for applied mechanics (see [Clapeyron, 1833; Lamé, 1852]) in the form of the theorem of Clapeyron (see eq. 6-22), and the implementation in the system of knowledge developing in theory of structures;
- the theory of linear-elastic trussed frameworks published by Maxwell in 1864 (see section 6.3.2.1);
- the trussed framework theory developed by Mohr in 1874/75 from the general work theorem (see section 6.3.2.1);
- the energy theorems summarised by Castigliano in 1879 (see section 6.4.1);
- the extension of the trussed framework theory of Maxwell and Mohr undertaken by Müller-Breslau for linear-elastic trusses (see section 6.3.2.2);
- Müller-Breslau's application of the energy theorems of Castigliano to linear-elastic trusses (see section 6.4.2).

The argument between Mohr and Müller-Breslau concerning the foundation of classical theory of structures lasting from 1883 to 1889 took place in trade journals and was continued in the appendices to their scientific writings. This scientific feud quickly spread to other areas of theory of structures such as the calculation of spatial frameworks (see section 8.1.3).

They did not managed to achieve an all-embracing, clear-cut breakdown of the principles of theory of structures into the principle of virtual displacements, principal of virtual forces, general work theorem and energy principle at this time.

## 11.2.9 The war of positions

Whereas the scientific controversy between 1883 and 1889 can in the first instance be allocated to type I (foundation) and type II (competition), after the end of the discipline-formation period of structural theory in the first decade of the 20th century it became a priority dispute between Mehrtens (see Fig. 6-41) and Weingarten (see Fig. 6-44) on the one side and Hertwig (see Fig. 6-60), Föppl (see Fig. 6-45) and Weyrauch (see Fig. 6-47) on the other (type III scientific controversy). The special feature of this scientific controversy was that the aforementioned persons were each arguing for the priority claims of Mohr or Müller-Breslau.

*Georg Christoph Mehrtens (1843–1917):*

- Leading German bridge-builder around 1900
- 1895: appointed professor of bridge-building and theory of structures at Dresden TH, Mohr's successor after 1900
- Three-volume main work *Vorlesungen über Statik der Baukonstruktionen und Festigkeitslehre* (lectures on theory of structures and strength of materials) [Mehrtens, 1903, 1904, 1905] in which he gives an outline of the history of theory of structures in one section and formulates Mohr's priority claims comprehensively as this concerns the development of theory of structures: influence lines, general work theorem, principle of virtual forces, kinematic theory of trusses, etc.

Furthermore, following Mohr's lead, he rules out the theorems of Castigliano as the foundation of classical theory of structures.

*August Hertwig (1872–1955):*
- 1902: appointed professor of theory of structures at Aachen RWTH on the recommendation of Müller-Breslau
- 1906: publication of a paper covering the development of some principles in theory of structures and the lectures on theory of structures and strength of materials by C. G. Mehrtens [Hertwig, 1906]. Priority issues concerning the kinematic theory of trusses, influence lines, substitute member method; differences in the theories of statically indeterminate systems by Maxwell, Mohr and Castigliano; criticism of Mehrtens' depiction of the theory of the elastic arch and the theory of secondary stresses;
- 1924: successor to Müller-Breslau at Berlin TH
Besides important work on theory of structures and subsoils research, Hertwig added important writings on the history of construction engineering.

*August Föppl (1854–1924):*
- 1894: Bauschinger's successor at Munich TH
- 1897: discussed the theorems of Castigliano in the *Festigkeitslehre* (strength of materials) volume of his *Vorlesungen über Technische Mechanik* (lectures in applied mechanics) [Föppl, 1897] using the example of the calculation of a system with one degree of static indeterminacy.

*Julius Weingarten (1836–1910):*
- 1874–1903: professor for mechanics at the Berlin Building Academy (Berlin TH after 1879), semi-retirement at the University of Freiburg after 1903
- "Rubbished" Föppl's Castigliano calculation in the form of a review in order to argue that the theorems of Castigliano involve a principle of elastic theory; Weingarten debated this aspect not only with Föppl, but also with Hertwig and Weyrauch (see section 6.4.3.3).

*Johann Jakob Weyrauch (1845–1917):*
- 1890: appointed professor for mechanical heat theory, aerostatics and aerodynamics plus some aspects of engineering mechanics at Stuttgart TH
- Polemic with Weingarten in which he stood up for including the thermal effect in the theorems of Castigliano (see section 6.4.3.3).

In terms of its content, the Berlin school of structural theory therefore essentially followed the geometric view of statics, whereas the Dresden school of applied mechanics was dedicated to the kinematic view of statics (see section 2.2.7).

**11.2.10 Until death do us part: Fillunger versus Terzaghi**

This scientific controversy in the new scientific discipline of soil mechanics between Karl von Terzaghi (Fig. 11-14) and Paul Fillunger (Fig. 11-15) and concerning the consolidation of deformable porous soils had a fatal end for Fillunger and his wife (see [Boer, 1990 & 2005]).

FIGURE 11-14
Karl von Terzaghi [Boer, 2005, p. 235]

FIGURE 11-15
Paul Fillunger [Boer, 2005, p. 159]

FIGURE 11-16
Cover of *Theorie der Setzung von Tonschichten* by Terzaghi and Fröhlich

*Karl von Terzaghi (1883–1963):*
- The founder of soil mechanics
- Born in Prague
- Studied mechanical engineering at Graz TH
- Afterwards, he worked for various building contractors in Austria, Russia and elsewhere; intrigued by cases of damage, he decided, while still young, to explore the frontier between geology and civil engineering.
- 1916–25: professor in Constantinople
- 1925: publication of his book *Erdbaumechanik auf bodenphysikalischer Grundlage* (soil mechanics based on the physics of soils) [Terzaghi, 1925]
- 1925–29: professor at MIT, Boston, USA
- The spring of 1936 saw him publish together with Fröhlich the book entitled *Theorie der Setzung von Tonschichten* (theory of settlement of clay strata) [Terzaghi & Fröhlich, 1936] (Fig. 11-16)
- 1929–38: professor at Vienna TH
- 1939: professor for engineering geology and soil mechanics at Havard University, Cambridge, USA

*Paul Fillunger (1883–1937):*
- Studied mechanical engineering at Vienna TH
- Engineering practice
- Teaching work at the Technological Trade & Industry Museum, Vienna
- Head of the testing laboratory at the Trade & Industry Museum
- 1923: appointed professor for applied mechanics at Vienna TH
- Publications on various branches of mechanics, especially elastic theory
- Early December 1936 saw the publication of his controversial treatise *Erdbaumechanik?* (soil mechanics?) [Fillunger, 1936] (Fig. 11-17)
- He and his wife Margarethe (née Gregorowitsch) committed suicide in the night of 6/7 March 1937.

The derivation of the – later very famous – partial differential equation of Terzaghi for the excess pore water pressure $w$ ($k$ and $a$ designate material parameters, $z$ is the downward coordinate and $t$ is the time)

$$\frac{\partial w(z,t)}{\partial t} = \left(\frac{k}{a}\right) \cdot \frac{\partial^2 w(z,t)}{\partial z^2} \qquad (11\text{-}8)$$

happened such that all considerations regarding the total body – solid body with fluid – were carried out and coupling mechanisms were introduced ad hoc. The derivation basically follows Fourier's heat conduction equation, but remained very obscure.

Fillunger criticised the consolidation theory of Terzaghi and Fröhlich in his controversial treatise *Erdbaumechanik?* with extraordinary vehemence and in solving the consolidation problem rigorously assumed a two-phase system. Fillunger's approach corresponds to the current state of the art in the theory of porous media for pure mechanical loading actions.

Although Gerhard Heinrich and Kurt Desoyer extended Fillunger's theory, the work of the Viennese professor became forgotten, which resulted in the wheel being re-invented later in the USA! Fillunger's personal attack on Terzaghi and Fröhlich resulted in disciplinary proceedings; the technical side of his offensive was checked by a scientific commission, which came to the conclusion that differential equation 11-8 is correct and is included in Fillunger's theory as a special case. The commission established that Fillunger himself had made a mistake when investigating the alleged mathematical errors of his rivals. Nevertheless, Terzaghi and Fröhlich could not refute Fillunger's theory. Fillunger gradually realised his mistakes, wrote a further pamphlet and had to reckon with being suspended. He became severely depressed and both he and his wife took their own lives in the night of 6/7 March 1937.

The conflicts between Terzaghi and Fröhlich on one side and Fillunger on the other helped to advance the development of soil mechanics as a discipline (type II scientific controversy).

### 11.2.11 "In principle, yes …": the dispute about principles

The themes of the dispute between Mohr and Müller-Breslau and their followers concerning the foundations of theory of structures became the objects of scientific disputes between 1936 and 1938, i.e. in the invention phase of structural theory (1925 – 50). In essence, this concerned the principle of virtual displacements and the variational principles of elastic theory. The discussion was opened by Pöschl in 1936 with his article on the minimal principle of elastic theory in the journal *Der Bauingenieur* [Pöschl, 1936]. In that article he comes to the conclusion that the principle of virtual displacements leads to fundamentally different conclusions "depending on whether the problem is a standard one involving elastic equilibrium or one involving buckling" (cited in [Schleusner, 1938/3]). In his reply, Domke proves that this conclusion is untenable [Domke, 1936] and refers to his paper published in 1915 [Domke, 1915] in which he clarifies the war of positions regarding the foundations of theory of structures (see section 11.2.9). Marguerre, too, writing in the journal *Zeitschrift für Angewandte Mathematik und Mechanik* in 1938, proved that Pöschl's conclusion could not be upheld, but from totally different perspectives [Marguerre, 1938/1]. One year prior to that, Kammüller had published his fundamental observations regarding the principle of virtual displacements in the journal *Beton und Eisen* [Kammüller, 1937], which provoked Arno Schleusner (Fig. 11-18) to contradict this [Schleusner, 1938/1] and led to Kammüller's reply [Kammüller, 1938/1], Schleusner's reply [Schleusner, 1938/2] and finally to Kammüller's reply to Schleusner's reply [Kammüller, 1938/2]! The debate in which Schleusner used the expression "principle of virtual forces" in the German language [Schleusner, 1938/1, p. 253] for the first time was terminated by the editors of the journal *Beton und Eisen*. Thereupon, Schleusner arranged for his freelance assistant Klaus Zweiling (Fig. 11-19) to compose the manuscript *Das Prinzip der virtuellen Verrückungen und die Variationsprinzipien der Elastizitätstheorie* (the principle of virtual displacements and the variational principles of elastic

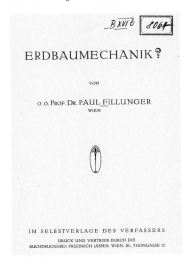

FIGURE 11-17
Cover of Fillunger's *Erdbaumechanik?*

theory), which Schleusner published under his name with the agreement of the author in the journal *Der Stahlbau* in 1938 [Schleusner, 1938/3].

In the above work a clear distinction is made between the principle of virtual displacements, the principle of virtual forces and the general work theorem (see sections 2.2.2, 2.2.3 and 2.2.4). It also establishes that the principle of virtual displacements represents the most general principle of elastic theory (see Fig. 10-25). Therefore, Zweiling and Schleusner took an important step towards formulating theory of structures in the language of the calculus of variations.

*Arno Schleusner (1882–1951):*
– Studied engineering sciences at Berlin TH and subsequently studied mathematics and physics at the universities of Berlin, Rostock and Jena.
– Engineering practice in various structural steelwork, reinforced concrete and aircraft construction companies (after 1916).
– 1930: set up an independent consulting engineering practice in Berlin (structures at Berlin-Tempelhof airport and other projects).
– Schleusner published on stability theory and the principles of structural analysis in *Beton und Eisen* (now *Beton- und Stahlbetonbau*), *Der Stahlbau* and other journals and wrote several monographs.

*Klaus Zweiling (1900–1968):*
– Studied mathematics and physics in Göttingen and afterwards was a laboratory physicist at Lorenz AG in Berlin.
– 1924–33: editor of workers' newspapers, arrested by the Gestapo and imprisoned, prohibited from publishing until 1945.
– From 1937 onwards, Zweiling worked at times for Schleusner, who acted like a fatherly figure to him.
– Zweiling wrote scientific works for Schleusner which appeared under the latter's name. The most important of these was the theoretical explanation and formulation of the principles of structural theory (principle of virtual displacements, principal of virtual forces, general work theorem) with the help of the calculus of variations. At the head of theory of structures stands the principle of virtual displacements, from which the principle of virtual forces can be derived under severely restrictive conditions; both principles, again under restrictive conditions, can be combined to form the general work theorem.
– After 1945 he enjoyed a career in publishing, and following that he was the leading Marxist theorist in the GDR.

In his book *Gleichgewicht und Stabilität* (equilibrium and stability) published in 1953 [Zweiling, 1953], he summarises those works that were published under the name of his great friend Schleusner because Zweiling was forbidden from publishing during the years of the Third Reich. Three years before that, Marguerre published a clear presentation of the duality of the principle of virtual displacements and the principle of virtual forces on the level of small displacements [Marguerre, 1950, pp. 70–90], but then works out that the principle of virtual displacements includes the principle of virtual forces in the case of large displacements (see

**FIGURE 11-18**
Arno Schleusner [Mayer, 1952, p. 152]

**FIGURE 11-19**
Klaus Zweiling (University of Leipzig archives)

Fig. 10-25). Marguerre was able to realise this because back in 1938 he had opened up a breach in the dominance of the linear in theory of structures with his theory of the curved plate with large deformation [Marguerre, 1938/2].

In the first place, the dispute surrounding the principle of virtual displacements is one concerning the development of the discipline of structural theory (type II scientific controversy).

## 11.2.12 Elastic or plastic? That is the question.

Stüssi (Fig. 11-20) and Thürlimann (Fig. 11-21) became involved in an argument about the ultimate load method in 1961/62; Thürlimann published the kinematic approach to the plastic hinge method, based on the principle of virtual displacements, in order to calculate the ultimate load.

*Fritz Stüssi (1901–1981):*
- After studying at Zurich ETH, he worked as a professorial assistant and then as a senior engineer in a structural steelwork company.
- 1930: doctorate at Zurich ETH and subsequently bridge-building practice under O. H. Amann in the USA.
- 1935: habilitation thesis at Zurich ETH and discovery of the paradox of the ultimate load method.
- 1936: active assistance at the "Olympic Games for structural engineering" in Berlin (see section 7.4.3).
- Rejected the ultimate load method from the very outset and disputed it again and again after the mid-1950s.

*Bruno Thürlimann:*
- 1946: completion of civil engineering studies at Zurich ETH
- 1951: doctorate at Lehigh University, USA
- 1953–60: professor at Lehigh University
- 1960–90: professor at Zurich ETH
- Research work on the ultimate load method in structural steelwork, on reinforced concrete shells and on prestressed concrete

FIGURE 11-20
Fritz Stüssi (Zurich ETH library)

FIGURE 11-21
Bruno Thürlimann (Institute for Theory of Structures & Design, Zurich ETH)

In 1935 Stüssi and Kollbrunner analysed the continuous beam with two degrees of static indeterminacy and discovered the paradox of ultimate load theory (see section 2.10.3). Some 17 years later, Symonds and Neal were able to explain the paradox of ultimate load theory. Nevertheless, Stüssi attacked the ultimate load theory in 1962. The outcome was the scientific controversy with Thürlimann described in detail in section 2.10.4.4.

The scientific controversy between Stüssi and Thürlimann can be attributed to type I and type II. In the more far-reaching context of breaching the dominance of the linear in theory of structures, the aim is to replace elastic theory by plastic theory, i.e. the foundation of theory of structures from the viewpoint of the material law (type I). At the same time, the controversy between Stüssi and Thürlimann was a competition between two structural theory approaches and therefore remains within the framework of the development of theory of structures as a discipline (type II).

## Résumé 11.3

Twelve scientific controversies from the history of mechanics and theory of structures have been investigated (Fig. 11-22). Three of those are type I (foundation), five are type II (competition), two are type I/II, one is type II/III (hybrid forms) and one is type III (priority). Further, it was shown that scientific controversy types I, II and III in mechanics and theory of structures follow one another in the historico-logical sense: I, I/II, II, II/III and III. Consequently, the history of mechanics and theory of structures can also be understood as the history of their controversies.

**FIGURE 11-22**
Overview of the 12 scientific controversies in mechanics and theory of structures

| Nature of controversy | Type of controversy | | |
|---|---|---|---|
| | **Foundation** *type I* | **Competition** *type II* | **Priority** *type III* |
| Galileo *(1632)* | Astronomy | | |
| Galileo *(1638)* | Strength of materials Dynamics | | |
| True measure of force *(1686–1749)* | | Search for the conservation laws in dynamics | |
| Principle of least action *(1696–1744)* | | | Variational problems in mechanics |
| St. Peter's Dome *(1743–1748)* | | Competing masonry arch theories | |
| Elastic theory *(1821–1889)* | Establishment of the continuum hypothesis | | |
| Graphical statics v. graphical analysis *(1871–1875)* | | Mathematical foundation of graphical methods | |
| Classical theory of structures *(1883–1886)* | | Linear-elastic theory of trusses | |
| Dispute among the followers *(1900–1910)* | | | Review of classical theory of structures |
| Soil mechanics *(1936/1937)* | | Consolidation theory | |
| Dispute about the principle of virtual displacements *(1936–1938)* | | Principles of theory of structures as a variational problem | |
| Dispute about the ultimate load method *(1935–1962)* | | Creation of a structural plastic theory | |

# TECHNICAL LITERATURE FOR CIVIL ENGINEERS

G. Rombach
**FEM in Concrete Construction**

2006. published in German
320 pages, 250 fig.
Softcover
€ 59,–
ISBN 978-3-433-01701-2

R. Kindmann, M. Stracke
**FEM in Steel Construction**

2007. published in German
382 pag. 200 fig.
Softcover
€ 55,–
ISBN 978-3-433-01837-8

U. Krüger
**Steel Construction I**
Part 1: Fundamentals

2007. published in German
349 pag. 148 fig. 41 tab.
€ 55,–
ISBN 978-3-433-01869-9

H. Kramer
**Structural Dynamics in Practice**

2006. published in German
286 pag. 188 fig 13 tab.
€ 55,–
ISBN 978-3-433-01823-1

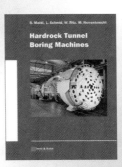

B. Maidl, L. Schmid, W. Ritz, M. Herrenknecht, D.S. Sturge
**Hardrock Tunnel Boring Machines**

2008. approx. 350 pag.
255 fig. Hardcover.
€ 89,–
ISBN 978-3-433-01676-3

Arbeitsausschuss „Ufereinfassungen" der HTG e. V. (ed.)
**Recommendations of the Committee for Waterfront Structures – Harbours and Waterways (EAU 2004)**

8th revised edition. 2005.
636 pages. Hardcover.
€ 119.–
ISBN: 978-3-433-01666-4

---

**Selected sample chapters from our books under
www.ernst-und-sohn.de**

---

008538026_my

**Ernst & Sohn**
Verlag für Architektur und
technische Wissenschaften GmbH & Co. KG

www.ernst-und-sohn.de

For order and customer service:
**Verlag Wiley-VCH**
Boschstraße 12
69469 Weinheim
Deutschland

Telefon: +49(0) 6201 / 606-400
Telefax: +49(0) 6201 / 606-184
E-Mail: service@wiley-vch.de

\* Prices are valid in Germany, exclusively, and subject to alterations.
Prices incl. VAT. Books excl. shipping. Journals incl. shipping.

ba

# CHAPTER 12

# Perspectives for theory of structures

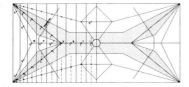

The author's difficulties with the current reduction of the computational activities of the structural engineer to mere manipulation of symbols led to his attempt to create an inherent aesthetic value for structural calculation. The lectures of Alfred Gotthold Meyer entitled *Eisenbauten. Ihre Geschichte und Ästhetik* (iron structures – history and aesthetics) (1907), Fritz Neumeyer's work on Friedrich Gilly (1997), Terry Eagleton's *The Ideology of the Aesthetic* (1994) and Friedrich Hölderlin's *Hyperion*, formed the backcloth to this work. It was at the 1998 *Vom Schönen und Nützlichen* (of beauty and utility) conference conceived and organised by Andreas Kahlow of Potsdam Polytechnic and held in Paretz Palace near Potsdam that the author was able to present a summary of his results. And finally, within the scope of the commemorative volume published on the occasion of the 60th birthday of Friedrich Führer (1998), it was Rolf Gerhardt who encouraged the author to work out his plea for the historico-genetic teaching of theory of structures. The concept of computer-assisted graphical analysis developed by the author in that same year later underwent further development and practical implementation in a computer laboratory at the University of the Arts in Berlin with the help of Prof. Gerhard Pichler. Sadly, Prof. Pichler passed away on 1 April 2004. In the meantime, Prof. John A. Ochsendorf at the MIT has taken charge of an international group of researchers that has notched up successes in the real-time analysis of masonry structures. The MIT research group has thus erected a milestone in the formation of computer-assisted graphical analysis.

What is it that holds together the engineering sciences, in particular those related to building, at their very core? Is it "the calculable reason" [Krämer, 1991/3] in the form of non-interpretive operations with written symbols for the purpose of problem-solving, which in historical terms turned into an upheaval in the symbolic foundation of the scientific perception of the modern age, like Sybille Krämer has shown? Leibniz' substitution of the philosophical pursuit of truth by the philosophical pursuit of correctness would later have an effect on the historico-logical development of classical theory of structures and its supplanting by modern structural mechanics. Since the 1880s, the theory and practice of structural analysis had formed the core of administration-type engineering activities because only structural analysis based on a scientific footing was able to achieve proof of the "correctness" of structural designs in both the physical and the social sense. This synthesis of analysis and building to create calculated building meant that the practical use of symbols in structural engineering became the norm. As was shown in the chapter 10, the history of modern theory of structures can be regarded as a history of the creation of model worlds which become increasingly far removed from building and are dominated more and more by the non-interpretive use of symbols. Formalisation in mathematics and in modern theory of structures consists of "expressing a problem with the help of an artificial language in such a way that the problem-solving steps can be set up as a step-by-step reformulation of symbolic expressions where the rules of this successive reformulation are exclusively related to the syntactic form of the symbols, but not to that for which the symbols 'stand', what they mean" [Krämer, 1991/3, p. 1]. Calculated building reduces the structural engineer to a mere symbols shifter unaware of the question regarding the consequences of his actions.

Did not Friedrich Hölderlin, 200 years ago, complain through his *Hyperion* that "pure intellect has never produced anything intelligent, nor pure reason anything reasonable" [Hölderlin, 1990, p. 92]? Did not Max Horkheimer and Theodor W. Adorno recognise in 1947 that the concept of enlightening thinking "is nothing less than the concrete historical forms, the institutions of society in which it is entwined, [and] already contains the germ of that backward step happening today everywhere" [Horkheimer & Adorno, 1994, p. 3]?

The deconstruction of the reasonable by mere reason, the intelligent by mere intellect, also takes place today in the form of a crisis of knowledge; knowledge in a materialised form "that only exists in the abstractions of models, which are not constructions of reality, but rather its deconstruction in the pluralism of the knowledge of fragments of reality based on division of labour," like Hans Jörg Sandkühler notes, concluding that "by separating knowledge from self-reflective experience, … the knowledge [is] divorced from the reality" [Sandkühler, 1988, p. 205]: incarnate instrumental reason. The social discourse about the anticipation as a target definition of the cognition is at risk. The following sections should be understood as a contribution to the history of the theory of structures for the discussion about the reasonable and the intelligent in any fundamental engineering science discipline.

**Theory of structures and aesthetics**

## 12.1

In the following it will be explained that
- beauty and utility in building are compatible.
- the chance for aesthetics is embodied in structural analysis.
- computer-aided graphical analysis could help to reduce the animosity between architects and structural engineers in the design of the load-bearing structure.

**The schism of architecture**

### 12.1.1

Aesthetics began with the publication of Alexander Baumgarten's *Aesthetica* in 1750. "Aesthetics is born as a discourse of the body," writes literary theorist Terry Eagleton at the start of his book *The Ideology of the Aesthetic* [Eagleton, 1990]. According to Baumgarten, aesthetic cognition is an intermediary between the generalities of reason and the specifics of the senses. Nearly a century after Baumgarten, Karl Marx called for the re-invention of aesthetics in his *Economic and Philosophical Manuscripts of 1844*: "*Sense-perception* (see Feuerbach) must be the basis of all science. Only when it proceeds from sense-perception in the two-fold form of *sensuous* consciousness and *sensuous* need – is it *true* science" [Marx, 1968, p. 543]. Only sensual cognition renders possible the steady state between the generalities of reason and the specifics of the senses. The historical process of separating the useful from the fine arts in building had already been settled by the turn of the 19th century. Denis Diderot (1713–84) can tell you plenty about the useful arts in his *Encyclopédie*. The mathematician and engineer Franz Joseph Ritter von Gerstner (1756–1832) impressed by the Josephinian enlightenment carries out the schism of architecture radically in his *Einleitung in die statische Baukunst* (introduction to structural engineering) published in 1789 (see section 5.5.2). In programmatic terms he formulates the hegemonic claim of a "mechanics derived from the nature of building itself" in architecture [Gerstner, 1789, p. 4]. In the first volume of the *Handbuch der Mechanik* (manual of mechanics) dating from 1831, structural engineering is embedded in the theory formation of construction and mechanical engineering with the help of mathematics and mechanics. According to Gerstner, structural engineering "obeys those rules on which the strength of the building with respect to the strength of the individual parts and their assembly are based. ... we have already set up the laws for the strength of the individual parts of a building, and all that remains for structural engineering are primarily the laws for the assembly" [Gerstner, 1833, p. 385].

In Gerstner's case there exists only a superficial relationship between theory of structures and strength of materials; a deeper relationship was to appear during the establishment phase of the discipline-formation period of structural theory. Gerstner had demonstrated paradigmatically the "laws for the assembly" of loadbearing system elements in the shape of pin-ended members to form the loadbearing system of the masonry arch in 1789, and now the discovery and the use of the "laws for the assembly" of the individual loadbearing systems to form a model of the structural system of the entire structure was the object of theory of structures. Hence, architecture, if we are to understand it like Bruno Taut as the art of proportions, i.e. the art of subdividing the whole (see [Taut, 1977]), faces an opposing concept in the form of a contradiction which today still forms the crux of the animosity between structural engineers and architects. When using the "laws for the assembly" of the loadbearing systems for the purpose of the loadbearing system synthesis, differential and integral calculus plays a key role. Gerstner uses the relationship between formalisation and mechanisation, discovered by Leibniz and expressed by him as *automate spirituel ou formel*, in a very singular way: he transforms the formation laws for loadbearing systems into the language of differential and integral calculus and thus generates mechanical character strings with Leibniz' "calculus" – character strings that are divorced from the concrete loadbearing system and only through specific re-interpretation represent a construction theory object. It is those three strides of transformation, formalised operations and specific re-interpretation that have characterised the business of theory of structures since the time of Gerstner. The heart of this is the "non-interpretive use of symbols" [Krämer, 1988, p.1] in which the *epistēmē* becomes the *tekhnē*. This programme of the founding father of theory of structures would become reality in the 19th century through the work of Navier, Clapeyron, Saint-Venant, Rankine, Culmann, Schwedler, Maxwell, Cremona, Castigliano, Mohr, Winkler, Müller-Breslau and Kirpichev. Nevertheless, Gerstner's plea for a mechanical architecture consists not only of the mere reflection of Diderot's separation of art into a mechanical and a beautiful art. His plea conceals the schism of construction theory: here structural engineering (theory of structures), there architectural theory. In this sense, Gerstner created a double separation between architecture and construction theory.

## 12.1.2 Beauty and utility in architecture – a utopia?

Can we overcome the schism in architecture? Can beauty and utility coexist in architecture? What is it that holds together construction theory at its very core? In his sensitive but likewise learned introductory article to the work *Ältestes Systemprogramm des deutschen Idealismus* (oldest system programme of German idealism) – undoubtedly the most important printed agenda of German philosophy in the 1790s – co-authored by that genius of building Friedrich Gilly (1772–1800), Friedrich Wilhelm Schelling (1775–1854), Friedrich Hölderlin (1770–1843) and Georg Friedrich Wilhelm Hegel (1770–1831) in 1795/96, the architecture theorist Fritz Neumeyer has surely defined the quintessence of the German intellectuals in the spiritual steady state between the classic and the romantic: a redefinition of the conditions for a reunification between poetry and philosophy as well as art and science – or put more succinctly: "Science must become sensual and poesy scientific" [Neumeyer, 1997, p. 87].

Friedrich Gilly obviously had in mind such a reunification between architecture and construction theory. This is revealed in his paper expressing his "thoughts on the need for the various parts of architecture to be

**FIGURE 12-1**
Cover of the first volume of the journal *Sammlung nützlicher Aufsätze und Nachrichten, die Baukunst betreffend* [Kahlow, 1998, p. 107]

united from a scientific and practical viewpoint" (reprinted in [Neumeyer, 1997, pp. 178–186]) which was published in 1799 in the first German-language journal for the building trade *Sammlung nützlicher Aufsätze und Nachrichten, die Baukunst betreffend* (collection of useful papers and bulletins on architecture) (Fig. 12-1) which had first appeared two years previously.

Whereas Eytelwein, in the same issue of this journal, was only calling for the unification of theory and practice in building [Eytelwein, 1799], Gilly's aim was to abolish the double separation between architecture and construction theory on the one hand, theory and practice on the other. Both Eytelwein and Friedrich Gilly were involved in devising the teaching plan for the Building Academy founded in Berlin in 1799, where Friedrich Gilly gave lessons in "optics and perspectives … also in architectural and technical drawing" [Neumeyer, 1997, p. 93], Eytelwein lessons in theory of structures, hydrostatics, mechanical engineering, dyke-building and river-training works, and David Gilly (1748–1808), Friedrich's father, lessons in lock-, bridge- and harbour-building plus building theory.

Eytelwein understands "building knowledge as an object of science and art [whose] application for accomplishing certain purposes of social life" [Eytelwein, 1799, p. 28] is useful. For Eytelwein, construction theory and construction engineering still appear as one entity in the form of building knowledge. But less than 10 years later, Eytelwein would help to break down that unity in building knowledge through its mathematicisation; and hence the encyclopaedic presentation of building knowledge in the tradition of the encyclopaedias of the German *Sturm und Drang* (storm and stress) movement were also obsolete. For Friedrich Gilly, the advantage of building knowledge characterised by the "grouping or connection of certain subjects" results from the "common points of their general principles rather than according to essential and direct relationship" [Neumeyer, 1997, p. 179]. In the end it is the fault of the thinking induced by manufacturing that the "certain subjects" were added to building knowledge, which is made up of the "features common to their general principles only – just like the additive assembly of the part-works in the end product in manufacturing". The consideration of these general principles of building knowledge "according to essential and direct relationship" later took on a mathematics-mechanics form in Eytelwein's works, especially for hydraulic structures, structural engineering subjects and mechanical engineering. And that instigated the divorce proceedings between beauty and utility in architecture. The divorce proceedings were also expressed in organisational terms by the separation of the Building Academy from the Academy of Arts in April 1824. The Building Academy under the directorship of Eytelwein from 1824 to 1831 would from now on be responsible for the technical side of building and oversee the training of proficient surveyors and master-builders for the provinces [Dobbert, 1899, p. 42]. Whereas the building students still continued to attend lectures in architectural subjects at the Academy of Arts, the Building Academy was alone responsible for the construction engineering subjects.

The "genuine architecture" on the other hand, writes Friedrich Gilly (Fig. 12-2), "already requires a very special diversity in its characteristic areas, not merely in the individual objects, but also in the purposes, claims and investigations. The study of this, like its practice, is highly varied in the particular views and their connections and must therefore be taken into account – but from separate viewpoints – when assessing the essential considerations. These considerations then come together to form a whole for comparison when one observes them according to the things they have in common, which are required in the execution, and this relationship in the consideration is necessary to the extent that the purposes and requirements themselves are necessarily interconnected" [Neumeyer, 1997, p. 179]. According to Friedrich Gilly, this includes not only the "more intensive science" and the "school of mathematics" [Neumeyer, 1997, p. 181], which qualify the master-builder in "secure foundations, long-lasting assembly and the execution of building", and enable him "to identify, to check the nature and durability of their building materials, and fasteners, and to observe effects of diverse kinds" [Neumeyer, 1997, p. 181], but also the theories he has been given through building practice. But that was not enough.

By referring to Karl Heinrich Heydenreich's (1763–1801) 1798 article on the "new concept of architecture as a beautiful art", Gilly hopes "that architecture can still be rescued from its exile" [Neumeyer, 1997, pp. 183–184]. In his article, Heydenreich devises "a new dialectic of purpose and idea, towards which Gilly was sympathetic" [Neumeyer, 1997,

FIGURE 12-2
Bust of Friedrich Gilly by Johann Gottfried Schadow
[Reelfs, 1984, p. 223]

p. 85], but distinguished the "natural purpose" of the building, to provide protection from the weather, from the "higher purpose" of the building, which he expressed as "churches, assembly buildings for state matters, arsenals, buildings for the cultivation of science and art, country houses and so on" [Neumeyer, 1997, p. 85]. Heydenreich would base the reason for this splitting of the purposes on the differentiation of architecture into a beautiful and a useful architecture. Nevertheless, both forms of art embody an inner relationship in the form of the physical purpose, the natural purpose, the means to the higher purpose. He therefore refers indirectly to the *tekhnē*, the most general and most comprehensive initial object of the poetic of Aristotle. For *tekhnē* is for Aristotle the epitome of all human capabilities of being able to accomplish something: through work, artistry and dexterity; but the poet has to say what could be or what is possible [Friemert, 1990, pp. 919–920].

The fact that *tekhnē* can belong to poiesis was examined by Gilly in terms of form and content in his description of the Bagatelle country house near Paris (reprinted in [Neumeyer, 1997, pp. 152–162]). To today's reader, this text still appears as a steady state between beauty and utility, as a concrete utopia of a sensualised science and science-based poesy. Bagatelle country house (Fig. 12-3), which was built for Marie Antoinette in 1777 (in just two months!) amid the delightful woodland of the Bois de Boulogne near Paris by François-Joseph Bélanger (1744–1818), whom the Count of Artois commissioned to carry out the work, is described by Gilly from the point of view of the roving observer who makes reflections in his poetic wanderings as a "beautiful and agreeable memorial to artistic assiduity" [Neumeyer, 1997, p. 155].

Embedded in the most famous *jardin anglais* of the era, Gilly reveals "an architectural wonder": there is the wonder of the dialectic of the simultaneity and non-simultaneity of the building process, which already looks beyond the change from the chronological succession to the spatial simultaneousness of the production in the contemporary organic manufacture, which anticipates the master-builder in a "demiurgic act" [Neumeyer, 1997, p. 71]. "It is our own imagination," Gilly writes, "that urges us in such undertakings, when one with this unusual exertion thinks of everything in mutual activity, like everyone considers, completes and adds his work to the whole for himself as his own law without understanding something about the interactions of the individual who has in mind solely the ordering sense of the inventor in every detail right up to completion" [Neumeyer, 1997, pp. 154–155]. There is the wonder of the aesthetic encounter with the subject, moulded into poesy in the text as the boundless play of his perceptions. Like Hölderlin's *Hyperion*, who in Calaurea encounters "the name of that which is one and all": "Its name is Beauty" [Hölderlin, 1990, p. 58]. Friedrich Gilly, akin in spirit to Friedrich Hölderlin; Friedrich Gilly, the Hölderlin of architecture.

But the continental drift of beauty and utility in architecture had already begun with the change from art to technology and the mathematicisation of the sciences. The schism of architecture would determine the

FIGURE 12-3
View, plan and location plan of Bagatelle House, after Friedrich Gilly [Kahlow, 1998, p. 116]

building of the 19th century. So the unity of beauty and utility in architecture remains a utopia to this very day. Despite this, we must ask one question: Did the fundamental construction theory discipline of structural analysis, which developed in close interaction with iron construction, finally seal the separation between beauty and utility in architecture, in particular in iron construction?

### 12.1.3 Alfred Gotthold Meyer's *Eisenbauten. Ihre Geschichte und Ästhetik*

It was in an essay that the mechanical engineering theorist Franz Reuleaux first posed the question "Can iron bridges be beautiful?" in the sense of tectonics. Alfred Gotthold Meyer (1864–1904), professor of history of arts and crafts at Berlin-Charlottenburg TH, investigated the style-forming and style-inhibiting forces of iron in his monograph *Eisenbauten. Ihre Geschichte und Ästhetik* (iron structures – history and aesthetics), which was published posthumously in 1907 and is now available in a new edition [Meyer, 1997]; it was intended to form part of his major work entitled *Das neunzehnte Jahrhundert in der Stilgeschichte* (the 19th century in the history of style), but this ambitious project remained only an outline. Meyer was not writing for the engineers of iron structures. Instead, his aim was to "introduce [such] structures – created with a new building material, with in some cases new means of construction and according to new methods for purposes previously unknown in some instances – into the history of building styles" [Meyer, 1997, p. 2]. His *Eisenbauten*.

*Ihre Geschichte und Ästhetik* covers a new object in the history of art. In contrast to the majority of his contemporaries, Meyer does not rule out the fact that the structural engineer of splendid talent is able to unite the artistic sense with the scientific and technical side of his profession. Although, with few exceptions, engineers and architects use different forces to achieve the common victory over theory and so march along different roads – "here more understanding, there more fantasy" [Meyer, 1997, p. 4] –, it is still the case that "whoever some day writes the 'history of building' in the 19th century will have to dedicate a major part to 'engineered structures' if he wishes to devise a complete and correct historical portrayal" [Meyer, 1997, p. 4]. For Meyer, it is not so much the form intentions of the creative personality that are decisive for the aim of formulating the style-forming and style-inhibiting characteristics of iron construction, but rather the forces that themselves act together in iron construction, among which the "new and at the same time most important [is] structural analysis" [Meyer, 1997, p. 4].

In chapter II "Rechnen und Bauen" (calculation and building) in the first book, dedicated to the fundamentals of iron construction, Meyer develops the history of the use of symbols in construction theory, which reached an interim climax in classical theory of structures. To this day, this competent presentation of this object from the history of art viewpoint has not been surpassed in this discipline.

Like every great work, Meyer's book contains forward-looking ideas. In his story of the development of theory of structures, the importance of the formal operations for the design and construction of iron structures comes to light. Calculation for him is more than just numerical computation – it also embraces the algebra that began with Vieta (1540-1603), the underlying idea of which according to Sybille Krämer is "to separate the manipulation of series of symbols from their interpretation" with the aim of "relieving the mind from the exertions of interpretation" [Krämer, 1988, p. 176]. Formal operations are not only linked with the written use of symbols, and therefore include the linearisation of perception; the schematic use of symbols is also inherent to formal operations: "While we form and reform series of characters," writes Sybille Krämer, "we have to behave as if we were a machine" [Krämer, 1988, p. 178]. This is the real crux of the aesthetic criticism of the iron structures created through calculation.

Meyer showed impressively that theory of structures does not necessarily lead to calculated building. The inner eye of the structural engineer already sees the (real) loadbearing structure in the structural system. "But in the legal framework of style, [the inner view] occurs [first] … as soon as it begins its real life as a tangible, physical, three-dimensional entity. Only then does it act upon the sensory organs, which are able to grasp the concept of 'size' as a property of a 'structure' and not only assess the means employed, but perceive in the *aisthesis*" [Meyer, 1997, p. 46]. The use of the Greek word *aisthesis* means that Meyer is in the first instance not relating aesthetics to art, but, like Baumgarten, to the whole field of human perception and sensation. But Meyer expresses his ideas boldly: he

FIGURE 12-4
Bridge over the Rhine between Mannheim and Ludwigshafen, built in 1999: the continuity effect of the three-span arch bridge can be seen through the "infill frames" above the intermediate piers [Stiglat, 2000, p. 35]

realises that structural analysis in iron construction can become a style-forming force. How can that be? At the phase transition "where the numbers and lines before the 'inner view' change into the realistic image of the iron framework, that 'realistic image' of the inner view at any point that no longer belongs to just the rational, necessary construction initially can have a modifying reactive effect on its means, i. e. the calculation" [Meyer, 1997, p. 47]. For only the mind is able to "place numbers and lines clearly in sheer endless rows and columns – but in the presence of the sensual power of imagination the rows and columns easily become a maze. And when this is regarded as undesirable, when the calculation from this standpoint is renewed in order to make the construction clearer, calmer, freer, prettier, in order to add a rhythmic interruption to the uniformity, in order to compensate for sharp contrasts, in order to emphasize the essentials more precisely and to mark the auxiliary forms – for the appearance, too – merely as inconsequential, then even the 'calculation' in iron construction itself becomes a style-forming force" [Meyer, 1997, p. 47].

So the steady state heaving back and forth between structural modelling and the active view of the structure through the inner eye of the structural engineer is able to produce structures that can be perceived aesthetically.

The twin-track railway bridge over the Rhine between Mannheim and Ludwigshafen, built in 1999, will be used as an example of this. The continuous polygonal arch (each span measures 91.30 m) embodies regularity, symmetry, harmony and proportion (Fig. 12-4). The loads transferred via the hangers into the arches become visible through the nodes and are not hidden like in a rounded arch [Stiglat, 1999, pp. 839/40].

Contrasting with this, those four internal moments on which the artistic value of the form rests are of a purely formal character in the TGV railway bridge over the Donzère Canal south of Montélimar (Fig. 12-5). By separating the design from the calculation and construction, the style-forming force of the structural analysis remained unused and turned into the opposite: the harmony and proportion of the play of forces between

**FIGURE 12-5**
TGV bridge over the Donzère Canal south of Montélimar, built in 1999: the loadbearing function of the superimposed middle arch can be calculated only, and not visualised [Stiglat, 1999, p. 840]

the two arches has congealed into a superficial flirtation with forms because the loadbearing function of the superimposed middle arch is unclear. The engineers who were confronted with this design by the architect Marc Mimram obviously had no chance to alter the concept; they write: "The architectural concept of the engineering structure ... and its large span for a TGV viaduct led in combination to an extraordinary loadbearing structure. These two aspects did not ease the work of the engineer thinking in terms of the structure" [Cremer et al., 1999, p. 408]. But the structural analysis should start to become apparent at the outset of the draft design and should be available without delay as the structure starts to take shape, sometimes moving ahead, sometimes following. The results of the structural analysis limit the draft design to fixed, confined limits, enable flights of design fantasy, but inhibit them too. Then and only then can the style-forming force of structural analysis be liberated, then and only then can the aesthetically possible be realised in structural forms.

**The aesthetics in the dialectic between building and calculation**

### 12.1.4

Meyer's bold idea regarding the possibility of the metamorphosis of structural analysis into the style-forming force in iron construction is more topical than ever today because the dialectic between calculation and building is an essential element that crops up every day in the working relationship between the structural engineer and the architect. Electronic calculation has painfully focused this dialectic. Apart from a few exceptions, the structural engineer is regarded by the general public as a building technician, even as a number cruncher who assists the architect. This downright Hegelian-sounding legend of master and servant cannot please structural engineers – but architects neither. How did we arrive at this situation?

In 1910, in his detailed review of Meyer's book in the first year of publication of the journal *Der Eisenbau*, Franz Czech wrote the following remark: "And now to make structural analysis the complete foundation of aesthetics; for what is the 'structural feel' other than the theory understood intellectually and transferred to flesh and blood, working from the subconscious" [Czech, 1910, p. 405]. This indeed means that Czech has understood aesthetics as a discourse about the body! How should we interpret that?

Müller-Breslau, who completed classical theory of structures, paved the way for the use of calculus in theory of structures, which pushed aside the real loadbearing structure as an object of structural modelling in the consolidation phase after 1900 in favour of the search for the most rational method of calculation. Only in this way could the $\delta$-symbols of the Berlin school of structural theory become the breeding ground for the technical realisation of the symbolic machine in the form of the computer of Konrad Zuse. The starting point for this technical development by the Berlin-based Zuse was his resolution of the calculations for a bridge framework with nine degrees of static indeterminacy into basic arithmetic operations and their processing in a computing plan with which the structural analysis could be carried out without knowledge of the structural theory. Zuse generalised this computing plan, which was still based on member analysis, to become the "computing plan or program" method. His computing plan formed the starting point for the world's first working program-controlled computing machine: the Zuse Z3 of 1941. The Leibniz concept of *automate spirituel ou formel* had thus been realised. In computer mechanics, calculation remains unconsidered: the civil or structural engineer can transform and manipulate strings of symbols without having been instructed in the meaning of the symbols. This abandonment of structural analysis is a historical product of a growing conditioning of the mental and physical capabilities of the structural engineer for structural methods. Nevertheless, it offered him – still in the classical phase of structural theory – the chance to create beautiful iron structures (Fig. 12-6). We could even say that Culmann's graphical statics not only rationalised the design work of the structural engineer, but at the same time also aestheticised it because the force diagrams and construction drawings occur in the twin form of both the sensual consciousness and the sensual need (Fig. 12-7). This development reached its zenith in the 1880s and 1890s; a prominent example of this is the Eiffel Tower analysed with the methods of graphical statics by Koechlin, a student of Culmann.

However, the increase in the rationalisation of engineering work quickly led to the decline of graphical statics into individual methods, to graphical analysis, whose methods represent merely graphical recipes. Like in the computer mechanics of a later day, even graphical analysis caused the structural engineer to become distanced from the draft design work insofar as he divested himself of graphical analysis recipes and allowed them to become merely the means of intellectual engineering. Here, too, we see the relinquishment of graphical calculations, i.e. the renunciation from the mental and physical sphere of the structural engineer in the form of the functional modelling to create the external artificial organ system of his activities – precisely as a technical means, as *tekhnē*. So the aesthetic component in the draft design work of the structural engineer

FIGURE 12-6
The Kaisersteg footbridge over the River Spree at Berlin-Oberschöneweide, designed by Müller-Breslau and built in 1897/98

also dwindled. Computer mechanics brought to a head the trend towards calculated building.

Nevertheless, modern information and communications technology is today already indicating a systematic design practice that could enable the structural engineer to regain lost design competence and the architect to regain lost construction competence, both on a higher level. For example, 20 years ago Ekkehard Ramm was already pointing out the visualisation and animation of mechanical relationships and processes; that would have been a decisive step towards the evolution of constructions and loadbearing system synthesis [Ramm et al., 1987]. Another example might be the computer-aided display of engineered structures [Klooster & Permantier, 1998]. A third way of departing from calculated building could be the creation of computer-aided graphical analysis, which could be placed in the phase transition between structural engineering studies and structural theory (Fig. 12-8).

This must have a modular structure so that every individual construction in graphical statics, e.g. the funicular polygon, can be mobilised by the user in the form of specially developed graphic editors, displayed on the screen and coupled with other graphic editors to form concrete con-

**FIGURE 12-7**
Graphical treatment of a steel plate crane after Ritter [Ritter, 1888]

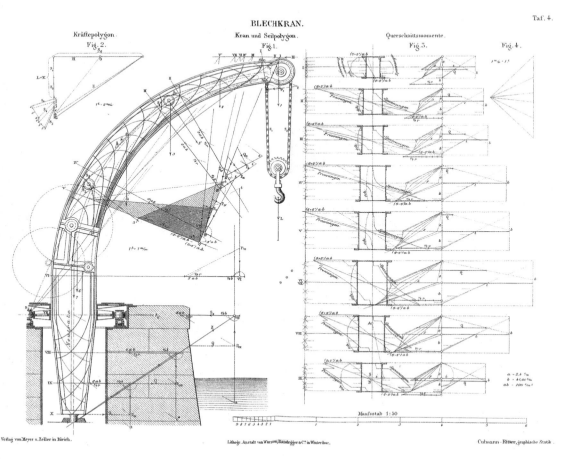

structions in the sense of graphical analysis. The use of such graphic editors and dimensioning modules based on approximation methods would enable the results of the graphical analysis synthesis to be transferred directly to the level of the loadbearing system. Computer-aided graphical analysis would be, so to speak, a hinge between conception and design because it goes beyond a pure morphology of the building fabric. And last but not least, at university level it would be a suitable means for bringing together students of structural engineering and architecture in the loadbearing system synthesis because the virtual building would increasingly link the two professional groups in teaching, research and practice.

Using numerous examples from the teaching of theory of structures and engineering practice, Gerhard Pichler, Karen Eisenloffel and Marko Ludwig demonstrate how, even today, graphical statics can become effective as a "language of the engineer" [Culmann, 1875, p. 5] with the help of the computer [Pichler et al., 1998]. Their approach is especially apparent in the structural investigation of the wide-span rib vault over the main staircase of Berlin's famous City Hall (Fig. 12-9). The plastered rib vault of clay brickwork exhibited cracks, the cause of which had to be ascertained because the room above the rib vault was due to be converted into a hall for festive events: "The investigations had to establish whether the damage was due to the vault being overloaded and whether the structure could withstand the high demands expected to be placed on it in the future. Owing to the pattern of damage, the initial idea was to install a steel beam grid to relieve the vault. However, a graphical analysis of the vault revealed that a 20 cm thick layer of ballast (sand) could influence the course of the line of thrust to such an extent that an imposed load of 5.0 kN/m$^2$ as well as 3.5 kN/m$^2$ on one side would be permissible in the assembly room above. An additional beam grid would be superfluous. The determination of the lines of thrust in rib vaults is an iterative process including a variable assumption of the horizontal force. The drawing work required was minimised by using a CAD program" [Pichler et al., 1998,

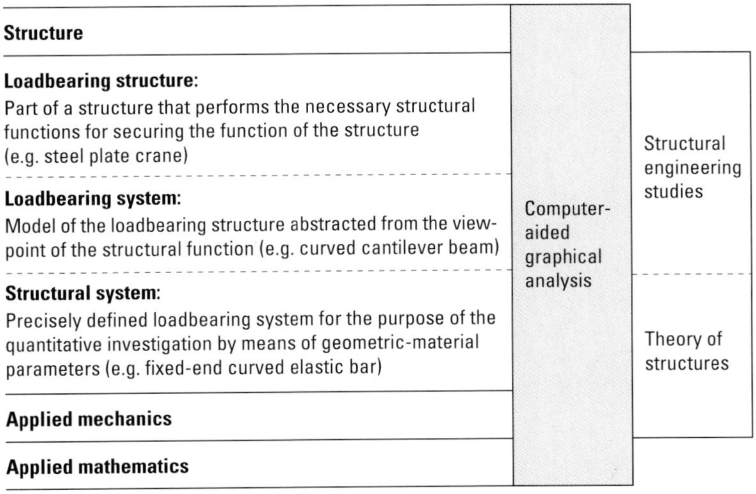

**FIGURE 12-8**
Object domain and disciplinary position of computer-aided graphical analysis [Kurrer, 1998, p. 209]

| Structure | | |
|---|---|---|
| **Loadbearing structure:** Part of a structure that performs the necessary structural functions for securing the function of the structure (e.g. steel plate crane) | Computer-aided graphical analysis | Structural engineering studies |
| **Loadbearing system:** Model of the loadbearing structure abstracted from the viewpoint of the structural function (e.g. curved cantilever beam) | | |
| **Structural system:** Precisely defined loadbearing system for the purpose of the quantitative investigation by means of geometric-material parameters (e.g. fixed-end curved elastic bar) | | Theory of structures |
| **Applied mechanics** | | |
| **Applied mathematics** | | |

p. 227]. Further, the authors demonstrate that coupling CAD with graphical analysis is not only useful for investigating vector- and form-active loadbearing structures, but also for mass-active loadbearing structures such as suspended floor slabs. They therefore make up the elements of a modern rules of proportion for the structural engineer.

Further elements were able to be added from the kinematic theory of structural analysis developed in the 1880s (see section 6.3.3). For example, the graphical method derived from the influence line theory created by Mohr, Winkler, Weyrauch, Müller-Breslau and, in particular, Land in the 1870s and 1880s would enable an adequate mapping on adapted graphic software. Land's theorem (see section 6.3.3.3) would be a nice example for the rediscovery of the pictorial representation of design analysis; its graphic representation by the symbolic machine in the shape of the computer could significantly improve the structural feel of the user for how moving actions influence the force and displacement variables.

Led by John A. Ochsendorf, MIT set up a research group for the real-time analysis of masonry structures in 2005 which in the meantime has published several papers (e.g. [Block & Ochsendorf, 2005; Block et al., 2006]). Based on the MSc dissertation by Philippe Block [Block, 2005] (supervised by Ochsendorf), "interactive THRUST" has now made it possible to determine the thrust lines in masonry structures by means of an online interactive graphical analysis. Through the synthesis of kinematic

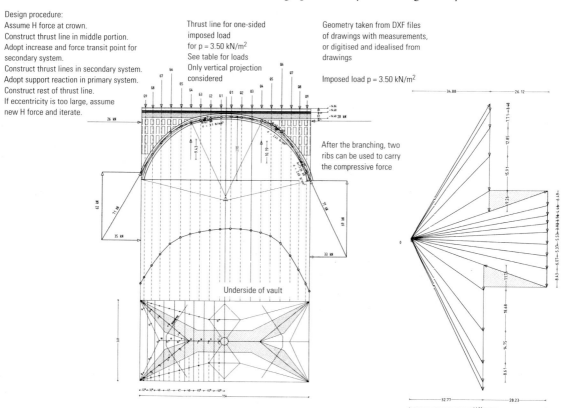

**FIGURE 12-9**
Computer-aided graphical analysis investigation of the rib vault over the main staircase of Berlin City Hall [Pitz, 2000, p. 85]

and static methods, the researchers have managed to combine the kinematic and geometric views of statics (see section 2.2.7) in the form of user-friendly computer programs with sophisticated graphic interfaces. The authors anticipate that the integration of these programs into the theory of structures studies of architects and engineers will result in considerable benefits. The work on the structural analysis and synthesis of masonry structures at MIT forms a nucleus of crystallisation for computer-aided graphical analysis.

The aesthetic component in the conception and design work could become real again with the help of computer-aided graphical analysis; it could transform the lost structural feel at the interface between conception and design into the flesh and blood of the user again. Science would become sensual and poesy scientific. And that in turn would open up a breach in the well-established role-play of structural engineers and architects. The number cruncher insult would also lose its meaning, likewise speaking of architects who shine in structural design through their incompetence.

## 12.2 A plea for the historico-genetic teaching of theory of structures

The historico-genetic concept thoroughly abolishes the breakdown of teaching into the traditional disciplines that materialised at the start of the previous century (structural design, constructional disciplines, planning disciplines, etc.). The interdisciplinary subsets of the aforementioned subjects would become the focal point of teaching. What this means for the historico-genetic teaching of the theory of structures is that the modelling process in practice is simulated in the teaching – from the historical structure (which could also be structures of the modern movement) and from the planned structure via the loadbearing system or the loadbearing structures to the loadbearing systems and the structural systems (Fig. 12-10). One intention of this inductive process from the concrete to the abstract is to make it clear to students of building that the design of the structure should not be considered from the point of view of loadbearing behaviour only, but that the structure can have other functions such as space-dividing, insulating, radiating (of electromagnetic waves), aesthetics, etc. And thus we can refer again and again to the interdisciplinary nature of the practice of architects and structural engineers by simulating it in teaching work.

Besides the modelling skills gained in cooperation with the constructional subjects, future architects and structural engineers will be provided with a historico-logical classification of loadbearing systems. In doing so, the history of design together with the history of design theory will play a major role; it should also provide insights into the fact that our present understanding of design and its theory is the result of a long constructional-technical and construction theory evolution, which therefore places students of building in the position of being able to identify current developmental tendencies and implement them effectively in practice in the future. The general content of these targets is therefore to enable the students to recognise the process character of design evolution in build-

| Object domain | Definition of object domain | Disciplines |
|---|---|---|
| Structure | | Constructional, planning and historical disciplines |
| Loadbearing structure | Part of a structure that performs the necessary structural functions for securing the function of the structure (e.g. brick arch, reinforced concrete arch) | Structural engineering studies |
| Loadbearing system | Model of the loadbearing structure abstracted from the viewpoint of the structural function (e.g. models of pre-elastic masonry arch theories) | Structural engineering studies, theory of structures |
| Structural system | Precisely defined loadbearing system for the purpose of the quantitative investigation by means of geometric-material parameters (e.g. fixed-end elastic arch) | Theory of structures, applied mechanics, applied mathematics |

FIGURE 12-10
Theory of structures and structural engineering studies in the context of the object domains and disciplines

ing and its construction theory basis together with its integration into the interrelated conditions of building. But these development processes were once the objects of an exciting search, a stimulating act – namely, back then as they were created.

**Historico-genetic methods for teaching of theory of structures**

### 12.2.1

The methodical implementation of the historico-genetic concept can take place as shown below depending on the themes and didactic requirements:

*Historico-logical longitudinal section analysis*: This form of presentation is suitable for reconstructing a series of design history steps in a clearly defined period of time. In doing so, the transitions between the steps should be processed especially intensively. This method has an overview and orientation character and at the same time should introduce basic theory of structures concepts. One example is the history and theory of beam constructions – the development from Leonardo da Vinci via Galileo and right up to the modelling of the beam loadbearing structure for the loadbearing system of the "beam" based on Navier's theory and still used today. This form of presentation is less suitable for the development of masonry arch structures and their theoretical analysis because the development history of the masonry arch spans more than 4000 years and there are still considerable gaps in that history to be researched.

*Historico-logical transverse section analysis*: This form of presentation is a momentary historical snapshot in which the complex fabric of conditions for a constructional-technical or constructional-theoretical revolution is investigated from various disciplinary perspectives. One example of this would be the report on the stability of the dome to St. Peter's in Rome written in 1742/43 by Tommaso Le Seur, Francesco Jacquier and Ruggiero Giuseppe Boscovich in which the horizontal thrust was calculated based on an analysis of the damage and cracks and was then used as the quantitative basis for the refurbishment of the dome construction with the help of circular iron ties. Using this (first-ever) damage report for the dome construction to establish the condition of the structure, the damage, the stability and the proposed methods of securing the structure, the three authors anticipated paradigmatically the task of the structural engineer in the preservation of historically important structures. If this example is suitable for illustrating the genuine tasks of the structural engineer in connection with the preservation of historically important structures, then the student can properly absorb the foundations of the modelling and the structural investigation of statically determinate hinged trussed frameworks from the presentation of trussed framework forms and their theoretical investigation by Culmann, Schwedler, Whipple, Zhuravsky and others around the middle of the 19th century.

*Historico-logical comparison*: This form of presentation allows two relevant developmental stages from the history of design and its theory to be compared with one another. Here, the comparative analysis of the differences and similarities between the two stages with the help of design history source materials can reveal the underlying developmental logic. The modelling of historical masonry arch structures with the help of elastic theory still practised today could be assessed critically by unearthing forgotten ultimate load theory approaches in the masonry arch analyses of the 18th century, which could then be modernised and further developed into alternative, practical and simple means of modelling historical masonry arch structures.

### 12.2.2 Content, aims, means and characteristics of the historico-genetic teaching of theory of structures

The detailed content of the historico-genetic teaching of theory of structures can be structured in four steps (Fig. 12-11).

### 12.2.3 Outlook

The reason and purpose behind the historico-genetic approach to teaching theory of structures for an integrated course of study in architecture and structural engineering will be developed. This will be followed by proposals for the specific configuration of the historico-genetic teaching of theory of structures. The plea for the historico-genetic teaching of theory of structures should be understood as part of the historical theory of structures concept proposed in section 4.8.4, which is characterised by the transformation of historical into engineering science knowledge and a historical appreciation of engineering science knowledge.

**FIGURE 12-11**
Four-step, historico-genetic teaching of theory of structures

|  | **First step** | **Second step** | **Third step** | **Fourth step** |
|---|---|---|---|---|
| **Content** | – Introduction to theory of structures<br>– Qualitative encounter with the structure<br>– Elementary structural theory in the construction history context | – Elementary strength of materials theory in the construction history context (historico-logical longitudinal section analysis)<br>– Quantitative analysis of loadbearing system elements such as cantilever beam, simply supported beam, etc.<br>– Analogy of beam and rope statics plus definition of thrust line (historico-logical longitudinal section analysis and structural analysis of historical masonry arch structures including a comparison with current theoretical approaches)<br>– Qualitative confrontation with principal types of loadbearing structures (beam structures, comparison of truss and trussed framework, gravity structures, masonry arch structures, cable structures, etc.) through the simple synthesis of loadbearing system elements | – Confrontation with the specific loadbearing systems of the built environment<br>– Principles of classical theory of structures and introduction to the theory of statically indeterminate systems in the historical context taking into account structural steelwork in particular<br>– Structure formation laws of loadbearing systems and development of a historical and logical loadbearing system classification | – Quantitative encounter with the structure<br>– Simulation of engineering practice in the preservation and securing of structures plus structural engineering for new structures<br>– Insight into the nature of non-linear theories in structural mechanics<br>– Graphical introduction to the loadbearing system analysis of shell structures<br>– In-depth study of technical codes of practice |
| **Aims** | – Nature and aims of structural theory in building in conjunction with strength of materials, the constructional and planning disciplines<br>– Use of examples to clarify the concepts of structure, design, loadbearing system, loadbearing structure, structural system, loadbearing behaviour, loadbearing function, loadbearing quality, loadbearing system analysis/synthesis, static determinacy/indeterminacy<br>– Direct and conscious cognizance of the built environment | – Quantitative assessment of stresses and deformations in beam structures<br>– Loadbearing system analysis of simple (historical) carpenter-built structures<br>– Knowledge of the loadbearing behaviour of structures and the loadbearing quality of loadbearing systems on the basis of equilibrium and simple deformation considerations<br>– Development of the ability to infer the nature of the loadbearing quality<br>– Construction of simple loadbearing systems from loadbearing system elements (loadbearing system synthesis) | – Quantitative assessment of internal force distribution and influence lines for internal force and deformation variables in statically indeterminate systems according to first-order theory<br>– Qualitative insight into the nature of stability problems (through discussion of historical building accidents attributable to loss of stability due to buckling)<br>– Mastery of the modelling processes from the loadbearing structure to the loadbearing system through classification of the loadbearing system and determination of loadbearing behaviour | – Acquisition of design competence in structural engineering<br>– Ability to assess possible securing measures for existing structures in need of refurbishment<br>– Insight into the necessity for interdisciplinary cooperation in the process of creating loadbearing systems |

|  | **First step** | **Second step** | **Third step** | **Fourth step** |
|---|---|---|---|---|
| **Aims** (contd.) | – Development of the ability to derive the nature of the loadbearing system from the appearance of the structure<br>– Quantitative assessment of the equilibrium phenomenon through the historico-logical comparison of lever principle and parallelogram of forces and the principle of virtual displacements (historico-logical comparison)<br>– Structural analysis of simple trussed frameworks, understanding the modelling from trussed framework structures to statically determinate hinged trussed frameworks (historico-logical transverse section analysis) | – Gaining awareness of the role of architect and structural engineer in the preservation of historically important structures, simple quantitative loadbearing system analyses of known, historically important structures (dome to St. Peter's in Rome, load-carrying characteristics of Gothic churches) | – Development of skills in assessing the influence of parameter changes on the internal force and deformation conditions<br>– Gaining awareness of the revolutionary influence that reinforced concrete had on the development of loadbearing systems (rigid frame structures, plane and curved shell structures, wide-span arch structures) using examples from the history of building<br>– Creating the conditions for developing modelling fantasy |  |
| **Means** | Photos, sketches, schemes, comparison of various constructional solutions for a functional problem, calculation and drawing aids | Photos, system sketches, historical source material, calculation and drawing aids, provision of principle statements regarding the internal forces and deformation conditions of statically determinate systems, models | Photos, system sketches, calculation aids, diagrams, simple computer programs, classical fundamental literature | Photos, damage records, crack patterns, building records, special literature, archive material, design aids, drawings, program systems, measuring instruments |
| **Characteristics** | Inductive, historical, use of examples, qualitative, discourse, interdisciplinary, analytical | Inductive-deductive, historico-logical, use of examples, quantitative-qualitative, discourse, interdisciplinary, analytic-synthetic | Deductive-inductive, logical-historical, theoretical, quantitative, analytic-synthetic, literary | Inductive, historical, use of examples, qualitative-quantitative, discourse, interdisciplinary, analytical, systematic, anticipatory, experimental |

# BRIEF BIOGRAPHIES

**AIRY, SIR GEORGE BIDDELL**
*27.7.1801 Alnwick, UK, †2.1.1892 Greenwich, UK

George Biddell Airy was the eldest child of William Airy – a farmer, later a tax collector – and his wife Ann Biddell. He attended Colchester Grammar School, where he learned mathematics and physics in addition to Ancient Greek and Latin. After leaving school, Airy's wealthy uncle recommended that he study mathematics in Cambridge. With the financial support of his uncle, Airy was able to complete his studies and in 1823 graduated with distinction from Trinity College as bachelor of arts. He was awarded the Smith Prize, the highest mathematics award the university can bestow, and also became senior wrangler. One year later he was elected a fellow of Trinity College, where he had already given lectures in mathematics and astronomy, and by 1826 had gained a master of arts, to which he added the Lucasian professorship for mathematics in the same year. On the advice of William Herschel (1738–1822), Airy applied for an astronomy professorship in Dublin, but William Rowan Hamilton (1805–65) was given the post instead. However, this did nothing to thwart his amazing scientific career; in 1828 he gained the Plumian professorship for astronomy and became director of the new observatory in Cambridge, and from 1835 to 1881 he was director of the Greenwich observatory. Airy's scientific and organisational activities manifest themselves in more than 500 publications on astronomy and other areas of science. In all his work he showed a special interest in the direct application of mathematics, particularly for engineering science problems; and this is what caused Airy to become involved in a heated debate with Arthur Cayley (1821–95), an advocate of pure mathematics. Airy became an honorary member of the Institution of Civil Engineers in 1842 and was a much-sought-after adviser to the British construction and railway engineering fraternity. He had close connections to Thomas Bouch (1822–80), William Henry Barlow (1812–1902), Isambard Kingdom Brunel (1806–59), William Fairbairn (1789–1874) and Robert Stephenson (1803–59). Airy gave impetus to the building of the Britannia Bridge (1846–50) by investigating the strains inside beams, which was the subject of a lecture he gave to the Royal Society in London in December 1862; this lecture was later printed in the Society's *Philosophical Transactions* [Airy, 1863]. Airy considered the beam to be an elastic plate and founded the theory of stress functions which was continued by Maxwell and completed by Michell. In 1870 Maxwell called the stress function $F(x,y)$ of the plane stress state the "Airy stress function", a designation which since then has been an important concept in elastic theory.

- Main contributions to structural analysis: *On the Strains in the Interior of Beams* [1863]
- Further reading: [Airy, 1896]; [Raack, 1977]; [Meleshko, 2003]
- Photo courtesy of [Raack, 1977, p. 3]

**ARGYRIS, JOHN HADJI**
*19.8.1913 Volos, Greece, †2.4.2004 Stuttgart, Federal Republic of Germany

Argyris stemmed from a Greek Orthodox family. His father was a direct descendant of a fighter in the battle for Greek independence, whereas his mother came from an old Byzantine family which had given birth to politicians, poets and scientists, e.g. Constantin Carathéodory (1873–1950), who in 1924 had been appointed professor of mathematics at the University of Munich [Hughes et al., 2004, p. 3763]. Argyris attended a classical grammar school in Athens and subsequently studied civil engineering at the technical universities in Athens and Munich; he graduated with distinction from Munich TH in 1936. After a brief period as an assistant to Günter Worch (1895–1981), who had been appointed to the chair of theory of structures and structural steelwork one year before, he decided on a practical career as a designer, structural engineer and then as chief project engineer at J. Gollnow & Sohn, the renowned steel fabricators based in Szczecin (Poland). Argyris' work there involved the calculation of guyed masts, which were appearing in great numbers for radio antennas during the 1930s. For instance, his first paper, published in the journal *Der Stahlbau*, is a precise analysis of a three-guy mast for a wind loading case that had not been considered hitherto [Argyris, 1940]. It was customary to apply the wind load in the plane in which the guy ropes support the mast. However, Argyris proved that the wind load perpendicular to this plane should also be investigated. He developed a method whereby this loading case could be considered and using the approach of the classical structural theory of Müller-Breslau and others modelled the mast as an elastically bedded beam. It is no coincidence that Christian Petersen, a student of Worch, would be left to deal with this theme exhaustively [Petersen, 1970]. But the highly talented Argyris could not resist the attraction of aircraft construction, a field that in the 1930s was more important than structural steelwork in terms of theory development in structural mechanics and as such was firmly embedded in the rearmament programme of the Third Reich. It was this field that Argyris had been studying at Berlin TH since 1939. Following the invasion of Greece by the German armed forces in early 1941, Argyris was arrested and accused of betraying research findings to the Allies. The head of the intelligence service of the German armed forces Wilhelm Franz Canaris (1887–1945), who was a member of the

AIRY

ARGYRIS

ARON

ASPLUND

conservative resistance against Hitler, arranged Argyris' escape: under the pretence that Argyris was to be executed outside the concentration camp, the guards let them pass by. Shortly afterwards, Argyris, holding his passport between his teeth, managed to swim across the Rhine during a night-time air raid [Hughes et al., 2004, p. 3764]. He resumed his studies in 1942 at the Zurich ETH Institute for Aerodynamics under Jakob Ackeret (1898–1981) and gained a further academic qualification. In 1943 he settled down in the UK, where until 1949 he was employed in the Research Department of the Royal Aeronautical Society. "Looking back," Erwin Stein says of the work of Argyris, "these were probably his most fruitful scientific years, if one places the fundamental ideas in the foreground" [Stein, 1985, p. 9]. As early as August 1943, Argyris had developed triangular elements for the structural mechanics analyses of sweptback wings. In doing so, he was able to make use of the shear field theory devised by Ebner and Köller [Ebner & Köller, 1937/2; 1939], which anticipated the notion of discretisation (see section 10.4.2.1). In his three secret memorandums, which he presented to the Research Council in 1943, 1947 and 1949, he also laid down in writing his bold idea of the triangular TRIM plate elements; but these works were regarded as "nonsense" by the Council and not approved for publication [Stein, 1985, p. 10]. The year 1949 saw Argyris appointed to the post of senior lecturer at Imperial College, London, and one year later reader in theory of aeronautical structures, and by 1955 he had become professor and director of the sub-department for aeronautical studies. During this period, Argyris published his brilliant matrix algebra reformulation of structural mechanics (see section 10.4.3), a watershed in non-classical engineering sciences which can be regarded as the first stage in the scientific revolution in structural mechanics as Thomas S. Kuhn saw it. He remained at Imperial College until 1975. Argyris became professor and director of the newly created Institute for Statics & Dynamics of Aerospace Structures at Stuttgart TH in 1959, a post he held until 1984; from then until 1994 he was in charge of the Institute of Computer Applications. It was within this organisational framework that Argyris pushed forward the development of FEM with every resource at his disposal, including the founding of the journal *Computer Methods in Applied Mechanics and Engineering* in 1970 and the monographs that Argyris wrote in conjunction with Hans Peter Mlejnek [Argyris & Mlejnek, 1986–88; 1991; 1997] to mention just two factors. Instead of concluding this brief biography with a list of the honours Argyris received, a quote from the obituary by Thomas J. R. Hughes, J. Tinsely Oden and Manolis Papadrakakis seems more pertinent: "His geometrical spirit, the elegance of his writings, his deep appreciation and understanding of classical ideas, his creativity and his epochal vision of the future initiated and defined the modern era of Engineering Analysis and set us all on life's path of discovery" [Hughes et al., 2004, p. 3766].

• Main contributions to structural analysis: *Untersuchung eines besonderen Belastungsfalles bei dreiseitig abgespannten Funkmasten* [1940]; *Structural Analysis* [1952]; *Energy Theorems and Structural Analysis* [1954/1955/1960]; *Die Matrizentheorie der Statik* [1957]; *Modern Fuselage Analysis and the Elastic Aircraft* [1963]; *Recent Advances on Matrix Methods in Structural Analysis* [1964]; *The Computer shapes the theory* [1965]; *Die Methode der Finiten Elemente* [1986–88]; *Dynamics of Structures* [1991]; *Computerdynamik der Tragwerke* [1997]

• Further reading:
[Stein, 1985]; [Hughes et al., 2004]; [Doltsinis, 2004]; [Phocas, 2005]

• Photo courtesy of University of Stuttgart archives

### ARON, HERMANN

*1.10.1845 Kempen, Posen province, Prussia (today: Poland), †29.8.1913 Bad Homburg, German Empire

Hermann Aron was the son of a trader and cantor who wished him to become a Jewish scholar. However, wealthy relations were of the opinion that the talented youngster should study. So at the age of 16 Aron went to the Kölln secondary modern school in Berlin, where classical education was combined with the modern natural sciences. After passing his university entrance examination in 1867, he initially studied at the University of Berlin and then at the University of Heidelberg. During those years he attended the lectures of Gustav Kirchhoff, worked as an assistant in the physics laboratory of the Berlin Industrial Academy (today: Berlin TU), gained a doctorate at the University of Berlin and took up the post of physics teacher at the Berlin Combined Artillery & Engineering School. It was in this period that Aron published the first bending theory of elastic shells with any curvature for static and dynamic loading cases [Aron, 1874], making use of the differential geometry approach employed in virtuoso fashion by Clebsch in his book *Theorie der Elasticität fester Körper* [Clebsch, 1862]. But Aron's shell theory was not adopted; Love formulated a bending theory for shells some 14 years later [Love, 1888], which finally became known through his famous textbook [Love, 1892/93]. "I often think and conceive slower than others," said Aron about himself, "but my understanding is more long-lasting, more thorough" (cited in [Förster, 2005, p. 15]). This self-appraisal fits in completely with his work on shell theory. Aron wrote his habilitation thesis at the University of Berlin in 1876, taking as his theme electrical measurement systems; he then taught physics and chemistry as a private lecturer at that university. In 1879 he and others founded the Berlin Electrical Engineering Society. Aron became famous through his inventions in the field of electrical engineering. For instance, in 1884 he designed the "Aron clock meter", the first electricity meter, which became the principal product of his successful business (Aron-Werke Elektricitäts-Gesellschaft). The company was renamed Heliowatt Werke Elektricitäts-AG in 1933 and during the years of the Third Reich was Aryanised – a verb of the *Lingua Tertii Imperii* (LTI – Victor Klemperer), the language of the Third Reich, which stood for the compulsory seizure of Jewish property. Hermann Aron's children saved themselves by emigrating to the USA and UK.

• Main contributions to structural analysis: *Ueber das Gleichgewicht und die Bewegung einer unendlich dünnen, beliebig gekrümmten elastischen Schale* [1874]

• Further reading:
[Harand, 1935, pp. 234–236]; [Förster, 2005]

• Photo courtesy of the archives of Electrosuisse

### ASPLUND, SVEN OLOF

*16.6.1902 Skön near Sundsvall, Sweden, †8.7.1984 Gothenburg, Sweden

Following his examinations as a structural engineer at Stockholm's Royal Institute of Technology, Asplund worked in the USA from 1925 to 1931.

It was there that he was able to play an active part in the upswing in the market for large bridges in the USA at the Melan Arch Bridge Construction Co. and the American Bridge Co.; he also spent time with Robinson & Steinman in New York designing and calculating suspension bridges. After returning to Sweden, he set up a contracting company in Örebro specialising in temporary and permanent suspension bridges with small and medium spans for hydroelectric power stations. In 1943 Asplund completed his dissertation *On the Deflection Theory of suspension Bridges* in which he applied the Green function $\Pi(\varepsilon)$ to this non-linear problem. He generalised this elegant approach later for other loadbearing structures, making use of matrix formulation [Asplund, 1958]. From 1949 until his retirement in 1967, Asplund worked as professor of bridge-building at Chalmers University of Technology in Gothenburg; in 1950 he was appointed to the vacant chair of theory of structures following the death of Sven Hultin. In his period as professor of bridge-building and structural mechanics it is his scientific work on the matrix formulation of structural mechanics, with a view to its programming on the computer, that is especially noteworthy. By the second half of the 1940s, he had already formulated a method for analysing pile groups with the help of matrices, which he later generalised [Asplund, 1956]. Asplund had already been working with a mainframe computer at home even before a computer was installed at his university at the end of the 1950s. It is therefore no surprise that he was very active in introducing computers for scientific calculations in general and structural mechanics calculations in particular at Chalmers as well. He was able to draw on his experiences as a visiting professor at universities in California, Florida and Pennsylvania. In 1966 Asplund crowned his scientific life's work with the monograph *Structural Mechanics: classical and matrix methods* [Asplund, 1966]. One year later, the suspension bridge designed by Asplund over the River Göta in Gothenburg, with a central span of 417.60 m, was opened. The roof structure to the Gothenburg Ullevi Stadium and the Stockholm TV tower (Kaknästornet, opened in 1967) can also be counted among Asplund's ingenious creations. Asplund can be regarded as the founding father of computer-assisted structural mechanics in Sweden, which also gained international acclaim. A number of his doctoral students became successful professors in structural mechanics and related fields in Sweden and abroad, e.g. Alf Samuelsson (1930 – 2005) [Wiberg et al., 2006]. Asplund was a member of the New York Academy of Sciences, the Swedish Academy for Engineering and an active member of the International Association for Bridge & Structural Engineering (IABSE).

- Main contributions to structural analysis:
*On the Deflection Theory of suspension Bridges* [1945]; *Generalized Elastic Theory for Pile Groups* [1956]; *Column-Beams and suspension bridges analyzed by 'Green's matrix'* [1958]; *An unified Analysis of indeterminate structures* [1961]; *Practical Calculation of suspension bridges* [1963]; *Structural Mechanics: classical and matrix methods* [1965]
- Further reading:
[Samuelsson, 1984]; [Samuelsson, 2004]
- Photo courtesy of [Samuelsson, 2004, p. 36]

### BAKER, SIR JOHN FLEETWOOD

(Baron Baker), *19.3.1901, Liscard, Cheshire, UK, †9.9.1985 Cambridge, UK

After completing his studies at the University of Cambridge in 1923, John Fleetwood Baker was employed by the Air Ministry on the structural design of airships. When he was 28 he was diagnosed as having tuberculosis and the doctors recommended that he give up work. However, financial reasons forced him to ignore the doctors' advice – a wise, if risky, decision that would later prove to be fortuitous for the development of structural analysis in its invention and innovation phases. In 1929 he was appointed Technical Officer on the Steel Structures Research Committee (SSRC), with the challenging task of revising the British structural steelwork standards. His measurements on steel structures revealed that the analysis principles led to results that deviated significantly from real loadbearing behaviour. The solution of this dilemma through the introduction of plastic methods of design formed Baker's scientific life's work. From 1933 to the start of World War 2 he was professor of engineering at the University of Bristol, where he was able to continue his ultimate-load tests successfully. He earned great praise as the scientific adviser to the British Home Defence thanks to his design of air-raid shelters. His analysis of buildings damaged by bombs formed the basis of a critical examination of existing air-raid shelters and led to the development of small steel shelters which could be erected in houses and which were designed using Baker's ultimate-load theory. More than 1.25 million of these so-called Morrison shelters (named after the then Minister responsible) saved countless lives. After Hitler conceded defeat in the Battle of Britain, and the tide had turned in favour of the Allies, Baker responded to the University of Cambridge's request (1943) to set up a Structural Research Laboratory at the University, which would later become famous for its pioneering contributions to the field of structural engineering: plastic hinge method, inherent stresses, brittle failure and fatigue problems in steel. The welded construction of a building for the Welding Institute (a spin-off of the University of Cambridge) built in 1946 near Cambridge was the first steel-frame building designed according to plastic theory; three more buildings belonging to Cambridge University's Faculty of Engineering followed in the early 1950s. He served his Alma Mater as dean of the Faculty of Engineering until being granted emeritus status in 1968. He was elected a member of the Royal Society 1956, knighted in 1961 and granted a life peerage in 1977. Lord Baker deserves to belong to the Hall of Fame of great British structural and civil engineering personalities together with famous names like Telford and Rankine.

- Main contributions to structural analysis:
*The Mechanical and Mathematical Analysis of Steel Building Frames* [1932]; *The Rational Design of Steel Building Frames* [1935/36]; *Modern Methods of Structural Design* [1936/37]; *The Steel Skeleton 1: Elastic Behaviour and Design* [1954]; *The Steel Skeleton 2: Plastic Behaviour and Design* [1956]; *Plastic Design of Frames 1: Fundamentals* [1969]
- Further reading:
[Heyman, 1985]
- Photo courtesy of Dr. W. Addis

### BAN, SHIZUO

*16.11.1896, Tokyo, Japan, †13.10.1989, Osaka, Japan

Following completion of his education at the 1st state grammar school in Tokyo, Shizuo Ban studied at the Imperial University in Tokyo from 1918 to 1921. From 1922 to 1933 he was associate professor at the Imperial University in Kyoto, and gained his doctorate in 1928 at the Imperial University in Tokyo with his dissertation on the analysis and design of reinforced concrete structures. From June 1931 to August 1933 Ban also studied at Karlsruhe TH and Zurich ETH. Upon returning to Japan he was appointed professor at the Imperial University in Kyoto, where he remained until November 1959. He was nominated honorary professor by his Almer Mater, and after being granted emeritus status he served as director of the General Building Research Corporation in Osaka (1964 – 74) and afterwards as its chairman (1974 – 83).

- Main contributions to structural analysis:
*Buckling of a rectangular plate under varying loading* [1935]; *Deformation of a hyperbolic paraboloid shell* [1953]
- Further reading:
[Yokoo, 1980]
- Photo courtesy of Prof. Dr. M. Yamada

### BARLOW, PETER

*October 1776, Norwich, Norfolk, UK, †1.3.1862, Woolwich, London, UK

Right from the start of his working life, Peter Barlow pursued his passion for mathematics and other sciences in his spare time. This led him to found a scientific society in which young people could discuss a wide range of scientific topics. He later became a teacher, and served on the teaching staff of the Woolwich Royal Military

 BAKER
 BAN
 BARLOW
 BAŽANT
 BÉLIDOR
 BELTRAMI

Academy from 1802 until his retirement in 1847. Barlow worked in various fields, e. g. mathematics, magnetism, strength of materials, performance of locomotives. His most important contribution to the strength of materials [Barlow, 1817] was the first British monograph on this subject and had a profound influence on the science of strength of materials during the constitution and establishment phases of structural theory. Barlow's book represents a convincing description of mathematical analyses and the strength tests he carried out with great care in his laboratory in Woolwich. He used his brilliant intellect to solve significant technical problems of the day, whether as consultant for Telford's suspension bridge over the Menai Strait (1817) or in overcoming errors in ships' compasses (1825). In just a few years he became the leading experimenter in the field of electromagnetism. He was elected to the Royal Society in 1823. His analysis of the load and pressure relationships of Bramah's large hydraulic press led in 1825 to reliable and safe design principles. He received inquiries from all over Europe concerning problems in shipbuilding, military matters and countless other scientific and technical fields.

- Main contributions to structural analysis:
*An Essay on the Strength and Stress of Timber* [1817]; *Experiments on the transverse strength and other properties of malleable iron with reference to its uses for railway bars; and a report founded on the same, addressed to the directors of the London and Birmingham Railway Company* [1835]
- Further reading:
[Barlow, 1867]; [Timoshenko, 1953]; [Smith, 2002]
- Photo courtesy of Dr. W. Addis

### BAŽANT, ZDENĚK
*25.11.1879 Prostějov, Mähren, Austria-Hungary (today: Czech Republic), †1.9.1954 Nové Město, Mähren, Czechoslovakia (today: Czech Republic)
Zdeněk Bažant studied at the Faculty of Civil Engineering at the Czech TH in Prague from 1896 to 1902. In 1901 he was appointed assistant to Prof. J. Šolín and was commissioned to design steel bridges, market halls and cable-car systems by various steel fabricators. He gained his doctorate at the Czech TH in Prague in 1904 with a dissertation on the subject of statically determinate continuous trussed girders and began his lectures on structural mechanics, theory of structures and strength of materials. In 1906 he completed his habilitation thesis entitled *Theory of influence lines* at the same university. He was promoted to lecturer in 1907, then associate professor (1909) and finally (1917) to full professor at the Czech TH in Prague. Bažant served his Alma Mater for 44 years and was its rector on two occasions. He had a great impact on the development of civil engineering in the former Czechoslovakia, supplying important contributions to systemising the theory of influence lines, to frame theory and to graphical methods in the theory of structures. In those fields he achieved international acclaim as editor and author of several textbooks and manuals, and published numerous papers in Czech, German and French. Bažant was a member of several learned societies such as the International Association for Bridge & Structural Engineering (IABSE) as well as the Czechoslovakian and Polish academies of sciences.

- Main contributions to structural analysis:
*Teorie příčinkových čar* (theory of influence lines) [1909/1910/1912]; *Statika stavebních konstrukcí* (manual of the theory of structural design) [1917/1946]; *Stavebná mechanika* (structural mechanics) [1918/1920/1946/1950]
- Further reading:
[Hořejší & Pirner, 1977, pp. 18 – 22]
- Photo courtesy of Prof. Dr. L. Frýba

### BÉLIDOR, BERNARD FOREST DE
*1697/98? Catalonia, Spain, †1761 Paris, France
Bélidor's father, Spanish cavalry officer Jean-Baptiste Forest de Bélidor, and his mother, Marie Hébert, both died when he was just five months old. Bernard Forest de Bélidor was therefore brought up by his godfather, De Fossiébourg, an artillery officer. Up until 1718, the young Bélidor assisted in the Meridian measurements between Paris and the English Channel coast led by Jacques Cassini and Philippe de La Hire, the results of which were published in 1720. The Duke of Orléans became aware of Bélidor's talents and set him up as professor of mathematics at the newly founded Artillery Academy in La Fère. During this period he published manuals entitled *La science des ingénieurs* and *Architecture hydraulique*, the very first civil engineering textbooks of the modern age, and which were to remain influential until the early 19th century. Bélidor's manuals inspired Lazare Carnot, Coulomb, Poncelet and Navier – the latter republished them with a comprehensive, critical commentary [Bélidor, 1813; 1819].

- Main contributions to structural analysis:
*La science des ingénieurs dans la conduite des travaux de fortification et d'architecture civile* [1729]; *Architecture hydraulique* [1737 – 53]
- Further reading:
[Gillispie, 1970]
- Photo courtesy of École Nationale des Ponts et Chaussées

### BELTRAMI, EUGENIO
*16.11.1835 Cremona, Austria-Hungary (today: Italy), †18.2.1900 Rome, Italy
Eugenio Beltrami was born into a family of artists. His father was a respectable painter of miniatures who as a result of political turmoil emigrated to Paris in 1848 where he became the curator of an art gallery. Eugenio Beltrami began studying mathematics at the University of Pavia, but had to interrupt his studies for financial reasons and took a job in the offices of the Lombardy–Venice Railway. At the age of 25 he was able to resume his studies, guided by the very assured advice of Francesco Brioschi (1824 – 97). Beltrami published his first mathematics paper in 1861 and others followed in quick succession. By 1862 he was able to resign from his job in the railway offices and at the age of 27 was appointed associate professor of algebra and analytical geometry at the University of Bologna. "From then on, his life was cheerful and relaxed, dedicated exclusively to caring for his family, his students and his favourite studies" [Pascal, 1903, p. 67]. That energetic and untiring promoter of young scholars, Enrico Betti, was responsible for Beltrami's appointment as professor at the University of Pisa, where Cremona was one of his colleagues and where he worked as a teacher and researcher from 1863 to 1866; thereafter, he moved back to the University of Bologna to take up a post as professor of theoretical mechanics until 1873. After Rome had become the capital of Italy in 1870, a plan was drafted to establish the largest university in the kingdom there staffed by the best scientists. And that is what happened. World-famous thanks to his contributions to differential geometry, Beltrami

was also invited to join the new university in Rome. Between 1873 and 1876, he lectured on the wide-ranging field of theoretical mechanics and higher analysis. During those years he drifted further and further away from his original areas of research and turned more and more to mathematical physics, which resulted in him being appointed professor for this subject at the University of Pavia, where he worked extremely successfully until 1891. During this second period of creativity, "he went through almost all the fields of mathematical physics in a series of 60 treatises: electricity, magnetism, potential theory, light, heat and elasticity" [Pascal, 1903, pp. 68–69]. In four treatises Beltrami managed to establish the mathematical foundations of linear elastic theory, in the forefront of which stands the tensorial differential equation (eq. 10-5, see section 10.2.1.1) named after him and John Henry Michell. Beltrami returned to the University of Rome in 1891, "where they had been trying to get him to return for some time; and it was a Rome that he died on 18 February 1900, just as he had lived – with the cheerful composure of an ancient philosopher" [Pascal, 1903, p. 69].

• Main contributions to structural analysis:
*Sulle equazioni generali dell'elasticità* [1881]; *Sulle condizioni di resistenzia die corpi elastici* [1885]; *Sull'interpretazione meccanica delle formole di Maxwell* [1886]; *Note fisico-matematiche (2ᵃ parte)* [1889/1]; *Sur la théorie de la déformation infiniment petite d'un milieu* [1889/2]; *Opere matematiche* [1902–20]
• Further reading:
[Pascal, 1903]; [Struik, 1970]
• Photo courtesy of [Beltrami, 1902]

## BENVENUTO, EDOARDO

*11.12.1940 Genoa, Italy, †27.11.1998 Genoa, Italy

After completing his education at the humanistic grammar school (*maturità classica*) in 1958, Edoardo Benvenuto went on to study piano at the Paganini Conservatorium. After successful completion of those studies, he switched to engineering and in 1965 graduated in civil engineering at the University of Genoa. From 1965 to 1974 Benvenuto worked on urban planning studies at the Istituto Ligure Ricerche Economiche e Sociali (ILRES). It was during this period that he began his outstanding academic career at the Faculty of Engineering at the University of Genoa: 1969–75 lectures on bridge-building, 1970 teaching certificate for structural dynamics and afterwards, until 1975, visiting professor of structural mechanics. In 1975 he was appointed to the chair of structural mechanics at the newly founded Faculty of Architecture in Genoa, on which he had a profound influence and managed extremely successfully during his years as dean from 1979 to 1997. His teaching experience and research work into the history of structural mechanics formed the basis for his 900-page monograph entitled *La Scienza delle Costruzioni e il suo sviluppo storico*, which was published in 1981. He thus became the founder of this branch of scientific history. Together with S. Di Pasquale and A. Giuffrè, Benvenuto initiated a research programme into the history of building theory and construction engineering, the results of which were intrinsic to a two-volume work published in 1991: *An Introduction to the History of Structural Mechanics*, which prompted that congenial spirit of rational mechanics C. A. Truesdell to write in his introduction: "This book is one of the finest I have ever read. To write a foreword for it is an honor, difficult to accept" [Benvenuto, 1991/1, p. VII]. And it is worth noting here that the Italian original dating from 1981 (reprinted in 2006) deals with the subject in considerably more detail. At the Faculty of Architecture in Genoa, Benvenuto became the spiritus rector of a course of study in which architecture and structural mechanics were integrated in an unprecedented creative symbiosis – clearly devoted to the retention of historical structures in the urban context. In 1993, together with the Belgian Bernoulli researcher Patricia Radelet-de Grave, he initiated the series of international conferences entitled *Between Mechanics and Architecture*, which advanced to become the itinerary for a school and since Benvenuto's early death has been continued by the Associazone Edoardo Benvenuto under its honorary president Jacques Heyman. Just four results of this itinerary will be mentioned here:
– the compendium *Towards a History of Construction* (2002) published by Becchi, Corradi, Foce and Pedemonte;
– *Degli archi e delle volte* (2002) by Becchi and Foce, with a very expertly annotated bibliography on the structural and geometrical analysis of past and present arches;
– the compendium on the status of the history of structural engineering *Construction History. Research Perspectives in Europe* (2004) published by Becchi, Corradi, Foce and Pedemonte;
– the reprint of *La Scienza delle Costruzioni e il suo sviluppo storico* (2006), Edoardo Benvenuto's main work, with an 18-page introduction by Becchi, Corradi and Foce.
At the same time, Benvenuto transcended the broad spectrum of building theory knowledge: scientific theory, philosophy and theology. For instance, as early as 1974 he published a book on materialism and scientific thinking, and from 1977 to 1980 taught part-time comparative cultural sciences at the Theological Faculty of Northern Italy. In 1997 he published a brilliant philosophical study on the genesis of the doctrine of Christian social teaching. And with a like-minded group of colleagues he founded the journal *Bailamme*. In 1996 Benvenuto was elected to the UNESCO Committee for Exact Sciences, and he also worked on the Committee for Culture for the same organisation. One year before his death, Benvenuto was elected president of the Ligurian Academy of Science and Literature in Genoa.

• Main contributions to structural analysis: *La Scienza delle Costruzioni e il suo sviluppo storico* [1981, 2006]; *An Introduction to the History of Structural Mechanics* [1991]; *Entre Méchanique et Architecture. Between Mechanics and Architecture* [1995]
• Further reading:
[Augusti, 1999]; [Becchi, A. et al., 2002]
• Photo courtesy of Università di Genova

## BETTI, ENRICO

*21.10.1823 near Pistoia, Italy, †11.8.1892 Pisa, Italy

After his father died when Enrico was very young, his mother brought him up alone. He later successfully completed studies in mathematics and physics at the University of Pisa. Under Mossotti's leadership he fought in the battles of Curtatone and Montanara for Italian independence. Betti taught mathematics at Pistoia Grammar School and was invited to become professor at the University of Pisa. He remained loyal to his Alma Mater until his death and also served as rector; he also headed the Pisa Teacher Training College. He became a member of parliament in 1862 and was a senator from 1884 onwards. For a few months in 1874 he worked as second permanent secretary in the Ministry for Public Education. However, his talents brought forth more fruits in other fields: he was a skilled teacher and scientist and so he played a important role in mathematics and elastic theory after the Risorgimento. His reciprocal theorem formulated in 1872 proved critical for the development of elastic theory and structural theory: "If in an elastic, homogeneous body two displacement systems are in equilibrium with two groups of loads applied to the surface, then the sum of the products of the force components of the first system multiplied by the displacement components at the same point of the second system is equal to the sum of the products of the force components of the second system multiplied by the displacement components of the first system at the same point" [Betti, 1872]. This theorem, which represents a generalisation of Maxwell's theorem [Maxwell, 1864/2], had already been incorporated by Mohr [Mohr, 1868]; in 1887 Robert Land had developed the reciprocal theorem independently from the work theorem, recognised its fundamental importance to classical structural theory and based his kinematic theory on this [Land 1887/2]. Betti's contributions to algebra and the theory of functions were equally important.

• Main contributions to structural analysis:
*Teorema generale intorno alle deformazioni che fanno equilibrio a forze che agiscono alla*

BENVENUTO        BETTI        BEYER        BISPLINGHOFF

*superficie* [1872]; *Sopra l'equazioni d'equilibrio dei corpi elastici* [1873–75]; *Teoria della elasticità* [1913]; *Opere matematiche* [1903–13]
- Further reading:
[Brioschi, 1892]; [Carruccio, 1970];
[Bottazzini, 1977]; [Bottazzini, 1977/1978];
[Benvenuto, 1981]; [Charlton, 1982];
[Biermann, 1983]; [Benvenuto, 1991/2]
- Photo courtesy of Università di Genova

### BEYER, KURT
*27.12.1881 Dresden, German Empire,
†9.5.1952 Dresden, German Democratic Republic
Kurt Beyer completed his civil engineering studies at Dresden TH in 1905. His dissertation, supervised by Georg Christoph Mehrtens, concerned the optimisation of bridge systems. Afterwards, he was employed for several years as a section engineer for the Siamese State Railways, which came to an end with the outbreak of World War 1. In 1919 he was appointed to the newly created professorship of theory of structures and applied mechanics for civil engineers at Dresden TH. His monograph *Statik im Eisenbetonbau* (1927), which was later reprinted more than once in considerably expanded editions under the title of *Statik im Stahlbetonbau*, became the most important standard work in the German language during the consolidation period of structural theory and influenced many engineers outside Germany as well. Besides the "Beyer Bible", as his work is often called, he published – together with Heinrich Spangenberg – the third, revised edition of Otto Mohr's *Abhandlungen aus dem Gebiete der Technischen Mechanik*. Without doubt, Beyer is, after Mohr, the most important advocate of the Dresden school of applied mechanics. As a consulting engineer, he devoted much of his time to structures for open-cast lignite mining (e. g. conveyor bridges), steel buildings, bridges and steel hydraulic structures. After the founding of the German Democratic Republic in 1949, his work on the theory of civil engineering was honoured by his appointments to several academies, including the German Academy of Sciences.
- Main contributions to structural analysis:
*Eigengewicht, günstige Grundmaße und geschichtliche Entwicklung des Auslegerträgers* [1908]; *Die Statik im Eisenbetonbau* [1927]; *Die Statik im Stahlbetonbau* [1956]
- Further reading:
[Koch et al., 1992]; [Möller & Graf, 2002]; [Stroetmann, 2007]
- Photo courtesy of Dresden TU archives

### BISPLINGHOFF, RAYMOND LEWIS
*7.2.1917 Hamilton, Ohio, USA,
†5.3.1985 Boston, Massachusetts, USA
Without doubt, Bisplinghoff belongs to the leading scientists and organisers of the USA's aerospace industry in the second half of the 20th century. Bisplinghoff, the son of a miller, attended the University of Cincinnati and graduated in the subjects of aircraft construction and physics. He chose a theme from physics for his dissertation project, but had to abandon this because of World War 2. Following service as an engineering officer in the US Navy Bureau of Aeronautics, he resumed his academic career as assistant professor for aeronautical engineering at the MIT in 1946. It was during this period that he completed his pioneering work on the structural dynamics behaviour of sweptback aircraft wings (see section 10.4.2.2 and [Bisplinghoff et al., 1955, pp. 51–56]). Hubert I. Flomenhoft, one of Bisplinghoff's colleagues at that time, remembers him well: "Soon after arriving at MIT, he sent a proposal to the office which he had left [US Navy Bureau of Aeronautics – the author] and was awarded a contract to study a range of problems in structural dynamics. This was completed in 1948, and the final report was a spiral-bound document about three centimeters in thickness with a striking blue cover. This became known as 'the big blue Bible', and was a basic reference for many later authors. The report also became the basis for about half of the book, *Aeroelasicity* [see [Bisplinghoff et al., 1955] – the author]" [Flomenhoft, 2007]. After two years, Bisplinghoff was promoted to associate professor, became full professor in 1953 and in 1957 deputy director of the MIT's Department of Aeronautics; in that same year Bisplinghoff was finally able to conclude his doctorate at Zurich ETH. The scientific output of his research group at MIT was summarised in three extraordinary monographs which he wrote together with Holt Asley, Robert L. Halfmann, James W. Mar and Theodore H. H. Pian [Bisplinghoff et. al., 1955; 1961; 1965], and which today are still among the standard works of structural mechanics. As an assistant administrator at NASA, Bisplinghoff was responsible for organising the scientific work. In 1966 he was nominated head of the Department of Aeronautics & Astronautics, and in 1968 dean of the MIT School of Engineering. During this same period, Bisplinghoff was in charge of the NASA Research & Technology Council and was therefore directly involved in preparing, organising and evaluating the scientific and technical work of the Apollo 8, 9, 10, 11 and 12 missions, the spectacular climax of which was the manned moon landing of 20 July 1969. One year after that, Bisplinghoff became deputy director of the National Science Foundation and in 1974 chancellor of the University of Missouri. He concluded his career as senior vice-president of research at Tyco Laboratories in Exeter, New Hampshire. "When Raymond L. Bisplinghoff died in 1985," Flomenhoft writes, "he was recognized by the United States Congress as a distinguished aeronautical engineer, renowned for his teaching, research, engineering, and writing, and for his leadership in education, government, and industry" [Flomenhoft, 1997, p. 2].
- Main contributions to structural analysis:
*Some results of sweepback wing structural studies* [1951]; *Aeroelasticity* [1955]; *Principles of Aeroelasticity* [1961]; *Statics of Deformable Solids* [1965]; *History of Aeroelasticity* [1997]
- Further reading:
[Stever, 1989]; [Flomenhoft, 1997]
- Photo courtesy of [Stever, 1989, p. 36]

### BLEICH, FRIEDRICH
*2.10.1878 Vienna, Austria-Hungary,
†17.2.1950 New York, USA
Friedrich Bleich studied civil engineering at Vienna TH (1896–1902); from 1902 to 1906 he was employed in the Prag-Bubno bridge-building company and afterwards as a senior engineer at Waagner-Biro A.G. in Vienna until he founded his own consultancy in 1910. In that same year he became co-founder of the journal *Der Eisenbau*, the scientific orientation of which occupied him extensively until publication was discontinued during the period of runaway inflation. Over the years 1914–16 he deputised as lecturer for bridge-building at Vienna TH. He gained his doctorate in 1917 with his thesis *Der Viermomentensatz und seine Anwendung auf die Berechnung statisch unbestimmter Tragwerke* supervised by Friedrich Hartmann and Anton Zschetzsche. He acted as an outstandingly

successful scientific secretary to the International Association for Bridge & Structural Engineering (IABSE) from its foundation in 1938. Hitler's annexation of Austria and the rise of the Third Reich forced Bleich to flee to Zurich in 1938. He emigrated to the USA in 1941, where he was employed by Albert Kahn, Associated Architects & Engineers, in Detroit; later he was offered a post at the US Institute for Steelwork. From 1947 until his death Bleich worked for Frankland & Lienhard, a firm of consulting engineers in New York. He worked on bridge designs, carried out research into the cause of the collapse of the Tacoma Narrows Bridge and – on behalf of the US Navy Department – compiled information on buckling and warping stresses in metal components. He was the first structural engineer to employ the finite difference method for the analysis of continuous beams (1904) and together with Ernst Melan developed a generalisation of this method [Bleich & Melan, 1927]. He also made vital contributions to the development of the theory of stability for steel buildings during the consolidation period of structural theory; over that period, his theorem of four moments enjoyed great popularity among practising engineers. Bleich's monograph [Bleich, 1924] on the theory and analysis of steel bridges gradually superseded Josef Melan's bridge books from the late 1920s onwards. Bleich was an excellent representative of that group of consulting engineers who made contributions to practical design and construction but also the formation of structural theory during the consolidation period.

- Main contributions to structural analysis:
*Die Berechnung statisch unbestimmter Tragwerke nach der Methode des Viermomentensatzes* [1918]; *Theorie und Berechnung der eisernen Brücken* [1924]; *Die gewöhnlichen und partiellen Differenzengleichungen der Baustatik* [1927]; *Buckling Strength of Metal Structures* [1952]
- Further reading:
[Girkmann, 1952]

### BLEICH, HANS HEINRICH

*24.3.1909 Vienna, Austria-Hungary,
†8.2.1985 New York, USA

Hans, the son of Friedrich Bleich, was awarded a diploma in structural engineering by Vienna TH in 1931 and three years later gained his doctorate at the same university. During his student days, he contributed a chapter to the book *Die gewöhnlichen und partiellen Differenzengleichungen der Baustatik*, which was published by his father and Ernst Melan in 1927. So in terms of his career he followed in the footsteps of his father. In 1932 Bleich drew attention to himself with a revolutionary publication in which the principle for the shake-down of steel structures was formulated for the first time, and which Ernst Melan generalised for structures in bending in 1938. Shake-down occurs when the plastic dissipation work remains limited in all parts of a loadbearing structure under the action of alternating loads. The Bleich-Melan shake-down principle forms an important cornerstone of plastic theory. Up until 1939, Bleich worked as an engineer for the Vienna-based contractor A. Porr AG and, following his emigration in that same year, for Braithwaite & Co. Ltd. in London. He started a new life in the USA in 1945 and it was there that he was able to develop his talents as both structural engineer and scientist ideally – as a research engineer at Chance-Vaught Aircraft in Stratford, and as a bridges engineer at Hardesty & Hanover in New York. From 1957 until his death he was active as a consultant for Weidlinger Associates in New York, where he played a major role in the conception, design and analysis of innovative structural systems for high-rise buildings, exhibition buildings and special structures like the Mount Wilson Observatory. It was in 1947 that he began teaching and researching at Columbia University, initially in the field of structural engineering, later aircraft construction (he became director of the Guggenheim Institute of Air Flight Structures in 1954). He resigned from his Alma Mater in 1975 as professor of structural engineering. He worked on and edited his father's book *Buckling Strength of Metal Structures*, which appeared two years after the latter's death. In total, Bleich published 68 papers and reports on theory of structures and applied mechanics. "His technical reports dealt with the gamut of those applied mechanics problems that are of practical significance in the field of dynamics and, particularly, in the interactions between fluids and elastic and plastic bodies [Salvadori, 1989, p. 47]. Understandably, he made major contributions to important technical directives such as the ASCE manual *Design of Cylindrical Shell Roofs* (1952), the *Guide for the Analysis of Ship Structures* (1960) published by the US Department of Commerce, and the report entitled *Support and Testing of Astronomical Mirrors* (1968) published by the Kitt Peak National Observatory in Arizona. His amazingly successful creativity was rewarded by renowned associations and societies, the ASCE Kármán Medal being just one example.

- Main contributions to structural analysis:
*Über die Bemessung statisch unbestimmter Stahltragwerke unter Berücksichtigung des elastisch-plastischen Verhaltens des Baustoffes* [1932]; *Die Berechnung verankerter Hängebrücken* [1935]
- Further reading:
[Salvadori, 1989]
- Photo courtesy of [Salvadori, 1989, p. 44]

### BOSCOVICH (BOŠKOVIĆ), RUGGERO GIUSEPPE

*18.5.1711 Dubrovnik, Croatia,
†13.2.1787 Milan, Italy

The son of the merchant Nikola Boscovich and Paula Bettera, the daughter of a merchant from Bergamo, Boscovich attended the Jesuit College of his home town and afterwards Sant' Andrea in Rome. He continued his studies in theology at the Collegium Romanum and from 1735 onwards adopted Newton's *Opticks* and *Principia*, and by 1740 he had been promoted to professor of mathematics at the Collegium Romanum. Two years later he worked with Tommaso Le Seur and Francesco Jacquier to compile a report on the damage to the dome of St. Peter's, which in the literature of civil engineering is today still regarded as the first theory of structures in the world. Based on an analysis of damage and cracks, the *tre mattematici* calculated the horizontal thrust with the help of the work theorem, which formed the quantitative basis for the refurbishment of the dome construction with the help of iron ties in the form of a ring beam. Boscovich was appointed by the Vatican to provide further reports, e. g. drainage of the Pontifical marshes, river training works for the Tiber, extensions to the harbours at Rimini and Savona. In 1758 Boscovich published his *Theoria philosophiae naturalis*, an atomic theory according to which the world consists not of continuous matter, but instead of countless "point-like structures"; the ultimate elements of matter are indivisible points – "atoms" – which are centres of force and this force varies in proportion to distance. From his revolutionary discovery that "every material point [links] a point in space and a moment in time", he surmised some 150 years before Ernst Mach, Henri Poincaré and Albert Einstein the relativity principle of modern physics from the criticism of the understanding of inertia in traditional mechanics. In the 19th century his atomic theory formed the basis of the rari-constant theory (discontinuum hypothesis, molecular hypothesis) of elastic theory, which, however, could not displace the multi-constant theory (continuum hypothesis), but later found favour again in the modern atomic theory of Max Born [Foce, 1993; 1995]. Criticism of Boscovich's *Theoria philosophiae naturalis* from the viewpoint of Aristotelian natural philosophy was not long in coming. The inflammatory discussion was fuelled by the fact that Boscovich's ideas for reforming studies were aimed directly at the scholastic teaching traditions of the universities influenced by the Catholic Church, which hindered the evolution of the exact sciences. In 1759 Boscovich left Rome via Paris. He was elected a member of the Royal Society in London in 1761. Boscovich was now a traveller with a scientific and political mission: he carried the torch of enlightenment across Europe like no other. He became an outstanding organiser and promoter of scientific activities on a European scale. For instance, he inspired a series of meridian curve measurements (measurement of geographical longitude) which proved that the Earth did not have the "form of a distended melon"

BLEICH, H. H.

BOSCOVICH

BOUGUER

BRESSE

(Voltaire), but instead confirmed Newton's hypo-thesis of an ellipsoid flattened at the poles. Newton's theory of gravity had finally found its way into geodesy. Boscovich's work was widely acclaimed in the 20th century. At EXPO 2000 in Hannover, Germany, a replica of the Boscovich monument in Zagreb formed the centrepiece of the Croatian exhibition pavilion – immersed in a multimedia, seemingly transcendental simulation of the universe. Even today, his prophetic worldly wisdom thus works like a universal promise of the coming spirituality of the human race.

- Main contributions to structural analysis:
*Parere di tre mattematici sopra i danni, che si sono trovati nella cupola di S. Pietro sul fine dell'Anno MDCCXLII, dato per ordine di Nostro Signore Papa Bendetto XIV* [1743]; *De motu corporis attracti in centrum immobile viribus decrescentibus in ratione distantiarum reciproca duplicata in spatiis non resistentibus. Dissertatio* [1743]; *Elementorum Universae Matheseos* [1754]; *De Lege Virium in Natura existentium, Dissertatio* [1755]; *Philosophiae naturalis theoria redacata ad unicam legem virium in natura existentium* [1758]; *Scrittura sulli danni osservati, nell'edificio delle Biblioteca Cesarea di Vienna, e loro riparazione, composta in occasione de' Sovrani commandi di Sua Maestà l'Imperatrice Regina Maria Teresa, e umiliata a suoi piedi pel felicissimo anniversario (13 Mai 1763) della sua nascita* [1763]; *Sentimento sulla solidità della nuova Guglia del Duomo di Milano o si consideri in se stessa, o rispetto al rimanente del vasto Tempio, esposto a richiesta del Nobilissimo e Vigilantissimo Capitolo che sopraintende alla sua gran fabbrica* [1764]

- Further reading:
[Rota, 1763]; [Gil, 1791]; [Thomson, 1907]; [Pighetti, 1964]; [Marković, 1970]; [Paoli, 1988]; [Becchi, 1988]; [Benvenuto, Corradi & Foce, 1996]; [Wapenhans & Richter, 2001]; [Wapenhans & Richter, 2002]
- Photo courtesy of Università di Genova

### BOUGUER, PIERRE
*16.2.1698 Le Croisic, Brittany, France,
†15.8.1758 Paris, France
Bouguer studied at the Jesuit College in Vannes, the capital of Morbihan Departement. In geodesy and geophysics he is primarily known for the free-air anomaly that bears his name and which is important in connection with the research into the exact shape of the Earth – a sphere flattened at the poles (oblate spheroid). His *Mémoire* [Bouguer, 1736] presented to the Paris Academy of Sciences in 1734 and published in 1736 was the first treatise on the theory of the dome. In this work, Bouguer investigates possible forms for stable domes while neglecting the friction, i. e. with compressive forces always acting in the direction of the meridian and at the midsurface of the dome. A. F. Frézier was able to continue this work. Frézier imagined the dome to be assembled from a series of masonry arch elements (see Fig. 4-32). This strip method enabled him to compare the thrust of the dome with the thrust of barrel vaults with the same profile which were known to him [Frézier, 1737 – 39]. The combination of the frictionless approach of Bouguer and the strip method of Frézier paved the way for the pure mathematical research into the loadbearing behaviour of domes. Such mathematical dome theories are characteristic of the application phase of structural theory (1700 – 75).

- Main contributions to structural analysis:
*Sur les lignes courbes qui sont propres à former les voûtes en dôme* [1736]; *Manœuvre des vaisseaux ou Traité de méchanique et de dynamique; Dans lequel on réduit à des solutions très-simples les problèmes de marine les plus difficiles, qui ont pour objet le mouvement du navire* [1757]

- Further reading:
[Kertz, 1999]; [Manzanares, 2000]; [Radelet-de Grave, 1999]; [Institut Culturel de Bretagne, 2002]
- Photo courtesy of [Institut Culturel de Bretagne, 2002]

### BOW, ROBERT HENRY
*27.1.1827 Alnwick, UK, †17.2.1909 Edinburgh, UK
Bow is an enigmatic figure among the British structural engineering fraternity. Many older engineers link his name with the elegant graphical methods of analysing trussed frameworks. In the early 1870s, Bow translated Maxwell's difficult-to-use graphical statics into serviceable graphical analysis, which formed an important intellectual resource for practical engineers for the next century. However, little is known about the life of Robert Henry Bow. We suppose that he was present at lectures at Edinburgh University in the mid-1840s. He first became known through his book on trussed bridges published in 1851 and various articles on iron roofs in the early 1850s. From 1854 to 1864 he worked as a designer for (or with) the bridge-builder Thomas Bouch, whose bridge over the Firth of Tay (1879) collapsed as a result of high winds. Following this period, Bow concentrated on the design and construction of trussed roofs, many of which were required for railway stations in particular. At various times in between he worked as a consulting engineer. His second book deals with trussed girders, breaking them down into 136 different types and classifying them as statically determinate (class I), kinematically determinate (class II), statically indeterminate (class III) and others (class IV), for which he draws the associated dual polygons of forces. Bow's classification and practical application of dual polygons of forces formed a crucial element in the rationalisation of engineering in the classical phase of structural theory, and led to these diagrams being directly attributed to Bow in some German publications [Scholz, 1989, p. 200], although the majority ascribed this method to Cremona, who had furnished the theoretical basis. Whereas Bow's second book [Bow, 1873] uses a classification based on economic use of materials, his first book [Bow, 1851] provides a typology of structures; Schwedler also published such a typology in 1851 (see [Hertwig, 1930/2]).

- Main contributions to structural analysis:
*A Treatise on Bracing* [1851]; *The Economics of Construction in Relation to Framed Structures* [1873]
- Further reading:
[Scholz, 1989]

### BRESSE, JAQUES ANTOINE CHARLES
*9.10.1822 Vienne, Isère, France,
†22.5.1883 Paris, France
Following his studies at the École Polytechnique and the École des Ponts et Chaussées, Bresse worked as assistant to J. B. Belanger in applied mechanics at the École des Ponts et Chaussées and was nominated his successor in 1853. One year later Bresse published his first monograph on the elastic theory of arches [Bresse, 1854], which represented a significant advancement on Navier's approaches: formulation of the concept of the middle-third of a cross-section, consideration of the strain stiffness and temperature de-

formations. He worked out the principle of superposition and demonstrated that it applies to Hookean bodies with small displacements only. Therefore, alongside Winkler, Bresse made the most important contribution to the establishment of the elastic arch theory in the analysis of masonry arches in the establishment phase of structural theory. His three-volume work *Cours de mécanique appliquée*, which was re-issued three times (1859, 1866, 1880), is a comprehensive and independent presentation of applied mechanics, which in terms of scientific originality is far superior to Weisbach's mechanics. For example, the third volume of the first edition (1865) contains a comprehensive theory of continuous beams. Bresse's contributions to theory of structures illustrate the high standard of civil engineering theory in France at the transition between the establishment and classical phases of structural theory. The Académie des Sciences awarded him the Poncelet Prize for his scientific work in 1873; he became a member of the Académie des Sciences in 1880.
- Main contributions to structural analysis: *Recherches analytiques sur la flexion et la résistance des pièces courbées* [1854]
- Further reading:
[Saint-Hardouin, 1883]; [n. n., 1898/1]; [Timoshenko, 1953]
- Photo courtesy of École Nationale des Ponts et Chaussées

## BUBNOV, IVAN GRIGORIEVICH

*18.1.1872 Nizhny Novgorod, Russia, †13.3.1919 St. Petersburg, Russia

After completing his secondary education, he entered the School of Naval Engineering and completed his studies there successfully in 1891; in 1896 Bubnov graduated from the Naval Academy, where in 1910 he was appointed full professor. From 1903 to 1913 he taught at the St. Petersburg Polytechnic Institute – from 1909 onwards as full professor. The year 1911 saw Bubnov formulate the general calculus of variations which outside Russia in particular is known as the Galerkin Method; Bubnov published his method in 1913 in the compendium of the St. Petersburg Institute of Engineers of Ways of Communication (Bubnov-Galerkin Method). Besides his university posts, he was also involved in projects for warships and submarines for the St. Petersburg naval harbour and the Baltic shipbuilding yard. From 1908 to 1914 Bubnov was in charge of the navy's Shipbuilding Testing Institute as successor to A. N. Krylov. He was promoted to major-general in 1912 and awarded the Order of the Holy Stanislav, 1st class, in 1915. Bubnov died of typhoid.
- Main contributions to structural analysis: *Otzyv o sochinenii professora Timoshenko "Ob ustoichivosti uprugikh sistem", udostoennom premii D. I. Zhuravskogo* (appraisal of Prof. Timoshenko's Werk "On the stability of elastic systems", which was awarded the D. I. Zhuravski Prize) [1913]; *Stroitel'naya mekhanika korablya (mechanics of shipbuilding)* [1912–1914]; *Trudy po teorii plastin* (works on plate theory) [1953]; *Izbrannye trudy* (selected works) [1956]
- Further reading:
[Vol'mir, 1953]; [Belkin, 1973]; [Grigolyuk, 1996]; [Rassol, 1999]; [Lehmann, 1999]
- Photo courtesy of Prof. Dr. G. Mikhailov

## BÜRGERMEISTER, GUSTAV

*10.5.1906 Woken, Bohemia, Austria-Hungary, †30.7.1983 Essen, Federal Republic of Germany

Following his school education in Aussig, Gustav Bürgermeister studied civil engineering at the German Technical University in Prague and became an assistant to J. Melan and J. Wanke. He gained his doctorate in 1940 and wrote his habilitation thesis on beam grids and the deck grids of steel bridges in 1942 and held the vacant chair of steel and timber engineering at his Alma Mater for a number of semesters. After that, he worked on diverse structural steelwork projects in the office of Prof. Richard Guldan (Prague) until 1945. After the war, following a period of internment, he was able to re-establish himself as a structural engineer and later as head of the design office of Beuchelt & Co. in Könnern a.d. Saale; it was in this capacity that he took charge of the (re)building of numerous new steel bridges between 1946 and 1952. He was appointed to the chair of structural engineering and steelwork at Dresden TH, the direct successor to Kurt Beyers, a post he held with great dedication to teaching and research from 1952 to 1971. Together with his colleagues Herbert Steup and Horst Kretzschmar, Bürgermeister published his two-volume *Stabilitätstheorie*, which was widely acclaimed by those involved in the theory and practice of steel structures both in the German Democratic Republic and abroad. He also contributed to the success of the multi-volume *Ingenieurtaschenbuch Bauwesen* through his work as editor and co-author of the two volumes on structural engineering. He was chairman of the GDR committee of the International Association for Bridge & Structural Engineering (IABSE), a member of the German Committee for Structural Steelwork (DASt) from 1955 to 1971, and in 1961 was nominated full member of the German Building Academy. In the GDR Bürgermeister was awarded several medals and honorary titles for his scientific and engineering achievements.
- Main contributions to structural analysis: *Stabilitätstheorie mit Erläuterungen zu DIN 4114* [1959]; *Stabilitätstheorie mit Erläuterungen zu den Knick- und Beulvorschriften* [1963]; *Ingenieurtaschenbuch Bauwesen. Konstruktiver Ingenieurbau. Grundlagen der Bauweisen* [1968]; *Ingenieurtaschenbuch Bauwesen. Konstruktiver Ingenieurbau. Entwurf und Ausführung* [1970]
- Further reading:
[Kinze, 1971]; [Roik, 1981]; [Knittel, 1984]
- Photo courtesy of Dresden TU archives

## CASTIGLIANO, ALBERTO

*8.11.1847 Asti, Italy, †25.10.1884 Milan, Italy

Alberto Castigliano grew up in poor circumstances. After completing his studies at the newly founded Istituto Industriale, Sezione di Meccanica e Costruzioni in Asti, he obtained an engineering diploma from the Reale Istituto Industriale e Professionale in Turin with the financial support of wealthy citizens. In 1873 he graduated with distinction in civil engineering from the Reale Scuola d'Applicazione degli Ingegneri, despite the difficult circumstances of his life. In the course of a legal dispute with Luigi Federico Menabrea (1809–96), which instigated Castigliano's diploma thesis *Intorno ai sistemi elastici* (on elastic systems), he wrote the extensive essay *Nuova teoria intorno all'equilibrio dei sistemi elastici* (new theory of equilibrium of elastic systems) in 1875, which was to become the core of his main work *Théorie de l'Équilibre des Systèmes Élastiques et ses Applications* published in 1879. Following his studies, it was not long before he became head of the design office of the Italian Railway Company, where as member of the board of directors he reorganised the pension fund. Unfortunately, he was unable to complete his planned multi-volume *Manuale pratico per gli ingegneri* (practical manual for engineers) before his death.
- Main contributions to structural analysis: *Intorno ai sistemi elastici* [1875/1]; *Intorno all'equilibrio dei sistemi elastici* [1875/2]; *Nuova teoria intorno all'equilibrio dei sistemi elastici* [1875/3]; *Théorie de l'Équilibre des Systèmes Élastiques et ses Applications* [1879]; *Intorno ad una proprietà dei sistemi elastici* [1882]; *Theorie des Gleichgewichtes elastischer Systeme und deren Anwendung* [1886]; *The Theory of Equilibrium of Elastic Systems and its Applications* [1966]
- Further reading:
[Crotti, 1884]; [Winkler, 1884]; [Oravas & McLean 1966/1]; [Oravas, 1966/2]; [Castigliano, 1935]; [Timoshenko, 1953]; [Benvenuto, 1981]; [Castigliano, 1984]; [Nascè, 1984]; [Benvenuto, 1991/2]
- Photo courtesy of Università di Genova

## CAUCHY, AUGUSTIN-LOUIS

*21.8.1789 Paris, France, †22.5.1857 Sceaux near Paris, France

Born the son of a senior civil servant of the *ancien régime* just one month after the storming of the Bastille, Augustin-Louis Cauchy was to remain influenced by the French Revolution and the political turmoil it unleashed throughout Europe. He grew up in the religious atmosphere of a very pious Catholic family in the village of Arcueil, where they were able to remain

 BUBNOV
 BÜRGERMEISTER
 CASTIGLIANO
 CAUCHY
 CHRISTOPHE

hidden from revolutionary attacks and the young Cauchy was taught the classical subjects admirably by his father. Later, he studied at the École Polytechnique and the École des Ponts et Chaussées (1805–09). It is alleged that as Cauchy was en route from Paris to Cherbourg in order to assist in the building of the naval harbour (1810–13), he carried with him *Mécanique céleste* (Pierre Simon Laplace, 1749–1827) and *Traité des fonctions analytiques* (Joseph Louis Lagrange, 1736–1813), among other works. At the start of his structural engineering activities he sent two essays on masonry arch theory to Gaspard Prony (1755–1839) in Paris [Cauchy, 1809 & 1810], but the latter lost them and they were never published. Before his return to Paris in 1813, Cauchy turned more and more to mathematics; he is supposedly one of the most productive mathematicians ever to have lived [Novy, 1978]. His influential father tried to use his official position to secure his highly talented son vacant posts in the Académie des Sciences (1813 and 1814), but without success. It was not until the restoration of the Bourbon monarchy and the expulsion of revolutionary sympathisers among the scientists at the Académie – such as Lazare Carnot and Gaspard Monge – that Cauchy was offered a post at, but not elected to, the Académie (1816). He never disputed the fact that he was the successor to the expelled Monge. That launched Cauchy's career, which was initially interrupted by the July Revolution of 1830 because he refused to declare allegiance to the Orleanist Louis Philippe ("King of the French") and instead followed the overthrown Bourbon King Charles X into exile. Cauchy was professor at the École Polytechnique and the Sorbonne as well as a member of the Collège de France. Encouraged by Laplace and Poisson, Cauchy wrote *Cours d'Analyse de l'École Polytechnique*, which became a work that placed differential and integral calculus on a new footing by assuming a precisely defined consistent concept of the infinitesimal, which formed the framework for his mechanics based on the continuum hypothesis in the early 1820s. In his work *Recherches sur l'équilibre et le mouvement intérieur des corps solides ou fluides, élastiques ou non élastiques* [Cauchy, 1823], presented to the Académie in 1822 and published in 1823, Cauchy explains continuum mechanics and presents a valid definition of the stress concept in extraordinarily clear language without resorting to equations [Cauchy, 1823, p. 10]. He introduced the stress tensor [Cauchy, 1827/1] and the strain tensor [Cauchy, 1827/2] in 1827. He generalised the Hookean law in 1828/29, which enabled him to generate the set of equations for elastic theory based on the molecular hypothesis in the (implicit) mathematical form of tensor calculus [Herbert, 1991]. It was not until 1837 that George Green was able to prove with the help of the law of conservation of energy that a consistent elastic theory for isotropic materials requires two elasticity constants and not just one, as would result from the molecular hypothesis [Green, 1839]. Cauchy returned to France at the end of 1838 and one year later was elected to the Bureau of Longitude. It was only after the oath of allegiance was abolished by the February Revolution of 1848 that Cauchy was able to take up his academic functions again without having to deviate from his active clericalism. The Catholic Church used his devotion as an example of the reconcilability of faith and science. As Cauchy fell ill in May 1857, clerics gathered around his deathbed and the Cardinal of Paris administered the last rites.

- Main contributions to structural analysis: *Mémoire sur les ponts en pierre, par A. L. Cauchy, élève des Ponts et Chaussées* [1809]; *Second mémoire sur les ponts en pierre, théorie des voûtes en berceau, par A. L. Cauchy, élève des Ponts et Chaussées* [1810]; *Recherches sur l'équilibre et le mouvement intérieur des corps solides ou fluides, élastiques ou non élastiques* [1823]; *De la pression ou tension dans un corps solide* [1827/1]; *Sur la condensation et la dilatation des corps solides* [1827/2]
- Further reading: [Freudenthal, 1971]; [Novy, 1978]; [Belhoste, 1991]
- Photo courtesy of École Nationale des Ponts et Chaussées

### CHRISTOPHE, PAUL

*25.7.1870 Verviers, Belgium, †1935? Liège?, Belgium

Following his studies at the School of Engineering in Ghent, Christophe worked from 1892 onwards as a civil servant in the Belgian road- and bridge-building authority, part of the Brussels Ministry of Public Works. At the start of his career in the civil service, he supervised the building of a number of large bridges in Liège and was quickly promoted to the post of vice-secretary of the Central Committee for Public Works in Brussels, where in 1898 he was entrusted with the experimental testing of bridges, "which gave him the chance to carry out much scientific work" [Emperger, 1905/3, p. 1]. For example, together with A. Lambin he published an analysis of the loadbearing behaviour of Vierendeel's steel bridge in Tervueren in the journal *Annales des Travaux publics de Belgique* [Lambin & Christophe, 1898]. One year later, he was dispatched to the international congress on reinforced concrete that Hennebique had organised on the occasion of the Paris World Exposition (1900) – for which he carried out careful preliminary studies. Later that year, the journal *Annales des Travaux publics de Belgique* published his lengthy report [Christophe, 1899], which shortly afterwards appeared as a monograph with the title *Le béton armé et ses applications* [Christophe, 1902]. At that time, the book was acknowledged "as the best-known and the best compendium in this field", and its translation into several languages, e.g. Russian (1903), German [Christophe, 1905], "made certainly the most important contribution to ensuring that reinforced concrete became as widespread as it is today" [Emperger, 1905/3, p. 2]. In the book, Christophe develops a design theory for reinforced concrete based on experiments (see section 9.2.2.2), a theory that Mörsch published in his book in 1902 and was soon to become the standard method of design. So Christophe chose a design method that relied neither on hard-to-swallow theories nor on unacceptable rules of thumb à la Hennebique. Working in the Brussels Ministry of Public Works, Christophe rose from Ingénieur to Ingénieur Principal and finally to Directeur Général des Ponts et Chaussées.

- Main contributions to structural analysis: *Le pont Vierendeel. Rapport sur les essais jusqu'à la rupture effectués au parc de Tervueren, par M. Vierendeel, sur un pont métallique de 31$^m$.50 de portée avec poutres à arcades de son système* [1898]; *Le béton armé et ses applications* [1899 & 1902]; *Der Eisenbeton und seine Anwendung im Bauwesen* [1905]
- Further reading: [Emperger, 1905/3]
- Photo courtesy of [Emperger, 1905/3, p. 1]

## CHWALLA, ERNST

*28.10.1901, Vienna, Austria-Hungary, †1.6.1960 Graz, Austria

After completing his civil engineering studies at Vienna TH, Chwalla worked as assistant to the chair of bridge-building at the same establishment. He gained his doctorate in 1926 with a dissertation on the lateral rigidity problems of open bridges and completed his habilitation thesis in 1928 with a contribution to the theory of stability. Both these works, written in Vienna under the direction of Prof. Friedrich Hartmann, helped Chwalla to create the foundation for his pioneering work on the theory of stability. When he was just 28 years old he was offered the chair of structural mechanics at the German Technical University in Brno, a post that had been held before by Joseph Melan and Paul Neumann. Together with Jokisch, he succeeded in integrating the stability problem into the displacement method in 1941. In the same year Chwalla presented his introduction to the theory of structures, which by 1954 had reached its third edition; this work is characterised by systematic inclusion of the higher sciences of strengths and mechanics of materials. During the 1940s he had a decisive influence on the drafts for German stability standards in structural steelwork (DIN 4114). At the end of the war Chwalla was taken prisoner by the Russians and sent to a Czech work camp. With the help of Czech professors, who were aware of his fate, a police investigation was initiated that finally led to his release from the camp in June 1945. From there he went on to set up a new career for himself in Vienna, with the help of Karl Girkmann and others. In 1947 he was certified as an engineering consultant for buildings; his advice on matters of structural theory in conjunction with the building of numerous large hydroelectric plants in Austria was highly regarded. In addition, he taught at the University for Soil Cultivation in Vienna. Graz TH appointed him professor of theory of structures in 1955. A year prior to that, Berlin TU had awarded him an honorary doctorate in recognition of his important scientific work in the field of stability theory and his great achievements in structural steelwork [Chwalla, 1954/2]. Shortly before his death, Chwalla became a member of the Austrian Academy of Sciences in Vienna. Over the years 1925 to 1960 he influenced the theory of stability in steel buildings quite unlike any other person in the German-speaking world, and hence contributed substantially to establishing the science of steelwork in the invention phase of structural theory. His last great work was the comprehensive integration of second-order theory into the displacement method.

• Main contributions to structural analysis: *Das Problem der Stabilität gedrückter Rahmenstäbe* [1934/1]; *Drei Beiträge zur Frage des Tragvermögens statisch unbestimmter Stahltragwerke* [1934/2]; *Über das ebene Knickproblem des Stockwerkrahmens* [1941]; *Einführung in die Baustatik* [1954/1]; *Ansprache und Vortrag anläßlich der Ehrenpromotion an der Technischen Universität Berlin-Charlottenburg* [1954/2]; *Die neuen Hilfstafeln zur Berechnung von Spannungsproblemen der Theorie zweiter Ordnung und von Knickproblemen* [1959]

• Further reading: [Chwalla, 1954/2]; [Sattler, 1960]; [Beer, 1960]

• Photo courtesy of [Sattler, 1960, p. 275]

## CLAPEYRON, BENOÎT-PIERRE-EMILE

*26.1.1799 Paris, France, †28.1.1864 Paris, France

In 1820, after finishing their studies at the École Polytechnique and the École des Mines, Clapeyron and his friend Gabriel Lamé left Paris to teach pure and applied mechanics, chemistry and design theory at the St. Petersburg Institute of Engineers of Ways of Communication for a period of 10 years. Together with Lamé he acted as a consulting engineer on numerous projects, including St. Isaac Cathedral and the Alexander Column in St. Petersburg as well as suspension bridges and the Schlüsselburg locks. Following the Paris July Revolution of 1830, Clapeyron returned to France and quickly rose to be a leading railway engineer. In 1844 he was appointed professor of steam engine construction at the École des Ponts et Chaussées. The theoretical consequences of his practical experience in bridge-building found their way into his famous *Mémoire* (1857) in the form of calculations for continuous beams. This book and another *Mémoire* on energy conservation in elastic theory (Clapeyron's theorem) published one year later earned him membership of the Académie des Sciences in Paris, where he succeeded Augustin-Louis Cauchy.

• Main contributions to structural analysis: *Mémoire sur la stabilité des voûtes* [1823]; *Mémoire sur la Construction des Polygones Funiculaires* [1828]; *Note sur un théorème de mécanique* [1833]; *Mémoire sur l'équilibre intérieur des corps solides homogènes* [1833]; *Calcul d'une poutre élastique reposant librement sur des apppuis inégalement espacés* [1857]; *Mémoire sur le travail des forces élastiques dans un corps solide élastique déformé par l'action de forces extérieures* [1858]

• Further reading: [Bradley, 1981]; [Gouzévitch, 1993]; [Tazzioli, 1995]

• Photo courtesy of Collection École Nationale des Ponts et Chaussées

## CLEBSCH, RUDOLF FRIEDRICH ALFRED

*19.1.1833 Königsberg, Prussia (today: Kaliningrad, Russia), †7.11.1872 Göttingen, German Empire

The son of a regimental doctor, Alfred Clebsch attended grammar school and in 1850 began studying at the University of Königsberg, where his studies in mathematics and physics under Franz Ernst Neumann (1798–1895), Friedrich Julius Richelot (1808–75) and Ludwig Otto Hesse (1811–74) were very successful. He passed his state examinations in mathematics and physics, and Neumann supervised his doctorate (a dissertation on the motion of an ellipsoid in an incompressible fluid [Clebsch, 1854]). Clebsch subsequently attended the teacher training college under the directorship of Karl Heinrich Schellbach (1805–92), which was associated with the Friedrich Wilhelm Grammar School in Berlin, and afterwards worked at various schools in Berlin as a mathematics teacher. He wrote his habilitation thesis in 1858 at the Berlin University for Mathematical Physics. It was in that same year that he was appointed professor of analytical mechanics at Karlsruhe Polytechnic. In his book *Theorie der Elasticität fester Körper* [1862], Clebsch manages to achieve a synthesis between the second-order theories of Saint-Venant [Saint-Venant, 1855 & 1856] and Kirchhoff [Kirchhoff, 1858], and to develop these further. And he presents Kirchoff's plate theory on a wider mathematical footing thanks to his masterly differential geometry approach – a method that for Hermann Aron (1845–1913) would become a model for the first systematic treatment of elastic shells [Aron, 1874]. The book also contains Clebsch's fundamental ideas for the displacement method (see section 10.2.1.2), which have been acclaimed by Zienkiewicz and Samuelsson (see [Zienkiewicz, 2004, p. 4; Samuelsson & Zienkiewicz, 2006, p. 150]). Saint-Venant was so enthusiastic about this book that he and Alfred Flamant (1839–1914) translated it into French and added a valuable commentary [Clebsch, 1883]; a reprint appeared in 1966 [Clebsch, 1966]. But the engineering community ignored his book on elastic theory. Clebsch was appointed professor of mathematics at the University of Gießen in 1863, whereupon he shifted his scientific interests to pure mathematics – five years later he was appointed professor of pure mathematics at the University of Göttingen. It was in that same year that he founded the journal *Mathematische Annalen* together with Carl Gottfried Neumann (1832–1925). Unfortunately, Clebsch died of diphtheria at the age of just 39.

• Main contributions to structural analysis: *Analytische Mechanik: nach Vorträgen, geh. an d. polytechn. Schule Carlsruhe, lithographiert* [1859/1]; *Elementar-Mechanik: nach Vorträgen, geh. an d. polytechn. Schule Carlsruhe von 1858–1859, lithographiert* [1859/2]; *Theorie der Elasticität fester Körper* [1862]; *Théorie de l'élasticité des corps solides de Clebsch* [1883]

• Further reading: [Husemann, 1876]; [Szabó, 1976]; [Shafarevich, 1983]

• Photo courtesy of [Szabó, 1976, p. 114]

CHWALLA   CLAPEYRON   CLEBSCH   COLONNETTI   COULOMB

### COLONNETTI, GUSTAVO
*8.11.1886 Turin, Italy, †20.3.1968 Turin, Italy

He concluded his civil engineering studies under Camillo Guidi in 1908 and his mathematics studies under Corrado Segre in 1911 at the Politecnico di Torino. In 1910 he was certified to teach engineering sciences at universities and one year later was appointed to teach applied mechanics at the Shipbuilding University in Genoa. The year 1914 saw him take over the chair of applied mechanics and engineering sciences at the School of Engineering in Pisa, where in 1918 he became director. Colonnetti was appointed to teach higher applied mechanics at the Politecnico di Torino in 1920, and was head of this establishment from 1922 to 1925. He succeeded Camillo Guidi in the chair of engineering sciences at the Politecnico di Torino in 1928. From 1936 onwards he was a member of the Papal Academy of Sciences. As an active member of the Catholic movement, Colonnetti rejected membership of the Fascist Party and in September 1943 fled to Switzerland, where together with other intellectuals he founded and took charge of a college for Italian students in exile at the University of Lausanne. During his period of exile he wrote political and cultural articles for the magazine *Gazzetta ticinese* under the pseudonym "Etegonon". Colonnetti returned to Italy in December 1944. In the post-war years he was active for the Democrazia Cristiana. He founded the Metrological Institute of the National Research Council of Italy, serving as president from 1946 to 1956. Afterwards, Colonnetti was responsible for engineering sciences at the Politecnico until being granted emeritus status in 1962. He was a member of several academies at home and abroad, and the universities of Toulouse, Lausanne, Poitiers and Liège awarded him doctorates. Among his many scientific writings, his contributions to elastic theory are most prominent. For example, more than once he worked on applications and possible extensions of the theorems of Menabrea, Castigliano and Betti; the theorem that bears his name is closely linked to these [Colonnetti, 1955]: the first theorem of elastic theory (Betti's reciprocal theorem) is paired with the second theorem of elastic theory (Colonnetti's theorem). Therefore, Colonnetti can be counted among the leading advocates of structural mechanics in the transition from the invention to the innovation phase of structural theory.
- Main contributions to structural analysis: *L'équilibre des corps déformables* [1955]
- Further reading:
[n. n., 1960–2001]; [n. n., 1968]; [Supino, 1969]; www.museovirtuale.polito.it
- Photo courtesy of Politecnico di Torino

### COULOMB, CHARLES AUGUSTIN
*14.6.1736 Angoulême, France, †23.8.1806 Paris, France

Charles Augustin Coulomb attended lectures at the Collège Mazarin and the Collège de France; in 1757 he became an associate member of the Société des Sciences de Montpellier, to which he contributed several articles on astronomy and mathematics. Afterwards he studied at the École du Génie de Mézière, from where he graduated in 1761 with the rank of lieutenant en premier des Corps du Génie. It was while studying here that he established his lifelong friendship with his mathematics teachers Jean Charles Borda and Abbé Charles Bossut. His first activities as an officer in the engineering corps were in Brest and the French colony of Martinique (1764–72), where he was in charge of building fortifications. Coulomb transformed his practical experiences into a book of theory which he presented to the Académie des Sciences in 1773; after favourable reviews by the academy members Borda and Bossut, his theories were published in 1776 under the title *Mémoires de mathématique et de physique présentés à l'académie royale des sciences par divers savants* [Coulomb, 1773/1776]. In his book, Coulomb solved Galileo's beam problem and developed a forward-looking earth pressure and masonry arch theory. Coulomb's 40-page *Mémoire* rounds off the preparatory period in structural theory quite conclusively and to some extent anticipates the theories of the coming consolidation period. However, the ideas of his *Mémoire* were not adopted until 40 years later. Coulomb also led the way in theories on electricity, magnetism and friction. In the meantime, Coulomb had been promoted to lieutenant-colonel, and in 1776 his proposal to restructure the Corps du Génie into a "Corps à talent" (Coulomb) was backed by the reform plans of the Turgot government. In 1781 he became a member of the Académie des Sciences and had a decisive influence on the profile of that institution until it was abolished in August 1793. He was critical of the French Revolution: prudently, he withdrew to his small estate near Blois in 1792. Only after the downfall of the Jacobins' "Reign of Terror" did he return to Paris and from 1795 onwards was responsible for experimental physics as an elected member of the newly founded Institut de France. In 1801 Coulomb became president of this highly respected scientific establishment. From 1802 until his death he was general inspector of all public education and in this capacity contributed significantly to creating the French system of Lycées.
- Main contributions to structural analysis: *Essai sur une application des règles des Maximis et Minimis à quelques Problèmes de statique relatifs à l'Architecture* [1773/1776]
- Further reading:
[Gillmor, 1971]; [Heyman, 1972/1]; [Radelet de Grave, 1994]
- Photo courtesy of [Szabó, 1996, p. 386]

### COUPLET, PIERRE
*unknown, †23.12.1743 Paris, France

Pierre Couplet was the son of Claude-Antoine Couplet (1642–1722), the treasurer of the Académie Royale des Sciences in Paris. The young Couplet entered the Académie in 1696 as a pupil of his father. Afterwards, Couplet went to Lisbon, where he learned Portuguese, and then signed on for a two-and-a-half-year astronomy research expedition to Brazil. In 1699 he became a member of the reformed Académie and in 1700 took part in the campaign to measure geographical longitude led by Cassini II. Later, Couplet was promoted to Professeur Royal de Mathématiques des Pages de la Grande Écurie and in 1717 to treasurer of the Académie – posts that his father had held previously. He presented several essays on astronomy, earth pressure theory, masonry arch theory, mansard roofs, pipe hydraulics and mechanics of carriages and sledges to the Académie; most significant among these was his second essay on masonry arch theory, which began the tradition of analysing the collapse mechanisms of arches. In Heyman's historical reconstruction of the history of masonry arch theories from the viewpoint of the ultimate load method, Couplet's theory therefore plays a very important role [Heyman, 1982]. After Couplet's death, the Académie refrained from the Éloge customary for its members. His

work on masonry arch theory was therefore gradually forgotten: "It is only unfortunate that the work was slowly forgotten, so that fifty years later Coulomb, in seeming ignorance of Couplet's contribution, had to rediscover much of the theory" [Heyman, 1976, pp. 35 – 36].
• Main contributions to structural analysis:
*De la poussée des voûtes* [1729/1731]; *Seconde partie de l'éxamen de la poussée des voûtes* [1730/1732]
• Further reading:
[Heyman, 1976]

## CREMONA, ANTONIO LUIGI GAUDENZIO GIUSEPPE

*7.12.1830 Pavia, Italy; †10.6.1903 Rome, Italy

Immediately after completing his education in his home town in 1848, Cremona joined the "Free Italy Battalion" in the fight against the Austrian rulers and took part in the defence of Venice, which ended in capitulation on 24 August 1849. In that same year he began studying civil engineering at the University of Pavia, from where he graduated with a doctor's degree in 1853. Following various teaching posts in Pavia, Cremona and Milan, Cremona became a professor at the University of Bologna in 1860. This was followed by a professorship in higher geometry at Milan Polytechnic (1867 – 73); it was during this period that he established the mathematical basis of his graphical statics, based on the Maxwell duality, which formed the practical foundation of Cremona's graphical analysis and was very quickly adopted in the teaching of the theory of structures and engineering practice: "Cremona therefore furnished a reason for constructing dual force diagrams for the graphical analysis of certain trussed frameworks which in general terms left no stone unturned" [Scholz, 1989, p. 198]. Together with Culmann and Maxwell, Cremona represents the graphical statics enhanced by projective geometry. In 1873 Minister Sella nominated him founding rector of the newly formed Technical University in Rome per decree, where he was responsible for graphical statics until 1877. Afterwards, he was professor of mathematics at the University of Rome until his death. He joined the Senate in 1879 and was one of its most highly respected members. The universities of Berlin, Stockholm and Oxford (among others) awarded him honorary doctorates for his pioneering work in geometry.
• Main contributions to structural analysis:
*Le figure reciproche nella statica grafica* [1872]; *Opere matematiche* [1914 – 17]
• Further reading:
[n. n., 1903]; [Loria, 1904]; [White, 1917/1918]; [Gabba, 1954]; [Greitzer, 1971]; [Arrighi, 1976]; [Benvenuto, 1981]; [Scholz, 1989]; [Maurer, 1998]
• Photo courtesy of Università di Genova

## CROSS, HARDY

*10.2.1885 Nansemond County, USA, †11.2.1959 Virginia Beach, USA

In 1902 Hardy Cross gained a Bachelor of Arts in English and in 1903 a Bachelor of Science at Hampden-Sydney College. For the next three years he taught English and mathematics at Norfolk Academy. At the age of just 23, he gained a Bachelor of Science in civil engineering at the Massachusetts Institute of Technology (MIT) and subsequently worked for two years in the bridge-building department of the Missouri Pacific Railroad in St. Louis. In 1911 Havard University awarded him the academic grade of Master of Civil Engineering. After that, Cross worked as assistant professor at Brown University and then, following a short period in practice, was promoted to professor of civil engineering at the University of Illinois in 1921. From 1937 until he was granted emeritus status in 1951, he taught and performed research at Yale University and was head of the Department of Civil Engineering. In his 10-page contribution to the *Proceedings of the American Society of Civil Engineers* (ASCE) in 1930, Cross solved the Gordian knot of the consolidation period of the theory of structures. His stroke of genius was to calculate statically indeterminate systems by iterative means using the simplest form of arithmetic [Cross, 1930]. The Cross method was admirably suited to analysing systems with a high degree of static indeterminacy, as is common in the design of high-rise buildings, for example. With one fell swoop, Cross ended the search which had characterised the application phase of structural theory – the hunt for suitable methods of calculation for solving systems with a high degree of static indeterminacy by rational means. The Cross method initiated not only an algorithmisation of structural theory, which was without precedent in the 20th century, but also raised the rationalisation of structural calculations to a new level. It is therefore not surprising that in the wake of his work a flood of lengthy discussion articles appeared in the *Transactions* of the ASCE [Cross, 1932]. His ingenious iterative method provoked countless engineers – well into the innovation phase of structural theory – to describe the Cross method and develop it further. Indeed, so much has been written that it would easily fill the medium-sized private library of any academic! And the Cross method was not just confined to theory of structures; it was also quickly accepted in disciplines such as shipbuilding and aircraft design. Cross himself transferred the basic idea of his iterative method to calculations of steady-state flows in pipework – the Hardy Cross method – and there, too, achieved a phenomenal breakthrough. The honours he received are too numerous to mention.
• Main contributions to structural analysis:
*Analysis of continuous frames by distributing fixed-end moments* [1930]; *Analysis of continuous frames by distributing fixed-end moments* [1932/1]; *Continuous Frames of Reinforced Concrete* [1932/2]; *Analysis of continuous frames by distributing fixed-end moments* [1949]; *Engineers and Ivory Towers* [1952]; *Arches, Continuous Frames, Columns and Conduits: Selected Papers of Hardy Cross* [1963]
• Further reading:
[Eaton, 2001]
• Photo courtesy of University of Illinois

## CULMANN, KARL

*10.7.1821 Bergzabern, Bavaria, †9.12.1881 Zurich, Switzerland

After attending Wissembourg Collège (1835 – 36) he moved to Metz, where his uncle, Friedrich Jakob Culmann (1787 – 1849), was a professor at the Artillery School; this awakened in him an interest in a career in engineering. From 1838 to 1841 he studied at Karlsruhe Polytechnic and was subsequently employed on public building works by Bavarian State Railways until 1855. With the help of his superior, Friedrich August von Pauli (1802 – 83), Culmann spent the years 1849 – 51 abroad in England, Ireland and the USA; his experiences were published in two travelogues which contain the theory of frameworks. After leaving Bavarian State Railways, he became full professor of engineering sciences at Zurich ETH, where he gave lectures on graphical statics from 1860 onwards; he gained his doctorate there in 1880. Culmann placed graphical statics on a sound footing and made a great contribution to the establishment phase of structural theory. Although his reasoning behind graphical statics was in the end rendered obsolete by projective geometry, together with Mohr he was the greatest structural engineer of the 19th century in the German-speaking world.
• Main contributions to structural analysis:
*Der Bau der hölzernen Brücken in den Vereinigten Staaten von Nordamerika* [1851]; *Der Bau der eisernen Brücken in England und Amerika* [1852]; *Die graphische Statik* [1864 & 1866].
• Further reading:
[Tetmajer, 1882]; [Stüssi, 1951]; [Stüssi, 1971]; [Charlton, 1982]; [Scholz, 1989]; [Maurer, 1998]; [Lehmann & Maurer, 2006]
• Photo courtesy of Zurich ETH library

## DAŠEK, VÁCLAV

*18.2.1882 Slavětín, Bohemia, Austria-Hungary (today: Czech Republic), †12.8.1970, Prague, Czechoslovakia (today: Czech Republic)

Following his secondary education, Václav Dašek studied civil engineering (1905 – 10) at the Czech TH in Prague. Afterwards, he completed his dissertation *Neue Methoden für die Berechnung der statisch unbestimmten Konstruktionen* at Prague TH. Dašek worked for various building contractors in Bohemia, Serbia, Switzerland and Yugoslavia as well as the municipal bridge

 CREMONA
 CROSS
 CULMANN
 DAŠEK
 DIJKSTERHUIS
 DINNIK

authority in Prague. He resumed his work at Prague TH in 1927 and one year later obtained his venia legendi with the groundbreaking habilitation thesis *Berechnung der Rahmenkonstruktionen mit Hilfe von Tensoren und Deformationsellipsen*. Dašek first applied tensor calculus to structural mechanics and prepared it for matrix analysis. His scientific accomplishments earned him a nomination as associate professor in 1929, and in 1934 he became a full professor at Prague TH's Faculty of Civil Engineering. The theory of sway frame systems and beam grids formed the focus of his research work. Dašek is regarded as an extremely modest scientist. Nevertheless, he received numerous honours: state prizes, honorary medals and election to full membership of the Czechoslovakian Academy of Sciences.
- Main contributions to structural analysis: *Výpočet rámových konstrukcí pomocí tensorů a elips deformačních* (calculation of frames by means of tensors and deformation ellipses) [1930]; *Výpočet rámových konstrukcí rozdělováním sil a momentů* (calculation of frames by means of force and moment distribution) [1943/1951/1966]; *Řešení trámových roštů metodou harmonického zatížení* (solving the beam grid problem using the method of harmonic loads) [1953]
- Further reading:
[Hořejší & Pirner, 1997, pp. 40–43]
- Photo courtesy of Prof. Dr. L. Frýba

### DIJKSTERHUIS, EDUARD JAN

*28.10.1892 Tilburg, Netherlands, †18.5.1965 Bilthoven, Netherlands
Eduard Jan Dijksterhuis was the son of the headmaster of the Willem II school in Tilburg, Berend Dijksterhuis, and his wife, Gezina Eerkes. He left his father's school in 1910 and went on to study mathematics at the University of Groningen, where he gained his doctorate in 1918 with a dissertation entitled *Bijdragen tot de kennis der meetkunde van het platte schroevenvlak*. He was employed as a teacher of mathematics and physics at the Willem II school in Tilburg from 1919 to 1953. He became interested in the early history of mathematics and mechanics as early as the 1920s. Initially influenced by Pierre Duhem's work on the history of science, Dijksterhuis made a major contribution to giving a professional status to the history of science in general and the history of the exact sciences in particular. For example, he published a history of mechanics from Aristotle to Newton in 1924, a commentary on Euclid's *Elemente* in 1929–30, a study on Archimedes in 1938 and one on Stevin in 1943. Dijksterhuis set standards in the historical study of the sciences in 1950 with his work *De mechanisering van het wereldbeeld*, which later appeared in German (1956) and English (*The Mechanization of the World Picture*, 1961); this helped the history of science gain a high social standing as an independent discipline on both national and international levels. He was awarded the P. C. Hooft Prize – one of the foremost state prices for literature in the Netherlands – for this work by the Royal Netherlands Academy of Arts & Sciences in 1952. Although his time as a private lecturer at the universities of Amsterdam and Leiden in the 1930s represents less successful years, from 1953 onwards, Dijksterhuis lectured in the history of mathematics and natural sciences at the universities of Utrecht and Leiden with more success. He spent his final years as a full professor for these scientific disciplines at the University of Utrecht (1960–63). In the Netherlands, Klaas van Berkel in particular has continued Dijksterhuis' scientific work, and his biography of Dijksterhuis is a worthy tribute to the old Dutch master of the history of science [Berkel, 1996].
- Main contributions to structural analysis: *Die Mechanisierung des Weltbildes* [1956]; *The Mechanization of the World Picture* [1961]; *Simon Stevin: Science in the Netherlands around 1600* [1970]; *Archimedes* [1987]
- Further reading:
[Hooykaas, 1967]; [Berkel, 1996]
- Photo courtesy of Boerhaave Museum, Leiden

### DINNIK, ALEKSANDR NIKOLAEVICH

*31.1.1876 Stavropol, Russia, †22.9.1950 Kiev, USSR (today: Ukraine)
A. N. Dinnik was the son of a physics teacher and in 1886 he started his secondary education in the humanities department of the grammar school in the town of his birth. He passed his school-leaving examinations with distinction in 1894 and began his studies at the faculty of mathematics and physics at the universities of Odessa and Kiev. Upon completion of his studies in 1899, Dinnik worked at the chair of physics at the Polytechnic Institute in Kiev and later switched to the chair of strength of materials. Dinnik gained his doctorate in 1909 with a dissertation on contact mechanics. He spent the following year studying at Munich TH under professors Arnold Sommerfeld and August Föppl. The year 1911 saw Dinnik appointed professor at the Polytechnic Institute in Novocherkask, and he was awarded the title Dr.-Ing. in 1912 following work on planar plate and shell structures at Danzig (Gdańsk) TH with Prof. G. Lorenz. Just one year later he was appointed to the chair of theoretical mechanics at the Mining Institute in Jekaterinoslav and in 1915 he completed his dissertation *Die Anwendung der Bessel-Funktionen bei Aufgaben der Elastizitätstheorie* at the University of Kharkov. Dinnik was elected a full member of the Academy of the Ukrainian Soviet Republic in 1929 and from 1946 onwards enjoyed the same status in the USSR Academy of Sciences. Unfortunately, he did not live to see the publication of the results of his creative work in the field of structural mechanics.
- Main contributions to structural analysis: *Izbrannye trudy* (selected works) [1952–56]
- Further reading:
[Grishkova & Georgievskaya, 1956]; [Malinin, 2000, p. 177]
- Photo courtesy of Prof. Dr. G. Mikhailov

### DI PASQUALE, SALVATORE

*27.11.1931 Naples, Italy, †2.11.2004 Florence, Italy
Salvatore Di Pasquale graduated from the faculty of architecture of the University of Naples in 1955. He lectured in descriptive geometry at the same establishment between 1961 and 1968 and after that bridges and large structures. Di Pasquale became a professor at the University of Naples as early as 1964 and finally became a full professor for construction theory at the faculty of architecture of the University of Florence in 1973, where he had begun work in 1971 and where he continued to work until his transfer to emeritus status in 1997. He served as dean of the faculty of architecture at his Alma Mater from 1986 to 1992 and was also head of the Construction Institute there from 1983 to 1995. Di Pasquale continued to give lectures as emeritus professor, e. g. in theory of structures and construction theory, at the faculty of architecture of the University of Catania, where

he also served as dean. In addition, from 1976 until his death he taught structural theory and the stability of monuments in the foundation studies course on the restoration of monuments at the Federico II University in Naples, and the structural analysis of masonry structures at the Centre d'études pour la conservation du patrimoine architectural et urbain R. Lemaire in Leuven (Belgium); Di Pasquale was also a visiting professor at the architecture faculties of the universities in Pescara, Ferrara, Venice and Milan. He was for a long time a member of the scientific advisory committee for the international journal *Meccanica* published by the Associazione Italiana Meccanica Teorica e Applicata (AIMETA) and the journal *Palladio* (journal of history of architecture and restoration). With more than 200 publications on modern theory of structures to his name (also with special emphasis on historically important structures), Di Pasquale can be counted among Italy's leading construction engineering theorists. That is why his name together with that of Giuffrè and Benvenuto represents the exploration of the historical dimensions of loadbearing systems. Salvatore Di Pasquale, Antonio Giuffrè and Edoardo Benvenuto form the triumvirate behind a history of construction engineering containing theory of structure elements, the foundations of which were principally created by them and whose productive energy they have shown in the sensitive refurbishment of historically important structures.

• Main contributions to structural analysis: *On the elastic problem of the non-homogeneous anisotropic body resting on lattice* [1967]; *On the elastic problem of the orthotropic plate* [1968]; *Energy forms in the finite element techniques* [1972]; *Scienza delle costruzioni. Introduzione alla progettazione strutturale* [1975]; *Metodi di calcolo per le strutture spaziali* [1978]; *Questions concerning the mechanics of masonry* [1988]; *New trends in the analysis of masonry structures* [1992]; *On the art of building before Galilei* [1995]; *L'arte del costruire. Tra conoscenza e scienza* [1996]; *Brunelleschi. La costruzione della cupola di Santa Maria del Fiore* [2002]
• Photo courtesy of Giovanni Di Pasquale

## DISCHINGER, FRANZ
*8.10.1887 Heidelberg, German Empire, †9.1.1953 Berlin, Federal Republic of Germany
He completed his studies in civil engineering at Karlsruhe TH in 1911. There he was influenced by the mathematician Karl Heun and the structural engineer Friedrich Engesser. From 1912 to 1932 he was a structural engineer with Dyckerhoff & Widmann, becoming a director before he left. In 1923 he developed methods of building and analysing shells, and gained his doctorate on this subject under Kurt Beyer at Dresden TH in 1928. He was full professor for reinforced concrete at Berlin TH from 1933 to 1945 and held the same post at Berlin TU until 1951 (apart from one year). Besides his work on the analysis of shells, Dischinger published articles on problems with reinforced and prestressed concrete bridges during his time in Berlin. His work in these fields lent great impetus to the establishment of reinforced concrete for structural purposes in the inter-war years. The Edward Longstreath Medal of the Franklin Institute in Philadelphia was awarded to Dyckerhoff & Widmann and Zeiss-Jena in 1938, and specifically mentioned Walter Bauersfeld, Ulrich Finsterwalder, Hubert Rüsch, Wilhelm Flügge and Franz Dischinger. Dischinger was awarded honorary doctorates by Karlsruhe TH (1948), Aachen RWTH (1949) and Istanbul TH (1952).

• Main contributions to structural analysis: *Schalen und Rippenkuppeln* [1928/1]; *Eisenbeton-Schalendächer System Dywidag* [1928/2]; *Die Theorie der Vieleckkuppeln und die Zusammenhänge mit den einbeschriebenen Rotationsschalen* [1929]; *Die weitere Entwicklung der Schalenbauweise „Zeiss-Dywidag"* [1932]; *Untersuchungen über die Knicksicherheit, die elastische Verformung und das Kriechen des Betons bei Bogenbrücken* [1937]; *Elastische und plastische Verformungen der Eisenbetontragwerke und insbesondere der Bogenbrücken* [1939]
• Further reading:
[Günschel, 1966]; [Specht, 1987]
• Photo courtesy of [Specht, 1987]

## DRUCKER, DANIEL C.
*3.6.1918 New York, USA, †1.9.2001 Gainesville, Florida, USA
Daniel C. Drucker studied at Columbia University, where he became interested in the conception, design and analysis of bridges. However, Raymond D. Mindlin suggested he write his dissertation on the subject of photoelasticity, which he completed in 1940. Afterwards, he lectured at Cornell University until 1943. Following military service, he worked for a short time at the Illinois Institute of Technology before transferring to Brown University, where he worked as a teacher and researcher from 1947 to 1967. It was at this university that William Prager founded the world-famous school of applied mathematics and mechanics in the 1940s and it was at this university that Drucker carried out his pioneering work on plastic theory. For example, based on the energy criterion for the stability of the elastic equilibrium (see section 10.4.5.1), he introduced the concept of material stability [Drucker, 1951], which today in the form of Drucker's stability postulate enjoys an established place in the literature. Material stability – especially the stability of the infinitesimal dimensions – is crucial for dealing with the shake-down of loadbearing structures (Bleich-Melan shake-down principle) and the formulation of the stress-strain relationships in plastic theory. Drucker became dean of the faculty of engineering at the University of Urbana-Champaign in 1968. From 1984 until his retirement in 1994, he worked as a graduate research professor at the University of Florida, and he was editor of the *Journal of Applied Mechanics* for 12 years. Drucker's work has been honoured with numerous awards, e.g. Lehigh, Brown, Northwestern and Urbana-Champaign universities plus the Haifa Technicon have all awarded him honorary doctorates, and the American Society of Mechanical Engineers (ASME) inaugurated its Daniel C. Drucker Medal in 1997. Charles E. Taylor wrote the following memorable words about Daniel C. Drucker: "In all of the thousands of hours we spent together, I never heard him utter a single swear word. He had a great sense of humor, but he never told a joke and he never spread gossip. I have never met a more honest man or pure person. Dan Drucker was the kind of person that we all try to be" [Taylor, 2003, p. 159].

• Main contributions to structural analysis: *A more fundamental approach to stress-strain relations* [1951]; *Coulombs friction, plasticity and limit loads* [1953]; *On uniqueness in the theory of plasticity* [1956]; *A definition of stable inelastic material* [1959]; *On Structural Concrete and the Theorems of Limit Analysis* [1961]
• Further reading:
[Taylor, 2003]
• Photo courtesy of [Taylor, 2003, p. 158]

## DUHEM, PIERRE MAURICE MARIE
*10.6.1861 Paris, France, †14.9.1916 Cabrespine, France
Duhem studied physics at the École Normale Supérieure in Paris. In his dissertation he applied thermodynamic concepts to chemistry and the theory of electricity and attacked theses that Marcellin Berthelot (1827–1907) – at that time the all-powerful secretary of the Paris Academy of Sciences and member of the Examinations Commission – had been working on for many years. That was too much for Berthelot, who declared that "this young man shall never teach in Paris" (Schäfer in: [Duhem, 1978, p. IX*]). And Duhem never did teach in Paris! His dissertation, which reveals him to be one of the co-founders of physical chemistry from the history of science perspective, was not acknowledged until much later. In a second attempt involving magnetism, Duhem proved to be the best doctor candidate of that year. Following short periods in Rennes and Lille, he worked as professor of theoretical physics at the University of Bordeaux from 1895 until his death. Physics was not the only subject in which Duhem was amazingly successful – his research into the history of science and scientific theory are also worthy of note. His works *L'évolution de la méchanique* [Duhem, 1903] and *Les origines de la statique* [Duhem, 1905/1906] are of extra-

DI PASQUALE　　DISCHINGER　　DRUCKER　　DUHEM　　EDDY　　EMPERGER

ordinary importance for the historical study of science in general and theory of structures in particular. In those works, Duhem single-handedly destroys the cliché originating from the Enlightenment that following the decline of the Hellenistic sciences, it was the Renaissance that brought liberation from the spiritual servitude of the Middle Ages and hence at the same time introduced the constitution of the sciences of the modern age. Duhem discovered the direct forefathers of the physicists of the 17th century in the shape of the impetus theory in the works of Nikolaus von Oresme (1330? – 82), Albert von Sachsen (c. 1316 – 90) and Johannes Buridan (c. 1300 – 58?), rector of the Sorbonne around 1327. "Through researching these sources, Duhem lent the history of science a whole new momentum" (Schäfer in: [Duhem, 1978, p. IX*]). So his two works on the evolution of mechanics contribute to a deeper understanding of the historico-logical sources of the orientation phase of structural theory (1575 – 1700). The German translation (1908) of his monograph on scientific theory *La théorie physique, son objet et sa structure* (*The aim and structure of physical theory*, 1954) [Duhem, 1906] had a long-lasting influence on the logical empiricism of the Viennese circle around Rudolph Carnap (1891 – 1970), Otto Neurath (1882 – 1945), Philipp Frank (1884 – 1966) and Hans Hahn (1879 – 1934). Duhem's work on the history of science was continued by Anneliese Maier (1905 – 71), Ernest Moody (1903 – 75), Alexandre Koyré (1892 – 1964), Marshall Clagett (1916 – 2005) and Eduard Jan Dijksterhuis (1892 – 1965) in their research into the period prior to the sciences of the modern age.

- Main contributions to structural analysis:
*L'évolution de la méchanique* [1903]; *Les origines de la statique* [1905/1906]; *La théorie physique, son objet et sa structure* [1906]; *The aim and structure of physical theory* [1954]; *Ziel und Struktur der physikalischen Theorien* [1978]
- Further reading:
Schäfer in: [Duhem, 1978]
- Photo courtesy of [Szabó, 1996; p. 534]

### EDDY, HENRY TURNER
*9.6.1844 Stoughton, USA,
†11.12.1921 Minneapolis, USA
After completing his mathematics studies at Yale University in 1867, Eddy attended engineering courses at Sheffield Scientific School. By 1868 he was already teaching mathematics and Latin at the University of East Tennessee in Knoxville and one year later became assistant professor for mathematics and civil engineering at Cornell University, where he also gained his doctorate. His academic career continued in Princeton, where in 1874 he was appointed professor of mathematics, astronomy and civil engineering, a post he held until 1890. It was during this period that he studied at the University of Berlin and the Sorbonne in Paris (1878/79), and also published his books on graphical statics [Eddy, 1877; 1878; 1880]. His books contain a method of graphical analysis for determining the meridian and hoop forces for the membrane stress state in domes [Eddy, 1878; 1880]. He was president of the University of Cincinnati in 1890 and one year later at Rose Polytechnic Institute, Terre Haute, Indiana. Eddy taught engineering and mechanics at the College of Engineering at the University of Minnesota from 1894 until his retirement in 1912. Afterwards, he collaborated with the consulting civil engineer Claude Allen Porter Turner (1869 – 1955) (see section 9.3.1.2) and published his theory of reinforced concrete slabs [Eddy, 1913; 1914]. Eddy therefore made a decisive contribution to reinforced concrete theory during the accumulation phase of structural theory (1900 – 25). The University of Minnesota acknowledged Eddy's scientific work as follows: "His ability as a mathematician won him an international reputation and his high general scholarship and Christian character endeared him to all with whom he came in contact. He was an educator of the highest type, an inspiration to his students and intimate associates, and a wise, sympathetic counsellor in the faculty conferences" [F., J. J., 1922, p. 13].

- Main contributions to structural analysis:
*New constructions in graphical statics* [1877]; *Researches in graphical statics* [1878]; *Neue Constructionen aus der graphischen Statik* [1880]; *The theory of the flexure and strength of rectangular flat plates applied to reinforced concrete floor slabs* [1913]; *Concrete-steel construction* [1914]
- Further reading:
[F., J. J., 1922]
- Photo courtesy of the University of Minnesota archives

### EMPERGER, FRITZ (FRIEDRICH IGNAZ) EDLER VON
*11.1.1862 Beraun, Bohemia, Austria-Hungary (today: Czech Republic), †7.2.1942 Vienna, Austria
Fritz von Emperger (or more properly Friedrich Ignaz Edler von Emperger) came from an old family of Austrian aristocrats. After attending Prague secondary modern school (1872 – 79), he studied civil engineering at Vienna TH and Prague TH, graduating from the latter in 1884. His main interest was bridge-building. At first he worked as an assistant to Prof. Friedrich Steiner at the chair of bridge-building at Prague TH, and was later employed by the Buschtěhrader Bridge Company in Falkenau (Bohemia) and the Prague-based bridge-building company Ruston & Co. The Monier form of construction Emperger saw at the Paris world exposition of 1889 left a deep impression on him. One year later, Emperger accepted a job offer at the Jackson Ironworks in New York, but by 1892 had already set up his own engineering consultancy on Broadway. He enthusiastically adopted the reinforced concrete system with rigid reinforcement patented by Joseph Melan in that same year in Austria-Hungary (Melan system) and in 1893 was commissioned to plan the Edenpark Bridge project in Cincinnati (Ohio), which he designed according to the Melan system. Owing to the cost of and other difficulties involved with the procurement of Portland cement from Germany, the majority of American building contractors were not inclined to build larger concrete structures. Emperger therefore founded the Melan Arch. Constr. Company in 1894, which built numerous bridges, underground railway systems and high-rise buildings in reinforced concrete and within a few years had provided enough impetus for the emergence of an American cement industry [Zesch, 1962, pp. 158/159]. It was in that same year that Emperger gave his famous presentation on reinforced concrete road bridges at the ASCE [Emperger, 1894]. He returned to Austria in 1896 and fought for the theoretical foundation of reinforced concrete construction: "All theory is dull, but practice without theory would be really dreadful" (cited in [Zesch, 1962, p. 159]). He worked as an honorary lecturer for the whole gamut of engineering sciences at Vienna TH from 1898 to 1902 and in 1903 was awarded a doctorate in engineering sciences for

his dissertation on reinforced concrete beams with top and bottom reinforcement [Emperger, 1903]. However, Emperger's greatest achievement was the founding of the journal *Beton und Eisen* (today: *Beton- und Stahlbetonbau*) in 1901, which evolved out of the reports he wrote for the Austrian Engineers & Architects Society concerning the Paris World Exposition of 1900 and represented the first independent technical/scientific journal for the civil engineering industry [Kurrer, 2001, pp. 214/215]. He agreed with the publisher Georg Ernst (1880–1950) that *Beton und Eisen* would be published by Wilhelm Ernst & Sohn from 1905 onwards. It was in that same year that Emperger produced the *Beton-Kalender* yearbook for the first time, and two years later he began releasing the four volumes of the *Handbuch für Eisenbetonbau* [Kurrer, 2005/2], which by the time it reached its third edition (1921–31) had grown to 14 volumes embracing nearly 8000 pages and approx. 12 000 illustrations [Kierdorf, 2007]. The *Handbuch* published by Wilhelm Ernst & Sohn can therefore be called an "encyclopaedia of reinforced concrete construction" [Kurrer, 1999, p. 48]. Emperger thus progressed to become the most successful author and editor of reinforced concrete literature in the first half of the 20th century and certainly helped Wilhelm Ernst & Sohn to become the leading German publishing house with an international reputation in the field of civil and structural engineering. The creation of a scientific footing for reinforced concrete construction was therefore really focused on the publication system of the journal *Beton und Eisen*, the *Beton-Kalender* yearbook and the *Handbuch für Eisenbetonbau* [Kurrer, 2005/2, pp. 796–98]. This publication system is the heart of what became known in the USA at that time as the "Germanization of Reinforced Concrete" (cited in [Kierdorf, 2007, p. 732]). Together with Joseph Melan (Melan system), Emperger can be regarded as a pioneer of composite construction [Eggemann, 2003/3]. For example, as early as the first decade of the 20th century, Emperger developed his Emperger column (composite column), which he patented in 1911 and used successfully in the USA (see section 7.5.2.1). Three years before his death, Emperger proposed casting in conventional reinforcement and prestressing tendons (he preferred two-core cables of high-strength steel) next to one another. The idea behind this partial prestressing was to increase the permissible steel stress and reduce the cracking compared to conventional reinforced concrete [Emperger, 1939]. Emperger's technical/scientific life's work enjoys considerable acclaim not only in Austria and Germany. He was, for example, an honorary member of the British Concrete Institute (today: Institution of Structural Engineers), the American Concrete Institute and the Masaryk Academy in Prague, and a corresponding member of the Polish Academy of Engineering Sciences for the final seven years of his life. He died of a heart attack on 7 February 1942 while preparing for a lecture tour. "A kindly fate," wrote Erwin Zesch, "spared him the miseries of a longer illness that would have condemned him to inactivity" [Zesch, 1962, pp. 166, 167]. Fritz von Emperger was buried with honours in the family's mausoleum in Vienna's main cemetery. At the request of the Austrian Engineers & Architects Society and the Austrian Concrete Society, a commemorative plaque was mounted on the house at Liechtensteinstraße 59 where Emperger had carried out much of his work while in Vienna on the occasion of the 100th anniversary of his birth.

- Main contributions to structural analysis: *The development and recent improvement of concrete-iron highway bridges* [1894]; *Neuere Bauweisen und Bauwerke aus Beton und Eisen. IV. Theil. Die Durchbiegung und Einspannung von armierten Betonbalken und Platten* [1902/4]; *Über die Berechnung von beiderseitig armierten Betonbalken, mit einem Anhang: einige Versuche über die Würfelfestigkeit von armiertem Beton* [1903]; *Die Rolle der Haftfestigkeit in den Verbundbalken* [1905/2]; *Die Abhängigkeit der Bruchlast vom Verbunde und die Mittel zur Erhöhung der Tragfähigkeit von Balken aus Eisenbeton* [1906]; *Versuche mit Säulen aus Eisenbeton und mit einbetonierten Säulen* [1908]; *Eine neue Verwendung des Gusseisens bei Säulen und Bogenbrücken* [1912]; *Neuere Bogenbrücken aus umschnürtem Gusseisen* [1913]; *Der Beiwert n = 15 und die zulässigen Biegespannungen* [1931]; *Stahlbeton mit vorgespannten Zulagen aus höherwertigem Stahl* [1939]

- Further reading:
[Kleinlogel, 1932]; [Brausewetter, 1942]; [Kleinlogel, 1942]; [Zesch, 1962]; [Ricken, 1992]; [Pauser, 1994]; [Kurrer, 1999]; [Kurrer, 2001]; [Eggemann, 2003/1; 2003/2; 2003/3]; [Kurrer, 2005/2]; [Eggemann & Kurrer, 2006]; [Kierdorf, 2007]

- Photo courtesy of the archives of Vienna Technical Museum

### ENGESSER, FRIEDRICH
*12.2.1848 Weinheim, Baden, †29.8.1931 Aachen, Germany
Friedrich Engesser studied at Karlsruhe Polytechnic from 1865 to 1869. Afterwards, he first worked on several structures for Black Forest Railways and then functioned as a central inspector for Baden State Railways in Karlsruhe. In 1885 he succeeded Prof. Hermann Sternberg at Karlsruhe TH, where he remained for 30 years as lecturer and researcher in structural engineering and theory of structures. For example, in 1889 he worked out the difference between deformation energy $\Pi$ and deformation complementary energy $\Pi^*$ and in doing so opened up the theory of structures for the quantitative dominance of non-linear material behaviour. In the same year Engesser published an article explaining the mechanical cause behind the deviation of buckling trials from the Euler curve. However, it was not until 1895 – after he had concluded trials to verify the buckling theory of Considère (1889) – that he specified his modified Euler equation for the non-elastic buckling zone within the scope of a discussion with Jasiński published in a Swiss building journal, and frankly admitted he had made a mistake in 1889 [Nowak, 1981, pp. 147–148]. His contributions to the theory of secondary stresses and theory of frameworks also had a lasting effect on structural theory in the consolidation period. Together with Müller-Breslau and Mohr, Engesser forms the triple star in the firmament of classical structural theory. He contributed to the underlying theories of structural steelwork more than any other. His outstanding achievements were rewarded in many ways, including a doctorate from Braunschweig TH.

- Main contributions to structural analysis: *Über statisch unbestimmte Trägersysteme bei beliebigem Formänderungsgesetz und über den Satz von der kleinsten Ergänzungsarbeit* [1889]; *Über die Knickfestigkeit gerader Stäbe* [1889]; *Die Zusatzkräfte und Nebenspannungen eiserner Fachwerkbrücken* [1892/1893]; *Über die Berechnung auf Knickfestigkeit beanspruchter Stäbe aus Schweiß- und Gußeisen* [1893]; *Über Knickfragen* [1895]; *Über die Knickfestigkeit von Stäben mit veränderlichem Trägheitsmoment* [1909]; *Die Berechnung der Rahmenträger mit besonderer Rücksicht auf ihre Anwendung* [1913]

- Further reading:
[Kriemler, 1918]; [Probst, 1931]; [Gaber, 1931]; [Steinhardt, 1949]

- Photo courtesy of University of Karlsruhe archives

### EYTELWEIN, JOHANN ALBERT
*31.12.1764 Frankfurt a. M., Holy Roman Empire, †18.8.1849 Berlin, Prussia
In 1779 Eytelwein was promoted to bombardier in the 1st Artillery Regiment in Berlin, where his superior at a later date was to be General von Tempelhoff – an officer who awakened in Eytelwein an understanding for the need to link engineering practice and theory. Following seven years of military service, he completed his surveying examinations. In 1790 he sat an examination for the Regional Building Department founded in 1770. Thereafter, he was employed as an inspector of dyke-building, and in 1794 was appointed Privy Senior Government Building Surveyor in the Regional Building Department in Berlin. From 1797 to 1806 he and other colleagues in the Regional Building Department published the first building journal in Germany *Sammlung nützlicher Aufsätze und Nachrichten, die Baukunst betreffend*. In 1799 he was one of the three founders of the Berlin Building Academy and was responsible for the mechanics

ENGESSER

EYTELWEIN

FAIRBAIRN

FALTUS

of solid bodies, hydraulics and mechanical engineering. Eytelwein's contribution to the theory of structures forms a bridge between the initial and constitution phases of structural theory. He became director of the Regional Building Department in 1809 and one year later rose to become First Secretary in the Prussian Ministry of Trade & Commerce, and in 1816 director of the Regional Building Department. He retired from Prussian state service in 1830 and was followed at the Building Academy by Peter Christian Beuth (1781–1853) and as director of the Regional Building Department by Karl Friedrich Schinkel (1781–1841). He was awarded numerous honours, including a doctorate by Berlin University's Faculty of Philosophy in 1811.

- Main contributions to structural analysis: *Handbuch der Mechanik fester Körper und Hydraulik* [1801, 1823, 1842]; *Handbuch der Statik fester Körper* [1808, 1832].
- Further reading: [Encke, 1851]; [Rühlmann, 1883]; [Scholl, 1990]
- Photo courtesy of [Scholl, 1990, p. 49]

### FAIRBAIRN, SIR WILLIAM

*19.2.1789 Kelso, Roxburghshire, UK,
†18.10.1874 Moore Park, Surrey, UK

Born to a farmer in the Scottish town of Kelso, Fairbairn completed an apprenticeship as a miner in the vicinity of North Shields, a port on the River Tyne important for shipping coal to London. It was there that William Fairbairn became friends with the young George Stephenson and helped him become a crane operator, unloading ballast from the incoming coal freighters. It was in 1850 that Fairbairn was granted a patent for a crane design which consisted of four riveted curving iron plates forming a hollow-box jib with a cross-section tapering to match the bending moment diagram and was therefore much better suited to loading and unloading operations than cranes with straight jibs; this so-called Fairbairn crane (see Fig. 12-7) became a characteristic feature of the ports of the second half of the 19th century. Following his apprenticeship, Fairbairn worked as a mill-builder in Newcastle Upon Tyne. He used his leisure time for learning. In Manchester he managed to obtain employment as the production engineer in a cotton mill and after five years was able to open a machine shop financed through his own savings and a bank loan. Fairbairn quickly gained a good reputation in the building of multi-storey factories, especially for the textiles industry; his turnkey fireproof factories were exported to Russia, Switzerland and Turkey, for example. It is therefore no surprise that Fairbairn – who signed his name "Fairbairn C.E." (civil engineer) – was elected to the Institution of Civil Engineers as early as 1830. As a protest against the conservatism of the Royal Society, he and a number of others founded the British Association for the Advancement of Science in 1831. During the 1830s, Fairbairn extended his technical and business activities to the building of locomotives and ships. His iron ship *Lord Dundas* entered the history books as the first serviceable iron steamship. Encouraged by his successes, Fairbairn set up a shipyard for building iron ships on the River Thames at Millwall in London, which, however, did not really attract the orders he had envisaged so he quickly took on orders for iron bridges as well [Lehmann, 1999, p. 130]. One highlight in Fairbairn's output is his collaboration with Robert Stephenson, Eaton Hodgkinson, Francis Thompson and Edwin Clark in the design, calculations and erection of the Britannia Bridge over the Menai Strait (1846–50) – not far from Telford's suspension bridge. The box section made from riveted wrought-iron plates had no precedent in bridge-building but was inspired by the most advanced iron ship designs of I. K. Brunel and Fairbairn. Fairbairn understood how to exploit the knowledge of structural theory and strength of materials for bridges and ships, although it must be said that he always preferred engineering science experimentation over theory. He was responsible for the first theoretical attempt to determine the longitudinal strength of ships with the help of beam theory (1860). Fairbairn published the first monograph on iron ships in 1865. Besides publications on shipbuilding and bridge-building, he also wrote articles about steam engines, material behaviour and riveting. William Fairbairn is the embodiment of the ingenious spirit of the "workshop of the world", as the United Kingdom was known as in those days. When Fairbairn died, that industrious island lost its last great engineering hero. More than 50 000 people turned out in Manchester to pay their respects at his funeral!

- Main contributions to structural analysis: *An Experimental Enquiry into the Strength and other Properties of Cast Iron, from Various Parts of the United Kingdom* [1838]; *On tubular girder bridges* [1849/1]; *An Account of the Construction of the Britannia and Conway Tubular Bridges with a complete History of their Progress, from the Conception of the Original Idea, to the Conclusion of the Elaborate Experiments which Determined the Exact Form and Mode of Construction ultimately Adopted* [1849/2]; *On the Application of Cast and Wrought Iron to Building Purposes* [1854]; *On the Application of Cast and Wrought Iron to Building Purposes. To which is added a short treatise on wrought iron bridges* [1857/58]; *Treatise on Iron Shipbuilding* [1865]
- Further reading: [Pole, 1877]; [Peters, 1996]; [Rennison, 1998]; [Lehmann, 1999, pp. 129–130]
- Photo courtesy of [Pole, 1877]

### FALTUS, FRANTIŠEK

*5.1.1901 Vienna, Austria-Hungary,
†6.10.1989 Prague, Czechoslovakia

František Faltus studied civil engineering from 1918 to 1923 at Vienna TH, where he gained his doctorate with the dissertation *Beitrag zur Berechnung statisch unbestimmter Tragwerke*. He started his professional career in 1923 with the steelwork fabricator Waagner-Biró-AG in Vienna and continued it in 1926 at the Škoda plant in Plzeň, Czechoslovakia. He was an enthusiastic advocate of the new method of jointing by welding and introduced it into steel bridges and other steel structures. Faltus thus became a pioneer of welding in structural steelwork. He designed the first welded trussed framework bridge in 1930 (in Plzeň), which was followed by the first fully welded arch bridge – also in Plzeň. His continuous steel bridge with a composite road deck erected in 1952 in Bytča, Czechoslovakia, also set new standards in structural steelwork. As early as 1938 he had been offered the chair of steelwork by the Czech TH in Prague, but owing to Germany's occupation of Czechoslovakia was unable to take up this post until 1945. Faltus quickly progressed to become a leading personality in the field of steelwork in the former Czechoslovakia. He made important scientific contributions to the theory of designing welded structures, but also to buckling and arch theory. In doing so, he understood the need to couple his national and international consulting activities closely with scientific analyses. On the occasion of its 50th anniversary, the International Association for Bridge & Structural Engineering (IABSE) elected Faltus (one of its founding members) honorary member. Other honours followed: an honorary doctorate from Dresden TU, a state prize, corresponding membership of the Czechoslovakian Academy of sciences, and numerous medals.

• Main contributions to structural analysis: *Svařování* (welding) [1947/1955]; *Prvky ocelových konstrukcí* (steel elements) [1951/1962]; *Joints with Fillet Welds* [1985]
• Further reading: [Hořejší & J. Pirner, 1997, pp. 57 – 61]
• Photo courtesy of Prof. Dr. L. Frýba

### FLÜGGE, WILHELM
*18.3.1904 Greiz, German Empire, †19.3.1990 Los Altos, California, USA
Wilhelm Flügge was just 21 when he graduated as a civil engineer from Dresden TH, and two years later he had already gained his doctorate at the same university under Kurt Beyer. But he was outdone by his younger brother, Siegfried Flügge (1912 – 97), who gained his doctorate with Max Born in Göttingen at the tender age of 21, at 25 was already professor of theoretical physics at the Kaiser Wilhelm Institute for Chemistry headed by Otto Hahn, and quickly went on to establish himself as a nuclear physicist of international repute. But back to Wilhelm Flügge. He worked for Dyckerhoff & Widmann from 1927 to 1930 and it was there that he became familiar with the most highly developed form of reinforced concrete construction: the emerging design language of shell construction. Franz Dischinger, Hubert Rüsch and Ulrich Finsterwalder were also at the company during those years. He left Dyckerhoff & Widmann to take up a post at the University of Göttingen, where he wrote his habilitation thesis in 1932 and co-founded the journal *Zentralblatt für Mechanik* in 1933. The following year saw the publication of Flügge's book *Statik und Dynamik der Schalen* [Flügge, 1934], the English translation of which (*Stresses in Shells*) appeared in 1960 [Flügge, 1960]; it gradually became the standard work on shell theory throughout the world. During the years of the Third Reich, Flügge was not appointed to any chair because he was regarded as politically unreliable. But this changed when Hermann Göring started to place more value on technical knowledge than political and racial purity while building up the air force and appointed Flügge to a senior position at the German Aviation Testing Authority (DVL) in 1938, where he helped with the analysis of shell structures for aircraft. After the war, Flügge and his wife, Irmgard Flügge-Lotz (1903 – 74), were ordered to carry out research work in Paris at the Office National d'Études et de Recherches Aéronautique (ONERA). That was in 1947. Following an inquiry from Timoshenko, the two scientists left France illegally and in 1949 started work at Stanford University, where they worked as teachers and researchers for 20 years, supervising more than 70 dissertations over those years. After he retired, Flügge published monographs on subjects like *Tensorial Methods in Continuum Mechanics* [Flügge, 1972] and *Viscoelasticity* [Flügge, 1975]. He understood perfectly how to combine rigorous derivations with interesting presentations.

• Main contributions to structural analysis: *Statik und Dynamik der Schalen* [1934]; *Stresses in Shells* [1960]; *Handbook of Engineering Mechanics* [1962]; *Tensorial Methods in Continuum Mechanics* [1972]; *Viscoelasticity* [1975]
• Further reading: [Duddeck, 1990]; [Gere et al., 2004]; [Stiglat, 2004, p. 147]
• Photo courtesy of [Stiglat, 2004, p. 147]

### FÖPPL, AUGUST
*25.1.1854 Groß-Umstadt near Darmstadt, Hesse, †12.10.1924 Ammerland (Bavaria), German Republic
August Föppl started his civil engineering studies in 1869 at Darmstadt TH and continued them at Stuttgart TH, where the excellent lectures of Otto Mohr developed his interest in mechanics and strength of materials. After Mohr's departure from Stuttgart in 1874, Föppl switched universities again and completed his civil engineering studies at Karlsruhe TH in 1874. He admitted not understanding the lectures of Franz Grashof in Karlsruhe; it was not until many years later that he referred to Grashof when carrying out his own research [Föppl, A. & Föppl, O., 1925, pp. 84 – 85]. Following his studies he worked for a short time in the Baden Highways Department, just enough to satisfy military service obligations, and then in the autumn of 1876 took up a temporary teaching post at the Building Trades School in Holzminden. From 1877 to 1894 he taught at the Municipal Industrial School in Leipzig. This period was his most fruitful in terms of contributions to the theory of structures. For example, he wrote four sections for the appendix of the 1878 German 2nd edition of Navier's *Mechanik der Baukunst* [Navier, 1878]: *Anwendungen der Elastizitätsgesetze auf die Berechnung der Baukonstruktionen* (pp. 506 – 522), *Theorie des Fachwerks* (pp. 523 – 556), *Die Theorie der Tonnengewölbe* (pp. 557 – 581) and *Dimensions-Berechnung der Eisen- und Stahl-Constructionen* (pp. 582 – 589). Of these four contributions, the second, *Theorie des Fachwerks*, is particularly noteworthy for its clear definitions, and appeared in expanded form as a book in 1880. This was followed one year later by the *Theorie der Gewölbe*, likewise much expanded. He combined both monographs under the title *Mathematische Theorie der Baukonstruktionen*, which in 1886 was acknowledged by the University of Leipzig as a dissertation. Föppl was the first to present a self-contained theory of space frames (1892). These books constitute a significant contribution to the consummation of theory of structures. In that same year he was appointed regular associate professor for cultivation and mechanical engineering in agriculture at the University of Leipzig. His studies in electrical engineering under Gustav Wiedemann (1826 – 99) at the University's Physics Institute were crowned in 1894 with his monograph *Einführung in die Maxwellsche Theorie der Elektricität*; it was here that he revealed for the first time the theoretical and practical strengths of vector analysis, which Föppl qualified as "the mathematical sign language of the physics of the future" [Hiersemann, 1990, p. 60]. In 1894 he succeeded Johann Bauschinger (1834 – 94) in the chair of applied mechanics and as head of the Mechanical-Technical Laboratory at Munich TH. He regarded the aim of materials testing more as "establishing the behaviour of complete construction parts, assembled constructions too, rather than the materials used themselves" [Prinz, 1924, p. 1]. This meant he was the first to establish a new understanding of materials testing, which later would become a successful reference model for engineering science experimentation. It was during his period in Munich that Föppl wrote what was the most influential textbook of applied mechanics in Germany up to that time: *Vorlesungen über Technische Mechanik* (1898 – 1910) – six volumes which appeared in numerous editions and were translated into several languages; more than 100 000 copies had been sold by 1925. His prolific output was complemented in 1920 by the two-volume work *Drang und Zwang* written in conjunction with his son Ludwig. In 1917 he succeeded in expanding the St. Venant torsion theory, work which his student Constantin Weber continued later. Föppl's most important student was Ludwig Prandtl, who in 1899 gained his doctorate under Föppl with a dissertation on lateral buckling phenomena of beams with slender rectangular cross-sections. A history of the teaching of applied mechanics in the German-speaking countries would have to analyse the textbooks of Franz Joseph Ritter von Gerstner, Julius Weisbach, August Föppl and Istvan Szabó. But one thing is certain: only August Föppl's books on mechanics became best-sellers. He thus created the most influential school of applied mechanics and was the first engineer to be accepted as a full member of the Bavarian Academy of Sciences. Both Munich TH and Darmstadt TH awarded him honorary doctorates.

• Main contributions to structural analysis: *Theorie des Fachwerks* [1880]; *Theorie der Gewölbe* [1881]; *Das Fachwerk im Raum* [1892]; *Vorlesungen über Technische Mechanik* [1898 – 1910]; *Über den elastischen Verdrehungs-*

FLÜGGE  FÖPPL  FRAEIJS DE VEUBEKE  FRÄNKEL  FREUDENTHAL

winkel eines Stabes [1917]; Drang und Zwang [1920]
- Further reading:
[Schlink, 1923]; [Föppl et al., 1924]; [Prinz, 1924]; [Föppl, L., 1924]; [Föppl, A. & Föppl, O., 1925]; [Hirsemann, 1990]; [Dittmann, 1994]
- Photo courtesy of [Föppl et al., 1924]

### FRAEIJS DE VEUBEKE, BAUDOUIN M.

*3.8.1917 Ramsgate, UK, †16.9.1976 Liège, Belgium

Following his studies at the universities of Leuven and Liège, Fraeijs de Veubeke joined the Belgian section of the Royal Air Force in 1944. Two years later he joined the aircraft design office of the Belgian aircraft company Avions Fairey; in 1948 he became a lecturer at the University of Leuven and in 1952 was finally appointed professor at the University of Liège. It was there that Fraeijs de Veubeke was in charge of the Laboratoire de Techniques Aeronautiques et Spatiales and held the chair of continuum mechanics. Like his colleague Charles Massonnet, his pioneering research work into the formation of structural mechanics theories during the innovation phase of structural theory (1950–75) gave the University of Liège an international reputation. For example, Guy Sanders (1937–87) was one of his students. Fraeijs de Veubeke's scientific works earned him not only membership of the Belgian Academy of Sciences, but also further honours in Belgium and abroad. A focal point of his scientific work was providing a foundation for FEM through general variational theorems (see section 10.4.5.4). Unfortunately, although this work had a great significance on structural mechanics, it did not receive the acclaim it was due. One reason for this is certainly Fraeijs de Veubeke's matter-of-fact style of presentation, which means that the reader has to prise out the jewels of the knowledge very carefully – Fraeijs de Veubeke never liked to present his findings as if they were propaganda. Felippa acknowledges Fraeijs de Veubeke as follows: "An aristocrat by birth and gentleman by nature, de Veubeke never displayed greed for priority and recognition" [Felippa, 2002, p. 10]. This becomes obvious in the example of Fraeijs de Veubeke's formulation of the general variational theorem: he never claimed to have discovered this variational theorem before Washizu and Hu. Fraeijs de Veubeke displayed human greatness based on honesty. Unfortunately, this honourable scientist passed away all too soon, and so a summary of his outstanding research studies never found its way into manuals.

- Main contributions to structural analysis:
Diffusion des inconnues hyperstatiques dans les voilures à longeron couplés [1951]; Matrix Method of Structural Analysis [1964]; Displacement and equilibrium models in the finite element method [1965, 2001]; Strain energy bounds in finite element analysis [1967]; A new variational principle for finite elastic displacements [1972]; Variational principles and the patch test [1974]; The dynamics of flexible bodies [1976]; B. M. Fraeijs de Veubeke Memorial Volume of Selected Papers [1980]
- Further reading:
[Géradin, 1980]; [Felippa, 2002]; [Zienkiewicz, 2001, pp. 287–289]
- Photo courtesy of [Zienkiewicz, 2001, p. 287]

### FRÄNKEL, WILHELM

*1.1.1841 Odessa, Russia, †13.4.1895 Dresden, German Empire

As the adopted son of Oskar Schlöhmilch (1823–1903), professor of mathematics at Dresden Polytechnic, Wilhelm Fränkel's CV was more or less preordained: studies in civil engineering at the Polytechnic School in Dresden, where from 1865 he served as assistant to Prof. Johann Andreas Schubert and carried out work for Saxony State Railways. Concurrently with Winkler and Mohr, he worked on the principles of the theory of influence lines (1867). In 1870 he was offered the chair of theory of structures and bridge-building (established in 1868) and in 1871 the Polytechnic School in Dresden was raised to the status of a Polytechnic. Fränkel remained here until he died; his successors were: Georg Christoph Mehrtens (1895–1912), Willy Gehler (1913–45), Kurt Beyer (1919–52) and Gustav Bürgermeister (1952–71). It may have been Winkler's theorem of the line of thrust in masonry arches [Winkler, 1879/1880, p. 128] that inspired Fränkel to derive (1882) – independently of Menabrea and Castigliano – the principle of Menabrea on a plane frame with n degrees of static indeterminacy and linear-elastic continuum from the principle of virtual forces and furnish proof of Winkler's theorem. He was aware that his "principle of the least work of elastic systems … would permit a consistent understanding of a whole series of structural theory problems" [Fränkel, 1882, p. 63]. Unfortunately, Fränkel's groundwork in the dispute over the fundamentals of the theory of structures (1883–89) was not fully adopted by either Müller-Breslau or Mohr – they could have helped clarify matters. T. M. Charlton paid tribute to Fränkel's pioneering contributions to classical structural theory [Charlton, 1982]. Fränkel invented the strain gauge named after him which was used for the experimental investigation of iron bridges; he later added the deflection gauge as well as the horizontal and vertical vibration gauges. He had thus created the instruments required to verify bridge engineering theories and classical structural theory. His chapter on moving bridges in the Handbuch der Ingenieurwissenschaften [Schäffer & Sonne, 1888/1] remained unequalled for two decades.

- Main contributions to structural analysis:
Berechnung eiserner Bogenbrücken [1867]; Zur Theorie der elastischen Bogenträger [1869]; Vorträge über Schiebebühnen und Drehscheiben [1872]; Anwendung der Theorie des augenblicklichen Drehpunktes auf die Bestimmung der Formänderung von Fachwerken – Theorie des Bogenfachwerkes mit zwei Gelenken [1875]; Über die ungünstigste Einstellung eines Systems von Einzellasten auf Fachwerkträger mit Hilfe von Influenzkurven [1876]; Das Princip der kleinsten Arbeit der inneren Kräfte elastischer Systeme und seine Anwendung auf die Lösung baustatischer Aufgaben [1882]
- Further reading:
[n. n., 1895]
- Photo courtesy of Dresden TU archives

### FREUDENTHAL, ALFRED MARTIN

*12.2.1906 Stryj, Austria-Hungary (today: Ukraine), †27.9.1977 Maryland, USA

Freudenthal studied civil engineering at Prague TH and Lemberg TH. He gained his doctorate in engineering sciences in 1930 at the German Technical University in Prague with a dissertation on plastic theory. During the 1930s he published a series of research findings concerning the plastic behaviour of reinforced concrete structures. In doing so, he considered the influence of creep in concrete on the internal forces of long-span arch structures [Freudenthal, 1935/1; 1936/2] before Dischinger [Dischinger, 1937; 1939]. His great interest in materials research, which he would later expand with great success, is even evident in these early works. He emigrated to Palestine in 1935. It was here that he was involved in the planning, design and administration of the new port at Tel Aviv from 1936 to 1946. Over those 10 years, Freudenthal also worked at the University of Haifa, initially as a lecturer and later as professor of bridge-building at the Hebrew Institute of Technology. Following the publication of this work on the statistical aspects of fatigue, he was appointed visiting professor for theoretical and applied mechanics at the University of Illinois in 1947. Two years later, Columbia University appointed him professor of civil engineering, and he remained at that university, teaching and researching, for 20 years. He was professor of civil and materials engineering from 1969 until his retirement at George Washington University and during that same period director of the Institute for the Study of Fatigue & Structural Reliability. This institute had been founded at Columbia University in 1962 and later transferred to George Washington University. Under the leadership of Freudenthal, the institute carried out pioneering work which enabled Freudenthal to become known as the father of structural reliability. Freudenthal was awarded many honours for these achievements, including the ASCE's Norman and Kármán Medals. Freudental published his scientific work in seven books and about 150 papers.

- Main contributions to structural analysis: *Beitrag zur Berechnung von Wälzgelenken aus Beton und Stein* [1933/1]; *Wirkung der Windkräfte auf Eisenbetonskelette mit Berücksichtigung des räumlichen Tragvermögens* [1933/2]; *Plastizitätstheoretische Methoden bei der Untersuchung statisch unbestimmter Tragwerke aus Eisenbeton* [1934]; *Die Änderung des Spannungszustandes weitgespannter, flacher Eisenbetonbogen durch die plastische Dauerverformung des Betons* [1935/1]; *Einfluß der Plastizität des Betons auf die Bemessung außermittig gedrückter Eisenbetonquerschnitte* [1935/2]; *Die plastische Dauerverformung des versteiften Stabbogens* [1936/1]; *Théorie des grandes voûtes en béton et en béton armé* [1936/2]; *The inelastic behaviour of engineering materials* [1950]; *Selected Papers by Alfred M. Freudenthal: Civil Engineering Classics* [1981]
- Further reading:
[Liebowitz, 1979]
- Photo courtesy of [Liebowitz, 1979, p. 62]

### GALERKIN, BORIS GRIGORIEVICH

*4.3.1871 Prudki near Polotsk (Vitebsk), Russia, †12.7.1945 Moscow, USSR
Following his primary education at the Jewish primary school in Polotsk (1885) and success in external examinations for the secondary school in Minsk, he studied at the St. Petersburg Technology Institute from 1893 onwards, from where he was expelled in 1899 for his revolutionary activities; nevertheless, he still graduated from the Institute as an external student in that same year. Following his engineering studies he worked in the locomotive factory in Kharkov and then in a mechanical workshop and boiler factory in St. Petersburg. He was imprisoned for a few months in 1905 for revolutionary activities and in 1907 was again sent to prison, this time for 18 months. Starting in 1909 he taught at the St. Petersburg Polytechnic Institute and in 1920 was appointed to the chair of structural mechanics. During this period Galerkin was closely linked with the Geneva Polytechnic Institute. From 1924 to 1929 he was professor of elastic theory at the Leningrad Institute for Highway Construction and professor of structural mechanics at the University of Leningrad. Galerkin organised and chaired consultations on the major structures of the USSR. The general calculus of variations, which Bubnov had already anticipated in 1911 (Bubnov-Galerkin method), is linked to his name. He became director of the Institute of Mechanics at the Russian Academy of Sciences in 1938 (he had been a full member since 1935). In 1942 he was awarded the Stalin Prize and was nominated Lieutenant-General of Engineering.

- Main contributions to structural analysis: *Sterzhni i plastinki: Ryady v nekotorykh voprosakh uprugogo ravnovesiua sterzhnei i plastinok* (bars and slabs: series for some problems of elastic equilibrium of bars and slabs) [1915]; *Uprugie tonkie plity* (thin elastic slabs) [1933]; *Sobranie sochinenii* (collected works) [1952–53]
- Further reading:
[PMM: Angewandte Mathematik & Mechanik, 1941]; [Galerkin, 1952–1953]; [Grigorian, 1978]; [Matviishin, 1989]
- Photo courtesy of Prof. Dr. G. Mikhailov

### GALLAGHER, RICHARD HUGO

*17.11.1927 New York City, USA, †30.9.1997 Tucson, Arizona, USA
Richard Hugo Gallagher was born into a Catholic family. His father was of Irish descent, but his mother had been born in Bohemia – the attributes of a true American! Following his secondary education at the Cardinal Hayes High School and military service in the US Navy, Gallagher studied structural engineering at New York University was awarded a bachelor degree in 1950. For the next five years he worked as a field engineer for the US Department of Commerce and thereafter as a design engineer in the New York offices of the Texas Corporation. During this period he studied structural engineering as an external student at New York University and upon completion was awarded a masters degree. Gallagher worked in the Structural Systems Department of Bell Aero Systems in Buffalo from 1955 to 1967, initially as assistant chief engineer. It was here that Gallagher continued to develop the FEM Turner and others had been working on at Boeing. "The opportunities offered by the finite element method fired his imagination and led to much creative research" [Zienkiewicz et al., 1997, p. 904]. Gallagher, working with J. Padlog and P. P. Bijlaard, wrote a paper in which tetrahedron elements appear as early as 1962 [Gallagher et al., 1962] – the very first publication concerning three-dimensional elements. He completed his doctorate at Buffalo University in 1966 with a dissertation on curved (finite) elements for thin shells, and from 1967 to 1978 was professor for civil and environmental engineering at Cornell University, teaching and researching alongside the well-known structural steelwork professor George Winter (1907–82); Gallagher took over from him as chairman of the faculty of engineering in 1969. It was at Cornell University that Gallagher was able to continue the work on stability theory and shell theory he had begun at Bell Aero Systems, extend this work considerably and apply FEM to fluid mechanics. He used his visiting professorship at the University of Tokyo in the autumn of 1973 and his research semester at University College, Swansea (UK), to prepare his main work, *Finite Element Analysis Fundamentals* [Gallagher, 1975], which was translated into German (1976), French (1977), Chinese (1979), Russian (1985) and Turkish (1994) as well as other languages. In historico-logical terms, this monograph can be regarded as a brilliant introduction to the diffusion phase of structural mechanics (1975 to date). Together with J. T. Oden, T. H. H. Pian, E. L. Wilson and O. C. Zienkiewcz, Gallagher visited China in 1981 in order to consolidate scientific contacts in the field of FEM (Gallagher was to return to China on a number of occasions); Shanghai Technical University awarded him an honorary doctorate in 1992 for his services to the development of scientific cooperation with China. Gallagher served as dean of the College of Engineering at the University of Arizona from 1978 to 1984, thereafter, until 1988, he was provost and vice-president of academic issues at Worcester Polytechnic Institute, and he rounded off his academic career as president of Clarkson University in 1995. The ASME Medal (1993) and honorary membership of ASCE are just two of the many honours and awards he received.

- Main contributions to structural analysis: *The stress analysis of heated complex shapes* [1962]; *A Correlation Study of Matrix Methods of Structural Analysis* [1964]; *Theory and Practice in Finite Element Structural Analysis* [1973]; *Finite Element Analysis Fundamentals* [1975]; *Finite Elements for Thin Shells and Curved Members* [1976]; *Introductory Matrix Structural Analysis* [1979]; *Optimum Structural Design* [1979]; *New Directions in Optimum Structural Design* [1984]

GALERKIN  GALLAGHER  GERSTNER  GIRARD  GIRKMANN

- Further reading:
[Zienkiewicz et al., 1997];
[Zienkiewicz & Taylor, 1998]
- Photo courtesy of [Zienkiewicz et al., 1997, p. 903]

## GERSTNER, FRANZ JOSEPH RITTER VON

*23.2.1756 Komotau, Bohemia, Austria, †25.6.1832 Mladějov, Bohemia, Austria
During his secondary education the young Gerstner made himself known to the local tradesmen in his home town. From 1773 to 1779 he studied at the University of Prague where he heard lectures on philosophy, theology, Greek, Hebrew, elementary mathematics, astronomy and higher mathematics. Some of his earnings came through playing the organ and private mathematics and physics pupils. Despite his speech impediment, he undertook two public doctor examinations on astronomy and Newton's *Principia*, which he passed with flying colours. Afterwards, he served as a surveyor on a royal commission which was set up by Emperor Joseph II in the course of abolishing serfdom. Inspired by the Emperor's vision, Gerstner studied medicine but still retained a serious interest in questions of astronomy and mathematics. In 1789 he published his introduction to structures and architecture, which became an important document in the initial phase of structural theory. That same year saw him appointed professor of higher mathematics at the University of Prague, where he was soon able to raise the numbers of students attending lectures from three or four to 70 or 80. One of those who attended was the later mathematician and philosopher Bernard Bolzano (1781–1848), whose genius was recognised by Gerstner and who did everything he could to help him. Five years after the death of his mentor, Bolzano wrote a biography as a tribute [Bolzano, 1837]. Gerstner's interests focused more and more on "advancing the commerce of the Fatherland through scientific teaching" [Gerstner, 1833, p. V]. With this in mind he founded the Prague Polytechnic in 1806 on behalf of the Bohemian Parliament and remained its director until 1822, responsible for mathematics and mechanics. In 1811 the Emperor appointed him director of hydraulic engineering of Bohemia; Gerstner became a valued adviser for countless engineering projects in his country of birth. In the 1820s he built the first railway line in continental Europe – the horse-drawn railway between Linz and Budweis – together with his son Franz Anton. His son also helped him with the publishing of his three-volume *Handbuch der Mechanik*, although this was not completed until after his death. Gerstner's *Handbuch* was the first independent work on applied mechanics in the German language and represents the constitution phase of this fundamental discipline of engineering science; not until many decades later did the theory of structures achieve emancipation from this in the German-speaking countries.

- Main contributions to structural analysis:
*Einleitung in die statische Baukunst* [1789]; *Handbuch der Mechanik* [1831–1834]
- Further reading:
[Bolzano, 1837]; [Wurzbach, 1859]; [Karmarsch, 1879]; [Gerstner, 1932]; [Rudolff, 1932]; [Mechtler, 1964]
- Photo courtesy of [Gerstner, 1833]

## GIRARD, PIERRE-SIMON

*4.11.1765 Caen, France, †30.11.1836 Paris, France
Pierre-Simon Girard came from a Protestant family. Girard's scientific talent was recognised at an early age, which enabled him to study at the École des Ponts et Chaussées, leaving in 1789 as a roads and bridges engineer. During the 1790s, Girard was involved in waterways and he published his first monograph on strength of materials in 1798 [Girard, 1798], which, however, was restricted to beam theory. In the first part of his book he presents the history of beam theory, but shows a preference for Euler's non-linear theory, which is why Girard's work contains complicated mathematical equations that are of little use in practice. Eytelwein would be the first to work with the linearised differential equation for bending in 1808 (see eq. 5-51), later Navier (see section 5.6.3.2), who based his practical bending theory on this. The second part of Girard's book deals with a problem that was popular in the 18th century – the beam of constant strength. But the third part of his book is original and covers bending and buckling tests. Girard's monograph on strength of materials is a focused demonstration of the initial phase of structural theory (1775–1825) and as such forms the historico-logical precursor to Navier's synthesis of statics and strength of materials to form theory of structures. In the same year his book on strength of materials was published, Girard together with other engineers and scientists accompanied Napoleon on his expedition to Egypt. After his return, Girard dedicated himself to problems and projects in hydraulic engineering, especially canal-building.

- Main contributions to structural analysis:
*Traité analytique de la résistance des solides, et des solides d'égale résistance, auquel on a joint une suite de nouvelles expériences sur la force, et l'élasticité spécifique des bois de chêne et de sapin* [1798]
- Further reading:
[Richet, 1933]; [Timoshenko, 1953/1, pp. 42–43, 58]
- Photo courtesy of Bibliothèque Nationale, Paris

## GIRKMANN, KARL

*22.3.1890 Vienna, Austria, †14.7.1959 Vienna, Austria
Karl Girkmann studied civil engineering at Vienna TH. Afterwards, he worked in industry, concentrating in particular on the mechanics problems of overhead lines, which were spreading fast in the 1920s. He gained his doctorate at Vienna TH in 1925 with a dissertation entitled *Eisengittermaste belastet durch wechselseitig wirkende Seilzüge*, followed by a habilitation thesis with the title *Beiträge zur Berechnung von Behältern*, which had been inspired by his work in the design office of the Vienna-based steelwork company Waagner-Biró-AG. Girkmann laid the foundations for the ultimate load method for frames in 1931–32. In 1938 came the appointment as full professor of applied mechanics and election to the board of the chair of elasticity and strength of materials at Vienna TH. His book *Flächentragwerke*, which appeared in numerous editions and various languages, became a milestone in the development of multi-dimensional loadbearing structures in the theory of structures. He was elected a member of the Academy of Sciences in Vienna in 1950 and awarded an honorary doctorate in engineering sciences by Graz TH in 1955.

- Main contributions to structural analysis:
*Bemessung von Rahmentragwerken unter Zugrundelegung eines ideal plastischen Stahles*

[1931]; *Über die Auswirkung der „Selbsthilfe" des Baustahls in rahmenartigen Stabwerken* [1932]; *Die Hochspannungs-Freileitungen* [1938]; *Flächentragwerke* [1949]
- Further reading:
[Chwalla et al., 1950]; [Beer, 1959]; [Karas, 1960]
- Photo courtesy of [Karas, 1960, p. 32]

### GIUFFRÈ, ANTONINO
\*17.1.1933 Messina, Sicily, Italy,
†27.11.1997 Rome, Italy
Antonino Giuffrè completed his structural engineering studies in Rome in 1957. Beginning in 1962, he worked at the faculty of architecture at the University of Rome I, where he became involved with the mechanical behaviour of reinforced concrete construction under seismic actions (1982). The damage caused to the Sant' Angelo dei Lombardi Cathedral and other historically important structures as a result of the Irpinia earthquake of November 1980 inspired Giuffrè to continue developing his concept for a method of restoration matched to the loadbearing quality and form of historical structures in regions prone to earthquakes (1988). Whereas the structural engineering design methods based on elastic theory resulted in a form of restoration that, using steel and reinforced concrete, represented a major intervention into the original structural system of the historical construction, Giuffrè explored the historical dimensions of the damaged structures to ensure a more appropriate form of refurbishment. He therefore investigated the kinematic view of masonry arch theory popular in the 18th century in his analysis of damaged masonry arches and generalised this approach to form his concept of the "modes of damage" of structures subjected to seismic actions, from which he devised his "masonry grammar". Like Heyman, Giuffrè was in favour of developing plastic methods of design and applying these to historic masonry arches. The difference was that Heyman based his work on a statics theorem, Giuffrè a kinematics theorem, and so Heyman in the end pursued the geometrical, Giuffrè the kinematic view of theory of structures.
- Main contributions to structural analysis:
*Analisi matriciale delle strutture. Statica, dinamica, dinamica aleatoria.* [1982/1]; *La risposta non lineare delle strutture in cemento armato* [1982/2]; *La meccanica nell'architettura* [1986]; *Monumenti e terremoti: aspetti statici del restauro* [1988]; *Un progetto in itinere: il restauro della cattedrale di S. Angelo dei Lombardi* [1988]; *Seismic response of mechanism of masonry assemblages* [1990]; *Letture sulla Meccanica delle Murature Storiche* [1991]; *Statics and dynamics of historical masonry buildings* [1992]; *Codice di pratica per la sicurezza e la conservazione dei Sassi di Matera* [1997]
- Further reading:
[Augusti, 1999]; [Piccarreta & Sguerri, 2001]; [Sorrentino, 2002]
- Photo courtesy of Dr. Luigi Sorrentino, Rome

### GRASHOF, FRANZ
\*11.7.1826 Düsseldorf, Prussia,
†26.10.1893 Karlsruhe, German Empire
When he was just 15 years old, Franz Grashof left school, initially in order to work for a locksmith. Later, he attended the industrial school in Hagen and then the secondary school in Düsseldorf. He studied at the Berlin Industrial Academy from 1844 to 1847. During the revolution of 1848, Grashof was forced into military service, whereupon he decided to join the German merchant fleet, but returned two-and-a-half years later to resume his studies at the Industrial Academy. At the same Academy he became a teacher of mathematics and mechanics in 1854 and one year later director of the Royal Office of Weights & Measures in Berlin. Grashof founded the Verein Deutscher Ingenieure (VDI, Association of German Engineers) in 1856 together with other members of "Die Hütte", the academic society of the Industrial Academy, and one year later the engineering pocketbook *Hütte*. In the role of its director, Grashof was to engineer the rise of the VDI to its position as Europe's premier engineering association. After the death of Ferdinand Redtenbacher in 1863, Grashof succeeded him as professor at Karlsruhe Polytechnic, teaching strength of materials, hydraulics, thermodynamics and mechanical engineering theory. His work *Theorie der Elasticität und Festigkeit mit Bezug auf ihre Anwendungen in der Technik* (1878) influenced applied mechanics in the period between Julius Weisbach and August Föppl. He prepared St. Venant's bending and torsion theories as well as Clebsch's *Theorie der Elasticität fester Körper* (1862) for the practical needs of engineers although he was an advocate of the deductive method. Grashof always paid attention to the academic presentation. This makes sense when we remember that he was a successful champion of pure academic expansion and the free constitution of the technical polytechnics; he saw that as a prerequisite for equality with universities. Grashof's name stands for the emancipation of the engineering science disciplines from mathematics and theoretical physics at the transition from the establishment to the classical phase. His achievements were widely acclaimed, e. g. an honorary doctorate from the University of Rostock and, posthumously, the Grashof Memorial Medal, awarded annually by the VDI.
- Main contributions to structural analysis:
*Die Festigkeitslehre mit besonderer Rücksicht auf die Bedürfnisse des Maschinenbaues. Abriss von Vorträgen an der Polytechnischen Schule zu Carlsruhe* [1866]; *Theorie der Elasticität und Festigkeit mit Bezug auf ihre Anwendungen in der Technik* [1878]
- Further reading:
[Plank, 1926]; [Lorenz, 1926]
- Photo courtesy of University of Karlsruhe archives

### GRIFFITH, ALAN ARNOLD
\*13.6.1893 London, UK,
†13.10.1963 Farnborough, UK
Following studies in mechanical engineering at the University of Liverpool, Alan Griffith worked in the Royal Aircraft Factory (later: Royal Aircraft Establishment, RAE) in Farnborough from 1915 onwards. One of his projects here involved the soap bubble analogy for torsion. His scientific interests covered a broad spectrum: applied mathematics and mechanics – especially aerodynamics and thermodynamics. In 1921 he published his famous work on fracture mechanics. Although Karl Wieghardt's theoretical work 13 years before had shown that that stresses at the tips of cracks propagate across all boundaries [Rossmanith, 1990, pp. 535–537], Griffith's great achievement was to consider the crack under load in the sense of an energy balance: in the equilibrium condition he equated the reduction in the elastic deformation energy stored in the material as the crack propagates with the rise in the surface energy as the area of the crack increases. According to Griffith, crack growth occurs when the deformation energy exceeds the energy required to form new surfaces. Today, his ingenious energy-based crack model still forms the basis of fracture mechanics, which are used to assess steel structures subjected to fatigue, for example. Griffith worked on aerodynamic and thermodynamic problems at the RAE until 1939. He was elected to the Royal Society in 1941. From 1939 to 1960, as a senior research engineer at Rolls-Royce Aeroengines in Derby, he made great contributions to the development of aircraft engines. His name is inextricably linked with the development of VTOL aircraft, and he was able to experience proof of their viability three months after he retired.
- Main contributions to structural analysis:
*The phenomenon of rupture and flow in solids* [1921]; *The theory of rupture* [1924]
- Further reading:
[Gordon, 1968]; [Rossmanith, 1990]

### GUIDI, CAMILLO
\*24.7.1853 Rome, Italy, †30.10.1941 Rome, Italy
He completed studies in engineering sciences at the School of Engineering in Rome in 1853. By 1882 he had become visiting professor of graphical statics at the Politecnico di Torino and from 1887 to 1928 held the chair of graphical statics and engineering sciences at the same establishment. In addition, from 1893 onwards

 GIUFFRÈ  GRASHOF  GUIDI  GVOZDEV  HAUPT  HAYASHI

he was also responsible for the theory of bridge-building and served as director of the materials testing unit at the Politecnico di Torino. Guidi established the theory of the elastic arched beam in massive structures and was principally responsible for developing reinforced concrete design theory in Italy. His lectures on engineering sciences (*Lezioni sulla scienza delle costruzioni*) were collected together in five volumes and passed through numerous editions; these exerted a permanent influence on structural engineering in Italy from the classical phase to the end of the consolidation period of structural theory. Guidi's work can be compared with that of Mörsch.

- Main contributions to structural analysis:
*Sugli archi elastici* [1883]; *Sulla curva delle pressioni negli archi e nelle volte* [1885]; *L'arco elastico* [1888]; *Lezioni sulla scienza delle costruzioni, Parte II, Elementi delle costruzioni, Statica delle costruzioni civili* [1896]; *L'arco elastico senza cerniere* [1903]; *Lezioni sulla scienza delle costruzioni, Parte IV, Teoria dei ponti* [1905]; *Influenza della temperatura sulle costruzioni murarie* [1906]; *Lezioni sulla scienza delle costruzioni, Parte I, Nozioni di statica grafica* [1933]; *Lezioni sulla scienza delle costruzioni, Parte II, Teoria dell'elasticità e resistenza die materiali* [1934/1]; *Lezioni sulla scienza delle costruzioni, Parte III, Elementi delle costruzioni, statica delle costruzioni civili* [1934/2]; *Lezioni sulla scienza delle costruzioni, Appendice: Le costruzioni in beton armato* [1935]; *Lezioni sulla scienza delle costruzioni, Parte IV, Teoria dei ponti* [1938]

- Further reading:
[n. n., 1970]; www.museovirtuale.polito.it
- Photo courtesy of Politecnico di Torino

### GVOZDEV, ALEKSEI ALEKSEEVICH

*8.5.1897 Bogucharovo, Tula Province, Russia,
†22.8.1986 Moscow, USSR

Upon completion of his studies at the Moscow Institute of Communication Engineers in 1922, Gvozdev gave lectures there in structural mechanics. He began working in the laboratory for plain and reinforced concrete at the Central Science Research Institute for Industrial Buildings in 1927, continued his career at the Moscow Civil Engineering Institute and in 1933 was appointed professor of structural mechanics at the Moscow Military Engineering Academy. Three years later he was awarded the title of doctor of engineering sciences. Gvozdev published his displacement method independently of Ludwig Mann [Mann, 1927]; his monograph [Gvozdev, 1927] is the first Russian publication on this important object of structural analysis. The book provides "a clear treatment of the nature of this method and the character of the basic system on which it rests, and sets out the properties of the coefficients of the canonical equations, the grouping of unknowns, etc. Aside from that, an extremely general viewpoint is maintained on the possible methods of formation of basic systems and the choice of unknowns" [Rabinovich, 1960, p. 16]. The fundamental theorems of ultimate load theory had already been discovered by Gvozdev in 1936, but they were published only in Russian and then only in the proceedings of the Moscow Academy of Sciences [Gvozdev, 1938/1960], and thus – like his pioneering work on the displacement method – remained essentially unknown to the international scientific community outside the USSR. However, Gvozdev's publications represent a decisive contribution to theory developments during the invention phase of structural theory (1925 – 50).

- Main contributions to structural analysis:
*Obshchii metod rascheta slozhnykh staticheski neopredelimykh system* (general method of analysis of complex statically indeterminate systems) [1927]; *The determination of the value of the collapse load for statically indeterminate systems undergoing plastic deformation* (in Russian) [1938/1960]

- Further reading:
[Malinin, 2000, p. 136]
- Photo courtesy of Prof. Dr. G. Mikhailov

### HAUPT, HERMAN

*26.3.1817 Philadelphia, USA,
†14.12.1905 Jersey City (New Jersey), USA

Herman Haupt graduated from West Point Military Academy in 1835 and later that year began work as a civil engineer (see section 2.3.6). He participated in bridge and tunnel projects for the Norristown Railroad. He was granted a patent for a bridge girder in 1839, which later became known as the Haupt Truss. This structural system consists of several superimposed propped beams, i. e. is a preliminary form of the trussed framework. One year later, Haupt was appointed professor of mathematics and engineering at Pennsylvania College (today: Gettysburg College), but he returned to railways in 1847. Within two years he had risen to the position of general superintendent of the Pennsylvania Railroad, and 1851 marked the year of the publication of his book on bridge-building [Haupt, 1851], which explains the design of structural elements in detail and was the first of its kind in the USA. Haupt's book is the "bedrock" of American bridge theory and marks the start of the establishment phase of structural theory (1850 – 75) in the USA. He was employed as chief engineer of the Southern Railroad of Mississippi from 1851 to 1853, and then returned to the Pennsylvania Railroad to take up the same post there, where he remained until 1856. The engineering works for the five-mile-long Hoosac Tunnel were his responsibility until shortly before the outbreak of the American Civil War in 1861. As an experienced bridge-builder, his work during the war played a considerable part in the Union's victory over the Confederate States. Before the end of the war, he retired from active service with the rank of brigadier-general and promptly published a book on military bridges [Haupt, 1864]. He continued his career successfully with various railway companies and his final post was as president of the Dakota & Great Southern Railroad (1885 – 86). When Herman Haupt died of heart failure at the stately age of 89, the USA lost one of its greatest civil engineers of the second half of the 19th century.

- Main contributions to structural analysis:
*Hints on Bridge Building by an Engineer* [1841]; *General Theory of Bridge Construction* [1851]; *Military Bridges* [1864]

- Further reading:
[Tyrrell, 1911]; [Parcel & Maney, 1926]
- Photo courtesy of http://en.wikipedia.org/wiki/Herman_Haupt

### HAYASHI, KEIICHI

*2.7.1879 Takada in Niigata, Japan,
†2.8.1957, Tokyo, Japan

Keiichi Hayashi was educated at the 1st state grammar school in Tokyo and afterwards studied at the Imperial University in Kyoto from 1900 to 1903. He gained his first practical engineering experience as a mining engineer at the Sumitomo-Bessi Mine. In 1912 he was awarded a doctorate by the Imperial University in Kyoto for his dissertation *On Beams on an*

*Elastic Bed*. From 1912 to 1917 he was visiting professor at the Imperial Kyushu University in Fukuoka and during this period undertook study tours of Germany, the UK and the USA. January 1917 saw him appointed professor of the 5th chair of building materials and building operations at the Faculty of Civil Engineering at the Imperial Kyushu University in Fukuoka. He was given the title of honorary professor at the same university in 1939. Hayashi's investigation (1912) of the beam on an elastic bed and its application in civil engineering was the first of its kind. His mathematical tables of higher functions proved invaluable to civil engineers during the invention and innovation phases of structural theory. However, physics – and not just the theory of structures – also benefited. For example, in his guest appearance at the Imperial Kyushu University in Fukuoka in November 1922, Albert Einstein paid thanks to Prof. Hayashi, saying that the hyperbola functions had been very useful to him.

- Main contributions to structural analysis:
*Theory of beams on an elastic bed and their application to civil engineering* [1921/1]; *Five-figure tables of trigonometric and hyperbolic functions as well as the functions $e^x$ and $e^{-x}$ using natural numbers as an argument* [1921/2]; *Five-figure tables of trigonometric and hyperbolic functions* [1921/3]; *Seven- and multi-figure tables of trigonometric and hyperbolic functions* [1926]; *Tables of Bessel, theta, spherical and other functions* [1930/1]; *Five-figure function tables, trigonometric, inverse trigonometric, exponential, hyperbolic, spherical, Bessel, elliptical functions, theta zero values, natural logarithms, gamma functions* [1930/2]; *Tables of Bessel functions* [1936]

- Further reading:
[Iseki, 1930, p. 143]; [Hayashi, 1951]; [Schleicher & Mehmel, 1957]; [Naruoka, 1974]; [Mizuno, 1983]; [Naruoka, 1999]

- Photo courtesy of Prof. Dr. M. Yamada

## HENCKY, HEINRICH

*2.11.1885 Ansbach, German Empire,
†6.7.1951 Gries, Austria

Heinrich Hencky was the son of a Bavarian treasury civil servant and completed his structural engineering studies at Munich TH in 1908. Following military service, Hencky worked on civil engineering and structural steelwork projects for the local authorities before becoming scientific assistant for theory of structures and bridge-building at Darmstadt TH for the two years prior to World War 1. It was while he was at Darmstadt TH that he was awarded a doctorate for his work on the theory of rectangular plates [Hencky, 1913]. He expanded his dissertation, which was conferred a "distinction", to cover the theory of circular plates [Hencky, 1915]. He was given a senior management post in a reinforced concrete company in Kharkov in 1914. Following the outbreak of World War 1,

Hencky was interned somewhere in the Ural Mountains (1915–18); this is where he met his future wife, Alexandra Yuditskaya. In the spring of 1918, Hencky was able to escape via Kharkov and return to Munich; his wife later followed him. After a short period working as an engineer in the materials testing department of the seaplane trials unit in Warnemünde, he wrote his habilitation thesis on structural mechanics at Darmstadt TH [Hencky, 1920], where he gained a post in the mechanical engineering department on the recommendation of Prof. Ludwig Föppl from Dresden TH. He then worked as a lecturer for applied mechanics at Delft TH from 1922 to 1929, where professors J. M. Burger and C. B. Biezeno were on the teaching and research staff. Hencky's years of creativity in Delft are characterised by his fundamental work on plastic theory [Hencky, 1923] and rheology [Hencky, 1925 & 1929], work which shows Biezeno's influence. His strength hypothesis [Hencky, 1923] was very important for modern structural mechanics. This pioneering work can be placed alongside the strength hypotheses of H. Tresca [Tresca, 1964], M. T. Huber [Huber, 1904/1] and R. v. Mises [Mises, 1913]. Nevertheless, during his time at Delft TH he did not manage to attain the post of professor – his original intention – because he was more a researcher than a teacher; apart from that, the "chemistry" between Hencky and Biezeno was not right. At the invitation of the MIT president at that time, S. Stratton, Hencky was appointed associate professor of mechanics in the MIT's department of mechanical engineering (1930–33). It was there that he supervised the dissertations *Yield conditioning of plates and shells by Mises-Hencky Criterion* (1931) by Theodore H. H. Pian and *Stress field of a plate reinforced by a longitudinal guide and subject to tension* (1932) by Robert Conrad. After Stratton's death in 1932, Hencky worked as a consulting engineer in Lisbon (New Hampshire). He was therefore pleased to except Galerkin's offer to take over the chair of applied mechanics at the Chemistry & Technology Institute in Kharkov in 1936. Despite good working conditions, Hencky became the victim of the deteriorating political situation between the USSR and Hitler's Germany. The authorities cancelled his work permit at the end of 1937 and he and his family were instructed to leave the USSR within 24 hours. Supported by the director of MAN, Richard Reinhardt, Hencky began work as a structural engineer responsible for special projects at MAN's Gustavsburg Works (Mainz) on 1 January 1938. The local SS authority mistrusted Hencky and demanded that he not have access to secret documents. However, Reinhardt was able to neutralise this dangerous situation. Therefore, Hencky, protected by Reinhardt, was able to rise to the post of senior engineer at MAN (1941) and was also given responsibility for the materials testing department

of the Gustavsburg Works. Only a fragment of his 1943 work *Neuere Verfahren der Festigkeitslehre* was able to be published [Hencky, 1951] because the original manuscript sent to the Oldenbourg Verlag in Munich was destroyed in an air raid. Hencky retired from MAN at the end of 1950, but just 18 months later he died in an accident in the Alps while pursuing his favourite pastime of mountain climbing.

- Main contributions to structural analysis:
*Über den Spannungszustand in rechteckigen ebenen Platten bei gleichmäßig verteilter und bei konzentrierter Belastung* [1913]; *Über den Spannungszustand in kreisrunden Platten mit verschwindender Biegesteifigkeit* [1915]; *Über die angenäherte Lösung von Stabilitätsproblemen im Raum mittels der elastischen Gelenkkette* [1920]; *Kippsicherheit und Achterbildung an angeschlossenen Kreisringen* [1921]; *Stabilitätsprobleme der Elastizitätstheorie* [1922]; *Über einige statisch bestimmte Fälle des Gleichgewichts in plastischen Körpern* [1923]; *Zur Theorie plastischer Deformationen und der hierdurch im Material hervorgerufenen Nachspannungen* [1924]; *Die Bewegungsgleichungen beim nichtstationären Fliessen plastischer Massen* [1925]; *Das Superpositionsgesetz eines endlich deformierten relaxationsfähigen elastischen Kontinuums und seine Bedeutung für eine exakte Ableitung der Gleichungen für zähe Flüssigkeit in der Eulerschen Form* [1929]; *Über die Berücksichtigung der Schubverzerrung in ebenen Platten* [1947]; *Neuere Verfahren in der Festigkeitslehre* [1951]

- Further reading:
[R. I. Tanner & E. Tanner, 2003]

- Photo courtesy of [R. I. Tanner & E. Tanner, 2003, p. 97]

## HERTWIG, AUGUST

*20.3.1872 Mühlhausen, German Empire,
†14.4.1955 Berlin, Federal Republic of Germany

August Hertwig, the son of a factory owner, began studying architecture at Berlin TH in late 1890, but switched to civil engineering in his first year there. Following the first state examination in 1894, Hertwig worked under Karl Bernhard at the bridge-building authority in Berlin, produced books on bridges for the Oldenburg Railway Company, worked as an assistant to Guidio Hauck, Siegmund Müller and Heinrich Müller-Breslau, and passed the second state examination (governmental building officer) in the summer of 1898. Thereafter, he was responsible for the structural calculations and design of the glasshouses for the botanic gardens in Berlin. Hertwig was appointed professor of structural analysis and steel bridges at Aachen RWTH in 1902 on the recommendation of Müller-Breslau and therefore became the youngest professor in Germany; and he had not yet published anything! He worked as a teacher and researcher at Aachen RWTH until he was

HENCKY     HERTWIG

awarded an honorary doctorate by Darmstadt TH in 1924, whereupon he transferred to Berlin TH to become Müller-Breslau's successor. He remained at Berlin TH until 1937, where he became a sort of father figure, the head of the many branches of the Berlin school of structural theory (see section 6.6.4.1). He was also chief editor of the journal *Der Stahlbau* from 1928 to 1938, and co-editor of the journal *Ingenieur-Archiv*. Hertwig also published articles on the history of the theory of structures [Hertwig, 1906; 1941], construction engineering [Hertwig, 1934/1; 1934/2; 1935] and the education of technical subjects at university level, and was also in favour of dynamic soil investigations. Shortly after World War 2, Hertwig took over the chair of structural steelwork at Berlin TU from Ferdinand Schleicher, who was forced to depart due to his membership of the Nazi party. Klöppel stressed not only Hertwig's excellent conduct, but also his brave and open fight to preserve justice and professionalism irrespective of a person's status (see [Klöppel, 1955, p. 122]).

• Main contributions to structural analysis:
*Beziehungen zwischen Symmetrie und Determinanten in einigen Aufgaben der Fachwerktheorie* [1905]; *Die Entwicklung einiger Prinzipien in der Statik der Baukonstruktionen und die Vorlesungen über Statik der Baukonstruktionen und Festigkeitslehre von G. C. Mehrtens* [1906]; *Über die Berechnung mehrfach statisch unbestimmter Systeme und verwandte Aufgaben in der Statik der Baukonstruktionen* [1910]; *Die Lösung linearer Gleichungssysteme durch unendliche Reihen und ihre Anwendung auf die Berechnung hochgradig unbestimmter Systeme* [1912]; *Die Hochschulreform* [1930/1]; *Johann Wilhelm Schwedler. Sein Leben und sein Werk* [1930/2]; *Das „Kraftgrößenverfahren" und das Formänderungsgrößenverfahren" für die Berechnung statisch unbestimmter Gebilde* [1933]; *Aus der Geschichte der Straßenbautechnik* [1934/1]; *Aus der Geschichte der Gewölbe. Ein Beitrag zur Kulturgeschichte* [1934/2]; *Die Eisenbahn und das Bauwesen* [1935]; *Die Entwicklung der Statik der Baukonstruktionen im 19. Jahrhundert* [1941]; *Lebenserinnerungen von August Hermann Adalbert Hertwig* [1947]; *Leben und Schaffen der Brückenbauer der Deutschen Reichsbahn Schwedler, Zimmermann, Labes und Schaper. Eine kurze Entwicklungsgeschichte des Brückenbaus* [1950]; *Rede, gehalten bei der Gedenkfeier für Müller-Breslau am 15. Juni 1925 in der Aula der Technischen Hochschule Charlottenburg* [1951]

• Further reading:
[DStV, 1943]; [Sattler, 1955/1]; [Klöppel, 1955]; [Kurrer, 2004/2]

• Photo courtesy of [Klöppel, 1955, p. 121]

### HODGKINSON, EATON

*26.2.1789 Anderton near Northwich, Cheshire, UK, †18.6.1861 Eglesfield House, Higher Broughton, Manchester, UK

The offspring of a poor farmer's family, Eaton Hodgkinson's intelligence was luckily discovered by John Dalton, an important scientist from Manchester. Dalton taught Hodgkinson mathematics and introduced him to the works of Euler, Lagrange and the Bernoullis. Hodgkinson's research into beams began in the early 1820s and the carefully recorded, comprehensive results were published as early as 1824 [Hodgkinson, 1824]. He gained even more recognition with his essay on the optimum form of beams [Hodgkinson, 1830]. By theoretical and experimental means, Hodgkinson showed that the cross-section of cast iron beams should be asymmetric and that the area of the tension zone must be about six times that of the compression zone. In 1840 he published the results of his buckling tests on cast- and wrought-iron circular columns and devised an empirical buckling formula; in doing so, he discovered two modes of failure: flexural (buckling) and fracture [Nowak, 1981, pp. 61 – 70]. His research into the optimisation of beams was carried out with the leading ironwork engineer of his time – William Fairbairn, who made use of the results in the building of the Conway and Britannia Bridges (1848 – 49). In his capacity as professor of mechanical engineering at University College, London, Hodgkinson was elected to the Royal Society and became vice-president of the Manchester Society. Hodgkinson's work made a major contribution to the empirical nature of the theory of structures during its constitution phase.

• Main contributions to structural analysis:
*On the Transverse Strain and Strength of Materials* [1824]; *Theoretical and Experimental Researches to Ascertain the strength and best forms of Iron beams* [1830]; *On the strength of pillars of cast iron and other materials* [1840]; *Experimental researches of the Strength and other properties of cast iron* [1846]

• Further reading:
[n. n., 1863]; [n. n. 1898/2]; [Timoshenko, 1953]

### HOOKE, ROBERT

*18.7.1635 Freshwater, Isle of Wight, UK, †3.3.1703 London, UK

After the death of his father, the 13-year-old Hooke was educated by the painter Peter Lely and attended Westminster School in London, where he studied Latin, Greek, Hebrew and the mathematical works of Euclid. In 1653 he went to Christ Church College in Oxford as a servant in order to pay for his scholarship by carrying out duties for more wealthy students. Hooke left the College in 1662 with a Master of Arts degree. One year later he was elected a member of the Royal Society and in 1665 became professor of geometry at Gresham College, where the Royal Society was holding its weekly meetings at that time. This illustrious circle dedicated to Francis Bacon's (1561 – 1626) concept of inductive sciences was made up of England's leading natural science researchers: John Wilkins, Thomas Willis, Seth Ward, William Petty, John Wallis, Christopher Wren, Robert Boyle and others. However, Hooke's pioneering achievements in almost all areas of natural science and technology (he invented the microscope) were founded methodically on the idea of formulating a hypothesis which was then proved right or wrong through experimentation. For example, from 1662 to 1664 he investigated the strength of timber beams and metal wires in order to verify Galileo's hypotheses [Galilei, 1638], but achieved little in the way of conclusive results. On the other hand, in 1675 Hooke was able to formulate verbally the form of the catenary arch [Hooke, 1675]. In his diary entry for 5 May 1675, Hooke notes, with a critical undertone, that his friend Wren had arranged for his masonry arch principles to be used to modify the design of the dome of St. Paul's Cathedral [Addis, 2002/1, p. 337]. Considerably more important for the strength of materials in the orientation phase of structural theory was the formulation and experimental verification of the theory of springs and springy bodies, later known as Hooke's law of elasticity [Hooke, 1678], which in historical terms could not unfold its logical potential until the discipline-formation period of elastic theory and structural theory. After the Great Fire of London in 1666, Hooke was involved in the rebuilding of the city; Bill Addis has been able to uncover a number of structures on which Hooke supplied the architectural input [Addis, 2002, pp. 336 – 37]. In terms of his social life, Hooke was regarded as a difficult person who always felt cheated (priority disputes with Isaac Newton and Christiaan Huygens), mistrusted others and, having become cynical and bitter, elected to spend his final years in solitude. "Hooke was a difficult man in an age of difficult men" [Westfall, 1972, p. 487]. He died in the room at Gresham College in which he had lived for 37 years.

- Main contributions to structural analysis: *A description of helioscopes, and some other instruments* [1675]; *Lectures de potentia restitutiva, or of spring explaining the power of springing bodies* [1678]
- Further reading:
[Westfall, 1972]; [Hambly, 1987]; [Addis, 2002/1]

### HRENNIKOFF (KHRENNIKOV), ALEKSANDR PAVLOVICH

*11.11.1896 Moscow, Russia, †31.12.1984 Vancouver, Canada

After studying at the Moscow Institute of Communication Engineers, Hrennikoff emigrated to Canada and completed a masters degree at the University of British Columbia in 1933. He gained a doctorate at the MIT in 1940 with his dissertation entitled *Plane stress and bending of plates by method of articulated framework* [Hrennikoff, 1940], the main results of which were summarised by Hrennikoff in a much-quoted series of papers [Hrennikoff, 1941] in which he develops the gridwork method with which, in particular, two-dimensional elastic continua can be modelled as a trussed framework system. Hrennikoff remained at the University of British Columbia as professor of structural engineering until his death. He was presented with the ASCE Moisseiff Award in 1949. Hrennikoff's gridwork method was generalised by himself and others, e. g. for the analysis of the stability of plates and shells. It became very important in the middle of the innovation phase of structural theory (1950 – 75) as powerful computers started to become available (see section 10.4.1.3). The gridwork method therefore formed the basis of the first computer-assisted structural analyses in the automotive industry (see section 10.4.1.4).

- Main contributions to structural analysis: *Plane stress and bending of plates by method of articulated framework* [1940]; *Solution of problems in elasticity by the framework method* [1941]; *Three-Dimensional bar cell for elastic stress analysis* [1971]; *Stability of plates using rectangular bar cells* [1972]; *Trapezoidal bar cells in plane stress* [1975]
- Photo courtesy of University of British Columbia archives

### HUBER, MAKSYMILIAN TYTUS

*4.1.1872 Krościenko near Nowy Sącz (Galicia), Austria (today: Poland), †9.12.1950 Kraków, Poland

He passed the entrance examination for Lemberg Polytechnic in 1889 and just one year later presented his first scientific publication. He passed his diploma examination with distinction in 1894 and, following military service and a one-year scholarship at the University of Berlin (reading mathematics and astronomy), returned to Lemberg in 1898 in order to take up the post of assistant to the chair of mathematics at the Polytechnic line at Lwów. During his teaching activities in Kraków training colleges, he studied mechanics and published his first strength hypothesis and his impact theory (both in 1904); the latter work formed the basis for his doctorate thesis. R. v. Mises and H. Hencky presented adequate strength hypotheses in 1913 and 1923 respectively without any knowledge of Huber's hypothesis. In 1908 he was appointed to the chair of applied mechanics at Lemberg Polytechnic. He was captured by the Russians in World War 1 and later met Timoshenko and Galerkin during a period at the University of Kazan. Following the restoration of the Polish state, he was able to resume his teaching and research activities – which he did with great success – at his old Alma Mater, which was now known under the name of Politechnika Lwowska. It was here that he devised his theory of the orthogonal-anisotropic slab (Huber's continuum). Huber became the "Grand Old Man" of applied mechanics and theory of structures in Poland. For instance, with the help of Wacław Olszak and Witold Nowacki he laid the foundation for the internationally acclaimed Polish school of applied mechanics. Like many surviving members of the Polish intelligentsia, he helped to maintain the scientific and cultural life of Poland (illegally) during the German occupation. Following the liberation of Poland, he became head of the chair of stereo mechanics at the Politechnika Gdańska, formerly Danzig TH.

- Main contributions to structural analysis: *Właściwa praca odkształcenia jako miara wytężenia materiału* (actual deformation work as a measure of material strain) [1904/1]; *Zur Theorie der Berührung fester elastischer Körper* [1904/2]; *Teoria płyt prostokątnie różnokierunkowych wraz z technicznymi zastosowaniami* (theory and technical applications of orthotropic plates) [1921]; *Die Theorie der kreuzweise bewehrten Eisenbetonplatten nebst Anwendungen auf mehrere bautechnisch wichtige Aufgaben über rechteckige Platten* [1923 – 26]; *Probleme der Statik technisch wichtiger orthotroper Platten* [1929]
- Further reading:
[Olszak, 1949]; [n. n., 1951]; [Olesiak, 1972]; [Olesiak, 1992]
- Photo courtesy of Prof. Dr. Z. Cywiński

### IGUCHI, SHIKAZO

*8.7.1889 Shizuoka, Japan, †13.3.1956, Muroran, Japan

Shikazo Iguchi studied at the Imperial University in Tokyo from 1912 to 1915. Following that, he was employed as an engineer in the River Training Works Department of the Japanese Interior Ministry. In 1930 he was appointed professor of hydraulic engineering at the Imperial Hokkaido University in Sapporo, and from 1930 to 1932 spent time studying in Germany. After his return to Japan he gained his doctorate at the Imperial University in Tokyo with a dissertation on the bending theory of orthotropic plates (1932). From 1949 until his death he was rector of the Muroran State Technical University. Iguchi contributed to the further development of plate theory in the invention phase of theory of structures. For instance, with the help of Robert Kirchhoff's (1824 – 87) plate theory he provided a complete mathematical analysis of Chladni's figures (1797) [Iguchi, 1953].

- Main contributions to structural analysis: *A solution for the calculation of a flexible rectangular plate* [1933]; *A general solution for the buckling of octagonal plates* [1936]; *The vibrations due to bending stress in rectangular plates encastré on four sides* [1937]; *The buckling of a rectangular plate due to shear forces* [1938]; *The natural oscillations and acoustic patterns of simply supported rectangular plates* [1953]
- Further biographical reading:
[Naruoka, 1970]; [Nomachi, 1983]
- Photo courtesy of Prof. Dr. M. Yamada

### JAEGER, THOMAS

*5.7.1929 Breslau, Germany, †21.8.1980 Berlin (West), Federal Republic of Germany

After completing his school education, Thomas Jaeger studied civil engineering at Dresden TH, graduating in 1956 with his diploma thesis on the ultimate load analysis of rigid steel structures. In that same year Jaeger published this work as a journal paper with the title *Grundzüge der Tragberechnung* [Jaeger, 1956], which was translated into five languages. The editor of the journal *Der Bauingenieur*, Ferdinand Schleicher, encouraged the young graduate engineer to translate the book *The Plastic Methods of Structural Analysis* by B. G. Neal (1956) into German, which he did within a very short time [Neal, 1958]. During the same period he translated papers on ultimate load theory by Horne and Baker for the journal *Bauplanung – Bautechnik*. By the time he was 30 his name had become known both inside and outside Germany, and he thus rose to become one of the advocates of the ultimate load method in Germany. He gained his doctorate at Berlin TU in 1963 with a dissertation entitled *Untersuchung zur Grenztragfähigkeit von Stahlbetonplatten*; in well over 100 tests he proved that the plastic hinge line theory was a suitable method for quantifying the

 HRENNIKOFF
 HUBER
 IGUCHI
 JAEGER
 JASIŃSKI
 JENKINS

limit state capacity of reinforced concrete slabs. In the same year he published, together with Antoni Sawczuk, *Grenztragfähigkeitstheorie der Platten*, which was to become a standard reference book. Jaeger's professional career lay on the boundary between the construction of nuclear power stations and structural engineering and was to culminate in structures for nuclear engineering: he was to become the founder and primus inter pares of this demanding field of work. In 1964 he introduced this new discipline into lectures at Berlin TU, where he also wrote his habilitation thesis in 1970. He had already founded the journal *Nuclear Engineering and Design* back in 1965. From 1968 to his early death he was director and professor at the Federal Materials Testing Institute (BAM) in Berlin. While in this position he used the full weight of his personality to set up an organisation for nuclear engineering structures; 1971 saw the founding of the International Association for Structural Mechanics in Reactor Technology (ASMiRT) and Jaeger managed the first five conferences on Structural Mechanics in Reactor Technology (SMiRT). The genesis of the discipline of nuclear engineering structures, which was primarily the work of Jaeger, revealed with a rare clarity the change from the scientific to the technological paradigms in the integration period of structural theory.

• Main contributions to structural analysis: *Grundzüge der Tragberechnung* [1956]; *Grenztragfähigkeitstheorie der Platten* [1963]

• Further reading:
[Brandes, Jaeger, Matthees, 1985]

• Photo courtesy of [Brandes, Jaeger, Matthees, 1985, p. 13]

### JASIŃSKI (YASINSKY), FELIKS
*15.9.1856 Warsaw, Russia (today: Poland), †18.11.1899 St. Petersburg, Russia

When he was 16 he passed his university entrance examination in Warsaw and in 1877 passed examinations for the St. Petersburg Institute of Engineers of Ways of Communication. From 1877 to 1890 Jasiński worked as an engineer on the St. Petersburg–Warsaw Railway in Pskow, thereafter in Vilna (today: Vilnius) and from 1888 onwards in St. Petersburg, where in 1890 he took charge of the Technical Building Department of the St. Petersburg–Moscow Railway. While in this position he greatly influenced the construction of the permanent way, bridges and stations and was held in high esteem by railway engineers. At the same time, he was looking into the theory of structures. In 1894 his contribution to buckling theory earned him a scientific degree comparable with a doctorate, and after 1895 he worked part-time at the St. Petersburg Institute of Engineers of Ways of Communication where one year later he was appointed visiting professor. A dispute between Jasiński and Engesser about inelastic buckling was published in a Swiss building journal in 1895 [Nowak, 1981, pp. 147–148]. Four years later Jasiński was planning to return to Warsaw when tuberculosis took its toll. He therefore died in St. Petersburg but was buried in the town of his birth. According to the writings of Timoshenko, Jasiński's lectures at the Institute of Engineers of Ways of Communication were original, comprehensible, interesting and masterly [Timoshenko, 2006].

• Main contributions to structural analysis: *Opyt razvitiya teorii prodol'nogo izgiba* (research into buckling theory) [1893]; *O soprotivlenii prodol'nomu izgibu* (on buckling strength) [1894]; *Recherches sur la flexion des pièces comprimées* [1894]; *Sobranie sochinenii* (collected works) [1902–04]; *Pisma* [1961]

• Further reading:
[Mitinsky, 1957]; [Mutermilch & Olszewski, 1964]; [Liepfeld, 1992]

• Photo courtesy of Prof. Dr. G. Mikhailov

### JENKINS, RONALD STEWART
*8.12.1907 Sutton, Surrey, UK, †27.12.1975 London, UK

Ronald Jenkins studied engineering from 1928 to 1931 at the City & Guilds School in London and afterwards worked with Oscar Faber. It was there that he met Ove Arup, who at the time was working in the same building for Christiani & Nielsen. Arup recognised Jenkins's analytical abilities and invited him to join him, initially with Kier and later in his own practice, which he set up with his cousin. Jenkins understood how the methods of applied mathematics and mechanics devised by scientists could be exploited for everyday structural engineering. He carried out invaluable work on shells in single and double curvature as well as prestressed concrete during the invention and innovation phases of structural theory. The shell roofs to the Bank of England printing works (Debden) and Smithfield Market (London), also the polygonal shells of the rubber works in Brwnmawr (South Wales) and, last but definitely not least, the world-famous roof to Sydney Opera House, which has become *the* landmark of that city. Furthermore, Jenkins also worked as a teacher in his company; he conveyed his technical and scientific knowledge to a whole generation of engineers who in turn advanced the theory, design, analysis and construction of shells and prestressed loadbearing structures. Ove Arup wrote of Jenkins: "Not only my firm, but the whole engineering profession and indeed the whole country owe him a debt for having contributed to the improvement of engineering education and the raising of engineering standards." Ronald Stewart Jenkins's professional career was a rare and happy synthesis of teaching, research and practical structural engineering.

• Main contributions to structural analysis: *Theory and design of cylindrical shell structures* [1947]; *Theory of new forms of shell* [1952]; *The design of a reinforced concrete factory at Brynmawr, South Wales* (1953/1); *Matrix analysis applied to indeterminate structures* [1953/2]; *Linear analysis of statically indeterminate structures* [1954]; *Design and Construction of the Printing Works at Debden* [1956]; *The evolution of the design of the concourse at the Sydney Opera House* [1968]

• Further reading:
[Arup, 1976]; [Ronald Jenkins Memorial Issue, 1976]; [Hobbs, 1985]

• Photo courtesy of Dr. W. Addis

### KANI, GASPAR
*16.10.1910 Franztal near Semlin, Austria-Hungary, †29.9.1968 Lake Simcoe near Toronto, Canada

After studying at the Technical Faculty of the University of Zagreb, Kani worked for a short time in the design office of the Königshütte Steelworks in Silesia, but very quickly obtained a post as assistant to the chair of structural theory, strength of materials and materials testing at the University of Zagreb. It was at this university that he became a lecturer for structural theory in 1942 and one year later became assistant to Otto Graf, thereafter Richard Deininger, at Stuttgart TH. After the war, he worked for four years as a building contractor.

It was during these years that he developed the iteration method of calculating systems with a high degree of static indeterminacy – a method which was later to bear his name. He gained his doctorate at Stuttgart TH in 1951 and afterwards worked primarily on prestressed concrete structures, initially as an industrial engineer and later as a consulting civil engineer. In 1960 he was offered a post at the University of Toronto, Canada, where he researched the causes of shear failures in reinforced concrete beams. At its conference in April 1969, the American Concrete Institute (ACI) awarded him posthumously the Wason Medal for the most important *ACI Journal* paper of the year 1967.
• Main contributions to structural analysis: *Die Berechnung mehrstöckiger Rahmen* [1949]; *Spannbeton in Entwurf und Ausführung* [1955]
• Further reading:
[Holzapfel, 1969]
• Photo courtesy of University of Toronto

## KAZINCZY, GÁBOR VON
*19.1.1889 Szeged, Austria-Hungary (today: Hungary), †23.5.1964 Motala, Sweden

The pioneer of the ultimate load method, Gábor Kazinczy, came from a family of Hungarian intellectuals. For example, his great-grandfather, Ferenc Kazinczy (1759 – 1831), had played a leading role in literary life, in the Hungarian enlightenment movement and in the reformation of the Hungarian language in the late 18th and early 19th centuries. Gábor Kazinczy completed his structural engineering studies at Budapest TH in 1911, then took charge of the Materials & Structural Testing Laboratory of the Budapest authorities and later became deputy head of the city's building department; his career in Budapest ended in 1943 as engineering chief counsellor. Kazinczy's first publication, which was to initiate the development of the ultimate load method (see section 2.10.1), had appeared in 1914 [Kazinczy, 1914]. The loading tests carried out on the structure for the house for Klinger Zsigmond (see Fig. 2-89), performed on behalf of the Budapest city authorities in 1913, supplied results that Kazinczy could not interpret with elastic theory. He therefore concluded that three cross-sections must become plastic (see Fig. 2 – 90c). Kazinczy's introduction of the fundamental concept of the plastic hinge, backed up by practical trials, marks the first and most important step in the direction of the plastic hinge method, which, however, would not take shape until the invention phase of structural theory (1925 – 50), albeit with the help of Kazinczy. Following a personal discussion with Maier-Leibnitz, a publication on his ultimate load trials appeared [Maier-Leibnitz, 1928; 1929], which in terms of method correspond with the ultimate load concept of Kazinczy. Kazinczy gained a doctorate at Budapest TH in 1931 with his dissertation *Design of clamped end steel beams with regard to the residual deformations*. Two years later, Kazinczy reported on ultimate load trials on continuous reinforced concrete beams [Kazinczy, 1933] – experiments that paved the way for the development of the ultimate load method in reinforced concrete, too. His habilitation thesis *Safety of structures* was written at Budapest TH in 1939. After World War 2, Kazinczy and his family moved to Denmark, where he worked for the Swedish Kooperativa Förbundets Arkitektkontor company from 1947 until his retirement in 1959. It was here that he prepared the designs and calculations for demanding engineering structures such as grain silos, long-span shells and prestressed suspended floors. Kazinczy published his findings and experiences in a total of 92 publications. Together with Maier-Leibnitz and John Fleetwood Baker, Kazinczy made a major contribution to providing a sound footing for the plastic hinge method during the invention phase of structural theory (1925 – 50).
• Main contributions to structural analysis: *Versuche mit eingespannten Trägern* [1914]; *Statisch unbestimmte Tragwerke unter Berücksichtigung der Plastizität* [1931]; *Die Weiterentwicklung der Plastizitätslehre* [1931]; *Die Plastizität des Eisenbetons* [1933]; *Kritische Betrachtungen zur Plastizitätstheorie* [1938]
• Further reading:
[Kaliszky, 1984]; [Lenkei, 2000]
• Photo courtesy of [Lenkei, 2000, p. 16]

## KHAN, FAZLUR RAHMAN
*3.4.1929 Faridpur, India (today: Bangladesh), †27.3.1982 Jeddah, Saudi Arabia

It was 1950 when Fazlur Rahman Khan graduated as a bachelor of science from the Bengal Engineering College of the University of Dhakar as the best student of his year. Benefiting from a Fulbright scholarship, he gained a masters degree in structural engineering and applied mechanics at the University of Illinois in 1952, and three years later was awarded a doctorate in structural engineering at the same university. He joined the Chicago-based architectural practice of Skidmore, Owings & Merill (SOM) later in 1955 and remained there for the rest of his career apart from a two-year interlude (1957 – 59). He was promoted to participating partner as early as 1961, associate partner in 1966 and, finally, general partner in 1970. During those one-and-a-half decades, Khan developed innovative structural systems for high-rise buildings: framed tube, tube-in-tube system, bundled tubes and diagonalised tube (see [Sobek, 2002]; [Mufti & Bakht, 2002]). The basic concept of the framed tube idea is a structural system consisting of a fixed-based vertical tube braced by the horizontal floor plates, which maximises the internal lever arm of the high-rise building cross-section. One example of this structural system was the World Trade Center in New York, which was completed in 1973 but unfortunately totally destroyed following the all-too-well-known terrorist attack of 11 September 2001. As a result of the shear lag effect, under horizontal loading those parts of the framed tube parallel to the plane of loading behave like a giant lattice frame. "The columns at the corners attract considerably more load than the columns in the middle of the sides, which means they therefore carry heavier loads than would be expected according to the principles of the practical bending theory" [Sobek, 2002, p. 425]. Like Eric Reissner had shown in 1941 that the shear lag effect governs the design of the thin-wall box beams of aircraft wings [Reissner, 1941], this is also the case for the framed tube, but in a totally different order of magnitude – and that is precisely the challenge of an innovative loadbearing system development. Khan first increased the efficiency of the framed tube by coupling the service core (inner tube) with the outer tube by means of the floor plates (tube-in-tube system) and increased it still further by using the multi-cell high-rise building cross-section (bundled tube). For example, the cross-section of Sears Tower in Chicago (442 m high, completed in 1975) consists of a group of nine individual cross-sections; in structural terms, this skyscraper can be modelled as a vertical cantilever beam with a nine-cell cross-section. Khan, together with B. Graham, realised Myron Goldsmith's design for the John Hancock Center (344 m high, completed in 1970) [Sobek, 2002, p. 430]. The prominent German structural engineer and architect Werner Sobek has acknowledged Khan's innovations in loadbearing structures with an impressive quick look at the history of skyscrapers and has called him the "vanguard of the 2nd Chicago School" [Sobek, 2002]. Khan's objective was to combine architecture and structural engineering. He was a conscious thinker and active citizen, and set up the Bangladesh Liberation Movement in the USA as early as 1971. He was awarded many honours, including honorary doctorates from Northwestern University (1973), Lehigh University (1980) and Zurich ETH (1980). In an obituary of this great engineering personality written in the journal *Engineering News Record* it says: "The consoling facts are that his structures will stand for years, and his ideas will never die" (cited in [Sobek, 2002, p. 433]). In this sense, his daughter, Yasmin Sabina Khan, has provided us with a living memorial in the shape of her book *Engineering Architecture: the vision of Fazlur R. Khan* [Khan, 2004].
• Main contributions to structural analysis: *Computer design of 100-story John Hancock Center* [1966/1]; *On some special problems of analysis and design of shear wall structures*

KANI   KAZINCZY   KHAN   KIRPICHEV   KLÖPPEL   KOITER

[1966/2]; *100-story John Hancock Center in Chicago – a case study of the design process* [1972]; *New structural systems for tall buildings and their scale effects on cities* [1974]
- Further reading:
[Beedle, 1984]; [Sobek, 2002]; [Mufti & Bakht, 2002]; [Khan, 2004]
- Photo courtesy of [Beedle, 1984, p. 152]

### KIRPICHEV, VIKTOR LVOVICH
*8.10.1845 St. Petersburg, Russia,
†20.10.1913 St. Petersburg, Russia
He completed his military engineering education with the Cadet Corps in Polotsk (1862) and at the Michaelis Artillery Academy in St. Petersburg (1868), where he subsequently taught. He left military service in 1870 and taught at the St. Petersburg Technological Institute (as professor from 1876 onwards). From 1885 to 1897 he was director of the newly founded Technological Institute in Kharkov and thereafter (until 1902) director of the newly founded Polytechnic Institute in Kiev. Owing to a serious conflict between the students and the administration of the Russian technical universities on the one hand, and the tsarist regime on the other, Kirpichev resigned to take up the post of chairman of the Building Commission of the St. Petersburg Polytechnic Institute in 1903, where he worked until the end of his life as professor of strength of materials and structural mechanics. Timoshenko was especially impressed by his book on the theory of statically indeterminate systems (1903) in which Kirpichev presented the whole theory of such systems in an extremely compact form, but at the same time with great clarity [Timoshenko, 2006, pp. 89 – 90]. Kirpichev undertook several scientific/technical study tours of Western Europe and the USA where he visited numerous universities and exhibitions.
- Main contributions to structural analysis: *Stroitel'naya mekhanika* (structural mechanics) [1874]; *Lishnie neizvestnye v stroitel'noi mekhanike: Raschet staticheski-neopredelimykh sistem* (supernumeration in structural mechanics: the analysis of statically indeterminate systems) [1903]; *Sobranie sochinenii* (collected works) [1917]
- Further reading:
[Radtsig, 1917]; [Timoshenko, 1958]; [Chekanov, 1982]
- Photo courtesy of Prof. Dr. G. Mikhailov

### KLÖPPEL, KURT
*15.9.1901 Aue, German Empire,
†13.8.1985 Darmstadt, Federal Republic of Germany
After completing his school education and an apprenticeship as a machine fitter, he went on to study mechanical engineering at the State Academy for Engineering in Chemnitz. That was followed by employment with Bleichert, a steel fabricator in Leipzig, and in the steelwork department of the Schichau shipbuilding yard in Danzig (Gdańsk); at the same time he studied civil engineering at Dresden TH, where Reinhold Krohn was on the teaching staff. Klöppel passed the diploma examination with distinction at Danzig (Gdańsk) TH in 1929. In that same year he took charge of the Technical-Scientific Section of the German Steelwork Association (DStV) and became managing director of the German Committee for Trials in Iron Construction (which later became the German Committee for Structural Steelwork, DASt). Klöppel gained his doctorate at Breslau TH in 1933. He was offered the chair of bridge-building (which also meant being responsible for the engineering laboratory) at Darmstadt TH in 1938 and was appointed chief editor of the journal *Der Stahlbau*, a job he performed until 1981. He lectured in theory of structures from 1939 onwards. After the war, Klöppel and his students contributed decisively to the international renown of (West) German steel construction. He fought consistently to raise the status of engineering sciences on a scientific policy level without neglecting his teaching and research duties as a university lecturer. Klöppel and his Darmstadt Institute for Theory of Structures & Structural Steelwork set standards in the field of structural theory, plastic theory and stability theory with respect to steel buildings, welding technology, mechanics of materials and structural steelwork as a whole. Without doubt, Klöppel was the No. 1 authority on German steel construction and the very incarnation of steel construction theory.
- Main contributions to structural analysis: *Hängebrücken mit besonderer Stützbedingung des Versteifungsträgers* [1940]; *Berechnung von Hängebrücken nach der Theorie II. Ordnung unter Berücksichtigung der Nachgiebigkeit der Hänger* [1941]; *Formänderungsgrößenverfahren in der Statik* [1947]; *Rückblick und Ausblick auf die Entwicklung der wissenschaftlichen Grundlagen des Stahlbaues* [1948]; *Aufgaben und Ziele der Zeitschrift „Der Stahlbau"* [1951/1]; *Die Theorie der Stahlverbundbauweise in statisch unbestimmten Systemen unter Berücksichtigung des Kriecheinflusses* [1951/2]; *Zur Einführung der neuen Stabilitätsvorschriften* [1952/3]; *Geheimrat Prof. Dr.-Ing. E. h. August Hertwig* [1955]; *Die Einheit der Wissenschaft und der Ingenieur* [1958]; *Unmittelbare Ermittlung von Einflußlinien mit dem Formänderungsgrößen-Verfahren* [1952/2]; *Beulwerte ausgesteifter Rechteckplatten* [1960/1968]; *Systematische Ableitung der Differentialgleichungen für ebene anisotrope Flächentragwerke* [1960]; *Zur Berechnung von Netzkuppeln* [1962]; *Übersicht über Berechnungsverfahren für Theorie II. Ordnung* [1964]; *Die Statik im Zeichen der Anpassung an elektronische Rechenautomaten* [1965]
- Further reading:
[Ramm, 1986]; [Schardt, 1986]; [Scheer, 1986]; [Scheer, 2001]
- Photo courtesy of Darmstadt TU archives

### KOITER, WARNER TJARDUS
*16.6.1914 Amsterdam, Netherlands,
†2.9.1997 Delft, Netherlands
After completing his studies in mechanical engineering at Delft TU in 1936, Koiter worked at the Dutch National Aeronautical Research Institute in Amsterdam (today: National Aerospace Laboratory, NLR). It was here that he carried out air-worthiness tests on aircraft structures and structural analyses of two-spar wings with a shear-resistant skin. Following a brief period spent at the patents office in 1938, Koiter moved to the Government Civil Aviation Office in 1939, where he quickly became director of the engineering division. During World War 2, he carried out research work at the NLR and gained his doctorate in 1945 at the faculty of mechanical engineering at Delft TU with his ground-breaking dissertation entitled *Over de Stabiliteit van het Elastisch Evenwicht* [Koiter, 1945]. He completed this dissertation while his country was still occupied by the German armed forces. "Professor Flügge had been sent from Germany to Delft to cover the Rector chair. According to the occupant law, Ph.D. students who were willing to discuss their thesis were obliged to take an oath of allegiance to the Nazi government. Koiter's thesis on the stability of elastic equilibrium was ready at that

time, but the author, refusing this imposition, waited for the liberation of his country. The thesis thus appeared only in 1945" [Pignataro, 1998, pp. 605/606]. Four years later, Koiter was appointed to the chair of applied mechanics at Delft TU. NASA published his dissertation in English in 1960 (*On the stability of elastic equilibrium*). The chair of theory of strength and stability of structures created specially for Koiter at his Alma Mater in 1973 remained his workplace until his transfer to emeritus status in 1979. His research into the stability of imperfect elastic structures in the supercritical range was a great help in overcoming the dominance of linear thinking during the innovation phase of structural mechanics (1950–75). Koiter's 150 or so research reports were acknowledged by the universities of Glasgow, Bochum, Ghent and Liège with honorary doctorates, the ASCE's Von Kármán Medal and the ASME's Timoshenko Medal. And as a further honour, the ASME inaugurated the Warner T. Koiter Medal in 1996, which was awarded to its namesake in the same year. The citation reads: "… for his fundamental work in stability of structures, for his diligence in the effective application of these theories, his international leadership in mechanics, and his effectiveness as a teacher and a researcher" (cited in [Pignataro, 1998, p. 605]). When Koiter died on 2 September 1997, the scientific world of structural mechanics lost not only one of its leading figures, but also an extraordinary personality: "Much less known has been, and perhaps still is to many people, the unparalleled human figure which was hidden behind the scientific one. His impeccable moral integrity was incompatible with any deviating compromise. 'Samurai in a world of Pharisees', but at the same time he was cordial, generous and deprived of arrogance" [Pignataro, 1998, p. 605]. The science director at the Koiter Institute at Delft TU, Prof. René de Borst, has provided us with a wonderful memorial to this magnificent researcher and teacher of structural mechanics in the shape of a biography *Warner Tjardus Koiter. Het instabiele hanteerbaar* [de Borst, 2002].

- Main contributions to structural analysis:
*Over de Stabiliteit van het Elastisch Evenwicht* [1945]; *Stress-strain relations, uniqueness and variational theorems for elastic-plastic materials with a singular yield surface* [1953]; *General theorems for elastic-plastic solids* [1960/1]; *A consistent first approximation in the general theory of thin elastic shells* [1960/2]; *On the nonlinear theory of thin elastic shells* [1966]; *On the foundations of the linear theory of thin elastic shells* [1970]; *Stijfheid en Sterkte 1: Grondslagen* [1972]

- Further reading:
[Koiter, 1979]; [Besseling & Heijden, 1979]; [Pignataro, 1998]; [Arbocz, 2000]; [Elishakoff, 2000]; [de Borst, 2002]

- Photo courtesy of [de Borst, 2002, p. 222]

### KOLOUŠEK, VLADIMÍR

*16.3.1909 Brünn, Mähren, Austria-Hungary (today: Czech Republic), †21.9.1976 Prague, Czechoslovakia (today: Czech Republic)
Following his school education, Vladimír Koloušek studied civil engineering at the Czech TH in Prague from 1927 to 1934; at the same time he passed an examination to study for two semesters in the Faculty of Natural Sciences at Karl University in Prague. He worked for the Vítkovice Steelworks from 1934 to 1937 and afterwards designed numerous reinforced concrete and steel structures for Czechoslovakian State Railways. It was during this period that he converted important F-functions to tabular form for dynamic analyses with a hand-operated calculating machine. Confronted by the vibration problems of radio masts and railway bridges, he was prompted to investigate structural dynamics, a field in which he was to become a pioneer. He published a forward-looking paper on the dynamics of frames in the *Ingenieur-Archiv* as early as 1941; here, he was able to integrate frame dynamics organically into the language of the displacement method. Before the advent of the computer, his dynamic displacement method was the most effective way of calculating natural frequencies. His doctorate came in 1946 at Prague TH with his dissertation *Statische und dynamische Lösung der abgespannten Antennenmasten*; just one year later he was able to complete his habilitation thesis on structural dynamics. After a period as a private lecturer at Prague TH, he worked as full professor of structural mechanics and dynamics at the Railway Academy in Prague from 1953 to 1962, a job he interrupted in 1954 to take on a professorship at the Transport Academy in Žilina. From 1963 to 1976 he was responsible for structural dynamics at Prague TH. Koloušek is regarded as the founder of structural dynamics. He always compared his theoretical findings with measurements. Many of his books have been translated. Although he was seen as an introverted scientist, many young engineers were attracted to him and that helped him found the internationally renowned Czechoslovakian school of structural dynamics. Koloušek was elected corresponding member of the Czechoslovakian Academy of Sciences and received numerous honorary medals and a state prize.

- Main contributions to structural analysis:
*Anwendung des Gesetzes der virtuellen Verschiebungen und des Reziprozitätssatzes in der Stabwerksdynamik* [1941]; *Dynamika stavebních konstrukcí* (dynamics of building structures) [1950/1954/1956/1967/1980]; *Baudynamik der Durchlaufträger und Rahmen* [1953]; *Schwingungen der Brücken aus Stahl und Stahlbeton* [1956]; *Efforts dynamiques dans les ossatures rigides* [1958]; *Dynamik der Baukonstruktionen* [1962]; *Dinamika strojitělnych konstrukcij* (structural dynamics) [1965];
*Dynamics in Civil Engineering Structures* [1973]; *Wind Effects on Civil Engineering Structures* [1983]

- Further reading:
[Hořejší & Pirner, 1997, pp. 115–118]

- Photo courtesy of Prof. Dr. L. Frýba.

### KROHN, REINHOLD

*25.11.1852 Hamburg, †29.6.1932 Danzig-Langfuhr (Gdańsk)
From 1869 to 1873 Reinhold Krohn studied civil engineering at Karlsruhe Polytechnic, where Hermann Sternberg (1825–85) was professor and introduced him to bridge-building, along with Friedrich Engesser, Hermann Zimmermann and August Föppl. After a number of years spent working for various consulting engineers and authorities in Hamburg, he became an assistant at Aachen RWTH in 1876 and was soon giving lectures on moving bridges, the theory of statically indeterminate trussed frameworks and graphical statics. He became a professor in 1881 and at the same time acted as a consulting engineer. It was in this capacity that he was introduced to the balanced cantilever method of building bridges at the former Southern Germany Bridge-Building Company in Gustavsburg. Since Krohn's publication on Maxwell's exchange theorem (1884), this theorem had become established as the theorem of the reciprocity of displacements (Maxwell's reciprocal theorem) in the classical phase of structural theory. From 1884 to 1886 he worked as an engineer, becoming familiar with American methods of bridge-building and learning about the much-admired truss bridge with pin joints, initially with C. C. Schneider, then with G. S. Morison and finally as senior engineer at the Pencoyd Steelworks. Upon his return to Germany, he joined the Gutehoffnung Foundry. Krohn adapted the new US industrial methods of fabrication to suit German bridge-building techniques. For example, under his leadership the Sterkrade Bridge-Building Company became the largest bridges operation in Germany with the highest export quota. Krohn paved the way for the building of large arch bridges in Germany. Together with Mehrtens and Kintzlé, he replaced wrought iron by mild steel. He founded the Association of German Bridge & Iron Fabricators in 1904 (renamed the German Iron Construction Association in 1913 and the German Steelwork Association in 1928). In the same year, Krohn was appointed professor of theory of structures and bridge-building at the newly established Danzig (Gdańsk) TH, where he taught and carried out research until being granted emeritus status. One of his last students was Kurt Klöppel. Krohn received numerous honours, e. g. election to the Prussian Parliament's Upper Chamber as the representative of Danzig TH, an honorary doctorate from Aachen RWTH.

 KOLOUŠEK
 KROHN
 KRYLOV
 LA HIRE
 LAMÉ

- Main contributions to structural analysis:
*Resultate aus der Theorie des Brückenbaus und deren Anwendung* [1882/1883]; *Der Satz von der Gegenseitigkeit der Verschiebungen und Anwendung desselben zur Berechnung statisch unbestimmter Fachwerkträger* [1884]
- Further reading:
[Bohny, 1923]; [Kusenberg, 1932]
- Photo courtesy of [Kusenberg, 1932, p. 422]

### KRYLOV, ALEKSEI NIKOLAEVICH

\*15.8.1863 Visyaga (Simbirsk), Russia, †26.10.1945 Leningrad, USSR

"The name Krylov," writes Lehmann, "stands for one of the most important theoreticians in international shipbuilding whose work is still relevant today" [Lehmann, 1999, p. 250]. Furthermore, Krylov also had a significant influence on structural mechanics and theory of structures. For example, the Krylov functions play a key role in the theory of the elastically bedded beam, and have been used with great success in shipbuilding theory, railway engineering, structural engineering and geotechnics. After his school education in France and Germany, Krylov returned to Russia in 1878 and studied shipbuilding in St. Petersburg. Following graduation in 1884, he was employed in the Hydrographic Institute in the city and in 1888 attended the Naval Academy (where he also taught) while simultaneously studying mathematics at the University of St. Petersburg. As early as 1898, Krylov presented a general theory of ship movements in waves, which made him one of the great figures in the hydromechanics of ships alongside Froude, Havelock, Michell and Thomson (Lord Kelvin) [Lehmann, 1999, p. 250]. He was quickly promoted to professor of shipbuilding theory at the Polytechnic and the Naval Academy in St. Petersburg; at the same time, he was in charge of the navy's Shipbuilding Testing Institute from 1900 to 1908, and until 1910 was the principal shipbuilding inspector and chairman of the Navy's Technical Committee. In addition, he lectured at the St. Petersburg Institute of Engineers of Ways of Communication as visiting professor from 1911 to 1913. He was awarded the Order of the Holy Vladimir, 2nd class, in 1915, and one year later was appointed Admiral of the Imperial Navy. In 1914 Krylov became a corresponding and in 1916 a full member of the St. Petersburg Academy of Sciences. After the October Revolution, he was chief of the Naval Academy from 1919 to 1921. He subsequently worked outside Russia until he was appointed professor at the Naval Academy in 1927. From 1928 to 1934 he was director of the Institute of Physics & Mathematics at the Academy of Sciences. Soviet Russia established a large merchant fleet and navy under Krylov's leadership. He became president of the Russian Society of Shipbuilders & Ship Engine Builders in 1933 and editor of the renowned shipbuilding journal *Sudostrojenie*. His scientific contributions to shipbuilding theory, applied mathematics and mechanics as well as the history of mechanics were highly acclaimed at home and abroad: Gold Memorial medal of the British Institution of Naval Architects (1898), honorary member of the Institution of Naval Architects (1941), Stalin Prize, 1st class (1941), Hero of Socialist Labour (1943). His final resting place is in the St. Petersburg Volkov Cemetery in the vicinity of the graves of Mendeleyev and Pavlov [Lehmann, 1999].

- Main contributions to structural analysis:
*O raschete balok, lezhashchikh na uprugom osnovanii* (on the analysis of elastically supported beams) [1931/1]; *O formakh ravnovesiya szhatykh stoek pri prodol'nom izgibe* (on the equilibrium forms of columns subject to buckling) [1931/2]; *Sobranie trudov* (collected works) [1936–56]
- Further reading:
[Krylov, 1942]; [Shtraikh, 1956]; [Khanovich, 1967]; [Grigorian, 1973]; [Lehmann, 1999]
- Photo courtesy of Prof. Dr. G. Mikhailov

### LA HIRE, PHILIPPE DE

\*18.3.1640 Paris, France, 21.4.1718 Paris, France

As the eldest son of Laurent de La Hire, artist, professor and founder of the Académie Royale de Peinture et de Sculpture, and pupil of Gérard Desargues (1593–1661), the young Philippe grew up among artists who sought to give their art a theoretical basis and who awakened his interest in painting, drawing, perspective and practical mechanics at an early age. After the death of his father, the devastated Philippe de La Hire went to Venice in 1660 to study art and mathematics for four years. Upon his return, he became friends with Abraham Bosse, Desargues's last student. It was during this period that La Hire became interested in the theory of conic sections and the cutting of stones, also geodetic and astronomy problems. In 1682 he became the successor to Gilles Personne de Roberval (1602–75) at the chair of mathematics (which had been vacant for seven years) at the Collège Royal. Five years later he succeeded Nicolas François Blondel (1618–86) at the Académie Royale d'Architecture, where he gave lectures on the theory of architecture and the theory of cutting of stones until 1717. Here, La Hire dealt with the problem of masonry arches with a clear interest for the *règles de l'art*, searching for a scientific basis that would explain the intuition behind the design and building of arches. These aspects, ignored by La Hire in his two *Mémoires* presented to the Académie des Sciences in 1695 and 1711, have not yet been analysed in their historical context [Becchi & Foce, 2002]. La Hire's masonry arch theory added one of building's most significant objects to the theory of mechanics. He therefore filled in the last piece of the jigsaw in the orientation phase of structural theory and at the same time laid the foundation stone for the application phase.

- Main contributions to structural analysis:
*Traité de la coupe des pierres* [1687–1690]; *Remarques sur l'épaisseur qu'on doit donner aux pieds droits des voutes et aux murs des dômes ou voutes de four* [1692/1912]; *Traité de mécanique, où l'on explique tout ce qui est nécessaire dans la pratique des Arts, et les propriétés des corps pesants lesquelles ont eu plus grand usage dans la Physique* [1695]; *Architecture civile* [1698]; *Remarques sur la forme de quelques arcs dont on se sert dans l'Architecture* [1702/1720]; *Sur la construction des voûtes dans les édifices* [1712/1731]; *Traité de mécanique, où l'on explique tout ce qui est nécessaire dans la pratique des Arts* [1730]
- Further reading:
[Taton, 1973]; [Buti & Corradi, 1981]; [Becchi & Foce, 2002]
- Photo courtesy of University of St. Andrew

### LAMÉ, GABRIEL

\*22.7.1795 Tours, France, †1.5.1870 Paris, France
Gabriel Lamé studied at the École Polytechnique and the École des Mines from 1813 to 1820. Together with his friend Clapeyron, Lamé worked as an engineer and scientist in St. Petersburg, where he was head of the Institute of Engineers of Ways of Communication and taught analysis,

mechanics, physics and chemistry. Both he and Clapeyron provided advice for important building projects in and around St. Petersburg. Following the French July Revolution of 1830, Lamé and Clapeyron left Russia because the tsarist government had taken the anti-revolutionary side. During the 1830s Lamé acted as a consulting engineer for railways but he was not able to develop his mathematical talents until later – as professor of physics at the École Polytechnique (1831), member of the Académie des Sciences (1843), examiner (1844) and professor (1851 – 62) of mathematical physics and probability theory at the University of Paris. Lamé excelled in elastic theory thanks to fundamental contributions such as the introduction of curvilinear coordinates, the stress ellipsoid, and the primary equations and constants of elastic theory. His book Leçons sur la théorie mathematique de l'élasticité des corps solides (1852) – the first monograph on elastic theory – rounded off the constitution phase of mathematical elastic theory. Lamé tried to find a synthesis between the theories of heat, electricity and light on the basis of the ether theory. He called the elastic medium ether "le véritable roi de la nature physique" [Lamé, 1861] and used elastic theory as his reference point in his attempt to find a synthesis. Maxwell managed to achieve a synthesis of the theories of electricity and light in his electrodynamics, but even he started with the ether hypothesis. Einstein's Theory of Relativity finally displaced ether from the throne of physics and enthroned the field concept.
- Main contributions to structural analysis:
Mémoire sur la stabilité des voûtes [1823]; Mémoire sur la construction des polygones funiculaires [1828]; Mémoire sur l'équilibre intérieur des corps solides homogènes [1833]; Mémoire sur la propagation de la chaleur dans les polyèdres, et principalement dans le prisme triangulaire régulier [1833/1]; Mémoire sur les lois de l'équilibre de l'éther dans les corps diaphanes [1833/2]; Mémoire sur les lois de l'équilibre du fluide éthéré [1834]; Mémoire sur le principe général de la Physique [1842]; Leçons sur la théorie mathematique de l'élasticité des corps solides [1852]; Course de physique mathématique rationelle. Discours préliminaire [1861]
- Further reading:
[Greitzer, 1973]; [Bradley, 1981]; [Gouzévitch, 1993]; [Tazzioli, 1995]
- Photo courtesy of École Nationale des Ponts et Chaussées

### LAND, ROBERT

*21.7.1857 Althammer, Prussia,
†late June 1899 Kamaran, Turkey
Land is one of the great unknowns in the classical phase of structural theory. He finished his secondary education in Görlitz in 1876 and began studying engineering sciences at Dresden TH, from where he graduated in 1880. Thereafter, he worked in the central offices of Berlin City Railways, but then returned to Dresden TH for teacher training until 1883. From 1883 to 1888 he was employed at the Imperial Hydraulic Engineering Office in Alsace-Lothringen. He derived the exchange theorem from the general work theorem – independently of Betti – in 1887 and based on this created a self-contained kinematic theory of frames within a very short time. Land recognised the duality of the principle of virtual displacements and the principle of virtual forces on the level of linear structural theory – an idea that does not feature in the works of his great teacher Otto Mohr nor his rival Heinrich Müller-Breslau. Land was finally able to realise his pedagogic aims as an assistant teacher in 1888 teaching mathematics at the State Technical Schools in Chemnitz. Two years later the Turkish government nominated him professor at the School of Engineering in Constantinople, where he worked until his untimely death. He left behind a wife and three children. But he also left behind the kinematic perspective of structural theory – which soon became forgotten – as an alternative to an energy-based theory of structures. It was not until the 1930s that articles in the journal Der Stahlbau approached step by step the problem of the distinctive displacement jumps in continuum elements. This is where Castigliano's theorem fails, as Hartmann was able to show [Hartmann, 1985], because limited deformation complementary energy $\Pi^*$ cannot be specified for displacement jumps in continuum elements, but the internal work certainly can be. Theoretically, Land could have specified the latter as well [Land, 1887/2], but he didn't do this because he was too committed to the kinematic machine model of structural theory, and so could imagine displacement jumps only through corresponding hinge mechanisms. This meant that the internal work in the unit displacement jump of a continuum element was converted into external work at the hinge mechanism; both formulations of this work theorem led to the same result. Klöppel therefore called this theorem-based influence lines relationship recognised by Land quite rightly the Land theorem [see Ramm, 1986, pp. 48 – 49]. Ebel expanded this theorem to provide an elegant way of determining influence lines for moving deformation variables [Ebel, 1979].
- Main contributions to structural analysis:
Über die statische und geometrische Bestimmtheit der Träger, insbesondere der Fachwerkträger (Zugleich ein Beitrag zur Kinematik der Stabwerke) [1887/1]; Über die Gegenseitigkeit elastischer Formänderungen als Grundlage einer allgemeinen Darstellung der Einflußlinien aller Trägerarten, sowie einer allgemeinen Theorie der Träger überhaupt [1887/2]; Kinematische Theorie der statisch bestimmten Träger [1887/3]; Kinematische Theorie der statisch bestimmten Träger [1888]; Eeitrag zur Ermittlung der Biegungslinien ebener elastischer Gebilde [1889]
- Further reading:
[n. n., 1899]

### LÉVY, MAURICE

*28.2.1838 Ribeauvillé, France, †30.9.1910 Paris, France
Maurice Lévy studied at the École Polytechnique and the École des Ponts et Chaussées from 1856 to 1861. Thereafter, he spent several years in the provinces before gaining his doctorate in Paris in 1867 and working at the École Polytechnique from 1862 to 1883 as a mechanics tutor. It was in this period that Lévy wrote his chief work on the theory of structures: his contribution to the elastic theory foundation of structural theory (1873) and the first edition of his graphical statics (1874), which followed on directly from Cremona's work and which became a standard reference book in the German-speaking world like Wilhelm Ritter's Anwendungen der graphischen Statik. Lévy's four-volume work on graphical statics (2nd edition) (1886 – 88) played the same role in the French-speaking world. In 1875 he was appointed professor of structural mechanics at the École Centrale des Arts et Manufactures and in 1885 professor of analytical mechanics and celestial mechanics at the Collège de France, where he was also promoted to the post of general inspector of bridges and highways. From 1883 onwards he was a member of the Académie des Sciences.
- Main contributions to structural analysis:
Note sur un système particulier de ponts biais [1869]; Application de la théorie mathématique de l'élasticité à l'étude de systèmes articulés [1873]; La statique graphique et ses applications aux constructions [1886 – 88]; Notes sur les diverses manières d'appliquer la règle du trapèze au calcul de la stabilité des barrages en maçonnerie [1897]
- Further reading:
[Koppelmann, 1973]; [Becchi, 1998]; [Maurer, 1998]
- Photo courtesy of École Nationale des Ponts et Chaussées

### LONG, STEPHEN HARRIMAN

*30.12.1784 Hopkinton (New Hampshire), USA, †4.9.1864 Alton (Illinois), USA
Stephen Harriman Long was one of the 10 children born to Moses and Lucy Harriman Long. At the age of 21 he began studying at Dartmouth College, from where he graduated in 1809. He then worked for several years as a teacher in New Hampshire and Pennsylvania. Influenced by General Joseph Swift, chief of the Engineering Corps of the US Army at that time, Long joined the army in 1815 as a second lieutenant, taught mathematics at West Point Military Academy for one year (see section 2.3.6) and was promoted to the rank of major. During the following decade, Long undertook

LÉVY    LONG    LOVE    LURIE

five great expeditions in western USA, including one in the Rocky Mountains and another in the Yellowstone region. He produced maps of these areas which would later prove very important for the building of railways. After Congress passed a law in 1824 permitting military engineers to work on the building of civilian infrastructure, especially railways, Long was put in charge of building the Baltimore & Ohio Railroad in 1827. Long had already had papers on railway engineering published in journals and was well known in this field. He had turned to bridge-building by 1829, and in 1830 designed the Jackson Bridge near Baltimore, a timber trussed framework bridge (see Fig. 2-32/top), for which he was granted a patent in that same year and which was presented in a paper [Long, 1830/1]. He returned to the Jackson Bridge in a second paper and investigated the loadbearing behaviour of his trussed framework system [Long, 1830/2]. "Long was the first person to understand the engineering principles of preload and to apply them to a bridge truss" [Griggs & DeLuzio, 1995, p. 1353]. It was in 1836 that Long provided tables of his trussed framework system for 20 spans between 55 and 300 feet with which the depth of truss and cross-sectional areas of members could be determined [Long, 1836; Gasparini & Provost, 1989, p. 30]. In the two patents of 1839, Long develops the trussed framework concept even further (see, for example, [Long, 1839/1]). No less a person than Karl Culmann revealed his admiration for Long's trussed framework bridges in his travelogue [Culmann, 1851; 1852], on which Timoshenko comments as follows: "Long's system of trusses was similar to that of Palladio, but he evidently knew a valid method of calculating stresses in truss members and, in his work [see [Long, 1839/1] – the author], gave very reasonable proportions for all the members of structures of various spans. After describing some of Long's bridges, Culmann remarks that he was unable to find out if Long was still alive and, if so, what he was doing" [Timoshenko, 1953/1, p. 191]. From 1838 onwards, Long worked in the newly formed US Corps of Topographical Engineers, and by 1861 had become chief topographical engineer to the federal government.

• Main contributions to structural analysis:
*Observations on wooden, or frame bridges* [1830/1]; *Description of the Jackson Bridge together with directions to builders of wooden framed bridges* [1830/2]; *Description of Col. Long's bridges together with a series of directions to bridge builders* [1836]; *Improved brace bridge* [1839/1]; *Specifications of a brace bridge and of a suspension bridge* [1839/2]

• Further reading:
[Wood, 1966]; [Gasparini & Provost, 1989]; [Griggs & DeLuzio, 1995]

• Photo courtesy of the US Corps of Topographical Engineers

### LOVE, AUGUSTUS EDWARD HOUGH

*17.4.1863 West-Super-Mare, UK,
†5.6.1940 Oxford, UK

"All you need is love" is the title of the famous song by the Beatles recorded in 1967. "All you need is Love" was Koiter's title for his contributions to the linear theory of thin shells. (E. Ramm and W. A. Wall have devised an approachable, easy-to-understand summary of modern shell theory [Ramm & Wall, 2004].) Koiter's use of the capital "L" for Love shows that he means not love in general, but rather A. E. H. Love and his shell theory [Love, 1888]. Who was A. E. H. Love? He was the second-eldest son of a dentist, John Henry Love, and attended Wolverhampton Grammar School, where he benefited from the mathematics lessons of the Rev. Henry Williams and, following his scholarship examination in 1881, was awarded a scholarship for St. John's College. He began his studies there in 1882 and finally gained distinction in 1885 as second wrangler of parts I and II of the Mathematics Tripos. One year later, Love became a fellow of St. John's College and in 1887 was awarded the Smith Prize, the highest accolade for mathematics at the University of Cambridge. His famous shell theory was published just one year later [Love, 1888] and triggered a scientific controversy with Lord Rayleigh and others (see [Calladine, 1988, pp. 4 & 5]). Love wrote his two-volume work on mathematical elastic theory in 1892 and 1893 [Love, 1892/1893], which compiled the knowledge of this branch of science up until the end of the 19th century in a synthesis hitherto unsurpassed. The second edition [Love, 1906] was translated into German [Love, 1907] and formed the scientific system of coordinates for the Göttingen school around Felix Klein in the field of elastic theory. Even today, the "historical introduction" to his work on elastic theory is still an excellent record of the historical study of the exact sciences. Love was elected a fellow of the Royal Society at the age of 30 and in 1899 was appointed to the Sedleian chair of natural philosophy at the University of Oxford, a post he held until his death. His monograph on geodynamics [Love, 1911] remains among the classic works on geophysics, seismology, foundation engineering dynamics and earthquake engineering. It was in this book that he developed a mathematical model for the surface wave which bears his name and which during earthquakes leads to horizontal ground movements perpendicular to the direction of propagation of the waves and which cause the greatest destruction. Love's work on elastic theory played a major role during the consolidation period of structural theory, or rather structural mechanics (1900 – 50), especially in the formation of the theory of plate and shell structures.

• Main contributions to structural analysis:
*The small free vibrations and deformation of a thin elastic shell* [1888]; *A treatise on the mathematical theory of elasticity* [1892/1893 & 1906]; *Lehrbuch der Elastizität* [1907]; *Some problems of geodynamics* [1911]

• Further reading:
[Bullen, 1973]; [Calladine, 1988]

• Photo courtesy of the University of St. Andrew

### LURIE, ANATOLY ISAKOVICH

*19.6.1901 Mogilev, Russia (today: Belarus),
†12.2.1980 Leningrad, USSR

After completing his grammar school education, Lurie studied at the faculty of physics and mechanics at Leningrad Polytechnic Institute and in 1925 became an assistant to the chair of theoretical mechanics at that establishment, later taking over there (1936 – 41). Although he had written no dissertation, Lurie was awarded the academic title of doctor of engineering sciences in 1939. Following the liberation of Leningrad, he returned to his Alma Mater, where he was director of the chair of dynamics and machine strength (later the chair for mechanics and control processes) from 1944 to 1977. At the same time, he worked as a consultant for industry. Lurie was the acknowledged head of the Leningrad school of mechanics, where L. G. Loitsyansky, G. I. Dzhanelidze and Y. G. Panovko were also active. He was elected a corresponding member of the USSR

745

Academy of Sciences in 1960. Lurie excelled in the fields of elastic theory, the theory of non-linear oscillations and the theory of plate and shell structures in particular.
• Main contributions to structural analysis:
*Statika tonkostennykh uprugikh obolochek* (statics of thin-walled elastic shells) [1947]; *Prostranstvennye zadachi teorii uprugosti* (three-dimensional problems of the theory of elasticity) [1955]; *Statics of Thin-walled Elastic Shells* [1959]; *Analiticheskaya mekhanika* (analytical mechanics) [1961]; *Räumliche Probleme der Elastizitätstheorie* [1963]; *Three-dimensional problems of the theory of elasticity* [1964]; *Teoriya uprugosti* [1970]; *Nelineinaya teoriya uprugosti* (nonlinear theory of elasticity) [1980]; *Nonlinear theory of elasticity* [1990]; *Analytical Mechanics* [2002]; *Theory of elasticity* [2005]
• Further reading:
[n. n., 1980]; [Malinin, 2000, p. 170];
[Lurie, K. A., 2001]
• Photo courtesy of Prof. Dr. G. Mikhailov

## MAGNEL, GUSTAVE
*14.9.1889 Essen, Belgium, †5.7.1955 Ghent, Belgium
Gustave Magnel studied civil engineering at the University of Ghent from 1907 to 1912. He left Belgium in 1914 and worked as a civil engineer at the D. G. Somerville & Co. Contractor Company until 1917, finally becoming chief engineer there. After he returned to Belgium in 1919, he joined the Strength of Materials Laboratory at the University of Ghent. He first started lecturing at the university in 1922 and went on to become a lecturer in 1927 and finally (1937) full professor of concrete and reinforced concrete construction as well as director of the Laboratory for Reinforced Concrete Construction. The French Association des Ingénieurs Docteurs honoured Magnel with its Grande Médaille, and he received the Frank P. Brown Medal from the Franklin Institute of Philadelphia. Magnel was elected a member of the Belgian Academy of Sciences and represented his country at UNESCO from 1945 to 1946. Further honours include Commander of the Yugoslavian St. Sava Order and Chevalier de la Légion d'Honneur. Magnel was one of the pioneers of prestressed concrete and the inventor of a prestressing anchorage system named after him and Blaton. He supported Freyssinet during the foundation of the Fédération Internationale de la Précontrainte (FIP) and was its first vice-president. In his capacity as a structural engineer, too, he was acclaimed by his contemporaries. For instance, he was the first person to use prestressed continuous beams – in his design for Sclayn Bridge; and it was Magnel's Walnut Lane Memorial Bridge that inaugurated the use of prestressed concrete in the USA.

• Main contributions to structural analysis:
*Calcul pratique des poutres Vierendeel* [1933]; *Stabilité des Constructions* [1948/1]; *Prestressed concrete* [1948/2]; *La pratique du béton armé* [1949]
• Further reading:
[Riessauw, 1960]
• Photo courtesy of Prof. Dr. R. Maquoi

## MAHAN, DENNIS HART
*2.4.1802 New York City, USA,
†16.9.1871 near Stony Point (New York), USA
Mahan spent his youth in Norfolk (Virginia) and studied at West Point Military Academy, where he was best in his class and graduated in 1824. After two years of lecturing in mathematics at West Point, Mahan was sent to France by the Department of War for a three-year study tour in which he was to investigate state building projects and military establishments. Sponsored by the French government, Mahan spent more than 12 months at the École d'application de l'Artillerie et du Génie in Metz, where Poncelet – an outstanding engineering personality who fused together science and engineering to form applied mechanics – was on the teaching and research staff. In Paris, Mahan was a regular guest of the legendary French general and liberal politician La Fayette (1757–1834). Mahan returned to West Point in 1830 and was released by the Engineering Corps initially for an interim period so that he could take up a post as acting professor, and finally (1832) as full-time professor, for civil and military engineering, a post he held until his death. Like no other during that period, Mahan, through his teaching and his books, had an influence on the self-image of the American civil engineer that endured into the final decades of the 19th century, a period that was determined by the tasks of internal colonisation through the military-like provision of infrastructure. To some extent, the US Army Engineering Corps and West Point adapted Vauban's Corps des Ingénieurs militaires and the emerging engineering schools of 18th-century France on a higher level. For example, Mahan published not only the first American book on fortifications (1836), but also the first American book on civil engineering, which in the first place was aimed at the West Point cadets [Mahan, 1837], but later became a standard work for civil engineers in the USA. The book includes numerous sample calculations from Navier's *Résumé des Leçons* [Navier, 1826]. In his introduction, Mahan recommends that "the best counsel that the author could give to every young engineer is to place in his library every work of science to which M. Navier's name is in any way attached" [Mahan, 1837]. Mahan republished Moseley's 1843 monograph *The mechanical principles of engineering and architecture* in 1856 [Moseley, 1856] and enriched it with numerous original ideas; in Germany, Hermann Scheffler adopted Moseley's "mechanical principles" in a creative way around the same time [Scheffler, 1857]. Mahan therefore implanted the structural theory of Navier and Moseley into the knowledge canon of the American civil engineer. In historical terms, Mahan's contribution to the development of the principles of civil engineering in the USA bridges the transition from the constitution phase (1825–50) to the establishment phase (1850–75) of structural theory.
• Main contributions to structural analysis:
*An elementary course of civil engineering* [1837]; *A treatise on fortification drawing and stereotomy* [1865]; *Descriptve geometry, as applied to the drawing of fortification and stereotomy* [1870]; *A treatise on civil engineering* [1878]
• Further reading:
[Abbot, 1886]
• Photo courtesy of the United States Military Academy Library, Special Collections

## MAILLART, ROBERT
*6.2.1872 Bern, Switzerland, †5.4.1940 Geneva, Switzerland
Robert Maillart grew up in a Calvinist family in Bern and his mathematical and drawing talents were already apparent during his years at grammar school. He studied structural engineering at Zurich ETH from 1890 to 1894, lectures by Wilhelm Ritter on graphical statics forming part of the curriculum. After graduating, Maillart worked for Pümpin & Herzog (Bern), the civil engineering department of the City of Zurich and Froté & Westermann. It was while he was employed in this latter company that he had his first spark of ingenuity: for the reinforced concrete arch bridge in Zuoz, completed in 1901, Maillart combined the road deck with the arch in such a way that a two-cell hollow box was formed. One year later, he founded his own company. Maillart designed a gasometer pit for the town of Sankt Gallen in 1903 and for the first time took into account the bending moments in the graphical analysis calculation of the internal forces for the cylindrical reinforced concrete shell fixed at the ground slab (see [Schöne, 1999, p. 71]). Later that year, Maillart observed long vertical cracks in the web in the vicinity of the abutments to the reinforced concrete arch bridge in Zuoz. These led to triangular cut-outs in the abutment elements and finally, in 1905, to the three-pin arch bridge spanning 51 m over the Rhine at Tavanasa. More than any other, Maillart gave the design language of reinforced concrete a valid architectural expression during the consolidation period of structural theory (1900–50). Not only his stiffened polygonal arches, e.g. over the Landquart at Klosters on the Chur–Davos railway line, or the Salginatobel Bridge at Schiers completed in 1930, but also his flat slab developed in 1908 according to the two-strip system are classic examples of the steady state between beauty and utility in

 MAGNEL  MAHAN  MAILLART  MANN  MASCHERONI  MASSONNET

the art of structural engineering. "Maillart was an engineer in the truest sense of the word. He placed theory and scientific findings entirely at the disposal of architecture: the first was his means, the other his goal. He saw experience and scientific knowledge as equal partners" [Roš, 1940, p. 224]. Maillart began his activities in Russia in 1912, but two years later he was caught unawares by the outbreak of World War 1 and had to be evacuated from Riga to Kharkov. He designed massive industrial structures for AEG and others in Kiev. Following the death of his wife and the outbreak of the October Revolution, Maillart returned to Switzerland with his three children penniless. Nevertheless, during his second period of creativity (1920–40), Maillart was able to complete 160 structures that embody the rigorous logic and artistic will of their creator. His most important contribution to theory of structures was the introduction of the concept of the shear centre and the clear formulation of its underlying theory in the early 1920s (see section 7.3.2.4). When Robert Maillart died on 5 April 1940, reinforced concrete construction lost a "concrete virtuoso" [Marti, 1996] and a genius of building. In his obituary, Mirko Gottfried Roš (1879–1972) writes: "You were both engineer and artist because your credo was the harmony between size, beauty and truth" [Roš, 1940, p. 226].

- Main contributions to structural analysis:
*Zur Frage der Biegung* [1921/1]; *Bemerkungen zur Frage der Biegung* [1921/2]; *Ueber Drehung und Biegung* [1922]; *Der Schubmittelpunkt* [1924/1]; *Zur Frage des Schubmittelpunktes* [1924/1 & 1924/3]; *Zur Entwicklung der unterzugslosen Decke in der Schweiz und in Amerika* [1926]; *Einige neuere Eisenbetonbrücken* [1936]
- Further reading:
[Roš, 1940]; [Kleinlogel, 1940]; [Bill, 1949]; [Günschel, 1966]; [Billington, 1979; 1980; 1990; 1997]; [Laffranchi & Marti, 1997]; [Gesellschaft für Ingenieurbaukunst, 1998]; [Billington, 2003]
- Photo courtesy of [Billington, 2003, p. 31]

### MANN, LUDWIG

*1.9.1871 Cologne, German Empire, †27.2.1959 Leipzig, German Democratic Republic
Following his education at the humanist grammar school in Wiesbaden, Mann studied civil engineering at Hannover TH, studies which he completed in 1895 as a student of Müller-Breslau at the Berlin-Charlottenburg TH. After a period of military service, he worked in an engineering practice in Berlin. In 1900 he was commissioned by the Berlin Magistrate to take charge of the office for the planning and design, structural calculations and site supervision of the building of the Tegel Gasworks. It was from the calculations of the 76 m-span pavilion roof that Mann developed his theory of cyclically symmetrical structural systems and their application to the theory of spatial structures [Mann, 1911]. He was promoted to senior engineer by Müller-Breslau in 1906 and appointed by him to take charge of the Theory of Structures Laboratory and to carry out earth pressure tests. One year later he gained his doctorate with distinction. His work on the analysis of rigid quadrilateral grids with the help of differential calculus [Mann, 1909] earned him an appointment to the chair of mechanics at Breslau TH (founded in 1910). One year after Ostenfeld's displacement method [Ostenfeld, 1926], Mann himself published a work on the same method, but in an expanded form based it faithfully on the principle of virtual displacements [Mann, 1927]. Besides his professorship, from 1928 onwards Mann also advised the German lignite industry on the use of large-scale machinery for open-cast mining, work which he continued in the German Democratic Republic after the war. The theoretical fruits of these engineering activities can be found in the form of completely new theory of structures methods which permitted fast and accurate three-dimensional calculations of loadbearing systems; it was within this context that he also developed his three-dimensional displacement method [Mann, 1939]. He formulated clearly the dual nature of structural theory based on the duality concept of projective geometry [Mann, 1957]. In this way, Mann placed a milestone in the invention phase (1925–50) of structural theory. Breslau TH appointed him dean of the Faculty of General Sciences and he was twice elected *rector magnificus*. He gained his doctorate at Dresden TH in 1957.

- Main contributions to structural analysis:
*Statische Berechnung steifer Vierecksnetze* [1909]; *Über zyklische Symmetrie in der Statik mit Anwendungen auf das räumliche Fachwerk* [1911]; *Theorie der Rahmenwerke auf neuer Grundlage* [1927]; *Grundlagen zu einer Theorie räumlicher Rahmentragwerke* [1939]; *Vergleich der Prinzipien und Begriffe für die Entwicklung der Kraft- und Deformationsmethoden in der Statik* [1958]
- Further reading:
[Rudolph, 1959]
- Photo courtesy of [Rudolph, 1959, p. 109]

### MASCHERONI, LORENZO

*13.5.1750 Castagneta near Bergamo, Italy, †14.7.1800 Paris, France
Ordained as a priest in 1767, three years later he was teacher of rhetoric and from 1778 onwards teacher of mathematics and physics at the teacher-training college in his home town. His work on masonry arches *Nuove ricerche sull'equilibrio delle volte* (1785) brought him an appointment as professor of algebra and geometry at the University of Pavia, of which he was principal in 1789 and 1793. From 1788 to 1791 he was president of the Accademia degli Affidati; he was also a member of the academies of Padua and Mantua, and the Società Italiana delle Scienze. In masonry arch theory, Mascheroni analysed collapse mechanisms with the help of the principle of virtual displacements, thereby finishing off in a historico-logical sense the tradition of analysing collapse mechanisms of arches that stretched back to Leonardo da Vinci [Sinopoli, 2002]. He was also active as a mathematician, poet and diplomat. For example, the legislative body of Milan sent him to Paris in 1797 in order to study France's new system of currency and metric measurement; however, the Austrian occupation of Milan prevented his return and he died in Paris following a brief illness.

- Main contributions to structural analysis:
*Nuove ricerche sull'equilibrio delle volte* [1785]; *Nuove ricerche sull'equilibrio delle volte* [1829]
- Further reading:
[Landi, 1829]; [Benvenuto, 1981]; [Seidenberg, 1981]; [Benvenuto, 1991/2]; [Becchi & Foce, 2002]
- Photo courtesy of Università di Genova

### MASSONNET, CHARLES

*14.3.1914 Arlon, Belgium, †4.4.1996 Liège, Belgium
He began studying civil engineering at the University of Liège in 1931 and graduated with distinction in 1936. Following a year of military service, he was employed on research work at the Fédération National Recherche Scientifique

(FNRS) under Ferdinand Campus, one of the founders of the International Association for Bridge & Structural Engineering (IABSE). However, Massonnet's research was rudely interrupted by five years in a prisoner-of-war camp (1940–45). During his years as a prisoner he learned German and studied the basic literature on theory of structures and strength of materials – especially the monographs of Timoshenko. From 1946 to 1949 he was assistant professor of strength of materials and elastic theory at the University of Liège, and from 1950 to 1983 he was responsible for theory of structures and steel construction at his Alma Mater. Like Klöppel, he represented the symbiosis between theory of structures and steel construction. For example, Massonnet carried out pioneering work in the fields of stability theory, ultimate load theory and the theory of boundary elements. As a leading member of important international scientific and technical associations, he made vital contributions to transferring the results of research into European steelwork standards. The research focus he established at the University of Liège continues successfully under his successor René Maquoi. Massonnet received numerous honours for his research work: honorary doctorates from Chalmers University of Technology (1976) and Zurich ETH, and corresponding membership of several academies, including the National Academy of Engineering (USA); he was appointed to the highly regarded Polish Association of Theoretical and Applied Mechanics in 1982.

- Main contributions to structural analysis: *Calcul plastique des constructions. Vol. 1: Structures planes* [1961]; *Calcul plastique des constructions. Vol. 2: Structures spatiales* [1963]; *Le calcul des grillages de poutres et dalles orthotropes selon la méthode Guyon-Massonnet-Bares* [1966]
- Further reading: [Maquoi et al., 1984]; [Maquoi, 1996]
- Photo courtesy of Prof. Dr. R. Maquoi

### MAXWELL, JAMES CLERK

*13.6.1831 Edinburgh, UK,
†5.11.1879 Cambridge, UK

James Clerk Maxwell was the son of a Scottish estate owner and lawyer, and studied in Edinburgh and Cambridge. From 1856 to 1860 he was professor of physics at Marischal College in Aberdeen and held the same post at King's College, London, until 1865. Poor health forced him to give up his professorship, move back to the estate he had in the meantime inherited and devote himself to writing his epochal treatise on electricity and magnetism. Nevertheless, Maxwell could not resist the offer of the newly created chair of experimental physics and the Cavendish Laboratory at the University of Cambridge in 1871 [Wiederkehr, 1986, p. 235]. Maxwell is regarded as the greatest physicist between Newton and Einstein. The Cavendish Laboratory evolved into the world's leading centre for physics. Maxwell's fame in physics overshadows his pioneering work in the theory of structures which, at best, gets only a cursory remark in the history-of-science literature. For instance, in a six-page article he formulated a comprehensive theory of statically indeterminate frames which contains the principle of virtual forces and the reciprocity theorem that Müller-Breslau named after him (1864/2). It was not until the first half of the 1880s that this spectacular work was noticed by the structural engineering community – the principles of the theory of statically indeterminate frames had by then already been established by Mohr. Nevertheless, Maxwell's theory was still able to unfold its effects in the second half of the classical phase of structural theory; together with Maxwell's theorem generalised by Müller-Breslau, this evolved into a theory for statically indeterminate frames. Thus, Maxwell's frame theory gained some fame, at least in academic circles. His contributions to graphical statics on the other hand, were adopted directly by advocates such as Cremona, and in 1869 Rankine included them in the second edition of his book *A Manual of Applied Mechanics* and hence ensured their dissemination among engineers. Based on Rankine's theorem, Maxwell explained the duality relation of frame geometry and the polygon of forces in 1864 and 1867 and created a theory of reciprocal diagrams [Scholz, 1989, pp. 187–91], to which Cremona made direct additions and generalised them. The Maxwell–Cremona duality of frame geometry and polygon of forces was called by Mohr in 1875 the "showpiece of graphical statics" [Scholz, 1989, p. 201]. Without doubt, Maxwell made the biggest contribution to frame theory during the establishment and classical phases of structural theory.

- Main contributions to structural analysis: *On reciprocal figures and diagrams of forces* [1864/1]; *On the calculation of the equilibrium and stiffness of frames* [1864/2]; *On the application of the theory of reciprocal polar figures to the construction of diagrams of forces* [1867]; *On reciprocal figures, frames, and diagrams of forces* [1870]; *On Bow's method of drawing diagrams in graphical statics with illustrations from Peaucellier's linkage* [1876]; *Scientific papers* [1890/1965]
- Further reading: [Niles, 1950]; [Benvenuto, 1981]; [Charlton, 1982]; [Wiederkehr, 1986]; [Scholz, 1989]; [Benvenuto, 1991/2]
- Photo courtesy of [Simonyi, 1990, p. 346]

### MEHRTENS, GEORG CHRISTOPH

*31.5.1843 Bremerhaven, Prussia,
†9.1.1917 Dresden, German Empire

At the age of 18, Mehrtens started studying at Hannover TH. From 1867 to 1894 he worked primarily for Prussian State Railways. In addition, during the early 1880s he was employed as Emil Winkler's assistant and as a private lecturer at Berlin TH; he also came into contact with Schwedler in Berlin. The highlight of his structural engineering career was supervising the Technical Office of the Royal Railways Department in Bromberg for the building of new bridges over the Weichsel at Dirschau and Nogat near Marienburg (1888–94). During the building of the Weichsel bridges, Mehrtens was able to avoid wrought iron completely and use mild steel exclusively, half in Martin steel and – the first time in bridge-building – half in basic Bessemer (Thomas) steel. He thus became the leading bridge-builder in Kaiser Wilhelm's reign. In 1894 he was approached by Aachen RWTH and one year later by Dresden TH, where he lectured in bridge-building and theory of structures until 1913, and – after the departure of Otto Mohr – strength of materials as well. His *Vorlesungen über Statik der Baukonstruktionen und Festigkeitslehre* formed the first modern textbook of theory of structures in the consolidation period. It was in this book that he developed the first ideas about a systematic history of structural theory. But his articles on the history of bridge-building, which comply with the scientific criteria for writings on the history of science and engineering (as promoted primarily by Conrad Matschoß in the first decade of the 20th century), are even more important.

- Main contributions to structural analysis: *Der deutsche Brückenbau im XIX. Jahrhundert. Denkschrift bei Gelegenheit der Weltausstellung 1900 in Paris* [1900]; *Vorlesungen über Statik der Baukonstruktionen und Festigkeitslehre* [1903–05]; *Eisenbrückenbau* [1908]
- Further reading: [Bleich, 1913]; [Bleich, 1917]; [Ricken, 1997]
- Photo courtesy of Dresden TU archives

### MELAN, ERNST

*16.11.1890 Brünn, Austria-Hungary (today: Brno, Czech Republic), †10.12.1963 Vienna, Austria

Ernst Melan was the son of Joseph Melan and spent his youth in Prague, where he studied at the German Technical University and gained his doctorate in 1917 with a work on the torsion of bodies of revolution. Afterwards, he turned to practical engineering at k. k. Statthalterei in Graz, moved to Waagner Biró AG, then took up a post as a production engineer in the strength of materials laboratory at Berlin-Charlottenburg TH and by 1921 had become a departmental head at Waagner Biró AG. He wrote his habilitation thesis during these years and was appointed to the chair of elastic theory at Vienna TH. His nomination as associate professor for theory of structures and strength of materials at the German Technical University in Prague followed in 1923, and two years after that he was appointed to the chair of theory of structures at Vienna TH, to which steel and timber engineering were added in 1939. Ernst Melan remained at Vienna

 MAXWELL
 MEHRTENS
 MELAN, E.
 MELAN, J.
 MENABREA

TH as a highly regarded teacher and researcher until 1962. The monograph *Die gewöhnlichen und partiellen Differenzengleichungen der Baustatik*, written together Friedrich Bleich, appeared in 1927 [Bleich & Melan, 1927], and this became a standard work for numerical methods in higher theory of structures during the invention phase of structural theory (1925–50). In the analysis of member buckling, he points out the possibility of using integral equations for an approximate determination of eigenvalues (see [Bargmann, 1990, p. 557]). Ernst Melan was an outstanding figure in the "paradigm articulation" (Kuhn) from elastic to plastic theory. This was already apparent in his analysis of a trussed framework with one degree of static indeterminacy using the ideal-elastic and the ideal-plastic material law [Melan, 1932]. In his most important publication in the field of plastic theory, he generalises the shake-down principle formulated by Hans Heinrich Bleich in 1932 [Melan, 1938]. Like his father, Ernst Melan was involved with bridges, as can be seen from his book entitled *Die genaue Berechnung von Trägerrosten*, co-written with Robert Schindler [Melan & Schindler, 1942], but also in the publication of the two-volume work on bridge-building [Melan, 1948 & 1950]. Together with Heinz Parkus (1909–82), Ernst Melan wrote the first comprehensive theoretical presentation on thermal stresses due to stationary temperature fields [Melan & Parkus, 1953]. His services to science were acknowledged by membership of the Austrian and Polish Academies of Science, among other honours.

- Main contributions to structural analysis:
*Zur Bestimmung des Sicherheitsgrades einfach statisch unbestimmter Fachwerke* [1932]; *Die gewöhnlichen und partiellen Differenzengleichungen der Baustatik* [1927]; *Zur Plastizität des räumlichen Kontinuums* [1938]; *Die genaue Berechnung von Trägerrosten* [1942]; *Einführung in die Statik* [1942]; *Einführung in die Festigkeitslehre für Studierende des Bauwesens* [1947]; *Der Brückenbau* [1948 & 1950/1]; *Einführung in die Baustatik* [1950/2]; *Wärmespannungen infolge stationärer Temperaturfelder* [1953]

- Further reading:
[Reinitzhuber, 1960]; [Chmelka, 1964]
- Photo courtesy of [Reinitzhuber, 1960, p. 390]

### MELAN, JOSEF
*18.11.1853 Vienna, Austrian Empire,
†6.2.1941 Prague, Czechoslovakia
Josef Melan studied civil engineering at Vienna TH from 1869 to 1874 and thereafter was assistant to Emil Winkler at the chair of railway engineering and bridge-building. He wrote his habilitation thesis entitled *Theorie des Brücken- und Eisenbahnbaues* at the same university in 1880 and remained on the teaching staff there until 1886. It was during this period that he worked in the design offices of the Ignaz Gridl bridge-building company and for the building contractor Gaertner – both based in Vienna. He was appointed associate professor of structural mechanics and graphical statics at Brünn TH in 1886, where he was promoted to full professor at the same chair in 1890 before switching to the chair of bridge-building in 1895. He was head of the chair of bridge-building at the German TH in Prague from 1902 until his transfer to emeritus status in 1923. Josef Melan was the outstanding authority on the theory and practice of bridge-building in Austria during the transition from the discipline-formation period to the consolidation period of structural theory. The Melan System, which links steel and concrete construction, won a significant market-share in European and American bridge-building as early as the 1890s and was awarded a gold medal at the World Exposition in Paris in 1900. Melan had published his work on concrete arches in conjunction with iron arches in 1893. However, it was not only in composite construction, but also in the field of steel bridge-building that Melan set standards. In 1888 he was the first person to quantify the effects of second-order theory. His books on bridges enjoyed international popularity. For example, in 1913 the American bridge-builder Steinman translated Melan's theory of arch and suspension bridges [Melan, 1913]. Melan also verified the calculations for the Williams Bridge on behalf of the New York Bridge-Building Department and the Hellgate Bridge for the New-York-based Lindenthal bridge design office. His influence on the theory and practice of large bridges in the USA during the first two decades of the 20th century is without precedent.

- Main contributions to structural analysis:
*Beitrag zur Berechnung statisch unbestimmter Stabsysteme* [1884]; *Theorie der eisernen Bogenbrücken und der Hängebrücken* [1888/2]; *Theorie des Gewölbes und des Eisenbetongewölbes im besonderen* [1908]; *Der Brückenbau. Nach Vorträgen, gehalten an der deutschen Technischen Hochschule in Prag* [1910, 1911, 1917]; *Theory of Arches and Suspension Bridges* [1913]; *Plain and reinforced concrete arches* [1915]

- Further reading:
[Nowak, 1923]; [Kluge & Machaczek, 1923]; [Bortsch, 1924]; [Fritsche, 1941]; [Emperger, 1941]; [Eggemann & Kurrer, 2006]
- Photo courtesy of [Nowak, 1923]

### MENABREA, LUIGI FEDERICO
*4.9.1809 Chambéry, Savoy, France,
†24.5.1896 St. Cassin near Chambéry, France
Luigi Federico Menabrea studied engineering and mathematics at the University of Turin. When Cavour relinquished his supreme command of the army in 1831 under King Albert, Menabrea was nominated his successor at the Bardo Alpine fortress. He left this post soon after in order to accept a job as professor of mathematics and engineering at the Military Academy of the Kingdom of Sardinia and the University of Turin. In 1842 he published a study of Charles Babbage's automatic calculating machine. It was around this time that his political career began, which was to govern his life until 1892: member of parliament, collaborator in Lombardy with Garibaldi in 1859 with the rank of major-general, vice-general of the Military Engineering Corps. In 1858 he published his theorem of minimum elastic work (deformation energy), which played a leading role in the classical phase of structural theory and bears his name. He was promoted to lieutenant-general in the 1860s and became a senator, Minister of the Navy and finally Minister of Public Works. In 1866 in Vienna he signed the peace treaty with Austria and Hungary as the representative of Italy. Afterwards he became chairman of the Italian Council of Ministers, head of the government and Foreign Minister. The year 1875 saw controversy arise with Castigliano over the energy-based theorems of structural theory. As a friend of and adviser to King Victor Emmanuel II and as ambassador in London and Paris, he influenced Italian foreign policy in the late 1870s and throughout the 1880s. He was a member of the academies of science in Paris, Rome and Turin.

- Main contributions to structural analysis:
*Nouveau principe sur la distribution des tensions dans les systèmes élastiques* [1858]; *Sul principio di elasticità, delucidazioni di L. F. M.* [1870]; *La determinazione delle tensioni e delle pressioni nei sistemi elastici* [1875]
- Further reading:
[Cavallari Murat, 1957]; [Bulferetti, 1969]; [Briguglio & Bulferetti, 1971]; [Boley, 1981]; [Benvenuto, 1981]; [Benvenuto, 1991/2]
- Photo courtesy of Università di Genova

## MERRIMAN, MANSFIELD

*27.3.1848 Southington, Connecticut, USA, †7.6.1925 New York, USA

Mansfield Merriman, the son of a farmer, completed his civil engineering studies at Yale University in 1871. He then worked for two years as an assistant engineer in the US Corps of Engineers. During his time as an instructor of civil engineering at Yale University (1875–78), Merriman was awarded a doctor title (1876) with the first American dissertation on the subject of statistics. His dissertation, published in 1877 [Merriman, 1877], would quickly become the standard work on statistics. The success of this dissertation enabled Merriman to work as a lecturer in astronomy at the same university from 1877 to 1878. Merriman published a paper on continuous bridge beams as early as 1876 [Merriman, 1876], and this earned him an appointment as professor of civil engineering at Lehigh University in 1878, where he worked as a very successful teacher until 1907. Over those years, Merriman wrote numerous textbooks on theory of structures, strength of materials, hydraulics, geodesy and statistics. To crown his literary life's work in the field of civil engineering, he took on the post of editor of that catechism of American civil engineering during the consolidation period of structural theory (1900–50): *The American Civil Engineer's Pocket-Book* [Merriman, 1911] (see section 3.2.3.2). "By the time he died in New York on June 7, 1925, 340,000 copies of his works had been published. He is said to have been one of the greatest engineering teachers of his day" [Stigler, 2004].

- Main contributions to structural analysis:
*On the theory and calculation of continuous bridges* [1876]; *Elements of the Method of least squares* [1877]; *A text-book on the mechanics of materials and of beams, columns, and shafts* [1885]; *A text-book on roofs and bridges. Part I. Stresses in simple trusses* [1888]; *A text-book on roofs and bridges. Part II. Graphic Statics* [1890]; *A text-book on retaining walls and masonry dams* [1892]; *A text-book on roofs and bridges. Part III. Bridge Design* [1894]; *Strength of materials. A text-book for manual training schools* [1897]; *A text-book on roofs and bridges. Part IV. Higher Structures* [1898]; *The principle of least work in mechanics and its use in investigations regarding the ether of space* [1903]; *Elements of mechanics. Forty lessons for beginners in engineering* [1905]; *The American Civil Engineer's Pocket-Book* [1911]
- Further reading:
[Stigler, 2004]
- Photo courtesy of Lehigh University archives

## MICHELL, JOHN HENRY

*26.10.1863 Maldon, Victoria, Australia, †3.2.1940 Camberwell, Victoria, Australia

John Henry Michell was the eldest son of the miner John Michell and his wife Grace (née Rowse). His energetic, adventurous parents, who had emigrated from Devonshire (UK) to Australia in 1854, had a great respect for learning and quickly recognised the intellectual talents of their two sons John and George. The parents moved to Melbourne in 1877 so their gifted sons could attend Wesley College. Afterwards, John Henry studied mathematics and natural philosophy at the University of Melbourne, from where he graduated in 1884 as a bachelor of arts (with first-class honours). He continued his mathematical studies at the University of Cambridge, where he was awarded a master of arts in 1887, then the Smith Prize in 1889 (the university's highest award for mathematics) and became senior wrangler. Michell became a fellow of Trinity College in 1890, and one year later was appointed professor of mathematics at the University of Melbourne, where he worked until the end of 1928. His 23 scientific publications are primarily concerned with applied mechanics. His papers published between 1898 and 1902 – elastic theory (1899–1902) and ship hydromechanics (1898) – are pioneering works. For example, it was in 1899 that Michell managed to derive the differential equation for the stress function $F(x,y)$ of the plane stress state in an elastic, isotropic body $\Delta\Delta F = 0$. He thus completed the work on the theory of the stress function started by Airy and continued by Maxwell, and together with Beltrami made a significant contribution to the force method of mathematical elastic theory. The six simultaneous partial differential equations of elastic theory are therefore known as the Beltrami-Michell formulation (BMF). Michell was also the first person to formulate the theory of thin elastic plates without the need for questionable assumptions. In recognition of his further development of mathematical elastic theory, Michell was elected a member of the Royal Society in 1902.

- Main contributions to structural analysis:
*On the direct determination of stress in an elastic solid, with application to the theory of plates* [1899]
- Further reading:
[Cherry, 1986]; [Raack, 1977]
- Photo courtesy of the University of St. Andrew

## MINDLIN, RAYMOND DAVID

*17.9.1906 New York, USA, †22.11.1987 Hanover, New Hampshire, USA

Mindlin spent the whole of his student and professional life – from 1924 to 1975 – at Columbia University, New York; with one exception: three years (1942–45) at the Applied Physics Laboratory, Silverspring (Maryland), where he was the person primarily responsible for developing the proximity fuse, one of the supreme achievements of the scientific wartime efforts of the USA, and for which Mindlin was awarded the President Medal for Merit (the highest US civilian award of World War 2) by Harry S. Truman in 1946. After he had gained three academic degrees in succession at Columbia University by 1932, Mindlin visited the summer courses organised by Timoshenko at the Department of Engineering Mechanics at the University of Michigan (where L. Prandtl, R. V. Southwell and H. M. Westergaard were among the teaching staff) in 1933, 1934 and 1935 as part of his doctorate studies. Westergaard was also Mindlin's unofficial doctorate supervisor, and the dissertation concerned the generalisation of the classic work of J. V. Boussinesq on the elastic half-space [Boussinesq, 1885], [Mindlin, 1936/1]. Based on this, it was possible to find, for example, analytical solutions to soil mechanics problems [Mindlin, 1939]. After spending two years as a research assistant, he was appointed instructor in 1938, and two years later lecturer, for structural engineering. By 1945 Mindlin had been promoted to associate professor and in 1947 he finally became a full professor for structural engineering. His theory of the shear-flexible plate was shared with the engineering community four years later [Mindlin, 1951]; E. Reissner (1945 & 1947), H. Hencky (1947) and L. Bollé (1947) published works on this theory. The theory of the shear-flexible plate marks the transition from the invention (1925–50) to the innovation phase (1950–75) of structural theory. In 1955 Mindlin summarised his plate theory to form his classic monograph [Mindlin, 1955], which has since been republished by J. Yang [Mindlin & Yang, 2007]. Among Mindlin's many honours are the Research Prize (1958) and the Kármán Medal (1961) of the ASCE plus the Timoshenko Medal (1964).

- Main contributions to structural analysis:
*Force at a point in the interior of a semi-infinite solid* [1936/1]; *Note on the Galerkin and Papkovitch stress functions* [1936/2]; *Stress distribution around a tunnel* [1939]; *Influence of rotatory inertia and shear on flexural motions of isotropic, elastic plates* [1951]; *An introduction to the mathematical theory of vibrations of elastic plates* [1955 & 2007]; *Effects of couple-stresses in linear elasticity* [1962]; *On Reissner's equations for sandwich plates* [1980]; *Flexural vibrations of rectangular plates*

MERRIMAN    MICHELL    MINDLIN    MISE    MOHR    MÖRSCH

with free edges [1986]; *The Collected Papers of Raymond D. Mindlin, Vol. I + II* [1989]
• Further reading:
[Herrmann, 1974]; [Deresiewicz et al., 1989]
• Photo courtesy of Prof. Peter C. Y. Lee

### MISE, KOZABURO
*8.3.1886 Ohzu in Ehime, Japan, †4.2.1955 Japan
He studied at the Imperial University in Tokyo from 1908 to 1911 and thereafter was a lecturer at the Imperial Kyushu University in Fukuoka. He started studying at the University of Illinois, USA, in 1912 and was appointed associate professor there (1915–18). He was awarded an honorary doctorate by the rector of the Imperial Kyushu University in Fukuoka and during that same year Mise was appointed to the chair of theory of structures at the university, where he was responsible for bridge-building from 1923 to 1946. He was nominated honorary professor of the university in 1946. Mise turned systematically to matrix analysis in theory of structures as early as the 1920s [Mise, 1922]. Like no other before him, Mise advanced the use of formalised theory in structural analysis. It was only with the technical realisation of the symbolic machine in the form of the computer that matrix formulation became valid on the disciplinary scale of structural theory. The first comprehensive work on matrix theory in theory of structures appeared shortly after he died [Argyris, 1957]. Mise therefore anticipated significant elements of the innovation phase of structural theory.
• Main contributions to structural analysis: *Elastic Distortions of Framed Structures* [1922]; *Elastic Distortions of Rigidly Connected Frames* [1927]; *General Solution of Secondary Stresses* [1929]; *Universal Solution of Framed Structures* [1946]
• Further reading:
[Iseki, 1930, p. 313]; [Reports of the Research Institute for Elasticity Engineering, Kyushu Imperial University, Fukuoka, 1946], see also [Mise, 1946]; [Naruoka, 1961; 1974]; [Oota, 1984]; [Naruoka, 1999]
• Photo courtesy of Prof. Dr. M. Yamada

### MOHR, OTTO
*8.10.1835 Wesselburen, Holstein, †2.10.1918 Dresden, German Empire
During his father's period of office as the local mayor, the young Otto met Friedrich Hebbel, who was later to become famous as an author but at that time was the 14-year-old scribe employed in his father's office. At the age of 16 Mohr went to the Polytechnic School in Hannover. After completing his studies he worked for Hannover State Railways and afterwards Oldenburg State Railways. Around 1860 he is supposed to have developed the method of sections for analysing a statically determinate frame (attributed to August Ritter) in working on a design for the first iron bridge with a simple triangulated frame at Lüneburg. A little later the young Mr. Mohr gained attention among his profession by publishing a paper on the consideration of displacements at the supports during the calculation of internal forces in continuous beams. But his work didn't stop there: he introduced influence lines at the same time as Winkler in 1868 and discovered the analogy since named after him, which gave graphical statics an almighty helping hand. He was appointed professor of structural mechanics, route planning and earthworks at Stuttgart Polytechnic in 1867. Six years later, Mohr accepted a post at Dresden Polytechnic as successor to Claus Köpcke (1831–1911) and taught graphical statics plus railway and hydraulic engineering there until 1893. After the departure of Gustav Zeuner in 1894, he took on the subjects of structural mechanics and strength of materials in conjunction with graphical statics. Mohr gave up teaching in 1900, but continued working on the development of structural mechanics and theory of structures. His work on the fundamentals of theory of structures based on the principle of virtual forces (1874/75) meant that he – alongside the work of Maxwell [Maxwell, 1864/2] – made the greatest contribution to classical theory of structures. Through his work, Mohr, like no other, provided impetus to the classical period of the discipline-formation period and the first half of the consolidation period of structural theory. Mohr argued with Müller-Breslau over the foundations of structural theory and later over priority issues regarding essential definitions, theorems and methods in the theory of structures. Numerous personalities from the world of science and engineering, e. g. Robert Land, Georg Christoph Mehrtens, Willy Gehler, Kurt Beyer and Gustav Bürgermeister, were influenced by the founder of the Dresden school of structural mechanics. Hannover TH awarded him a doctorate. After lengthy deliberations, Mohr accepted the post of Working Privy Counsellor with the title "Excellency", which he had been awarded by the Saxony government.
• Main contributions to structural analysis: *Beitrag zur Theorie der Holz- und Eisenkonstruktionen* [1868]; *Beitrag zur Theorie der Bogenfachwerksträger* [1874/1]; *Beitrag zur Theorie des Fachwerks* [1874/2]; *Beiträge zur Theorie des Fachwerks* [1875]; *Die Berechnung der Fachwerke mit starren Knotenverbindungen* [1892/93]; *Abhandlungen aus dem Gebiete der Technischen Mechanik* [1906, 1914, 1928]
• Further reading:
[Gehler, 1916]; [Grübler, 1918]; [Steiding, 1985]
• Photo courtesy of Hebbel Museum, Wesselburen

### MÖRSCH, EMIL
*30.4.1872 Reutlingen, German Empire, †29.12.1950 Weil im Dorfe near Stuttgart, Federal Republic of Germany
Emil Mörsch studied civil engineering at Stuttgart TH from 1890 to 1894. Upon graduating he worked as a senior civil servant and superintendent in the Ministerial Department for Highways & Waterways, and afterwards was employed in the bridge unit of Württemberg State Railways. He joined the Wayss & Freytag company in Neustadt, Palatinate, in early 1901 and it was here, commissioned by the company, that he published the first edition of his book *Der Betoneisenbau. Seine Anwendung und Theorie*, which later underwent numerous reprints (substantially enlarged) under the title of *Der Eisenbeton. Seine Theorie und Anwendung*. This book set standards in reinforced concrete writing during the consolidation period of structural theory. Its theory based on practical trials made it *the* standard work of reference in reinforced concrete construction for more than half a century. In 1904 Mörsch was appointed professor of theory of structures, bridge-building and reinforced concrete construction at Zurich ETH. However, four years later he returned to the board of Wayss & Freytag AG. From 1916 onwards, Mörsch worked as professor of theory of structures, reinforced concrete construction and masonry arch bridges at Stuttgart TH. He adhered rigorously to elastic theory for designing reinforced concrete components right up until his death. Among his numerous honours

are honorary membership of the Concrete Institute (today: Institution of Structural Engineers) (1913) and honorary doctorates from Stuttgart TH (1912) and Zurich ETH (1929).
- Main contributions to structural analysis:
*Der Betoneisenbau. Seine Anwendung und Theorie* [1902]; *Berechnung von eingespannten Gewölben* [1906]; *Der Eisenbetonbau. Seine Theorie und Anwendung* [1920]
- Further reading:
[Graf, 1951]; [Bay, 1990]
- Photo courtesy of University of Stuttgart archives

## MÜLLER-BRESLAU, HEINRICH

*13.5.1851 Breslau, Prussia, †23.4.1925 Berlin-Grunewald, Germany

Following service in the Franco-Prussian War of 1870–71, the young Müller, who a few years later was to change his name to Müller-Breslau, left the place of his birth to study at the Berlin Building Academy. However, the birth of a son in December 1872, who was also christened Heinrich (1872–1962), forced him to start earning money. He tutored his fellow students at the Building Academy in theory of structures in readiness for the dreaded state examination set by Schwedler, although he himself did not sit the examination. Müller-Breslau, however, turned duty into a virtue by publishing his theory of structures notes as a book in 1875 and setting himself up as an independent civil engineer. In October 1883 he was appointed assistant and lecturer at Hannover TH and in April 1885 professor of civil engineering at the same establishment before succeeding Emil Winkler in the chair of theory of structures, building and bridge design at Berlin TH in October 1888. Taking the theorems of Castigliano and Maxwell's frame theory as his starting point, Müller-Breslau worked out a consistent theory of statically indeterminate frames between 1883 and 1888 which, just a few years later, officially became the force method. Müller-Breslau's completion of classical theory of structures brought to a close the discipline-formation period of structural theory. During the 1880s, the dispute between Mohr and Müller-Breslau over the fundamentals of theory of structures led to the formation of the Dresden school of applied mechanics and the Berlin school of structural theory, which also gained international recognition. Müller-Breslau's appointment as a full member of the Prussian Academy of Sciences in 1901 demonstrates the high status accorded to theory of structures and iron bridge-building – indeed engineering sciences on the whole – by Imperial Germany.
- Main contributions to structural analysis:
*Die neueren Methoden der Festigkeitslehre und der Statik der Baukonstruktionen* [1886, 1893, 1904, 1913]; *Die Graphische Statik der Baukonstruktionen* [1887, 1903, 1892, 1908]
- Further reading:
[Hertwig & Reissner, 1912]; [Bernhard, 1925]; [Hertwig, 1951]; [Hannover TH, 1956]; [Hees, 1992]
- Photo courtesy of [Hertwig, 1951, p. 53]

## MUSKHELISHVILI, NIKOLAI IVANOVICH

*16.2.1891 Tiflis, Russia (today: Georgia), †16.7.1976 Tiflis, USSR (today: Georgia)

After leaving grammar school, Muskhelishvili, the son of an engineering officer, studied at the faculty for mathematics and physics at the University of Petrograd (today: St. Petersburg) from 1909 to 1914. After completing his studies, he switched to theoretical mechanics and worked at the university as a lecturer until 1920. Muskhelishvili subsequently worked at the polytechnic faculty of the State University of Tiflis and from 1922 onwards was a professor there until his faculty became independent in 1928. He was a member of the faculty for mathematics and physics at the State University of Tiflis from 1928 to 1936 and was also involved with setting up the teaching organisation of the polytechnic institute (1928–30). Muskhelishvili, more than any other, influenced the structure of the scientific network of Georgia. He became a corresponding member of the USSR Academy of Sciences in 1933 and from that date onwards had a significant influence on the development of the Georgian branch of that Academy, which became the independent Georgian Academy of Sciences in 1941, where Muskhelishvili served as president (1941–72) and director (1941–76) of the Mathematics Institute. He was awarded a doctor title in 1934, became director of the Department of Theoretical Mechanics at the State University of Tiflis in 1938 and one year later a full member of the USSR Academy of Sciences. Muskhelishvili received the Stalin Prize in 1941 and 1946 for his monographs on the mathematical foundations of elastic theory [Muskhelishvili, 1933] and the integral equation methods of mathematical physics [Muskhelishvili, 1946]; both of these standard works enjoyed worldwide acclaim after they were translated into several languages. Muskhelishvili was president of the USSR National Committee on Theoretical & Applied Mechanics, an important scientific body, from its foundation in 1956 until 1976. He established the Georgian school of mathematics, a scientific school of international repute.
- Main contributions to structural analysis:
*Nekotorye osnovnye zadachi matematicheskoi teorii uprugosti* (some basic problems of the mathematical theory of elasticity) [1933]; *Singulyarnye integral'nye uravneniya* (singular integral equations: boundary problems of function theory and their application to mathematical Physics) [1946]; *Some basic problems of the mathematical theory of elasticity* [1963]; *Singuläre Integralgleichungen. Randwertprobleme der Funktionentheorie und Anwendungen auf die mathematische Physik* [1965]; *Einige Grundaufgaben zur mathematischen Elastizitätstheorie* [1971]; *Singular Integral Equations: Boundary Problems of Function Theory and Their Application to Mathematical Physics* [1992]
- Further reading:
[Vekua, 1991]; [Khvedelidze & Manjavidze, 1993]
- Photo courtesy of Prof. Dr. G. Mikhailov

## MUSSCHENBROEK, PETRUS VAN

*14.3.1692 Leiden, Netherlands, †19.9.1761 Leiden, Netherlands

The second son of the famous instrument-maker Johan Musschenbroek (1660–1707), Petrus van Musschenbroek studied medicine at the University of Leiden. After gaining his doctorate in 1715, he undertook a study tour to London. Back in Leiden, Musschenbroek worked as a doctor, and together with his elder brother Jan enjoyed the friendship and philosophy of 'sGravesande. In 1719 he became doctor of philosophy and professor of mathematics and philosophy at the University of Duisburg, where from 1721 onwards he also worked as associate professor of medicine. He held the chair of natural philosophy and mathematics at the University of Utrecht from 1723 to 1740 and took on the chair of astronomy as well in 1732. Following the death of 'sGravesande in 1742, Musschenbroek became his successor in Leiden and taught experimental natural philosophy in the tradition of Newton and 'sGravesande; in doing so he often used the instruments of his brother Jan. His lectures published in Latin were translated into Dutch, English, French and German. During his time in Utrecht, Musschenbroek published his experiments on strength of materials [Musschenbroek, 1729], to which Boscovich, Le Seur and Jacquier referred in their report on the dome of St. Peter's in 1742. The systematic introduction of experimentation during the application phase of structural theory is indebted to the work of Musschenbroek.
- Main contributions to structural analysis:
*Physicae, experimentales et geometricae* [1729]
- Further reading:
[Struik, 1981]
- Photo courtesy of www.polytechphotos.dk

## NAGHDI, PAUL MANSOUR

*29.3.1924 Teheran, Iran, †9.7.1994 Berkeley, California, USA

Searching for freedom and education, Naghdi emigrated and found his way – via an adventurous route – to the USA in 1943. He was awarded a bachelor degree in mechanical engineering at Cornell University in 1946. Following a short period of service in the US Army, Naghdi studied engineering mechanics at the University

MÜLLER-BRESLAU  MUSKHELISHVILI  MUSSCHENBROEK  NAGHDI  NAVIER

of Michigan, where he was awarded a masters degree in 1948 and a doctorate in 1951. It was during this time that he became a citizen of the USA (1948) and started lecturing in engineering mechanics (1949–51). Naghdi worked as an assistant professor from 1951 to 1954 and thereafter was promoted to full professor. As professor of engineering science, Naghdi switched to the University of California, Berkeley, in 1958, where he was to play a leading role in the establishment of the applied mechanics division of the mechanical engineering department, which he headed from 1964 to 1969. This researcher and teacher, so enthusiastic about democratic ideals and the democratic process, played an active part in the academic life of his faculty at Berkeley. In 1991 he was appointed to the Roscoe & Elizabeth Hughes chair of mechanical engineering and held this newly created professorship at the graduate school in Berkeley until he died of lung cancer in 1994. Naghdi succeeded in formulating a rigorously non-linear shell theory, the so-called Cosserat surface theory [Naghdi, 1972], and the theory of the elastic-plastic material behaviour in the region of large deformations [Green & Naghdi, 1965]; furthermore, he published works on viscoelasticity, continuum thermodynamics, mixture theory and the micromechanical aspects of plasticity. His pioneering research work in the field of plasticity and shell theory earned him the ASME Timoshenko Medal in 1980. The National University of Ireland, the Université Catholique de Louvain and the University of California, Berkeley, awarded Naghdi honorary doctorates in 1987, 1992 and 1994 respectively.

- Main contributions to structural analysis:
*On the theory of thin elastic shells* [1957]; *Foundations of elastic shell theory* [1963]; *A general theory of an elastic-plastic continuum* [1965]; *The theory of plates and shells* [1972]; *A brief history of the Applied Mechanics Division of ASME* [1978]; *Non-linear elasticity and theoretical mechanics: In honour of A. E. Green* [1994]; *Theoretical, Experimental and Numerical Contributions to the Mechanics of Fluids and Solids. A Collection of Papers in Honour of Paul M. Naghdi* [1998]
- Further reading:
[Nordgren, 1996]
- Photo courtesy of [Nordgren, 1996, p. 154]

**NAVIER, CLAUDE-LOUIS-MARIE-HENRI**
*10.2.1785 Dijon, France, †21.8.1836 Paris, France

After losing his father – a lawyer in Dijon – at the age of just 14, the young Navier was cared for by his uncle and his wife, Emil and Marie Gauthey. Emil Gauthey taught part-time at the École des Ponts et Chaussées and in 1791 was appointed general inspector of the Bridges & Highways Corps. Navier's uncle therefore became his role model. He studied at the École Polytechnique and École des Ponts et Chaussées from 1802 to 1806 and afterwards, in addition to practical employment in bridge-building, dedicated himself to preparing a new edition of Gauthey's *Traité des ponts* (1813) and Bélidor's engineering manuals [Bélidor, 1813], [Bélidor, 1819]. In 1819 he was appointed professeur suppléant at the École des Ponts et Chaussées, which resulted in his *Leçons* [Navier, 1820]. During the early 1820s, Navier established the principles of elastic theory together with Cauchy and Lamé. In May 1821 Navier submitted a paper to the Académie des Sciences in which he derived the basic equations of elastic theory (to be named after him and Lamé) from the discontinuum (molecular) hypothesis; an extract from this paper was published in 1823 [Navier, 1823/3], but publication of the complete work had to wait until 1827 [Navier, 1827]. The year 1828 was marked by a dispute between Navier and Denis Poisson (1781–1840) in the journal *Annales de Chimie et de Physique* concerning the principles of elastic theory, which, however, did not supply any clarification because both based their ideas on the molecular hypothesis. Navier was commissioned by the government to travel to England and Scotland in order to find out about the construction of chain suspension bridges; his findings were published in his famous *Rapport*, which contained the first theory of suspension bridges [Navier, 1823/1]. Although this publication earned him membership of the Académie in 1824, its implementation in practice resulted in numerous difficulties for Navier in connection with his failed Pont des Invalides suspension bridge project [Stüssi, 1940, p. 204]. At the same time, the 1820s can be seen as his most creative years. His *Résumé des Leçons* [Navier, 1826] made Navier the founder of theory of structures; this work was to challenge great minds in the establishment phase of structural theory – like Saint-Venant, who obtained a copy of the third edition and improved on it by adding a grandiose historico-critical commentary [Navier, 1864]. In Germany, Moritz Rühlmann in particular is credited with establishing Navier's theory of structures [Navier, 1851/1878]. Navier became Cauchy's successor at the chair of analysis and mechanics at the École Polytechnique, Chevalier de la Légion d'Honneur and section inspector of the Bridges & Highways Corps – all in 1831. In sociological terms, Navier – like Clapeyron and other prominent scientists and engineers – was committed to the ideas of Saint-Simon and his followers. Therefore, Navier nominated Auguste Comte as his assistant at the École Polytechnique and played an active part in events in Raucourt de Charleville's Institut de la Morale Universelle [McKeon, 1981, p. 2]. In this way the classical engineering sciences established in France at that time – in the first place theory of structures and applied mechanics – experienced an implicit sociological significance on which the, as it were, natural positivism of the engineering scientist dedicated to the "scientific paradigms" [Ropohl, 1999, pp. 20–23] could draw sustenance.

- Main contributions to structural analysis:
*Leçons données à l'École Royale des Ponts et Chaussées sur l'Application de la Mécanique* [1820]; *Rapport et Mémoire sur les Ponts suspendus* [1823/1]; *Extrait des recherches sur la flexion des planes élastiques* [1823/2]; *Sur les lois de l'équilibre et du mouvement des corps solides élastiques* [1823/3]; *Résumé des Leçons données à l'École Royale des Ponts et Chaussées sur l'Application de la Mécanique à l'Etablissement des Constructions et des Machines. 1er partie: Leçons sur la résistance des matériaux et sur l'établissement des constructions en terre, en maçonnerie et en charpente* [1826]; *Mémoire sur les lois de l'équilibre et du mouvement des corps solides élastiques* [1827]; *Résumé des Leçons données à l'École des Ponts et Chaussées sur l'Application de la Mécanique à l'Etablissement des Constructions et des Machines. 2. Aufl., Vol. 1: Leçons sur la résistance des matériaux et sur l'établissement des constructions en terre, en maçonnerie et en charpente, revues et corrigées. Vol. 2: Leçons sur le mouvement et la résistance des fluides, la conduite et la distribution des eaux. Vol. 3: Leçons sur l'établissement des machines* [1833/1836]; *Mechanik der Baukunst (Ingenieur-*

*Mechanik) oder Anwendung der Mechanik auf das Gleichgewicht von Bau-Constructionen* [1833/1878]; *Résumé des leçons données à l'École des Ponts et Chaussées sur l'application de la mécanique à l'établissement des constructions et des machines, avec des Notes et des Appendices par M. Barré de Saint-Venant* [1864]
- Further reading:
[n. n., 1837]; [Stüssi, 1940]; [McKeon, 1981]; [Charlton, 1982]; [Hänseroth, 1985]; [Picon, 1995]
- Photo courtesy of École Nationale des Ponts et Chaussées

### NEWMARK, NATHAN MORTIMORE

*22.9.1910 Plainfield, New Jersey, USA, †25.1.1981 Urbana, Illinois, USA
After completing his structural engineering studies at Rutgers University in 1930, Newmark attended the University of Illinois in Urbana, where Hardy Cross, Harold M. Westergaard and Frank E. Richart were among his tutors. He was awarded the academic title master of science in 1932, gained his doctorate two years later and by 1943 had become research professor for structural engineering. Even at an early date, Newmark made a name for himself internationally through his dynamic analyses of the actions on loadbearing structures due to impact, waves, explosion and earthquakes. He remained true to his Alma Mater until his retirement in 1976. Newmark was in charge of the university laboratory for digital computers for 10 years (1947 – 57) and was responsible for the development of one of the first mainframe computers (ILLIAC II). "This activity led to the university's eminent status in developing computer science in engineering" [Hall, 1984, p. 218]. Newmark also played a significant role in the development of the Minute Man and MX missile systems and was dean of the structural engineering faculty of his university from 1956 to 1973. But he was also extremely successful in engineering practice. For example, in the late 1940s and early 1950s he collaborated on the building of the Latino Americana Towers in Mexico City as a consultant for seismic issues. This high-rise structure survived a severe earthquake in 1957 without damage. He summarised his experiences in the monograph *Design of Multi-Story Reinforced Concrete Buildings for Earthquake Motion*, written together with J. A. Blume and L. H. Corning [Blume et al., 1961]. The output of his scientific research and engineering practice can be read in more than 200 publications. Of his many honours, just two will be mentioned here: the National Medal of Science presented to him personally by Lyndon B. Johnson, US President at that time (1968), and the 16th Gold Medal of the British Institution of Structural Engineers (1980).

- Main contributions to structural analysis: *Design of Multi-Story Reinforced Concrete Buildings for Earthquake Motion* [1961]; *Fundamentals of Earthquake Engineering* [1971]
- Further reading:
[Hall, 1984]
- Photo courtesy of the University of Illinois

### NOVOZHILOV, VALENTIN VALENTINOVICH

*18.5.1910 Lublin, Russia (today: Poland), †14.6.1987 Leningrad, USSR
Following his graduation from the Leningrad Polytechnic Institute (at that time known as the Physics-Mechanics Institute) in 1931, Novozhilov worked in various research institutes in Leningrad, and from 1949 onwards in the Central Research Institute for Shipbuilding (later known as the Krylov Institute). He was appointed professor at the University of Leningrad in 1945. Novozhilov carried out research into shell theory and worked on the principles of non-linear elastic theory. He gained his doctorate in 1946, and became a corresponding (1958) and then a full (1966) member of the USSR Academy of Sciences. His awards include Hero of Socialist Labour (1969) and the Lenin Prize (1984).
- Main contributions to structural analysis: *Teoriya tonkikh obolochek* (theory of thin shells) [1947 – 65]; *Osnovy nelineinoi teorii uprugosti* (principles of non-linear elastic theory) [1948 – 53]
- Photo courtesy of Prof. Dr. G. Mikhailov

### NOWACKI, WITOLD

*20.7.1911 Zakrzewo near Gniezno, German Empire (today: Poland), †23.8.1986 Warsaw, Poland
Witold Nowacki studied civil engineering at Danzig (Gdańsk) TH from 1929 to 1934. After completing his military service he worked as an engineer on numerous structures in northern Poland. Following the defence of Warsaw in September 1939, Nowacki was captured by the Germans and was not released until January 1945. As a prisoner-of-war in the Oflag II C Woldenberg concentration camp, he taught his fellow POWs theory of structures and structural engineering; he also prepared a number of lengthier works on the structural theory of grillages and the application of the displacement method to the stability and dynamics of frames and plates while serving as a prisoner. He used these works to gain his doctorate in September 1945 and write his habilitation thesis in December of that same year for the Politechnika Warszawska. Six months after being released he had already been appointed chair of theory of structures and strength of materials at the Politechnika Gdańska. In 1952 he was appointed to the post of chair of structural mechanics at the Politechnika Warszawska and became a full member of the Polish Academy of Sciences, of which he was president from 1978 to 1980. From 1956 onwards he held the chair of theory of elasticity and plasticity at the University of Warsaw. Nowacki wrote numerous fundamental Polish textbooks dealing with modern theory of structures. Besides his scientific contributions to theory of structures and mechanics – of international significance –, Nowacki had an influence on the establishment and development of relevant post-war journals in Poland: *Archiwum Mechaniki stosowanej* (archives of applied mechanics) (first published in Gdańsk in 1949 and continued from 1953 in Warsaw under the title of *Archives of Mechanics*), *Rozprawy Inżynierskie* (engineering proceedings), *Mechanika Teoretyczna i Stosowana* (theoretical and applied mechanics) and *Archiwum Inżynierii Lądowej* (archives of civil engineering). Nowacki's students include internationally renowned personalities such as M. Bieniek (emeritus professor of Columbia University, N.Y.), Z. Cywiński (emeritus professor of Politechnika Gdańska and University of Tokyo), R. Dąbrowski (emeritus professor of Politechnika Gdańska, †2004), S. Kaliski (professor at the WAT Military Academy of Engineering and Minister of Higher Education, †1979), Z. Kączkowski (emeritus professor of Politechnika Warszawska) and many others.
- Main contributions to structural analysis: *The state of stress in a thin plate due to the action of sources of heat* [1956]; *Dynamics of elastic systems* [1963]; *Theorie des Kriechens – lineare Viskoelastizität* [1965]; *Dynamic problems of thermoelasticity* [1966]; *Theory of asymmetric elasticity* [1981]
- Further reading:
[Litwiniszyn, 1981]; [Nowacki, 1985]; [n. n., 1995]
- Photo courtesy of Prof. Dr. Z. Cywiński

### OLSZAK, WACŁAW

*24.10.1902 Karwina near Cieszyn, Austria (today: Czech Republic), †8.12.1980 Udine, Italy
Wacław Olszak was an outstanding student of civil engineering at Vienna TH (1920 – 25) and progressed to postgraduate studies in continuum and fluid mechanics at the Sorbonne in Paris. He collaborated on the planning of numerous structures in south-west Poland and in 1927 visited the countries of the Middle East; afterwards, there followed a period of severe illness. He gained a doctorate at Vienna TH in 1933 and one year later another at the Politechnika Warszawska. Olszak wrote his habilitation thesis at the Kraków Mining Academy in 1937. During World War 2 he worked in a factory and as a truck driver. He refused offers of chairs at Munich, Dresden and Vienna THs, but in 1946 responded to an offer of the chair of strength of materials at the Kraków Mining Academy. From

NEWMARK    NOVOZHILOV    NOWACKI    OLSZAK    OSTENFELD    PAPKOVICH

1952 onwards he held the chair of strength of materials at the Faculty of Industrial Structures at the Politechnika Warszawska, which he later reorganised into the chair of elastic and plastic theory. Olszak's important contributions to plastic, viscoelasticity and prestressed concrete theory earned him numerous national and international honours. Together with Nowacki, he organised the Centre International des Sciences Mécaniques in Udine (Italy).
- Main contributions to structural analysis: *The method of inversion in the theory of plates* [1956]; *Plasticity under nonhomogeneous conditions* [1962]; *Recent trends in the development of the theory of plasticity* [1963]; *Inelastic behaviour in shells* [1967]
- Further reading: [n. n., 1972]
- Photo courtesy of Prof. Dr. Z. Cywiński

## OSTENFELD, ASGER SKOVGAARD
*13.10.1866 Jutland, Denmark,
†17.9.1931 Copenhagen, Denmark
Following completion of a civil engineering degree at Copenhagen TH in 1890, Ostenfeld practised for six years as an engineer while at the same time acting as assistant for highway and waterway construction at Copenhagen TH, where he became a lecturer in 1894 and served as professor of applied mechanics from 1900 to 1904. Ostenfeld's professorship was given the title theory of structures and steel construction in 1905. After being granted emeritus status, he took charge of the Theory of Structures Laboratory at Copenhagen TH from 1926 until his death. Ostenfeld became known through his original publications in the areas of theory of structures as well as steel and reinforced concrete construction. For instance, a long period in Vienna resulted in the publication of several papers on reinforced concrete construction – which was new at the time – in the *Zeitschrift des Österreichischen Ingenieur- und Architekten-Vereins* between 1896 and 1900. He transformed his autographed lecture notes into textbooks, which were characterised by their clarity, high level of scientific knowledge and practicality: *Technische Elastizitätslehre* (1898), *Technische Statik* (1900/1903), *Eisenkonstruktionen im Hoch- und Brückenbau* (1906/1909/1912), *Eisenbetonbrücken* (1917). Ostenfeld therefore advanced to become the founding father of theory of structures in Denmark. Among his scientific works it is his book *Die Deformationsmethode*, published by the Julius Springer publishing house in 1926, that stands out, indeed, can definitely be considered as *the* most pioneering contribution during the consolidation period of structural theory. Ostenfeld served as a juror for international bridge competitions in Scandinavia on several occasions. His honours include a doctorate from the German TH in Prague, membership of the Prague Mazaryk-Akademie and the Stockholm Ingeniørvetenskabsakademie, and honorary membership of the Danish Engineering Association (1930).
- Main contributions to structural analysis: *Berechnung statisch unbestimmter Systeme mittels der „Deformationsmethode"* [1921]; *Die Deformationsmethode* [1926]
- Further reading: [Suenson, 1931]; [Emperger, 1932]
- Photo courtesy of www.polytechphotos.dk

## PAPKOVICH, PETR FEDOROVICH
*5.4.1887 Brest-Litovsk, Russia,
†3.4.1946 Leningrad, USSR
Papkovich completed his school education at the classical grammar school in Samara with a gold medal. He went on to study in the Shipbuilding Department of the St. Petersburg Polytechnic Institute, from where he graduated in 1911. He joined the Navy and remained true to the Navy throughout his life, being awarded the Order of the Holy Stanislav, 3rd class, in 1915, promoted to staff captain of the Corps of Shipbuilding Engineers in 1916, and serving in the great shipbuilding yards of St. Petersburg from 1911 to 1929. Papkovich taught at the Polytechnic Institute from 1916 to 1930 and in 1925 was appointed professor of shipbuilding mechanics. From then until 1939 he worked for the Leningrad Shipbuilding Institute, an offshoot of the Polytechnic Institute. From 1920 onwards he also taught at the Naval Academy and in 1934 was appointed to the chair of shipbuilding mechanics. The end of his career saw him working in the Shipbuilding Research Institute from 1929 to 1939. Papkovich succeeded in specifying general solutions for the homogeneous Lamé-Navier displacement differential equations for the static case as early as 1932. Neuber specified solutions independently in 1934 (three-function arrangement according to Papkovich-Neuber). Papkovich's awards and honours: nominated corresponding member of the Russian Academy of Sciences (1933), doctorate (1935) (without oral assessment), promotion to engineering rear admiral (1940), Order of Lenin (1943 and 1946), Worthy Scientist & Engineer of the Russian Federation (1944), Stalin Prize, 1st class (1946).
- Main contributions to structural analysis: *Solution génerale des équations différentielles fondamentales d'élasticité, exprimée par trois fonctions harmoniques* [1932/1]; *Expressions génerales des composantes des tensions, ne renfermant comme fonctions arbitraires que des fonctions harmonique* [1932/2]; *Teoriya uprugosti* (elastic theory) [1939]; *Stroitel'naya mekhanika korablya* (shipbuilding mechanics) [1941–47]; *Trudy po stroitel'noi mekhanike* (works on structural mechanics) [1962–63]
- Further reading: [Kurdyumov, 1952]; [Slepov, 1984; 1991]
- Photo courtesy of Prof. Dr. G. Mikhailov

## PERRONET, JEAN-RODOLPHE
*8.10.1708 Surennes near Paris, France,
†27.2.1794 Paris, France
On completion of his studies at the École du Génie Militaire, he worked in the Bureau Central des Dessinateurs at the Ponts et Chaussées. In 1747 he became director of the École des Ponts et Chaussées (the oldest school of civil engineering in the world) founded by Daniel Charles Trudaine (1703–69) in that same year. In 1763 he was promoted to Premier Ingénieur du Corps at the Ponts et Chaussées. His masonry arch bridges, e. g. Pont de Neuilly (1768–74) and Pont de la Concorde (1787–91), set standards in stone bridge-building that remained valid until well into the 19th century. Perronet's publications specify empirically based design guidelines for masonry arch bridges, and he recommends using the results of strength of materials experiments.
- Main contributions to structural analysis: *Déscription des projets et de la construction des ponts de Neuilly, de Nantes, d'Orléans, etc.* [1782/1783]
- Further reading: [Dartein, 1906]; [Picon, 1985]
- Photo courtesy of [Dartein, 1906]

## PISARENKO, GEORGY STEPANOVICH

*12.11.1910 Poltava district, Russia (today: Ukraine), †9.1.2001 Kiev, Ukraine

Pisarenko was born into a family of Cossacks. He completed his studies at the faculty of shipbuilding at the Industrial Institute in Gorki in 1936 with a distinction. When he started his postgraduate studies at Kiev Polytechnic Institute in 1939, he was also working simultaneously at the Institute for Structural Mechanics at the Ukrainian Academy of Sciences, to which Pisarenko belonged for 62 years and which, under his leadership (1966–88), gained an international reputation – and today carries his name. Following his doctorate dissertation in 1948, he turned to research into strength problems under extreme conditions, a field in which he would record spectacular achievements and for which he would found a special academic institute in 1966 (Institute of Fastness Issues), serving as president until 1988. Pisarenko was in charge of the Department of Strength of Materials at Kiev Polytechnic Institute from 1952 to 1984. He was elected a corresponding member (1957) and later (1964) a full member of the Ukrainian Academy of Sciences, where he later served as general secretary (1962–66) and vice-president (1970–78). He published more than 800 scientific works, mostly together with his numerous students, and in 1969 founded the journal *Issues of Strength*, which today appears under the name of *Strength of Materials*. So Pisarenko created a scientific school of international renown with his Ukrainian school of mechanics. In recognition of his scientific and organisational accomplishments, he received the State Prize of the Ukraine in 1969 and 1980, and the State Prize of the USSR in 1982.

- Main contributions to structural analysis: *Prochnost' materialov pri vysokikh temperaturakh* (high-temperature strength of materials) [1966]; *High-Temperature Strength of Materials* [1969]; *Prochnost' materialov i elementov konstruktsii v ekstremal'nykh usloviyakh* (strength of materials and structure elements under extremal conditions) [1980]
- Further reading: [Pisarenko, 1994]; [Troshchenko, 2005]
- Photo courtesy of Prof. Dr. V. T. Troshchenko

## POLENI, GIOVANNI

*23.8.1683 Venice, Italy, †14.11.1761 Padua, Italy

Giovanni Poleni worked as professor at the University of Padua from 1708 until his death, initially in the chair of astronomy and meteorology, changing to the chair of physics in 1715, succeeding Nicolaus Bernoulli (1687–1759) in the chair of mathematics in 1719/20 and finally holding the chair of experimental philosophy from 1738 onwards. In his role as technical adviser to the Republic of Venice, he was of course intensively involved in waterways construction and hydraulic engineering. Furthermore, Poleni was called in to advise on the refurbishment of the dome of St. Mark's Church in Venice (1729), and contributed to numerous similar projects in Brescia, Padua, Vicenza and elsewhere in Venice. His report (commissioned by Pope Benedict XIV) on the damage to the dome of St. Peter's in Rome was published in 1748. Here, he drew a thrust line that was the result of a "thought experiment". Together with the report published in 1743 by the *tre mattematici* Thomas Le Seur (1703–70), François Jacquier (1711–88) and Ruggero Giuseppe Boscovich, Poleni's publication forms the last milestone in the preparatory period of structural theory. He was honoured with membership of numerous academies, including those of Bologna, Padua, London (1710), Berlin (1715) and St. Petersburg (1724).

- Main contributions to structural analysis: *Memorie istoriche della gran cupola del tempio Vaticano e de'danni di essa, e de'ristoramenti loro* [1748]
- Further reading: [Passadore, 1963]; [Cavallari-Murat, 1963]; [Benvenuto, 1981]; [Franke, 1983]; [Soppelsa, 1988]; [Corradi, 1989]; [Benvenuto, 1991/2]
- Photo courtesy of Museo di Storia della Fisica, Padova

## PONCELET, JEAN-VICTOR

*1.7.1788 Metz, France, †22.12.1867 Paris, France

Poor and therefore excluded from good schooling, Poncelet was nevertheless far more talented than any of his fellow pupils and so quickly rose to be top of his class. He was therefore able to attend the Lycée impériale in Metz as an external student and enrol at the École Polytechnique in 1807. It was here that his teachers were Ampère, Fourier, Lacroix, Legendre, Poinsot and Poisson. Poncelet quickly moved on to the École d'Application de Metz in 1810, but was ordered to leave for fortification works on Walchem Island in February 1812. After that he took part in Napoleon's field campaign against Russia and was taken prisoner by the Russians in November 1812. During his years as a prisoner, he extended the *Géométrie descriptive* of Monge [Monge, 1794/95] and turned the principles into his famous book on projective geometry [Poncelet, 1822]. Following his return in September 1814, he worked as an engineering officer on various military engineering projects, finally becoming responsible for the fortifications in Metz. In 1824 Poncelet was appointed professor at the École d'application de l'Artillerie et du Génie in Metz and taught – with elegance, simplicity and clarity – the fundamentals of an applied mechanics based on machines to the officers attending the famous course *Mécanique appliquée aux machines* from 1825 to 1834. Furthermore, between 1827 and 1830 he presented the popular evening lectures on applications of geometry and mechanics in industry to workers and industrialists (*Cours de mécanique industrielle*). Lithographed editions of both courses were published, but these were for the most part not edited by Poncelet himself. Like Navier's *Résumé des Leçons* forms the principal work of the constitution phase of structural theory (1825–50), Poncelet's courses made a decisive contribution to the constitution phase of applied mechanics (1825–50). Based on his work on projective geometry [Poncelet, 1822], he also solved problems in masonry arch theory [Poncelet, 1835] and earth pressure theory. For example, Poncelet's *Mémoire sur la stabilité des revêtements et de leurs fondation* (1840), which was translated into German and embellished by J. W. Lahmeyer [Poncelet, 1844], contains the graphical determination of the earth pressure acting on retaining walls. Poncelet became a member of Metz City Council, Secrétair du conseil général du département de la Moselle (1830), a member of the Paris Academy of Sciences (1834) and from 1838 to 1848 was engaged as a professor at the Faculté des sciences in Paris. His military career is impressive, too: he reached the rank of brigadier-general in 1848 and in that same year was appointed commander of the École Polytechnique, and in this capacity he was appointed commander-in-chief of the National Guard of the Seine Departement. Poncelet retired at the end of October 1850. The French government sent Poncelet to serve on the juries at the World Expositions in London (1851) and Paris (1855), and he reported on these in detail in his books. Rühlmann called Poncelet the "Euler of the 19th century" because, like Euler, Poncelet was a "creator of totally new theories, a promoter of abstracts and empirical sciences … He was blessed with being able to take part in the most important period of the emergence and development of industry, building and machine mechanics … Like Euler, though, Poncelet was also an excellent teacher who, with the simplest presentations and with moderate thoroughness knew how to captivate his students and make them enthusiastic for science" [Rühlmann, 1885, pp. 387–389].

- Main contributions to structural analysis: *Traité des propriétés projectives des figures* [1822, 1865/1866]; *Cours de mécanique appliquée aux machines* [1826]; *Solution graphique des principales questions sur la stabilité des voûtes* [1835]; *Introduction à la mécanique industrielle* [1840]; *Über die Stabilität der Erdbekleidungen und deren Fundamente* [1844]; *Examen critique et historique des principales théories ou solutions concernant l'équilibre des voûtes* [1852]
- Further reading: [Taton, 1981]
- Photo courtesy of [Rühlmann, 1885, p. 398]

PERRONET

PISARENKO

POLENI

PONCELET

PRAGER

PRANGE

### PRAGER, WILLIAM
*23.5.1903 Karlsruhe, German Empire,
†16.3.1980 Zurich, Switzerland

Prager studied at Darmstadt TH and gained his doctorate there in 1926. In 1929 he became assistant to Ludwig Prandtl in Göttingen and his deputy at the Institute of Applied Mechanics at the University of Göttingen. He was not quite 30 years old when he was appointed professor of applied mechanics at Karlsruhe TH. During his time in Göttingen he wrote the book *Dynamik der Stabwerke* together with Hohenemser and carried out fundamental work on plastic theory. Germany's youngest professor gained an international reputation. Prager left Germany in protest against the way in which the National Socialists took power and until 1941 served as professor of theoretical mechanics at the University of Istanbul. As Hitler's troops advanced to within 200 km of Istanbul, Prager and his family fled by road via the Middle East and Pakistan to India, from where they boarded a ship and finally reached New York. Prager managed, within a very short time, to establish his world-famous school of applied mathematics and mechanics at Brown University. By 1943 he had founded the Journal *Quarterly of Applied Mathematics*. He was involved in teaching and research work at Brown University until 1973, except for the years 1963–68. Prager knew how to present difficult problems with an amazing degree of simplicity – whether in discussions, lectures, presentations, papers or books. Following his transfer to emeritus status, he and his wife settled in Savognin, Switzerland. Prager is the author of extraordinary contributions in the fields of fluid mechanics, elastic theory, plastic theory, dynamics, numerics, transport technology and structural optimisation. This latter field was researched particularly intensively by Prager in his final period of creativity. In terms of theory of structures, it was his work – in cooperation with the group of researchers around J. F. Baker from the University of Cambridge – on the fundamentals of the ultimate load method that proved crucial because it initiated the paradigm change from elastic to plastic methods of design worldwide. Prager therefore also had a lasting influence on the style of theory in the innovation phase of structural theory. He was awarded countless honours for his scientific work: honorary doctorates from the universities of Brussels, Hannover, Liège, Manchester, Milan, Stuttgart, etc., membership of scientific academies, prizes and medals too many to mention.

- Main contributions to structural analysis: *Beitrag zur Kinematik des Raumfachwerkes* [1926]; *Dynamik der Stabwerke. Eine Schwingungslehre für Bauingenieure* [1933]; *Theory of perfectly plastic solids* [1951]; *On Limit Design of Beams and Frames* [1951/1952]; *Theorie ideal plastischer Körper* [1954]; *Probleme der Plastizitätstheorie* [1955]; *An Introduction to Plasticity* [1959]
- Further reading:
[Rozvany, 1989; 2000]; [Drucker, 1984]
- Photo courtesy of [Drucker, 1984, p. 232]

### PRANGE, GEORG
*1.1.1885 Hannover, German Empire,
†3.2.1941 Hannover, Germany

Georg Prange studied mathematics and physics at the University of Göttingen, but due to severe illness had to interrupt his studies from 1906 to 1910 and was not able to sit his examination for teaching in higher education until 1912. From 1912 to 1921 he was assistant for mathematics at Hannover TH. It was during this time that Prange became interested in the relationships between applied mechanics and mathematics, in particular with the variational principles of elastic theory. He gained his doctorate at the University of Göttingen in 1914, and two years later wrote his habilitation thesis *Das Extremum der Formänderungsarbeit* at Hannover TH. He published extracts from this in 1919. The complete thesis was, however, not published until 1999 (by Klaus Knothe). Prange established the calculus of variations fundamentals of structural theory and recognised their dual nature quite clearly. He therefore anticipated the style of theory closely linked with the method of finite elements during the integration period of structural theory. Prange was appointed professor of higher mathematics at Hannover TH in 1921.

- Main contributions to structural analysis: *Das Extremum der Formänderungsarbeit* [1916]; *Die Theorie des Balkens in der technischen Elastizitätslehre* [1919]
- Further reading:
[Hannover TH, 1956]; [Knothe, 1999]
- Photo courtesy of University of Hannover archives

### PUGSLEY, SIR ALFRED GRENVILLE
*13.5.1903 Wimbledon, UK, †9.3.1998 Bristol, UK

Alfred Pugsley studied engineering at Battersea Polytechnic at a time when such courses still had a high practical content. After graduating, he spent a period working as a civil engineering student in the Royal Arsenal, Woolwich (London). This is where he became familiar with a wide range of engineering research and in 1926 started work in the R&D team of the Royal Airship Works in Cardington near Bedford. It was this that started him out on his future career – Pugsley spent the next two decades in aircraft engineering. The design, analysis and construction of airships presented an exciting challenge for young engineers and therefore Pugsley was proud to work on the great R101 airship. During his design activities in Cardington, he learned that the dynamic loadbearing behaviour of such structures is essentially influenced by high, constant loads. He moved to the Royal Aircraft Establishment (Farnborough) in 1931, where he was primarily responsible for investigating the dynamic behaviour of the wings and ailerons of military aircraft, but was also introduced to new types of lightweight metal alloys, which would later find their way into structural engineering too. After World War 2, Pugsley worked as professor of civil engineering at the University of Bristol until his transfer to emeritus status in 1968. It was here that he continued his research into dynamic structural behaviour and established the safety concept in structural engineering [Pugsley, 1951 & 1966]; the ideas of the latter work found their way into automotive and ship design, engineering structures, aircraft design and the design of suspension bridges [Pugsley, 1957]. During his long period as professor of civil engineering, Pugsley still remained closely linked with aircraft design, e.g. as head of the Aeronautical Research Council (1952–57) and adviser to the Air Registration Board (1955–65). His ground-breaking contributions to modern engineering science during the integration period were acknowledged with numerous awards: election to the Royal Society (1952), a knighthood (1956), Gold Medal of the Institution of Structural Engineers (1968).

- Main contributions to structural analysis: *Concepts of Safety in structural engineering* [1951]; *The Theory of Suspension Bridges* [1957];

*The Safety of Structures* [1966]; *The Engineering Climatology of Structural Accidents* [1969]; *James Forrest Lecture 1978. Statics in Engineering hands* [1979]; *The works of Isambard Kingdom Brunel* [1980]; *The nonlinear behaviour of a suspended cable* [1983]
• Further reading:
[Bulson et al., 1983]

## RABICH, REINHOLD
*12.1.1902 Gotha, German Empire, †7.11.1974 Dresden, German Democratic Republic

After completing his school education in Gotha, Rabich initially worked as a draughtsman and design engineer in the railway works of his home town before moving to Dresden to study civil engineering. He passed his diploma examination with flying colours and was awarded the Francius Badge and the Engels Memorial Medal. Right from the outset of his work on the chapters on slabs, plates and shells in Kurt Beyer's book *Die Statik im Stahlbetonbau* (1933/34), Rabich demonstrated his skills in transforming the theory of plate and shell structures into the practical world of structural analysis, which he introduced independently during the 1950s. During his time as design and structural engineer with Dyckerhoff & Widmann KG in Berlin (1934–45), he designed various types of shell structure and carried out research in this field together with Franz Dischinger and Ulrich Finsterwalder. He was a self-employed checking engineer in Gotha from 1945 to 1949 and afterwards was in charge of the Dresden I Design Office for Industrial Buildings. Under his leadership, shell structures were widely used for industrial buildings in the former GDR. He gained his doctorate in 1953 with his work *Die Membrantheorie der einschalig hyperbolischen Rotationsschale* and five years later was appointed to the chair of reinforced concrete and monolithic bridges at Cottbus University of Building. In 1962 he took over a similar chair at Dresden TU. Rabich knew how to express the assumptions of shell theory with respect to the bending theory of circular cylindrical shells – which are not easy to formulate in mathematical terms – as clear, understandable concepts for everyday engineering purposes.
• Main contributions to structural analysis:
*Die Membrantheorie der einschalig hyperbolischen Rotationsschale* [1953]; *Einführung in die Statik der Schalenträger mit kreisförmigen Querschnittsteilen* [1954]; *Die Statik der Schalenträger* [1955]; *Die Statik der Schalenträger. Die Berechnung der Randstörung am Randträger* [1956]; *Statik der Platten, Scheiben, Schalen* [1964]; *Leitfaden Berechnung von Kreiszylinderschalen mit Randgliedern* [1965]
• Further reading:
[Hoyer, 1967]; [Zerna, 1967]; [Eckold, 2002]
• Photo courtesy of Dresden TU archives

## RABINOVICH, ISAAK MOISEEVICH
*23.1.1886 Mogilev, Russia, †28.4.1977 Moscow, USSR

He completed his education in Mogilev and then in 1904 passed a university entrance examination for Moscow Imperial Technical University, from where he was expelled in 1911 for his revolutionary activities. At the same time, he was permanently barred from studying at any university in the country and banished to the government district of Olonetz. Therefore, Rabinovich was not able to complete his studies until after the October Revolution. He worked in the Scientific-Experimental Institute for Highway Construction from 1918 to 1932 and also taught at various universities in Moscow. From 1932 onwards he held the chair of structural mechanics at the Military Engineering Academy and from 1933 to 1955 also held the same chair at the Moscow Civil Engineering Institute. Dr. sc. techn. Rabinovich was part of the USSR's military apparatus from 1932 to 1966. He was nominated engineering major-general (1943), Worthy Scientist & Engineer of the Russian Federation (1944), and corresponding member of the Russian Academy of Sciences (1946). Following his retirement in 1966, he was awarded the title Hero of Socialist Labour.
• Main contributions to structural analysis:
*Metody rascheta ram* (methods of frame calculation) [1934–37]; *Dostizheniya stroitel'noi mekhaniki sterzhnevykh sistem v SSSR* (achievements in the structural mechanics of frames in the USSR) [1949]; *Structural Mechanics in the USSR 1917–1957* [1960]; *Hängedächer* [1966]; *Stroitel'naya mekhanika v SSSR 1917–1967* (structural mechanics in the USSR 1917–67) [1969]
• Further reading:
[Umansky, 1966]; [n. n., 1976]; [Rabinovich, 1984]
• Photo courtesy of Prof. Dr. G. Mikhailov

## RANKINE, WILLIAM JOHN MACQUORN
*5.7.1820 Edinburgh, UK, †24.12.1872 Glasgow, UK

William Rankine attended Ayr Academy (1828–29) and Glasgow High School (1830). Illness forced him to leave the latter and for six years he was taught by his father, David Rankine, a respected railway engineer in the Edinburgh & Dalkeith Railway and later the Caledonian Railway Company. When he was 14, his father gave him a copy of Newton's *Principia*. The young Rankine absorbed this work written in Latin and thus laid the foundation for his knowledge of higher mathematics, dynamics and physics. He began studying chemistry, natural history, botany and natural philosophy at the University of Edinburgh in 1836, but had to break off after two years for personal reasons – to assist his father in his work for the Edinburgh & Dalkeith Railway. After that, Rankine worked for several years under Sir John MacNeill on the railways and canals of Ireland. He returned to Edinburgh in 1842 and worked for railway companies and consulting engineers and later in the London office of Lewis Gordon, who at that time held the chair of civil engineering and mechanics set up in 1840 at the University of Glasgow. However, it was not until 1855 that Rankine succeeded him in this position. In his inaugural lecture, *The harmony between theory and practice in engineering*, "Rankine distinguished between theoretical science, which is concerned with what we are to think, and practical science, where the question is what we are to do, often in situations where scientific theory and existing data can be insufficient" [Sutherland, 1999, p. 183]. Rankine was active in both fields, adding valuable contributions to thermodynamics, elastic theory and hydrodynamics from 1848 onwards. Maxwell even went so far as to say that Rankine was one of the three founding fathers of thermodynamics. On the other hand, Rankine pushed back the boundaries of engineering science disciplines decisively, primarily through his *Manuals* of applied mechanics [Rankine, 1858] and civil engineering [Rankine, 1862], both of which enjoyed numerous editions and became standard works of reference in the process of creating a scientific basis for engineering and technology. In 1872, after a long battle, the authorities agreed to Rankine's proposal of awarding the academic degree bachelor of science for the study of engineering at British universities – the first step on the way to raising the status of engineering sciences in universities. During the formation period of structural theory, Rankine came to the fore through his contributions to earth pressure theory [Rankine, 1857], masonry arch theory [Rankine, 1865] and graphical statics [Rankine, 1858, 1864 & 1870]. For example, he formulated the equilibrium criterion of kinematic and statically determinate plane frames – Rankine's theorem [Maxwell, 1864/1] – and found by implication the dual relationship between frame geometry and the polygon of forces. Rankine did not explore this duality further and published a supplement to his space frames theorem [Rankine, 1864] without any proof. Rankine had a great affect on the establishment phase of structural theory – both in the UK and continental Europe. And his achievements in the creation of scientifically based ship theory and mechanical engineering remain undisputed. He was therefore in favour of combining the individual engineering sciences. Rankine was elected to the Royal Society in 1853, and in 1857 he became co-founder and first president of the Scottish Institution of Engineers & Shipbuilders. Interestingly, he was only an associate member of the Institution of Civil Engineers. He composed songs, wrote fables and humorous poems such as *The Mathematician in Love*. His biographer, H. B. Suther-

 RABICH  RABINOVICH  RANKINE  REISSNER, E.  REISSNER, H.

land (emeritus professor of every chair that Rankine held!), paid tribute to him with the following words: "Rankine was a wonderful combination of the man of genius and of humour. How much more pleasant and effective is the contribution, scientific or otherwise, when you know behind it lies a man capable of having a twinkle in his eyes. As Clerk Maxwell wrote, Rankine's death at the early age of 52 was 'as great a loss to the diffusion of science as to its advancement'" [Sutherland, 1999, p. 187].
• Main contributions to structural analysis:
*On the stability of loose earth* [1857]; *A Manual of Applied Mechanics* [1858]; *A Manual of Civil Engineering* [1862]; *Principle of the equilibrium of polyhedral frames* [1864]; *Graphical measurement of elliptical and trochtoidal arcs, and the construction of a circular arc nearly equal to a given straight line* [1865]; *Einige graphische Constructionen* [1866/2]; *Diagrams of forces in frameworks* [1870]
• Further reading:
[Parkinson, 1981]; [Charlton, 1982]; [Scholz, 1989]; [Sutherland, 1999]
• Photo courtesy of [Sutherland, 1999, p. 186]

### REISSNER, ERIC
*5.1.1913 Aachen, German Empire,
†1.11.1996 La Jolla (California), USA
The son of Hans and Josephine Reissner grew up in Berlin. After completing his schooling, Eric Reissner began studying applied physics at Berlin TH in 1931, but switched to applied mathematics after two years. Georg Hamel made a deep impression on him, and acquainted him with all the aspects of theoretical and applied mechanics. He also had fond memories of E. Jacobsthal (theory of complex functions), R. Fuchs (differential equations) and R. Becker (theoretical physics). In 1934 Reissner spent one semester at Zurich ETH, where he listened to lectures by E. Meissner, W. Pauli and G. Pólya; later that year he published his first paper on the calculation of T-beams [Reissner, 1934]. He extended his diploma thesis into a dissertation in 1935/36. As the official campaign of the Nazis against German Jews took on ever more threatening proportions, Reissner explored the chance of career opportunities in the USA, and found the MIT receptive to his approach, taking him on as a research assistant in aeronautics (1937–39).

He gained a doctorate in mathematics in 1938 on a theme from aircraft construction, was promoted to assistant professor in 1942, associate professor in 1946 and finally full professor for applied mathematics in 1949. The years 1940 to 1950 saw the publication of Reissner's work on plate theory [Reissner, 1944] and a hybrid variational theorem of elastic theory [Reissner, 1950] (see section 10.4.5.1), which was to have a profound influence on the innovation phase of structural mechanics (1950–75) and for which he received a prize from the American Institute of Aeronautics & Astronautics in 1984. In the 1950s Reissner made a decisive contribution to the solution of structural dynamic problems in the development of the Atlas and Polaris rockets. Reissner remained true to the MIT until 1970 and thereafter worked as a professor for applied mechanics at the University of California in San Diego until his retirement. He published nearly 300 papers in scientific and technical journals. Reissner received many honours and awards over the years, including the ASCE Kármán (1964) and Timoshenko (1973) Medals.
• Main contributions to structural analysis:
*Über die Berechnung von Plattenbalken* [1934]; *Least work of shear lag problems* [1941]; *On the theory of bending of elastic plates* [1944]; *Analysis of shear lag in box beams by the principle of minimum potential energy* [1946]; *Note on the method of complementary energy* [1948]; *On a variational theorem in elasticity* [1950]; *On a variational theorem for finite elastic deformations* [1953]; *Selected Works in Applied Mechanics and Mathematics* [1996]
• Further reading:
[Reissner, 1996]; [Fung et al., 2001]
• Photo courtesy of [Fung et al., 2001, p. 242]

### REISSNER, HANS
*18.1.1874 Berlin, German Empire,
†2.10.1967 Berlin, Federal Republic of Germany
He completed his civil engineering studies at Berlin-Charlottenburg TH in 1898 and afterwards left for a two-year period of research in the USA. He gained his doctorate under Müller-Breslau with a dissertation on structural dynamics (1902) while working as a consulting engineer in Berlin (1900–04) and attending the lectures of Max Planck and H. A. Schwartz at the University of Berlin. It was during this time that he spent another period studying in

the USA, which Reissner wrote about in the series of articles entitled *Nordamerikanische Eisenbauwerkstätten* in *Dinglers Polytechnisches Journal* (1905/06). Reissner eventually started his academic career in 1904 and became assistant to Müller-Breslau, and two years later succeeded Arnold Sommerfeld in the chair of mechanics at Aachen RWTH, where he established the flow laboratory and the first wind tunnel. After switching to the chair of mechanics at Berlin-Charlottenburg TH, his successor in Aachen was Theodore von Kármán. Both founded the engineering science disciplines of aerodynamics and aircraft construction. During his time in Berlin, Reissner carried out research into fundamental issues of physics and mechanics, strength of materials, plastic theory, gas dynamics and aircraft construction, and also continued his work on the theory of structures. He was awarded a doctorate by Aachen RWTH in 1929. Until 1933 Reissner was vice-president of the Society for Applied Mathematics & Mechanics and chairman of the German Aircraft Committee. He and his family left Hitler's Germany in 1938 "as the principles and ideals under which Germany had placed its youth and its best creative years were renounced and ridiculed" [Szabó, 1959, p. 82]. He was a professor in Chicago at the Illinois Institute of Technology from 1938 to 1943, and it was here that he published a work on bridge dynamics in connection with the collapse of the Tacoma Narrows Bridge. In 1954 he became professor of aerodynamics and aircraft construction at the Polytechnic Institute of Brooklyn, and Reissner's period of creativity in the USA concentrated on those two engineering sciences. During the consolidation period of structural theory, his main contributions were in the areas of structural dynamics, shell theory and stability theory.
• Main contributions to structural analysis:
*Zur Dynamik des Fachwerks* [1899]; *Anwendungen der Statik und Dynamik monozyklischer Systeme auf die Elastizitätstheorie* [1902]; *Schwingungsaufgaben aus der Theorie des Fachwerks* [1903]; *Über die Spannungsverteilung in zylindrischen Behälterwänden* [1908]; *Über die Knicksicherheit ebener Bleche* [1909]; *Spannungen in Kugelschalen (Kuppeln)* [1912]; *Energiekriterium der Knicksicherheit* [1925]; *Neuere Probleme der Flugzeugstatik* [1926]; *Theorie der Biegeschwingungen*

*frei aufliegender Rechteckplatten unter dem Einfluß beweglicher, zeitlich periodisch veränderlicher Belastungen* [1932]; *Formänderung und Spannungen einer dünnwandigen, an den Rändern frei aufliegenden beliebig belasteten Zylinderschale. Eine Erweiterung der Navierschen Integrationsmethode* [1933]; *Spannungsverteilung in der Gurtplatte einer Rippendecke* [1934]; *Oscillations of Suspension Bridges* [1943]
• Further reading:
[Szabó, 1959]; [Reissner, 1977]
• Photo courtesy of Aachen RWTH archives

### RITTER, AUGUST
*11.12.1826 Lüneburg, Hannover,
†26.2.1908 Lüneburg, German Empire
August Ritter ended his schooling at an early age in order to become a sailor. After two years training as ship's boy, he continued his education at the Polytechnic School in Hannover (1843 – 46) and the School of Mining in Nienburg. His subsequent studies at the University of Göttingen were completed with the title Dr. phil. His career then reads as follows: mechanical engineer in Rome and Naples from Easter 1854 to summer 1855; lecturer in applied mathematics and mechanical engineering at the Polytechnic School in Hannover from September 1855 to October 1856; after mechanical engineering was dropped, teaching of higher mechanics (formerly: mechanics of architecture) in addition to mechanics (formerly: applied mathematics) from July 1859 onwards; appointment as professor at Hannover Polytechnic in December 1868; professor of mechanics at Aachen RWTH from October 1870 to October 1899. Ritter's book on iron roof and bridge structures (1862), which contains the method (named after him) of determining member forces in statically determinate frameworks (method of static moments, Ritter's method of sections), represents the level of practical theory of structures at the transition from the establishment to the classical phase. The idea behind this method, which Ritter had published one year before in the *Zeitschrift des Arch.- u. Ing.-Vereins zu Hannover* allegedly stems from Otto Mohr. In the late 1870s and throughout the 1880s Ritter developed the first comprehensive theory of the structure and evolution of stars, but scientific recognition for this work was not forthcoming. He was awarded a doctorate by Dresden TH in 1903 "in honour of his fundamental and outstanding work in the fields of applied mechanics and statics of structural engineering."
• Main contributions to structural analysis: *Über die Berechnung eiserner Dach- und Brücken-Constructionen* [1861]; *Elementare Theorie und Berechnung eiserner Dach- und Brücken-Construktionen* [1862]

• Further reading:
[Walther, 1925]; [Hannover TH, 1956]; [Schwarz, 1993]
• Photo courtesy of University of Hannover archives

### RITTER, WILHELM
*14.4.1847 Liestal, Switzerland,
†18.10.1906 Remismühle, Switzerland
Following an outstanding result in his diploma examination at Zurich ETH in 1868, Wilhelm Ritter worked for a year as a civil engineer on railways in Hungary before becoming assistant to Karl Culmann. By 1870 he had already written his habilitation thesis as a private lecturer in engineering sciences at Zurich ETH. His publication on the use of Mohr's continuous beam analogy, which Maurice Koechlin translated into French in 1886, stems from this period. Ritter was professor of engineering sciences at the Polytechnic in Riga from 1873 to 1881. After Culmann's death, the Swiss Education Council decided to split Culmann's subjects and asked Ritter to take over the newly created professorship of graphical statics and bridge-building at Zurich ETH. This is where he wrote his chief work, the four-volume *Anwendungen der graphischen Statik*. This represented a continuation of Culmann's planned second volume, fragmentary drafts of which were discovered after his death. Nevertheless, Ritter's work, the fourth volume of which was published by his son Hugo Ritter in 1906, is still an independent work because the graphical statics are presented to the practising engineer in the form of graphical analysis methods without the need for an understanding of projective geometry. At the transition from the discipline-formation to the consolidation period of theory of structures, Ritter's book formed the most influential and most comprehensive work on graphical analysis in the German language. Like his famous predecessor, Culmann, Ritter submitted numerous reports. The year 1899 saw Ritter publish a pioneering article on the design of reinforced concrete beams. He also took on a number of prestigious but arduous positions, e. g. director of Zurich ETH (1887 – 91). He was president of the Naturforschende Gesellschaft in Zurich from 1893 to 1896 and served outstandingly during the 150th anniversary of that body; it was for this reason that the University of Zurich appointed him doctor honoris causa. From 1902 until his death, Ritter was unfortunately confined to the Remismühle Sanatorium.
• Main contributions to structural analysis: *Statische Berechnung der Versteifungsfachwerke der Hängebrücken* [1883]; *Anwendungen der graphischen Statik* [1888 – 1906]; *Die Bauweise Hennebique* [1899]
• Further reading:
[Meister, 1906]
• Photo courtesy of Zurich ETH library

### RÜHLMANN, CHRISTIAN MORITZ
*15.2.1811 Dresden, Saxony,
†16.1.1896 Hannover, German Empire
The first stages of Rühlmann's education took place in his home town: Kreuzschule (1825 – 26), School of Building and Surgical Academy (1827 – 28), and then the technical training school (today: Dresden TU) from 1829 onwards; Rühlmann departed from the latter as a technician and was employed there as an assistant mathematics teacher in 1835. His special interest in classical languages, philosophy, physics, chemistry and mathematics would later mean that he was predestined, for example, to emphasize the cultural value of applied mechanics in his monograph on the history of that subject [Rühlmann, 1885]; the division into humanistic culture on the one side and natural sciences/engineering culture on the other was alien to him. With the help of the Saxony State Government, Rühlmann undertook study trips to Austria, Prussia, Belgium, France, Styria and Switzerland. He lectured in mathematics, descriptive geometry, machine drawings, machine design, mechanical engineering and mechanical technology at Chemnitz Industrial School (today: Chemnitz TU) from 1836 to 1840. He gained a doctorate from the University of Jena in 1840 and transferred to the Hannover Higher Industrial School (today: Hannover TU), where he worked as professor of mechanics and mechanical engineering until his retirement. *Mechanik*, the first volume of his *Die technische Mechanik und Maschinenlehre*, was published in 1840, the second edition of which followed in 1845 and 1847; the work was eventually completed by the addition of the second volume on hydromechanics in 1853. The third, fully revised and expanded edition of the first volume appeared in 1860 under the title of *Grundzüge der Mechanik im Allgemeinen und der Geostatik im Besondern* [Rühlmann, 1860]. The second, improved and expanded edition of his *Hydromechanik* did not appear until 1879 [Rühlmann, 1879]. Rühlmann published his four-volume work *Allgemeine Maschinenlehre* between 1862 and 1874; this is an encyclopaedia of mechanical engineering which unfolds the historico-technical development of the entirety of mechanical engineering with clarity. Rühlmann's books provided with copious historico-critical notes were finally summarised by him in 1885 in the form of a monograph entitled *Vorträge über Geschichte der Technischen Mechanik* [Rühlmann, 1885]. Rühlmann's contribution to applied mechanics with respect to theory of structures can be summarised in four points: 1. He published the first German textbooks on applied mechanics, adopting, in particular, Poncelet creatively. 2. He used Navier's *Résumé des Leçons* [Navier, 1826] in his lectures after 1845, encouraging his student G. Westphal to translate the work into German under the title

RITTER, A.   RITTER, W.   RÜHLMANN   RÜSCH   RVACHEV

of *Mechanik der Baukunst* [Navier, 1851] and hence ensured a broader adoption of this classic work of structural analysis in the German-speaking countries, a fact that provided considerable impetus for the development of structural theory in the first half of its establishment phase (1850 – 75). 3. As the spiritus rector of the *Zeitschrift des Architekten- und Ingenieur-Vereins zu Hannover* during the 1850s, he ensured that this journal became the most important German engineering journal in this field during the second half of the establishment phase of structural theory (1850 – 75); it was in this journal that, for example, Mohr published his pioneering work on continuous beam theory [Mohr, 1860 & 1868] and trussed framework theory [Mohr, 1874 & 1875]. 4. Finally, Rühlmann published the first monograph on the history of applied mechanics from its beginnings to the end of the establishment phase of applied mechanics (1850 – 75) [Rühlmann, 1885]. He thus founded the historical approach to the teaching of applied mechanics and made a decisive contribution to the self-image of this fundamental engineering science discipline. Without doubt, this erudite and understandable book marks the pinnacle of the scientific life's work of Rühlmann and is as yet unsurpassed. Rühlmann was awarded many honours, his nomination as an "officer of public education in France" being just one important example.
- Main contributions to structural analysis: *Grundzüge der Mechanik im Allgemeinen und der Geostatik im Besondern* [1860]; *Vorträge über Geschichte der Technischen Mechanik* [1885]
- Further reading:
[Hoyer, 1907]; [Heymann & Naumann, 1985]
- Photo courtesy of the university archives of Chemnitz Technical University, No. UAC 502/176

### RÜSCH, HUBERT
*13.12.1903 Dornbirn, Austria-Hungary, †17.10.1979 Munich, Federal Republic of Germany

Hubert Rüsch studied civil engineering at Munich TH from 1922 to 1926, where Ludwig Föppl (applied mechanics) and Heinrich Spangenberg (reinforced concrete construction) awakened his special liking for the control of dynamics and the imaginative shaping of reinforced concrete constructions [Knittel & Kupfer, p. VII]. He joined Dyckerhoff & Widmann and it was here that he developed shell construction, together with Franz Dischinger and Ulrich Finsterwalder, to the point of being ready for prefabrication. He gained his doctorate under Ludwig Föppl in 1930 and thereafter was in charge of the design office of Dyckerhoff & Widmann's Buenos Aires branch until 1933. After that he worked on the further development of shell methods of construction for industrial buildings. He became head of the Industrial Building Department at the headquarters of Dyckerhoff & Widmann in Berlin in 1939. His appointment to the chair of masonry and concrete structures and as manager of the Building Materials Testing Laboratory at Munich TH followed in 1948. Rüsch made substantial contributions to the first prestressed concrete standard in the world and his article in *Beton-Kalender 1954* [Rüsch & Kupfer, 1954], written jointly with Herbert Kupfer, became the first textbook on prestressed concrete design. During the innovation phase of structural theory, Rüsch published several pioneering contributions on reinforced concrete theory. Like Fritz Leonhardt, Hubert Rüsch worked internationally and established schools. He was granted emeritus status at his own request in 1969. He was a founding member of the Comité Européen du Béton (CEB) in 1953, was awarded the 1957 Emil Mörsch Memorial Medal by the German Concrete Association, was given a doctorate by Dresden TH in 1959, won the 1962 Wason Medal of the American Concrete Institute, served as vice-president of the International Association for Bridge & Structural Engineering (IABSE) from 1962 to 1966, in 1965 was visiting professor at Cornell University (USA), became an honorary member of the American Concrete Institute and the Réunion Internationale des Laboratoires d'Essais et des Recherches sur les Matériaux et les Constructions (RILEM) (both in 1968), was president of the CEB from 1969 to 1971, and in 1973 was visiting professor at Texas University (USA).
- Main contributions to structural analysis: *Die Großmarkthalle in Leipzig, ein neues Kuppelbausystem, zusammengesetzt aus Zeiss-Dywidag-Schalengewölben* [1929]; *Theorie der querversteiften Zylinderschalen für schmale, unsymmetrische Kreissegmente* [1931]; *Shedbauten in Schalenbauweise, System Zeiss-Dywidag* [1936]; *Die Hallenbauten der Volkswagenwerke in Schalenbauweise, System Zeiss-Dywidag* [1939]; *Bemessung von Spannbetonbauteilen* [1954]; *Fahrbahnplatten von Straßenbrücken* [1957]; *Berechnungstafeln für schiefwinklige Fahrbahnplatten von Straßenbrücken* [1967]; *Stahlbeton – Spannbeton. Bd. 1: Die Grundlagen des bewehrten Betons unter besonderer Berücksichtigung der neuen DIN 1045. Werkstoffeigenschaften und Bemessungsverfahren* [1972]; *Stahlbeton – Spannbeton. Bd. 2: Die Berücksichtigung der Einflüsse von Kriechen und Schwinden auf das Verhalten der Tragwerke* [1976]
- Further reading:
[Knittel & Kupfer, 1969]; [Kupfer, 1979]
- Photo courtesy of [Kupfer, 1979]

### RVACHEV, VLADIMIR LOGVINOVICH
*21.10.1926 Chigirin, USSR (today: Ukraine), †26.4.2005 Kharkov, Ukraine

Rvachev, the son of a teacher, began studying at the Polytechnic Institute in Kharkov in 1943, but the occupation of his home town by the German armed forces forced him to flee and sign on for military service. Not until after the war was Rvachev able to resume his studies at the faculty of mathematics and physics at the University of Lemberg, from where he graduated in 1952 and where three years later attained his first doctorate with a work on elastic theory. Thereafter, he was in charge of the department of higher mathematics at Berdyansk Pedagogic Institute until 1963. During this time he completed his dissertation on three-dimensional contact problems in elastic theory at the Institute for Problems in Mechanics at the USSR Academy of Sciences and at the age of just 35 was appointed professor. Rvachev was in charge of the Computational Mathematics Department at the Kharkov Institute of Radioelectronics from 1963 to 1967, and afterwards was head of the Department of Applied Mathematics & Computer methods at the Institute for Problems in Mechanical Engineering at the Ukrainian Academy of Sciences until he retired. With his theory of R-functions, Rvachev founded a mathematical theory in which the mathematical logic was linked with classical methods of mathematics and modern cybernetics as early as 1963. He summarised his findings in 1982 in a monograph [Rvachev, 1982], which did not gain international recognition until V. Shapiro's English edition appeared six years later [Shapiro, 1988]. "With R-functions there appears the possibility of creating a constructive mathematical tool which incorporates the capabilities of classical continuous analysis and logic algebra. This allows one to overcome the main obstacle which hinders the use of variational methods when solving boundary problems in domains of complex shape with complex boundary conditions, this obstacle being connected with the construction of so-called coordinate sequences. In contrast to widely used methods of the network type (finite difference, finite and boundary elements), in the R-functions method all the geometric information present in the boundary value problem statement is reduced to analytical form, which allows one to search for a solution in the form of formulae called solution structures containing some indefinite functional components" [Rvachev & Sheiko, 1995, p. 151]. Rvachev therefore gave computational mechanics a new form of mathematical foundation with considerable heuristic potential. Together with his students, he published more than 500 scientific papers and 17 monographs. He became a corresponding (1972) and later a full (1978) member of the Ukrainian Academy of Sciences. He was awarded numerous honours in recognition of his pioneering findings.

- Main contributions to structural analysis: *Teoriya R-funktsii i nekotorye ee prilozheniya* (theory of R-functions and some applications) [1982]; *Metod R-funktsii v teorii uprugosti i plastichnosti* (R-functions method in elasticity and plasticity theory) [1990]; *R-functions in boundary value problems in mechanics* [1995]
- Further reading: [Shidlovsky, 1988]
- Photo courtesy of Prof. Dr. G. Mikhailov

## SAAVEDRA, EDUARDO

*27.2.1829 Tarragona, Spain, †12.2.1912 Madrid, Spain

Eduardo Saavedra studied civil engineering at the Escuela de Ingenieros de Caminos in Madrid from 1846 to 1851, and was also a professor at this establishment in the periods 1854 – 62 and 1867 – 70. During those years Saavedra wrote several books, contributed articles to journals and translated important civil engineering books – including those of Fairbairn (1857 & 1859) and Michon (1860) – into Spanish; he added a comprehensive commentary to the latter work. His work set standards and raised the status of Madrid's Escuela de Ingenieros de Caminos considerably. Saavedra excelled in his work as civil engineer, architect, historian and archaeologist. In the Spain of the second half of the 19th century, he represented the engineer influenced by the spirit of humanism, a not uncommon manifestation. This is why the royal academies of history (from 1861), sciences (from 1869) and language (from 1874) could count him among their members. The burden of official posts did not hinder him and he continued to provide important contributions in the aforementioned fields, especially engineering and theory of structures. One of the main things we have to thank him for is the way he spread the knowledge of the engineering sciences through his books, translations and papers; the latter appeared mainly in the journals *Revista de Obras Públicas* and *Anales de la construcción y de la Industria*. In the field of theory of structures, his structural analysis of masonry arches based on elastic theory is particularly noteworthy. Saavedra was the first to explore this new field. He also ensured the dissemination of the ideas of Yvon Villarceau on the design and calculation of arched buttressing.

- Main contributions to structural analysis: *Teoría de los puentes colgados* [1856]; *Lecciones sobre la resistencia de los materiales* [1859]; *Nota sobre el coeficiente de estabilidad* [1859]; *Nota sobre la determinación del problema del equilibrio de las bóvedas* (equilibrium of the arch) [1860]; *Experimento sobre los arcos de máxima estabilidad* [1866]; *Teoría de los contrafuertes* [1868]
- Further reading: [Mañas Martínez, 1983]; [Sáenz Ridruejo, 1990]
- Photo courtesy of Prof. Dr. S. Huerta

## SAINT-VENANT, ADHÉMAR JEAN CLAUDE BARRÉ DE

*23.8.1797 Villiers-en-Bière, Seine et-Marne, France, †6.1.1886 St.-Ouen, Loir-et-Cher, France

Napoleon lost the battle of Leipzig in 1813 and Paris faced its downfall. It was in this year that Saint-Venant started studying at the École Polytechnique and all the students were mobilised to help defend Paris. The 17-year-old Saint-Venant refused to take part, saying: "My conscience forbids me to fight for a usurper" [Benvenuto, 1997, p. 4]. The young conscientious objector was forced to quit the École Polytechnique and work as an assistant in the Service des Poudres et Salpêtres (gunpowder factories). It was not until 1823 that the government permitted him to resume his studies at the École des Ponts et Chaussées, which he completed in 1825. He worked for the Service des Ponts et Chaussées until 1848 and later as Professeur du génie rurale at the Agricultural Institute in Versailles, where he was involved with typical civil engineering duties. Showing a high awareness of social responsibility, Saint-Venant committed himself to improving the miserable living conditions in the countryside through the targeted application and further development of hydraulic engineering for agriculture (land improvement, irrigation and rational use of ponds). In 1842 the authority discharged him from his duties and until 1848 he had to make himself available to the authority but on a reduced salary and without any responsibilities. It was during this period that Saint-Venant made his main contributions to the further evolution of structural mechanics [Saint-Venant, 1844], in particular torsion theory [Saint-Venant, 1847]. As the social issue was dramatically revealed in the Paris Revolution of 1848 and the ruling powers struck back with the military, Saint-Venant took sides: a military solution would not improve the social injustices, which mainly affected the unemployed. More charity would be much better for improving the working and living conditions of the lower classes. A short time later a nephew of Napoleon would use the lower classes to set himself up in power and then declare himself Emperor Napoleon III. On the whole, Saint-Venant's remarks were based on the ideology of religiously motivated socialism, which fought to improve the living conditions of rural populations who had been made "superfluous" by the Industrial Revolution [Benvenuto, 1997, pp. 6 – 7]. In 1852 Saint-Venant was promoted to Ingénieur en chef and in 1868 was elected Poncelet's successor in the Mechanics Section of the Académie des Sciences. During the 1850s and 1860s, Saint-Venant formulated the semi-inverse method [Saint-Venant, 1855] within the scope of his torsion theory and expanded the practical bending theory of Navier [Saint-Venant, 1856]. He published Navier's *Résumé des leçons* in a third edition with a comprehensive historico-critical commentary [Navier, 1864]. In his presentation to the Paris *Société Philomathique* on 28 July 1860, Saint-Venant formulated the compatibility conditions of elastic theory for the first time [Saint-Venant, 1860] and hence completed their set of equations: equilibrium conditions, material equations, kinematic relationships and compatibility conditions. Three years before his death, he published *Theorie der Elasticität fester Körper* [Saint-Venant, 1862] together with Flamant, Clebsch, which contained an extensive appendix [Clebsch, 1883]. Also pioneering are Saint-Venant's contributions to the theory of viscous fluids, structural dynamics, plastic theory and vector calculus. His work on structural mechanics did not become evident until the structural theory of the consolidation period.

- Main contributions to structural analysis: *Mémoire sur les pressions qui se développent à l'intérieur des corps solides lorsque les déplacements de leurs points, sans altérer l'élasticité, ne peuvent cependant pas être considérés comme très petits* [1844/1]; *Mémoire sur l'équilibre des corps solides, dans les limites de leur élasticité, et sur les conditions de leur résistance, quand les déplacements éprouvés par leurs points ne sont pas très-petits* [1844/2]; *Note sur l'état d'équilibre d'une verge élastique à double courbure lorsque les déplacements éprouvés par ses points, par suite l'action des forces qui la sollicitent, ne sont pas très-petits* [1844/3]; *Deuxième note: Sur l'état d'équilibre d'une verge élastique à double courbure lorsque les déplacements éprouvés*

 SAAVEDRA  SAINT-VENANT  SAVIN  SCHWEDLER  SERENSEN

par ses points, par suite l'action des forces qui la sollicitent, ne sont pas très-petits [1844/4]; *Mémoire sur l'équilibre des corps solides, dans les limites de leur élasticité, et sur les conditions de leur résistance, quand les déplacements éprouvés par leurs points ne sont pas très-petits* [1847/1]; *Mémoire sur la torsion des prismes et sur la forme affectée par leurs sections transversales primitivement planes* [1847/2]; *Suite au Mémoire sur la torsion des prismes* [1847/3]; *Mémoire sur la Torsion des Prismes, avec des considérations sur leur flexion, ainsi que sur l'équilibre intérieur des solides élastiques en général, et des formules pratiques pour le calcul de leur résistance à divers efforts s'exerçant simultanément* [1855]; *Mémoire sur la flexion des prismes, sur les glissements transversaux et longitudinaux qui l'accompagnent lorsqu'elle ne s'opère pas uniformément ou en arc de cercle, et sur la forme courbe affectée alors par leurs sections transversales primitivement planes* [1856]; *Sur les conditions pour que six fonctions des coordonées x, y, z des points d'un corps élastiques représentent des composantes de pression s'exercant sur trois plans rectangulaires à l'intérieur de ce corps, par suite de petits changements de distance de ses parties* [1860]; *Résumé des leçons données à l'École des Ponts et Chaussées sur l'application de la mécanique à l'établissement des constructions et des machines, avec des Notes et des Appendices par M. Barré de Saint-Venant* [1864]; *Théorie de l'élasticité des corps solides de Clebsch. Traduite par M. M. Barré de Saint-Venant et Flamant, avec des Notes étendues de M. de Saint-Venant* [1883]
• Further reading:
[Boussinesq & Flamant, 1886]; [Pearson, 1889]; [Itard, 1981]; [Benvenuto, 1997]
• Photo courtesy of École Nationale des Ponts et Chaussées

### SAVIN, GURY NIKOLAEVICH
*1.2.1907 Vesiegonsk, Russia,
†28.10.1975 Kiev, USSR (today: Ukraine)
It was in 1928 that Savin, the son of a teacher, started studying at the faculty of mathematics and physics at the University of Dnepropetrovsk. After completing his studies, Savin worked at the chair for structural mechanics at the Dnepropetrovsk Civil Engineering Institute until 1941 – in the end as professor. He was head of the Institute for Rock Mechanics at the Academy of Sciences of the Ukrainian Soviet Republic from 1942 to 1945, becoming a corresponding (1945) and then a full (1948) member of that Academy. It was during this period that Savin also headed the Lemberg branch of the Academy of Sciences of the Ukrainian Soviet Republic. Afterwards, he was dean of the Ivan Franko University in Lemberg until 1952. During his period of creativity in Lemberg, Savin managed to gather around him a group of researchers from the field of elastic theory. He was vice-president and director of the Institute for Mathematics at the Academy of Sciences of the Ukrainian Soviet Republic from 1952 to 1957, and after 1959 he headed the Institute of Mechanics at the same Academy, where he and his students made substantial progress in the development of mathematical elastic theory and rheology. Together with physicists and chemists, he researched the material behaviour of polymers in the mid-1960s. Following the example of his teacher, Dinnik, Savin built up a school of applied mechanics with an international reputation in Kiev. He supervised numerous doctor and habilitation theses, published about 300 journal articles, 13 monographs and 9 textbooks. Savin founded the mechanics journal *Prikladna Mekhanika* (applied mechanics), which since 1992 has been published under the title of *International Applied Mechanics*. He also served as co-editor of renowned scientific journals and was a member of the presidium of the USSR Academy of Sciences. Savin was awarded numerous honours and prizes for his outstanding achievements in the fields of theoretical and applied mechanics.
• Main contributions to structural analysis: *Kontsentratsiia napriazhenii okolo otverstii* (stress concentration around holes) [1951]; *Spannungserhöhung am Rande von Löchern* [1956]; *Stress concentration around holes* [1961]; *Ocherki razvitiya nekotorykh fundamental'nykh problem mekhaniki* (essays in the advancement of some fundamental problems of mechanics) [1964]; *Plastinki I obolochki s rebrami zhestkosti* (plates and shells with stiffening ribs) [1964]; *Rib-reinforced plates and shells* [1967]; *Raspredelenie napryazhenii okolo otverstii* (stress distribution around holes) [1968]; *Stress distribution around holes* [1970/1]; *Savin, G. N., 1970/2. Mekhanicheskoe podobie konstruktsii iz armirovannogo materials* (mechanical similarity of structures made of reinforced materials) [1970/2]; *Mekhanika deformiruemykh tel: Izbrannye trudy* (mechanics of deformable bodies: selected works) [1979]
• Further reading:
[n. n., 1976]; [Khoroshun & Rushchitsky, 2007]; [Guz & Rushchitsky, 2007]; [Panasyuk, 2007]
• Photo courtesy of [Guz, 2007, p. 1]

### SCHWEDLER, JOHANN WILHELM
*28.6.1823 Berlin, Prussia, †9.6.1894 Berlin, German Empire
Schwedler attended the Friedrich Werder Training College in Berlin and the Berlin Building Academy. He passed a building inspector examination in 1852 and thereafter was employed in various post in the Prussian State Building Authority, finally becoming Privy Senior Building Counsellor (1873) and as such the final authority on all larger engineering structures in Prussia. Schwedler designed important engineering structures and developed new types of loadbearing assemblies – Schwedler's cupola and Schwedler's truss – which had an influence on German iron structures in the last third of the 19th century. As an active member of the Berlin Architecture Society, a member of the editorial staff of the *Zeitschrift für Bauwesen* and the teaching staff of the Building Academy (1864 – 73), he supplied notable contributions to the theory of structures. His greatest achievement in the realm of structural theory is, however, his truss theory, published in 1851, which together with that of Culmann became the "signature" of the establishment phase of structural theory.
• Main contributions to structural analysis: *Theorie der Brückenbalkensysteme* [1851]; *Die Construction der Kuppeldächer* [1866]
• Further reading:
[Deutsche Bauzeitung, 1894]; [Hertwig, 1930/2]; [Ricken, 1994/1]; [Lorenz, 1997]
• Photo courtesy of [Hertwig, 1930/2]

### SERENSEN, SERGEI VLADIMIROVICH
*29.8.1905 Khabarovsk, Russia,
†2.5.1977 Moscow, USSR
After completing his studies in mechanical engineering at Kiev Industrial Institute (today: Polytechnic Institute) in 1926, Serensen started training for a university lectureship at the chair of agricultural mechanics of the people's cooperative for education of the Ukrainian Soviet Republic and completed this in 1929

with a dissertation on anisotropic beam grids. Between 1928 and 1934, Serensen worked as a scientific assistant at the Institute for Structural Mechanics at the Academy of Sciences of the Ukrainian Soviet Republic, finally becoming deputy director (1934–36) and then director (1936–40) of the Institute. He was elected a corresponding member of the Academy of Sciences of the Ukrainian Soviet Republic as early as 1934 and a full member in 1939. After 1940, Serensen spent two years as head of the fatigue strength of structural elements department in the Institute for Structural Mechanics at the Academy of Sciences of the Ukrainian Soviet Republic und the strength department at the engine works in Ufa. After that, he worked with great success in the field of the fatigue strength and thermodynamics of engines at the P. I. Baranov Central Science Research Institute for Aircraft Engines until 1967. His final post was in the laboratory for fatigue strength and thermomechanics in the Institute of Machine Control at the Academy of Sciences of the Ukrainian Soviet Republic. Besides his research activities, Serensen taught at the chair for strength of materials at Kiev Polytechnic Institute and from 1943 onwards was head of the chair of strength of materials at the Moscow Aviation Technology Institute. Serensen was awarded the USSR State Prize in 1949 for his research into the fatigue strength and thermomechanics of engines, and Prague TH awarded him an honorary doctorate in 1965.
- Main contributions to structural analysis: *Izbrannye trudy* (selected works) [1985]
- Further reading: [n. n., 1977]; [Pisarenko, 1993]; [Malinin, 2000, p. 210]
- Photo courtesy of Prof. Dr. G. Mikhailov

## ŠOLÍN, JOSEF

*4.3.1841 Trhová Kamenice, Mähren, Austrian Empire (today: Czech Republic), †19.9.1912 Prague, Austria-Hungary (today: Czech Republic)
Following his secondary education, he studied at the Prague Polytechnic Institute (1860–64) and at the Faculty of Philosophy at Karl University in Prague. He worked as an assistant for descriptive geometry at the Prague Polytechnic Institute from 1864 to 1868. Afterwards, he was a teacher of mathematics and descriptive geometry at various secondary schools in Prague until 1870. From 1870 to 1873 he was lecturer in structural mechanics, then associate professor and, finally (1876), full professor of theory of structures and later also strength of materials at the Czech University in Prague. Šolín founded the Czech school of structural mechanics. He served as rector of his Alma Mater three times. He was a member of the Royal Scientific Society and the first president of the Czech Engineering Association, which published technical books. He was appointed Royal & Imperial Privy Counsellor in 1904.
- Main contributions to structural analysis: *Zur Theorie des continuirlichen Trägers veränderlichen Querschnittes* [1885]; *Bemerkungen zur Theorie des Erddruckes* [1887]; *Über das allgemeine Momentenproblem des einfachen Balkens bei indirekter Belastung* [1889]; *Stavebná mechanika* (structural mechanics) [1889/1893]; *Nauka o pružnosti a pevnosti* (strength of materials) [1893/1902/1904]
- Further reading: [Hořejší & Pirner, 1997, pp. 210–213]
- Photo courtesy of Prof. Dr. L. Frýba

## SOUTHWELL, RICHARD VYNNE

*2.7.1888 Norwich, UK, †9.12.1970 near Nottingham, UK
Richard Southwell studied mechanical engineering (up to 1910) and applied mathematics at the University of Cambridge. However, the outbreak of World War 1 initially prevented him from finishing his studies. During the war he worked on the development of non-rigid airships and switched to the newly created Royal Aircraft Establishment (RAE), where he served in the aerodynamics department until 1919. Afterwards, Southwell completed his studies in Cambridge and gained a post as head of the Flying Department of the National Physical Laboratory, where he worked until 1925, mainly on the development of rigid airships. He was elected to the Royal Society in 1925, and from that year until 1928 was a lecturer in Cambridge and thereafter professor of engineering sciences at the University of Oxford (until 1942). He was rector of the Imperial College of Science & Technology in London from 1942 to 1948, but afterwards returned to Cambridge. Southwell's relaxation method established not only the structural theory iteration method of the invention phase in mathematical terms, but also advanced the practical application of the structural analysis of systems with a high degree of static indeterminacy. He therefore had a lasting influence on modern structural mechanics, which was becoming established in the innovation phase of structural theory. Today, the relaxation method is used with great success in the analysis of cable and membrane structures (e. g. form-finding). Here is a selection of his honours: honorary doctorates from the universities of St. Andrew (1939), Brussels (1949), Bristol (1949), Belfast (1952), Glasgow (1957) and Sheffield (1958); numerous medals, including the Timoshenko Medal of the American Society of Mechanical Engineers (1959).
- Main contributions to structural analysis: *The General Theory of Elastic Stability* [1914]; *Stress determination in braced frameworks I* [1921]; *Stress determination in braced frameworks II, III, IV* [1922]; *Stress calculation for the hulls of rigid airships* [1926]; *On the calculation of stresses in braced frameworks* [1933]; *Introduction to the theory of elasticity for engineers and physicists* [1934]; *Stress calculation in frameworks by the method of 'systematic relaxation of constraints'. Part I & II* [1935]; *Relaxation Methods in Engineering Science* [1940]
- Further reading: [Christopherson, 1972]; [n. n., 1989]
- Photo courtesy of Dr. W. Addis

## STEINMAN, DAVID BARNARD

*11.6.1886 Khomsk, Brest, Russia (today: Belarus), †21.8.1960 New York, USA
David Barnard Steinman came from a large family of Russian migrants and grew up in the shadow of the Brooklyn Bridge in New York. He studied at the City College of New York from 1902 to 1906 and passed his bachelor of science degree with distinction. After that, Steinman continued his structural engineering studies at New York's Columbia University and was awarded an engineering doctorate there in 1911. He had already begun working as a structural engineer during his period of study at Columbia University. Steinman worked as a consulting engineer from 1910 to 1914 and was at the same time professor of structural engineering at the University of Idaho. It was during this period that he published his book *Suspension Bridges and Cantilevers. Their Economic Proportions and Limiting Spans* [Steinman, 1911] and translated Joseph Melan's book *Theorie der eisernen Bogenbrücken und der Hängebrücken* [Melan, 1888/2] into English, which appeared with the title *Theory of Arches and Suspension Bridges* [Melan, 1913]. He began his work on the Sciotoville Bridge in 1914 under Gustav Lindenthal (1850–1935). Othmar Ammann (1879–1965) was another Lindenthal employee at this time, working on the Hell Gate Bridge (opened in 1917). Steinman was to remain Ammann's greatest rival throughout his life: "Bringing Lindenthal, Steinman and Ammann together meant that the USA's three greatest bridge-builders were all working in one office" [Dicleli, 2006/1, p. 73]. Steinman's practical activities were accompanied by scientific studies in the tried-and-tested way. For example, he published his translation of Joseph Melan's *Theorie des Gewölbes und des Eisenbetongewölbes im besonderen* [Melan, 1908] in 1915 under the title of *Plain and reinforced concrete arches* [Melan, 1915] for the first volume of the *Handbuch für Eisenbetonbau* edited by Emperger. It is here that Melan provides an overview of the calculation of plain and reinforced concrete arches with special emphasis on the method of construction he had invented in 1892. The Melan method makes use of steel arch ribs, mostly in the form of a trussed arch, that are

ŠOLÍN   SOUTHWELL   STEINMAN   STEVIN   STRAUB   STÜSSI

connected with concrete to form a concrete arch with rigid reinforcement (Melan system). According to an estimate made by Heinrich Spangenberg (1879–1936), more than 5000 Melan bridges were constructed in the USA up until 1924 [Spangenberg, 1924]; even today, they still play an important role in the building of large bridges, especially in Japan, [Eggemann & Kurrer, 2006]. At the start of the 1920s, Steinman founding an engineering firm together with Holton D. Robinson which was to be active in the international market for large bridges over the next 20 years. Steinman expanded his translation of Melan's work on arch and suspension bridges [Melan, 1913] in 1922, publishing it under his name as a monograph entitled *Suspension Bridges, their Design, Construction and Erection* [Steinman, 1922]; the second edition appeared in 1929 with the title *A Practical Treatise on Suspension Bridges* and became the "bible" of suspension bridge construction in the first half of the 20th century. Steinman was to supervise the design and construction of more than 400 bridges on five continents! Together with the Swiss bridge-builder Othmar Ammann (who became a US citizen in 1924), Steinman represents the emancipation of the large American bridge from its European influences. And more besides: starting in the 1920s, both set standards for the building of bridges with very long spans. Steinman crowned his life's work with the design and construction of the Mackinac Strait Bridge (1158 m main span, completed in 1957). "But he was never able to achieve his dream of building a suspension bridge with the world's longest span" [Páll, 1960, p. 391].

- Main contributions to structural analysis: *Suspension Bridges and Cantilevers. Their Economic Proportions and Limiting Spans* [1911]; *Suspension Bridges, their Design, Construction and Erection* [1922]; *A Practical Treatise on Suspension Bridges* [1929]; *Deflection theory for continuous suspension bridges* [1934]; *The builders of the bridge: The story of John Roebling and his son* [1945]; *Der Entwurf einer Brücke von Italien nach Sizilien mit der größten Spannweite der Welt* [1951]; *Famous bridges in the world* [1953]
- Further reading:
[Páll, 1960]; [Petroski, 1996]
- Photo courtesy of [Páll, 1960]

**STEVIN, SIMON**
*1548 Bruges, Netherlands (today: Belgium),
†March 1620 The Hague, Netherlands
The illegitimate son of prosperous parents, Stevin's career began as a merchant in the Financial Administration of Antwerp and Bruges. He undertook trips to Poland, Russia and Norway in the years 1571–77 and did not start studying until 1583 (at the University of Leiden). He was then appointed professor of mathematics in The Hague and finance officer to Prince Moritz of Orange, the Governor of the Netherlands. In 1604 Stevin was promoted to general quartermaster of the Dutch army, became inspector of dyke-building and senior waterways engineer. It was always his goal to apply science in practice: mathematics, mechanics, astronomy, navigation, geodesy, military science, engineering, accounting, building theory. His work, *De Beghinselen der Weeghconst* (1586), stands at the start of the orientation phase of structural theory and was only surpassed by Galileo's *Discorsi* (1638). *De Beghinselen der Weeghconst* includes his books *De Weeghdaet* and *De Beghinselen des Waterwichts*, the first book that went beyond the hydrostatics of Archimedes. As a co-founder and organiser of Renaissance culture in the Netherlands, Stevin contributed substantially to the rise of his country.

- Main contributions to structural analysis: *De Beghinselen der Weeghconst* [1586]
- Further reading:
[Dijksterhuis, 1961; 1970;] [Minnaert, 1981]; [Grabow, 1985]; [Berkel, 1985]
- Photo courtesy of [Szabó, 1996, p. 146]

**STRAUB, HANS**
*30.11.1895 Berg (Thurgau), Switzerland,
†24.12.1962 Winterthur, Switzerland
Hans Straub was the son of a priest and studied civil engineering and later architecture at Zurich ETH from 1914 to 1919. He toured Italy for the first time in 1920, an undertaking that would determine his professional and spiritual development. After a short time as a university assistant, he worked as a civil engineer for a building contractor in Rome. He very soon took responsibility for numerous engineering structures, especially in the field of hydraulic engineering. In addition, he became interested in the history of civil engineering. While working in Rome, he discovered the report on the structural analysis of the dome of St. Peter's compiled by the *tre mattematici* Boscovich, Jacquier and Le Seur in 1743, the purpose of which had been to establish the causes of damage and propose remedial measures. Since that discovery, the year 1743 has marked the birth of modern structural engineering. Straub's most important work is his book *Die Geschichte der Bauingenieurkunst* (1949), the fourth, revised and expanded edition of which was published in 1992 by Peter Zimmermann; it contains numerous passages relevant to the history of structural theory.

- Main contributions to structural analysis: *Die Geschichte der Bauingenieurkunst* [1949]; *Zur Geschichte des Bauingenieurwesens* [1960/1]; *Drei bisher unveröffentlichte Karikaturen zur Frühgeschichte der Baustatik* [1960/2]
- Further reading:
[Halász, 1963]
- Photo courtesy of Prof. Dr. E. Straub

**STÜSSI, FRITZ**
*3.1.1901 Wädenswil, Switzerland,
†15.3.1981 Wädenswil, Switzerland
After attending a humanist grammar school in Zurich, Fritz Stüssi studied civil engineering at Zurich ETH, from where he graduated with distinction in 1923. He then worked as an assistant at Zurich ETH and took charge of the technical office of the Döttingen steel fabrication shop of Conrad Zschokke AG. He gained his doctorate in 1930 and travelled to the USA, where he worked for the bridge-builder O. H. Ammann on the Kill van Kull arch bridge. After returning to Switzerland in 1934, he was appointed senior engineer in the Swiss Railways Company and two years later founded an engineering practice. He wrote his habilitation thesis in 1935 and was appointed professor of theory of structures, structural and bridge engineering in steel and timber at Zurich ETH in 1937, to which Stüssi remained loyal until his transfer to emeritus status. It was during this period that he rose to become the top authority on steel construction in Switzerland, which gained him an excellent international reputation; his contributions to the theory of bridges, but also the extraordinarily clear structure of his lectures on theory of structures helped in this respect. In addition, Stüssi worked very dedicatedly in the management of the International Association for

Bridge & Structural Engineering (IABSE). Nevertheless, his theory of structures never lost touch with the straightforwardness of classical methods, although he also suggested methods in which graphical statics were combined with numerics in a highly original way. Despite his vigorous campaign against the ultimate load method in the 1950s and early 1960s, the fight was already lost. Stüssi was one of the few members of his profession who was familiar with the history of theory of structures and also contributed to the writings on this subject. His great work in bridge-building was rewarded with numerous honours, including eight honorary doctorates!

- Main contributions to structural analysis: *Beitrag zum Traglastverfahren* [1935]; *Vorlesungen über Baustatik* [1946/1954]; *Ausgewählte Kapitel aus der Theorie des Brückenbaues* [1955]; *Gegen das Traglastverfahren* [1962]; *Über die Entwicklung der Wissenschaft im Brückenbau* [1964]
- Further reading: [Steinhardt, 1981]; [Cosandey, 1981]
- Photo courtesy of Zurich ETH library

## SWAIN, GEORGE FILLMORE

*2.3.1857 San Francisco, USA,
†1.7.1931 New Hampshire, USA

George Fillmore Swain's father was very successful in the shipping and commission business, gradually became a leading businessman in San Francisco and was president of the Chamber of Commerce in that city; Swain's mother was a woman of refined literary tastes, but unfortunately died early. The house of the Swains was visited by important personalities of the city's cultural life, e. g. the outstanding violinist Camella Urso, the clergyman Dr. Bellows and the author Mark Twain. After the death of his father in 1872, George Fillmore Swain's much-travelled and art-loving uncle, the doctor Charles Wesley Fillmore, decided on the route the young man's education should take. George Fillmore Swain was hardly 16 years old when he enrolled at MIT, but he became the best-ranking student of his class! He attended Prof. George H. Howison's course on logical thinking, expanded his literary education and showed a liking for the writings of Shakespeare. Swain was also profoundly influenced by the lectures of the mathematically skilled civil engineering professor John B. Henck. "His influence on Professor Swain's later teaching must have been very great" [Hovgaard, 1937, p. 334]. Swain attained a bachelor of science in civil engineering and typography in 1877. At the suggestion of his uncle, he then undertook a three-year course of study in Europe. Swain enrolled at the Berlin Building Academy, where he studied bridge-building and theory of structures with Winkler, railway engineering with Goering and hydraulic engineering with Hagen. Although Swain did not take any final examinations, he returned to the USA with flattering recommendations from Winkler and Goering. Swain presented the fruits of his theory of structure studies in several journal articles, one of them being his excellent paper on the application of the principle of virtual velocities to trussed frameworks [Swain, 1883], much quoted in the literature, also in the German-speaking countries. In Germany, Swain also improved on his musical talents; in later years, Swain's sensitive piano accompaniments and his playing of the Beethoven sonatas would delight the visitors to his uncle's musical soirées. Swain was appointed instructor for structural engineering at the MIT in 1881, and two years later he was promoted to assistant professor, then to associate professor in 1886 and finally to professor of structural engineering in 1887. He was senior engineer at the Railroad Commission of Massachusetts from 1887 to 1914 and in this position was responsible for checking more than 2000 bridges. The year 1909 brought a change as he moved to Havard University as professor of structural engineering, where he remained until his transfer to emeritus status in 1929. Swain served the ASCE as director (1909 – 10), vice-president (1908 – 09) and president (1913), the first time the business of the ASCE had been under the auspices of a university professor. With his all-round education and talents, Swain considerably helped to introduce historical aspects into the teaching of engineers [Swain, 1917 & 1922]. His three-volume monumental work *Structural Engineering* [Swain, 1924/1; 1924/2; 1927] formed the final element in the accumulation phase of structural theory (1900 – 25) in the USA. After Swain had been elected an honorary member of the ASCE in 1929, Hardy Cross characterised him at the AGM of January 1930 in New York as follows: "As a scholar he has been honest and accurate in detail and broad in vision. As an engineer he has shown discrimination in the choice of appropriate tools of thought and resourcefulness in application to engineering work. As a teacher he has been preeminent in his ability to inspire men and to train them. This man, this teacher, is no mere academic pedagogue. He was a man who never permitted mazes of mathematics and mazes of statistics to befog the vision of the men who studied under him. He was a prophet, a priest of clear individual thought and aggressive individual judgment" (cited in [Hovgaard, 1937, p. 342]).

- Main contributions to structural analysis: *On the Application of the Principle of Virtual Velocities to the Determination of the Deflection and Stresses of Frames* [1883]; *How to Study* [1917]; *The Young Man and Civil Engineering* [1922]; *Structural Engineering. The Strength of Materials* [1924/1]; *Structural Engineering. Fundamental Properties of Materials* [1924/2]; *Structural Engineering. Stresses, Graphical Statics and Masonry* [1927]
- Further reading: [Hovgaard, 1937]
- Photo courtesy of [Hovgaard, 1937]

## SZABÓ, ISTVÁN

*13.12.1906 Orosháza, Austria-Hungary;
†21.1.1980 Berlin, Federal Republic of Germany

After passing his university entrance examination in 1926, István Szabó studied physics at Berlin-Charlottenburg TH. His professional career was influenced by professors R. Rothe (mathematics), R. Becker (physics), G. Hertz (physics) and G. Hamel (mathematics, mechanics). While he was still a student, he worked in the Berlin research laboratories of Osram (1930 – 33), where he was also employed (1935 – 39) after completing his studies. He gained his doctorate in 1943 at Berlin-Charlottenburg TH under W. Schmeidler and G. Hamel with a dissertation entitled *Die Strömung um eine Fläche von elliptischem Umriß*. As soon as his Alma Mater reopened in 1946 under the name of Berlin Technical University, he became senior engineer to Ernst Mohr's chair of mathematics at the Faculty of General Engineering Sciences, where he wrote his habilitation thesis *Das Temperaturfeld in der Anode einer Röntgenröhre* in 1947. He was appointed to the chair of mechanics (which had been vacant since 1945) at the Faculty of Civil Engineering at Berlin TU in 1948. This post had been held earlier by F. Kötter (1901 – 12), H. Reissner (1913 – 34) and F. Tölke (1937 – 45). Within just a few years Szabó was able to present this chair, founded in 1901 and steeped in tradition, in a new light. The fruits of his scientific activities were his textbooks *Einführung in die Technische Mechanik* (1954) and *Höhere Technische Mechanik* (1956), which for more than three decades set the standards for the teaching of applied mechanics in German universities. The last 15 years of his life were dedicated to researching the history of mechanics, the results of which were published in 31 papers and a book entitled *Geschichte der mechanischen Prinzipien* (1977). P. Zimmermann and E. A. Fellmann were responsible for the third edition of this magnificent history book in 1996 and enriched it with less accessible writings, Szabó's obituary [Zimmermann, 1981] and an annotated list of his works on the history of mechanics and applied mathematics [Szabó, 1996, pp. 555 – 557]. Therefore, Szabó not only established a school in the field of mechanics, but also created the foundations for a modern history of mechanics in the German language.

- Main contributions to structural analysis: *Einführung in die Technische Mechanik* [1954]; *Höhere Technische Mechanik* [1956]; *Geschichte der mechanischen Prinzipien* [1977];

 SWAIN
 SZABÓ
 TAKABEYA
 TANABASHI
 TETMAJER

*Einige Marksteine in der Entwicklung der theoretischen Bauingenieurkunst* [1980]
- Further reading:
[Raack, 1971]; [Zimmermann, 1981]
- Photo courtesy of [Raack, 1971, p. 1]

### TAKABEYA, FUKUHEI

*9.9.1893 Okazaki near Nagoya, Japan,
†24.4.1975 Kamakura, Japan
After completing his education at the 8th state grammar school in Nagoya, Fukuhei Takabeya studied at the Imperial Kyushu University in Fukuoka (1916–19), where he afterwards served as a lecturer until 1921. He gained his doctorate at the Imperial Kyushu University in 1922 with the dissertation *On the calculation of a beam encastré at both ends taking special account of the axial force*. Over the years 1921 to 1925 he was an associate professor at the Imperial Kyushu University and in May 1925 he was appointed professor at the Imperial Hokkaido University in Sapporo, but returned to the Imperial Kyushu University in 1947. He was professor at the Japanese Defence Academy from 1954 to 1966 and afterwards worked at the University of Tokai until 1972. He became an honorary member of the Japanese Society of Civil Engineers (JSCE) in 1963, and honorary professor of the Japanese Defence Academy in 1966. Takabeya developed the iteration methods of Cross and Kani further to form a highly effective computational algorithm at the transition between the invention and innovation phases of structural theory [Takabeya, 1965 & 1967], which was especially useful for analysing systems with a high degree of static indeterminacy in high-rise buildings.
- Main contributions to structural analysis:
*On the calculation of a beam encastré at both ends taking special account of the axial force* [1924]; *Frame tables* [1930/1]; *On the calculation of stresses in plane, encastré flat metal sheets* [1930/2]; *Multistorey frames* [1965 & 1967]
- Further reading:
[Uchida, 1983]; [Naruoka, 1999]
- Photo courtesy of Prof. Dr. M. Yamada

### TANABASHI, RYO

*2.3.1907 Shizuoka, Japan, †5.5.1974 Kyoto, Japan
Like Takabeya, Ryo Tanabashi also attended the 8th state grammar school in Nagoya, but afterwards studied at the Imperial University in Kyoto (1926–29) instead. He was a lecturer at Kobe Technical University from 1929 to 1931, thereafter a lecturer at the Imperial University in Kyoto before becoming an associate professor there in 1933. He gained his doctorate at the Technical Faculty of the Imperial University in Kyoto in 1936 with a dissertation entitled *Investigations into the structural theory of building structures*. His appointment as professor at the Imperial University in Kyoto came in 1945. He was made director of the Research Institute for the Prevention of Natural Catastrophes at the University of Kyoto when it was founded in 1951. He was given emeritus status and made an honorary professor of the University of Kyoto in 1970. From then until his death he worked at Kinki University in Osaka. His revolutionary proposal (dating from 1937) to define and estimate the seismic resistance of structures as the energy absorption until failure of the structure at the plastic limit state [Tanabashi, 1937] was verified in 1956 by Prof. G. W. Housner (USA) by using the example of the failure of holding-down bolts at an oil refinery after the Arasuka earthquake. Today, Tanabashi's method forms the basis for the design of structures to withstand earthquakes.
- Main contributions to structural analysis:
*On the resistance of structures to earthquake shocks* [1937/1]; *Approximate method for determining the maximum and minimum bending moments of frame structures* [1937/2]; *Systematic resolution of elasticity equations for a multi-storey frame and the question of the influence line* [1938]; *Studies on the nonlinear vibration of structures subjected to destructive earthquakes* [1956]; *Analysis of statically indeterminate structures in the ultimate state* [1958]
- Further reading:
[Housner, 1959]; [Yamada, 1980]; [Kobori, 1980]
- Photo courtesy of Prof. Dr. M. Yamada

### TETMAJER, LUDWIG VON

*14.7.1850 Krompach, Austrian Empire,
†31.1.1905 Vienna, Austria-Hungary
Born in Hungary, the son of the director of the Krompach-Hernader ironworks, Tetmajer studied at Zurich Polytechnic (Zurich ETH) from 1868 to 1872 and then worked for the Swiss Railways Company. He had already written his habilitation thesis by 1873 and went on to assist Culmann and teach his method of graphical statics. In the late 1870s he turned to materials testing, which at that time was still in its infancy. Upon Culmann's death he was appointed professor of theory structures and technology of building materials at Zurich ETH. At the same time he also took charge of the Swiss Building Materials Testing Institute. Thus began the systematic establishment of materials testing in Switzerland, which very soon enjoyed a good international reputation. In the early years of his professorship he formulated the forward-thinking hypothesis that the mechanical material quality is characterised quantitatively by the internal work; the energy doctrine becoming established in the classical phase of structural theory thus appears in materials testing as well. Tetmajer published two reports (1883/84) on buckling tests carried out on timber members where for the first time the buckling load was presented as a function of the slenderness of the member and the beginnings of the Tetmajer Line (as it was later called) can be seen [Nowak, 1981, pp. 112–113]. His numerous buckling tests and buckling theory deliberations were brought together analytically in 1890 in the form of an empirical formula (Tetmajer Line) for all the materials he had investigated. In 1895, following the death of Johann Bauschinger, Tetmajer developed the Bauschinger Conferences on materials testing into the Internationaler Verband für die Materialprüfungen der Technik (now known as the New International Association for the Testing of Materials), and became its first president. He taught and carried out research at Vienna TH from 1901 onwards, and also served as rector there. He unfortunately did not live to see the founding of a central laboratory for technical materials testing in Austria [Rossmanith, 1990, p. 530]; he died while giving a lecture.
- Main contributions to structural analysis:
*Methoden und Resultate der Prüfung der schweizerischen Bauhölzer* [1884];
*Die angewandte Elastizitäts- und Festigkeitslehre* [1889]; *Methoden und Resultate der Prüfung der Festigkeitsverhältnisse des Eisens und anderer Metalle* [1890]; *Die Gesetze der Knickfestigkeit der technisch wichtigsten Baustoffe* [1896]; *Die Gesetze der Knickungs- und der zusammengesetzten Druckfestigkeit der technisch*

wichtigsten Baustoffe [1903]; *Die angewandte Elastizitäts- und Festigkeitslehre* [1904]
- Further reading:
[Schüle, 1905]; [n. n., 1905]; [Emperger, 1905]; [Rös, 1925]
- Photo courtesy of Zurich ETH library

### TIMOSHENKO, STEPAN PROKOFIEVICH

*22.12.1878 Shpotovka (Poltava), Ukraine, Russia, †29.05.1972 Wuppertal, Federal Republic of Germany

Timoshenko finished his secondary education at Romenski School in 1896 and five years later he had completed his studies at the Institute of Engineers of Ways of Communication in St. Petersburg; he taught here and at the Petersburg Polytechnic Institute from 1902 onwards. Krylov's method of analysing engineering problems mathematically using differential equations was to have a considerable influence on Timoshenko's scientific career. He was a professor at the Faculty of Civil Engineering at the Polytechnic Institute in Kiev from 1906 to 1911, serving as dean there from 1909. He was dismissed in February 1911 in connection with the dispute between the universities and the tsarist government. But in that same year he was awarded the Zhuravsky Gold Medal of the St. Petersburg Polytechnic Institute for his treatise *On the stability of elastic systems*. It was at this latter institute that he served as professor of theoretical mechanics from 1913 to 1917 in the place of A. N. Krylov. According to his former friend A. F. Joffe, Timoshenko joined the group around Maxim Gorki in the years leading up to the revolution and "was probably the most left-wing of the Russian professors" [Joffe, 1967, p. 106]. In 1918 Timoshenko became a professor of the Polytechnic Institute in Kiev as well as director and founder of the Mechanics Research Institute at the reorganised Ukraine Academy of Sciences. Following an interlude as professor in Zagreb (1921–22), Timoshenko became involved in a wide range of activities in the USA, first at the Westinghouse company, then as a professor at Michigan University (1928–35) and afterwards at Stanford University. He influenced applied mechanics in the 20th century like no other and that had a decisive knock-on effect for theory of structures. For example, in 1921 he published his theory of the shear-flexible beam, which was later named after him (Timoshenko beam). The American Society of Mechanical Engineers was founded in 1957 and Timoshenko was awarded the society's first gold medal. Following a leg injury in 1964, he moved to Wuppertal, Federal Republic of Germany, to be with his daughter. He undertook extensive travels in the USSR in 1958 and 1967.
- Main contributions to structural analysis:
*On the question of the strength of rails* [1915/2001]; *On the correction for shear of the differential equation for transverse vibration of prismatic bars* [1921]; *History of strength of materials. With a brief account of the history of theory of elasticity and theory of structures* [1953/1]; *The collected papers* [1953/2]; *Ustoichivost' sterzhnei, plastin i obolochek: Izbrannye raboty* (stability of bars, plates and shells: collected works) [1971]; *Prochnost i kolebaniya elementov konstruktsii* (strength and vibrations of construction elements) [1975/1]; *Staticheskie i dinamicheskie zadachi teorii uprugosti* (static and dynamic problems of elastic theory) [1975/2]
- Further reading:
[Timoshenko, 1968]; [Frenkel, 1990]; [Pisarenko, 1991]; [Timoshenko, 2006]
- Photo courtesy of [Joffe, 1967, p. 107]

### TREDGOLD, THOMAS

*22.8.1788 Brandon near Durham, UK, †28.1.1829 London, UK

Tredgold was neither a mathematician nor a scientist, but fought with spirit and passion for a deeper understanding of the loadbearing capacity of timber and iron structures. His genius bridged in a unique way the gap between the emerging engineering sciences and the needs of practical engineers at the time of the Industrial Revolution. Tredgold's book on carpentry [Tredgold, 1820] was still being printed in the 1930s, but his book on cast iron [Tredgold, 1822] was regarded as out of date within 50 years. Both books helped the practising engineer to understand theory of structures and strength of materials through timber and cast iron applications. It was Tredgold who introduced the term "neutral axis" into beam theory [Tredgold, 1820]. From him stem two quotations that even today are imprinted on the minds of many civil engineers: "Civil engineering is the art of directing the great sources of power in Nature for the use and convenience of mankind" and "The stability of a building is inversely proportional to the science of the builders". Notwithstanding, let it not be said that Tredgold was averse to science; indeed, providing a scientific basis for building technology formed a thread in his life's work. Tredgold was apprenticed to a carpenter and in 1813 settled London, where he worked in an architectural practice. He soon became a much-sought-after engineer skilled in designing timber and cast iron structures. From 1817 onwards he published articles in the *Philosophical Magazine*. His contributions to theory of structures and strength of materials can be grouped together with the structural problems of the practising engineer. It is for this reason that his influence on practical engineering was greater than that on the constitution phase of theory of structures. Unfortunately, the Tredgold family was unable to maintain the prosperity it had earned in the early days and after Thomas' untimely death they became poor and had to scrape a living by selling many books.
- Main contributions to structural analysis:
*Elementary Principles of Carpentry* [1820]; *A Practical essay on the strength of cast iron ...* [1822]
- Further reading:
[Booth, 1979/1980/1]; [Booth, 1979/1980/2]; [Booth, 1979/1980/3]; [Sutherland, 1979/1980]; [Booth, 2002]
- Photo courtesy of Dr. W. Addis

### TREFFTZ, ERICH IMMANUEL

*21.2.1888 Leipzig, German Empire, †21.1.1937 Dresden, Germany

Erich Immanuel Trefftz was the son of a Leipzig businessman. He attended the Thomas School in Leipzig for three years and then the Kaiser Wilhelm Grammar School in Aachen (1900–06). After passing his university entrance examination, he studied mechanical engineering at Aachen RWTH, but switched to the faculty of mathematics after one year, continued his studies at the University of Göttingen, where he worked as assistant to his uncle, Carl Runge, and also studied at the Ludwig Prandtl Institute for Applied Mechanics. Trefftz completed his studies in mathematics with a dissertation on a flow mechanics problem under Richard von Mises at the University of Strasbourg. Following a period as an assistant at Aachen RWTH, he volunteered for military service in World War 1, but after being wounded towards the end of the war, he was transferred to the derelict Aerodynamics Institute at Aachen RWTH, where he became professor of applied mathematics in 1919. Just one year later, Trefftz was appointed professor of applied mechanics at Dresden TH, where he worked until his early death. In 1929 Stuttgart TH awarded Trefftz an honorary doctorate – he was only 41 years old! After Richard von Mises had been forced to leave the country by the Nazis, Trefftz took over the editor's chair at the journal *Zeitschrift für Angewandte Mathematik und Mechanik* in 1934. Trefftz attained long-lasting achievements in hydromechanics, applied mathematics, vibration theory and elastic theory. In this latter field, this highly skilled and morally upright scientist excelled in his publications on fundamentals, torsion theory, shear centre, shell theory and stability theory. According to R. Grammel, the scientific work of Trefftz has three traits: "… great courage – almost always tackling and overcoming precisely the most difficult problems; secure mastery when handling the mathematics required to solve the task; a strong sense of self-criticism which only allowed him to accept that for which he could find no more objections" [Grammel, 1938, p. 10].
- Main contributions to structural analysis:
*Über die Torsion prismatischer Stäbe von polygonalem Querschnitt* [1921]; *Über die Wirkung einer Abrundung auf die Torsionsspannungen in der inneren Ecke eines Winkeleisens* [1922];

 TIMOSHENKO  TREDGOLD  TREFFTZ  TRUESDELL  VIANELLO

*Über die Spannungsverteilung in tordierten Stäben bei teilweiser Überschreitung der Fließgrenze* [1925]; *Ein Gegenstück zum Ritzschen Verfahren* [1926]; *Über die Ableitung der Stabilitätskriterien des elastischen Gleichgewichtes aus der Elastizitätstheorie endlicher Deformationen* [1930]; *Zur Theorie der Stabilität des elastischen Gleichgewichts* [1933]; *Ableitung der Schalenbiegungsgleichungen nach dem Castiglianoschen Prinzip* [1935/1]; *Über den Schubmittelpunkt in einem durch eine Einzellast gebogenen Balken* [1935/2]; *Die Bestimmung der Knicklast gedrückter, rechteckiger Platten* [1935/3]; *Graphostatik* [1936]; *Die Bestimmung der Schubbeanspruchung beim Ausbeulen rechteckiger Platten* [1936]; *Über die Tragfähigkeit eines längsbelasteten Plattenstreifens nach Überschreiten der Beullast* [1937]

• Further reading:
[Prandtl, 1937]; [Grammel, 1938];
[Riedrich, 2003]

• Photo courtesy of Dresden TU archives photo collection

### TRUESDELL, CLIFFORD AMBROSE
*18.2.1919 Los Angeles, USA,
†14.1.2000 Baltimore, USA

After completing his education at the Polytechnic High School in Los Angeles, Truesdell spent two years in Europe mastering the German, French and Italian languages perfectly and improving his knowledge of Latin and Greek. It was during this period that he developed his lifelong love of the literature, art and music of the Baroque period. He graduated as Bachelor of Science in mathematics and physics and also as Master of Science in mathematics at the California Institute of Technology; he gained his PhD in Princeton. After various posts at the Massachusetts Institute of Technology and state laboratories, he became professor – at 30! – of mathematics in Indiana. From 1961 until being given emeritus status he taught and carried out research as professor of rational mechanics at the John Hopkins University in Baltimore. Truesdell not only had a lasting influence on modern rational mechanics and thermodynamics in the second half of the 20th century; his unconventional work in mathematics, natural philosophy and the history of mechanics is also exceptional. In 1960 he founded the journal *Archive for the History of Exact Sciences*, and in that same year provided a commentary on essential elements of Euler's *Opera Omnia* from the viewpoint of rational mechanics. He studied the entire history of mechanics with relentless clarity in order to reconstruct it rationally; scientific knowledge concerning the history of mechanics was considered by Truesdell to be an intrinsic part of the research process in the field of rational mechanics. His scientific work received worldwide acclaim. He was awarded honorary doctorates by Milan TH (1965) and the universities of Tulane (1976), Uppsala (1979) and Ferrara (1992); the USSR Academy of Sciences twice awarded him its Euler Medal (1958, 1983); he was a member of many academies.

• Main contributions to structural analysis: *The rational Mechanics of flexible or elastic bodies 1638–1788. Introduction to Leonhardi Euleri opera omnia vol. 10 et 11 seriei secundae* [(1960)]; *Die Entwicklung des Drallsatzes* [1964]; *Essays in the History of Mechanics* [1968]

• Further reading:
[Capriz, 2000]

• Photo courtesy of Università di Genova

### VARIGNON, PIERRE
*1654 Caen, France, †23.12.1722 Paris, France

As one of the three sons of a less prosperous architect, Pierre Varignon chose a career in the church, and studied theology and philosophy at the Jesuit College of his home town; his ordination followed in 1683. His interest in geometry can be traced back to Euclid's *Elemente*, which he discovered in a second-hand bookshop and read zealously. Thanks to funds provided by a former fellow student, the abbot of Saint-Pierre, the pair of them were able to leave Caen in 1686 and continue their philosophical and mathematical studies in Paris. Just one year later he published his book *Projet d'une nouvelle mécanique* (1687), which was dedicated to the Académie des Sciences and earned him membership of the Académie and a job as professor of mathematics at the newly founded Collège Mazarin. *Nouvelle Mécanique ou Statique* was produced from his lectures and published posthumously in 1725, and was to influence the development of theory of structures in three ways: firstly, compared to the parallelogram rule, the lever principle does not have any particular significance in the history of theory of structures (and so Varignon finally eliminated the special role of simple machines in ancient mechanics); secondly, a clear recognition of the relationship between the polygon of forces and the funicular polygon; thirdly, he explained the equilibrium of forces with the orthogonality between resultant and possible displacement. Varignon therefore came very close to the principle of virtual displacements. Graphical statics was given a great boost through Varignon's book on mechanics, Lagrange was also inspired. Therefore, Varignon's work had a substantial influence on the discipline-formation period of structural theory.

• Main contributions to structural analysis: *Projet d'une nouvelle mécanique* [1687]; *Nouvelle Mécanique ou Statique* [1725]

• Further reading:
[Rühlmann, 1885]; [Costabel, 1981]

### VIANELLO, LUIGI
*29.9.1862 Treviso, Italy, †16.7.1907 Berlin, German Empire

Luigi Vianello was the son of a highly respected public notary in Treviso near Venice. After attending the local grammar school, he studied mathematics for two years at the University of Padua and following a further three years of study at Turin Technical University was awarded his diploma. He gained his first practical experience as an engineer in machine factories, railway workshops and railway engineering in Venice, Treviso and Milan between 1885 and 1892. After that, he moved to Germany, where he first worked at the Egestorff locomotive plant in Linden (Hannover) until 1895 before moving on to the bridge-building office of the Gutehoffnung foundry in Sterkrade for two years. It was here that he worked with Reinhold Krohn, laying the foundations for his later important work in structural steelwork and structural analysis. At Siemens & Halske (1897–1902) he was responsible for the technical side of many of the most difficult structures of Berlin's elevated railway, e. g. the structures around the Gleisdreieck junction. After completing these structures, he joined the Monorail Department of the Continentale Gesellschaft für elektrische Unternehmungen – the company that built and operated the unique monorail in Wuppertal. Based on the model of the Wuppertal monorail, erected between 1898 and 1901, the company under the leadership of Richard Petersen, co-founder of the VDI and its chairman from 1868 to 1872 [Brandt & Poser, 2006, p. 271], offered this spectacular form of transport to other cities, e. g. Hamburg, Berlin; much of the design work, the structural calculations and cost estimates for a monorail in Berlin which, however, was never realised, are the work of Vianello. His 700-page monograph *Der Eisenbau* (1905) was the first comprehensive standard work on structural steelwork in the German language. It enjoyed several editions and was still a worthy favourite in design offices in the second half of the consolidation period of structural theory (1900 – 50). Among Vianello's contributions to structural analysis, especially worthy of note is his graphical iteration method for calculating the buckling load of straight members (1898), with which even members of varying bending stiffness can be analysed easily. "There were certainly hardly any technical problem of significance," Vianello's fatherly friend and boss Petersen wrote in his heart-warming obituary, "with which he was not familiar – in the natural sciences, in chemistry –, it was a delight to debate scientific problems with him" [Petersen, 1907, p. 2034]. Besides his native Italian, Vianello could speak and write perfectly in German, English and French, and he could make himself understood in Spanish, Russian and Norwegian, too! Extremely modest and obliging in his conduct towards others, Vianello hated lies, hated appearance for the sake of it, and always called a spade a spade. "He repelled some with this attitude, but his friends appreciated the integrity of his character – there was no guile in him" [Petersen, 1907, p. 2034]. His painful hip joint ailment prevented him from enjoying his leisure time in the Alps, and his failing sight increasingly hindered the progress of his scientific work. "In order to avoid a long, sad invalidism, he took his own life calmly and collectedly" [Oder, 1907, p. 452]. So Vianello, like his congenial fellow countryman Castigliano, died shortly before his 45th birthday. As he had no relations in Germany, Vianello's last employer, the Continentale Gesellschaft für elektrische Unternehmungen, was happy to deal with the formalities of his death. The Berlin district association of the VDI honoured the memory of one of its highly skilled members at a meeting. The dignified funeral, the worthy memorial meeting and the obituary were basically all the work of his friend Richard Petersen, as a tribute to the structural steelwork engineer Luigi Vianello.

- Main contributions to structural analysis: *Der kontinuirliche Balken mit Dreiecks- oder Trapezlast* [1893]; *Der Kniehebel* [1895]; *Die Doppelkonsole* [1897]; *Graphische Untersuchung der Knickfestigkeit gerader Stäbe* [1898]; *Die Konstruktion der Biegungslinie gerader Stäbe und ihre Anwendung in der Statik* [1903]; *Der durchgehende Träger auf elastisch senkbaren Stützen* [1904]; *Der Eisenbau* [1905]; *Knickfestigkeit eines dreiarmigen ebenen Systems* [1906]; *Der Flachträger. Durchgehender räumlicher Träger auf nachgiebigen Stützen* [1907]
- Further reading: [Petersen, 1907]; [Oder, 1907]
- Photo courtesy of [Petersen, 1907, p. 2034]

### VIERENDEEL, ARTHUR

*10.4.1852 Leuven, Belgium, †8.11.1940 Uccle, Belgium

Arthur Vierendeel graduated in civil and mine engineering at the Catholic University of Leuven in 1874. Afterwards, he worked as an engineer in the offices of the Ateliers Nicaise et Delcuve company in La Louvière (Belgium). In 1876 Vierendeel won the tender for the Royal Circus in Brussels, at that time one of the largest steel structures in Belgium, which triggered a controversial public debate owing to its extraordinary lightness. Vierendeel was appointed director of the newly founded Technical Department of the West Flanders Ministry of Public Works in 1885. Four years later he became professor of strength of materials, theory of structures and history of building at the Catholic University in Leuven. He retired from his director's post in 1927 and was granted emeritus status in 1935. The idea of the girder that bears his name became known as early as 1895 and was implemented as a frame bridge without diagonals in 1896 – 97. Vierendeel had this 31.5 m-span bridge built at his own cost within the scope of the Brussels World Exposition in Tervueren (Belgium) and loaded it to failure in order to verify the agreement between calculations and measurements. As a result, many bridges based on Vierendeel girders were built in Belgium – especially for Belgian State Railways – but also abroad. For instance, the first Vierendeel bridge in the USA was erected as early as 1900. The journal *Der Eisenbau* published a debate on the structural pros and cons of the Vierendeel girder in 1912; numerous articles on the structural analysis of the Vierendeel girder appeared, which promoted the development of the displacement method. The Vierendeel girder thus presented a challenge for the structural theory of the accumulation phase. Even today, the Vierendeel girder is still used frequently, e. g. for the Qian Lin Xi Bridge in China (1989 – 9 spans each of 100 m and 12.5 m wide), or in the Commerzbank Headquarters in Frankfurt a. M. (1996) in the form of huge Vierendeel bays.

- Main contributions to structural analysis: *L'architecture du Fer et de l'Acier* [1897]; *Théorie générale des poutres Vierendeel* [1900]; *La construction architectureale en fonte, fer et acier* [1901]; *Der Vierendeelträger im Brückenbau* [1911]; *Einige Betrachtungen über das Wesen des Vierendeelträgers* [1912]; *Cours de stabilité des constructions* [1931]
- Further reading: [Lederer, 1970]; [Radelet de Grave, 2002]
- Photo courtesy of Prof. Dr. R. Maquoi

### VLASOV, VASILY ZACHAROVICH

*24.02.1906 Kareevo near Tarussa, Russia, †7.8.1958 Moscow, USSR

Following three years at the local village school, Vlasov attended a secondary school which he left in 1924 to attend various universities in Moscow. He completed his civil engineering studies at the University of Civil Engineering in 1930 (known as Moscow Civil Engineering Institute after 1932), which was the former Faculty of Civil Engineering of Moscow TH. Vlasov taught at this newly formed university and became I. M. Rabinovich's successor in 1956. During this same period he also worked at the Scientific Research Institute for Industrial Building (1932 – 46), initially under the leadership of A. A. Gvozdev, and at the Military Engineering Academy (1932 – 42). He gained his doctorate and was appointed professor in 1937. The Stalin Prize, 1st class (1941) and 2nd class (1950), followed. Vlasov was in charge of the Structural Mechanics Section at the Institute of Mechanics at the USSR Academy of Sciences from 1946 onwards, and in 1953 became a corresponding member of the latter academy. Vlasov achieved a consistent formulation of the theory of thin-wall elastic bars based on the theory of shells and folded plates.

- Main contributions to structural analysis: *Novyi method rascheta tonkostennykh prizmaticheskikh skladchatykh pokrytii i obolochek* (new method of calculation for thin-wall prismatic folded-plate roofs and shells) [1933]; *Tonkostennye uprugie sterzhni* (thin-wall elastic bars) [1940]; *Obshchaya teoriya obolochek i ee prilozheniya v tekhnike* [1949]; *Allgemeine Schalentheorie und ihre Anwendung in der Technik* [1958]; *Izbrannye trudy* (selected works) [1962 – 64]; *Dünnwandige elastische Stäbe* [1964/65]
- Further reading: [n. n., 1962]; [Leont'ev, 1963]; [Stel'makh & Vlasov, 1982]
- Photo courtesy of Prof. Dr. G. Mikhailov

### WASHIZU, KYUICHIRO

*12.3.1921, Owari-Ichinomiya in Aichi, Japan, †25.11.1981, Tokyo, Japan

Following his education at the 3rd state grammar school in Kyoto, Kyuichiro Washizu studied at the Imperial University in Tokyo (1940 – 42) and was thereafter lecturer at the Faculty of Aircraft Construction at the University of Tokyo, serving as associate professor there from 1947 to 1958. Washizu studied and carried out research at the Massachusetts Institute of Technology (MIT) from 1953 to 1955 and gained his doctorate at the University of Tokyo in 1957 with a dissertation entitled *Approximate solutions in*

VIERENDEEL　　VLASOV　　WASHIZU　　WEBER　　WEISBACH

elastomechanics. In 1958 he was appointed professor of aircraft construction at the University of Tokyo, where he worked until being granted emeritus status in 1981; he was appointed honorary professor of his Alma Mater later that same year. His subsequent appointment as professor of mechanical engineering at the Faculty of Fundamentals of Engineering Sciences at the University of Osaka was cut short after just a few months through sudden heart failure. Washizu's contributions to the calculus of variation fundamentals in structural mechanics had a profound influence on the finite element method during the integration period of structural theory. His professional career was therefore of an international character: MIT research member (1960), visiting professor at the University of Washington (1962), distinguished visiting professor at the Georgia Institute of Technology (1979); President of the Japanese Institute of Aerospace Research in Tokyo (1979–80).

- Main contributions to structural analysis: *Boundary Value Problems in Elasticity* [1953]; *On the Variational Principles of Elasticity and Plasticity* [1955]; *Bounds for Solution of Note on the Principle of Stationary Complementary Energy Applied to Free Vibration of an Elastic Body* [1966]; *Variational Methods in Elasticity and Plasticity* [1968]
- Further reading: [Yamamoto, 1983]
- Photo courtesy of Prof. Dr. M. Yamada

## WEBER, CONSTANTIN HEINRICH

*14.8.1885 Bärenwalde near Zwickau, German Empire, †14.8.1976 Hannover, Federal Republic of Germany

Weber spent the first 10 years of his life in the village of his birth, a small place in Saxony's Erz Mountains; afterwards, the family moved to Riga, where Constantin Heinrich Weber completed his schooling in May 1904. He studied mechanical engineering at Braunschweig TH from 1906 to 1911 and completed his diploma with a distinction. His diploma thesis was a design for a mobile slewing jib crane with electric drive. Apart from a time spent in military service and his years as an infantry officer during World War 1, Weber worked from 1912 to 1926 as a mechanical engineer in industry, gaining experience with the design and construction of water turbines, milking machines and, in particular, cranes. His work with cranes introduced him to torsion problems, which he analysed in a short paper [Weber, 1921]. He would later publish further notable writings on torsion theory. Weber gained his doctorate as an external student in 1923 at the mechanical engineering department of Braunschweig TH under Otto Föppl. His dissertation, *Biegung, Schub und Drehung von Balken* (see also [Weber, 1924]), earned him a distinction and the title Dr.-Ing., and brought to an end the debate about the shear centre (see section 7.3.2.5). He taught at a mechanical engineering school in Dortmund from 1926 to 1928 before being appointed professor of mechanics and strength of materials in the Mechanics Department of Dresden TH in autumn 1928. He remained in Dresden until autumn 1945. In 1942 he also collaborated on the development of the V2 rocket in Peenemünde. After World War 2, Weber carried out scientific activities at the Institute for Machine Elements at Braunschweig TH under Prof. G. Niemann, and from 1948 to 1951 lectured in special areas of mechanics. His Alma Mater awarded him an honorary doctorate in 1950. Weber crowned his scientific life's work with a book on torsion theory, which he wrote together with his colleague Wilhelm Günther [Weber & Günther, 1958]. The focus of this is the method behind the calculation of the torsion stiffness of complex cross-sections – as are very common in mechanical engineering (sections with notches, multiple coupled cross-sectional areas) – by way of two-dimensional harmonic functions and conformal mapping.

- Main contributions to structural analysis: *Die Lehre der Drehungsfestigkeit* [1921]; *Bisherige Lösungen des Torsionsproblems* [1922]; *Biegung und Schub in geraden Balken* [1924]; *Übertragung des Drehmomentes in Balken mit doppelflanschigem Querschnitt* [1926]; *Veranschaulichung und Anwendung der Minimalsätze der Elastizitätstheorie* [1938]; *Halbebene mit Kreisbogenkerbe* [1940]; *Eingrenzung von Verschiebungen und Zerrungen mit Hilfe der Minimalsätze* [1942]; *Zur nichtlinearen Elastizitätstheorie* [1948]; *Torsionstheorie* [1958]
- Further reading: [Petschel, 2003, p. 1016]
- Photo courtesy of Dresden TU archives, photo collection

## WEISBACH, JULIUS

*10.8.1806 Mittelschmiedeberg near Annaberg, Saxony, †24.2.1871 Freiberg, German Empire (Saxony)

Julius Weisbach was the eighth child of a foreman at the Mittelschmiedeberg iron forging works and demonstrated his exceptional intellect at an early age. Following his education at Annaberg Grammar School, he attended the Freiberg School of Mining in 1820. He then went on to study at the Freiberg Mining Academy from 1822 to 1826 and rounded off his education with mathematical and natural science studies in Göttingen and Vienna. In 1830 he undertook a six-month miner's journey (on foot) through Austria and Hungary before returning to Freiberg. He earned a living as a private tutor of mathematics before being appointed to the chair of applied mathematics and mechanical engineering of mining at Freiberg Mining Academy in 1833; the subjects of mine surveying, crystallography, descriptive geometry and mechanical engineering were added to his remit later. Weisbach bequeathed lasting achievements in each of these fields. Like Franz Joseph Ritter von Gerstner before him and August Föppl afterwards, Weisbach can be called a leader in the didactics of engineering sciences. His book *Lehrbuch der Ingenieur- und Maschinen-Mechanik* influenced structural mechanics and made it popular in the middle of the discipline-formation period in an encyclopaedic way. He was the first person to investigate systematically the combined actions of bending and axial forces in frames (1848). Weisbach was a corresponding member of the academies in St. Petersburg, Stockholm and Florence. The University of Leipzig awarded him an honorary doctorate in 1859, and one year later he became the first honorary member of the newly formed (1856) Association of German Engineers (VDI).

- Main contributions to structural analysis: *Lehrbuch der Ingenieur- und Maschinen-Mechanik* [1845–1887]; *Die Theorie der zusammengesetzten Festigkeit* [1848/1]; *Der Ingenieur. Sammlung von Tafeln, Formeln und Regeln der Arithmetik, Geometrie und Mechanik* [1848/2]
- Further reading: [Beck, 1956]; [Höffl & Grabow, 1982]
- Photo courtesy of the Bergakademie Freiberg Photographs Department

## WESTERGAARD, HAROLD MALCOLM

*9.10.1888 Copenhagen, Denmark,
†22.6.1950 Cambridge, Massachusetts, USA

Westergaard came from a family of scholars. His grandfather was professor for oriental languages at the University of Copenhagen, and his father was professor for economics and statistics at the same university. Westergaard studied at Copenhagen TH, where he worked under Ostenfeld, and completed his engineering studies in 1911. He maintained contact with Ostenfeld until the death of this great structural engineer in 1931. Following practical experience in reinforced concrete construction in Copenhagen, Hamburg and London, he inhaled the spirit of the Göttingen school around Felix Klein, studied at that university under Ludwig Prandtl and by 1915 had prepared the written edition of his dissertation at Munich TH with the help of August Föppl. However, World War 1 prevented him from finishing his doctorate work in Munich; the oral examination could not take place until September 1921, with Sebastian Finsterwalder and Ludwig Föppl, and the written edition of Westergaard's dissertation did not appear until 1925 [Westergaard, 1925], which meant that his Dr.-Ing. title could not be recognised until after that date. But by then, Westergaard had already made a name for himself in structural engineering in the USA. With the help of a research scholarship provided by the American Scandinavian Foundation, he was able to attain a PhD at the University of Illinois in Urbana in 1916 and on the strong recommendation of his mentor, Prof. Ostenfeld, was appointed lecturer for theoretical and applied mechanics at that university. He became assistant professor there in 1921, associate professor in 1924 and full professor in 1927. And Westergaard would not disappoint the university. Together with W. A. Slater, he published a paper on the theory of reinforced concrete plates as early as 1921 [Westergaard, 1921], which earned him the Wason Medal of the American Concrete Institute (ACI) in that same year. One year later, he published a paper on buckling theory [Westergaard, 1922], which he would later expand. In 1926 he investigated the mechanical behaviour of concrete pavements for roads [Westergaard, 1926], which would soon become the basis of the relevant regulations. The first systematic work on the history of the theory of structures appeared in 1930 [Westergaard, 1930/1] and that started the historical study of the theory of structures in the USA. He worked as an adviser to the US States Bureau of Reclamation during the building of the Hoover Dam and also published a much-quoted paper on this [Westergaard, 1933], with his original scientific findings, also acted as an adviser to the US Navy Bureau of Yards & Docks and the US Bureau of Public Roads (see, for example, [Westergaard, 1930/2]). In 1936 he was appointed to the Gordon McKay professorship for structural engineering at Havard University and one year later became dean of the Graduate School of Engineering, a post he held until 1946. It was during these years that he turned more and more to the fundamentals, his work on fracture mechanics [Westergaard, 1939] bearing witness to this. He served as a commander in the Civil Engineer Corps of the US Navy during World War 2 and was a member of a commission set up to assess the effects on structures of the atomic bombs dropped on Hiroshima and Nagasaki. It was in the spring of 1949 that Westergaard began to summarise his scientific life's work spread over nearly 40 papers. He fought hard against his severe illness, but was only able to complete the first part of his manuscript on elastic theory, which was published posthumously [Westergaard, 1952]. His death meant the loss of a first-class personality in American structural analysis during its invention phase (1925 – 50). "Westergaard was a striking figure, intellectually brilliant and physically strong … He loved art and music, and although somewhat shy, he was warm and thoughtful of others" [Newmark, 1974, p. 874].

- Main contributions to structural analysis:
*Moments and stresses in slabs* [1921]; *Buckling of elastic structures* [1922]; *Anwendung der Statik auf die Ausgleichsrechnung* [1925]; *Stresses in concrete pavements computed by theoretical analysis* [1926/1]; *Computation of stresses in concrete roads* [1926/2]; *One hundred years advance in structural analysis* [1930/1]; *Computation of Stresses in bridge slabs due to wheel loads* [1930/2]; *Water pressure on dams during earthquake* [1933]; *General solution of the problem of elastostatics of an n-dimensional homogeneous isotropic solid in an n-dimensional space* [1935]; *Bearing pressures and cracks* [1939]; *Theory of Elasticity and Plasticity* [1952]
- Further reading:
[Fair, 1950]; [Newmark, 1974]
- Photo courtesy of the University of Illinois

## WEYRAUCH, JOHANN JAKOB

*8.10.1845 Frankfurt a. M., Holy Roman Empire,
†13.2.1917 Stuttgart, German Empire

After losing his parents, Weyrauch was brought up in the home of a Protestant priest in Frankfurt. He attended Frankfurt Higher Training College and studied engineering sciences at Zurich ETH from 1864 to 1867 while at the same time studying for a teaching diploma in mathematics and natural sciences. He gained his doctorate at the University of Zurich in 1868 before taking part in military service and then becoming an engineer on the new Berlin Railway. Following active service in the Franco-Prussian War of 1870 – 71, he furthered his education. He lived in Brussels and Paris, where he attended lectures at the Sorbonne, and travelled through Belgium, England, Germany and Austria. In 1873 he published his book on beams in which he uses the method of influence lines (the name stems from him). One year later, he completed his habilitation thesis at Stuttgart TH as a private lecturer in pure and applied mathematics and published the first historico-critical analysis of graphical statics. He was appointed associate professor (1876) and full professor (1880) of mechanical heat theory, aerostatics and aerodynamics as well as a special engineering mechanics chapter at Stuttgart TH. From 1880 onwards, Weyrauch's interest shifted more and more towards theory of structures and elastic theory. For instance, he published the writings of the Heilbronn-based doctor Robert Mayer, and he opposed the prevailing view that it was not Robert Mayer but rather Hermann Helmholtz who discovered the law of energy conservation. Hence, Weyrauch started the scientific history research surrounding Mayer. Typical of the scientific approach of Weyrauch "is his desire to work out fundamental principles and comprehensive viewpoints and place them at the top, e. g. the law of conservation of energy, the so-called second theorem of heat theory, the principle of virtual displacements; the special case is then dealt with using general principles" (M. Enslin in Weyrauch, 1922, p. 26). In doing so, he criticised the monistic *weltanschauung*, which reduced everything to mechanics, but also distanced himself from dogmatic Christianity. Weyrauch's deductive method was a major instrument in transforming the energy doctrine into the energy-based theory of structures in the classical phase of structural theory. The British Institution of Civil Engineers awarded him the Telford Prize on two occasions (1881, 1883) for his outstanding scientific achievements. In Germany he was awarded neither an honorary doctorate nor membership of any academies; this may well be because of his political criticism fed by his liberal thinking. Despite tempting offers, he remained loyal to Stuttgart TH and served there as rector for two periods (1889 – 92, 1899 – 1902).

- Main contributions to structural analysis:
*Allgemeine Theorie und Berechnung der kontinuierlichen und einfachen Träger* [1873]; *Die graphische Statik. Historisches und Kritisches* [1874]; *Theorie der elastischen Bogenträger* [1879]; *Theorie elastischer Körper* [1884]; *Die elastischen Bogenträger. Ihre Theorie und Berechnung entsprechend den Bedürfnissen der Praxis. Mit Berücksichtigung von Gewölben und Bogenfachwerken* [1897]; *Über statisch unbestimmte Fachwerke und den Begriff der Deformationsarbeit* [1908]; *Über den Begriff der Deformationsarbeit in der Theorie der Elastizität fester Körper* [1909]
- Further reading:
[Weyrauch, 1922]; [Böttcher & Maurer, 2008, pp. 101 – 110]
- Photo courtesy of University of Stuttgart archives

WESTERGAARD  WEYRAUCH  WHIPPLE  WIEGMANN  WIERZBICKI

## WHIPPLE, SQUIRE
*16.9.1804 Hardwick (Massachusetts), USA,
†15.3.1888 Albany (New York), USA
Squire Whipple, the son of a farmer, was confronted with engineering at an early age. In the years from 1811 to 1817, his father designed, built and operated a cotton mill near Greenwich (Massachusetts). Afterwards, the Whipple family moved to Otsego County (New York), where the father worked as a farmer once again. Squire Whipple attended nearby Hartwick Academy and Fairfield Academy. After just one year, he graduated from Union College in Schenectady (New York) in 1830, and over the following decade worked on various railway and canal projects, filling the periods of unemployment between projects by selling mathematical instruments he had made himself. His experience gained while working on the extension to the Erie Canal led him to the conclusion that the timber bridges crossing the old canal were unsuitable for the new, widened waterway, and iron bridges would be the only option. In 1841 Whipple was granted a patent for his bowstring truss made from cast and wrought iron (see [Griggs & DeLuzio, 1995, pp. 1356, 1357]), and he formulated two patent claims in which he specifies in clear terms the division of work between cast iron for compression members and wrought iron for tension members. He was able to erect such a structure for a bridge over the Erie Canal at Utica that same year, and six more were to follow in New York and Erie. Whipple published his findings on trussed framework and beam theory in 1847 [Whipple, 1847] (see Fig. 2-54). Whipple's modest book inaugurated the emancipation of theory of structures formation in the USA from that of Europe. F. E. Griggs, Jr. and A. J. DeLuzio have examined this pioneering work in detail [Griggs & DeLuzio, 1995, pp. 1358 – 60]. Whipple was the first person to develop the simple trussed framework theory using both graphical and trigonometrical methods. Based on tests, he specifies an equation for sizing cast iron struts. In addition, Whipple derives the simple beam theory and describes the phenomenon of the elastic-plastic behaviour of cast iron beams in bending (see Fig. 2 – 96). Finally, he provides a summary of fatigue behaviour, but without using the word "fatigue"! So Whipple laid the foundation for the establishment phase of structural theory

(1850 – 75) in the USA. F. E. Griggs, Jr. therefore quite rightly describes Whipple, the bridge-building practitioner and theorist, as the "Father of Iron Bridges" [Griggs, 2002, p. 146].
• Main contributions to structural analysis:
*A work on bridge building* [1847]; *An appendix to Whipple's bridge building* [1869]; *An elementary and practical treatise on bridge building* [1873]
• Further reading:
[Tyrrell, 1911]; [Parcel & Maney, 1926]; [Gasparini & Provost, 1989]; [Griggs & DeLuzio, 1995]; [Griggs, 2002]
• Photo courtesy of [Griggs & DeLuzio, 1995, p. 1356]

## WIEGMANN, RUDOLF
*16.4.1804 Adensen near Hannover (today: Nordstemmen), Hannover, †17.4.1865 Düsseldorf, Prussia
Wiegmann lost his father at the age of 11, an army officer who fell at the Battle of Waterloo. Nevertheless, he was still able to enjoy a decent upbringing, and studied in Hannover with senior building engineer Wedekind, in Darmstadt with Georg Moller and at the University of Göttingen with Otfried Müller. The subsequent study tour of Italy (1828 – 32) had a profound influence on his later creative output. In 1835 Wiegmann moved from Hannover to Düsseldorf, where in 1838 he was appointed professor of architecture and teacher of perspective drawing at the local academy of arts. Wiegmann's paper on trusses published in 1839 had actually been written three years before and had been inspired by an iron roof construction for a theatre designed by Heinrich Hübsch [Schädlich, 1967, p. 84]. Wiegmann designed a trussed beam with wrought-iron chains and joined together two such beams with a tie to form a roof-type truss, which he also suggested should be built entirely in wrought iron; he even attempted to calculate the member forces from the equilibrium of the joints. Priority for this trussed framework construction later known as the Polonceau truss was claimed not only by Wiegmann, but by the French engineers Polonceau and Emy as well. In 1909 Max Förster proposed renaming the Polonceau truss the Wiegmann-Polonceau truss [Schädlich, 1967, p. 87]. However, Wiegmann is primarily remembered for his work as an architect and author on art. For example, from 1843 onwards he served as secretary of the Art

Society for Rhineland & Westphalia; he was later attacked for this work "in such a vigorous manner that the trouble gave him consumption, whereupon he resigned from office and died shortly afterwards" [Daelen, 1897, p. 391].
• Main contributions to structural analysis:
*Über die Konstruktion von Kettenbrücken nach dem Dreiecksysteme und deren Anwendung auf Dachverbindungen* [1839]
• Further reading:
[Daelen, 1897]; [Schädlich, 1967]
• Photo courtesy of Malkasten Art Society archives, Düsseldorf

## WIERZBICKI, WITOLD
*26.1.1890 Warsaw, Russia (today: Poland), †30.1.1965 Warsaw, Poland
He gained a diploma in 1916 at the St. Petersburg Institute of Engineers of Ways of Communication and in the same year was awarded the Rippas Prize for his research into applied mechanics. After World War 1, he returned to Poland and until 1929 was employed in the Ministry of Transport redesigning the Warsaw railway junction. At the same time, he made scientific contributions to the theory of structures. He gained his doctorate in 1925 and wrote his habilitation thesis in 1926, both at the Faculty of Civil Engineering at the Politechnika Warszawska. He was in charge of the chair of forestry engineering and surveying in the Faculty of Forestry at the Agricultural University from 1929 to 1935, and thereafter professor of theory of structures at the Politechnika Warszawska until he died. Wierzbicki's textbook *Mechanika budowli* (structural mechanics) published in 1929 passed through several editions and became the standard work for the teaching of theory of structures in Poland. His greatest scientific achievement was the creation of the fundamentals for the reliability theory of structures. During World War 2 he served as a teacher at various technical education establishments in Warsaw, in the later years of the war doing so illegally. After the war, Wierzbicki played a key role in the restoration of the scientific and technical fabric of his country.
• Main contributions to structural analysis:
*Bezpieczeństwo budowli jako zagadnienie prawdopodobieństwa* (safety of structures as a probability problem) [1936]; *La securité des constructions considerée comme problème*

*de probabilité* [1946]; *Probabilistic and semi-probabilistic method for the investigation of structure safety* [1957]
- Further reading:
[Mutermilch & Janiczek, 1959];
[Mutermilch, 1965]
- Photo courtesy of Prof. Dr. Z. Cywiński

## WINKLER, EMIL
*18.4.1835 Falkenberg near Torgau, Saxony,
†27.8.1888 Friedenau near Berlin, German Empire
Following his apprenticeship as a bricklayer, Emil Winkler attended the Building Trades School in Holzminden and then studied at Dresden Polytechnic (Dresden TH) from 1854 to 1858. He then became an assistant engineer in the Saxony Waterways Department. He gained his doctorate in 1861 at the Faculty of Philosophy at the University of Leipzig with a dissertation entitled *Über den Druck im Inneren von Erdmassen*. He was employed as an assistant at Dresden Polytechnic from 1861 to 1865, where he taught design of engineering structures from 1863 onwards under Prof. Johann Andreas Schubert. He was appointed professor of engineering structures at Prague Polytechnic (Prague TH) in the autumn of 1865, and three years later full professor of railway engineering and the structural elements of bridge-building at the Polytechnic Institute in Vienna (Vienna TH). November 1877 saw him accept an appointment as professor of structural theory of buildings and bridges at the Berlin Building Academy (renamed Berlin-Charlottenburg TH in 1879). He remained active in this post until his death following a stroke. Shortly before died he was awarded an honorary doctorate by the University of Bologna.
- Main contributions to structural analysis:
*Die Lehre von der Elasticität und Festigkeit* [1867]; *Vorträge über Brückenbau* [1872–86]; *Die Lage der Stützlinie im Gewölbe* [1879/1880].
- Further reading:
[Melan, 1888/1]; [Zentralblatt der Bauverwaltung, 1888/1]; [Deutsche Bauzeitung, 1888/2]; [Kurrer, 1988]; [Knothe & Tausendfreund, 2000]; [Knothe, 2004]
- Photo courtesy of [Stark, 1906]

## WÖHLER, AUGUST
*22.6.1819 Soltau, Hannover,
†21.3.1914 Hannover, German Empire
The mathematical talents of this teacher's son became evident at an early date and – financed through an annual scholarship of 100 thaler – was able to attend the Hannover Higher Industrial School (today: Hannover TU), starting in 1835. Wöhler joined the Borsig company in Berlin (1840), worked as a locomotive driver for Hannover Railways (1843) and was senior machine master of the Lower Silesia–Märkisch Railway in Frankfurt a.d. Oder (1847), where he remained for the next 23 years. It was in this period that Wöhler was to achieve his greatest accomplishments as a research engineer. In 1855 he became the first person to publish the correct equations for calculating the deflection of continuous iron bridge beams with lattice sides [Wöhler, 1855]; in the paper he also recommends constructing the supports as rocker bearings with rollers. As early as 1854, sudden axle failures, which led to catastrophic railway accidents, were linked with the word "fatigue" in a presentation at the Institution of Civil Engineers. Wöhler and Prof. Schwarz from the Berlin Building Academy were therefore appointed to perform tests to assess the effects of rail impacts on the wheel axles of locomotives and rolling stock. After 1856 Wöhler carried on the tests alone, and discovered the relationship between stress difference and number of load repetitions (Wöhler diagram) [Wöhler, 1858–70]; he therefore founded the experimental research into fatigue. Although Wöhler's fatigue tests initially went largely unnoticed, they were continued after 1870 at the Berlin Industrial Academy (which merged with the Building Academy in 1879 to form Berlin TH) on the test apparatus available there. That work led to the creation of the mechanical-technical testing facility from which the Prussian Materials Testing Authority developed (today: Federal Institute for Materials Research & Testing). Wöhler became director at the Norddeutsche Aktiengesellschaft für Eisenbahnbedarf in Berlin in 1869 and from 1874 until his retirement in 1889 was a railway director and member of the General Directorate of German Railways in Strasbourg. Berlin TH awarded him an honorary doctorate in 1901 for his services to materials research.
- Main contributions to structural analysis:
*Theorie rechteckiger eiserner Brückenbalken mit Gitterwänden und mit Blechwänden* [1855]; *Bericht über die Versuche, welche auf der Königl. Niederschlesisch-Märkischen Eisenbahn mit Apparaten zum Messen der Biegung und Verdrehung von Eisenbahnwagen-Achsen während der Fahrt, angestellt wurden* [1858]; *Versuche zur Ermittlung der auf die Eisenbahnwagen-Achsen einwirkenden Kräfte und der Widerstandsfähigkeit der Wagen-Achsen* [1860]; *Über die Versuche zur Ermittlung der Festigkeit von Achsen, welche in den Werkstätten der Niederschlesisch-Märkischen Eisenbahn zu Frankfurt a. d. O. angestellt sind* [1863]; *Resultate der in der Centralwerkstatt der Niederschlesisch-Märkischen Eisenbahn zu Frankfurt a. d. O. angestellten Versuche über die relative Festigkeit von Eisen, Stahl und Kupfer* [1866]; *Über die Festigkeits-Versuche mit Eisen und Stahl* [1870]
- Further reading:
[Blaum, 1918]; [Kahlow, 1987]; [Knothe, 2003/3]
- Photo courtesy of [Ruske, 1971, p. 100]

## WREN, SIR CHRISTOPHER JULIUS
*20.10.1632 East Knoyle, Wiltshire, UK,
†25.2.1723 London, UK
Following the early death of his mother, Wren was looked after by his older sisters and attended Westminster School in London from 1641 to 1646, where he proved to be an outstanding scholar of Latin, mathematics and the natural sciences. Upon completing his education, he helped Charles Scarburgh in his preparations for lectures on anatomy. At the age of 17 Wren entered Wadham College in Oxford as a fellow commoner, which entitled him to eat at the same table as the fellows. He left the College in 1653 as Master of Arts. He was appointed professor of geometry at Gresham College in London in 1657 and four years later became professor of astronomy at the University of Oxford. At the age of 30 he found himself co-founder and primus inter pares of a scientific society that was to have a profound influence on the course of natural sciences: The Royal Society of London for the Promotion of Natural Knowledge. In the age of the Scientific Revolution, the Royal Society provided an adequate stage for the emerging sciences – from Copernicus to Galileo to Newton. Wren served as president of this focus of new scientific knowledge from 1680 to 1682. His architectural career had begun back in 1663, and after the Great Fire of London in 1666, Wren became the spiritus rector of the rebirth of the City of London, and was regarded as England's most important architect until well into the 20th century. Above all else, Wren is remembered for St. Paul's Cathedral, which represents a milestone in the history of building. He realised that buildings must be designed according to structural theory and not geometrical rules: "The design … must be regulated by the art of staticks, or invention of the centers of gravity, and the duly poising all parts to equiponderate; without which, a fine design will fail and prove abortive. Hence I conclude, that all designs must, in the first place, be brought to this test, or rejected" [Addis, 2002/2, p. 802]. Criticising the (geometrical) rule of Blondel, Wren turned to structural theory to determine the relationship of the clear span of a semicircular arch to the thickness required for its abutments [Addis, 1990, pp. 144–146]. To do this, he combined the self-weight of half of the arch plus the imposed load at the centre of gravity of force $F_1$ and compared them with the dead load of the abutment plus imposed load (force $F_2$). Both forces act at the ends of an unequal lever arm (see fig. 2-2c), whose centre of rotation is positioned at the inner face of the springing a (distance between centre of gravity and force $F_1$ is $l_1$, distance between centre of gravity and force $F_2$ is $l_2$). $F_2$ and hence the thickness of the abutment then follow from equilibrium condition 2-1 for the unequal lever. The influence of horizontal thrust is missing in

 WINKLER
 WÖHLER
 WREN
 YOUNG
 ZERNA
 ZHURAVSKY

Wren's approach. Later, Wren would use Hooke's discovery of the catenary arch to revise his design for St. Paul's Cathedral. Nevertheless, Wren was the first to formulate the task of theory of structures thoroughly and rounded off the orientation phase of structural theory.
- Main contributions to structural analysis: *Parentalia or Memoirs of the Family of Wrens* [1750]; *Memoirs of the Life and Works of Sir Christopher Wren* [1823]
- Further reading: [Hamilton, 1933/1934]; [Addis, 2002/2]
- Photo courtesy of [Szabó, 1996, p. 443]

## YOUNG, THOMAS

*13.6.1773 Milverton, Somerset, UK,
†10.5.1829 London, UK
Thomas Young studied medicine at universities in London, Edinburgh and Göttingen. Following completion of his studies (1796), he continued his research at the University of Cambridge. Young discovered the interference of light in 1801 and helped the wave theory of light (which can be traced back to Huygens) to achieve a breakthrough. He was elected to the Royal Society in that same year and appointed professor of natural philosophy (natural sciences) at the Royal Institution in London in 1802. Young was not a good teacher and resignedly gave up his position after just one year in order to publish his lectures [Young, 1807]. This work with its comprehensive bibliography became a guideline for the technical literature of that period – strength of materials in particular. In the book, Young introduces the concept of the modulus of elasticity and realises that the design of loadbearing elements should be based on the yield point of a material in addition to its tensile strength. His masonry arch theory published in the *Bridge* article in the supplement to the fourth edition of *Encyclopaedia Britannica* in 1817 was several decades ahead of its time; unfortunately, Young's arch theory was not adopted [Huerta, 2005 & 2006]. From 1811 onwards he worked as a doctor at St. George's Hospital in London. He was also the Secretary for External Relations of the Royal Society, Secretary of the Royal Commission on Weights & Measures and a scientific adviser to the Admiralty. His excellent command of languages was certainly one of the reasons why he was asked to help decipher the cuneiform writing on the Rosetta Stone. Young's findings in the field of strength of materials formed an important foundation for the structural theory of the discipline-formation period.
- Main contributions to structural analysis: *A course of lectures in Natural Philosophy and the Mechanical Arts* [1807]; *Encyclopaedia Britannica (Supplement) (Carpentry & Bridge articles)* [1817]
- Further reading: [Peacock, 1855]; [Timoshenko, 1953]; [Kahlow, 1994]; [Beal, 2000]; [James, 2002]; [Huerta, 2005]; [Huerta, S., 2006]; [Beal, 2007]
- Photo courtesy of Dr. W. Addis

## ZERNA, WOLFGANG

*11.10.1916 Berlin, German Empire,
†14.11.2005 Celle, Germany
Zerna studied structural engineering at Berlin TH, where his tutors included Franz Dischinger, August Hertwig, Ferdinand Schleicher and Arnold Agatz. After gaining his diploma in June 1940, there followed four years of military service and subsequent internment after the war before he was able to return to Germany in 1946. He then took a post as assistant to Alf Pflüger at the chair of theory of structures and structural steelwork at Hannover TH. He completed his doctorate in June 1947 with a dissertation on the membrane theory of general shells of revolution [Zerna, 1949/2]. Zerna finished his habilitation thesis in mechanics in September 1948 with a work on the basic equations of elastic theory [Zerna, 1950]. Afterwards, he was a guest lecturer at the Department of Mathematics at the University of Durham, UK (October 1948 to September 1949), where together with A. E. Green he wrote the monograph *Theoretical Elasticity* [Green & Zerna, 1954], which was to become the standard work of reference for elastic theory based on tensor analysis during the innovation phase of structural theory (1950–75). After his return from England, he first worked at Polensky & Zöllner (Cologne) and then at Ph. Holzmann AG (Frankfurt a. M.), where in the end he became responsible for all prestressed concrete works. He held the chair of concrete and masonry construction at Hannover TH from 1957 to 1967 and after that was professor for concrete and masonry construction at the Institute of Structural Engineering at the Ruhr University in Bochum (founded in 1963), where he remained until being transferred to emeritus status in 1983. Zerna's chair became the cornerstone of the Institute of Structural Engineering and "a highly acclaimed research institute with what was at that time a new style of theoretical-numerical-experimental research" [Krätzig et al., 2006, p. 177]. Zerna and his colleagues became known for their pioneering work on the design of large natural-draught cooling towers and nuclear engineering structures; in particular, Zerna's research work had a major influence on the design and construction of prestressed concrete reactor pressure vessels. Furthermore, Zerna inspired the establishment of two highly successful consulting engineering practices in Bochum. The honorary doctorates awarded to him by the universities of Stuttgart and Essen are just two of the many honours he received for his services to science.
- Main contributions to structural analysis: *Beitrag zur allgemeinen Schalenbiegetheorie* [1949/1]; *Zur Membrantheorie der allgemeinen Rotationsschalen* [1949/2]; *Grundgleichungen der Elastizitätstheorie* [1950]; *Theoretical Elasticity* [1954]; *Rheologische Beschreibung des Werkstoffes Beton* [1967]; *Beuluntersuchungen an hyperbolischen Rotationsschalen* [1974]; *Kriterien zur Optimierung des Baues von Großnaturzugkühltürmen im Hinblick auf Standsicherheit, Bauausführung und Wirtschaftlichkeit* [1978]
- Further reading: [Krätzig et al., 1976]; [Krätzig et al., 2006]
- Photo courtesy of Prof. Dr. W. B. Krätzig

## ZHURAVSKY (JOURAVSKI), DMITRY IVANOVICH

*29.12.1821 Beloye (Kursk), Russia,
†30.11.1891 St. Petersburg, Russia
After attending Nezhin secondary school, he graduated from the St. Petersburg Institute of Engineers of Ways of Communication in 1842. He subsequently worked on bridges and railways, where the Howe truss was very widespread; Zhuravsky analysed the truss theoretically and calculated the shear stresses in beams. In 1865 he received the Demidov Prize of the St. Petersburg Academy of Sciences for his frame theory. He was in charge of the conversion of the top of the tower to the SS Peter & Paul Cathedral in St. Petersburg from 1857 to 1858, designing and providing the structural

calculations for an iron space frame. He was appointed colonel of the Corps of Engineers of Ways of Communication in 1859, and from 1877 to 1884 he was director of the Railway Department in the Ministry of Public Works. After that he was head of the Technical Department of the Council of Ministries. He retired as a Privy Counsellor in 1889 and was awarded the Order of the Holy Alexander Nevsky, 1st class.

- Main contributions to structural analysis: *Remarques sur la résistance d'un corps prismatique et d'une pièce composée en bois ou en tôle de fer à une force perpendiculaire à leur longueur* [1856]; *Remarques sur les poutres en treillis et les poutres pleines en tôle* [1860]; *O slozhnoi sisteme, predstavlyayushchei soedinenie arki s raskosnoyu sistemoyu* (on a complex system of an arch braced by a system of struts) [1864]
- Further reading: [Gersevanov, 1897]; [Timoshenko, 1950]; [Csonka, 1960]; [Rakcheev, 1984]
- Photo courtesy of Prof. Dr. G. Mikhailov

## ZIMMERMANN, HERMANN

*17.12.1845 Langensalza, Saxony,
†3.4.1935 Berlin, Germany

He served as a sailor from 1862 to 1869 before studying mechanical and civil engineering at Karlsruhe Polytechnic, where he was especially influenced by Grashof and Sternberg. He gained his doctorate at the University of Leipzig in 1874 with a subject from contact kinematics. After that, he was employed as a civil engineer in the head office of German Railways in Strasbourg and from 1881 to 1891 in the administration offices of German Railways in Berlin. During this period he published a book on the theory of railway permanent ways which remained the basis for calculating the stresses in rails and sleepers and the pressure on ballast until the 1960s. He became Schwedler's successor at the Prussian Ministry of Public Works in 1891 and was therefore the most senior technical civil servant of Prussian State Railways. He designed the dome over the plenary hall of the new German parliament building after the architect Paul Wallot failed to deliver a solution. Zimmermann published his findings on the analysis of space frames in 1901. In that same year, Mohr used the structural analysis of the Zimmermann dome to start a dispute with Müller-Breslau about the priority of the equivalent member method; many other structural engineers took part in this debate on the theory of space frames. After Müller-Breslau, Zimmermann was the second engineer to be elected a full member of the Prussian Academy of Sciences (1904). Before very long, he was using the minutes of the academy's meetings to investigate questions of buckling theory. He retired in 1911. Zimmermann was awarded numerous state and academic honours.

- Main contributions to structural analysis: *Die Berechnung des Eisenbahnoberbaues* [1888]; *Über Raumfachwerke, neue Formen und Berechnungsweisen für Kuppeln und sonstige Dachbauten* [1901]; *Die Lehre vom Knicken auf neuer Grundlage* [1930]
- Further reading: [Beyer, 1925]; [Kulka, 1930]; [Hertwig, 1950]
- Photo courtesy of [Kulka, 1930, p. 881]

## ZWEILING, KLAUS

*18.2.1900 Berlin, German Empire,
†18.11.1968 Berlin, German Democratic Republic

He was born in Berlin-Moabit, the son of Adolf Zweiling, who was later to become Privy Counsellor and Principal of the German Patents Office, and his wife Olara Zweiling (née Gosselmann). He passed his university entrance examination in 1917, but then had to serve on the land in Falkenberg/Mark before becoming a draughtsman with the Artillery Testing Commission in Berlin (October 1917 to September 1918) and serving as a soldier (September to December 1918). Afterwards, he studied mathematics and physics in Berlin und Göttingen, finishing in Göttingen with the dissertation *Eine graphische Methode zur Bestimmung von Planeten- und Kometenbahnen* (published in 1923). During his time at university he attended lectures on philosophy, history, art history, economics and ancient languages; he was fluent in Latin, Greek, French, English and Italian. From 1918 onwards he read Marxist literature – especially that of Karl Marx, Friedrich Engels, Franz Mehring, August Bebel, Rosa Luxemburg and Vladimir I. Lenin. He worked as a laboratory physicist and deputy laboratory manager at G. Lorenz AG in Berlin-Tempelhof from January 1923 to May 1924, where he was reprimanded for taking part in the May Day celebrations of 1924. As Zweiling was now on the blacklist of the Society of Berlin Metalworking Industrialists, he could no longer pursue his career. From September 1924 onwards he worked as an editor for various workers' newspapers in Münster (*Volkswille*), Zwickau (*Sächsisches Volksblatt*) and Plauen (chief editor of the *Volkszeitung*) and found ample work as a teacher of Marxism for the workers' parties. He co-founded the Socialist Workers Party (SAP) in 1931 – a left-wing SPD splinter group – and became editor of the Berlin-based *Sozialistische Arbeiter-Zeitung* – the mouthpiece of the SAP. Zweiling lost his job again in 1932 and worked without pay in the research laboratory of Dr. Otto Emersleben, investigating high-frequency vibrations. When Hitler came to power in 1933, the workers' movement was quickly disbanded and that abruptly put an end to Zweiling's activities as the "organischer Intellektueller" (Antonio Gramsci) of the workers' movement. He was arrested by the Gestapo on 22 August 1933 and sentenced to three years imprisonment for incitement to high treason. Following his release from prison on 1 September 1936 he was unemployed. He acted as a freelance mathematician advising university professors and checking engineers on questions of the application of mathematics. Until 1943 he thus scraped a living as a semi-legal intellectual prevented from publishing. But Zweiling's misfortune to be a victim of Nazi persecution had already borne creative fruits in the late 1930s in the form of a more rigorous theoretical basis for the theory of structures. The checking engineer Arno Schleusner entrusted Zweiling with a comprehensive stability investigation of the compression chords of the long-span cantilevering steel girder construction at Berlin-Tempelhof airport. Zweiling solved the stability problem of the multi-span elastically supported members on the basis of calculus of variations [Graße & Steup, 1999, pp. 5–6]. He also used this method for his theoretical explanation of the specific differences between the principle of virtual displacements and the principle of virtual forces. Zweiling verified that the latter, in contrast to the former, cannot be directly shown physically, but is merely a mathematical reformulation of the former under the restrictive condition of minor displacements [Graße & Steup, 1999, pp. 3–4]. As Zweiling was not allowed to publish, Arno Schleusner declared his readiness to publish the findings under his name in the journal *Der Stahlbau* [Schleusner, 1938]. In a manuscript that has unfortunately never been published, the Dresden-based professors Wolfgang Graße and Herbert Steup (of the Gustav Bürgermeister school) pay tribute to Zweiling's contributions to the treatment of the plate problems of elastic theory with the help of the theory of biharmonic polynomes and to the convergence of the iteration methods of Vianello [Vianello, 1898] and Engesser [Engesser, 1909] for determining the buckling load on struts with varying second moment of area [Graße & Steup, 1999]. During World War 2, Zweiling was at first declared unfit for service, but in the end had to serve on the eastern front. Following the Allies' victory over Hitler, Zweiling resumed his editorial activities until he was appointed manager and chief editor of *Verlag Technik* (Berlin) in 1949. The German Democratic Republic (GDR) was founded in the same year and Zweiling became a Marxist theorist and academic teacher in the field of the relationship between philosophy and natural sciences – his habilitation thesis on materialism and natural sciences (1948) had already indicated this (linked to a lectureship in philosophy at Berlin's Humboldt University). He worked as a lecturer in philosophy at that university from 1949 to 1955 and during that period brought together his articles on the theory of structures in two monographs [Zweiling, 1952; 1953]. Afterwards, he was appointed professor of dialectic and historical

ZIMMERMANN     ZWEILING

materialism with responsibility for the study of philosophy at Berlin's Humboldt University. He took charge of the Institute of Philosophy at Karl Marx University in Leipzig from 1960 to 1965. Besides Georg Klaus, Hermann Ley and, later, Herbert Hörz, Zweiling made major contributions to the philosophical analysis of natural sciences from the viewpoint of the Marxist thinking in the GDR. His monograph *Gleichgewicht und Stabilität* had a lasting influence on the successful theory style of Gustav Bürgermeister and his students.

- Main contributions to structural analysis: *Das Prinzip der virtuellen Verrückungen und die Variationsprinzipien der Elastizitätstheorie* [1938]; *Grundlagen einer Theorie der biharmonischen Polynome* [1952]; *Gleichgewicht und Stabilität* [1953]
- Further reading:
[Zweiling, 1948]; [Cadre Department, Karl Marx University Leipzig, 1968]; [Dickhoff, 1982]; [Graße & Steup, 1999]
- Photo courtesy of University of Leipzig archives

# BIBLIOGRAPHY

Abbot, H. L., 1886. Memoir of Dennis Hart Mahan 1802–1871. Read before the National Academy, Nov. 7, 1878. In: *Biographical Memoirs of National Academy of Sciences*, vol. ii, pp. 31–37. Washington: National Academy of Sciences.

Abeles, C., 1907. Statische Untersuchung einiger im Eisenbetonbau häufig vorkommenden Aufgaben. *Beton und Eisen*, vol. 6, No. 5, pp. 128–131 & No. 6, pp. 154–157.

Addis, W. (Bill), 1990. *Structural Engineering. The Nature of Theory and Design*. Chichester: Ellis Horwood.

Addis, W. (Bill), 2002/1. Hooke, Robert. In: *A Biographical Dictionary of Civil Engineers in Great Britain and Ireland, vol. 1: 1500–1830*. Ed. A. W. Skempton, M. M. Chrimes, R. C. Cox, P. S. M. Cross-Rudkin, R. W. Rennison & E. C. Ruddock, pp. 334–337. London: ICE & Thomas Telford.

Addis, W. (Bill), 2002/2. Wren, Sir Christopher. In: *A Biographical Dictionary of Civil Engineers in Great Britain and Ireland, vol. 1: 1500–1830*. Ed. A. W. Skempton, M. M. Chrimes, R. C. Cox, P. S. M. Cross-Rudkin, R. W. Rennison & E. C. Ruddock, pp. 799–802. London: ICE & Thomas Telford.

Addis, W. (Bill), 2007. Building: *3000 Years of Design Engineering and Construction*. London/New York: Phaidon.

Affan, A., Calladine, C. R., 1986. Structural mechanics of double-layer space grids. In: *Proceedings, IASS Symposium, Osaka, 1986*, vol. 3, ed. K. Heki, pp. 41–48. Amsterdam: Elsevier Science Publishers B.V.

Affan, A., Calladine, C. R., 1989. Initial bar tensions in pin-jointed assemblies. *International Journal of Space Structures*, vol. 4, No. 1, pp. 1–16.

Aimond, F., 1936. Étude statique des voiles minces en paraboloïde hyperbolique travaillant sans flexion. In: *Proceedings of the International Association for Bridge & Structural Engineering (IABSE)*, vol. 4, ed. L. Karner & M. Ritter, pp. 1–112. Zurich: IABSE secretariat.

Airy, G. B., 1863. On the strains in the interior of beams. In: *Philosophical Transactions of the Royal Society of London*, vol. 153, pt. 1, pp. 49–80.

Airy, W., 1896. *Autobiography of Sir George Biddell Airy*. Cambridge: At the University Press.

Albert, J., Herlitzius, E., Richter, F., 1982. *Entstehungsbedingungen und Entwicklung der Technikwissenschaften*. Leipzig: VEB Deutscher Verlag für Grundstoffindustrie.

Anders, G., 1985. *Die Antiquiertheit des Menschen. Erster Band: Über die Seele im Zeitalter der zweiten industriellen Revolution*. 7th ed. (1st ed. 1956). Munich: C. H. Beck.

Andreaus, U. A., Ruta, G. C., 1998. A review of the problem of the shear centre(s). *Continuum Mechanics and Thermodynamics*, vol. 10, pp. 369–380.

Andrée, W. L., 1908. *Die Statik des Kranbaues*. Munich: R. Oldenbourg.

Andrée, W. L., 1913. *Die Statik des Kranbaues mit Berücksichtigung der verwandten Gebiete Eisenhoch-, Förder- und Brückenbau*, 2nd ed. Munich: R. Oldenbourg.

Andrée, W. L., 1917. *Die Statik des Eisenbaues*. Munich: R. Oldenbourg.

Andrée, W. L., 1919/1. *Zur Berechnung statisch unbestimmter Systeme. Das B-U-Verfahren*. Munich: R. Oldenbourg.

Andrée, W. L., 1919/2. *Die Statik der Schwerlastkrane. Werft- und Schwimmkrane und Schwimmkranpontons*. Munich: R. Oldenbourg.

Angerer, T., Rubin, D., Taus, M., 1999. Verbundstützen und Querkraftanschlüsse der Verbundflachdecken beim Millennium Tower. *Stahlbau*, vol. 68, No. 8, pp. 641–646.

Arbocz, J., 2000. Preface to W. T. Koiter Commemorative Issue on Stability, Strength and Stiffness in Materials and Structures. *International Journal of Solids and Structures*, vol. 37, pp. 6773–6775.

Ardant, P., 1847. Theoretische und auf Erfahrung gegründete Studien über die Einrichtung der Zimmerungen von großer Spannung. *Zeitschrift für Praktische Baukunst*, pp. 59–112 & 145–210.

Argyris, J. H., 1940. Untersuchung eines besonderen Belastungsfalles bei dreiseitig abgespannten Funkmasten. *Der Stahlbau*, vol. 13, No. 8/9, pp. 33–38.

Argyris, J. H., Dunne, P. C., 1952. Structural Analysis. In: *Structural Principles and Data, Part 2. Handbook of Aeronautics No. 1*, 4th ed. London: Sir Isaac Pitman & Sons, Ltd.

Argyris, J. H., 1954. Energy Theorems and Structural Analysis. Part I. General Theory. *Aircraft Engineering*, vol. 26, pp. 347–387 & 394.

Argyris, J. H., 1955. Energy Theorems and Structural Analysis. Part I. General Theory. *Aircraft Engineering*, vol. 27, pp. 42–58, 80–94, 125–134 & 145–158.

Argyris, J. H., Kelsey, S., 1954. Energy Theorems and Structural Analysis. Part II. Applications to Thermal Stress Problems and St. Venant Torsion. *Aircraft Engineering*, vol. 26, pp. 410–422.

Argyris, J. H., 1957. Die Matrizentheorie der Statik. *Ingenieur-Archiv*, vol. 25, pp. 174–192.

Argyris, J. H., Kelsey, S., 1960. *Energy theorems and structural analysis*. London: Butterworth.

Argyris, J. H., Kelsey, S., 1963. *Modern Fuselage Analysis and the Elastic Aircraft*. London: Butterworth.

Argyris, J. H., 1964. *Recent Advances on Matrix Methods in Structural Analysis*. Oxford: Pergamon Press.

Argyris, J. H., 1965. The Computer shapes the theory. *Journal of the Royal Aeronautical Society*, vol. 65, p. XXXII.

Argyris, J. H., Mlejnek, H.-P., 1986. *Die Methode der Finiten Elemente. Band I*. Braunschweig: Vieweg.

Argyris, J. H., Mlejnek, H.-P., 1987. *Die Methode der Finiten Elemente. Band II*. Braunschweig: Vieweg.

Argyris, J. H., Mlejnek, H.-P., 1988. *Die Methode der Finiten Elemente. Band III*. Braunschweig: Vieweg.

Argyris, J. H., Mlejnek, H.-P., 1991. *Dynamics of Structures*. Amsterdam: Elsevier.

Argyris, J. H., Mlejnek, H.-P., 1997. *Computerdynamik der Tragwerke*. Braunschweig: Vieweg.

Aron, H., 1874. Über das Gleichgewicht und die Bewegung einer unendlich dünnen, beliebig gekrümmten elastischen Schale. *Journal für die reine und angewandte Mathematik*, vol. 78, pp. 136–174.

Arrighi, G., 1976. Nuovo contributo al carteggio Guido Grandi – Tommaso Narducci. Storia di una polemica – una memoria sconosciuta del matematico di Cremona 'Sopra le curve geometriche, o meccaniche'. *Physis – Rivista Internazionale di Storia della Scienza*. vol. 18, pp. 366–382.

Arup, O., Jenkins, R. S., 1968. The Evolution of the design of the concourse at the Sydney Opera House. *Proc. Instn. Civ. Engrs.*, vol. 39, April, pp. 541–565.

Arup, O., 1976. Ronald Jenkins, *Newsletter*, No. 94, Jan/Feb, pp. 281–282.

Asplund, S. O., 1945. *On the Deflection Theory of suspension Bridges*. Stockholm: General Staff Litho. Dept.

Asplund, S. O., 1956. Generalized Elastic Theory for Pile Groups. In: *Proceedings of the International Association for Bridge & Structural Engineering (IABSE)*, vol. 16, ed. F. Stüssi & P. Lardy, pp. 1–22. Zurich: IABSE secretariat.

Asplund, S. O., 1958. Column-Beams and suspension bridges analyzed by 'Green's matrix'. Gothenburg: Gumperts.

Asplund, S. O., 1961. A unified Analysis of indeterminate structures. Gothenburg: Gumperts.

Asplund, S. O., 1963. *Practical Calculation of suspension bridges*. Gothenburg: Gumperts.

Asplund, S. O., 1966. *Structural Mechanics: classical and matrix methods*. New Jersey: Prentice-Hall.

Aster, R., 1968. Stahlbetonbestimmungen der Deutschen Demokratischen Republik. In: *Beton-Kalender*, vol. 57, pt. II, pp. 416–535. Berlin: Wilhelm Ernst & Sohn.

Atrek, E., Gallagher, R. H., Ragsdell, K. M., Zienkiewicz, O. C. (ed.), 1984. *New Directions in Optimum Structural Design*. New York: Wiley.

Augusti, G., Sinopoli, A., 1992. Modelling the dynamics of large block structures. *Meccanica*, vol. 27, No. 3, pp. 195–211.

Augusti, G., 1999. Un ricordo di Antonino Giuffrè ed Edoardo Benvenuto. *IX Convegno Nazionale L'Ingegneria Sismica in Italia, Torino, 20–23 Settembre*. Torino: CD-ROM.

Autenrieth, E., 1887/1. Über die Berechnung der Anker, welche zur Befestigung von Platten an ebenen Flächen dienen. *ZVDI*, vol. 31, No. 17, pp. 341–345.

Autenrieth, E., 1887/2. Letter. *ZVDI*, vol. 31, No. 22, p. 460.

Babuška, I., 1987. Comments on the development of computational mathematics in Czechoslovakia and in the USSR. In: *Proceedings of the ACM conference on History of scientific and numeric computation*, ed. G. E. Crane, pp. 95a–95g. New York: ACM Press.

Bach, C., 1889/1890. *Elastizität und Festigkeit*. Berlin: Julius Springer.

Bach, C., 1890. Versuche über die Widerstandsfähigkeit ebener Platten. *Zeitschrift des Vereines deutscher Ingenieure*, vol. 34, pp. 1041–1048, 1080–1086, 1103–1111 & 1139–1144.

Bach, C., 1909. Versuche über die tatsächliche Widerstandskraft von Balken mit U-förmigem Querschnitt. *Zeitschrift des Vereines deutscher Ingenieure*, vol. 53, pp. 1790–1795.

Bach, C., 1910. Versuche über die tatsächliche Widerstandskraft von Trägern mit U-förmigem Querschnitt. *Zeitschrift des Vereines deutscher Ingenieure*, vol. 54, pp. 382–387.

Bach, C., Graf, O., 1911/1. *Versuche mit Eisenbeton-Balken zur Ermittlung der Widerstandsfähigkeit verschiedener Bewehrung gegen Schubkräfte. Erster Teil*. Deutscher Ausschuß für Eisenbeton, No. 10. Berlin: Wilhelm Ernst & Sohn.

Bach, C., Graf, O., 1911/2. *Versuche mit Eisenbeton-Balken zur Ermittlung der Widerstandsfähigkeit verschiedener Bewehrung gegen Schubkräfte. Zweiter Teil*. Deutscher Ausschuß für Eisenbeton, No. 12. Berlin: Wilhelm Ernst & Sohn.

Bach, C., Graf, O., 1915. *Versuche mit allseitig aufliegenden quadratischen und rechteckigen Eisenbetonplatten*. Deutscher Ausschuß für Eisenbeton, No. 15. Berlin: Wilhelm Ernst & Sohn.

Bach, C., 1920. *Elastizität und Festigkeit*. 8th ed., with Prof. R. Baumann. Berlin: Julius Springer.

Bach, C., 1926. *Mein Lebensweg und meine Tätigkeit. Eine Skizze*. Berlin: Julius Springer.

Bachmann, O., Cohrs, H.-H., Whiteman, T., Wislicki, A., 1997. *Krantechnik von der Antike zur Neuzeit*. Isernhagen: Giesel Verlag für Publizität GmbH.

Badr, I. E. A., 1962. *Vom Gewölbe zum räumlichen Tragwerk*. Dissertation, Zurich ETH. Zurich.

Baer, 1929. Karl Bernhard zu seinem 70. Geburtstag. *Der Bauingenieur*, vol. 10, No. 45, pp. 794–795.

Baker, J. F., 1932. The Mechanical and Mathematical Analysis of Steel Building Frames. *J. Instn. Civ. Engrs.*, Sept, No. 131.

Baker, J. F., 1935/36. The Rational Design of Steel Building Frames. *J. Instn. Civ. Engrs.*, vol. 3, pp. 127–210.

Baker, J. F., 1936. A new method for the design of steel building frames. In: *Proceedings of the International Association for Bridge & Structural Engineering (IABSE)*, vol. 4, ed. L. Karner & M. Ritter, pp. 113–129. Zurich: IABSE secretariat.

Baker, J. F., 1936/37. Modern Methods of Structural Design. *J. Instn. Civ. Engrs.*, vol. 6, pp. 297–324.

Baker, J. F., 1949. The design of steel frames. *The Structural Engineer*, vol. 27, pp. 397–431.

Baker, J. F., 1954. *The steel skeleton 1: Elastic behaviour and design*. Cambridge: Cambridge University Press.

Baker, J. F., Horne, M. R., Heyman, J., 1956. *The steel skeleton 2: Plastic behaviour and design*. Cambridge: Cambridge University Press.

Baker, J. F., Heyman, J., 1969. *Plastic Design of Frames 1: Fundamentals*. Cambridge: Cambridge University Press.

Baldi, B., 1621. *In mechanica Aristotelis problemata exercitationes. Adiecta succinta narratione de autoris vita et scriptis*. Moguntiae: Viduae Ioannis Albini.

Balet, J. W., 1908. *Analysis of elastic arches three-hinged, two-hinged and hingeless of steel, masonry and reinforced concrete*. New York: The Engineering News Publishing Company. London: Archibald Constable & Co. Ltd.

Ban, S., 1935. Knickung der rechteckigen Platte bei veränderlicher Randbelastung. In: *Proceedings of the International Association for Bridge & Structural Engineering (IABSE)*, vol. 3, pp. 1–18. Zurich.

Ban, S., 1953. Formänderung der hyperbolischen Paraboloidschale. In: *Proceedings of the International Association for Bridge & Structural Engineering (IABSE)*, vol. 13, pp. 1–16. Zurich.

Bandhauer, C. G. H., 1829. *Verhandlungen über die artistische Untersuchung des Baues der Hängebrücke über die Saale bei Mönchen-Nienburg*. Leipzig: C. H. F. Hartmann.

Bandhauer, C. G. H., 1831. *Bogenlinie des Gleichgewichts oder der Gewölbe und Kettenlinien*. Leipzig: Hartmann'sche Buchhandlung.

Banse, G., 1976. Philosophische Fragen der technischen Wissenschaften – Probleme und Ergebnisse. *Dt. Ztschr. f. Philos.*, vol. 24, No. 3, pp. 307–318.

Banse, G., 1978: Technikwissenschaften. In: H. Hörz, R. Löther, S. Wollgast (ed.): *Philosophie und Naturwissenschaften. Wörterbuch zu den philosophischen Fragen der Naturwissenschaften*, pp. 904–907. Berlin: Dietz.

Banse, G., Wendt, H. (ed.), 1986: *Erkenntnismethoden in den Technikwissenschaften. Eine methodologische Analyse und philosophische Diskussion der Erkenntnisprozesse in den Technikwissenschaften*. Berlin: VEB Verlag Technik.

Banse, G., Grunwald, A., König, W., Ropohl, G. (ed.), 2006. *Erkennen und Gestalten. Eine Theorie der Technikwissenschaften*. Berlin: edition sigma.

Bargmann, H., 1990. Heinz Parkus und die Wiener Schule der Mechanik. *Österreichische Ingenieur- und Architekten-Zeitschrift (ÖIAZ)*, vol. 135, No. 10, pp. 554–560.

Barlow, P., 1817. *An Essay on the Strength and Stress of Timber.* London.

Barlow, P., 1835. *Experiments on the transverse strength and other properties of malleable iron with reference to its uses for railway bars; and a report founded on the same, Addressed to the directors of the London and Birmingham Railway company.* London.

Barlow, P., 1867. *An Essay on the Strength and Stress of Timber.* 6th ed., London.

Barlow, W. H., 1846. On the Existence (practically) of the line of equal Horizontal Thrust in Arches, and the mode of determining it by Geometrical Construction. *Minutes and Proceedings of the Institution of Civil Engineers,* vol. 5, pp. 162–182.

Barthel, R., 1993/1. Tragverhalten gemauerter Kreuzgewölbe. In: *Aus Forschung und Lehre,* No. 26. Karlsruhe: University of Karlsruhe, 1993.

Barthel, R., 1993/2. Tragverhalten und Berechnung gemauerter Kreuzgewölbe. *Bautechnik,* vol. 70, No. 7, pp. 379–391.

Barthel, R., 1994. Tragverhalten und Berechnung gemauerter Kreuzgewölbe. In: *Erhalten historisch bedeutsamer Bauwerke,* Jahrbuch 1992, ed. F. Wenzel, pp. 21–40. Berlin: Ernst & Sohn.

Başar, Y., Krätzig, W. B., 2001. *Theory of Shell Structures.* 2nd ed. Research Reports, VDI series 18, No. 258. Düsseldorf: VDI-Verlag.

Bathe, K.-J., Tinsley, J., Wunderlich, W. (ed.), 1977. *Formulations and Computational Algorithms in Finite Element Analysis.* U. S.–Germany Symposium 1976, Cambridge, Mass. MIT.

Bauschinger, J., 1871. *Elemente der graphischen Statik.* Munich: R. Oldenbourg.

Bauschinger, J., 1884. Untersuchungen über die Elasticität und Festigkeit der wichtigsten natürlichen Bausteine in Bayern. In: *Mittheilungen aus dem mechanisch-technischen Laboratorium der K. Technischen Hochschule in Munich,* Nr. 10. Munich.

Bay, H., 1990. Emil Mörsch. Erinnerungen an einen großen Lehrmeister des Stahlbetonbaus. In: *Wegbereiter der Bautechnik,* ed. VDI-Gesellschaft Bautechnik, pp. 47–66. Düsseldorf: VDI-Verlag.

Bažant, Z., 1909/1910/1912. *Teorie příčinkových čar* (theory of influence lines), 3 vols. Prague: SPI (in Czech).

Bažant, Z., 1917/1946. *Statika stavebních konstrukcí* (manual of theory of structures). Prague: ČMT (in Czech).

Bažant, Z., 1918/1920/1946/1950. *Stavebná mechanika* (structural mechanics), 4 vols. Prague: ČMT & VTN (in Czech).

Beal, A. N., 2000. Who invented Young's Modulus? *The Structural Engineer,* vol. 78, No. 14, pp. 27–32.

Beal, A. N., 2007. Thomas Young and the theory of structures 1807–2007. *The Structural Engineer,* vol. 85, No. 23, pp. 43–47.

Becchi, A., 1988. *Radici storiche della teoria molecolare dell'elasticità con particolare riguardo alla Theoria Philosophiae Naturalis di R. G. Boscovich.* Tesi di laurea, University of Genoa.

Becchi, A., 1994/1. *I criteri di plasticità: cent anni di dibattito (1864–1964).* Doctorate thesis, University of Florence.

Becchi, A., Corradi, M., Foce, F., 1994/2. Dibattiti e interpretazioni sul compartamento statico delle Strutture voltate in Muratura. *Scienza e beni Culturali,* vol. X, pp. 243–250.

Becchi, A., 1998. M. Lévy versus J. de La Gournerie: a debate about skew bridges. In: *Proceedings of the Second International Arch Bridge Conference,* ed. A. Sinopoli, pp. 65–72. Rotterdam: Balkema.

Becchi, A., 2002. *Zwischen Mechanik und Architektur: die Gewölbetheorie an der Architekturakademie in Paris (1687–1718).* Lecture of 24 Oct 2002 at the German Science Museum, Berlin. Berlin: "Engineering History" study group of the VDI Berlin-Brandenburg District Society.

Becchi, A., Corradi, M., Foce, F., Pedemonte, O. (ed.), 2002. *Towards a History of Construction.* Dedicated to Edoardo Benvenuto. Basel: Birkhäuser.

Becchi, A., Foce, A., 2002. *Degli archi e delle volte. Arte del costruire tra meccanica e stereotomia.* Venice: Marsilio Editori.

Becchi, A., 2003. Before 1695: The statics of arches between France and Italy. In: *Proceedings of the First International Congress on Construction History,* ed. S. Huerta, vol. I, pp. 353–364. Madrid: Instituto Juan de Herrera.

Becchi, A., Corradi, M., Foce, F., Pedemonte, O. (ed.), 2003. *Essays in the History of Mechanics.* Basel: Birkhäuser.

Becchi, A., 2004/1. *Q. XVI. Leonardo, Galileo e il caso Baldi: Magonza, 26 marzo 1621.* Traduzione die testi latini, note e glossario a cura di Sergio Aprosio. Venice: Marsilio Editori.

Becchi, A., Corradi, M., Foce, F., Pedemonte, O. (ed.), 2004/2. *Construction History. Research Perspectives in Europe.* Florence: Kim Williams Books.

Becchi, A., 2005. Vaults in the air: Signor Fabritio's English theory. In: *Essays in the history of theory of structures,* ed. S. Huerta, pp. 45–59. Madrid: Instituto Juan de Herrera.

Beck, W., 1956. Julius Weisbach. Gedenkschrift zu seinem 150. Geburtstag. *Freiberger Forschungshefte Reihe D,* No. 16, ed. dean of Freiberg Mining Academy. Berlin: Akademie-Verlag.

Becker, F., 1930. *Die Industrialisierung im Eisenbetonbau.* Dissertation, Karlsruhe TH.

Beedle, L. S., 1984. Fazlur Rahman Khan. In: *Memorial Tributes.* National Academy of Engineering, vol. 2, pp. 152–156. Washington, D.C.: National Academy Press.

Beer, H., 1959. †Prof. Karl Girkmann. *Der Bauingenieur,* vol. 34, No. 10, pp. 414–415.

Beer, H., 1960. †Prof. Dr. techn. Dr.-Ing. E. h. Ernst Chwalla. *Der Stahlbau,* vol. 29, No. 7, pp. 223–224.

Belanger, J. B., 1858. *Théorie de la résistance et la flexion plane des solides.* Paris.

Belanger, J. B., 1862. *Théorie de la résistance et la flexion plane des solides.* 2nd ed. Paris.

Belhoste, B., 1991. *Augustin-Louis Cauchy. A Biography.* Berlin: Springer.

Belhoste, B., Dahan Dalmedico, A., Picon, A. (ed.), 1994. *La formation polytechnicienne 1794–1994.* Paris: Dunod.

Belhoste, B., 2003. *La Formation d'une Technocratie. L'École polytechnique et ses élèves de la Révolution au Second Empire.* Paris: Belin.

Bélidor, B. F. de, 1725. *Nouveau cours de Mathématique a l'Usage de l'Artillerie et du Génie.* Paris: Chez Charles-Antoine Jombert.

Bélidor, B. F. de, 1729. *La science des ingénieurs dans la conduite des travaux de fortification et d'architecture civile.* Paris: C.-A. Jombert.

Bélidor, B. F. de, 1737–1753. *Architecture hydraulique, ou l'art de conduire, d'élever et de ménager les eaux pour les différens besoins de la vie.* Paris: C.-A. Jombert.

Bélidor, B. F. de, 1729/1757. *Ingenieur-Wissenschaft bey aufzuführenden Vestungswerken und bürgerlichen Gebäuden, 1. Teil.* Trans. from the French. Nuremberg: Christoph Weigels.

Bélidor B. F. de, 1813. *La Science des ingénieurs dans la conduite des travaux de fortification et d'architecture civile, par Bélidor. Nouvelle édition avec des notes par M. Navier.* Paris: F. Didot.

Bélidor, B. F. de, 1819. *Architecture hydraulique, ou l'art de conduire, d'élever et de ménager les eaux pour les différens besoins de la vie. Nouvelle édition avec des notes par M. Navier.* Paris: F. Didot.

Belkin, V. P., 1973. Znamenityi korablestroitel' i vydayushchiisya uchenui I. G. Bubnov (I. G. Bubnov, famous shipbuilder and outstanding scholar). In: *Problemy stroitel'noi mekhaniki korablya* (problems of shipbuilding mechanics), pp. 3–27. Leningrad: Sudostroenie (in Russian).

Beltrami, E., 1881. Sulle equazioni generali dell'elasticità. *Ann. di mat.,* series II, vol. X.

Beltrami, E., 1885. Sulle condizioni di resistenzia die corpi elastici. *Rend. Ist. Lomb.,* series II, vol. XVIII.

Beltrami, E., 1886. Sull'interpretazione meccanica delle formole di Maxwell. *Mem. Dell'Accad. di Bologna,* series IV, vol. VII.

Beltrami, E., 1889/1. Note fisico-matematiche (2ª parte). *Rend. del Circ. mat. di Palermo,* vol. III.

Beltrami, E., 1889/2. Sur la théorie de la déformation infiniment petite d'un milieu. *Comptes rendus de l'acad. de Paris,* vol. CVIII.

Beltrami, E., 1902–20. *Opere matematiche.* vol. 4, Milan: Hoepli.

Bendixsen, A., 1914. *Die Methode der Alpha-Gleichungen zur Berechnung von Rahmenkonstruktionen.* Berlin: Verlag von Julius Springer.

Bendixsen, A., 1915. Die Berechnung von Rippenkuppeln mit oberem und unterem Ringe. *Armierter Beton,* vol. 8, pp. 45–49, 76–80, 95–101 & 114–119.

Benjamin, W., 1983. Paris, Hauptstadt des 19. Jahrhunderts. In: W. Benjamin: *Das Passagenwerk,* ed. R. Tiedemann. Frankfurt a. M.: Suhrkamp.

Benjamin, W., 1989. Das Kunstwerk im Zeitalter seiner technischen Reproduzierbarkeit. In: *Walter Benjamin. Gesammelte Schriften, Band VII/1 (Nachträge),* ed. R. Tiedemann & H. Schweppenhäuser. Frankfurt a. M.: Suhrkamp.

Benvenuto, E., 1981. *La Scienza delle Costruzioni e il suo sviluppo storico.* Florence: Sansoni.

Benvenuto, E., 1991/1. *An Introduction to the History of Structural Mechanics. Part. I: Statics and Resistance of Solids.* Berlin: Springer.

Benvenuto, E., 1991/2. *An Introduction to the History of Structural Mechanics. Part. II: Vaulted Structures and Elastic Systems.* Berlin: Springer.

Benvenuto, E., Radelet-de Grave, P. (ed.), 1995. *Entre Méchanique et Architecture. Between Mechanics and Architecture.* Basel: Birkhäuser.

Benvenuto, E., Corradi, M., Foce, F., 1996. Metaphysical Roots of the nineteenth Century Debate on the Molecular Theory of Elasticity. In: R. C. Batra & M. F. Beatty (ed.), *Contemporary research in the mechanics and mathematics of materials, CIMNE,* Barcelona, pp. 79–91.

Benvenuto, E., 1997. Adhémar Jean-Claude Barré de Saint-Venant: man, scientist and engineer. In: *Entre mécanique et architecture IV, Louvain la Neuve, les 28 et 29 juillet 1997. Résumés des communications,* pp. 2–18. Louvain la Neuve.

Benvenuto, E., 2006. *La scienza delle Costruzioni e il suo sviluppo storico.* Rome: Edizioni di storia e letteratura.

Berg, P., 2001. Kurt Klöppel – Sein Bild von der Brücke. *Stahlbau,* vol. 70, No. 9, pp. 591–611.

Berkel, K. van, 1985. *In het voetspoor van Stevin: geschiedenis van de natuurwetenschap in Nederland 1580–1940.* Amsterdam: Boom.

Berkel, K. van, 1996. *Dijksterhuis: een biografie.* Amsterdam: Uitgeverij Bert Bakker.

Bernhard, K., 1925. †Müller-Breslau. *Die Bautechnik,* vol. 3, No. 20, pp. 261–262.

Bernhard, K., 1930. Über die Verwindungssteifigkeit von zweigleisigen Eisenbahn- Fachwerkbrücken. *Der Stahlbau,* vol. 3, No. 8, pp. 85–92.

Bernshtein, S. A., 1957. *Ocherki po istorii stroitel'-noi mekhaniki* (essays in the history of structural mechanics). Moscow: Stroiizdat (in Russian).

Bernshtein, S. A., 1961. *Izbrannye trudy po stroitel'noi mekhanike* (selected works on structural mechanics). Moskva: Stroiizdat (in Russian).

Bertram, D., 1994. Stahl im Bauwesen. In: *Beton-Kalender,* vol. 83, pt. I, ed. J. Eibl, pp. 139–259. Berlin: Ernst & Sohn.

Besseling, J. F., Heijden, A. M. A. van der, 1979. *Trends in solid mechanics: Proceedings of the symposium dedicated to the 65th birthday of W. T. Koiter.* Delft University of Technology, June 13–15, 1979. Delft: Delft University Press.

Betti, E., 1872. Teorema generale intorno alle deformazioni che fanno equilibrio a forze che agiscono alla superficie. *Il Nuovo Cimento,* series 2, vol. 7–8, pp. 87–97.

Betti, E., 1873–1875. Sopra l'equazioni d'equilibrio dei corpi elastici. *Annali di matematica pura ed applicata,* series II, vol. VI, pp. 101–111.

Betti, E., 1874/1913. Sopra l'equazioni d'equilibrio dei corpi elastici. In: *Opere matematiche,* vol. II, pp. 379–390. Milan: Hoepli. (In *Annali di matematica pura ed applicata,* series II, vol. VI, 1874, pp. 101–111).

Betti, E., 1903–13. *Opere matematiche.* vol. 2, Milan: Hoepli.

Betti, E., 1913. Teoria della elasticità. In: *Opere matematiche,* vol. II, pp. 291–378. Milan: Hoepli. (In *Il nuovo cimento,* series II, vol. VII).

Beyer, K., 1908. Eigengewicht, günstige Grundmaße und geschichtliche Entwicklung des Auslegerträgers. Leipzig: Engelmann.

Beyer, K., 1925. Über die Bedeutung Zimmermanns als Forscher. *Der Bauingenieur,* vol. 6, No. 37, pp. 1013–1015.

Beyer, K., 1927. *Die Statik im Eisenbetonbau.* Stuttgart: Wittwer.

Beyer, K., 1933. *Die Statik im Eisenbetonbau.* Berlin: Springer.

Beyer, K., 1956. *Die Statik im Stahlbetonbau.* 2nd reprint of 2nd, rev. ed. Berlin: Springer.

Biermann, K.-R., 1983. Die Wahlvorschläge für Betti, Brioschi, Beltrami, Casorati und Cremona zu Korrespondierenden Mitgliedern der Berliner Akademie der Wissenschaften. *Bollettino di Storia delle Scienze Matematiche,* vol. 3, pp. 127–136.

Bill, M., 1949. *Robert Maillart.* Zurich: Verlag für Architektur.

Billington, D. P., 1979. *Robert Maillart's Bridges.* Princeton: Princeton University Press.

Billington, D. P., 1980. Wilhelm Ritter: Teacher of Maillart and Ammann. *Journal of the Structural Division (ASCE),* May 1980, pp. 1103–1116.

Billington, D. P., 1990. *Robert Maillart and the Art of Reinforced Concrete.* Cambridge, Mass.: The MIT Press.

Billington, D. P., 1997. *Robert Maillart.* Cambridge: Cambridge University Press.

Billington, D. P., 2003. *The Art of Structural Design: A Swiss Legacy.* New Haven: Yale University Press.

Birnstiel, C., 2005. The Nienburg cable-stayed bridge collapse: an analysis eighteen decades later. In: *Proccedings of the 5th International Conference on Bridge Management,* ed. G. A. R. Parke & P. Disney, pp. 179–186. London: Thomas Telford.

Bisplinghoff, R. L., Ashley, H., Halfman, R. L., 1955. *Aeroelasticity.* Reading (Mass.): Addison-Wesley.

Bisplinghoff, R. L., Ashley, H., 1961. *Principles of Aeroelasticity.* New York: Wiley & Sons.

Bisplinghoff, R. L., Mar, J. W., 1965. *Statics of Deformable Solids.* Reading (Mass.): Addison-Wesley.

Bisplinghoff, R. L., 1997. History of Aeroelasticity. In: *The Revolution in Structural Dynamics,* ed. H. I. Flomenhoft, pp. 3–33. Palm Beach Gardens: Dynaflo Press.

Björnstad, E., 1909. *Die Berechnung von Steifrahmen nebst andern statisch unbestimmten Systemen.* Berlin: Julius Springer.

Blaum, R., 1918. August Wöhler. In: *Beiträge zur Geschichte der Technik und Industrie,* vol. 8, ed. C. Matschoss, pp. 35–55. Berlin: Springer.

Bleich, F., 1913. G. C. Mehrtens zum siebzigsten Geburtstag. *Der Eisenbau,* vol. 4, No. 5, pp. 155–157.

Bleich, F., 1917. †G. C. Mehrtens. *Der Eisenbau,* vol. 8, No. 5, pp. 123–124.

Bleich, F., 1918. *Die Berechnung statisch unbestimmter Tragwerke nach der Methode des Viermomentensatzes.* Berlin: Springer.

Bleich, F., 1924. *Theorie und Berechnung der eisernen Brücken.* Berlin: Springer.

Bleich, F., Melan, E., 1927. *Die gewöhnlichen und partiellen Differenzengleichungen der Baustatik.* Berlin: Springer.

Bleich, F., 1952. *Buckling Strength of metal structures.* New York: McGraw-Hill.

Bleich, H. H., 1932. Über die Bemessung statisch unbestimmter Stahltragwerke unter Berücksichtigung des elastisch-plastischen Verhaltens des Baustoffes. *Der Bauingenieur,* vol. 13, No. 19/20, pp. 261–267.

Bleich, H. H., 1935. *Die Berechnung verankerter Hängebrücken.* Vienna: Springer.

Bletzinger, K.-U., Ziegler, R., 2000. Theoretische Grundlagen zur numerischen Formfindung von Membrantragwerken. In: *Beton-Kalender,* vol. 89, pt. 2, ed. J. Eibl, pp. 441–456. Berlin: Ernst & Sohn.

Block, P., 2005. *Equilibrium System. Studies in Masonry Structure.* Dissertation ms., Dept. of Architecture, MIT.

Block, P., Ochsendorf, J. A., 2005. Interactive Thrust Line Analysis for Masonry Structures. In: *Theory and Practice of Construction: Knowledge, Means, and Models,* Ravenna, Italy, ed. G. Mochi, pp. 473–483.

Block, P., Ciblac, T., Ochsendorf, J. A., 2006. Real-time Limit Analysis of Vaulted Masonry Buildings. *Computers and Structures,* vol. 84, No. 29–30, pp. 1841–1852.

Blume, J. A., Newmark, N. M., Corning, L. H., 1961. *Design of Multi-Story Reinforced Concrete Buildings for Earthquake Motion.* Chicago: Portland Cement Association.

Blumenthal, O., 1913. Über die asymptotische Integration von Differentialgleichungen mit Anwendung auf die Berechnung von Spannungen in Kugelschalen. *Zeitschrift für Mathematik und Physik,* vol. 63, pp. 342–362.

Blumtritt, O., 1988. Genese der Technikwissenschaften: Ein Resümee methodologischer Konzepte. *Technikgeschichte,* vol. 55, No. 2, pp. 75 – 86.

Boer, R. d., 1990. Wiener Beitrag zur Theorie poröser Medien und zur theoretischen Bodenmechanik. *Österreichische Ingenieur- und Architekten-Zeitschrift (ÖIAZ),* vol. 135, No. 10, pp. 546 – 554.

Boer, R. d., 2005. *The Engineer and the Scandal. A Piece of Science History.* Berlin/Heidelberg: Springer-Verlag.

Bogolyubov, A. N., 1983. *Matematiki, mekhaniki: Biograficheský spravochnik* (mathematicians, mechanicians: biographical dictionary). Kiev: Naukova Dumka (in Russian).

Böhm, T., Dorn, S., 1988. Bestimmungsgrundlagen des Automatisierungsbegriffs. *Dresdener Beiträge zur Geschichte der Technikwissenschaften,* No. 16, pp. 9 – 12.

Bohny, F., 1923. Reinhold Krohn zum siebzigsten Geburtstage. *Der Bauingenieur,* vol. 4, No. 1, pp. 1 – 2.

Boistard, L. C., 1810. Expériences sur la stabilité des voûtes. In: *Recueil de divers mémoires extraits de la bibliothèque impériale des ponts et chaussées a l'usage de MM. les ingénieurs,* ed. P. Lesage, vol. 2, pp. 171 – 217. Paris: Chez Firmin Didot.

Bolenz, E., 1991. *Vom Baubeamten zum freiberuflichen Architekten. Technische Berufe im Bauwesen (Preußen/Deutschland, 1799 – 1931).* Frankfurt a. M.: Peter Lang.

Boley, B. A., 1981. Menabrea, Luigi Federico. In: *Dictionary of Scientific Biography,* vol. 9. C. C. Gillispie (ed.), pp. 267 – 268. New York: Charles Scribner's Sons.

Boll, K., Vogel, U., 1969. Die Stahlkernstütze und ihre Bemessung. *Die Bautechnik,* vol. 46, No. 8, pp. 253 – 262 & No. 9, pp. 303 – 309.

Bollé, L., 1947. Contribution au problème linéaire de flexion d'une plaque élastique. *Bull. Techn. Suisse Romande,* vol. 73, pp. 281 – 285 & 293 – 298.

Bolzano, B., 1837. *Leben Franz Joseph Ritter von Gerstner.* Prague: Gottlieb Haase Söhne.

Booth, L. G., 1979/1980/1. Tredgold 1788 – 1829: Some aspects of his work Part 1: His Life. *Transactions of the Newcomen Society,* vol. 51, pp. 57 – 64.

Booth, L. G., 1979/1980/2. Tredgold 1788 – 1829: Some aspects of his work Part 2: Carpentry. *Transactions of the Newcomen Society,* vol. 51, pp. 65 – 71.

Booth, L. G., 1979/1980/3. Tredgold 1788 – 1829: Some aspects of his work Part 5: His Publications. *Transactions of the Newcomen Society,* vol. 51, pp. 86 – 92.

Booth, L. G., 2002/2. Tredgold, Thomas. In: *A Biographical Dictionary of Civil Engineers in Great Britain and Ireland,* vol. 1: *1500 – 1830.* Ed. A. W. Skempton. M. M. Chrimes, R. C. Cox, P. S. M. Cross-Rudkin, R. W. Rennison & E. C. Ruddock, pp. 716 – 722. London: ICE & Thomas Telford.

Boothby, T. E., 2001. Analysis of masonry arches and vaults. *Progress in Structural Engineering and Materials,* vol. 3, pp. 246 – 256.

Borelli, J. A., 1927. *Die Bewegung der Tiere.* Translation with commentary by Max Mengeringhausen. Ostwald's Klassiker der exakten Wissenschaften, No. 221. Leipzig: Akademische Verlagsgesellschaft.

Born, J., 1954. *Faltwerke. Ihre Theorie und Berechnung.* Stuttgart: Konrad Wittwer.

Born, M., 1906. *Untersuchungen über die Stabilität der elastischen Linie in Ebene und Raum unter verschiedenen Grenzbedingungen.* Dissertation, University of Göttingen.

Bornscheuer, F. W., 1948. Baustatik. In: *FIAT Review of German Science 1939 – 1946. Applied Mathematics, Part II. Mechanics of rigid and elastic bodies,* ed. A. Walther, pp. 13 – 52. Wiesbaden: Dieterich'sche Verlagsbuchhandlung.

Bornscheuer, F. W., 1952/1. Systematische Darstellung des Biege- und Verdrehvorganges unter besonderer Berücksichtigung der Wölbkrafttorsion. *Der Stahlbau,* vol. 21, No. 1, pp. 1 – 9 & No. 12, pp. 225 – 232.

Bornscheuer, F. W., 1952/2. Beispiel und Formelsammlung zur Spannungsberechnung dünnwandiger Stäbe mit wölbbehindertem Querschnitt. *Der Stahlbau,* vol. 21, No. 12, pp. 225 – 232 & vol. 22, No. 2, pp. 32 – 44.

Bornscheuer, F. W., 1961. Tafeln der Torsionskenngrößen für die Walzprofile der DIN 1025 – 1027. *Der Stahlbau,* vol. 30, No. 3, pp. 81 – 82.

Borrmann, M., 1992. *Historische Pfahlgründungen.* Institute of History of Building, University of Karlsruhe.

Borst, R. de, 2002. Warner Tjardus Koiter – Het instabiele hanteerbaar. In: *Delfts Goud – Leven en werk van 18 markante hoogleraren.* K. F. Wakker, B. Herbergs, M. v. d. Sanden (ed.), pp. 222 – 231. Delft: Beta Imaginations.

Bortsch, R., 1924. Zum 70. Geburtstage von Josef Melan. *Der Bauingenieur,* vol. 5, No. 1, pp. 9 – 10.

Bosc, J.-L., Chauveau, J.-M., Clément, J., Degenne, J., Marrey, B., Paulin, M., 2001. *Joseph Monier et la naissance du ciment armé.* Paris: Éditions du Linteau.

Bosch, J. B., 1905. Die Berechnung der Eisenbetonplatte. *Beton und Eisen,* vol. 4, No. 7, pp. 177 – 180 & No. 11, p. 281.

Boscovich, R. G., 1743. *De motu corporis attracti in centrum immobile viribus decrescentibus in ratione distantiarum reciproca duplicata in spatiis non resistentibus.* Dissertation. Rome.

Boscovich, R. G., 1754. *Elementorum Universae Matheseos.* vol. I & II, Tomae; vol. III, Venice.

Boscovich, R. G., 1755. *De Lege Virium in Natura existentium.* Dissertation. Rome.

Boscovich, R. G., 1758. *Philosophiae naturalis theoria redacata ad unicam legem virium in natura existentium.* Vienna.

Boscovich, R. G., 1763. *Scrittura sulli danni osservati, nell'edificio delle Biblioteca Cesarea di Vienna, e loro riparazione, composta in occasione de' Sovrani comandi di Sua Maestà l'Imperatrice Regina Maria Teresa, e umiliata a suoi piedi pel felicissimo anniversario (13 Mai 1763) della sua nascita.* Vienna.

Boscovich, R. G., 1764. *Sentimento sulla solidità della nuova Guglia del Duomo di Milan o si consideri in se stessa, o rispetto al rimanente del vasto Tempio, esposto a richiesta del Nobilissimo e Vigilantissimo Capitolo che sopraintende alla sua gran fabbrica.* Milan.

Bossenmayer, H.-J., 2004. Der Deutsche Ausschuß für Stahlbau, DASt. *Stahlbau,* vol. 73, No. 10, pp. 830 – 837.

Bottazzini, U., 1977. The mathematical papers of Enrico Betti in the Scuola Normale Superiore of Pisa. *Historia Mathematica.* vol. 4, pp. 207 – 209.

Bottazzini, U., 1977/1978. Riemanns Einfluß auf E. Betti und F. Casorati. *Archive for History of Exact Sciences.* vol. 18, pp. 27 – 37.

Böttcher, K.-H., Maurer, B., 2008. Stuttgarter Mathematiker. Geschichte der Mathematik an der Universität Stuttgart von 1829 bis 1945 in Biographien. *Publications of the University of Stuttgart archives,* vol. 2. Ed. N. Becker. Stuttgart: Wais & Partner.

Bouguer, P., 1736. Sur les lignes courbes qui sont propres à former les voûtes en dôme. *Mémoires de l'Académie Royale des Sciences,* 1734, pp. 149 – 166. Paris.

Bouguer, P., 1757. *Manœuvre des vaisseaux ou Traité de méchanique et de dynamique; Dans lequel on réduit à des solutions très-simples les problèmes de marine les plus difficiles, qui ont pour objet le mouvement du navire.* Paris: H. L. Guerin & L. F. Delatour.

Boussinesq, J. V., 1885. *Application des potentiels à l'étude de l'équilibre et du mouvement des solides élastiques.* Lille: Imprim. L. Danel.

Boussinesq, J. V., Flamant, A., 1886. Notice sur la vie et les travaux de M. de Saint-Venant. *Annales des ponts et chaussées,* vol. 12, pp. 557 – 595.

Bow, R. H., 1851. *A Treatise on Bracing, with its application to bridges and other structures of wood or iron.* Edinburgh: Black.

Bow, R. H., 1873. *The Economics of Construction in relation to Framed Structures.* London: Spon.

Bradley, M., 1976. Scientific Education for a New Society. The École Polytechnique 1795 – 1830. *History of Education,* vol. 5, No. 1, pp. 11 – 24.

Bradley, M., 1981. Franco-Russian Engineering Links: The Careers of Lamé and Clapeyron, 1820 – 1830. *Annals of Science,* vol. 38, pp. 291 – 312.

Brandes, K., Jaeger, B., Matthees, W., 1985. Stationen des Lebens – Ziele und Wirkungen.

In: *Thomas A. Jaeger. Ein Leben im Spannungsfeld zwischen Technik und Risiko*. Ed.: International Association for Structural Mechanics in Reactor Technology, Federal Institute for Materials Research & Testing, Klaus Brandes in close cooperation with Brunhilt Jaeger & Wolfgang Matthees, pp. 27 – 106. Berlin: Federal Institute for Materials Research & Testing (BAM).

Brandt, S., Poser, S. (ed.), 2006. *Zukunft des Ingenieurs – Ingenieure der Zukunft. 150 Jahre VDI Berlin-Brandenburg*. Berlin: VDI Bezirksverein Berlin-Brandenburg e.V.

Braun, H.-J., 1975. Allgemeine Fragen der Technik an der Wende zum 20. Jahrhundert. Zum Werk von P. K. Engelmeyers. *Technikgeschichte*, vol. 42, No. 4, pp. 306 – 326.

Braun, H.-J., 1977. Methodenprobleme der Ingenieurwissenschaft, 1850 bis 1900. *Technikgeschichte*, vol. 44, No. 1, pp. 1 – 18.

Brausewetter, B., 1942. Zum 80. Geburtstag von unserem Oberbaurat Dr.-Ing. e. h. Fritz v. Emperger. *Beton und Eisen*, vol. 41, No. 1/2, pp. 1 – 2.

Bravais, A., 1890. *Abhandlungen über symmetrische Polyeder*. Trans. in conjunction with P. Groth, ed. C. & E. Blasius. Ostwald's Klassiker der exakten Wissenschaften, No. 17. Leipzig: Verlag von Wilhelm Engelmann.

Bredt, R., 1878. Maschine zur Prüfung der Elastizität und Festigkeit von Stahl und Eisen. *Organ für die Fortschritte des Eisenbahnwesens in technischer Beziehung*, vol. 33, new series, vol. 15, pp. 138 – 140.

Bredt, R., 1883. Hydraulische Krahne für Stahlwerke. *Stahl und Eisen*, vol. 3, No. 12, pp. 673 – 74.

Bredt, R., 1885/1. Freistehender hydraulischer Krahn von 3000 kg Tragkraft. *Stahl und Eisen*, vol. 5, No. 6, p. 285.

Bredt, R., 1885/2. Berechnung der Kettenhaken. *Zeitschrift des Vereines deutscher Ingenieure (ZVDI)*, vol. 29, No. 15, pp. 283 – 285.

Bredt, R., 1886. Zerknickungsfestigkeit und excentrischer Druck. *ZVDI*, vol. 30, No. 29, pp. 621 – 625.

Bredt, R., 1887. Berechnung von Fundamentplatten. *ZVDI*, vol. 31, No. 22, pp. 459 – 460.

Bredt, R., 1893. Festigkeit von Röhren und Kugelschalen mit innerem und äußerem Druck. *ZVDI*, vol. 37, No. 30, pp. 903 – 905 & No. 31, pp. 935 – 938.

Bredt, R., 1894. Studien über Zerknickungsfestigkeit. *ZVDI*, vol. 38, No. 27, pp. 810 – 816, No. 28, pp. 844 – 847 & No. 29, pp. 875 – 879.

Bredt, R., 1894/1895. *Krahn-Typen der Firma Ludwig Stuckenholz Wetter a. d. Ruhr*. Düsseldorf: L. Schwann.

Bredt, R., 1895. Elastizität und Festigkeit krummer Stäbe. *ZVDI*, vol. 39, No. 35, pp. 1054 – 1059 & No. 36, pp. 1074 – 1078.

Bredt, R., 1896/1. Kritische Bemerkungen zur Drehungselastizität. *ZVDI*, vol. 40, No. 28, pp. 785 – 790 & No. 29, pp. 813 – 817.

Bredt, R., 1896/2. Letter of 24 July 1896 to the editor of ZVDI. *ZVDI*, vol. 40, No. 33, pp. 943 – 944.

Bredt, R., 1898. Das Elastizitätsgesetz und seine Anwendung für praktische Rechnung. *ZVDI*, vol. 42, No. 25, pp. 694 – 699.

Bredt, R., 1899. Zur Frage der Ingenieurausbildung. *ZVDI*, vol. 43, No. 22, pp. 662 – 664.

Brendel, G., 1958. Johann Andreas Schubert als Bauingenieur. *Bauplanung-Bautechnik*, vol. 12, No. 7, pp. 321 – 322.

Bresse, J. A. C., 1848. Etudes théoriques sur la résistance des arcs employés dans les ponts en fonte ou en bois. *Annales des Ponts et Chaussées*, vol. 25, pp. 150 – 193.

Bresse, J. A. C., 1854. *Recherches analytiques sur la flexion et la résistance des pièces courbées, accompagnées de tables numeriques pour calculer la poussée des arcs chargés de poids d'une manière quelconque et leur pression maximum sous une charge uniformément répartie*. Paris: Mallet-Bachelier.

Bresse, J. A. C., 1859. *Cours de mécanique appliquée*. pt. 2. Paris: Mallet-Bachelier.

Bresse, J. A. C., 1865. *Cours de mécanique appliquée*. pt. 2, 2nd ed. Paris: Mallet-Bachelier.

Breymann, G. A., 1854. *Allgemeine Bau-Constructions-Lehre mit besonderer Beziehung auf das Hochbauwesen. Ein Leitfaden zu Vorlesungen un zum Selbstunterrichte. Theil 3: Construction in Metall*. Stuttgart: Hoffmann.

Briguglio, L., L Bulferetti, L. (ed.), 1971. *Memorie*. Florence: Giunti.

Brioschi, F., 1892. Enrico Betti. *Annali di matematica pura ed applicata*, vol. 20, p. 256.

Bronneck, H. v., 1913. *Einführung in die Berechnung der im Eisenbetonbau gebräuchlichen biegungsfesten Rahmen*. Berlin: Wilhelm Ernst & Sohn.

Broszko, M., 1932. Beitrag zur allgemeinen Lösung des Knickproblems. In: *Proceedings of the International Association for Bridge & Structural Engineering (IABSE)*, vol. 1, ed. IABSE; pp. 1 – 8. Zurich: A.-G. Gebr. Leemann & Co.

Brizzi, E., 1951. Architettura statica e geometria classica del Ponte a S. Trinitá. *Nuova Cittá*, Numero speciale, Luglio.

Bubnov, I. G., 1913. *Otzyv o sochinenii professora Timoshenko "Ob ustoichivosti uprugikh sistem", udostoennom premii D. I. Zhuravskogo* (report on Prof. Timoshenko's work "On the stability of elastic systems", which was awarded the D. I. Zhuravsky Prize). Sbornik SPb. Instituta inzhenerov putei soobshcheniya (compendium of the St. Petersburg Institute of Engineers of Ways of Communication), No. 81, pp. 33 – 36 (in Russian).

Bubnov, I. G., 1912 – 1914. *Stroitel'naya mekhanika korablya* (mechanics of shipbuilding). 2 parts. St. Petersburg: Marine Ministry (in Russian).

Bubnov, I. G., 1953. *Trudy po teorii plastin* (works on plate theory). Moscow: Gostekhizdat (in Russian).

Bubnov, I. G., 1956. *Izbrannye trudy* (selected works). Leningrad: Sudpromgiz (in Russian).

Buchheim, G., 1984. Zu einigen Fragen der Periodisierung der Technikwissenschaften. *Dresdener Beiträge zur Geschichte der Technikwissenschaften*, No. 9, pp. 72 – 83.

Buchheim, G., Sonnemann, R., 1986. Die Geschichte der Technikwissenschaften – eine neue wissenschaftshistorische Disziplin. *Wiss. Ztschr. Techn. Univers. Dresden*, vol. 35, No. 2, pp. 135 – 142.

Buchheim, G., Sonnemann, R. (ed.), 1990. *Geschichte der Technikwissenschaften*. Leipzig: Edition Leipzig.

Bulferetti, L., 1969. Luigi Federico Menabrea e i suoi inediti souvenirs. *Physis*. vol. 11, pp. 89 – 99.

Bullen, K. E., 1973. Love, Augustus Edward Hough. In: *Dictionary of Scientific Biography*, vol. VIII. C. C. Gillispie (ed.), pp. 516 – 517. New York: Charles Scribner's Sons.

Bulson, P. S., Caldwell, J. B., Severn, R. T., 1983. *Engineering Structures: Developments in the Twentieth Century*. Bristol: University of Bristol Press. (A collection of essays to mark the eightieth birthday of Sir Alfred Pugsley).

Buonopane, St. G., 2006. The Technical Writings of John A. Roebling and his Contributions to Suspension Bridge Design. In: *John A. Roebling. A bicentennial celebration of his birth*, ed. Theodore Green, pp. 21 – 36. Pub. by the American Society of Civil Engineers (ASCE).

Bürgermeister, G., Steup, H., 1959. *Stabilitätstheorie mit Erläuterungen zu DIN 4114*. Pt. I, 2nd, rev. ed. Berlin: Akademie-Verlag.

Bürgermeister, G., Steup, H., Kretzschmar, H., 1963. *Stabilitätstheorie mit Erläuterungen zu den Knick- und Beulvorschriften*. Pt. II. Berlin: Akademie-Verlag.

Bürgermeister, G. (ed.), 1968. *Ingenieurtaschenbuch Bauwesen. Band II: Konstruktiver Ingenieurbau. Teil 1: Grundlagen der Bauweisen*. Leipzig: Edition Leipzig.

Bürgermeister, G. (ed.), 1970. *Ingenieurtaschenbuch Bauwesen. Band II: Konstruktiver Ingenieurbau. Teil 2: Entwurf und Ausführung*. Leipzig: B. G. Teubner Verlagsgesellschaft.

Burmester, L., 1880. Über die momentane Bewegung ebener kinematischer Ketten. *Zivilingenieur*, vol. 26, pp. 247 – 290.

Busemann, A., 1935. Aerodynamischer Auftrieb bei Überschallgeschwindigkeit. *Luftfahrtforschung*, vol. 12, No. 6, pp. 210 – 220.

Busse, R., 1912. Entwurf einer Rahmenbrücke über die Ems. *Der Eisenbau*, vol. 3, No. 6, pp. 214 – 219.

Buti, A., Corradi, M., 1981. I Contributi di un Matematico del XVII Secolo ad un Problema di Architettura: Philippe De la Hire e la statica degli Archi. *Estratto degli Atti dell'Accademia Ligure di Scienze e Lettere*, vol. XXXVIII, pp. 3 – 23, Genoa.

Büttner, O., Hampe, E., 1977. *Bauwerk – Tragwerk – Tragstruktur, Bd. 1*. Berlin: VEB Verlag für Bauwesen.

Büttner, O., Hampe, E., 1984. *Bauwerk – Tragwerk – Tragstruktur, Bd. 2*. Berlin: VEB Verlag für Bauwesen.

Čališev, K. A., 1922. Izračunavanje višestruko statici neodredenik sistemy pomoču postepnik aproksimacija. *Tehnički List*, No. 4, pp. 1 – 6.

Čališev, K. A., 1923. Izračunavanje višestruko statici neodredenik sistemy pomoču postepnik aproksimacija. *Tehnički List*, No. 5, pp. 125 – 127, 141 – 143, 151 – 154 & 157 – 158.

Čališev, K. A., 1936. Die Methode der sukzessiven Annäherung bei der Berechnung von vielfach statisch unbestimmten Systemen. In: *Proceedings of the International Association for Bridge & Structural Engineering (IABSE)*, vol. 4, ed. L. Karner & M. Ritter, pp. 199 – 215. Zurich: IABSE secretariat.

Calladine, C. R., 1978. Buckminster Fuller's 'Tensegrity' Structures and Clerk Maxwell's rules for the construction of stiff frames. *International Journal of Solids and Structures*, vol. 14, pp. 161 – 172.

Calladine, C. R., 1988. The theory of thin shell structures 1888 – 1988. *Proceedings of the Institution of Mechanical Engineers*, vol. 202, No. 42, pp. 1 – 9.

Calladine, C. R., 2006. Letter of 2 April 2006.

Cambournac, L., 1932. Poutrelles en acier enrobées de béton (beams of encased rolled steel). In: *Proceedings of the International Association for Bridge & Structural Engineering (IABSE)*, vol. 1, ed. L. Karner & M. Ritter, pp. 25 – 34. Zurich: A.-G. Gebr. Leemann & Co.

Capriz, G., 2000. Obituary: Clifford Ambrose Truesdell (1919 – 2000). *Meccanica*, vol. 35, No. 5, pp. 463 – 466.

Carnot, S., 1824. *Réflexions sur la puissance motrice du feu et sur les machines propres à développer cette puissance*. Paris: Bachelier, Libraire.

Caruccio, E., 1970. Betti, Enrico. In: *Dictionary of Scientific Biography*, vol. II. C. C. Gillispie (ed.), pp. 104 – 106. New York: Charles Scribner's Sons.

Castigliano, C. A. P., 1875/1. *Intorno ai sistemi elastici*. Dissertazione presentata da Castigliano Alberto alla Commissione Esaminatrice della R. Scuola d'applicazione degli Ingegneri in Torino per ottenere la Laurea di ingegnere civile. Torino: V. Bona.

Castigliano, C. A. P. 1875/2. Intorno all'equilibrio dei sistemi elastici. *Atti della Reale Accademia delle scienze di Torino*, (2), 2, pp. 201 – 221.

Castigliano, C. A. P., 1875/3. Nuova teoria intorno all'equilibrio dei sistemi elastici. *Atti della Reale Accademia delle scienze di Torino*, (2), 11, pp. 127 – 286

Castigliano, C. A. P., 1879. *Théorie de l'Équilibre des Systèmes Élastiques et ses Applications*. Turin: A. F. Negro.

Castigliano, C. A. P., 1882. Intorno ad una proprietà dei sistemi elastici. *Atti della Reale Accademia delle scienze di Torino*, (2), 17, pp. 705 – 713.

Castigliano, C. A. P., 1886. *Theorie des Gleichgewichtes elastischer Systeme und deren Anwendung*. Translation from the French by E. Hauff. Vienna: Verlag von Carl Gerold's Sohn.

Castigliano, C. A. P., 1935. *Selecta*. A cura di Gustavo Colonnetti. Turin: Luigi Avalle.

Castigliano, C. A. P., 1966. *The Theory of Equilibrium of Elastic Systems and its Applications*. Translated by Ewart S. Andrews with a new Introduction and Biographical Portrait section by Gunhard Æ. Oravas. New York: Dover.

Castigliano, C. A. P., 1984. *Selecta 1984*. A cura di Edoardo Benvenuto e Vittorio Nascé. Turin: Levrotto Bella.

Cauchy, A. L., 1809. *Mémoire sur les ponts en pierre, par A. L. Cauchy, élève des Ponts et Chaussées*. In: Centre de Documentation de l'École Nationale des Ponts et Chaussées, Manus. 1982. Paris.

Cauchy, A. L., 1810. *Second mémoire sur les ponts en pierre, théorie des voûtes en berceau, par A. L. Cauchy, élève des Ponts et Chaussées*. In: Centre de Documentation de l'École Nationale des Ponts et Chaussées, Manus. 1982. Paris.

Cauchy, A. L., 1823. Recherches sur l'équilibre et le mouvement intérieur des corps solides ou fluides, élastiques ou non élastiques. In: *Bulletin des Sciences de la Société Philomathique de Paris*, pp. 9 – 13.

Cauchy, A. L., 1827/1. De la pression ou tension dans un corps solide. *Exercices de Mathématiques*, vol. 2, pp. 42 – 56.

Cauchy, A. L., 1827/2. Sur la condensation et la dilatation des corps solides. *Exercices de Mathématiques*, vol. 2, pp. 60 – 69.

Cavallari-Murat, A. (ed.), 1957. Rimembranze e divagazioni a proposito del centenario teorema di Menabrea. pp. 1 – 11. Turin: Graziano.

Cavallari-Murat, A., 1963. *Giovanni Poleni e la costruzione architettonica*. In: Giovanni Poleni (1683 – 1761) nel bicentenario della morte. Accademia Patavina di Scienze Lettere ed Arti, pp. 95 – 118. Padova.

Charlton, T. M., 1976. Contributions to the science of bridge-building in the nineteenth century by Henry Moseley, Hon. L.l.D., F.R.S. & William Pole, D. Mus., F.R.S. *Notes and Records of the Royal Society of London*, vol. 30, pp. 169 – 179.

Charlton, T. M., 1982. *A history of theory of structures in the nineteenth century*. Cambridge: Cambridge University Press.

Chatzis, K., 1997. Die älteste Bauingenieurschule der Welt – die École des Ponts et chaussées (1747 – 1997). *Bautechnik*, vol. 74, No. 11, pp. 776 – 789.

Chatzis, K., 2004. La réception de la statique graphique en France durant le dernier tiers du XIXe siècle. *Revue d'histoire des mathématiques*, vol. 10, pp. 7 – 43.

Chatzis, K., 2007. La mécanique dans l'Encyclopédie des sciences mathématiques pures et appliquées. In: *D'une encyclopédie à l'autre. Felix Klein et l'Encyclopédie des sciences mathématiques pures et appliquées*, H. Gispert & C. Goldstein (ed.). Paris.

Chekanov, A. A., 1982. V. L. Kirpichev 1845 – 1913. Moscow: Nauka (in Russian).

Cherry, T. M., 1986. Michell, John Henry (1863 – 1940). In: *Australian Dictionary of Biography*, vol. 10, pp. 494 – 495. Melbourne: Melbourne University Press.

Cheung, Y. K., 2005. Research in Computational Mechanics with Prof. O. C. Zienkiewicz during the Picneering Days (1962 – 1967). *iacm expressions*, 18/05, pp. 20 – 23.

Chew, K., Wilson, A., 1993. *Victorian science and engineering*. Dover: Alan Sutton Publishing.

Chmelka, F., Melan, E., 1942. *Einführung in die Statik*. Vienna: Springer.

Chmelka, F., Melan, E., 1947. *Einführung in die Festigkeitslehre für Studierende des Bauwesens*. Vienna: Springer.

Chmelka, F., 1964. †Ernst Melan. *Der Stahlbau*, vol. 33, No. 4. p. 127.

Christophe, P., 1899. Le béton armé et ses applications. *Annales des Travaux publics de Belgique*, vol. 56, 2nd series, vol. IV, pp. 429 – 538, 647 – 678 & 961 – 1124.

Christophe, P., 1902. *Le béton armé et ses applications*. 2nd ed. Paris/Liège: Librairie Polytechnique, C. Béranger.

Christophe, P., 1905. *Der Eisenbeton und seine Anwendung im Bauwesen*. Berlin: Verlag Tonindustrie-Zeitung.

Christopherson, D. G., 1972. *Biographical Mem. Fellows Roy. Soc.*, vol. 18, pp. 549 – 565.

Chwalla, E., 1934/1. Das Problem der Stabilität gedrückter Rahmenstäbe. In: *Proceedings of the International Association for Bridge & Structural Engineering (IABSE)*, vol. 2, ed. L. Karner & M. Ritter, pp. 80 – 95. Zurich: IABSE secretariat.

Chwalla, E., 1934/2. Drei Beiträge zur Frage des Tragvermögens statisch unbestimmter Stahltragwerke. In: *Proceedings of the International Association for Bridge & Structural Engineering (IABSE)*, vol. 2, ed. L. Karner & M. Ritter, pp. 96 – 125. Zurich: IABSE secretariat.

Chwalla, E., Jokisch, F., 1941. Über das ebene Knickproblem des Stockwerkrahmens. *Der Stahlbau*, vol. 14, No. 8/9, pp. 33 – 37 & No. 10/11, pp. 47 – 51.

Chwalla, E., 1943. *Knick-, Kipp-, und Beulvorschriften für Baustahl. DIN 4114 E. Entwurf 4a der Beratungsunterlage vom Oktober 1943*. Berlin: Fachgruppe Stahlbau, im N.S.-Bund deutscher Technik.

Chwalla, E., Leon, A., Parkus, H., Reinitzhuber, F. (ed.), 1950. *Federhofer-Girkmann-Festschrift. Beiträge zur angewandten Mechanik*. Vienna: Deuticke.

Chwalla, E., 1954/1. *Einführung in die Baustatik.* Reprint of 2nd rev. ed., 1944. Cologne: Stahlbau-Verlags-GmbH.

Chwalla, E., 1954/2. Ansprache und Vortrag anläßlich der Ehrenpromotion an der Technischen Universität Berlin-Charlottenburg. *Veröffentlichungen des Deutschen Stahlbau-Verbandes No. 3/54.* Cologne: Stahlbau-Verlags-GmbH.

Chwalla, E., 1959. Die neuen Hilfstafeln zur Berechnung von Spannungsproblemen der Theorie zweiter Ordnung und von Knickproblemen. *Der Bauingenieur,* vol. 34, No. 4, pp. 128–37, No. 6, pp. 240–255 & No. 8, pp. 299–309.

Clapeyron, E., 1833. Note sur un théorème de mécanique. *Annales des Mines,* III, pp. 63–70.

Clapeyron, E., Lamé, G., 1833. Mémoire sur l'équilibre intérieur des corps solides homogènes. In: *Mém. Sav. Ètrang.* IV, pp. 463–562.

Clapeyron, E., 1834. Sur la puissance motrice de la chaleur. *École Polyt. Journ.,* vol. 14, No. 23, pp. 153–190.

Clapeyron, E., 1837. Sur la puissance motrice de la chaleur. *Taylor, Scient. Mem.* I, pp. 347–376.

Clapeyron, E., 1842. Mémoire sur le réglement des tiroirs dans les machines a vapeur. *Comptes Rendus,* XIV, pp. 632–633.

Clapeyron, E., 1843. Sur la puissance motrice de la chaleur. *Poggend. Annal.* LIX, pp. 446–450.

Clapeyron, E., 1857. Calcul d'une poutre élastique reposant librement sur des appuis inégalement espacés. *Comptes Rendus,* XLV, pp. 1076–1080.

Clapeyron, E., 1858. Mémoire sur le travail des forces élastiques dans un corps solide élastique déformé par l'action de forces extérieures, *Comptes Rendus,* XLVI, pp. 208–212.

Clapeyron, E. & Lamé, G., 1828. Mémoire sur la construction des polygones funiculaires. *Journal Génie Civil,* I, pp. 496–504.

Clapeyron, E., 1926. *Abhandlung über die bewegende Kraft der Wärme.* Trans. & ed. K. Schreber. Leipzig: Akademische Verlagsgesellschaft.

Clark, E., 1850. *The Britannia and Conway tubular bridges.* 2 vols. & atlas. London: for the author by Day & Son & John Weale.

Clebsch, A., 1854. *De motu ellipsoidis in fluido incompressibili viribus quibuslibet impulsi.* Dissertatio Inauguralis Physico-Mathematica, Universität Königsberg: Königsberg: Regimontani.

Clebsch, A., 1859/1. *Analytische Mechanik: nach Vorträgen, geh. an d. polytechn. Schule Carlsruhe, lithographiert.* Karlsruhe: Litho. Dept. of L. Geißendörfer.

Clebsch, A., 1859/2. *Elementar-Mechanik: nach Vorträgen, geh. an d. polytechn. Schule Carlsruhe von 1858–1859, lithographiert.* Karlsruhe: Litho. Dept. of L. Geißendörfer.

Clebsch, A., 1862. *Theorie der Elasticität fester Körper.* Leipzig: B. G. Teubner.

Clebsch, A., 1883. *Théorie de l'élasticité des corps solides de Clebsch.* Trans. by M. M. Barré de Saint-Venant et Flamant, with notes by M. de Saint-Venant. Paris: Dunod.

Clebsch, A., 1966. *Théorie de l'élasticité des corps solides de Clebsch.* Trans. by M. M. Barré de Saint-Venant et Flamant, with notes by M. de Saint-Venant. New York/London: Johnson Reprint Cooperation.

Clough, R. W., 1960. The finite element method in plane stress analysis. In: *Proceedings of the Second ASCE Conference on Electronic Computation,* Pittsburgh, PA, 8–9 September 1960.

Clough, R. W., 1965. The Finite Element Method in Structural Mechanics. Philosophy of the Finite Element Procedure. In: *Stress Analysis.* Ed. O. C. Zienkiewicz & G. S. Holister, pp. 85–119. New York: John Wiley & Sons.

Clough, R. W., 2004. Early history of the finite element method from the viewpoint of a pioneer. *International Journal for Numerical Methods in Engineering,* vol. 60, pp. 283–287.

Coignet, E., Tedesco, N. de, 1894. Le Calcul des ouvrages en ciment avec ossature métallique. *Mémoires de la Société des Ingénieurs civils de France,* p. 282ff. Paris.

Coignet, F., 1861. *Bétons agglomérés appliqués à l'art de construire notamment à l'état de pierres artificielles.* Paris: E. Lacroix.

Collar, A. R., 1978. The first fifty years of aeroelasticity. *Aerospace,* Feb. pp. 12–20.

Collins, M. P., Mitchell, D., 1991. *Prestressed Concrete Structures.* Englewood Cliffs: Prentice Hall.

Colonnetti, G., 1955. *L'équilibre des corps déformables.* Paris: Dunod.

Como, M., 1992. Equilibrium and collapse analysis of masonry bodies. *Meccanica,* vol. 27, No. 3, pp. 185–194.

Conrad, D., Hänseroth, T., 1995. Johann Andreas Schubert (1808–1870). *Bautechnik,* vol. 72, No. 10, pp. 671–675 & No. 11, pp. 756–765.

Considère, A., 1898/1899. Influence des armatures métalliques sur les propriétés des ciments et bétons. *Le Génie civil,* vol. XXXIV, p. 213ff.

Cook, W. D., Mitchell, D., 1988. Studies of Disturbed Regions near Discontinuities in Reinforced Concrete Members. *ACI Structural Journal,* vol. 85, No. 2, pp. 206–216.

Cornelius, W., 1952. Die Berechnung der ebenen Flächentragwerke mit Hilfe der Theorie der orthogonal-anisotropen Platte. *Der Stahlbau,* vol. 21, No. 2, pp. 21–24, No. 3, pp. 43–48 & No. 4, pp. 60–64.

Corradi, L., 1984. On the developments of computational mechanics. *Meccanica,* vol. 19, pp. 76–85.

Corradi, M., 1989. Contributi italiani alla statica delle volte e delle cupole. In: *Contributi italiani alla Scienza delle Costruzioni* (ed. E. Benvenuto, M. Corradi & G. Pigafetta), pub. in volume *Storia sociale e culturale d'Italia,* vol. V. *La cultura filosofica e scientifica,* vol. II: *La Storia delle Scienze* (ed. C. Maccagni & P. Freguglia), pp. 938–944. Busto Arsizio: Bramante.

Corradi, M., 2005. The mechanical sciences in Antiquity. In: *Essays in the history of theory of structures,* ed. S. Huerta, pp. 103–116. Madrid: Instituto Juan de Herrera.

Courant, R., 1943. Variational methods for the solution of problems of equilibrium and vibration. *Bulletin of the American Mathematical Society,* vol. 49, pp. 1–23.

Courant, R., 1994. Variational methods for the solution of problems of equilibrium and vibration. *International Journal for Numerical Methods in Engineering,* vol. 37, pp. 2161–2162.

Cosandey, M., 1981. Nachruf auf Fritz Stüssi. *Der Stahlbau,* vol. 50, No. 7, pp. 222–223.

Cosserat, E., Cosserat, F., 1909. *Théorie des corps deformables.* Paris: Hermann.

Costabel, P., 1981. Varignon, Pierre. In: *Dictionary of Scientific Biography,* vol. 13. C. C. Gillispie (ed.), pp. 584–587. New York: Charles Scribner's Sons.

Coste, A., 1995. La méthode des éléments finis appliqué à la restauration de la cathédrale de Beauvais. In: *Entre Mécanique et Architecture,* ed. P. Radelet-de Grave & E. Benvenuto, pp. 348–360. Basel: Birkhäuser.

Coulomb, C. A., 1773/1776. Essai sur une application des règles des Maximis et Minimis à quelques Problèmes de statique relatifs à l'Architecture. In: *Mémoires de mathématique & de physique, présentés à l'Académie Royale des Sciences par divers savans,* vol. 7, 1773, pp. 343–382. Paris.

Couplet, P., 1729/1731. De la poussée des voûtes. In: *Mémoires de l'Académie Royale des Sciences,* 1729, pp. 79–117. Paris.

Couplet, P., 1730/1732. Seconde partie de l'examen de la poussée des voûtes. In: *Mémoires de l'Académie Royale des Sciences,* 1730, pp. 117–141. Paris.

Cousinéry, B.-É., 1839. *Le calcul par le trait, ses éléments et ses applications à la mesure des lignes, des surfaces et des cubes, à l'interpolation graphique et à la détermination, sur l'épure, de l'épaisseur des murs de soutènement et des murs de culées des voutes.* Paris: Carillan-Gœury & Dalmont.

Cowan, H. J., 1977. *The master builders: A history of structural and environmental design from ancient egypt to the nineteenth century.* London/ New York: Wiley-Interscience.

Craemer, H., 1929/1. Scheiben und Faltwerke als neue Konstruktionselemente im Eisenbetonbau. *Beton und Eisen,* vol. 28, No. 13, pp. 254–257 & No. 14, pp. 269–272.

Craemer, H., 1929/2. Reply. *Beton und Eisen,* vol., No. 18, pp. 338–339.

Craemer, H., 1929/3. Further reply. *Beton und Eisen,* vol. 28, No. 18, p. 340.

Craemer, H., 1930. Allgemeine Theorie der Faltwerke. *Beton und Eisen,* vol. 29, No. 15, pp. 276–281.

Crane, G. E. (ed.), 1987. *Proceedings of the ACM conference on History of scientific and numeric computation.* Princeton, New Jersey, 13–15 May 1987. New York: ACM Press.

Cremer, J.-M. et al., 1999. Der TGV-Viadukt bei Donzère. *Stahlbau,* vol. 68, No. 6, pp. 399–408.

Cremona, L., 1872. *Le figure reciproche nella statica grafica.* Milan: Hoepli.

Cremona, L., 1914–1917. *Opere matematiche.* vol. 3, Milan: Hoepli.

Cross, H., 1930. Analysis of continuous frames by distributing fixed-end moments. *Proceedings of the American Society of Civil Engineers,* vol. 56, pp. 919–928.

Cross, H., 1932/1. Analysis of continuous frames by distributing fixed-end moments. *Transactions of the American Society of Civil Engineers,* vol. 96, pp. 1–10.

Cross, H., Morgan, N. D., 1932/2. *Continuous Frames of Reinforced Concrete.* New York: Wiley.

Cross, H., 1949. Analysis of continuous frames by distributing fixed-end moments. In: *Numerical Methods of Analysis in Engineering (Successive Corrections),* L. E. Grinter (ed.), pp. 1–12. New York: Macmillan & Co.

Cross, H., 1952. *Engineers and Ivory Towers.* New York: Ayer.

Cross, H., 1963. *Arches, Continuous Frames, Columns and Conduits: Selected Papers of Hardy Cross.* Introduction by N. M. Newmark. Urbana, Illinois.

Crotti, F., 1884. Commemorazione di Alberto Castigliano, *Il Politecnico.* vol. 32 (11/12), p. 597.

Csonka, S., 1960. Dimitri Ivanovich Jourawski. *Acta techn. Acad. scient. hung.,* vol. 28, p. 423–439 (in English).

Culmann, K., 1851. Der Bau der hölzernen Brücken in den Vereinigten Staaten von Nordamerika. *Allgemeine Bauzeitung,* vol. 16, pp. 69–129.

Culmann, K., 1852. Der Bau der eisernen Brücken in England und Amerika. *Allgemeine Bauzeitung,* vol. 17, pp. 163–222.

Culmann, K., 1864/1866. *Die graphische Statik.* Zurich: Verlag von Meyer & Zeller.

Culmann, K., 1875. *Die Graphische Statik.* 2nd, rev. ed. Zurich: Meyer & Zeller.

Cywiński, Z., Kurrer, K.-E., 2005. Reinhold Krohn (1852–1932): Director of Gutehoffnung Foundry & Prof. of Bridge & Structural Engineering at Danzig TH. In: *Proceedings of the International Conference Heritage of Technology – Gdańsk Outlook 4,* pp. 55–61. Gdańsk University of Technology 2005.

Czech, F., 1910. Eisenbauten. Ihre Geschichte und Ästhetik. *Der Eisenbau,* vol. 1, No. 10, pp. 405–407.

Czech, F., 1912. Der Vierendeelträger in der Geschichte des Eisenbaues. *Der Eisenbau,* vol. 3, No. 3, pp. 104–113.

Czerwenka, G., Schnell, W., 1967. *Einführung in die Rechenmethoden des Leichtbaus I.* Mannheim: Bibliographisches Institut.

Czerwenka, G., 1983. Vom Biegebalken zur Stabschale. In: *Humanismus und Technik, Jahrbuch 1983,* pp. 41–67. Berlin: Berlin TU library/Publications Dept.

Daelen, E., 1897. Wiegmann: Rudolf W. In: *Allgemeine Deutsche Biographie,* Band 42. Ed. Historical Commission, Royal Academy of sciences at the request and with the support of the King of Bavaria. Leipzig: Verlag von Duncker & Humblot.

DAfStb (ed.), 1982. *Festschrift 75 Jahre Deutscher Ausschuß für Stahlbeton.* Berlin/Munich: Wilhelm Ernst & Sohn.

Danilecki, W., 1954. Stahlbetonkonstruktionen mit steifen, tragenden Formstählen. *Bauplanung – Bautechnik,* vol. 8, No. 7, pp. 308–312.

Danusso, A., 1913. Beitrag zur Berechnung der kreuzweise bewehrten Eisenbetonplatten und deren Aufnahmeträger. Forscherarbeiten auf dem Gebiete des Eisenbetons, No. 21. Berlin: Wilhelm Ernst & Sohn.

Danyzy, A. A. H., 1732/1778. Méthode générale pour déterminer la résistance qu'il faut opposer à la poussée des voûtes. *Histoire de la Société Royale des Sciences établie à Montpellier* 2 (1718–1745), pp. 40-56 & 203–205. Montpellier: Jean Martel.

Dartein, F. de, 1906. La vie et les travaux de Jean-Rodolphe Perronet. *Annales des Ponts et Chaussées,* vol. 76, 8th series, vol. XXIV, pp. 5–87.

Dašek, V., 1930. *Výpočet rámových konstrukcí pomocí tensorů a elips deformačních* (calculation of frame constructions by means of tensors and deformation ellipses). Prague: MAP (in Czech).

Dašek, V., 1943/1951/1966. *Výpočet rámových konstrukcí rozdělováním sil a momentů* (calculation of frame constructions by means of force and moment distribution). Prague: TVV (in Czech).

Dašek, V., 1953. *Řešení trámových roštů metodou harmonického zatížení* (solution of the beam grid problem using the method of harmonic loading). Prague: ČSAV (in Czech).

Delhumeau, G., 1999. L'invention du béton armé – Hennebique 1890–1914. Paris: Norma Editions.

Denfert-Rochereau, 1859. Mémoire sur les voûtes en berceau portant une surcharge limitée à un plan horizontal. *Revue de l'Architecture et des Travaux Publics,* vol. 17, col. 114–124, 158–179, 207–224 & 257–268.

Denke, P. H., 1956. The matrix solution of certain nonlinear problems in structural analysis. *Journal of the Aeronautical Sciences,* vol. 23, No. 2, pp. 231–236.

Deresiewicz, H., Bieniek, M. P., DiMaggio, F. L. (ed.), 1989. *The Collected Papers of Raymond D. Mindlin,* vol. I+II. New York/Berlin: Springer.

Dernedde, W., Barbré, R., 1961. *Das Cross'sche Verfahren zur schrittweisen Berechnung durchlaufender Träger und Rahmen.* 4th ed. Berlin: Verlag von Wilhelm Ernst & Sohn.

Deutsche Reichsbahn (ed.), 1922. *Vorschriften für Eisenbauwerke. Grundlagen für das Entwerfen und Berechnung eiserner Eisenbahnbrücken (vorläufige Fassung).* Berlin: Verlag von Wilhelm Ernst & Sohn.

Deutsche Reichsbahn-Gesellschaft (ed.), 1925. *Vorschriften für Eisenbauwerke. Berechnungsgrundlagen für eiserne Eisenbahnbrücken (BE) vom 25. Februar 1925.* Berlin: Verlag von Wilhelm Ernst & Sohn.

Deutscher Beton-Verein (ed.), 1903. *Bericht über die VI. Haupt-Versammlung des Deutschen Beton-Vereins am 20. und 21. Februar 1903.* Berlin: Verlag Tonindustrie-Zeitung.

Deutscher Beton-Verein (ed.), 1907. *Bericht über die VI. Haupt-Versammlung des Deutschen Beton-Vereins am 22. und 23. Februar 1907.* Berlin: Verlag Tonindustrie-Zeitung.

DStV (ed.), 1929. *Denkschrift zum 25jährigen Bestehen des Deutschen Stahlbau-Verbandes 1904–1929.* Essen: Graphische Anstalt der Friedrich Krupp Aktiengesellschaft.

DStV (ed.), 1942. *Stahlbau-Kalender,* ed. G. Unold. Berlin: Wilhelm Ernst & Sohn.

DStV (ed.), 1943. Forschungsheft aus dem Gebiete des Stahlbaus. Beiträge zur Baustatik, Elastizitätstheorie, Stabilitätstheorie, Bodenmechanik, No. 6. Berlin: Springer-Verlag.

DStV (ed.), 1950. Stahlbau-Tagung in Braunschweig 1949. *Abhandlungen aus dem Stahlbau,* No. 8. Bremen-Horn: Industrie- & Handelsverlag Walter Dorn.

DStV (ed.), 1952. Stahlbau-Tagung in Munich 1952. *Abhandlungen aus dem Stahlbau,* No. 12. Bremen-Horn: Industrie- & Handelsverlag Walter Dorn.

DStV (ed.), 1954. *Deutscher Stahlbau-Verband 1904–1954. Betrachtungen zum 50jährigen Verbandsjubiläum am 17. September 1954.* Oldenburg: Gerhard Stalling AG.

DStV (ed.), 1964. Stahlbau-Tagung in Aachen 1964. *Veröffentlichungen des DStV,* No. 18. Cologne: Stahlbau-Verlags-GmbH.

Dickhoff, M., 1982. Zweiling, Klaus. In: *Philosophenlexikon.* Ed. E. Lange & D. Alexander, pp. 964–966. Berlin: Dietz Verlag.

Dicleli, C., 2000. Karl Bernhard. Die Durchdringung von Kunst und Technik. *deutsche bauzeitung,* vol. 134, No. 6, pp. 116–120.

Dicleli, C., 2006/1. Othmar Ammann. Brückenbauer des 20. Jahrhunderts in den USA. *deutsche bauzeitung,* vol. 140, No. 2, pp. 72–76.

Dicleli, C., 2006/2. Ulrich Finsterwalder. Ingenieur aus Leidenschaft. *deutsche bauzeitung,* vol. 140, No. 10, pp. 76–80.

Diderot, D., 1751. Art. In: Encyclopédie ou dictionnaire raisonné des sciences, des arts et des métiers, ed. D. Diderot & J. L. R. d'Alembert. Paris: Le Breton.

Dietrich, R. J., 1998. Faszination Brücken: Baukunst – Technik – Geschichte. Munich: Callwey.

Dijksterhuis, E. J., 1956. *Die Mechanisierung des Weltbildes*. Berlin: Springer.

Dijksterhuis, E. J., 1961. *The Mechanization of the World Picture*. Oxford: Clarendon Press.

Dijksterhuis, E. J., 1970. *Simon Stevin: Science in the Netherlands around 1600*. The Hague: Martinus Nijhoff.

Dijksterhuis, E. J., 1987. *Archimedes*. Princeton: Princeton University Press.

Dimitrov, N., 1971. Festigkeitslehre. In: *Beton-Kalender*, vol. 60, ed. G. Franz, pp. 237–309. Berlin/Munich/Düsseldorf: Verlag von Wilhelm Ernst & Sohn.

Dinnik, A. N., 1952–1956. *Izbrannye trudy* (selected works), vol. 3, Kiev: AN USSR (in Russian).

Di Pasquale, S., 1967. On the elastic problem of the non-homogeneous anisotropic body resting on lattice. *Meccanica*, n. 3, pp. 153–157.

Di Pasquale, S., 1968. On the elastic problem of the orthotropic plate. *Meccanica*, n. 2, pp. 111–120.

Di Pasquale, S., Leggeri, B., Nencioni, S., 1972. Energy forms in the finite element techniques. In: *Proceedings of the International Conference on Variationals Methods in Engineering*, vol. I, ed. C. Brebbia and H. Tottenham, pp. 4.33–4.43. Old Woking, Surrey: Unwin Brothers Ltd., The Gresham Press.

Di Pasquale, S., 1975. *Scienza delle costruzioni. Introduzione alla progettazione strutturale*. Milan: Tamburini – Masson.

Di Pasquale, S., 1978. *Metodi di calcolo per le strutture spaziali*. Milan: CISIA.

Di Pasquale, S., 1988. Questions concerning the mechanics of masonry. In: *Stable-unstable?* pp. 249–264, ed. R. M . Lemaire, K. Van Balen, Leuven. Centre for the Conservation of Historic Towns & Buildings. Leuven: University Press.

Di Pasquale, S., 1992. New trends in the analysis of masonry structures. *Meccanica*, vol. 27, No. 3, pp. 173–184.

Di Pasquale, S., 1995. On the art of building before Galilei. In: *Entre Mécanique et Architecture – Between Mechanics and Architecture*, pp. 102–121, ed. P. Radelet-de Grave & E. Benvenuto. Basel: Birkhäuser.

Di Pasquale, S., 1996. *L'arte del costruire. Tra conoscenza e scienza*. Venice: Marsilio.

Di Pasquale, S., 2002. *Brunelleschi. La costruzione della cupola di Santa Maria del Fiore*, Venice: Marsilio.

Dirichlet, P. G. L., 1846. Über die Stabilität des Gleichgewichts. *Journal für die reine und angewandte Mathematik*, vol. 32, pp. 85–88.

Dirschmid, W., 1969. Methode zur Auslegung einer Fahrzeug-Karosserie hinsichtlich optimaler Gesamtsteifigkeit. *ATZ Automobiltechnische Zeitschrift*, vol. 71, No. 1, pp. 11–14.

Dirschmid, W., 2007. Erster Einsatz der Karosserie-Optimierung mit FEM am Audi 100, Baujahr 1968. Presentation at *ANSYS Conference & 25th CADFEM User's Meeting, 21–23 November 2007*, Dresden.

Dischinger, F., 1925. Fortschritte im Bau von Massivkuppeln. *Der Bauingenieur*, vol. 6, No. 10, pp. 362–366.

Dischinger, F., 1928/1. Schalen und Rippenkuppeln. In: *Handbuch für Eisenbetonbau*, 3rd ed., vol. 12, pt. II, ed. F. v. Emperger, pp. 151–371, Berlin: Wilhelm Ernst & Sohn.

Dischinger, F., Finsterwalder, U., 1928/2. Eisenbeton-Schalendächer System Dywidag. *Der Bauingenieur*, vol. 9, No. 44, pp. 807–812, No. 45, pp. 823–827 & No. 46, pp. 842–846.

Dischinger, F., 1929. Die Theorie der Vieleckkuppeln und die Zusammenhänge mit den einbeschriebenen Rotationsschalen. *Beton und Eisen*, vol. 28, No. 5, pp. 100–107, No. 6, pp. 119–122, No. 8, pp. 150–156 & No. 9, pp. 169–175.

Dischinger, F., Rüsch, H., 1929. Die Großmarkthalle in Leipzig – ein neues Kuppelsystem, zusammengesetzt aus Zeiss-Dywidag-Schalengewölben. *Beton und Eisen*, vol. 28, No. 18, pp. 325–329, No. 19, pp. 341–346, No. 23, pp. 422–429 & No. 24, pp. 437–442.

Dischinger, F., Finsterwalder, U., 1932. Die weitere Entwicklung der Schalenbauweise "Zeiss-Dywidag". *Beton und Eisen*, vol. 33, No. 7/8, pp. 101–108, No. 10, pp. 149–155, No. 11, pp. 165–170, No. 12, pp. 181–184, No. 14, pp. 213–220, No. 15, pp. 229–235 & No. 16, pp. 245–247.

Dischinger, F., 1937. Untersuchungen über die Knicksicherheit, die elastische Verformung und das Kriechen des Betons bei Bogenbrücken. *Der Bauingenieur*, vol. 18, No. 33/34, pp. 487–520, No. 35/36, pp. 539–552 & No. 39/40, pp. 595–621.

Dischinger, F., 1939. Elastische und plastische Verformungen der Eisenbetontragwerke und insbesondere der Bogenbrücken. *Der Bauingenieur*, vol. 20, No. 5/6, pp. 53–63, No. 21/22, pp. 286–294, No. 31/32, pp. 426–437 & No. 47/48, pp. 563–572.

Dittmann, F., 1994. August Föppl: Einführung in die Maxwellsche Theorie der Elektrizität. *Wiss. Z. Techn. Univers. Dresden*, vol. 43, No. 6, pp. 89–91.

Dobbert, E., 1899. Bauakademie, Gewerbeakademie und Technische Hochschule bis 1884. In: *Chronik der Königlichen Technischen Hochschule zu Berlin 1799–1899*. Berlin: Wilhelm Ernst & Sohn.

Doltsinis, I., 2004. Obituary of John Argyris. *Communications in Numerical Methods in Engineering*, vol. 20, pp. 665–669.

Domke, O., 1915. Über Variationsprinzipien der Elastizitätslehre nebst Anwendungen auf die technische Statik. *Zeitschrift für Mathematik und Physik*, vol. 63, pp. 174–192.

Domke, O., 1936. Zum Aufsatz "Über die Minimalprinzipie der Elastizitätstheorie" von T. Pöschl. *Der Bauingenieur*, vol. 17, No. 41/42, p. 459ff.

Dorn, H. I., 1970. *The Art of Building and the Science of Mechanics. A Study of the Union of Theory and Practice in the Early History of Structural Analysis in England*. PhD diss.: Princeton University.

Dörnen, A., 1955. Stahlüberbau der Weserbrücke Porta. *Der Stahlbau*, vol. 24, No. 5, pp. 97–101.

Dougall, J., 1903/1904. An analytical theory of the equilibrium of an isotropic elastic plate. *Edinburgh Royal Society, Transactions*, vol. 41, pp. 129–227.

Drucker, D. C., 1951. A more fundamental approach to stress-strain relations. In: *Proceedings of the 1st U.S. National Congress of Applied Mechanics*, pp. 487–491.

Drucker, D. C., 1953. Coulombs friction, plasticity and limit loads. *Transactions of the American Society of Mechanical Engineers*, vol. 21, pp. 71–74.

Drucker, D. C., 1956. On uniqueness in the theory of plasticity. *Quarterly Applied Mathematics*, vol. 14, pp. 35–42.

Drucker, D. C., 1959. A definition of stable inelastic material. *Journal of Applied Mechanics*, vol. 26, pp. 101–106.

Drucker, D. C., 1961. On Structural Concrete and the Theorems of Limit Analysis. In: *Proceedings of the International Association for Bridge & Structural Engineering (IABSE)*. vol. 21, pp. 49–59. Zurich.

Drucker, D. C., 1984. William Prager. In: *Memorial Tributes. National Academy of Engineering*, vol. 2, pp. 232–235. Washington, D.C.: National Academy Press.

Du Bois, A. J., 1875. *The new method of graphical statics* (extract from engineering magazine). New York: Van Nostrand.

Du Bois, A. J., 1875. The elements of graphical statics, and their applications to framed structures. New York: Wiley.

Duddeck, H., 1972. Traglasttheorie der Stabtragwerke. In: *Beton-Kalender*, vol. 61, pt. II, pp. 621–676. Berlin: Wilhelm Ernst & Sohn.

Duddeck, H., 1990. Wilhelm Flügge verstorben. *Der Bauingenieur*, vol. 65, p. 486.

Duddeck, H., Ahrens, H., 1994. Statik der Stab-tragwerke. In: *Beton-Kalender*, vol. 83, pt. I, ed. J. Eibl, pp. 261–376. Berlin: Ernst & Sohn.

Duddeck, H., 1996. Und machet Euch die Erde untertan…? *Der Bauingenieur*, vol. 71, No. 6, pp. 241–248.

Duddeck, H., Ahrens, H., 1998. Statik der Stabtragwerke. In: *Beton-Kalender*, vol. 87, pt. I, ed. J. Eibl, pp. 339–454. Berlin: Ernst & Sohn.

Duddeck, H., 1999. Die Sprachlosigkeit der Ingeniure. In: *Die Sprachlosigkeit der Ingenieure*, ed. H. Duddeck & J. Mittelstraß, pp. 1–17. Opladen: Leske + Budrich.

Dugas, R., 1950. *Histoire de la Mechanique*. Neuchâtel: ed. du Griffon.

Duhem, P., 1903. *L'évolution de la méchanique*. Paris: A. Joanin.

Duhem, P., 1905/1906. *Les origines de la Statique*. vol. 2. Paris: Hermann.

Duhem, P., 1906/2007. *La théorie physique, son objet et sa structure*. Paris: J. Vrin (reprint).

Duhem, P., 1954. *The aim and structure of physical theory*. Trans. by Wiener with a preface by Louis de Broglie. Princeton: Princeton University Press.

Duhem, P., 1978. *Ziel und Struktur der physikalischen Theorien*. Auth. trans. by Friedrich Adler with a preface by Ernst Mach, with intro. & bibl., ed. Lothar Schäfer. Hamburg: Felix Meiner Verlag.

Duleau, M. M., 1820. *Essai théorique et experimental sur la résistance du fer forgé*. Paris.

Duncan, W. J., Collar, A. R., 1934. A method for the solution of oscillation problems by matrices. *Philosophical Magazine*, vol. 15 (series 7), pp. 865–885.

Duncan, W. J., Collar, A. R., 1935. Matrices applied to the motions of damped systems. *Philosophical Magazine*, vol. 19 (series 7), pp. 197–214.

Dunn, W., 1904. Notes on the Stresses in Framed Spires and Domes. *Journal of the Royal Institute of British Architects*, 3rd series, 11 (Nov 1903 – Oct 1904), pp. 401–412.

Dupuit, J., 1870. *Traité de l'équilibre des voûtes et de la construction des ponts en maçonnerie*. Paris: Dunod.

Durand-Claye, A., 1867. Note sur la vérification de la stabilité des voûtes en maçonnerie et sur l'emploi des courbes de pression. *Annales des Ponts et Chaussées*, vol. 13, pp. 63–93.

Dürbeck, A., 1931. Bauliche Fragen der Weiterentwicklung der amerikanischen Wolkenkratzer. *Der Stahlbau*, vol. 4, No. 20, pp. 236–239.

Dürrenmatt, F., 1980. *Die Physiker*. Zurich: Diogenes.

Dyckerhoff & Widmann, 1929/1. Letter concerning plates and folded plates as new construction elements in structural steelwork. *Beton und Eisen*, vol. 28, No. 14, p. 276.

Dyckerhoff & Widmann, 1929/2. Letter concerning plates and folded plates as new construction elements in structural steelwork. *Beton und Eisen*, vol. 28, No. 18, pp. 337–338.

Dyckerhoff & Widmann, 1929/3. Reply. *Beton und Eisen*, vol. 28, No. 18, p. 339.

Eagleton, T., 1994. *Ästhetik. Die Geschichte ihrer Ideologie*. Stuttgart/Wiemar: J. B. Metzler.

Eaton, L. K., 2001. The Engineering Achievements of Hardy Cross. *Nexus Network Journal*, vol. 3, No. 2, pp. 15–24.

Ebel, H., 1979. Zur Berücksichtigung von Verformungslastfällen in den Reziprozitätssätzen von Betti, Maxwell und Krohn-Land. *Der Stahlbau*, vol. 48, No. 5, pp. 137–140.

Eberhard, F., Ernst, E., Wolf, W., 1958. 50 Jahre deutscher Ausschuß für Stahlbau. In: *Deutscher Ausschuß für Stahlbau, 1908–1958*, pp. 9–46. Cologne: Stahlbau-Verlags-GmbH.

Eberlein, H., 1970. Beitrag zur geometrischen Ableitung der zweilagigen Stabwerkroste aus dem cubischen Gitter. Berlin: Dissertation, Berlin TU.

Ebner, H., 1929/1. Zur Berechnung räumlicher Fachwerke im Flugzeugbau. Berlin: Dissertation, Berlin TH.

Ebner, H., 1929/2. Zur Berechnung räumlicher Fachwerke im Flugzeugbau. In: *Jahrbuch 1929 der Deutschen Versuchsanstalt für Luftfahrt e.V.*, Berlin-Adlershof, pp. 371–414. Munich: Oldenbourg.

Ebner, H., 1931. Die Berechnung regelmäßiger, vielfach statisch unbestimmter Raumfachwerke mit Hilfe von Differenzengleichungen. In: *Jahrbuch 1931 der Deutschen Versuchsanstalt für Luftfahrt e.V.*, Berlin-Adlershof, pp. 246–288. Munich: Oldenbourg.

Ebner, H., 1932. Zur Berechnung statisch unbestimmter Raumfachwerke (Zellwerke). *Der Stahlbau*, vol. 5, No. 1, pp. 1–6 & No. 2, pp. 11–14.

Ebner, H., 1933. Die Beanspruchung dünnwandiger Kastenträger auf Drillung bei behinderter Querschnittsverwölbung. *Zeitschrift für Flugtechnik und Motorluftschiffahrt*, vol. 24, No. 23, pp. 645–655 & No. 24, pp. 684–692.

Ebner, H., 1937/1. Theorie und Versuche zur Festigkeit von Schalenrümpfen. In: *Luftfahrtforschung*, vol. 14, pp. 93–115.

Ebner, H., 1937/2. Zur Festigkeit von Schalen- und Rohrholmflügeln. In: *Jahrbuch 1937 der Deutschen Versuchsanstalt für Luftfahrt e.V.*, Berlin-Adlershof, pp. 221–232.

Ebner, H., Köller, H., 1937/1. Über die Einleitung von Längskräften in versteifte Zylinderschalen. In: *Jahrbuch 1937 der Deutschen Versuchsanstalt für Luftfahrt e.V.*, Berlin-Adlershof, pp. 243–252.

Ebner, H., Köller, H., 1937/2. Zur Berechnung des Kraftverlaufs in versteiften Zylinderschalen. In: *Luftfahrtforschung*, vol. 14, pp. 607–626.

Ebner, H., Köller, H., 1938/1. Über den Kraftverlauf in längs- und querversteiften Scheiben. In: *Luftfahrtforschung*, vol. 15, pp. 527–542.

Ebner, H., Köller, H., 1938/2. *Calculation of load distribution in stiffened cylindrical shells*. N.A.C.A. T.M., No. 866, June 1938.

Ebner, H., Köller, H., 1939. Über den Kraftverlauf in längs- und querversteiften Scheiben. In: *Jahrbuch 1939 der Deutschen Versuchsanstalt für Luftfahrt e.V.*, Berlin-Adlershof, p. 121–136.

Eckold, K., 2002. Der Statik von Schalen verschrieben. *Dresdener Universitätsjournal* 1/2002.

Eddy, H. T., 1877. *New constructions in graphical statics*. New York: Van Nostrand.

Eddy, H. T., 1878. *Researches in graphical statics*. New York: Van Nostrand.

Eddy, H. T., 1880. *Neue Constructionen aus der graphischen Statik*. Leipzig: Teubner.

Eddy, H. T., 1913. *The theory of the flexure and strength of rectangular flat plates applied to reinforced concrete floor slabs*. Minneapolis: Rogers & company.

Eddy, H. T., Turner, C. A. P., 1914. *Concrete-steel construction*. Minneapolis: Heywood mfg. company.

Efimow, N. W., 1970. *Höhere Geometrie II. Grundzüge der projektiven Geometrie*. Braunschweig/Basel: Vieweg & C. F. Winter.

Egerváry, E., 1956. Begründung und Darstellung einer allgemeinen Theorie der Hängebrücken mit Hilfe der Matrizenrechnung. In: *Proceedings of the International Association for Bridge & Structural Engineering (IABSE)*. vol. 16, ed. F. Stüssi & P. Lardy, pp. 149–184. Zurich: IABSE secretariat.

Eggemann, H., 2003/1. *Vereinfachte Bemessung von Verbundstützen im Hochbau*. Dissertation, Aachen RWTH.

Eggemann, H., 2003/2. Development of composite columns. Emperger's effort. In: *Proceedings of the First International Congress on Construction History*, ed. S. Huerta, vol. I, pp. 787–797. Madrid: Instituto Juan de Herrera.

Eggemann, H., 2003/3. Fritz von Emperger – Verbundbaupionier. *Beton- und Stahlbetonbau*, vol. 98, No. 11, pp. 701–705.

Eggemann, H., 2006. Simplified Design of Composite Columns, Based on a Comparative Study of the Development of Building Regulations in Germany and the United States. In: *Proceedings of the Second International Congress on Construction History*, ed. M. Dunkeld, J. Campbell, H. Louw, M. Tutton, W. Addis & R. Thorne, vol. I, pp. 1023–1041. Cambridge. Construction History Society.

Eggemann, H., Kurrer, K.-E., 2006. Zur internationalen Verbreitung des Systems Melan. *Beton- und Stahlbetonbau*, vol. 101, No. 11, pp. 911–922.

Eggenschwyler, A., 1918. Über die Drehungsbeanspruchung rechteckiger Kastenquerschnitte. *Der Eisenbau*, vol. 9, No. 3, pp. 45–54.

Eggenschwyler, A., 1920/1. Zur Festigkeitslehre (letter). *Schweizerische Bauzeitung*, vol. 76, No. 18, pp. 206–208.

Eggenschwyler, A., 1920/2. Zur Festigkeitslehre. *Schweizerische Bauzeitung*, vol. 76, No. 23, p. 266.

Eggenschwyler, A., 1921. Über die Drehungsbeanspruchung von dünnwandigen symmetrischen ⌶-förmigen Querschnitten. *Der Eisenbau*, vol. 12, No. 9, pp. 207–215.

Eggenschwyler, A., 1924. Zur Frage des Schubmittelpunktes. *Schweizerische Bauzeitung*, vol. 83, No. 22, pp. 259–261.

Eggert, H., Henke, G., 2007. Stahlbaunormen. Kommentierte Stahlbauregelwerke. In: *Stahlbau-Kalender*, ed. U. Kuhlmann, pp. 1–243. Berlin: Ernst & Sohn.

Ehlers, G. 1930/1. Ein neues Konstruktionsprinzip. *Der Bauingenieur*, vol. 11, No. 8, p. 125–132.

Ehlers, G. 1930/2. Die Spannungsermittlung in Flächentragwerken. *Beton und Eisen*, vol. 29, No. 15, pp. 281–286 & No. 16, pp. 291–296.

Eiffel, G., 1889. *Mémoire sur le Viaduc de Garabit.* Paris: Librairie Polytechnique, Baudry et C$^{ie}$, Editeurs.

Eiffel, G., 1900. *La tour de trois mètres.* Paris: Société des Imprimeries Lemercier.

Eiffel, G., 2006. *La tour de trois mètres.* Reprint with commentary by Bertrand Lemoine. All texts in German, English, Spanish, Italian, Portuguese, Dutch & Japanese. Cologne: Taschen.

Eiselin, O., 1938. Stahlbrücken mit Plattengurtungen. *Die Bautechnik,* vol. 16, No. 4, pp. 41–43, No. 6, p. 76 & No. 16, pp. 205–207.

Elbern, 1920. Über deutsche Eisenbauanstalten und ihre Arbeitsweisen. *Zeitschrift für Bauwesen,* vol. 70, col. 281–330.

Elishakoff, I., 2000. Elastic stability: From Euler to Koiter there was none like Koiter. *Meccanica,* vol. 35, pp. 375–380.

Elmes, J., 1823. *Memoirs of the Life and Works of Sir Christopher Wren.* London.

Emde, H., 1977. *Geometrie der Knoten-Stab-Tragwerke.* Würzburg: Strukturforschungszentrum e.V.

Emerson, W., 1754. *The principles of mechanics.* London: W. Ynnys, J. Richardson.

Emperger, F. v., 1894. The development and recent improvment of concrete-iron highway bridges. *Transactions ASCE,* vol. 31, pp. 437–457.

Emperger, F. v., 1901/1. Neuere Bauweisen und Bauwerke aus Beton und Eisen, nach dem Stande bei der Pariser Weltausstellung 1900. *Zeitschrift des Österreichischen Ingenieur- und Architekten-Vereines,* vol. 53, No. 7, pp. 97–103, No. 8, pp. 117–124, No. 43, pp. 713–719 & No. 46, pp. 765–770.

Emperger, F. v., 1901/2. *Neuere Bauweisen und Bauwerke aus Beton und Eisen, nach dem Stande bei der Pariser Weltausstellung 1900 mit einem Anhang über Stiegenbauten.* Vienna: Self-pub.

Emperger, F. v., 1902/1. *Eine Belastungsprobe mit Massivdecken nach dem System Hennebique.* Vienna: Lehmann & Wentzel (Paul Krebs).

Emperger, F. v., 1902/2. Neuere Bauweisen und Bauwerke aus Beton und Eisen, nach dem Stande bei der Pariser Weltausstellung 1900. *Zeitschrift des Österreichischen Ingenieur- und Architekten-Vereines,* vol. 54, No. 24, pp. 441–446, No. 25, pp. 453–458 & No. 30, pp. 518–523.

Emperger, F. v., 1902/3. *Neuere Bauweisen und Bauwerke aus Beton und Eisen. III. Theil. Fortsetzung des Berichtes über den Stand der Pariser Ausstellung aus dem Gebiete des Wasserbaues mit einem Anhang über den Hauseinsturz in Basel.* Vienna: von Lehmann & Wentzel (Paul Krebs).

Emperger, F. v., 1902/4. *Neuere Bauweisen und Bauwerke aus Beton und Eisen. IV. Theil. Die Durchbiegung und Einspannung von armierten Betonbalken und Platten.* Vienna: Lehmann & Wentzel (Paul Krebs).

Emperger, F. v., 1903. *Über die Berechnung von beiderseitig armierten Betonbalken, mit einem Anhang: einige Versuche über die Würfelfestigkeit von armiertem Beton.* Dissertation, Vienna TH. Berlin: Wilhelm Ernst & Sohn.

Emperger, F. v., 1905/1. †H. Ludwig v. Tetmajer. *Beton und Eisen,* vol. 4, No. 3, p. 72.

Emperger, F. v., 1905/2. *Die Rolle der Haftfestigkeit in den Verbundbalken.* Forscherarbeiten auf dem Gebiete des Eisenbetons, No. 3. Berlin: Wilhelm Ernst & Sohn.

Emperger, F. v., 1905/3. Zur Geschichte des Eisenbetons in Belgien. *Beton und Eisen,* vol. 4, No. 1, pp. 1–2.

Emperger, F. v., 1906. *Die Abhängigkeit der Bruchlast vom Verbunde und die Mittel zur Erhöhung der Tragfähigkeit von Balken aus Eisenbeton.* Forscherarbeiten auf dem Gebiete des Eisenbetons, No. 5. Berlin: Wilhelm Ernst & Sohn.

Emperger, F. v., 1908. *Versuche mit Säulen aus Eisenbeton und mit einbetonierten Säulen.* Forscherarbeiten auf dem Gebiete des Eisenbetons, No. 8. Berlin: Wilhelm Ernst & Sohn.

Emperger, F. v., 1912. *Eine neue Verwendung des Gusseisens bei Säulen und Bogenbrücken.* Berlin: Wilhelm Ernst & Sohn.

Emperger, F. v., 1913. *Neuere Bogenbrücken aus umschnürtem Gusseisen.* Berlin: Wilhelm Ernst & Sohn.

Emperger, F. v., 1931. Der Beiwert n = 15 und die zulässigen Biegespannungen. *Beton und Eisen,* vol. 30, pp. 340–49.

Emperger, F. v., 1932. †Prof. Ostenfeld. *Beton und Eisen,* vol. 31, No. 3, p. 51.

Emperger, F. v., 1939. *Stahlbeton mit vorgespannten Zulagen aus höherwertigem Stahl.* Berlin: Wilhelm Ernst & Sohn.

Emperger, F. v., 1941. †Josef Melan. *Beton und Eisen,* vol. 40, p. 110.

Emy, A. R., 1837. *Traité de l'art de la charpenterie.* vol. 1. Paris.

Emy, A. R., 1841. *Traité de l'art de la charpenterie.* vol. 2. Paris.

Encke, J. F., 1851. Gedächtnisrede auf Eytelwein. In: *Abhandlungen der Königlichen Akademie der Wissenschaften zu Berlin.* 1849, pp. XV–XXXIV. Berlin: Dümmler's Buchhandlung.

Engelmann, H., 1929. Experimentelle und theoretische Untersuchungen zur Drehfestigkeit der Stäbe. *Zeitschrift für angewandte Mathematik und Mechanik,* vol. 9, No. 5, pp. 386–401.

Engels, F., 1962. Dialektik der Natur. In: *Karl Marx-Friedrich Engels-Werke, Bd. 20.* Berlin: Dietz.

Engesser, F., 1879. Über die Durchbiegung von Fachwerkträgern und die hierbei auftretenden zusätzlichen Spannungen. *Zeitschrift für Baukunde,* vol. 2, pp. 590–602.

Engesser, F., 1880. Ueber die Lage der Stützlinie in Gewölben. *Deutsche Bauzeitung,* vol. 14, pp. 184-186, pp. 210 & 243.

Engesser, F., 1889/1. Über statisch unbestimmte Träger bei beliebigem Formänderungs-Gesetze und über den Satz von der kleinsten Ergänzungsarbeit. *Zeitschrift d. Arch.- & Ing.-Vereins zu Hannover,* vol. 35, pp. 733–744.

Engesser, F., 1889/2. Über die Knickfestigkeit gerader Stäbe. *Zeitschrift d. Arch.- & Ing.-Vereins zu Hannover,* vol. 35, pp. 455–462.

Engesser, F., 1892/1893. *Die Zusatzkräfte und Nebenspannungen eiserner Fachwerkbrücken.* Berlin: Springer.

Engesser, F., 1893. Über die Berechnung auf Knickfestigkeit beanspruchter Stäbe aus Schweiß- und Gußeisen. *Zeitschrift d. Österr. Ing. & Arch.-Vereins,* vol. 45, pp. 506–508.

Engesser, F., 1895. Über Knickfragen. *Schweizerische Bauzeitung,* vol. 26, No. 4, p. 24.

Engesser, F., 1909. Über die Knickfestigkeit von Stäben mit veränderlichem Trägheitsmoment. *Zeitschrift d. Österr. Ing. & Arch.-Vereins,* vol. 61, pp. 544–548.

Engesser, F., 1913/1. Ueber Rahmenträger und ihre Beziehung zu den Fachwerkträgern. *Zeitschrift für Architektur und Ingenieurwesen.* vol. 59 (vol. 18 of new series), col. 67–86.

Engesser, F., 1913/2. *Die Berechnung der Rahmenträger.* Berlin: Wilhelm Ernst & Sohn.

Ernst, A., 1883/1. *Die Hebezeuge. Theorie und Kritik ausgeführter Konstruktionen. Textband.* Berlin: Julius Springer.

Ernst, A., 1883/2. *Die Hebezeuge. Theorie und Kritik ausgeführter Konstruktionen, Atlas.* Berlin: Julius Springer.

Ernst, A., 1886. Letter. *ZVDI,* vol. 30, No. 47, pp. 1030–1032.

Ernst, A., 1899. *Die Hebezeuge. Theorie und Kritik ausgeführter Konstruktionen. Band 1,* 3rd, rev. ed. Berlin: Julius Springer.

Ersch, J. S., Gruber, J. G. (ed.), 1857. Gewölbe. In: *Allgemeine Encyklopädie der Wissenschaften und Künste, 66. Teil.* Leipzig: F. A. Brockhaus.

Eßlinger, M., 1952. *Statische Berechnung von Kesselböden.* Berlin: Springer-Verlag.

Euler, L., 1744. *Methodus inveniendi lineas curvas maximi minimive proprietate gaudentes, sive solutio problematis isoperimetrici latissimo sensu accepti.* Lausanne/Geneva.

Ewert, S., 2002. *Brückensysteme.* Berlin: Ernst & Sohn.

Eytelwein, J. A., 1799. Nachrichten von der Errichtung der Königlichen Bauakademie Berlin. *Sammlung nützlicher Aufsätze und Nachrichten, die Baukunst betreffend,* vol. 3, No. 2, pp. 28–40.

Eytelwein, J. A., 1801. *Handbuch der Mechanik fester Körper und der Hydraulik.* Berlin: Lagarde.

Eytelwein, J. A., 1808. *Handbuch der Statik fester Körper.* 3 vols. 1st ed. Berlin: Realschulbuchhandlung.

Eytelwein, J. A., 1823. *Handbuch der Mechanik fester Körper und der Hydraulik.* 2nd ed. Leipzig: Koechly.

Eytelwein, J. A., 1832. *Handbuch der Statik fester Körper.* 3 vols. 2nd ed. Berlin: Reimer.

Eytelwein, J. A., 1842. *Handbuch der Mechanik fester Körper und der Hydraulik.* 3rd ed. Leipzig: Koechly.

Fair, G. M., 1950. Harold Malcolm Westergaard (1888–1950). *Year Book of the American Philosophical Society*, pp. 339–42. Philadelphia: American Philosophical Society.

Fairbairn, W., 1838. *An Experimental Enquiry into the Strength and other Properties of Cast Iron, from Various Parts of the United Kingdom.* Manchester Memoirs, vol. 6, New Series. Manchester: Printed by Francis Looney.

Fairbairn, W., 1849/1. On tubular girder bridges. *Minutes of the Proceedings of the Institution of Civil Engineers*, vol. 9, pp. 233–241. Discussion, pp. 242–287.

Fairbairn, W., 1849/2. *An Account of the Construction of the Britannia and Conway Tubular Bridges with a complete History of their Progress, from the Conception of the Original Idea, to the Conclusion of the Elaborate Experiments which Determined the Exact Form and Mode of Construction ultimately Adopted.* London: John Weale and Longman, Brown, Green & Longmans, Paternoster Row.

Fairbairn, W., 1854. *On the Application of Cast and Wrought Iron to Building Purposes.* London: John Weale.

Fairbairn, W., 1857/58. *On the Application of Cast and Wrought Iron to Building Purposes. To which is added a short treatise on wrought iron bridges.* London: John Weale.

Fairbairn, W., 1865. *Treatise on Iron Shipbuilding.* London: Longman.

Falk, S., 1951. Ein übersichtliches Schema für die Matrizenmultiplikation. *Zeitschrift für angewandte Mathematik und Mechanik (ZAMM)*, vol. 31, No. 4/5, pp. 152–153.

Falk, S., 1956. Die Berechnung des beliebig gestützten Durchlaufträgers nach dem Reduktionsverfahren. *Ingenieur-Archiv*, vol. 24, pp. 216–232.

Falkenheiner, H., 1950. Calcul systématique des charactéristiques élastiques des systèmes hyperstatiques. *La Recherche Aéronautique*, No. 17, p. 17ff.

Falkenheiner, H., 1951. La systématisation du calcul hyperstatiques d'après l'hypothèse du 'schéma du champ homogène'. *La Recherche Aéronautique*, No. 2, p. 61ff.

Falkenheiner, H., 1953. Systematic Analysis of Redundant Elastic Structures by Means of Matrix Calculus. *Journal of the Aeronautical Sciences*, vol. 20, No. 4, p. 293ff.

Falter, H., 1994. Die Sicherheit im bautechnischen Gestaltungsprozeß, eine Frage des Erkennens und der Darstellung. *Dresdener Beiträge zur Geschichte der Technikwissenschaften*, No. 23/1, pp. 53–61.

Falter, H., 1999. *Untersuchungen historischer Wölbkonstruktionen – Herstellverfahren und Werkstoffe.* Stuttgart: Dissertation, University of Stuttgart.

Falter, H., Kahlow, A., Kurrer, K.-E., 2001. Vom geometrischen Denken zum statisch-konstruktiven Ansatz im Brückenentwurf. *Bautechnik*, vol. 78, No. 12, pp. 889–902.

Faltus, F., 1947/1955. *Svařování.* (welding). Prague: ČMT (in Czech).

Faltus, F., 1951/1962. *Prvky ocelových konstrukcí.* (structural steelwork elements). Prague: TVV (in Czech).

Faltus, F., 1985. *Joints with Fillet Welds.* Amsterdam: Elsevier.

Federhofer, K., 1909. Graphisches Verfahren für die Ermittlung der Spannungsverteilung in zylindrischen Behälterwänden. *Beton und Eisen*, vol. 8, No. 16, pp. 387–388.

Federhofer, K., 1910. Graphisches Verfahren für die Ermittlung der Spannungsverteilung in zylindrischen Behälterwänden. *Beton und Eisen*, vol. 9, No. 2, pp. 40–41 & 45.

Fedorov, S. J., 2000. *Wilhelm von Traitteur. Ein badischer Baumeister als Neuerer in der russischen Architektur.* Berlin: Ernst & Sohn.

Fedorov, S. G., 2005. *Carl Friedrich von Wiebeking und das Bauwesen in Russland.* Munich/Berlin: Deutscher Kunstverlag.

Feldtkeller, R., 1937. *Einführung in die Vierpoltheorie der elektrischen Nachrichtentechnik.* Leipzig: Hirzel.

Felippa, C. A., 1994. An appreciation of R. Courant's 'Variational Methods for the solution of problems of equilibrium and vibrations', 1943. *International Journal for Numerical Methods in Engineering*, vol. 37, pp. 2159–2161.

Felippa, C. A., 2000. On the original publication of the general canonical functional of linear elasticity. *Journal of Applied Mechanics*, vol. 67/1, pp. 217–219.

Felippa, C. A., 2001. A historical outline of matrix structural analysis: a play in three acts. *Computers & Structures*, vol. 79, pp. 1313–1324.

Felippa, C. A., 2002. Fraeijs de Veubeke: Neglected Discoverer of the 'Hu-Washizu Functional'. *iacm expressions*, 12/02, pp. 8–10.

Felippa, C. A., 2005. The Amusing History of Shear Flexible Beam Elements. *iacm expressions*, 17/05, pp. 15–19.

Ferroni, N., 1808. Della vera curva degli archi del ponte a S. Trinità di Firenze. *Società It. delle Scienze*, vol. 14. Verona.

Fillunger, P., 1936. *Erdbaumechanik?* Vienna: Self-pub.

Fischer, M., Kleinschmidt, C. (ed.), 2002. *Stahlbau, in Dortmund.* Essen: Klartext-Verlag.

Fischmann, H., 1915. Letter to Carstanjen of October 1915. From: *Handakte "Versuchsausschuß" des DStV*.

F., J. J., 1922. Henry Turner Eddy. *Science*, vol. 55, Issue 1410, pp. 12–13.

Fleckner, S., 2003. Gotische Kathedralen – Statische Berechnungen. *Bauingenieur*, vol. 78, No. 1, pp. 13–23.

Fliegel, E., 1938. Die Elastizitätsgleichungen zweiter Art der Stabwerksdynamik. *Ingenieur-Archiv*, vol. 9, pp. 20–38.

Flomenhoft, H. I., 1997. *The Revolution in Structural Dynamics.* Palm Beach Gardens: Dynaflo Press.

Flomenhoft, H. I., 2007. Letter to the author from Dr. Hubert I. Flomenhoft dated 21 Nov 2007.

Flower, F. A., 1905. *Edwin McMasters Stanton. The autocrat of rebellion, emancipation, and reconstruction.* Akron: The Saalfield Publishing Co.

Flügge, W., 1934. *Statik und Dynamik der Schalen.* Berlin: Springer.

Flügge, W., 1957. *Statik und Dynamik der Schalen.* 2nd, rev. ed. Berlin: Springer.

Flügge, W., 1960. *Stresses in Shells.* Berlin: Springer.

Flügge, W., 1962. *Handbook of Engineering Mechanics.* New York: McGraw-Hill.

Flügge, W., 1972. *Tensorial Methods in Continuum Mechanics.* New York: Springer.

Flügge, W., 1975. *Viscoelasticity.* New York: Springer.

Foce, F., 1993. *La teoria molecolare dell'elasticità dalla fondazione ottocentesca ai nuovi sviluppi del XX secolo.* Doctorate thesis, University of Florence.

Foce, F., 1995. The Theory of Elasticity between Molecular and Continuum Approach. In: *Entre Mécanique et Architecture*, ed. P. Radelet-de Grave & E. Benvenuto, pp. 301–314. Basel: Birkhäuser.

Foce, F., Aita, D., 2003, The masonry arch between 'limit' and 'elastic' analysis. A critical re-examination of Durand-Claye's method. In: *Proceedings of the First International Congress on Construction History*, ed. S. Huerta, vol. II, pp. 895–908. Madrid: Instituto Juan de Herrera.

Foce, F., 2005. On the safety of the masonry arch. Different formulations from the history of structural mechanics. In: *Essays in the history of theory of structures*, ed. S. Huerta, pp. 117–142. Madrid: Instituto Juan de Herrera.

Föppl, A., 1880. *Theorie des Fachwerks.* Leipzig: Verlag von Arthur Felix.

Föppl, A., 1881. *Theorie der Gewölbe.* Leipzig: Verlag von Arthur Felix.

Föppl, A., 1887. Zur Fachwerkstheorie. *Schweizerische Bauzeitung*, vol. 9, No. 7, pp. 42–43.

Föppl, A., 1888. Über das räumliche Fachwerk. *Schweizerische Bauzeitung*, vol. 11., No. 17, pp. 115–117.

Föppl, A., 1891/1. Über das räumliche Fachwerk. *Schweizerische Bauzeitung*, vol. 17, No. 13, pp. 77–80.

Föppl, A., 1891/2. Über das Flechtwerk. *Schweizerische Bauzeitung*, vol. 17, No. 16, pp. 95–96.

Föppl, A , 1891/3. Die Theorie des räumlichen Fachwerks und der Brückeneinsturz bei Mönchenstein. *Schweizerische Bauzeitung*, vol. 18, No. 3, pp. 15–17.

Föppl, A., 1892. *Das Fachwerk im Raume.* Leipzig: B. G. Teubner.

Föppl, A., 1894. Einführung in die Maxwellsche Theorie der Elektricität. Leipzig: B. G. Teubner.

Föppl, A., 1896. Letter. *ZVDI*, vol. 40, No. 33, p. 943.

Föppl, A., 1897. *Vorlesungen über Technische Mechanik, Dritter Band: Festigkeitslehre.* Leipzig: B. G. Teubner.

Föppl, A., 1898. *Vorlesungen über Technische Mechanik, Erster Band: Einführung in die Mechanik.* Leipzig: B. G. Teubner.

Föppl, A., 1899. *Vorlesungen über technische Mechanik. Vierter Band. Dynamik.* Leipzig: Verlag von B. G. Teubner.

Föppl, A., 1900. *Vorlesungen über Technische Mechanik, Zweiter Band: Graphische Statik.* Leipzig: B. G. Teubner.

Föppl, A., 1901/1. Entgegnung auf das Referat des Herrn Weingarten (Weingarten, 1901). *Archiv der Mathematik und Physik, III. Reihe, 1. Bd.*, pp. 352 – 357. Leipzig: B. G. Teubner.

Föppl, A., 1901/2. *Vorlesungen über Technische Mechanik, Vierter Band: Dynamik.* 2nd ed. Leipzig: B. G. Teubner.

Föppl, A., 1901/3. Zeichnerische Behandlung der Zimmermannschen Kuppel. *Zentralblatt der Bauverwaltung*, vol. 21, No. 79, pp. 487 – 488.

Föppl, A., 1907. *Vorlesungen über Technische Mechanik, Fünfter Band: Die wichtigsten Lehren der höheren Elastizitätstheorie.* Leipzig: B. G. Teubner.

Föppl, A., 1910. *Vorlesungen über Technische Mechanik, Sechster Band: Die wichtigsten Lehren der höheren Dynamik.* Leipzig: B. G. Teubner.

Föppl, A., 1911. *Vorlesungen über technische Mechanik, Zweiter Band: Graphische Statik.* 3rd ed. Leipzig: B. G. Teubner.

Föppl, A., 1917/1. Über den elastischen Verdrehungswinkel eines Stabes. In: *Sitzungsberichte der mathematisch-physikalischen Klasse der Bayerischen Akademie der Wissenschaften zu Munich*, Munich, pp. 5 – 31.

Föppl, A., 1917/2. Der Drillungswiderstand von Walzeisenträgern. *ZVDI*, vol. 60, No. 33, pp. 694 – 695.

Föppl, A., Föppl, L., 1920. *Drang und Zwang*, 2 vols. Munich: Oldenbourg.

Föppl, A., 1922. Versuche über die Verdrehungssteifigkeit der Walzeisenträger. In: *Sitzungsberichte der mathematisch-physikalischen Klasse der Bayerischen Akademie der Wissenschaften zu Munich*, Munich, pp. 295 – 313.

Föppl, A., Föppl, O., 1923. *Grundzüge der Festigkeitslehre.* Leipzig/Berlin: B. G. Teubner.

Föppl, A., Föppl, O., 1925. *Lebenserinnerungen. Rückblick auf meine Lehr- und Aufstiegsjahre.* Munich: Oldenbourg.

Föppl, L., Föppl, O., Prandtl, L., Thoma, H., (ed.), 1924. *Beiträge zur Technischen Mechanik und Technischen Physik. August Föppl zum siebzigsten Geburtstag am 25. Januar 1924.* Berlin: Julius Springer.

Föppl, L., 1924. †August Föppl. *Zeitschrift für angewandte Mathematik und Mechanik,* vol. 4, No. 6, pp. 530 – 531.

Förster, H. C., 2005. Was wir vergessen können, war nie unser. *TU internal memo*, No. 10, p. 15.

Förster, M., 1908. Die Grundzüge der geschichtlichen Entwicklung des Eisenbetonbaus. In: Emperger, F. v. (ed.): *Handbuch für Eisenbetonbau I. Band*, pp. 1 – 52. Berlin: Wilhelm Ernst & Sohn.

Förster, M., 1909. *Die Eisenkonstruktionen der Ingenieur-Hochbauten.* 4th, rev. ed. Leipzig: Engelmann.

Förster, M., 1911. Die Gründe des Einsturzes des großen Gasbehälters am Großen Grasbrook zu Hamburg vom 7. Dez. 1909. *Der Eisenbau*, vol. 2, No. 4, pp. 178 – 182.

Förster, M., 1921. Die geschichtliche Entwicklung des Eisenbetonbaus. In: Emperger, F. v. (ed.): *Handbuch für Eisenbetonbau I. Band*, 3rd, rev. ed. pp. 1 – 52. Berlin: Wilhelm Ernst & Sohn.

Förster, M., 1929. Matthias Koenen, der geistige Vater des Eisenbetonbaus. *Bauingenieur*, vol. 10, No. 48, pp. 583 – 586.

Fox, L., 1987. Early numerical Analysis in the United Kingdom. In: *Proceedings of the ACM conference on History of scientific and numeric computation,* ed. G. E. Crane, pp. 21 – 39. New York: ACM Press.

Fraeijs de Veubeke, B., 1951. Diffusion des inconnues hyperstatiques dans les voilures à longerons couplés. *Bulletin Serv. Technique de L'Aéronautique No. 24.* Brussels: Imprimerie Marcel Hayez.

Fraeijs de Veubeke, B., 1964. Upper and lower bounds in matrix structural analysis. In: *AGARDograph 72: Matrix Methods of Structural Analysis,* ed. B. M. Fraeijs de Veubeke, pp. 165 – 2016. Oxford: Pergamon Press.

Fraeijs de Veubeke, B., 1965. Displacement and equilibrium models in the finite element method. In: *Stress Analysis,* ed. O. C. Zienkiewicz & G. S. Holister, pp. 145 – 197. New York: John Wiley & Sons.

Fraeijs de Veubeke, B., Zienkiewicz, O. C., 1967. Strain energy bounds in finite element analysis. *Journal of Strain Analysis,* vol. 2, pp. 265 – 271.

Fraeijs de Veubeke, B., 1972. A new variational principle for finite elastic displacements. *International Journal of Engineering Science,* vol. 10, pp. 745 – 763.

Fraeijs de Veubeke, B., 1974. Variational principles and the patch test. *International Journal for Numerical Methods in Engineering,* vol. 8, pp. 783 – 801.

Fraeijs de Veubeke, B., 1976. The dynamics of flexible bodies. *International Journal of Engineering Science,* vol. 14, pp. 895 – 913.

Fraeijs de Veubeke, B., 2001. Classic Reprints Series: Displacement and equilibrium models in the finite element method by B. Fraeijs de Veubeke, chap. 9, pp. 145 – 197 of *Stress Analysis,* ed. O. C. Zienkiewicz & G. S. Holister, pub. by John Wiley & Sons, 1965. Reprint in *International Journal for Numerical Methods in Engineering,* vol. 52, pp. 287 – 342.

Fränkel, W., 1867. Berechnung eiserner Bogenbrücken. *Zivilingenieur*, vol. 13, pp. 57 – 67.

Fränkel, W., 1869. Zur Theorie der elastischen Bogenträger. *Zeitschrift d. Arch.- & Ing.-Vereins zu Hannover*, vol. 15, pp. 115 – 131.

Fränkel, W., 1872. Vorträge über Schiebebühnen und Drehscheiben. In: *Vorträge über Eisenbahnbau von E. Winkler.* Prague: Verlag von Dominicus.

Fränkel, W., 1875. Anwendung der Theorie des augenblicklichen Drehpunktes auf die Bestimmung der Formänderung von Fachwerken – Theorie des Bogenfachwerkes mit zwei Gelenken. *Zivilingenieur*, vol. 21, pp. 515 – 538.

Fränkel, W., 1876. Über die ungünstigste Einstellung eines Systems von Einzellasten auf Fachwerkträger mit Hilfe von Influenzkurven. *Zivilingenieur*, vol. 22, pp. 446 – 454.

Fränkel, W., 1882. Das Princip der kleinsten Arbeit der inneren Kräfte elastischer Systeme und seine Anwendung auf die Lösung baustatischer Aufgaben. *Zeitschrift d. Arch.- & Ing.-Vereins zu Hannover*, vol. 28, pp. 63 – 76.

Frank, 1906. Der Einfluß veränderlichen Querschnitts auf die Biegungsmomente kontinuierlicher Träger, unter besonderer Berücksichtigung von Betoneisenkonstruktionen. *Beton und Eisen*, vol. 5, No. 12, pp. 315 – 318.

Franke, P.-G., 1983. Giovanni Poleni (1683 – 1761). *Die Bautechnik*, vol. 60, No. 8, pp. 261 – 262.

Frazer, R. A., Duncan, W. J., Collar, A. R., 1938. *Elementary Matrices and some Applications to Dynamics and Differential Equations.* Cambridge: The University Press.

Frenkel, V. J., 1990. Timoshenko, S. P. In: *Dictionary of Scientific Biography,* vol. 18, pp. 926 – 927. New York: Charles Scribner's Sons.

Freudenthal, A. M., 1933/1. Beitrag zur Berechnung von Wälzgelenken aus Beton und Stein. *Beton und Eisen,* vol. 32, No. 9, pp. 139 – 144 & No. 10, pp. 157 – 162.

Freudenthal, A. M., 1933/2. Wirkung der Windkräfte auf Eisenbetonskelette mit Berücksichtigung des räumlichen Tragvermögens. *Beton und Eisen,* vol., 32 No. 18, pp. 285 – 291.

Freudenthal, A. M., 1934. Plastizitätstheoretische Methoden bei der Untersuchung statisch unbestimmter Tragwerke aus Eisenbeton. In: *Proceedings of the International Association for Bridge & Structural Engineering (IABSE).* vol. 2, ed. L. Karner & M. Ritter, pp. 180 – 192. Zurich: IABSE secretariat.

Freudenthal, A. M., 1935/1. Die Aänderung des Spannungszustandes weitgespannter, flacher Eisenbetonbogen durch die plastische Dauerverformung des Betons. *Beton und Eisen,* vol. 34, No. 11, pp. 176 – 184.

Freudenthal, A. M., 1935/2. Einfluß der Plastizität des Betons auf die Bemessung außermittig gedrückter Eisenbetonquerschnitte. *Beton und Eisen*, vol. 34, No. 21, pp. 335–338.

Freudenthal, A. M., 1936/1. Die plastische Dauerverformung des versteiften Stabbogens. *Beton und Eisen*, vol. 35, No. 12, pp. 206–209.

Freudenthal, A. M., 1936/2. Théorie des grandes voûtes en béton et en béton armé. In: *Proceedings of the International Association for Bridge & Structural Engineering (IABSE)*. vol. 4, ed. L. Karner & M. Ritter, pp. 249–264. Zurich: IABSE secretariat.

Freudenthal, A. M., 1950. *The inelastic behaviour of engineering materials*. New York: Wiley.

Freudenthal, A. M., 1981. *Selected Papers by Alfred M. Freudenthal: Civil Engineering Classics*. New York: ASCE.

Freudenthal, H., 1971. Cauchy, Augustin-Louis. In: *Dictionary of Scientific Biography*, vol. III. C. C. Gillispie (ed.), pp. 131–148. New York: Charles Scribner's Sons.

Freyssinet, E., 1939. *Une révolution dans les techniques du béton*. Librairie de l'enseignement technique. Paris: ed. Léon Eyrolles.

Frézier, A. F., 1737–1739. *La théorie et la pratique de la coupe des pierres*, 3 vols., Strasbourg/ Paris: Charles-Antoine Jombert.

Friedrichs, K. O., 1929. Ein Verfahren der Variationsrechnung, das Minimum eines Integrals als das Maximum eines anderen Ausdruckes darzustellen. *Nachrichten d. Gesellschaft d. Wissenschaften zu Göttingen. Math.-phys. Klasse*, pp. 13–20.

Friemert, C., 1990. Kunst, Künste. In: *Europäische Enzyklopädie zu Philosophie und Wissenschaften*, Band 2, ed. H. J. Sandkühler, pp. 919–939. Hamburg: Felix Meiner.

Fritsche, J., 1930. Die Tragfähigkeit von Balken aus Stahl mit Berücksichtigung des plastischen Verformungsvermögens. *Der Bauingenieur*, vol. 11, No. 49, pp. 851–855, No. 50, pp. 873–874 & No. 51, pp. 888–893.

Fritsche, J., 1936. Grundsätzliches zur Plastizitätstheorie. *Der Stahlbau*, vol. 9, No. 9, pp. 65–68.

Fritsche, J., 1941. †Prof. J. Melan. *Der Bauingenieur*, vol. 22, No. 9/10, pp. 89–90.

Fritz, B., 1934. *Theorie und Berechnung vollwandiger Bogenträger bei Berücksichtigung des Einflusses der Systemverformung*. Berlin: Julius Springer.

Fritz, B., 1950/1. Vereinfachtes Berechnungsverfahren für Stahlträger mit einer Betondruckplatte bei Berücksichtigung des Kriechens und Schwindens. *Die Bautechnik*, vol. 27, No. 2, pp. 37–42.

Fritz, B., 1950/2. Vorschläge für die Berechnung durchlaufender Träger in Verbund-Bauweise. A. Verbund-Vollwandträger. B. Verbund-Fachwerkträger. *Der Bauingenieur*, vol. 25, No. 8, pp. 271–277.

Fritz, B., 1961. *Verbundträger-Berechnungsverfahren für die Brückenbaupraxis*. Berlin: Springer-Verlag.

Fröhlich, H., 1949. Einfluß des Kriechens auf Verbundträger. *Der Bauingenieur*, vol. 24, No. 10, pp. 300–307.

Fröhlich, H., 1950. Theorie der Stahlverbund-Tragwerke. *Der Bauingenieur*, vol. 25, No. 3, pp. 80–87; Discussion: pp. 91–93.

Fuchssteiner, W., 1954. "Wenn Navier noch lebte…" Theorie und Praxis in der Statik. *Beton- und Stahlbetonbau*, vol. 49, No. 1, pp. 15–21.

Fürst, A., Marti, P., 1997. Robert Maillart's Design Approach for Flat Slabs. *Journal of Structural Engineering*, vol. 123, No. 8, pp. 1102–1110.

Fuhrke, H., 1955. Bestimmung von Balkenschwingungen mit Hilfe des Matrizenkalküls. *Ingenieur-Archiv*, vol. 23, pp. 329–348.

Fung, Y. C., Penner, S. S., Seible, F., Williams, F. A., 2001. Eric Reissner. In: *Memorial Tributes*. National Academy of Engineering, vol. 9, pp. 242–245. Washington, D.C.: National Academy Press.

Gabba, A., 1954. Le trasformazioni cremoniane in una lettera di Luigi Cremona a Giovanni Schiaparelli. *Istituto Lombardo di Scienze e Lettere Rendiconto della Classe di Scienze Matematiche e Naturali*, pp. 290–294.

Gaber, E., 1931. †Friedrich Engeßer. *Beton und Eisen*, vol. 30, No. 20, p. 368.

Galerkin, B. G., 1915. *Sterzhni i plastinki: Ryady v nekotorykh voprosakh uprugogo ravnovesiua sterzhnei i plastinok* (bars and plates: series in some problems of the elastic equilibrium of bars and plates). Petrograd (in Russian).

Galerkin, B. G., 1933. *Uprugie tonkie plity* (thin elastic plates). Leningrad/Moscow: Gosstroiizdat (in Russian).

Galerkin, B. G., 1952–1953. *Sobranie sochinenii* (collected works). 2 Bände. Moscow: AN SSSR (in Russian).

Galilei, G., 1638. *Discorsi e dimostrazioni matematiche, intorno a due nuove scienze*. Leiden: Elsevier.

Galilei, G., 1638/1964. *Unterredungen und mathematische Demonstrationen über zwei neue Wissenszweige, die Mechanik und die Fallgesetze betreffend*. Reprint of trans. by Arthur v. Oettingen, 1890, 1891 & 1904. Darmstadt: Wissenschaftliche Buchgesellschaft.

Galilei, G., 1987. *Schriften, Briefe, Dokumente*, vol. 1. Berlin.

Gallagher, R. H., Padlog, J., Bijlaard, P. P., 1962. The stress analysis of heated complex shapes. *American Rocket Society Journal*, vol. 32, No. 5, pp. 700–707.

Gallagher, R. H., 1964. *A Correlation Study of Matrix Methods of Structural Analysis*. Oxford: Pergamon.

Gallagher, R. H., Yamada, Y. (ed.), 1973. *Theory and Practice in Finite Element Structural Analysis*. Tokyo: University of Tokyo Press.

Gallagher, R. H., 1975. *Finite Element Analysis Fundamentals*. New Jersey: Prentice-Hall.

Gallagher, R. H., Ashwell, D. (ed.), 1976. *Finite Elements for Thin Shells and Curved Members*. New York: Wiley.

Gallagher, R. H., McGuire, W., 1979. *Introductory Matrix Structural Analysis*. New York: Wiley.

Gallagher, R. H., Zienkiewicz, O. C. (ed.), 1979. *Optimum Structural Design*. New York: Wiley.

Gargiani, R. (ed.), 2008. *Nouvelle Histoire de la Construction. La Colonne*. Edition monographique, EPFL LTH3. Lausanne: PPUR – Presses Polytechniques et Universitaires Romandes.

Gasparini, D. A., 2002. Contributions of C. A. P. Turner to Development of Reinforced Concrete Flat Slabs 1905–1909. *Journal of Structural Engineering*, vol. 128, No. 10, pp. 1243–1252.

Gasparini, D. A., Provost, C., 1989. Early Nineteenth Century Developments in Truss Design in Britain, France and the United States. *Construction History*, vol. 5, pp. 21–33.

Gasparini, D. A., Vermes, W., 2001. C. A. P. Turner and reinforced concrete flat-slab bridges. In: *Proceedings of the 7th Historic Bridges Conference*, pp. 12–27. Cleveland.

Gebbeken, N., 1988. *Eine Fließgelenktheorie höherer Ordnung für räumliche Stabtragwerke (zugleich ein Beitrag zur historischen Entwicklung)*. Mitteilungen des Institutes für Statik der Universität Hannover, ed. H. Rothert. Hannover: Institute for Theory of Structures, University of Hannover.

Geckeler, J. W., 1926. *Über die Festigkeit achsensymmetrischer Schalen*. Forschungsarbeiten auf dem Gebiete des Ingenieurwesens, No. 276. Berlin: VDI-Verlag.

Gehler, W., 1910. *Die Entwicklung der Nebenspannungen eiserner Fachwerkbrücken und das praktische Rechnungsverfahren nach Mohr*. Berlin: Wilhelm Ernst & Sohn.

Gehler, W., 1913. *Der Rahmen*. Berlin: Wilhelm Ernst & Sohn.

Gehler, W. (ed.), 1916. *Otto Mohr zum achtzigsten Geburtstage*. Berlin: Wilhelm Ernst & Sohn.

Gehler, W., 1916. Rahmenberechnung mittels der Drehwinkel. In: *Otto Mohr zum achtzigsten Geburtstage*, ed. W. Gehler, pp. 88–123. Berlin: Wilhelm Ernst & Sohn.

Géradin, M. (ed.), 1980. *B. M. Fraeijs de Veubeke Memorial Volume of Selected Papers*. Alphen aan den Rijn: Sitthoff & Noordhoff.

Gere, J., Herrmann, G., Steele, C. R., 2004. *Memorial Resolution Wilhelm Flügge (1904–1990)*. http://histsoc.stanford.edu/pdfmem/FluggeW.pdf.

Gerhardt, R., 1989. *Experimentelle Momenten-Darstellung. Darstellung der Biegebeanspruchung ebener Tragwerke durch Seillinien*. Dissertation, Aachen RWTH. Aachen: J. Mainz.

Gerhardt, R., Rickert, N., Rust, J, Thiemann, C., 2003. *Der Knochen. Betrachtungen zur Knochenstatik*. Aachen: Shaker Verlag.

Gersevanov, M. N., 1897. Chestvovanie pamyati inzhenera D. I. Zhuravskogo (commemorative event for the engineer D. I. Zhuravsky). *Izvestiya Sobraniya inzhenerov putei soobshcheniya* (bulletin of the association of communication engineers), vol. 17, No. 5, pp. 65–76 (in Russian).

Gerstner, F. A. v., 1826. *Bericht über Linz–Budweiser Bahn 1825.* Vienna Technical Museum, archive No. HSS 128/4, fol. 58.

Gerstner, F. A. v., 1839. *Berichte aus den Vereinigten Staaten von Nordamerica, über Eisenbahnen, Dampfschiffahrten, Banken und andere öffentliche Unternehmungen.* Leipzig: C. P. Melzer.

Gerstner, F. A. v., 1842. *Die innern Communicationen der Vereinigten Staaten von Nordamerica.* 1st vol., ed. L. Klein. Vienna: Förster's artistische Anstalt.

Gerstner, F. A. v., 1843. *Die innern Communicationen der Vereinigten Staaten von Nordamerica.* 2nd vol., ed. L. Klein. Vienna: Förster's artistische Anstalt.

Gerstner, F. A. v., 1997. *Early American Railroads. Franz Anton Ritter von Gerstner's 'Die innern Communicationen' 1842–1843.* Ed. Frederick C. Gamst. Trans. by David J. Diephouse & John C. Decker. Palo Alto: Stanford University Press.

Gerstner, F. J. v., 1789. *Einleitung in die statische Baukunst.* Prague: publications of the k. k. Normalschul-Buchdruckerei.

Gerstner, F. J. v., 1813. *Zwei Abhandlungen über Frachtwägen und Straßen.* Prague.

Gerstner, F. J. v., 1831. *Handbuch der Mechanik, Band 1,* 1st ed., ed. F. A. v. Gerstner. Prague: Johann Spurny.

Gerstner, F. J. v., 1832. *Handbuch der Mechanik, Band 2,* ed. F. A. v. Gerstner. Prague: Johann Spurny.

Gerstner, F. J. v., 1833. *Handbuch der Mechanik, Band 1,* 2nd ed. (1st ed. 1831), ed. F. A. v. Gerstner. Prague: Johann Spurny.

Gerstner, F. J. v., 1834. *Handbuch der Mechanik, Band 3,* ed. F. A. v. Gerstner. Vienna: J. P. Sollinger.

Gerstner, F. J. v., 1932. Gerstner's autobiography. *HDI-Mitteilungen,* vol. 21, pp. 254–259.

Gesellschaft für Ingenieurbaukunst (ed.), 1998. *Robert Maillart – Betonvirtuose.* Exhibition cat., Institute for Theory of Structures & Design at Zurich ETH. 2nd ed. Zurich: vdf Hochschulverlag AG, Zurich ETH.

Giencke, E., 1958. Die Berechnung von durchlaufenden Fahrbahnplatten. *Der Stahlbau,* vol., 27 No. 9, pp. 229–237, No. 11, pp. 291–298 & No. 12, pp. 326–332.

Giencke, E., 1960. Die Berechnung der Hohlrippenplatten. *Der Stahlbau,* vol. 29, No. 1, pp. 1–11 & No. 2, pp. 47–59.

Giencke, E., 1961. Einfluß der Steifen-Exzentrizität auf Biegung und Stabilität orthotroper Platten. In: *Beiträge aus Statik und Stahlbau,* pp. 35–61. Cologne: Stahlbau-Verlags-GmbH.

Giencke, E., 1967. Ein einfaches und genaues finites Verfahren zur Berechnung von orthotropen Scheiben und Platten. *Der Stahlbau,* vol. 36, No. 9, pp. 260–268 & No. 10, pp. 303–315.

Giencke, E., Petersen, J., 1970. Ein finites Verfahren zur Berechnung schubweicher orthotroper Platten. *Der Stahlbau,* vol. 39, No. 6, pp. 161–166 & No. 7, pp. 202–207.

Gil, M. G., 1791. *Theoria Boscovichiana vindicata, et defensa ab impugnationibus.* Fulginae.

Gillispie, C. C., 1970. Bélidor, Bernard Forest de. In: *Dictionary of Scientific Biography,* vol. I. C. C. Gillispie (ed.), pp. 581–582. New York: Charles Scribner's Sons.

Gillmor, C. S., Freudenthal, H., 1971. Coulomb, Charles Augustin. In: *Dictionary of Scientific Biography,* vol. III. C. C. Gillispie (ed.), pp. 439–447. New York: Charles Scribner's Sons.

Girard, P. S. 1798. *Traité analytique de la résistance des solides, et des solides d'égale résistance, auquel on a joint une suite de nouvelles expériences sur la force, et l'élasticité spécifique des bois de chêne et de sapin.* Paris: Firmin Didot-Du Pont.

Girkmann, K., 1931. Bemessung von Rahmentragwerken unter Zugrundelegung eines ideal plastischen Stahles. In: *Sitzungsberichte der Akademie der Wissenschaften in Vienna, math.-naturw. Klasse Abt. IIa,* 140, pp. 679–628. Vienna.

Girkmann, K., 1932. Über die Auswirkung der "Selbsthilfe" des Baustahls in rahmenartigen Stabwerken. *Der Stahlbau,* vol. 5, pp. 121–127.

Girkmann, K., Königshofer, E., 1938. *Die Hochspannungs-Freileitungen.* Vienna: Springer.

Girkmann, K., 1946. *Flächentragwerke.* Vienna: Springer.

Girkmann, K., 1952. †Dr.-Ing. Friedrich Bleich. *Der Stahlbau,* vol. 21, No. 1, pp. 18–19.

Gispen, K., 2006. Der gefesselte Prometheus: Die Ingenieure in Großbritannien und in den Vereinigten Staaten 1750–1945. In: *Geschichte des Ingenieurs,* ed. W. Kaiser & W. König, pp. 127–177. Munich/Vienna: Carl Hanser.

Giuffrè, A., 1982/1. *Analisi matriciale delle strutture. Statica, dinamica, dinamica aleatoria.* Milan: Masson.

Giuffrè, A., Giannini, R., 1982/2. La risposta non lineare delle strutture in cemento armato. In: *Progettazione e particolari costruttivi in zona sismica,* ed. G. Grandori, pp. 175–238. Rome: Edilstampa.

Giuffrè, A., 1986. *La meccanica nell'architettura.* Rome: NIS.

Giuffrè, A., 1988. *Monumenti e terremoti: aspetti statici del restauro.* Rome: Multigrafica.

Giuffrè, A., Marconi, P., 1988. Un progetto in itinere: il restauro della cattedrale di S. Angelo dei Lombardi. *Ricerche di storia dell'arte,* XIII, 35, pp. 37–54.

Giuffrè, A., Baggio, C., Mariani, R., 1990. Seismic response of mechanism of masonry assemblages. In: *Proceedings of the 9th European conference on earthquake engineering (Moscow),* vol. 5, pp. 221–230.

Giuffrè, A., 1991. *Letture sulla Meccanica delle Murature Storiche.* Rome: Kappa.

Giuffrè, A., Carocci, A., 1992. Statics and dynamics of historical masonry buildings. In: *International Workshop on Structural Restoration of Historical Buildings in Old City Centres,* Heraclion (Crete), pp. 35–95.

Giuffrè, A., Carocci, A., 1997. *Codice di pratica per la sicurezza e la conservazione dei Sassi di Matera.* Matera: La Bautta.

Gizdulich, R., 1957. La ricostruzione del Ponte a S. Trinità. *Comunità* No. 51, July.

Goethe, J. v., 1808. *Zur Farbenlehre. Didaktischer Teil.* Wiemar: Verlag Hermann Böhlaus.

Gold, P., 2006. *Architektur als Wissenschaft – Wissenschaftstheoretische Betrachtungen zu Architektur und Architekturtheorie.* Dissertation, University of Vienna.

Gold, P., 2007. Wissenschaftstheoretische Überlegungen zur Entwicklung der Baustatik. *Stahlbau,* vol. 76, No. 12, pp. 924–933.

Goldbeck, O., 2004. Vom Stahlbauer zum Generalunternehmer. *Stahlbau,* vol. 73, No. 10, pp. 818–821.

Gollert, H. J., 2007. Vom Rechenzentrum zum Premium-Lösungsanbieter im Bereich Baumanagement. *transparent,* No. 33, pp. 19–21.

Gordon, J., 1968. *The New Science of Strong Materials.* Harmondsworth: Penguin.

Gorbunov, B. N., Krotov, Y. V., 1936. *Osnovy rascheta prostranstvennykh ram* (fundamentals of the analysis of three-dimensional frames). Moscow/Leningrad: Glavn. red. stroit. lit (in Russian).

Gorokhov, V. G., 2001. *Technikphilosophie und Technikfolgenabschätzung in Russland.* Graue Reihe Nr. 26. Europäische Akademie zur Erforschung von Folgen wissenschaftlich-technischer Entwicklungen Bad Neuenahr-Ahrweiler GmbH (Director: Prof. Dr. Carl Friedrich Gethmann). Bad Neuenahr-Ahrweiler: Europäische Akademie.

Göschel, H. (ed.), 1976. Technik. In: *Meyers Neues Lexikon, Band 13,* 2nd ed., pp. 463–465. Leipzig: VEB Bibliographisches Institut.

Göschel, H. (ed.), 1980: Technik. In: *Meyers Universal-Lexikon, Band 4,* p. 275. Leipzig: VEB Bibliographisches Institut.

Gouzévitch, I., Gouzévitch, D., 1993. Les contacts franco-russes dans le domaine de l'enseignement supérieur technique et de l'art de l'ingénieur. *Cahier du Monde russe et soviétique,* vol. 34, No. 3, pp. 345–368.

Grabow, R., 1985. *Simon Stevin.* Leipzig: BSB B. G. Teubner Verlagsgesellschaft.

Graefe, R., 1986. Zur Formgebung von Bögen und Gewölben. *architectura – Zeitschrift für Geschichte der Baukunst,* vol. 16, No. 1, pp. 50–67.

Graefe, R., 1990. Netzdächer, Hängedächer und Gitterschalen. In: *Vladimir G. Šuchov,*

1853–1939. Die Kunst der sparsamen Konstruktion. Ed. R. Graefe, M. Gapoëv, M. & O. Pertschi, pp. 28–53. Stuttgart: Deutsche Verlagsanstalt.

Graf, O., 1937. Über Leichtfahrbahnen für stählerne Straßenbrücken. Der Stahlbau, vol. 10, No. 14, pp. 110–12 & No. 16, pp. 123–127.

Graf, O., 1938. Aus Untersuchungen mit Leichtfahrbahndecken zu Straßenbrücken. Berichte des Deutschen Ausschusses für Stahlbau. ed. B, No. 9. Ed. Deutschen Stahlbau-Verband. Berlin: Springer.

Graf, O., 1951. †Emil Mörsch. Die Bautechnik, vol. 28, No. 2, p. 41.

Graf, W., Vassilev, T., 2006. Einführung in computerorientierte Methoden der Baustatik. Berlin: Ernst & Sohn.

Grammel, R., 1938. Das wissenschaftliche Werk von Erich Trefftz. Zeitschrift für angewandte Mathematik und Mechanik (ZAMM), vol. 18, No. 1, pp. 1–11.

Grashof, F., 1866. Die Festigkeitslehre mit besonderer Rücksicht auf die Bedürfnisse des Maschinenbaues. Abriss von Vorträgen an der Polytechnischen Schule zu Carlsruhe. Berlin: Rudolph Gaertner.

Grashof, F., 1878. Theorie der Elasticität und Festigkeit mit Bezug auf ihre Anwendungen in der Technik. 2nd, rev. ed. Berlin: Rudolph Gaertner.

Graße, W., Steup, H., 1999. Zum 100. Geburtstag von Klaus Zweiling. Unpub. ms. Dresden.

Grasser, E., 1971. Bemessung der Stahlbetonbauteile. Bemessung auf Biegung mit Längskraft (ohne Knickgefahr), Schub und Torsion. In: Beton-Kalender, vol. 60, pt. I, pp. 513–630. Berlin: Wilhelm Ernst & Sohn.

Grasser, E., 1987. Bemessung der Stahlbetonbauteile. Bemessung für Biegung mit Längskraft, Schub und Torsion. In: Beton-Kalender, vol. 76, pt. I, pp. 455–570. Berlin: Wilhelm Ernst & Sohn.

Grasser, E., 1997. Bemessung der Stahlbetonbauteile. Bemessung für Biegung mit Längskraft, Schub und Torsion nach DIN 1045. In: Beton-Kalender, vol. 86, pt. I, pp. 363–477. Berlin: Wilhelm Ernst & Sohn

Graubner, C.-A., Jäger, W., 2007. Tragfähigkeit von unbewehrtem Mauerwerk unter zentrischer und exzentrischer Druckbeanspruchung nach DIN 1053-100. Mauerwerk, vol. 11, No. 1, pp. 19–26.

Green, A. E., Zerna, W., 1950/1. Theory of Elasticity in General Coordinates. In: Philosophical Magazine 41, p. 313ff.

Green, A. E., Zerna, W., 1950/2. The equilibrium of thin elastic shells. Quarterly Journal of Mechanics and Applied Mathematics, vol. III, pt. 1, pp. 9–22.

Green, A. E., Zerna, W., 1954. Theoretical Elasticity. Oxford: Clarendon Press.

Green, A. E., Nagdhi, P. M., 1965. A general theory of an elastic-plastic continuum. Archive for Rational Mechanics and Analysis, vol. 18, pp. 251–281.

Green, G., 1839. On the Laws of Reflection and Refraction of Light at the common Surface of two non-crystallized Media. In: Transactions of the Cambridge Philosophical Society (1838–1842), vol. 7, pp. 1–24.

Green, T. (ed.), 2006. John A. Roebling. A bicentennial celebration of his birth. Published by the American Society of Civil Engineers.

Greenberg, H. J., Prager, W., 1951/1952. On Limit Design of Beams and Frames. Transactions ASCE, 1951, vol. 117, pp. 447–484. Discussion: Proc. ASCE, 1952, vol. 78, pp. 459–484.

Gregory, D., 1697. Catenaria. Philosophical Transactions of the Royal Society, vol. 19, No. 231, pp. 637–652.

Greitzer, S. L., 1971. Cremona, Antonio Luigi Gaudenzio Giuseppe. In: Dictionary of Scientific Biography, vol. III. C. C. Gillispie (ed.), pp. 467–469. New York: Charles Scribner's Sons.

Greitzer, S. L., 1973. Lamé, Gabriel. In: Dictionary of Scientific Biography, vol. VII. C. C. Gillispie (ed.), pp. 601–602. New York: Charles Scribner's Sons.

Griffith, A. A., 1921. The phenomenon of rupture and flow in solids. Philosophical Transactions of the Royal Society of London, A, vol. 221, pp. 163–198.

Griffith, A. A., 1924. The theory of rupture. Proceedings of First International Congress of Applied Mechanics, pp. 55–63. Delft.

Griggs, F. E., 2002. Squire Whipple – Father of Iron Bridges. Journal of Bridge Engineering, vol. 7, Issue 3, pp. 146–155.

Griggs, F. E., DeLuzio A. J., 1995. Stephen H. Long and Squire Whipple: The first American Structural Engineers. Journal of Structural Engineering, vol. 121, No. 9, pp. 1352–1361.

Grigolyuk, E. I., 1996. Metod Bubnova: Istoki, formulirovka, razvitie (Bubnov's method: sources, formulation, development). Moscow: Institute for Mechanics, University of Moscow (in Russian).

Grigorian, A. T., 1973. Krylov, A. N. In: Dictionary of Scientific Biography, vol. 7, pp. 513–514. New York: Charles Scribner's Sons.

Grigorian, A. T., 1978. Galerkin, Boris Grigorievich. In: Dictionary of Scientific Biography, vol. 15, pp.164–165. New York: Charles Scribner's Sons.

Grimm, J., Grimm, W., 1854. Anmut. In: Deutsches Wörterbuch. I. Band, ed. J. Grimm & W. Grimm, col. 409–410. Leipzig: Verlag von S. Hirzel.

Grimm, J., Grimm, W. (ed.), 1860. Bogen. In: Deutsches Wörterbuch, II. Band, ed. J. Grimm & W. Grimm. Leipzig: Verlag von S. Hirzel.

Grimm, J., Grimm, W., 1973. Gewölbe. In: Deutsches Wörterbuch, IV. Band, I. Abt. 4. Teil 1. Lieferung, ed. GDR Academy of Sciences. Leipzig: S. Hirzel Verlag.

Grinter, L. E., 1932. Discussion to paper by H. Cross. Transactions of the American Society of Civil Engineers, vol. 96, pp. 11–20.

Grinter, L. E., 1933. Wind stress analysis simplified. Transactions of the American Society of Civil Engineers, vol. 59, pp. 3–27.

Grinter, L. E. (ed.), 1949. Numerical Methods of Analysis in Engineering (Successive Corrections). New York: Macmillan Co.

Grinter, L. E. (Chair), 1955. Report on the Committee on Evaluation of Engineering Education. Journal of Engineering Education, vol. 46, Sept., pp. 25–60.

Grishkova, N. P., Georgievskaya, V. V., 1956. Aleksandr Nikolaevich Dinnik. Kiev: AN USSR (in Russian).

Groh, C., 1999. The old Dirschau Bridge in its technical detail. In: Proceedings of the International Conference Preservation of the Engineering Heritage – Gdańsk Outlook 2000, ed. Z. Cywiński & W. Affelt, pp. 103–109. Gdańsk: Technical University of Gdańsk.

Grote, J., Marrey, B., 2000. Freyssinet. Prestressing in Europe 1930–1945. Paris: Éditions du Linteau.

Groth, P., 2000. Alfred Zimmer – Pionier der FEM-Anwendung. ATZ Automobiltechnische Zeitschrift, vol. 102, No. 5, pp. 338–341.

Groth, P., 2002. FEM-Anwendungen. Statik-, Dynamik- und Potenzialprobleme mit professioneller Software lösen. Berlin/Heidelberg: Springer.

Grübler, M., 1883. Allgemeine Eigenschaften der zwangsläufigen ebenen kinematischen Ketten. Zivilingenieur, vol. 29, pp. 167–199.

Grübler, M., 1918. †Otto Mohr. Der Eisenbau, vol. 9, pp. 296–297.

Grüning, G., 1937. Versuche zur Bestimmung der Verbundwirkung von Eisenbeton- und Massivdecken mit darin einbetonierten Walzträgern bei schwingenden Beanspruchungen. Deutscher Ausschuss für Eisenbeton No. 84. Berlin: Wilhelm Ernst & Sohn.

Grüning, M., 1912. Theorie der Baukonstruktionen I: Allgemeine Theorie des Fachwerks und der vollwandigen Systeme. In: Encyklopädie der mathematischen Wissenschaften, vol. IV/2, pp. 419–534. Leipzig: B. G. Teubner.

Grüning, M., 1925. Die Statik des ebenen Tragwerkes. Berlin: Julius Springer.

Grüning, M., 1926. Die Tragfähigkeit statisch unbestimmter Tragwerke aus Stahl bei beliebig häufig wiederholter Belastung. Berlin: Springer.

Gubler, J., 2004. Les Beautés du Béton Armé. In: Häuser aus Beton. Vom Stampfbeton zum Großtafelbau, ed. U. Hassler & H. Schmidt, pp. 27–39. Tübingen/Berlin: Ernst Wasmuth Verlag.

Guidi, C., 1883. Sugli archi elastici. Memorie della Reale Accademia delle Scienze di Torino, series II, vol. 36, pp. 181–197.

Guidi, C., 1885. Sulla curva delle pressioni negli archi e nelle volte. Memorie della Reale Accademia delle Scienze di Torino, series II, vol. 37, 1885, pp. 625–642.

Guidi, C., 1888. *L'arco elastico*. Turin: Tip. e Lit. Commerciale.

Guidi, C., 1896. *Lezioni sulla scienza delle costruzioni, Parte III, Elementi delle costruzioni, Statica delle costruzioni civili*. Turin: Camilla e Bertolero.

Guidi, C., 1903. L'arco elastico senza cerniere. *Memorie della Reale Accademia delle Scienze di Torino*, series II, vol. 52, pp. 294–314.

Guidi, C., 1905. *Lezioni sulla scienza delle costruzioni, Parte IV, Teoria dei ponti*. Turin: V. Bona.

Guidi, C., 1906. Influenza della temperatura sulle costruzioni murarie. *Atti della Reale Accademia delle Scienze di Torino*, vol. 41, pp. 359–370.

Guidi, C., 1933. Lezioni sulla scienza delle costruzioni, Parte I, Nozioni di statica grafica. (reprint). Turin.

Guidi, C., 1934/1. Lezioni sulla scienza delle costruzioni, Parte II, Teoria dell'elasticità e resistenza die materiali. (reprint). Turin.

Guidi, C., 1934/2. Lezioni sulla scienza delle costruzioni, Parte III, Elementi delle costruzioni, statica delle costruzioni civili. (reprint). Turin.

Guidi, C., 1935. Lezioni sulla scienza delle costruzioni, Appendice: Le costruzioni in beton armato. (reprint). Turin.

Guidi, C., 1938. Lezioni sulla scienza delle costruzioni, Parte IV, Teoria dei ponti. (reprint). Turin.

Günschel, G., 1966. Große Konstrukteure 1. Freyssinet – Maillart – Dischinger – Finsterwalder. Berlin: Ullstein.

Guntau, M., 1978. Zur Herausbildung wissenschaftlicher Disziplinen in der Geschichte. *Rostocker Wissenschaftshistorische Manuskripte*, No. 1, pp. 11–24.

Guntau, M., 1982. Gedanken zur Herausbildung wissenschaftlicher Disziplinen in der Geschichte und zu Problemen der Disziplingenese in der Wissenschaftsgeschichtsschreibung. *Rostocker Wissenschaftshistorische Manuskripte*, No. 8, pp. 19–49.

Güntheroth, N., Kahlow, A., 2005. Johann August Röbling – John Augustus Roebling (1806–1869). In: Bundesingenieurkammer (ed.), *Ingenieurbaukunst in Deutschland 2005/2006*, pp. 124–137. Hamburg: Junius Verlag.

Güntheroth, N., Kahlow, A., 2006 (ed.). Von Mühlhausen in die Welt – Der Brückenbauer J. A. Röbling (1806–1869). Mühlhausen: Mühlhäuser Beiträge, special issue 15.

Guralnick, S. A., LaFraugh, R. W., 1963. Laboratory study of a 45 foot square flat plate structure. *ACI Journal*, vol. 60, pp. 1107–1185.

Gurtin, M. E., 1983. The linear theory of elasticity. In: *Mechanics of Solids* vol. II, ed. C. A. Truesdell, pp. 1–296. Berlin: Springer-Verlag.

Guz, A. N., 1979. Mechanics of a deformable body at the Academy of Sciences of the Ukrainian SSR. *Soviet Applied Mechanics*, pp. 901–910.

Guz, A. N., 2007. On the 100th anniversary of the birth of academian G. N. Savin (1907–2007). *International Applied Mechanics*, vol. 43, No. 1, p. 1.

Guz, A. N., Rushchitsky, J. J., 2007. Monographic heritage of G. N. Savin. *International Applied Mechanics*, vol. 43, No. 1, pp. 28–34.

Gvozdev, A. A., 1927. *Obshchii metod rascheta slozhnykh staticheski neopredelimykh system* (general method of analysis of complex statically indeterminate systems). Moscow: Mosk. inst. inzh. zhel.-dor. transp (in Russian).

Gvozdev, A. A., 1938/1960. The determination of the value of the collapse load for statically indeterminate systems undergoing plastic deformation (in Russian). In: *Proceedings of the Conference on Plastic Deformations, December 1936*, p. 19. Moscow/Leningrad: Akademiia Nauk SSSR 1938. In English: *International Journal of Mechanical Sciences*, vol. 1, pp. 322–335.

Haase, H., 1882–1885. Zur Theorie der parabolischen und elliptischen Gewölbe. *Allgemeine Bauzeitung*, vol. 47, pp. 89–103, vol. 48, pp. 75–79 & 89–92, vol. 49, pp. 12–15, 17–20, 25–27 & 41–43, vol. 50, pp. 44–46, 49–54, 57–73 & 77–82.

Hacker, 1888. Statische Bestimmung der Spannungen des Fachwerks im Raume bei schiefer Belastung. *Zeitschrift für Bauwesen*, vol. 38, col. 43–82.

Hadamard, J., 1908. Sur le problème d'analyse relatif à l'équilibre des plaques élastiques encastrées. Institute de France, Académie des Sciences, Mémoires presentés par divers savants, pt. 33, No. 4.

Hagen, G., 1862. *Form und Stärke gewölbter Bogen*. Berlin: Verlag von Ernst & Korn.

Hager, K., 1914. *Theorie des Eisenbetons*. Munich: Oldenbourg.

Halász, R., v., 1963. †Hans Straub. *Die Bautechnik*, vol. 40, No. 7, p. 217.

Hall, W. J., 1984. Nathan Mortimore Newmark. In: *Memorial Tributes. National Academy of Engineering*, vol. 2, pp. 216–220. Washington, D.C.: National Academy Press.

Haller, H.-W., Hörenbaum, C., Osterrieder, P., Saal, H., 2004. Produktmodelle zur Optimierung der Projektabwicklung. *Stahlbau*, vol. 73, No. 3, pp. 196–204.

Hambly, E. C., 1987. Robert Hooke, the City's Leonardo. In: *The Cities Universities*, vol. 2, No. 2, July.

Hamel, G., 1912. *Elementare Mechanik*. Leipzig/Berlin: B. G. Teubner.

Hamel, G., 1949. *Theoretische Mechanik*. Berlin/Göttingen/Heidelberg: Springer-Verlag.

Hamilton, S. B., 1933/1934. The Place of Sir Christopher Wren in the history of structural engineering. *Trans. Newcomen Society Transactions*, vol. 14, pp. 27–42.

Hamilton, S. B., 1952. The Historical Development of Structural Theory. *Proceedings of the Institution of Civil Engineers*, vol. 1, pp. 374–419.

Hannover TH (ed.), 1956. Catalogus Professorum. Der Lehrkörper der Technischen Hochschule Hannover 1831–1956. Hannover.

Hänseroth, T., 1980. Zur Vorgeschichte der Baumechanik. *Dresdener Beiträge zur Geschichte der Technikwissenschaften*. No. 1, pp. 35–72.

Hänseroth, T., 1982. Der deutsche Brückenbau am Scheideweg: Johann Andreas Schubert und die Göltzschtalbrücke. *Dresdener Beiträge zur Geschichte der Technikwissenschaften*, No. 5, pp. 1–10.

Hänseroth, T., 1985. Die 'Mechanik der Baukunst'. Zum 150. Todestag des Begründers der Baumechanik L. M. H. Navier (1785–1836). *NTM-Schriftenr. Gesch. Naturwiss., Technik, Med.*, vol. 22, No. 2, pp. 33–41.

Hänseroth, T., 1987. Zur Rezeption und Fortbildung bau- und hydromechanischer Theorien in Deutschland im 18. und frühen 19. Jh. In: *Proceedings ICOHTEC Dresden*, ed. R. Sonnemann & K. Krug. Berlin: VEB Deutscher Verlag der Wissenschaften.

Hänseroth, T., Mauersberger, K., 1987. Technisches Wissen um Bauwerke und Maschinen – Quellen und programmatische Ansätze der Technikwissenschaften. In: *Beiträge zur Wissenschaftsgeschichte. Wissenschaft im Mittelalter und Renaissance*, ed. G. Wendel, for the History of Science Study Group, GDR Ministry of Higher Education, pp. 205–41. Berlin: VEB Deutscher Verlag der Wissenschaften.

Hänseroth, T., Mauersberger, K., 1989. Lazare Carnots Werk im Lichte der Herausbildung der Bau- und Maschinenmechanik in Frankreich. *Wissenschaftliche Zeitschrift der Technischen Universität "Otto von Guericke" Magdeburg*, vol. 33, No. 2, pp. 45–54.

Hänseroth, T., 1991. Willy Gehler (1876–1953) – Zur Dichotomie eines Bauingenieurs und Ingenieurwissenschaftlers. *Dresdener Beiträge zur Geschichte der Technikwissenschaften*, No. 19, pp. 65–75.

Hänseroth, T., 2003/1. Die Konstruktion 'verwissenschaftlichter' Praxis: Zum Aufstieg eines Paradigmas in den Technikwissenschaften des 19. Jahrhunderts. In: *Wissenschaft und Technik. Studien zur Geschichte der TU Dresden*, ed. T. Hänseroth, pp. 15–36. Cologne/Wiemar/Vienna: Böhlau Verlag.

Hänseroth, T., 2003/2. Gehler, Willy (Gustav). In: *Die Professoren der TU Dresden 1828–2003*, ed. D. Petschel, pp. 255–257. Cologne: Böhlau.

Hantschk, H., 1990. Johann Joseph Prechtl und das Wiener Polytechnische Institut. *Österreichische Ingenieur- und Architekten-Zeitschrift (ÖIAZ)*, vol. 135, No. 10, pp. 487–494.

Häntzschel, G., 1889. Das Verhalten der Gleisbettung in statischer Beziehung nach den Versuchen der Reichseisenbahnen. *Organ für*

*Fortschritte des Eisenbahnwesens,* No. 26, pp. 141–144, 194–198, 227–229 & plates XX, XXI & XXXIV.

Happel, H., 1914. Über das Gleichgewicht rechteckiger Platten. *Nachrichten d. Kgl. Ges. d. Wiss. zu Göttingen. Math.-phys. Klasse,* pp. 37–62.

Harand, I., 1935. 'Sein Kampf'. *Antwort an Hitler.* Vienna: Self-pub.

Harris, J. G., DeLoatch, E. M., Grogan, W. R., Peden, I. C., Whinnery, J. R., 1994. Education Round Table: Reflections on the Grinter Report. *Journal of Engineering Education,* vol. 83, No. 1, pp. 69–94.

Hartmann, Friedel, 1982. Castigliano's Theorem and its Limits. *Zeitschrift für angewandte Mathematik und Mechanik (ZAMM),* vol. 62, No. 12, pp. 645–650.

Hartmann, Friedel, 1983/1. Castiglianos Theorem bei Scheiben und Körpern. *Die Bautechnik,* vol. 60, No. 10, pp. 299–303.

Hartmann, Friedel, 1983/2. Wann gilt Castiglianos Theorem? *Die Bautechnik,* vol. 60, No. 10, pp. 358–359.

Hartmann, Friedel, 1985. *The Mathematical Foundation of Structural Mechanics.* Berlin: Springer.

Hartmann, Friedrich, 1913. *Die statisch unbestimmten Systeme des Eisen- und Eisenbetonbaues.* Berlin: Wilhelm Ernst & Sohn.

Haupt, H., 1841. *Hints on Bridge Building by an Engineer.*

Haupt, H., 1851. *General Theory of Bridge Construction.* New York: D. Appleton & Company.

Haupt, H., 1864. *Military Bridges.* New York: D. Van Nostrand.

Hayashi, K., 1921/1. *Theorie des Trägers auf elastischer Unterlage und ihre Anwendung auf den Tiefbau.* Berlin: Springer.

Hayashi, K., 1921/2. *Fünfstellige Tafeln der Kreis- und Hyperbelfunktionen sowie der Funktionen $e^x$ und $e^{-x}$ mit den natürlichen Zahlen als Argument.* Berlin: Walter de Gruyter.

Hayashi, K. 1921/3. *Fünfstellige Tafeln der Kreis- und Hyperbelfunktionen.* Berlin: Springer.

Hayashi, K., 1926. *Sieben- und mehrstellige Tafeln der Kreis- und Hyperbelfunktionen.* Berlin: Springer.

Hayashi, K., 1930/1. *Tafeln der Besselschen, Theta-, Kugel- und anderer Funktionen.* Berlin: Springer.

Hayashi, K., 1930/2. *Fünfstellige Funktionentafeln, Kreis-, zyklometrische, Exponential-, Hyperbel-, Kugel-, Besselsche, elliptische Funktionen, Thetanullwerte, natürlicher Logarithmus, Gammafunktionen.* Berlin: Springer.

Hayashi, K., 1936. *Tables of Bessel Functions.* London: British Assoc. Math. Tables VI.

Hayashi, K., 1951. Review of my publications in Germany. In: *Gakuto,* No. 4, vol. 48, pp. 6–9. Tokyo: Verlag Maruzen (in Japanese).

Heck, O. S., 1937. Über die Berechnung versteifter Scheiben und Schalen. In: *Jahrbuch 1937 der Deutschen Versuchsanstalt für Luftfahrt e. V.,* Berlin-Adlershof, pp. 233–42.

Hees, G., 1991. Heinrich Müller-Breslau. In: *VDI Bau. Jahrbuch 1991,* ed. VDI-Gesellschaft Bautechnik. pp. 324–370. Düsseldorf: VDI-Verlag.

Heinle, E., Schlaich, J., 1996. *Kuppeln aller Zeiten – aller Kulturen.* Stuttgart: Deutsche Verlags-Anstalt.

Heinrich, B., 1979. *Am Anfang war der Balken. Zur Kulturgeschichte der Steinbrücke.* Deutsches Museum. Munich.

Heinrich, B., 1983. *Brücken. Vom Balken zum Bogen.* Hamburg: Rowohlt.

Heinrich, E., 1957–1971. Gewölbe. In: *Reallexikon der Assyriologie und Vorderasiatischen Archäologie,* 3. Band, ed. E. Weidner & W. v. Soden, pp. 323–340. Berlin: Walter de Gruyter.

Heinzerling, W., 2002. Flügelpeilung und Flächenregel – zwei grundlegende Patente der Flugzeugaerodynamik. In: *Neuntes Kolloquium Luftverkehr an der Technischen Universität Darmstadt,* ed. Aviation Study Group, Darmstadt TU, pp. 1–44.

Hellinger, E., 1913. Die allgemeinen Ansätze der Mechanik der Kontinua. In: *Encyklopädie der mathematischen Wissenschaften,* vol. IV/2, ed. F. Klein & C. H. Müller, p. 601–694. Leipzig: B. G. Teubner.

Hencky, H., 1913. *Über den Spannungszustand in rechteckigen ebenen Platten bei gleichmäßig verteilter und bei konzentrierter Belastung.* Munich: Oldenbourg.

Hencky, H., 1915. Über den Spannungszustand in kreisrunden Platten mit verschwindender Biegesteifigkeit. *Zeitschrift für Mathematik und Physik,* vol. 63, pp. 311–317.

Hencky, H., 1920. Über die angenäherte Lösung von Stabilitätsproblemen im Raum mittels der elastischen Gelenkkette. *Der Eisenbau,* vol. 11, pp. 437–451.

Hencky, H., 1921. Kippsicherheit und Achterbildung an angeschlossenen Kreisringen. *Zeitschrift für angewandte Mathematik und Mechanik (ZAMM),* vol. 1, pp. 451–454.

Hencky, H., 1922. Stabilitätsprobleme der Elastizitätstheorie. *Zeitschrift für angewandte Mathematik und Mechanik (ZAMM),* vol. 2, pp. 58–66.

Hencky, H., 1923. Über einige statisch bestimmte Fälle des Gleichgewichts in plastischen Körpern. *Zeitschrift für angewandte Mathematik und Mechanik (ZAMM),* vol. 3, pp. 241–251.

Hencky, H., 1924. Zur Theorie plastischer Deformationen und der hierdurch im Material hervorgerufenen Nachspannungen. *Zeitschrift für angewandte Mathematik und Mechanik (ZAMM),* vol. 4, pp. 323–335.

Hencky, H., 1925. Die Bewegungsgleichungen beim nichtstationären Fliessen plastischer Massen. *Zeitschrift für angewandte Mathematik und Mechanik (ZAMM),* vol. 5, pp. 144–146.

Hencky, H., 1929. Das Superpositionsgesetz eines endlich deformierten relaxationsfähigen elastischen Kontinuums und seine Bedeutung für eine exakte Ableitung der Gleichungen für zähe Flüssigkeit in der Eulerschen Form. *Annalen der Physik,* vol. 394, No. 6, pp. 617–630.

Hencky, H., 1947. Über die Berücksichtigung der Schubverzerrung in ebenen Platten. *Ingenieur-Archiv,* vol. 16, pp. 72–76.

Hencky, H., 1951. *Neuere Verfahren in der Festigkeitslehre.* Munich: Verlag R. Oldenbourg.

Henneberg, L., 1886. *Lehrbuch der technischen Mechanik. 1. Teil: Statik der starren Systeme.* Darmstadt: Verlag von Arnold Bergsträsser.

Henneberg, L., 1894. Über die Entwicklung und die Hauptaufgaben der Theorie der einfachen Fachwerke. In: *Jahresbericht der Deutschen Mathematiker-Vereinigung,* vol. 3, pp. 567–599. Berlin: Verlag von Georg Reimer.

Henneberg, L., 1902. Die sog. Methode des Ersatzstabes. *Zentralblatt der Bauverwaltung,* vol. 23, No. 60, pp. 377–378.

Henneberg, L., 1903. Die graphische Statik der starren Körper. In: *Enzyklopädie der mathematischen Wissenschaften. Vierter Band: Mechanik, erster Teilband,* ed. F. Klein & C. Müller, pp. 345–434. Leipzig: Verlag von B. G. Teubner.

Henneberg, L., 1911. *Die graphische Statik der starren Systeme.* Leipzig: Verlag von B. G. Teubner.

Heppel, J. M., 1860. On a method of computing the strains and deflections of continuous beams under various conditions of load. *Minutes of the Proceedings of the Institution of Civil Engineers,* vol. 19, pp. 625–643.

Heppel, J. M., 1870. On the theory of continuous beams (with commentary by W. J. M. Rankine). *Proceedings of the Royal Society of London,* vol. 19, pp. 56–71.

Herbert, D., 1991. *Die Entstehung des Tensorkalküls. Von den Anfängen in der Elastizitätstheorie bis zur Verwendung in der Baustatik.* Stuttgart: Franz Steiner Verlag.

Herbst, 1928. Das Stahlskelett im heutigen amerikanischen Hochhausbau. *Der Stahlbau,* vol. 1, No. 19, pp. 231–232.

Herrbruck, J., Groß, J.-P., Wapenhans, W., 2001. Gewölbebrücken: Ersatz der linearen 'Kaputt-rechnung'. *Bautechnik,* vol. 78, No. 11, pp. 805–814.

Herrmann, G. (ed.), 1974. *R. D. Mindlin and Applied Mechanics.* Amsterdam: Elsevier.

Herrmann, L. R., 1965/1. Elasticity equations for nearly incompressible materials by a variational theorem. *AIAA Journal,* vol. 3, pp. 1896–2000.

Herrmann, L. R., 1965/2. A bending analysis for plates. In: *Proceedings of 1st Conference on Matrix Methods in Structural Mechanics,* AFFDL-TR-66-80, pp. 577–604. Dayton: Air Force Institute of Technology.

Hertwig, A., 1905. Beziehungen zwischen Symmetrie und Determinanten in einigen Aufgaben der Fachwerktheorie. In: *Festschrift Adolph Wüllner,* pp. 194–213. Leipzig: B. G. Teubner.

Hertwig, A., 1906. Die Entwicklung einiger Prinzipien in der Statik der Baukonstruktionen und die Vorlesungen über Statik der Baukonstruktionen und Festigkeitslehre von G. C. Mehrtens. *Zeitschrift für Arch.- & Ing.-Wesen,* pp. 493–516.

Hertwig, A., 1907/1. Zu den Bemerkungen des Herrn Geheimen Regierungsrats Prof. Dr. Weingarten über meine Abhandlung: 'Die Entwicklung einiger Prinzipien der Statik der Baukonstruktionen'. *Zeitschrift für Arch.- & Ing.-Wesen,* pp. 374–378.

Hertwig, A., 1907/2. Entgegnung auf die Bemerkungen des Herrn Mehrtens 'In eigener Sache'. *Zeitschrift für Arch.- & Ing.-Wesen,* pp. 372–374.

Hertwig, A., 1910. Über die Berechnung mehrfach statisch unbestimmter Systeme und verwandte Aufgaben in der Statik der Baukonstruktionen. *Zeitschrift für Bauwesen,* vol. 60, pp. 109–120.

Hertwig, A., Reissner, H. (ed.) 1912. *Festschrift Heinrich Müller-Breslau.* Leipzig: Alfred Kröner.

Hertwig, A., 1912. Die Lösung linearer Gleichungssysteme durch unendliche Reihen und ihre Anwendung auf die Berechnung hochgradig unbestimmter Systeme. In: *Festschrift Heinrich Müller-Breslau,* ed. A. Hertwig & H. Reissner, pp. 37–59. Leipzig: Alfred Kröner Verlag.

Hertwig, A., 1930/1. *Die Hochschulreform.* Berlin: Wilhelm Ernst & Sohn.

Hertwig, A., 1930/2. *Johann Wilhelm Schwedler. Sein Leben und sein Werk.* Berlin: Wilhelm Ernst & Sohn.

Hertwig, A., 1933. Das "Kraftgrößenverfahren" und das "Formänderungsgrößenverfahren" für die Berechnung statisch unbestimmter Gebilde. *Der Stahlbau,* vol. 6, No. 19, pp. 145–149.

Hertwig, A., 1934/1. Aus der Geschichte der Straßenbautechnik. *Beiträge zur Geschichte der Technik und Industrie. Technikgeschichte,* vol. 23, pp. 1–5.

Hertwig, A., 1934/2. Aus der Geschichte der Gewölbe. Ein Beitrag zur Kulturgeschichte. *Beiträge zur Geschichte der Technik und Industrie. Technikgeschichte,* vol. 23, pp. 86–93.

Hertwig, A., 1935. Die Eisenbahn und das Bauwesen. *Beiträge zur Geschichte der Technik und Industrie. Technikgeschichte,* vol. 24, pp. 29–37.

Hertwig, A., 1941. Die Entwicklung der Statik der Baukonstruktionen im 19. Jahrhundert. *Beiträge zur Geschichte der Technik und Industrie. Technikgeschichte,* vol. 30, pp. 82–98.

Hertwig, A., 1947. *Lebenserinnerungen von August Hermann Adalbert Hertwig.* Typewritten ms.

Hertwig, A., 1950. *Leben und Schaffen der Brückenbauer der Deutschen Reichsbahn Schwedler, Zimmermann, Labes und Schaper. Eine kurze Entwicklungsgeschichte des Brückenbaus.* Berlin: Wilhelm Ernst & Sohn.

Hertwig, A., 1951. Presentation held at the commemorative event for Müller-Breslau on 15 June 1925 at Charlottenburg Technical University. *Der Stahlbau,* vol. 20, No. 5, pp. 53–54 (in German)

Heyman, J., 1966. The Stone Skeleton. *International Journal of Solids and Structures,* vol. 2, pp. 249-279.

Heyman, J., Baker, J. F., 1966. Plastic theory and design, Chap. 23. In: *The Steel Designers' Manual,* 3rd Ed., London.

Heyman, J., 1967. Westminster Hall Roof. *Proceedings of the Institution of Civil Engineers,* vol. 37, pp. 137-162.

Heyman, J., 1969. The safety of masonry arches. *International Journal of Mechanical Sciences,* vol. 11, pp. 363–385.

Heyman, J., 1972/1. *Coulomb's memoir on statics: An essay in the history of civil engineering.* Cambridge: Cambridge University Press.

Heyman, J., Padfield, C. J., 1972/2. Two masonry bridges: Clare College Bridge. *Proc. Instn. Civ. Engrs.,* vol. 52, pp. 305–318.

Heyman, J., Threlfall, B. D., 1972/3. Two masonry bridges: Telford's bridge at Over. *Proc. Instn. Civ. Engrs.,* vol. 52, pp. 319–330.

Heyman, J., 1976. Couplet's engineering memoirs, 1726–33. In: *History of technology, First annual volume,* A. Rupert Hall & N. Smith (ed.), pp. 21–44. London: Mansell.

Heyman, J., 1980/1. The estimation of the strength of masonry arches. *Proc. Instn. Civ. Engrs.,* Pt. 2, vol. 68, pp. 921–37.

Heyman, J., Hobbs, N. B., Jermy, B. S., 1980/2. The rehabilitation of Teston bridge. *Proc. Instn. Civ. Engrs.,* Pt. 1, vol. 68, pp. 489–497.

Heyman, J., 1982. *The masonry arch.* Chichester: Ellis Horwood Ltd.

Heyman, J., 1985. John Fleetwood Baker, Baron Baker of Windrush. *Royal Society Biographical Memoirs.* London.

Heyman, J., 1995/1. *The Stone Skeleton. Structural Engineering of Masonry Architecture.* Cambridge: Cambridge University Press.

Heyman, J., 1995/2. *Teoría, historia y restauración de estructuras de fábrica. Colección de ensayos,* ed. S. Huerta. Madrid: Instituto Juan de Herrera.

Heyman, J., 1998/1. *Structural Analysis. A historical approach.* Cambridge: Cambridge University Press.

Heyman, J., 1998/2. Hooke's cubico-parabolical conoid. *Notes and Records of the Royal Society of London,* vol. 52, pp. 39-50.

Heyman, J., 1999. *El esqueleto de piedra. Mecánica de la arquitectura de fábrica,* ed. S. Huerta. Madrid: Instituto Juan de Herrera.

Heyman, J., 2004. *Análisis de Estructuras. Un estudio histórico,* ed. S. Huerta. Madrid: Instituto Juan de Herrera.

Heymann, J., Naumann, F. 1985. Christian Moritz Rühlmann (1811–1896) – Der erste Lehrer an der Königlichen Gewerbeschule zu Chemnitz und bedeutender Technikwissenschaftler. *Wiss. Zeitschrift der TH Karl-Marx-Stadt,* vol. 27, No. 5, pp. 673–682.

Hiersemann, L., 1990. Der Leipziger Gewerbeschullehrer August Föppl und die Theoretisierung der Technik. *Technische Mechanik,* vol. 11, No. 1, pp. 55–64.

Higgins, T. J., 1949. A Survey of the Approximate Solution of Two-Dimensional Physical Problems by Variational Methods and Finite Difference Procedures. In: *Numerical Methods of Analysis in Engineering (Successive Corrections),* L. E. Grinter (ed.), pp. 169–198. New York: Macmillan Co.

Hilbert, D., 1901. Über das Dirichlet'sche Prinzip. In: *Festschrift zur Feier des 150jährigen Bestehens der Kgl. Ges. der Wiss. Zu Göttingen.* Berlin: Weidenhammer.

Hilbert, D., 1906. Zur Variationsrechnung. *Mathematische Annalen,* vol. 62, pp. 351–370.

Hilz, H., 1993. *Eisenbrückenbau und Unternehmertätigkeit in Süddeutschland.* Stuttgart: Franz Steiner Verlag.

Hobbs, R., 1985. Ronald Stewart Jenkins: engineer and mathematician. *The Arup Journal,* vol. 20, No. 2, pp. 9–13.

Hobsbawm, E., 1994. *The Age of Extremes: The Short Twentieth Century 1914–1991.* London: Penguin.

Hobsbawm, E., 1999. Das Zeitalter der Extreme. *Weltgeschichte des 20. Jahrhunderts.* Trans. from English by Y. Badal, 3rd ed. Munich: Deutscher Taschenbuch Verlag.

Hodgkinson, E., 1824. On the Transverse Strain and Strength of Materials. *Mem. Proc. Manchester Literary and Philosophical Society.*

Hodgkinson, E., 1830. Theoretical and Experimental Researches to Ascertain the strength and best forms of Iron beams, *Mem. Proc. Manchester Literary and Philosophical Society,* pp. 407–544.

Hodgkinson, E., 1840. On the strength of pillars of cast iron and other materials. *Phil. Trans. Part II,* pp. 385–456.

Hodgkinson, E., 1846. *Experimental researches of the Strength and other properties of cast iron.* London.

Höffl, K., Grabow, G., 1982. Julius Ludwig Weisbach (1806 bis 1871). *Freiberger Forschungshefte Reihe D,* No. 152, ed. dean of Freiberg Mining Academy. Leipzig: VEB Deutscher Verlag für Grundstoffindustrie.

Hölderlin, F., 1990. *Hyperion oder der Eremit in Griechenland.* Stuttgart: Philipp Reclam.

Hölderlin, F., 1992. *Gedichte.* Stuttgart: Philipp Reclam.

Holzapfel, M., 1969. †Prof. Dr.-Ing. G. Kani. *Beton- und Stahlbetonbau,* vol. 64, No. 1, p. 23.

Holzer, St. M., 2006. Vom Einfluß des Analysewerkzeugs auf die Modellbildung. Zur statischen Analyse des Wiegmann-Polonceau-Trägers in der 2. Hälfte des 19. Jahrhunderts. *Bautechnik,* vol. 83, No. 6, pp. 428–434.

Homberg, H., 1949. *Einflußflächen für Trägerroste*. Dahl: Self-pub.

Homberg, H., 1951. *Kreuzwerke: Statik der Trägerroste und Platten*. Berlin: Springer.

Homberg, H., 1952. Über die Lastverteilung durch Schubkräfte, Theorie des Plattenkreuzwerks. *Der Stahlbau*, vol. 21, No. 3, pp. 42–43, No. 4, pp. 64–67, No. 5, pp. 77–79 & No. 10, pp. 190–192.

Homberg, H., Weinmeister, J., 1956. *Einflußflächen für Kreuzwerke*. Berlin: Springer.

Homberg, H., Trenks, H., 1962. *Drehsteife Kreuzwerke*. Berlin: Springer.

Hooke, R., 1675. *A description of helioscopes, and some other instruments*. London: John & Martin Printer to the Royal Society.

Hooke, R., 1678. *Lectures de potentia restitutiva, or of spring explaining the power of springing bodies*. London: John & Martin Printer to the Royal Society.

Hooykaas, R., 1967. Eduard Jan Dijksterhuis (1892–1965). *Isis*, vol. 58, pp. 223–225.

Hopkins, H. G., 1980. Obituary of Professor William Prager 23 May 1903 – 17 March 1980. *International Journal of Mechanical Science*, vol. 22, pp. 393–394.

Hořejší, J., Pirner, M. 1997. *Osobnosti stavební mechaniky* (personalities in structural mechanics). Prague: ÚTAM (in Czech).

Horkheimer, M., Adorno, T. W., 1994. *Dialektik der Aufklärung. Philosophische Fragmente*. Frankfurt a. M.: Fischer Taschenbuch Verlag.

Hossdorf, H., 1976. *Modellstatik*. Wiesbaden: Bauverlag.

Housner, G. W., 1959. Behavior of Structures During Earthquake. *Proceedings of American Society of Civil Engineers*, EM4, pp. 109–129.

Hovgaard, W., 1937. Biographical Memoir of George Fillmore Swain. In: *National Academy of Sciences of the United States of America. Biographical Memoirs*, vol. XVII, pp. 331–350. Washington: National Academy of Sciences.

Hoyer, E. v., 1907. Rühlmann, Christian Moritz. In: *Allgemeine Deutsche Biographie, Band 53*. Ed. Historical Commission, Royal Academy of sciences at the request and with the support of the King of Bavaria, pp. 587–593. Leipzig: Verlag von Duncker & Humblot.

Hoyer, W., 1967. Prof. Dr.-Ing. Reinhold Rabich 65 Jahre. *Wiss. Z. Techn. Univers. Dresden*, vol. 16, No. 1, p. 41.

Hrennikoff, A. P., 1940. *Plane stress and bending of plates by method of articulated framework*. Sc. D. thesis, MIT, Boston.

Hrennikoff, A. P., 1941. Solution of problems in elasticity by the framework method. *Transactions of the American Society of Mechanical Engineers. Journal of applied mechanics*, vol. 8, pp. A169–A175.

Hrennikoff, A. P., Gantayat, A., 1971. Three-Dimensional bar cell for elastic stress analysis. *Journal of the Engineering Mechanics Division*, vol. 97, No. 3, pp. 1021–1024.

Hrennikoff, A. P., Mathew, C. I., Sen, R., 1972. Stability of plates using rectangular bar cells. In: *Proceedings of the International Association for Bridge & Structural Engineering (IABSE)*, vol. 32. Ed. IABSE, pp. 109–123. Zurich.

Hrennikoff, A. P., Agrawal, K. M., 1975. Trapezoidal bar cells in plane stress. In: *Proceedings of the International Association for Bridge & Structural Engineering (IABSE)*, vol. 35. Ed. IABSE, pp. 65–87. Zurich.

Hruban, I., 1982. Kettenbrücken von damals. In: *Neue Prager Presse*, 29.10.1982, No. 43.

Hruban, K., 1921. Zur Berechnung der Pilzdecke. *Beton und Eisen*, vol. 20, No. 16, pp. 187–188 & No. 17/18, pp. 200–202.

Hu, H.-C., 1955. On some variational methods on the theory of elasticity and the theory of plasticity. *Scientia Sinica*, vol. 4, pp. 33–54.

Huber, G., Rubin, D., 1999. Verbundrahmen mit momententragfähigen Knoten beim Millennium Tower. *Stahlbau*, vol. 68, No. 8, pp. 612–622.

Huber, G., Obholzer, A., 1999. Verbundflachdecken beim Millennium Tower. *Stahlbau*, vol. 68, No. 8, pp. 623–630.

Huber, M. T., 1904/1. *Właściwa praca odkształcenia jako miara wytężenia materiału* (true deformation work as a measure of material strain). Lwów: Czasopismo Techniczne (in Polish).

Huber, M. T., 1904/2. Zur Theorie der Berührung fester elastischer Körper. *Annalen der Physik*, vol. 14, pp. 153–163.

Huber, M. T., 1914. Die Grundlagen einer rationellen Bemessung der kreuzweise bewehrten Eisenbetonplatten. *Zeitschrift des Österreichischen Ingenieur- und Architekten-Vereines*, vol. 66, No. 30, pp. 557–564.

Huber, M. T., 1921. *Teoria płyt prostokątnie różnokierunkowych wraz z technicznymi zastosowaniami* (theory and technical applications of orthropic plates). Lwów: Archiwum Towarzystwa Naukowego (in Polish).

Huber, M. T., 1923. Die Theorie der kreuzweise bewehrten Eisenbetonplatten nebst Anwendungen auf mehrere bautechnisch wichtige Aufgaben über rechteckige Platten. *Der Bauingenieur*, vol. 4, No. 12, pp. 354–360 & No. 13, pp. 392–395.

Huber, M. T., 1924. Über die Biegung einer rechteckigen Platte von ungleicher Biegungssteifigkeit in der Längs- und Querrichtung bei einspannungsfreier Stützung des Randes (Mit besonderer Berücksichtigung der kreuzweise bewehrten Betonplatten). *Der Bauingenieur*, vol. 5, No. 9, pp. 259–263 & No. 10, pp. 305–306.

Huber, M. T., 1925/1. Über die Biegung einer sehr langen Eisenbetonplatte. *Der Bauingenieur*, vol. 6, No. 1, pp. 1–19 & No. 2, pp. 46–53.

Huber, M. T., 1925/2. Über die genaue Biegungsgleichung einer orthropenen Platte in ihrer Anwendung auf kreuzweise bewehrte Betonplatten. *Der Bauingenieur*, vol. 6, No. 30, pp. 878–879.

Huber, M. T., 1926. Vereinfachte strenge Lösung der Biegungsaufgabe einer rechteckigen Eisenbetonplatte bei geradliniger freier Stützung aller Ränder. *Der Bauingenieur*, vol. 7, No. 7, pp. 121–127, No. 8, pp. 152–154 & No. 9, pp. 170–175.

Huber, M. T., 1929. *Probleme der Statik technisch wichtiger orthroper Platten*. Warsaw: Akademia Nauk Technicznych (in Polish).

Huberti, G., 1964. *Vom Caementum zum Spannbeton. Band II/Teil B: Die erneuerte Bauweise*. Wiesbaden/Berlin: Bauverlag.

Huerta, S., 2003. The mechanics of timbrel vaults: a historical outline. In: *Essays in the History of Mechanics*, ed. A. Becchi, M. Corradi, F. Foce & O. Pedemonte, pp. 89-133. Basel: Birkhäuser.

Huerta, S., Foce, F., 2003. Vault theory in Spain between XVIIIth and XIXth century: Monasterio's unpublished manuscript 'Nueva teorica sobre el empuje de bovedas'. In: *Proceedings of the First International Congress on Construction History*, ed. S. Huerta, vol. II, pp. 1155–1166. Madrid: Instituto Juan de Herrera.

Huerta, S., 2004. *Arcos, bóvedas y cúpulas. Geometría y equilibrio en el cálculo tradicional de estructuras de fábrica*. Madrid: Instituto Juan de Herrera.

Huerta, S., 2005. Thomas Young's theory of the arch. His analysis of Telford's design for an iron arch of 600 feet over the Thames in London. In: *Essays in the history of the theory of structures. In honour of Jacques Heyman*, ed. S. Huerta, pp. 189–233. Madrid: Instituto Juan de Herrera.

Huerta, S., 2006. The first thermal analysis of an arch bridge: Thomas Young 1817. In: *Proceedings of the First International Conference on Advances in Bridge Engineering*, ed. A. Kumar, C. J. Brown, L. C. Wrobel, pp. 18–29. Uxbridge: Brunel University Press.

Huerta, S., Kurrer, K.-E., 2008. Zur baustatischen Analyse gewölbter Steinkonstruktionen. In: *Mauerwerk-Kalender*, vol. 33, ed. W. Jäger, pp. 373–422. Berlin: Ernst & Sohn.

Hughes, T. J. R., Oden, J. T., Papadrakakis, M., 2004. In Memoriam to Prof. John H. Argyris: 19 August 1913 – 2 April 2004. *Computer Methods in Applied Mechanics and Engineering*, vol. 193, pp. 3763–3766.

Husemann, T., 1876. Clebsch: Rudolf Friedrich Alfred C. In: *Allgemeine Deutsche Biographie, Band 4*. Ed. Historical Commission, Royal Academy of sciences at the request and with the support of the King of Bavaria, pp. 299–300. Leipzig: Verlag von Duncker & Humblot.

Hutton, C., 1772. *The Principles of Bridges*. Newcastle-upon-Tye: Kincaird and Creech.

Hutton, C., 1812. *Tracts on mathematical and philosophical subjects comprising, among numerous important articles, the Theory of Bridges*. London: Wilkie and Robinson.

Iguchi, S., 1933. *Eine Lösung für die Berechnung der biegsamen rechteckigen Platten*. Berlin: Springer.

Iguchi, S., 1936. Allgemeine Lösung der Knickungsaufgabe für rechteckige Platten. *Ingenieur-Archiv*, vol. 7, pp. 207–215.

Iguchi, S., 1937. Die Biegungsschwingungen der vierseitigen eingespannten rechteckigen Platte. *Ingenieur-Archiv*, vol. 8, pp. 11–25.

Iguchi, S., 1938. Die Knickung der rechteckigen Platte durch Schubkräfte. *Ingenieur-Archiv*, vol. 9, pp. 1–12.

Iguchi, S., 1953. Die Eigenschwingungen und Klangfiguren der vierseitig freien rechteckigen Platte. *Ingenieur-Archiv*, vol. 21, pp. 303–322.

Ingerslev, A., 1921. Om en elementaer Beregningsmade af krydsarmerede Plader. *Ingeniøren*, vol. 30, pp. 507–515.

Institut Culturel de Bretagne (ed.), 2002. *Pierre Bouguer, un savant breton au XVIIIe siècle.* Vannes.

IABSE – International Association for Bridge & Structural Engineering (ed.), 1936/1. *2nd Congress, Berlin/Munich, 1–11 October 1936. Preliminary Report.* Berlin: Verlag von Wilhelm Ernst & Sohn 1936.

IABSE – International Association for Bridge & Structural Engineering (ed.), 1936/2. *Proceedings, vol. 4.* Zurich: A.-G. Gebr. Leemann & Co. 1936.

IABSE – International Association for Bridge & Structural Engineering (ed.), 1938. *2nd Congress, Berlin/Munich, 1–11 October 1936. Final Report.* Berlin: Verlag von Wilhelm Ernst & Sohn 1938.

Iori, T., 2001. *Il cemento armato in Italia dalle origini alla seconda guerra mondiale.* Rome: EDILSTAMPA.

Iseki, K. R., 1930. Who's Who in "Hakushi" in Great Japan. vol. V, p. 143, p. 313. Tokyo: Hattensha.

Itard, J., 1981. Saint-Venant, Adhémar Jean Claude Barré de. In: *Dictionary of Scientific Biography*, vol. 12, C. C. Gillispie (ed.), pp. 73–74. New York: Charles Scribner's Sons.

Iwanow, B. I. et al., 1980. *Spezifik der technischen Wissenschaften.* Trans. from the Russian by H. Reichel & G. Homuth. Moscow, Leipzig: Verlag MIR & VEB Fachbuchverlag.

Iwanow, B. I., Tscheschew, W. W., 1982. *Entstehung und Entwicklung der technischen Wissenschaften.* Moscow/Leipzig: MIR & VEB Fachbuchverlag.

Jacobi, C. G. J., 1891. Über die Pariser Polytechnische Schule. In: *Collected Works vol. 7,* ed. K. Weierstraß, pp. 355–370. Berlin.

Jaeger, T. A., 1956. Grundzüge der Tragberechnung. *Der Bauingenieur*, vol. 31, No. 8, pp. 273–268.

Jaeger, T. A., Sawczuk, A., 1963. *Grenztragfähigkeitstheorie der Platten.* Berlin: Springer.

Jäger, K., 1937. *Festigkeit von Druckstäben aus Stahl.* Vienna: Verlag von Julius Springer.

Jäger, W., Pflücke, T., Schöps, P., 2006. Kommentierte Technische Regeln für den Mauerwerksbau. Teil 1: DIN 1053-100: Mauerwerk – Berechnung auf der Grundlage des semiprobabilistischen Sicherheitskonzepts – Kommentare und Erläuterungen. In: *Mauerwerk-Kalender,* vol. 31, pp. 363–431, ed. H.-J. Irmschler, W. Jäger & P. Schubert. Berlin: Ernst & Sohn.

Jagfeld, M., Barthel, R., 2004. Zur Gelenkbildung in historischen Tragsystemen aus Mauerwerk. *Bautechnik,* vol. 81, No. 2, pp. 96–102.

James, F. A. J. L., 2002. Young, Thomas. In: *A Biographical Dictionary of Civil Engineers in Great Britain and Ireland, vol. 1: 1500–1830.* Ed. A. W. Skempton, M. M. Chrimes, R. C. Cox, P. S. M. Cross-Rudkin, R. W. Rennison & E. C. Ruddock, pp. 818–819. London: ICE & Thomas Telford.

Jasiński (Yasinsky), F. S., 1893. *Opyt razvitiya teorii prodol'nogo izgiba* (research into buckling theory). St. Petersburg: Erlich (in Russian).

Jasiński (Yasinsky), F. S., 1894. *O soprotivlenii prodol'nomu izgibu* (on buckling theory). St. Petersburg (in Russian). 1894 French ed.: Recherches sur la flexion des pièces comprimées. *Annales des ponts et chaussées,* vol. 8, pp. 233–364.

Jasiński (Yasinsky), F. S., 1902–1904. *Sobranie sochinenii* (collected works). 3 vols. St. Petersburg: Institute of Engineers of Ways of Communication (in Russian).

Jasiński (Yasinsky), F. S., 1952. *Izbrannye raboty po ustoichivosti szhatykh sterzhnei* (selected works on the stability of struts). Moscow/Leningrad: Gostekhizdat (in Russian).

Jasiński (Yasinsky), F., 1961. *Pisma.* 2 vols. Warsaw: PWN (in Polish).

Jelinek, C., 1856. *Das ständisch-polytechnische Institut zu Prag.* Prague: Gottlieb Haase Söhne.

Jenkins, R. S., 1947. *Theory and design of cylindrical shell structures.* London: Arup.

Jenkins, R. S., 1952. Theory of new forms of shell. *Symposium on shell roof construction* No. 1, July, pp. 127–148.

Jenkins, R. S., 1953/1. The design of a reinforced concrete factory at Brynmawr, South Wales. *Proc. Instn Civ. Engrs.,* vol. 2, No. 3, Dec, pp. 345–397.

Jenkins, R. S., 1953/2, *Matrix analysis applied to indeterminate structures.* London: Ove Arup & Partners.

Jenkins, R. S., 1954. *Linear analysis of statically indeterminate structures.* Note No. 1, Euler Society.

Jesberg, P., 1996. *Die Geschichte der Ingenieurbaukunst aus dem Geist des Humanismus.* Stuttgart: Deutsche Verlags-Anstalt.

Joffe, A. F., 1967. *Begegnungen mit Physikern.* Basel: Pfalz-Verlag.

Johanson, K. W., 1932. Bruchmomente der kreuzweise bewehrten Platten. In: *Proceedings of the International Association for Bridge & Structural Engineering (IABSE).* vol. 1. Ed. IABSE; pp. 277–296. Zurich: A.-G. Gebr. Leemann & Co.

Johanson, K. W., 1949/1951. *Pladeformler.* 2 vols. Copenhagen: Polyteknisk Forenig.

Jones, R. E., 1964. A generalization of the direct-stiffness method of structural analysis. *AIAA Journal,* vol. 2, pp. 821–826.

Jones, R. T., 1945. *Wing Plan Forms for High-Speed Flight.* NACA Report No. 863, 23 June 1945.

Jouravski (Zhuravsky), D., 1856. Remarques sur la résistance d'un corps prismatique et d'une pièce composée en bois ou en tôle de fer à une force perpendiculaire à leur longueur. *Annales des ponts et chaussées,* vol. 12, pp. 328–351.

Jouravski (Zhuravsky), D., 1860. Remarques sur les poutres en treillis et les poutres pleines en tôle. *Annales des ponts et chaussées,* vol. 20, pp. 113–134 (excerpts from both articles already appeared between 1850 and 1859 in Russian publications).

Jungwirth, D., 1968. *Elektronische Berechnung des in einem Stahlbetonbalken in gerissenem Zustand auftretenden Kräftezustandes unter besonderer Berücksichtigung der Querkraftbereichs.* Dissertation, Munich TH.

Jürges, T., 2000. *Die Entwicklung der Biege-, Schub- und Verformungsbemessung im Stahlbetonbau und ihre Anwendung in der Tragwerkslehre.* Dissertation, Aachen RWTH.

Kaderabteilung der Karl-Marx-Universität Leipzig, 1968. Kurzbiographie von Prof. Dr. Klaus Zweiling. Typewritten ms.

Kafka, F., 1970. *Sämtliche Erzählungen,* ed. P. Raabe, pp. 148–149. Frankfurt a. M.: Fischer Verlag.

Kahlow, A., 1987. August Wöhler und die Entwicklung der Festigkeits- und Materialforschung im 19. Jahrhundert. In: *Dresdener Beiträge zur Geschichte der Technikwissenschaften,* No. 14, pp. 5–67.

Kahlow, A., 1990. Thomas Young und die Herausbildung des Begriffs Elastizitätsmodul. *NTM-Schriftenr. Gesch. Naturwiss., Technik, Med.,* vol. 27, No. 1, pp. 13–26.

Kahlow, A., Kurrer, K.-E. (ed.), 1994. Zeichnung, Grafik, Bild in Technikwissenschaften und Architektur. Materials for a conference at Potsdam Polytechnic, 4–5 Dec 1993. In: *Dresdener Beiträge zur Geschichte der Technikwissenschaften,* No. 23/1 & 23/2.

Kahlow, A., 1994. Wissenschaft und Ätherglaube. Der Elastizitäts-Modul von Thomas Young und seine Wellentheorie des Lichts. *Kultur & Technik,* No. 3, pp. 40–45.

Kahlow, A., 1995/1996. Bautechnikgeschichte zwischen Geschichte der Technikwissenschaften, Technikgeschichte, und Baugeschichte. *Blätter für Technikgeschichte,* 57/58, pp. 65–70.

Kahlow, A., Schendel, A. (ed.), 1998. *Vom Schönen und Nützlichen.* Potsdam: Stiftung Preußische Schlösser & Gärten Berlin-Brandenburg.

Kaiser, C., 2003. The Fleischbrücke in Nuremberg: A stone arch bridge as an object for researching the History of Building Technology. In: *Proceedings of the First International Congress on Construction History*, ed. S. Huerta, vol. II, pp. 1189–1200. Madrid: Instituto Juan de Herrera.

Kaiser, C., 2005. *Die Fleischbrücke in Nürnberg 1596–1598*. Cottbus: Dissertation, Faculty for Architecture, Civil Engineering & Urban Planning, Brandenburg Technical University.

Kaliszky, S., 1984. Gábor Kazinczy 1889–1964. *Periodica Polytechnica (Civil Engineering)*, vol. 28, pp. 75–76.

Kammüller, K., 1937. Das Prinzip der virtuellen Verschiebungen. Eine grundsätzliche Betrachtung. *Beton und Eisen*, vol. 36, No. 22, pp. 363–365.

Kammüller, K., 1938/1. Entgegnung zu Schleusner 1938/1. *Beton und Eisen*, vol. 37, No. 16, p. 271.

Kammüller, K., 1938/2. Entgegnung zu Schleusner 1938/2. *Beton und Eisen*, vol. 37, No. 16, pp. 271–272.

Kani, G., 1949. *Die Berechnung mehrstöckiger Rahmen*. Stuttgart: Wittwer.

Kani, G., 1955. *Spannbeton in Entwurf und Ausführung*. Stuttgart: Wittwer.

Kann, F., 1932. Rechnerische Untersuchung über die Größe des Fließbereiches in stählernen Durchlaufbalken unter Berücksichtigung des Momentenausgleiches. *Der Stahlbau*, vol. 5, No. 14, pp. 105–109.

Kant, I., 1746. *Gedanken von der wahren Schätzung der lebendigen Kräfte und Beurtheilung der Beweise derer sich Herr von Leibnitz und andere Mechaniker in dieser Streitsache bedienet haben, nebst einigen vorhergehenden Betrachtungen welche die Kraft der Körper überhaupt betreffen, durch Immanuel Kant*. Königsberg: Martin Eberhard Dorn.

Kapp, E., 1877. *Grundlinien einer Philosophie der Technik. Zur Entstehungsgeschichte der Cultur aus neuen Gesichtspunkten*. Braunschweig: Westermann.

Kappus, R., 1937. Drillknicken zentrisch gedrückter Stäbe mit offenem Profil im elastischen Bereich. In: *Luftfahrt-Forschung*, vol. 14, pp. 444–457.

Kappus, R., 1953. Zentrisches und exzentrisches Drehknicken von Stäben mit offenem Profil. *Der Stahlbau*, vol., 22 No. 1, pp. 6–12.

Karas, K., 1960. †Prof. Karl Girkmann. *Der Stahlbau*, vol. 29, No. 1, p. 32.

Karmarsch, K., 1879. Franz Joseph Ritter von Gerstner. In: *Allgemeine Deutsche Biographie*, 9. Bd., pp. 67–69. Leipzig: Duncker & Humblot.

Karner, L., 1938. Die Bedeutung der Zähigkeit des Stahles für die Berechnung und Bemessung von Stahlbauwerken, insbesondere von statisch unbestimmten Konstruktionen (general report on session I). In: *International Association for Bridge & Structural Engineering (ed.). 2nd Congress, Berlin/Munich, 1–11 October 1936. Final Report*, pp. 29–32. Berlin: Verlag von Wilhelm Ernst & Sohn.

Karner, L., Ritter, M., 1938. Die Bedeutung der Zähigkeit des Stahles für die Berechnung und Bemessung von Stahlbauwerken, insbesondere von statisch unbestimmten Konstruktionen (conclusions & proposals from session I). In: *International Association for Bridge & Structural Engineering (ed.). 2nd Congress, Berlin/Munich, 1–11 October 1936. Final Report*, pp. 933–934. Berlin: Verlag von Wilhelm Ernst & Sohn, 1938.

Kaufmann, G., 1906. Kontinuierliche Balken und statisch unbestimmte Systeme im Eisenbetonbau. *Beton und Eisen*, vol. 5, No. 5, pp. 125–128, No. 6, pp. 154–156 & No. 7, pp. 175–178.

Kazinczy, G. v., 1914. Trials with fixed-end beams (in Hungarian). *Betonszemle*, vol. 2, pp. 68–71, 83–87 & 101–104.

Kazinczy, G. v., 1931/1. Statisch unbestimmte Tragwerke unter Berücksichtigung der Plastizität. *Der Stahlbau*, vol. 4, No. 5, pp. 58–59.

Kazinczy, G. v., 1931/2. Die Weiterentwicklung der Plastizitätslehre. *Technika*, vol. 12, Nos. 5–7, pp. 168–172.

Kazinczy, G. v., 1933. Die Plastizität des Eisenbetons. *Beton und Eisen*, vol. 32, No. 5, pp. 74–80.

Kazinczy, G. v., 1938. Kritische Betrachtungen zur Plastizitätstheorie. In: *International Association for Bridge & Structural Engineering (ed.). 2nd Congress, Berlin/Munich, 1–11 October 1936. Final Report*, pp. 56–69. Berlin: Ernst & Sohn.

Kedrow, B. M., 1975. *Klassifizierung der Wissenschaften. Band 1*. Cologne: Pahl-Rugenstein Verlag.

Keith, 1962. Emperger-Gedenkfeier in Vienna. *Allgemeine Bauzeitung*, No. 671, 14 February.

Kertz, W., 1999. *Geschichte der Geophysik*. Hildesheim: Georg Olms.

Kesselring, F., 1954. *Technische Kompositionslehre*. Berlin: Springer-Verlag.

Khan, F. R., Iyengar, S. H., Colaco, J. P., 1966/1. Computer design of 100-story John Hancock Center. *Journal of the Structural Division, ASCE*, vol. 92, pp. 55–74.

Khan, F. R., 1966/2. On some special problems of analysis and design of shear wall structures. *Proceedings on Tall Buildings*, University of Southampton, April, pp. 321–344.

Khan, F. R., 1972. 100-story John Hancock Center in Chicago – a case study of the design process. *IABSE Journal*, vol. 16, pp. 27–34.

Khan, F. R., 1974. New structural systems for tall buildings and their scale effects on cities. *Proceedings on Tall Building, Planning, Design and Construction Symposium*, Nov 14–15, Nashville, Tenn., pp. 99–128.

Khan, Y. S., 2004. *Engineering Architecture: the vision of Fazlur R. Khan*. New York: W. W. Norton & Company.

Khanovich, I. G., 1967. *Akademik A. N. Krylov*. Leningrad: Nauka (in Russian).

Khoroshun, L. P., Rushchitsky, J. J., 2007. Gurii Nikolaevich Savin: Highlights from his life, and from his research and teaching career. *International Applied Mechanics*, vol. 43, No. 1, pp. 2–9.

Khvedelidze, B., Manjavidze, G., 1993. A survey of N. I. Muskhelishvili's scientific heritage. In: *Continuum Mechanics and Related Problems of Analysis: Proc. Internat. Symposium* (Tiflis, 1991), pp. 11–67. Tiflis: Metsniereba.

Kierdorf, A., 2006. Early Mushroom Slab Construction in Switzerland, Russia and the U.S.A. – A Study in Parallel Technological Development. In: *Proceedings of the Second International Congress on Construction History*, ed. M. Dunkeld, J. Campbell, H. Louw, M. Tutton, W. Addis & R. Thorne, vol. II, pp. 1793–1807. Cambridge. Construction History Society.

Kierdorf, A., 2007. 100 Jahre 'Handbuch für Eisenbetonbau'. *Beton- und Stahlbetonbau*, vol. 102, No. 10, pp. 725–732.

Killer, J., 1959. *Die Werke der Baumeister Grubenmann*. 2nd ed., Zurich: Verlag Leemann.

Kinze, W., 1971. Prof. Dr.-Ing. habil. Gustav Bürgermeister zum 65. Geburtstag. *Wiss. Z. Techn. Univers. Dresden*, vol. 20, No. 2, p. 519.

Kirchhoff, G. R., 1850. Ueber das Gleichgewicht und die Bewegung einer elastischen Scheibe. *Journal für die reine und angewandte Mathematik*, vol. 40, pp. 51–88.

Kirchhoff, G. R., 1858. Ueber das Gleichgewicht und die Bewegung eines unendlich dünnen elastischen Stabes. *Journal für die reine und angewandte Mathematik*, vol. 56, pp. 285–313.

Kirchhoff, G. R., 1876. *Vorlesungen über mathematische Physik. Mechanik*. Leipzig: B. G. Teubner.

Kirpichev, V. L., 1874. *Stroitel'naya mekhanika* (structural mechanics). 2 vols. St. Petersburg: Sakhatov (in Russian).

Kirpichev, V. L., 1903. *Lishnie neizvestnye v stroitel'noi mekhanike: Raschet staticheski-neopredelimykh sistem* (redundancy in structural mechanics: calculation of statically indeterminate systems). Kiev: Kul'zhenka (in Russian).

Kirpichev, V. L., 1917. *Sobranie sochinenii* (collected works), vol. 1. Petrograd: Petrograd Polytechnic Institute (in Russian).

Kirpichev, V. L., 1934. *Lishnie neizvestnye v stroitel'noi mekhanike: Raschet staticheski-neopredelimykh sistem* (redundancy in structural mechanics: calculation of statically indeterminate systems). Moscow (in Russian).

Kirsch, E. G., 1863. Die Fundamentalgleichungen der Theorie der Elasticität fester Körper, hergeleitet aus der Betrachtung eines Systems von Punkten, welche durch elastische Streben verbunden sind. *Zeitschrift des Vereines deutscher Ingenieure*, vol. 12, No. 8, col. 481–488, No. 9, col. 553–570 & No. 10, col. 631–638.

Kist, N. C., 1920. Die Zähigkeit des Materials als Grundlage für die Berechnung von Brücken, Hochbauten und ähnlichen Konstruktionen aus Flußeisen. *Der Eisenbau,* vol. 11, No. 23, pp. 425–428.

Klein, F., 1889. Referate zu Castiglianos Théorie de l'Équilibre des Systèmes Élastiques et ses Applications (1879) und Müller-Breslaus Die neueren Methoden der Festigkeitslehre und der Statik der Baukonstruktionen (1886). In: *Jahrbuch über die Fortschritte der Mathematik Bd. 18 (1886),* pp. 947–955. Berlin: Verlag von Georg Reimer.

Klein, F., Wieghardt, K., 1904. Über Spannungsflächen und reziproke Diagramme, mit besonderer Berücksichtigung der Maxwellschen Arbeiten. *Archiv der Mathematik und Physik,* 3rd series, vol. 8, pp. 1–10 & 95–119.

Kleinlogel, A., 1914. *Rahmenformeln.* Berlin: Wilhelm Ernst & Sohn.

Kleinlogel, A., 1932. Zum 70. Geburtstage F. von Empergers. *Beton und Eisen,* vol. 31, No. 1, pp. 1–4.

Kleinlogel, A., 1940. †Robert Maillart. *Beton und Eisen,* vol. 39, No. 13, pp. 182–183.

Kleinlogel, A., 1942. †Fritz v. Emperger. *Beton und Eisen,* vol. 41, No. 5/6, p. 41.

Kleinlogel, A., 1952. *Rigid frame formulas.* Trans. from the German *Rahmenformeln,* 11th ed. by F. S. Morgenroth. New York: Frederick Ungar Publishing Co.

Kleinlogel, A., Haselbach, W., 1993. *Rahmenformeln.* 17th ed., Berlin: Ernst & Sohn.

Klemm, F. 1977. Naturwissenschaften und Technik in der Französischen Revolution. *Deutsches Museum. Abhandlungen und Berichte,* vol. 45, No. 1.

Klemm, F., 1979. *Zur Kulturgeschichte der Technik. Aufsätze und Vorträge.* Munich: Deutsches Museum.

Klimke, H., 1976. *Berechnung der Traglast statisch unbestimmter räumlicher Gelenkfachwerke unter Berücksichtigung der überkritischen Reserve der Druckstäbe.* Dissertation, University of Karlsruhe.

Klimke, H., 1983. Zum Stand der Entwicklung der Stabwerkskuppeln. *Der Stahlbau,* vol. 52, No. 9, pp. 257–262.

Klooster, T., Permantier, M., 1998. Rechnen und Bauen. Über den Einsatz des Computers im Bauwesen. *Stahlbau,* vol. 67, No. 3, pp. 178–182.

Klöppel, K., Lie, K., 1940. Hängebrücken mit besonderer Stützbedingung des Versteifungsträgers. *Der Stahlbau,* vol. 13, No. 21/22, pp. 109–116.

Klöppel, K., Lie, K., 1941. Berechnung von Hängebrücken nach der Theorie II. Ordnung unter Berücksichtigung der Nachgiebigkeit der Hänger. *Der Stahlbau,* vol. 14, No. 19/20, pp. 85–93.

Klöppel, K., Lie, K., 1947. Formänderungsgrößenverfahren in der Statik. *Die Technik,* vol. 2, No. 10, pp. 445–452.

Klöppel, K., 1948. Rückblick und Ausblick auf die Entwicklung der wissenschaftlichen Grundlagen des Stahlbaues. In: *Abhandlungen aus dem Stahlbau, H. 2, Stahlbau-Tagung Hannover 1947.* Bremen-Horn: Industrie- und Handelsverlag Walter Dorn.

Klöppel, K., 1951/1. Aufgaben und Ziele der Zeitschrift "Der Stahlbau". *Der Stahlbau,* vol. 20, No. 1, pp. 1–3.

Klöppel, K., 1951/2. Die Theorie der Stahlverbundbauweise in statisch unbestimmten Systemen unter Berücksichtigung des Kriecheinflusses. *Der Stahlbau,* vol. 20, No. 2, pp. 17–23.

Klöppel, K., 1952/1. Geheimrat Hertwig 80 Jahre. *Der Stahlbau,* vol. 21, No. 3, pp. 37–38.

Klöppel, K., 1952/2. Unmittelbare Ermittlung von Einflußlinien mit dem Formänderungsgrößen-Verfahren. *Der Stahlbau,* vol. 21, No. 8, pp. 132–136 & No. 9, p. 176.

Klöppel, K., 1952/3. Zur Einführung der neuen Stabilitätsvorschriften. In: *Abhandlungen aus dem Stahlbau.* No. 12. Ed. Deutscher Stahlbau-Verband, Stahlbau-Tagung Munich 1952, pp. 84–143. Bremen-Horn: Industrie- und Handels-Verlag Walter Dorn.

Klöppel, K., 1955. Geheimrat Prof. Dr.-Ing. E. h. August Hertwig. Schriftleiter der Zeitschrift "Der Stahlbau" von 1928 bis 1938. *Der Stahlbau,* vol. 24, No. 6, pp. 121–122.

Klöppel, K., 1958. Die Einheit der Wissenschaft und der Ingenieur. *Deutsches Museum Munich. Abhandlungen und Berichte,* vol. 26, No. 2.

Klöppel, K., Scheer, J., 1960. *Beulwerte ausgesteifter Rechteckplatten.* Berlin: Ernst & Sohn.

Klöppel, K., Schardt, R., 1960. Systematische Ableitung der Differentialgleichungen für ebene anisotrope Flächentragwerke. *Der Stahlbau,* vol. 29, No. 2, pp. 33–43.

Klöppel, K., Schardt, R., 1962. Zur Berechnung von Netzkuppeln. *Der Stahlbau,* vol. 31, No. 6, pp. 129–136.

Klöppel, K., Friemann, H., 1964. Übersicht über Berechnungsverfahren für Theorie II. Ordnung. *Der Stahlbau,* vol., 33 No. 9, pp. 270–277.

Klöppel, K., Friemann, H., 1965. Die Statik im Zeichen der Anpassung an elektronische Rechenautomaten. *VDI-Z,* vol. 107, No. 3, pp. 1603–1607.

Klöppel, K., Möller, K. H., 1968. *Beulwerte ausgesteifter Rechteckplatten. II. Band.* Berlin: Ernst & Sohn.

Kluge, K., Machaczek, F., 1923. Melan als Lehrer der Praxis. In: *Joseph Melan zum siebzigsten Geburtstage,* pp. 151–190. Leipzig/Vienna: Deuticke.

Knauer, K. H., 1983. Die Pferdeeisenbahn Linz-Budweis. Erste Eisenbahn des europäischen Kontinents. Vienna: Vienna Technical Museum.

Knebel, K., Sanchez, J., Zimmermann, St., 2001. Das Eden-Projekt. Konstruktion, Fertigung und Montage des größten Gewächshauses der Welt. *Stahlbau,* vol. 70, No. 8, pp. 513–525.

Knittel, G., Kupfer, H., 1969. *Stahlbetonbau. Berichte aus Forschung und Praxis. Festschrift Rüsch.* Berlin: Wilhelm Ernst & Sohn.

Knittel, G., 1984. Prof. Dr.-Ing. habil. Gustav Bürgermeister gestorben. *Stahlbau,* vol. 53, No. 1, p. 30.

Knobloch, E., 2000. Mathematik an der Technischen Hochschule und der Technischen Universität Berlin. In: *1799–1999. Von der Bauakademie zur Technischen Universität Berlin. Geschichte und Zukunft,* ed. Karl Schwarz for the president of Berlin TU, pp. 394–398. Berlin: Ernst & Sohn.

Knobloch, E. (ed.), 2004. *"The shoulders on which we stand" – Wegbereiter der Wissenschaft. 125 Jahre Technische Universität.* Berlin/Heidelberg/New York: Springer-Verlag.

Knopf, J. (ed.), 2001. *Brecht-Handbuch. Band 2: Gedichte.* Stuttgart: J. B. Metzler.

Knothe, K., 1983. *Finite Methoden zur Konstruktionsberechnung II (WS 1983/84). Modifizierte Variationsprinzipien – insbesondere für Stabwerke – und ihre Verwendbarkeit für Finite-Element-Verfahren.* Berlin: Aeorspace Institute, Berlin TU.

Knothe, K., Wessels, H., 1999. *Finite Elemente. Eine Einführung für Ingenieure.* 3rd, rev. ed. Berlin/Heidelberg/New York: Springer-Verlag.

Knothe, K. (ed.), 1999. *Das Extremum der Formänderungsarbeit. Habilitationsschrift. Technische Hochschule Hannover 1916. Georg Prange.* ALGORISMUS. Studien zur Geschichte der Mathematik und der Naturwissenschaften, No. 31. Munich: Institute for the History of the Natural Sciences.

Knothe, K. & Tausendfreund, D., 2000. Emil Oskar Winkler (1835–1888). Begründer der Statik der Baukonstruktionen an der TH Berlin. Leben und Werk. In: *1799–1999. Von der Bauakademie zur Technischen Universität Berlin. Geschichte und Zukunft,* ed. Karl Schwarz for the president of Berlin TU, pp. 164–178. Berlin: Ernst & Sohn.

Knothe, K., 2001. *Gleisdynamik.* Berlin: Ernst & Sohn.

Knothe, K., 2003/1. Emil Winkler und die Oberbauforschung. *ZEVrail Glasers Annalen,* vol. 127, No. 2, pp. 97–100.

Knothe, K., 2003/2. Schwedler, Zimmermann und Timoshenko: Die Verbreitung und Erweiterung der Winklerschen Theorie. *ZEVrail Glasers Annalen,* vol. 127, No. 3/4, pp. 188–191.

Knothe, K., 2003/3. August Wöhler (1819–1914). Dauerfestigkeit und Langzeitverhalten, gestern und heute. *ZEVrail Glasers Annalen,* vol. 127, No. 8, pp. 384–386.

Knothe, K., 2003/4. 150 Jahre Geschichte bahntechnischer Forschung. Von J. A. Schubert bis C. T. Müller. *ZEVrail Glasers Annalen,* vol. 127, No. 9, pp. 448–451.

Knothe, K., 2004. *Fiedlerbriefe und Bibliographie Emil Winklers.* ALGORISMUS. Studien zur Geschichte der Mathematik und der Naturwissenschaften, No. 48. Munich: Institute for the History of the Natural Sciences. Augsburg: Dr. Erwin Rauner Verlag.

Knothe, K., 2005. Baustatik und Flugzeugbau. Etappen im Leben von Alfred Teichmann (1902–1971). *Stahlbau,* vol. 74, No. 5, pp. 373–378.

Kobori, T. (ed.), 1980. *Collected works of Prof. Tanabashi. Commemorative publication.*

Koch, M., Franz, G., Steup, H., 1992. Kurt Beyer. In: *VDI Bau. Jahrbuch 1992,* ed. VDI-Gesellschaft Bautechnik, pp. 354–393. Düsseldorf: VDI-Verlag.

Koenen, M., 1882. Vereinfachung der Berechnung continuirlicher Balken mit Hilfe des Satzes von der Arbeit. *Wochenblatt für Architekten und Ingenieure,* pp. 402–403 & 416–418.

Koenen, M., 1886. Berechnung der Stärke der Monierschen Cementplatten. *Zentralblatt der Bauverwaltung,* vol. 6, No. 47, p. 462.

Koenen, 1902. *Grundzüge für die statische Berechnung der Beton- und Betoneisenkonstruktionen.* Berlin: Wilhelm Ernst & Sohn.

Koenen, M., 1905. Über die gefährlichen Abscherflächen in Beton eingebetteter Eisenstäbe. *Beton und Eisen,* vol. 4, No. VI, pp. 148–149.

Koenen, M., 1906. *Grundzüge für die statische Berechnung der Beton- und Eisenbetonbauten.* 3rd, rev. ed. Berlin: Wilhelm Ernst & Sohn.

Koenen, M., 1921. Zur Entwicklungsgeschichte des Eisenbetons. *Der Bauingenieur,* vol. 2, No. 13, pp. 347–349.

Kögler, F., 1915. *Berichte des Ausschusses für Versuche im Eisenbau.* ed. B, No. 1. Zur Einführung – Bisherige Versuche. Berlin: Verlag von Julius Springer.

Kogon, E., 1946. *Der SS-Staat – Das System der deutschen Konzentrationslager.* Munich: Alber.

Koiter, W. T., 1945. *Over de Stabiliteit van het Elastisch Evenwicht.* Proefschrift (thesis) Technische Hogeschool Delft. Amsterdam: H. J. Paris.

Koiter, W. T., 1953. Stress-strain relations, uniqueness and variational theorems for elastic-plastic materials with a singular yield surface. *Quarterly of Applied Mathematics,* vol. 11, pp. 350–354.

Koiter, W. T., 1960/1. General theorems for elastic-plastic solids. In: *Progress in Solid Mechanics 6,* ed. I. N. Sneddon & R. Hill, pp. 165–221. Amsterdam: North-Holland.

Koiter, W. T., 1960/2. A consistent first approximation in the general theory of thin elastic shells. In: *Proceedings of IUTAM symposium on the theory of thin elastic shells,* ed. W. T. Koiter, pp. 12–33. Amsterdam: North Holland.

Koiter, W. T., 1966. On the nonlinear theory of thin elastic shells. I-III. In: *Proc. Kon. Ned. Ak. Wet.,* vol. 69, pp. 1–54.

Koiter, W. T., 1970. On the foundations of the linear theory of thin elastic shells. In: *Proc. Kon. Ned. Ak. Wet.,* vol. 73, pp. 169–195.

Koiter, W. T., 1972. *Stijfheid en Sterkte 1: Grondslagen.* Haarlem: Scheltema & Holkema.

Koiter, W. T., 1979. Omzien in verwondering, maar niet in wrok. In: *Trends in solid mechanics: Proceedings of the symposium dedicated to the 65th birthday of W. T. Koiter,* ed. J. F. Besseling & A. M. A. van der Heijden, pp. 237–246. Delft: Delft University Press.

Koloušek, V., 1941. Anwendung des Gesetzes der virtuellen Verschiebungen und des Reziprozitätssatzes in der Stabwerksdynamik. *Ingenieur-Archiv,* vol. 12, pp. 363–370.

Koloušek, V., 1950/1954/1956/1967/1980. *Dynamika stavebních konstrukcí* (dynamics of building structures), 3 vols. Prague: SNTL (in Czech).

Koloušek, V., 1953. *Baudynamik der Durchlaufträger und Rahmen.* Leipzig: Fachbuchverlag.

Koloušek, V., 1956. Schwingungen der Brücken aus Stahl und Stahlbeton. In: *Proceedings of the International Association for Bridge & Structural Engineering (IABSE).* vol. 16, ed. F. Stüssi & P. Lardy, pp. 301–332. Zurich: IABSE secretariat.

Koloušek, V., 1958. *Efforts dynamiques dans les ossatures rigides.* Paris: Dunod.

Koloušek, V., 1962. *Dynamik der Baukonstruktionen.* Berlin: Verlag für Bauwesen.

Koloušek, V., 1965. *Dinamika strojitělnych konstrukcij* (structural dynamics). Moscow: Strojizdat (in Russian).

Koloušek, V., 1973. *Dynamics in Civil Engineering Structures.* London: Butterword.

Koloušek, V., Pirner, M., Fischer, O., Náprstek, J., 1983. *Wind Effects on Civil Engineering Structures.* Amsterdam: Elsevier.

König, W., 1999. Die Akademie und die Technikwissenschaften. Ein unwillkommenes Geschenk. In: *Die Königlich Preußische Akademie der Wissenschaften zu Berlin im Kaiserreich,* ed. J. Kocka, R. Hohlfeld & P. T. Walther, pp. 381–398. Berlin: Akademie-Verlag.

König, W., 2006/1. Vom Staatsdiener zum Industrieangestellten: Die Ingenieure in Frankreich und Deutschland 1750–1945. In: *Geschichte des Ingenieurs,* ed. W. Kaiser & W. König, pp. 179–231. Munich/Vienna: Carl Hanser.

König, W., 2006/2. Struktur der Technikwissenschaften. In: *Erkennen und Gestalten. Eine Theorie der Technikwissenschaften,* ed. G. Banse, A. Grunwald, W. König & G. Ropohl, pp. 37–44. Berlin: edition sigma.

Kooharian, A., 1953. Limit analysis of voussoir (segmental) and concrete arches. *Proc. Am. Concr. Inst.,* vol. 49, pp. 317–28.

Köpcke, C., 1856. Über die Dimensionen von Balkenlagen. *Zeitschrift des Architekten- und Ingenieur-Vereins zu Hannover,* vol. 2, col. 82–91.

Köpcke, W., 1968. Prof. Dr.-Ing. A. Teichmann 65 Jahre. *Der Bauingenieur,* vol. 43, p. 70.

Koppelmann, E., 1973. Lévy, Maurice. In: *Dictionary of Scientific Biography,* vol. VIII. C. C. Gillispie (ed.), pp. 287–288. New York: Charles Scribner's Sons.

Kordina, K., Quast, U., 1971. Bemessung der Stahlbetonbauteile. Bemessung von schlanken Bauteilen – Knicksicherheitsnachweis. In: *Beton-Kalender,* vol. 60, pt. I, pp. 631–729. Berlin: Wilhelm Ernst & Sohn.

Kordina, K., Quast, U., 1997. Bemessung der Stahlbetonbauteile. Bemessung von schlanken Bauteilen – Knicksicherheitsnachweis. In: *Beton-Kalender,* vol. 86, pt. I, pp. 479–575. Berlin: Wilhelm Ernst & Sohn.

Körner, C., 1901. *Gewölbte Decken. Handbuch der Architektur.* 3rd pt., vol. 2, No. 3b. Stuttgart: Arnold Bergsträsser Verlagsbuchhandlung.

Krämer, S., 1982. *Technik, Gesellschaft und Natur. Versuch über ihren Zusammenhang.* Frankfurt/New York: Campus.

Krämer, S., 1988. *Symbolische Maschinen. Die Idee der Formalisierung in geschichtlichem Abriß.* Darmstadt: Wissenschaftliche Buchgesellschaft.

Krämer, S., 1989. Über das Verhältnis von Algebra und Geometrie in Descartes' "Géomètrie". *Philosophia naturalis,* vol. 26, No. 1, pp. 19–40.

Krämer, S., 1991/1. Denken als Rechenprozedur: Zur Genese eines kognitionswissenschaftlichen Paradigmas. *Kognitionswissenschaft,* No. 2, pp. 1–10.

Krämer, S., 1991/2. Zur Begründung des Infinitesimalkalküls durch Leibniz. *Philosophia naturalis,* vol. 28, No. 2, pp. 117–146.

Krämer, S., 1991/3. *Berechenbare Vernunft. Kalkül und Rationalismus im 17. Jahrhundert.* Berlin/New York: Walter de Gruyter.

Krämer, S., 1993. Operative Schriften als Geistestechnik. Zur Vorgeschichte der Informatik. In: *Informatik und Philosophie,* ed. P. Schefe, H. Hastedt, Y. Dittrich, G. Keil, pp. 69–83. Mannheim: BI-Wissenschaftverlag.

Krämer, S., 1994. Geist ohne Bewußtsein? Über einen Wandel in den Theorien vom Geist. In: *Geist – Gehirn – künstliche Intelligenz,* ed. S. Krämer, pp. 88–110. Berlin: Walter de Gruyter.

Kranakis, E., 1996. *An Exploration of Engineering Culture, Design and Research in the Nineteenth-Century France and America.* Cambridge, Mass.: The MIT Press.

Krankenhagen, G., Laube, H., 1983. *Werkstoffprüfung. Von Explosionen, Brüchen und Prüfungen.* Reinbek bei Hamburg: Rowohlt Taschenbuch Verlag.

Krätzig, W. B., Roik, K., Kotulla, B. (ed.), 1976. *Konstruktiver Ingenieurbau in Forschung und Praxis: Festschrift Wolfgang Zerna und Institut KIB.* Düsseldorf: Werner-Verlag.

Krätzig, W. B., Schmidt-Schleicher, H., Stangenberg, F., 2006. Wolfgang Zerna verstorben. *Stahlbau,* vol. 75, No. 2, pp. 177–178.

Kraus, I., 2004. *Dějiny technických věd a vynálezů v českých zemích* (Geschichte der technischer Wissenschaften und Erfindungen in Böhmen). Prague: Academia.

Kraus, K., Wisser, S., Knöfel, D. Über das Löschen von Kalk vor der Mitte des 18. Jhdt. In: *Arbeitsblätter, Nr. 1, Gruppe 6, Stein,* pp. 206–221. Siegen University. Siegen.

Krauß, K., 1985. Vom Materialwissen und den Bautechniken der alten Baumeister. *Denkmalpflege in Baden-Württemberg.* No. 4, pp. 218–223.

Kriemler, C. J., 1918. Friedrich Engesser zum 70. Geburtstag dem 12. Februar 1918. *Der Eisenbau,* vol. 9, No. 2, pp. 21–25.

Krohn, R., 1882. *Resultate aus der Theorie des Brückenbaus und deren Anwendung. I. Abteilung, Balkenbrücken.* Leipzig: Baumgärtners Buchhandlung.

Krohn, R., 1883. *Resultate aus der Theorie des Brückenbaus und deren Anwendung. Zweiter Teil. Bogenbrücken.* Leipzig: Baumgärtner's Verlagsbuchhandlung.

Krohn, R., 1884. Der Satz von der Gegenseitigkeit der Verschiebungen und Anwendung desselben zur Berechnung statisch unbestimmter Fachwerkträger. *Zeitschrift des Arch.- & Ing.-Vereins Hannover,* vol. 30, No. 4, pp. 269–274.

Kromm, A., Marguerre, K., 1938. Verhalten eines von Schub- und Druckkräften beanspruchten Plattenstreifens oberhalb der Beulgrenze. In: *Jahrbuch 1938 der Deutschen Versuchsanstalt für Luftfahrt e.V.,* Berlin-Adlershof, pp. 263–275.

Kron, G., 1939. *Tensor Analysis of Networks.* New York: John Wiley & Sons.

Kruck, G. E., 1934. *Beitrag zur Berechnung statisch unbestimmter, biegungsfester Tragwerke.* Dissertation, Zurich ETH.

Krug, K., Meinicke, K.-P., 1989. Nicolas Léonard Sadi Carnot (1796–1832). Pionier der Technischen Thermodynamik. In: *Lebensbilder von Ingenieurwissenschaftlern,* ed. G. Buchheim & R. Sonnemann, pp. 116–126. Leipzig: VEB-Fachbuchverlag.

Krylov, A. N., 1931/1. *O raschete balok, lezhashchikh na uprugom osnovanii* (on the calculation of elastically bedded beams). 2nd ed. Leningrad: AN SSSR (in Russian).

Krylov, A. N., 1931/2. O formakh ravnovesiya szhatykh stoek pri prodol'nom izgibe (on the equilibrium forms of columns in buckling). *Izvestiya Akademii nauk SSSR, OMEN (Nachrichten der Akademie der Wissenschaften der UdSSR, Abt. Math. und Naturwiss.),* No. 7, pp. 963–1012 (in Russian).

Krylov, A. N., 1936–1956. *Sobranie trudov* (collected works). 12 vols. (in 18 half-vols.). Moscow/Leningrad: AN SSSR-Nauka. (mainly in Russian).

Krylov, A. N., 1942. *Moi vospominaniya* (memoirs). Moscow/Leningrad: AN SSSR (in Russian).

Kuhlmann, U., 1996. Perspektiven im Verbundbrückenbau. *Stahlbau,* vol. 65, No. 10, pp. 331–337.

Kuhn, T. S., 1962. *The Structure of Scientific Revolutions.* Chicago.

Kuhn, T. S., 1979. *Die Struktur wissenschaftlicher Revolutionen.* Frankfurt a. M.: Suhrkamp.

Kulka, H., 1930. Hermann Zimmermann zum fünfundachtzigsten Geburtstag. *Der Bauingenieur,* vol. 11, No. 51, p. 881.

Kunert, K., 1955. *Beitrag zur Berechnung der Verbundkonstruktionen.* Dissertation, Berlin TU.

Kupfer. H., 1964. Erweiterung der Mörsch'schen Fachwerkanalogie mit Hilfe des Prinzips vom Minimum der Formänderungsarbeit. *CIB-Bulletin.* No. 40.

Kupfer, H., 1979. *Zum Gedenken an Hubert Rüsch.* Munich: Wilhelm Ernst & Sohn.

Kurdyumov, A. A., 1952. P. F. Papkovich. *Trudy Leningradskogo korablestroitel'nogo instituta* (proceedings of the Leningrad Shipbuilding Institute), No. 10, pp. 15–30 (in Russian).

Kurrer, K.-E., 1985/1. Das Verhältnis von Bautechnik, und Statik. *Bautechnik,* vol. 62, No. 1, pp. 1–4.

Kurrer, K.-E., 1985/2. Zur Geschichte der Theorie der Nebenspannungen in Fachwerken. Otto Mohr zum 150. Geburtstag. *Bautechnik,* vol. 62, No. 10, pp. 325–330.

Kurrer, K.-E., 1987. Beitrag zur Entwicklungsgeschichte des Kraftgrößenverfahrens. *Bautechnik,* vol. 64, No. 1, pp. 1–8.

Kurrer, K.-E., 1988. Der Beitrag Emil Winklers zur Herausbildung der klassischen Baustatik. In: *Humanismus und Technik, Jahrbuch 1987,* pp. 11–39. Berlin: Berlin TU Library/Publications Dept.

Kurrer, K.-E., 1990/1. Technik. In: *Europäische Enzyklopädie zu Philosophie und Wissenschaften, Band 4,* ed. H. J. Sandkühler, pp. 534–550. Hamburg: Felix Meiner.

Kurrer, K.-E., 1990/2. Das Technische in der Mechanik des F. J. Ritter v. Gerstner (1756–1832). *Österreichische Ingenieur- und Architekten-Zeitschrift (ÖIAZ),* vol. 135, No. 10, pp. 501–508.

Kurrer, K.-E., 1991/1. Zur Entstehung der Stützlinientheorie. *Bautechnik,* vol. 68, No. 4, pp. 109–117.

Kurrer, K.-E., 1991/2. Auf der Suche nach der wahren Stützlinie in Gewölben. In: *Humanismus und Technik, Jahrbuch 1990,* pp. 20–54. Berlin: Berlin TU Library/Publications Dept.

Kurrer, K.-E., 1992. Von der Theorie des Gewölbes zur Theorie des elastischen Bogens (Teil I). Das Verhältnis der Gewölbe- zur Bogentheorie zwischen 1826 und 1860. In: *Humanismus und Technik, Jahrbuch 1991,* pp. 16–47. Berlin: Berlin TU Library/Publications Dept.

Kurrer, K.-E., 1994/1. Von der graphischen Statik zur Graphostatik. Die Rezeption des Theorieprogramms Culmanns durch die klassische Baustatik. *Dresdener Beiträge zur Geschichte der Technikwissenschaften,* No. 23/1, pp. 79–86.

Kurrer, K.-E., 1994/2. Von der graphischen Statik zur Graphostatik. Die Rezeption des Theorieprogramms Culmanns durch die klassische Baustatik. *Dresdener Beiträge zur Geschichte der Technikwissenschaften,* No. 23/2, pp. 102–113.

Kurrer, K.-E., 1995. Comment la théorie de l'élasticité s'est imposée à l'analyse de la structure portante des voûtes dans les pays germanophones de 1860 à 1900. In: *Entre Mécanique et Architecture,* ed. P. Radelet-de Grave et E. Benvenuto, pp. 331–347. Basel: Birkhäuser.

Kurrer, K.-E., 1997/1. Stahl + Beton = Stahlbeton? Stahl + Beton = Stahlbeton! Die Entstehung der Triade von Verwaltung, Wissenschaft und Industrie im Stahlbetonbau in Deutschland. *Beton- und Stahlbetonbau,* vol. 92, No. 1, pp. 13–18 & No. 2, pp. 45–49.

Kurrer, K.-E., 1997/2. Zur Entwicklungsgeschichte der Gewölbetheorien von Leonardo da Vinci bis ins 20. Jahrhundert. *architectura,* vol. 27, No. 1, pp. 87–114.

Kurrer, K.-E., 1998/1. Baustatik und Ästhetik. *Stahlbau,* vol. 67, No. 3, pp. 205–210.

Kurrer, K.-E., 1998/2. Zur Debatte um die Theoreme von Castigliano in der klassischen Baustatik. *Bautechnik,* vol. 75, No. 5, pp. 311–322.

Kurrer, K.-E., Kahlow, A., 1998. Arch and vault from 1800 to 1864. In: *Arch Bridges. History, analysis, assessment, maintenance and repair,* ed. A. Sinopoli, pp. 37–42. Rotterdam/Brookfield: Balkema.

Kurrer, K.-E., 1999. Zur Entwicklung der deutschsprachigen Literatur auf dem Gebiet des Stahlbetonbaus bis 1920. In: Zur Geschichte des Stahlbetonbaus – Die Anfänge in Deutschland 1850 bis 1910, ed. Hartwig Schmidt. *Beton- und Stahlbetonbau, Spezial* (special issue), pp. 42–50.

Kurrer, K.-E., 2000/1. Die Berliner Schule der Baustatik. In: *1799–1999. Von der Bauakademie zur Technischen Universität Berlin. Geschichte und Zukunft,* ed. Karl Schwarz for the president of Berlin TU, pp. 152–163. Berlin: Ernst & Sohn.

Kurrer, K.-E., 2000/2. Von der Kunst zur Automation des statischen Rechnens. In: *1799–1999. Von der Bauakademie zur Technischen Universität Berlin. Geschichte und Zukunft,* ed. Karl Schwarz for the president of Berlin TU, pp. 188–199. Berlin: Ernst & Sohn.

Kurrer, K.-E., 2001. 100 Jahre Zeitschrift 'Beton- und Stahlbetonbau'. *Beton- und Stahlbetonbau,* vol. 96, No. 4, pp. 219–222.

Kurrer, K.-E., 2002. Accueil de la théorie de la torsion de Saint-Venant dans la littérature technique allemande jusqu'en 1950. *Construction Métallique,* vol. 39, No. 1, p. 5–16.

Kurrer, K.-E., 2003. The Development of the Deformation Method. In: *Essays on the History of Mechanics,* ed. A. Becchi, M. Corradi, F. Foce & O. Pedemonte, pp. 57–86. Basel: Birkhäuser.

Kurrer, K.-E., 2004/1. Zur Komposition von Raumfachwerken von Föppl bis Mengeringhausen. *Stahlbau,* vol. 73, No. 8, pp. 603–623.

Kurrer, K.-E., 2004/2. August Herwigs Lebenserinnerungen (1947): Rechenschaftsbericht einer konservativen Ingenieurpersönlichkeit zwischen Scylla und Charybdis zweier Weltkriege. In: *Technik und Verantwortung im Nationalsozialismus,* ed. W. Lorenz & T. Meyer, pp. 109–142. Münster: Waxmann Verlag.

Kurrer, K.-E., 2005/1. Hermann Maier-Leibnitz (1885–1962): Wegbereiter des Industriebaus der klassischen Moderne. *Stahlbau*, vol. 74, No. 8, pp. 623–634.

Kurrer, K.-E., 2005/2. 100 Jahre Beton-Kalender. *Beton- und Stahlbetonbau*, vol. 100, No. 9, pp. 795–811.

Kurrer, K.-E., 2006. Zum 75. Jahrgang von STAHLBAU. *Stahlbau*, vol. 75, No. 4, pp. 249–256.

Laffranchi, M., Marti, P., 1997. Robert Maillart's Curved Concrete Arch Bridges. *Journal of the Structural Division (ASCE)*, October 1997, pp. 1280–1286.

La Hire, P. de, 1687–1690. *Traité de la coupe des pierres*. Paris: Bibliothèque de l'Institut, ms. 1596; Bibliothèque de l'École Nationale des Ponts et Chaussées, ms. 228. Rennes: Bibliothèque municipale, ms. 224. Langres: Bibliothèque municipale, ms. 108.

La Hire, P. de, 1692/1912. Remarques sur l'épaisseur qu'on doit donner aux pieds droits des voutes et aux murs des dômes ou voutes de four. In: *Procès-verbaux de l'Académie Royale d'Architecture (1671–1793)*, vol. 2 (1682–1696), ed. H. Lemonnier. Paris: J. Schemit.

La Hire, P. de, 1695. *Traité de mécanique, où l'on explique tout ce qui est nécessaire dans la pratique des Arts, et les propriétés des corps pesants lesquelles ont eu plus grand usage dans la Physique*. Paris: Imprimerie Royale.

La Hire, P. de, 1698. *Architecture civile*. Paris 1698 (London: Royal Institute of British Architects Library, ms. 725).

La Hire, P. de, 1702/1720. Remarques sur la forme de quelques arcs dont on se sert dans l'Architecture. In: *Mémoires de l'Académie Royale des Sciences*, 1702, pp. 100–103. Paris.

La Hire, P. de, 1712/1731. Sur la construction des voûtes dans les édifices. In: *Mémoires de l'Académie Royale des Sciences*, 1712, pp. 69–77. Paris.

La Hire, P. de, 1730. Traité de mécanique, où l'on explique tout ce qui est nécessaire dans la pratique des Arts. *Mémoires de l'Académie Royale des Sciences*, vol. 9, pp. 1–331. Paris.

Lafaille, B., 1935. Mémoire sur l'étude général des surfaces gauches minces. In: *Proceedings of the International Association for Bridge & Structural Engineering (IABSE)*. vol. 3, ed. L. Karner & M. Ritter, pp. 295–332. Zurich: IABSE secretariat.

Lagrange, J.-J., 1788. *Mécanique analytique*. Paris: Chez La Veuve Desaint, Libraire.

Laissle, A., Schübler, A., 1869. *Der Bau der Brückenträger mit besonderer Rücksicht auf Eisen-Constructionen. Erster Teil*, 3rd, rev. ed. Stuttgart: Paul Neff.

Laitko, H., 1977. Der Begriff der wissenschaftlichen Schule – theoretische und praktische Konsequenzen seiner Bestimmung. In: *Wissenschaftliche Schulen, Band 1*. Ed. Mikulinskij, S. R., Jaroševskij, M. G., Kröber, G., Steiner, H., pp. 257–290. Berlin: Akademie-Verlag.

Laitko, H., 1995. Letter to the author from Prof. Hubert Laitko dated 4 April 1995.

Lamarle, E., 1855. Note sur un moyen très-simple d'augmenter, dans une proportion notable, la résistance d'une pièce prismatique chargée uniformément. *Bulletin de l'Académie Royale Belgique*, vol. 22, pt. 1, pp. 232–252 & 503–525.

Lambin, A., Christophe, P., 1898. Le pont Vierendeel. Rapport sur les essais jusqu'à la rupture effectués au parc de Tervueren, par M. Vierendeel, sur un pont métallique de $31^m.50$ de portée avec poutres à arcades de son système. *Annales des Travaux publics de Belgique*, vol. 55, 2nd series, vol. III, pp. 53–139.

Lamé, G., Clapeyron, E., 1823. Mémoire sur la stabilité des voûtes. *Annales des mines*, vol. VIII, pp. 789–818.

Lamé, G., Clapeyron, E., 1828. Mémoire sur la construction des polygones funiculaires. *Journal du génie civil*, vol. I, pp. 496–504.

Lamé, G., 1833/1. Mémoire sur la propagation de la chaleur dans les polyèdres, et principalement dans le prisme triangulaire régulier. *Journal de l'École Polytechnique*, vol. 14, pp. 194–251.

Lamé, G., 1833/2. Mémoire sur les lois de l'équilibre de l'éther dans les corps diaphanes. *Annales de Chimie et de Physique*, vol. 55, pp. 322–335.

Lamé, G., 1834. Mémoire sur les lois de l'équilibre du fluide éthéré. *Journal de l'École Polytechnique*, vol. 15, pp. 191–288.

Lamé, G., 1842. Mémoire sur le principe général de la Physique. In: *Comptes Rendus hebdomadaires des séances de l'académie des Sciences*, vol. 14, pp. 35–37.

Lamé, G., 1852. *Leçons sur la théorie mathematique de l'élasticité des corps solides*. Paris: Bachelier.

Lamé, G., 1854. Mémoire sur l'équilibre de l'élasticité des envenloppes sphériques. *Journal de Mathématique (Liouville)*, vol 19, pp. 51–87.

Lamé, G., 1861. *Course de physique mathématique rationelle. Discours préliminaire*. Paris.

Land, R., 1887/1. Über die statische und geometrische Bestimmtheit der Träger, insbesondere der Fachwerkträger (Zugleich ein Beitrag zur Kinematik der Stabwerke). *Zentralblatt der Bauverwaltung*, vol. 7, pp. 363–370.

Land, R., 1887/2. Über die Gegenseitigkeit elastischer Formänderungen als Grundlage einer allgemeinen Darstellung der Einflußlinien aller Trägerarten, sowie einer allgemeinen Theorie der Träger überhaupt. *Wochenblatt für Baukunde*, pp. 14–16, 24–25 & 33–35.

Land, R., 1887/3. Kinematische Theorie der statisch bestimmten Träger. *Schweizerische Bauzeitung*, vol. 21, pp. 157–160.

Land, R., 1888. Kinematische Theorie der statisch bestimmten Träger. *Zeitschrift des Österreichischen Ingenieur- und Architekten-Vereins*, vol. 40, pp. 11–39 & 162–181.

Land, R., 1889. Beitrag zur Ermittlung der Biegungslinien ebener elastischer Gebilde. *Zeitschrift des Österreichischen Ingenieur- und Architekten-Vereins*, vol. 41, pp. 157–161.

Landi, F., 1829. Elogio dell'Abate Lorenzo Mascheroni, prefazione a Mascheroni, L., 1829. *Nuove ricerche sull'equilibrio delle volte*. Milan: Silvestri.

Lang, A. L., Bisplinghoff, R. L., 1951. Some results of sweepback wing structural studies. *Journal of the Aeronautical Sciences*, vol. 18, pp. 705–717.

Lang, G., 1890. *Zur Entwicklungsgeschichte der Spannwerke des Bauwesens*. Riga: Verlag von N. Kymmel.

Lange, P., 1986. Die Herausbildung der Silikattechnik als eine Disziplin der Technikwissenschaften (theses). *Dresdener Beiträge zur Geschichte der Technikwissenschaften*, No. 12, pp. 73–81.

Langefors, B., 1952. Analysis of elastic structures by matrix transformation with special regard to semimonocoque structures. *Journal of the Aeronautical Sciences*, vol. 19, pp. 451–458.

Lauenstein, R., Bastine, P., 1913. *Die graphische Statik. Elementares Lehrbuch für den Schul- und Selbstunterricht sowie zum Gebrauch in der Praxis*. 12th ed. Leipzig: Alfred Kröner.

Lavoinne, J., 1872. Sur la résistance des parois planes des chaudières à vapeur. *Annales des Ponts et Chaussées*, vol. 3, pp. 267–303.

Lederer, A., 1970. Arthur Vierendeel. In: *Biographie nationale, Tome 35, supplément Tome VII* (facs. 2). Brussels.

Lederer, F., 1972. *Fachwerk- und Rostplatten*. Düsseldorf: Werner-Verlag 1972.

Lehmann, C., Maurer, B., 2006. *Karl Culmann und die graphische Statik. Zeichnen, die Sprache des Ingenieurs*. Berlin: Ernst & Sohn.

Lehmann, D., 2000. Berechnung des nichtlinearen Tragverhaltens gezogener vorgespannter L-Flansche. *Stahlbau*, vol. 69, No. 1, pp. 35–54.

Lehmann, E., 1970/1. Übersicht über die Berechnung schiffbaulicher Konstruktionen mit Hilfe der Methode der finiten Elemente. *HANSA – Schiffahrt – Schiffbau – Hafen*, vol. 107, trade fair special issue, pp. 589–596.

Lehmann, E., 1970/2. Finite Berechnungsmethoden und ihre Anwendung auf die Berechnung schiffbaulicher Konstruktionen. *HANSA – Schiffahrt – Schiffbau – Hafen*, vol. 107, special issue, pp. 1895–1902.

Lehmann, E., 1999. *100 Jahre Schiffbautechnische Gesellschaft e.V. Biografien zur Geschichte des Schiffbaus*. Berlin: Springer.

Lehmann, E., 2001. Festigkeitsanalysen schiff-baulicher Konstruktionen. In: *100 Jahre Schiff-bautechnische Gesellschaft e.V. Festveranstaltung vom 25. bis 29. Mai in Berlin*, ed. H. Keil, pp. 288–303. Berlin: Springer.

Lehmann, E., 2004. *Schiffbautechnische Forschung in Deutschland*. Hamburg: Seehafen Verlag.
Leipholz, H., 1968. *Einführung in die Elastizitätstheorie*. Karlsruhe: G. Braun.
Leitz, H., 1914. *Die Berechnung der frei aufliegenden, rechteckigen Platten*. Forscherarbeiten auf dem Gebiete des Eisenbetons, No. 22. Berlin: Wilhelm Ernst & Sohn.
Lenk, H., Moser, S., Schönert, K. (ed.), 1981. *Technik zwischen Wissenschaft und Praxis. Technikphilosophische und techniksoziologische Schriften aus dem Nachlaß von Hans Rumpf*. Düsseldorf: VDI-Verlag.
Lenk, K., 1937. Herstellung und Anwendung von Spannbeton. *Beton und Eisen*, vol. 36, No. 10, pp. 161–169.
Lenkei, P., 2000. Introduction. In: *Experiments with fixed end bars*. Instituto Técnico de Materiales y Construcciones (INTEMAC), pp. 15–19. Madrid: INTEMAC.
Leonardo da Vinci, 1974/1. *Codices Madrid (Codex Madrid I)*, ed. L. Reti, German facs. ed. Frankfurt a.M.: S. Fischer-Verlag.
Leonardo da Vinci, 1974/2. *Codices Madrid (Codex Madrid II)*, ed. L. Reti, German facs. ed. Frankfurt a.M.: S. Fischer-Verlag.
Leonhardt, F., 1937/1938. *Die vereinfachte Berechnung zweiseitig gelagerter Trägerroste*. Dissertation, Stuttgart.
Leonhardt, F., 1940. TH Leichtbau – eine Forderung unserer Zeit. Anregungen für den Hoch- und Brückenbau. *Die Bautechnik*, vol. 18, No. 36/37, pp. 413–423.
Leonhardt, F., 1950. Die neue Straßenbrücke über den Rhein von Cologne nach Deutz. In: special issue of *Die Bautechnik*. Berlin: Wilhelm Ernst & Sohn.
Leonhardt, F., Andrä, W., 1950. *Die vereinfachte Trägerrostberechnung*. Stuttgart: Julius Hoffmann Verlag.
Leonhardt, F., 1955. *Spannbeton für die Praxis*. Berlin: Verlag von Wilhelm Ernst & Sohn.
Leonhardt, F., 1956. Der Stuttgarter Fernsehturm. *Beton- und Stahlbetonbau*, vol. 51, No. 4, pp. 73–85 & No. 5, pp. 104–111.
Leonhardt, F., 1964. *Prestressed Concrete. Design and Construction*. 2nd, rev. ed. Berlin: Verlag von Wilhelm Ernst & Sohn.
Leonhardt, F., 1965. Die verminderte Schubabdeckung bei Stahlbetontragwerken. Begründung durch Versuchsergebnisse mit Hilfe der erweiterten Fachwerkanalogie. *Der Bauingenieur*, vol. 40, No. 1, pp. 1–15.
Leonhardt, F., Baur, W., 1971. Erfahrungen mit dem Taktschiebeverfahren im Brücken- und Hochbau. *Beton- und Stahlbetonbau*, vol. 66, No. 7, pp. 161–167.
Leonhardt, F., 1973. Das Bewehren von Stahlbetontragwerken. In: *Beton-Kalender*, vol. 62, pt. II, pp. 331–433. Berlin: Wilhelm Ernst & Sohn.
Leont'ev, N. N., 1963. *V. Z. Vlasov*. Moscow: Gosstroiizdat (in Russian).

Le Seur, T., Jacquier, F., Boscovich, R. G., 1742. *Parere di tre matematici sopra i danni, che si sono trovati nella cupola di S. Pietro sul fine dell'Anno MDCCXLII, dato per ordine di Nostro Signore Papa Bendetto XIV*. Romae.
Le Seur, T., Jacquier, F., Boscovich, R. G., 1743. *Riflessioni de' Padri Tommaso Le Seur, Francesco Jacquieur dell'Ordine de' Minimi, e Ruggero Giuseppe Boscovich della Compagnia di Gesù sopra alcune difficoltà spettanti i danni, e i risarcimenti della cupola di S. Pietro*. Rome.
Leuprecht, O., 1907. Beitrg zur Berechnung steifer Rahmenkonstruktionen. *Beton und Eisen*, vol. 6, No. 9, pp. 233–235 & No. 10, pp. 258–262.
Lévy, M. 1869. Note sur un système particulier de ponts biais. In: *Comptes rendus de l'Académie des Sciences de Paris*, vol. 69, No. 22, pp. 1132–1133.
Lévy, M., 1873. *Application de la théorie mathématique de l'élasticité à l'étude de systèmes articulés*. Paris.
Lévy, M., 1874. *La statique graphique et ses applications aux constructions*. Paris: Gauthier-Villars.
Lévy, M., 1886–1888. *La statique graphique et ses applications aux constructions*. 4 vols., ed. II, Paris: Gauthier-Villars.
Lévy, M., 1897. Notes sur les diverses manières d'appliquer la règle du trapèze au calcul de la stabilité des barrages en maçonnerie. *Annales des Ponts et Chaussées*, IV, pp. 5–19.
Lévy, M., 1899. Sur l'équilibre élastique d'une plaque rectangulaire. *Comptes rendus de l'Académie des Sciences de Paris*, vol. 129, pp. 535–539.
Levy, S., 1947. Computation of influence coefficients for aircraft structures with discontinuities and sweepback. *Journal of the Aeronautical Sciences*, vol. 14, No. 10, pp. 547–560.
Levy, S., 1953. Structural analysis and influence coefficients for delta wings. *Journal of the Aeronautical Sciences*, vol. 20, No. 7, pp. 449–454.
Lewe, V., 1920. Die Lösung des Pilzdeckenproplems durch Fouriersche Reihen. *Der Bauingenieur*, vol. 1, No. 22, pp. 631–636.
Lewe, V., 1922. Beitrag zur strengen Lösung des Pilzdeckenproplems durch Fouriersche Reihen, Streifenlast und Stützenkopfeinspannung. *Der Bauingenieur*, vol. 3, No. 4, pp. 111–112.
Lewe, V., 1923. Die statische Berechnung der Flüssigkeitsbehälter. In: *Handbuch für Eisenbetonbau*, 3rd ed., vol. 5, ed. F. v. Emperger, pp. 71–181, Berlin: Wilhelm Ernst & Sohn.
Lewe, V., 1926/1. *Pilzdecken und andere trägerlose Eisenbetonplatten*. Berlin: Wilhelm Ernst & Sohn.
Lewe, V., 1926/2. Rezension zu "Die elastische Platten" von Nádai. *Die Bautechnik*, vol. 4, No. 10, p. 118.
Lie, K., 1941. Praktische Berechnung von Hängebrücken nach der Theorie II. Ordnung. Einfeldrige und durchlaufende Versteifungsträger mit konstantem und veränderlichem Trägheitsmoment. *Der Stahlbau*, vol. 14, No. 14/15; pp. 65–69 & No. 16/18, pp. 78–84.
Liebowitz, H., 1979. Alfred Martin Freudenthal. In: *Memorial Tributes. National Academy of Engineering* vol. 1, pp. 62–65. Washington, D.C.: National Academy Press.
Liepfeld, A., 1992. *Polacy na szlakach techniki, wyd* (Poland on the road of technology). Wydawnictwa Szkolne i Pedagogiczne. Warsaw (in Polish).
Lindner, J., Scheer, J., Schmidt, H. (ed.), 1993. *Stahlbauten. Erläuterungen zu DIN 18800 T. 1 bis T. 4*. Berlin: Beuth und Ernst & Sohn.
Lindner, J., Scheer, J., Schmidt, H. (ed.), 1998. *Stahlbauten. Erläuterungen zu DIN 18800 T. 1 bis T. 4*. 3rd ed. Berlin: Ernst & Sohn.
Litwiniszyn, J., 1981. Działalność naukowa Witolda Nowackiego (scientific activities of Witold Nowacki). *Nauka Polska*, Nos. 11–12, p. 118.
Livesley, R. K., 1956. The Application of an Electronic Digital Computer to some Problems of Structural Analysis. *The Structural Engineer*, vol. 34, No. 1, pp. 1–12.
Livesley, R. K., 1957. *An Introduction to Digital Computers*. Cambridge: Cambridge University Press.
Livesley, R. K., 1964. *Matrix Methods of Structural Analysis*. Oxford: Pergamon Press.
Livesley, R. K., 1992. A computational model for the limit analysis of three-dimensional masonry structures. *Meccanica*, vol. 27, No. 3, pp. 161–171.
Long, S. H., 1830/1. Observations on wooden, or frame bridges. *Journal of the Franklin Institute*, pp. 252–55.
Long, S. H., 1830/2. *Description of the Jackson Bridge together with directions to builders of wooden framed bridges*. Baltimore: Sands & Neilson.
Long, S. H., 1836. *Description of Col. Long's bridges together with a series of directions to bridge builders*. Concord: John P. Brown.
Long, S. H., 1839/1. *Improved brace bridge*. Specification of a patent for an improved bridge, granted to Lt. Col. S. H. Long.
Long, S. H., 1839/2. *Specifications of a brace bridge and of a suspension bridge*. Philadelphia.
Lorenz, H., 1926. Die wissenschaftlichen Leistungen F. Grashofs. *Beiträge zur Geschichte der Technik und Industrie*, vol. 15, pp. 1–12.
Lorenz, W., 1990. Die Entwicklung des Dreigelenksystems im 19. Jahrhundert. *Stahlbau*, vol. 59, No. 1, pp. 1–10.
Lorenz, W., 1997. 200 Jahre eisernes Berlin. *Stahlbau*, vol. 66, No. 6, pp. 291–310.
Lorenz, W., 2005. Archäologie des Konstruierens. In: Bundesingenieurkammer (ed.), *Ingenieurbaukunst in Deutschland 2005/2006*, pp. 172–181. Hamburg: Junius Verlag.

Loria, G., 1904. Luigi Cremona et son œuvre mathématique. *Bibliotheca mathematica.* pp. 125–195.

Love, A. E. H., 1888. The small free vibrations and deformation of a thin elastic shell. *Philosophical Transactions,* A 179, pp. 491–546.

Love, A. E. H., 1892/1893. *A treatise on the mathematical theory of elasticity,* 2 vols. Cambridge: Cambridge University Press.

Love, A. E. H., 1906. *A treatise on the mathematical theory of elasticity.* 2nd ed. Cambridge: Cambridge University Press.

Love, A. E. H., 1907. *Lehrbuch der Elastizität.* Trans. from English by A. Timpe. Leipzig: B. G. Teubner.

Love, A. E. H., 1911. *Some problems of geodynamics.* Cambridge: Cambridge University Press.

Lurie, A. I., 1947. *Statika tonkostennykh uprugikh obolochek* (statics of thin-walled elastic shells). Leningrad/Moscow: Gostekhizdat (in Russian).

Lurie, A. I., 1955. *Prostranstvennye zadachi teorii ugrugosti* (three-dimensional problems of the theory of elasticity). Moscow: Gostekhizdat (in Russian).

Lurie, A. I., 1959. *Statics of thin-walled elastic shells.* Trans. AEC-tr–3798. International Atomic Energy Agency.

Lurie, A. I., 1961. *Analiticheskaya mekhanika* (analytical mechanics). Moscow: Fizmatgiz (in Russian).

Lurie, A. I., 1963. *Räumliche Probleme der Elastizitätstheorie.* Translation from the Russian by H. Göldner, G. Landgraf & H. Kirchhübel. Berlin: Akademie-Verlag.

Lurie, A. I., 1964. *Three-dimensional problems of the theory of elasticity.* New York: Interscience Publishers.

Lurie, A. I., 1970. *Teoriya uprugosti* (theory of elasticity). Moscow: Nauka (in Russian).

Lurie, A. I., 1980. *Nelineinaya teoriya uprugosti* (non-linear theory of elasticity). Moscow: Nauka (in Russian).

Lurie, A. I., 1990. *Nonlinear theory of elasticity.* Amsterdam: North-Holland.

Lurie, A. I., 2002. *Analytical Mechanics.* Berlin/New York: Springer.

Lurie, A. I., 2005. *Theory of elasticity.* Berlin/New York: Springer.

Lurie, K. A., 2001. A. I. Lurie: Rannie gody (A. I. Lurie: early years). *Nauchno-tekhnicheskie vedomosti SPbGTU* (Sci.-techn. Transactions, St. Petersburg Technical University), No. 4, pp. 169–184 (in Russian).

Mach, E., 1912. *Die Mechanik in ihrer Entwicklung.* 7th, rev. ed. Leipzig: F. A. Brockhaus.

McHenry, D., 1943. A Lattice Analogy for the Solution of Plane Stress Problems. *Journal of the Institution of Civil Engineers,* vol. 21, pp. 59–82.

McKeon, R. M., 1981. Navier, Claude-Louis-Marie-Henri. In: *Dictionary of Scientific Biography,* vol. X. C. C. Gillispie (ed.), pp. 1–5. New York: Charles Scribner's Sons.

Mader, F. W., 1957. Die Berücksichtigung der Diskontinuität bei der Berechnung orthotroper Platten. *Der Stahlbau,* vol. 26, No. 10, pp. 283–289.

Magnel, G., 1933. *Calcul pratique des poutres Vierendeel.*

Magnel, G., 1948/1. *Stabilité des Constructions.* vols. I & II, 3rd ed., vols. III & IV, 2nd ed.

Magnel, G., 1948/2. *Prestressed concrete.* 1st ed.

Magnel, G., 1948/3. *Le béton précontraint.* Ghent: Editions Fecheyr.

Magnel, G., 1949. *La pratique du béton armé.* Parts I & II, 5th ed.

Mahan, D. H., 1837. *An elementary course of civil engineering.* New York: Wiley.

Mahan, D. H., 1865. *A treatise on fortification drawing and stereotomy.* New York: Wiley.

Mahan, D. H., 1870. *Descriptve geometry, as applied to the drawing of fortification and stereotomy.* New York: Wiley.

Mahan, D. H., Wood, D. V., Mahan, F. A., 1878. *A treatise on civil engineering.* New York: Wiley.

Maier-Leibnitz, H., 1928. Beitrag zur Frage der tatsächlichen Tragfähigkeit einfacher und durchlaufender Balkenträger aus Baustahl St 37 und Holz. *Die Bautechnik,* vol. 6, No. 1, pp. 11–14 & No. 2, pp. 27–31.

Maier-Leibnitz, H., 1929. Versuche mit eingespannten Balken von I-Form aus Baustahl St 37. *Die Bautechnik,* vol. 7, No. 20, pp. 313–318.

Maier-Leibnitz, H., 1936. Versuche zur weiteren Klärung der Frage der tatsächlichen Tragfähigkeit durchlaufender Träger aus Baustahl St 37. *Der Stahlbau,* vol. 9, No. 20, pp. 153–160.

Maier-Leibnitz, H., 1938. Die Beziehungen $M_{st}(P)$ und $M_F(P)$ beim durchlaufenden Balken mit drei Öffnungen, belastet durch $P$ im Mittelfeld. In: *International Association for Bridge & Structural Engineering (ed.). 2nd Congress, Berlin/Munich, 1–11 October 1936. Final Report,* pp. 70–73. Berlin: Verlag von Wilhelm Ernst & Sohn.

Maier-Leibnitz, H., 1941. Zusammenwirken von I-Trägern mit Eisenbetondecken. *Die Bautechnik,* vol. 19, No. 25, pp. 265–270.

Maillart, R., 1921/1. Zur Frage der Biegung. *Schweizerische Bauzeitung,* vol. 77, No. 18, pp. 195–197.

Maillart, R., 1921/2. Bemerkungen zur Frage der Biegung. *Schweizerische Bauzeitung,* vol. 78, No. 2, pp. 18–19.

Maillart, R., 1922. Ueber Drehung und Biegung. *Schweizerische Bauzeitung,* vol. 79, No. 20, pp. 254–257.

Maillart, R., 1924/1. Der Schubmittelpunkt. *Schweizerische Bauzeitung,* vol. 83, No. 10, pp. 109–111.

Maillart, R., 1924/2. Zur Frage des Schubmittelpunktes. *Schweizerische Bauzeitung,* vol. 83, No. 15, pp. 176–177.

Maillart, R., 1924/3. Zur Frage des Schubmittelpunktes. *Schweizerische Bauzeitung,* vol. 83, No. 22, pp. 261–262.

Maillart, R., 1926. Zur Entwicklung der unterzugslosen Decke in der Schweiz und in Amerika. *Schweizerische Bauzeitung,* vol. 87, No. 21, pp. 263–266.

Maillart, R., 1936. Einige neuere Eisenbetonbrücken. *Schweizerische Bauzeitung,* vol. 107, No. 15, pp. 157–162.

Mainstone, R. J., 1968. Structural Theory and Design before 1742. *Architectural Review,* vol. 143, pp. 303–310.

Mainstone, R. J., 2003. Saving the dome of St Peter's. *Construction History,* vol. 19, pp. 3–18.

Mairle, L., 1933/1935. *Die Entwicklung der Berechnung statisch unbestimmter massiver Bogen und Gewölbe unter dem Einfluß der Elastizitätslehre.* Braunschweig: Dissertation, Technical University of Carolo-Wilhelmina zu Braunschweig.

Makowski, Z. S., 1963. *Räumliche Stabwerke aus Stahl.* Düsseldorf: Verlag Stahleisen.

Malberg, A., 1857/1859. Historisch-kritische Bemerkungen über Kettenbrücken. *Zeitschrift für Bauwesen,* vol. 7, col. 226–238 & 560–574, vol. 9, col. 397–412 & 547–570.

Malinin, N. N., 2000. *Kto jest' kto v soprotivlenii materialov* (who's who in strength of materials). Moscow: N.-E. Baumann.

Mañas Martínez, J., 1983. *Eduardo Saavedra, Ingeniero y Humanista.* Madrid: Colegio de Ingenieros de Caminos, Canales y Puertos/Ediciones Turner.

Manderla, H., 1880. Die Berechnung der Sekundärspannungen, welche im einfachen Fachwerk in Folge starrer Knotenverbindungen auftreten. *Allgemeine Bauzeitung,* vol. 45, pp. 27–43.

Manegold, K. H., 1970. *Universität, Technische Hochschule und Industrie.* Berlin: Duncker & Humblodt.

Mang, H., Hofstetter, G., 2004. *Festigkeitslehre.* Mit einem Beitrag von Josef Eberhardsteiner. 2nd ed. Vienna/New York: Springer-Verlag.

Mann, L., 1909. Statische Berechnung steifer Vierecksnetze. *Zeitschrift für Bauwesen,* vol. 59, p. 539.

Mann, L., 1911. Über zyklische Symmetrie in der Statik mit Anwendungen auf das räumliche Fachwerk. *Der Eisenbau,* vol. 2, No. 1, pp. 18–27.

Mann, L., 1927. *Theorie der Rahmenwerke auf neuer Grundlage.* Berlin: Verlag von Julius Springer.

Mann, L., 1939. Grundlagen zu einer Theorie räumlicher Rahmentragwerke. *Der Stahlbau,* vol. 12, No. 19/20, pp. 145–149 & No. 21/22, pp. 153–158.

Mann, L., 1958. Vergleich der Prinzipien und Begriffe für die Entwicklung der Kraft- und Deformationsmethoden in der Statik. *Bauplanung-Bautechnik,* vol. 12, No. 1, pp. 12–15.

**M**anzanares, G. L., 2000. La forma ideal de las cúpulas: el ensayo de Bouguer. In: *Actas del III Congreso Nacional de Historia de la Construcción, Sevilla 26 – 28 octubre de 2000,* pp. 603 – 613. Madrid: Instituto Juan de Herrera.

Manzanares, G. L., 2005. Boscovich's contribution to the theory of domes: the expertise for the Tiburio of Milan cathedral. In: *Essays in the history of theory of structures,* ed. S. Huerta, pp. 273 – 301. Madrid: Instituto Juan de Herrera.

**M**aquoi, R. et al., 1984. Anniversary volume, *Verba volant, scripta manent.* Liège: Imprimerie Cerès.

Maquoi, R., 1996. Charles Massonnet gestorben. *Stahlbau,* vol. 65, No. 6, pp. 229 – 230.

**M**arcus, H., 1912. Beitrag zur Theorie der Rippenkuppel. *Der Eisenbau,* vol. 3, No. 11, pp. 387 – 397.

Marcus, H., 1919. Die Theorie elastischer Gewebe und ihre Anwendung auf die Berechnung elastischer Platten. *Armierter Beton,* vol. 12, pp. 107 – 112, 129 – 135, 164 – 170, 181 – 190, 219 – 229, 245 – 250 & 281 – 289.

Marcus, H., 1924. *Die Theorie elastischer Gewebe und ihre Anwendung auf biegsame Platten.* Berlin: Julius Springer.

Marcus, H., 1925. *Die vereinfachte Berechnung biegsamer Platten.* Berlin: Julius Springer.

**M**arguerre, K., Trefftz, E., 1937. Über die Tragfähigkeit eines längsbelasteten Plattenstreifens nach Überschreiten der Beullast. In: *Jahrbuch 1937 der Deutschen Versuchsanstalt für Luftfahrt e.V.,* Berlin-Adlershof, pp. 212 – 220.

Marguerre, K., 1938/1. Über die Behandlung von Stabilitätsproblemen mit Hilfe der energetischen Methode. *Zeitschrift für angewandte Mathematik und Mechanik,* vol. 18, No. 1, pp. 57 – 73.

Marguerre, K., 1938/2. Zur Theorie der gekrümmten Platte großer Formänderung. In: *Proceedings of the Fifth International Congress for Applied Mechanics,* pp. 93 – 101.

Marguerre, K., 1938/3. Über die Anwendung der energetischen Methode auf Stabilitätsprobleme. In: *Jahrbuch 1938 der Deutschen Versuchsanstalt für Luftfahrt e.V.,* Berlin-Adlershof, pp. 252 – 262.

Marguerre, K., 1940. Torsion von Voll- und Hohlquerschnitten. *Der Bauingenieur,* vol. 21, No. 41/42, pp. 317 – 322.

Marguerre, K. (ed.), 1950. *Neuere Festigkeitsprobleme des Ingenieurs.* Berlin: Springer-Verlag.

**M**ark, R., 1982. *Experiments in Gothic Structure.* Cambridge (Mass.): The MIT Press.

Marković, Ž., 1970. Bošković, Rudjer J. In: *Dictionary of Scientific Biography,* vol. II. C. C. Gillispie (ed.), pp. 326 – 332. New York: Charles Scribner's Sons.

**M**arks, R. W., 1960. *The Dymaxion World of Buckminster Fuller.* New York: Reinhold Publishing Corporation.

**M**arti, P., 1985. Basic Tools of Reinforced Concrete Beam Design. *ACI Structural Journal,* vol. 82, No. 1, pp. 46 – 56.

Marti, P., 1986. Staggered Shear Design of Simply Supported Concrete Beams. *ACI Structural Journal,* vol. 83, No. 1, pp. 36 – 42

Marti, P. (ed.), 1996. *Robert Maillart – Beton-Virtuose.* Gesellschaft für Ingenieurbaukunst, vol. 1. Zurich: vdf Hochschulverlag AG.

**M**atheson, J. A. L., 1959. *Hyperstatic Structures. An Introduction to the Theory of Statically Indeterminate Structures. Volume I* (with chapters by N. W. Murray & R. K. Livesley). London: Butterworth's Scientific.

Matheson, J. A. L., Francis, A. J., 1960. *Hyperstatic Structures. An Introduction to the Theory of Statically Indeterminate Structures. Volume II* (with chapters by N. W. Murray & R. K. Livesley). London: Butterworth's Scientific.

**M**autner, K. W., 1936. Spannbeton nach dem Freyssinet-Verfahren. *Beton und Eisen,* vol. 35, No. 10, pp. 320 – 324.

**M**arx, K., 1968. Ökonomisch-philosophische Manuskripte. In: *Karl Marx-Friedrich Engels-Werke, Ergänzungsband, Zweiter Teil, Bd. 40.* Berlin: Dietz.

Marx, K., 1979. Das Kapital. Kritik der politischen Ökonomie, Erster Band. In: *Karl Marx-Friedrich Engels-Werke, Bd. 23.* Berlin: Dietz.

Marx, K., 1981. Grundrisse der Kritik der politischen Ökonomie. In: *Marx-Engels-Gesamtausgabe, Abt. II/Bd. 1, T. 2.* Berlin: Dietz.

**M**ascheroni, L., 1785. *Nuove ricerche sull'equilibrio delle volte.* Bergamo.

Mascheroni, L., 1829. *Nuove ricerche sull'equilibrio delle volte.* Milan: Silvestri.

**M**assonnet, C., Save, M., 1961. *Calcul plastique des constructions. vol. 1: Structures planes.* Brussels/New York: CBLIA and Blaisdell Publ. Co.

Massonnet, C., 1963. Kritische Betrachtungen zum Traglastverfahren. *VDI-Z,* vol. 105, No. 23, pp. 1057 – 1063.

Massonnet, C., Save, M., 1963. *Calcul plastique des constructions. vol. 2: Structures spatiales.* Brussels: CBLIA.

Massonnet, C., Bares, R., 1966. *Le calcul des grillages de poutres et dalles orthotropes selon la méthode Guyon-Massonnet-Bares.* Paris: Dunod.

Massonnet, C., 1976. Die europäischen Empfehlungen (EKS) für die plastische Bemessung von Stahltragwerken. *Acier-Stahl-Steel,* vol. 32, pp. 146 – 156.

**M**atschoss, C., 1919. *Ein Jahrhundert deutscher Maschinenbau. Von der mechanischen Werkstätte bis zur deutschen Maschinenfabrik 1819 – 1919.* Berlin: Julius Springer.

**M**atviishin, Y. A., 1989. *Novye dannye k biografii B. G. Galerkina i ego pervye raboty po mekhanike* (new details on the biography of B. G. Galerkin and his first works in mechanics). In: *Differentsial'nye uravneniya s parametrom* (differential equations with one parameter). Kiev: Institute of Mathematics, pp. 68 – 79 (in Russian).

**M**auersberger, K., 1983. Studienmaterial zur Vorlesung "Geschichte der Technikwissenschaften/Maschinenwesen". Teil 1: Von den Anfängen bis zur Herausbildungsperiode (1850). *Dresdener Beiträge zur Geschichte der Technikwissenschaften,* No. 7.

Mauersberger, K., 2003. Das wissenschaftliche Maschinenwesen im Spannungsfeld methodischer Auseinandersetzungen. In: *Wissenschaft und Technik. Studien zur Geschichte der TU Dresden,* ed. Thomas Hänseroth, pp. 37 – 66. Cologne/Wiemar/Vienna: Böhlau Verlag.

**M**aurer, B., 1998. *Karl Culmann und die graphische Statik.* Berlin/Diepholz/Stuttgart: Verlag für Geschichte der Naturwissenschaft und der Technik.

**M**axwell, J. C., 1864/1. On reciprocal figures and diagrams of forces. *Philosophical Magazine,* vol. 27, pp. 250 – 261.

Maxwell, J. C., 1864/2. On the calculation of the equilibrium and stiffness of frames. *Philosophical Magazine,* vol. 27, pp. 294 – 299.

Maxwell, J. C., 1867. On the application of the theory of reciprocal polar figures to the construction of diagrams of forces. *Engineering,* vol. 24, p. 402.

Maxwell, J. C., 1870. On reciprocal figures, frames, and diagrams of forces. *Transactions of the Royal Society of Edinburgh,* vol. 26, pp. 1 – 47.

Maxwell, J. C., 1876. On Bow's method of drawing diagrams in graphical statics with illustrations from Peaucellier's linkage. *Proceedings of the Cambridge Philosophical Society,* vol. 2, pp. 407 – 414.

Maxwell, J. C., 1890/1965. *Scientific papers.* Ed. W. Niven, 2 vols. Cambridge. Reprint: New York, Dover.

**M**ayer, M., 1912. Die trägerlose Eisenbetondecke. *Deutsche Bauzeitung, Zementbeilage,* vol. 46, No. 21, pp. 162 – 166.

Mayer, R., 1951. †Dr.-Ing. Arno Schleusner. *Der Stahlbau,* vol. 20, No. 12, p. 152.

Mayor, B., 1910. *Statique graphique des systèmes de l'espace.* Lausanne: F. Rouge & Cie, Librairie und Paris: Librairie Gauthier-Villars.

Mayor, B., 1926. *Introduction à la statique graphique des systèmes de l'espace.* Lausanne: Librairie Payot & Cie.

**M**echtler, P., 1964. Franz Joseph Ritter von Gerstner. In: *Neue Deutsche Biographie, 6. Bd.,* pp. 328 – 329. Berlin: Duncker & Humblot.

**M**ehlhorn, G., Hoshino, M., 2007. Brückenbau auf dem Weg vom Altertum zum modernen Brückenbau. In: *Handbuch Brücken. Entwerfen, Konstruieren, Berechnen, Bauen und Erhalten,* ed. G. Mehlhorn, pp. 1 – 101. Berlin/Heidelberg: Springer.

**M**ehrtens, G. C., 1885. Fortschritte im Bau von Brückengewölben. *Zentralblatt der Bauverwaltung,* vol. 5, pp. 473 – 475, 490 – 492 & 517 – 519.

Mehrtens, G. C., 1893. Zur Baugeschichte der alten Eisenbahnbrücken bei Dirschau und

Marienburg. *Zeitschrift für Bauwesen,* vol. 43, pp. 97–122.

Mehrtens, G. C., 1900. *Der deutsche Brückenbau im XIX. Jahrhundert. Denkschrift bei Gelegenheit der Weltausstellung 1900 in Paris.* Berlin: Springer.

Mehrtens, G. C., 1903. *Vorlesungen über Statik der Baukonstruktionen und Festigkeitslehre. Erster Band: Einführung in die Grundlagen.* Leipzig: Engelmann.

Mehrtens, G. C., 1904. *Vorlesungen über Statik der Baukonstruktionen und Festigkeitslehre. Zweiter Band: Statisch bestimmte Systeme.* Leipzig: Engelmann.

Mehrtens, G. C., 1905. *Vorlesungen über Statik der Baukonstruktionen und Festigkeitslehre. Dritter Band: Formänderungen und statisch unbestimmte Träger.* Leipzig: Engelmann.

Mehrtens, G. C., Mohr, O., 1907. In eigener Sache. *Zeitschrift für Arch.- & Ing.- Wesen,* pp. 367–372.

Mehrtens, G. C., 1908. Vorlesungen über Ingenieurwissenschaften. Zweiter Teil: Eisenbrückenbau. Leipzig: Engelmann.

Mehrtens, G. C., 1909. Vorlesungen über Ingenieur-Wissenschaften. Erster Teil: Statik und Festigkeitslehre; Erster Band: Einführung in die Grundlagen. 2nd, rev. ed. Leipzig: Wilhelm Engelmann.

Mehrtens, G. C., 1912. *Vorlesungen über Ingenieur-Wissenschaften I. Teil: Statik und Festigkeitslehre, Dritter Band II. Hälfte.* 2nd, rev. ed. Leipzig: Verlag von Wilhelm Engelmann.

Meinicke, K.-P., 1988. Die historische Entwicklung der Ähnlichkeitstheorie. *Dresdener Beiträge zur Geschichte der Technikwissenschaften,* No. 15, pp. 3–54.

Meissner, E., 1913. Das Elastizitätsproblem für dünne Schalen von Ringflächen-, Kugel- und Kegelform. *Physikalische Zeitschrift,* vol. 14, pp. 343–349.

Meister, E., 1906. Prof. Dr. Wilhelm Ritter. *Verhandlungen der schweiz. Naturf. Gesellschaft,* offprint from Obituaries supplement, pp. 1–15. St. Gallen.

Melan, E., 1932. Zur Bestimmung des Sicherheitsgrades einfach statisch unbestimmter Fachwerke. *Zeitschrift für angewandte Mathematik und Mechanik (ZAMM),* vol. 12, No. 3, pp. 129–136.

Melan, E., 1938. Zur Plastizität des räumlichen Kontinuums. *Ingenieur-Archiv,* vol. 9, pp. 116–126.

Melan, E., Schindler, R., 1942. *Die genaue Berechnung von Trägerrosten.* Vienna: Springer.

Melan, E. (ed.), 1948. *Der Brückenbau, Band II.* Vienna: Deuticke.

Melan, E. (ed.), 1950/1. *Der Brückenbau, Band III.* Vienna: Deuticke.

Melan, E., 1950/2. *Einführung in die Baustatik.* Vienna: Springer.

Melan, E., Parkus, H., 1953. *Wärmespannungen infolge stationärer Temperaturfelder.* Vienna: Springer.

Melan, J., 1884. Beitrag zur Berechnung statisch unbestimmter Stabsysteme. *Zeitschrift des Österreichischen Ingenieur- und Architekten-Vereines,* vol. 36, pp. 100–108.

Melan, J., 1888/1. Prof. Dr. Emil Winkler. *Zeitschrift des Österreichischen Ingenieur- und Architekten-Vereines,* vol. 40, pp. 186–187.

Melan, J., 1888/2. Theorie der eisernen Bogenbrücken und der Hängebrücken. In: *Der Brückenbau. Handbuch der Ingenieurwissenschaften II. Band. Vierte Abteilung. Eiserne Bogenbrücken und Hängebrücken,* ed. J. Melan & T. Schäffer, ed. T. Schäffer & E. Sonne, 2nd, rev. ed., pp. 1–144. Leipzig: Engelmann.

Melan, J., 1890. Zur rechnungsmäßigen Ermittelung der Biegespannungen in Beton- und Monier-Constructionen. *Wochenschrift des österreichischen Ingenieur- und Architekten-Vereins,* vol. 15, pp. 223–226.

Melan, J., 1908. Theorie des Gewölbes und des Eisenbetongewölbes im besonderen. In: *Handbuch für Eisenbetonbau. Erster Band: Entwicklungsgeschichte und Theorie des Eisenbetons,* ed. F. v. Emperger, pp. 387–449. Berlin: Wilhelm Ernst & Sohn.

Melan, J., 1910, 1911, 1917. Der Brückenbau. Nach Vorträgen, gehalten an der deutschen Technischen Hochschule in Prague. Leipzig/Vienna: Deuticke.

Melan, J., 1913. *Theory of Arches and Suspension Bridges.* English trans. by Prof. Dr. Steinman. Chicago: The Myron Clark Publ. Co.

Melan, J., 1915. *Plain and reinforced concrete arches.* English trans. by Prof. Dr. Steinman. New York: Wiley & Sons.

Meleshko, V. V., 2003. Selected topics in the history of the two-dimensional biharmonic problem. *Applied Mechanics Reviews,* vol. 56, No. 1, pp. 33–85.

Melosh, R. J., 1961. A stiffness matrix for the analysis of thin plates in bending. *Journal of the Aeronautical Sciences,* vol. 28, pp. 34–40.

Melosh, R. J., 1963. Bases for the derivation of matrices for the direct stiffness method. *AIAA Journal,* vol. 1, pp. 1631–1637.

Menabrea, L. F., 1858. Nouveau principe sur la distribution des tensions dans les systèmes élastiques. *Comptes Rendus,* 46, pp. 1056–1060.

Menabrea, L. F., 1870. Sul principio di elasticità, delucidazioni di L. F. M. *Atti della Reale Accademia delle scienze di Torino,* p. 687.

Menabrea, L. F., 1875. La determinazione delle tensioni e delle pressioni nei sistemi elastici. *Atti della reale Accademia dei Lincei,* series 2a, 2, pp. 201–221.

Mengeringhausen, M., 1928. *Die Entwicklung der Schienenfabrikation in Deutschland.* Dissertation, Munich TH.

Mengeringhausen, M., 1942. *Die MERO-Bauweise.* Berlin: Self-pub.

Mengeringhausen, M., 1943. Verbindung von Rohrstäben und knotenbildenden Verbindungsstücken, insbesondere für zerlegbare Fachwerkkonstruktionen. German Patent of 12 March 1943 & FRG Patent of 12 March 1953 (No. 874657).

Mengeringhausen, M., 1944. *Die MERO-Bauweise.* 2nd ed. Berlin: Self-pub.

Mengeringhausen, M., 1962. *Komposition im Raum. Einführung in die Konstruktion und Anwendung von Raum-Fachwerken für das Bauwesen.* Würzburg: MERO Rohrkonstruktionen und Geräte (Self-pub.).

Mengeringhausen, M., 1967. Kompositionslehre räumlicher Stab-Fachwerke. In: *Proceedings of the International Conference on Space Structures,* ed. R. M. Davies, Paper R 7. Oxford/Edinburgh: Blackwell Scientific Publications.

Mengeringhausen, M., 1975. *Komposition im Raum. Raumfachwerke aus Stäben und Knoten.* Wiesbaden: Bauverlag.

Mengeringhausen, M., 1978. Lecture of 10 March 1978 at VDI, Schweinfurt. Typewritten ms.

Mengeringhausen, M., 1979. Der Weg zu einer neuen Generation von Raumfachwerken. *Bauwelt,* No. 38, pp. 1627–1634.

Mengeringhausen, M., 1983. *Komposition im Raum. Die Kunst individueller Baugestaltung mit Serienelementen.* Gütersloh: Bertelsmann-Fachzeitschriften.

Merriman, M., 1876. *On the theory and calculation of continous bridges.* New York: D. Van Nostrand.

Merriman, M., 1877. *Elements of the Method of least squares.* London: Macmillan.

Merriman, M., 1885 *A text-book on the mechanics of materials and of beams, columns, and shafts.* New York: Wiley.

Merriman, M., 1888. *A text-book on roofs and bridges. Part I. Stresses in simple trusses.* New York: Wiley.

Merriman, M., 1890. *A text-book on roofs and bridges. Part II. Graphic Statics.* New York: Wiley.

Merriman, M., 1892. *A text-book on retaining walls and masonry dams.* New York: Wiley.

Merriman, M., 1894. *A text-book on roofs and bridges. Part III. Bridge Design.* New York: Wiley.

Merriman, M., 1897. *Strength of materials. A text-book for manual training schools.* New York: Wiley.

Merriman, M., 1898. *A text-book on roofs and bridges. Part IV. Higher Structures.* New York: Wiley.

Merriman, M., 1903. The principle of least work in mechanics and its use in investigations regard-ing the ether of space. *Proceedings of the American Philospical Society,* vol. 42, No. 173, pp. 162–165.

Merriman, M., 1905. *Elements of mechanics. Forty lessons for beginners in engineering.* New York: Wiley.

Merriman, M. (ed.), 1911. *The American Civil Engineer's Pocket-Book.* New York: John Wiley & Sons.

Méry, E., 1840. Mémoire sur l'equilibre des voûtes en berceau. *Annales des Ponts et Chaussées*, vol. 19, pp. 50 – 70, pp. 133 – 134.

Meyer, 1980. Technik. In: H. Göschel (ed.): *Meyers Universal-Lexikon, Band 4*, p. 275. Leipzig: VEB Bibliographisches Institut.

Meyer, A. G., 1997. *Eisenbauten. Ihre Geschichte und Ästhetik*. With an epilogue by T. J. Heinisch. Berlin: Gebr. Mann.

Michaux, B., 1990. Epistemologie. In: *Europäische Enzyklopädie zu Philosophie und Wissenschaften, Bd. 1*, ed. H. J. Sandkühler, pp. 757 – 761. Hamburg: Felix Meiner Verlag.

Michell, J. H., 1899. On the direct determination of stress in an elastic solid, with application to the theory of plates. In: *Proc. of the London Mathematical Society*, vol. 31, pp. 100 – 124.

Michl, T., 1999. Gebrauchstauglichkeitsuntersuchungen beim Millennium Tower. *Stahlbau*, vol. 68, No. 8, pp. 631 – 640.

Michon, F., 1848. Tables et formules pratiques pour l'établissement des voûtes cylindriques. *Mémorial de l'Officier du Génie*, vol. 15, pp. 7-117.

Mikulinskij, S. R., Jaroševskij, M. G., Kröber, G., Steiner, H. (ed.), 1977. *Wissenschaftliche Schulen, Band 1*. Berlin: Akademie-Verlag.

Mikulinskij, S. R., Jaroševskij, M. G., Kröber, G., Steiner, H., Winkler, R.-L., Altner, P. (ed.), 1979. *Wissenschaftliche Schulen, Band 2*. Berlin: Akademie-Verlag.

Milankovitch, C., 1907. Theorie der Druckkurven. *Zeitschrift für Mathematik und Physik*, vol. 55, pp. 120 – 128.

Mindlin, R. D., 1936/1. Force at a point in the interior of a semi-infinite solid. *Physics*, vol. 7, pp. 195 – 202.

Mindlin, R. D., 1936/2. Note on the Galerkin and Papkovitch stress functions. *Bulletin of the American Mathematical Society*, vol. 42, pp. 373 – 376.

Mindlin, R. D., 1939. Stress distribution around a tunnel. *Proceedings of the American Society of Civil Engineers*, vol. 65, pp. 619 – 642.

Mindlin, R. D., 1951. Influence of rotatory inertia and shear on flexural motions of isotropic, elastic plates. *Journal of Applied Mechanics*, vol. 18, pp. 31 – 38.

Mindlin, R. D., 1955. *An introduction to the mathematical theory of vibrations of elastic plates*. New Jersey: US Army Signal Corps Eng. Lab., Fort Monmouth, New Jersey.

Mindlin, R. D., Tiersten, H. F.,1962. Effects of couple-stresses in linear elasticity. *Archive for Rational Mechanics and Analysis*, vol. 11, pp. 415 – 448.

Mindlin, R. D., 1980. On Reissner's equations for sandwich plates. In: *Mechanics Today*, vol. 5 (E. Reissner 65th Birthday Volume), ed. S. Nemat-Nasser, pp. 315 – 328. Oxford: Pergamon Press.

Mindlin, R. D., 1986. Flexural vibrations of rectangular plates with free edges. *Mechanics Research Communications*, vol. 13, pp. 349 – 357.

Mindlin, R. D., Yang, J., 2007. *An introduction to the mathematical theory of vibrations of elastic plates*. Singapur: World Scientific Publishing Co. Pte. Ltd.

Minnaert, M. G. J., 1981. Stevin, Simon. In: *Dictionary of Scientific Biography*, vol. 13. C. C. Gillispie (ed.), pp. 47 – 51. New York: Charles Scribner's Sons.

Minten, J., Sedlacek, G., Paschen, M., Feldmann, M., Geßler, A., 2007. SPS – ein neues Verfahren zur Instandsetzung und Ertüchtigung von stählernen orthotropen Fahrbahnplatten. *Stahlbau*, vol. 76, No. 7, pp. 438 – 454.

Mise, K., 1922. Elastic Distortions of Framed Structures. In: *Memoirs of the College of Engineering, Kyushu Imperial University*, pp. 175 – 212. (1961: application by M. Naruoka in an IABSE publication).

Mise, K., 1927, Elastic Distortions of Rigidly Connected Frames. In: *Memoirs of the College of Engineering, Kyushu Imperial University*. Fukuoka.

Mise, K., 1929. General Solution of Secondary Stresses, Paper No. 763, *World Engineering Congress*, Tokyo, Japan.

Mise, K., 1946. Universal Solution of Framed Structures. In: *Reports of the Research Institute for Elasticity Engineering, Kyushu Imperial University*. Fukuoka, Commemoration No. Abstract, pp. 1 – 2.

Mises, R. v., 1913. Mechanik der festen Körper im plastischen deformablen Zustand. *Nachrichten d. Kgl. Ges. d. Wiss. zu Göttingen. Math.-phys. Klasse*, pp. 582 – 592.

Mises, R. v., 1917. Graphische Statik räumlicher Kräftesysteme. *Zeitschrift für Mathematik und Physik*, vol. 64, pp. 209 – 232.

Mises, R. v., 1924/1. Motorrechnung, ein neues Hilfmittel der Mechanik. *Zeitschrift für angewandte Mathematik und Mechanik*, vol. 4, No. 2, pp. 155 – 181.

Mises, R. v., 1924/2. Anwendungen der Motorrechnung. *Zeitschrift für angewandte Mathematik und Mechanik*, vol. 4, No. 4, pp. 193 – 213.

Mitinsky, A. N., 1957. *F. S. Yasinsky: Ocherk zhizni i nauchno-inzhenernoi deyatel'nosti* (F. S. Yasinsky: portrait and scientific and technical work). Moscow: Gostekhizdat (in Russian).

Mizuno, T., 1983. Biography of Prof. Hayashi. *JSCE*, vol. 68, Aug., pp. 38 (in Japanese).

Möbius, A. F., 1837. *Lehrbuch der Statik*. Leipzig: G. J. Göschen'sche Verlagsbuchhandlung.

Mohr, O., 1860. Beitrag zur Theorie der Holz- und Eisen-Constructionen. *Zeitschrift des Architekten- und Ingenieur-Vereins zu Hannover*, vol. 6, No. 2/3, pp. 323 – 346 & No. 4, pp. 407 – 442.

Mohr, O., 1862. Beitrag zur Theorie der Holz- und Eisen-Constructionen. *Zeitschrift des Architekten- und Ingenieur-Vereins zu Hannover*, vol. 8, No. 3/4, pp. 245 – 280.

Mohr, O., 1868. Beitrag zur Theorie der Holz- und Eisen-Constructionen. *Zeitschrift des Architekten- und Ingenieur-Vereins zu Hannover*, vol. 14, No. 1, pp. 19 – 51.

Mohr, O., 1874/1. Beitrag zur Theorie der Bogenfachwerksträger. *Zeitschrift des Arch.- & Ing.-Vereins zu Hannover*, vol. 20, pp. 223 – 238.

Mohr, O., 1874/2. Beitrag zur Theorie des Fachwerks. *Zeitschrift des Architekten- und Ingenieur-Vereins zu Hannover*, vol. 20, pp. 509 – 526.

Mohr, O., 1875. Beiträge zur Theorie des Fachwerks. *Zeitschrift des Architekten- und Ingenieur-Vereins zu Hannover*, vol. 21, pp. 17 – 38.

Mohr, O., 1876. Über die Zusammensetzung der Kräfte im Raume. *Zivilingenieur*, 30 (new series, vol. 22), col. 121 – 130.

Mohr, O., 1885. Beitrag zur Theorie des Fachwerkes. *Zivilingenieur*, vol. 31, pp. 289 – 310.

Mohr, O., 1886. Über die Elasticität der Deformationsarbeit. *Zivilingenieur*, vol. 32, pp. 395 – 400.

Mohr, O., 1892. Die Berechnung der Fachwerke mit starren Knotenverbindungen. *Zivilingenieur*, vol. 38, pp. 577 – 594.

Mohr, O., 1893. Die Berechnung der Fachwerke mit starren Knotenverbindungen. *Zivilingenieur*, vol. 39, pp. 67 – 78.

Mohr, O., 1902/1. Beitrag zur Theorie des Raumfachwerkes. *Zentralblatt der Bauverwaltung*, vol. 22, No. 34, pp. 205 – 208.

Mohr, O., 1902/2. Zur Berechnung der Raumfachwerke. *Zentralblatt der Bauverwaltung*, vol. 22, No. 102, pp. 634 – 636.

Mohr, O., 1903. Zur Berechnung der Raumfachwerke. *Zentralblatt der Bauverwaltung*, vol., 23 No. 38, pp. 237 – 239, No. 64, pp. 402 – 403 & No. 102, pp. 641 – 642.

Mohr, O., 1906. *Abhandlungen aus dem Gebiete der Technischen Mechanik*. Berlin: Wilhelm Ernst & Sohn.

Mohr, O., 1912. Die Berechnung der Pfostenträger (Vierendeelträger). *Der Eisenbau*, vol. 3, No. 3, pp. 85 – 96.

Molinos, L., Pronnier, C., 1857. *Traité théoretique et pratique de la construction des ponts métalliques*. Paris: Editeur Inconnu.

Möller, B., Graf, W., 2002. Kurt Beyer (1881 – 1952) – Erinnerungen an einen bedeutenden Statiker und Bauingenieur. *Bautechnik*, vol. 79, No. 5, pp. 335 – 339.

Monasterio, J., o. J. *Nueva teorica sobre el empuje de bovedas*. Ms. Biblioteca de la Escuela Técnica Superior de Ingenieros de Caminos, Canales y Puertos.

Monge, G., 1794/95. *Géometrié descriptive*. Paris: Baudouin.

Mörsch, E., 1901. Theorie der Betoneisenkonstruktionen. *Süddeutsche Bauzeitung*, No. 47, pp. 1 – 20.

Mörsch, E., 1906/1. Berechnung von eingespannten Gewölben. *Schweizerische Bauzeitung*, vol. 47, pp. 83 – 85 & 89 – 91.

Mörsch, E., 1906/2. *Der Eisenbetonbau. Seine Theorie und Anwendung*. 2nd, rev. ed., Stuttgart: Wittwer.

Mörsch, E., 1907. Versuche über die Schubwirkungen bei Eisenbetonträgern. *Deutsche Bauzeitung*, vol. 41, Mitteilungen über Zement, Beton- und Eisenbetonbau, vol. 4, No. 30, pp. 207–212, No. 32, pp. 223–228 & No. 35, pp. 241–243.

Mörsch, E., 1908. *Der Eisenbetonbau. Seine Theorie und Anwendung.* 3rd, rev. ed., Stuttgart: Wittwer.

Mörsch, E., 1920. *Der Eisenbetonbau. Seine Theorie und Anwendung.* vol. I, 1st half-vol., 5th, rev. ed., Stuttgart: Wittwer.

Mörsch, E., 1922. *Der Eisenbetonbau. Seine Theorie und Anwendung.* vol. I, 2nd half-vol., 5th, rev. ed., Stuttgart: Wittwer.

Mörsch, E., 1943. *Der Spannbetonträger. Seine Herstellung, Berechnung und Anwendung.* Stuttgart: Wittwer.

Mörsch, E., 1947. *Statik der Gewölbe und Rahmen. Teil A.* Stuttgart: Wittwer.

Mörsch, E., 2000. Gewölbte Brücken. In: *Beton-Kalender*, vol. 89, pt. II, pp. 1–53. Berlin: Ernst & Sohn.

Monier, J., 1880. *Verfahren zur Herstellung von Gegenständen verschiedener Art aus einer Verbindung von Metallgerippen mit Zement.* German Patent No. 14673 of 22 Dec 1880.

Morse, E. W., 1981. Young, Thomas. In: *Dictionary of Scientific Biography*, vol. 14. C. C. Gillispie (ed.), pp. 562–572. New York: Charles Scribner's Sons.

Moseley, H., 1833/1. On a new principle in statics, called the principle of least pressure. *The London, Edinburgh and Dublin Philosophical Magazine*, series 3, vol. 3, pp. 285–288.

Moseley, H., 1833/2. On the theory of resistances in statics. *The London, Edinburgh and Dublin Philosophical Magazine*, series 3, vol. 3, pp. 431–436.

Moseley, H., 1835. On the equilibrium of the arch. *Cambridge Philosophical Transactions*, vol. 5, pp. 293–313.

Moseley, H., 1838. On the theory of the equilibrium of bodies in contact. *Cambridge Philosophical Transactions*, vol. 6, pp. 463–491.

Moseley, H., 1843. *The mechanical principles of engineering and architecture.* London: Longman.

Moseley, H., 1856. *The mechanical principles of engineering and architecture.* Ed. by D. H. Mahan. New York: Wiley.

Moser, S., 1958: *Metaphysik einst und jetzt. Kritische Untersuchungen zu Begriff und Ansatz der Ontologie.* Berlin: De Gruyter.

Muess, H., Sauerborn, N., Schmitt, J., 2004. Höhepunkte im modernen Verbundbau – eine beispielhafte Entwicklungsgeschichte. *Stahlbau*, vol. 73, No. 10, pp. 791–800.

Mufti, A. A., Bakth, B., 2002. Fazlur Khan (1929–1982). *Canadian Journal of Civil Engineering*, vol. 29, No. 2, pp. 238–245(8).

Müllenhoff, A., 1922. Versuche über die Verdrehungsfestigkeit der Walzeisen-Träger von A. Föppl. *Der Eisenbau*, vol. 13, No. 12, pp. 269–272.

Müller, J., 1967. Philosophische Probleme der technischen Wissenschaften. In: M. Guntau, H. Wendt (ed.): *Naturforschung und Weltbild. Eine Einführung in Philosophische Probleme der modernen Naturwissenschaften*, 2nd, rev. ed., pp. 341–369. Berlin: VEB Deutscher Verlag der Wissenschaften.

Müller, P., 1978. *Plastische Berechnung von Stahlbetonscheiben und -balken.* Institute of Theory of Structures & Design, Zurich ETH, Report No. 83. Basel: Birkhäuser Verlag.

Müller, R. K., Haas, E., 1971. *Handbuch der Modellstatik.* Berlin: Springer.

Müller, S., 1907. Zur Berechnung mehrfach statisch unbestimmter Tragwerke. *Zentralblatt der Bauverwaltung*, vol. 27, No. 4, pp. 23–27.

Müller, S., 1908. *Technische Hochschulen in Nordamerika.* Leipzig: Verlag von B. G. Teubner.

Müller-Breslau, H., 1883/1. Elasticitätstheorie der Tonnengewölbe. *Zeitschrift für Bauwesen*, vol. 33, pp. 35–52 & 211–228.

Müller-Breslau, H., 1883/2. Über die Anwendung des Princips der Arbeit in der Festigkeitslehre. *Wochenblatt für Architekten und Ingenieure*, pp. 87–89 & 96–98.

Müller-Breslau, H., 1884. Der Satz von der Abgeleiteten der ideellen Formänderungs-Arbeit. *Zeitschrift d. Arch.- & Ing.-Vereins zu Hannover*, vol. 30, pp. 211–214.

Müller-Breslau, H., 1886. *Die neueren Methoden der Festigkeitslehre und der Statik der Baukonstruktionen.* Leipzig: Baumgärtner's Buchhandlung.

Müller-Breslau, H., 1887/1. *Die Graphische Statik der Baukonstruktionen, Band I.* Leipzig: Baumgärtner's Buchhandlung.

Müller-Breslau, H., 1887/2. Beitrag zur Theorie des ebenen Fachwerkes. *Schweizerische Bauzeitung*, vol. 21, pp. 121–123.

Müller-Breslau, H., 1887/3. Beitrag zur Theorie der ebenen Träger. *Schweizerische Bauzeitung*, vol. 21, pp. 129–131.

Müller-Breslau, H., 1891/92. Beitrag zur Theorie des räumlichen Fachwerks. *Zentralblatt der Bauverwaltung*, vol. 11, No. 44, pp. 437–440, vol. 12, No. 19, pp. 201–207, No. 21, pp. 225–227, No. 23, pp. 244–246 & No. 24, pp. 256–259.

Müller-Breslau, H., 1892. *Die graphische Statik der Baukonstruktionen, Band II. Erste Abteilung.* 2nd, rev. ed. Leipzig: Baumgärtner's Buchhandlung.

Müller-Breslau, H., 1893. *Die neueren Methoden der Festigkeitslehre und der Statik der Baukonstruktionen.* 2nd, rev. ed. Leipzig: Baumgärtner's Buchhandlung.

Müller-Breslau, H., 1898. Zur Theorie der Kuppel- und Turmdächer und verwandter Konstruktionen. *Z-VDI*, vol. 42, No. 44, pp. 1205–1213 & No. 45, pp. 1233–1241.

Müller-Breslau, H., 1899/1. Zur Theorie der Kuppel- und Turmdächer. *Zeitschrift des Vereines deutscher Ingenieure*, vol. 43, No. 14, pp. 385–389.

Müller-Breslau, H., 1899/2. Die Berechnung achtseitiger Turmpyramiden. *Zeitschrift des Vereines deutscher Ingenieure*, vol. 43, No. 37, pp. 1126–1134.

Müller-Breslau, H., 1900. Kaisersteg über die Spree bei Oberschöneweide. *Zeitschrift für Bauwesen*, vol. 50, pp. 65–76.

Müller-Breslau, H., 1902. Über räumliche Fachwerke. *Zentralblatt der Bauverwaltung*, vol. 22, No. 8, pp. 49–51, No. 10, pp. 61–63, No. 70, pp. 429–32 & No. 82, pp. 501–503.

Müller-Breslau, H., 1903/1. *Die Graphische Statik der Baukonstruktionen, Band II. Erste Abtheilung.* 3rd, rev. ed. Leipzig: Baumgärtner's Buchhandlung.

Müller-Breslau, H., 1903/2. Bemerkungen zur Berechnung des Raumfachwerks. *Zentralblatt der Bauverwaltung*, vol. 23, No. 10, pp. 65–66.

Müller-Breslau, H., 1903/3. Zur Berechnung räumlicher Fachwerke. *Zentralblatt der Bauverwaltung*, vol. 23, No. 48, pp. 298–300.

Müller-Breslau, H., 1903/4. Zur Berechnung der Raumfachwerke. *Zentralblatt der Bauverwaltung*, vol. 23, No. 82, pp. 509–512, No. 84, pp. 523–524 & No. 102, pp. 642–643.

Müller-Breslau, H., 1904. *Die neueren Methoden der Festigkeitslehre und der Statik der Baukonstruktionen.* 3rd, rev. ed. Leipzig: Baumgärtner's Buchhandlung.

Müller-Breslau, H., 1906. *Erddruck auf Stützmauern.* Stuttgart: Alfred Kröner Verlag.

Müller-Breslau, H., 1908. *Die Graphische Statik der Baukonstruktionen, Band II. Zweite Abtheilung.* Berlin: Alfred Kröner Verlag.

Müller-Breslau, H., 1913. *Die neueren Methoden der Festigkeitslehre und der Statik der Baukonstruktionen.* 4th, rev. ed. Leipzig: Alfred Kröner.

Müller-Seidel, W. (ed.), 1993. *Exzentrische Bahnen. Ein Hölderlin-Brevier.* Munich: Deutscher Taschenbuch Verlag.

Muskhelishvili, N. I., 1933. *Nekotorye osnovnye zadachi matematicheskoi teorii uprugosti* (some basic problems of the mathematical theory of elasticity). Leningrad: AN SSSR (in Russian).

Muskhelishvili, N. I., 1946. *Singulyarnye integral'nye uravneniya* (singular integral equations: boundary problems of function theory and their application to mathematical physics). Moscow/Leningrad: Gostekhizdat (in Russian).

Muskhelishvili, N. I., 1963. *Some basic problems of the mathematical theory of elasticity.* Groningen: P. Noordhoff.

Muskhelishvili, N. I., 1965. *Singuläre Integralgleichungen. Randwertprobleme der Funktionentheorie und Anwendungen auf die mathematische Physik.* Berlin: Akademie-Verlag.

Muskhelishvili, N. I., 1971. *Einige Grundaufgaben zur mathematischen Elastizitätstheorie.* Munich: Hanser.

Muskhelishvili, N. I., 1992. *Singular Integral Equations: Boundary Problems of Function*

Theory and Their Application to Mathematical Physics. New York: Dover.

Musschenbroek, P. v., 1729. *Physicae, experimentales et geometricae.* Utrecht: Samuelem Luchtmans.

Mutermilch, J., Janiczek, J., 1959. *Księga Jubileuszowa.dla uczczenia zasług naukowych profesora doktora inżyniera Witolda Wierzbickiego* (commemorative publication in honour of the scientific work of Prof. Dr.-Ing. Witold Wierzbicki). Warsaw: Polska Akademia Nauk (in Polish).

Mutermilch, J., Olszewski, E., 1964. Jasiński Feliks. In: *Polski Słownik Biograficzny. 11. Band*, pp. 38–39 (in Polish).

Mutermilch, J., 1965. Profesor Witold Wierzbicki. *Inżynieria i Budownictwo*, vol. 22, No. 3. pp. 111–112 (in Polish).

Muttoni, A., Schwartz, J., Thürlimann, B., 1996. *Bemessung von Betontragwerken mit Spannungsfeldern.* Basel: Birkhäuser Verlag.

Nádai, A. L., 1923. Der Beginn des Fließvorganges in einem tordierten Stab. *Zeitschrift für angewandte Mathematik und Mechanik*, vol. 3, No. 6, pp. 442–454.

Nádai, A. L., 1925. *Die elastischen Platten.* Berlin: Julius Springer.

Nádai, A. L., 1968. *Die elastischen Platten.* Berlin: Springer.

Naghdi, P. M., 1957. On the theory of thin elastic shells. *The Quarterly of Applied Mathematics*, vol. 14, pp. 369–380.

Naghdi, P. M., 1963. Foundations of elastic shell theory. In: *Progress in Solid Mechanics*, vol. IV, ed. L. N. Sneddon & R. Hill, pp. 1–90. Amsterdam: North-Holland Publishing.

Naghdi, P. M., 1972. The theory of plates and shells. In: *Handbuch der Physik*, vol. 6a/2, ed. C. A. Truesdell, pp. 425–640. Berlin: Springer.

Naghdi, P. M., 1978. A brief history of the Applied Mechanics Division of ASME. *Journal of Applied Mechanics*, vol. 46, pp. 723–794.

Naghdi, P. M., Spencer, A. J., England, A. H., 1994. *Non-linear elasticity and theoretical mechanics: In honour of A. E. Green.* Oxford: Oxford University Press.

Naghdi, P. M. (author), Casey, J. (ed.), Crochet, M. J. (ed.), 1998. *Theoretical, Experimental and Numerical Contributions to the Mechanics of Fluids and Solids. A Collection of Papers in Honour of Paul M. Naghdi.* Basel: Birkhäuser.

Napolitani, P. D., 1995. Le modèle de la théorie des proportions. In: *Entre Mécanique et Architecture*, ed. P. Radelet-de Grave & E. Benvenuto, pp. 69–86. Basel: Birkhäuser.

Naruoka, M., 1961. *Publ. IABSE.*

Naruoka, M., 1970. *Theory of structures, vol. II, plates*, p. 207. Tokyo: Maruzen (in Japanese).

Naruoka, M., 1974. *Selected chapters on theory of structures.* Tokyo: Maruzen (in Japanese).

Naruoka, M., 1999. *Lectures on exemplary structural analysis.* Osaka, 4 Dec 1995.

Osaka: Kansai-Doro-Kenkyukai (in Japanese).

Nascè, V., 1984. Alberto Castigliano, railway engineer: his life and times. *Meccanica*, vol. 19, Special issue 'Alberto Castigliano 1847–1884', pp. 5–14.

Navier, C. L. M. H., 1820. *Leçons données à l'Ecole Royale des Ponts et Chaussées sur l'Application de la Mécanique.* Paris.

Navier, C. L. M. H., 1823/1. *Rapport et Mémoire sur les Ponts suspendus.* Paris: Carilian-Goeury.

Navier, C. L. M. H., 1823/2. Extrait des recherches sur la flexion des planes élastiques. In: *Bulletin des Sciences de la Société Philomathique de Paris*, pp. 92–102.

Navier, C. L. M. H., 1823/3. Sur les lois de l'équilibre et du mouvement des corps solides élastiques. In: *Bulletin des Sciences de la Société Philomathique de Paris*, pp. 177–181.

Navier, C. L. M. H., 1826. Résumé des Leçons données à l'Ecole Royale des Ponts et Chaussées sur l'Application de la Mécanique à l'Etablissement des Constructions et des Machines. 1$^{er}$ partie: Leçons sur la résistance des materiaux et sur l'établissement des constructions en terre, en maçonnerie et en charpente. Paris: Firmin Didot père et fils.

Navier, C. L. M. H., 1827. Mémoire sur les lois de l'équilibre et du mouvement des corps solides élastiques. In: *Mémoires de l'Institut*, vol. 7, pp. 375–393.

Navier, C. L. M. H., 1833/1838. Résumé des Leçons données à l'Ecole des Ponts et Chaussées sur l'Application de la Mécanique à l'Etablissement des Constructions et des Machines. 2nd ed., vol. 1: Leçons sur la résistance des materiaux et sur l'établissement des constructions en terre, en maçonnerie et en charpente, revues et corrigées. vol. 2: Leçons sur le mouvement et la résistance des fluides, la conduite et la distribution des eaux. vol. 3: Leçons sur l'établissement des machines. Paris: Carilian-Goeury.

Navier, C. L. M. H., 1839. Résumé des Leçons données à l'Ecole des Ponts et Chaussées sur l'Application de la Mécanique à l'Etablissement des Constructions et des Machines. New edition. Brussels: Société Belge de Libraire, etc.

Navier, C. L. M. H., 1864. Résumé des leçons données à l'Ecole des Ponts et Chaussées sur l'application de la mécanique à l'établissement des constructions et des machines, avec des Notes et des Appendices par M. Barré de Saint-Venant. 3rd ed. Paris: Dunod.

Navier, C. L. M. H., 1833/1878. *Mechanik der Baukunst (Ingenieur-Mechanik) oder Anwendung der Mechanik auf das Gleichgewicht von Bau-Constructionen.* Trans. from 2nd French ed. of 1833 (vol. 1) by G. Westphal (1st ed. trans. 1851) with an appendix by G. Westphal & A. Föppl, 2nd ed., Hannover: Helwing'sche Verlags-Buchhandlung.

Neal, B. G., 1956. *The plastic methods of structural analysis.* London: Chapman & Hall.

Neal, B. G., 1958. *Die Verfahren der plastischen Berechnung biegesteifer Stahlstabwerke.* Trans. from English by T. A. Jaeger. Berlin: Springer.

Nemat-Nasser, S., 1972. General variational principles in nonlinear and linear elasticity with applications. In: *Mechanics Today*, vol. II, pp. 214–261. New York/Toronto/Oxford/Sydney/Braunschweig: Pergamon Press.

Neufert, E., 1936. *Bau-Entwurfslehre.* Berlin: Bauwelt-Verlag.

Neumann, P., 1890. Ueber die Berechnung der Monier-Constructionen. *Wochenschrift des österreichischen Ingenieur- und Architekten-Vereins*, vol. 15, pp. 209–212.

Neumeyer, F. (ed.), 1997. *Friedrich Gilly. Essays zur Architektur 1796–1799.* Berlin: Ernst & Sohn.

Newby, F. (ed.), 2001. *Early Reinforced Concrete. Studies in the History of Civil Engineering*, vol. 11. Aldershot/Burlington/Singapore/Sydney: Ashgate.

Newmark, N. M., Rosenblueth, E., 1971. *Fundamentals of Earthquake Engineering.* New Jersey: Prentice-Hall.

Newmark, N. M., 1974. Westergaard, Harald Malcolm. In: *Dictionary of American Biography*, supplement 4, pp. 873–874. New York: Scribner.

Newton, I., 1999. *Die mathematischen Prinzipien der Physik.* Trans. & ed. V. Schüller. Berlin: Walter de Gruyter.

Nielsen, N. J., 1920. *Bestemmelse af Spændinger i Plader ved Anvendelse af Differensligninger.* Kopenhagen.

Niles, A. S., 1950. Clerk Maxwell and the theory of indeterminate structures. *Engineering*, vol. 170, pp. 194–198.

n.n., 1837. Ludwig Navier. *Allgemeine Bauzeitung*, vol. 2, No. 39, pp. 325–328.

n.n., 1849/1. Die atmosphärische Eisenbahn von Paris nach St. Germain und die hölzerne Bogenbrücke derselben über die Seine. *Allgemeine Bauzeitung*, vol. 14, pp. 171–175.

n.n., 1849/2. *An account of the grand flotation of one of the monster tubes over the Menai Straits, Britannia bridge, June 20th, 1849, with an engraving.* Carnarvon: Printed by James Rees and sold by the booksellers

n.n., 1853. Chronique, Pont d'Asnières. *Annales des Ponts et Chaussées*, pp. 375–880.

n.n., 1863. Hodgkinson, Eaton. In: *J. C. Poggendorf. Biographisch-Literarisches Handwörterbuch der exakten Naturwissenschaften, Erster Band: A – L*, pp. 1117–1118. Leipzig: Verlag von Johann Ambrosius Barth.

n.n., 1877. *Engineering*, Aug. 3, p. 87.

n.n., 1881. *Engineering*, Feb. 25, pp. 192–194.

n.n., 1884/1. *Engineering*, Feb. 22, pp. 166–167.

n.n., 1884/2. *Engineering*, April 11, pp. 308 & 315.

n.n., 1888/1. †Prof. Emil Winkler. *Zentralblatt der Bauverwaltung*, vol. 8, pp. 387–388.

n.n., 1888/2. †Prof. Emil Winkler. *Deutsche Bauzeitung*, vol. 22, pp. 434–436.

n.n., 1894. Gedächtnisfeier für Johann Wilhelm Schwedler. *Deutsche Bauzeitung*, vol. 28, pp. 585–586.

n.n., 1895. †Prof. Dr. W. Fränkel. *Zentralblatt der Bauverwaltung*, vol. 15, p. 168.

n.n., 1898/1. Bresse, Jacques Antoine Charles. In: *J. C. Poggendorf. Biographisch-Literarisches Handwörterbuch der exakten Naturwissenschaften, Dritter Band, I. Abteilung: A – L*, p. 188. Leipzig: Verlag von Johann Ambrosius Barth.

n.n., 1898/2. Hodgkinson, Eaton. In: *J. C. Poggendorf. Biographisch-Literarisches Handwörterbuch der exakten Naturwissenschaften, Dritter Band, I. Abteilung: A – L*, p. 642. Leipzig: Verlag von Johann Ambrosius Barth.

n.n., 1899. †Prof. Robert Land in Constantinople. *Zentralblatt der Bauverwaltung*, vol. 19, p. 364.

n.n., 1903. †Luigi Cremona. *Schweizerische Bauzeitung*, vol. 37, pp. 11–12.

n.n., 1905. †Ludwig v. Tetmajer. *Zeitschrift des Österreichischen Ingenieur- und Architekten-Vereins*, vol. 57, pp. 85–87.

n.n., 1924. Zur Frage des Schubmittelpunktes. *Schweizerische Bauzeitung*, vol. 83, No. 25, p. 297.

n.n., 1941/1. K 70-letiyu so dnya rozhdeniya i 45-letiyu nauchnoi deyatel'nosti B. G. Galerkina (On the Occasion of the 70th Birthday of B. G. Galerkin). *Prikladnaya matematika i mekhanika (PMM: Applied Mathematics & Mechanics)*, vol. 5, pp. 331–344 (Russian & English).

n.n., 1941/2. Generalbevollmächtigter für die Regelung der Bauwirtschaft, 1941. 13. Anordnung des Generalbevollmächtigten für die Regelung der Bauwirtschaft vom 21.5.1940. In: *Stahlbau-Kalender 1941*. Ed. Deutscher Stahlbau-Verband; ed. Georg Unold, pp. 576–577. Berlin: Verlag von Wilhelm Ernst & Sohn.

n.n., 1943. Fachgruppe Stahlbau, im N.S.-Bund Deutscher Technik (ed.). *Stahlbau-Kalender 1944*; ed. Georg Unold. Berlin: Verlag von Wilhelm Ernst & Sohn.

n.n., 1950/1. Special issue "Stahlverbund-Bauweise". Conference on 8 Dec 1949 in Hannover. *Der Bauingenieur*, vol. 25, No. 3, pp. 73–112.

n.n., 1950/2. Special issue "Verbund-Bauweise". Conference on 20–21 April 1950 in Hannover. *Der Bauingenieur*, vol. 25, No. 8, pp. 269–324.

n.n., 1951. Dr. M. T. Huber, Applied Mechanics Authority is Dead. *Mechanical Engineering, ASME*, vol. 73, p. 265.

n.n., 1952. B. G. Galerkin i obzor ego nauchnykh trudov. (B.G. Galerkin and an overview of his scientific work). In: *B. G. Galerkin, collected works*, vol. 1, pp. 5–20 (in Russian).

n.n., 1954/1. Vorwort. In: *Beton-Kalender*, vol. 43, pt. I, p. III. Berlin: Wilhelm Ernst & Sohn.

n.n., 1954/2. Spannbeton. Richtlinien für die Bemessung und Ausführung – DIN 4227. In: *Beton-Kalender*, vol. 43, pt. I, pp. 639–66. Berlin: Wilhelm Ernst & Sohn.

n.n., 1960–2001. *Dizionario biografico degli italiani*. Istituto della Enciclopedia Italiana, Rome, vol. 27, pp. 464–466.

n.n., 1962. Vasily Zakharovich Vlasov i obzor ego nauchnoi deyatel'nosti. (V. Z. Vlasov and an overview of his scientific work). In: *V. Z. Vlasov, collected works*, vol. 1, pp. 3–13 (in Russian).

n.n., 1968. Colonnetti, Gustavo. In: *J. C. Poggendorf. Biographisch-Literarisches Handwörterbuch der exakten Naturwissenschaften, Band VIIb, Teil 2: C – E*, pp. 871–75. Berlin: Akademie-Verlag.

n.n., 1969. Draft of DIN 1045. In: *Beton-Kalender*, vol. 58, pt. II, pp. 415–566. Berlin: Wilhelm Ernst & Sohn.

n.n., 1970. Guidi, Camillo. In: *J. C. Poggendorf. Biographisch-Literarisches Handwörterbuch der exakten Naturwissenschaften, Band VIIb, Teil 3: F – Hem*, pp. 1778. Berlin: Akademie-Verlag.

n.n., 1971. DIN 1045 neu. In: *Beton-Kalender*, vol. 60, pt. I, pp. 1147–1292. Berlin: Wilhelm Ernst & Sohn.

n.n., 1972. *Mechanika Teoretyczna i Stosowana*, vol. 10, No. 2 (in Polish).

n.n., 1974. Richtlinien für die Bemessung und Ausbildung von Verbundträgern. June 1974 ed. Berlin: Deutsches Institut für Normung e.V.

n.n., 1976. Gurii Nikolaevich Savin. *Soviet Applied Mechanics*, pp. 1348–1349.

n.n., 1976. I. M. Rabinovich. *Prikladnaya matematika i mekhanika (PMM: Applied Mathematics and Mechanics)*, vol. 40, pp. 387–398 (in Russian).

n.n., 1977. Sergei Vladimirovich Serensen (1905–1977). *Strength of Materials* (Springer) vol. 9, No. 6, pp. 779–781.

n.n., 1980. Pamyati A. I. Lurie. (A. I. Lurie – in memorium). *Mekhanika Tverdogo Tela* (Proc. USSR Acad. Sci.: Mechanics of Solids), No. 3, pp. 162–169 (in Russian).

n.n., 1989. Southwell, Sir Richard Vynne. In: *J. C. Poggendorf. Biographisch-Literarisches Handwörterbuch der exakten Naturwissenschaften, Band VIIb, Teil 8: Sn – Vl*, pp. 5028–5030. Berlin: Akademie-Verlag.

n.n., 1995. Pięćdziesiąt lat Wydziału Budownictwa Lądowego 1945–1995 (50 years of the Faculty of Civil Engineering). Gdańsk: Księga Jubileuszowa, Politechnika Gdańska (in Polish).

Nomachi, S., 1984. Biography of Prof. Iguchi, *JSCE*, vol. 68, Aug, p. 42.

Norris, C. H., Wilbur, J. B., 1960. *Elementary structural analysis*. 2nd ed. New York: McGraw-Hill.

Novozhilov, V. V., 1947–1965. *Teoriya tonkikh obolochek* (the theory of thin shells). Leningrad: Marine Academy. 1951 New edition. 1962: 2nd ed. Leningrad: Sudpromgiz (in Russian). 1959: English edition *The theory of thin shells*, and 1965: *Thin shell theory*. Groningen: Noordhoff.

Novozhilov, V. V., 1948–1943. *Osnovy nelineinoi teorii uprugosti* (Foundation of the non-linear theory of elasticity). Leningrad/Moscow: Gostekhizdat (in Russian). 1953: English edition *Foundation of the non-linear theory of elasticity*. Rochester/New York: Graylock.

Novy, L., 1978. Augustin Louis Cauchy (1789 bis 1857). In: *Biographien bedeutender Mathematiker*, ed. H. Wußing & W. Arnold, pp. 334–344. Cologne: Aulis Verlag Deubner & Co. KG.

Nordgren, R. P., 1996. Paul M. Naghdi. In: *Memorial Tributes*. National Academy of Engineering, vol. 8, pp. 154–158. Washington, D.C.: National Academy Press.

Nowack, B., 1981. *Die historische Entwicklung des Knickstabproblems und dessen Behandlung in den Stahlbaunormen*. Publications of the Institute of Theory of Structures & Structural Steelwork, Darmstadt TU, No. 35. Darmstadt.

Nowacki, W., 1956. The State of Stress in a Thin Plate due to the Action of Sources of Heat. In: *Proceedings of the International Association for Bridge & Structural Engineering (IABSE)*. vol. 16, ed. F. Stüssi & P. Lardy, pp. 373–398. Zurich: IABSE secretariat.

Nowacki, W., 1963. *Dynamics of elastic systems*. New York: Wiley.

Nowacki, W., 1965. *Theorie des Kriechens – lineare Viskoelastizität*. Vienna: Deuticke.

Nowacki, W., 1966. *Dynamic problems of thermoelasticity*. Leyden: Noordhoff.

Nowacki, W., 1981. *Theory of asymmetric elasticity*. Warsaw: PWN.

Nowacki, W., 1985. *Notatki autobiograficzne* (autobiographical notes). Warsaw: Państwowe Wydawnictwo Naukowe (PWN) (in Polish).

Nowak, A., 1923. Joseph Melan. In: *Joseph Melan zum siebzigsten Geburtstage*. pp. VII–XIV. Leipzig/Vienna: Deuticke.

Ochsendorf, J. A., 2002. *Collapse of Masonry Structures*. PhD dissertation, Cambridge University.

Ochsendorf, J. A., 2006. The masonry arch on spreading supports. *The Structural Engineer*, vol. 84, pp. 29–35.

Ockleston, A. J., 1955. Load tests on a three-storey reinforced concrete building in Johannesburg. *Structural Engineer*, vol. 33, pp. 304–322.

Oden, J. T., 1969. A general theory of finite elements. *International Journal of Numerical Methods in Engineering*, vol. 1, pp. 205–221 & 247–259.

Oden, J. T., Reddy, J. N., 1974. On dual complementary variational principles in mathematical physics. *International Journal of Engineering Science*, vol. 12, pp. 1–29.

Oden, J. T., 1987. Historical comments on Finite Elements. In: *Proceedings of the ACM conference on History of scientific and numeric computation*, ed. G. E. Crane, pp. 125–130. New York: ACM Press.

Oder, 1907. †Luigi Vianello. *Zentralblatt der Bauverwaltung*, vol. 27, No. 68, p. 452.

Olesiak, Z., 1972. Setna rocznica urodzin profesora Maksymiliana Tytusa Hubera. (100th anniversary of the birth of Prof. Maksymilian Tytus Huber). *Mechanika Teoretyczna i Stosowana,* vol. 10, pp. 350–354 (in Polish).

Olesiak, Z. S., 1992. Maksymilian Tytus Huber – w 120-lecie Urodzin. (Maksymilian Tytus Huber – on his 120th birthday). *Nauka Polska,* No. 5/6, pp. 231–249 (in Polish).

Olszak, W., 1949. 50-lecie pracy naukowej M. T. Hubera. (50 years of scientific work by M. T. Huber). *Archiwum Mechaniki Stosowanej,* vol. 1, pp. 265–270 (in Polish).

Olszak, W., Mróz, Z., W., 1956. The Method of Inversion in the Theory of Plates. In: *Proceedings of the International Association for Bridge & Structural Engineering (IABSE).* vol. 16, ed. F. Stüssi & P. Lardy, pp. 399–424. Zurich: IABSE secretariat.

Olszak, W., Rychlewski, J., Urbanowski, W., 1962. Plasticity under nonhomogeneous conditions. In: Advances in Applied Mechanics, vol. VII. Acad. Press.

Olszak, W., Mróz, Z., Perzyna, P., 1963. *Recent trends in the development of the theory of plasticity.* Oxford: Pergamon Press & Warsaw: PWN.

Olszak, W., Sawczuk, A., 1967. *Inelastic behaviour in shells.* Groningen: Noordhoff.

Onat, E. T., Prager, W., 1953. Limit Analysis of Arches. *Journal of the Mechanics and Physics of Solids,* vol. 1, pp. 77–89.

Oota, 1984. Biography of Prof. Mise. *JSCE,* vol. 69, June, p. 92 (in Japanese).

Oravas, G. Æ., McLean, L., 1966/1. Historical Development of Energetical Principles in Elastomechanics. *Applied Mechanics Reviews,* vol. 19, No. 8, pp. 647–658 & No. 11, pp. 919–933.

Oravas, G. Æ. (ed.), 1966/2. *The Theory of Equilibrium of Elastic Systems and its Applications by Carlo Pio Castigliano.* New ed. of E. S. Andrews' trans. of *Théorie de l'Équilibre des Systèmes Élastiques et ses Applications* dating from 1919. New York: Dover Publications.

Ostenfeld, A., 1898. Zur Berechnung von Monier-constructionen. *Zeitschrift des österreichischen Ingenieur- und Architekten-Vereines,* vol. 50, pp. 22–25.

Ostenfeld, A., 1921. Berechnung statisch unbestimmter Systeme mittels der "Deformationsmethode". *Der Eisenbau,* vol. 12, No. 11, pp. 275–289.

Ostenfeld, A., 1926. *Die Deformationsmethode.* Berlin: Springer.

Pahl, P. J., Damrath, R., 2000. *Mathematische Grundlagen der Ingenieurinformatik.* Berlin/Heidelberg/New York: Springer-Verlag.

Pahl, P. J., Damrath, R., 2001. *Mathematical Foundations of Computational Engineering. A handbook.* Trans. from German by Felix Pahl. Berlin/Heidelberg/New York: Springer-Verlag.

Pahl, P. J., 2006. Bauwesen. Praxis und Wissenschaft des Bauingenieurwesens. In: *Erkennen und Gestalten. Eine Theorie der Technikwissenschaften,* ed. G. Banse, A. Grunwald, W. König & G. Ropohl, pp. 278–289. Berlin: edition sigma.

Páll, G., 1960. †Dr. David B. Steinman. *Der Stahlbau,* vol. 29, No. 12, pp. 390–391.

Panasyuk, V. V., 2007. G. N. Savin and the Lviv school of mechanics. *International Applied Mechanics,* vol. 43, No. 1, pp. 10–27.

Panetti, 1906. *Giornale del Genio civile,* Anno XLIV, Marzo.

Petri, R., Kreutz, J.-S., 2004. Der Kettensteg in Nürnberg – die älteste erhaltene eiserne Hängebrücke Kontinentaleuropas. *Stahlbau,* vol. 73, No. 5, pp. 308–311.

Paoletti, P., 1987. *Il Ponte a Santa Trinità.* Florence.

Paoli, G., 1988. *Ruggiero Giuseppe Boscovich nella scienza e nella storia del 1700.* Rome: Accademia Nazionale delle Scienze detta dei XL.

Papkovich, P. F., 1932/1. Solution générale des équations différentielles fondamentales d'élasticité, exprimée par trois fonctions harmoniques. *Comptes rendus,* vol. 195, pp. 513–515.

Papkovich, P. F., 1932/2. Expressions générales des composantes des tensions, ne renfermant comme fonctions arbitraires que des fonctions harmonique. *Comptes rendus,* vol. 195, pp. 754–756.

Papkovich, P. F., 1939. *Teoriya uprugosti* (elastic theory). Leningrad/Moscow: Oborongiz (in Russian).

Papkovich, P. F. 1941–1947. *Stroiteľnaya mekhanika korablya* (mechanics of ships). 3 vols. Leningrad: Morskoi transport-Sudpromgiz (in Russian).

Papkovich, P. F., 1962–1963. *Trudy po stroiteľnoi mekhanike* (works on structural mechanics). 4 vols. Leningrad: Sudpromgiz (in Russian).

Parcel, J. I., Maney, G. A., 1926. *An Elementary Treatise of Statically Indeterminate Stresses.* New York: Wiley.

Parkinson, E. M., 1981. Rankine, William John Macquorn. In: Dictionary of Scientific Biography, vol. 11. C. C. Gillispie (ed.), pp. 291–295. New York: Charles Scribner's Sons.

Pascal, E., 1903. Eugenio Beltrami. *Mathematische Annalen,* vol. 57, No. 1, pp. 65–107.

Passadore, G., 1963. Indice bibliografico Poleniano. In: *Giovanni Poleni (1683–1761) nel bicentenario della morte. Accademia Patavina di Scienze Lettere ed Arti,* pp. 95–118. Padova.

Pasternak, H., Rabiega, J., Biliszczuk, J., 1996. 200 Jahre eiserne Brücken auf dem europäischen Kontinent – auf Spurensuche in Schlesien und der Lausitz. *Stahlbau,* vol. 65, No. 12, pp. 542–546.

Pasternak, P., 1922. Beiträge zur Berechnung vielfach statisch unbestimmter Stabsysteme. *Der Eisenbau,* vol. 13, No. 11, pp. 239–254.

Pasternak, P., 1926. *Der abgekürzte Gauss'sche Algorithmus als eine einheitliche Grundlage in der Baustatik.* Dissertation, Zurich ETH.

Paulling, J. R., 1963. Finite element analysis of ship structures. Report No. 63–145, *Det Norske Veritas Research Department.* Oslo.

Pauser, A., 1994. *Eisenbeton 1850–1950: Idee, Versuch, Bemessung, Realisierung; unter Berücksichtigung des Hochbaus in Österreich.* Vienna: Manz.

Pauser, A., 2003. *Gemauerte Kreisbogengewölbe. Bewertung des Tragvermögens von Bestandsobjekten.* Gutachten für die Österreichischen Bundesbahnen.

Pauser, A., 2005. *Brücken in Wien. Ein Führer durch die Baugeschichte.* Vienna: Springer.

Peacock, G. (ed.), 1855. *Miscellaneous works of the late Thomas Young.* London: John Murray.

Pearson, K., 1889. *The Elastical Researches of Barré de Saint-Venant.* Cambridge: At the University Press.

Pechstein, K., 1975. *Allerlei Visierungen und Abriß wegen der Fleischbrücken 1595.* Nuremberg: Anzeiger des Germanischen Nationalmuseums 1975.

Pelikan, W., Eßlinger, M., 1957. Die Stahlfahrbahn. Berechnung und Konstruktion. *MAN-Forschungsheft Nr. 7.* Augsburg: K. G. Kieser.

Pelke, E., Ramm, W., Stiglat, K., 2005. *Geschichte der Brücken. Zeit der Ingenieure.* Germersheim: Deutsches Straßenmuseum.

Pelke, E., 2007. Entwicklung der Spannbetonbrücken in Deutschland – der Beginn. *Bauingenieur,* vol. 82, No. 6, pp. 262–269.

Pellegrino, S., Calladine, C. R., 1986. Matrix analysis of statically and kinematically indeterminate frameworks. *International Journal of Solids and Structures,* vol. 22, pp. 409–428.

Perino, A. M. S., 1995. Un monument du XIX siècle à Turin: Le Pont Mosca sur la Doire. In: *Entre Mécanique et Architecture,* ed. P. Radelet-de Grave et E. Benvenuto, pp. 274–287. Basel: Birkhäuser.

Perrodil, F., 1872. Applications des équations du problème général de la résistance des matériaux au problème de la stabilité d'une voûte d'épaisseur variable traitée comme un monolithe homogène. *Annales des Ponts et Chaussées,* vol. 4, sem. II, pp. 42–83.

Perrodil, F., 1876. Théorie de la stabilité des voûtes. Applications des équations de la résistance des matériaux au problème de la stabilité des voûtes. *Annales des Ponts et Chaussées,* vol. 11, sem. I, pp. 178–222.

Perrodil, F., 1879. *Résistance des voûtes et arcs métalliques employés dans la construction des ponts.* Paris: Gauthier-Villars.

Perrodil, F., 1880. Résistance des voûtes et des arcs métalliques. *Annales des Ponts et Chaussées,* vol. 19, sem. I, pp. 212–232.

Perrodil, F., 1882. Arc d'expérience en maçonnerie de brique et ciment de Portland. *Annales des Ponts et Chaussées,* vol. 4, pp. 111–139.

Perronet, J.-R., 1782/1783. *Déscription des projets et de la construction des ponts de Neuilly, de Mantes, d'Orléans, etc.* Paris: Imprimerie Royale.

Perronet, J. R., Chezy, A., 1810. Formule générale pour déterminer l'épaisseur des piles et culées des arches des ponts, soit qu'elles soient en plein cintre ou surbaissées. In: *Recueil de divers mémoires extraits de la bibliothèque impériale des ponts et chaussées a l'usage de MM. les ingénieurs*, ed. P. Lesage, vol. 2, pp. 243-273. Paris: Chez Firmin Didot.

Pesciullesi, C., Rapallini, M., 1995. The analogy between equilibrium of threads and thin masonry structures. In: *Entre Mécanique et Architecture,* ed. P. Radelet-de Grave & E. Benvenuto, pp.123–139. Basel: Birkhäuser.

Pestel, E. C., Leckie, A., 1963. *Matrix Methods in Elastomechanics.* New York: McGraw-Hill.

Peters, T. F., 1987. *Transitions in Engineering. Guillaume Henri Dufour and the Early 19th Century Cable Suspension Bridges.* Basel/Boston: Birkhäuser.

Peters, T. F., 1996. *Building the Nineteenth Century.* Cambridge, Mass.: The MIT Press.

Petersen, C. R., 1907. Luigi Vianello. *Zeitschrift des Vereines deutscher Ingenieure,* vol. 51, No. 51, pp. 2033–2034.

Petersen, C., 1970. *Abgespannte Maste – Statik und Dynamik.* Berlin/Munich: Wilhelm Ernst & Sohn.

Petersen, C., 1977. Der wahrscheinlichkeitstheoretische Aspekt der Bauwerkssicherheit im Stahlbau. In: *4/1977 Berichte aus Forschung und Entwicklung. Beiträge zum Tragverhalten und zur Sicherheit von Stahlkonstruktionen, Vorträge aus der Fachsitzung III des Deutschen Stahlbautages Stuttgart 1976,* ed. DASt, pp. 26–42. Cologne: Stahlbau-Verlags-GmbH.

Petersen, C., 1980. *Statik und Stabilität der Baukonstruktionen.* Braunschweig/Wiesbaden: Vieweg.

Petersen, C., 1988. *Stahlbau – Grundlagen der Berechnung und baulichen Ausbildung von Stahlbauten.* Braunschweig/Wiesbaden: Vieweg.

Petersen, C., 1996. *Dynamik der Baukonstruktionen.* Braunschweig/Wiesbaden: Vieweg.

Petit, 1835. Mémoire sur le calcul des voûtes circulaires. *Mémorial de l'Officier du Génie,* vol. 12, pp. 73-150.

Petropavlovskaja, I. A., 1990. Der Sendeturm für die Radiostation Šabolovka in Moscow. In: *Vladimir G. Šuchov, 1853–1939. Die Kunst der sparsamen Konstruktion.* Ed. R. Graefe, M. Gapoev, M. & O. Pertschi, pp. 92–103. Stuttgart: Deutsche Verlagsanstalt.

Petroski, H., 1996. *Engineers of Dreams. Great Bridge Builders and the Spanning of America.* New York: Random House.

Petschel, D. (ed.), 2003. *Die Professoren der TU Dresden 1828–2003.* Cologne: Böhlau.

Petzold, H., 1992. *Moderne Rechenkünstler. Die Industrialisierung der Rechentechnik in Deutschland.* Munich: Verlag C. H. Beck.

Pflüger, A., 1947. Zum Beulproblem der anisotropen Rechteckplatte. *Ingenieur-Archiv,* vol. 16, pp. 113–120.

Pflüger, A., 1950. *Stabilitätsprobleme der Elastostatik.* 2nd ed. 1964, 3rd ed. 1975. Berlin: Springer.

Pflüger, A., 1957. *Elementare Schalenstatik.* 2nd ed. 1960, 3rd ed. 1967. Berlin: Springer.

Phocas, M. C., 2005. John Argyris and his decisive contribution in the dverelopment of lightweight structures. Form follows force. In: *Proceedings of the 5th GRACM International Congress on Computational Mechanics,* ed. G. Georgiou, P. Papanastasion & M. Papadrakakis, pp. 97–103. Nicosia: Kantzilaris Publications.

Pian, T. H. H., 1964. Derivation of element stiffness matrices by assumed stress distributions. *AIAA Journal,* vol. 2, pp. 1333–1336.

Pian, T. H. H., Tong, P., 1969. Basis of finite element methods for solid continua. *International Journal of Engineering Science,* vol. 1, pp. 3–28.

Pian, T. H. H., 1973. Finite element methods by variational principles with relaxed continuity requirements. In: *Variational Methods in Engineering,* ed. C. A. Brebbia & H. Tottenham, pp. 3/1–3/24. Southampton: Southampton University Press.

Pian, T. H. H., 1979. A historical note about 'hybrid elements'. *International Journal of Engineering Science,* vol. 12, pp. 891–892.

Pian, T. H. H., 2000. Some notes on the early history of hybrid stress finite element method. *International Journal of Engineering Science,* vol. 47, pp. 419–425.

Piccarreta, F., Sguerri, L., 2001. I protagonisti della progettazione strutturale. Le esperienze progettuali, scientifiche e didattiche dalla fine della guerra agli inizi degli anni '90. In: *La Facoltà di Architettura dell'Università 'La Sapienza' dalle Origini al Duemila. Discipline, Docenti, Studenti,* ed. V. Franchetti Pardo, pp. 497–541. Rome: Gangemi Editore.

Pichler, G. et al., 1998. Die graphische Statik in Studium und Praxis. In: *Brückenschläge. Festschrift Herrn Univ.-Prof. Dr.-Ing. Wilhelm Friedrich Führer zur Vollendung des 60. Lebensjahres.* Ed. staff of the Chair of Structural Engineering, Aachen RWTH, pp. 223–231. Aachen: Self-pub.

Picon, A., 1985. Die Vorstellungskraft eines Konstrukteurs: Jean-Rodolphe Perronet und das Modell einer Brücke. *Daidalos,* No. 15, pp. 74–87.

Picon, A., 1988. Navier and the introduction of suspension bridges in France. *Construction History,* vol. 4, pp. 21–34.

Picon, A., 1995. Entre science et art de l'ingénieur. L'enseignement de Navier à l'Ecole des Ponts et Chaussées. In: *Entre Mécanique et Architecture,* ed. P. Radelet-de Grave & E. Benvenuto, pp. 257–273. Basel: Birkhäuser.

Picon, A. (ed.), 1997. *L'Art de l'ingénieur.* Paris: Éditions du Centre Pompidou.

Pieper, K., 1983. *Sicherung historischer Bauten.* Berlin: Wilhelm Ernst & Sohn.

Pighetti, C. 1964. Discutendo del newtonianesimo di R. G. B. In: *Physis,* VI, 15.

Pignataro, M., 1993. W. T. Koiter (1914–1997). Meccanica, vol. 33, No. 6, Dec, pp. 605–606

Pippard, A. J. S., Tranter, E., Chitty, L., 1936/1937. The mechanics of the voussoir arch. *Journal of the Institution of Civil Engineers,* vol. 4, pp. 281–306 & vol. 6, pp. 4–26 (discussion).

Pippard, A. J. S., Ashby, R. J., 1938. An experimental study of the voussoir arch. *Journal of the Institution of Civil Engineers,* vol. 10, pp. 383–404.

Pippard, A. J. S., Baker, J. F., 1943. *The analysis of engineering structures.* 2nd ed. London: E. Arnold.

Pippard, A. J. S., 1961. Elastic theory and engineering structures. *Proceedings of the Institution of Civil Engineers,* vol. 19, pp. 129–156.

Pirlet, J., 1910. Die Berechnung statisch unbestimmter Systeme. *Der Eisenbau,* vol. 1, No. 9, pp. 331–349.

Pisarenko, G. S., Rudenko, V. N., Tret'yachenko, G. N., 1966. *Prochnost' materialov pri vysokikh temperaturakh* (high-temperature strength of materials). Kiev: Naukova Dumka (in Russian).

Pisarenko, G. S., Rudenko, V. N., Tret'yachenko, G. N., 1969. *High-Temperature Strength of Materials.* Jerusalem: Sci. Translations.

Pisarenko, G. S., et al., 1980. *Prochnost' materialov i elementov konstruktsii v ekstremal'nykh usloviyakh* (strength of materials and structural elements under extremal conditions), vol. 2, Kiev: Naukova Dumka (in Russian).

Pisarenko, G. S., 1991. *Stepan Prokof'evich Timoshenko.* Moscow: Nauka (in Russian).

Pisarenko, G. S., 1993. *Sergei Vladimirovich Serensen.* Kiev: Naukova Dumka (in Russian).

Pisarenko, G. S., 1994. *Vospominaniya i razmyshleniya* (reminiscences and reflections). Kiev: Naukova Dumka (in Russian).

Pitz, H., 2000. Das Rote Rathaus. In: *Tragwerkstatt Gerhard Pichler. Entwürfe, Bauten, Konstruktionen.* Ed. I. Ermer & K. Eisenloffel, pp. 84–86. Berlin: Gebr. Mann.

Planat, P., 1887. *Pratique de la mécanique appliquée à la resistance des matériaux.* Paris: La Construction Moderne.

Plank, R., 1926. Franz Grashof als Lehrer und Forscher. *ZVDI,* vol. 70, No. 28, pp. 933–938.

Plato, 1994. *Sämtliche Werke. Band 4: Timaios, Kritias, Minos, Nomoi.* Trans. H. Müller & F. Schleiermacher. Reinbek: Rowohlt Taschenbuch Verlag.

Pohlmann, G., 1956. Balken auf elastischer Unterlage als Teil einer Konstruktion. In: *Hütte III. Bautechnik,* 28th ed. Berlin: Ernst & Sohn.

Poisson, S. D., 1829. Mémoire sur l'équilibre et le movement des corps élastiques. *Mémoires de l'Académie Paris,* vol. 8, pp. 357–570.

Pole, W., 1877. *The Life of Sir William Fairbairn, Bart.* London: Longmans, Green & Co.

Poleni, G., 1748. *Memorie istoriche della gran cupola del tempio Vaticano e de' danni di essa, e de' ristoramenti loro.* Padova: Stamperia del Seminario.

Polónyi, S., 1995. Sicherheit – sich absichern. *Bautechnik*, vol. 72, No. 3, pp. 199–205.

Polonceau, C., 1840/1. Notice sur un nouveau système de charpente en bois et en fer. *Revue générale de l'Architecture*, vol. 1, col. 27–32.

Polonceau, C., 1840/2. Neues Dachkonstrukzionssystem aus Holz und Eisen. *Allgemeine Bauzeitung*, vol. 5, pp. 273–280.

Poncelet, J. V., 1822. *Traité des propriétés projectives des figures.* Paris.

Poncelet, J. V., 1826. *Cours de mécanique appliquée aux machines.* Metz.

Poncelet, J. V., 1835. Solution graphique des principales questions sur la stabilité des voûtes. *Mémorial de l'Officier du Génie*, vol. 12, pp. 151–213.

Poncelet, J. V., 1840. *Introduction à la mécanique industrielle.* Metz.

Poncelet, J. V., 1844. *Über die Stabilität der Erdbekleidungen und deren Fundamente.* Trans. from the French and with an appendix by J. W. Lahmeyer. Braunschweig: Verlag von G. C. E. Meyer sen.

Poncelet, J. V., 1852. Examen critique et historique des principales théories ou solutions concernant l'équilibre des voûtes. *Comptes Rendus de l'Académie des Sciences*, vol. 35, pp. 494–502, 532–540 & 577–587.

Poncelet, J. V., 1865/66. *Traité des propriétés projectives des figures.* vols. 1–2, 2nd ed. Paris: Gauthier-Villars.

Porstmann, W., 1917. *Normenlehre. Grundlagen, Reform und Organisation der Maß- und Normensysteme dargestellt für Wissenschaft, Unterricht und Wirtschaft.* Leipzig: Schulwissenschaftlicher Verlag A. Haase.

Pöschl, T., Terzaghi, K. v., 1913. *Berechnung von Behältern nach neueren analytischen und graphischen Methoden.* Berlin: Springer.

Pöschl, T., 1936. Über die Minimalprinzipie der Elastizitätstheorie. *Der Bauingenieur*, vol. 17, No. 17/18, pp. 160ff.

Potterat, L., 1920/1. Zur Festigkeitslehre. *Schweizerische Bauzeitung*, vol. 76, No. 13, pp. 141–143.

Potterat, L., 1920/2. Zur Festigkeitslehre (statement). *Schweizerische Bauzeitung*, vol. 76, No. 18, p. 208.

Pottgiesser, H., 1985. *Eisenbahnbrücken aus zwei Jahrhunderten.* Basel: Birkhäuser.

Prager, W., 1926. Beitrag zur Kinematik des Raumfachwerkes. *Zeitschrift für angewandte Mathematik und Mechanik*, vol 6, No. 5, pp. 341–355.

Prager, W., 1927. Die Formänderungen von Raumfachwerken. *Zeitschrift für angewandte Mathematik und Mechanik*, vol. 7, No. 6, pp. 421–424.

Prager, W., Hohenemser, K. H., 1933. *Dynamik der Stabwerke. Eine Schwingungslehre für Bauingenieure.* Berlin: Springer.

Prager, W., Hodge, P. G., 1951. *Theory of perfectly plastic solids.* New York: Wiley.

Prager, W., Hodge, P. G., 1954. *Theorie ideal plastischer Körper.* Vienna: Springer.

Prager, W., 1955. *Probleme der Plastizitätstheorie.* Basel: Birkhäuser.

Prager, W., 1959. *An Introduction to Plasticity.* London: Addison-Wesley Publishing Company.

Prandtl, L., 1937. †Erich Trefftz. *Zeitschrift für Angewandte Mathematik und Mechanik*, vol. 17, No. 1, pp. I–IV.

Prange, G., 1915. *Die Hamilton-Jacobische Theorie der Doppelintegrale (mit einer Übersicht der Theorie für einfache Integrale).* Dissertation, University of Göttingen. Göttingen: Universitäts-Buchdruck.

Prange, G., 1916. *Das Extremum der Formänderungsarbeit.* Habilitationsschrift. Technische Hochschule Hannover, typewritten ms.

Prange, G., 1919. Die Theorie des Balkens in der technischen Elastizitätslehre. *Zeitschrift für Architektur- und Ingenieurwesen*, vol. 65, col. 83–96 & 121–150.

Prato, C., 1968. *A mixed finite element method for thin shell analysis.* PhD thesis, Dep. of Civil Engineering, MIT.

Prinz, C., 1924. A. Föppl als Forscher und Lehrer. In: *Beiträge zur Technischen Mechanik und Technischen Physik. August Föppl zum siebzigsten Geburtstag am 25. Januar 1924*, ed. L. Föppl, O. Föppl, L. Prandtl & H. Thoma, p. 1–3. Berlin: Julius Springer.

Probst, E., 1931. †Friedrich Engesser. *Der Bauingenieur*, vol. 12, No. 39.

Prony, G. de. 1823. Rapport fait à l'Académie royale des sciences, le 26 mai 1823, sur le Mémoire de Lamé et Clapeyron. *Annales des mines*, vol. VIII, pp. 818–836.

Prony, G. de. 1839. Notice biographique sur Navier. In: *Résumé des Leçons données à l'Ecole des Ponts et Chaussées sur l'Application de la Mécanique à l'Etablissement des Constructions et des Machines*, pp. x-lj. Brussels: Société Belge de Libraire, etc.

Proske, D., Lieberwirth, P. van Gelde, P., 2006. *Sicherheitsbeurteilung historischer Bogenbrücken.* Dresden: Dirk Proske Verlag 2006.

Puchta, S., 2000. Die antimathematische Bewegung in den Technikwissenschaften am Ende des 19. Jahrhunderts. In: *1799–1999. Von der Bauakademie zur Technischen Universität Berlin. Geschichte und Zukunft*, ed. Karl Schwarz for the president of Berlin TU, pp. 130–136. Berlin: Ernst & Sohn.

Pugsley, A. G., 1951. Concepts of Safety in structural engineering. *Jnl. Instn. Civ. Engrs.*

Pugsley, A. G., 1957. *The Theory of Suspension Bridges.* London: Arnold.

Pugsley, A. G., 1966. *The Safety of Structures.* London: Arnold.

Pugsley, A. G., 1969. The Engineering Climatology of Structural Accidents. *Proceedings of the International Conference on Structural Safety and Reliability*, pp. 335–340. Washington.

Pugsley, A. G., 1979. James Forrest Lecture 1978. Statics in Engineering Hands. *ICE Proceedings*, vol. 66, No. 2, pp. 159–168.

Pugsley, A. G. (Ed.), 1980. *The works of Isambard Kingdom Brunel.* Cambridge: Cambridge University Press.

Pugsley, A. G., 1983. The nonlinear behaviour of a suspended cable. *The Quarterly Journal of Mechanics and Applied Mathematics*, vol. 36, No. 2, pp. 157–162.

Quade, W., 1940. Matrizenrechnung und elektrische Netze. *Archiv für Elektrotechnik*, vol. 34, No. 10, pp. 545–567.

Raack, W., 1971. István Szabó 65 Jahre. In: *Aus Theorie und Praxis der Ingenieurwissenschaften. Mathematik Mechanik Bauwesen. Festschrift zum 65. Geburtstag von Prof. Dr.-Ing. István Szabó*, ed. R. Trostel & P. Zimmermann, pp. 1–3. Berlin: Wilhelm Ernst & Sohn.

Raack, W., 1977. Zur Airyschen Spannungsfunktion. *Die Bautechnik*, vol. 54, No. 1, pp. 1–7 & No. 4, pp. 138–142.

Rabich, R., 1953. Die Membrantheorie der einschalig hyperbolischen Rotationsschale. *Bauplanung-Bautechnik*, No. 7, pp. 310–318 & 320.

Rabich, R., 1954. Einführung in die Statik der Schalenträger mit kreisförmigen Querschnittsteilen. *Bauplanung-Bautechnik*, No. 9, pp. 389–395.

Rabich, R., 1955. Die Statik der Schalenträger. *Bauplanung-Bautechnik*, No. 3, pp. 115–167.

Rabich, R., 1956. Die Statik der Schalenträger. Die Berechnung der Randstörung am Randträger. *Bauplanung-Bautechnik*, No. 1, pp. 1–15.

Rabich, R., 1964. Statik der Platten, Scheiben, Schalen. In: *Ingenieurtaschenbuch Bauwesen. Band I: Grundlagen des Ingenieurbaus*, ed. G. Grüning & A. Hütter, pp. 861–1120. Leipzig: Edition Leipzig.

Rabich, R., 1965. *Leitfaden Berechnung von Kreiszylinderschalen mit Randgliedern.* Berlin: VEB Verlag für Bauwesen.

Rabinovich, I. M., 1934–1937. *Metody rascheta ram (methods of frame analysis).* 3 vols. 3rd ed. Moscow/Leningrad: Gosstroiizdat (in Russian).

Rabinovich, I. M., 1949. *Dostizheniya stroitel'noi mekhaniki sterzhnevykh sistem v SSSR (achievements in the structural mechanics of trusses in the USSR).* Moscow: Akad. Arkhitektury SSSR (in Russian).

Rabinovich, I. M. (ed.), 1960. *Structural Mechanics in the USSR 1917–1957.* New York: Pergamon Press.

Rabinovich, I. M. (ed.), 1966. *Hängedächer.* Wiesbaden/Berlin: Bauverlag.

Rabinovich, I. M. (ed.), 1969. *Stroitel'naya mekha-nika v SSSR 1917–1967* (structural mechanics in the USSR, 1917–67). Moscow: Stroiizdat (in Russian).

Rabinovich, I. M., 1984. *Vospominaniya 1904–1974* (memoirs, 1904–74). Moscow: Nauka (in Russian).

Radelet-de Grave, P., 1994. Etude de l'Essai sur une application des règles de Maximis et Minimis à quelques Problèmes de statique relatifs à l'Architecture, par Coulomb. In: *Science et technique en perspective*, vol. 27, Université de Nantes, pp. 2–22.

Radelet-de Grave, P., 1995. Le 'De Curvatura Fornicis' de Jacob Bernoulli ou l'Introduction des infiniment Petits dans le Calcul des Voûtes. In: *Entre Mécanique et Architecture*, ed. P. Radelet-de Grave & E. Benvenuto, pp. 142–163. Basel: Birkhäuser.

Radelet-de Grave, P., 1999. La théorie des voûtes de Pierre Bouguer: jeu mathématique et enjeu pratique. *Sciences et Techniques en perspective*, 2nd series, fasc. 2, pp. 397–421.

Radelet-de Grave, P., 2002. Arthur Vierendeel (1852–1940), pour une architecture du fer. In: *Towards a History of Construction*, ed. A. Becchi, M. Corradi, F. Foce & O. Pedemonte, pp. 417–435. Basel: Birkhäuser.

Radtsig, A. A., 1917. V. L. Kirpichev. In: *V. L. Kirpichev, collected works*, vol. 1, pp. iii–xxxvi. Petrograd: Petrograd Polytechnic Institute (in Russian).

Rakcheev, E. N., 1984. *D. I. Zhuravsky*. Moscow: Nauka (in Russian).

Ramm, E., Andelfinger, U., Höcklin, H., Kimmich, S., 1987. *Baustatik und Computer – Entwicklungen und Tendenzen*. 3rd Conference on Theory & Practice of Structural Engineering, Stuttgart, pp. 20.1–20.18. Stuttgart.

Ramm, E., 1990. Der Behälterbau. In: *Vladimir G. Šuchov, 1853–1939. Die Kunst der sparsamen Konstruktion*. Ed. R. Graefe, M. Gapoev, M. & O. Pertschi, pp. 120–27. Stuttgart: Deutsche Verlagsanstalt.

Ramm, E., Reitinger, R., 1992. Force follows form in shell design. In: *IASS-CSCE International Congress*, Toronto, 13–17 July 1992, pp. 1–17.

Ramm, E., Hofmann, T. J., 1995. Stabtragwerke. In: *Der Ingenieurbau. Grundwissen in 9 Bänden: Baustatik/Baudynamik*, ed. G. Mehlhorn, pp. 1–349. Berlin: Ernst & Sohn.

Ramm, E., 2000. Entwicklung der Baustatik von 1920 bis 2000. *Bauingenieur*, vol. 75, No. 8, pp. 319–331.

Ramm, E., Wall, W. A., 2004. Shell structures – a sensitive interrelation between physics and numerics. *International Journal for Numerical Methods in Engineering*, vol. 60, pp. 381–427.

Ramm, W., 1986. Würdigung des Werkes in dem Gebiet der Statik. In: *Kurt-Klöppel-Gedächtnis-Kolloquium TH Darmstadt 15. bis 16. September 1986. THD Schriftenreihe Wissenschaft und Technik*, vol. 31, ed. President of Darmstadt TH, pp. 37–84. Darmstadt.

Ramm, W., 1998. Matthias Koenen (1849–1924). Schöpfer der ersten Biegebemessung für Eisenbetonplatten und Mitbegründer der Eisenbetonbauweise in Deutschland. In: *VDI Bau. Jahrbuch 1998*, ed. VDI-Gesellschaft Bautechnik, pp. 349–379. Düsseldorf: VDI-Verlag.

Ramm, W., 1999. History of the Vistula Bridges in Tczew. In: *Proceedings of the International Conference Preservation of the Engineering Heritage – Gdańsk Outlook 2000*, ed. Z. Cywiński & W. Affelt, pp. 195–204. Gdańsk: Technical University of Gdańsk.

Ramm, W., 2001. Über die Geschichte des Eisenbaus und das Entstehen des Konstruktiven Ingenieurbaus. *Stahlbau*, vol. 70, No. 9, pp. 628–641.

Ramm, W. (ed.), 2004. *Zeugin der Geschichte: Die Alte Weichselbrücke in Dirschau [Świadek przeszłości: Dawny most przez Wisłę w Tczewie]*. Kaiserslautern: Kaiserslautern TU (in Geman and Polish).

Ramm, W., 2007. Über die faszinierende Geschichte des Betonbaus vom Beginn bis zur Zeit nach dem 2. Weltkrieg. In: *Gebaute Visionen. 100 Jahre Deutscher Ausschuss für Stahlbeton 1907–2007*, Deutscher Ausschuss für Stahlbeton im DIN Deutsches Institut für Normung e.V. (ed.), pp. 27–130. Berlin: Beuth Verlag.

Ramme, W., 1939. *Über die geschichtliche Entwicklung der Statik in ihren Beziehungen zum Bauwesen*. Braunschweig: Waisenhausbuchdruck.

Rankine, W. J. M., 1857. On the stability of loose earth. *Philosophical Transactions of the Royal Society*. vol. 147, pp. 9–27.

Rankine, W. J. M., 1858. *A Manual of Applied Mechanics*. London: C. Griffin & Co.

Rankine, W. J. M., 1859. *A Manual of the Steam Engine and other Prime Movers*. London: C. Griffin & Co.

Rankine, W. J. M., 1862. *A Manual of Civil Engineering*. London: C. Griffin & Co.

Rankine, W. J. M., 1864. Principle of the equilibrium of polyhedral frames. *Philosophical Magazine*. series 4, vol. 27, p. 92.

Rankine, W. J. M., 1865. Graphical measurement of elliptical and trochtoidal arcs, and the construction of a circular arc nearly equal to a given straight line. *Philosophical Magazine*. Series 4, vol. 29, pp. 22–25.

Rankine, W. J. M., 1866/1. *Useful Rules and Tables Relating to Mensuration, Engineering, Structures, and Machines*. London: C. Griffin & Co.

Rankine, W. J. M., 1866/2. Einige graphische Constructionen. *Zivilingenieur*, vol. 12, pp. 223–224.

Rankine, W. J. M., 1870. Diagrams of forces in frameworks. *Proceedings of the Royal Society of Edinburgh*, vol. 7, pp. 171–172.

Rankine, W. J. M., 1872. *A Manual of Civil Engineering*. 8th ed. London: C. Griffin & Co.

Rasch, J. J., 1989. Die Kuppeln in der römischen Architektur. Entwicklung, Formgebung, Konstruktion. In: *Zur Geschichte des Konstruierens*, ed. R. Graefe, pp. 17–37. Stuttgart: Deutsche Verlags-Anstalt.

Rassol, I. R., 1999. *Ivan Grigorievich Bubnov 1872–1919*. Moscow: Nauka & St. Petersburg: Elmor (in Russian).

Rayleigh, J. W. S., 1877. *The Theory of Sound*. vol. I. New York.

Rayleigh, J. W. S., 1878. *The Theory of Sound*. vol. II. New York.

Rayleigh, J. W. S., 1879. *Die Theorie des Schalles. Erster Band*. Trans. from the English by F. Neesen. Braunschweig: Vieweg.

Rayleigh, J. W. S., 1880. *Die Theorie des Schalles. Zweiter Band*. Trans. from the English by F. Neesen. Braunschweig: Vieweg.

Rayleigh, J. W. S., 1894. *The Theory of Sound*, vol. I., 2nd, rev. ed. London/New York: Macmillian & Co.

Rayleigh, J. W. S., 1896. *The Theory of Sound*. vol. II, 2nd, rev. ed. London/New York: Macmillian & Co.

Reckling, K.-A., 1957. *Plastizitätstheorie und ihre Anwendung auf Festigkeitsprobleme*. Berlin: Springer.

Redtenbacher, F., 1852. *Prinzipien der Mechanik und des Maschinenbaus*. Mannheim: Bassermann.

Reelfs, H., 1984. Konrad Levezow, Denkschrift auf Friedrich Gilly, königlichen Architecten und Professor der Academie der Baukunst in Berlin. In: *Friedrich Gilly (1772–1800) und die Privatgesellschaft junger Architekten* (exhibition catalogue), pp. 217–242. Berlin: Verlag Willmuth Arenhövel.

Reich, E., 1907. Beitrag zur Berechnung zylindrischer Reservoire. *Beton und Eisen*, vol. 6, No. 10, pp. 257–258

Rein, W., 1930. *Berichte des Ausschusses für Versuche im Stahlbau*. Ed. B. No. 4. Versuche zur Ermittlung der Knickspannungen für verschiedene Baustähle. Berlin: Verlag von Julius Springer.

Reinitzhuber, F., 1941. Die statische Wirkungsweise der Hohlplatten. *Die Bautechnik*, vol. 19, No. 6, pp. 66–67.

Reinitzhuber, F., 1960. Prof. Dr. Ernst Melan 70 Jahre. *Der Stahlbau*, vol. 29, No. 12, p. 390.

Reissner, E., 1934. Über die Berechnung von Plattenbalken. *Der Stahlbau*, vol. 7, No. 26, pp. 206–208.

Reissner, E., 1941. Least work of shear lag problems. *Journal of the Aeronautical Sciences*, vol. 8, pp. 284–291.

Reissner, E., 1944. On the theory of bending of elastic plates. *Journal of Mathematics and Physics*, vol. 23, pp. 184–191.

Reissner, E., 1946. Analysis of shear lag in box beams by the principle of minimum potential energy. *The Quarterly of Applied Mathematics*, vol. 4, pp. 268–278.

Reissner, E., 1948. Note on the method of complementary energy. *Journal of Mathematics and Physics*, vol. 27, pp. 159–160.

Reissner, E., 1950. On a variational theorem in elasticity. *Journal of Mathematics and Physics*, vol. 29, pp. 90–95.

Reissner, E., 1953. On a variational theorem for finite elastic deformations. *Journal of Mathematics and Physics*, vol. 32, pp. 129–135.

Reissner, E., 1977. Hans Reissner, Engineer, Physicist and Engineering Scientist. *The engineering science*, vol. 2, No. 4, Dec, pp. 97–104.

Reissner, E., 1996. *Selected Works in Applied Mechanics and Mathematics*. Sudbury: Jones & Barlett Publishers.

Reissner, H., 1899. Zur Dynamik des Fachwerks. *Zeitschrift für Bauwesen*, vol. 49, pp. 478–484.

Reissner, H., 1902. Anwendungen der Statik und Dynamik monozyklischer Systeme auf die Elastizitätstheorie. *Annalen der Physik*, vol. 314, 4th series, vol. 9, pp. 44–79.

Reissner, H., 1903. Schwingungsaufgaben aus der Theorie des Fachwerks. *Zeitschrift für Bauwesen*, vol. 53, pp. 138–162.

Reissner, H., 1905. Nordamerikanische Eisenbauwerkstätten. *Dinglers Polytechnisches Journal*, vol. 86 (320), No. 38, pp. 593–598, No. 39, pp. 609–613, No. 40, pp. 628–632, No. 41, pp. 643–649, No. 42, pp. 662–666, No. 46, pp. 726–731 & No. 47, pp. 741–746.

Reissner, H., 1906. Nordamerikanische Eisenbauwerkstätten. *Dinglers Polytechnisches Journal*, vol. 87 (321), No. 3, pp. 33–35, No. 4, pp. 54–59, No. 5, pp. 65–70, No. 6, pp. 88–92, No. 7, pp. 97–100, No. 9, pp. 131–135, No. 10, pp. 145–149, No. 11, pp. 164–168, No. 12, pp. 182–185, No. 14, pp. 214–218 & No. 15, pp. 230–234.

Reissner, H., 1908. Über die Spannungsverteilung in zylindrischen Behälterwänden. *Beton und Eisen*, vol. 7, No. 6, pp. 150–155.

Reissner, H., 1909. Über die Knicksicherheit ebener Bleche. *Zentralblatt der Bauverwaltung*, vol. 29, pp. 93–96 & 151.

Reissner, H., 1912. Spannungen in Kugelschalen (Kuppeln). In: *Festschrift Heinrich Müller-Breslau*, ed. A. Hertwig & H. Reissner, pp. 181–193. Leipzig: Alfred Kröner.

Reissner, H., 1925. Energiekriterium der Knicksicherheit. *Zeitschrift für Angewandte Mathematik und Mechanik*, vol. 5, pp. 475–478.

Reissner, H., 1926. Neuere Probleme aus der Flugzeugstatik. *Zeitschrift für Flugtechnik und Motorluftschiffahrt*, vol. 17, No. 7, pp. 137–146, No. 9, pp. 179–185 & No. 18, pp. 384–393.

Reissner, H., 1927. Neuere Probleme aus der Flugzeugstatik. *Zeitschrift für Flugtechnik und Motorluftschiffahrt*, vol. 18, No. 7, pp. 153–158.

Reissner, H., 1932. Theorie der Biegeschwingungen frei aufliegender Rechteckplatten unter dem Einfluß beweglicher, zeitlich periodisch veränderlicher Belastungen. *Ingenieur-Archiv*, vol. 3, pp. 668–673.

Reissner, H., 1933. Formänderung und Spannungen einer dünnwandigen, an den Rändern frei aufliegenden beliebig belasteten Zylinderschale. Eine Erweiterung der Navierschen Integrationsmethode. *Zeitschrift für Angewandte Mathematik und Mechanik*, vol. 13, pp. 133–138.

Reissner, H., 1934. Spannungsverteilung in der Gurtplatte einer Rippendecke. *Zeitschrift für Angewandte Mathematik und Mechanik*, vol. 14, pp. 312–313.

Reissner, H., 1943. Oscillations of Suspension Bridges. *Transactions of the American Society of Mechanical Engineers*, vol. 65, pp. A23–A32.

Renn, J., Damerow, P., Rieger, S., 2000. Hunting the White Elephant: When and How did Galileo Discover the Law of Fall? *Science in Context*, vol. 13, Nos. 3–4, pp. 299–419.

Rennert, K., 1999. Rudolph Bredt (1842–1900). In: *Ingenieure im Ruhrgebiet, Rheinisch-Westfälische Wirtschaftsbiographien Bd. 17*, ed. W. Weber, pp. 57–73. Münster: Aschendorffsche Verlagsbuchhandlung.

Rennison, R. W., 1998. The influence of William Fairbairn on Robert Stephenson's bridge designs: four bridges in north-east England. *Industrial Archaeology Review*, vol. XX, pp. 37–48.

Résal, J., Degrand, J., 1887. *Ponts en maçonnerie. I. Stabilité des voûtes*. Paris.

Reuleaux, F., 1875. *Theoretische Kinematik. Grundzüge einer Theorie des Maschinenwesens*. Braunschweig: Friedrich Vieweg und Sohn.

Richardson, L. F., 1909. The approximate arithmetical solution by finite differences of physical problems involving differential equations, with an application to stresses in masonry dams. *Philosophical Transactions*, A 210, p. 307

Richet, C., 1933. Pierre Simon Girard. *Comtes rendus de l'Académie des sciences*, vol. 197, pp. 1481–1486.

Ricken, H., 1992. Erinnerung an Fritz Edlen von Emperger. *Bautechnik*, vol. 69, No. 2, pp. 97–99.

Ricken, H., 1994/1. Johann Wilhelm Schwedler. In: *VDI Bau. Jahrbuch 1994*, ed. VDI-Gesellschaft Bautechnik. pp. 320–367. Düsseldorf: VDI-Verlag.

Ricken, H., 1994/2. *Der Bauingenieur: Geschichte eines Berufes*. Berlin: Verlag für Bauwesen.

Ricken, H., 1995. Erinnerungen an Richard Buckminster Fuller (1895–1983). *Bautechnik*, vol. 72, No. 7, pp. 460–466.

Ricken, H., 1997. Georg Christoph Mehrtens (1843–1917). *Bautechnik*, vol. 74, No. 2, pp. 117–120.

Ricken, H., 2001. Erinnerung an Emil Mörsch (1872–1950). *Bautechnik*, vol. 78, No. 1, pp. 46–51.

Ricken, H., 2003. Erinnerungen an Wladimir Grigorjewitsch Schuchow (1853–1939). *Bautechnik*, vol. 80, No. 8, pp. 542–547.

Ricketts, J. T., Loftin, M. K., Merritt, F. S. (ed.), 2003. *Standard Handbook for Civil Engineers*. 5th ed. New York: McGraw-Hill.

Riedel, W., 1927. Beiträge zur Lösung des ebenen Problems eines elastischen Körpers mittels der Airyschen Spannungsfunktion. *Zeitschrift für angewandte Mathematik und Mechanik (ZAMM)*, vol. 7, No. 3, pp. 169–188.

Riedrich, T., 2003. Trefftz, Erich Immanuel. In: *Die Professoren der TU Dresden 1828–2003*, ed. D. Petschel, pp. 973–974. Cologne: Böhlau.

Riessauw, F., 1960. *Liber Memorialis*. Rikjuniversiteit Gent, Fac. Wetenschappen en Fac. Toegepaste Wetenschappen, pp. 364–371. Ghent

Rippenbein, Ya., M., 1933. *Ramy i fermy prostanstvennye i ploskiye (On the analysis of two- and three-dimensional statically indeterminate systems, in collection two- and three-dimensional frames and trusses)*. Moscow: Gosstroiizdat (in Russian).

Ritter, A., 1861. Über die Berechnung eiserner Dach- und Brücken-Constructionen. *Zeitschrift des Arch.- & Ing.-Vereins Hannover*, vol. 7, No. 4, pp. 410–426.

Ritter, A., 1862. *Elementare Theorie und Berechnung eiserner Dach- und Brücken-Constructionen*. Hannover: Carl Rümpler.

Ritter, A., 1883. Statische Berechnung der Versteifungsfachwerke der Hängebrücken. *Schweizerische Bauzeitung*, vol. 1, No. 1, pp. 6–38.

Ritter, W., 1888. *Anwendungen der graphischen Statik. Nach Professor C. Culmann. Erster Teil.: Die im Inneren eines Balkens wirkenden Kräfte*. Zurich: Verlag von Meyer & Zeller.

Ritter, W., 1890. *Anwendungen der graphischen Statik. Nach Professor C. Culmann. Zweiter Teil.: Das Fachwerk*. Zurich: Verlag von Meyer & Zeller.

Ritter, W., 1899. Die Bauweise Hennebique. *Schweizerische Bauzeitung*, vol. 33, No. 5, pp. 41–43, No. 6, pp. 49–52 & No. 7, pp. 59–61.

Ritter, W., 1900. *Anwendungen der graphischen Statik. Nach Professor C. Culmann. Dritter Teil: Der kontinuierliche Balken*. Zurich: A. Raustein.

Ritter, W., 1906. *Anwendungen der graphischen Statik. Nach Professor C. Culmann. Erster Teil: Der Bogen*. Zurich: A. Raustein.

Rittershaus, T., 1875. Zur heutigen Schule der Kinematik. *Zivilingenieur*, vol. 21, pp. 425–450.

Ritz, W., 1909. Über eine neue Methode zur Lösung gewisser Variationsprobleme der mathematischen Physik. *Journal für die reine und angewandte Mathematik*, vol. 135, No. 1, pp. 1–61.

Roberts, V. L., Trent, I., 1991. *Bibliotheca Mechanica*. New York: Jonathan A. Hill.

Robertson, H., Arup, O., Jenkins, R. S., Rosevear, R. S., 1956. Design and Construction of the Printing Works at Debden. *The Structural Engineer*, vol. 34, No. 4, April, pp. 137–151.

Robison, J., 1801. Arch. In: *Supplement to the third edition of the Encyclopaedia Britannica*. Edinburgh: Thomson Bonar.

Rohn, A., 1924. Zur Frage des Schubmittelpunktes. *Schweizerische Bauzeitung*, vol. 83, No. 12, pp. 131–132.

Roik, K., Bode, H., Haensel, J., 1975. *Erläuterungen zu den "Richtlinien für die Bemessung und Ausführung von Stahlverbundträgern". Anwendungsbeispiele.* Ruhr University Bochum. Technische-Wissenschaftliche Mitteilungen, No. 75-11. Bochum.

Roik, K., 1981. Prof. Dr.-Ing. habil. Gustav Bürgermeister 75 Jahre. *Der Stahlbau,* vol. 50, p. 159.

Ronald Jenkins Memorial Issue, 1976. *The Arup Journal,* vol. 11, No. 1, London.

Ropohl, G., 1979. *Eine Systemtheorie der Technik. Zur Grundlegung der Allgemeinen Technologie.* Munich/Vienna: Hanser.

Ropohl, G., 1999/1. *Allgemeine Technologie. Eine Systemtheorie der Technik.* Munich/Vienna: Hanser.

Ropohl, 1999/2. Der Paradigmenwechsel in den Technikwissenschaften. In: *Die Sprachlosigkeit der Ingenieure,* ed. H. Duddeck & J. Mittelstraß, pp. 19–32. Opladen: Leske + Budrich.

Roš, M., 1925. Ludwig von Tetmajer 1850–1905. In: *Eidg. Materialprüfungsanstalt an der E.T.H. Zürich,* pp. 1–7. Zurich.

Roš, M., 1940. †Robert Maillart zum Gedächtnis. *Schweizerische Bauzeitung,* vol. 115, No. 19, pp. 224–226.

Rossmanith, H.-P., 1990. Die österreichischen Wegbereiter der Bruchmechanik und der Materialprüfung. *Österreichische Ingenieur- und Architekten-Zeitschrift (ÖIAZ),* vol. 135, No. 10, pp. 526–537.

Rota, A., 1763. *Ragionamento su la teoria fisico-matematica del p. R. G. Boscovich.* Rome.

Rothe, A., 1967. *Statik der Stabtragwerke. Band 2.* Berlin: VEB Verlag für Bauwesen.

Rothert, H., Gebbeken, N., 1988. Zur Entwicklung des Traglastverfahrens. In: *Festschrift Professor Dr.-Ing. Heinz Duddeck zu seinem sechszigsten Geburtstag,* ed. J. Scheer, H. Ahrens & H.-J. Bargstädt, pp. 15–42. Berlin: Springer.

Rozvany, G. I. N., 1989. *Structural design via optimality criteria: The Prager approach to structural optimization.* Dordrecht: Kluwer.

Rozvany, G. I. N., 2000. William Prager (1903–1980). *Structural Multidisciplinary Optimization,* vol. 19, pp. 167–168.

Ruddock, T., 1979. *Arch Bridges and their Builders 1735–1835.* Cambridge: Cambridge University Press.

Ruddock, T., 1999/2000. Galashiels wire suspension bridge, 1816. *Transactions of the Newcomen Society,* vol. 71, No. 1, pp. 103–113.

Rudeloff, M., 1920. *Berichte des Ausschusses für Versuche im Eisenbau.* ed. A, No. 2. Versuche zur Prüfung und Abnahme der 3000-t-Maschine. Berlin: Verlag von Julius Springer.

Rudolff, O., 1932. Gerstners Persönlichkeit im Rahmen seiner Zeit und seiner Heimat. *HDI-Mitteilungen,* vol. 21, pp. 266–270.

Rudolph, C. L., 1959. †Prof. Dr.-Ing. Ludwig Mann. *Der Stahlbau,* vol. 28, No. 4, pp. 109–110.

Rueb, B., 1914. *Der Einfluß der Längs- und Querkräfte auf statisch unbestimmte Bogen- und Rahmentragwerke.* Berlin: Wilhelm Ernst & Sohn.

Ruff, F., 1903. *Auskunftsbuch für statische Berechnungen (Schnellstatiker).* Frankfurt a. M.: Verlag des "Auskunftsbuch für statische Berechnungen".

Ruh, M., Klimke, H., 1981. Darstellung eines Konstruktions-Systems zur Erzeugung der Geometrie von Knoten-Stab-Tragwerken. In: *IKOSS – Intern. FEM Congress,* pp. 107–122. Baden-Baden.

Rühlmann, M., 1860. *Grundzüge der Mechanik im Allgemeinen und der Geostatik im Besondern.* Leipzig: Arnoldische Buchhandlung.

Rühlmann, M., 1879. *Hydromechanik oder die Technische Mechanik flüssiger Körper.* Hannover: Hahn'sche Buchhandlung.

Rühlmann, M., 1883. Leben und Wirken Eytelwein's. *Zeitschrift des Arch.- & Ing.-Vereins Hannover,* vol. 29, No. 5, pp. 301–304.

Rühlmann, M., 1885. *Vorträge über Geschichte der Technischen Mechanik.* Leipzig: Baumgärtner's Buchhandlung.

Rumpf, H., 1966. Struktur der Zerkleinerungswissenschaft. *Aufbereitungs-Technik,* vol. 7, No. 8, pp. 421–435.

Rumpf, H., 1973. Gedanken zur Wissenschaftstheorie der Technikwissenschaften. In: H. Lenk, S. Moser (ed.): *Techne, Technik, Technologie. Philosophische Perspektiven.* Pullach bei Munich: Verlag Documentation.

Runge, C., 1904. Über die Formänderung eines zylindrischen Wasserbehälters durch den Wasserdruck. *Zeitschrift für Mathematik und Physik,* vol. 51, p. 254ff.

Rüsch, H., 1929. Die Großmarkthalle in Leipzig, ein neues Kuppelbausystem, zusammengesetzt aus Zeiss-Dywidag-Schalengewölben. *Beton und Eisen,* vol. 28, No. 18, pp. 325–329, No. 19, pp. 341–346, No. 23, pp. 422–429 & No. 24, pp. 437–442.

Rüsch, H., 1931. *Theorie der querversteiften Zylinderschalen für schmale, unsymmetrische Kreissegmente.* Dissertation, Munich TH.

Rüsch, H., 1936. Shedbauten in Schalenbauweise, System Zeiss-Dywidag. *Beton und Eisen,* vol. 35, No. 10, pp. 159–165.

Rüsch, H., 1939. Die Hallenbauten der Volkswagenwerke in Schalenbauweise, System Zeiss-Dywidag. *Der Bauingenieur,* vol. 20, No. 9/10, pp. 123–129.

Rüsch, H., 1953. *Spannbeton, Erläuterungen zu den Richtlinien für Bemessung und Ausführung.* Berlin: Verlag von Wilhelm Ernst & Sohn.

Rüsch, H., Kupfer, H., 1954. Bemessung von Spannbetonbauteilen. In: *Beton-Kalender,* vol. 43, pt. I, pp. 401–468. Berlin: Wilhelm Ernst & Sohn.

Rüsch, H., 1957. Fahrbahnplatten von Straßenbrücken. Deutscher Ausschuß für Stahlbeton, No. 106. Berlin: Wilhelm Ernst & Sohn.

Rüsch, H., 1964. Über die Grenzen der Anwendbarkeit der Fachwerkanalogie bei der Berechnung der Schubfestigkeit von Stahlbetonbalken. In: Festschrift F. Campus "Amici et Alumni", Université de Liège.

Rüsch, H., 1967. *Berechnungstafeln für schiefwinklige Fahrbahnplatten von Straßenbrücken.* Deutscher Ausschuß für Stahlbeton, No. 166. Berlin: Wilhelm Ernst & Sohn.

Rüsch, H., 1972. *Stahlbeton – Spannbeton. vol. 1: Die Grundlagen des bewehrten Betons unter besonderer Berücksichtigung der neuen DIN 1045. Werkstoffeigenschaften und Bemessungsverfahren.* Düsseldorf: Werner-Verlag.

Rüsch, H., Jungwirth, D., 1976. *Stahlbeton – Spannbeton. Bd. 2: Die Berücksichtigung der Einflüsse von Kriechen und Schwinden auf das Verhalten der Tragwerke.* Düsseldorf: Werner-Verlag.

Rüsch, H., Kupfer, H., 1980. Bemessung von Spannbetonbauteilen. In: *Beton-Kalender,* vol. 69, pt. I, pp. 989–1086. Berlin: Wilhelm Ernst & Sohn.

Ruske, W., 1971. *100 Jahre Materialprüfung in Berlin.* Berlin: Bundesanstalt für Materialprüfung. Berlin: Bundesanstalt für Materialprüfung.

Rvachev, V. L., 1982. *Teoriya R-funktsii i nekotorye ee prilozheniya* (theory of R-functions and some applications). Kiev: Naukova Dumka, 1982 (in Russian).

Rvachev, V. L., Sinekop, N. S., 1990. *Metod R-funktsii v teorii uprugosti i plastichnosti* (R-functions method in elastic and plastic theory). Kiev: Naukova Dumka (in Russian).

Rvachev, V. L., Sheiko, Sinekop, T. I., 1995. R-functions in boundary value problems in mechanics. *Applied Mechanics Reviews,* vol. 48, No. 4, pp. 151–188.

Saavedra, E., 1856. *Teoría de los puentes colgados.* Madrid: Imprenta de José Peña.

Saavedra, E., 1859/1. *Lecciones sobre la resistencia de los materiales.* 2nd ed. Madrid: Escuela Superior de Ingenieros de Caminos, Canales y Puertos.

Saavedra, E., 1859/2. Nota sobre el coeficiente de estabilidad. *Revista de Obras Públicas,* vol. 7, pp. 277–281.

Saavedra, E., 1860. Nota sobre la determinación del problema del equilibrio de las bóvedas. *Revista de Obras Públicas,* vol. 8, pp. 101–104. (published in the same year in German with the title "Gleichgewicht der Gewölbe". *Zeitschrift des Arch.- & Ing.-Vereins Hannover,* vol. 6, pp. 459–461).

Saavedra, E., 1866. Experimento sobre los arcos de máxima estabilidad. *Revista de Obras Públicas,* vol. 14, pp. 13–21.

Saavedra, E., 1868. Teoría de los contrafuertes. *Revista de Obras Públicas,* vol. 8, pp. 92–95.

Sáenz Ridruejo, F., 1990. *Los Ingenieros de Caminos del siglo XIX.* Madrid: Colegio Oficial de Ingenieros de Caminos, Canales y Puertos.

Saint-Guilhem, P., 1859. Mémoire sur l'établissement des arches de pont aussujetties aux conditions du maximum de stabilité. *Annales des Ponts et Chaussées*, vol. 17, pp. 83–106.

Saint-Venant, A. J. C. B. de, 1844/1. Mémoire sur les pressions qui se développent à l'intérieur des corps solides lorsque les déplacements de leurs points, sans altérer l'élasticité, ne peuvent cependant pas être considérés comme très petits. *L'Institut*, pp. 26–28, 10 April.

Saint-Venant, A. J.-C. B. de, 1844/2. Mémoire sur l'équilibre des corps solides, dans les limites de leur élasticité, et sur les conditions de leur résistance, quand les déplacements éprouvés par leurs points ne sont pas très-petits. *Société Philomatique de Paris*, 26 March.

Saint-Venant, A. J.-C. B. de, 1844/3. Note sur l'état d'équilibre d'une verge élastique à double courbure lorsque les déplacements éprouvés par ses points, par suite l'action des forces qui la sollicitent, ne sont pas très-petits. *Comptes Rendus de l'Académie des Sciences*, vol. 19, II, pp. 36–44, 1 July.

Saint-Venant, A. J.-C. B. de, 1844/4. Deuxième note: Sur l'état d'équilibre d'une verge élastique à double courbure lorsque les déplacements éprouvés par ses points, par suite l'action des forces qui la sollicitent, ne sont pas très-petits. *Comptes Rendus de l'Académie des Sciences*, vol. 19, II, pp. 181–187, 15 July.

Saint-Venant, A. J.-C. B. de, 1847/1. Mémoire sur l'équilibre des corps solides, dans les limites de leur élasticité, et sur les conditions de leur résistance, quand les déplacements éprouvés par leurs points ne sont pas très-petits. *Comptes Rendus de l'Académie des Sciences*, vol. 24, pp. 260–263, 22 Feb.

Saint-Venant, A. J.-C. B. de, 1847/2. Mémoire sur la torsion des prismes et sur la forme affectée par leurs sections transversales primitivement planes. *Comptes Rendus de l'Académie des Sciences*, vol. 24, pp. 485–488, 22 March.

Saint-Venant, A. J.-C. B. de, 1847/3. Suite au Mémoire sur la torsion des prismes. *Comtes Rendus de l'Académie des Sciences*, vol. 24, pp. 847–49, 10 May.

Saint-Venant, A. J.-C. B. de, 1855. Mémoire sur la Torsion des Prismes, avec des considérations sur leur flexion, ainsi que sur l'équilibre intérieur des solides élastiques en général, et des formules pratiques pour le calcul de leur résistance à divers efforts s'exerçant simultanément. *Mémoires présentés par divers savants à l'Académie des Sciences de l'Institut Impérial de France. Recueil des savants étrangers*, vol. 14, pp. 233–560.

Saint-Venant, A. J.-C. B. de, 1856. Mémoire sur la flexion des prismes, sur les glissements transversaux et longitudinaux qui l'accompagnent lorsqu'elle ne s'opère pas uniformément ou en arc de cercle, et sur la forme courbe affectée alors par leurs sections transversales primitivement planes. *Journal de Mathématiques pures et appliquées de Liouville*, II, 1, pp. 88–189.

Saint-Venant, A. J.-C. B. de, 1860. Sur les conditions pour que six fonctions des coordonées $x, y, z$ des points d'un corps élastiques représentent des composantes de pression s'exerçant sur trois plans rectangulaires à l'intérieur de ce corps, par suite de petits changements de distance de ses parties. Paris: L'Institut, vol. 28.

Salvadori, M. G., 1989. Hans Heinrich Bleich. In: *Memorial Tributes*. National Academy of Engineering, vol. 3, pp. 44–48. Washington, D.C.: National Academy Press.

Samuelsson, A., 1984. In Memoriam S. O. Asplund. *International Journal for Numerical Methods in Engineering*, vol. 20, p. 2325.

Samuelsson, A., 2002. Computational Mechanics – 50 Years. *iacm expressions* 12/02, pp. 6–7.

Samuelsson, A., 2004. Professorn som skrew kompendier på vers. *Chalmers magasin*, 2-2002, pp. 36–37.

Samuelsson, A., Zienkiewicz, O. C., 2006. History of the stiffness method. *International Journal for Numerical Methods in Engineering*, vol. 67, pp. 149–157.

Sanchez, J., 1980. *FORMEX Formulation of Structural Configurations*. PhD thesis, University of Surrey, Dept. of Civil Engineering.

Sanders, A., 1898. Theorie des Eisenbetons. *De Ingenieur*, p. 187ff.

Sandkühler, H. J., 1988. Dialektik, Krise des Wissens, Enzyklopädie und Emanzipation. Das Enzyklopädische Wörterbuch zu Philosophie und Wissenschaft. In: *DIALEKTIK 16. Enzyklopädie und Emanzipation. Das Ganze wissen*. Ed. H. J. Sandkühler, H.-H. Holz and L. Lambrecht, pp. 203–224. Cologne: Pahl-Rugenstein.

Sandkühler, H. J., 1990. *Enzyklopädie, Krise des Wissens, Emanzipation*. In: Europäische Enzyklopädie zu Philosophie und Wissenschaften, vol. 1, ed. H. J. Sandkühler, pp. 746–757. Hamburg: Felix Meiner.

Sattler, K., 1953. *Theorie der Verbundkonstruktionen*. Berlin: Wilhelm Ernst & Sohn.

Sattler, K., 1955/1. Academic commemorative event for Privvy Counsellor August Hertwig, 7 June 1955, Berlin-Charlottenburg TU. *Humanismus und Technik*, vol. 3, No. 1, 30 June, pp. 3–10.

Sattler, K., 1955/2. Ein allgemeines Berechnungsverfahren für Tragwerke mit elastischem Verbund. Cologne: Stahlbau-Verlags-GmbH.

Sattler, K., 1959. Theorie der Verbundkonstruktionen. Band 1: Theorie. Band 2: Zahlenbeispiele. Berlin: Wilhelm Ernst & Sohn.

Sattler, K., 1960. †Prof. Dr.-Ing. E. h. Dr.-Ing. Ernst Chwalla. *Der Bauingenieur*, vol. 35, No. 7, pp. 275–278.

Sattler, K., 1962. Betrachtungen über neuere Verdübelungen im Verbundbau. *Der Bauingenieur*, vol. 37, No. 1, pp. 1–8 & No. 2, pp. 60–67.

Sauer, R., 1940. Graphische Statik räumlicher Kräftesysteme mit Hilfe der dualen Kräfteabbildung. *Zeitschrift für angewandte Mathematik und Mechanik*, vol. 20, No. 3, pp. 174–180.

Savin, G. N., 1951. *Kontsentratsiia napriazhenii okolo otverstii* (stress concentration around holes). Moscow: Gostekhizdat (in Russian).

Savin, G. N., 1956. *Spannungserhöhung am Rande von Löchern*. Berlin: VEB Verlag Technik.

Savin, G. N., 1961. *Stress concentration around holes*. Oxford: Pergamon Press.

Savin, G. N., Putyata, T. V., Fradlin, B. N., 1964. *Ocherki razvitiya nekotorykh fundamental'nykh problem mekhaniki* (essays in the advancement of some fundamental problems of mechanics). Kiev: Naukova Dumka (in Russian).

Savin, G. N., Fleishman, N. P., 1964. *Plastinki I obolochki s rebrami zhestkosti* (plates and shells with stiffening ribs). Kiev: Naukova Dumka (in Russian).

Savin, G. N., Fleishman, N. P., 1967. *Rib-reinforced plates and shells.* Jerusalem: JHS.

Savin, G. N., 1968. *Raspredelenie napryazhenii okolo otverstii* (stress distribution around holes). Kiev: Naukova Dumka (in Russian).

Savin, G. N., 1970/1. *Stress distribution around holes.* Technical Translation NASA TT F-607. Washington: NASA.

Savin, G. N., 1970/2. *Mekhanicheskoe podobie konstruktsii iz armirovannogo materials* (mechanical similarity of structures made of reinforced materials). Kiev: Naukova Dumka (in Russian).

Savin, G. N., 1979. *Mekhanika deformiruemykh tel: Izbrannye trudy* (mechanics of deformable bodies: selected works). Kiev: Naukova Dumka (in Russian).

Schade, H. H., 1938. Bending Theory of Ship Bottom Structure. *Transactions: Society of Naval Architects and Marine Engineers*, vol. 46.

Schade, H. H., 1940. The Orthogonally Stiffened Plate under Uniform Lateral Load. Transactions of the American Society of Mechanical Engineers. *Journal of Applied Mechanics*, vol. 7, A143–A146.

Schade, H. H., 1953. The Effective Breadth Concept in Ship-Structural Design. *Transactions: Society of Naval Architects and Marine Engineers*, vol. 61.

Schadek v. Degenburg, R., Demel, K., 1915. *Hilfsmittel zur einfachen Berechnung von Formänderungen und von statisch unbestimmten Trägern*. Berlin: Wilhelm Ernst & Sohn.

Schädlich, C., 1967. *Das Eisen in der Architektur des 19. Jahrhunderts. Beitrag zur Geschichte eines neuen Baustoffs*. Habilitation thesis at the Weimar Polytechnic for Architecture & Construction (typewritten ms.).

Schaechterle, K., 1912. *Beiträge zur Berechnung der im Eisenbetonbau üblichen elastischen Bogen und Rahmen*. Forscherarbeiten auf dem Gebiete des Eisenbetons, No. 17. Berlin: Wilhelm Ernst & Sohn.

Schaechterle, K., 1914. *Beiträge zur Berechnung der im Eisenbetonbau üblichen elastischen Bogen*

und Rahmen. 2nd, rev. ed. Berlin: Wilhelm Ernst & Sohn.
Schaechterle, K., 1929. Von den allgemeinen Grundlagen der Festigkeitsrechnung. *Der Stahlbau,* vol. 2, No. 11, pp. 125–130 & No. 12, pp. 135–142.
Schaechterle, K., 1934. Neue Fahrbahnkonstruktionen für stählerne Straßenbrücken. *Die Bautechnik,* vol. 12, No. 37, pp. 479–483 & No. 42, pp. 564–567.
Schaechterle, K., Leonhardt, F., 1936. Stahlbrücken mit Leichtfahrbahnen. Versteifte Tonnenbleche, Versuche und Ausführungen. *Die Bautechnik,* vol. 14, No. 43, pp. 626–630 & No. 45, pp. 659–662.
Schaechterle, K., Leonhardt, F., 1938. Fahrbahnen der Straßenbrücken. Versteifte Erfahrungen, Versuche und Folgerungen. *Die Bautechnik,* vol. 16, No. 23/24, pp. 306–324.
Schäfer, J., Hilsdorf, H. K., 1991. Struktur und mechanische Eigenschaften von Kalkmörteln. In: *Jahrbuch des Sonderforschungsbereichs 315 Universität Karlsruhe 1991,* pp. 65–76. Berlin: Ernst & Sohn.
Schäffer, T., Sonne, E. (ed.), 1886. *Handbuch der Ingenieurwissenschaften. Zweiter Band: Der Brückenbau. Erste Abteilung,* 2nd, rev. ed. Leipzig: Wilhelm Engelmann.
Schäffer, T., Sonne, E. (ed.), 1888/1. *Handbuch der Ingenieurwissenschaften. Zweiter Band: Der Brückenbau. Dritte Abteilung,* 2nd, rev. ed. Leipzig: Wilhelm Engelmann.
Schäffer, T., Sonne, E. (ed.), 1888/2. *Handbuch der Ingenieurwissenschaften. Zweiter Band: Der Brückenbau. Vierte Abteilung,* 2nd, rev. ed. Leipzig: Wilhelm Engelmann.
Schäffer, T., Sonne, E. (ed.), 1889. *Handbuch der Ingenieurwissenschaften. Zweiter Band: Der Brückenbau. Fünfte Abteilung,* 2nd, rev. ed. Leipzig: Wilhelm Engelmann.
Schäffer, T., Sonne, E., Landsberg, T. (ed.), 1890. *Handbuch der Ingenieurwissenschaften. Zweiter Band: Der Brückenbau. Zweite Abteilung,* 2nd, rev. ed. Leipzig: Wilhelm Engelmann.
Schaper, G., 1935/1. †Hermann Zimmermann. *Die Bautechnik,* vol. 13, No. 18, p. 255.
Schaper, G., 1935/2. Deutscher Ausschuß für Stahlbau. *Die Bautechnik,* vol. 13, No. 55, p. 762.
Schaper, G., 1935/3. Leichte Fahrbahndecken und leichte Fahrbahntafeln für stählerne Straßenbrücken. *Die Bautechnik,* vol. 13, No. 4, pp. 47–49.
Schapitz, E., Fellner, H., Köller, H., 1939. Experimentelle und rechnerische Untersuchung eines auf Biegung belasteten Schalenflügelmodells. In: *Jahrbuch 1939 der Deutschen Versuchsanstalt für Luftfahrt e.V.,* Berlin-Adlershof, pp. 137–150.
Schardt, R., 1986. Würdigung der Persönlichkeit Kurt Klöppel. In: *Kurt-Klöppel-Gedächtnis-Kolloquium TH Darmstadt 15. bis 16. September 1986. THD Schriftenreihe Wissenschaft und Technik, Bd. 31,* ed. President of Darmstadt TH, pp. 13–36. Darmstadt.

Scheer, J., 1958. Benutzung programmgesteuerter Rechenautomaten für statische Aufgaben, erläutert am Beispiel der Durchlaufträgerberechnung. *Der Stahlbau,* vol. 27, No. 9, pp. 225–229 & No. 10, pp. 275–280.
Scheer, J., 1986. Würdigung des Werkes auf dem Gebiet der Stabilitätstheorie. In: Kurt-Klöppel Memorial Colloquium, Darmstadt TH, 15–16 Sept 1986. THD "Wissenschaft und Technik" series, vol. 31, ed. President of Darmstadt TH, pp. 85–111. Darmstadt.
Scheer, J., 1992. Entwicklungen im Stahlbau, im Rückblick und Ausblick eines Universitätsprofessors. In: *18/1992 Bericht aus Forschung, Entwicklung und Normung. Neues aus Forschung, Entwicklung und Normung,* ed. DASt, pp. 5–16. Cologne: Stahlbau-Verlagsgesellschaft mbH.
Scheer, J., 1998. Letter from Prof. Joachim Scheer to the author dated 29 Nov 1998.
Scheer, J., 2000. *Versagen von Bauwerken. Ursachen, Lehren. Band 1: Brücken.* Berlin: Verlag Ernst & Sohn.
Scheer, J. (ed.), 2001. *Stahlbau,* vol. 70, No. 9.
Scheer, J., 2003. *Einige Informationen zur Entwicklung des MERO-Raumtragwerkes.* Letter to the author dated 9 Aug 2003.
Scheffler, H., 1857. *Theorie der Gewölbe, Futtermauern und eisernen Brücken.* Braunschweig: Verlag der Schulbuchhandlung.
Scheffler, H., 1858/1. Festigkeits- und Biegungsverhältnisse eines über mehrere Stützpunkte fortlaufenden Trägers. *Der Civilingenieur,* vol. 4, pp. 62–73.
Scheffler, H., 1858/2. Continuirliche Brückenträger. *Der Civilingenieur,* vol. 4, pp. 142–146.
Scheffler, H., 1858/3. *Theorie der Festigkeit gegen das Zerknicken nebst Untersuchungen über verschiedenen inneren Spannungen gebogener Körper und über andere Probleme der Biegungstheorie mit praktischen Anwendungen.* Braunschweig.
Scheffler, H., 1860. Continuirliche Brückenträger. *Der Civilingenieur,* vol. 6, pp. 129–202.
Schellbach, K. H., 1851. Probleme der Variationsrechnung. *Journal für die reine und angewandte Mathematik,* vol. 41, No. 4, pp. 293–363.
Schischkoff, G., 1982. Technik. In: G. Schischkoff (ed.): *Philosophisches Wörterbuch,* 21st ed., pp. 1209–1210. Stuttgart: Alfred Kröner.
Schlegel, R., Rautenstrauch, K., 2000. Ein elastoplastisches Berechnungsmodell zur räumlichen Untersuchung von Mauerwerkstrukturen. *Bautechnik,* vol. 77, No. 6, pp. 426–436.
Schlegel, R., Will, J., Fischer, D., Rautenstrauch, K., 2003. Tragfähigkeitsbewertung gemauerter Brückenbauwerke mit modernen Berechnungsmethoden am Beispiel der Göltzschtalbrücke. *Bautechnik,* vol. 80, No. 1, pp. 15–23.
Schlegel, R., 2004. *Numerische Berechnung von Mauerwerksstrukturen in homogenen und diskreten Modellierungsstrategien.* Dissertation, Bauhaus University, Wiemar.

Schlegel, R., Konietzky, H., Rautenstrauch, K., 2005. Mathematische Beschreibung von Mauerwerk unter statischer und dynamischer Beanspruchung im Rahmen der Diskontinuumsmechanik mit Hilfe der Distinkt-Element-Methode. *Mauerwerk,* vol. 9, No. 4, pp. 143–151.
Schleicher, F. (ed.), 1943. *Taschenbuch für Bauingenieure,* 1st ed. Berlin: Springer-Verlag.
Schleicher, F., 1949. Lebenslauf von Prof. Dr.-Ing. habil. Ferdinand Schleicher. Typewritten ms.
Schleicher, F. (ed.), 1955. *Taschenbuch für Bauingenieure,* 2nd, rev. ed. Berlin: Springer-Verlag.
Schleicher, F., Mehmel, A., 1957. Nachruf zu Prof. Hayashi. *Der Bauingenieur,* vol. 32, No. 9, p. 367.
Schleusner, A., 1938/1. Zum Prinzip der virtuellen Verschiebungen. *Beton und Eisen,* vol. 37, No. 15, pp. 252–254.
Schleusner, A., 1938/2. Erwiderung zu Kammüller 1938/1. *Beton und Eisen,* vol. 37, No. 16, p. 271.
Schleusner, A., 1938/3. Das Prinzip der virtuellen Verrückungen und die Variationsprinzipien der Elastizitätstheorie. *Der Stahlbau,* vol. 11, No. 24, pp. 185–192.
Schierk, H.-F., 1994. Die Talbrücke bei Müngsten. Vor 100 Jahren begann der Bau dieses Meisterwerks der Ingenieurbaukunst. In: *VDI Bau Jahrbuch 1994,* ed. VDI-Gesellschaft Bautechnik, pp. 369–381. Düsseldorf: VDI-Verlag.
Schlaich, J., Schäfer, K., 1984. Konstruieren im Stahlbetonbau. In: *Beton-Kalender,* vol. 73, pt. II, pp. 787–1005. Berlin: Wilhelm Ernst & Sohn.
Schlaich, J., Schäfer, K., 1989. Konstruieren im Stahlbetonbau. In: *Beton-Kalender,* vol. 78, pt. II, pp. 563–715. Berlin: Wilhelm Ernst & Sohn.
Schlaich, J., Schäfer, K., 1993. Konstruieren im Stahlbetonbau. In: *Beton-Kalender,* vol. 82, pt. II, pp. 327–486. Berlin: Wilhelm Ernst & Sohn.
Schlaich, J., Schäfer, K., 1998. Konstruieren im Stahlbetonbau. In: *Beton-Kalender,* vol. 87, pt. II, pp. 721–895. Berlin: Wilhelm Ernst & Sohn.
Schlaich, J., Schäfer, K., 2001. Konstruieren im Stahlbetonbau. In: *Beton-Kalender,* vol. 90, pt. II, pp. 311–492. Berlin: Wilhelm Ernst & Sohn.
Schlaich, M., 2003. Der Messeturm in Rostock – ein Tensegrityrekord. *Stahlbau,* vol. 72, No. 10, pp. 697–701.
Schlink, W., 1904. Über räumliche Dachfachwerke. *Zeitschrift für Arch.- & Ing.-wesen,* vol. 50 (new series, vol. 9), col. 183–198.
Schlink, W., 1907/1. Über Stabilitätsuntersuchungen von Raumfachwerken. In: *Jahresbericht der Deutschen Mathematiker-Vereinigung,* vol. 16, pp. 46–53. Leipzig: Verlag von B. G. Teubner.
Schlink, W., 1907/2. *Statik der Raumfachwerke.* Leipzig: Verlag von B. G. Teubner.
Schlink, W., 1923. August Föppl. *Zeitschrift für angewandte Mathematik und Mechanik,* vol. 3, No. 6, pp. 481–483.
Schlink, W., Dietz, H., 1946. *Technische Statik. Ein Lehrbuch zur Einführung ins technische*

Denken. 2nd & 3rd, rev. ed. Berlin: Springer-Verlag.

Schmidt, H., 2005. Von der Steinkuppel zur Zeiss-Dywidag-Schalenbauweise. *Beton- und Stahlbetonbau,* vol. 100, No. 1, pp. 79 – 92.

Schmiedel, K., 1994. *Konstruktion und Gestalt. Ein Vierteljahrhundert Stahlbauarchitektur.* Berlin: Ernst & Sohn.

Schnabel, F., 1925. Die Anfänge des technischen Hochschulwesens. In: *Festschrift anläßlich des 100jährigen Bestehens der Technischen Hochschule Fridericiana Karlsruhe,* pp. 1 – 44. Berlin: VDI-Verlag.

Schneider, R., 1886. Letter. *ZVDI,* vol. 30, No. 47, p. 1030.

Schober, H., Kürschner, K., Jungjohann, H., 2004. Neue Messe Mailand – Netzstruktur und Tragverhalten einer Freiformfläche. *Stahlbau,* vol. 73, No. 8, pp. 541 – 551.

Scholl, L. U., 1990. Johann Albert Eytelwein. In: *Berlinische Lebensbilder. Techniker,* ed. W. Treue & W. König, pp. 47 – 63. Berlin: Colloquium Verlag 1990.

Scholz, E., 1989. *Symmetrie, Gruppe, Dualität. Zur Beziehung zwischen theoretischer Mathematik und Anwendungen in Kristallographie und Baustatik des 19. Jahrhunderts.* Basel: Birkhäuser.

Schöne, L., 1999. Kuppelschale und Rippenkuppel – Zur Entwicklung von zwei frühen Eisenbeton-Konstruktionsarten. In: Zur Geschichte des Stahlbetonbaus – Die Anfänge in Deutschland 1850 bis 1910, ed. Hartwig Schmidt. *Beton- und Stahlbetonbau, Spezial* (special issue), pp. 66 – 74.

Schönhöfer, R., 1907. Review of Love's textbook on elasticity. *Beton und Eisen,* vol. 6, No. 11, p. 296.

Schrader, K. H., 1969. *Die Deformationsmethode als Grundlage einer problemorientierten Sprache.* Mannheim: Bibliographisches Institut.

Schreyer, C., Ramm, H., Wagner, W., 1967/1. *Praktische Baustatik. Teil 2.* 10th, rev. ed. Stuttgart: B. G. Teubner.

Schreyer, C., Ramm, H., Wagner, W., 1967/2. *Praktische Baustatik. Teil 3.* 5th, rev. ed. Stuttgart: B. G. Teubner.

Schröder, H., 1962. Baustatik. Formeln und Tabellen. In: *Beton-Kalender,* vol. 51, pt. I, pp. 225 – 321. Berlin: Wilhelm Ernst & Sohn.

Schubert, J. A., 1845. *Über freie und vorgeschriebene Stützlinien.* Dresden: Druck der Teubnerschen Officin.

Schubert, J. A., 1847/1848. *Theorie der Construction steinerner Bogenbrücken,* 2 vols. Dresden/Leipzig: Arnoldische Buchhandlung.

Schüle, F., 1905. †L. von Tetmajer. *Schweizerische Bauzeitung,* vol. 39, p. 65.

Schulz, F. J. E., 1808. *Versuch einiger Beiträge zur hydraulischen Architektur.* Königsberg: Friedrich Ricolovius.

Schwarz, K. (ed.), 1979. *100 Jahre Technische Universität Berlin 1879 – 1979.* Exhibition catalogue. Berlin: Berlin TU library/Publications Dept.

Schwarz, O., 1993. August Ritter und die erste Theorie des Aufbaus und der Entwicklung von Fixsternen als konvektive Gaskugeln. *Naturwissenschaft Technik Medizin (NTM).* New series vol. 1. No. 3, pp. 137 – 145.

Schwedler, J. W., 1851. Theorie der Brückenbalkensysteme. *Zeitschrift für Bauwesen,* vol. 1, col. 114 – 123, 162 – 173 & 265 – 278.

Schwedler, J. W., 1863. Zur Theorie der Kuppelgewölbe. *Zeitschrift für Bauwesen,* vol. 13, col. 535ff.

Schwedler, J. W., 1865. Resultate über die Konstruktion der eisernen Brücken. *Zeitschrift für Bauwesen,* vol. 15, col. 331 – 340.

Schwedler, J. W., 1866. Die Construction der Kuppeldächer. *Zeitschrift für Bauwesen,* vol. 16, col. 7 – 34.

Schwedler, J. W., 1868. Die Stabilität des tonnenförmigen Kappengewölbes. *Deutsche Bauzeitung,* vol. 2, pp. 153 – 155.

Schwedler, J. W., 1869/1. Eiserne Dachconstructionen über Retortenhäusern der Gas-Anstalten zu Berlin. *Zeitschrift für Bauwesen,* vol. 19, col. 65 – 70, atlas sht. 24 – 27.

Schwedler, J. W., 1869/2. Dach- und Decken-Construction über dem Festsaale des neuen Rathauses zu Berlin. *Zeitschrift für Bauwesen,* vol. 19, col. 387 – 390, atlas sht. 54 – 55.

Schwedler, J. W., 1869/3. Dach- und Decken-Construction über dem Stadtverordnetensaale im neuen Rathause zu Berlin. *Zeitschrift für Bauwesen,* vol. 19, col. 389 – 392, atlas sht. 56.

Schwedler, J. W., 1882. Discussion on iron permanent way. In: C. Wood. Iron permanent way. *Minutes of Proceedings of the Institution of Civil Engineers,* vol. 67, pp. 95 – 118.

Schweizerischer Ingenieur- und Architekten-Verein (ed.), 1994. *Tragwerksnormen 1892 – 1956. Eine Sammlung der in der Schweiz zwischen 1892 und 1956 erlassenen Verordnungen, Vorschriften und Normen für Tragwerke aus Stahl, Beton, Mauerwerk und Holz.* Zurich: Self-pub.

Schwerin, E., 1919. Über Spannungen in symmetrisch belasteten Kugelschalen (Kuppeln) insbesondere bei Belastung durch Winddruck. *Armierter Beton,* vol. 12, pp. 25 – 37, 54 – 63 & 81 – 88.

Seidel, H., 1980. *Von Thales bis Platon. Vorlesungen zur Geschichte der Philosophie.* Cologne: Pahl-Rugenstein Verlag.

Seidenberg, A., 1981. Mascheroni, Lorenzo. In: *Dictionary of Scientific Biography,* vol. 9. C. C. Gillispie (ed.), pp. 156 – 158. New York: Charles Scribner's Sons.

Séjourné, P., 1913 – 1916. *Grandes Voûtes.* 6 vols. Bourges: Imprimerie Vve Tardy-Pigelet et Fils.

Serensen, S. V., 1985. *Izbrannye trudy* (selected works), vol. 3. Kiev: Naukova Dumka (in Russian).

Sganzin, J. M., 1840 – 1844. *Programme ou résumé des leçons d'un cours de constructions, avec des applications tirées specialement de l'art de l'ingenieur des ponts et chaussées.* 5th ed. Liège: Dominique Avanzo.

Shanley, F. R., 1944. *Basic Structures.* New York: Wiley & Sons.

Shanley, F. R., 1946. The column paradox. *Journal of the Aeronautical Sciences,* vol. 13, p. 648.

Shanley, F. R., 1947/1. Inelastic column theory. *Journal of the Aeronautical Sciences,* vol. 14, pp. 261 – 267.

Shanley, F. R., 1947/2. Cardboard-Box Wing Structures. *Journal of the Aeronautical Sciences,* vol. 14, pp. 713 – 715.

Shanley, F. R. (ed.), 1952. *Weight-Strength Analysis of Aircraft Structures.* New York: McGraw-Hill.

Shafarevich, I. R., 1983. Zum 150. Geburtstag von Alfred Clebsch. *Mathematische Annalen,* vol. 266, pp. 135 – 140.

Shapiro, V., 1988. *Theory of R-functions and applications: A primer.* Technical Report CPA88 – 3, Cornell Programmable Automation. Ithaca/New York: Sibley School of Mechanical Engineering.

Shidlovsky, A. K. (ed.). 1988. Vladimir Logvinovich Rvachev. In: *Bibliography of scientists of the Ukrainian Soviet Socialist Republic.* Kiev: Naukova Dumka (in Russian).

Shtraikh, S. Y., 1956. *Aleksei Nikolaevich Krylov.* Moscow: Voenizdat (in Russian).

Siebke, H., 1977. Grundgedanken zur Bemessung stählerner Eisenbahnbrücken auf Betriebsfestigkeit vor wahrscheinlichkeitstheoretischem Hintergrund. In: *4/1977 Berichte aus Forschung und Entwicklung. Beiträge zum Tragverhalten und zur Sicherheit von Stahlkonstruktionen, Vorträge aus der Fachsitzung III des Deutschen Stahlbautages Stuttgart 1976,* ed. DASt, pp. 43 – 47. Cologne: Stahlbau-Verlags-GmbH 1977.

Siebke, H., 1990. Geheimrat Gottwalt Schaper. Wegbereiter für den Stahlbrückenbau vom Nieten zum Schweißen. In: *Wegbereiter der Bautechnik.* Ed. VDI-Gesellschaft Bautechnik, pp. 103 – 127. Düsseldorf: VDI-Verlag.

Sievers, H., Görtz, W., 1956. Der Wiederaufbau der Straßenbrücke über den Rhein zwischen Duisburg-Ruhrort und Homberg (Friedrich-Ebert-Brücke). Statische Grundlagen und Konstruktion. *Der Stahlbau,* vol. 25, No. 4, pp. 77 – 88.

Sigrist, V., 2005. Zur Querkraftbemessung von Stahlbetonträgern. *Beton- und Stahlbetonbau,* vol. 100, No. 5, pp. 390 – 397.

Silberschlag, J. E., 1773. *Ausführliche Abhandlung der Hydrotechnik oder des Wasserbaues.* pt. 2. Leipzig: Caspar Fritsch.

Simonnet, C., 2005. *Le Béton. Histoire d'un Matériau.* Marseille: Éditions Parenthèses.

Simonyi, K., 1990. *Kulturgeschichte der Physik.* Frankfurt a. M./Leipzig: Verlag Harri Deutsch, Thun & Urania-Verlag.

**S**inopoli, A., Corradi, M., Foce, F., 1997. Modern formulation for preelastic theories on masonry arches. *Journal of Engineering Mechanics*, vol. 123, No. 3, pp. 204–213.

Sinopoli, A., Corradi, M., Foce, F., 1998. Lower and upper bound theorems for masonry arches as rigid systems with unilateral contacts. In: *Proceedings of the Second International Arch Bridge Conference*, ed. A. Sinopoli, pp. 99–108. Rotterdam: Balkema.

Sinopoli, A., 2002. A re-examination of some theories on vaulted structures: the role of geometry from Leonardo to de La Hire. In: *Towards a History of Construction*, ed. A. Becchi, M. Corradi, F. Foce & O. Pedemonte, pp. 601–624. Basel: Birkhäuser.

**S**lepov, B. I. (ed.), 1984. *Vospominaniya o P. F. Papkoviche* (reminiscences of P. F. Papkovich). Leningrad: Nauka (in Russian).

Slepov, B. I., 1991. *P. F. Papkovich 1887–1946*. Leningrad: Nauka (in Russian).

**S**mith, D., 2002. Barlow, Peter. In: *A Biographical Dictionary of Civil Engineers in Great Britain and Ireland*, vol. 1: 1500–1830. Ed. A. W. Skempton, M. M. Chrimes, R. C. Cox, P. S. M. Cross-Rudkin, R. W. Rennison & E. C. Ruddock, pp. 39–41. London: ICE & Thomas Telford.

**S**nell, G., 1846. On the Stability of Arches, with practical methods for determining, according to the pressures to which they will be subjected, the best form of section, or variable depth of voussoir, for any given extrados or intrados. *Minutes and Proceedings of the Institution of Civil Engineers*, vol. 5, pp. 439–476.

**S**obek, W., 2002. Fazlur Khan (1929–1982). Wegbereiter der 2. Chicago School. In: *VDI Bau. Jahrbuch 2002*, ed. VDI-Gesellschaft Bautechnik, pp. 417–434. Düsseldorf: VDI-Verlag.

**Š**olín, J., 1885. Zur Theorie des continuirlichen Trägers veränderlichen Querschnittes. *Civilingenieur*, vol. 31

Šolín, J., 1887. Bemerkungen zur Theorie des Erddruckes. *Allgemeine Bauzeitung*, vol. 54

Šolín, J., 1889. Über das allgemeine Momentenproblem des einfachen Balkens bei indirekter Belastung. *Civilingenieur*, vol. 35

Šolín, J., 1889/1893. *Stavebná mechanika* (structural mechanics), pt. I & II. Prague: Prague TH (in Czech).

Šolín, J., 1893/1902/1904. *Nauka o pružnosti a pevnosti* (strength of materials). Prague: ČMT (in Czech).

**S**onnemann, R. (ed.), 1978. *Geschichte der Technischen Universität Dresden*. Berlin: VEB Deutscher Verlag der Wissenschaften.

Sonnemann, R., Krug, K. (ed.), 1987. *Technology and Technical Sciences in History*. Proc. of the ICOHTEC Symposium, Dresden, 25–29 Aug 1986, Berlin: Deutscher Verlag der Wissenschaften.

**S**ontag, H. J., 1951. *Beitrag zur Ermittlung der zeitabhängigen Eigenspannungen von Verbundträgern*. Dissertation, Karlsruhe TH.

**S**oppelsa, M. L. (ed.), 1988. *Giovanni Poleni, idraulico, matematico, architetto, filologo (1683–1761)*. Atti della giornata di studi, Padua 15 March 1986. Padua: Grafice Erredici.

**S**or, S., 1905. Die Berechnung der Eisenbetonplatte. *Beton und Eisen*, vol. 4, No. 10, pp. 256–257.

**S**orrentino, L., 2002. Remembering another master: Antonio Giuffrè (1933–1997). In: *Towards a History of Construction*, ed. A. Becchi, M. Corradi, F. Foce & O. Pedemonte, pp. 625–643. Basel: Birkhäuser.

**S**outhwell, R. V., 1914. The General Theory of Elastic Stability. *Phil. Trans. Royal Soc. London, A*, vol. 213, pp. 187–245.

Southwell, R. V., 1921. Stress determination in braced frameworks I. *Aeronaut. Res. Com., Rep. and Memo.* No. 737.

Southwell, R.V., 1922. Stress determination in braced frameworks II, III, IV. *Aeronaut. Res. Com., Rep. and Memo.* Nos. 790, 792, 819.

Southwell, R. V., 1926. Stress calculation for the hulls of rigid airships. *Aeronaut. Res. Com., Rep. and Memo.* No. 1057.

Southwell, R. V., 1933. On the calculation of stresses in braced frameworks. *Proc. R. Soc. London, A.* vol. 139, pp. 475–508.

Southwell, R. V., 1934. *Introduction to the theory of elasticity for engineers and physicists*. Oxford: Oxford University Press.

Southwell, R. V., 1935. Stress calculation in frameworks by the method of 'systematic relaxation of constraints'. Pt. I & II. *Proceedings of the Royal Society of London, A.* vol. 151, pp. 56–95.

Southwell, R. V., 1940. *Relaxation Methods in Engineering Science*. Oxford: Oxford University Press.

**S**päth, J. L., 1811. *Statik der hölzernen Bogen-Brücken nach der Construction des Königlich-Baierischen Geheimen-Raths und General-Directors des Wasser-, Brücken- und Strassen-Baues, Herrn Carl Friedrich von Wiebeking*. Munich: Verlag der Fleischmann'schen Buchhandlung.

**S**pangenberg, H., 1924. Eisenbetonbogenbrücken für große Spannweiten. *Der Bauingenieur*, vol. 5, pp. 461–468 & 503–512.

**S**pecht, M. (ed.), 1987. *Spannweite der Gedanken. Zur 100. Wiederkehr des Geburtstages von Franz Dischinger*. Berlin: Springer.

**S**pierig, S., 1963. *Beitrag zur Lösung von Scheiben-, Platten- und Schalenproblemen mit Hilfe von Gitterrostmodellen*. Dissertation, Hannover TH.

**S**pitzer, J. A., 1896. Berechnung der Monier-Gewölbe. Wissenschaftliche Verwertung der Versuchsergebnisse bei dem Purkersdorfer Probegewölbe nach System Monier. *Zeitschrift des österreichischen Ingenieur- und Architekten-Vereines*, vol. 48, pp. 305–320.

Spitzer, J. A., 1898. Träger aus Materialien von veränderlichen Formänderungs-Coefficienten. *Zeitschrift des österreichischen Ingenieur- und Architekten-Vereines*, vol. 50, pp. 270–273 & 286–289.

Spitzer, J. A., 1908. Versuche mit Gewölben. In: *Handbuch für Eisenbetonbau, Bd.1*, ed. F. v. Emperger, pp. 302–386. Berlin: Verlag von Wilhelm Ernst & Sohn.

**S**pur, G., Fischer, W. (ed.), 2000. *Georg Schlesinger und die Wissenschaft vom Fabrikbetrieb*. Munich/Vienna: Carl Hanser.

**S**tamm, K., Witte, H., 1974. *Sandwichkonstruktionen*. Vienna/New York: Springer-Verlag.

**S**tark, F. (ed.), 1906. *Die Deutsche Technische Hochschule in Prague 1806–1906*. Prague: German TU.

**S**taudt, K. G. C. v., 1847. *Geometrie der Lage*. Nuremberg: Verlag von Bauer und Raspe – Fr. Korn.

**S**teiding, I., 1985. Prof. Dr.-Ing. h.c. Christian Otto Mohr zum 150. Geburtstag. *Bauplanung-Bautechnik*, vol. 39, No. 9, pp. 395–397.

**S**tein, E., 1985. Speech in praise of Prof. Drs. h. c. J. H. Argyris on the occasion of the award of his honorary doctorate (Dr.-Ing. E. h.) by the Faculty of Civil Engineering & Surveying, University of Hannover on 22 April 1983. In: *Forschungs- und Seminarberichte aus dem Bereich der Mechanik der Universität Hannover*, Bericht-Nr. F 85/1, CRT-K – 2/84, pp. 6–16. Hannover.

Stein, E., de Borst, R., Hughes, T. J. R. (ed.), 2004/1. *Encyclopedia of Computational Mechanics. vol. 1: Fundamentals*. Chichester: John Wiley & Sons.

Stein, E., de Borst, R., Hughes, T. J. R. (ed.), 2004/2. *Encyclopedia of Computational Mechanics. vol. 2: Solids and Structures*. Chichester: John Wiley & Sons.

Stein, E., de Borst, R., Hughes, T. J. R. (ed.), 2004/3. *Encyclopedia of Computational Mechanics. vol. 3: Fluids*. Chichester: John Wiley & Sons.

**S**teinhardt, O., 1949. *Friedrich Engesser*. Karlsruhe: C. F. Müller.

Steinhardt, O., 1981. Fritz Stüssi zum 80. Geburtstag am 3. Januar 1981. *Schweizerische Bauzeitung*, vol. 99, No. 1, pp. 4–5.

**S**teinman, D. B., 1911. *Suspension Bridges and Cantilevers. Their Economic Proportions and Limiting Spans*. New York: D. Van Nostrand Co.

Steinman, D. B., 1922. *Suspension Bridges, Their Design, Construction and Erection*. New York: Wiley & Sons.

Steinman, D. B., 1929. *A Practical Treatise on Suspension Bridges*. New York: Wiley & Sons.

Steinman, D. B., 1934. Deflection theory for continuous suspension bridges. In: *Proceedings of the International Association for Bridge & Structural Engineering (IABSE)*. vol. 2, ed. L. Karner & M. Ritter, pp. 400–51. Zurich: IABSE secretariat.

Steinman, D. B., 1945. *The builders of the bridge: The story of John Roebling and his son.* New York: Hartcourt, Brace and Co.

Steinman, D. B., 1951. Der Entwurf einer Brücke von Italien nach Sizilien mit der größten Spannweite der Welt. *Der Stahlbau,* vol. 20, No. 3, pp. 29–32.

Steinman, D. B., 1953. *Famous bridges in the world.* New York: Random House.

Stelling, E. G., 1929. Verdrehfeste, dreiflächige Brückenträger mit Ausführungsbeispielen an neueren Bauten der Hamburger Hochbahn. *Der Stahlbau,* vol. 2, No. 7, pp. 75–84.

Stel'makh, S. I., Vlasov, V. V., 1982. *V. Z. Vlasov i ego vklad v sozdanie sovremennoi stroitel'noi mekhaniki tonkostennykh konstruktsy* (V. Z. Vlasov and his contribution to the formation of the modern mechanics of thin-walled constructions). Moscow: Stroiizdat (in Russian).

Stephan, S., Sánchez-Alvarez, J., Knebel, K., 2004. Stabwerke auf Freiformflächen. *Stahlbau,* vol. 73, No. 8, pp. 562–572.

Steuermann, M., 1936. Die Versuchspilzdecke in Baku. *Beton und Eisen,* vol. 35, No. 21, pp. 357–363 & No. 22, pp. 374–376.

Stever, H. G., 1989. Raymond L. Bisplinghoff. In: *Memorial Tributes. National Academy of Engineering,* vol. 3, pp. 36–42. Washington, D.C.: National Academy Press.

Stevin, S., 1586. *De Beghinselen der Weeghconst.* Leiden: Raphelinghen.

Stevin, S., 1987. *L'art pondéraire ou la statique* (vol. 4) ed. A. Girard 1634 (reprint). Paris: ACL-éditions.

Stickforth, J., 1986. Über den Schubmittelpunkt. *Ingenieur-Archiv,* vol. 56, pp. 438–452.

Stiegler, K., 1969. Einige Probleme der Elastizitätstheorie im 17. Jahrhundert. *Janus,* vol. 56, pp. 107–122.

Stiglat, K., Wippel, H., 1966. *Platten.* Berlin/Munich: Wilhelm Ernst & Sohn.

Stiglat, K., 1997. *Brücken am Weg. Frühe Brücken aus Eisen und Beton in Deutschland und Frankreich.* Berlin: Ernst & Sohn.

Stiglat, K., 1999. Muß ein Bogen immer rund sein? Zur neuen Eisenbahnbrücke über den Rhein zwischen Mannheim und Ludwigshafen. *Stahlbau,* vol. 68, No. 10, pp. 839–842.

Stiglat, K., 2000. INGENIEURGRUPPE BAUEN 1965–2000, ed. INGENIEURGRUPPE BAUEN. Karlsruhe: Self-pub.

Stiglat, K., 2004. *Bauingenieure und ihr Werk.* Berlin: Ernst & Sohn.

Stocker, J., 1987. *Zum Tragverhalten der Zwischenkuppel von St. Geneviève in Paris.* Diploma thesis, Institute for Theory of Structures (Prof. E. Ramm), University of Stuttgart.

Stodola, A., 1910. *Die Dampfturbinen.* 4th ed. Berlin: Springer.

Strassner, A., 1912. Beitrag zur Berechnung mehrstöckiger Rahmen mit Rücksicht auf die Veränderlichkeit des Trägheitsmomentes. Forscherarbeiten auf dem Gebiete des Eisenbetons, No. 18. Berlin: Wilhelm Ernst & Sohn.

Strassner, A., 1916. *Neuere Methoden zur Statik der Rahmentragwerke und der elastischen Bogenträger.* Berlin: Wilhelm Ernst & Sohn.

Straub, H., 1949. *Die Geschichte der Bauingenieurkunst.* Basel: Birkhäuser.

Straub, H., 1992. *Die Geschichte der Bauingenieurkunst.* 4th, rev. ed., ed. Peter Zimmermann. Basel: Birkhäuser.

Straub, H. & Halász, R. v., 1960/1. Zur Geschichte des Bauingenieurwesens. *Die Bautechnik,* vol. 37, No. 4, pp. 121–122.

Straub, H. & Halász, R. v., 1960/2. Drei bisher unveröffentlichte Karikaturen zur Frühgeschichte der Baustatik. *Die Bautechnik,* vol. 37, No. 10, pp. 365–366.

Strecker, F., Feldtkeller, R., 1929. Grundlagen der Theorie des allgemeinen Vierpols. *Elektrische Nachrichten-Technik,* vol. 6, No. 2, pp. 93–112.

Striebing, L., 1966. Theorie und Methodologie der technischen Wissenschaften aus philosophischer Sicht. *Wiss. Ztschr. Techn. Univ. Dresden,* vol. 15, No. 4.

Striebing, L., Herlitzius, E., Teichmann, D., Sonnemann, R., 1978. "Phil.-Hist. 78" – Rückschau und Ausblick. *Wiss. Ztschr. Techn. Univers. Dresden,* vol. 28, No. 4, pp. 1037–1044.

Stigler, St. M., 2004. Merriman, Mansfield. In: *Encyclopedia of Statistical Sciences.* New York: Wiley.

Stroetmann, R., 2007. In memoriam Kurt Beyer – zum 125. Geburtstag. *Stahlbau,* vol. 76, No. 5, pp. 347–353.

Struik, D. J., 1970. Beltrami, Eugenio. In: *Dictionary of Scientific Biography,* vol. II. C. C. Gillispie (ed.), pp. 599–600. New York: Charles Scribner's Sons.

Struik, D. J., 1981. Musschenbroek, Petrus van. In: *Dictionary of Scientific Biography,* vol. 9. C. C. Gillispie (ed.), pp. 594–597. New York: Charles Scribner's Sons.

Study, E., 1903. *Geometrie der Dynamen.* Leipzig: B. G. Teubner.

Stüssi, F., Kollbrunner, C. F., 1935. Beitrag zum Traglastverfahren. *Die Bautechnik,* vol. 13, No. 21, pp. 264–267.

Stüssi, F., 1938. Zur Auswertung von Versuchen über das Traglastverfahren. In: *International Association for Bridge & Structural Engineering* (ed.). 2nd Congress, Berlin/Munich, 1–11 October 1936. Final Report, pp. 74–76. Berlin: Verlag von Wilhelm Ernst & Sohn.

Stüssi, F., 1940. Baustatik vor 100 Jahren – die Baustatik Naviers. *Schweizerische Bauzeitung,* vol. 58, No. 18, pp. 201–204.

Stüssi, F., 1946. *Vorlesungen über Baustatik. Band I. Statisch bestimmte Systeme, Spannungsberechnung, Elastische Formänderungen, Stabilitätsprobleme, Seile.* Basel: Birkhäuser.

Stüssi, F., 1951. Karl Culmann und die graphische Statik. *Schweizerische Bauzeitung* vol. 69, No. 1, pp. 1–3.

Stüssi, F., 1954. *Vorlesungen über Baustatik. Band II. Statisch unbestimmte Systeme.* Basel: Birkhäuser.

Stüssi, F., 1955. Ausgewählte Kapitel aus der Theorie des Brückenbaues. In: *Taschenbuch für Bauingenieure. Erster Band.* 2nd, rev. ed., ed. F. Schleicher, pp. 905–963. Berlin: Springer.

Stüssi, F., 1962. Gegen das Traglastverfahren. *Schweizerische Bauzeitung,* vol. 80, No. 4, pp. 53–57.

Stüssi, F., 1964. Über die Entwicklung der Wissenschaft im Brückenbau. In: *Neujahrsblatt,* ed. Naturforschende Gesellschaft in Zurich.

Stüssi, F., 1971. Zum 150. Geburtstag von Karl Culmann. *Schweizerische Bauzeitung,* vol. 89, pp. 694–97.

Šuchov, F. V., 1990. Erinnerungen an meinen Großvater V. G. Šuchov. In: *Vladimir G. Šuchov, 1853–1939. Die Kunst der sparsamen Konstruktion.* Ed. R. Graefe, M. Gapoëv, M. & O. Pertschi, pp. 20–21. Stuttgart: Deutsche Verlagsanstalt.

Suenson, E., 1931. †Asger Skovgaard Ostenfeld. *Der Bauingenieur,* vol. 12, pp. 838–839.

Supino, G., 1969. Necrologio. *Celebrazioni Lincee,* No. 20.

Sutherland, H. B., 1999. Prof. William John Macquorn Rankine. *Proc. Instn. Civ. Engrs., Civ. Engng.* vol. 132, Nov, pp. 181–187.

Sutherland, R. J. M., 1979/1980. Tredgold 1788–1829: Some aspects of his work Part 3: Cast iron. *Transactions of the Newcomen Society,* vol. 51, pp. 71–82.

Swain, G. F., 1883. On the Application of the Principle of Virtual Velocities to the Determination of the Deflection and Stresses of Frames. *Journal of the Frankline Institute,* vol. 115 (3rd series), pp. 102–115.

Swain, G. F., 1917. *How to Study.* New York: McGraw-Hill Book Company.

Swain, G. F., 1922. *The Young Man and Civil Engineering.* New York: Macmillan.

Swain, G. F., 1924/1. *Structural Engineering. vol. 1. The Strength of Materials.* New York: McGraw-Hill Book Company.

Swain, G. F., 1924/2. *Structural Engineering. vol. 2. Fundamental Properties of Materials.* New York: McGraw-Hill Book Company.

Swain, G. F., 1927. *Structural Engineering. vol. 3. Stresses, Graphical Statics and Masonry.* New York: McGraw-Hill Book Company.

Symonds, P. S., Neal, B. G., 1952. The Interpretation of Failure Loads in the Plastic Theory of Continuous Beams and Frames. *Journal of the Aeronautical Sciences,* vol. 19, No. 15, pp. 15–22.

Szabó, I., 1954. *Einführung in die Technische Mechanik.* Berlin: Springer.

Szabó, I., 1956. *Höhere Technische Mechanik.* Berlin: Springer.

Szabó, I., 1959. Prof. Dr.-Ing. Dr.-Ing. E. h. Hans J. Reissner 85 Jahre. *Der Stahlbau,* vol. 28, No. 3, pp. 81–82.

Szabó, I., 1976. Die Entwicklung der Elastizitätstheorie im 19. Jahrhundert nach Cauchy. *Die Bautechnik*, vol. 53, No. 4, pp. 109–116.

Szabó, I., 1977. *Geschichte der mechanischen Prinzipien und ihrer wichtigsten Anwendungen*. Basel: Birkhäuser.

Szabó, I., 1980. Einige Marksteine in der Entwicklung der theoretischen Bauingenieurkunst. In: *Beiträge zur Bautechnik. Robert von Halász zum 75. Geburtstag gewidmet*, ed. J. Bauer, C. Scheer & E. Cziesielski, pp. 1–20. Berlin: Wilhelm Ernst & Sohn.

Szabó, I., 1984. *Einführung in die Technische Mechanik nach Vorlesungen von Istvan Szabó*. Berlin/Heidelberg/New York: Springer-Verlag.

Szabó, I., 1987. *Geschichte der mechanischen Prinzipien und ihrer wichtigsten Anwendungen*. 2nd, rev. ed., ed. P. Zimmermann & E. A. Fellmann. Basel: Birkhäuser

Szabó, I., 1996. *Geschichte der mechanischen Prinzipien und ihrer wichtigsten Anwendungen*. 3rd, rev. ed., ed. P. Zimmermann & E. A. Fellmann. Basel: Birkhäuser.

Szilard, R. 1982. *Finite Berechnungsmethoden der Strukturmechanik. Band 1: Stabwerke*. Berlin/Munich: Ernst & Sohn.

Szilard, R. 1990. *Finite Berechnungsmethoden der Strukturmechanik. Band 2: Flächentragwerke im Bauwesen*. Berlin/Munich: Ernst & Sohn.

Takabeya, F., 1924. *Zur Berechnung des beiderseits eingemauerten Trägers unter besonderer Berücksichtigung der Längskraft*. Berlin: Springer.

Takabeya, F., 1930/1. *Rahmentafeln*. Berlin: Springer.

Takabeya, F., 1930/2. Zur Berechnung der Spannungen in ebenen eingespannten Flachblechen. In: *3rd International Congress of Theoretical and Applied Mechanics*, vol. II, pp. 71–73.

Takabeya, F., 1965. *Multi-story Frames*. Berlin: Verlag von Wilhelm Ernst & Sohn.

Takabeya, F., 1967. *Mehrstöckige Rahmen*. Berlin: Verlag von Wilhelm Ernst & Sohn.

Tanabashi, R. 1937/1. On the Resistance of Structures to Earthquake Shocks. *Memoirs of the College of Engineering, Kyoto Imperial University*, vol. IX, No. 4, Kyoto.

Tanabashi, R., 1937/2. Angenähertes Verfahren zur Ermittlung der maximalen und minimalen Biegemomente von Rahmentragwerken. In: *Memoirs of the College of Engineering, Kyoto Imperial University*, vol. IX, No. 4, Kyoto.

Tanabashi, R., 1938. Systematische Auflösung der Elastizitätsgleichungen des Stockwerkrahmens und die Frage der Einflußlinie. In: *Memoirs of the College of Engineering*, vol. X, No. 4. Kyoto.

Tanabashi, R., 1956. Studies on the Nonlinear Vibration of Structures Subjected to Destructive Earthquakes. *Proc., WCEE*, Berkeley, Calif.

Tanabashi, R., 1958. Analysis of Statically Indeterminate Structures in the Ultimate State. Bulletin, 20, Disaster Prevention Research Institute, Kyoto University, Kyoto.

Tanner, R. I., Tanner, E., 2003. Heinrich Hencky: a rheological pioneer. *Rheologica Acta*, vol. 42, pp. 93–101.

Taton, R., 1973. La Hire, Philippe de. In: *Dictionary of Scientific Biography*, vol. VII. C. C. Gillispie (ed.), pp. 576–579. New York: Charles Scribner's Sons.

Taus, M., 1999. Verbundkonstruktion beim Millennium Tower. *Stahlbau*, vol. 68, No. 8, pp. 647–651.

Taut, B., 1977. *Architekturlehre. Grundlagen, Theorie und Kritik, Beziehungen zu den anderen Künsten und zur Gesellschaft*. Ed. T. Heinisch & G. Peschken. Hamburg: Verlag für das Studium der Arbeiterbewegung (VSA).

Taylor, C. E., 2003. Dr. Daniel Drucker 1918–2001. Graduate Research Professor Emeritus University of Florida. *Journal of Applied Mechanics*, vol. 70, No. 1, pp. 158–159.

Taylor, R. L., 2002. Ritz and Galerkin: the road to the Finite Element Method. *iacm expressions* 12/02, pp. 2–5.

Tazzioli, R., 1995. Construction engineering and natural philosophy: the work by Gabriel Lamé. In: *Entre Mécanique et Architecture*, ed. P. Radelet-de Grave et E. Benvenuto, pp. 317–329. Basel: Birkhäuser.

Technische Akademie Esslingen (ed.), 1970. Ankündigungsblatt zu den Lehrgängen, *Statik- und Festigkeitsprobleme der Praxis auf Digitalrechnern*. Esslingen.

Teichmann, A., 1931. *Zur Berechnung auf Knickbiegung beanspruchter Flugzeugholme*. Munich: Oldenbourg.

Teichmann, A., 1939. Gedankengänge zur Flatterberechnung. In: *Jahrbuch 1939 der Deutschen Versuchsanstalt für Luftfahrt e.V.*, Berlin-Adlershof, pp. 224–249.

Teichmann, A., 1942. Das Flattern von Trag- und Leitwerken. Lecture from 12 Dec 1941. In: *Schriften der Deutschen Akademie der Luftfahrtforschung*, No. 49, pp. 1–49. Munich/Berlin: Oldenbourg.

Teichmann, A., Pohlmann, G., 1957. *Umdrucke zur Statik der Baukonstruktion, Teil II*. Berlin: Alleinvertrieb durch die Buchhandlung Kiepert.

Teichmann, A., 1958. *Statik der Baukonstruktionen III. Statisch unbestimmte Systeme*. Berlin: Walter de Gruyter.

Terzaghi, K. v., 1925. *Erdbaumechanik auf bodenphysikalischer Grundlage*. Leipzig/Vienna: Franz Deuticke.

Terzaghi, K. v., Fröhlich, O. K., 1936. *Theorie der Setzung von Tonschichten*. Leipzig/Vienna: Franz Deuticke.

Thalau, K., Teichmann, A., 1933. *Aufgaben aus der Flugzeugstatik*. Berlin: Springer.

Thode, D., 1975. *Untersuchungen zur Lastabtragung in spätantiken Kuppelbauten*. Studien zur Bauforschung, No. 9, ed. Koldewey-Gesellschaft. Darmstadt.

Teßmann, K., 1965. Die experimentelle Methode in den technischen Wissenschaften. In: University of Rostock (ed.): *Struktur und Funktion der experimentellen Methode*. Rostock.

Tetmajer, L., 1882. *Über Culmanns bleibende Leistungen*. A lecture held at the Naturforschende Gesellschaft in Zurich.

Tetmajer, L., 1884. *Methoden und Resultate der Prüfung der schweizerischen Bauhölzer*. Bulletin No. 2 of the Materials Testing Institute, Zurich ETH.

Tetmajer, L., 1889. *Die angewandte Elasticitäts- und Festigkeitslehre*. Zurich: Verlag von Züricher & Furrer.

Tetmajer, L., 1890. *Methoden und Resultate der Prüfung der Festigkeitsverhältnisse des Eisens und anderer Metalle*. Bulletin No. 4 of the Materials Testing Institute, Zurich ETH. Zurich.

Tetmajer, L., 1896 *Die Gesetze der Knickfestigkeit der technisch wichtigsten Baustoffe*. Bulletin No. 8 of the Materials Testing Institute, Zurich ETH.

Tetmajer, L., 1903. *Die Gesetze der Knickungs- und der zusammengesetzten Druckfestigkeit der technisch wichtigsten Baustoffe*. 3rd, rev. ed. Leipzig/Vienna: Deuticke.

Teut-Nedeljkov, A., 1979. Zwischen Revolution und Reform: in Preußen entsteht das erste deutschsprachige Polytechnikum. Präliminarien zur Entstehungsgeschichte. In: *100 Jahre Technische Universität Berlin 1879*. 1979, Exhibition catalogue, ed. K. Schwarz. Berlin: Berlin TU library/Publications Dept.

Thul, H., 1978. Entwicklungen im Brückenbau. *Beton- und Stahlbetonbau*, vol. 73, No. 1, pp. 1–7.

Thullie, M. R. v., 1896. Über die Berechnung der Biegespannungen in den Beton- und Monier-Constructionen. *Zeitschrift des österreichischen Ingenieur- und Architekten-Vereines*, vol. 48, pp. 365–369.

Thullie, M. R. v., 1897. Über die Berechnung der Monierplatten. *Zeitschrift des österreichischen Ingenieur- und Architekten-Vereines*, vol. 49, pp. 193–197.

Thullie, M. R. v., 1902. Beitrag zur Berechnung der Monierplatten. *Zeitschrift des österreichischen Ingenieur- und Architekten-Vereines*, vol. 54, pp. 242–244.

Thürlimann, B., 1961. Grundsätzliches zu den plastischen Berechnungsverfahren. *Schweizerische Bauzeitung*, vol. 79, No. 48, pp. 863–869 & No. 49, pp. 877–881.

Thürlimann, B., 1962. Richtigstellungen zum Aufsatz "Gegen das Traglastverfahren". *Schweizerische Bauzeitung*, vol. 80, pp. 123–136.

Thürlimann, B., Marti, P., Pralong, J., Ritz, P., Zimmerli, B., 1983. *Anwendungen der Plastizitätstheorie auf Stahlbeton. Vorlesung zum Fortbildungskurs für Bauingenieure*. Institute of Theory of Structures & Design, Zurich ETH.

Timoshenko, S. P., 1910. Einige Stabilitätsprobleme der Elastizitätstheorie. *Zeitschrift für Mathematik und Physik*, vol. 58, pp. 337–385.

Timoshenko, S. P., 1915/2001. Zur Frage der Festigkeit von Schienen. Trans. from Russian by H. Elsner & ed. by K. Knothe & A. Duda. In: K. Knothe, *Gleisdynamik,* pp. 5 – 44. Berlin: Ernst & Sohn.

Timoshenko, S. P., 1921. On the correction for shear of the differential equation for transverse vibration of prismatic bars. *Philosophical Magazine,* vol. 41, pp. 744 – 746.

Timoshenko, S. P., 1950. D. I. Jouravski and his contribution to the theory of structures. In: *Beiträge zur angewandten Mechanik (Federhofer-Girkmann Festschrift),* ed. E. Chwalla et al., pp. 115 – 123. Vienna: Deuticke.

Timoshenko, S. P., 1953/1. *History of strength of materials. With a brief account of the history of theory of elasticity and theory of structures.* New York: McGraw-Hill.

Timoshenko, S. P., 1953/2. *The collected papers.* New York: McGraw-Hill.

Timoshenko, S. P., 1958. *Kruzhok imeni V. L. Kirpicheva* (V. I. Kirpichev circle). Munich: Einheit (in Russian).

Timoshenko, S. P., 1959. *Engineering education in Russia.* New York: McGraw-Hill.

Timoshenko, S. P., 1968. *As I remember: The autobiography.* Princeton: Van Nostrand.

Timoshenko, S. P., 1971. *Ustoichivost' sterzhnei, plastin i obolochek: Izbrannye raboty* (stability of bars, plates and shells: selected works). Moscow: Nauka (in Russian).

Timoshenko, S. P., 1975/1. *Prochnost i kolebaniya elementov konstruktsii* (strength and vibrations of construction elements). Moscow: Fizmatlit (in Russian).

Timoshenko, S. P., 1975/2. *Staticheskie i dinamicheskie zadachi teorii uprugosti* (static and dynamic problems of eleastic theory). Kiev: Naukova Dumka (in Russian).

Timoshenko, S. P., 2006. *Erinnerungen Stepan P. Timoshenko: Eine Autobiographie.* Trans. from the Russian by Albert Duda. Berlin: Ernst & Sohn.

Todhunter, I., Pearson, K., 1886. *A History of the Theory of Elasticity and of the Strength of Materials. From Galilei to Lord Kelvin. vol. I. Galilei to Saint-Venant 1639 – 1850.* Cambridge: At the University Press.

Todhunter, I., Pearson, K., 1893/1. *A History of the Theory of Elasticity and of the Strength of Materials. From Galilei to Lord Kelvin. vol. II. Saint-Venant to Lord Kelvin. Part I.* Cambridge: At the University Press.

Todhunter, I., Pearson, K., 1893/2. *A History of the Theory of Elasticity and of the Strength of Materials. From Galilei to Lord Kelvin. Saint-Venant to Lord Kelvin. vol. II. Part II.* Cambridge: At the University Press.

Tobies, R., 1981. *Felix Klein.* Leipzig: B. G. Teubner.

Tomlow, J., 1993. Die Kuppel des Gießhauses der Firma Henschel in Kassel (1837). Eine frühe Anwendung des Entwurfsverfahrens mit Hängemodellen. *architectura,* vol. 23, pp. 151 – 172.

Torroja, E., 1948. Rapport sur les voiles minces construits en Espagne. In: *3rd Congress, Liège. Final Report,* pp. 575 – 584. Zurich: IABSE secretariat.

Trautwine, J. C., 1872. *Civil Engineer's Pocket-Book.* New York: Wiley & Son.

Trautz, M., Tomlow, J., 1994. Alessandro Antonelli (1798 – 1888) – Lightweight masonry structures in neoclassical architecture. In: *3rd International Symposium on Special Area of Research 230,* Stuttgart, 4 – 7 Oct 1994, pp. 119 – 124.

Trautz, M., 1998. Zur Entwicklung von Form und Struktur historischer Gewölbe aus der Sicht der Statik. Dissertation, University of Stuttgart.

Trautz, M., 2001. Formunvollendet. Eine Trilogie von der Suche nach der Form statisch idealer Kuppeln. *deutsche bauzeitung,* vol. 135, No. 11, pp. 105 – 111.

Trautz, M., 2002. Maurice Koechlin. *deutsche bauzeitung,* vol. 136, No. 4, pp. 105 – 110.

Tredgold, T., 1820. *Elementary Principles of Carpentry.* (12th ed. 1919). London: Taylor.

Tredgold, T., 1822. *A Practical essay on the strength of cast iron* (5th ed. 1861 – 62). London: Taylor.

Trefftz, E., 1921. Über die Torsion prismatischer Stäbe von polygonalem Querschnitt. *Mathematische Annalen,* vol. 82, pp. 97 – 112.

Trefftz, E., 1922. Über die Wirkung einer Abrundung auf die Torsionsspannungen in der inneren Ecke eines Winkeleisens. *Zeitschrift für Angewandte Mathematik und Mechanik,* vol. 2, No. 4, pp. 263 – 267.

Trefftz, E., 1925. Über die Spannungsverteilung in tordierten Stäben bei teilweiser Überschreitung der Fließgrenze. *Zeitschrift für Angewandte Mathematik und Mechanik,* vol. 5, No. 1, pp. 64 – 73.

Trefftz, E., 1926. Ein Gegenstück zum Ritzschen Verfahren. In: *Verhandlungen des zweiten internationalen Kongresses für technische Mechanik,* pp. 131 – 137. Zurich: Füssli.

Trefftz, E., 1930. Über die Ableitung der Stabilitätskriterien des elastischen Gleichgewichtes aus der Elastizitätstheorie endlicher Deformationen. In: *Proceedings of the 3rd International Congress for Applied Mechanics,* vol. 3, pp. 44 – 50. Stockholm.

Trefftz, E., 1933. Zur Theorie der Stabilität des elastischen Gleichgewichts. *Zeitschrift für Angewandte Mathematik und Mechanik,* vol. 13, No. 2, pp. 160 – 165.

Trefftz, E., 1935/1. Ableitung der Schalenbiegungsgleichungen nach dem Castiglianoschen Prinzip. *Zeitschrift für Angewandte Mathematik und Mechanik,* vol. 15, No. 1/2, pp. 101 – 108.

Trefftz, E., 1935/2. Über den Schubmittelpunkt in einem durch eine Einzellast gebogenen Balken. *Zeitschrift für Angewandte Mathematik und Mechanik,* vol. 15, No. 5, pp. 220 – 225.

Trefftz, E., 1935/3. Die Bestimmung der Knicklast gedrückter, rechteckiger Platten. *Zeitschrift für Angewandte Mathematik und Mechanik,* vol. 15, No. 6, pp. 339 – 344.

Trefftz, E., 1936. *Graphostatik.* Leipzig/Berlin: B. G. Teubner.

Trefftz, E., Willers, F. A., 1936. Die Bestimmung der Schubbeanspruchung beim Ausbeulen rechteckiger Platten. *Zeitschrift für Angewandte Mathematik und Mechanik,* vol. 16, No. 6, pp. 336 – 344.

Tresca, H., 1864. Sur l'écoulement des Corps solides soumis à des fortes pressions. *Comptes Rendus de l'Académie des Sciences,* vol. 59, pp. 754 – 756.

Treue, W., 1989/1990. D'Alemberts Einleitungen zur Enzyklopädie – heute gelesen. *Blätter für Technikgeschichte,* 51/52, pp. 173 – 186.

Troshchenko, V. T. (ed.), 2005. *G. S. Pisarenko – uchenyi, pedagog i organizator nauki* (G. S. Pisarenko – scientist, teacher and organiser of science). Kiev: Akademperiodika (in Russian).

Trost, H., 1966. Spannungs-Dehnungs-Gesetz eines viskoelastischen Festkörpers wie Beton und Folgerungen für Stabwerke aus Stahlbeton und Spannbeton. *Beton,* vol. 16, No. 6, pp. 233 – 248.

Trost, H., 1967. Auswirkungen des Superpositionsprinzips auf Kriech- und Relaxationsprobleme des Beton und Spannbeton. *Beton- und Stahlbetonbau,* vol. 62, No. 10, pp. 230 – 238 & No. 11, pp. 261 – 269.

Trost, H., 1968. Zur Berechnung von Stahlverbundträgern im Gebrauchszustand auf Grund neuerer Erkenntnisse des viskoelastischen Verhaltens des Betons. *Der Stahlbau,* vol. 37, No. 11, pp. 321 – 331.

Truesdell, C. A., 1960. *The rational Mechanics of flexible or elastic bodies 1638 – 1788. Introduction to Leonhardi Euleri opera omnia vol. 10 et 11 seriei secundae.* Zurich: Orell Füssli.

Truesdell, C. A., 1964. Die Entwicklung des Drallsatzes. *Zeitschrift für Angewandte Mathematik und Mechanik,* vol. 44, No. 4/5, pp. 149 – 158.

Truesdell, C. A., 1968. *Essays in the History of Mechanics.* Berlin: Springer.

Tschemmernegg, F., 1999. Innsbrucker Mischbautechnologie im Wiener Millennium Tower. *Stahlbau,* vol. 68, No. 8, pp. 606 – 611.

Turner, M. J., Clough, R. W., Martin, H. C., Topp, L. J., 1956. Stiffness and deflection analysis of complex structures. *Journal of the Aeronautical Sciences,* vol. 23, No. 9, pp. 805 – 823, 854.

Turner, M. J., 1959. *The direct stiffness method of structural analysis.* Structural & Materials Panel Paper, AGARD Meeting, Aachen.

Turner, M. J., Dill, E. H., Martin, H. C., Melosh, R. J., 1960. Large deflection analysis of complex structures subjected to heating and external loads. *Journal of the Aeronautical Sciences,* vol. 27, No. 2, pp. 97 – 106, 127.

Turner, M. J., Martin, H. C., Weikel, R. C., 1962/1964. Further development and applications of the stiffness method AGARD Structures Materials Panel, Paris, July 1962. In: *AGARDograph 72: Matrix Methods of Structural Analysis*, ed. B. M. Fraeijs de Veubeke, pp. 203–266. Oxford: Pergamon Press.
Tyrrell, H. G., 1911. *History of Bridge Engineering*. Chicago.

Uchida, I., 1983. Biography of Prof. Takabeya. *JSCE*, vol. 68, Aug, p. 49 (in Japanese).
Umansky, A. A., 1966. Kratkii obzor nauchnykh trudov I. M. Rabinovicha (brief overview of the scientific works of I. M. Rabinovich). In: *Stroitel'naya mekhanika* (structural mechanics). Moscow: Stroiizdat, pp. 5–31 (in Russian).
Ungewitter, G., 1890. *Lehrbuch der gotischen Konstruktionen*. 3rd ed. rev. by K. Mohrmann. 2 vols. Leipzig: successor to T. O. Weigel.

Vainberg, D. V., Chudnovskii, V. G., 1948. *Prostranstvennye ramnye karkasy inzhenernykh sooruzhenii* (three-dimensional frameworks in engineering structures). Kiev: Gostekhizdat Ukrainy (in Russian).
Valeriani, S., 2006. *Kirchendächer in Rom. Beiträge zu Zimmermannskunst und Kirchenbau von der Spätantike bis zur Barockzeit*. Petersberg: Michael Imhof Verlag.
Varignon, P., 1687. *Projet d'une nouvelle mécanique*. Paris.
Varignon, P., 1725. *Nouvelle mécanique ou statique*. 2 vols. Paris: C. Jombert.
V.d.B.u.E.-F. in Berlin (ed.), 1910. *Geschäftsbericht. 6. Geschäftsjahr 1909–1910*. Berlin.
V.d.B.u.E.-F. in Berlin (ed.), 1912. *Geschäftsbericht. 8. Geschäftsjahr 1911–1912*. Berlin.
Vekua, I. N., 1991. *Academician N. Muskhelishvili: Short Biography and Survey of Scientific Works*. Tiflis: Metsniereba.
Vianello, L., 1893. Der kontinuirliche Balken mit Dreiecks- oder Trapezlast. *Zeitschrift des Vereines deutscher Ingenieure*, vol. 37, No. 13, pp. 361–364.
Vianello, L., 1895. Der Kniehebel. *Zeitschrift des Vereines deutscher Ingenieure*, vol. 39, No. 9, pp. 253–257.
Vianello, L., 1897. Die Doppelkonsole. *Zeitschrift des Vereines deutscher Ingenieure*, vol. 41, No. 45, pp. 1275–1278.
Vianello, L., 1898. Graphische Untersuchung der Knickfestigkeit gerader Stäbe. *Zeitschrift des Vereines deutscher Ingenieure*, vol. 42, No. 52, p. 1436–1443.
Vianello, L., 1903. Die Konstruktion der Biegungslinie gerader Stäbe und ihre Anwendung in der Statik. *Zeitschrift des Vereines deutscher Ingenieure*, vol. 47, No. 3, pp. 92–97.
Vianello, L., 1904. Der durchgehende Träger auf elastisch senkbaren Stützen. *Zeitschrift des Vereines deutscher Ingenieure*, vol. 48, No. 4, pp. 128–132 & No. 5, pp. 161–166.

Vianello, L., 1905. *Der Eisenbau*. Munich/Berlin: R. Oldenbourg.
Vianello, L., 1906. Knickfestigkeit eines dreiarmigen ebenen Systems. *Zeitschrift des Vereines deutscher Ingenieure*, vol. 50, No. 43, pp. 1753–1754.
Vianello, L., 1907. Der Flachträger. Durchgehender räumlicher Träger auf nachgiebigen Stützen. *Zeitschrift des Vereines deutscher Ingenieure*, vol. 51, No. 42, pp. 1661–1666.
Vierendeel, A., 1897. *L'architecture du Fer et de l'Acier*. Brussels/Paris: Ep. Lyon-Claesen.
Vierendeel, A., 1900. *Théorie générale des poutres Vierendeel*. Mémoires et comptes rendus des Travaux de la société des ingénieurs civils de France. Paris.
Vierendeel, A., 1901. *La construction architectureale en fonte, fer et acier*. Louvain: Upstpruyst. Paris: Dunod.
Vierendeel, A., 1911. Der Vierendeelträger im Brückenbau. *Der Eisenbau*, vol. 2, No. 10, pp. 381–385.
Vierendeel, A., 1912. Einige Betrachtungen über das Wesen des Vierendeelträgers. *Der Eisenbau*, vol. 3, No. 6, pp. 242–244.
Vierendeel, A., 1931. *Cours de stabilité des constructions*. vol. I, 5th ed. Louvain: Librairie universitaire. Paris: Dunod.
Villarceau, A. J. F. Y., 1844. Équilibre des voûtes en berceaux cylindriques. *Revue de l'Architecture et des Travaux Publics*, vol. 5, col. 57–82, 256–266 & 393–419.
Villarceau, A. J. F. Y., 1853. *Sur l'établissement des arches de pont envisagé au point de vue de la plus grande stabilité*. Paris: Mallet-Bachelier.
Villarceau, A. J. F. Y., 1854. *Sur l'établissement des arches de pont*. Mémoires présentées par divers savants à l'Académie des Sciences de l'Institut de France, vol. 12, pp. 503–522.
Virlogeux, M., 2006. Der Viadukt über das Tarntal bei Millau. *Bautechnik*, vol. 83, No. 2, pp. 85–107.
Vitruvius, P. M., 1981. *Zehn Bücher über Architektur*. Trans. with commentary by Curt Fensterbusch. 3rd ed. Darmstadt: Wissenschaftliche Buchgesellschaft.
Viviani, R., 1957. Riccardo Gizdulich – La ricostruzione del Ponte a S. Trinità. *Comunità* No. 51 – Luglio (review). Colletino Technico, Dec 1957, pp. 22–23.
Vlasov, V. Z., 1933. *Novyi metod rascheta tonkostennykh prizmaticheskikh skladchatykh pokrytii i obolochek* (new method of calculation for thin-walled prismatic folded-plate roofs and shells). Moscow: Gosstroiizdat (in Russian).
Vlasov, V. Z., 1940. *Tonkostennye uprugie sterzhni* (thin-walled elastic bars). Moscow/Leningrad: Gosstroiizdat (in Russian).
Vlasov, V. Z., 1949. *Obshchaya teoriya obolochek i ee prilozheniya v tekhnike* (general theory of shells and its application in engineering). Moscow/Leningrad: Gostekhizdat (in Russian).

Vlasov, V. Z., 1958. *Allgemeine Schalentheorie und ihre Anwendung in der Technik*. Berlin: Akademie-Verlag.
Vlasov, V. Z., 1962–1964. *Izbrannye trudy* (selected works). 3 vols. Moscow: AN SSSR-Nauka (in Russian).
Vlasov, V. Z., 1964/65. *Dünnwandige elastische Stäbe*. 2 vols. Berlin: VEB Verlag für Bauwesen.
Völker, P., 1908. Theorie des Eisenbetons. In Emperger, F. v. (ed.): *Handbuch für Eisenbetonbau I. Band*, pp. 214–301. Berlin: Wilhelm Ernst & Sohn.
Vogel, U., 1965. *Die Traglastberechnung stählerner Rahmentragwerke nach der Plastizitätstheorie II. Ordnung*. Structural steelwork research pamphlet No. 15. Berlin: Springer.
Vol'mir, A. S., 1953. Ocherk zhizni i deyatel'nosti I. G. Bubnova (outline of the life and works of I. G. Bubnov). In: *I. G. Bubnov, Trudy po teorii plastin* (works on plate theory), pp. 309–393 (in Russian).
Voltaire, F. M. A., 1927. *Der Doktor Akakia und sein Schildknappe. Zwei Streitschriften aus der Zeit Friedrichs des Großen*. Ed. and with epilogue by C. Diesch. Berlin: Berliner Bibliophilenbund.
Voormann, F., 2005. Von der unbewehrten Hohlsteindecke zur Spannbetondecke. Massivdecken zu Beginn des 20. Jahrhunderts. *Beton- und Stahlbetonbau*, vol. 100, No. 9, pp. 836–846.
Vorländer, K., 2004. *Immanuel Kant. Der Mann und das Werk. Drittes Buch: Die Höhezeit. Zweiter Teil: Der Mensch*. 3rd ed. Wiesbaden: Matrix Verlag.

Wachsmann, K., 1989. *Wendepunkt im Bauen*. Reprint of the 1959 ed. (Krausskopf-Verlag, Wiesbaden) with a foreword by O. Patzelt. Dresden: VEB Verlag der Kunst.
Waddell, J. A. L., 1916. *Bridge Engineering*. 2 vols. New York: John Wiley & Sons.
Wagner, H., 1928. Über räumliche Flugzeugfachwerke. Die Längsstabkraftmethode. *Zeitschrift für Flugtechnik und Motorluftschiffahrt*, vol. 19, No. 15, pp. 337–347.
Wagner, H., 1929/1. Verdrehung und Knickung von offenen Profilen. In: *Festschrift 25 Jahre T. H. Danzig*, pp. 329–344. Danzig: Verlag A. W. Kefermann.
Wagner, H., 1929/2. Ebene Blechwandträger mit sehr dünnem Stegblech. *Zeitschrift für Flugtechnik und Motorluftschiffahrt*, vol. 20, No. 8, pp. 200–207, No. 9, pp. 227–233, No. 10, pp. 256–262, No. 11, pp. 279–284 & No. 12, pp. 306–314.
Wagner, H., Kimm, G., 1940. *Bauelemente des Flugzeuges*. Munich/Berlin: Oldenbourg.
Wagner, R., Egermann, R., 1987. *Die ersten Drahtkabelbrücken*. Düsseldorf: Werner-Verlag.
Wagner, W., Sauer, R., Gruttmann, E., 1999. Tafeln der Torsionskenngrößen von Walzprofilen unter Verwendung von FE-Diskretisierungen. *Stahlbau*, vol. 68, No. 2, pp. 102–111.

Walbrach, K. F., 2006. Ohne ihn gäbe es keinen Eiffelturm. Maurice Koechlin (1856–1946). *Bautechnik*, vol. 83, No. 4, pp. 271–284.

Walthelm, U., 1989. Beitrag zur Auffindung von Traglastmodellen. *Beton- und Stahlbetonbau*, vol. 84, No. 2, pp. 38–40.

Walther, A., 1952. Moderne mathematische Maschinen und Instrumente und ihre Anwendungsmöglichkeiten auf Probleme des Stahlbaus. In: Munich Steelwork Conference 1952. *Abhandlungen aus dem Stahlbau*, ed. Deutscher Stahlbau-Verband, No. 12, pp. 144–197. Bremen-Horn: Industrie- und Handelsverlag Walter Dorn.

Walther, A., Barth, W., 1961. Elektronisches Rechnen im Stahlbau. In: *Beiträge aus Statik und Stahlbau*, pp. 151–156. Cologne: Stahlbau-Verlags-GmbH.

Walther, A., 1925. Ritter, Georg Dietrich August. In: *Männer der Technik. Ein biographisches Handbuch*, ed. C. Matschoss, pp. 227–228. Berlin: VDI-Verlag.

Wapenhans, W., 1992. *Zur Entwicklung des Stahlverbundbaus in Deutschland bis 1992*. Dresden: Wapenhans und Richter.

Wapenhans, W., Richter, J., 2001. *Die erste Statik der Welt von 1742 zur Peterskuppel in Rom*. Dresden: Wapenhans und Richter (see also: http://www.wundr.com/html/05_06.htm)

Wapenhans, W., Richter, J., 2002. Die erste Statik der Welt vor 260 Jahren. *Bautechnik*, vol. 79, No. 8, pp. 543–553.

Washizu, K., 1953. Bounds for Solutions of Boundary Value Problems in Elasticity. *Journal of Mathematics and Physics*, vol. 32, No. 2–3, Jul–Oct, pp. 119–128.

Washizu, K., 1955. *On the Variational Principles of Elasticity and Plasticity*. Aeroelastic and Structures Research Laboratory, MIT, Technical Report 25-18, March.

Washizu, K., 1966. Note on the Principle of Stationary Complementary Energy Applied to Free Vibration of an Elastic Body. *International Journal of Solids and Structures*, vol. 2, No. 1, Jan, pp. 27–35.

Washizu, K., 1968. *Variational Methods in Elasticity and Plasticity*. Oxford: Pergamon Press.

Wasiutynski, A., 1899. Beobachtungen über die elastischen Formänderungen des Eisenbahngleises. *Organ für Fortschritte des Eisenbahnwesens* (supplement), No. 54, pp. 293–327 & plates XXXVIII–XLIV.

Wayss & Freytag AG (ed.), Mörsch, E., 1902. *Der Betoneisenbau. Seine Anwendung und Theorie*. Neustadt a. d. H.: Wayss & Freytag AG.

Wayss & Freytag AG (ed.), 1925. *Festschrift aus Anlaß des fünfzigjährigen Bestehens der Wayss & Freytag A.-G.* Stuttgart: Konrad Wittwer.

Wayss, G. A. (ed.), 1887. *Das System Monier in seiner Anwendung auf das gesammte Bauwesen*. Berlin/Vienna.

Weber, C., 1921. *Die Lehre der Drehungsfestigkeit*. Forschungsarbeiten auf dem Gebiete des Ingenieurwesens, No. 249. Berlin: Verlag des Vereines deutscher Ingenieure.

Weber, C., 1922. Bisherige Lösungen des Torsionsproblems. *Zeitschrift für angewandte Mathematik und Mechanik*, vol. 2, No. 4, pp. 299–302.

Weber, C., 1924. Biegung und Schub in geraden Balken. *Zeitschrift für angewandte Mathematik und Mechanik*, vol. 4, No. 4, pp. 334–348.

Weber, C., 1926. Übertragung des Drehmomentes in Balken mit doppelflanschigem Querschnitt. *Zeitschrift für angewandte Mathematik und Mechanik*, vol. 6, No. 2, pp. 85–97.

Weber, C., 1928. Letter conc.: Beiträge zur Lösung des ebenen Problems eines elastischen Körpers mittels der Airyschen Spannungsfunktion von W. Riedel. *Zeitschrift für angewandte Mathematik und Mechanik*, vol. 8, No. 2, pp. 159–160.

Weber, C., 1938. Veranschaulichung und Anwendung der Minimalsätze der Elastizitätstheorie. *Zeitschrift für angewandte Mathematik und Mechanik*, vol. 18, No. 6, pp. 375–379.

Weber, C., 1940. Halbebene mit Kreisbogenkerbe. *Zeitschrift für angewandte Mathematik und Mechanik*, vol. 20, No. 5, pp. 262–270.

Weber, C., 1942. Eingrenzung von Verschiebungen und Zerrungen mit Hilfe der Minimalsätze. *Zeitschrift für angewandte Mathematik und Mechanik*, vol. 22, No. 3, pp. 130–136.

Weber, C., 1948. Zur nichtlinearen Elastizitätstheorie. *Zeitschrift für angewandte Mathematik und Mechanik*, vol. 28, No. 6, pp. 189–190.

Weber, C., Günther, W., 1958. *Torsionstheorie*. Braunschweig: Friedrich Vieweg.

Weber, M. v., 1869. *Die Stabilität des Gefüges der Eisenbahn-Gleise*. Wiemar: Verlag B. F. Voigt.

Weber, M., 1993. *Die protestantische Ethik und der "Geist" des Kapitalismus*. Bodenheim: Athenäum Hain Haustein.

Weber, W., 1999. *Die gewölbte Eisenbahnbrücke mit einer Öffnung. Begriffserklärungen, analytische Fassung der Umrißlinien und ein erweitertes Hybridverfahren zur Berechnung der oberen Schranke ihrer Grenztragfähigkeit, validiert durch ein Großversuch*. Dissertation, Munich TU.

Wehle, L. B., Lansing, W., 1952. A method for reducing the analysis of complex redundant structures to a routine procedure. *Journal of the Aeronautical Sciences*, vol. 19, pp. 677–684.

Weichhold, A., 1968. *Johann Andreas Schubert – Lebensbild eines bedeutenden Hochschullehrers und Ingenieurs aus der Zeit der industriellen Revolution*. Leipzig/Jena: Urania-Verlag.

Weingarten, J., 1901. Review of (Föppl, 1898/1). *Archiv der Mathematik und Physik, III. Reihe, 1. Bd.*, pp. 343–352. Leipzig: B. G. Teubner.

Weingarten, J., 1902. Über den Satz vom Minimum der Deformationsarbeit. *Archiv der Mathematik und Physik, III. Reihe, 2. Bd.*, pp. 233–39. Leipzig: B. G. Teubner.

Weingarten, J., 1905. Über die Lehrsätze Castiglianos. *Archiv der Mathematik und Physik, III. Reihe, 8. Bd.*, pp. 183–192. Leipzig: B. G. Teubner.

Weingarten, J., 1907/1. Zu der Abhandlung des Herrn Professor A.Hertwig: Die Entwicklung einiger Prinzipien der Statik der Baukonstruktionen. *Zeitschrift für Arch.-u. Ing.-Wesen*, pp. 107–110.

Weingarten, J., 1907/2. Zur Theorie der Wirkung der ungleichen Erwärmung auf elastische Körper in Beziehung auf Fachwerke. *Zeitschrift für Arch.- & Ing.- Wesen*, pp. 453–462.

Weingarten, J., 1909/1. Über das Clapeyronsche Theorem in der technischen Elastizitätstheorie. *Zeitschrift für Arch.- & Ing.- Wesen*, pp. 515–520.

Weingarten, J., 1909/2. Über den Begriff der Deformationsarbeit in der Theorie der Elasticität fester Körper. *Nachrichten d. Kgl. Ges. d. Wiss. zu Göttingen. Math.-phys. Klasse*, pp. 64–67.

Weisbach, J., 1845. *Lehrbuch der Ingenieur- und Maschinen-Mechanik. Erster Teil: Theoretische Mechanik*, 1st ed. Braunschweig: Friedrich Vieweg.

Weisbach, J., 1846. *Lehrbuch der Ingenieur- und Maschinen-Mechanik. Zweiter Teil.: Praktische Mechanik*, 1st ed. Braunschweig: Friedrich Vieweg.

Weisbach, J., 1847/48. *Principles of mechanics of machinery and engineering*, ed. L. Gordon. London: Hippolyte Ballière Publisher.

Weisbach, J., 1848/1. Die Theorie der zusammengesetzten Festigkeit, *Der Ingenieur*, vol. 1, pp. 252–265.

Weisbach, J., 1848/2. *Der Ingenieur. Sammlung von Tafeln, Formeln und Regeln der Arithmetik, Geometrie und Mechanik*. Braunschweig: Friedrich Vieweg.

Weisbach, J., 1850. *Lehrbuch der Ingenieur- und Maschinen-Mechanik. Erster Teil.: Theoretische Mechanik*, 2nd ed. Braunschweig: Friedrich Vieweg.

Weisbach, J., 1851. *Lehrbuch der Ingenieur- und Maschinen-Mechanik. Zweiter Teil.: Statik der Bauwerke und Mechanik der Umtriebsmaschinen*, 2nd ed. Braunschweig: Friedrich Vieweg.

Weisbach, J., 1855. *Lehrbuch der Ingenieur- und Maschinen-Mechanik. Erster Teil.: Theoretische Mechanik*, 3rd ed. Braunschweig: Friedrich Vieweg.

Weisbach, J., 1857. *Lehrbuch der Ingenieur- und Maschinen-Mechanik. Zweiter Teil.: Statik der Bauwerke und Mechanik der Umtriebsmaschinen*, 4th ed. Braunschweig: Friedrich Vieweg.

Weisbach, J., 1863/1. *Lehrbuch der Ingenieur- und Maschinen-Mechanik. Erster Teil.: Theoretische Mechanik, 1. Hälfte, 1. Abt.: Hilfslehren, Bewegungslehre, Statik fester Körper; 2. Abt.: Elastizitäts- und Festigkeitslehre*, 4th ed. Braunschweig: Friedrich Vieweg.

Weisbach, J., 1863/2. *Lehrbuch der Ingenieur- und Maschinen-Mechanik. Erster Teil: Theoretische Mechanik, 2. Hälfte, 3. Abt.: Dynamik der festen Körper. Statik der flüssigen Körper. Dynamik der flüssigen Körper. Schwingungen*, 4th ed. Braunschweig: Friedrich Vieweg.

Weisbach, J., 1863/3. *Lehrbuch der Ingenieur- und Maschinen-Mechanik. Zweiter Teil.: Statik der Bauwerke und Mechanik der Umtriebsmaschinen,* 4th ed. Braunschweig: Friedrich Vieweg.

Weisbach, J., 1875. *Lehrbuch der Ingenieur- und Maschinen-Mechanik. Erster Teil: Theoretische Mechanik,* 5th ed. Braunschweig: Friedrich Vieweg.

Weisbach, J., 1882. *Lehrbuch der Ingenieur- und Maschinen-Mechanik. Zweiter Teil, 1. Abt.: Statik der Bauwerke,* 5th ed., ed. G. Herrmann. Braunschweig: Friedrich Vieweg.

Weisbach, J., 1887. *Lehrbuch der Ingenieur- und Maschinen-Mechanik. Zweiter Teil, 2. Abt.: Mechanik der Umtriebsmaschinen,* 5th ed., ed. G. Herrmann. Braunschweig: Friedrich Vieweg.

Weitz, F. R., 1975. *Entwurfsgrundlagen und Entscheidungskriterien für Konstruktionssysteme im Großbrückenbau unter besonderer Berücksichtigung der Fertigung.* Dissertation, Darmstadt TH.

Wenzel, F. (ed.), 1987–1997. *Erhalten historisch bedeutsamer Bauwerke, Jahrbücher.* Berlin: Ernst & Sohn.

Werner, E., 1974. *Die ersten eisernen Brücken (1777–1859).* Dissertation, Munich TU.

Werner, E., 1980. *Technisierung des Bauens. Geschichtliche Grundlagen moderner Bautechnik.* Düsseldorf: Werner-Verlag.

Werner, F., Seidel, J., 1992. *Der Eisenbau. Vom Werdegang einer Bauweise.* Berlin/Munich: Verlag für Bauwesen.

Westergaard, H. M., Slater, W. A., 1921. Moments and stresses in slabs. *Proceedings of the American Concrete Institute,* vol. 17, No. 2, pp. 415–538.

Westergaard, H. M., 1922. Buckling of elastic structures. *Transactions of the American Society of Civil Engineers,* vol. 85, pp. 576–634.

Westergaard, H. M., 1925. *Anwendung der Statik auf die Ausgleichsrechnung.* Dissertation, Munich TH.

Westergaard, H. M., 1926/1. Stresses in concrete pavements computed by theoretical analysis. *Public Roads,* vol. 7, No. 2, pp. 25–35.

Westergaard, H. M., 1926/2. Computation of stresses in concrete roads. *Proceedings, Highway Research Board,* vol. 5, No. 1, pp. 90–112.

Westergaard, H. M., 1930/1. One hundred years advance in structural analysis. *Transactions of the American Society of Civil Engineers,* vol. 94, pp. 226–246.

Westergaard, H. M., 1930/2. Computation of Stresses in bridge slabs due to wheel loads. *Public Roads,* vol. 11, No. 1, pp. 1–23.

Westergaard, H. M., 1933. Water pressure on dams during earthquakes. *Transactions of the American Society of Civil Engineers,* vol. 98, pp. 418–472.

Westergaard, H. M., 1935. General solution of the problem of elastostatics of an $n$-dimensional homogeneous isotropic solid in an $n$-dimensional space. *Bulletin of the American Mathematical Society,* vol. 41, No. 10, pp. 695–699.

Westergaard, H. M., 1939. Bearing pressures and cracks. *Journal of Applied Mechanics,* vol. 6, pp. 49–53.

Westergaard, H. M., 1952. *Theory of Elasticity and Plasticity.* Cambridge: Havard University Press.

Westfall, R. S., 1972. Hooke Robert. In: *Dictionary of Scientific Biography,* vol. VI. C. C. Gillispie (ed.), pp. 481–488. New York: Charles Scribner's Sons.

Weyrauch, J., 1873. *Allgemeine Theorie und Berechnung der kontinuierlichen und einfachen Träger.* Leipzig: B. G. Teubner.

Weyrauch, J. J., 1874. Die graphische Statik. Historisches und Kritisches. *Zeitschrift für Mathematik und Physik,* vol. 19, pp. 361–390.

Weyrauch, J. J., 1879. *Theorie der elastischen Bogenträger.* Munich: Theodor Ackermann.

Weyrauch, J. J., 1884. *Theorie elastischer Körper. Eine Einleitung zur mathematischen Physik und technischen Mechanik.* Leipzig: B. G. Teubner.

Weyrauch, J. J., 1897. *Die elastischen Bogenträger. Ihre Theorie und Berechnung entsprechend den Bedürfnissen der Praxis. Mit Berücksichtigung von Gewölben und Bogenfachwerken.* 2nd, rev. ed. Munich: Theodor Ackermann.

Weyrauch, J. J., 1908. Über statisch unbestimmte Fachwerke und den Begriff der Deformationsarbeit. *Zeitschrift für Arch.- & Ing.- Wesen,* pp. 91–108.

Weyrauch, J. J., 1909. Über den Begriff der Deformationsarbeit in der Theorie der Elastizität fester Körper. *Nachrichten d. Kgl. Ges. d. Wiss. zu Göttingen. Math.-phys. Klasse,* pp. 242–246.

Weyrauch, R., 1922. Weyrauch, Jakob, ord. Professor der Technischen Hochschule Stuttgart. In: *Württembergischer Nekrolog für 1917,* pp. 22–31. Stuttgart.

Whipple, S., 1847. *A work on bridge building.* Utica: H. H. Curtiss.

Whipple, S., 1869. *An appendix to Whipple's bridge building.* Albany.

Whipple, S., 1873. *An elementary and practical treatise on bridge building.* New York: D. Van Nostrand.

White, H. S., 1917/1918. Cremona's works. *Bull. Amer. Math. Soc.,* pp. 238–243.

Wiberg, N.-E., Oñate, E., Zienkiewicz, O. C., Taylor, R., 2006. Obituary of Alf Samuelsson (1930–2006). *International Journal for Numerical Methods in Engineering,* vol. 67, pp. 158–159.

Wiederkehr, K. H., 1986. Maxwell. In: *Große Naturwissenschaftler. Biographisches Lexikon.* 2nd, rev. ed., ed. F. Krafft, pp. 235–237. Düsseldorf: VDI-Verlag.

Wieghardt, K., 1906. Über einen Grenzübergang der Elastizitätslehre und seine Anwendung auf die Statik hochgradig statisch unbestimmter Fachwerke. *Verhandlungen des Vereines zur Beförderung des Gewerbefleißes,* vol. 85, pp. 139–176.

Wiegmann, R., 1839. *Über die Konstruktion von Kettenbrücker nach dem Dreiecksysteme und deren Anwendung auf Dachverbindungen.* Düsseldorf: J. H. C. Schneider.

Wilhelm Ernst & Sohn (ed.), 1967. *Wilhelm Ernst & Sohn. Verlag für Architektur und technische Wissenschaften* (company chronology). Berlin/Munich: Wilhelm Ernst & Sohn.

Williamson. F., 1979. An historical note on the finite element method. *International Journal for Numerical Methods in Engineering,* vol. 15, pp. 930–934.

Williot, V.-J., 1877. Notions pratiques sur la statique graphique. *Annales du Génie civil,* 2nd series, vol. 6, Nos. 10 & 12, Oct & Dec.

Williot, V.-J., 1878. *Notions pratiques sur la statique graphique.* Paris: E. Lacroix.

Winkler, E., 1858. Formänderung und Festigkeit gekrümmter Körper, insbesondere der Ringe. *Civilingenieur,* vol. 4, pp. 232–246.

Winkler, E., 1860. Festigkeit der Röhren, Dampfkessel und Schwungringe. *Civilingenieur,* vol. 6, pp. 325–362 & 427–462.

Winkler, E., 1862. Beiträge zur Theorie der continuierlichen Brückenträger. *Civilingenieur,* vol. 8, pp. 136–182.

Winkler, E., 1867. *Die Lehre von der Elasticität und Festigkeit.* Prague: H. Dominicus.

Winkler, E., 1868/1869. Vortrag über die Berechnung der Bogenbrücken. *Mitteilungen des Architekten- und Ingenieurvereins Böhmen,* 1868, pp. 6–12 & 1869, pp. 1–7.

Winkler, E., 1871. Abriss der Geschichte der Elasticitätslehre. *Technische Blätter,* vol. 3, No. 1, pp. 22–33 & No. 3, pp. 232–245.

Winkler, E., 1872/1. Theorie der continuierlichen Träger. *Zeitschrift des österreichischen Ingenieur- und Architekten-Vereines,* vol. 24, pp. 27–32 & 61–65.

Winkler, E., 1872/2. *Die Gitterträger und Lager gerader Träger eiserner Brücken.* Vienna: Carl Gerold's Sohn.

Winkler, E., 1872/3. *Neue Theorie des Erddruckes nebst einer Geschichte der Theorie des Erddruckes und der hierüber angestellten Versuche.* Vienna R. v. Waldheim.

Winkler, E., 1879/1880. Die Lage der Stützlinie im Gewölbe. *Deutsche Bauzeitung,* vol. 13, pp. 117–119 & 127–130, vol. 14, pp. 58–60.

Winkler, E., 1881/1. *Theorie der Brücken. Theorie der gegliederten Balkenträger.* 2nd ed. Vienna: Carl Gerold's Sohn.

Winkler, E., 1881/2. Die Sekundär-Spannungen in Eisenkonstruktionen. *Deutsche Bauzeitung,* vol. 14, pp. 110–111, 129–130 & 135–136.

Winkler, E., 1883. *Vorträge über Statik der Baukonstruktionen, I. Nr.: Festigkeit gerader Stäbe. Teil I,* 3rd ed. Berlin: printed lecture notes.

Winkler, E., 1884. Alberto Castigliano. *Deutsche Bauzeitung,* vol. 18, pp. 570–573.

Wittek, K. H., 1964. *Die Entwicklung des Stahlhochbaus.* Düsseldorf: VDI-Verlag.

Wittfoht, H., 1986. Wohin entwickelt sich der Spannbeton-Brückenbau? Analysen und Tendenzen. *Beton- und Stahlbetonbau*, vol. 81, No. 2, pp. 29–35.

Wittmann, W., 1879. Zur Theorie der Gewölbe. *Zeitschrift für Bauwesen*, vol. 29, col. 61–74.

Wöhler, A., 1855. Theorie rechteckiger eiserner Brückenbalken mit Gitterwänden und mit Blechwänden. *Zeitschrift für Bauwesen*, vol. 5, col. 122–166.

Wöhler, A., 1858. Bericht über die Versuche, welche auf der Königl. Niederschlesisch-Märkischen Eisenbahn mit Apparaten zum Messen der Biegung und Verdrehung von Eisenbahnwagen-Achsen während der Fahrt, angestellt wurden. *Zeitschrift für Bauwesen*, vol. 8, col. 642–652.

Wöhler, A., 1860. Versuche zur Ermittlung der auf die Eisenbahnwagen-Achsen einwirkenden Kräfte und der Widerstandsfähigkeit der Wagen-Achsen. *Zeitschrift für Bauwesen*, vol. 10, col. 584–616.

Wöhler, A., 1863. Über die Versuche zur Ermittlung der Festigkeit von Achsen, welche in den Werkstätten der Niederschlesisch-Märkischen Eisenbahn zu Frankfurt a. d. O. angestellt sind. *Zeitschrift für Bauwesen*, vol. 13, col. 233–258.

Wöhler, A., 1866. Resultate der in der Centralwerkstatt der Niederschlesisch-Märkischen Eisenbahn zu Frankfurt a. d. O. angestellten Versuche über die relative Festigkeit von Eisen, Stahl und Kupfer. *Zeitschrift für Bauwesen*, vol. 16, col. 67–84.

Wöhler, A., 1870. Über die Festigkeits-Versuche mit Eisen und Stahl. *Zeitschrift für Bauwesen*, vol. 20, col. 74–106.

Wolossewitsch, O. M., et al., 1980. *Spezifik der technischen Wissenschaften*. Moscow/Leipzig: MIR & VEB Fachbuchverlag.

Wood, R. G., 1966. *Stephen Harriman Long, 1784–1864: Army engineer, explorer, inventor*. Glendale: Arthur H. Clark & Co.

Wood, R. H., 1961. *Plastic and elastic design of slabs and plates*. London: Thames & Hudson.

Worch, G., 1962. Lineare Gleichungen in der Baustatik. In: *Beton-Kalender*, vol. 51, pt. II, pp. 310–384. Berlin: Verlag von Wilhelm Ernst & Sohn.

Worch, G., 1965. Elektronisches Rechnen in der Baustatik. Eine Einführung. In: *Beton-Kalender*, vol. 54, pt. II, pp. 319–384. Berlin: Verlag von Wilhelm Ernst & Sohn.

Wren, S., 1750. *Parentalia or Memoirs of the Family of Wrens*. London.

Wuczkowski, R., 1907. Geschäftshaus des Architekten C. Hofmeier, Vienna I, Kärntnerstraße 27. *Beton und Eisen*, vol. 6, No. 6, p. 150, No. 7, pp. 166–168 & No. 9, p. 244.

Wuczkowski, R., 1912. *Zur Statik der Stockwerkrahmen*. 2nd, rev. ed. Berlin: Wilhelm Ernst & Sohn.

Wurzbach, C. v., 1859. Franz Joseph Ritter von Gerstner. In: *Biographisches Lexikon des Kaisertums Österreich*, pt. 5, pp. 160–163. Vienna: Verlag der typogr.-literar.-artist. Anstalt.

Wußing, H., 1958. Die École Polytechnique – eine Errungenschaft der Französischen Revolution. *Pädagogik*, vol. 13, No. 9, pp. 646–662.

Wußing, H., Arnold, W. (ed.), 1978. *Biographien bedeutender Mathematiker*. Cologne: Aulis Verlag Deubner & Co. KG.

Wußing, H., 1979. *Vorlesungen zur Geschichte der Mathematik*. Berlin: VEB Deutscher Verlag der Wissenschaften.

Wußing, H. (ed.), 1983. *Geschichte der Naturwissenschaften*. Cologne: Aulis-Verlag Deubner.

Wyklicky, H., 1984. *200 Jahre Allgemeines Krankenhaus*. Vienna: Facultas.

Wyss, T., 1923. *Beitrag zur Spannungsuntersuchung an Knotenblechen eiserner Fachwerke. Forschungsarbeiten auf dem Gebiete des Ingenieurwesens*, No. 262. Berlin: Springer.

Yamada, M., 1980. Erdbebensicherheit von Hochbauten. *Stahlbau*, vol. 49, pp. 225–231.

Yamamoto, Y., 1966. *A formulation of matrix displacement method*. MIT, Dept. of Aeronautics & Astronautics.

Yamamoto, Y., 1983. Obituary of Prof. Washizu. In: *Proceedings zum Symposium "Analyse mittels Matrizen"*, ed. Japanese Society of Steel Construction (JSSC). Tokyo (in Japanese).

Yokoo, Y. (ed.), 1980. *60 Jahre Fakultät für Architektur*, Kyoto University. Kyoto.

Young, D. M., 1987. A historical review of iterative methods. In: *Proceedings of the ACM conference on History of scientific and numeric computation*, ed. G. E. Crane, pp. 117–124. New York: ACM Press.

Young, T., 1807. *A Course of Lectures on Natural Philosophy and the Mechanical Arts*. 2 vols. London: Joseph Johnson.

Young, T., 1817. *Encyclopaedia Britannica (Supplement)* (Entries on Carpentry and Bridge). London.

Young, T., 1824. Bridge. In: *Supplement to the fourth, fifth and sixth editions of the Encyclopaedia Britannica*, vol. 2, pp. 497–520, plates 42–44. Edinburgh: Archibald Constable.

Young, T., 1855. *Miscellaneous Works*, ed. G. Peacock. London: John Murray.

Zammattio, C., 1974. Naturwissenschaftliche Studien. In: *Leonardo. Künstler – Forscher – Magier*, ed. L. Reti. Frankfurt a. M.: S. Fischer.

Zedler, J. H. (ed.), 1735. Gewölbe. In: *Grosses vollständiges Universal-Lexikon*, vol. 10. Halle/Leipzig: Verlag Johann Heinrich Zedler.

Zerna, W., 1949/1. Beitrag zur allgemeinen Schalenbiegetheorie. *Ingenieur-Archiv*, vol. 17, pp. 149–164.

Zerna, W., 1949/2. Zur Membrantheorie der allgemeinen Rotationsschalen. *Ingenieur-Archiv*, vol. 17, pp. 223–232.

Zerna, W., 1950. Grundgleichungen der Elastizitätstheorie. *Ingenieur-Archiv*, vol. 18, pp. 211–220.

Zerna, W., 1953. Zur neueren Entwicklung der Schalentheorie. *Beton- und Stahlbetonbau*, vol. 48, No. 4, pp. 88–89.

Zerna, W., 1967. Das wissenschaftliche Werk von Prof. Dr.-Ing. Reinhold Rabich. *Wiss. Z. Techn. Univers. Dresden*, vol. 16, No. 1, pp. 43–44.

Zerna, W., Trost, H., 1967. Rheologische Beschreibung des Werkstoffes Beton. *Beton- und Stahlbetonbau*, vol. 62, No. 7, pp. 165–170.

Zerna, W., 1974. *Beuluntersuchungen an hyperbolischen Rotationsschalen*. Opladen: Westdeutscher Verlag.

Zerna, W., 1978. *Kriterien zur Optimierung des Baues von Großnaturzugkühltürmen im Hinblick auf Standsicherheit, Bauausführung und Wirtschaftlichkeit*. Opladen: Westdeutscher Verlag.

Zesch, E., 1962. Friedrich Emperger, ein Pionier der Stahlbetonbauweise. *Blätter für Technikgeschichte*, vol. 24, pp. 157–167.

Zhuravsky (Jouravski), D. I., 1864. O slozhnoi sisteme, predstavlyayushchei soedinenie arki s raskosnoyu sistemoyu (on a complex system for an arch stiffened by a system of struts). *Zhurnal Glavnogo upravleniya putei soobshcheniya i publichnykh zdanii* (journal of the principal authority for public works), No. 4, pp. 165–194 (in Russian).

Ziegler, R., 1985. *Die Geschichte der geometrischen Mechanik im 19. Jahrhundert*. Stuttgart: Franz Steiner Verlag.

Zienkiewicz, O. C., Cheung, Y. K., 1964. The finite element method for analysis of elastic isotropic and orthotropic slabs. *Proceedings of the Institution of Civil Engineers*, vol. 28, pp. 471–488.

Zienkiewicz, O. C., 1965. Finite element procedures in the solution of plate and shell problems. In: *Stress Analysis*. Ed. O. C. Zienkiewicz & G. S. Holister, pp. 120–144. New York: John Wiley & Sons.

Zienkiewicz, O. C., Cheung, Y. K., 1967. *The Finite Element Method in Structural and Continuum Mechanics*. London/New York/Toronto/Sydney: McGraw-Hill.

Zienkiewicz, O. C., Gallagher, R. H., Lewis, R. W., 1994. International Journal for Numerical Methods in Engineering: The first 25 years and the future. *International Journal for Numerical Methods in Engineering*, vol. 37, pp. 2151–2158.

Zienkiewicz, O. C., 1995. Origins, milestones and directions of the finite element method. *Archives of Computational Methods in Engineering*, vol. 2, No. 1, pp. 1–48.

Zienkiewicz, O. C., Lewis, R. W., Carey, G. F., 1997. In Memorian to a great engineering scientist and educator. Prof. Richard Hugo Gallagher 17 November 1927 – 30 September 1997. *Communications in Numerical Methods in Engineering*, vol. 13, pp. 903–907.

Zienkiewicz, O. C., Taylor, C., 1998. In Memorian to a great engineering scientist and educator.

Prof. Richard Hugo Gallagher 17 November 1927 – 30 September 1997. *International Journal for Numerical Methods in Fluids*, vol. 27, pp. 1 – 11.

Zienkiewicz, O. C. (ed.), 2001. Classic Reprints Series: Displacement and equilibrium models in the finite element method by B. Fraeijs de Veubeke chap. 9, pp. 145 – 197 of Stress Analysis, ed. O. C. Zienkiewicz and G. S. Holister, Published by John Wiley & Sons, 1965. Reprinted in *International Journal for Numerical Methods in Engineering*, vol. 52, pp. 287 – 342.

Zienkiewicz, O. C., 2004. The birth of the finite element method and of computational mechanics. *International Journal for Numerical Methods in Engineering*, vol. 60, pp. 3 – 10.

Zilch, K., Rogge, A., 1999. Bemessung von Stahlbeton- und Spannbetonbauteilen nach EC 2 – pt. II. In: *Beton-Kalender*, vol. 88, pt. 1, ed. J. Eibl, pp. 341 – 454. Berlin: Ernst & Sohn.

Zilch, K., Diederichs, C. J., Katzenbach, R., 2002. *Handbuch für Bauingenieure*. Berlin: Springer-Verlag.

Zimmer, A., 1969. *Digitales Verfahren für die wirtschaftliche Lösung von Festigkeitsproblemen und anderen Aufgaben der Elastizitätslehre*. Dissertation, Clausthal TU.

Zimmer, A., Eiselt, H. P., 1969. ESEM – Elastostatik-Elementmethode. Teil 1: Lösung von Kerbspannungsproblemen. *IBM-Nachrichten*, vol. 19, No. 195, pp. 701 – 709.

Zimmer, A., Nolting, H., 1969. ESEM – Elastostatik-Elementmethode. Teil 2: Organisation und Datenablauf für ESEM-Eingabe. *IBM-Nachrichten*, vol. 19, No. 196, pp. 779 – 785.

Zimmer, A., Bausinger, R., 1969. ESEM – Elastostatik-Elementmethode. Teil 3: Berechnung von Lagerreaktionen und Biegeflächen von verschieden gelagerten Platten. *IBM-Nachrichten*, vol. 19 No. 197, pp. 860 – 866.

Zimmer, A., Groth, P., 1970. *Elementmethode der Elastostatik – Programmierung und Anwendung*. Munich/Vienna: Oldenbourg.

Zimmermann, H., 1888. *Die Berechnung des Eisenbahnoberbaues*. Berlin: Ernst & Sohn.

Zimmermann, H., 1889. *Rechentafel*. Berlin: Verlag von Wilhelm Ernst & Sohn.

Zimmermann, H., 1901/1. *Über Raumfachwerke, neue Formen und Berechnungsweisen für Kuppeln und sonstige Dachbauten*. Berlin: Ernst & Sohn.

Zimmermann, H., 1901/2. Das Raumfachwerk der Kuppel des Reichstagshauses. *Zentralblatt der Bauverwaltung*, vol. 21, No. 33, pp. 201 – 203 & No. 34, pp. 209 – 214.

Zimmermann, H., 1922/1. Anleitung zur Berechnung der Fehlerhebel aus den Neigungsänderungen der Stabenden – Letter to DASt dated 22 June 1922. From: *Handakte "Versuchsausschuß" des DStV*.

Zimmermann, H., 1922/2. Die Lagerung bei Knickversuchen und ihre Fehlerquellen. *Sitzungsberichte der Preußischen Akademie der Wissenschaften*, No. 12.

Zimmermann, H., 1922/3. Die Knickfestigkeit vollwandiger Stäbe in neuer einheitlicher Darstellung. *Zentralblatt der Bauverwaltung*, vol. 42, pp. 34 – 39.

Zimmermann, H., 1930. *Die Lehre vom Knicken auf neuer Grundlage*. Berlin: Ernst & Sohn.

Zimmermann, P., 1981. István Szabó (1906 – 1980). *Verhandlungen der Naturforschenden Gesellschaft in Basel*, vol. 91, pp. 77 – 78.

Zipkes, S., 1906/1. Eisenbetonbrücken mit versenk-ter Fahrbahn. *Beton und Eisen*, vol. 5, No. 6, pp. 140 – 144, No. 7, pp. 164 – 166 & No. 8, pp. 197 – 198.

Zipkes, S., 1906/2. Fachwerkträger aus Eisenbeton. *Beton und Eisen*, vol. 5, No. 10, pp. 244 – 247 & No. 11, pp. 281 – 284.

Zöllner, G., 1956. Lebensbild Julius Weisbachs. In: *Julius Weisbach. Gedenkschrift zu seinem 150. Geburtstag. Freiberger Forschungshefte Reihe D*, No. 16. Ed. dean of Freiberg Mining Academy, pp. 11 – 61. Berlin: Akademie-Verlag.

Zöllner, E., 1970. *Geschichte Österreichs*. 4th ed. Munich: Oldenbourg.

Zschetzsche, A. F., 1901. Die Kuppel des Reichstagshauses. *Zeitschrift des Österr. Ing.- & Arch.-Vereines*, vol. 53, No. 4, pp. 52 – 60, No. 5, pp. 65 – 70 & No. 6, pp. 81 – 87.

Zschimmer, E., 1925. Zur Erkenntniskritik der technischen Wissenschaft. In: *Festschrift anläßlich des 100jährigen Bestehens der Technischen Hochschule Fridericiana Karlsruhe*, pp. 531 – 542. Berlin: VDI-Verlag.

Zumpe, G., 1975. *Angewandte Mechanik, Band 1: Bildung und Beschreibung von Modellen*. Berlin: Akademie Verlag.

Zurmühl, R., 1950. *Matrizen. Eine Darstellung für Ingenieure*. Berlin: Springer-Verlag.

Zuse, K., 1936. *Die Rechenmaschine des Ingenieurs*. Berlin: Typewritten ms. dated 30 Jan 1936.

Zuse, K., 1993. *Der Computer. Mein Lebenswerk*. 3rd ed. Berlin: Springer-Verlag.

Zuse, 1995. Conversation between the author and Konrad Zuse on 16 Jan 1995 in Hünfeld.

Zweiling, K., 1948. *Lebenslauf*. Appendix to application by Dr. Klaus Zweiling upon approval for habilitation thesis. Typewritten ms.

Zweiling, K., 1952. *Grundlagen einer Theorie der biharmonischen Polynome*. Berlin: Verlag Technik.

Zweiling, K., 1953. *Gleichgewicht und Stabilität*. Berlin: Verlag Technik.

# NAME INDEX

**A**beles 530
Ackeret 713
Addis 26, 136
Adorno 693
Affan, A. 495
Agatz 775
Aimond 552
Aiken 617
Airy 77, *712*, 750
Aita 210, 219
Albert 749
Alberti 138f., 199, 205, 209, 260
Albert von Sachsen 727
Alexander I 58, 310f.
Allievi 358
Ammann 764f.
Ammannati 190, 192
Ammontons 203
Ampère 352, 756
Anders, G. 572
Andrä, W. 457
Andreaus 440
Andrée 410, *431ff.*
Anthemios 255
Antoinette 698
Archimedes *41ff.*, 46, 255f., 725, 765
Ardant 223
Argyris 39, 380, 394, 597, 600, 610, 619, 630, 635, *640ff.*, 654, 662, *712f.*
Aristotle 43, 46, 137, 194, 204f., 675, 698, 725
Armstrong, W. G. 420
Aron 548, *713*, 722
Aronhold 354
Arons, J. 511
Arup, O. 738
Asimont 587
Asley 717
Asplund *713f.*
Audoy 212
Augusti 240
Autenrieth 425, 522

**B**abbage 749
Bach 62, *155f.*, 334, *405ff.*, 410, *436f.*, 438f., 523, 526, 535, 539, 540, 564
Bachelard 243
Bachot 190
Bacon 737
Baker, J. F. *130ff.*, 646, *714*, 740, 757
Baldi 70, 194, *204ff.*, 258
Balet 230
Ban *714*
Bandhauer 67, *221f.*, 239
Bänsch 101
Banse 149, 151
Bargellini 190
Barlow 217, 712, *714f.*
Baron, F. 605
Barthel 235, 239, *243f.*, 325
Başar 554
Bastine 326
Bauby 82
Bauersfeld 474, 475, *486ff.*, 487, 551
Baumeister 180
Baumgarten, A. 694
Bauschinger 220, 224, 324, 407, 502, *682f.*, 730, 767
Bay, H. 560, *566f.*
Bazaine 312
Bažant *715*
Beatles 745
Becchi 24, 186, 201, 240, 396, 716
Bechert, H. 560
Becker, R. 759, 766
Beedle, L. S. 133
Belanger 305, 698, 719
Bell, A. G. 474, 475, *486f.*
Bella 196
Bélidor 33, 189, *201ff.*, 207f., 244, 283f., 292, 296, *715*
Beltrami 583, *715f.*
Bendixsen 581, *593ff.*
Benedetti 258
Benedict XIV 756

Benjamin 574
Benvenuto 24, 246, 396, 581, 584, *716*, 726
Berkel 725
Bernal 150
Bernhard, K. 123, *388*, 736
Bernhard, R. 441
Bernoulli, D. 677
Bernoulli, Jakob 90, 214, 258, 272, 293, 298, 533
Bernoulli, Jakob II 534
Bernoulli, Johann 43, *678f.*
Bernoulli, N. 756
Berry, A. 587
Berthelot 726
Bertot 305, 317
Bessemer 420
Bétancourt 58, 312
Betti 355, 579, 715, *716f.*, 723, 744
Beuth 729
Beyer, K. 123, 527, 552, *717*, 720, 726, 730, 751, 758
Biadego 358
Bieniek, M. 754
Biezeno 736
Bijlaard 732
Bisplinghoff *637ff.*, 641, *717*
Bismarck 50
Björnstad 530
Blaton 746
Bleich, F. 449, 540, 543, 596, *717f.*, 749
Bleich, H. H. *718*, 749
Bletzinger 674
Block, P. 706
Blondel 31, 743, 774
Blume, J. A. 754
Blumenthal, O. 550
Bohny 123
Boistard 209, 217
Böll, H. 450
Boll, K. 464, 560
Bollé, L. 750
Bolzano 733
Boothby 233, 240f.

Borda 723
Born, J. 547
Born, M. 611, 654, 718, 730
Bornemann 170
Bornscheuer 115, 435, 443, 646
Borsig 586
Borst, R. 742
Bosch, J. B. 535
Boscovich 679f., 708, 718f., 752, 756
Bossut 211, 217, 723
Bouch 712, 719
Bouguer 719
Bouniceau 502
Boussinesq 750
Bow 320, 322, 719
Boyer 82
Boyle 737
Braun, H.-J. 144
Bravais 493
Bredt 411ff., 443
Brecht 250, 543 f.
Bresse 82, 89, 95, 223, 305, 336f., 719f.
Breymann 48
Brik 180, 228, 529
Brioschi 715
Brindley 277
Brissat 52
Brizzi 191
Bronneck 530, 539
Broszko 451
Brückmann 170
Brunel, I. K. 712
Brunner 127, 451
Bubnov 720
Buchheim 306 f.
Buckminster Fuller 474 f., 486, 489 f.
Buffon 276, 296
Buonopane 69
Burger 736
Bürgermeister 720, 731, 751
Buridan, J. 727
Burmester 353 f., 356
Burr 70
Busch, W. 659
Busemann, A. 638
Büttner, O. 528
Byernadskii, N. M 603

Čališev, K. A. 603, 605
Calladine, C. R. 240, 320, 495
Cambournac 465
Canaris 712
Candela 238, 552
Canevazzi 358
Carathéodory 712
Cardano 46, 258
Carl, P. 195
Carnap 727
Carnot, L. 52, 308 f., 312, 721
Carnot, N. L. S. 154, 312 ff.
Carstanjen 446
Cassini II 723
Cassini, J. 715

Castigliano 34, 36, 89, 113, 187, 224, 307, 347,
 348, 351, 356, 358ff., 362, 364 f., 367 ff., 374,
 378 f., 382 f., 397, 408, 655, 685, 695, 720, 723,
 731, 744, 752
Cauchy 22, 95, 270, 301, 317, 331, 399, 681,
 720f., 722
Cavour 749
Cayley 610, 712
Ceradini 358
Chamberlain, H. S. 622
Charles X 721
Charlton 23, 731
Chatzis 319
Cheung, Y. K. 654
Chezy 208
Chitty 230
Chladni 533, 738
Christie, J. 96
Christoffel 381
Christophe 498 f., 515, 518ff., 524 ff., 535, 563,
 591, 721
Chudnovskii, V. G. 608
Chwalla 116, 443 f., 451, 453 f., 597, 722
Cichocki 115
Clagett 726
Clairaut 52
Clark, E. 77ff., 93, 304, 728
Clark, W. T. 67
Clapeyron 95, 172, 301, 305, 308ff., 342 f.,
 346 f., 350, 548, 682, 695, 722, 743 f., 753
Clausius 313
Clebsch 414, 429, 533, 581, 584f., 713, 722,
 734, 762
Clericetti 358
Clough 25, 40, 249, 640, 643, 648ff., 653 ff., 669
Cohn, B. 639
Coignet, F. 499, 515, 519
Colbert 273
Coleridge 174
Collar, A. R. 615f., 637
Collins, M. P. 568
Colonnetti 723
Como 240
Condorcet 53
Conrad, R. 736
Considère, A. 498, 514 f., 519, 728
Cook, W. D. 568
Copernicus 143, 774
Coriolis 161
Cornelius 457, 461
Corning, L. H. 754
Corradi, M. 24, 42, 210, 240, 246, 396, 716
Cort, H. 278
Cosimo I 190
Cosserat, E. 548
Cosserat, F. 548
Coste, A. 243
Coulomb 33, 90, 95, 203, 205, 211f., 217 f.,
 223, 246, 301, 311, 402, 715, 723, 724
Couplet 189, 203, 208f., 213, 221, 233, 237,
 723f.
Courant 655
Cousinéry 323

Craemer 527, 545
Cremona 307, 317, 320ff., 358, 695, 719, 724,
 744, 748
Cross 37 ff., 603ff, 724, 751, 766 f.
Cubitt 73
Culmann 34, 36, 48, 80, 82, 91ff., 95 f., 144,
 219, 307, 317ff., 329, 351, 397, 576, 577, 585,
 668, 682, 695, 703, 709, 724, 745, 760, 767
Cywiński, Z. 754
Czech, F. 591, 702
Czerwenka 637

Dąbrowski, R. 754
D'Alembert 270, 677 f.
Dalton 737
Damerow 194
Damrath 40, 152, 156 f.
Danusso 539
Danyzy 204 f., 209
Da Ponte 195
Darby II, A. 277
Dašek 614, 724f.
De Aranda 201
De Borde 211
De Catelan 677
De Hontañon 201
Deininger 738
DeLuzio, A. J. 773
De Macas 514
Del Monte 46, 192ff., 258
De Montemor 273
Demel 362
Denfert-Rochereau 227
Denke, P. H. 641
Derand 201
Desargues 743
Descartes 194, 251, 273, 677
Desoyer 687
De Tedesco 515, 519
Diderot 52, 157, 174, 181, 246, 694f.
Dijksterhuis 250, 725, 727
Dill, E. H. 652
Dimitrov 253
Dinnik 725, 763
Di Pasquale 240, 716, 725f.
Dirichlet 642, 655ff., 665
Dirschmid, W. 627
Dischinger 326 f., 463, 467 f., 474, 486ff., 527,
 547, 551f., 555, 558, 726, 730 f., 758
Domke, O. 662 f., 687
Dorn, H. I. 23
Dougall 534
Drucker 232, 568, 726
Du Bois 329
Duddeck 28, 562
Dufour 66
Duhem 23, 250, 258, 725, 726f.
Duleau 400
Duncan, W. J. 615f., 637
Dunn 326, 327
Dupuit 227
Durand-Claye 219, 227
Dürbeck, R. 603

Dürer 190
Dürrenmatt 130
Dzhanelidze, G. I. 745

**E**agleton 692, 694
Ebel 744
Eberlein 490, 493
Ebner 442, 630, 640, 713
Eddy 326 f., 329, 540, *727*
Edwards, W. 210
Efimow 318
Eggemann 396, 464
Eggenschwyler *437 f.*, 439, 634
Ehlers 527, 545
Eiffel *80 ff.*, 94 f., 227, 307, 325, 487
Einstein 621, 718, 744, 748
Eiselin 442
Eisenloffel 705
Emde, H. 490, 494 f.
Emerson, W. 214
Emperger, F. v. *463 f.*, 516, 524, 529, *727 f.*, 764
Emy, A. R. 47
Engelmeyer 144
Engesser 363, 447, 451, 530, 581, 587, 682, 726, *728*, 738, 742
Ersch 187
Ernst, A. 425, 427
Ernst, G. 728
Etzel, K. v. 314
Euclid 769
Euler 32, 45, 90, 107, 137, 203, 270, 274, 277, 293, 296, 298, 335, 377, 443, 447, 453, 479, 733, 737, 756
Ewert 63
Eytelwein 33, 171, 250, 279, 281, 290, *294 ff.*, 302, 657, 696 f., *728 f.*, 733
Eyth 144

**F**aber, O. 738
Fabri 31, 194, *206 f.*, 209, 358
Fairbairn 421, 712, *729*, 762
Faiss, H. 628 ff.
Falk, S. 617
Falkenheiner 614
Falleni 190
Falter, H. 186, 198
Faltus *729 ff.*
Favaro 358
Favero 358
Federhofer 548 f.
Fedorov 57
Feldmann, E. A. 766
Feldtkeller, R. 614
Felippa 25, 621, 640, 653 f., 672, 731
Ferradin-Gazan 310
Ferroni 191
Figari 358
Fillunger 451, *685 ff.*
Finley 64
Finsterwalder, S. 327, 547
Finsterwalder, U. 547, 551, 555, 560, 726, 730, 758, 761, 772

Fischmann 446
Flachat, E. 49, 95, 312
Flamant 533, 722, 762
Fleckner 255
Fleming, R. 602
Flomenhoff 639 f., 717
Flügge, S. 730
Flügge, W. 327, 552, 726, *730*
Flügge-Lotz 730
Foce 24, 186, 210, 212, 219, 240, 246, 396, 716
Föppl, A. 157 f., *177 ff.*, 227, 254, 326 f., 329, *367 ff.*, 402, 411, 429, 433, 440, 474 ff., *478 ff.*, 485, 534, 548, 631, 685, 725, 730 f., 734, 742, 771 f.
Föppl, L. 177, 441, 548, 736, 761, 772
Föppl, O. 440
Förster, M. 48, 503, 773
Föttinger 456
Fourcroy 53
Fourier 95, 756
Fraeijs de Veubeke 665 ff., 668, *731*
Frank, P. 727
Franz II 66, 280
Fränkel 36, 103, 180, 362, *731*
Frazer, R. A. *615 f.*, 637
Freudenthal, G. 250
Freudenthal, A. M. *731 f.*
Freyssinet 487, *554 f.*, 746
Freytag, C. 500, 513, 523, 526
Frézier *204 f.*, 719
Friedrich II 678
Friedrichs, K. O. 666
Friemann 117, 647
Fritsche, J. 123, 451
Fritz, B. 115, 467
Frobenius 611
Frocht, M. M. 605
Fröhlich, H. 467
Fröhlich, O. K. 686
Froude 743
Fuchs, R. 759
Fuchssteiner 567
Führer, F. 692
Fuhrke, H. 617
Fürst, A. 538

**G**aber 123
Gaddi 189
Galerkin 720, *732*, 738
Galileo 31, 46, 90, 192 f., 211, 250 f., 253, 257, *260 ff.*, 301, 358, *675 ff.*, 708, 723, 737, 765, 774
Gallagher 463, 653, 671, *732 f.*
Gargiani 26
Garibaldi 749
Gassendi 273
Gaudi 238
Gauß 534
Gauthey, E. 291 f., 753
Gebbeken 125 ff., 129
Geckeler, J. 550 f.
Gehler 453, 465, 530, 589 ff., *594*, 731, 751
Gerhardt, R. 27, 692
Germain, S. 533 f.

Gerstner, F. J. v. 33, 55, 59, 67, 157 f., *161 ff.*, 172, 177, *216 ff.*, 247, 250 f., 275 f., 278 ff., 282, *283 ff.*, 332, 570, 574, 694 f., 730, 733, 771
Gerstner, F. A. v. 161, 282 f., 733
Ghezzi 680
Giencke 462
Gilly, D. 295, 696
Gilly, F. 295, 692, 695
Girard 276 f., 296, *733*
Girkmann *124 ff.*, 131, 452, 722, *733 f.*
Gispen, K. 61
Giuffrè 716, 726, *734*
Gizdulich 191
Goering 766
Goethe 31, 581, 662
Göring, H. 478
Goffin 560
Gold 90
Goldsmith, M. 740
Gorbunov, B. N. 608
Gordon, L. 758
Gorki 768
Gorokhov 145
Gottl-Ottlilienfeld 148
Graf, O. 457, 539 f., 564, 738
Graham, B. 740
Grandi 31, 358
Grashof 332, 414, 425, 429, 460, 534 ff., 539 f., 548, 730, *734*
Grasser, E. 561 f.
Graubner 222, 239
Green, A. E. 24, 528, *553 f.*, 621
Green, G. 176, 360, 534, 642, 654, *665 ff.*, *681 f.*, 721, 775
Gregory 90, 237
Greiner-Mai 306
Griffith *734*
Griggs, F. E. 773
Grimm, J. 139, 187
Grimm, W. 139, 187
Grinter, L. E. 26, 603, 605
Groh 80
Gropius 474, 491
Groß 241
Groth, P. 627 ff.
Grothe, J. 555
Grubenmann, H. U. 70
Gruber 187
Grübler 354
Grüning, G. 465
Grüning, M. 38, 122 f., 372, 578 f.
Grunsky 68
Grunwald 151
Gruttmann, F. 435
Guastavino Jr. 326
Günther, W. 771
Güntheroth 68
Guidi 358, 723, *734 f.*
Guldan 720
Guntau 306 f.
Guralnick 539
Gvozdev 133, 607, *735*, 770

Haase, H. 227
Haberkalt 519
Hacker 478
Hadamard 534
Hagen 227, 281, 766
Hager, K. 540
Hahn, H. 727
Hahn, O. 730
Hahn, V. 646
Halfmann 717
Hambly 214
Hamel, G. 655, 662 f., 766
Hamilton, R. W. 660
Hamilton, S. B. 23
Hamilton, W. R. 712
Hampe, E. 528
Hänseroth 53, 261, 306
Häntzschel 106
Happel 534
Hargreaves 277
Hartmann, Friedel 373, 744
Hartmann, Friedrich 115, *443f.*, 451, 530, 717, 722
Hauck 736
Haupt, H. 60, *735*
Havelock 743
Hayashi 107, *735f.*
Hebbel, F. 751
Heck, O. S. 636
Hees 30
Hegel 695
Heinrich, B. 188
Heinrich, E. 188
Heinrich, G. 687
Heinzerling 180 f.
Heisenberg 611
Hellinger *658ff.*, 665
Helmholtz 313, 374, 677, 772
Henck, J. B. 766
Hencky 540, *736*, 738, 750
Henneberg 476, 482 ff.
Hennebique 509, 513, *515ff.*, 523 f., 536, *563*, 591, 721
Henri, A. 310
Heppel 305
Herbert, D. 24
Hermann, J. 678
Hermite 611
Heron 46, 256
Herrbruck 241
Herrmann, G. 23, 607
Herrmann, L. R. 670
Herschel 712
Hertel, H. 442
Hertwig, A. 23, 115, 146, 255, 366 f., *371f.*, 380, *387ff.*, 397, 596, 601, 662, 685, *736f.*, 775
Hertwig, R. 391
Hertz, G. 766
Hertz, H. 534
Herzog, M. 560
Hesse 722
Heun 726

Heydenreich, K. H. 697 f.
Heyman 24, 121, 131, 186, 209 f., 214, *232ff.*, 244, 246, 248, 311, 326, 568, 716, 723, 734
Higgins, T. J. 605
Hilbert 657 f.
Hinrich 181
Hitler 443, 450, 452, 664, 713 f., 757
Hodgkinson, E. 76, 92, 729, 737
Hoening 115
Hoff, N. 637
Hofmann, T. J. 41
Hofstetter, G. 255
Hölderlin 29, 471, 692 f., 695, 698
Holl 195
Holle 527
Holzer, S. M. 48
Homberg, H. 457, 618
Hooke 31, 90, 213 f., *271f.*, 296, *737f.*
Horn, F. 456
Horne 133
Housner, G. W. 767
Howe, W. 64, 72
Hrennikoff 493, 562, *624*, 627, *738*
Hruban, K. 540
Hu 665
Huber, M. T. *459ff.*, 543, 736, *738*
Hübsch 773
Huerta 1, 24, 87 f., 186, 212, 217, 246
Hughes 713
Hultin 714
Humboldt, A. v. 312
Hutton, C. 214, 240
Huygens 32, 90, 272, 738, 775
Hyatt 499

Iguchi *738*
Ingerslev 539
Isherwood 456
Ivanov 307

Jacobi, K. G. J. 662
Jacobsthal, E. 759
Jacquier *679f.*, 708, 718, 752, 756
Jaeger, T. *738f.*
Jäger, K. *443f.*, 451
Jäger, W. 222, 239
Jagfeld 235, 239
Jagschitz 123
Jasiński 728, *739*
Jenkin, F. 218
Jenkins *738*
Joffe, A. F. 768
Johanson 539
Johnson, B. 754
Johnson, L. 754
Jokisch 116, 597, 722
Jones, R. E. 670
Jones, R. T. 639
Jordan, P. 611
Joseph II 279
Joule 313, 677
Jung 358
Jungbluth, O. 469

Jünger, F. G. 446
Jürges, T. 518
Jungwirth, D. 568

Kączkowski, Z. 754
Kafka 28
Kahlow 68 f., 186, 692
Kahn 718
Kaiser, C. 195, 199, 241, 260
Kaliski, S. 754
Kammüller 687
Kani 604 f., *739f.*, 767
Kann, F. 127
Kant 662, 677 f.
Kapp, E. 143
Kappus 439
Kármán 447, 451, 539, 759
Karmarsch 56
Kaufmann, G. 529
Kazinczy, F. 740
Kazinczy, G. v. *121f.*, 127, 129, *740*
Kelsey, S. 643
Kennard 73
Kepler 137
Kesselring 493
Khan, F. R. *740f.*
Khan, Y. S. 740
Kier 738
Kierdorf, A. 538
Kinzlé 742
Kirchhoff 460, *533f.*, 541, 543, 657, 664, 713, 722, 738
Kirkwood-Dodds 554 f.
Kirpichev 34, 36, 307, 374, *377ff.*, 606, 695, *741*
Kirsch, E. G. 493, 562, *622f.*
Kist, N. C. 122, 451
Klein, F. 372, 414, 582, 600, 623, 654, *657f.*, 745, 772
Klein, L. 60
Kleinlogel 464, 530 f., 591
Klemm, F. 53, 261
Klimke, H. 470, 474, 490, 494
Klöppel 115, 117, 146, 390, 398, 451 ff., *455f.*, 458, 462, 469, 618, 646 f., 737, *741*, 742, 748
Knebel, K. 470
Knothe 106, 392, 657 f., 666 ff., 757
Koechlin 81 f., 84, 94, 307, 703, 760
Koenen 362, 382, 465, 498 f., *501ff.*, *507ff.*, 512 ff., 522, 526
Kögler, F. 444, 446
Koiter *741f.*
Kollbrunner 127, 689
Köller 635 f., 640, 713
Koloušek *742*
Konietzky 242
König, J. S. *678*
König, W. 151
Kooharian 232
Köpcke, C. 47, 224, 305, 352, 751
Kordina 561 f.
Körner 326
Kötter, F. 409, 766

Koyré 727
Krabbe 115, 123
Krämer 570, 573, 580, 598, 620 f., 693, 700
Kranakis 293
Krätzig 554
Krauß, K. 259, 260
Kretzschmar 720
Krohn, G. 614
Krohn, R. 98, 444, 741, *742 f.*
Kromm, A. 657
Krotov, Y. V. 608
Kruck 596
Krug, K. 314
Krylov 720, *743*, 768
Ktesibios 256
Kudriaffsky, J. v. 66
Kuhlmann, U. 468
Kuhn 749
Kuhn, T. S. 136, 450, 674, 713
Kunert, K. 467
Kulka 123
Kupfer 559, 567, 761
Kuppler, J. G. 68
Kurrer 25

Lacroix 756
Lafaille, B. 552
La Fayette 746
LaFraugh 539
Lagrange 42, *45 f.*, 95, 253, 377, 533 f., 595, 681, 721, 737
La Hire 189, *201 f.*, *207 f.*, 215, 217, 259, 715, *743*
Lahmeyer, J. W. 756
Laitko 306, 380 f.
Laissle 101
La Lande 280
Lamarle 305
Lamb, H. 548
Lambin, A. 591, 721
Lambot 499
Lamé 95, 301, *309 ff.*, 317, 399, 548, 583, 682, 722, *743 f.*, 753
Land 46, 101, 350, 353 f., *355 ff.*, 606, 706, *744*
Landau, E. 654
Landsberg 158, 180, 581
Lang, A. N. 641
Lang, G. 22
Langefors 614, 641
Lansing, W. 641
Laplace 401, 721
Lauenstein 326
Lavoinne 534
Lees 64
Lefèvre 250
Legendre 756
Leibniz 32, 90, 249, 272, 279, 283, 384, 598, 655, *677 f.*, 693, 695, 703
Leckie, A. 643
Lehmann, C. 27
Lehmann, D. 108
Lehmann, E. 743
Lemoine 94

Lenk, H. 148
Lenk, K. 554
Lentze, C. 79
Leonhardt 442, 457, *555 ff.*, 560, 561 ff., 567, 646, 761
Leonardo da Vinci 46, 134, 186, 200, 202, 253, *256 ff.*, 265, 708, 747
Le Seur *679 f.*, 708, 718, 752, 756
Lévy, M. 329, 534, *744*
Levy, S. 640 f.
Lewe, V. 527 f., 540, 543, 548
Lewis, R. W. 671
Lie 115
Lieberwirth 240
Lienau 123
Lincoln 60
Lindenthal 114, 749, 764
Livesley, R. K. 240, 643, *644 ff.*
Loitsyansky, L. G. 745
Loleit, A. F. 537 f.
Long, S. H. 64, 70, *744 f.*
Lord Kelvin (Thomson, W.) 313, 434, 682, 743
Lorenz, G. 725
Lorenz, W. 72, 388, 496, 586
Louis Philippe 721
Love 532, 534, 548, *582 f.*, 713, *745*
Ludwig, M. 705
Lurie *745 f.*

MacAdam 277
Mach 718
McHenry 624
McLean 663
MacNeill, J. 758
Mader 462
Magnel 555, *746*
Mahan *746*
Maier, A. 727
Maier-Leibnitz *122 f.*, 129, 131, 463, 465, 740
Maillart 238, *438 ff.*, 487, 532, 538, 548 f., *746 f.*
Mainstone 255
Malinin 25
Maltzahn, C. v. 195
Manderla 108, 581, *587 ff.*
Maney, G. A. 594
Mang, H. 255
Mann, L. 112, 410, 582, *595 f.*, 606, 735, *747*
Maquoi 748
Marchetti 31, 358
Marcus 410, 527, *541 ff.*
Mar 717
Marg 474
Marguerre 442, 654, 658, 687 ff.
Mariotte 90, 273
Mark, R. 241
Marrey, B. 555
Marti, P. 538
Martin, H. C. 39, 249, 640 f., *649 ff.*, 652 f., 669
Marx 245, 246 f., 249, 313, 352, 413, 694
Mascheroni 213, 246, *747*
Massonnet 135, 731, *747 f.*
Matheson, J. A. L. 594
Matschoß 146, 389, 432, 748

Mauersberger 53, 261, 306
Maupertuis *678*
Maurer, B. 25, 27
Mautner 554 f.
Maxwell 44, 307, 313 f., *319 ff.*, *342 ff.*, 345 f., 348 ff., 355 f., 359, 382, 397, 577, 579, 589, 683, 685, 695, 712, 716, 719, 724, 742, 744, *748*, 750 ff., 758
Mayer, M. 540
Mayer, R. 313, 476, 677, 772
Mayor 476, 485
Medici, G. 191
Mehrtens 64, 80, 93, *365 ff.*, 371 f., 389, 409, 588, 684, 717 f., *748*, 751
Meinicke, K.-P. 262, 314
Meissner, E. 550, 759
Melan, E. 543, 718, *748 f.*
Melan, J. *113 f.*, 116, 181, 183, 463, 488, 513 ff., 519, 718, 720, 722, 727, 748, *749*, 764
Mellin 79
Melosh, R. J. 652, 670
Melzi 258
Menabrea 226, 247, 348, 360, 362, 370, 408, 567, 658, 665, 720, 731, *749 f.*
Mengeringhausen 139 f., 474 f., 486, *490 ff.*
Menzel 412 f.
Merriman 170, *750*
Mersenne 272 f.
Méry 217, 227
Metzler 123
Meyer, A. G. 692, 699
Meyer, E. 124
Michaux 244
Michelangelo 190, 192, 680
Michell 583, 623, 716, 743, *750*
Michon 220
Mikhailov 306
Milankovitch 209
Mindlin 726, *750 f.*
Minkowski 654
Mise *751*
Mises 476, 485, 553, 608, *612 ff.*, 736, 738, 751, 768
Mitchell, D. 568, 712
Mitis, I. v. 66
Möbius, A. F. 475
Modigliani 358
Mohr 34, 36, 44, 46, 83, 99, 103, 109, 122 f., 177, 305 ff., 327, 343, *344 ff.*, 347 f., 350 ff., 359, *362 ff.*, 371, 379, *382 f.*, 386, 389, 397, 475 f., 484, 541 f., 562, 581, 588 f., 595, 606, 623, 683, 685, 687, 695, 706, 716 f., 728, 730 f., 744, 748, *751 f.*, 760 f.
Mohrmann 325 f.
Moiseiff 114
Molinos 305
Moller, G. 773
Monasterio 212
Monge 51, 53, 95, 279, 281, 308, 721, 756
Monier *499 ff.*, 504 f.
Monzoni 73
Moody 727
Morison, G. S. 742

Mörsch 228, 229f., 467, 498, 509, 513, 515, *522ff.*, 562, 564, 566, 735, *751f.*
Moseley, H. 77, 217, *218f.*, 297, 304, 746
Moser, S. 146f.
Müllenhoff, A. 434
Müller, C. H. 659
Müller, J. 150
Muller, J. 555
Müller, O. 773
Müller, S. 61, 391, 601, 736
Müller-Breslau 34, 36, 46, 62, 183, 227, 306f., 328ff., 343, *348ff.*, 353ff., *362ff.*, 371, 377, 379, *381ff.*, 389, 397, 408f., 431, 445ff., 476, 481ff., 500, 531, 548, 570, 577ff., 581, 589, 592, 599ff., 608, 683, 687, 695, 703, 706, 712, 728, 736f., 744, 747f., 751, *752*, 759
Muskhelishvili *752*
Musschenbroek 203, 274f., 296, *752*

**N**achtigall 474
Nádai 441, 527, 543
Naghdi *752f.*
Nagel 331
Napoleon 58, 309, 733
Napoleon III 762
Napolitani 261
Navier 22, 34, 66, 88, *90*, 95, 101, 113, 134, 161, *220ff.*, 250, 270, 278, *290ff.*, 296, 301ff., 307, 348, 397, 402, 507, 534, 548, 583, 598, 657, 681, 695, 708, 715, 719, 730, 733, 746, 753f., 756, 760
Neal, B. G. 130f., 133, 135, 689, 738
Nebenius 57
Nemorarius 43, 46
Nervi 238
Neuber 755
Neufert 490
Neukirch 115
Neumann, C. G. 722
Neumann, F. E. 722
Neumann, P. 513, 515, 519, 722
Neumeyer, F. 692, 695
Néville 64, 72, 73
Newcomen 277
Newmark 605, *754*
Newton 45, 143, 214, 251, 255, 265, 279, 384, 718f., 725, 738, 748, 752, 758, 774
Niemann, G. 771
Nietzsche 448
Nikolaus I 311, 312
Nikolaus von Oresme 727
Nouguier 84
Novozhilov *754*
Nowacki 738, *754*
Nowak, B. 451

**O**chsendorf 209, 236, 240, 692, 706
Ockleston 539
Oden 671, 713, 732
Olszak 738, *754f.*
Onat 232
Oravas 663

Ostenfeld 109, 514f., 519, 581, *594f.*, 596, 599, 606, 625, 647, 747, *755*, 772
Ostwald, W. 313, 452
Ott, K. v. 332
Otto, K. 492

**P**adlog 732
Pahl 40, 152, 156f.
Palladio 194f., 745
Panetti 548
Panovko, Y. G. 745
Papadrakakis 713
Papin 677
Papkovich *755*
Parent 90, 296
Parigi 191
Parkus 749
Parnell 278
Pascal 273
Pasternak 596
Pauli, F. A. v. 75, 724
Pauli, W. 759
Pauser 66, 239, 518
Peacock 86
Pearson 22, 277, 398, 400, 532
Pedemonte 24, 716
Peirce, B. 611
Peirce, C. S. 611
Pellegrino, S. 495
Perrodil 187, 225
Perronet 51, 191, 208, *755*
Peters, T. F. 76
Petersen, C. 398, 712
Petersen, J. 462
Petersen, R. 769f.
Petit 209, 220
Petty 737
Petzold 617
Pflüger 462, 775
Pian 638, 664, 670, 717, 732
Pichler, G. 692, 705
Pieper, K. 241
Piobert 399
Pippard *230f.*
Pirlet 601
Pisarenko 756
Planat 325
Planck, M. 759
Plato 46, *136ff.*, 581
Podszus 195
Pohl, K. 391f., 608
Pohlmann 108, 393
Poincaré 718
Poinsot 95, 756
Poisson 95, 301, 533f., 753, 756
Pole, W. 77f., 304
Poleni 215f., 247, 680f., *756*
Polonceau, C. 47
Polonceau, A.-R. 95
Pólya 759
Poncelet 95, 161, 220, *223*, 225, 318f., 399, 574, 715, 746, *756*
Porstmann 490

Pöschl, T. 548, 687
Potterat *437*
Prager *130ff.*, 233, 476, 485, *757*
Prandtl 177, 553, 750, 772
Prange *658ff.*, 665, *757*
Prato, C. 670
Prechtl 55
Pronnier 305
Prony, G. 95, 291
Proske 240
Pugsley *757f.*

**Q**uade, W. 614
Quast, U. 561f.

**R**abich *758*
Rabinovich 23, *758*, 770
Radelet-de Grave 24, 396, 716
Ramm, E. 25, 41, 186, 704, 745
Ramm, W. 80, 586
Ramme 23
Ramsbottom, J. 414f., 419
Rankine 89, 157f., *173ff.*, 187, 223, 237, 307, *320ff.*, 397, 575, 695, 714, 748, *758f.*
Raschdorff 385
Rathenau, E. 99
Raucourt 310
Rautenstrauch 242f.
Rayleigh *374ff.*, 377, 534, 548, 745
Rebhann 305, 332
Redtenbacher 56, *155*, 414, 427, 734
Reich, E. 549
Reichenbach, G. v. *85f.*, 90
Rein, W. 447, 451
Reinhardt, R. 736
Reinitzhuber 442
Reissner, E. 543, 549f., *663f.*, 665, 740, 750, *759*
Reissner, H. 95f., 380, *387f.*, 456, 548, 634, *759f.*, 766
Renn 194
Rennert, K. 423
Résal 227
Reuleaux 317, 347, 350, *351f.*, 356, 427, 699
Reuter, W. 429
Ricardo, D. 312
Richardson 620
Richardson, L. F. 540
Richart, F. E. 754
Richelot 722
Richter, J. 680
Ricken, H. 487
Rickert, H. 145
Riedel, H.-K. 295
Riedel, W. 623f.
Riedler 61, 144, 386, 427
Rieger 194
Rippenbein, Y. M. 608
Ritter, A. 751, *760*
Ritter, H. 760
Ritter, W. 113, *324f.*, 439, 480, 509, 513ff., 519, 531, 563, 581, 588, 601, 704, 744, *760*
Ritter, W. E. 459

Rittershaus 353
Ritz, W. *375f.*, 658
Roberval 45f., 743
Robinson, J. 215
Röbling, J. A. *68ff.*
Roebling, E. W. 69
Roebling, W. A. 69
Rohn, A. 440
Roik 468
Ropohl *148f.*, 151
Rös, M. G. 127, 451, 747
Rothert 125ff., 129
Röting 170
Rottenhan 280
Rousseau 52
Rueb, B. 530
Ruh, M. 494
Rühlmann 34, 261, 310, 752, 756, *760f.*
Rumjancev 57, 58
Rumpf *147f.*
Runge, C. 548f., 654, 768
Rüsch 327, *547*, 551, 555, *559*, 561, 567, 726, 730, *761*
Ruta 440
Rvachev *761f.*

**S**aavedra 187, 223, *762*
Saint-Guilhem 227
Saint-Simon 309f., 312
Saint-Venant 22, 335, *398ff.*, 406ff., 410, 428f., 437, 443, 533, 695, 722, *762f.*
Samuelsson 25, 604, 614, 714, 722
Sanders, A. 514f., 519
Sanders, G. 731
Sandkühler 157, 693
Sanchez, J. 494
Sattler 467f.
Sauer, R. 435, 476
Savart 400
Savin *763*
Saviotti 358
Sawczuk 739
Sayno 358
Schade, H. H. 456
Schadek 362
Schädlich 48
Schaechterle, K. 441, 456f., 530
Schäfer, K. 496, *568f.*
Schäffer 158, 180
Schaim 124
Schairer, S. 639
Schaper, G. 390, 397, 447, *448f.*, 457
Schardt 462
Scheer, J. 480, 493, 618
Scheffler, H. 219, 223, 227, 297, 305, *575f.*, 746
Schellbach 655, 722
Schelling 695
Schifkorn 64
Schindler, R. 749
Schinkel 278, 729
Schinz 80, 93
Schlaich, J. 470, 496, 561, *568f.*

Schlegel, R. 242f.
Schleicher *391f.*, 737f., 775
Schleusner 608, 663, *687f.*
Schlink 476, 485
Schmidt, H. 496
Schnabel 51, 55, 146, 309
Schneider, C. C. 96, 742
Schneider, R. 425
Schnirch 67
Schober, H. 470
Schönert 148
Schönfließ 389
Scholz, E. 24, 324
Schopenhauer 136, 662
Schröder, H. 647
Schubert, J. A. *74f.*, 331, 731, 774
Schübler 101
Schüle, F. 524
Schulz, F. J. E. 204, 206, 208
Schuster, R. 513
Schwartz, H. A. 759
Schwartzkopff 105
Schwedler 47, 50, 80, *91*, 93, 98, 101, 106, 223, 224, 341, 382, 474, *475ff.*, 483f., 550, *585f.*, 588, 695, 709, 719, 748, *763*
Schwerin, E. 550
Sedlmayr, H. 472
Seger, H. 511
Seguin 66, 95
Seifert, L. 445
Séjourné 228, 230
Selberg 115
Senefelder 574
Serensen *763f.*
Serlio 190
Seyrig 82, 227
Sganzin 208
S'Gravesande 274, 752
Shapiro 762
Shaw, S. E. 605
Shukhov 474f., 486, *487f.*, 548
Silberschlag 206, 295
Sinopoli 210, 240, 246
Slaby 386
Slater, W. A. 538, 772
Smith, A. 312
Sobek 474, 740
Sobolev 373
Šolín 715, *764*
Sommerfeld 725, 759
Sonne 158, 180
Sonnemann 306
Sontag, H. J. 467
Sor, S. 535
Southwell 604f., 750, *764*
Spangenberg 717, 761, 765
Späth, J. L. 220
Spierig, S. 624f., 627
Spitzer, J. A. 513, 515
Stamm 469
Standfuß, F. 560
Staudt, K. G. C. v. 318
Stein, E. 644

Stein, P. 469
Steiner, F. 180, 182ff., 727
Steinman 114, 749, *764f.*
Stelling, E. G. 440f.
Stephan, S. 470
Stephenson, R. *76ff.*, 315f., 712, 729
Steppling 279
Sternberg, H. 223, 728
Steuermann 539
Steup 720
Stevin 32, *44f.*, 46, 193, 259, 725, *765*
Stickforth 440
Stiglat 496, 543, 560
Stirling 215
Stodola, A. 550
Stokes 682
Strassner, A. 530f.
Stratton, S. 736
Straub, H. 253, 261, *765*
Strecker, F. 614
Striebing 150
Stuckenholz, G. 414
Stuckenholz, L. 414, *417ff.*
Study, E. 611f.
Stüssi *127ff.*, *134f.*, 294, *689*, *765f.*
Sutherland, H. B. 758
Swain 26, 99, 232, 237, *766*
Swift, J. 744
Sylvester, J. J. 610
Symonds 130f., 135, 689
Szabó 177, 253, 532, *766f.*
Szilard, R. 543

**T**acola 194
Tait, P. G. 434
Takabeya *767*
Tanabashi *767*
Tartaglia 46
Taton 52
Taut, B. 695
Taylor, C. H. 726
Teichmann *116f.*, 135, 380, *392ff.*, 597, 645
Telford 65, 87, 89, 164, 278, 714
Tempelhof 295, 728
Terzaghi 548, *685ff.*
Tessanek 279, 289
Tetmajer, L. 325, 447f., *767f.*
Teßmann 150
Thabit ibn Qurra 46
Thalau 116
Theimer 115
Thompson, F. 729
Thomson, W. 313, 434, 682, 743
Thul, H. 560
Thullie 513ff., 519
Thürlimann *134f.*, 568, *689*
Timoshenko 63, 106, 177f., 261, 336, 399, 532, 672, 730, 738, 745, 750, *768*
Timpe, A. 582
Todhunter 22, 277, 397, 400, 532
Todt 392, 451
Tölke, F. 766

837

Tong, P. 670
Topp, L. J. 39, 249, *640f.*, *649ff.*, 669
Torroja, E. 238, 529, 552, 555
Town, I. 64, 70 f.
Traitteur, W. v. 66
Tranter 230
Trautwine 169 f.
Trautz 80
Tredgold *768*
Trefftz 654, *768 f.*
Tresca 95, 736
Trost 463, 468
Trostel 250
Trudaine 51, 295, 755
Truesdell 716, *769*
Tschauner 115
Turner, C. A. P. 532, *536 ff.*, 727
Turner, M. J. 39, 249, *640 f.*, *648 ff.*, *652 ff.*, 658, 669
Turley, E. 522
Twain 766

Udet, E. 491
Ungewitter 325
Unold 115

Vainberg, D. V. 608
VanGelde 240
Vauban 50, 746
Varignon 41, 43, 45 f., 202, 319, *769*
Vasari 190
Vène 296, 297
Vèrges 292
Vermeulen 414
Vianello 409 f., *769 f.*
Vicat 95
Vierendeel 529, 591, 721, *770*
Vieta 573, 700
Villarceau 227
Vitruvius 46, *138 f.*
Vlaslov *770*
Vogel, U. 135, 464
Vogeler 504 f.
Voigt, W. 426, 682
Völker, P. 507, 515
Voltaire 52, 678, 718

Wadell, J. A. L. 603
Wagner, H. 439, 617, 630, 634
Wagner, W. 435
Wall, W. A. 745
Wallis 737
Walther, A. 616 f., 646
Walthelm 567
Wanke 720
Wapenhans 241, 466, 680
Ward 737
Warren 64, 73, 341
Washizu *665 ff.*, *770 f.*
Wasiutynski 105
Watt 154, 277
Wayss, G. A. 498, 500, *503 ff.*, 509, 512 f., 548 f.
Weber, C. H. 177, 439 ff., 730, *771*

Weber, M. 413, 428, 456
Weber, M. M. v. 105
Weber, W. 188, *238 ff.*
Wedekind 773
Wedler, B. 560 f.
Wehle, L. B. 641
Weierstraß 381
Weikel, R. C. 652
Weingarten 366, *367 ff.*, *371 ff.*, 658, 685
Weisbach 157 f., *166 ff.*, *402 ff.*, 407 f., 572, 575, 734, *771*
Wellmann 421
Wendt, H. 151
Werner, E. 261
Werner, G. 69
Westergaard 23, 538, 750, 754, *772*
Westphal, G. 305, 760
Weyrauch 103, 224, 305, 307, 366, *372 f.*, 658, 685, 706, *772*
Whewell 167
Whipple 92, 98, 709, *773*
Victor Emmanuel II 749
Wiebeking 57 f.
Wiedemann, G. 730
Wieghardt, K. 623, 734
Wiegmann 47, *773*
Wierzbicki *773 f.*
Wiley, W. H. 170
Wilhelm II 379, 386
Wilke, R. 74
Wilkins 737
Williot 329
Willis 737
Wilson, E. L. 732
Windelband 145
Winkler 22, 34, 89, *102 ff.*, 182, 187, 217, *223 ff.*, 232, 244, 275, 307, 330, *331 ff.*, 344, 350, 379, 397, 414, 426, 481, 500, 548, 581, 586, 588, 706, 731, 748 f., 751 f., 774
Wippel 543
Witte, H. 469
Wittfoth 555, 560
Wittmann 325 f.
Wöhler 424, *774*
Wöhlert, F. 414
Wölfel, P. L. 96
Wolff, C. 284
Wolff, F. 505
Wolff, J. 195 f.
Wolff, M. 250
Woltmann 281
Wood 539
Worch 647, 712
Wren 214, 255, 737, *774 f.*
Wuczkowski 530
Wußing 51
Wydra 279
Wyss, T. 566

Yamamoto, Y. 670
Yang, J. 750
Young *86 ff.*, 221 ff., 247, *775*

Zerna, W. 24, 528, *553 f.*, 561, 621, *775*
Zesch 728
Zeuner 170, 414, 427, 751
Zhuravsky 92, 709, *775 f.*
Zienkiewicz 25, 40, 462, 604, 620, *653 f.*, 668, *671 f.*, 722, 732
Zimmer, A. 625
Zimmermann, H. *105 f.*, 439, *443 ff.*, 447 f., 474, *476 ff.*, 481, 484, 597, 742, 776
Zimmermann, P. 376, 765 f.
Zipkes 529
Zöllner 170
Zott, R. 380
Zschetzsche 484, 717
Zschimmer 145, 147
Zurmühl 615, 617
Zuse 39, 393, 486, 570, 580, 597, 599, *606 ff.*, 617, 647, 703
Zweiling *687 ff.*, *775 f.*

# SUBJECT INDEX

**A**butment thickness  206, 208, 212, 220
Accumulation phase  *37*, 106, 109, 121, 526, 531 f., 543, 547, 550, 562, *571 f.*
Acoustics  374
Active earth pressure  211
Additional boundary condition  668
Addition principle  463
Administration and science  511
Aerodynamics  394, 611
Ageing theory  467, 558
Aircraft design  442
Aircraft construction  596, 664
Aircraft fuselage  635, 638
Aircraft industry  469
Aircraft spar  392
Aircraft structure  630
Aircraft wings  635, 637 f.
Airship and aircraft construction  384
Airship hanger  432
Airy stress function  623
Algebraic scholastics  572
Ammonite shell  491
Analytical mechanics  659
Analytical theory of science  148
Anchor  522
Anchorages for splices in prestressing tendons  556
Angle of rotation  407
Angle of rupture  208 f.
Angle of twist  400
Angle lever  265 f.
Anisotropic shell  462
Annular plate  628
Antenna systems  491
Anti-mathematics movement  62
Application phase  *32*, 54
Applied thermodynamics  277, 414
Approximation method  587
Approximation function  249, 376, 669
Aqueduct  180
Arch profile  89
Arch thickness  205, 207, 575

Aristotelian motion theorem  46
Aristotelian natural philosophy  675, 677
Aristotelian tradition  45
Arithmetical operation  607
Assembly language  625
Asymmetric cross-section  441
Asymmetric load  550
Asymmetric masonry arch  575
Automatic System for Kinematics Analysis (ASKA)  630
Automation  156, 161, 248
Automation of engineering calculation  646
Automatic calculating machine  607
Automation of structural calculations  608
Automotive engineer  628
Automotive industry  *625 ff.*
Average compressive stress  221
Average stress  88
Axial force  169
Axial force diagram  120
Axial stress  88, 108
Axially loaded elastic extensible bar  655
Axially symmetric shell  476

**B**alanced cantilever  420, 573
Band matrix  316
Band structure  632
Bar end moments  109 f., 120
Bar cell  624
Bar element  639 f.
Base coordinates  595
Based rule  139
Basic Bessemer steel  97
Basic symbol  600
Basket arch  189, 199
Beam grid  455, 457, 539, 573, 624, 705
Beam grid model  539
Beam models  76
Beam on elastic supports  103, 335
Beam on two supports  269
Beam oscillation  617
Beam trussed from above  49

Beam trussed from below  49
Beauty and law  *137 ff.*
Beauty and utility  692, *695 ff.*
Beltrami-Michell stress differential equation  583
Bending about two axes  336
Bending deflection  257
Bending elasticity  334
Bending failure  276
Bending failure problem  46, 251, 264, 273
Bending line  300, 303
Bending moment  171, 301, 303, 521
Bending moment diagram  80, 112, 126, 171 f., 368, 385, 580
Bending stiffness  109, 115, 298, 300, 316, 335, 337, 375, 460, 587
Bending strength  171
Bending stress  113
Bending stress equation  520
Bending test  123, 275 f., 438
Bending theory for cylindrical shells  549
Bending theory for rotational shells  549, 551
Bending theory for shells  548
Bending theory of straight beams  179
Bendixsen's method  594
Bent-up reinforcing bars  564 ff.
Berlin school of structural theory  *379 ff.*, 572, 579, 592, 685, 703
Bernoulli hypothesis  438, 519 f., 548
Bessel function  534
Bessemer method  35
Bessemer steel  397
Betti's theorem  607
Beyer bible  552
Biharmonic equation  534, 623
Bleich's theorem of four moments  596
Bodywork calculation  627
Bodywork optimisation procedure  627
Boiler formula  549
Bologna process  63
Bolted pinned joint  98
Boolean algebra  574

Boolean localisation matrix  653
Boundary element method (BEM)  619
Boundary value problem  658
Box beam  638, 664
Braced sway frame  581, 593
Brachistochrone  655, 679
Bredt's theorem  443
Breaking length  264
Bubnov-Galerkin method  642
Buckling coefficient  126
Buckling of columns  294
Buckling load  274, 376, 446
Buckling matrix  619
Buckling problem  93
Buckling strength  179, 253
Buckling strength test  302
Buckling theory  171, 277, *443 ff.*, 618
Buckling test  274 f., 443
Building industry  511
Burr's system  70

**C**able-stayed bridge  95
Cableways  431
CAD program  705
Calcomp plotter  628
Calculus of variations  375, 608, 621, 655, 657, 659, 661 f., 669, 678 f., 688
Canal bridge  180
Canonic transformation  659 ff.
Canonic variational theorem  662
Canonic variational theorem of Hellinger and Prange  658 ff.
Cantilever  171, 276
Cantilever beam  48, 117, 265, 267 ff., 335, 337, 573
Cantilever method  602 ff.
Carry-over factor  604
Carry-over method  *617 ff.*, 621
Cartesianism  52
Castigliano's energy theorem  640
Castigliano's first theorem  *360*, 378, 607, 642
Castigliano's second theorem  113, 348, *360*, 378, 606 f., 642, 658
Castigliano's theorems  355, 358, 366 f., *373 f.*, 379, 661
Castigliano's third theorem  *360*
Cast-iron arch  85
Cast-iron arch bridge  87 f., 90
Cast-iron bridge  64 f., 72, 76
Cast-iron tube  85
Catenary  192
Catenary arch  214, 221, 290
Catenary model  235
Cause-effect relationship  163, 261, 290, 579
Cell tube  *630 ff.*, 635, 637
Cellular hollow section  433
Cement  522
Cement research  511
Centre-of gravity axis  88
Chain suspension bridge  66
Chaos and fraction  671
Characteristic compressive strength  222
Chemical energy  154

Civil engineering laboratory  390 f.
Civil engineering school  51
Clapeyron's theorem  342 f., 346, 607
Clapeyron's theorem of three moments  *314 ff.*, 632
Classical civil engineering theory  161
Classical engineering  156
Classical engineering form  153
Classical fundamental engineering science disciplines  322 f., 572, 647
Classical phase  36, 46, 84, 105 f., 108, 143, 182, 225, *307*, 317 f., 324 f., 327, 329 f., 475 f., 531, 571, *576 f.*, 623, 703
Close-mesh lattice girder  85
Coefficient matrix  632
Cold-formed hollow section  109
Collapse  444 ff., 479, 524
Collapse mechanism  238, 262, 266
Collapse mechanism analysis  245
Complementary energy formulation  668
Complementary energy principle  668
Complementary work  642
Composite beam  465
Composite column  463
Composite floor slab  465
Composite frame  465
Composite strength  169, 170 f.
Composite system  49, 72, 585 f.
Composition law  136, 478, 481, 485, 491
Compression test  219, 525
Compression zone  507
Compressive strength  173, 209, 233 f., 301
Compressive strength test  302
Compressive stress  220 f., 224
Compressive stress curve  525
Compressive stress diagram  221
Compressive stress distribution  221 f.
Computational engineering  40, 570
Computer Aided Design (CAD)  470, 472, 627
Computer Aided Engineering (CAE)  627
Computer-aided framework topology  494
Computer-aided graphical analysis  694, 704 f., 707
Computer-aided structural analyses  *625 ff.*
Computer Aided Testing (CAT)  627
Computational Fluid Dynamics (CFD)  672
Computational Solid Mechanics (CSM)  672
Computer-assisted graphical analysis  692
Computer network  41
Computer statics  570
Computing plan  580, 608 f., 703
Computing sheet  608
Concentric joint  588
Concept of the truss model  568
Concrete arch  85
Concrete arch bridge  562
Concrete dome  551
Concrete stress  521
Concrete compressive stress  509, 520 f.
Conical shell  550
Conjugated potential function  666
Conservation laws in dynamics  690
Conservation of energy law  314, 677, 683

Conservative system  377
Conservative vibration problem  615
Consolidation of deformable porous soils  685
Consolidation period  22, 24, *37*, 115, 144, 158, 173, 180, 230, 307, 318, 323, 358, 373, 379, 387, 433, 476, 485, 570 f., 573, *578 ff.*, 596, 600, 605 f., 615, 623, 642
Consolidation problem  686
Consolidation theory  690
Constitution phase  22, *34*, 75, 90, 305, *307 f.*, 533, 548, 574
Constitutive relationship  682
Construction language  72, 545, 560
Constructional discipline  708
Construction engineer  560, 597
Continuous arch  573
Continuous stiffening girder  115
Continuous beam  50, 76, 77, 80, 103, 122 f., 127, 129, 179, *296 ff.*, 316, 327, 335, 362, 369, 385, 452, 540, 542, 557, 573, 580, 617, 632, 698
Continuous beam theory  317, 587
Continuous frame  38
Continuous multi-storey frame  531
Continuous longitudinal stiffener  457
Continuous polygonal arch  701
Continuous two-span beam  48
Continuous theory  462
Continuum analysis  455, 459
Continuum hypothesis  270, 690
Continuum mathematical model  621
Continuum mechanics  440, 458, 554, 612, 642
Continuum model  426
Continuum physics  648
Control system  672
Cooling tower  432
Copernican planetary system  675
Copperplate engravings  571
Corbelled arch  188
Corps of military engineers  33
Corpuscular model  426
Counterweight  420
Crane-building  411, 562
Crane column  421, 424, 426
Crane hook  425
Crane engineer  432
Crane manufacture  432
Crane rail  431
Crane rail beam  431
Cranked lever  43, 208
Creep and shrinkage  556 f.
Creep and shrinkage of concrete  467
Cremona diagram  50, 317, 321 f.
Crystallography  493
Critical factor  132
Critical load  121, 376 f.
Critical load increase factor  121
Critical loading case  223
Critical rationalism  146
Cross-girder  456 ff.
Cross method  *603 f.*, 605
Cubic grids  493
Curved arched beam  337

Curved bar  334 f., 426
Curved body  331
Curved cantilever beam  705
Curved elastic arched beam  337
Curved elastic trussed girder  81
Curved plate  335
Curvilinear coordinate  553
Curvilinear integral  534
Cylindrical shell  179, 546, 549, 551, 624
Cylindrical water tank  508, 549

D'Alembert principle  374
Deck plate  458
Deflection curve  327, 579 f.
Deflection measurement  220, 538
Deflection theory  114
Deformation complementary energy  364, 378, 642
Deformation energy  247, 257, 358, 364, 367, 375, 378, 408, 410, 642, 683
Deformation equation  401
Deformation figure  112
Deformation measurement  93, 220, 224
Deformation parameter  592, 594
Deformation work  179, 348
Deformed system  117 f.
Degree of freedom  624, 626
Degree of prestress  556
Delta wing  640, 648
Denominator  120
Desktop computer  41
Descriptive geometry  51, *53 f.*, 308, 352, 354, 475, 572
Design equation  502
Design language  586
Design method  514, 518
Design model  525
Design office  97, 600
Design theory  503, 514, 529
Determinant theory  37, 389, 601
Deterministic safety concept  472
Diagram analysis  575
Differential calculus  32, 39, 54, 164, 179, 213 f., 216, 247, 270, 289, 383, 681, 695
Differential equation for the elastic membrane  541
Differential equation for plates  533
Differential geometry  43, 188
Diffusion phase  *41*, 117, 398, 528, 654, 662
Dirichlet's energy criterion  657
Dirichlet's stability theorem  655
Direct stiffness method  621, 647, *652 ff.*, 654, 658, 669
Discipline-formation period  22 f., *34 ff.*, 45 f., 84, 91, 94, 113 f., 158, 166, 168, 220, 251, 260, 290, 297, *306 ff.*, 577, 585, 589, 596, 642, 684
Disk memory  628
Discontinuous element  624
Discontinuum mechanics  243
Discrete mathematical model  621
Displacement equations  593
Displacement jump  594, 599

Displacement method  37 f., 107, *108 ff.*, 116, 358, 570 f., *581 ff.*, 602, 606 f., 621, 625, 641 f., 645, 652, 656
Displacement parameter  595
Displacement variable  581, 599, 606
Displacement vector  625, 664 f.
Dissipation work  126
Disturbed region  568
Double-layer frame grid  493, 495
Double strut system  564
Dockyard crane  431
Dodecahedron  137
Domain decomposition  671
Dome  325
Dome theory  475
Domical rib vault  244
Double Fourier series  540
Double trigonometric series  533 f.
Drawing office  517
Dresden school of applied mechanics  383, 592, 594, 684
Dresden school of kinematics  *351 ff.*, 382
Dual diagrams  321 f.
Dual nature of structural theory  596
Dual nature of theory of structures  606, 644
Dual polygons of forces  322
Dynamic boundary conditions  376, 583

Earth pressure  251, 294
Earth pressure theory  33, 182
Eccentric plastic hinge  126
Eclecticism  385, 389
Effect-cause relationship  579
Effective torsional stiffness  461
Effective width  456
Eigenvalue analysis  615 f.
Eigenvalue problem  637
Einstein's summation convention  594
Elastic arch beam  294
Elastic arch theory  82, 102, 115, 185, 187, *220 ff.*
Elastic continuum  661, 681
Elastic frame theory  531
Elastic/inelastic zone  449
Elastic in-plane-loaded plate  540, 566
Elastic limit state  125, 127, 276
Elastic line  293, 296 f.
Elastic membrane  294, 542
Elastic modulus  173, 198, 220, 226, 228, 257, 301 f., 347, 403, 519, 583, 622, 681
Elastic plate  528, 623
Elastic pole  601
Elastic-plastic torsion of bars  441
Elastic slab  294, 528
Elastic theory for masonry arches  *246*
Elastic ultimate load  266
Elastic weight  540, 542
Elasticity conditions  226, 558
Elasticity equations  234, 349, 365, 558
Elasticity equations of the second order  112
Electricity pylon  487
Electronic computation  644
Electro magnetism  672
Elasto-plastic deformation  123

Elasto-Statics Element Method (ESEM)  628
Electrical engineering  179, 386, 411, 605
Electrical engineering theory  614
Electrical networks  602, 605, 614, 646
Electrical vibrations  374
Elementary frame  632
Elementary mathematics  179
Element axial force  548
Element bending moment  548
Element shear force  548
Element stiffness matrix  625, 649, 651 f., 653 f., 669
Element stress matrix  654
Element torsion moment  548
Element trust force  548
Elementwise linear approximation  655
Elliptical arch  190
Emperger column  464
Empirical rule  210
Encyclopaedia of applied mechanics  *166 ff.*
Encyclopaedia of bridge-building  332 f.
Energy conservation law  375 f.
Energy conservation principle  247
Energy dissipation  314
Energy doctrine  *354 ff.*, 367, 373, 379, 382 f., 589
Energy method  607, 618
Energy principle  42, 172, 176, 294, 348, 359, 460
Energy theorems of Castigliano  684
Engesser's complementary work  661
Engineering aesthetics  388
Engineering education  26
Engineering hydromechanics  163
Engineering manual  168
Engineering model  534
Engineering office  523
Engineering philosophy  143, 146, 148 f., 151
Engineering science history  146
Engineering science theory  149
Engineering sociology  151
Enlightenment  52, 174, 181, 204, 272, 277, 284, 693, 694
Entropy  154
Epistēmē  40, 158, 351 f., 576 f., 580, 611, 695
Epistemology  142, 145
Equilibrium conditions  45, 120, 337, 401, 478, 509, 567, 583 f., 574, 593, 596, 606, 622, 633, 659, 661
Equilibrium equations  520
Equilibrium principle  45
Equivalent trussed framework system  623
Erection procedure  77, 84
Eccentric joint  588
Establishment phase  22, *35 f.*, 46, 50, 72, 79, 89, 91, 105, 168, 176, 305, *307 f.*, 324, 398, 531, *574 f.*, 593, 622, 694
Euler buckling cases  335
Euler buckling theory  425, 443
Euler cases  *377*
Euler curve  447
Euler differential equation  656, 658
Euler safety factor for buckling  453 f.

Euler's polyhedron theorem  137, 479
Experimental mathematics  646
Experimental physics  177
External memory  627
External pressure  548
External torsion moment  402
External virtual work  342, 683
External work  187, 247, 353, 376
Extra-symbolic meaning  574 f.
Extremum of deformation work  659
Eye-bar shop  98

Factory organisation  422
Failure load  539
Failure mechanism  217, 239
Failure mode  121, 217
Failure moment  86
Fairbairn foundation  421
Falk scheme  619
False arch  188
Fatigue  418
FE main program  654
FEM textbook  654
Fictitious forces  117 ff.
Field problem  619
Finite bar segment  647
Finite difference method (FDM)  493, 540 ff., 619 f., 631
Finite elements  249
Finite element analysis  248 f.
Finite element mapping (FEMAP)  628
Finite element method (FEM)  40, 241, 249, 435, 462, 493, 562, 570 f., 619 ff., 648
Finite procedure  646
Fire resistance  505
Fire test  509
First boundary value problem  583
First Bredt equation  412, 429
First formation law of trussed framework theory  47, 487
First fundamental law of thermodynamics  682
First-order tensor  612
First-order theory  119, 588
First prime task of thrust line theory  214 f., 217
First stage of formal operations  574, 577
Fixed arch  223, 335, 338
Fixed bar  595
Fixed-based frame  573
Fixed-end arch  115, 528, 573
Fixed-end arch system  210
Fixed-end beam  122 f., 253, 263
Fixed-end column  258
Fixed-end curved elastic bar  705
Fixed-end elastic arch  241
Fixed-end moment  267, 269, 301
Fixed three-pin arch  230
Fixed two-pin arch  230
Flat plate  335
Flat steel plate  469
Fluid mechanics  612, 671
Flexibility  643
Flexibility matrix  613, 642 f.
Flexibility method  641

Flexibility of complete structure  644
Flexibility of element  644
Flexural buckling  392
Floating crane  431
Flutter Bible  615
Flutter calculation  393, 645
Folded plate  460, 528, 544, 545 ff.
Folded shell  528, 544
Force-deformation behaviour  220
Force-deformation diagram  172
Force diagram  36, 703
Force method  37 f., 107, 116 f., 340 ff., 349 f., 383, 467, 483, 558, 570, 581 ff., 592 ff., 602, 604, 606 f., 614, 621, 632, 635 ff., 641 ff., 654
Form-active loadbearing structure  706
Forging shop  97
Formalisation idea  598
Formalisation of structural calculations  571
Formalised theory  620, 641, 645, 662
Formal language  599
Formal operation  574, 700
Formalised language  574, 579
Formation law  136 ff., 141, 478, 481, 485, 490 f., 493, 495, 695
Form-finding  191, 194
FORTRAN  628
FORTRAN IV  654
Foundation anchor  425
Foundation dynamics  390
Four-bar linkage  352 f.
Fourier's heat conduction equation  686
Four-span continuous beam  78
Frame  544
Frame analysis  591
Frame equation  531 f.
Frame table  602
Frame theory  595
Framework  71
Framework cell  631 f.
Framework cell discretisation  630
Framework theory  37
Frictionless wedges  202
Full-size test  539
Fully plastic cross-section  123, 125, 266
Functional  664
Functional analysis  655
Fundamental engineering science disciplines  28, 37, 84, 142 ff., 277, 313, 348, 359, 498, 548, 585, 600, 612, 614, 617, 621, 641
Fundamental theory  33, 571
Funicular force  164
Funicular curve  165, 540
Funicular polygon  179, 193, 201 f., 206, 311, 318 ff., 325, 329, 577, 704
Fuselage  630

Gantry crane  491
Gas tank  549
Gaussian curvature  533
General bending theory of shells  553
General dynamics  612
General technology  148, 151
General theory of linear elastic trusses  36

General theory of trusses  347, 498
General law of work  43 f.
General work  688
General work theorem  355, 367, 539, 579, 606, 679 f., 683 f.
Generalised coordinates  374, 377
Generalised displacement  377 f., 643 f.
Generalised flexibility  642
Generalised force  374, 377 f., 643 f.
Generalised stiffness  644
Geodesic dome  489 f.
General variational theorem  665 f., 668
Geometrical imperfections  447
Geometrical mechanics  46
Geometrical mechanics of rigid bodies  611
Geometrical theory of proportion  251
Geometrically determinate  109, 111, 120
Geometrically determinate system  119
Geometrical factor of safety  236
Geometrically indeterminate  109, 120
Geometric boundary conditions  583, 656
Geometric composition theory  139
Geometric proportioning rule  139
Geometric series  491 ff.
Geometric similarity  262, 268
Geometric view of statics  46, 135, 218, 681, 685, 707
German idealism  695
German Standards Committee (DIN)  527
Glider model  486
Global displacement state  625
Global system of coordinates  626, 653
Golden rule of mechanics  412
Graph theory  152, 156
Graphic editor  704 f.
Graphic interface  707
Graphic monitor  628 f.
Graphic software  706
Graphical analysis  50, 182, 317 ff., 355, 484, 531, 572, 576 f., 682, 703, 705
Graphical integration machine  317, 322
Graphical method of fixed points  601
Graphical statics  35 f., 48, 50, 84, 94, 144, 179, 307, 311, 317 ff., 347, 355, 484, 572, 576, 682, 703, 705
Grashof's grillage method  535, 538, 540
Gridwork method  624 f.
Gridwork model  628
Grillage method  534, 539
Groin vault  244
Group theory  156
Guyed cantilever  84

Hamilton-Jacobi theory  659, 662 ff.
Hammerhead crane  430
Hanging chain  192 f.
Harmony  701
Heavy-duty crane  432
Hemispherical shell  488
Hemispherical spatial framework  488
Hennebique's system  515 ff., 523
Hennebique's T-beam system  514
Hennebique's trussed framework model  563

Heterogeneous manufacture  245
Hexahedron  137
Higher dynamics  180
Higher engineering education  *50 ff.*
Higher technical education  55, 58
High-rise buildings  560, 592
High-speed aerodynamic  637
High-strength concrete  555
Historical disciplines  708
Historical engineering science  25, 186, *248*
Historical epistemology  243 f.
Historical theory of structures  25, 186, *248*
Historicity of structures  25
Historico-genetic teaching  27, 692, *707 f.*
Historico-logical comparison  709
Historico-logical longitudinal section analysis  708
Historico-logical transverse section analysis  708
History of art  700
History of elastic theory  335
History of engineering  145 f., 151
History of engineering science  248, 389
History of engineering science knowledge  248
Holistic design  *568 f.*
Hollow box  76, 316
Hollow-rib plate  462
Hollow section  442
Homogeneous quadratic function  681
Hookean body  313, 576
Hookean relationship  650
Hooke's law  32, 104, 121, 173, *271 ff.*, 390, 519, 607, 682
Hoop force  326 f.
Hoop tension  679
Horizontal frame  593
Howe truss  72
Huber differential equation  460
Huber's plate theory  *459 f.*
Hybrid displacement model  670
Hybrid finite elements  664
Hybrid model  670
Hybrid stress model  670
Hybrid variational theorem  663 f.
Hydrostatic pressure  549, 681
Hydraulic crane  417, 420
Hydraulic cylinder  421
Hydraulic jack  78
Hyperbolic paraboloid  553
Hyperbolic paraboloid shells  552
Hyperboloid  487

Icosahedron  137, 488 f.
Idea of formalisation  573, 580, 620
Ideal buckling load  447
Ideal compressive stress  449
Ideal-elastic and ideal-plastic material law  122 f., 452
Ideal elastic material behaviour  252, 275, 277, 294
Ideal elastic modulus  467
Ideal plastic material behaviour  252, 275, 494
Imposed load  72, 82, 101, 165, 199, 223

Inclined pin-jointed bar  286
Inclined principal tensile stress  563
Incremental launching method  560
Industrial Revolution  21, 57, 64, 93, 101, 153, 155, 160, 163, 168, 180, 246, 251, 272, 277 f., 283, 289, 291, 296, 308, 313, 317, 330, 397, 414, 418, 421, 586, 597
Industrial standard  525
Infinitely convergent series  602
Infinitesimal calculus  40, 598, 648, 655
Influence coefficient  640
Influence line  74, *99 ff.*, 185, 229, 305, 329, 335, 338 ff., 348, 354 f., 379, 579 f., 580, 684 f.
Influence line theory  230, 431, 578, 706
Influence line concept  102
Initial phase  23, *33*, 90, 213, 277
Innovation phase  *40*, 107, 117, 121, 231, 398, 456, 469, 471, 475, 528, 539, 558, 567, 571, 619 ff., 641, 643, 647, 653, 662
In-plane-loaded plates  *545 ff.*
Integral calculus  32, 39, 54, 164, 179, 213 f., 216, 247, 270, 289, 383, 681, 695
Integral equation method  534
Integral plate  469
Integral tables  362
Integration machine  320, 322
Integration period  *39*, 121, 230, 553 f., 605, 608
Interactive THRUST  706
Internal bending work  187
Internal pressure  548
Internal secondary conditions  668
Internal virtual work  119, 131, 342, 355, 683
Intra-symbolic meaning  574, 620
Invention phase  *38*, 107, 121, 327, 398, 448, 527, 534, 543 f., 547, 551, 553 f., *596 f.*, 621, 624, 637, 687
Inverted catenary  67, 90, 189, 287, 289, 335, 681
Inverted funicular polygon  286 f.
Iron arch bridge  181, 183
Iron beam bridge  180
Irregularly formed roof  471
Isherwood system  456
Isochromatic lines  188
Isothermic deformation  372
Iteration method  394

Jacobian determinant  480, 482
Jacobi-Gauss iteration method  605
Joint angles of rotation  589, 593 f.
Joint displacement  593
Joint of rupture  311

Kern  88, 237, *335 ff.*
Kern point moment  229
Keystone  224
$k_h$-method  508
Kinematic chain  126 f.
Kinematic doctrine  *354 ff.*, 379, 382, 589
Kinematic machine  46, *350*
Kinematic machine model  683
Kinematic method  131 f., 328, 607, 706
Kinematic model  680
Kinematic pin-jointed system  591

Kinematic pin system  593
Kinematic relationship  583, 682
Kinematic theorem  103
Kinematic theory  103, 483, 706
Kinematic theory of beams  355 f.
Kinematic theory of trusses  353, 684 f.
Kinematic series  353 f.
Kinematic spatial framework theory  484
Kinematic trussed framework theory  327, 484
Kinematic ultimate load theorem  237
Kinematic view of statics  42, *46*, 135, 206, 218, 347, 685, 707
Kinematically determinate  47
Kinematically determinate system  119
Kinematically indeterminate system  286
Kinematically permissible hinge mechanism  235, 237
Kinetic energy  257, 374 f.
Kirchhoff differential equation for plates  540, 657
Kirchhoff's plate theory  460, 533

Lagrange equation  377 ff.
Lagrange formalism  606
Lagrange's generalised coordinates  595
Lamé elastic constants  622
Lamé-Navier displacement differential equation  583
Land's theorem  99, 355
Language of matrix algebra  645
Language of matrix theory  644, 653
Laplace differential equation  401
Laplace operator  461
Large deflections  664
Large deformations  629
Large displacements  688
Lateral buckling  177, 437, 453, 455, 504
Lattice dome  462, 478 f., 481, 485
Lattice girder  70 f., 113, 630
Lattice girder bridge  79
Lattice shell  487
Lattice structure  344
Law of friction  295
Law of materials  583
Legendre transformation  666
Leibnizians  677
Lever arm  274
Lever arm of the internal forces  507
Lever principle  42, 45, 265
Lifting bridge  431
Lifting gear  418
Lifting procedure  78
Light pen  628 f.
Lightweight steel construction  *469 ff.*
Lightweight steel road deck  457
Limitation principle  668
Linear algebra  183, 323, 577, 579, 645
Linear buckling analysis  624
Linear combination  601
Linear compressive stress distribution  525
Linear distribution of compressive stress  509
Linear elastic analysis  241
Linear elastic body  622

Linear elastic continuum  230, 362, 623 f.
Linear elastic FEM  243
Linear-elastic spring  296
Linear-elastic theory of truss  690
Linear-elastic trussed framework theory  364, 382
Linear electrical networks  614
Linearisation of perception  700
Linear programming  646
Linear theory of structures  607
Linear transformation  610, 612
Line diagram analysis  573, 575 ff., 600
Line of pressure  218
Line of resistance  218
Line of thrust  89 f., 192, 194 f., 201, 206, 209, 218 ff., 225 f., 234 f., 236 f., 287, 326, 335, 337, 567, 681, 705
Line of thrust method  223
Line of thrust theory  89, 213 ff., 246
Line of fracture  539
Lithography  574, 576
Load-carrying capacity  122, 124, 127, 129, 222, 239, 258, 454, 466, 506, 565
Load increase factor  131
Loading tests  509 f., 529
Load vector  120
Local approximation function  670
Local buckling  453, 455
Local system of coordinates  626
Localised Ritz approache  654
Locomotive crane  421
Longitudinal rib  456
Longitudinal stiffener  456, 458, 635
Long's system  71
Lower kern line  336
Lower kern point moments  601

**M**achine dynamics  175
Machine kinematics  175, 317, 414
Machine shop  97
Machine tool  352, 354, 413, 415
Macroelement  632, 635 f.
Main bridge shop  97
Mainframe computer  628
Main girder  456 ff.
Main routine  645
Manipulation of symbols  692
Manipulator of symbols  580
Marxism  149 ff., 312
Masonry arch theory  74, 87
Masonry arch collapse mechanisms  204
Mass-active loadbearing structure  706
Mass inertia tensor  612
Materials tests  519
Materials research  539
Materials testing  255, 512
Materials testing science  514
Materials laboratory  62
Mathematical elastic theory  179, 336, 532 ff., 543, 548, 582 f.
Mathematical physics  659
Mathematical shell theory  548
Mathematicisation of the science  698

Matrix algebra  574, 611, 620, 640 ff.
Matrix calculation  652
Matrix formulation  610
Matrix iteration  637
Matrix iteration procedure  615
Matrix multiplication  613, 619
Maximum bending moment  127, 508
Maximum line of thrust  219
Maximum principal stress  566
Maximum span moment  101
Maxwell-Betti reciprocal equation  658
Maxwell's theorem  343 f., 348 f., 607, 661
Means-purpose relationship  579
Mechanics of rigid bodies  484
Mechanical calculator  598, 599
Mechanical engineering  154, 167, 173, 180, 337, 352, 386, 405, 411, 459, 586, 596, 686
Mechanical networks  614
Melan arch bridge  464
Melan system  488, 537
Member end moments  589, 594 f.
Member angles of rotation  589, 593 f.
Member chord rotation  117 f., 120
Member stiffness  604
Membrane  374
Membrane stress condition  326 f., 548
Membrane theory  476, 479, 550
Membrane theory for rotational shells  550
Menabrea's principle of minimum deformation energy  222, 607, 658
Meridian force  327
MERO system  491 ff.
Mesh geometry  542
Method of fluxions  216, 383
Method of base coordinates  596
Method of successive approximation  39, 603 f.
Microprocessor  571
Middle-third rule  88, 223, 237
Military engineering corps  309
Military schools  51 ff.
Minimal area problem  655
Minimal line of thrust  219
Minimum arch thickness  208 f.
Minimum load increase factor  116
Minimum of total energy  661
Minimum thickness  224, 236 f.
Mobile shed structure  491
Modern structural mechanics  641
Mohr's analogy  83, 329, 349, 540, 542
Mohr's general work theorem  607
Molecular hypothesis  681
Möller's T-beam  514
Moment equilibrium  588 f.
Moment of resistance  410, 426
Monier-Broschüre  500 ff.
Monier slab  502
Monier system  498 ff.
Monier's patent  499 ff.
Monolithic reinforced concrete frame  516
Motor dyad  612
Motor matrix  612
Motor symbolism  612
Motor tensor  612

Movable counterweight  420
Moving bridge  180
Moving load  101
Multi-bay frame  573, 592
Multi-constant theory  623, 682
Multi-dimensional continua  647
Multi-storey frame  592 f., 602
Mushroom flat slab  536 ff., 540, 543

**N**apoleonic Wars  55
NASA Structural Analysis System (NASTRAN)  629 f.
Natural causality  153
Natural frequency  375
Natural philosophy of Aristotle  676
Navier's beam theory  598
Navier-type boundary conditions  534
NC machine  472
Neutral axis  88, 125, 507, 557, 603
Neutral fibre  521
Néville truss  73
Newton's law of force  143
Newton's law of inertia  143
Newton's law of reaction  143
Newton's second law  45
$n$-method  509
Nodal load  626
Node displacement  342
Node equilibrium  111
Node rotation  109
Non-central system of forces  321
Non-classical engineering mathematics  157
Non-classical engineering sciences  154 f., 156, 580
Non-elastic buckling theory  450
Non-Euclidean geometry  490
Non-interpretive operation  693
Non-isothermic deformation  372
Non-linear algebra  646
Non-linear analysis  627
Non-linear elastic analysis  241
Non-linear-elastic material behaviour  427
Non-linear elastic truss  363
Non-linear FEM  243
Non-linear procedure  571
Non-linear sets of equations  135
Non-linear stress-strain diagram  224
Non-linear stress-strain relationship  642
Non-$n$ cross-section design process  561
Non-rigid foundation  104
Non-sway frame  593
Non-sway system  603
Non-Uniform Rational B-Splines (NURBS)  470
Normal stress diagram  439
Number juggler  702, 707
Numerical engineering methods  602

**O**ctahedron  137
Octagonal dome  551
One-dimensional elastic continua  335, 666
One-sided uniformly distributed load  508 f.
Online interactive graphical analysis  706
Open-hearth furnace  97

Organic manufacture  245, 698
Organ projection  144
Orientation phase  *31 f.*
Orthogonal frames  593
Orthogonalisation methods  *601*
Orthogonally anisotropic plate  458
Orthotropic bridge deck  *456 ff.*, 469
Orthotropic plate  455 ff., 469, 543
Out-of-plane-loaded structure  *532 ff.*
Overhead moving crane  99
Overhead travelling crane  414 ff., 419, 422, 431
Overall stiffness matrix  651 ff.

**P**air of elements  354 ff.
Parabolic distribution of compressive stress  509
Parabolic two-pin arch  220
Paradigm change  451
Paradox of elastic theory  134
Paradox of the plastic hinge method  *127 ff.*
Paradox of the plastic hinge theory  133
Paradox of ultimate load theory  689
Parallelogram of forces  32, 42, *44 f.*, 575
Partial differential equation of Terzaghi  686
Partially plastic cross-section  125
Particle physics  672
Passive earth pressure  211
Pauli truss  294
Perforate plate  624
Perforated quadratic plate  628
Permanent way theory  105
Permissible concrete compressive stress  508
Permissible tensile stress of concrete  549
Permissible tensile stress of steel  521, 549
Permissible steel tensile stress  508
Permissible stress  334, 509
Perpetuum mobile  193
Philosophy of technology  143
Photoelastic experiment  188
Photoelastic measurement  567
Photoelastic test  566
Photogrammetric method  191
Piecewise linear approximation  655
Pinned beam  122
Pinned trussed framework model  48, 50, 91, 98, 108, 113
Pin-jointed bar  289
Pin-jointed framework  342, 529, 623
Pin-jointed truss  49
Pin-jointed trussed framework  585
Pin-jointed trussed framework model  529, 584
Pithead gear  431 f.
Plane of rupture  207 f.
Plane shell structure  566
Planetarium dome  488, 551
Planning disciplines  708
Plastic deformation  199, 260
Plastic design method  127
Plasticity condition  126, 131
Plastic hinge  121, 125 ff., 129, 210
Plastic hinge method  127, 134
Plastic hinge theory  122, 135, 266, 276
Plastic limit state  127
Plate  374

Plate bending stiffness  461
Plate buckling  456
Plate deflection  533
Plate stiffness  540
Plate theory  179, 460, 493, 532
Plates with in-plane loading  528, 544
Platonic bodies  *136 f.*, 479, 491
Plotter  629
Poisson's ratio  440, 533, 583, 622
Polar moment of inertia  408
Pole diagram  353, 356
Pole distance  320
Pole plan  354
Polygonal dome  551
Polygon of forces  202, *311, 318 ff.*, 325, 577
Polyhedron  319, 492
Political economics  52
Portal frame  431
Portal method  602 ff.
Portland cement  497, 511, 524
Potential energy  374, 377, 655
Potential equation  666
Powerful Efficient Reliable Mechanical Analysis System (PERMS)  630
Practical elastic theory  414
Practical beam theory  220
Practical bending theory  32, 34 f., 37, 90, 253, 293, 301 f., 307, 331, 364, 382, 433 f., 437 ff., 676
Practical torsion theory  433
Prange's variational theorem  663
Precambered leaf spring  628
Prefabrication  77
Prefabrication industry  555
Preparatory period  23, *31 ff.*
Pressure cylinder  424, 426
Pressure-volume diagram  313
Prestressed carriageway  557
Prestressed concrete  *554 ff.*
Prestressed concrete beam  467, 555
Prestressed concrete bridge  560
Prestressed concrete construction  496
Prestressed concrete engineer  556
Prestressed concrete research  555
Prestressed force  558
Prestressed ground anchors  557
Prestressed masts  557
Prestressed piles  557
Prestressed pipes  557
Prestressed ring flanges  108
Prestressed shells  557
Prestressed tanks  557
Primary boundary conditions  668
Primary stress  587
Prime mover  352 f., 413
Principal compressive stress  566
Principle of deformation minimum  614
Principle of Dirichlet  607
Principle of Engesser  607
Principle of extremum of potential energy  659
Principle of induction  167
Principle of least action  *678 f.*
Principle of Maxwell and Betti  579

Principle of Menabrea  230, 362 f., 368, 370, 379, 408, 567, 660 f., 683
Principle of minimum complementary potential energy  642
Principle of minimum deformation energy  183, 226, 640
Principle of minimum potential energy  654
Principle of minimum total potential  642
Principle of minimum loading  219
Principle of the stability of the elastic equilibrium  655
Principle of virtual displacements  37 f., 42, *43*, 45, 99, 119 f., 131, 133, 135, 236, 239, 294, 347 ff., 355 ff., 377, 413, 484, 533, 595 f., 606 ff., 642, 656, 659 f., 661, 681, 687 f.
Principle of virtual forces  36 ff., 42, *44*, 294, 342, 348 ff., 355 ff., 362 f., 378, 383, 577, 589, 596, 606 ff., 642, 658 f., 666, 683 f., 687 f.
Principle of virtual work  214
Principle of virtual velocities  43, 45, 345, 383
Principal tensile stress  566
Principal stress  566
Principal stress line  566
Principal stress trajectories  565 f.
Programmability  394
Program systems  609
Projective geometry  34, 318 ff., 322, 324 f., 329, 354, 576, 611, 682
Proportion  701
Propped cantilever  131 f.
Ptolemaic planetary system  675
Puddling furnace  35, 278
Pulley  251
Pure bending  436, 509, 520
Pure mathematics  582
Pure mechanics  175
Purpose-means relationship  *158*, 163, 261, 290

**Q**uadratic approximation function  669
Quadratic element  668
Quadratic trussed element  623
Quadripole theory  614
Quantum mechanics  611
Quantum theory  621

**R**adial stress  550
Radio mast  487
Rari-constant concept  426
Rari-constant theory  622, 682
Rationalisation of structural calculations  *600 f.*
Rationalisation movement  318, 324, 486, 571
Rational mechanics  596
Rayleigh-Ritz coefficient  *375 ff.*
Rayleigh-Ritz method  *375 ff.*, 642
Real-time analysis  692
Reciprocal diagram  320 ff.
Reciprocal figure  319
Reciprocal force polygon  319
Reciprocal funicular polygon  319
Reciprocal theorem  343, 374, 379, 612
Rectangular bar cell  625 f.
Rectangular element  649, 670
Rectangular plate  649

Rectangular reinforced concrete slabs 534 f.
Rectangular slab 533 f., 624
Rectangular triangular bar cell 624
Reduction method 618
Regula falsi 34
Regular polyhedra 136
Regularity 701
Reinforced concrete arch 509, 528
Reinforced concrete beam 520, 525, 528, 564
Reinforced concrete column 536
Reinforced concrete construction 498
Reinforced concrete plate 528
Reinforced concrete shell 488, 528, 547
Reinforced concrete shell structure 496
Reinforced concrete slabs 223, 499, 502, 515, 528
Reinforced concrete standard 521, 524
Reinforced concrete theory 499, 592
Reinforcing steel 555
Relaxation method 604 f.
Resolved strut 445
Rheological model concept 468
Restraint forces *111 ff.*, 119, 593, 595, 599, 606
Residual stresses 180
Reversible electric motor 419
Ribbed slab 628
Rib vault 244, 325 f., 705
Rigid body mechanics 612, 672
Rigid foundation 104
Rigid frames *528 ff.*, 573, 589
Rigid frame system 578
Rigid framework 578
Rigid joint 530
Rigid-jointed frame 585
Rigid-jointed framework 581, 590
Rigid-plastic material behaviour 222
Rigid-plastic material law 233
Rise/span ratio 196
Ritter's method of sections 50
Riveted construction 450
Riveted gusset plate 108
Riveted joint 49, 98, 342, 485, 529, 586
Riveted solid-web beams 85
Riveted steel truss 566
Riveting machine 97
Rolling shop 97
Rotationally symmetrical domes 326
Rotational restraints 109
Rotational shell 548

**S**afety factor for load-carrying capacity 454
Safety theorem 236
Sail vault 244
Saint-Venant torsion theory 177, 398, 405, 412, 433, 443
Sandwich construction 469
Schism of architecture *694 f.*
Schlink dome 485
Schwedler dome *475 ff.*, 481, 484 f., 550
Schwedler truss 294
Scientific revolution 143, 154, 251, 259, 674
Scientific revolution in structural mechanics 647

Scientific school *380 f.*
Scientific-technical revolution 155
Screw 251
Secondary condition 670
Secondary stresses 98, 108, 529 f., 566, 581, 587, 603
Second boundary value problem 584
Second Bredt equation 412, 429
Second fundamental law of thermodynamics 682
Second industrial revolution 512, 513, 522 f.
Second-order displacement theory 135
Second-order tensor 612
Second-order theory 38, *113 ff.*, 121, 126, 171, 184, 335, 380, 393, 581, 587, 596, 645 f.
Second prime task of thrust line theory 201 f., 214 f., *217*, 287, 320
Second principle of thermodynamics 154
Second stage of formal operations 574 ff., 579 f., 606
Second theorem of Castigliano 408, 683
Segmental arch 203, 206
Semicircular arch 202, 205 f., 209, 236 f.
Semi-graphical fixed-point method 531
Semi-inverse method 400, 404
Semi-octahedron 493
Semi-probabilistic safety concept 472
Semi-regular Archimedean bodies 492
Sensitivity to errors 601 f.
Shallow foundation 534
Shear centre *435 ff.*
Shear connector 467
Shear crack 565 f.
Shear design 524, 565
Shear elasticity 334
Shear field *633 ff.*, 638, 640
Shear field layout *630 ff.*, 636 f.
Shear field theory 635
Shear-flexible beam 672
Shear-flexible elastic plate 664
Shear-flexible orthotropic plate 462
Shear flow 429
Shear force 563 f., 633
Shear lag 664
Shear link 564
Shear measurement 568
Shear modulus 402 f.
Shear reinforcement 563, 565
Shear stiffness 335
Shear stress diagram 335
Shear stress vector 401
Shear strength 567
Shear studs 465, 467
Shear test 564, 567
Shell analysis 670
Shell construction 560
Shell roof 494
Shell plate 460
Shell structure 489
Shell theory 327, 527, *547 ff.*, 621
Shell-type spatial framework 476
Shell-type trussed framework 476
Shell wing model 637

Shift lift project 390
Shipbuilding 596
Shipbuilding engineering 456
Shipbuilding industry 469
Shipbuilding science 456
Shot-fired fixings 465
Shrinkage 519
Similarity theory 262
Siemens-Martin steel 97
Simple machines 46, 162, 200, 251, 295
Simply supported beam 47 f., 78, 116, 127, 129, 171, 253, 298, 528, 753
Simply supported beam with cantilever 573
Simply supported I-beam 123
Single-pin arch 573
Single-pin frame 573
Single strut system 564
Six-node triangle 668
Skew bridges 431
Slab bending moment 536
Slab strip 658
Slab test 540
Slab theory 179
Slab thickness 507 ff.
Slenderness 447
Slenderness ratio 448
Slewing jib crane 416 f., 431
Slipway frame 431 f.
Slope deflection method 594
Soap-bubble allegory 434
Sobolev's embedding theorem 373
Software company 646
Soil mechanics *685 f.*
Solid-web box beam 634
Solution vector 112
Space frame 631
Space truss model 622
Spatial bridge system 481
Spatial elastic continuum 493
Spatial elastic theory 553
Spatial framework theory *481 ff.*
Spatial trussed framework 562
Spatial truss system 596
Specific symbolic machine 643
Spherical shell 550
Spring constant 272
Square slab 535
Stable equilibrium condition 209
Stabilised pin-jointed system 590 f.
Stabilised pinned system 118 f., 593
Stability case 453
Stability of the elastic line 657
Stability problem 117, 120, 645
Stability theory 455, 605, 688
Standardisation theory 490
Statically determinate 47
Statically determinate trussed framework 91
Static law *136 ff.*, 139, *140 f.*, 478, 481, 485, 491, 495
Static method 131 f., 707
Statics of stone constructions 333
Static ultimate load 238
Statically equivalent node loads 650

Statically indeterminate  47
Statically indeterminate parameters  592
Stationary value of canonic functional  659
Steam boiler  414
Steam engine  173, 277, 313, 315, 415
Steel core  465
Steel plate crane  704 f.
Steel tensile stress  509
Steelwork industry  92 ff.
Stiffened cylindrical shell  635
Stiffened rectangular plates  618
Stiffened shell structure  635
Stiffening beam  114
Stiffness  643
Stiffness coefficient  119
Stiffness matrix  112, 120, 642 f., 649, 651 f., 672
Stiffness method  641
Stirrups  563
Stochastic  156
Stochastic computation  671
Stone bridge  180
Storage yard  97
Storey frame  604
Storey-height frame  628
Strain-displacement relationship  649
Strain distribution  258, 336
Strain stiffness  109, 115, 335, 337
Strain tensor  665
Strength test  35, 65
Stress analysis  113, 220, 304
Stress diagram  125
Stress distribution  266, 276, 515
Stress equation  507
Stress field  568
Stress pattern  566
Stress problem  117, 120
Stress proof  449
Stress-strain diagram  252, 313, 682
Stress-strain relationship  468, 525
Stress tensor  583, 612, 664, 665
Stress trajectory  144, 565
Stringer  635, 637
Strip method  324
Strip of slab  508, 536
Structural Analysis Program (SAP)  494
Structural engineering disciplines  498
Structural imperfections  447
Structural iteration methods  602 ff.
Structural law  136 ff., 139, 140, 478, 485, 487 490 f., 493, 495
Structural machine  478
Structural matrix analysis  380, 394, 570, 610 f.
Structural matrix method  617
Structural models  35
Structural shell theory  327
Structure of the computer  645 f.
Stüssi beam  129, 134
Subroutine  645
Substitute member method  366, 482 ff., 685
Supercritical buckling  658
Superposition equation  343
Superposition principle  588
Superposition theorem  112 ff., 120

Support moment  269
Suspension bridge  64 ff., 79, 181, 183, 292
Suspension bridge theory  65, 90, 115
Suspended platform  431
Sway frame  593
Sweptback wings  639
Swing bridge  431
Symbolic machine  573, 580, 598 f., 610, 620 f., 703, 706
Symmetry  701
System of classic engineering sciences  143, 153, 155, 157, 278, 598
Symbolic operations  642
System matrix  349 f., 362
System theory  142, 148 f., 151

Tabular calculation  394
Tangential stress  550
Teaching of engineering  427
T-Beam  521, 544, 565 f.
T-Beam section  562
Technical finality  153
Technical revolution  498, 512, 514, 522, 555
Tectonic  699
Tekhnē  40, 158, 351, 576 f., 579 f., 695, 698, 703
Template shop  97
Temporary bridge  491
Temperature stress  661
Tensile resistance  265, 267
Tensile strength  165, 172 f., 188, 261, 263, 275, 301
Tensile strength test  86, 225, 233 f., 302
Tensile stress  199, 209, 241
Tensile test  166, 219, 251, 252 f., 263, 265, 268, 270, 274
Tension field theory  634
Tensor algebra  609, 611
Tensor analysis  156, 553 f., 574, 608, 611
Tensor calculation  381
Tensor calculus  528
Test and model  115
Test beam  466, 565
Test model  638 f.
Testing facility  97
Testing machine  423, 445
Tetmajer straight line  448
Tetrahedron  137, 493
Tetrahedron framework  140 f., 480
Theorem of Castigliano  382
Theorem of Clapeyron  314, 682, 684 f.
Theorem of Maxwell  382
Theorem of minimum deformation energy  360
Theorem of three moments  308, 313
Theorem of ultimate load design  133
Theoretical kinematics  351
Theoretical mechanics  662
Theory of beam grids  457
Theory of continuous suspension bridges  114
Theory of curved elastic bars  425 f.
Theory of earth pressures  331, 333
Theory of engineering sciences  144 f., 148, 151 ff.
Theory of elastica  272
Theory of elastic arches  337, 366

Theory of elastic plates  77
Theory of elastic trussed frameworks  621, 684
Theory of folded plates  545 ff.
Theory of friction resistances  162
Theory of influence lines  366
Theory of linear-elastic truss systems  562
Theory of orthotropic plates  459 ff.
Theory of out-of-plane loaded structures  532
Theory of plates  527
Theory of plate and shell structures  543 f.
Theory of porous media  686
Theory of prestressed concrete  528
Theory of proportion  261
Theory of reinforced concrete  518
Theory of resistance  162
Theory of rigid frames  611
Theory of sandwich construction  469
Theory of secondary stresses  37, 109, 366, 581, 587 ff., 591, 685
Theory of sets of equations  600 f.
Theory of sets of linear equations  599, 610, 612
Theory of shell structures  37 f., 381
Theory of spatial frameworks  475 ff.
Theory of stability  596
Theory of statically indeterminate trussed framework  341 ff.
Theory of statically indeterminate systems  318
Theory of steam engine  168
Theory of suspension bridges  164 ff.
Theory of the arch plate  567
Theory of the beam on elastic supports  107 f.
Theory of the elastic line  32, 35, 298
Theory of thin elastic plates  533
Theory of trussed frameworks  294, 612
Theory of viscoelastic bodies  463, 468
Theory of warping torsion  433, 435, 439 f.
Thermal effect  82, 372
Thermal energy  373
Thermal expansion coefficient  501 f.
Thermal load case  363
Thermal stress  179
Thermodynamics  154, 168, 173, 308, 313 f., 672
Thin skin  637
Thin reinforced concrete shell  326
Thin shell  327
Thin-wall box beam  634
Thin-wall hollow-box beam  560
Thin-wall hollow section  412, 428, 442
Thin-wall open section  434
Thin-wall structure  470
Third boundary value problem  584
Third prime task of thrust line theory  214, 217
Third stage of formal operations  574
Thought experiment  265
Three-cell hollow cross-section  442
Three-centred arch  189 f., 196, 238
Three-dimensional continua  659
Three-dimensional displacement method  608
Three-dimensional elastic continua  666
Three-dimensional lattice girder  631 f.
Three-dimensional trussed framework  614
Three-hinge system  188
Three-moment equation  316

Three-pin arch  115, 199, 223, 335, 338, 357, 573
Three-pin arch plate  566
Three-pin frame  47, 573
Three-pin system  206, 285 f., 289
Three-pin truss  48
Three-span beam  124
Thrust line  74, 188 f., 575 f., 706
Timber arch  200, 224
Timber arch structure  223
Timber bridge  70, 180
Timoshenko beam  672
Tower crane  431
Torsional-flexural buckling  455
Torsional stiffness  433, 539
Torsion constant  402, 408, 411 f., *433 ff.*
Torsion moment  400, 403, 407, 410
Torsion of thin-wall sections  443
Torsion stiffness  402 f., 409, 441
Torsion test  405
Torsion theory  *428 f.*
Total potential energy  655
Total stiffness matrix  625
Trajectory diagram  566
Transistor technology  647
Transversely stiffened space frame  631
Transverse stiffener  637
Transcendental equation  165, 209
Translation equilibrium  259
Transmission mechanism  352 f.
Transmission tool  413, 415
Transporter cranes  431
Transverse bending  439
Transverse stiffener  456
Trapezoidal steel profile  469
Travelling gantry crane  418, 420 f.
Travelling jib crane  417, 419
Travelling slewing jib crane  417, 420, 430
Triad of industry, administration and science  522, 559
Triangle of forces  320, 322
Triangular bar cell  625 f.
Triangular element  649 f.
Triangular plate  649
Triple strut system  564
True measure of force  *677*
Truss dynamic  596
Truss model  562, 622
Trussed arch  81 f., 94
Trussed frame  573
Trussed framework  85, 589
Trussed framework bridge  73
Trussed framework model  71, 562, 564
Trussed framework structure  598
Trussed framework theory  34 ff., 49, 72, 80, 84, 91, 93, 98, 112, 139, 168, 313, 342, 347, 475, 564, 585, *591 ff.*, 596, 660
Trussed girder  573
Trussed rigid framework  578
Trussed shell  478 f., 481, 487, 631
Tubular bridge  77
Turbulence  671
Turner mushroom system  536
TV tower  560

Twin-arch bridge  196
Twisting theorem  45
Two-dimensional elastic continuum  335, 528, 623
Two-phase system  686
Two-span beam  124, 298, 303, 331
Two-span continuous beam  78
Two-strip system  538
Two-pin arch  81, 115, 335, 338
Two-pin frame  573
Two-pin trussed girder  345
Type 1 elasticity equations  596, 599, 601, 658
Type 2 elasticity equations  596, 599, 606 f., 656

Ultimate load  228, 464, 502
Ultimate load method  *121 ff.*, 450, 452, 466 ff., 689
Ultimate load ratio  267
Ultimate load theory  122, *131 ff.*, 186, 207, 210, 213, 216, 218, 232, 246, 562, 568
Ultimate load theory of slabs  539
Ultimate limit state  561
Ultimate moment  269
Ultimate strength of beam  261
Ultimate tensile force  267, 270
Unbraced nodes  604
Uniformly distributed load  109, 116, 121, 171, 285, 508, 534, 536
Unit displacement method  643
Unit load method  643
Universal symbolic machine  610, 641, 643
Unstable equilibrium  208, 237
Upper bound of ultimate load  237
Upper kern line  336
Upper kern point moments  601

Variation methods  621
Variation principles of elastic theory  596, 687
Variational problem  690
Variational theorems  *654 ff.*, 663, 667
Variational theorem of Dirichlet and Green  663, *655 ff.*, 658, 663, 665, 667, 669
Variational theorem of Fraeijs de Veubeke, Hu and Washizu  *665 ff.*
Variational theorem of Hellinger, Prange and Reissner  665 ff., 670
Variational theorem of Hu and Washizu  *665 ff.*
Variational theorem of Menabrea and Castigliano  663 ff., 667, 670
Varying elastic modulus  655
Vector-loadbearing structure  706
Vehicle axles  627
Vertical frame  592
Vibration mechanics  615
Vibration of ropes  374
Vibration of shells  374
Vibration theory  605
Vierendeel girder  529 f., 590 ff., 612
Virtual external work  119
Virtual internal work  408
Virtual work  119, 408
Viscoplastic deformation  199
Visintini beam  514

Visual communication  53
Voussoir rotation theory  *204 ff.*, 217 f., 220, 226, 231, 245, 247

Waffle slabs  539
Warpage function  400 f.
Warping normal stress  634
Warping torsion  443
Water tank  549
Water tower  487
Watt's steam engine  154
Wedge theory  *199 ff.*, 216, 245
Weighted residual method  619
Welded construction  450
Wheel on axle  251
Wiegmann-Poloncreau truss  48 ff., 320
Williot's displacement diagram  329
Wind load  602
Wind pressure  82
Winkler bedding  103
Wire rope  68
Wood engraving  575
Wrought-iron arch  224
Wrought-iron bridge  72, 79

Yield stress  113, 124, 126, 166, 266, 275, 464
Yield surface  233

Zeiss-Dywidag shell  489, 545, 547, *551 f.*
Zimmermann dome  *476 ff.*, 485